# The Neuroscience of Depression

Features, Diagnosis, and Treatment

# The Neuroscience of Depression

## Features, Diagnosis, and Treatment


Edited by

**Colin R. Martin**

Professor of Perinatal Mental Health, Institute of Clinical and Applied Health Research (ICAHR), University of Hull, UK

**Lan-Anh Hunter**

School of Primary Care, Thames Valley, Health Education England, Oxford, United Kingdom

**Vinood B. Patel**

School of Life Sciences, University of Westminster, London, United Kingdom

**Victor R. Preedy**

Professor of Nutritional Biochemistry, Department of Nutrition and Dietetics, Professor of Clinical Biochemistry, Department of Clinical Biochemistry; Director of the Genomics Centre, King's College, London, UK

**Rajkumar Rajendram**

King's College London, United Kingdom



ELSEVIER

ACADEMIC PRESS

An imprint of Elsevier

Academic Press is an imprint of Elsevier
125 London Wall, London EC2Y 5AS, United Kingdom
525 B Street, Suite 1650, San Diego, CA 92101, United States
50 Hampshire Street, 5th Floor, Cambridge, MA 02139, United States
The Boulevard, Langford Lane, Kidlington, Oxford OX5 1GB, United Kingdom



**Library of Congress Cataloging-in-Publication Data**
A catalog record for this book is available from the Library of Congress

**British Library Cataloguing-in-Publication Data**
A catalogue record for this book is available from the British Library

ISBN 978-0-12-817933-8

For information on all Academic Press publications
visit our website at https://www.elsevier.com/books-and-journals

*Publisher:* Nikki Levy
*Acquisitions Editor:* Joslyn T. Chaiprasert-Paguio
*Editorial Project Manager:* Timothy Bennett
*Production Project Manager:* Sreejith Viswanathan
*Cover Designer:* Mark Rogers

Typeset by SPi Global, India

# Contents

# Contributors

*Numbers in parenthesis indicate the pages on which the authors' contributions begin.*


**Fatemeh Abdollahi** (21), Department of Public Health, Faculty of Health, Mazandaran University of Medical Sciences, Sari, Iran

**Amani Ahmed** (229), Caen University Hospital, Emergency Department, Caen, France

**Muaweah Ahmad Alsaleh** (229), Psychological Studies, Neuropsychologist, Psychotherapist, Psychology Researcher, Behavioural Science, Psychotherapy Practice and Clinical Psychopathology, Psychology Biostatistics, CERReV Laboratory; Public Health-Health Ethics, INSERM & Adapted Physical Activities and Health, COMETE & Pain Management, University of Caen Normandy, Caen, & New Navarre Hospital, & University of Lorraine, Interpsy Laboratory, Nancy; Independent Psychotherapist, France Asylum Land & Reception Service for Unaccompanied Foreign Minors, Caen; Faculty of Education, Counseling Psychology, University of Aleppo, Aleppo, Syria; Psychological Courses Developer, University of the People, Pasadena, CA, United States

**Efruz Pirdogan Aydin** (251), Department of Psychiatry, University of Health Sciences, Sisli Etfal Teaching and Research Hospital, Istanbul, Turkey

**Susana Barbosa-Méndez** (401), Laboratorio de Neurofarmacología Conductual, Microcirugía y Terapéutica Experimental, Subdirección de Investigaciones Clínicas, Instituto Nacional de Psiquiatría "Ramón de la Fuente", México DF, México

**Philip J. Batterham** (423), Centre for Mental Health Research, Research School of Population Health, The Australian National University, Canberra, ACT, Australia

**Amir Sasan Bayani Ershadi** (505), Departments of Pharmacology and Toxicology, School of Pharmacy; Applied Pharmacology Research Center, Zanjan University of Medical sciences, Zanjan, Iran

**Luis Enrique Becerril-Villanueva** (401), Laboratorio de Psicoinmunología, Dirección de Neurociencias, Instituto Nacional de Psiquiatría "Ramón de la Fuente", México DF, México

**Olivia E. Bogucki** (239), Department of Psychiatry and Psychology, Mayo Clinic, Rochester, MN, United States

**Álvaro G. Bris** (345), Department of Pharmacology and Toxicology, School of Medicine, CIBERSAM-ISCIII, Universidad Complutense de Madrid, Madrid, Spain

**Alison L. Calear** (423), Centre for Mental Health Research, Research School of Population Health, The Australian National University, Canberra, ACT, Australia

**Andres Cardenas** (537), Division of Environmental Health Sciences, University of California, Berkeley, Berkeley, CA, United States

**Allison J. Carroll** (239), Department of Psychiatry and Behavioral Sciences, Northwestern University Feinberg School of Medicine, Chicago, IL, United States

**Javier R. Caso** (345), Department of Pharmacology and Toxicology, School of Medicine, CIBERSAM-ISCIII, Universidad Complutense de Madrid, Madrid, Spain

**Shigeyuki Chaki** (381), Research Headquarters, Taisho Pharmaceutical Co., Ltd., Saitama, Japan

**Sally Wai-Chi Chan** (487), Division of Global and Engagement and Partnership; Centre for Brain and Mental Health, The University of Newcastle, Callaghan, NSW, Australia

**Ankita Chattopadhyay** (547), Department of Psychiatry, All India Institute of Medical Sciences, New Delhi, India

**Mahboubeh Dadfar** (211), Department of Master for Public Health, School of Behavioral Sciences and Mental Health-Tehran Institute of Psychiatry, International Campus, School of Public Health, Student Committee of Education and Development Center (EDC), Iran University of Medical Sciences, Tehran, Iran

**Kenta Deriha** (369), Department of Neuropsychiatry, Graduate School of Medicine, Sapporo Medical University, Sapporo, Japan

**Crisanto Diez-Quevedo** (185), Department of Psychiatry and Legal Medicine, School of Medicine, Universitat

Autònoma de Barcelona, Hospital Universitari Germans Trias i Pujol, Badalona, Spain

**Daniel Teixeira dos Santos** (99), Department of Neurology, Hospital of Clinics of Porto Alegre, Porto Alegre, Rio Grande do Sul, Brazil

**Marco Aurélio dos Santos Carvalho** (99), School of Medicine, Universidade do Grande Rio, Rio de Janeiro, Brazil

**David J.A. Dozois** (219), Department of Psychology, The University of Western Ontario, London, ON, Canada

**Mohaddeseh Ebrahimi-Ghiri** (293), Department of Biology, Faculty of Sciences, University of Zanjan, University Blvd., Zanjan, Iran

**Tatjana Ewais** (413), Clinical Medicine, School of Medicine, Griffith University, Gold Coast; Mater Clinical School, Faculty of Medicine, University of Queensland, Brisbane, QLD, Australia

**Sabrina Faleschini** (537), School of Psychology, Laval University, Quebec, QC, Canada

**Andrea Fekete** (283), 1st Department of Pediatrics; SE Diabetes Research Group, Semmelweis University, Budapest, Hungary

**Renerio Fraguas** (121), Departamento e Instituto de Psiquiatria, Hospital das Clinicas, Faculdade de Medicina da Universidade de São Paulo, São Paulo, SP, Brazil

**B. García Bueno** (131), Department of Pharmacology and Toxicology, Faculty of Medicine, University Complutense, Cibersam, IUIN, Imas12, Madrid, Spain

**María Paz García-Portilla** (3), Department of Psychiatry, School of Medicine, University of Oviedo, Oviedo, Asturias; Biomedical Research Networking Center for Mental Health (CIBERSAM); FINBA (Fundación para la Investigación y la Innovación Biosanitaria del Principado de Asturias), Oviedo, Spain

**Jennifer C.P. Gillies** (219), Department of Psychology, The University of Western Ontario, London, ON, Canada

**Fernando S. Goes** (515), Department of Psychiatry and Behavioral Sciences, Johns Hopkins University School of Medicine, Baltimore, MD, United States

**Jacqueline K. Gollan** (205), Department of Psychiatry and Behavioral Sciences, Department of Obstetrics and Gynecology, Feinberg School of Medicine, Northwestern University, Chicago, IL, United States

**Clarice Gorenstein** (57, 155, 165), Department and Institute of Psychiatry, LIM-23, Hospital das Clínicas, University of São Paulo Medical School, São Paulo, SP; Department of Pharmacology, Institute of Biomedical Sciences, University of São Paulo, São Paulo, Brazil

**Isabel A. Gortner** (205), Latin School of Chicago, Chicago, IL, United States

**Francisco Guimarães** (459), Department of Pharmacology, School of Medicine of Ribeirão Preto (FMRP), University of São Paulo (USP), Ribeirão Preto, Brazil

**Girdhari Lal Gupta** (79, 89), Department of Pharmacology, Shobhaben Pratapbhai Patel School of Pharmacy & Technology Management, SVKM'S NMIMS, Mumbai, India

**Eri Hashimoto** (369), Department of Neuropsychiatry, Graduate School of Medicine, Sapporo Medical University, Sapporo, Japan

**Kenji Hashimoto** (393), Division of Clinical Neuroscience, Chiba University Center for Forensic Mental Health, Chiba, Japan

**Antonis Hatzigeorgiadis** (115, 477), Department of Physical Education and Sport Science, University of Thessaly, Trikala, Greece

**Elaine Henna** (155), Department and Institute of Psychiatry, LIM-23, Hospital das Clínicas, University of São Paulo Medical School, São Paulo, SP; Department of Medicine, Discipline of Psychiatry, College of Medical and Health Sciences, Pontifícial Catholic University of Sao Paulo, Sorocaba, Brazil

**Vinícius Medeiros Henriques** (99), School of Medicine, Universidade do Grande Rio, Rio de Janeiro, Brazil

**Lisa D. Hinz** (431), Art Therapy Doctoral Program, Norte Dame de Namur University, Belmont, CA, United States

**Nant Thin Thin Hmwe** (487), School of Nursing and Midwifery, The University of Newcastle, Callaghan, NSW, Australia

**Caroline J. Hollins Martin** (195), School of Health and Social Care, Edinburgh Napier University, Edinburgh, United Kingdom

**Mir-Jamal Hosseini** (505), Departments of Pharmacology and Toxicology, School of Pharmacy; Applied Pharmacology Research Center, Zanjan University of Medical sciences, Zanjan, Iran

**Bing Hu** (357), Department of Neurology, University of California, San Francisco, San Francisco, CA, United States

**Normala Ibrahim** (21), Department of Psychiatry, Faculty of Medicine and Health Sciences, Universiti Putra Malaysia, Serdang, Selangor, Malaysia

**Maria Iglesias-González** (185), Department of Psychiatry and Legal Medicine, School of Medicine, Universitat Autònoma de Barcelona, Hospital Universitari Germans Trias i Pujol, Badalona, Spain

**Samia Joca** (459), Department of BioMolecular Sciences, School of Pharmaceutical Sciences of Ribeirao Preto (FCFRP), University of São Paulo (USP), Ribeirão Preto, Brazil; Aarhus Institute of Advanced Studies (AIAS), Aarhus University, Aarhus, Denmark

**Chiaki Kawanishi** (369), Department of Neuropsychiatry, Graduate School of Medicine, Sapporo Medical University, Sapporo, Japan

**Fatemeh Khakpai** (293), Cognitive and Neuroscience Research Center (CNRC), Islamic Azad University of Medical Sciences, Tehran, Iran

**Ahmad Khosravi** (443), Center for Health Related Social and Behavioural Sciences Research, Department of Epidemiology, Shahroud University of Medical Sciences, Shahroud, Iran

**Jian Kong** (469), Department of Psychiatry, Massachusetts General Hospital, Harvard Medical School, Charlestown, MA, United States

**Ella Kurz** (423), Centre for Mental Health Research, Research School of Population Health, The Australian National University, Canberra, ACT, Australia

**Lorena de la Fuente-Tomás** (3), Department of Psychiatry, School of Medicine, University of Oviedo, Oviedo, Asturias; Biomedical Research Networking Center for Mental Health (CIBERSAM); FINBA (Fundación para la Investigación y la Innovación Biosanitaria del Principado de Asturias), Oviedo, Spain

**Chien-Han Lai** (271), Institute of Biophotonics, National Yang-Ming University, Taipei; PhD Psychiatry & Neuroscience Clinic, Taoyuan, Taiwan

**Moon-Soo Lee** (261), Department of Psychiatry, Korea University, College of Medicine, Seoul, Republic of Korea

**Lilla Lenart** (283), 1st Department of Pediatrics; SE Diabetes Research Group, Semmelweis University, Budapest, Hungary

**David Lester** (211), Psychology Program, Stockton University, Galloway, NJ, United States

**J.C. Leza** (131), Department of Pharmacology and Toxicology, Faculty of Medicine, University Complutense, Cibersam, IUIN, Imas12, Madrid, Spain

**Katarzyna A. Lisowska** (143), Department of Pathophysiology, Medical University of Gdańsk, Gdańsk, Poland

**Cheng Long** (337), School of Life Sciences, South China Normal University; South China Normal University-Panyu Central Hospital Joint Laboratory of Translational Medical Research, Panyu Central Hospital, Guangzhou, PR China

**Vija B. Lusebrink** (431), University of Louisville, Louisville, KY, United States

**Munn-Sann Lye** (21), Department of Population Medicine, Faculty of Medicine and Health Sciences, Universiti Tunku Abdul Rahman, Sungai Long Campus, Cheras, Kajang, Selangor, Malaysia

**K. MacDowell** (131), Department of Pharmacology and Toxicology, Faculty of Medicine, University Complutense, Cibersam, IUIN, Imas12, Madrid, Spain

**J.L.M. Madrigal** (131), Department of Pharmacology and Toxicology, Faculty of Medicine, University Complutense, Cibersam, IUIN, Imas12, Madrid, Spain

**Colin R. Martin** (195), Institute for Clinical and Applied Health Research, University of Hull, Hull, United Kingdom

**Donel Martin** (447), Black Dog Institute, School of Psychiatry, University of New South Wales, Randwick, NSW, Australia

**David Martín-Hernández** (345), Department of Pharmacology and Toxicology, School of Medicine, CIBERSAM-ISCIII, Universidad Complutense de Madrid, Madrid, Spain

**Yutaka J. Matsuoka** (497), Division of Health Care Research, Behavioral Science and Survivorship Research Group, Center for Public Health Sciences, National Cancer Center Japan, Tokyo, Japan

**Gabrielle A. Mesches** (205), Department of Psychiatry and Behavioral Sciences, Department of Obstetrics and Gynecology, Feinberg School of Medicine, Northwestern University, Chicago, IL, United States

**Kristin Mignogna** (515), Department of Mental Health, Bloomberg School of Public Health, Johns Hopkins University, Baltimore, MA; Department of Psychiatry and Behavioral Sciences, Johns Hopkins University School of Medicine, Baltimore, MD, United States

**Afzal Misrani** (337), School of Life Sciences, South China Normal University; School of Life Sciences, Guangzhou University, Guangzhou, PR China

**Ioannis D. Morres** (115, 477), Department of Physical Education and Sport Science, University of Thessaly, Trikala, Greece

**Seyed Abbas Mousavi** (443), Department of Psychiatry, Mazandaran University of Medical Sciences, Sari, Iran

**Joshua K.M. Nan** (431), Department of Social Work, Faculty of Social Science, Hong Kong Baptist University, Kowloon Tong, Kowloon, Hong Kong

**Stevan Nikolin** (447), Black Dog Institute, School of Psychiatry, University of New South Wales, Randwick, NSW, Australia

**Trevor R. Norman** (301), Department of Psychiatry, Austin Hospital, University of Melbourne, Heidelberg, VIC, Australia

**Ryo Okubo** (497), Department of Clinical Epidemiology, Translational Medical Center, National Center of Neurology and Psychiatry, Kodaira, Tokyo, Japan

**Nikita Patil** (79), Department of Pharmacology, Shobhaben Pratapbhai Patel School of Pharmacy and Technology Management, Mumbai, India

**C. Pellegrino** (309), Inmed, INSERM, Aix-Marseille University, Marseille, France

**Man-Man Peng** (69), Department of Social Work and Social Administration, Faculty of Social Sciences, University of Hong Kong, Hong Kong SAR, China

**Marta P. Pereira** (345), Department of Molecular Biology and Center of Molecular Biology Severo Ochoa, Universidad Autonoma de Madrid, Madrid, Spain

**Krzysztof Pietruczuk** (143), Department of Pathophysiology, Medical University of Gdańsk, Gdańsk, Poland

**Tanvi Pingale** (89), Department of Pharmacology, Shobhaben Pratapbhai Patel School of Pharmacy & Technology Management, SVKM'S NMIMS, Mumbai, India

**María Dolores Ponce-Regalado** (401), Laboratorio de investigaciones en Biociencias, Departamento de Ciencias de la Salud, Centro Universitario de los Altos, Universidad de Guadalajara, Guadalajara, Mexico

**Somayeh Ramezani** (443), Sexual Health and Fertility Research Center, Bahr Hospital, Shahroud University of Medical Sciences, Shahroud, Iran

**Mao-Sheng Ran** (69), Department of Social Work and Social Administration, Faculty of Social Sciences, University of Hong Kong, Hong Kong SAR, China

**Lubova Renemane** (175), Department of Psychiatry and Narcology, Riga Stradins University, Riga, Latvia

**A. Rezzag** (309), Inmed, INSERM, Aix-Marseille University, Marseille, France

**C. Rivera** (309), Inmed, INSERM, Aix-Marseille University, Marseille, France; Neuroscience Center, University of Helsinki, Helsinki, Finland

**Yasodha Maheshi Rohanachandra** (45), Department of Psychiatry, Faculty of Medical Sciences, University of Sri Jayewardenepura, Gangodawila, Sri Lanka

**Peijing Rong** (469), Institute of Acupuncture and Moxibustion, China Academy of Chinese Medical Sciences, Beijing, China

**Antonio Reis de Sá Junior** (57), Department of Internal Medicine, Health Sciences Center, Federal University of Santa Catarina, Florianopolis, SC, Brazil

**Elizabeth M. Sajdel-Sulkowska** (33), Department of Psychiatry, Harvard Medical School & Brigham and Women's Hospital, Boston, MA, United States

**Alberto Salazar-Juárez** (401), Laboratorio de Neurofarmacología Conductual, Microcirugía y Terapéutica Experimental, Subdirección de Investigaciones Clínicas, Instituto Nacional de Psiquiatría "Ramón de la Fuente", México DF, México

**Amanda Sales** (459), Department of Pharmacology, School of Medicine of Ribeirão Preto (FMRP), University of São Paulo (USP), Ribeirão Preto, Brazil

**Ariandra Sartim** (459), Department of BioMolecular Sciences, School of Pharmaceutical Sciences of Ribeirao Preto (FCFRP), University of São Paulo (USP), Ribeirão Preto, Brazil

**Maria Semkovska** (527), Department of Psychology, University of Southern Denmark, Odense M, Denmark

**Gabriela P. Silote** (459), Department of BioMolecular Sciences, School of Pharmaceutical Sciences of Ribeirao Preto (FCFRP), University of São Paulo (USP), Ribeirão Preto, Brazil; Department of Clinical Medicine, Translational Neuropsychiatry Unit (TNU), Aarhus University, Aarhus, Denmark

**Bruno Pinatti Ferreira de Souza** (121), Departamento e Instituto de Psiquiatria, Hospital das Clinicas, Faculdade de Medicina da Universidade de São Paulo, São Paulo, SP, Brazil

**Guilherme de Souza Paulo Filho** (99), School of Medicine, Universidade do Grande Rio, Rio de Janeiro, Brazil

**Łukasz P. Szałach** (143), Department of Pathophysiology, Medical University of Gdańsk, Gdańsk, Poland

**Monika Talarowska** (12), Department of Psychology and Individual Differences, Institute of Psychology, University of Lodz, Lodz, Poland

**M. Tessier** (309), Inmed, INSERM, Aix-Marseille University, Marseille, France

**Yin-Yee Tey** (21), In personal capacity

**Yannis Theodorakis** (115, 477), Department of Physical Education and Sport Science, University of Thessaly, Trikala, Greece

**Wataru Ukai** (369), Department of Neuropsychiatry, Graduate School of Medicine, Sapporo Medical University, Sapporo, Japan

**Cristina Ulecia-Morón** (345), Department of Pharmacology and Toxicology, School of Medicine, CIBERSAM-ISCIII, Universidad Complutense de Madrid, Madrid, Spain

**Ece Turkyilmaz Uyar** (251), Department of Psychiatry, Okmeydani Teaching and Research Hospital, Istanbul, Turkey

**Jelena Vrublevska** (175), Department of Psychiatry and Narcology, Riga Stradins University, Riga, Latvia

**Yuan-Pang Wang** (57, 155, 165), Department and Institute of Psychiatry, LIM-23, Hospital das Clínicas, University of São Paulo Medical School, São Paulo, SP, Brazil

**Gregers Wegener** (459), Department of Clinical Medicine, Translational Neuropsychiatry Unit (TNU), Aarhus University, Aarhus, Denmark

**Marta Weinstock** (325), Institute of Drug Research, School of Pharmacy, The Hebrew University Medical Centre, Ein Kerem, Jerusalem, Israel

**Georgia Wilson** (469), Department of Psychiatry, Massachusetts General Hospital, Harvard Medical School, Charlestown, MA, United States

**Jinzhong Xiao** (497), Next Generation Science Institute, Morinaga Milk Industry Co., Ltd., Zama City, Kanagawa, Japan

**Jun-ichi Yamaguchi** (381), Research Headquarters, Taisho Pharmaceutical Co., Ltd., Saitama, Japan

**Kai Zhang** (393), Department of Psychiatry, Chaohu Hospital, Anhui Medical University, Hefei, China

# Foreword

Depression is a significant mental health issue that is also surprisingly common. Pervasive beyond boundaries of class, status, ethnicity, and physical health, depression can affect anybody and its impact can be debilitating and in extreme cases, life limiting. The notion of melancholia and depression has been described in historical texts for millennia. Indeed, depression has been a focus for clinical intervention for more than 100 years, for example, by the use of psychoanalysis. Despite this it is somewhat surprising that the incidence and prevalence of depression and, in particular, the major depressive disorders, has only been explored systematically over the past 50 years. A key feature of such observations is that estimations of prevalence of depression have increased over time. Approaches to clinical research with regard to depression have also broadened, as attempts are made to understand this psychobiological gestalt with a myriad of potential contributory factors, including those that are of social, psychological, biological, and genetic origin. This represents a synergy and sensitivity to the underlying complexity of this condition. The goal of such research enterprise is of course, primarily, to foster good mental health by enhancing prevention and improving interventions, to improve outcomes notably the reduction of self-harm. The challenge of enhancing outcomes is influenced by many factors but key is contextualizing treatment interventions that are sensitive and tailored to the individual needs of the person concerned, informed, and justified by a robust and research-informed evidence base.

Another key factor is the consideration of the relationship of depression to other clinical presentations, whether they are circumscribed as primarily mental health or manifestations of physical disease states.

Importantly, understanding depression in a research-informed and person-centered manner is crucial for extrapolating from the neuroscience paradigm of understanding normative function by contrasting with dysfunction. Contemporary insights in the depression research literature thus then enable us not only to understand with greater clarity the processes and interactions thereof that lead to depression, but also the processes that promote positive mental health and well-being.

My experience working as a psychiatrist in clinical practice, as with others working clinically with people experiencing mental health problems, we are only too aware of the impact of the disorder on everyday functioning and well-being, not only on the individual themselves, but invariably on their family and loved ones. It operates not only on an individual level but on a societal one as well.

Though we are becoming ever more innovative and sophisticated in our approaches to patient care, we must always remember that the experience for each person with the diagnosis of major depression is unique.

Reflecting on the above, it is my great pleasure to offer this Foreword for this exciting new book on depression. Professors Preedy and Martin and Drs. Rajendram, Hunter, and Patel have done a significant service to clinicians and researchers alike in bringing together leading experts in depression research to produce *The Neuroscience of Depression: Features, Diagnosis, and Treatment*. Comprehensively addressing key research in depression across 70 chapters, this contemporary volume covers key "hot topics" ranging from neurodevelopmental theories and clinical staging models, through to screening and assessment, the use of novel neuromarkers and MRI, promising "horizon scanning" pharmacological interventions, and comorbidities with a range of physical disease states, all within the context of diagnosis and treatment. I believe this extensive, evidence-anchored, integrated, and contemporary account of depression will be an invaluable resource for those of us working in the field, irrespective of whether the venue be the clinical consultation or the research laboratory.

**Sandra Scott**
London, May 2020

**Dr. Sandra Scott** trained at the School of Medicine at St. Bartholomew's Hospital, London, United Kingdom, is a consultant Forensic Psychiatrist and a member of the

Royal College of Psychiatrists. By virtue of working in the NHS for over 19 years Dr. Scott has enjoyed a high profile in the media bringing psychological and psychiatric concepts to a broad public audience through her work on television productions such as *Confidence Lab*, *Celebrity Big Brother*, *Big Brother's 2,3,4, and 5* and *Tomorrow's World*. She has also assessed contributors and provided psychological support during and post filming for participants for a wide variety of television programs including *I'm a Celebrity, Get Me Out of Here!*, *Love Island Series 5 and 6*, *71 Degrees North*, *The Door*, *Supersize Superskinny*, *Dancing On Ice*, *Hell's Kitchen*, *Fat Teens Can't Hunt*, *Celebrity Fit Club*, *Unanimous*, *Temptation Island*, and *Dispatches*.

# Preface

*The Neuroscience of Depression* focuses on *Features, Diagnosis, and Treatment*. Depression, also called major depressive disorder or clinical depression, is a mood disorder, with a persistent feeling of sadness and loss of interest. The symptoms of depression are far ranging from mild to severe and encompass disturbed sleep, lack of energy, loss of libido, changes in appetite, continuous low mood, feeling anxious, having low esteem, no motivation in life and suicidal thoughts, and feeling helpless. Over time, this may result in avoiding contact with family and friends with difficulties to motivate oneself to work. Depression has many causes, but ultimately can affect anyone in their life, from childhood events, after giving birth, or due to other medical conditions such as cancer, heart disease, and Parkinson's disease. As discussed in these volumes, there are different causes of depression from cellular, molecular to genetic changes, as well as social and life events. Therefore, various diagnostic tools can be used, resulting in psychological, pharmacological, or alternate treatments for moderate-to-severe depression. New tools and approaches are constantly being developed, with the goal to support mental health and well-being. Furthermore, studies showing the beneficial effects of plant or natural extracts provide the foundation for further rigorous studies in clinical trials.

The book contains five sections. Part 1 presents coverage on General Aspects of depression, which includes chapters on clinical staging in depression, theories of depression, depression during and after pregnancy, depression in Alzheimer's disease, depression in college students due to obesity, disasters and traumatic events, and the relationship between depression and alcohol use among students. Part 2: Biomarkers and Diagnosis includes chapters on scoring tools for depression including, The Beck Depression Inventory, The Hamilton Depression Rating Scale, The Patient Health Questionnaire, Depression Anxiety Stress Scales, methods of neuroimaging in depression, Thioredoxin as a biomarker and neural MRI markers. Part 3: Pharmacological Treatments for Depression includes chapters on drugs used for depression such as citalopram, agomelatine, bumetanide, ketamine, mirtazapine, angiotensin receptor 1 blockade, and the role of cannabinoid CB1 receptors. In Part 4: Counselling, Psychotherapy and Behavioural Treatments for Depression includes chapters on art therapy, cognitive emotional training, online web-based programs, and mindfulness-based cognitive therapy. Part 5: Other Aspects of Treatment: Specific Groups, Monitoring and Novel Regimens includes chapters on exercise for depression, transcutaneous vagus nerve stimulation, acupressure and depression, bifidobacterium breve A-1 probiotics, Coenzyme Q use in depression, and gene expression and DNA methylation changes in depression.

**Each chapter has key facts, Summary Points and a mini-dictionary**.

The book is designed for psychologists, psychiatrist, behavioral scientists, neurologists, pathologist, counsellors, and specialists, as well as other health care workers and scientists. Contributions are from leading national and international experts including those from world renowned institutions. It is suitable for undergraduate to graduates, lecturers, senior lecturers and professors as well as libraries.


**Colin R. Martin**
University of Hull

**Lan-Anh Hunter**
School of Primary Care, Thames Valley,
Health Education England

**Vinood B. Patel**
University of Westminster

**Victor R. Preedy**
King's College, London

**Rajkumar Rajendram**
King's College, London

Part I

# Depression: Introductory chapters

Chapter 1

# Clinical staging in depression

Lorena de la Fuente-Tomás[a,b,c] and María Paz García-Portilla[a,b,c]

[a]*Department of Psychiatry, School of Medicine, University of Oviedo, Oviedo, Asturias, Spain;* [b]*Biomedical Research Networking Center for Mental Health (CIBERSAM), Oviedo, Spain;* [c]*FINBA (Fundación para la Investigación y la Innovación Biosanitaria del Principado de Asturias), Oviedo, Spain*

## List of abbreviations

| | |
|---|---|
| **AD** | antidepressants |
| **CBASP** | cognitive behavioral analysis system of psychotherapy |
| **CBT** | cognitive behavioral therapy |
| **DM-TRD** | Dutch measure for quantification of treatment resistance in depression |
| **DSM-IV** | Diagnostic and Statistical Manual of Mental Disorders, 4th edition |
| **ECT** | electroconvulsive therapy |
| **GAF** | Global Assessment of Functioning |
| **MDD** | major depressive disorder |
| **MGH-S** | Massachusetts General Hospital staging method |
| **MSM** | Maudsley staging method |
| **TAU** | treatment as usual |
| **TRD** | treatment-resistant depression |

## Introduction

Clinical staging is a system of classification for determining the position of an individual along a continuum of severity based on clinical phases, called stages. It is widely used in medicine, particularly in fields such as cardiology, endocrinology, and oncology. Clinical staging is a simpler and more refined system than conventional classifications, where the emphasis is on potential changes and the longitudinal nature of diseases. Furthermore, staging models are not just about symptom severity but also involve disease extension.

The concept of clinical staging incorporates five assumptions (Scott et al., 2013):

1. Treatment of earlier stages is associated with better initial response or prognosis.
2. Earlier treatments have a more favorable risk–benefit ratio than later treatments.
3. The impact of early intervention can be assessed against changes in the stage distribution of the disease over time.
4. The provision of stage-appropriate treatment modifies the individual's risk of disease progression.
5. As knowledge on the underlying disease mechanisms develops, more robust clinicopathological models of staging become achievable, within which "bio-signatures" may be characterized to either validate or redefine stages.

In psychiatry, clinical staging was first proposed by Fava and Kellner (1993) and was developed for schizophrenia, depression, mania, and panic disorders. Since then, there have been important developments in the literature (Cosci & Fava, 2013; de la Fuente-Tomas et al., 2019; McGorry, Hickie, Yung, Pantelis, & Jackson, 2006; Vieta, Reinares, & Rosa, 2011). It has been suggested that staging models may increase the number of patients treated early and adequately according to disease course, which may delay the onset of the disease or prevent its progression (Hetrick et al., 2008). Furthermore, staging models may guide the choice of treatment according to disease course (Guidi, Tomba, Cosci, Park, & Fava, 2017). In this chapter, we focus on proposals for clinical staging of depression.

## Clinical staging in depression

Despite the fact that some patients with major depressive disorders (MDD) experience only one lifetime depressive episode, many others exhibit illness characteristics consistent with a progressive disorder. For example, 20%–35% develop a chronic disorder over time (Boschloo et al., 2014). Several studies have pointed to MDD as a progressive condition, specifically based on the concept of allostatic load and kindling theory (Ferensztajn, Remlinger-Molenda, & Rybakowski, 2014). However, there is no conclusive evidence of this (Dodd, Berk, Kelin, Mancini, & Schacht, 2013; Verduijn et al., 2015) (Fig. 1).

Staging models of MDD have been proposed with two different purposes: (1) staging of disease progression (Cosci & Fava, 2013; Fava & Kellner, 1993; Fava & Tossani, 2007; Hetrick et al., 2008; Verduijn et al., 2015) and (2) staging of treatment resistance (Fava, 2003; Fekadu et al., 2009; Peeters et al., 2016; Thase & Rush, 1997; van Diermen et al., 2018).

The Neuroscience of Depression. https://doi.org/10.1016/B978-0-12-817933-8.00026-8

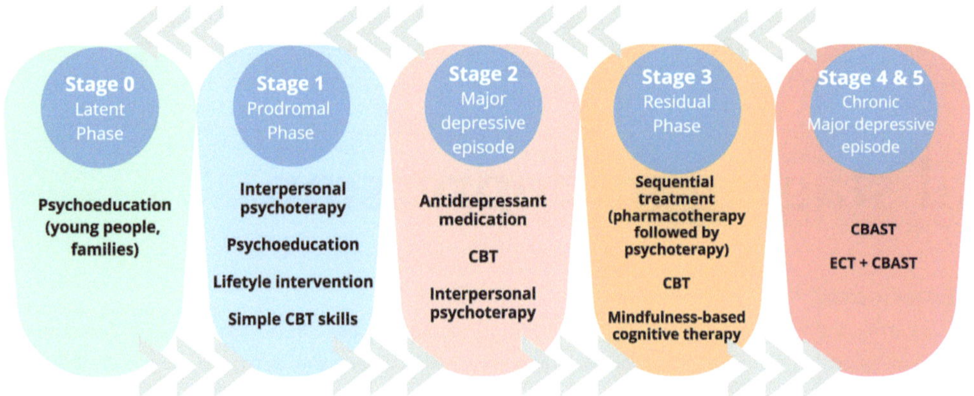


**FIG. 1** Interventions according to clinical stages of MDD. Treatment strategies suggested by different authors according to the clinical stages of MDD.

## Clinical staging and progression in depression

### First clinical staging model proposal for unipolar depression

The first model for MDD was developed in 1993 (Fava & Kellner, 1993) and was updated some years later (Cosci & Fava, 2013; Fava & Tossani, 2007). The latest version of the model divides the course of MDD into five clinical stages, which are defined by severity of affective psychopathology and number of previous episodes. The five stages of the model consist of one preclinical stage (prodromal phase) and four clinical stages (stage 2, first episode; stage 3, residual phase; stage 4, dysthymia disorder or recurrent depression; and stage 5, chronic major depressive episode).

The first stage (prodromal phase) is characterized by generalized anxiety, anhedonia, irritability, fatigue, and sleep disorders (initial or delayed insomnia). The inclusion of a prodromal stage is in line with the concept of allostatic load defined as "the cost of chronic exposure to fluctuating or heightened neural activation" (Fava & Tossani, 2007). Stage 1 is divided into 1a, characterized by the presence of risk factors without depressive symptoms, and 1b, characterized by subsyndromal depressive symptoms but not achieving the severity of a depressive episode. Stage 2 is when the first episode of depression occurs. Stage 3 is the residual phase, divided into two clinical stages according to Fava and Tossani (2007): remission with no residual symptoms (stage 3a) and a diagnosis of dysthymia (stage 3b), and divided into three clinical stages according to Cosci and Fava (2013): remission with no residual depressive symptoms (stage 3a), remission with residual mood symptoms (depressed mood, guilt, hopelessness) (stage 3b), and a diagnosis of dysthymia (stage 3c). Stage 4, divided into two stages, is characterized by recurrent MDD, consisting of at least two episodes at least 2 months apart before return to a phase of regular functioning (stage 4a) and a diagnosis of double depression if dysthymia was present in the residual phase (stage 4b). Finally, stage 5 refers to patients who do not reach full remission, which means a chronic course lasting at least 2 years (see Table 1).

### Staging model proposed by Hetrick

Building on the notion that there had been no interventions developed for the different depression stages, Hetrick et al. (2008) suggested a staging model based on McGorry's work (McGorry et al., 2006). This model consists of eight stages and takes into account cognition and functioning, starting with a latent phase with an increased risk of anxiety or depressive disorder but no current symptoms (stage 0). In stage 1, individuals may experience mild or nonspecific symptoms of anxiety or depression, including cognitive deficits and mild functional decline (stage 1a) or moderate but subthreshold symptoms of anxiety or depression, together with moderate cognitive and functional decline. In stage 2, subjects experience the first episode of MDD along with moderate to severe cognitive and functional decline. Then there may be a residual phase without complete remission from the first episode (stage 3a), recurrence or relapse with a period of remission (stage 3b), or multiple relapses (stage 3c). Finally, stage 4 refers to severe, persistent, or unremitting illness.

In 2015, a first and only attempt was made to empirically validate a clinical staging model for MDD (Verduijn, Milaneschi, van Hemert, et al., 2015). The authors tried to test the staging model in its eight stages, and participants with a broader range of conditions were therefore included (MDD, anxiety disorders, and psychotic disorders). Statistically significant differences were found across preclinical stages (stages: 0, 1a, 1b, 2) but not across more progressive stages of full-threshold MDD. It was concluded that MDD staging based on number of previous episodes seems to be less powerful than staging based on illness duration.

**TABLE 1** **Staging models for depression.**

| | Fava and Kellner (1993) | Fava and Tossani (2007) | Cosci and Fava (2013) | Hetrick et al. (2008) |
|---|---|---|---|---|
| Stage 0 | | | | *Latent phase*<br>Increased risk of anxiety or depressive disorder; asymptomatic |
| Stage 1 | *Prodromal phase* | *Prodromal phase* | *Prodromal phase* | *Prodromal phase* |
| Stage 1a | | No depressive symptoms | No depressive symptoms | Mild or nonspecific symptoms of anxiety or depression, neurocognitive deficits, and functional change |
| Stage 1b | | Minor depression | Minor depression | Ultrahigh risk: moderate but subthreshold symptoms with moderate cognitive and functional decline (GAF < 70) |
| Stage 2 | *First episode of MDD* | *First episode of MDD* | *First episode of MDD* | *First episode of MDD*<br>Moderate to severe symptoms, cognitive deficits, and functional decline (GAF 30–50) |
| Stage 3 | *Residual phase* | *Residual phase* | *Residual phase* | *Residual phase* |
| Stage 3a | | No depressive symptoms | No depressive symptoms | Incomplete remission from first episode |
| Stage 3b | | Dysthymia | Mood symptoms | Recurrence or relapse of depressive disorder |
| Stage 3c | | | Dysthymia | Multiple relapses |
| Stage 4 | | | | *Severe, persistent, or unremitting illness* |
| Stage 4a | Dysthymia | Recurrent depression | Recurrent depression | |
| Stage 4b | Recurrent major depression | Double depression | Double depression | |
| Stage 5 | *Chronic MDD* | *Chronic MDD* | *Chronic MDD* | |

Clinical staging models applied to depression over the last three decades.

## Potential interventions according to clinical stages

Some authors have suggested that staging could prevent or delay the progression of depression thanks to its potential to adapt therapeutic interventions to specific phases of the disease (Guidi et al., 2017; Hetrick et al., 2008). For the latent phase, it has been shown that there is a need to improve mental health literacy and psychoeducation for young people and their families in order to identify early signals (Hetrick et al., 2008).

In the prodromal phase, the importance of identifying individuals at high risk has been suggested (Malda et al., 2019; McGorry & Mei, 2018), and one study has proved the clinical benefits of early intervention in recurrent depression, showing that early treatment (a combination of pharmacotherapy and interpersonal psychotherapy) significantly shortened episodes by approximately 4–5 months (Kupfer, Frank, & Perel, 1989). The importance of developing psychometric instruments capable of measuring small changes has also been suggested (Guidi et al., 2017).

For stage 2, the joint use of psychotherapy and antidepressant drugs has been shown to have limited effectiveness in terms of relapse prevention compared with antidepressant drug treatment alone in the acute phase of a depressive episode (Biesheuvel-Leliefeld et al., 2015). However, there is evidence in favor of cognitive behavioral therapy (CBT) and interpersonal psychotherapy (Guidi et al., 2017).

Stage 3 is more challenging and involves multiples scenarios, including failure to achieve full remission or the occurrence of relapse that could lead to multiple episodes. Furthermore, it is a phase characterized by residual symptoms (see Table 1). A recent meta-analysis has shown that sequential administration of pharmacotherapy followed by psychotherapy seems to reduce relapses and recurrence in MDD (Guidi, Tomba, & Fava, 2016). The effectiveness of this sequential strategy appear to be associated with decreased residual symptoms and/or development of psychological well-being and coping skills (Guidi et al., 2016). Guidi et al. (2016) have proposed six steps for implementing this sequential treatment. First, they recommend beginning by assessing the patient 3 months after initiating antidepressant drug treatment, paying special attention to residual symptoms. Secondly, cognitive behavioral treatment for residual symptoms may be followed by mindfulness-based cognitive therapy, shown to be efficacious for relapse prevention, particularly in those with pronounced residual symptoms (Kuyken et al., 2016). The next two steps focus on tapering off antidepressant drugs as slowly as possible, and the last step consists of administering psychotherapy and carefully assessing the patient 1 month after drug discontinuation.

In stages 4 and 5, the depressive disorder becomes persistent (defined as a minimum duration of 2 years), including the four diagnostic groups: dysthymia, chronic major depression, recurrent major depression with incomplete remission between episodes, and double depression. There is supporting evidence that the cognitive behavioral analysis system of psychotherapy (CBASP) is effective in the treatment of chronic depression (Negt et al., 2016). Furthermore, it has been found to be effective after electroconvulsive therapy (ECT) for the treatment of severe persistent depressive disorder (Brakemeier et al., 2014). Finally, we found a recent systematic review that analyzed the effects of pharmacological and psychological treatments (alone or combined) in comparison with placebo or treatment as usual (TAU) for persistent depressive disorder. The beneficial effects of continued or maintenance pharmacotherapy are uncertain due to the clinical heterogeneity of the samples and the moderate or high risk of bias in the studies. For the rest of the comparisons, the body of evidence was too small, and the authors suggest that further high-quality trials should be conducted for psychological interventions (Machmutow et al., 2019).

## Clinical staging and treatment-resistant depression

The case of treatment-resistant depression (TRD) merits special focus, considering that it is a relatively common phenomenon that requires significant human resources and constitutes a public health problem (Fekadu et al., 2009).

Despite its clinical importance, more than 40 years after it was first defined, there is no universally accepted definition of TRD. In the last two decades, several authors have proposed the use of a dimensional description based on at least the number and type of failed treatments (in terms of dose and duration of the trial), rather than a categorical one, following a staging model approach. In fact, since 1997, several models have been developed to provide valid and reliable dimensional classifications of this phenomenon, as well as to improve the ability of doctors to predict treatment outcomes and prognosis.

In the following, we summarize the main characteristics of the best proposed models. The older models have the disadvantages of having been developed theoretically without empirical validation and being mainly unidimensional, based on treatment failure. On the contrary, the newer ones incorporate other dimensions along with treatment variables and have been at least partially validated (see Fig. 2 and Table 2). Each model represents an effort to improve our ability to allocate patients their best treatment options and to make a more accurate diagnosis and prognosis.

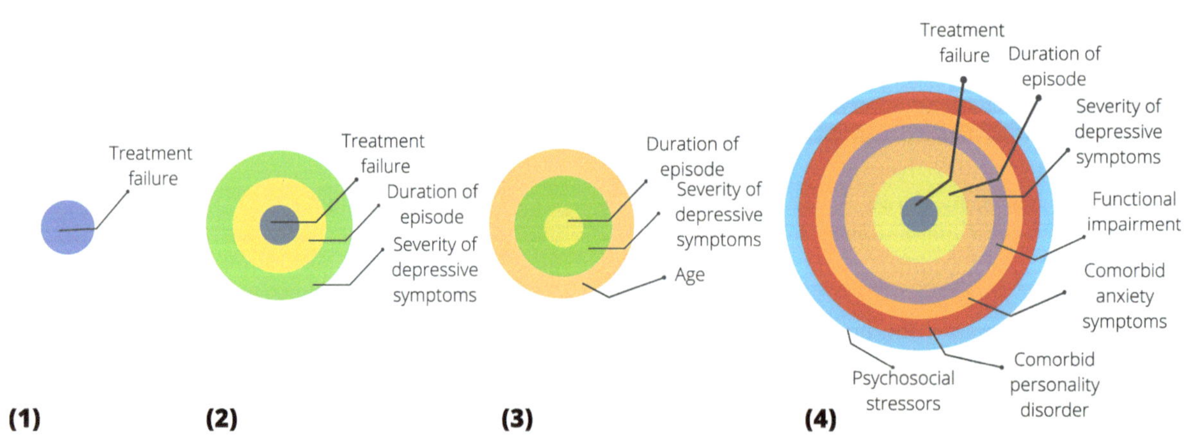


FIG. 2 (1) The Thase and Rush staging model, and he Maudsley staging method (MSM). (2) The Maudsley staging method (MSM). (3) The Electroconvulsive Therapy - Maudsley staging method (ECT-MSM ). (4) The Dutch measure for quantification of treatment resistance in depression (DM-TRD).

## The Thase and Rush staging model

In 1997, Thase and Rush proposed a hierarchical staging model based on the number of failed trials and the class of antidepressants that failed, implying that the antidepressants utilized at later stages had greater efficacy than those used in the initial phases of the illness (see Table 2). This model classifies patients into stages from I (failure of at least one adequate trial of one major class of antidepressant) to V (stage IV resistance as well as failure of bilateral ECT). Failure of a trial with a tricyclic antidepressant corresponds to stage III and a monoamine oxidase inhibitor to stage IV.

This model has been criticized by Fava (2003) and Fekadu et al. (2009) because its conceptual basis on the use of antidepressants highly differs from daily clinical practice, it neglects the role of optimization/augmentation strategies, and there is no clear evidence of the superiority of any of them over the others.

## The Massachusetts General Hospital staging method

In 2003, Fava proposed this model that considers the number of failed antidepressant trials and the intensity/optimization of each trial without making assumptions about antidepressant hierarchies (see Table 2). The Massachusetts General Hospital staging method (MGH-S) model confers one point to each failed adequate antidepressant trial and 0.5 point per trial per optimization/augmentation strategy. In addition, it confers special weight to ECT failure by adding three points to the overall score. The MGH-S generates a score that reflects the level of resistance to the treatment, the greater the score, the greater the treatment resistance.

## The Maudsley staging method

The multidimensional staging model of Fekadu et al. was developed by applying a theoretical framework to empirical data from patients with TRD. This model includes three factors: duration of the current depressive episode, severity at the onset of the episode, and types of treatment failure (see Table 2). These dimensions are scored according to a series of operational criteria, and their scores are added to provide a total TRD severity score (Fekadu et al., 2009). Thus, the Maudsley staging method (MSM) makes it possible to differentiate between different degrees of severity in the TRD group.

Concerning the duration of the episode, the MSM is differentiated into three levels, ranging from acute (≤12 months, score = 1) to chronic (>24 months, score = 3). With respect to the severity of the episode, the authors divide this into subsyndromal (score = 1) and syndromal [ratings from 2 to 5 based on the degree of the severity and the presence of psychotic symptoms, mild (2), moderate (3), severe without psychotic symptoms (4), and severe with psychotic symptoms (5)]. Finally, the dimension of treatment failures is made up of three therapeutic strategies: antidepressants, augmentation, and electroconvulsive therapy (ECT). The first one, antidepressants, refers to the number of antidepressants used in the episode and is rated in levels, from level 1 (1–2 antidepressants, score = 1) to level 5 (more than 10 antidepressants, score 5). The other two areas refer to the use of these two interventions (augmentation and ECT) and are scored identically, 0 = not used and 1 = used, without giving particular weight to ECT. Thus, the total score ranges from 3 to 15; the higher the score, the greater the severity of the TRD. Furthermore, the authors propose a categorical interpretation of the total score: mild TRD (scores = 3–6), moderate (scores = 7–10), and severe (scores = 11–15). Also, a descriptive characterization of the three dimensions can be given.

Although the original authors found reasonable face and predictive validities in a sample of inpatients with TRD, its predictive validity has been questioned in the case of patients receiving ECT (van Diermen et al., 2018). To avoid its limitations, the author proposes an adapted version for more accurate prediction of ECT outcome that includes age along with the duration of the episode and severity of depressive symptoms. The operational criteria for scoring age are: under 50 years = 2, between 50 and 65 years = 1, 65 years and older = 0. Duration of episode is scored as in the MSM (≤12 months, score = 1; 13–24 months, score = 2; >24 months, score = 3) while the scoring system for severity of depressive symptoms is reversed (severe with psychosis = 1, severe without psychosis = 2, moderate = 3). The total score of this adapted version ranges between 2 and 8; the lower the score, the better the effectiveness of the ECT.

Among the advantages of this model, different authors have pointed out its ease of use in daily clinical practice and its good flexibility, the fact that it takes into account the complexity and multidimensionality of TRD, as well as the possibility of identifying different degrees of severity in the TRD group, and the absence of distinction between classes of antidepressants when switching (Fekadu et al., 2009; van Diermen et al., 2018). One further advantage is an open-access paper published in 2018 (Fekadu, Donocik, & Cleare, 2018) that provides detailed guidelines for correctly using and completing the MSM. It includes the definition of TRD that underlies the model, the operational criteria to accurately rate each dimension, and the equivalence between scores on the most used instruments for rating depression and the severity categories obtained with the MSM.

**TABLE 2** Dimensions of the staging models proposed for treatment-resistant depression.

| Model | Thase and Rush | MGH-S | MSM | ECT-MSM | DM-TRD |
|---|---|---|---|---|---|
| **Authors** | **Thase and Rush (1997)** | **Fava (2003)** | **Fekadu et al. (2009)** | **van Diermen et al. (2018)** | **Peeters et al. (2016)** |
| Treatment failure | Number of failed AD trials | Number of failed AD trials | Number of failed AD trials | | Number of failed AD trials |
| | Class of AD that failed | Optimization/augmentation strategies | Augmentation strategies | | Augmentation/ combination strategies |
| | ECT | ECT | ECT | | ECT |
| | | | | | Psychotherapy |
| | | | | | Intensified treatment |
| Duration of episode | | | Acute | Acute | Acute |
| | | | Subacute | Subacute | Subacute |
| | | | Chronic | Chronic | Chronic |
| Severity of depressive symptoms | | | Subsyndromal | | Subsyndromal |
| | | | Syndromal<br>– Mild<br>– Moderate<br>– Severe without psychosis<br>– Severe with psychosis | Syndromal<br>– Moderate<br>– Severe without psychosis<br>– Severe with psychosis | Syndromal<br>– Mild<br>– Moderate<br>– Severe without psychosis<br>– Severe with psychosis |
| Age | | | | Under 50 years | |
| | | | | Between 50 and 65 | |
| | | | | Over 65 years | |
| Functional impairment | | | | | No impairment |
| | | | | | Impairment<br>– Mild<br>– Moderate<br>– Severe |

| | | | | | |
|---|---|---|---|---|---|
| Comorbid anxiety symptoms | | | | | Not present |
| | | | | | Present<br>– Not meeting DSM-IV criteria<br>– Meeting DSM-IV criteria $\geq 1$ anxiety disorder |
| Comorbid personality disorder | | | | | Not present |
| | | | | | Present<br>– Not based on a formal interview<br>– Based on a formal interview |
| Psychosocial stressors | | | | | None |
| | | | | | At least one psychosocial stressor |

Dimensions and profilers of the staging models proposed for treatment-resistant depression over the last two decades.

## The Dutch measure for quantification of treatment resistance in depression

In 2016, Peeters et al. developed the Dutch measure for quantification of treatment resistance in depression (DM-TRD) model by redefining the treatment failures dimension of the MSM model and adding the following new dimensions to improve its predictive validity for treatment outcome of TRD (see Table 2):

1. Maximum functional impairment during the current episode, divided into four categories according to GAF ratings (GAF: 90–100, score = 0; 60–90 = 1; 30–60 = 2; and < 30 = 3).
2. Comorbid anxiety symptoms: not present = 0; present, but without DSM-IV diagnosis = 0.5; and present with at least one DSM-IV anxiety disorder = 1.
3. Comorbid personality disorder: not present = 0; present, but not based on formal interview = 0.5; present and based on formal interview = 1.
4. Presence of psychosocial stressors based on Axis IV of DSM-IV: none = 0; at least one psychosocial stressor = 1.

The redefinition of the treatment failures dimension also consists of psychotherapy treatment (not used = 0; supportive therapy = 0.5; one empirically supported psychotherapy = 1; at least two empirically supported therapies = 2) and intensified treatment (not used = 0; day patient for at least 12 weeks, 3 days/week = 1; inpatient for at least 4 weeks = 2), as well as establishing four levels instead of two in the augmentation/combination area (level 0, not used = 0; level 1, 1–2 medications = 1; level 2, 3–4 medications = 2; level 3, 5–6 medications = 3). In addition, the use of ECT treatment is taken into consideration only if at least eight sessions were administered. All treatment criteria refer to the current episode. This model provides a total TRD score consisting of the sum of the scores on all dimensions, ranging from 2 to 27; the higher the score, the worse the treatment outcome.

The authors state that, after further validation with prospective and larger samples, this model may serve as a "starting point for a staging and profiling approach based on psychopathological and biological markers that may individually predict course and prognosis of the disorder" and help clinicians in treatment planning and appropriate allocation for patients with major depressive disorders.

In 2019, van Dijk et al. expanded the DM-TRD by adding one more item concerning the presence of severe adversity before the age of 16 years (no = 0; yes = 1); thus, scores of this new version range from 2 to 28. In the longitudinal validation study of the extended version, the authors demonstrate a strong relationship between scores on the DM-TRD and clinical course, that is, the higher the score, the poorer the treatment outcomes, concluding that the DM-TRD has good long-term predictability of clinical severity. However, counterintuitively, they found that the new item did not improve the predictive power of the old version (van Dijk et al., 2019).

## Conclusion

In this chapter, we have summarized the state of the art of the staging method applied to major depressive disorders, including treatment-resistant depression. From the initial proposals to the latest ones, staging models have become more sophisticated, including several dimensions of the illness, providing standardized operational criteria for their administration to improve inter-rater reliability, and being empirically tested. All of this has been made possible by advances in the knowledge of depression and its treatment strategies along with advances in statistical methods. Thus, at present, clinicians have available several easy-to-use staging models to assist them with diagnosis, treatment planning, and prognosis processes.

## Key facts

- Staging models are classification tools for diagnostic assistance and therapeutic and prognostic orientation.
- Staging models take into account the extent of the disease and determine the position of a person within a continuum of severity.
- The use of clinical staging is widespread in medicine, especially in oncology and cardiology.
- In psychiatry, clinical staging was first proposed by Fava and Kellner (1993) and mainly developed for schizophrenia, bipolar disorder, and depression.
- Several staging models have been developed for depression and treatment-resistant depression, but few of them have been validated.

## Summary points

- This chapter focuses on clinical staging for depression.
- Clinical staging is proposed as a more refined form of diagnosis where the longitudinal course of the disease is taken into account.
- Several studies have noted that major depressive disorder is a progressive condition that could benefit from a staging approach.
- Over the last three decades, four staging models were developed for depression, and only one has been empirically tested.
- Since 1997, five staging models have been proposed for treatment-resistant depression, but only three have been validated.
- Different therapeutic interventions for depression have been suggested according to clinical stages.

## Mini-dictionary of terms

**Antidepressants** Drugs used in the treatment of depression. There are several types, but there is no clear evidence of the superiority of any of them over the others.

**Clinical staging** A useful classification tool used in medical sciences, which may determine the extent of a disease.

**Electroconvulsive therapy** Electrical stimulation of the brain that induces seizures. It is indicated in the treatment of severe mental disorders, usually when other treatments fail.

**High-risk groups** Individuals who have a first-degree relative with a disease, most often a parent or sibling.

**Prodromal phase** A period of time characterized by mental features that represent a change in a person's premorbid functioning, which frequently indicates the onset of a disease.

**Psychotherapy** A modality of treatment based on psychological methods that help a person change behavior and get over problems.

**Treatment-resistant** A condition that does not respond to at least two appropriate treatment trials, in terms of dose and duration. In the case of depression, there is no clear consensus on this definition.

## References


Biesheuvel-Leliefeld, K. E., Kok, G. D., Bockting, C. L., Cuijpers, P., Hollon, S. D., van Marwijk, H. W., et al. (2015). Effectiveness of psychological interventions in preventing recurrence of depressive disorder: Meta-analysis and meta-regression. *Journal of Affective Disorders*, *174*, 400–410.

Boschloo, L., Schoevers, R. A., Beekman, A. T., Smit, J. H., van Hemert, A. M., & Penninx, B. W. (2014). The four-year course of major depressive disorder: The role of staging and risk factor determination. *Psychotherapy and Psychosomatics*, *83*, 279–288.

Brakemeier, E. L., Merkl, A., Wilbertz, G., Quante, A., Regen, F., Buhrsch, N., et al. (2014). Cognitive-behavioral therapy as continuation treatment to sustain response after electroconvulsive therapy in depression: A randomized controlled trial. *Biological Psychiatry*, *76*, 194–202.

Cosci, F., & Fava, G. A. (2013). Staging of mental disorders: Systematic review. *Psychotherapy and Psychosomatics*, *82*, 20–34.

de la Fuente-Tomas, L., Sanchez-Autet, M., Garcia-Alvarez, L., Gonzalez-Blanco, L., Velasco, A., Saiz Martinez, P. A., et al. (2019). Clinical staging in severe mental disorders; bipolar disorder, depression and schizophrenia. *Revista de Psiquiatría y Salud Mental—Journal of Psychiatry and Mental Health*, *12*, 106–115.

Dodd, S., Berk, M., Kelin, K., Mancini, M., & Schacht, A. (2013). Treatment response for acute depression is not associated with number of previous episodes: Lack of evidence for a clinical staging model for major depressive disorder. *Journal of Affective Disorders*, *150*, 344–349.

Fava, G. A. (2003). Diagnosis and definition of treatment-resistant depression. *Biological Psychiatry*, *53*, 649–659.

Fava, G. A., & Kellner, R. (1993). Staging: A neglected dimension in psychiatric classification. *Acta Psychiatrica Scandinavica*, *87*, 225–230.

Fava, G. A., & Tossani, E. (2007). Prodromal stage of major depression. *Early Intervention in Psychiatry*, *1*, 9–18.

Fekadu, A., Donocik, J. C., & Cleare, A. J. (2018). Standardisation framework for the Maudsley staging method for treatment resistance in depression. *BMC Psychiatry*, *18*, 100.

Fekadu, A., Wooderson, S., Donaldson, C., Markopoulou, K., Masterson, B., Ponn, L., et al. (2009). A multidimensional tool to quantify treatment resistance in depression: The Maudsley staging method. *The Journal of Clinical Psychiatry*, *70*, 177–184.

Ferensztajn, E., Remlinger-Molenda, A., & Rybakowski, J. (2014). Staging of unipolar affective illness. *Psychiatria Polska*, *48*, 1127–1141.

Guidi, J., Tomba, E., Cosci, F., Park, S. K., & Fava, G. A. (2017). The role of staging in planning psychotherapeutic interventions in depression. *The Journal of Clinical Psychiatry*, *78*, 456–463.

Guidi, J., Tomba, E., & Fava, G. A. (2016). The sequential integration of pharmacotherapy and psychotherapy in the treatment of major depressive disorder: A meta-analysis of the sequential model and a critical review of the literature. *The American Journal of Psychiatry*, *173*, 128–137.

Hetrick, S. E., Parker, A. G., Hickie, I. B., Purcell, R., Yung, A. R., & McGorry, P. D. (2008). Early identification and intervention in depressive disorders: Towards a clinical staging model. *Psychotherapy and Psychosomatics*, *77*, 263–270.

Kupfer, D. J., Frank, E., & Perel, J. M. (1989). The advantage of early treatment intervention in recurrent depression. *Archives of General Psychiatry*, *46*, 771–775.

Kuyken, W., Warren, F. C., Taylor, R. S., Whalley, B., Crane, C., Bondolfi, G., et al. (2016). Efficacy of mindfulness-based cognitive therapy in prevention of depressive relapse: An individual patient data meta-analysis from randomized trials. *JAMA Psychiatry*, *73*, 565–574.

Machmutow, K., Meister, R., Jansen, A., Kriston, L., Watzke, B., Harter, M. C., et al. (2019). Comparative effectiveness of continuation and maintenance treatments for persistent depressive disorder in adults. *The Cochrane Database of Systematic Reviews*, *5*, Cd012855.

Malda, A., Boonstra, N., Barf, H., de Jong, S., Aleman, A., Addington, J., et al. (2019). Individualized prediction of transition to psychosis in 1,676 individuals at clinical high risk: Development and validation of a multivariable prediction model based on individual patient data meta-analysis. *Frontiers in Psychiatry*, *10*, 345.

McGorry, P., Hickie, I. B., Yung, A. R., Pantelis, C., & Jackson, H. J. (2006). Clinical staging of psychiatric disorders: A heuristic framework for choosing earlier, safer and more effective interventions. *The Australian and New Zealand Journal of Psychiatry*, *40*(8), 616–622.

McGorry, P. D., & Mei, C. (2018). Ultra-high-risk paradigm: Lessons learnt and new directions. *Evidence-Based Mental Health*, *21*, 131–133.

Negt, P., Brakemeier, E. L., Michalak, J., Winter, L., Bleich, S., & Kahl, K. G. (2016). The treatment of chronic depression with cognitive behavioral analysis system of psychotherapy: A systematic review and meta-analysis of randomized-controlled clinical trials. *Brain and Behavior: A Cognitive Neuroscience Perspective*, *6*, e00486.

Peeters, F., Ruhe, H. G., Wichers, M., Abidi, L., Kaub, K., van der Lande, H. J., et al. (2016). The Dutch measure for quantification of treatment resistance in depression (DM-TRD): An extension of the Maudsley staging method. *Journal of Affective Disorders*, *205*, 365–371.

Scott, J., Leboyer, M., Hickie, I., Berk, M., Kapczinski, F., Frank, E., et al. (2013). Clinical staging in psychiatry: A cross-cutting model of diagnosis with heuristic and practical value. *The British Journal of Psychiatry*, *202*, 243–245.

Thase, M. E., & Rush, A. J. (1997). When at first you don't succeed: Sequential strategies for antidepressant nonresponders. *The Journal of Clinical Psychiatry*, *58*(Suppl 13), 23–29.

van Diermen, L., Hebbrecht, K., Schrijvers, D., Sabbe, B. C. G., Fransen, E., & Birkenhäger, T. K. (2018). The Maudsley staging method as predictor of electroconvulsive therapy effectiveness in depression. *Acta Psychiatrica Scandinavica*, *138*, 605–614.

van Dijk, D. A., van der Boogaard, T. M., Deen, M. L., Spijker, J., Ruhé, H. G., & Peeters, F. P. M. L. (2019). Predicting clinical course in major depressive disorder: The association between DM-TRD score and symptom severity over time in 1115 outpatients. *Depression and Anxiety*, *36*, 345–352.

Verduijn, J., Milaneschi, Y., Schoevers, R. A., van Hemert, A. M., Beekman, A. T., & Penninx, B. W. (2015). Pathophysiology of major depressive disorder: Mechanisms involved in etiology are not associated with clinical progression. *Translational Psychiatry*, *5*, e649.

Verduijn, J., Milaneschi, Y., van Hemert, A. M., Schoevers, R. A., Hickie, I. B., Penninx, B. W., et al. (2015). Clinical staging of major depressive disorder: An empirical exploration. *The Journal of Clinical Psychiatry*, *76*, 1200–1208.

Vieta, E., Reinares, M., & Rosa, A. R. (2011). Staging bipolar disorder. *Neurotoxicity Research*, *19*, 279–285.

Chapter 2

# Neurodevelopmental theory of depression

Monika Talarowska
*Department of Psychology and Individual Differences, Institute of Psychology, University of Lodz, Lodz, Poland*

## List of abbreviations

| | |
|---|---|
| **5-HTT** | serotonin transporter |
| **BDNF** | brain-derived neurotrophic factor |
| **CRF** | corticotropin-releasing hormone |
| **CRP** | C-reactive protein |
| **FKBP5** | binding protein 5 |
| **HDAC1** | histone deacetylase 1 |
| **HDAC2** | histone deacetylase 2 |
| **HPA axis** | hypothalamic–pituitary–adrenal axis |
| **Il1** | interleukin 1 |
| **Il6** | interleukin 6 |
| **Il10** | interleukin 10. |
| **Il12** | interleukin 12. |
| **miRNA/mRNA** | microRNA. |
| **MMPI-2** | Minnesota Multiphasic Personality Inventory 2. |
| **PTSD** | post-traumatic stress disorder. |
| **SLC6A4** | 5-hydroxy-tryptamine transporter. |
| **SSRI** | selective serotonin reuptake inhibitor. |
| **TNF-α** | tumor necrosis factor α. |
| **TRYCATs** | tryptophan catabolites. |

## Introduction

A new understanding of mood disorders, including depressive disorders, has emerged over the past decade. They began to be perceived as neuroprogressive disorders associated with neurodegenerative changes in the frontal cortex and the limbic system as a consequence of inflammatory factors, including cytokines, penetrating the blood–brain barrier (Pandey, 2017). In this chapter, the author presents a different view on the formation of depression, i.e., the so-called neurodevelopmental theory of depression. Its strength is the integration of previous approaches explaining the etiology of depression, both purely biological and those that refer in essence to the psychological understanding of this disease. This theory emphasizes the importance of the earliest stages of our lives (prenatal period, early childhood, and adolescence) for the development of personality traits that favor the occurrence of a depressive episode in adult life. The question arises about the universality of the abovementioned traits, which can be described as a depressive personality.

The presented theory combines the abovementioned elements into one whole. The common denominator is epigenetic mechanisms, which reflect innate changes in gene expression, unrelated to changes in DNA sequence. The determination of specific gene expression patterns is crucial for the morphological and functional differentiation of cells in the human body (Bakusic, Schaufeli, Claes, & Godderis, 2017). Epigenetic mechanisms are responsible for these expression patterns. The most important consequence of epigenetic changes is the appearance of different phenotypes of the same genome based on different epigenetic status. Disruption of the normal epigenetic environment at an early stage of development can have serious consequences.

The presented theory has been described as "neurodevelopmental" in order to emphasize the importance and impact of early stages of human life, including the prenatal period, on the occurrence and development of depressive disorders. The author of this chapter has attempted to find answers to the questions why this period plays such an important role in human life, what kind of biological mechanisms are activated then, and what aspects of further functioning are affected by these mechanisms.

## Early childhood experience and personality traits

Human personality is determined by the functioning of branched networks of nerve endings. It involves typical interpersonal behaviors, subjective reactions, feelings, and objectives we are striving after (Pingault, Falissard, Côté, & Berthoz, 2012).

Three development periods, i.e., *prenatal period, early childhood (until the age of 5–6), and adolescence*, have particular significance for the shaping of personality. The process of personality shaping is also affected by both genetic and environmental factors, which indicate the direction of the structural and functional development of the brain, the hormonal system, and the immune system

The Neuroscience of Depression. https://doi.org/10.1016/B978-0-12-817933-8.00010-4

(Gałecki & Talarowska, 2018). They may affect, either negatively or positively, each of the previously mentioned developmental stages; hence, reduce or increase our biological immunity as well as modify psychological coping skills.

A meta-analysis of 154 longitudinal studies has led to the conclusion that the personality of an adult person still changes in the direction of increasing cohesion, and this process reaches a plateau around the age of 50 (Roberts & DelVecchio, 2000). The stability indexes of the traits are no more than 0.31 in childhood; at the age of 30 they reach 0.64, and at the age of 50 reach 0.74. This means that life events in the early years of our lives have the greatest impact on the structure of our personality, which is particularly visible in the case of neuroticism.

Mainly limbic system structures with the amygdala, the hippocampus, and the prefrontal cortex, as well as the effectiveness of connections between them, have particular relevance in the process of personality shaping (Davis & Panksepp, 2011). The moment of maturation of those structures is convergent with the periods critical for the shaping of permanent personality traits of a human being and maturity of the immune system. The hippocampus region reaches maturity close to an adult person between week 13 and 20 of pregnancy. Further significant structural changes in these areas occur during the first year of life of the child, especially in the dentate gyrus and the entorhinal cortex. In functional terms, frontal lobes "mature" gradually as late as at the age of 20–25 (Morrison, Rodgers, Morgan, & Bale, 2014), reaching the highest specialization in this period.

Many authors stress the importance of early childhood experiences, especially those with trauma traits, for the development of mental disorders in adult life. Studies indicate a direct connection between trauma from childhood and abuse of psychoactive substances, psychosis, mood disorders, anxiety disorders and the risk of attempted suicide (Aas et al., 2016). These experiments indicate that trauma lead to changes in the reactivity of the hormonal system and immune system, changes in brain function (mainly in the frontal cortex and the hippocampus area), and at the psychological level to the persistence of non-adaptive ways of reacting to stressors (Nagy, Vaillancourt, & Turecki, 2018). The latter are based on neural connections reinforced by repeating sensory experiences, both those of a positive and a traumatic nature. Creating and reinforcing neural connections is a key task in the early stages of brain development and forms the foundations of personality.

A human being is a social animal. For each of us, a particularly important need is a relationship with another person. It is just as important to us as the need to satisfy hunger and thirst. Loneliness strongly motivates us to change this state (Cacioppo, Cacioppo, & Boomsma, 2014). Insufficient satisfaction of the need for proximity in the early stages of a child's life (e.g., due to social isolation, low parental skills, emotional rejection of the child by parents) leads to changes in neurobehavioral responses to experienced stress, which shapes patterns of our future relationships with people (Kinnally & Capitanio, 2015). At the same time, the hormonal and immune systems are deregulated through the network of mutual feedback in the HPA axis. Personality traits in the form of anxiety attitude, established by means of dysregulating the HPA axis, are, on the other hand, a source of constant pro-inflammatory activity of the immune system. Through excessive production of neurotoxic compounds (especially the so-called tryptophan catabolites, TRYCATs), this cascade of mutual feedback loops leads gradually to neurodegenerative processes, which are revealed among others in the form of depression (Talarowska & Gałecki, 2016).

Moreover, through epigenetic mechanisms, these patterns are passed on to future generations. This relationship was confirmed by Saavedra-Rodríguez and Feig (2013). In the study carried out by the authors mentioned previously, males and females of mice were subjected to social stress during their early childhood and adolescence in the form of instability and the need to fight for social position. These factors not only changed the behavior of the examined individuals in the form of increased anxiety reaction but were also passed on to the next three generations through epigenetic changes. In the studies by Koutra et al. (2017) it was demonstrated that the severity of postnatal depression symptoms in the mother and the degree of anxiety she experienced as a permanent trait of her personality were related to the quality of neuropsychological development in the children. Furthermore, intensification of emotional proximity between parents and children during early childhood was a factor significantly affecting frontal cortex volume in the offspring's frontal gyrus area and correlated with the personality traits conducive to depression in children.

## The affective and rational system—The basis for personality formation

Regulation of emotions takes place in three collaborating areas of the brain. The structures of the brain stem are responsible for the most elementary, innate, and unconscious impulsive reactions (excitation vs inhibition, autonomous reactions). The limbic system, including the hippocampus and the amygdala, modifies our emotional reactions depending on the incoming environmental stimuli. The prefrontal cortex is responsible for control over emotions and feelings (Hallam et al., 2015).

Negative emotional attitudes, typical of patients with symptoms of depression, are most likely to result from an imbalance between "emotional" (structures of the limbic system with the amygdala and the hippocampus) and "motivational/regulatory" brain regions (frontal lobes, mainly the area of the prefrontal cortex of the brain) (Penner et al., 2016). The "emotional" brain of the people affected by depression is hyperactive in response to negative stimuli, whereas it reacts too poorly to information characterized by a positive emotional charge. Meanwhile, the "motivational/regulatory" brain does not cope with the blocking of unwanted and unpleasant contents. The described dysfunctions seem to be a permanent feature of the cognitive and emotional functioning of patients with depression. They are also likely to cause a pessimistic style of information processing (as a permanent trait of personality) characteristic for people with depressive disorders, associated with numerous ruminations of a negative emotional nature (Hamlat et al., 2015). Thus, the cerebral cortex, by deciding how to deal with primary emotions coming from deeper structures, is responsible for the foundations of our personality.

## Epigenetics

Epigenetic mechanisms play an important role in the inheritance of neurobehavioral patterns and personality traits. These mechanisms (histone modification, DNA methylation, gene expression changes at the miRNA level) prepare our body to cope with changes in the environment. They increase the chances of survival of the species by significantly reducing the time needed to pass them on to future generations. However, if our children do not experience adverse events in the future that are the source of our fear and anxiety, then the inherited coping mechanism will do a lot of harm (Saavedra, Molina-Marquez, Saavedra, Zambrano, & Salazar, 2016).

When adult rats were starving during pregnancy, their offspring (group A) had a significantly lower birth weight than the offspring of the mothers who were not restricted in their access to food (group B). However, if adult rats from group A are raised in conditions that provide free access to any amount of food, they will become more obese than group B rats. The mechanism associated with the accumulation of energy reserves becomes pathological. The fear of a predicted shortage of food inherited from mothers will push them toward excessive accumulation of food (Rantala, Luoto, Krams, & Karlsson, 2018).

Early childhood experiences associated with severe stressors (considered a risk factor for depression in adult life) are linked with changes in gene expression. They include genes involved in a response to stress (activity of the hypothalamic–pituitary–adrenal axis), related to autonomic nervous system hyperactivity and cortical and subcortical processes of neuroplasticity and neurodegeneration. These are glucocorticoid receptor encoding gene, FK506-binding protein 5 (FKBP5) gene (Tozzi et al., 2018), arginine vasopressin and estrogen receptor alpha encoding gene, and 5-hydroxy-tryptamine transporter gene among others (SLC6A4) (Provenzi, Giorda, Beri, & Montirosso, 2016), as well as brain-derived neurotrophic factor encoding gene (BDNF) (Stonawski et al., 2019).

## Emotional immunity or immune emotionality?—The key to understanding depression

Dysregulation of the immune system as an etiological factor, and also affecting the course of depression, is no longer questionable (Euteneuer et al., 2017). D'Acquisto (2017) uses the term affective immunology. In his opinion, it means that the immune and affective systems are dynamic systems, subject to constant changes, but constituting the mirror reflection of one another. The interaction between the immune system and emotions is evidenced by the frequency of emotional disorders in patients with immune system diseases and deterioration of the immune system in patients with various groups of mental disorders (Maes et al., 2012). D'Acquisto stresses that the variability of the two systems is expressed in their plasticity, understood as the ability to change (adapt) under the influence of extrinsic factors. In the case of both the immune system and the affective system, by means of changes in the DNA chain, we obtain from our ancestors only a biological predisposition determining the risk of incidence of a given disease. Our ability to adapt (diet, lifestyle, but also our ability to cope) determines whether or not the disease manifests itself. D'Acquisto also introduces the concept of immunological personality, asking the question of its convergence with psychological personality. It seems that the answer to this question may be in the affirmative. The personality trait important for the activation of the immune system is neuroticism, which mediates the psychological response to stress stimuli.

An increased level of neuroticism, as a personality trait, combined with low conscientiousness and openness is linked not only with an elevated risk of attempting suicide (Isung et al., 2014), but also with a rise in the following indicators of an active inflammatory process: interleukin 6 (Il-6) and C-reactive protein (CRP) (Luchetti, Barkley, Stephan, Terracciano, & Sutin, 2014). A tendency to often experience the feeling of anger and hostility is accompanied by an increase in the level of CRP (Smith, Uchino, Bosch, & Kent, 2014) and TNF-α (Girard, Tardif, Boisclair Demarble, & D'Antono, 2016). Furthermore, a tendency to have an anxious approach when evaluating reality correlates positively with the level of CRP and negatively with the level of self-control (Henningsson et al., 2008). In our study (Gałecki & Talarowska, 2018) we found that the scales of the MMPI-2 (Minnesota Multiphasic Personality Inventory) personality questionnaire by S. Hathaway and J. McKinley associated with the intensification of anxiety symptoms (the scale of hypochondria, depression, hysteria, and the Welsh anxiety scale) correlated positively with expression at the mRNA level and at the protein level for Il-1, Il-10, and Il-12. Additionally, it turns out that a high level of neuroticism is a common feature for the people susceptible to depressive disorders and dementia (Montag & Panksepp, 2017), while personality changes in the form of intensive fear as a permanent personality trait turn out to be a predictor of dementia.

Table 1.

## Mother's fear as well as grandmother's fear as a source of depression

Emotions experienced by pregnant women have an impact on the developing organism through the impact of the hormonal system.

**TABLE 1** **Inflammatory process indicators and personality traits.**

| | |
|---|---|
| ↑ Il-1 | ↑ Hypochondria<br>↑ Hysteria<br>↑ Depression<br>↑ Welsh anxiety (Gałecki & Talarowska, 2018) |
| ↑ Il-6 | ↑ Neuroticism (Sutin et al., 2010; Turiano, Mroczek, Moynihan, & Chapman, 2013)<br>↓ Conscientiousness (Sutin et al., 2010)<br>↓ Openness to experience (Chapman et al., 2011)<br>↑ Impulsiveness (Isung et al., 2014) |
| ↑ Il-10 | ↑ Hypochondria<br>↑ Hysteria<br>↑ Depression<br>↑ Welsh anxiety (Gałecki & Talarowska, 2018) |
| ↑ Il-12 | ↑ Hypochondria<br>↑ Hysteria<br>↑ Depression<br>↑ Welsh anxiety (Gałecki & Talarowska, 2018) |
| ↑ CRP | ↑ Neuroticism (Sutin et al., 2010)<br>↓ Conscientiousness (Sutin et al., 2010)<br>↓ Openness to experience (Luchetti et al., 2014)<br>↑ Hostility (Smith et al., 2014)<br>↑ Impulsiveness (Isung et al., 2014)<br>↑ Anxious attitude in reality evaluation (Henningsson et al., 2008)<br>↓ Self-control (Henningsson et al., 2008) |
| ↑ TNF-α | ↑ Hostility (Boisclair Demarble, Moskowitz, Tardif, & D'Antono, 2014; Girard et al., 2016) |

Zheng, Fan, Zhang, and Dong (2016) underlined that prenatal stress experienced by mothers during pregnancy leads to an increased risk of symptoms and behavior similar to depression as well as anxiety in adulthood. The following features are associated with changes in gene expression: decreased expression of BDNF and AcH3K14 with enhanced expression of histone deacetylase (HDAC1 and HDAC2) in the hippocampus. Boulle et al. (2016) demonstrated that fetal exposure to selective serotonin reuptake inhibitors (SSRIs) alone has an impact on neuroplasticity and increases the likelihood of depression and anxiety-related behavior. This phenomenon seems to be stronger among women, although—as the authors themselves made clear—the number of studies carried out on this group is far from sufficient to draw far-reaching conclusions. Interestingly, the same team (Boulle et al., 2016) concluded that exposure to fluoxetine in the postnatal period leads to a change in anxiety behavior by strengthening the anxiety felt in non-stressed offspring and reducing anxiety in the offspring who came into contact with stress in the fetal period. The last of the cited studies refers only to male individuals.

Due to epigenetic changes, prenatal stress is considered to be the strongest factor affecting mental health in later stages of life (Babenko, Kovalchuk, & Metz, 2015). Can we speak, then, of intergenerational trauma, which is transmitted from generation to generation and thus gradually becomes a constant evolutionary feature? In connection with the abovementioned frontal lobe dysfunctions and hyperactivity of limbic structures, does it lead directly and inevitably to depression epidemics?

Importantly, the costs incurred by the offspring in the form of various pathological behavior patterns depend on their mothers' ability to cope with biological stressors (maternal immune responses) and psychological stressors (coping skills) (Bronson, Ahlbrand, Horn, Kern, & Richtand, 2011). These observations are in line with the conclusions that the hereditary element in severe depression is relatively important and that more moderate forms of the disease are increasingly dependent on accompanying environmental factors in life.

In the context of the hypothesis formulated above, the results of the study conducted by Pawluski et al. (2012) and Rayen, Gemmel, Pauley, Steinbusch, and Pawluski (2015) appear to be optimistic. Independently of each other, these two scientific teams showed that exposure to SSRI at early stages of life reverses the neurobiological effects of prenatal stress and thus has a protective effect on the human body. Nevertheless, Pawluski et al. (2012) stressed that the strengthening and protective effect of SSRI are only visible in the male individuals studied by the researchers. Similar results were achieved by (Ignácio et al. 2017). The authors confirmed the antidepressant action of quetiapine in motherless animals and demonstrated the protective effect of quetiapine in the reduction of epigenetic changes caused by stress in the early stages of life.

You can ask a different question, this time a more difficult one. Do antidepressants have a negative impact on our coping skills? Is it possible to "immunize" oneself to their action?

## Early childhood trauma

Experiences resembling early childhood trauma are associated with the risk of depression symptoms at later stages of life (Saavedra et al., 2016). Moreover, a reduced volume of the hippocampus and gray matter in the prefrontal cortex is associated not only with an episode of depression itself, but also with the emotionally difficult experiences that took place in the early stages of development. Additionally, both early childhood trauma and the presented structural changes may be the factors responsible for the low effectiveness of pharmacological treatment (Frodl, Reinhold, Koutsouleris, Reiser, & Meisenzahl, 2010). In response to traumatic experience, epigenetic modifications have proven to be important factors in long-term biological trajectories that lead to stress-related psychiatric disorders, reflecting both individual genetic predisposition and environmental influences. Therefore, the question should be asked about the effectiveness of psychotherapeutic measures taken in case of confirmed anatomical changes associated with early childhood trauma.

## Glucocorticoid cascade hypothesis—Epigenetics again?

Through the neurotoxic action of cortisol and neuroinflammatory processes, chronic or acute stress leads to anatomical and functional changes in the central nervous system. These changes affect mainly the hippocampus and the frontal lobes, i.e., structures crucial for the occurrence of depressive symptoms (Dantzer, O'Connor, Freund, Johnson, & Kelley, 2008).

Excessive activity of the HPA axis, preceding a depression episode, is a consequence of genetic factors (epigenetics is again of great importance) and contacts with averse stimuli at early stages of ontogenetic development or in adulthood (Lee & Sawa, 2014).

It is believed that the interaction between childhood negligence and the polymorphic area within the serotonin transporter encoding gene (5-HTT) is associated with increased levels of anxiety and the release of glucocorticoids during stress exposure, as well as with the risk of developing a severe form of depression (van der Doelen et al., 2015). van der Doelen et al. (2015) have reported that stress experienced at an early stage of life and the 5-HTTT genotype interact and influence DNA methylation of the corticotropin-releasing hormone (CRF) encoding gene promoter in the medial amygdala in adult male rats. The quoted authors further emphasized that DNA methylation of a specific area in the CRF promoter is significantly correlated with the levels of mRNA CRF in the medial part of the amygdala. Moreover, the expression of CRF at the mRNA level in the amygdala is related to the ability to cope with stress (van der Doelen et al., 2015). The above correlation was also confirmed on the basis of the results of research conducted by Dannlowski et al. (2014) devoted to the reduction in the volume of the hippocampus, as well as the study by Auxéméry (2012) on post-traumatic stress disorder (PTSD).

## Personality of the 21st century

The presence of the personality traits typical for the so-called Cluster C (avoidant, dependent, and obsessive–compulsive personality) reduces the likelihood of achieving remission of depressive disorders by 30% and increases the risk of relapse after the first episode by as much as 80% (Bukh, Andersen, & Kessing, 2016). In one of our previous study (Talarowska, Zboralski, Chamielec, & Gałecki, 2011), we also indicated that the premorbid personality structure (anxiety as a constant feature of emotional functioning) may have a significant importance for the effectiveness of applied antidepressant pharmacotherapy. It seems, therefore, that what binds all types of personalities, also from Clusters A and B, leading in each case to the development of depression, is anxiety. Perhaps in the classification of personality disorders it is worth considering the introduction of a separate unit, i.e., neurotic or depressive personality, being a risk factor for the occurrence of affective disorders.

To sum up or deliberations, it is worth considering whether the traits of our personality define our later diseases in an unchangeable way, being a kind of a life sentence we have no influence on. Are people with neurotic features doomed to evolutionary failure? It turns out that for thousands of years of the human race's development, the anxiety attitude has been conducive to survival. Our ancestors were more vigilant and focused on anticipating potential threats, which allowed them to avoid risks more effectively. Today, in an environment that is objectively assessed as low risk, a neurotic person will continue to be overly vigilant, consuming its immune resources pointlessly (Montag & Panksepp, 2017). In this pessimistic approach, however, it turns out that the level of intelligence is a mediator and a specific protective factor between the neurotic trait and the risk of depression. Therefore, the maturity of the frontal lobes, strengthening their development, and improvement of their functioning should be the therapeutic goal, regarding both pharmacotherapy and psychotherapy.

Fig. 1 illustrates a perfect summary of the considerations presented herein.

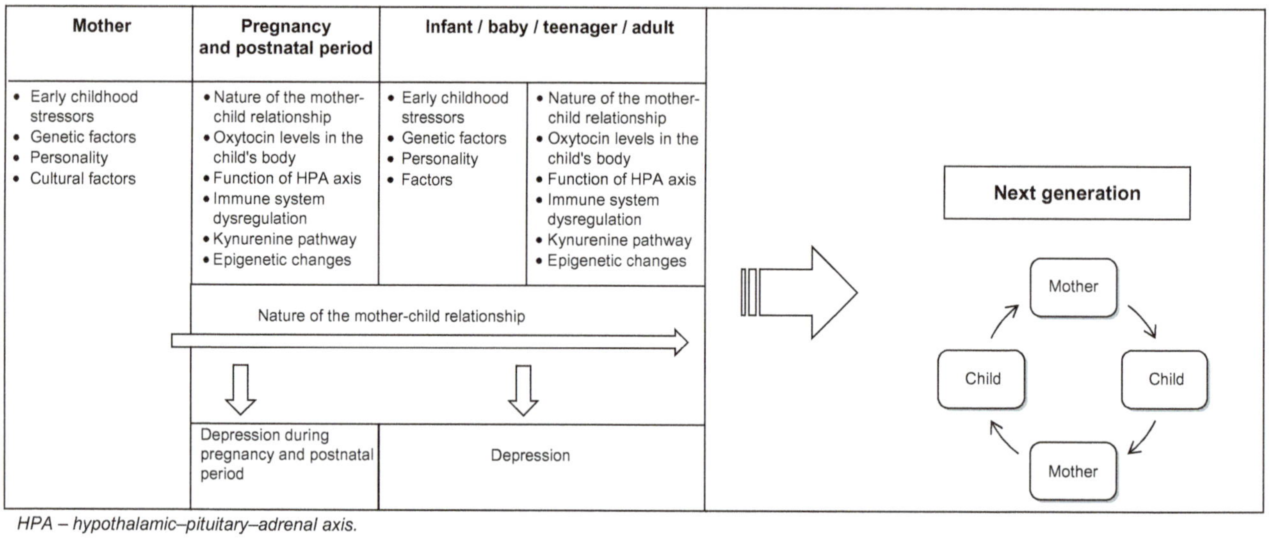


FIG. 1 Model summarizing the neurodevelopmental theory of depression. *HPA*—hypothalamic–pituitary–adrenal axis.

## Key facts

- Neurodevelopmental theory of depression emphasizes the importance of the earliest stages of our lives (prenatal period, early childhood, and adolescence) for the development of personality traits that favor the occurrence of a depressive episode in adult life.
- The affective (the limbic system with the amygdala and the hippocampus) and rational system of the brain (the front al lobes) are the basis for personality formation.
- Epigenetic mechanisms play an important role in the inheritance of neurobehavioral patterns and personality traits.
- Dysregulation of the immune system is an etiological factor in depression (emotional immunity/immune emotionality).
- The presence of the personality traits typical for the so-called Cluster C (avoidant, dependent, and obsessive–compulsive personality) is an etiological factor in depression.

## Summary points

- Neurodevelopmental theory of depression—what does it mean?
- Early childhood experience and personality traits.
- The affective and rational system—the basis for personality formation.
- Epigenetics.
- Emotional immunity or immune emotionality?—The key to understanding depression.
- Mother's fear as well as grandmother's fear as a source of depression.
- Personality of the 21st century.

## Mini-dictionary of terms

**Epigenetics** Mechanisms including DNA methylation, modifications of histones, and chromatin structures, as well as functions of non-coding RNA, are co-responsible for specific patterns of gene expression. Each of the three processes is not dependent on DNA sequence. These are rapid and reversible changes, which are affected to the largest extent by environmental factors.

**Personality** Typical for each of us, involves interpersonal behavior, subjective reactions, feelings, and goals to which we aspire.

**Amygdala, the hippocampus, the prefrontal cortex** Brain structures crucial for the development of our personality.

**Emotional brain** Structures of the limbic system with the amygdala and the hippocampus.

**Motivational brain regions** Frontal lobes, mainly the area of the prefrontal cortex of the brain.

**Affective immunology** It means that the immune and affective systems are dynamic systems, subject to constant changes, but constituting the mirror reflection of one another.

## References

Aas, M., Henry, C., Andreassen, O. A., Bellivier, F., Melle, I., & Etain, B. (2016). The role of childhood trauma in bipolar disorders. *International Journal of Bipolar Disorders*, *4*(1), 2.

Auxéméry, Y. (2012). Posttraumatic stress disorder (PTSD) as a consequence of the interaction between an individual genetic susceptibility, a traumatogenic event and a social context. *Encephale*, *38*(5), 373–380.

Babenko, O., Kovalchuk, I., & Metz, G. A. (2015). Stress-induced perinatal and transgenerational epigenetic programming of brain development and mental health. *Neuroscience & Biobehavioral Reviews*, *48*, 70–91.

Bakusic, J., Schaufeli, W., Claes, S., & Godderis, L. (2017). Stress, burnout and depression: A systematic review on DNA methylation mechanisms. *Journal of Psychosomatic Research*, *92*, 34–44. https://doi.org/10.1016/j.jpsychores.2016.11.005.

Boisclair Demarble, J., Moskowitz, D. S., Tardif, J. C., & D'Antono, B. (2014). The relation between hostility and concurrent levels of inflammation is sex, age, and measure dependent. *Journal of Psychosomatic Research*, *76*(5), 384–393.

Boulle, F., Pawluski, J. L., Homberg, J. R., Machiels, B., Kroeze, Y., Kumar, N., et al. (2016). Prenatal stress and early-life exposure to fluoxetine have enduring effects on anxiety and hippocampal BDNF gene expression in adult male offspring. *Developmental Psychobiology*, *58* (4), 427–438.

Bronson, S. L., Ahlbrand, R., Horn, P. S., Kern, J. R., & Richtand, N. M. (2011). Individual differences in maternal response to immune challenge predict offspring behavior: Contribution of environmental factors. *Behavioural Brain Research*, *220*(1), 55–64.

Bukh, J. D., Andersen, P. K., & Kessing, L. V. (2016). Personality and the long-term outcome of first-episode depression: A prospective 5-year follow-up study. *Joural of Clinical Psychiatry*, *77*(6), 704–710.

Cacioppo, J. T., Cacioppo, S., & Boomsma, D. I. (2014). Evolutionary mechanisms for loneliness. *Cognition & Emotion*, *28*, 3–21.

Chapman, B. P., van Wijngaarden, E., Seplaki, C. L., Talbot, N., Duberstein, P., & Moynihan, J. (2011). Openness and conscientiousness predict 34-week patterns of Interleukin-6 in older persons. *Brain, Behavior, and Immunity*, *25*(4), 667–673.

D'Acquisto, F. (2017). Affective immunology: Where emotions and the immune response converge. *Dialogues in Clinical Neuroscience*, *19* (1), 9–19.

Dannlowski, U., Kugel, H., Redlich, R., Halik, A., Schneider, I., Opel, N., et al. (2014). Serotonin transporter gene methylation is associated with hippocampal gray matter volume. *Human Brain Mapping*, *35*(11), 5356–5367.

Dantzer, R., O'Connor, J. C., Freund, G. G., Johnson, R. W., & Kelley, K. W. (2008). From inflammation to sickness and depression: When the immune system subjugates the brain. *Nature Reviews Neuroscience*, *9* (1), 46–56.

Davis, K. L., & Panksepp, J. (2011). The brain's emotional foundations of human personality and the affective neuroscience personality scales. *Neuroscience & Biobehavioral Reviews*, *35*(9), 1946–1958.

Euteneuer, F., Dannehl, K., Del Rey, A., Engler, H., Schedlowski, M., & Rief, W. (2017). Peripheral immune alterations in major depression: The role of subtypes and Pathogenetic characteristics. *Frontiers in Psychiatry*, *8*, 250.

Frodl, T., Reinhold, E., Koutsouleris, N., Reiser, M., & Meisenzahl, E. M. (2010). Interaction of childhood stress with hippocampus and prefrontal cortex volume reduction in major depression. *Journal of Psychiatr Research*, *44*(13), 799–807.

Gałecki, P., & Talarowska, M. (2018). Neurodevelopmental theory of depression. *Progress in Neuro-Psychopharmacology & Biological Psychiatry*, *80*(Pt C), 267–272.

Girard, D., Tardif, J. C., Boisclair Demarble, J., & D'Antono, B. (2016). Trait hostility and acute inflammatory responses to stress in the laboratory. *PLoS One*, *11*(6), e0156329.

Hallam, G. P., Webb, T. L., Sheeran, P., Miles, E., Wilkinson, I. D., Hunter, M. D., et al. (2015). The neural correlates of emotion regulation by implementation intentions. *PLoS One*, *10*(3), e0119500.

Hamlat, E. J., Connolly, S. L., Hamilton, J. L., Stange, J. P., Abramson, L. Y., & Alloy, L. B. (2015). Rumination and overgeneral autobiographical memory in adolescents: An integration of cognitive vulnerabilities to depression. *Journal of Youth and Adolescence*, *44*(4), 806–818.

Henningsson, S., Baghaei, F., Rosmond, R., Holm, G., Landén, M., Anckarsäter, H., et al. (2008). Association between serum levels of C-reactive protein and personality traits in women. *Behavioral and Brain Functions*, *4*, 16.

Ignácio, Z. M., Réus, G. Z., Abelaira, H. M., Maciel, A. L., de Moura, A. B., Matos, D., … Quevedo, J. (2017). Quetiapine treatment reverses depressive-like behavior and reduces DNA methyltransferase activity induced by maternal deprivation. *Behavioural Brain Research*, *320*, 225–232.

Isung, J., Aeinehband, S., Mobarrez, F., Nordström, P., Runeson, B., Asberg, M., et al. (2014). High interleukin-6 and impulsivity: Determining the role of endophenotypes in attempted suicide. *Translational Psychiatry*, *4*, 470.

Kinnally, E. L., & Capitanio, J. P. (2015). Paternal early experiences influence infant development through non-social mechanisms in rhesus macaques. *Frontiers in Zoology*, *12*(Suppl 1), 14.

Koutra, K., Roumeliotaki, T., Kyriklaki, A., Kampouri, M., Sarri, K., Vassilaki, M., et al. (2017). Maternal depression and personality traits in association with child neuropsychological and behavioral development in preschool years: Mother-child cohort (Rhea Study) in Crete, Greece. *Journal of Affective Disorders*, *217*, 89–98.

Lee, R. S., & Sawa, A. (2014). Environmental stressors and epigenetic control of the hypothalamic-pituitary-adrenal axis. *Neuroendocrinology*, *100*(4), 278–287.

Luchetti, M., Barkley, J. M., Stephan, Y., Terracciano, A., & Sutin, A. R. (2014). Five-factor model personality traits and inflammatory markers: New data and a meta-analysis. *Psychoneuroendocrinology*, *50*, 181–193.

Maes, M., Berk, M., Goehler, L., Song, C., Anderson, G., Gałecki, P., & Leonard, B. (2012). Depression and sickness behavior are Janus-faced responses to shared inflammatory pathways. *BMC Medicine*, *10*, 66. https://doi.org/10.1186/1741-7015-10-66.

Montag, C., & Panksepp, J. (2017). Emotional systems and personality: An evolutionary perspective. *Frontiers in Psychology*, *8*, 464.

Morrison, K. E., Rodgers, A. B., Morgan, C. P., & Bale, T. L. (2014). Epigenetic mechanisms in pubertal brain maturation. *Neuroscience*, *264*, 17–24.

Nagy, C., Vaillancourt, K., & Turecki, G. (2018). A role for activity dependent epigenetics in the development and treatment of major depressive disorder. *Genes, Brain and Behavior*, *17*(3), e12446.

Pandey, G. N. (2017). Inflammatory and innate immune markers of neuroprogression in depressed and teenage suicide brain. *Modern Trends in Pharmacopsychiatry*, *31*, 79–95.

Pawluski, J. L., Rayen, I., Niessen, N. A., Kristensen, S., van Donkelaar, E. L., Balthazart, J., et al. (2012). Developmental fluoxetine exposure differentially alters central and peripheral measures of the HPA system in adolescent male and female offspring. *Neurosciences*, *220*, 131–141.

Penner, J., Ford, K. A., Taylor, R., Schaefer, B., Théberge, J., Neufeld, R. W., et al. (2016). Medial prefrontal and anterior insular connectivity in early schizophrenia and major depressive disorder: A resting functional MRI evaluation of large-scale brain network models. *Frontiers in Human Neuroscience*, *10*, 132.

Pingault, J. B., Falissard, B., Côté, S., & Berthoz, S. (2012). A new approach of personality and psychiatric disorders: A short version of the affective neuroscience personality scales. *PLoS One*, *7*(7), e41489.

Provenzi, L., Giorda, R., Beri, S., & Montirosso, R. (2016). SLC6A4 methylation as an epigenetic marker of life adversity exposures in humans: A systematic review of literature. *Neuroscience & Biobehavioral Reviews*, *71*, 7–20.

Rantala, M. J., Luoto, S., Krams, I., & Karlsson, H. (2018). Depression subtyping based on evolutionary psychiatry: Proximate mechanisms and ultimate functions. *Brain, Behavior and Immunity*, *69*, 603–617.

Rayen, I., Gemmel, M., Pauley, G., Steinbusch, H. W., & Pawluski, J. L. (2015). Developmental exposure to SSRIs, in addition to maternal stress, has long-term sex-dependent effects on hippocampal plasticity. *Psychopharmacology*, *232*(7), 1231–1244.

Roberts, B. W., & DelVecchio, W. F. (2000). The rank-order consistency of personality from childhood to old age: A quantitative review of longitudinal studies. *Psychological Bulletin*, *126*, 3–25.

Saavedra, K., Molina-Márquez, A. M., Saavedra, N., Zambrano, T., & Salazar, L. A. (2016). Epigenetic modifications of major depressive disorder. *International Journal of Molecular Sciences*, *17*(8).

Saavedra-Rodríguez, L., & Feig, L. A. (2013). Chronic social instability induces anxiety and defective social interactions across generations. *Biological Psychiatry*, *73*(1), 44–53.

Smith, T. W., Uchino, B. N., Bosch, J. A., & Kent, R. G. (2014). Trait hostility is associated with systemic inflammation in married couples: An actor-partner analysis. *Biological Psychology*, *102*, 51–53.

Stonawski, V., Frey, S., Golub, Y., Rohleder, N., Kriebel, J., Goecke, T. W., et al. (2019). Associations of prenatal depressive symptoms with DNA methylation of HPA axis-related genes and diurnal cortisol profiles in primary school-aged children. *Development and Psychopathology*, *31* (2), 419–431.

Sutin, A. R., Terracciano, A., Deiana, B., Naitza, S., Ferrucci, L., Uda, M., et al. (2010). High neuroticism and low conscientiousness are associated with interleukin-6. *Psychological Medicine*, *40*(9), 1485–1493.

Talarowska, M., & Gałecki, P. (2016). Cognition and emotions in recurrent depressive disorders - the role of inflammation and the kynurenine pathway. *Current Pharmaceutical Design*, *22*(8), 955–962.

Talarowska, M., Zboralski, K., Chamielec, M., & Gałecki, P. (2011). The MMPI-2 neurotic triad subscales and depression levels after pharmacological treatment in patients with depressive disorders - clinical study. *Psychiatria Danubina*, *23*(4), 347–354.

Tozzi, L., Farrell, C., Booij, L., Doolin, K., Nemoda, Z., Szyf, M., et al. (2018). Epigenetic changes of FKBP5 as a link connecting genetic and environmental risk factors with structural and functional brain changes in major depression. *Neuropsychopharmacology*, *43*(5), 1138–1145.

Turiano, N. A., Mroczek, D. K., Moynihan, J., & Chapman, B. P. (2013). Big 5 personality traits and interleukin-6: Evidence for "healthy neuroticism" in a US population sample. *Brain, Behavior, and Immunity*, *28*, 83–89. https://doi.org/10.1016/j.bbi.2012.10.020.

van der Doelen, R. H., Arnoldussen, I. A., Ghareh, H., van Och, L., Homberg, J. R., & &Kozicz T. (2015). Early life adversity and serotonin transporter gene variation interact to affect DNA methylation of the corticotropin-releasing factor gene promoter region in the adult rat brain. *Development and Psychopathology*, *27*(1), 123–135.

Zheng, Y., Fan, W., Zhang, X., & Dong, E. (2016). Gestational stress induces depressive-like and anxiety-like phenotypes through epigenetic regulation of BDNF expression in offspring hippocampus. *Epigenetics*, *11*(2), 150–162.

Chapter 3

# Depression after pregnancy

Munn-Sann Lye[a], Normala Ibrahim[b], Fatemeh Abdollahi[c], and Yin-Yee Tey[d]
[a]Department of Population Medicine, Faculty of Medicine and Health Sciences, Universiti Tunku Abdul Rahman, Sungai Long Campus, Cheras, Kajang, Selangor, Malaysia; [b]Department of Psychiatry, Faculty of Medicine and Health Sciences, Universiti Putra Malaysia, Serdang, Selangor, Malaysia; [c]Department of Public Health, Faculty of Health, Mazandaran University of Medical Sciences, Sari, Iran; [d]In personal capacity

## List of abbreviations

| | |
|---|---|
| **5-HTT** | 5-hydroxytryptamine (serotonin)transporter |
| **5HTTLPR** | 5HTT-linked polymorphic region |
| **APA** | American Psychiatric Association |
| **BDI-II** | Beck Depression Inventory |
| **CNS** | central nervous system |
| **COMT** | catechol-*O*-methyl transferase |
| **CRH** | corticotrophin-releasing hormone |
| **CYP2D6** | cytochrome P4502D6 |
| **ECT** | electroconvulsive therapy |
| **EDPS** | Edinburgh Postnatal Depression Scale |
| **MAO-A** | monoamine oxidase-A |
| **PDSS** | Postpartum Depression Screening Scale |
| **PHQ-9** | Patient Health Questionnaire-9 |
| **PPD** | postpartum depression |
| **SSRI** | selective serotonin reuptake inhibitor |
| **TPH** | tryptophan hydroxylase |

## Introduction

### Preamble

Depression after pregnancy has been described by a variety of other terms, including postpartum depression (PPD), postnatal depression, and, in some recent publications, is subsumed under depressive disorders occurring during the antenatal period and in the 4 weeks after delivery (American Psychiatric Association (APA), 2014). Although PPD may resemble other forms of depression, it has its own uniqueness, happening at a time in which pregnancy and parturition play an inherently vital role in influencing its occurrence and sequelae. Mental health experts also distinguish between baby blues and postpartum depression, in which the former starts shortly after delivery, is usually milder, and lasts for a shorter period of time, resolving typically by the second week after delivery. In this chapter, we are more concerned with the more severe and longer lasting postpartum depression, which impacts on the mother's economic productivity, social functioning, quality of life and fertility, as well as on the infant's emotional and physical development, and its cognitive and behavioral health. In some cultures, signs and symptoms are often missed by health-care workers and where women silently struggle with their illness due to family politics and/or lack of intercultural communication in the health services (Babatunde & Moreno-Leguizamon, 2012).

### Definition

Depression after pregnancy is addressed in the Diagnostic and Statistical Manual of Mental Disorders, 5th edition (DSM-5) (APA, 2014) under the heading of depressive disorders, specified as "with peripartum onset" with or without psychotic symptoms. In this definition, DSM-5 included onset of depression during pregnancy ("50% of major depressive episodes actually begin prior to delivery") and limited onset to within 4 weeks after delivery. However, many studies did not strictly adhere to the 4-week interval, and may range from 2 months up to even a year after delivery.

The American Psychiatric Association describes symptoms that include "sluggishness, fatigue, feeling sad, hopeless, helpless, or worthless; difficulty sleeping/sleeping too much, changes in appetite, difficulty concentrating/confusion, crying for 'no reason,' lack of interest in the baby, not feeling bonded to the baby, or feeling very anxious about the baby, feelings of being a bad mother, fear of harming the baby or oneself, and loss of interest or pleasure in life." (APA, 2014).

## Epidemiology

### Magnitude of the problem

The occurrence of postpartum depression fluctuates depending on location and population. In a recent systematic review, metaanalysis and metaregression of 291 studies from

The Neuroscience of Depression. https://doi.org/10.1016/B978-0-12-817933-8.00005-0

56 countries that included low-, middle-, and high-income countries, Hahn-Holbrook, Cornwell- Hinrichs, and Anaya (2018) reported a pooled prevalence of 18% (CI: 16–19%) on a per capita basis. High-income countries, for example, Singapore, Netherlands, United States, and Australia tend to have lower prevalence of PPD compared to low- to middle-income countries (Fig. 1). According to the authors, some of the heterogeneity could be due to differences in the degree of cultural collectivism while there may be a degree of over-reporting of PPD due to self-reporting when using the Edinburgh Postpartum Depression Scale rather than clinical interview.

### Risk factors

Strong predictors of PPD include depression during pregnancy, various stressors, lack of social support, and a previous history of depression, among others. Depression during pregnancy, or antenatal depression, has been linked to postpartum depression (Leung et al., 2017; Mersky & Janczewski, 2018; Mohamad Yusuff, Tang, Binns, & Lee, 2015). The prevalence ranges from 4% to 81%, being more common in the lower income countries (Sheeba et al., 2019). Various stressors have been identified to be associated with increased risk of PPD; these include anxiety trait, feelings of guilt, or insecurity (American Psychiatric Association, 2014; Pham et al., 2018), exposure to stressful life events (Armstrong, Fraser, Dadds, & Morris, 1999; Mersky & Janczewski, 2018; Shwartz, Shoahm-Vardi, & Daoud, 2019), intimate partner violence (physical or sexual) (Mersky & Janczewski, 2018; Shwartz et al., 2019), and stressful maternity and childcare experiences (Chojenta, Lucke, Forder, & Loxton, 2016; Mohamad Yusuff et al., 2015). In addition, lack of social support (Shwartz et al., 2019) and a previous history of depression (Chojenta et al., 2016; Pham et al., 2018) have also been associated with an increased risk of PPD.

### Protective factors

On the other hand, wanted pregnancy (Shwartz et al., 2019), having more education, having a permanent job, being of the ethnic majority, and having a kind, trustworthy intimate partner (Fisher et al., 2012) tended to protect mothers from PPD. Other factors favorable to reduced risk of PPD include supportive behavior of husband who assisted with infant care and satisfaction with marital relationship (Mohamad Yusuff et al., 2015); partner being employed, pregnant mother who routinely attends prenatal sessions (Fan et al., 2020); and a high level of social support (Leung et al., 2017).

## Consequences of postpartum depression

In a systematic review of 122 studies, Slomian, Honvo, Emonts, Reginster, and Bruyère (2019) found that PPD affected the mother's physical and psychological health and relationships, and increased risky behaviors while infants suffered from impaired anthropometry, physical health, sleep, as well as motor, cognitive, language, emotional, social, and behavioral development. In addition, mother–child interactions, including bonding and breastfeeding, and the maternal role are all affected negatively. Netsi et al. (2018) showed a dose–response relationship between severity of mother's PPD and behavioral problems in their children at 3.5 years of age (Table 1). Mothers with PPD have been reported to have higher relationship discord with their partners while their partners have also experienced depressive symptoms, anxiety, and stress (Moore Simas et al., 2019).

There is also a dose–response relationship between mother's severity of PPD and depression in their children that creates an environment detrimental to the personal development of mothers or the optimal development of a child and it behooves health policy makers to develop and institute health programs that will detect PPD promptly to prevent the negative consequences to both mother and child (Netsi et al., 2018). In the United Kingdom, Bauer, Knapp, and Parsonage (2016) estimated the total lifetime costs of perinatal depression as £75,728 (£34,811) per woman with PPD, aggregated costs being £6.6 billion, with most of the costs associated with impact on children.

## Neurobiological basis of PPD

PPD is one of the psychiatric conditions that is amenable to treatment. However, as the etiology of PPD is multifactorial, therapy should be guided by our understanding of the underlying evidence not only on the neurobiology of the disease but also on psycho-socio-cultural factors (Fig. 2) that interact with the biology of the disease (Beck, 2002; Halbreich, 2005). An awareness of which theory contributes to PPD could assist clinicians in developing treatment plans.

### Neurobiological theories

The predominant theoretical perspective for PPD is the biological model. In this model, for each individual woman, a basic set of pathological and neurobiological factors act upon the supposedly passive individual (Beck, 2002; Yim, Tanner Stapleton, Guardino, Hahn-Holbrook, & Dunkel Schetter, 2015). Several hypotheses have been proposed and tested, including hormonal changes (Brummelte & Galea, 2016) and inflammatory processes (Kettunen, 2019) as well as genetic vulnerability (Serati, Redaelli, Buoli, & Altamura, 2016).

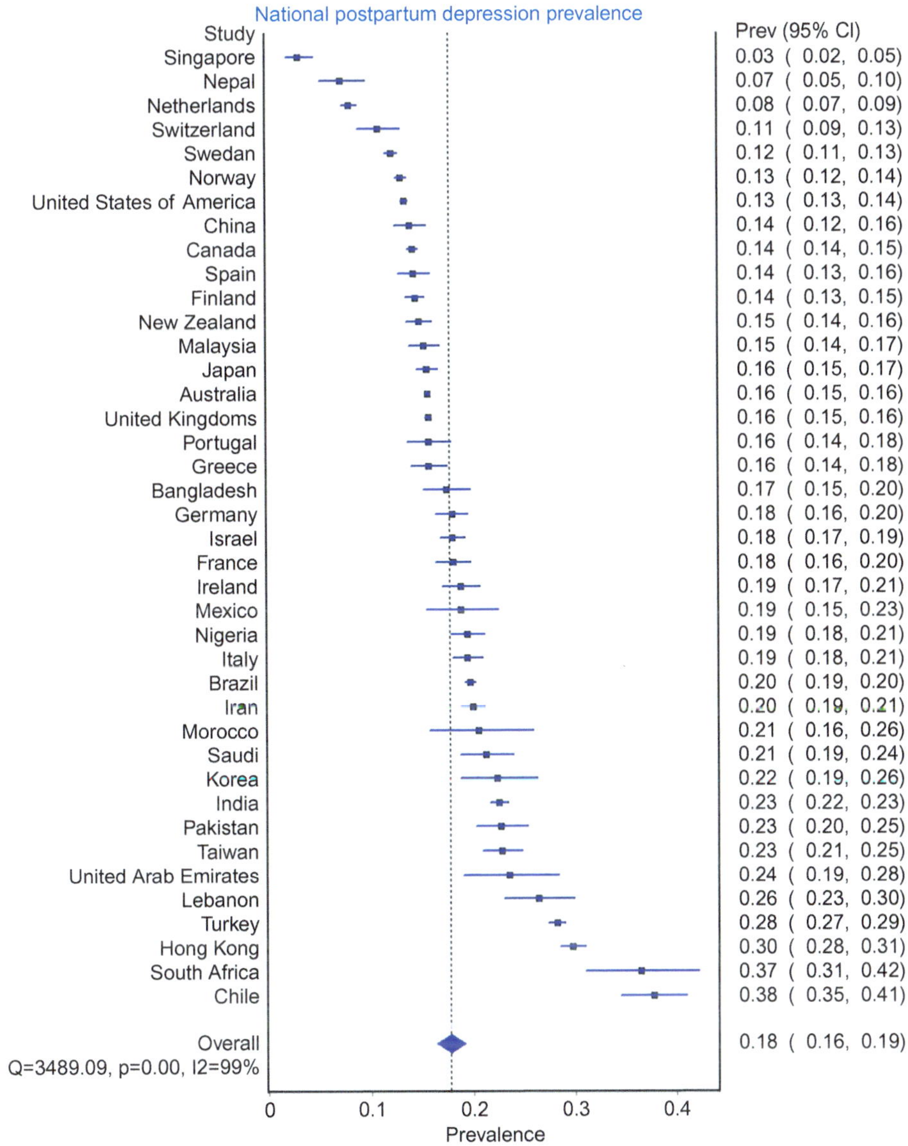


**FIG. 1** Meta-analytically derived postpartum depression prevalence in 40 countries. The forest plot shows that the meta-analytically derived PPD prevalence ranged from 3.1% in Singapore to 37.7% in Chile. *From: Hahn-Holbrook, J., Cornwell-Hinrichs, T., & Anaya, I. (2018). Economic and Health Predictors of National Postpartum Depression Prevalence: A Systematic Review, meta-analysis, and meta-Regression of 291 Studies from 56 countries.* Frontiers in Psychiatry. *Copyright: © 2018 Hahn-Holbrook, Cornwell-Hinrichs and Anaya. This is an open-access article distributed under the terms of the Creative Commons Attribution License (CC BY). The use, distribution or reproduction in other forums is permitted, provided the original author(s) and the copyright owner are credited and that the original publication in this journal is cited, in accordance with accepted academic practice. No use, distribution or reproduction is permitted which does not comply with these terms.*

There exist different theories regarding the pathophysiological hormonal effects that occur during puerperium in increasing the risk of PPD, including the withdrawal theory (Wang et al., 2020), interaction among the hypothalamic–pituitary–gonadal and the hypothalamic–pituitary–adrenal systems (Brummelte & Galea, 2016; Serati et al., 2016; Shadrina, Bondarenko, & Slominsky, 2018) (Fig. 3), and the changes in levels of gonadal hormones (Wang et al., 2020).

It has been hypothesized that the sharp decline in reproductive hormone levels that occurs during postpartum is the main cause of PPD (Serati et al., 2016). The fluctuation in the reproductive hormone levels has been postulated to be a trigger for changes in peripheral and central monoamine centers (Yim et al., 2015). Estrogen withdrawal after birth that results in reduced hippocampal glucocorticoid receptor (GR) expression was assumed to be the main mechanism causing PPD. Estrogen is involved in the

**TABLE 1 Logistic and ordinal logistic regressions investigating the association between postnatal depression (PND) and adverse child outcomes, controlling for maternal education.**

| Level of PND severity | Behavioral problems at 3.5 years (*n* = 7917)[a] | | Low GCSE mathematics grades at 16 years (*n* = 4941) | | Offspring depression at 18 years (*n* = 3486) | |
|---|---|---|---|---|---|---|
| | OR (95% CI) | *P* value | OR (95% CI) | *P* value | OR (95% CI) | *P* value |
| Below threshold[b] | **1 [Reference]** | NA | **1 [Reference]** | NA | **1 [Reference]** | NA |
| Moderate but not persistent[c] | 2.22 (1.74–2.83) | <0.001 | 1.14 (0.77–1.68) | 0.51 | 1.11 (0.51–2.44) | 0.79 |
| Marked but not persistent[d] | 1.91 (1.36–2.68) | <0.001 | 1.53 (0.89–2.63) | 0.13 | 2.34 (1.03–5.29) | 0.04 |
| Severe but not persistent[e] | 2.39 (1.78–3.22) | <0.001 | 1.40 (0.89–2.22) | 0.15 | 1.72 (0.77–3.82) | 0.18 |
| Moderate persistent[f] | 3.04 (2.10–4.38) | <0.001 | 1.65 (0.89–3.05) | 0.11 | 1.05 (0.32–3.42) | 0.94 |
| Marked persistent[g] | 2.84 (1.71–4.71) | <0.001 | 1.32 (0.60–2.90) | 0.46 | 2.30 (0.67–7.90) | 0.19 |
| Severe persistent[h] | 4.84 (2.94–7.98) | <0.001 | 2.65 (1.26–5.57) | 0.01 | 7.44 (2.89–19.11) | <0.001 |

[a]*Using the Rutter revised total problems scale.*
[b]*EPDS score of less than 13 points in the postnatal year.*
[c]*EPDS score of 13–14 points at 2 months and less than 13 points at 8 months.*
[d]*EPDS score of 15–16 points at 2 months and less than 15 points at 8 months.*
[e]*EPDS score of 17 or more points at 2 months and less than 17 points at 8 months.*
[f]*EPDS score of 13–14 points at 2 months and 13 or more points at 8 months.*
[g]*EPDS score of 15–16 points at 2 months and 15 or more points at 8 months.*
[h]*EPDS score of 17 or more points at 2 and 8 months.*
*Data shows that risks for adverse outcomes for children of women with persistent PND were higher compared to nonpersistent PND.*
*Abbreviations:* EPDS, *Edinburgh Postnatal Depression Scale;* GCSE, *General Certificate of Secondary Education;* NA, *not applicable;* OR, *odds ratio;* PND, *postnatal depression.*
From: Netsi, E., Pearson, R. M., Murray, L., Cooper, P., Craske, M.G., & Stein, A. (2018). Association of persistent and severe postnatal depression with child outcomes. *JAMA Psychiatry, 75*(3), 247–253. 

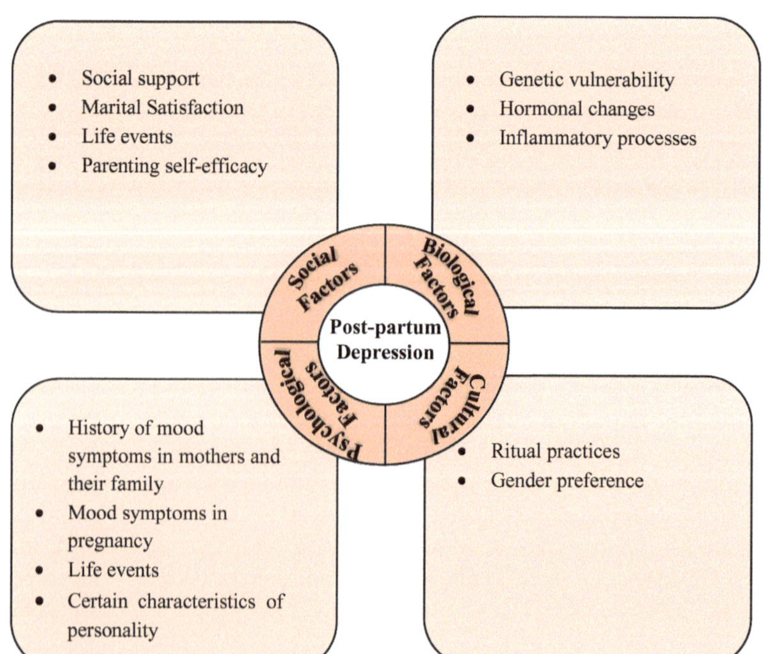


FIG. 2 Bio-psycho-socio-cultural theoretical concept of postpartum depression. The figure shows theoretical basis of PPD involving biological, psychological, social, and cultural factors. *Figure drawing Copyright: © Fatemeh Abdollahy.*

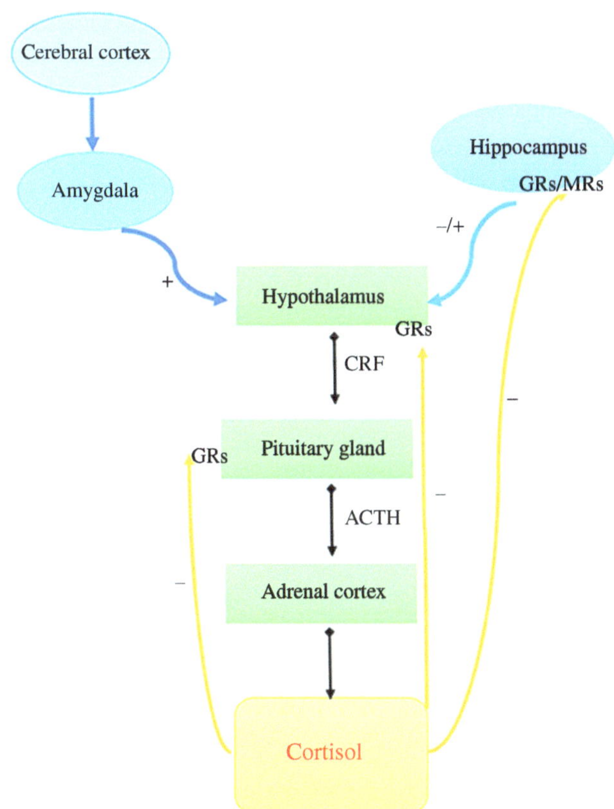


**FIG. 3** Schematic diagram of hypothalamic–pituitary–adrenal (HPA) axis. The diagram shows the HPA axis's response to stress and the negative feedback loops. *ACTH*, adrenocorticotropic hormone; *CRF*, corticotropin-releasing factor; *GRs*, glucocorticoid receptor; *MRs*, mineralocorticoid receptors
Secretion ⟷ ; stimulation ⟶+ ; inhibition ⟶− .
*From: Shadrina, M., Bondarenko, E. A., & Slominsky, P. A. (2018). Genetics factors in major depression disease.* Frontiers in Psychiatry. *Permission: Frontiers is fully compliant with open access mandates, by publishing its articles under the Creative Commons Attribution license (CC-BY). Funder mandates such as those by the Wellcome Trust (UK), National Institutes of Health (USA) and the Australian Research Council (Australia) are fully compatible with publishing in Frontiers. Authors retain copyright of their work and can deposit their publication in any repository. The work can be freely shared and adapted provided that appropriate credit is given and any changes specified.*

emotional and cognitive processes through its effect on neurotransmitters. It stabilizes serotonin levels and helps to retain more of these neurotransmitters in the brain, and also has an influence on adrenaline, norepinephrine, and serotonin receptors (Brummelte & Galea, 2016).

The fluctuation of other hormones transitioning from pregnancy to postpartum that is suspected to affect the development of PPD are beta-endorphin (Yim et al., 2010), cortisol (Taylor, Glover, Marks, & Kammerer, 2009), prolactin (Groer & Morgan, 2007), oxytocin (Skrundz, Bolten, Nast, Hellhammer, & Meinlschmidt, 2011), corticotropin-releasing hormone (Hahn-Holbrook, Schetter, Arora, & Hobel, 2013), and thyroid hormones (Yim et al., 2015).

Conclusive evidence, however, is lacking; even though many researchers have proposed the physiological fluctuation of certain hormones as the cause of mood disorders after birth, this has not been supported consistently by the literature and only a small subset of women experience PPD (Yim et al., 2015). Although it is suspected that the rapid reduction in reproductive hormones occurring after birth predisposes genetically susceptible women to PPD, Yim et al. (2015) opined that moderators of biological vulnerability such as a history of depression or life stressor events have not been examined sufficiently in these studies. Hence the available evidence needs to be interpreted with caution.

Some researchers have explored the inflammatory processes in the context of PPD. During pregnancy the immune system's function is complicated by the fetus which carries paternal antigens. It is identified as foreign by the maternal immune system but is however tolerated by the maternal immune system (Zenclussen, 2013) although the mechanism of action is unknown.

During pregnancy and postpartum a shift in the balance of inflammatory biomarkers and the innate immune system occurs (Corwin & Pajer, 2008). A complicated balance of pro-inflammatory cytokines such as interleukin (IL)-6, IL-1β, and tumor necrosis factor-alpha (TNF-α), which initiate the inflammatory response, and anti-inflammatory cytokines (e.g., IL-10) contribute to the regulation the immune system (Kronborg et al., 2011). Raised inflammatory markers accompanied by psychological or biological stressors is linked with depression in nonpuerperal women (Raison, Capuron, & Miller, 2006). During pregnancy and early postpartum period, the pro-inflammatory state is enhanced (Maes et al., 2000) and may be related to the development of PPD (Kendall-Tackett, 2007). Although this relationship was significant in some studies (Boufidou et al., 2009), other studies have not found this to be so (Okun et al., 2011). The reliance on nonspecific biomarkers for depression and small sample sizes in some of these studies warrant more detailed empirical investigations in this area (Yim et al., 2015).

Genes associated with the inflammation response may play a causal role in the pathogenesis of PPD; for instance, polymorphisms in functional allelic variants of the genes for IL-1β and TNF-α have been shown to increase risk of major depressive disorder (Raison et al., 2006), although this relationship has not been examined in relation to PPD risk. The genes linked to 5-hydroxytryptamine (serotonin) transporter (5-HTT), 5HTT-linked polymorphic region (5HTTLPR), tryptophan hydroxylase (TPH), monoamine oxidase-A (MAO-A), and catechol-*O*-methyl transferase (COMT) have been implicated in the context of PPD (Yim et al., 2015). It has also been hypothesized that polymorphisms in genes that regulate reproductive hormones may render some women susceptible to mood disorders (Yim et al., 2015). It is probable that effects of polymorphic variations in genes within the monoaminergic system, and in genes for the estrogen receptor, oxytocin peptide, glucocorticoid receptor and corticotrophin-releasing hormone (CRH) receptor 1 predispose women to increased risk of PPD (Yim et al., 2015). In one study, Josefsson (2003) found that the metabolic rate of women with polymorphisms in the CYP2D6 (Cytochrome P4502D6) gene as a biomarker of personality phenotype in depressed pregnant and postpartum women was more than that anticipated in the general population. Some researchers noted that the influence of genes by itself is insufficient to cause psychiatric disorders such as PPD (Yim et al., 2015) and development of these diseases depends on epigenetics and environmental triggers (Comasco, Sylvén, Papadopoulos, Oreland, et al., 2011; Comasco, Sylvén, Papadopoulos, Sundström-Poromaa, et al., 2011; Mitchell et al., 2011).

### Conclusion

There is no consensus regarding the etiological mechanisms of PPD. It may be possible that one or a combination of the theories discussed above may operate in the depressed postpartum woman (Abdollahi, Lye, & Zarghami, 2016). Halbreich (2005) developed a Bio-Psycho-Socio-Cultural model of postpartum depression and postulated that the essential causative factor may be related to genetic predisposition to the disorders of reproduction. This susceptibility may be due to the individual's hypersensitivity to fluctuation of gonadal hormones or other steroids leading to symptoms during other periods of hormonal instability such as in the days prior to menstruation, use of hormonal contraception, and during menopause. This susceptibility is combined with a tendency to dysregulation and unsuccessful adjustment mechanisms of the central nervous system (CNS). It should be noted that CNS susceptibility is not confined only to fluctuation in steroid levels. CNS vulnerability may be linked to impaired adjustment to other conditions, for example, external stressors, abrupt drug and hormonal withdrawal, and other psychobiological as well as social changes. However, only some exposed and susceptible individuals may show symptoms and illness (Halbreich, 2005).

## Management of PPD

Recognizing the illness at an early stage is key in the management of PPD; unrecognized illness leads to prolonged depression that affects mothers and infants. Delay in adequate treatment is the most significant factor influencing the duration of PPD (Beck, 2002). The approach to manage PPD should be comprehensive and multidisciplinary as well as holistic, as suffering mothers, infants, and their families are affected. Apart from ensuring that depressive symptoms are treated using pharmacological or psychotherapeutic approach, the safety and well being of mothers and infants need to be considered as well. Thomas, Gandhi, & Parel, 2018 have proposed an integrated approach by including a team that comprises a mental health professional, an obstetrician, community health worker, nurses, and trained midwives in managing perinatal mental illness including PPD (Fig. 4). The focus of care is mainly on mothers' ability to cope with the motherhood role and assisting them to care for and promote bonding with their infants, in addition to resolving their acute depressive symptoms.

In the pharmacological treatment of PPD, antidepressants such as selective serotonin reuptake inhibitors (SSRI) are the mainstay medication and their usage is similar to that for major depressive disorder in non-postpartum women (Sharma & Sharma, 2012). However, the lowest effective dosage of antidepressant should be prescribed in mothers who favor breastfeeding (Davanzo, Copertino, De Cunto, Minen, & Amaddeo, 2011). The potential risk of antidepressant medication exposure to infants also requires careful consideration in its prescription as the medications pass into the breast milk (Hallberg & Sjöblom, 2005). Therefore, mothers with PPD who receive antidepressants should be adequately informed on the benefits and risks of antidepressants and infants should be monitored regularly for symptoms and signs of antidepressant exposure such as persistent irritability, sedation, or failure to thrive (Sit & Wisner, 2005). As PPD commonly coexists with anxiety symptoms, benzodiazepine may be useful; intermittent insomnia may benefit from the use of non-benzodiazepine hypnotic medications (BC Reproductive Mental Health & Perinatal Services, 2014).

In managing PPD, careful consideration should be given to those who are at imminent risk of committing suicide or infanticide. Therefore, suicide and infanticide risk assessment is crucial in the evaluation of their mental status examination. Hospitalization should be considered in those who are at risk of harming themselves or their infants. Infanticide risk occurs when severely depressed postpartum women neglect and abuse the infants as being unwanted or are having marital discord that leads to seeking revenge against the partner. It was reported between 16% and 29% mothers who performed infanticide also have completed suicide (Krischer, Stone, Sevecke, & Steinmeyer, 2007). Electroconvulsive therapy (ECT) is a reasonable option for rapid improvement in symptoms and recommended for women with PPD who have severe depression and at risk of suicide, although it may give rise to minimal and short-lasting risks to breastfeeding infants due to medication given for anesthesia and muscle relaxation (Focht & Kellner, 2012; Centre of Perinatal Excellence, 2017). However, for treatment-naïve individuals, to convince them or their family to opt for ECT may be a challenge.

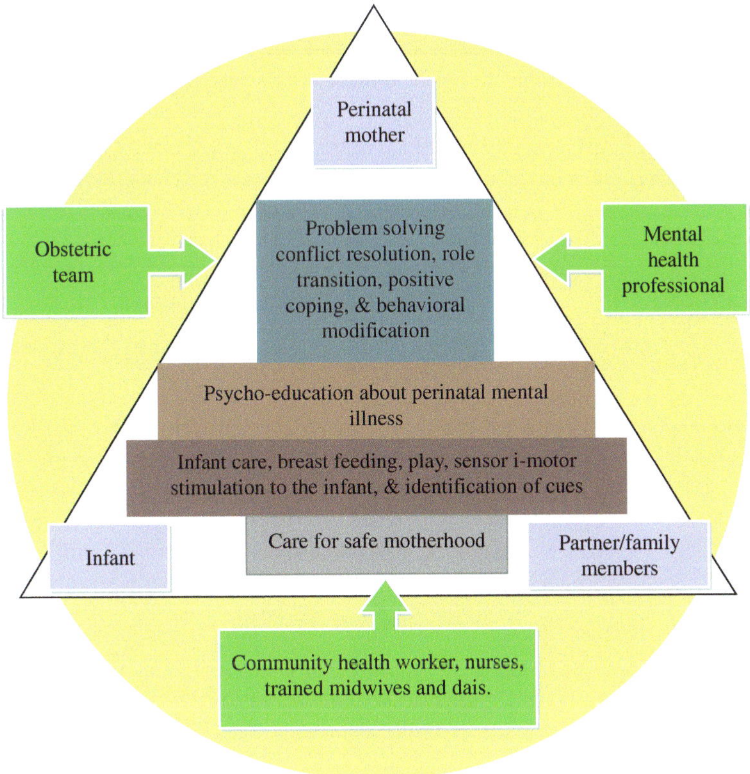


**FIG. 4** Proposed Care Model in managing mothers with perinatal mental illness. The diagram shows proposed care model as an approach for management of perinatal mental illness that includes PPD. *From: Thomas, L., Gandhi, S., & Parel, J. T. (2018) Interventions for mothers with postpartum depression: A systematic review.* International Journal of Depression and Anxiety, 1, *002. Copyright: © 2018 Thomas L, et al. This is an open-access article distributed under the terms of the Creative Commons Attribution License, which permits unrestricted use, distribution, and reproduction in any medium, provided the original author and source are credited.*

Psychotherapeutic approaches in treating PPD are well established as reported in the literature. Clinical practice guidelines recommend a combination of psychotherapy and pharmacological treatments for mild to moderate PPD (Centre of Perinatal Excellence, 2017; Ministry of Health Malaysia, 2019; National Collaborating Centre for Mental Health (UK), 2014). Specific psychotherapies, such as cognitive behavioral therapy, and brief dynamic interpersonal psychotherapy, have been documented to reduce symptoms of depression in controlled studies (Cooper, Murray, Wilson, & Romaniuk, 2003). A moderate positive effect in reducing symptoms of PPD in the short term, using psychosocial or psychological interventions, has also been reported in a meta-analysis of randomized controlled trials by Cuijpers, Brännmark, and van Straten (2008); however, the long-term effect is unknown. In a recent systematic review, concluded that psychotherapeutic approaches addressing coping skills, problem solving, and interpersonal skills that are culturally sound, either individually or group based, have a beneficial role in PPD treatment.

In conclusion, clinicians must be aware of a number of important issues in the management of PPD. Early recognition of symptoms is vital as this will facilitate early treatment and thus improve prognosis of the illness. Multidisciplinary approaches in managing PPD would assist mothers in fulfilling their roles on safe motherhood and in improving relationships with partners and families as well as in managing their physical and psychological health. Suicide and infanticide risk must be evaluated in all cases and the choice of treatment should be informed by risk–benefit analysis, as treatment especially pharmacotherapy may have adverse effects on breastfeeding mothers and infants. Surveillance of infants during treatment is important in order to prevent any harmful consequences on their development.

## Prevention of PPD

Interventions to alleviate the major risk factors that predispose women to develop PPD could be instituted at any stage of a woman's reproductive life, preceded by risk evaluation to identify risk factors. Women at risk of PPD can be identified during routine preconception, antenatal and postnatal care visits. Screening has been recommended at a woman's first contact with primary care and during early postpartum National Collaborating Centre for Mental Health (UK), 2014. In addition, the American Academy of Pediatrics recommends routine screening in mothers during well-child clinic visits at 1, 2, 4, and 6 months of age (Earls, Yogman, Mattson, & Rafferty, 2019). Various screening tools have been used over the past two decades as presented in Table 2.

**TABLE 2 Some screening tools for postpartum depression.**

| Screening tool | Description | Sensitivity[a] and Specificity[b] |
|---|---|---|
| Edinburgh Postnatal Depression Scale (EDPS) | Self-report instrument, with 10 questions about PPD symptoms for the past 7days that can be used during pregnancy and postpartum period. Each question is rated on a scale of 0–3, with total score range of 0–30 | Uses cut-off point of ≥12:<br>(a) 86% (b) 78% (Moraes, Lorenzo, Pontes, Montenegro, & Cantilino, 2017) |
| Beck Depression Inventory (BDI-II) | Self-report instrument, consisting of 21 items on symptoms of depression for the past 2weeks. Each item is scored 0 to 3 points for a total score range of 0–63 | 0 – 13: Minimal depression<br>14 – 19: Mild depression<br>20 – 28: Moderate depression<br>29 – 63: Severe depression<br>(a) 81% (b) 92% (Beck, Steer, & Brown, 1996; Hubley, 2014). |
| Patient Health Questionnaire-9 (PHQ-9) | Self-administered questionnaire comprising of 9 questions on symptoms of major depressive disorder for the past 2weeks. Scores for each question range from 0 to 3, with total score ranging from 0 to 27 | Uses cut-off point of >10: (a) 88% (b) 86% (Levis et al., 2019) |
| Postpartum Depression Screening Scale (PDSS) | Self-report questionnaire comprising of 35 items and 7 factors on depressive symptoms for the past 2 weeks. It uses Likert scale ranging from 1 to 5, with total score ranging from 35 to 175 | Uses cut-off point of 80: (a) 94% (b) 98% (Beck & Gable, 2001) |
| Center for Epidemiologic Studies Depression Scale (CES-D) | Self-administered questionnaire with 20 items. Scores for each question range from 0 to 3, with total score ranging from 0 to 60 | Uses cut-off point 16 or greater to identify individuals at risk for clinical depression (Radloff, 1977)<br>(a) 95% (b) 29% (Radloff, 1977) |
| Zung Self-Rating Depression Scale (Zung SDS) | Self-administered questionnaire with 20 items. Uses 4-point Likert scale, ranging from 1 to 4 with total score ranging from 20 to 80 | Scoring:<br><50 Normal<br>50–59 mild depression<br>60–69 marked depression<br>>70 severe to extreme major depression<br>(a) 83%<br>(b) 82%<br>(Dunstan & Scott, 2019) |

Characteristics of each screening tool are summarized in the table.

Among women with risk factors, educational initiatives to prepare women physically and emotionally to adapt to the changes that could potentially occur, exhibited enhanced effectiveness when they were given during both the antenatal and postpartum periods (Armstrong et al., 1999). Psychoeducation during antenatal and postpartum periods decreased the rate of depression among primiparous women who were at risk (Elliott et al., 2000). Psychotherapeutic approaches including interpersonal therapy among high-risk women during the antenatal period particularly in the second and third trimesters (Grote et al., 2009) and cognitive behavioral therapy (Kozinszky et al., 2012) have shown promising effects in the primary prevention of postpartum depression by reducing depression symptomology during the antenatal period. Continuous midwifery care and monitoring of women in vulnerable groups both antenatally and postpartum assisting them to engage in the treatment (Marks, Siddle, & Warwick, 2003) and providing telephone-based peer support to high-risk women (Dennis et al., 2009) have been shown to reduce depression symptomology as a secondary prevention measure.

## Key facts in screening for PPD

- Screening for PPD helps to detect women at risk early.
- Women at risk of PPD can be identified early during routine preconception, antenatal and postnatal care visits.
- Routine screening is recommended in mothers during well-child clinic visits at 1, 2, 4, and 6 months of age.

- Past personal and family history of psychiatric illness are strong risk factors of PPD and should be elicited during screening.
- Screening also serves to establish the woman's current state of mind and mood and could be useful as baseline information for follow-up and management.

## Summary points

- This chapter discusses the epidemiology, consequences, neurobiology, management, and prevention of PPD.
- Prevalence of PPD varies across countries and cultures, and is higher in low- and middle-income countries.
- A number of neurobiological mechanisms have been proposed for the causal mechanism of PPD.
- Management of women with PPD should preferably involve multidisciplinary teams, using pharmacological and psycho-social-cultural approaches.
- Preventive programs targeting women at risk could reduce the risk of PPD and depressive symptoms.

## Mini-dictionary of terms

**Cognitive behavioral therapy** A form of psychotherapy that targets cognition and behavior.

**Electroconvulsive therapy** A mode of treatment that uses electrical energy to provoke convulsions in patients with depressive symptoms.

**Interpersonal therapy** A form of psychotherapy that focuses on symptoms that arise from interpersonal difficulties.

**Neurobiology** A study of the biological mechanisms in the nervous system.

**Personality trait** A characteristic that reflects how an individual reacts to situations in life.

**Postpartum depression** A mental condition with onset within 4 weeks after delivery in a mother, displaying symptoms of low mood that includes feeling sad, loss of sense of self-worth, fear of harming baby or self, and loss of pleasure in life.

**Prevalence** A measure of the magnitude or extent of a health condition, usually computed by number of old and new cases of a disease per unit population.

**Risk factors** Variables that increase risk of disease or adverse health effects.

**Role stressor** Stress arising from the role of being a wife, a mother, a daughter-in-law, an employee, etc.

**Stressful life events** Acute incidents that occur in life that cause stress during pregnancy and postpartum.

## References


Abdollahi, F., Lye, M. S., & Zarghami, M. (2016). Perspective of postpartum depression theories: A narrative literature review. *North American Journal of Medical Sciences*, *8*(6), 232–236.

American Psychiatric Association. (2014). *Diagnostic and statistical manual of mental disorders* (5th ed.). Washington, DC: American Psychiatric Publishing.

Armstrong, K. L., Fraser, J. A., Dadds, M. R., & Morris, J. (1999). A randomized, controlled trial of nurse home visiting to vulnerable xfamilies with newborns. *Journal of Paediatrics and Child Health*, *35*(3), 237–244.

Babatunde, T., & Moreno-Leguizamon, C. J. (2012). Daily and cultural issues of postnatal depression in African women immigrants in South East London: Tips for health professionals. *Nursing Research & Practice*, *2012*.

Bauer, A., Knapp, M., & Parsonage, M. (2016). Lifetime costs of perinatal anxiety and depression. *Journal of Affective Disorders*, *192*, 83–90.

Beck, C. T. (2002). Theoretical perspectives of postpartum depression and their treatment implications. *The American Journal of Maternal Child Nursing*, *27*(5), 282–287.

Beck, C. T., & Gable, R. K. (2001). Further validation of the postpartum depression screening scale. *Nursing Research*, *50*(3), 155–164.

Beck, A. T., Steer, R. A., & Brown, G. K. (1996). *BDI-II: Beck depression inventory manual* (2nd ed.). San Antonio, TX: Psychological Corporation.

Boufidou, F., Lambrinoudaki, I., Argeitis, J., Zervas, I. M., Pliatsika, P., Leonardou, A. A., et al. (2009). CSF and plasma cytokines at delivery and postpartum mood disturbances. *Journal of Affective Disorders*, *115*(1–2), 287–292.

Brummelte, S., & Galea, L. A. (2016). Postpartum depression: Etiology, treatment and consequences for maternal care. *Hormones and Behavior*, *77*, 153–166.

Centre of Perinatal Excellence. (2017). *Mental health care in the perinatal period: Australian clinical practice guideline. [pdf file]. Melbourne: Centre of Perinatal Excellence*. Retrieved from https://cope.org.au/wp-content/uploads/2017/10/Final-COPE-Perinatal-Mental-Health-Guideline.pdf.

Chojenta, C. L., Lucke, J. C., Forder, P. M., & Loxton, D. J. (2016). Maternal health factors as risks for postnatal depression: A prospective longitudinal study. *PLoS One*, *11*(1), e0147246.

Comasco, E., Sylvén, S. M., Papadopoulos, F. C., Oreland, L., Sundström-Poromaa, I., & Skalkidou, A. (2011). Postpartum depressive symptoms and the BDNF Val66Met functional polymorphism: Effect of season of delivery. *Archives of Women's Mental Health*, *14*(6), 453–463.

Comasco, E., Sylvén, S. M., Papadopoulos, F. C., Sundström-Poromaa, I., Oreland, L., & Skalkidou, A. (2011). Postpartum depression symptoms: A case-control study on monoaminergic functional polymorphisms and environmental stressors. *Psychiatric Genetics*, *21*(1), 19–28.

Cooper, P. J., Murray, L., Wilson, A., & Romaniuk, H. (2003). Controlled trial of the short- and long-term effect of psychological treatment of post-partum depression. I. Impact on maternal mood. *The British Journal of Psychiatry*, *182*, 412–419.

Corwin, E. J., & Pajer, K. (2008). The psychoneuroimmunology of postpartum depression. *Journal of Women's Health*, *17*(9), 1529–1534.

Cuijpers, P., Brännmark, J. G., & van Straten, A. (2008). Psychological treatment of postpartum depression: A meta-analysis. *Journal of Clinical Psychology*, *64*(1), 103–118.

Davanzo, R., Copertino, M., De Cunto, A., Minen, F., & Amaddeo, A. (2011). Antidepressant drugs and breastfeeding: A review of the literature. *Breastfeeding Medicine*, *6*(2), 89–98.

Dennis, C. L., Hodnett, E., Kenton, L., Weston, J., Zupancic, J., Stewart, D. E., et al. (2009). Effect of peer support on prevention of postnatal depression among high risk women: Multisite randomised controlled trial. *BMJ*, *338*, a3064.

Dunstan, D. A., & Scott, N. (2019). Clarification of the cut-off score for Zung's self-rating depression scale. *BMC Psychiatry*, *19*, 177.

Earls, M. F., Yogman, M. W., Mattson, G., & Rafferty, J. (2019). AAP Committee on Psychosocial Aspects of Child and Family Health. Incorporating recognition and management of perinatal depression into pediatric practice. *Pediatrics*, *143*(1), e20183259.

Elliott, S. A., Leverton, T. J., Sanjack, M., Turner, H., Cowmeadow, P., Hopkins, J., et al. (2000). Promoting mental health after childbirth: A controlled trial of primary prevention of postnatal depression. *The British Journal of Clinical Psychology*, *39*(3), 223–241.

Fan, Q., Long, Q., De Silva, V., Gunarathna, N., Jayathilaka, U., Dabrera, T., et al. (2020). Prevalence and risk factors for postpartum depression in Sri Lanka: A population-based study. *Asian Journal of Psychiatry*, *47*, 101855.

Fisher, J., Cabral de Mello, M., Patel, V., Rahman, A., Tran, T., Holton, S., et al. (2012). Prevalence and determinants of common perinatal mental disorders in women in low- and lower-middle-income countries: A systematic review. *Bulletin of the World Health Organization*, *90*(2), 139G–149G.

Focht, A., & Kellner, C. H. (2012). Electroconvulsive therapy (ECT) in the treatment of postpartum psychosis. *The Journal of ECT*, *28*(1), 31–33.

Groer, M. W., & Morgan, K. (2007). Immune, health and endocrine characteristics of depressed postpartum mothers. *Psychoneuroendocrinology*, *32*(2), 133–139.

Grote, N. K., Swartz, H. A., Geibel, S. L., Zuckoff, A., Houck, P. R., & Frank, E. (2009). A randomized controlled trial of culturally relevant, brief interpersonal psychotherapy for perinatal depression. *Psychiatric Services*, *60*(3), 313–321.

Hahn-Holbrook, J., Cornwell-Hinrichs, T., & Anaya, I. (2018). Economic and health predictors of national postpartum depression prevalence: A systematic review, meta-analysis, and meta-regression of 291 studies from 56 countries. *Frontiers in Psychiatry*, *8*, 248.

Hahn-Holbrook, J., Schetter, C. D., Arora, C., & Hobel, C. J. (2013). Placental corticotropin-releasing hormone mediates the association between prenatal social support and postpartum depression. *Clinical Psychological Science: A Journal of the Association for Psychological Science*, *1*(3), 253–264.

Halbreich, U. (2005). Postpartum disorders: Multiple interacting underlying mechanisms and risk factors. *Journal of Affective Disorders*, *88*(1), 1–7.

Hallberg, P., & Sjöblom, V. (2005). The use of selective serotonin reuptake inhibitors during pregnancy and breast-feeding: A review and clinical aspects. *Journal of Clinical Psychopharmacology*, *25*(1), 59–73.

Hubley, A. M. (2014). Beck Depression Inventory. *Encyclopedia of Quality of Life and Well-Being Research*, 338–345. https://doi.org/10.1007/978-94-007-0753-5_156.

Josefsson, A. (2003). *Postpartum depression : Epidemiological and biological aspects (PhD dissertation)*. Linköping: Linköping University Electronic Press. Retrieved from http://urn.kb.se/resolve?urn=urn:nbn:se:liu:diva-5208.

Kendall-Tackett, K. (2007). A new paradigm for depression in new mothers: The central role of inflammation and how breastfeeding and anti-inflammatory treatments protect maternal mental health. *International Breastfeeding Journal*, *2*, 6.

Kettunen, P. (2019). *Postpartum depression: Time of onset, severity, symptoms, risk factors and treatment* (academic dissertation). Finland: Tampere University.

Kozinszky, Z., Dudas, R. B., Devosa, I., Csatordai, S., Tóth, E., Szabó, D., et al. (2012). Can a brief antepartum preventive group intervention help reduce postpartum depressive symptomatology? *Psychotherapy and Psychosomatics*, *81*(2), 98–107.

Krischer, M. K., Stone, M. H., Sevecke, K., & Steinmeyer, E. M. (2007). Motives for maternal filicide: Results from a study with female forensic patients. *International Journal of Law and Psychiatry*, *30*(3), 191–200.

Kronborg, C. S., Gjedsted, J., Vittinghus, E., Hansen, T. K., Allen, J., & Knudsen, U. B. (2011). Longitudinal measurement of cytokines in pre-eclamptic and normotensive pregnancies. *Acta Obstetricia et Gynecologica Scandinavica*, *90*(7), 791–796.

Leung, B. M., Letourneau, N. L., Giesbrecht, G. F., Ntanda, H., Hart, M., & Team, A. O. N. (2017). Predictors of postpartum depression in partnered mothers and fathers from a longitudinal cohort. *Community Mental Health Journal*, *53*(4), 420–431.

Levis, B., Benedetti, A., Thombs, B. D., & DEPRESsion Screening Data (DEPRESSD) Collaboration. (2019). Accuracy of Patient Health Questionnaire-9 (PHQ-9) for screening to detect major depression: Individual participant data meta-analysis. *BMJ*, *365*, l1476.

Maes, M., Lin, A. H., Ombelet, W., Stevens, K., Kenis, G., De Jongh, R., et al. (2000). Immune activation in the early puerperium is related to postpartum anxiety and depressive symptoms. *Psychoneuroendocrinology*, *25*(2), 121–137.

Marks, M. N., Siddle, K., & Warwick, C. (2003). Can we prevent postnatal depression? A randomized controlled trial to assess the effect of continuity of midwifery care on rates of postnatal depression in high-risk women. *The Journal of Maternal-Fetal & Neonatal Medicine*, *13*(2), 119–127. https://doi.org/10.1080/jmf.13.2.119.127.

Mersky, J. P., & Janczewski, C. E. (2018). Adverse childhood experiences and postpartum depression in home visiting programs: Prevalence, association, and mediating mechanisms. *Maternal and Child Health Journal*, *22*(7), 1051–1058.

Ministry of Health Malaysia. (2019). *Management of major depressive disorders, 2nd ed [pdf file]*. Malaysia: Ministry of Health. Retrieved from http://www.moh.gov.my/moh/resources/Penerbitan/CPG/1)_CPG_Management_Major_Depressive_Disorder_(Second_Edition).pdf.

Mitchell, C., Notterman, D., Brooks-Gunn, J., Hobcraft, J., Garfinkel, I., Jaeger, K., et al. (2011). Role of mother's genes and environment in postpartum depression. *Proceedings of the National Academy of Sciences of the United States of America*, *108*(20), 8189–8193.

Mohamad Yusuff, A. S., Tang, L., Binns, C. W., & Lee, A. H. (2015). Prevalence and risk factors for postnatal depression in Sabah, Malaysia: A cohort study. *Women and Birth*, *28*(1), 25–29.

Moore Simas, T. A., Huang, M. Y., Patton, C., Reinhart, M., Chawla, A. J., Clemson, C., et al. (2019). The humanistic burden of postpartum depression: A systematic literature review. *Current Medical Research and Opinion*, *35*(3), 383–393.

Moraes, G. P., Lorenzo, L., Pontes, G. A., Montenegro, M. C., & Cantilino, A. (2017). Screening and diagnosing postpartum depression: When and how? *Trends in Psychiatry and Psychotherapy*, *39*(1), 54–61.

National Collaborating Centre for Mental Health (UK). (2014). *Antenatal and postnatal mental health: Clinical management and service guidance*. (British Psychological Society).

Netsi, E., Pearson, R. M., Murray, L., Cooper, P., Craske, M. G., & Stein, A. (2018). Association of persistent and severe postnatal depression with child outcomes. *JAMA Psychiatry*, *75*(3), 247–253.

Okun, M. L., Luther, J., Prather, A. A., Perel, J. M., Wisniewski, S., & Wisner, K. L. (2011). Changes in sleep quality, but not hormones predict time to postpartum depression recurrence. *Journal of Affective Disorders*, *130*(3), 378–384.

Pham, D., Cormick, G., Amyx, M. M., Gibbons, L., Doty, M., Brown, A., et al. (2018). Factors associated with postpartum depression in women from low socioeconomic level in Argentina: A hierarchical model approach. *Journal of Affective Disorders*, *227*, 731–738.

Radloff, L. S. (1977). The CES-D scale: a self-report depression scale for research in the general population. *Applied Psychological Measurement*, *1*, 385–401.

Raison, C. L., Capuron, L., & Miller, A. H. (2006). Cytokines sing the blues: Inflammation and the pathogenesis of depression. *Trends in Immunology*, *27*(1), 24–31.

Reproductive Mental Health, B. C., & Perinatal Services, B. C. (2014). Best practice guidelines for mental health disorders in the perinatal period. *pdf file*. BC: Vancouver. Retrieved from http://www.perinatalservicesbc.ca/Documents/Guidelines-Standards/Maternal/MentalHealthDisordersGuideline.pdf.

Serati, M., Redaelli, M., Buoli, M., & Altamura, A. C. (2016). Perinatal major depression biomarkers: A systematic review. *Journal of Affective Disorders*, *193*, 391–404.

Shadrina, M., Bondarenko, E. A., & Slominsky, P. A. (2018). *Genetics factors in major depression disease*. (Front Psychiatry).

Sharma, V., & Sharma, P. (2012). Postpartum depression: Diagnostic and treatment issues. *Journal of Obstetrics and Gynaecology Canada*, *34* (5), 436–442.

Sheeba, B., Nath, A., Metgud, C. S., Krishna, M., Venkatesh, S., Vindhya, J., et al. (2019). Prenatal depression and its associated risk factors among pregnant women in Bangalore: A hospital based prevalence study. *Frontiers in Public Health*, *7*, 108.

Shwartz, N., Shoahm-Vardi, I., & Daoud, N. (2019). Postpartum depression among Arab and Jewish women in Israel: Ethnic inequalities and risk factors. *Midwifery*, *70*, 54–63.

Sit, D. K., & Wisner, K. L. (2005). Decision making for postpartum depression treatment. *Psychiatric Annals*, *35*(7), 577–584.

Skrundz, M., Bolten, M., Nast, I., Hellhammer, D. H., & Meinlschmidt, G. (2011). Plasma oxytocin concentration during pregnancy is associated with development of postpartum depression. *Neuropsychopharmacology*, *36*(9), 1886–1893.

Slomian, J., Honvo, G., Emonts, P., Reginster, J. Y., & Bruyère, O. (2019). Consequences of maternal postpartum depression: A systematic review of maternal and infant outcomes. *Womens Health (Lond)*, *15*. 1745506519844044.

Taylor, A., Glover, V., Marks, M., & Kammerer, M. (2009). Diurnal pattern of cortisol output in postnatal depression. *Psychoneuroendocrinology*, *34*(8), 1184–1188.

Thomas, L., Gandhi, S., & Parel, J. T. (2018). Interventions for mothers with postpartum depression: A systematic review. *International Journal of Depression and Anxiety*, *1:002*.

Wang, J., Yun, Q., Ma, S. F., Song, H. R., Guo, M. N., & Zhang, W. N. (2020). Inhibition of expression of glucocorticoids receptors may contribute to postpartum depression. *Biochemical and Biophysical Research Communications*, *523*(1), 159–164.

Yim, I. S., Glynn, L. M., Schetter, C. D., Hobel, C. J., Chicz-Demet, A., & Sandman, C. A. (2010). Prenatal beta-endorphin as an early predictor of postpartum depressive symptoms in euthymic women. *Journal of Affective Disorders*, *125*(1–3), 128–133.

Yim, I. S., Tanner Stapleton, L. R., Guardino, C. M., Hahn-Holbrook, J., & Dunkel Schetter, C. (2015). Biological and psychosocial predictors of postpartum depression: Systematic review and call for integration. *Annual Review of Clinical Psychology*, *11*, 99–137.

Zenclussen, A. C. (2013). Adaptive immune responses during pregnancy. *American Journal of Reproductive Immunology*, *69*(4), 291–303.

Chapter 4

# Modeling maternal depression during pregnancy: Rodent models of major depressive disorder with peripartum onset

Elizabeth M. Sajdel-Sulkowska

*Department of Psychiatry, Harvard Medical School & Brigham and Women's Hospital, Boston, MA, United States*

## List of abbreviations

**BDNF** model brain-derived neurotrophic factor
**CMS** chronic mild stress
**CORT** axiscorticosterone
**CRH** corticotrophin-releasing hormone
**CVD** cardiovascular disease
**DMS** statistical manual of mental disorders
**ED** embryonic day
**FSL** flinders sensitive line
**FST** forced swim test
**GABA** gamma aminobutyric acid
**GD** gestational day
**GR** glucocorticoid receptor
**HPA** hypothalamus-pituitary-adrenal
**ICD** International Classification of Diseases
**LH** learned helplessness
**MDD** major depressive disorder
**MGB** microbiota-gut-brain axis
**MIA** maternal immune activation
**PD** postnatal day
**SD** Spraque Dawley
**TST** tail suspension test
**UMCS** unpredictable mild chronic stress
**WHO** World Health Organization
**WKY** Wistar Kyoto

## Introduction

Maternal depression during pregnancy is a growing public health issue affecting 20% of women and exceeding the postpartum depression almost twofold. Furthermore, postpartum depression is frequently a continuation of depression during pregnancy (Verreault et al. 2014). It presents a significant risk factor for both mothers and their unborn children, with developmental health problems and long-term health issues increasing burden to the society.

Maternal depression during pregnancy is often comorbid with cardiovascular disease (CVD) involving the uteroplacental circulation that is a crucial player in early fetal development and impacts fetal cardiovascular system and fetal neurodevelopment. It is associated with increased levels of stress hormones, proinflammatory cytokines, and altered levels of neurotrophins, all of which contribute to altered maternal physiology and fetal environment. Importantly, the intensity of maternal depressive symptoms is positively associated with depressive symptoms in the offspring.

Furthermore, depression during pregnancy is a significant problem facing the expectant mothers and health professionals regarding the use of antidepressant treatment that may be a risk to the unborn children as data on antidepressant exposure during pregnancy are often conflicting (Singhal, Manjula, & Vijay Sagal, 2016). Some data indicate that exposure to antidepressants during pregnancy may have adverse neurodevelopmental consequences in one-third of the neonates (Pearlstein 2015). On the other hand, depression in pregnant mothers is related to changes in serotonin levels in the fetal brain (St-Pierre et al. 2016).

Questions regarding the mechanisms involved in maternal depression during pregnancy and the course of treatments cannot be answered through clinical studies and require experimental animal models to understand the causes, pathology, and modes of therapy. However, there are very few animal models of maternal depression, as compared to animal models of depression in men, and most of those are focused on the postpartum period. The most critical factor that contributed to the absence of clinically relevant animal models of maternal depression during pregnancy is the failure of clinical recognition of this crucial and distinct form of depression.

The American Psychiatric Association only recognized depression in women of childbearing age as a separate

The Neuroscience of Depression. https://doi.org/10.1016/B978-0-12-817933-8.00007-4

entity of Major Depressive Disorder (MDD) in the 1998 edition of the Statistical Manual of Mental Disorders (DMS)-IV with the specifier of the postpartum period. DMS-5 edited in 2013 recognized the depression during pregnancy with the crucial qualifier—the peripartum period. The International Statistical Classification of Diseases and Related Health Problems (ICD-10CM), the manual of WHO, recognized the pregnancy during the depression as part of MDD in 2017. The current definition of maternal depression during pregnancy is: "Major Depressive Disorder with a Peripartum Onset defined as a most recent episode during pregnancy as well as in the 4 weeks following delivery."

Thus the animal models of maternal depression before 2013 and the majority of the most recent ones focused on postpartum depression. Consequently, the animal models of maternal depression during pregnancy are just emerging, and there is a great need to discuss and agree on the uniform and clinically relevant models.

Several excellent reviews discuss the general criteria of the animal models. In general, there is an agreement that a model must fulfill the face validity representing a phenotype similar to the human disease, construct validity defined as processes that present risks in human pathology and the animal, and predictive validity defined as sensitivity to the treatment interventions effective in humans (Belzung & Lemoine, 2011; Wang et al. 2017).

## Characteristics of animal models of maternal depression

Animal models of the disease refer to the induction of human-like pathology in nonhuman animals. Depression is a disease that affects several functions such as the way one thinks, feels, and behaves. Moreover, while there is no way we can get animals to reproduce all symptoms, we can find similar symptomatic behaviors. For example, depression causes feelings of sadness or hopelessness while animals give up struggling in certain situations referred to as "learned helplessness" or "behavioral despair" and are incorporated into behavioral tests of "forced swim test" FST and 'tail suspension" test (TST).

To date, most of the animal models of depression focus on depression in males with more than twice as many studies as in females. The PubMed database searches with "depression animal model males," result in 8695 articles; "depression animal model females," 3725, "depression animal models postpartum depression," 171; and "depression animal models during pregnancy," one (Czarzasta et al, 2019). In part, the paucity of models of maternal depression during pregnancy reflects the relatively recent clinical recognition of maternal peripartum depression.

Early animal models of depression relied on dogs (Seligman, 1972) and learned helplessness (LH) to electric shocks similar to the human onset of depression in response to uncontrollable stress. Both for ethical and practical reasons, further research turned toward rodents. Current studies of maternal depression during pregnancy involve almost exclusively rodents: rats and mice, with studies involving rats far exceeding those with mice. Rodents, especially rats, show brain structure and connectivity similar to humans, reflect better human physiology and diseases, are more capable of learning tasks, better mimic human behavior, and are much more social (Bryda 2013).

Rats as model animals offer the advantage of relatively short pregnancy of around 4 weeks (rats 21–22.5 days, mice 19.5 days), they have a large number of pups. They have been long a species of choice in neuroanatomical, neurophysiological, and behavioral studies. Mice with a shorter gestation (19.5 days) have been involved mostly in genetic models (Clancy et al. 2007). Initial studies suggested that uncontrollable stress precipitates depression-like behaviors in rodents, fulfilling the construct validity. Furthermore, rodents exposed to uncontrollable stress showed several symptoms similar to those observed in human depression, such as decreased motor activity, weight loss, and reduced motivationfor reward - acore feature of anhedonia. These symptoms can be used as diagnostic tools of depression and provide a model with symptomatology resembling clinical depression (Wang et al. 2017).

### Several strains of rats proved amiable in studies of maternal depression

*Sprague–Dawley (SD)* rats are frequently selected for behavioral models, such as chronic mild stress (CMS), unpredictable mild chronic stress (UMCS), and corticosterone excess model (CORT). They have not been successfully adapted for the LH model. They are calm and easy to handle. However, there is some variability between this strain associated with individual vendors and internally inbred animals.

Flinders sensitive line (FSL) derived from SD strain has been widely used as a rat model of depression but not in the studies of maternal depression. Another strain derived from SD rats, the Holtzman strain, has been used in the learned helplessness model of depression. However, neither Holtzman nor Long Evans rat strains have been used in models of maternal depression.

Wistar–Kyoto (WKY) strain is a model of endogenous depression. Adult WKY rats demonstrate decreased bodyweight, increased immobility in the swim test, and a greater degree of anhedonia (assessed by sugar consumption). Furthermore, pregnant females have a heightened sensitivity to chronic stress (Schroeder et al. 2012), and antidepressant resistance and individual behavioral variability have also been reported (Nam, Clinton, Jackson, & Kerman, 2014).

### Mice strains frequently used in models of maternal depression

Based on the stress paradigms, CMS and UMCS include the C57BL/6N—most wildly used an inbred strain of laboratory mice, NMR1 mice—an outbred model with increased risk of spontaneous tumors and renal disease, and the CD1 an albino outbred strain (Table 1).

Also, several natural mouse mutations and transgenic mice with a deficit of serotonergic activity (5Tph2 −/− mice, (Krishnan & Nestler, 2008)), altered GABAnergic regulation (Gabrd −/− mice, Maguire and Mody 2016), and maternal immune activation model (MIA; C3H/He mice; Berger et al. 2018) have been used in studies of maternal depression.

## Face validity of animal models of maternal depression; symptoms and other physiological markers that resemble those found in humans and are used to verify the model

The majority of animal models of maternal depression use behavioral symptoms and other physiological markers observed in humans to validate the newly developed model. However, it must be noted that once the given experimental paradigm is established, validation of maternal depression in subsequent studies is seldom confirmed; the use of depression markers to monitor maternal depression is presented in Table 2.

Most frequently used tests of maternal depression include maternal weight gain during pregnancy (mice and rats), sucrose preference test (SPT); rats and mice) that measures the loss of pleasure (anhedonia) in terms of sucrose consumption (Kalueff, Gallagher, & Murphy, 2006), tail suspension test (TST; mostly mice) that measures the despair in terms of immobility during the brief tail suspension (Alexandre et al., 2006; El Yacoubi et al., 2003), the Porsolt forced swim test (FST; rats) that measures the helplessness in terms of immobility in an inescapable container of water (Porsolt, Le Pichon, & Jalfre, 1977), and the splash test developed in mice (Marocco et al. 2014; Santarelli et al. 2003) and adapted in rats (Czarzasta et al. 2019) that measures self-grooming behavior related to depressive-like behavior in terms of the length of time spent washing their sides after being splashed with sucrose solution.

Brain-derived neurotrophic factor (BDNF) is another marker of depression. Brain BDNF plays an essential role during neurodevelopment and has been implicated in several psychiatric disorders, including depression (Autry and Monteggia 2012). A decrease in hippocampal BDNF has been observed in human subjects with depression (Sen, Duman, & Sanacora, 2008), in a mouse model of maternal postpartum depression (Zhou et al. 2017) and in the rat offspring following maternal exposure to stress (Boulle et al. 2016a, 2016b).

### Time equivalence of human and animal pregnancy and the relative chronology of human and rat development concerning the embryonic and fetal periods (Fig. 1) is an important issue in animal models of maternal depression

Furthermore, the relative chronology concerns also individual organs and even individual parts of the given organ. Extrapolating brain development from humans to experimental animals is not that straight forward and has been based on the morphological, "rules of thumb," and "event-based' analyses. It has been suggested that in the case of brain development neuroinformatics, combining neuroscience, evolutionary science, and statistical modeling and computer science may provide the best approximation. Meantime, the customarily accepted conversion of rat brain at postnatal days (PD) 1–10 equating to the third trimester in humans, or PD7 equivalence of the human brain at birth, is based on the old data. Based on analyses of a growth spurt, PD1–PD10 corresponds to the last trimester in humans, and neurotransmitter GABA-based data suggests that PD 2–7 in rats corresponds to the human third trimester (Clancy et al. 2007).

Furthermore, individual neurodevelopmental events have a unique equivalence, such as the limbic system maturation on an embryonic day (ED 11) in rat corresponds to the first month of pregnancy in humans (Clancy et al. 2007). Accordingly, there is a need to establish uniform chronology as a guideline for a meaningful comparison of both maternal and embryonic and fetal equivalence for the particular organ under consideration. Fig. 1 illustrates the chronology of rat vs human pregnancy (20–22.5 days vs 42 weeks), developmental periods: embryonic (conception–17th day vs 0–11th week), fetal (17th day–birth vs 11th week–birth), heart (16th day vs third week), and brain (18th day vs ninth week). A separate seldom discussed problem lies in the extrapolation of the first postnatal week in rodents to the last trimester of human pregnancy is the difference in the impact of in utero vs extra-uterus environment on offspring development.

Very few of the existing animal models of maternal depression include early gestation periods in their stress paradigm. This period, before neural tube closure, is the critical period of fetal development, and most sensitive to the environmental impact; several pathologies have been

**TABLE 1 CMS and UCMS rodent models of maternal depression: nature of stressors and time of exposure.**

| Stress type, species and strain | Period of stress | Stress components and duration | References |
|---|---|---|---|
| CMS<br>Rat: Sprague–Dawley | GD14-GD21 | Restraint + light: 8 days, 2 h 15 min/day | Boulle et al. (2016a, 2016b) |
| CMS<br>Rat: Sprague–Dawley | GD14-GD21 | Restraint: 8 days, 2 h 15 min/day | Glombik et al. (2015)<br>Trojan et al. (2016) |
| CMS<br>Rat: Sprague–Dawley | GD11-GD20 | Restraint: 10 days, 2 h 15 min/day | Palacios-Garcia et al. (2015) |
| CMS<br>Rat: Sprague–Dawley | GD11-GD21 | Restraint + light: 11 days, 2 h 15 min | Marocco et al. (2014). |
| CMS<br>Rat: Sprague–Dawley | Pregestation: 5 weeks | Restraint: 35 days, 1 h/day | Czarzasta et al. (2019) |
| CMS<br>Rat: Wistar | GD15-GD21 | Restraint + light: 7 days, 2 h/day | Butkevich et al. (2019) |
| CMS<br>Rat: Sprague–Dawley | GD10-GD20 | Restraint + light: 7 days, 1 h/day | Smith, Secki, Evans, Costall, & Smythe, (2014) |
| CMS<br>Rat: Sprague–Dawley | GD1-GD7vs<br>GD8-GD14vs<br>GD15-GD21vs | Restraint: 7 days, 2 h 15 min/day<br>Restraint: 7 days, 2 h 15 min/day<br>Restraint: 7 days 15 min/day | Amugongo & Hlusko, (2013) |
| UMCS<br>Rat: Sprague–Dawley | Pregestation: 3 weeks | Daily: food and water deprivation or cage tilting or cold swim or electric shock or cage rocking or light<br>O/N or elevated temperature or restraint or soiled cage or cage change | Huang et al., (2010, 2012) |
| UMCS<br>Rat Sprague–Dawley | GD15-GD21 | Daily: food and water deprivation or cage tilting or cold swim or electric shock or cage rocking or light<br>O/N or elevated temperature or restraint or soiled cage or cage change | Rayen et al. (2011, 2013); Pawluski et al. (2012a, 2012b), Gemmel et al., (2017); Gemmel, Rayen, Lotus, et al., (2016); Gemmel, Rayen, van Donkelaar, et al., (2016) |
| UMCS<br>Rat: Sprague–Dawley | GD9-GD19 | Daily: restraint or cage tilt or damp bedding or noise or cage changes or O/N light | Bourke et al. (2013) |
| UMCS<br>Rat: Sprague–Dawley | GD9-GD21 | Daily: restraint or cage tilt or damp cage or cage change or noise or light | Ehrlich et al. (2015) |
| UMCS<br>Rat: Sprague–Dawley | GD1-GD10 vs.<br>GD11-GD20 | Daily: FST, restraint, water deprivation, restraint in cold, isolation, food deprivation, water deprivation, restraint in cold, food deprivation | Modir et al. (2014) |
| Wistar–Kyoto | GENETIC MODEL | NA | Millard et al. (2020) |
| CMS<br>Mouse | GD13-GD17 | Restraint +light + water: 5 days, 2 h 15 min/day | Behan et al. (2011);<br>Schmidt et al. (2018) |
| CMS<br>Mouse: NMRI | GD5-GD19 | Restraint +light: 15 days, 90 min/day | Salari et al. (2016) |
| CMS<br>Mouse: Balb/c | Pregestation: 3 weeks | Restraint +light: restraint 21 days, 6 h/day +12 x O/N light | Xia et al. (2016) |
| UMCS<br>Mouse:C57BL/6 | GD4-GD18 | Daily: food restriction or O/N light or paired housing<br>or foreign object or soiled cage or irregular ultrasound or white noise or restraint | Kiryanova et al. (2016) |

Nature of stressors and the time of exposure in individual paradigms of stress-based models of maternal depression. *CMS* (chronic mild stress), *GD* (gestational day), and *UCMS* (unpredictable chronic mild stress).

**TABLE 2** **Impact of maternal depression on preadolescent and adolescent offspring.**

| Stress type species and strain | Time of stress | Confirmation of maternal depression | Parameters measured in offspring | References |
|---|---|---|---|---|
| CMS<br>Rat: Sprague–Dawley | GD14-GD21 | None | PD21 males: decreased sucrose consumption, corticosterone | Glombik et al. (2015)<br>Trojan et al. (2016) |
| CMS<br>Rat: Sprague–Dawley | Pregestation: 5 weeks | Weight gain, splash test, corticosterone | Birth-PD21: neurodevelopmental changes, splash test<br>Adolescent: behavioral changes, brain, and heart | Czarzasta et al. (2019) |
| CMS<br>Rat: Wistar | GD15-GD21 | None | PD28: increased immobility on FST, increased corticosterone | Butkevich et al. (2019) |
| UMCS<br>Rat: Sprague–Dawley | GD9-GD21 | None | Adolescent females only: increased anxiety | Ehrlich et al. (2015) |
| UMCS<br>Rat: Sprague–Dawley | Pregestation: 3 weeks | Weight gain sucrose consumption, serum corticosterone (after stress & on GD20) | Adolescent (PD30) males and females: increased immobility on FST and corticosterone | Huang et al., (2010, 2012) |
| | | | Immobility on FST and corticosterone + brain BDNF level | Huang et al. (2012) |
| | | | Immobility on FST and corticosterone + serotonin synthesis | Huang et al., (2012) |
| UMCS<br>Rat: Sprague–Dawley | GD15-GD21 | None | Adolescent females: increased immobility on FST | Rayen et al. (2011) |
| | | None | Adolescent males: increased immobility on FST | Rayen et al. (2013) |
| | | None | Adolescent: increase in corticosterone | Pawluski et al. (2012a, 2012b) |
| | | Weight gain, corticosterone, FST | No results (maternal effect only) | Pawluski et al. (2012a, 2012b) |
| | | | PD 21: social behavior and hippocampal neurogenesis | Gemmel et al. (2017) |
| Wistar–Kyoto, rat | | | PD39 adolescent males; sex difference | Millard et al. (2020) |
| CMS<br>Mouse | GD13-GD17 | None | PD45 females: decreased sucrose consumption, increased immobility on FST; changes in the hippocampus | Behan et al. (2011) |
| CMS<br>Mouse: NMRI | GD5-GD15 | Weight gain, decreased sucrose consumption, increased immobility on FST | Adolescent males only: increased anxiety | Salari et al. (2016) |
| CMS<br>Mouse: CD1 | GD14-GD18 | None | Adolescent PD30-PD35: decreased sucrose consumption | Avitsur (2017) |
| CMS<br>Mouse | Pregestation: 3 weeks | Decreased sucrose consumption, increased immobility on TST | Early postnatal, adolescent female only: decrease sucrose consumption | Xia et al. (2016)<br>Xia et al. (2017) |

Summary of the studies of the impact of maternal depression on preadolescent and adolescent offspring: note the absence of maternal depression verification. *CMS* (chronic mild stress), *GD* (gestational day), and *UCMS* (unpredictable chronic mild stress).

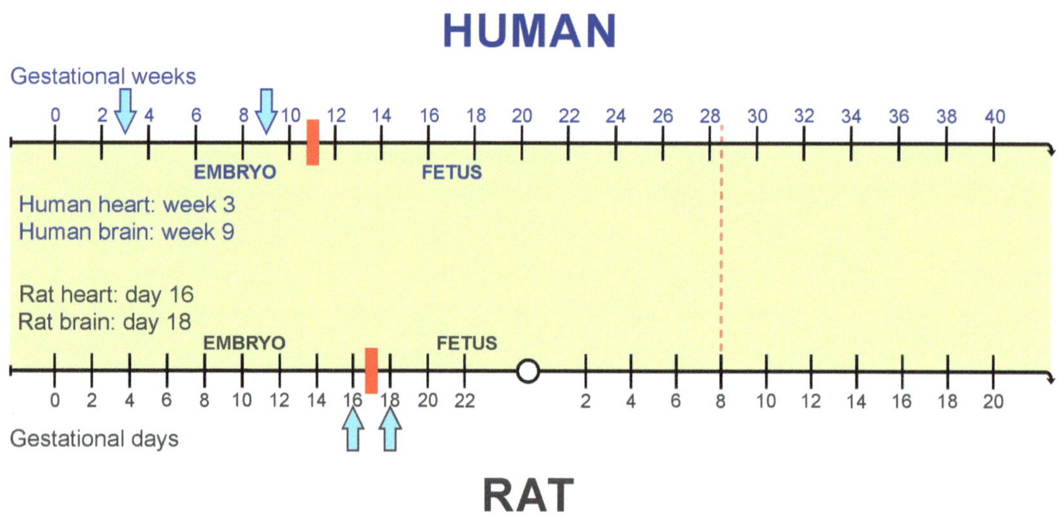


FIG. 1 Chronology of rat vs human pregnancy and development. Schematic representation of time correspondence between rat and human pregnancy, embryonic and fetal periods, and the time of heart and brain maturation. References: https://embryology.med.unsw.edu.au.embryology/index.Php/rat_-Development_Stages; neurodevelopment.org/human-rat-comparison

associated with environmental insults during the first trimester of human pregnancy (Walder et al. 2014). Furthermore, in the case of maternal depression during pregnancy, the process of pregnancy and the characteristics of different stages of pregnancy must be taken into consideration, as not only the first but also the last trimesters of pregnancy are more critical and more sensitive to manipulation than the middle.

The use of the stressors during the second part of pregnancy has been justified by the assumption that neurodevelopment in rats during the first 10 postnatal days corresponds to the third trimester of human pregnancy (Rice et al., 2010). With the inclusion of the peripartum period in the clinical definition of depression, future animal studies of maternal depression should include the early as well as late pregnancy periods, and optimally the maternal depression should be documented by both behavioral and physiological tests.

## The delay of physiological response to stress is a critical aspect of a stress-induced animal model of depression

Data from separate groups indicate that the development of the maternal depressive state is gradual and takes from 1 week (Conrad, Kerchner, & Mosher, 1999), 3 weeks (Gemmel et al. 2017; Huang et al. 2010) to 5 weeks (Czarzasta et al. 2019). It is possible that the induction of maternal depression by gestational stress is more rapid than by pregestational exposure, but further studies are needed to establish maternal depression in individual experiments.

## Construct validity of maternal models of depression: The mode of induction of depression

Construct validity defines processes that present risks in human pathology and can be applied to the animal (Wang et al. 2017). While establishing the animal model of maternal depression, one should consider plausible etiologies of human maternal depression that include genetic predisposition, stressful life events, and medical problems. It has been assumed that approximately one-third of MDD is of genetic origin and two-thirds due to the environmental factors. Genetics determines susceptibility to mood disorders, neurotransmitter balance, and response to drugs, and the environmental factors include stress, acquired medical problems, housing, and diet.

Currently, it is hypothesized that the mechanisms involved in animal depression are similar to humans and are based on the excess of corticosterone, corticotropin-releasing hormone (CRH), or glucocorticoid receptor (GR) resistance, and cytokines levels. Although so far, the behavioral symptoms of human depression cannot be reproduced in animal models, some behaviors such as anhedonia, despair, weight fluctuation, self-care, and neuroanatomical changes can be documented (Bogi et al. 2017; Czarzasta et al. 2019).

Stress is a significant factor contributing to depression, and animal models utilizing stress have been used in many studies. Furthermore, it has been shown that exposure of animals to chronic mild stress (CMS model) generates different effects when the stress is unpredictable mild chronic stress (UMCS; Schroeder et al. 2012). Maternal depression can be triggered by sustained physiological or psychological stress that results in increased secretion of several hormones, primarily glucocorticoids, to meet higher energy demands. However, chronic elevation of glucocorticoids in stress or depression is accompanied by increased levels of pro-inflammatory cytokines (Leonard 2010) associated with increased risk of preterm birth and neurodevelopmental pathologies (Carlo et al. 2011).

The existing rat models are based on the excessive stress, and naturally occurring genetic variants, while the majority of mice models are genetic variants with some based on the stress. In considering the construct validity of an animal model of maternal depression, one thus has to consider sensitivity to stress, reaction to antidepressants, as well as the nature of depression trigger, housing condition, and diet. Just like in humans, some rats or mice species are more sensitive to stress, and even stress of light or crowding may affect the overall response.

Stress is intimately related to depression. The stress response involves activation of the hypothalamus–pituitary–adrenal (HPA) axis resulting in the release of glucocorticoids, cortisol in humans and corticosterone in rodents, from the adrenal glands. Stress, and specifically chronic stress, triggers physiological responses in the body that involve activation of the HPA axis, which in turn affects the activity of other hormonal systems as well as neurotransmitters. While a low level of stress is a normal mobilizing mechanism, the excessive and chronic, and especially uncontrollable/unpredicted stress, results in excessive activation and prolonged activation of the HPA axis and impaired HPA feedback and hypersecretion of glucocorticoids. Prolonged elevation of cortisol/corticosterone is neurotoxic to the hippocampal neurons and inhibits neurogenesis, which leads to the development of depression in humans and rodents. Stress during pregnancy activates both the maternal and the fetal HPA axis; the excessive stress during the time of critical brain development affects fetal programming (Barker 2001), resulting in both short- and long-term consequences in the developing fetus.

*Rat models* of maternal depression include *the CMS* model, which is equivalent to everyday stresses, the *UMCS* model that, because of its unpredictability, represents the stress such as the one experienced during present pandemics, and *the CORT* model. The three models, while initially developed in rats, have been subsequently used in mice. An additional model, Wistar–Kyoto rat strain that displays behavioral and neurobiological phenotypes similar to that observed in MDD (Millard et al. 2020), has been used primarily to study maternal–offspring interactions during the postpartum period. Interestingly altered emotional behavior, such as abnormal separation-associated vocalization, and reluctance to seek proximity with their mothers, can be detected as early as postnatal day 10. WKY mothers exhibit a deficit in caregiver behavior reactions.

*Mouse models of maternal depression* rely less frequently on *CMS* and *UMCS* paradigms (Table 1), and mostly on the genetics-based paradigm. A knockout mouse model of maternal depression during pregnancy was based on the observation of altered GABAnergic system regulation during pregnancy (Maguire and Mody 2016).

*Very few CMS-based and UMCS-based studies have focused on the mid- to late-gestational periods*, thus excluding the initial and the most vulnerable period of pregnancy. A review of over 30 rodent studies revealed three reports that examined the impact of maternal stress during prepregnancy and the first part of pregnancy on the rat (Czarzasta et al. 2019; Huang et al., 2012, 2013) and mouse (Modir et al. 2014) development.

*Individual CMS and UMCS paradigms* used by different groups involve several modifications regarding nature, duration of the stressors, and timing concerning pregnancy period (pregestation, pregnancy, postnatal). Importantly, the majority of studies do not monitor routinely maternal depression and rely on the maternal depression markers established in the initial report of the model (Table 2). Furthermore, different forms of stress have been shown to elicit stress-specific effects in dams and their offspring. While the CMS model of maternal restraint may provide a temporary experience of hopelessness and despair, the UMCS results in unpredictability that is much more stressful, while extreme stress may result in aggression (Wei et al. 2017).

*CMS model* relies on daily restraint and involves exposure of dams to stress by placing them in a tube one to three times a day for several days. Dams undergoing this stress during the pregnancy or before have shown increased depressive-like behavior (Mairesse et al. 2014), while their offspring show reduced social play during adolescence and increased immobility during FST as adults (Mairesse et al. 2014). Importantly, it has been shown that (Conrad, et al. 1999) 1 5 weeks of pregestational stress exposure is required to produce maternal depression and obtain a statistically significant effect on dam's splash test and altered neuro- (Czarzasta et al., 2019) and cardiovascular (Czarzasta et al., 2020) development in the offspring (Czarzasta et al. 2019).

*UMCS*—the unpredictable chronic mild stress—model was developed in the early 1980s (Katz, Roth, & Carroll, 1981) and improved in the early 1990s (Willner, Golembiowska, Klimek, & Muscat, 1991). The original paradigm consisted of decreasing stressors, and the depressive-like behavior was assessed by anhedonia. The stressors included food and water deprivation, overnight illumination, cage tilt to develop a depressive state that was tested by a decrease in reward sensitivity and anhedonia. Maternal depression was measured with sugar consumption and weight loss. That particular model has been frequently used as a model of maternal depression. However, individual groups have frequently modified the nature, the number of stressors, their frequency, and duration, as well as developmental period making it difficult to compare the resulting data.

*CORT model* of maternal depression (Gregus, Wintink, Davis, & Kalynchuk, 2005) relies on the observation that increased corticosterone levels accompany stress response. It involves direct injections of high levels of corticosterone to pregnant Sprague–Dawley rat dam between 10 and 20 days of gestation or 2–21 days postpartum and results in age- and tissue-dependent accumulation of corticosterone in offspring (Brummelte et al. 2010). Rats repeatedly injected with corticosterone lost weight and showed longer immobility period on FST and lower sucrose consumption on the sucrose preference test (Dwivedi, 2016).

## Predictive validity of rodent models of maternal depression: Response to antidepressants

Animal models of depression are indispensable in assessing the impact of depression on both maternal and fetal health as well as the effect of antidepressants on the developing fetus. Clinical data regarding the safety of antidepressants for developing babies are inconclusive. Thus, depressed pregnant mothers face a difficult choice of whether "to use or not to use" antidepressant treatment. Untreated depression is associated with hormonal changes, such as elevated cortisol (corticosterone), which affect the developing CNS. It has been documented that the intensity of maternal depressive symptoms is positively associated with depressive symptoms in the offspring.On the other hand, antidepressants cross the placenta and may affect offspring neurodevelopment as well.

Rodent models are instrumental in solving thedilemma regarding the use of antidepressants during pregnancy. The effect of maternal antidepressant treatment has been evaluated in rats exposed to CMS (Butkevich et al. 2019; Ehrlich et al. 2015) and UCMS models (Pawluski et al. 2012a, 2012b; Rayen et al. 2011, 2013) and mice exposed to the CMS model (Avitsur 2017). However, some animal studies of antidepressants have examined their effect on nondepressed animals (Gobinath et al. 2017), while other studies failed to provide supportive depression data. The use of antidepressants in stressed versus unstressed animals is an important issue as the action of antidepressants in the depressed human/animals is different from that in nondepressed human/animals.

## Summary of studies addressing the impact of maternal depression on preadolescent and adolescent offspring

Rat models of maternal depression include two stress-based depression models, CMS and UMCS, and a natural genetic strain (Tables 1 and 2). This section focuses on the studies that examined the impact of maternal depression on newborn and adolescent offspring; the results are presented in Table 2.

*CMS model* involves exposure of dams to restraint stress by placing them in a tube/bag one to three times a day for several days. As can be observed, individual studies have introduced additional stressors, such as exposure to light or water (Tables 1 and 2), and applied stressors during a pregestational period or at different periods of pregnancy. Furthermore, the onset and duration of depressive behavior have not been examined. It has been observed that 5 weeks of a daily 1-h restraint before pregnancy was required to induce a decline in self-grooming behavior. However, the effect of stress persisted through pregnancy and lactation, as measured by a self-grooming splash test, and plasma and urine corticosterone levels. Offspring of these dams sustained neurodevelopmental, behavioral, and cardiac abnormalities (Czarzasta et al. 2019, 2020).

The original *UMCS model* consisted of the decreasing nature of stressors; the maternal depressive-like behavior was assessed by weight gain and sugar consumption. The stressors included food and water deprivation, overnight illumination, and cage tilt. The initially developed paradigm has been since modified in both the nature and the number of stressors, their frequency, duration, as well as developmental period making it difficult to compare the resulting data from different studies. While several groups used the same strain and breeder, the duration and the frequency of daily stress were variable, thus affecting the stress magnitude. Furthermore, different groups selected different gestational periods for exposure to stress. The unpredictability of the daily stressors increases their effect on the HPA axis; when used for 3 weeks before pregnancy results in significant activation of both maternal as well as fetal HPA (Huang et al. 2012) and serotonergic sex-dependent changes in the offspring (Huang et al. 2012, 2016). Furthermore, pregestational maternal stress resulted in decreased social memory, play behavior, and hippocampal neurogenesis in male, but not female offspring (Gemmel et al. 2017; Grunwald, Benitez, & Brunton, 2016).

Importantly, it appears that different modes of stress elicit different responses in both the dams and the offspring. The CMS may induce a daily 1-h period of hopelessness and despair enough to generate depressive-like behavior, more severe forms of stress have been shown evoke aggressive behavior (Wei et al. 2017).

Very few of both CMS-based and UMCS-based studies focused on the mid- to late-gestational periods (Tables 1 and 2), and excluded the initial and the most critical period of pregnancy (Table 1, Fig. 1). It has been reported that CMS during the first or second week of pregnancy results in lower maternal body weight gain as compared to CMS during the third week (Amugongo & Hlusko, 2013).

While establishing the initial model, the maternal depressive state is usually confirmed by weight gain, sugar consumptions, and other depression markers (Huang et al. 2010) only some groups verify maternal depression in subsequent studies (Huang et al., 2010, 2012). However,that particular practice is not routinely utilized, yet it is absolutely necessary to obtain meaningful data as it can be impacted by small environmental changes, such as laboratory temperature, humidity, noise and overcrowding, and even animal handling. In summary, data presented in Tables 1 and 2 illustrate a large discrepancy in the strain, nature of "depressive factor" timing, and duration of stress, all of which impact the results. More importantly, because of these differences, it is virtually impossible to generalize the findings and draw reproducible conclusions.

## Future animal models of maternal depression

Based on the developing concept of the vital link between gut microbiota and the brain, microbiota-gut-brain (MGB) axis and behavior (Sajdel-Sulkowska et al. 2019; Sajdel-Sulkowska and Zabielski 2013), and the critical connection between changes in gut microbiota and neuropsychiatric disorders, including depression, future animal models may be critical in the development of new treatments. According to the MGB axis theory, maternal microbiota determine the type of microbiota that first colonize the infant's gut. Thus changes in maternal microbiota associated with untreated vs antidepressant-treated depression during pregnancy may impact neurodevelopment in a distinct and sex-dependent manner. Indeed, human depression is related to changes in gut microbiota ( Jiang et al. 2015; Liang et al. 2018), and similar microbiota changes can be observed in stress-exposed animals (Liang et al. 2018). Furthermore, probiotic treatments reduce depression scores (Desbonnet et al., 2010; Desbonnet, Garret, Clarke, Bienenstock, & Dinan, 2008).

Maternal gut microbiota also undergoes programmed changes during pregnancy (Dunlop et al., 2015; Jasarevic et al. 2017) and is sensitive to maternal physiological and psychological factors, including stress. Future animal models may assess the effect of probiotics as an alternative or adjunct therapy to prevent the neurodevelopmental consequences of maternal depression during pregnancy.

An interesting aspect of antidepressants and their antimicrobial effects have been recently reviewed (Macedo et al., 2017) and appears to be relevant to the microbiota role in human depression. Existing data suggest that both depression in pregnant mothers and antidepressants induce perturbations in serotonin brain levels (St-Pierre et al. 2016) during the critical fetal developmental periods (Walder et al. 2014), and maternal pregnancy/antidepressants during this period are likely to impact the neurodevelopmental process.

Thus future studies aimed at establishing scientific bases for a potential therapeutic role of probiotic intervention in pregnancy-associated depression or supplementation of current ones with probiotics have, therefore, a significant clinical potential.

## Conclusions

Clinical recognition of maternal depression during pregnancy in 2013 calls for creating new rodent models that are necessary and indispensable to devise the optimal treatment for both the mother and her developing offspring. In creating these new models it is time to address several issues such as (1) relative chronology of rodent and human pregnancy, embryonic and fetal development, and structural and functional organization of individual organs; (2) sensitivity to and latency of particular stressors in CMS and UMCS models of depression; and (3) reevaluation of reliable diagnostic tools and careful monitoring of depressive state. Importantly, standardization of animal strains and experimental protocols across different laboratories would facilitate more rapid progress in understanding this critical disease. Future models of depression during pregnancy should be open to the concept of the vital role of microbiota-gut-brain (MGB) axis in mental health and disease and explore probiotic therapy in the treatment of MDD with Perinatal Onset.

## Key facts of maternal depression during pregnancy

- Maternal depression during pregnancy recognized by DSM-5 in 2013 as major depressive disorder with peripartum onset is a significant public health issue affecting up to 20% women exceeding the postpartum depression almost twofold.
- Existing rodent models of maternal depression seldom addressed critical periods of pregnancy and embryonic and fetal development, and provide conflicting data regarding the use of antidepressants.
- Revised animal models focusing on depression during pregnancy are necessary and indispensable to devise the optimal treatment for both the mother and her developing offspring and to help in solving the dilemma whether to use or not to use antidepressants during pregnancy.

- New rodent models of maternal depression require the revised status of the animal to the human chronology of pregnancy and development and methodology better standardized for animal species/strain, challenge agent, method of exposure, and the endpoints.
- Future rodent models of depression during pregnancy should be open to exploring the developing concept of the microbiota-gut-brain (MGB) and mental health, and probiotic therapy in depression.

## Summary points

- This chapter focuses on maternal depression during pregnancy that is recognized as a significant public health issue, affecting 20% of women, and exceeding the postpartum depression almost twofold.
- It was included in the DSM-5 in 2013 under Major Depressive Disorders with Peripartum Onset.
- Maternal depression during pregnancy affects both the mother and her offspring and presents a significant dilemma to the mothers whether to use the antidepressant or risk untreated depression, as existing models provide inconclusive data.
- Most of the existing animal models of maternal depression are in rats based on the stress paradigms chronic mild stress (CMS) or unpredictable mild chronic stress (UMCS) and the second part of pregnancy for exposure, thus neglecting the most critical part of pregnancy.
- Clinical recognition of maternal depression during pregnancy calls for the development of new animal models utilizing new criteria, especially the time of exposure to stress.
- Significant problems are the animal to the human chronology of pregnancy and offspring development, taking into consideration that relative chronology refers to individual structures such as brain or heart, and also parts of the organ such as brain hippocampus.
- Future animal models of maternal depression during pregnancy should explore the concept of the microbiota-gut-brain (MGB) and mental health, and probiotic therapy in depression that may provide an alternative to antidepressants.

## Mini-dictionary of terms

I do not find a need for a dictionary in this chapter.

## References


Alexandre, C, Popa, D, Fabre, V, Bouali, S, Venault, P, Lessh, KP, et al. (2006). Early life blokade of 5-hydroxytryptamine 1A receptors normalizes sleep and depression-like behavior in adult knock-out mice lacing the serotonin transporter. *Journal of Neuroscience*, *26*, 5554–5564.

Amugongo, S. K., & Hlusko, L. J. (2013). Impact of maternal prenatal stress on growth of the offspring. *Aging and Disease*, *5*, 1–16.

Autry, A. E., & Monteggia, L. M. (2012). Brain-derived neurotrophic factor and neuropsychiatric disorders. *Pharmacological Reviews*, *64*, 238–258.

Avitsur, R. (2017). Increased symptoms of illness following prenatal stress: Can it be prevented by fluoxetine? *Behavioural Brain Research*, *317*, 62–70.

Barker, D. J. P. (2001). Fetal and infant origins of adult disease. *Monatsschrift Kinderheilkunde*, *149*, 2–6.

Behan, A. T., van den Hove, D. L. A., Mueller, L., Jetten, M. J. A., Steinbusch, H. W. M., Cotter, D. R., et al. (2011). Evidence of female-specific glial deficit in the hippocampus in a mouse model of prenatal stress. *European Neuropsychopharmacology*, *21*, 71–79.

Belzung, C., & Lemoine, M. (2011). Criteria of validity for animal models of psychiatric disorders: Focus on anxiety disorders and depression. *Biology of Mood & Anxiety Disorders*, *1*, 9.

Berger, S., Ronovsky, M., Horvath, O., Berger, A., & Pollak, C. C. (2018). Impact of maternal immune activation on maternal care behavior, offspring emotionality and intergenerational transmission in C3H/He mice. *Brain, Behavior, and Immunity*, *70*, 131–140.

Bogi, E., Belovicova, K., Csaltlosova, K., & Dubovicky, M. (2017). Animal models of maternal depression for monitoring neurodevelopmental changes occurring in dams and offspring. *Interdisciplinary Toxicology*, *10*, 35–39.

Boulle, F., Pawluski, J. L., Homberg, J. R., Machiels, B., Kroeze, Y., Kumar, N., et al. (2016a). Prenatal stress and early–life fluoxetine have enduring effects on anxiety and hippocampal BDNF gene expression in adult male offspring. *Developmental Psychobiology*, *58*, 427–438.

Boulle, F., Pawluski, J. L., Homberg, J. R., Machiels, B., Kroeze, Y., Kumar, N. M., et al. (2016b). Developmental fluoxetine exposure increases behavioral despair and alters epigenetic regulation on the hippocampal BDNF gene in adult female offspring. *Hormones and Behavior*, *80*, 47–57.

Bourke, C. H., Capello, C. F., Rogers, S. M., Yu, M. L., Boss-Williams, K. A., Weiss, J. M., et al. (2013). Prenatal exposure to escitalopram and/or stress in rats: A prenatal stress model of maternal depression and its treatment. *Psychopharmacology*, *228*, 231–241.

Brummelte, S., Schmidt, K. L., Taves, M. D., & Soma, K. K. (2010). Elevated corticosterone levels in stomach milk, serum, and brain of male and female offspring after maternal corticosterone treatment in the rat. *Developmental Neurobiology*. https://doi.org/10.1002/dneu.20805.

Bryda, E. C. (2013). The mighty mouse: The impact of rodents on advances in biomedical research. *Missouri Medicine*, *110*, 207–2011.

Butkevich, I. P., Mikhailenko, V. A., Vershinina, E. A., & Barr, G. A. (2019). Differences between the prenatal effects of fluoxetine or buspirone alone or in combination on pain affective behaviors in prenatally stressed male and female rats. *Frontiers in Behavioral Neuroscience*, *13*, 125.

Carlo, W. A., McDonald, S. A., Tyson, J. E., Stoll, B. J., Ehrenkratz, R. A., Shankaran, S., et al. (2011). Cytokines and neurodevelopmental outcomes in extremely low birth weight infants. *The Journal of Pediatrics*, *159*, 919–925.

Clancy, B., Finlay, B. L., Darlington, R. B., & Anand, K. J. S. (2007). Extrapolating brain development from experimental species to humans. *Neurotoxycology*, *28*. https://doi.org/10.1016/j.neuro.2007.01.014.

Conrad, K. P., Kerchner, L. J., & Mosher, M. D. (1999). Plasma and 24-h (NO (x) and cGMP during normal pregnancy and preeclampsia in women on reduced NO(x) diet. *American Journal of Physiology*, *277*, F48–F57.

Czarzasta, K., Makowska, M., Kasarello, K., Skital, V. M., Tyszkowska, K., Matusik, K., et al. (2019). A rat model to study maternal depression during pregnancy and postpartum periods, its comorbidity with cardiovascular diseases and neurodevelopmental impact in the offspring. *Physiology & Behavior*, *199*, 258–264.

Czarzasta, K., Wojciechowska, M., Segiet-Swiecicka, A., Borodzicz-Jazdzyk, S., Niedziela, M., Sajdel-Sulkowska, E. M., et al. (2020). The effect of depressive-like behavior in pregnant rat dams on the cardiovascular system in their offspring. *Stress*, *in press*.

Desbonnet, L., Garret, L., Clarke, G., Bienenstock, J., & Dinan, T. G. (2008). The probiotic Bifidobacteria infantis:nAn assessment of potential antidepressant properties in rat. *Journal of Psychiatric Research*, *43*, 164–167.

Desbonnet, L., Garret, L., Clarke, G., Kiely, B., Cryan, J. F., & Dinan, T. G. (2010). Effects of the probiotics Bifidobacterium infantis in the maternal separation model of depression. *Neuroscience*, *170*, 1179–1188.

Dunlop, A. L., Mulle, J. G., Ferranti, E. P., Edwards, S., Dunn, A. B., & Corwin, E. J. (2015). Maternal microbiome and pregnancy outcomes that impact infant health: A review. *Advances in Neonatal Care*, *15*, 377–385.

Dwivedi, Y. (2016). Pathogenic nf therapeutic applications of microRNAs in major depressive disorder. *Progress in Neuro-Psychopharmacology & Biological Psychiatry*, *64*, 341–348.

Ehrlich, D. E., Neigh, G. N., Bourke, C. H., Nemeth, C. L., Hazra, R., Ryan, S. J., et al. (2015). Prenatal stress, regardless of concurrent escitalopram treatment, alters behavior and amygdala gene expression of adolescent female rats. *Neuropharmacology*, *97*, 251–258.

El Yacoubi, M., Bouali, S., Popa, D., Naudon, L., Lerou-Nicollet, I., Hamon, M., et al. (2003). Behavioral, neurochemical, and electrophysiological characterization of a genetic mouse model of depression. *Proceedings of the National Academy of Sciences of the United States of America*, *100*, 6227–6232.

Gemmel, M., Hazlett, M., Bogi, E., De Lacalle, S., Hill, L. A., Kokras, N., et al. (2017). Perinatal fluoxetine effects on social play, the HPA system, and hippocampal plasticity in pre-adolescent male and female rats: Interactions with pre-gestational maternal stress. *Psychoneuroendocrinology*, *84*, 159–171.

Gemmel, M., Rayen, I., Lotus, T., van Donkelaar, E., Steinbusch, H. W., De Lacalle, N., et al. (2016). Developmental fluoxetine and prenatal stress effects on serotonin, dopamine, and synaptophysin density in the PFC and hippocampus of offspring at weaning. *Developmental Psychobiology*, *58*, 315–327.

Gemmel, M., Rayen, I., van Donkelaar, E., Loftus, T., Steinbusch, H. W., Kokras, M., et al. (2016). Gestational stress and fluoxetine tratment differentially affect plascticity, methylation and serotonin levels in the PFC and Hippocampus of rat dams. *Neuroscience*, *327*, 3242.

Glombik, K., Stachowicz, A., Slusarczyk, J., Trojan, E., Budziszewska, B., Suski, M., et al. (2015). Maternal stress predict altered biogenesis and the profile of mitochondria proteins in the frontal cortex and hippocampus of adult offspring rats. *Psychoneuroendocrynology*, *60*, 151–162.

Gobinath, A. R., Workman, J. L., Chow, C., Lieblich, S., Galea, E., & L. A. M. (2017). Sex-dependent effects of maternal corticosterone and SSRI treatment on hippocampal neurogenesis across development. *Biology of Sex Differences*, *8*, 20.

Gregus, A., Wintink, A. J., Davis, A. C., & Kalynchuk, L. E. (2005). Effect of repeated corticosterone injections and restraint stress on anxiety and dpression-like behavior in male rats. *Behavioural Brain Research*, *156*, 105–114.

Grunwald, N. J., Benitez, D. P., & Brunton, P. J. (2016). Sex-dependent effects of prenatal stress on social memory in rats: a role for differential expressioncentral vasopressin-1a receptors. *Journal of Neuroendocrinology*, *28*. https://doi.org/10.1111/jne.12343.

Huang, Y., Shen, Z., Hu, L., Xia, F., Li, Y., Zhuang, J, et al. (2016). Exposure of mother rats to chronic unpredictable stress befor pregnancy alters the metabolism of gamma-aminobuteric acid and glutamate in the right hippocampus of offspring in early adolescence in a sexually dimorphic manner. *Psychiatry Research*, *246*, 236–245.

Huang, Y., Shi, X., Xu, H., Yang, H., Chen, T., Chen, S., et al. (2010). Chronic unpredictable stress before pregnancy reduces the expression of brain-derived neurotrophic factor and N-methyl-D-aspartate receptor in hippocampus of offspring rats associated with impairment of memory. *Neurochemical Research*, *35*, 1038–1049.

Huang, Y., Xu, H., Li, H., Yang, H., Chen, Y., & Shi, X. (2012). Pre-gestational stress reduces the ratio of 5-HIAA to 5-HT and the expression of 5-HT1A receptor and serotonin transporter in the brain of fetal rat. *BMC Neuroscience*, *13*, 22.

Jasarevic, E., Howard, C. D., Misic, A. M., Beiting, D. P., & Bale, T. L. (2017). Stress during pregnancy alters temporal and spatial dynamics of the maternal and offspring microbiome in a sex-specific manner. *Scientific Reports*, *7*, 44182.

Jiang, H., Ling, Z., Zhang, Y., Mao, H., Ma, Z., Yin, Y., et al. (2015). Altered fecal microbiota composition in patients with major depressive disorder. *Brain, Behavior, and Immunity*, *48*, 186–194.

Kalueff, A. V., Gallagher, P. S., & Murphy, D. L. (2006). Are serotonin transporter knockout mice "depressed"?: hypoactivity but not anhedonia. *NeuroReport*, *17*, 1347–1351.

Katz, R. J., Roth, K. A., & Carroll, B. J. (1981). Acute and chronic stress effects on open field activity in the rat: implications for a model of depression. *Neuroscience & Biobehavioral Reviews*, *5*, 247–251.

Kiryanova, V., Meunier, S. J., Vecchiarelli, H. A., Hill, M. N., & Dyck, R. H. (2016). Effects of maternal stress and perinatal fluoxetine exposure on behavioral outcomes of adult male offspring. *Neuroscience*, *320*, 281–296.

Krishnan, V., & Nestler, E. J. (2008). The molecular neurobiology of depression. *Nature*, *455*, 894–902.

Leonard, B. E. (2010). The concept of depression as a dysfunction of the immune system. *Current Immunology Reviews*, *6*, 205–212.

Liang, S., Wu, X., Hu, X., Wang, T., & Jin, F. (2018). Recognizing depression from the microbiota-gut-brain axis. *International Journal of Medical Sciences*, *19*, 1592.

Macedo, D., Filho, A. J. M. C., Soares de Sousa, C. N., Quevedo, J., Barichello, T., Junior, H. V. N., et al. (2017). Antodepressants, antimicrobials or both? Gut microbiota dysbiosis in depression and possible implications of the antimicrobial effects of antidepressant drugs for antidepressant effectiveness. *Journal of Affective Disorders*, *208*, 22–32.

Maguire, J., & Mody, J. (2016). Behavioral deficits in juveniles mediated by maternal stress hormones in mice. *Neural Plasticity*, *2016*, 2762518.

Mairesse, J., Silletti, V., Laloux, C., Zuena, A. R., Giovine, A., Consolazione, M., et al. (2014). Chronic agomelatine treatment corrects the abnormalities in the circadian rhythm of motor activity and sleep/wake cycle induced by prenatal restraint stress in adult rats. *The International Journal of Neuropsychopharmacology*, *16*, 323–338.

Marocco, J., Reynaert, M.-L., Gatta, E., Gabriel, C., Mocaer, E., Di Prisco, S., et al. (2014). The effects of antidepressant treatment in prenatally stressed rats support the glutaminergic hypothesis of stress-related disorders. *The Journal of Neuroscience*, *34*, 2015–2024.

Millard, S. J., Weston-Green, K., & Newell, K. A. (2020). The Wistar-Kyoto rat model of endogenous depression: a tool for exploring treatment resistance with an urgent need to focus on sex differences. *Progress in Neuropsychopharmacology & Biological Psychiatry*, *101*, 109908.

Modir, F., Elahdadi Salmani, M., Goudarzi, I., Lashkarboluki, T., & Abrari, K. (2014). Prenatal stress decreases spatial learning and memory retrieval of the adult male offspring of rats. *Physiology & Behavior*, *129*, 104–109.

Nam, H., Clinton, S. M., Jackson, N. L., & Kerman, I. A. (2014). Learned helplessnsess and social avoidance in Wistar Kyoto rat. *Frontiers in Behavioral Neuroscience*, *8*, 109.

Palacios-Garcia, I., Lara-Vasquez, A., Montiel, J. F., Diaz-Veliz, G. F., Sepulveda, H., Utreras, E., et al. (2015). Prenatal stress down-regulates Reelinexpressionby methylation of its promoter and induces adult behavioral impairments in rats. *PLoS One*, *10*, e0117680.

Pawluski, J. L., Charlier, T. D., Fillet, M., Houbart, V., Crispin, H. T., Steinbusch, H. W., et al. (2012a). Chronic fluoxetine treatment and maternal adversity differentially alter neurobehavioral outcomes in the rat dam. *Behavioural Brain Research*, *228*, 159–168.

Pawluski, J. L., Rayen, I., Niessen, N. A., Kristensen, S., van Donkelaar, E. L., Balthazart, J., et al. (2012b). Developmental fluoxetine exposure differentially alters central and peripheral measures of the HPA system in adolescent male and female offspring. *Neuroscience*, *220*, 131–141.

Pearlstein, T. (2015). Depression during pregnancy. *Best Practice & Research. Clinical Obstetrics & Gynaecology*, *29*, 754–764.

Rayen, I., van den Hove, D. L., Prickaerts, J., Steinbusch, H. W., & Pawluski, J. L. (2011). Fluoxetine during development reverses the effects of prenatal stress on depressive-like behavior and hippocampal neurogenesis in adolescence. *PLoS One*, *6*, e24003.

Porsolt, R. D., Le Pichon, M., & Jalfre, M. (1977). A NW model sensitive to antidepressant treatment. *Nature*, *266*, 730–732.

Rayen, I., Steinbusch, H. W., Charlier, T. D., & Pawluski, J. L. (2013). Developmental fluoxetine exposure and prenatal stress alter sexual differentiation of the brain and reproductive behavior in male rat offspring. *Psychoneuroendocrinology*, *38*, 1618–1629.

Sajdel-Sulkowska, E. M., & Zabielski, R. (2013). Gut microbiome and brain-gut axis in autism—Aberrant development of gut-brain communication and reward circuitry. *Recent Advances in Autism Spectrum Disorders*, *VI*.

Rice, F., Harold, G. T., Boivin, J., van den Bree, M., Hay, D. F., & Thapar, A. (2010). The links between prenatal stress and offspring development and psychopathology: disentangling environmental and inherited influences. *Psychological Medicine*, *40*, 335–345.

Sajdel-Sulkowska, E. M., Makowska-Zubrycka, M., Czarzasta, K., Kasarello, K., Aggarval, V., Bialy, M., et al. (2019). Common genetic variants link the abnormalities in the gut-brain axis in prematurity and autism. *Cerebellum*, *18*, 255–265.

Salari, A.-A., Fatehl-Gharehlar, L., Motayagheni, N., & Homberg, J. R. (2016). Fluoxetine normalizes the effects of prenatal maternal stress on depression- and anxiety-like behaviours in mouse dams and male offspring. *Behavioural Brain Research*, *311*, 354–367.

Santarelli, L., Saxe, M., Gross, C., Surget, A., Battaglia, F., Dulawa, S., et al. (2003). Requirement of hippocampal neurogenesis for the behavioral effects of antidepressants. *Science*, *301*, 805–809.

Schmidt, M., Braun, K., Brandwein, C., Rossetti, A. C., Guara Ciuara, S., Riva, M. A., et al. (2018). Maternal stress during pregnancy induces depressive-like behavior only in female offspring and correlates to their hippocampal Avp and Oxt receptor expression. *Behavioural Brain Research*, *353*, 1–10.

Schroeder, M., Sultany, T., & Weller, A. (2012). Prenatal stress effects on emotion regulation differ by genotype and sex in prepubertal rats. *Developmental Psychobiology*. https://doi.org/10.1002/dev.21010.

Seligman, M. E. (1972). Learned helplessness. *Annual Review of Medicine*, *23*, 407–412.

Sen, S., Duman, R., & Sanacora, G. (2008). Serum brain-derived neurotrophic factor, depression, and antidepressants medications: Meta-analyses and implications. *Biological Psychiatry*, *64*, 527–532.

Singhal, M., Manjula, M., & Vijay Sagal, K. J. (2016). Subclinical depressionin Urban Indian adolescents: prevalence, felt needs, andcorrelates. *Indian Journal of Psychiatry*, *58*, 394–402.

Smith, J. W., Secki, J. R., Evans, A. T., Costall, B., & Smythe, J. W. (2014). Gestational stress induces post-partum depression-like behavior and alters care in rats. *Psychoneuroendocrinology*, *29*, 227–244.

St-Pierre, J., Laurent, L., King, S., & Vaillancourt, C. (2016). Effects of prenatal maternal stress on serotonin and fetal development. *Placenta*, *48*(Suppl. 1), S66–S71.

Trojan, E., Glombik, K., Slusarczyk, J., Budziszewska, B., Kubera, M., Roman, A., et al. (2016). The beneficial impact of antidepressant drugs on prenatal stress-evoked malfunction of the insulin-like growth factor-1 (IGF-1) protein family of olfactory bulbs of adult rats. *Neurotoxicity Research*, *29*, 288–298.

Verreault, N., Da Costa, D., Marchand, A., Ireland, K., Dritsa, M., & Khalifé, S. (2014). Rates and risk factors associated with depressive symptoms during pregnancy and with postpartum onset. *Journal of Psychosomatic Obstetrics and Gynaecology*, *35*, 84–91.

Willner, P., Golembiowska, K., Klimek, V., & Muscat, R. (1991). Changes in mesolimbic dopamine may explain stress-induced anhedonia. *Psychobiology*, *19*, 79–84.

Xia, B., Chen, C., Zhang, H., Xue, W., Tang, J., Tao, W., et al. (2016). Chronic stress prior to pregnancy potentiated long-lasting postpartum depressive-like behavior, regulated by Akt-mTOR signaling in the hippocampus. *Scientific Reports*, *6*, 35042.

Xia, B., Chen, C., Zhang, H., Zhou, X., Tao, W., Wu, H., et al. (2017). Expression of disrupted-in schizophrenia 1 hippocampus in postpartum depression animal models induced by pre-pregnancy stress. *Acta Academiae Medicinae Sinicae*, *39*, 230–235.

Walder, D. J., Laplante, D. P., Sousa-Pires, A., Veru, F., Brunet, A., & King, S. (2014). Prenatal maternal stress predicts autism traits in 6½ year-old children: Project ice storm. *Psychiatry Research*, *219*, 353–360.

Wang, Q., Timberlake, M. A., II, Prall, K., & Dwivedi, Y. (2017). The recent progress in animal models of depression. *Progress in Neuro-Psychopharmacology & Biological Psychiatry*, *77*, 99–109.

Wei, S., Xiao, X., Wang, J., Shiguang, S. S., Li, Z., Xu, K., et al. (2017). Impact of anger emotional stress before pregnancy on adult male offspring. *Oncotarget*, *8*, 98837–98852.

Zhou, Y., Wang, X., Zhao, Y., Liu, A., Zhao, T., Zhang, Y., et al. (2017). Elevated thyroid peroxidase antibody increases risk of postpartum depression by decreasing prefrontal cortex BDNF and 5-HT levels in mice. *Frontiers in Cellular Neuroscience*, *10*, 307.

Chapter 5

# Depression in mothers and mental health in their children: Impact, risk factors, and interventions

Yasodha Maheshi Rohanachandra
*Department of Psychiatry, Faculty of Medical Sciences, University of Sri Jayewardenepura, Gangodawila, Sri Lanka*

## List of abbreviations

ADHD attention-deficit hyperactivity disorder
IQ intelligence quotient
PCIT parent-child interaction therapy

## Introduction

Children of depressed parents are at an increased risk for depression and other psychiatric and psychosocial problems compared to the general population (Beardslee, Bemporad, Keller, & Klerman, 1983). Early reviews found 40%–45% of children of parents with affective disorders having psychiatric disorders, which is 3–4 times higher than that in children of nondepressed mothers (Beardslee et al., 1983; Cummings & Davies, 1994). However, later studies have shown a lower prevalence of 11.5%–12.5% (Maselko et al., 2016; Woolhouse, Gartland, Mensah, Giallo, & Brown, 2016).

Depression in mothers, whether occurring antenatally or postnatally, is associated with many adverse outcomes in their offspring including impaired bonding, insecure attachment, sleep disturbances, development delays, externalizing and internalizing problems, and increased risk of self-harm (Goodman et al., 2011; Hammerton et al., 2015; Moehler, Brunner, Wiebel, Reck, & Resch, 2006; Murray & Cooper, 1997; Quevedo et al., 2012; Taylor et al., 2017).

Many of these effects are shown to persist from infancy, preschool age, middle childhood through to adolescence (Ahun et al., 2018; Herba et al., 2013; Murray et al., 2011; Taylor et al., 2017). Therefore, attention should be paid toward initiating preventive interventions to minimize the negative impacts on these children.

This chapter focuses on the effects of maternal depression on child mental health, the risk factors, resilient factors, and the evidence-based interventions to minimize the effects of maternal depression on offspring.

## Effects of maternal depression on mental health of children

### Mother-child bonding

Even mild or unrecognized depressive symptoms in the mother, during the first 4 months after birth can result in impaired mother-child bonding (Moehler et al., 2006). Depressive symptoms in the mother at 2 weeks, 6 weeks, and 4 months postpartum are strongly linked to impaired bonding from 2 weeks to 14 months of age. However, no relationship has been demonstrated between maternal depression at 14 months and mother-child bonding (Moehler et al., 2006). This suggests that there could be a "sensitive period" or a "window of opportunity" for optimal development of bonding. This highlights the importance of early identification and prompt treatment of depression in the mother, in order to minimize the effect on mother child bonding.

### Attachment

Maternal depression is associated with reduced secure attachment and increased risk of developing avoidant and disorganized attachment in infants (Martins & Gaffan, 2000). The magnitude of risk for developing insecure attachment and the risk of reduced secure attachment varies considerably between studies. However, the increase in disorganized attachment has been a consistent finding (Martins & Gaffan, 2000).

The impact on disrupted attachment is seen not only in infancy but also among preschool and early school-age children. In preschool and early school-age (5–7 years)

The Neuroscience of Depression. https://doi.org/10.1016/B978-0-12-817933-8.00045-1

children, maternal depression has been consistently associated with disorganized/controlling attachment styles (Badovinac et al., 2018).

Maternal depression is thought to exert its effects on attachment through its impact on maternal sensitive caregiving. Depressive symptoms can cause difficulty in regulating negative emotions and may limit the mothers' capacity to identify process and respond appropriately to their children's needs, which may in turn affect early attachment relationships.

## Child development

Maternal depression is associated with delay in gross motor, fine motor, language, social-emotional and cognitive development in children (Ali, Mahmud, Khan, & Ali, 2013; Kaplan et al., 2014; Quevedo et al., 2012; Vameghi, Amir Ali Akbari, Sajjadi, Sajedi, & Alavimajd, 2015) (Figs. 1 and 2).

Delay in development has been described in up to 11.8% of children of mothers with depression (Vameghi et al., 2015). Fine motor delay has been demonstrated to be increased by fourfold and delay in emotional development by sixfold, compared to children of nondepressed mothers (Ali et al., 2013).

Poor cognitive function has been reported in infancy and early childhood (Murray & Cooper, 1997). Moreover, longitudinal studies have found poor academic performance in 16-year-old boys whose mother had a history of postnatal depression. This suggests that the early cognitive deficits may persist (Murray et al., 2010).

Maternal depressive symptoms may interfere with her ability to provide early learning opportunities with diminished motivation to engage in play or to allow the child to interact with the environment in a manner that promotes cognitive growth, all of which may lead to delays in development (Black et al., 2007; O'Brien Caughy, Huang, & Lima, 2009).

## Sleep disturbances

Maternal depression, either occurring prenatally or postnatally, is associated with sleep disturbances in their offspring in early life and up to adolescence. Higher levels of prenatal depressive symptoms are linked with increased sleep problems in children at 18 and 30 months of age, independent of the mother's postnatal mood and psychosocial variables (O'Connor et al., 2007). Maternal postnatal depressive symptoms have significant effect on increased nocturnal awakenings at 18 months (Ystrom et al., 2017). However, no such association has been found between maternal depressive symptoms and nocturnal awakening in children at 6 months of age. This suggests that there may be developmentally specific associations and "sensitive periods" for environmental influences on infants' sleep pattern (Ystrom et al., 2017).

Maternal postnatal depression and increased risk of sleep problems in adolescents have also been demonstrated. Compared to 22% of adolescents of nondepressed mothers, 36% of adolescent offspring of depressed mothers reported having sleep problems at 18 years of age (Taylor et al., 2017) (Fig. 3). Several mechanisms have been postulated

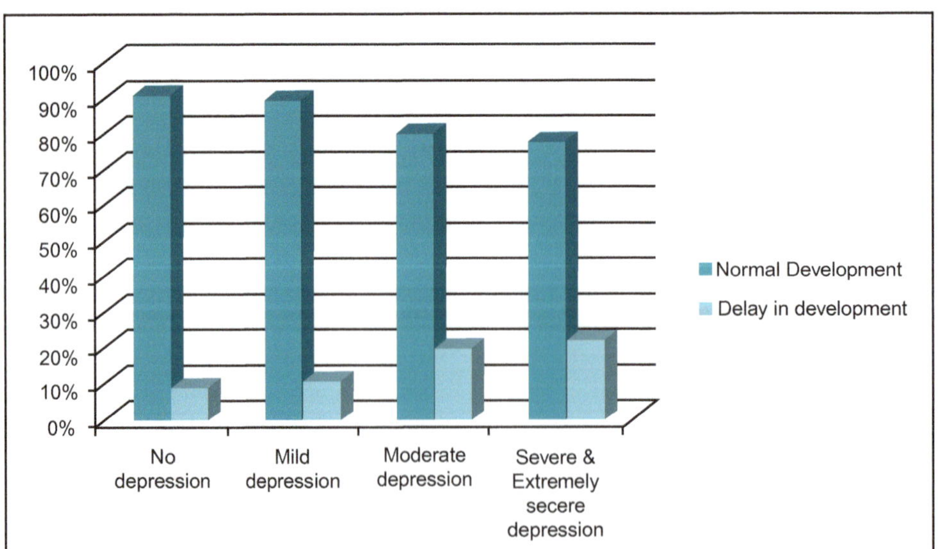


FIG. 1 Correlation between severity of mothers' depression and developmental delay in offspring. The $X$-axis shows the severity of depression, and the $Y$-axis shows the percentage of normal and delayed development. *Adapted from Vameghi, R., Amir Ali Akbari, S., Sajjadi, H., Sajedi, F., & Alavimajd, H. (2015). Correlation between mothers' depression and developmental delay in infants aged 6-18 months.* Global Journal of Health Science, 8 *(5), 11–18. https://doi.org/10.5539/gjhs.v8n5p11.*

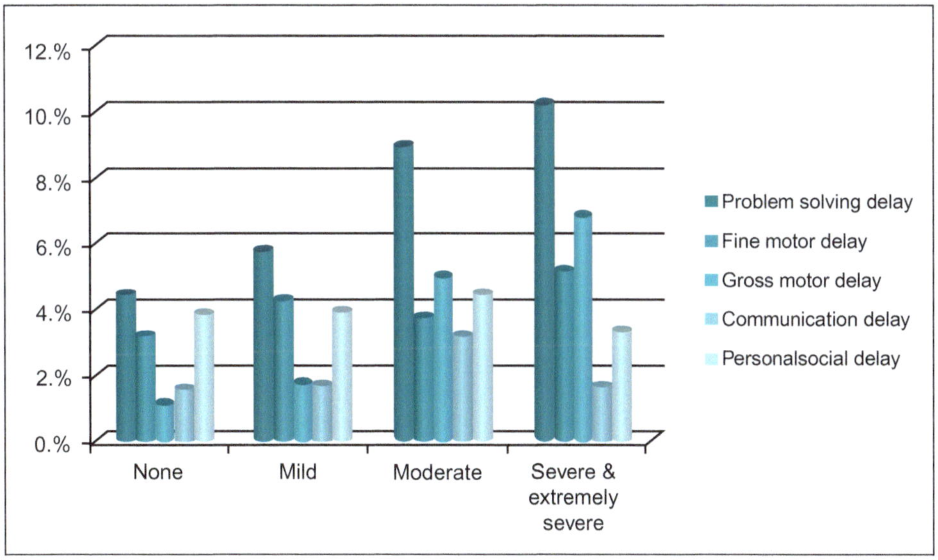


FIG. 2 Correlation between mother's depression and developmental delay in the five domains. The *X*-axis shows the severity of depression, and the *Y*-axis shows the percentage of normal and delayed development. *Adapted from Vameghi, R., Amir Ali Akbari, S., Sajjadi, H., Sajedi, F., & Alavimajd, H. (2015). Correlation between mothers' depression and developmental delay in infants aged 6-18 months. Global* Journal of Health Science, 8 *(5), 11–18. https://doi.org/10.5539/gjhs.v8n5p11.*

regarding the effects of maternal depression on sleep of offspring. First, depressed mothers may have difficulty in setting sleep-wake routines (Taylor et al., 2017), which may result in later bedtimes (Ystrom et al., 2017). Second, maternal negative cognitions may lead to increased maternal involvement at bedtime (Taylor et al., 2017; Ystrom et al., 2017), which may reduce the infant's ability to self-regulate and fall asleep (Taylor et al., 2017). The first 2 years of life is a time of rapid brain development and the sleep-wake rhythm undergoes substantial changes during this period. Thus, any disturbance occurring at this stage may have lasting effects on offspring's sleep (Sadeh, 2007).

## Externalizing problems

Children of depressed mothers are more likely to be physically aggressive when compared to their normal counterparts (Hay, Pawlby, Angold, Harold, & Sharp, 2003; Kim-Cohen, Moffitt, Taylor, Pawlby, & Caspi, 2005). The risk of aggression is increased independent of other risk

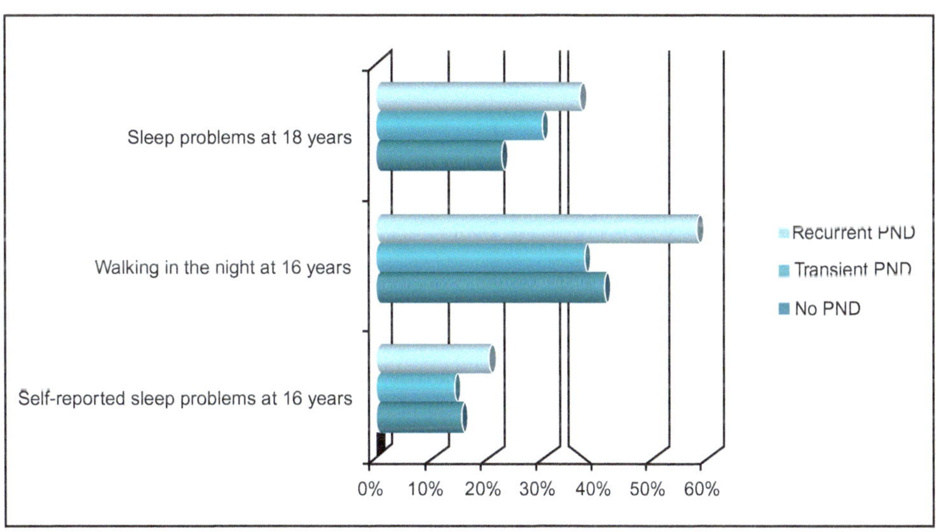


FIG. 3 Correlation of sleep problems in children of mothers with postnatal depression at 16 and 18 years of age, with the chronicity of mother's depression. The *X*-axis shows the percentage of problems, and the *Y*-axis shows the chronicity of maternal depression. *Adapted from Taylor, A. K., Netsi, E., O'Mahen, H., Stein, A., Evans, J., & Pearson, R. M. (2017). The association between maternal postnatal depressive symptoms and offspring sleep problems in adolescence.* Psychological Medicine, 47 *(3), 451–459. https://doi.org/10.1017/s0033291716002427.*

factors such as parental history of arrest, parental antisocial personality disorder, social class, family financial stress, and single parent households (Hay et al., 2003; Kim-Cohen et al., 2005). Increased risk of aggression in offspring may persist up to 15–20 years (Hay et al., 2003; Kim-Cohen et al., 2005). Several mechanisms have been suggested to account for this increased aggression. These include quality of parent-child relationship, interpersonal skill deficits, history of depression in the child (Keenan-Miller, Hammen, & Brennan, 2010), and low perception of parental warmth and acceptance (Hale, Van Der Valk, Engels, & Meeus, 2005). Symptoms of ADHD in children are also thought to contribute toward the aggression in children of depressed mothers (Hay et al., 2003).

Maternal depression is a predictor of later development of conduct problems in children with ADHD. A dose-response relationship has been described between the degree of conduct problems and the number of developmental periods during which the mother remained depressed (Agha, Zammit, Thapar, & Langley, 2017; Chronis et al., 2007).

Chronic or recently elevated depression in the mother also increases the risk of loud, rowdy behavior, alcohol use, and bullying in young male adolescents but the female adolescents did not exhibit elevated risks of such behavior. However, they were more likely to view alcohol as harmless (Flouri & Ioakeimidi, 2018). Perceived harm from alcohol is known to be a predictor in alcohol use and having a more positive attitude toward alcohol could predict a higher risk for the initiation of alcohol use (Maggs, Staff, Patrick, Wray-Lake, & Schulenberg, 2015).

### Internalizing problems

Maternal depression is associated with internalizing problems in the offspring from preschool years (Herba et al., 2013), through middle childhood (Ahun et al., 2018), adolescence (Murray et al., 2011), and up to early adulthood (Hammen, Hazel, Brennan, & Najman, 2012; Murray et al., 2011; Pearson et al., 2013). The risk of developing internalizing problems is similar to that of developing externalizing problems (Goodman et al., 2011). Maternal depressive symptoms during preschool years are associated with emotional problems, separation anxiety, and social withdrawal (Herba et al., 2013). In adolescents, maternal depression is associated with increased risk of depression and anxiety disorders (Murray et al., 2011). The lifetime risk of depression at 16 years of age in children of depressed mothers was over 40%, with the average age of onset being 14 years. Comorbid anxiety disorders are present in more than half of the depressed offspring (Murray et al., 2011).

There are several explanations for children of depressed mothers to have a higher risk of internalizing problems. The effects of depression on parenting seem to mediate these effects. Frequent display of psychological aggression (i.e., shouting at the child, name calling, threatening), maternal rejection, and low levels of positive interactions along with high levels of negative interactions, especially in positive context, were associated with higher levels of internalizing symptoms in the offspring (Kuckertz, Mitchell, & Wiggins, 2018; Olino et al., 2016; Zalewski, Thompson, & Lengua, 2017). Empathic sensitivity of the child has also been linked to increased incidence of internalizing problems in offspring. It is suggested that children who experience high levels of negative emotions in response to their mothers' distress, experience happiness when the mother stays happy and paying constant attention to their mothers' emotional status carry a higher risk of them developing internalizing problems (Tully & Donohue, 2017).

### Suicidal ideation

Depression in the mother is associated with increased suicidal ideation in the offspring, especially in mothers with chronic and severe depression (Hammerton et al., 2015; Hammerton, Zammit, Thapar, & Collishaw, 2016). The higher risk of suicidal ideation is only partially mediated by exposure to maternal suicidal attempts and offspring's psychiatric disorder (Hammerton et al., 2015; Hammerton et al., 2016). Impaired parent-child relationship also contributes to the increased risk of suicidal ideation, independent of child psychopathology (Hammerton et al., 2016). This highlights that the risk of suicide should be considered in the offspring of depressed mothers, even when the children themselves are not depressed (Hammerton et al., 2016) (Table 1).

## Risk factors

### Demographic factors

Socioeconomic status, ethnicity, age of the mother, and her education level are all associated with the risk of developing psychopathology in the offspring. Poverty is shown to be associated with a higher risk of psychopathology in the children of depressed mothers due to increased stressors and poor resourcefulness (Goodman et al., 2011).

Low education level in mothers who suffered from postnatal depression has shown to be associated with higher rates of depression among their offspring (Flynn, Chung, Ozer, & Fernald, 2017; Pearson et al., 2013). In contrast, the mothers with higher educational level have more positive home environments, more positive and sensitive parenting styles, and additional sources of social support, which may mitigate the effects of maternal depression on offspring (Pearson et al., 2013).

**TABLE 1 Mental health problems in children of mothers with depression.**

| Mental health outcomes in children |
|---|
| Impaired bonding |
| Attachment difficulties<br>– Reduced secure attachment<br>– Increased disorganized attachment |
| Developmental delay<br>– Gross motor<br>– Fine motor<br>– Language<br>– Social-emotional<br>– Cognitive development |
| Sleep disturbances |
| Externalizing problems<br>– Aggression<br>– Increased risk of rowdy behavior, alcohol use, and bullying in male adolescents |
| Internalizing problems<br>– Emotional problems, separation anxiety, and social withdrawal in preschoolers<br>– Increased risk of depression and anxiety disorders in adolescence |
| Suicidal ideation |

The age of the mother has also been shown to impact the rates of psychological problems in children, but the findings have been inconsistent. One study had demonstrated that children of younger mothers with depression have better language development when compared to older mothers. It was hypothesized that this was possibly because the older mothers were more tired and spoke less with their children (Quevedo et al., 2012). However, some studies have found that the psychological and behavioral problems are higher when the mothers are at the extremes of ages (less than 30 years or more than 50 years) when compared to children of mothers between 30 and 50 years (Rohanachandra, Prathapan, & Wijetunge, 2018). It was suggested that as the younger mothers are relatively immature and have inadequate coping skills, they find it more difficult to cope with the demands of parenting when depressed. As the older mothers are more likely to be less agile and possibly have more physical comorbidities, they would find it more difficult to cope with the demands of parenting with the added burden of tiredness due to depression (Rohanachandra et al., 2018). The children of mothers from ethnic minorities also experience higher rates of externalizing and internalizing problems when the mother is depressed (Goodman et al., 2011).

## Child factors

The age, gender, and personality traits of children influence the impact of maternal depression on children. Younger children are more affected by their mothers' depression than older children (Goodman et al., 2011). First, younger children are exclusively dependent on their mothers, while older children have a wider social network of other relatives and friends, which may mitigate the effects of maternal depression. Second, compared to younger children, older children have more cognitive maturity which provides them with a better understanding that their mothers' changed behavior is due to an illness. Third, the older children who get exposed to their mothers' illness for the first time have already experienced more years of normal development and normal mother-child interactions than younger children. Therefore, children who are at a younger age when exposed to their mothers' psychopathology may be more vulnerable to develop psychological consequences (Goodman et al., 2011).

The gender of the child also influences the types of problems experienced. The cognitive development in boys has shown to be more affected by maternal depression than that in girls (Murray et al., 2010), whereas girls are known to experience more internalizing problems (Goodman et al., 2011). No such gender difference has been demonstrated in externalizing problems and general psychopathology (Goodman et al., 2011). Several reasons have been postulated to explain the higher rates of internalizing problems among girls. First, the hereditability of depression is higher in girls than in boys (Kendler, Gardner, Neale, & Prescott, 2001). Second, girls experience more stressors, especially at interpersonal level which make them more vulnerable to develop internalizing problems (Hammen & Brennan, 2003). Third, the parenting styles used by depressed mothers may vary between the two sexes (Goodman et al., 2011).

Offspring personality traits especially neuroticism, conscientiousness, and agreeableness have also been demonstrated to mediate the risk of child behavior problems in mothers with depression (Allen, Oshri, Rogosch, Toth, & Cicchetti, 2019).

## Family factors

Increased externalizing problems are found in single parent households where the mother is depressed. However, there is no increase in internalizing problems or general psychopathology (Goodman et al., 2011).

Paternal psychopathology also plays a role in mediating the effects of maternal depression on children. Paternal alcoholism occurring together with maternal depression is associated with an increased risk of conduct disorder and oppositional defiant disorder in boys (Foley et al., 2001). Fathers' psychological status also contribute to their children's social and emotional competence (Goodman et al., 2011) and the presence of paternal psychopathology is thought to increase the risk of psychological problems through increased genetic predisposition and by adding to the adverse qualities of the family environment (Goodman et al., 2011).

A relationship between the parity of the mother and the language development of the child has also been demonstrated and families with more than two children are more likely to have children with language delays if the mother is depressed (Quevedo et al., 2012). In larger families there is relatively less time to be spent with each child, which may result in less stimulation and contribute to speech delay (Halpern, Giugliani, Victora, Barros, & Horta, 2002).

### Illness factors

There is consistent evidence that the chronicity of maternal depression is associated with higher risk of psychopathology in children (Brennan et al., 2000; Grace, Evindar, & Stewart, 2003; Hammerton et al., 2015). Most previous literature suggests that greater severity of symptoms also confers a higher risk of psychological problems in the offspring (Brennan et al., 2000; Hammen & Brennan, 2003; Hammerton et al., 2015). However, some suggest that the impact on children may be higher when the mother suffers from moderate depression, rather than severe depression (Rutter, 1990). One explanation for this is that when the mother has a severe depressive episode the child finds it easier to understand that the mother is ill and to depend on other caregivers and severe depression is more likely to be noticed by others, who would then intervene to offer more support to the child (Rutter, 1990).

Maternal depression occurring comorbid with other mental health problems increases the risk of developing psychopathology in the offspring. Therefore, in order to minimize the psychological impact on children, the focus should not only be on treating depression, but also addressing the comorbid mental health issues (Ahun et al., 2018; Kim-Cohen et al., 2005).

It has also been shown that the maternal suicidal attempts are associated with an increased risk of suicidal behavior in their adolescent offspring (Hammerton et al., 2016).

### Resilience factors

Not all children develop mental health problems as a result of their mothers' depression. Factors in the child that contribute to resilience include high IQ (Pargas, Brennan, Hammen, & Le Brocque, 2010) and more nonverbal communication at 15 months of age (Savage-McGlynn et al., 2015). Maternal factors such as high levels of maternal warmth, low levels of maternal over-involvement, low level of maternal psychological control, and mother's positive perception about her parenting have been shown to be protective against the impact of maternal depression on children (Pargas et al., 2010; Savage-McGlynn et al., 2015). Furthermore, co-parent support and quality of parental couple relationship have also been described to mitigate the effects of maternal depression on the offspring (Collishaw et al., 2016) (Table 2).

**TABLE 2** Factors effecting mental health outcomes in children.

| Risk factors | Resilient factors |
|---|---|
| Demographic factors<br>– Poverty<br>– Low educational attainment in the mother<br>– Ethnic minorities | Maternal factors<br>– High levels of maternal warmth<br>– Low levels of maternal over-involvement<br>– Low level of maternal psychological control<br>– Mother's positive perception about her parenting |
| Child factors<br>– Younger age<br>– Child's personality traits (neuroticism, conscientiousness, and agreeableness) | Child factors<br>– High IQ<br>– More nonverbal communication at 15 months of age |
| Family factors<br>– Single parent families<br>– Paternal alcoholism<br>– Paternal psychopathology<br>– Larger family size | Family factors<br>– Co-parent support<br>– Good parental couple relationship |
| Illness factors<br>– Higher chronicity and severity of depression in mother<br>– Comorbid mental illness | |

## Interventions to reduce the impact of maternal depression on children

### Screening of mother for PPD

Given the high prevalence of depression during pregnancy and postpartum period, it is recommended that all women be routinely screened for depression in the antenatal and perinatal care settings (Rishel, 2012) for early detection.

## Screening children for psychological well-being

In addition to screening mothers, screening of children for psychological well-being may be an important step in early identification of vulnerable children. If during routine visits, the pediatrician or family physician notices abnormal parent-child interactions or features suggestive of emotional problems in the child, both the child and the mother needs to be screened for depression (Rishel, 2012) (Table 3).

## Effective treatment of the mother

Vigorous and prompt treatment of the mother is the first step in minimizing the impact of maternal depression on children. Remission of mother's depressive symptoms is significantly associated with improvement in internalizing problems, externalizing problems, and total symptoms in the children, independent of the child's age, sex, baseline severity of the problems, socioeconomic status, and severity of the mothers' illness (Weissman et al., 2006). In contrast, the child's mental health problems were found to be increased if the mother remained depressed (Weissman et al., 2006).

## Parent-child interaction therapy (PCIT)

Maternal depression has a profound impact on the mother-child relationship (Rishel, 2012). Therefore, focusing on mother-child interactions is a key step in mitigating the effects of maternal depression on children. PCIT is a family-centered therapy designed for caregivers of abused or at-risk children between 2 and 8 years of age. PCIT has been recommended as a preventive intervention for children of depressed mothers (Rishel, 2012).

## High-quality childcare

There is ample evidence that children from disadvantaged backgrounds benefit from high-quality child care (Bradley & Vandell, 2007). This has also been demonstrated to be beneficial in buffering the impact of maternal depression on children. High-quality childcare characterized by warm and sensitive educators and stimulating material has a positive effect on child mental health, especially on externalizing problems (Charrois et al., 2017). Therefore, ensuring that such children have access to high-quality childcare is an important step in mitigating the adverse effects of maternal depression on children (Charrois et al., 2017).

## School-based interventions

School-based interventions also play a role in minimizing the effects of maternal depression on children. Children placed in classrooms with a warm and positive environment were shown to be less effected by their mothers' depression with regard to externalizing problems, social skills, and cognitive performance (Yan, Zhou, & Ansari, 2016). Positive classroom environment can improve children's coping skills (Yan et al., 2016), promote emotional regulation (Skinner & Belmont, 1993), strengthen children's perceived self-control (Wilson, Pianta, & Stuhlman, 2007), and increase their perceived social support through forming

**TABLE 3** List of Instruments used to measure well-being and depression in children.

| Child/parent rated instruments | Structured and semistructured interviews |
|---|---|
| • Beck Depression Inventory (BDI)<br>• Beck Youth Inventories of Emotional and Social Impairment (BYI)<br>• Children Depression Inventory (CDI)<br>• Children's Depression Rating Scale Revised (CDRS-R)<br>• Center for Epidemiological Studies Depression Scale for Children (CES-DC)<br>• Depression Self-Rating Scale (DSRS)<br>• Kutcher Adolescent Depression Scale (KADS)<br>• Mood and feelings questionnaire (MFQ)<br>• Multiscore Depression Inventory for Children (MDI-C)<br>• Pediatric Quality-of-Life Enjoyment and Satisfaction Questionnaire (PQ-LES-Q)<br>• Reynolds Child Depression Scale- 2nd Edition (RCSD2)<br>• Screen for Child Anxiety Related Disorders (SCARED)<br>• Spence Children's Anxiety Scale<br>• Strengths & Difficulties Questionnaire (SDQ) | • Anxiety Disorders Interview Schedule (ADIS-IV)<br>• Child and Adolescent Psychiatric Assessment (CAPA)<br>• Child Assessment Schedule (CAS)<br>• Children's Interview for Psychiatric Syndromes (ChIPS)<br>• Diagnostic Interview for Children and Adolescents (DICA)<br>• Diagnostic Interview Schedule for Children (DISC-IV)<br>• Interview Schedule for Children and Adolescents (ISCA)<br>• Kiddie Schedule for Affective Disorders and Schizophrenia (K-SADS)<br>• Preschool Age Psychiatric Assessment (PAPA) |

a strong emotional bond with the teacher (Wentzel, 2002) which can all contribute to alleviating the effects of maternal mental health. Therefore, the development of school-based intervention programs aimed at children of mothers with depression is a vital step in reducing the mental health risks in children.

### Home visitation

Previous literature has shown that, apart from a few exceptions, most studies do not show a beneficial effect of home visitation on the manifestation or course of maternal depressive symptoms (Ammerman, Putnam, Bosse, Teeters, & Van Ginkel, 2010).

## Conclusion

Given the impact of maternal depression on children's mental health, more attention should be given to establish evidence-based interventions to minimize the psychological consequences on children. This should not be considered only as a duty of the specialized mental health services, but rather a role of all health-care workers including primary care workers, family physicians, and pediatricians as well as educational services. Further research is also needed to develop evidence-based interventions for this vulnerable group of children.

## Key facts

Key facts about the effect on maternal depression on parent-child interactions.

- Lack of energy experienced by depressed mothers may impair their capacity to respond to their children's needs.
- Depressed mothers are less emotionally available, less attuned to their infants' emotional needs, less sensitive, and more hostile toward their children compared to non-depressed mothers.
- In managing conflicts, depressed mothers are known to express more guilt inducing, critical and harsh statements to their children.
- Depressed mothers are less engaging, less vocal, and less skillful at getting and sustaining their children's attention and are less sensitive about their child's levels of understanding, are uncomfortable in the teaching role, provide less stimulation, and engage in less structured teaching than nondepressed mothers.
- Depressed mothers make more negative appraisals about their children's behavior and are more likely to use physical punishment in disciplining their children.

## Summary points

- Maternal depression is associated with impaired bonding; insecure attachment; sleep disturbances; delays in gross motor, fine motor, language social-emotional and cognitive development; externalizing problems such as aggression, conduct problems and ADHD; internalizing problems such as separation anxiety, social withdrawal, and depression; and increased risk of self-harm in their offspring.
- Poor socioeconomic status, mother's low educational level, single parent households, younger age of the child, increased severity and chronicity of maternal depressive symptoms, presence of comorbid mental health problems in the mother, and presence of psychopathology in the father are associated with a higher risk of psychopathology in children.
- Screening of mother for postpartum depression, prompt treatment of the mother, PCIT, high-quality childcare, and school-based interventions are effective in minimizing the impact on children.

## Mini-dictionary of terms

**Bonding** Bonding is the special emotional attachment that the parents develop toward their children.

**Attachment** Attachment is the baby's tendency to seek proximity to their caregiver, especially when they feel threatened, and feel secure in their presence.

**Secure attachment** Secure attachment style is when a child has trust in their caregivers' availability and responsiveness and have positive expectations from their caregivers. They usually display positive interactions with their caregivers.

**Avoidant attachment** Avoidant attachment style occurs when a child believes that they cannot trust the caregivers' availability and responsiveness and cannot turn to their caregivers for comfort.

**Disorganized attachment** Disorganized attachment occurs in children of abusive caregivers. Such children may display negative behaviors and rejection toward their caregivers.

**Externalizing problems** Behaviors that are manifested in the children's outward negative behavior toward the external environment. These include aggression, hyperactivity, and disruptive behaviors.

**Internalizing problems** Behaviors that are manifested as children's negative behavior toward their internal psychological environment. These include withdrawal, anxiety, inhibition, and depression.

## References

Agha, S. S., Zammit, S., Thapar, A., & Langley, K. (2017). Maternal psychopathology and offspring clinical outcome: A four-year follow-up of boys with ADHD. *European Child & Adolescent Psychiatry*, *26*(2), 253–262. https://doi.org/10.1007/s00787-016-0873-y.

Ahun, M. N., Consoli, A., Pingault, J. B., Falissard, B., Battaglia, M., Boivin, M., et al. (2018). Maternal depression symptoms and internalising problems in the offspring: The role of maternal and family factors. *European Child & Adolescent Psychiatry*, *27*(7), 921–932. https://doi.org/10.1007/s00787-017-1096-6.

Ali, N. S., Mahmud, S., Khan, A., & Ali, B. S. (2013). Impact of postpartum anxiety and depression on child's mental development from two peri-urban communities of Karachi, Pakistan: A quasi-experimental study. *BMC Psychiatry*, *13*, 274. https://doi.org/10.1186/1471-244x-13-274.

Allen, T. A., Oshri, A., Rogosch, F. A., Toth, S. L., & Cicchetti, D. (2019). Offspring personality mediates the association between maternal depression and childhood psychopathology. *Journal of Abnormal Child Psychology*, *47*(2), 345–357. https://doi.org/10.1007/s10802-018-0453-3.

Ammerman, R. T., Putnam, F. W., Bosse, N. R., Teeters, A. R., & Van Ginkel, J. B. (2010). Maternal depression in home visitation: A systematic review. *Aggression and Violent Behavior*, *15*(3), 191–200. https://doi.org/10.1016/j.avb.2009.12.002.

Badovinac, S., Martin, J., Guerin-Marion, C., O'Neill, M., Pillai Riddell, R., Bureau, J. F., et al. (2018). Associations between mother-preschooler attachment and maternal depression symptoms: A systematic review and meta-analysis. *PLoS One*. *13*(10) e0204374. https://doi.org/10.1371/journal.pone.0204374.

Beardslee, W. R., Bemporad, J., Keller, M. B., & Klerman, G. L. (1983). Children of parents with major affective disorder: A review. *The American Journal of Psychiatry*, *140*(7), 825–832. https://doi.org/10.1176/ajp.140.7.825.

Black, M. M., Baqui, A. H., Zaman, K., McNary, S. W., Le, K., Arifeen, S. E., et al. (2007). Depressive symptoms among rural Bangladeshi mothers: Implications for infant development. *Journal of Child Psychology and Psychiatry*, *48*(8), 764–772. https://doi.org/10.1111/j.1469-7610.2007.01752.x.

Bradley, R. H., & Vandell, D. L. (2007). Child care and the well-being of children. *Archives of Pediatrics & Adolescent Medicine*, *161*(7), 669–676.

Brennan, P. A., Hammen, C., Andersen, M. J., Bor, W., Najman, J. M., & Williams, G. M. (2000). Chronicity, severity, and timing of maternal depressive symptoms: Relationships with child outcomes at age 5. *Developmental Psychology*, *36*(6), 759–766.

Charrois, J., Cote, S. M., Japel, C., Seguin, J. R., Paquin, S., Tremblay, R. E., et al. (2017). Child-care quality moderates the association between maternal depression and children's behavioural outcome. *Journal of Child Psychology and Psychiatry*, *58*(11), 1210–1218. https://doi.org/10.1111/jcpp.12764.

Chronis, A. M., Lahey, B. B., Pelham, W. E., Jr., Williams, S. H., Baumann, B. L., Kipp, H., et al. (2007). Maternal depression and early positive parenting predict future conduct problems in young children with attention-deficit/hyperactivity disorder. *Developmental Psychology*, *43*(1), 70–82. https://doi.org/10.1037/0012-1649.43.1.70.

Collishaw, S., Hammerton, G., Mahedy, L., Sellers, R., Owen, M. J., Craddock, N., et al. (2016). Mental health resilience in the adolescent offspring of parents with depression: A prospective longitudinal study. *Lancet Psychiatry*, *3*(1), 49–57. https://doi.org/10.1016/s2215-0366(15)00358-2.

Cummings, E. M., & Davies, P. T. (1994). Maternal depression and child development. *Journal of Child Psychology and Psychiatry*, *35*(1), 73–112. https://doi.org/10.1111/j.1469-7610.1994.tb01133.x.

Flouri, E., & Ioakeimidi, S. (2018). Maternal depressive symptoms in childhood and risky behaviours in early adolescence. *European Child & Adolescent Psychiatry*, *27*(3), 301–308. https://doi.org/10.1007/s00787-017-1043-6.

Flynn, E. P., Chung, E. O., Ozer, E. J., & Fernald, L. C. H. (2017). Maternal depressive symptoms and child behavior among Mexican women and their children. *International Journal of Environmental Research and Public Health*, *14*(12), 1566. https://doi.org/10.3390/ijerph14121566.

Foley, D. L., Pickles, A., Simonoff, E., Maes, H. H., Silberg, J. L., Hewitt, J. K., et al. (2001). Parental concordance and comorbidity for psychiatric disorder and associate risks for current psychiatric symptoms and disorders in a community sample of juvenile twins. *Journal of Child Psychology and Psychiatry*, *42*(3), 381–394.

Goodman, S. H., Rouse, M. H., Connell, A. M., Broth, M. R., Hall, C. M., & Heyward, D. (2011). Maternal depression and child psychopathology: A meta-analytic review. *Clinical Child and Family Psychology Review*, *14*(1), 1–27. https://doi.org/10.1007/s10567-010-0080-1.

Grace, S. L., Evindar, A., & Stewart, D. E. (2003). The effect of postpartum depression on child cognitive development and behavior: A review and critical analysis of the literature. *Archives of Women's Mental Health*, *6*(4), 263–274. https://doi.org/10.1007/s00737-003-0024-6.

Hale, W. W., 3rd, Van Der Valk, I., Engels, R., & Meeus, W. (2005). Does perceived parental rejection make adolescents sad and mad? The association of perceived parental rejection with adolescent depression and aggression. *The Journal of Adolescent Health*, *36*(6), 466–474. https://doi.org/10.1016/j.jadohealth.2004.04.007.

Halpern, R., Giugliani, E. R., Victora, C. G., Barros, F. C., & Horta, B. L. (2002). Fatores de risco para suspeita de atraso no desenvolvimento neuropsicomotor aos 12 meses de vida. *Revista Chilena de Pediatría*, *73*(5), 529–539.

Hammen, C., & Brennan, P. A. (2003). Severity, chronicity, and timing of maternal depression and risk for adolescent offspring diagnoses in a community sample. *Archives of General Psychiatry*, *60*(3), 253–258. https://doi.org/10.1001/archpsyc.60.3.253.

Hammen, C., Hazel, N. A., Brennan, P. A., & Najman, J. (2012). Intergenerational transmission and continuity of stress and depression: Depressed women and their offspring in 20 years of follow up. *Psychological Medicine*, *42*(5), 931–942. https://doi.org/10.1017/s0033291711001978.

Hammerton, G., Mahedy, L., Mars, B., Harold, G. T., Thapar, A., Zammit, S., et al. (2015). Association between maternal depression symptoms across the first eleven years of their Child's life and subsequent offspring suicidal ideation. *PLoS One*. *10*(7)e0131885 https://doi.org/10.1371/journal.pone.0131885.

Hammerton, G., Zammit, S., Thapar, A., & Collishaw, S. (2016). Explaining risk for suicidal ideation in adolescent offspring of mothers with depression. *Psychological Medicine*, *46*(2), 265–275. https://doi.org/10.1017/s0033291715001671.

Hay, D. F., Pawlby, S., Angold, A., Harold, G. T., & Sharp, D. (2003). Pathways to violence in the children of mothers who were depressed postpartum. *Developmental Psychology*, *39*(6), 1083–1094. https://doi.org/10.1037/0012-1649.39.6.1083.

Herba, C. M., Tremblay, R. E., Boivin, M., Liu, X., Mongeau, C., Seguin, J. R., et al. (2013). Maternal depressive symptoms and children's emotional problems: Can early child care help children of depressed mothers? *JAMA Psychiatry*, *70*(8), 830–838. https://doi.org/10.1001/jamapsychiatry.2013.1361.

Kaplan, P. S., Danko, C. M., Everhart, K. D., Diaz, A., Asherin, R. M., Vogeli, J. M., et al. (2014). Maternal depression and expressive communication in one-year-old infants. *Infant Behavior & Development*, *37*(3), 398–405. https://doi.org/10.1016/j.infbeh.2014.05.008.

Keenan-Miller, D., Hammen, C., & Brennan, P. A. (2010). Mediators of aggression among young adult offspring of depressed mothers. *Journal of Abnormal Psychology*, *119*(4), 836–849. https://doi.org/10.1037/a0021079.

Kendler, K. S., Gardner, C. O., Neale, M. C., & Prescott, C. A. (2001). Genetic risk factors for major depression in men and women: Similar or different heritabilities and same or partly distinct genes? *Psychological Medicine*, *31*(4), 605–616. https://doi.org/10.1017/s0033291701003907.

Kim-Cohen, J., Moffitt, T. E., Taylor, A., Pawlby, S. J., & Caspi, A. (2005). Maternal depression and children's antisocial behavior: Nature and nurture effects. *Archives of General Psychiatry*, *62*(2), 173–181. https://doi.org/10.1001/archpsyc.62.2.173.

Kuckertz, J. M., Mitchell, C., & Wiggins, J. L. (2018). Parenting mediates the impact of maternal depression on child internalizing symptoms. *Depression and Anxiety*, *35*(1), 89–97. https://doi.org/10.1002/da.22688.

Maggs, J. L., Staff, J., Patrick, M. E., Wray-Lake, L., & Schulenberg, J. E. (2015). Alcohol use at the cusp of adolescence: A prospective national birth cohort study of prevalence and risk factors. *The Journal of Adolescent Health*, *56*(6), 639–645. https://doi.org/10.1016/j.jadohealth.2015.02.010.

Martins, C., & Gaffan, E. A. (2000). Effects of early maternal depression on patterns of infant-mother attachment: A meta-analytic investigation. *Journal of Child Psychology and Psychiatry*, *41*(6), 737–746.

Maselko, J., Sikander, S., Bangash, O., Bhalotra, S., Franz, L., Ganga, N., et al. (2016). Child mental health and maternal depression history in Pakistan. *Social Psychiatry and Psychiatric Epidemiology*, *51*(1), 49–62. https://doi.org/10.1007/s00127-015-1143-x.

Moehler, E., Brunner, R., Wiebel, A., Reck, C., & Resch, F. (2006). Maternal depressive symptoms in the postnatal period are associated with long-term impairment of mother-child bonding. *Archives of Women's Mental Health*, *9*(5), 273–278. https://doi.org/10.1007/s00737-006-0149-5.

Murray, L., Arteche, A., Fearon, P., Halligan, S., Croudace, T., & Cooper, P. (2010). The effects of maternal postnatal depression and child sex on academic performance at age 16 years: A developmental approach. *Journal of Child Psychology and Psychiatry*, *51*(10), 1150–1159. https://doi.org/10.1111/j.1469-7610.2010.02259.x.

Murray, L., Arteche, A., Fearon, P., Halligan, S., Goodyer, I., & Cooper, P. (2011). Maternal postnatal depression and the development of depression in offspring up to 16 years of age. *Journal of the American Academy of Child and Adolescent Psychiatry*, *50*(5), 460–470. https://doi.org/10.1016/j.jaac.2011.02.001.

Murray, L., & Cooper, P. (1997). Effects of postnatal depression on infant development. *Archives of Disease in Childhood*, *77*(2), 99–101. https://doi.org/10.1136/adc.77.2.99.

O'Brien Caughy, M., Huang, K. Y., & Lima, J. (2009). Patterns of conflict interaction in mother-toddler dyads: Differences between depressed and non-depressed mothers. *Journal of Child and Family Studies*, *18*(1), 10–20. https://doi.org/10.1007/s10826-008-9201-6.

O'Connor, T. G., Caprariello, P., Blackmore, E. R., Gregory, A. M., Glover, V., & Fleming, P. (2007). Prenatal mood disturbance predicts sleep problems in infancy and toddlerhood. *Early Human Development*, *83*(7), 451–458. https://doi.org/10.1016/j.earlhumdev.2006.08.006.

Olino, T. M., McMakin, D. L., Nicely, T. A., Forbes, E. E., Dahl, R. E., & Silk, J. S. (2016). Maternal depression, parenting, and youth depressive symptoms: Mediation and moderation in a short-term longitudinal study. *Journal of Clinical Child and Adolescent Psychology*, *45*(3), 279–290. https://doi.org/10.1080/15374416.2014.971456.

Pargas, R. C. M., Brennan, P. A., Hammen, C., & Le Brocque, R. (2010). Resilience to maternal depression in young adulthood. *Developmental Psychology*, *46*(4), 805–814. https://doi.org/10.1037/a0019817.

Pearson, R. M., Evans, J., Kounali, D., Lewis, G., Heron, J., Ramchandani, P. G., et al. (2013). Maternal depression during pregnancy and the postnatal period: Risks and possible mechanisms for offspring depression at age 18 years. *JAMA Psychiatry*, *70*(12), 1312–1319. https://doi.org/10.1001/jamapsychiatry.2013.2163.

Quevedo, L. A., Silva, R. A., Godoy, R., Jansen, K., Matos, M. B., Tavares Pinheiro, K. A., et al. (2012). The impact of maternal post-partum depression on the language development of children at 12 months. *Child: Care, Health and Development*, *38*(3), 420–424. https://doi.org/10.1111/j.1365-2214.2011.01251.x.

Rishel, C. W. (2012). Pathways to prevention for children of depressed mothers: A review of the literature and recommendations for practice. *Depression Research and Treatment*, *2012*, 313689. https://doi.org/10.1155/2012/313689.

Rohanachandra, Y. M., Prathapan, S., & Wijetunge, S. (2018). Psychological problems and socioemotional wellbeing among children of mothers with depression and their association with sociodemographic factors in a Sri Lankan setting. *Psychiatry Journal*, *2018*, 3809384. https://doi.org/10.1155/2018/3809384.

Rutter, M. (1990). Commentary: Some focus and process considerations regarding effects of parental depression on children. *Developmental Psychology*, *26*(1), 60.

Sadeh, A. (2007). Consequences of sleep loss or sleep disruption in children. *Sleep Medicine Clinics*, *2*(3), 513–520.

Savage-McGlynn, E., Redshaw, M., Heron, J., Stein, A., Quigley, M. A., Evans, J., et al. (2015). Mechanisms of resilience in children of mothers who self-report with depressive symptoms in the first postnatal year. *PLoS One*. *10*(11)e0142898https://doi.org/10.1371/journal.pone.0142898.

Skinner, E. A., & Belmont, M. J. (1993). Motivation in the classroom: Reciprocal effects of teacher behavior and student engagement across the school year. *Journal of Educational Psychology*, *85*(4), 571.

Taylor, A. K., Netsi, E., O'Mahen, H., Stein, A., Evans, J., & Pearson, R. M. (2017). The association between maternal postnatal depressive symptoms and offspring sleep problems in adolescence. *Psychological Medicine*, *47*(3), 451–459. https://doi.org/10.1017/s0033291716002427.

Tully, E. C., & Donohue, M. R. (2017). Empathic responses to Mother's emotions predict internalizing problems in children of depressed mothers. *Child Psychiatry and Human Development*, *48*(1), 94–106. https://doi.org/10.1007/s10578-016-0656-1.

Vameghi, R., Amir Ali Akbari, S., Sajjadi, H., Sajedi, F., & Alavimajd, H. (2015). Correlation between Mothers' depression and developmental delay in infants aged 6-18 months. *Global Journal of Health Science*, *8*(5), 11–18. https://doi.org/10.5539/gjhs.v8n5p11.

Weissman, M. M., Pilowsky, D. J., Wickramaratne, P. J., Talati, A., Wisniewski, S. R., Fava, M., et al. (2006). Remissions in maternal depression and child psychopathology: A STAR*D-child report. *JAMA*, *295*(12), 1389–1398. https://doi.org/10.1001/jama.295.12.1389.

Wentzel, K. R. (2002). Are effective teachers like good parents? Teaching styles and student adjustment in early adolescence. *Child Development*, *73*(1), 287–301. https://doi.org/10.1111/1467-8624.00406.

Wilson, H. K., Pianta, R. C., & Stuhlman, M. (2007). Typical classroom experiences in first grade: The role of classroom climate and functional risk in the development of social competencies. *The Elementary School Journal*, *108*(2), 81–96.

Woolhouse, H., Gartland, D., Mensah, F., Giallo, R., & Brown, S. (2016). Maternal depression from pregnancy to 4 years postpartum and emotional/behavioural difficulties in children: Results from a prospective pregnancy cohort study. *Archives of Women's Mental Health*, *19*(1), 141–151. https://doi.org/10.1007/s00737-015-0562-8.

Yan, N., Zhou, N., & Ansari, A. (2016). Maternal depression and children's cognitive and socio-emotional development at first grade: The moderating role of classroom emotional climate. *Journal of Child and Family Studies*, *25*(4), 1247–1256.

Ystrom, E., Hysing, M., Torgersen, L., Ystrom, H., Reichborn-Kjennerud, T., & Sivertsen, B. (2017). Maternal symptoms of anxiety and depression and child nocturnal awakenings at 6 and 18 months. *Journal of Pediatric Psychology*, *42*(10), 1156–1164. https://doi.org/10.1093/jpepsy/jsx066.

Zalewski, M., Thompson, S. F., & Lengua, L. J. (2017). Parenting as a moderator of the effects of maternal depressive symptoms on preadolescent adjustment. *Journal of Clinical Child and Adolescent Psychology*, *46*(4), 563–572. https://doi.org/10.1080/15374416.2015.1030752.

Chapter 6

# The neuroscience of depression: Mechanisms and treatments

Yuan-Pang Wang[a], Antonio Reis de Sá Junior[b], and Clarice Gorenstein[a]
[a]Department and Institute of Psychiatry, University of São Paulo Medical School, São Paulo, SP, Brazil; [b]Department of Internal Medicine, Health Sciences Center, Federal University of Santa Catarina, Florianopolis, SC, Brazil

## List of abbreviations

| | |
|---|---|
| **BDI** | Beck Depression Inventory |
| **BPSL** | biopsychosocial and lifestyle model |
| **CES-D** | Center for Epidemiologic Studies for Depression |
| **DSM-5** | Diagnostic and Statistical Manual of Mental Disorders 5th Edition |
| **HADS** | Hospital Anxiety and Depression Scale |
| **HAM-D** | Hamilton Depression Rating Scale |
| **IPTS** | interpersonal-psychological theory of suicide |
| **MADRS** | Montgomery-Åsberg Depression Rating Scale |
| **NSSI** | non-suicidal self-injury |
| **PHQ-9** | Patient Health Questionnaire |
| **USPSTF** | US Preventive Services Task Force |

## Introduction

Most of the episodes of depression start during the college period when students face important circumstances that place them under a risk for depression. College students used to experience a wide range of hardships, precipitating and aggravating the liability to depression. Major complaints of students refer to unsatisfactory academic performance, psychological distress, and social isolation. Problematic life routines like sleeplessness, erratic eating habits, and sedentarism worsen student's health. Additional pressure with the future comes from financial difficulties, unsuccessful relationships, pressure to thrive after school, and postgraduation plans. During the college years, this complexity of factors is at stake, sometimes without enough support or a protective network.

Many features of the college period are especially stressful for several reasons. First, this is a transitional period, when adolescents leave high school to go to college. This represents a switch over period toward students' autonomy and independence, full of academic and psychological challenges to adjust. Second, these changes bring up an array of choices and decisions about their future professional life. In other words, this is a time for people to develop their own professional identity, to cultivate a reliable relationship network, and to make career choices. Third, this is a testing period for the desire of being successful and the frustration of failure. Expectations, academic demands, recognition, overcharging, self-recrimination unbalance the students' well-being.

Depression is a serious but common and treatable medical illness. The core characteristics of depression are pervasive unhappiness which manifests itself in a range of symptoms, including feelings of sadness, sleep disturbance, feelings of guilt, lack of self-care, poor concentration, lack of interest in everyday experiences, and may cause physical symptoms (DSM-5, American Psychiatric Association, 2013). This condition can harm the student's ability to perform schoolwork assignments, interact with peers, or take care of them. Investigations have revealed that depression is associated with huge morbidity and mortality. The price of depression is associated with negative societal, educational, financial, and health consequences.

In the present chapter, we review the epidemiological aspects of depression in college students, how to screen student's clinical manifestations, management strategies of depression, and depression-related risk behaviors like suicide, self-injury, and substance abuse. Particular features of this population may serve as a framework to set an effective prevention program and to ameliorate their mental health.

## Overview of depression in students

Depression is one of the most important contributors to disability in the world. The prevalence of depression in the general population is estimated at 4.4% in 2015, is more common among females than males (5.1% vs 3.6%) (World Health Organization, 2017). People with depression often seek help in both general practice and mental health settings, complaining somatic or emotional discomfort.

The Neuroscience of Depression. https://doi.org/10.1016/B978-0-12-817933-8.00030-X

Depression-related conditions account for approximately 50% of psychiatric consultations and 12% of all hospital admissions (Kuo, Tran, Shah, & Matorin, 2015) even though, only 20% of them obtain adequate treatment (Kessler et al., 2003).

The onset of depressive episodes used to start during college life and is associated with academic impairments. Cross-sectional studies have estimated the prevalence of depression in adolescents and young adults as high as 30% (Auerbach et al., 2018; Hill, Yaroslavsky, and Pettit, 2015). In the same direction, a systematic review of existing studies on college students (Ibrahim, Kelly, Adams, & Glazebrook, 2013) has quantified the pooled prevalence of depression as 30.6%, which is much higher than the general population. Female college students are more likely to report depressive symptoms than their male counterparts (American College Health Association, 2016; Ibrahim et al., 2013). It can be conjectured that the expansion of depression in students will gain far greater attention in the next decades (Center for Collegiate Mental Health, 2017).

Likewise, the increasing trend of depression worldwide in recent decades (GBD 2015 Disease and Injury Incidence and Prevalence Collaborators, 2016), some large-scale studies has also shown a mounting prevalence of depression in undergraduates when compared with previous age cohorts (American College Health Association, 2009; Center for Collegiate Mental Health, 2017). The rates of moderate-to-severe depression almost duplicated among the US undergraduates from 2007 to 2018, respectively, 23.2% and 41.1% (Duffy, Twenge, & Joiner, 2019). Anxiety, non-suicidal self-injury, suicidal ideation, and suicide attempts also markedly increased over the same period (Figs. 1 and 2). These findings suggest a broad worsening of students' mental health across the time. Hence, it is easy to forecast that more and more college students will require specialized attention and healthcare services in the coming years.

The escalation and the impact of depression on students' health call for urgent actions, to understand and mitigate predisposing and contributing factors to the onset or exacerbation of depression. Accurate screening and identification of depression in students is the first step to define those students who are or not in need of intervention. Following evidence-based programs of prevention and treatment of depression can be deployed cost-effectively. Efforts to manage depression-related detrimental consequences could involve coordinated actions across educational policy-makers, civil society, and private sectors.

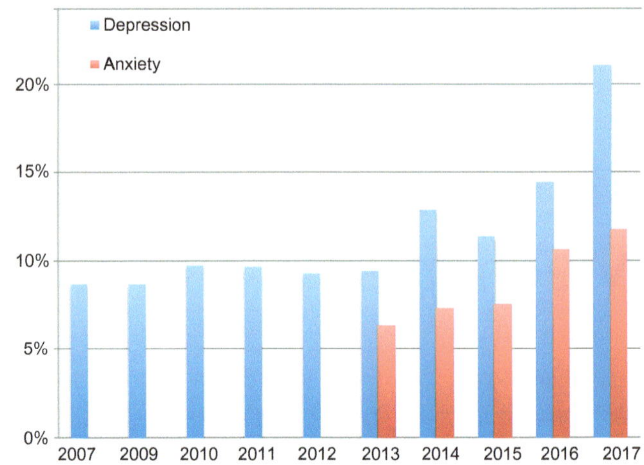


FIG. 1 Trends of depression and anxiety among the US college students (2007–2017). Data were from the National College Health Assessment ($n$ = 610,543) and Healthy Minds Study ($n$ = 177,692). Depression is defined following PHQ-9 score > 15 (severe depression) and anxiety with GAD score > 15 (severe anxiety). *Adapted from Duffy, M.E., Twenge, J.M., & Joiner, T.E. (2019). Trends in mood and anxiety symptoms and suicide-related outcomes among U.S. undergraduates, 2007–2018: Evidence from two National Surveys.* Journal of Adolescent Health, *65(5), 590–598. https://doi.org/10.1016/j.jadohealth.2019.04.033.*

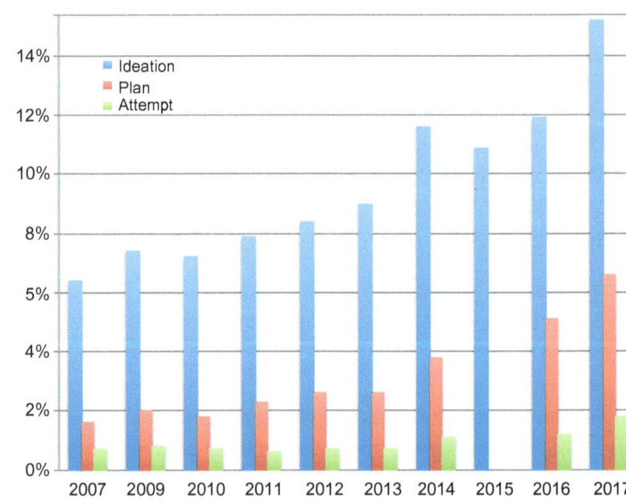


FIG. 2 Trends of suicidal ideation, plan, and attempt among the US college students (2007–2017). Data were from the National College Health Assessment ($n$ = 610,543) and Healthy Minds Study ($n$ = 177,692). Participants were asked, "In the past year, did you ever seriously think about attempting suicide?" "In the past year, did you make a plan for attempting suicide?" Those who answered yes were asked the follow-up question, "In the past year, did you attempt suicide?" *Adapted from Duffy, M.E., Twenge, J.M., & Joiner, T.E. (2019). Trends in mood and anxiety symptoms and suicide-related outcomes among U.S. undergraduates, 2007–2018: Evidence from two National Surveys.* Journal of Adolescent Health, *65(5), 590–598. https://doi.org/10.1016/j.jadohealth.2019.04.033.*

## Screening depression

A screening program purposes to detect vulnerable students and further provide appropriate evaluation, follow-up, prevention, and treatment. The US Preventive Services Task Force (USPSTF) recommends screening depression in all adults over 18 years regardless of risk factors (Siu et al., 2016). The reviewed evidence on the benefits and harms of screening for depression in adult populations without related signs or symptoms has indicated B recommendation to USPSTF's guideline. This means that there is a moderate-to-high certainty of net benefit about the effectiveness of specific preventive care services for patients.

The implementation of a screening program should assure an adequate infrastructure in place, to guarantee correct diagnosis, efficient treatment, and suitable follow-up. Major reasons for not screening included the shortage of mental health professionals, provider unwillingness, responsibility concerns, and lack of clinic space (English & Campbell, 2019).

Screening and rating scales of depression provide an objective measurement of symptoms and evidence of both reproducibility and validity. Reliable and validated self-report screening tools for the detection of depression allow fast identification of those students in need of help. Accordingly, follow-up evaluations, prevention programs, and intervention resources can be further provided (Garlow et al., 2008). This is an opportunity to examine students' functional outcomes and the success of the program in health centers, monitoring those students who will experience either transient or persistent symptoms of depression.

Common tools for screening depression among adult populations are the Center for Epidemiologic Studies for Depression (CES-D, Radloff, 1977), the Hospital Anxiety and Depression Scale (HADS, Zigmond & Snaith, 1983), the Patient Health Questionnaire (PHQ-9, Kroenke, Spitzer, & Williams, 2001), among others (Table 1). Typically, screening tools are brief, concise, and cost-effective. The self-report format is the usual method of administration. Reluctant respondents are more likely to accept the self-report questionnaires. Summarizing, an advantage of screening scales is its easy application in large student samples.

After the screening, some rating scales can be included to assess the severity of depression and follow-up. Examples are the Hamilton Depression Rating Scale (HAM-D, Hamilton, 1967), the Montgomery-Åsberg Depression Rating Scale (MADRS, Montgomery & Åsberg, 1979), Beck Depression Inventory (BDI, Beck et al., 1996) among others (Table 1). The MADRS and the HAM-D are anchored in a standard schedule and should be administered by a trained professional. It is also worth mentioning herein that a clinician should establish the definite diagnosis of depression, insofar that assessment tools (screening and rating scales) cannot replace a personal interview. For additional information, refer to Chapter 15 on "Assessment scoring tools of depression" and on specific chapters dedicated to BDI and HAM-D in the present volume.

Depression screening program holds the potential to increase identification of mental health concerns among college students (English & Campbell, 2019). Although the great majority of students might have complained at least one depressive symptom in the BDI-II (de Sá Junior, de Andrade, Andrade, Gorenstein, & Wang, 2018), the presence of some symptoms deserves more attention from clinicians and mentors (Fig.3). For example, the following items might indicate a higher level of severity: sadness, pessimism, past failure, punishment feelings, self-dislike, crying, suicidal thoughts, worthlessness, and loss of interest in sex. Overlooked depressive symptoms in the student population could delay their help-seeking for proper treatment. Also, low treatment rates of depression in college students can be partially explained

**TABLE 1** Screening and rating scales of depressive symptoms.

| Scale (Short name) | Purpose | Items | Item scoring | Method of application | Author, date |
|---|---|---|---|---|---|
| Center for Epidemiologic Studies for Depression (CES-D) | Screening | 20 | 0–3 | Self-report | Radloff (1977) |
| Hospital Anxiety and Depression Scale (HADS) | Screening | 14[a] | 0–3 | Self-report | Zigmond and Snaith (1983) |
| Patient Health Questionnaire (PHQ-9) | Screening | 9 | 0–3 | Self-report | Kroenke et al. (2001) |
| Hamilton Rating Scale for Depression (HAM-D) | Rating | 17 and 21 | 0–2 or 0–4 | Clinician | Hamilton (1967) |
| Montgomery-Åsberg Depression Rating Scale (MADRS) | Rating | 10 | 0–3 | Clinician | Montgomery and Åsberg (1979) |
| Beck Depression Inventory (BDI-II) | Rating | 21 | 0–3 | Self-report | Beck, Steer, and Brown (1996) |

[a]*Seven items for depression and seven items for anxiety.*

Increasing severity

| Severity | Symptoms |
|---|---|
| Mild | Changes in sleep, self-criticalness, tiredness or fatigue, loss of energy, changes in appetite |
| Moderate | Concentration difficulty, irritability, indecisiveness, guilty feeling, agitation, loss of interest |
| Severe | Sadness, pessimism, crying, self-dislike, worthlessness, past failure, punishment feelings, loss of interest in sex, suicide thoughts |

FIG. 3 Levels of the severity of depressive symptoms among college students. The severity classification of items is extracted from responses of college students ($n=12{,}711$) for the Beck Depression Inventory-II. *Adapted from de Sá Junior A.R., de Andrade A.G., Andrade L.H., Gorenstein C., & Wang Y.P. (2018). The response pattern of depressive symptoms among college students: What lies behind items of the Beck depression inventory-II?.* Journal of Affective Disorders, 234, *124–130. https://doi.org/10.3389/fpsyt.2019.00050.*

by its under-recognition and the stigma surrounding the label of depression.

A direct consequence of screening programs refers to the available resource on campus, because the demand for evaluation and intervention may increase in poor resource and overburdened college-based health centers. Many universities have indeed implemented universal depressive symptom screening programs, but the number of students in need of services significantly exceeds available mental health resources. Bearing in mind this scenario, college stakeholders should weigh the imbalance between the interest for mental health services and the ability of the campus to provide support for those in need of treatment before the decision to implement a universal screening program of depression.

## Treatment and management

On-campus counseling centers constitute the closest place for a college student to address their feelings, sadness, and academic failures. Because depression is prevalent among college students, counselors should be alert for its early signs. Additional tasks should develop students' coping competencies, controlling crises, or addressing specific needs. College campi must determine how best to provide appropriate advice and intervention services for students. Also, opportunities for prevention and intervention can be easily deployed in student communities. After all, many students did not even know they were suffering from depression.

The presence of mental health problems is an important driver for help-seeking behaviors. In general, the more negative consequence of mental health problems for individuals' lives increases, the more likely one will seek help. Nevertheless, not all students will seek help for their problems; many of them would prefer to handle eventual adversities (Cramer, 1999; Gurin, Veroff, & Feld, 1960). In a large cohort student study (Liu et al., 2017), male students reported higher levels of mental health problems than did female counterparts in the first college year. However, female students reported more help-seeking behaviors than did male peers in the following four college years. When problems are not obvious, help-seeking intentions and behavior are low; when problems are moderately severe, help-seeking behaviors are also high; when problems are extremely severe, help-seeking behaviors may decrease as sufferers give up. Possibly, the motivation to seek professional help due to mental health problems is also moderated by the stigma toward mental disorders.

Perceived stigma can represent a barrier to seeking treatment for depression in colleges (Eisenberg, Downs, Golberstein, & Zivin, 2009). In several societies, individuals with depression-related suicidal thoughts are stigmatized, even being excluded from social life. Some negative stereotypes, prejudice, and discrimination, either self-imposed or labeled by society, might hamper the help-seeking behaviors of students. Concerns over confidentiality could be related to the unwillingness to seek help. Many reluctant students fear the institutional banishment of undergraduates who experience depression, which can end up as an involuntary medical leave.

Additional reasons for low treatment rate might be linked to information-related and attitudinal issues, including the lack of knowledge about mental health services, and the wish to deal with problems by their own (Clement et al., 2015). For example, ethnic minorities, youth, men, religious people, and those in the military and health professions were more discouraged to seek help due to stigma (Clement et al., 2015; Eisenberg et al., 2009). In this perspective, key needs of students are (1) to obtain the help that brings into line with the perceived severity of the problem, (2) to structure a trustful relationship with support givers, and (3) to unfasten the tension between the need to disclose and the stigma related with disclosure. Developing coping strategies to deal with environmental adversities is a fundamental survival skill to prevent depression in students. Policy-makers of education should start an in-campus dialogue with students and carefully structure the student health service to prevent failures.

### The setting

College campi are an obvious environment for the implementation of the treatment of depression. Ingredients for starting a campus mental health centers should consider the structure, the staff, the provision of care, and the coordination with off-campus providers and primary care practitioners. The main tasks involve the assessment of the program of identification of mental health issues, access to mental health services, patient engagement, and follow-up of mental health conditions (Woltmann et al., 2012). Generally, most colleges have deficient financing resources and infrastructure for prevention and early intervention of depression. The availability of mental health resources in most college campi might be poor for detection, identification, and planned treatment following a rational protocol (Duffy et al., 2019).

Ideal integrated care for managing a standardized treatment plan should have both mental health and medical components. While many students possess health insurance through their family support, some of them seek help in university-sponsored care. Alternative care combination for students between the college and the primary care might be cost-effectively emphasized (Kates, McPherson-Doe, & George, 2011). Keeping data privacy without sharing personal information is mandatory (Turner et al., 2018), but allowing free access to the multidisciplinary team can strengthen students' trust and adherence in the treatment program. The model of integrated health care systems is a potential practice to increase the identification of mental health issues, availability to mental health services, and improved patient engagement (Kates et al., 2011; Woltmann et al., 2012). However, students less likely to use primary care when integrated care is adopted (Turner et al., 2018). The decision over the in-campus and integrated model should be set after the assessment of local resources.

## Targeting depressive symptoms

Less than half of college students with depression had access to any mental health services at a minimally adequate level (Eisenberg & Chung, 2012), had obtained either a prescription of antidepressants or counseling. Most of the students sought counseling than antidepressant medication. Even though, the vast majority (87%) of students with significant symptoms of depression had at least some contact with a health provider over a year. This suggests that the adequacy of depression care is a huge problem in college health. Greater availability of psychiatry care, better coordination between specialty and primary care using collaborative care models, and increased efforts to retain students in psychotherapy are some directions to pursue.

In general, both pharmacotherapy and psychotherapy have shown evidence of comparable benefits (Cuijpers et al., 2013). Pharmacotherapy was significantly more efficacious than nondirective counseling, and psychotherapy was significantly more efficacious than pharmacotherapy with tricyclic antidepressants (Cuijpers et al., 2013). As the response to treatments in the general population, some students will be more responsive to pharmacotherapy, others to psychotherapy. Conversely, many students may delay treatment seeking, arguing that depression is not urgent and their distress might be a transient stage of the college experience. Therefore, educational actions that increase knowledge about depression can expand students' help-seeking behavior.

## The biopsychosocial intervention

College health centers face multiple challenges such as scarcity of evidence-based interventions that are suitable across different profiles of college students. How to deliver psychotherapy (in terms of technique, number, length, frequency, and duration of sessions, etc.) to college students is not well established. Most interventions rest on the foundation of traditional rather than evidence-based practices. Different interventions for treatment or modification of unhealthy behaviors can be carried for college students on an individual or small group. Frequently, the number of therapy visits falls below the guidelines for minimal adequacy (Hansen & Lambert, 2003). Due to practical and economic constraints of rapid increases in demand, the feasibility of adopting short-term therapy in college counseling centers or specialized services is debatable.

The general objectives of treatment of any depressive disorder are to alleviate symptoms, mitigate the associated morbidity, limit the disability, and reduce self-harm risk. Recovery to the premorbid level of functioning with better health awareness and quality of life is the final purpose. From the standpoint of college health, a successful program should consider the multiple characteristics of students, such as physical illness, psychosocial stressors, and life events. For case formulation, the biopsychosocial model is a useful framework for individualized management and clinical care. The proposal of the biopsychosocial & lifestyle (BPSL) model considers contributors to the development of depression during the planning phase. It includes biological, psychological, and social perspectives alongside lifestyle factors (Malhi et al., 2015). Typically, depressive disorders arise from factors from multiple domains. This model identifies potential causes of depression in association with planned interventions. A comprehensive set of recommendations and interventions can be administered to students, following potential contributing dimensions (Table 2).

This BPSL is a holistic approach and explores different ways to embrace health promotion, illness prevention, and health maintenance. Students should be examined individually from multiple perspectives, taking into account the influence of people's behavior on their health status. In doing so, students can gradually develop and maintain healthy lifestyles to promote health and happiness at the biological, psychological, and social levels.

Although some experts think that the prevention of depression is possible, there is no well-accepted program for use in the college campus. Possibly, the best strategy

**TABLE 2 Comprehensive and individualized treatment alternatives for students with depression, based on biopsychosocial and lifestyle model (BPSL).**

| Biopsychosocial and lifestyle components | | | |
|---|---|---|---|
| **Biological** | **Psychological** | **Social** | **Lifestyle** |
| Antidepressants | Cognitive-behavioral therapy | Family psychoeducation | Exercise |
| Antipsychotics | Brief and formal interpersonal therapy | Family/friends | Diet |
| Mood stabilizers | Mindfulness | Support groups | Smoking cessation |
| Electroconvulsive therapy | Acceptance and commitment therapy | Community groups | Alcohol cessation |
| Transcranial magnetic stimulation | Psychodynamic therapy | Caregivers | Ceasing drugs |
| Anxiolytics | Counseling | Employment | Managing substance misuse |
| | | Housing | Sleep |

After obtaining the diagnosis of major depression, the professional should assess the severity and the contribution of biological, psychosocial, and lifestyle components in the precipitation and maintenance of clinical picture in an individual student. Thereafter, a suitable combination of treatment choices can be recommended.
Adapted from Malhi, G.S., Bassett, D., Boyce, P., Bryant, R., Fitzgerald, P.B., Fritz, K., et al. (2015). Royal Australian and New Zealand College of Psychiatrists clinical practice guidelines for mood disorders. *Australian and New Zealand Journal of Psychiatry, 49*(12), 1087–1206. https://doi.org/10.1177/0004867415617657

to avoid the increase of depression among students is its early detection and comprehensive management.

## Risk behaviors in students

The media coverage of suicide and other risk behaviors among college students has amplified the concern for mental health issues and campus safety. The experience of living in a college environment can lead to a chaotic situation because this is a highly sensitive period for many people. Students need to manage academic competition amid harmonizing different aspects of college life, including work, social activities, and personal relationships. Emerging unfavorable outcomes comprise somatization, eating disorders, self-harm, and addictive behaviors. Frequent visit to healthcare services or emergency rooms might represent the iceberg tip of the "cry for help," regardless of current major depression (Auerbach et al., 2019). Therefore, educators should not neglect these overt behaviors that students may show up (Duffy et al., 2019).

### Suicidal thoughts and attempt

Suicidal thoughts are potentially dangerous in themselves and it is a predictor of the pathway to suicidal attempts, varying from fleeting thoughts of being better off dead to persistent, ruminative thoughts about suicide.

International attention to the problem of suicide among college students has been growing steadily over the past years (Mortier et al., 2018). Administrators and policymakers are increasingly concerned with the prevention of suicide among students (Kuo et al., 2015). Episodes of suicidal thoughts and behaviors are the result of the interaction between psychosocial, psychological, environmental, and genetic risk factors. Suicide ideation among college students may be the combined result of the social and academic pressures.

Suicide is one of the leading causes of death among young adults, including college students (Giannetta et al., 2012). Although few college students report suicidal behaviors, between 7% and 10% of students present severe suicidal thoughts (American College Health Association, 2016; de Sá Junior et al., 2018) and 1.5% reported a suicide attempt. Suicide among the US students is increasing in the past decade (Fig.2). Globally, the past-year prevalence of suicidal ideation, plans, and attempts among college students is17.2%, 8.8%, and 1.0%, respectively (Mortier et al., 2018).Unpredictably, when compared prospectively over 30 years, the rate of death by self-harm among students in Oxford did not differ from that of young people from the general population (Hawton, Bergen, Mahadevan, Casey, & Simkin, 2012).

The level of depressive symptoms during the first year of college can predict persistent suicide ideation during college among students (Mortier et al., 2018). Moreover, female gender, low social support, high affective dysregulation, and high depressive symptoms are associated with increased risk of a suicide plan or attempt (Wilcox et al., 2010). Unfortunately, only 34% of depressed college students with suicidal ideation receive minimally adequate care and those with untreated depression or suicidal ideation report negative attitudes and beliefs about treatment (Eisenberg & Chung, 2012). Timely identification and proper referral of college students at risk of suicide are fundamental steps to stop self-inflicted deaths.

There is a scarcity of evidence-based intervention for the treatment of suicidal students. Few longitudinal and experimental designs have verified causal pathways, and whether trait vulnerabilities of depression and attachment interact with acute stressors to increase suicide risk. The attachment style and depression in students might contribute to suicidality as distal and proximal dimensions of suicide risk, respectively.

Psychotherapy can be beneficial for suicidal patients if the therapist does not reduce the intervention to symptomatic management and control; recognizes how the patient uses the planned death as part of an adaptation process; and prevents counter transference hazards (Hendin, 1981). Some psychosocial modalities for management and control were considered a highly effective method of treatment for suicidal students (Hawton et al., 2016). Traditional interventions include cognitive-behavioral and interpersonal therapy (Hetrick, Cox, Witt, Bir, & Merry, 2016), as well as psychoanalytic/psychodynamic psychotherapy (Briggs et al., 2019; Hendin, 1981).

A promising model to address taxing suicidal behaviors is the three component interpersonal psychological theory of suicide (IPTS; Joiner, 2005) (Fig.4). The first one—the perceived burdensomeness—refers to the feeling of being unable to contribute to, or burdens family, friends, and/or society. The second—the thwarted belongingness—indicates the sense of defeated desire to connect with others that are not an integral part of a family, the circle of friends, or other valued groups. The third the acquired ability for suicide—must be present, because burdensomeness and

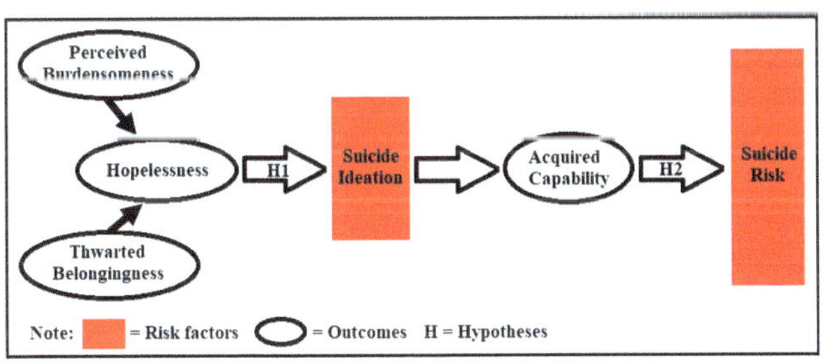


FIG. 4 The Interpersonal–Psychological Theory of Suicide (IPTS). *Adapted from Van Orden, K.A., Witte, T.K., Cukrowicz, K.C., Braithwaite, S.R., Selby, E.A, & Joiner, T.E. (2010). The interpersonal theory of suicide.* Psychological Review, 117*(2), 575–600. https://doi.org/10.1037/a0018697.*

belongingness might instill a desire for suicide, but are not sufficient to ensure that desire will lead to a suicide attempt.

In a subsequent series of Joiner's publications, it is shown the interactive nature of the IPTS, which by according to IPTS, depression and other risk factors contribute to suicidality because they affect components of perceived burdensomeness, thwarted belongingness, and acquired the ability for suicide. Increasing belongingness during the intervention is potentially lifesaving. Furthermore, it is suggested that changing even just one of the three components may substantially reduce the risk for suicide (Van Orden et al., 2010). Further investigations in rigorous randomized trials should be implemented to confirm the utility of IPTS in the management of suicide among students.

Effective suicide prevention efforts include limiting access to methods of suicide like firearms, drugs, and poisons; treating mental disorders and substance misuse; careful media reporting about suicide and improving economic conditions. Even though crisis hotlines are common, they have not been well studied in college samples.

## Non-suicidal self-injury

Non-suicidal self-injury (NSSI) behavior is deliberate and direct bodily harm but without suicidal intent. It comprises self-inflicted injuries of increasing severity, ranging from scratches, cuts, bites, to self-mutilation. The NSSI is often associated with dissatisfaction with one's existence, which can sometimes evolve into suicidal thoughts. Individuals who present NSSI also used to research suicide methods. Conceptually, NSSI is not classified as a mental disorder and it can occur particularly among those who lack the skills and coping strategies to manage psychological distress in adaptive ways. This condition may foster a desire to take one's life, which can be aggravated when accompanied by a suicidal plan. Although NSSI is a behavior distinct from a suicide attempt, it is not always associated with the risk of suicide.

The lifetime prevalence of NNSI was reported as 22.8% in Belgian and Australian universities (Kiekens et al., 2018), with higher rates among females than males (25.4% vs 19.5%). Regarding age brackets, the lifetime prevalence ranged from 17.2% for adolescents, 13.4% for young adults, and 5.5% for adults (Swannell, Martin, Page, Hasking, & St John, 2014). These rate variations can be attributed to different sampling, instrument, location, incentive for participation, etc. Suicidal behaviors and NSSI are associated with many factors such as mental problems, sociological, economic and cultural contexts, demographic characteristics, physical impairment, personality traits, and religiousness.

NSSI behavior is often depicted as a coping mechanism that is both difficult to stop and addictive. There is a plethora of NSSI-related content that is disturbingly available online. Contrary to support to the practice of self-harm, some NSSI websites also deliver positive messages regarding attitudes, psycho-educational discussions to increase, or even discussing first aid tips. Clinical assessment is an imperative component in the counseling process and the establishment of effective treatment among students. Interventions designed to reduce self-critical thoughts and bolster self-compassion may decrease the intensity of self-injury urges, thereby changing the frequency of NSSI episodes.

### Substance abuse

Alcohol and drug use are highly prevalent among college students, with annual prevalence estimates of 79% for alcohol use and 43% for any illicit substance use (Schulenberg et al., 2017). Between 26% and 37% of students have engaged in polydrug use in the past 12 months (Oliveira, Alberghini, Santos, & Andrade, 2013). Approximately 65% of college students drink alcohol monthly (Hingson & White, 2013), but marijuana is considered the most often used illicit substance in college and universities (American Academy of Pediatrics, et al., 2011).The nonmedical use of stimulants in the college setting is a clinical concern (Mochrie, Whited, Cellucci, Freeman, & Corson, 2018). This scenario is worrisome and should be readily tackled.

Substance abuse is associated with impaired academic performance, delayed college graduation, social-interpersonal difficulties, and engagement in risk behaviors (Arria, Allen, Caldeira, Vincent, & O'Grady, 2019; Phillips, Phillips, Lalonde, & Tormohlen, 2015; Skidmore, Kaufman, & Crowell, 2016). Often, depression co-occurs with substance use, challenging the diagnosis and treatment of this dual disorder. The self-medication hypothesis is one of the explanations of frequent co-occurrence of depression and substance use. Maladaptive students suffering from depression may self-medicate with substances to drown bad feelings, cope with problems, and relieve the symptoms. However, this is just a clumsy attempt to mask or alleviate their depressive symptoms.

The recreational use of substances, e.g., alcohol and marijuana, can facilitate social interactions. The purpose is to seek acceptance in groups that the student present personal identification. While the social life in the college may open the door to widespread substance use, many students will withdraw from interpersonal contact and fall into solitary substance use (Mason, Stevens, & Fleming, 2019).

Substance abuse is a severe problem among college students and it is considered as one of the main social challenges worldwide. Motivating a drug addict into treatment is a big challenge because many students do not view their substance use as a problem. Regardless of the seriousness of substance abuse, they rarely seek treatment and rather prefer resolving the problem on their own. Acknowledging a need for help may be an important first step toward seeking treatment. An applicable strategy to engage students to seek help is encouraging drug users by a knowledge-based approach. Also, brief counselor-delivered motivational interventions seem to be helpful (Dennhardt & Murphy, 2013).

Few studies have examined the prevention and treatment of substance abuse in college students, but parent-based and in-person brief motivational interventions appear to be promising. Differences in treatment response include patterns of substance use and help-seeking behaviors. Including stress management and treatment of concurrent depression are ingredients of personalized management (Dennhardt & Murphy, 2013).In the same direction, prevention programs should consider the social, cognitive, personality, and environmental features of students.

## Conclusion

The college period overlaps the transition from adolescence to adulthood. This is a thrilling period of opportunities and responsibilities, with plenty of transformations and emotional swings. However, college students experience a wide range of stressors that can result in several undesirable outcomes and devastating consequences for the student, family, and society at large. Depression, aside from causing suffering among college students, impairs academic achievement and can interfere with interpersonal relationships. Therefore, managing depression in this group is considered a global public health issue.

High prevalence of depression in combination with the suboptimal adherence and response rate to treatment requires effective interventions across different profiles of college students. Early management of depression may improve symptoms, increase academic performance and prevent college dropout. The mental health of college students is worrisome and requires greater attention from students, parents, mentors, education experts, and policymakers. There is a need for policy reformulation in colleges that allow increasing the access to health care, considering the adequacy and quality of the service for students with depression.

## Key facts

- College students experience higher rates of depression than those found in the general population.
- Early identification of depression among college students can be implemented through universal screening.

- Resources for mental health should be available before implementing screening.
- Counseling, psychotherapy, and psychopharmacotherapy are usual treatments for depression in college campus.
- Suicidal thoughts and attempts, non-suicidal self-injury, and substance abuse are comorbid outcomes of students with depression.
- The level of depressive symptoms during the first year of college can predict persistent suicide ideation during college among students.
- Early management of depression may improve symptoms, increase academic performance, and prevent college dropout.

## Summary points

- This chapter summarizes findings on depression that occurs among college students, from epidemiology, screening, to management.
- Depression among college students is estimated at around 30%, higher than the general population.
- Students with depression used to handle the problem by developing further risk behaviors: suicidal thought, self-harm, substance abuse, among others.
- Early identification of depression in students with brief assessment scales is the key to mapping the in-campus problem. Screening of depression can be deployed through brief assessment scales, such as the CES-D, PHQ-9, and HADS.
- The implementation of universal screening of depression is recommended to all college students, although most institutions often do not have financial resources, clinical facilities, and mental health professionals to assist the increased demand.
- The effectiveness of interventions is associated with student's suboptimal adherence and response rate to treatment. Many students with depression are reluctant to seek help. Some of them rather prefer managing their problems by themselves, fearing discrimination, or prejudice. Perceived stigma is a component of low adherence.
- High prevalence of depression is a great concern; therefore the access to health care should be improved, in terms of the adequacy and quality of the service offered to students with depression.

## Mini-dictionary of terms

**Screening tools** These assessment instruments are usually self-reported, short and brief scales to identify cases and non-cases of depression, for example. There are several rapid tools designed for the detection of depression in community and primary care. The utility of screening tools is associated with its capacity for early identification of target conditions, aiming for further treatment and prevention. Cost-effectiveness is an additional quality of its widespread adoption.

**Rating scale** This type of assessment tool is adapted to measure psychopathology severity and follow-up. The assessment covers specific constructs measured by the instrument, for example, dimensions of depressive disorders. Usually, rating scales are lengthier than screening tools, can be self-administered by the respondent or be rated by an interviewer. These scales can detect changes or improvements across several time-points.

**Interpersonal Psychological Theory of Suicide** This is a comprehensive theory of necessary components of suicide. The first one—the perceived burdensomeness—refers to the feeling of being unable to contribute to, or burdens family, friends, and/or society. The second—the thwarted belongingness—indicates the sense of defeated desire to connect with others that are not an integral part of a family, the circle of friends, or other valued groups. The third—the acquired ability for suicide—must be present because burdensomeness and belongingness might instill a desire for suicide, but are not sufficient to ensure that desire will lead to a suicide attempt. The combination of three components can result in high lethality. This model can be incorporated during the management of suicidal patients.

**Biopsychosocial and Lifestyle model** This is a framework for the case formulation of individualized management. Assuming that there are several causes and contributors for depression, this model considers the biological, psychosocial, and lifestyle of patients with depression during the planning stage. Following, meaningful treatments can be recommended following the affected dimensions. This model can be adapted for managing depression in students.

**Non-suicidal self-injury** This behavior is deliberate and direct bodily harm but without suicidal intent. It comprises self-inflicted injuries of increasing severity, ranging from scratches, cuts, bites, to self-mutilation. Non-suicidal self-injury is associated with dissatisfaction with one's existence. It is a distinct behavior from a suicide attempt, without a clear relationship with the risk of suicide. Individuals who present self-injury also used to research suicide methods and some scholars view these behaviors as a non-adaptive coping strategy to manage psychological distress. This condition may foster a desire to take one's life, which can be aggravated when accompanied by a suicidal plan.

**Suicide** This is an umbrella term to embrace a variety of cognitions and behaviors related to self-inflicted death. Suicide ranges from suicidal thoughts, desire to be dead, make a plan to die, suicidal attempt, and complete suicide. Conceptually, suicide overlaps with self-harm, self-injury, and life-threatening behaviors. Although suicide on its own is not labeled as a mental disorder, people who attempt suicide often present psychiatric conditions like depression, schizophrenia, and bipolar disorders.

## References

American Academy of Pediatrics, American Academy of Family Physicians, American College of Physicians, Transitions Clinical Report Authoring Group, Cooley, W. C., & Sagerman, P. J. (2011). Supporting the health care transition from adolescence to adulthood in the medical home. *Pediatrics*, *128*(1), 182–200. https://doi.org/10.1542/peds.2011-0969.

American College Health Association. (2009). American college health association-National College Health Assessment Spring 2008 reference group data report (abridged): The American college health association. *Journal of American College Health*, *57*(5), 477–488. https://doi.org/10.3200/JACH.56.5.469-480.

American College Health Association. (2016). *National College Health Assessment II: Reference group executive summary fall 2015*. Hanover, MD: American College Health Association.

American Psychiatric Association. (2013). *Diagnostic and statistical manual of mental disorders* (5th. edn). DC American Psychiatric Association: Washington.

Arria, A. M., Allen, H. K., Caldeira, K. M., Vincent, K. B., & O'Grady, K. E. (2019). Excessive drinking and drug use during college: Prospective associations with graduate school plans and attendance. *Journal of American College Health*, *14*, 1–7. https://doi.org/10.1080/07448481.2018.1535494.

Auerbach, R. P., Mortier, P., Bruffaerts, R., Alonso, J., Benjet, C., Cuijpers, P., et al. (2018). WHO world mental health surveys international college student project: Prevalence and distribution of mental disorders. *Journal of Abnormal Psychology*, *127*, 623e38. https://doi.org/10.1037/abn0000362.

Auerbach, R. P., Mortier, P., Bruffaerts, R., Alonso, J., Benjet, C., Cuijpers, P., et al. (2019). Mental disorder comorbidity and suicidal thoughts and behaviors in the World Health Organization world mental health surveys international college student initiative. *International Journal of Methods in Psychiatric Research*, *28*(2). https://doi.org/10.1002/mpr.1752.

Beck, A. T., Steer, R. A., & Brown, G. K. (1996). *Beck depression inventory-II manual*. San Antonio, TX: Psychological Corporation.

Briggs, S., Netuveli, G., Gould, N., Gkaravella, A., Gluckman, N. S., Kangogyere, P., et al. (2019). The effectiveness of psychoanalytic/psychodynamic psychotherapy for reducing suicide attempts and self-harm: Systematic review and meta-analysis. *British Journal of Psychiatry*, *214*(6), 320–328. https://doi.org/10.1192/bjp.2019.33.

Center for Collegiate Mental Health. (2017). *2016 Annual Report (Publication No. STA 17–74)*. Available at: http://sites.psu.edu/ccmh/files/2017/01/2016-Annual-Report-FINAL_2016_01_09-1gc2hj6.pdf (Accessed on 10 September 2019).

Clement, S., Schauman, O., Graham, T., Maggioni, F., Evans-Lacko, S., Bezborodovs, N., et al. (2015). What is the impact of mental health-related stigma on help-seeking? A systematic review of quantitative and qualitative studies. *Psychological Medicine*, *45*(1), 11–27. https://doi.org/10.1017/S0033291714000129.

Cramer, K. M. (1999). Psychological antecedents to help-seeking behavior: A reanalysis using path modeling structures. *Journal of Counseling Psychology*, *46*(3), 381–387. https://doi.org/10.1037/0022-0167.46.3.381.

Cuijpers, P., Sijbrandij, M., Koole, S. L., Andersson, G., Beekman, A. T., & Reynolds, C. F., 3rd. (2013). The efficacy of psychotherapy and pharmacotherapy in treating depressive and anxiety disorders: A meta-analysis of direct comparisons. *World Psychiatry*, *12*(2), 137–148. https://doi.org/10.1002/wps.20038.

de Sá Junior, A. R., de Andrade, A. G., Andrade, L. H., Gorenstein, C., & Wang, Y. P. (2018). The response pattern of depressive symptoms among college students: What lies behind items of the Beck depression inventory-II? *Journal of Affective Disorders*, *234*, 124–130. https://doi.org/10.3389/fpsyt.2019.00050.

Dennhardt, A. A., & Murphy, J. G. (2013). Prevention and treatment of college student drug use: A review of the literature. *Addictive Behaviors*, *38*(10), 2607–2618. https://doi.org/10.1016/j.addbeh.2013.06.006.

Duffy, M. E., Twenge, J. M., & Joiner, T. E. (2019). Trends in mood and anxiety symptoms and suicide-related outcomes among U.S. undergraduates, 2007–2018: Evidence from two National Surveys. *Journal of Adolescent Health*, *65*(5), 590–598. https://doi.org/10.1016/j.jadohealth.2019.04.033.

Eisenberg, D., & Chung, H. (2012). Adequacy of depression treatment among college students in the United States. *General Hospital Psychiatry*, *34*(3), 213–220. https://doi.org/10.1016/j.genhosppsych.2012.01.002.

Eisenberg, D., Downs, M. F., Golberstein, E., & Zivin, K. (2009). Stigma and help seeking for mental health among college students. *Medical Care Research and Review*, *66*(5), 522–541. https://doi.org/10.1177/1077558709335173.

English, I., & Campbell, D. G. (2019). Prevalence and characteristics of universal depression screening in U.S. college health centers. *Families, Systems & Health*, *37*(2), 131–149. https://doi.org/10.1037/fsh0000411.

Garlow, S. J., Rosenberg, J., Moore, J. D., Haas, A. P., Koestner, B., Hendin, H., et al. (2008). Depression, desperation, and suicidal ideation in college students: Results from the American Foundation for Suicide Prevention College screening project at Emory University. *Depression and Anxiety*, *25*(6), 482–488. https://doi.org/10.1002/da.20321.

GBD 2015 Disease and Injury Incidence and Prevalence Collaborators. (2016). Global, regional, and national incidence, prevalence, and years lived with disability for 310 diseases and injuries, 1990–2015: A systematic analysis for the global burden of disease study 2015. *Lancet*, *388*(10053), 1545–1602. https://doi.org/10.1016/S0140-6736(16)31678-6.

Giannetta, M. M., Betancourt, L. M., Brodsky, N. L., Wintersteen, M. B., Romer, D., Giannetta, J. M., et al. (2012). Suicidal ideation and self-harm behavior in a community sample of preadolescent youth: A case-control study. *Journal of Adolescent Health*, *50*(5), 524–526. https://doi.org/10.1016/j.jadohealth.2011.09.013.

Gurin, G., Veroff, J., & Feld, S. (1960). *Americans view their mental health: A nationwide interview survey*. New York, NY: Basic Books.

Hamilton, M. (1967). Development of a rating scale for primary depressive illness. *The British Journal of Social and Clinical Psychology*, *6*(4), 278–296. https://doi.org/10.1111/j.2044-8260.1967.tb00530.x.

Hansen, N. B., & Lambert, M. J. (2003). An evaluation of the dose-response relationship in naturalistic treatment settings using survival analysis. *Mental Health Service Research*, *5*(1), 1–12. https://doi.org/10.1023/a:1021751307358.

Hawton, K., Bergen, H., Mahadevan, S., Casey, D., & Simkin, S. (2012). Suicide and deliberate self-harm in Oxford University students over a 30-year period. *Social Psychiatry and Psychiatric Epidemiology*, *47*(1), 43–51. https://doi.org/10.1007/s00127-010-0310-3.

Hawton, K., Witt, K. G., Salisbury, T. L. T., Arensman, E., Gunnell, D., Hazell, P., et al. (2016). Psychosocial interventions following self-harm in adults: A systematic review and meta-analysis. *Lancet Psychiatry*, *3*(8), 740–750. https://doi.org/10.1016/S2215-0366(16)30070-0.

Hendin, H. (1981). Psychotherapy and suicide. *American Journal of Psychotherapy*, *35*(4), 469–480. https://doi.org/10.1176/appi.psychotherapy.1981.35.4.469.

Hetrick, S. E., Cox, G. R., Witt, K. G., Bir, J. J., & Merry, S. N. (2016). Cognitive behavioural therapy (CBT), third-wave CBT and interpersonal therapy (IPT) based interventions for preventing depression in children and adolescents. *Cochrane Database of Systematic Reviews*, *8*. https://doi.org/10.1002/14651858.CD003380.pub4.

Hill, R. M., Yaroslavsky, I., & Pettit, J. W. (2015). Enhancing depression screening to identify college students at risk for persistent depressive symptoms. *Journal of Affective Disorders*, *174*, 1–6. https://doi.org/10.1016/j.jad.2014.11.025.

Hingson, R. W., & White, A. (2013). Trends in extreme binge drinking among US high school seniors. *JAMA Pediatrics*, *167*(11), 996–998. https://doi.org/10.1001/jamapediatrics.2013.3083.

Ibrahim, A. K., Kelly, S. J., Adams, C. E., & Glazebrook, C. (2013). A systematic review of studies of depression prevalence in university students. *Journal of Psychiatric Research*, *47*(3), 391–400. https://doi.org/10.1016/j.jpsychires.2012.11.015.

Joiner, T. E. (2005). *Why people die by suicide*. Cambridge, MA: Harvard University Press.

Kates, N., McPherson-Doe, C., & George, L. (2011). Integrating mental health services within primary care settings: The Hamilton family health team. *Journal of Ambulatory Care Management*, *34*, 174–182. https://doi.org/10.1097/JAC.0b013e31820f6435.

Kessler, R. C., Berglund, P., Demler, O., Jin, R., Koretz, D., Merikangas, K. R., et al. (2003). National Comorbidity Survey Replication. The epidemiology of major depressive disorder: Results from the National Comorbidity Survey Replication (NCS-R). *JAMA*, *289*(23), 3095–3105. https://doi.org/10.1001/jama.289.23.3095.

Kiekens, G., Hasking, P., Boyes, M., Claes, L., Mortier, P., Auerbach, R. P., et al. (2018). The associations between non-suicidal self-injury and first onset suicidal thoughts and behaviors. *Journal of Affective Disorders*, *239*, 171–179. https://doi.org/10.1016/j.jad.2018.06.033.

Kroenke, K., Spitzer, R. L., & Williams, J. B. (2001). The PHQ-9: Validity of a brief depression severity measure. *Journal of General Internal Medicine*, *16*(9), 606–613. https://doi.org/10.1046/j.1525-1497.2001.016009606.x.

Kuo, D. C., Tran, M., Shah, A. A., & Matorin, A. (2015). Depression and the suicidal patient. *Emergency Medicine Clinics of North America*, *33*(4), 765–778. https://doi.org/10.1016/j.emc.2015.07.005.

Liu, F., Zhou, N., Cao, H., Fang, X., Deng, L., Chen, W., et al. (2017). Chinese college freshmen's mental health problems and their subsequent help-seeking behaviors: A cohort design (2005-2011). *PLoS One*, *12*(10). https://doi.org/10.1371/journal.pone.0185531.

Malhi, G. S., Bassett, D., Boyce, P., Bryant, R., Fitzgerald, P. B., Fritz, K., et al. (2015). Royal Australian and New Zealand College of Psychiatrists clinical practice guidelines for mood disorders. *Australian and New Zealand Journal of Psychiatry*, *49*(12), 1087–1206. https://doi.org/10.1177/0004867415617657.

Mason, W. A., Stevens, A. L., & Fleming, C. B. (2019). A systematic review of research on adolescent solitary alcohol and marijuana use in the United States. *Addiction*, *115*(1), 19–31. https://doi.org/10.1111/add.14697.

Mochrie, K. D., Whited, M. C., Cellucci, T., Freeman, T., & Corson, A. T. (2018). ADHD, depression, and substance abuse risk among beginning college students. *Journalof American College Health*, *26*, 1–5. https://doi.org/10.1080/07448481.2018.1515754.

Montgomery, S. A., & Åsberg, M. (1979). A new depression scale designed to be sensitive to change. *British Journal of Psychiatry*, *134*, 382–389. https://doi.org/10.1192/bjp.134.4.382.

Mortier, P., Auerbach, R. P., Alonso, J., Bantjes, J., Benjet, C., Cuijpers, P., et al. (2018). Suicidal thoughts and behaviors among first-year college students: Results from the WMH-ICS project. *Journal of American Academy of Child and Adolescent Psychiatry*, *57*(4). https://doi.org/10.1016/j.jaac.2018.01.018. 263–273.e1.

Oliveira, L. G., Alberghini, D. G., Santos, B., & Andrade, A. G. (2013). Polydrug use among college students in Brazil: A nationwide survey. *Brazilian Journal of Psychiatry*, *35*(3), 221–230. https://doi.org/10.1590/1516-4446-2012-0775.

Phillips, K. T., Phillips, M. M., Lalonde, T. L., & Tormohlen, K. N. (2015). Marijuana use, craving, and academic motivation and performance among college students: An in-the-moment study. *Addictive Behaviors*, *47*, 42–47. https://doi.org/10.1016/j.addbeh.2015.03.020.

Radloff, L. S. (1977). The CES-D scale: A self-report depression scale for research in the general population. *Applied Psychological Measurement*, *1*(3), 385–401. https://doi.org/10.1177/01466216 7700100306.

Schulenberg, J. E., Johnston, L. D., O'Malley, P. M., Bachman, J. G., Miech, R. A., & Patrick, M. E. (2017). *Monitoring the future: National Survey Results on drug use, 1975–2016: Volume 2, college students and adults ages 19–55*. Ann Arbor, MI: Institute for Social Research, The University of Michigan.

Siu, A. L., US Preventive Services Task Force (USPSTF), Bibbins-Domingo, K., Grossman, D. C., Baumann, L. C., Davidson, K. W., et al. (2016). Screening for depression in adults: US Preventive Services Task ForceRecommendation statement. *JAMA*, *315*(4), 380–387. https://doi.org/10.1001/jama.2015.18392.

Skidmore, C. R., Kaufman, E. A., & Crowell, S. E. (2016). Substance use among college students. *Child and Adolescent Psychiatric Clinics of North America*, *25*(4), 735–753. https://doi.org/10.1016/j.chc.2016.06.004.

Swannell, S. V., Martin, G. E., Page, A., Hasking, P., & St John, N. J. (2014). Prevalence of nonsuicidal self-injury in nonclinical samples: Systematic review, meta-analysis and meta-regression. *Suicide and Life-threatening Behavior*, *44*(3), 273–303. https://doi.org/10.1111/sltb.12070.

Turner, J. C., Keller, A., Wu, H., Zimmerman, M., Zhang, J., & Barnes, L. E. (2018). Utilization of primary care among college students with mental health disorders. *Health Psychology*, *37*(4), 385–393. https://doi.org/10.1037/hea0000580.

Van Orden, K. A., Witte, T. K., Cukrowicz, K. C., Braithwaite, S. R., Selby, E. A., & Joiner, T. E., Jr. (2010). The interpersonal theory of suicide. *Psychological Review*, *117*(2), 575–600. https://doi.org/10.1037/a0018697.

Wilcox, H. C., Arria, A. M., Caldeira, K. M., Vincent, K. B., Pinchevsky, G. M., & O'Grady, K. E. (2010). Prevalence and predictors of persistent suicide ideation, plans, and attempts during college. *Journal of Affective Disorders*, *127*(1–3), 287–294. https://doi.org/10.1016/j.jad.2010.04.017.

Woltmann, E., Grogan-Kaylor, A., Perron, B., Georges, H., Kilbourne, A. M., & Bauer, M. S. (2012). Comparative effectiveness of collaborative chronic care models for mental health conditions across primary, specialty, and behavioral health care settings: Systematic review and meta-analysis. *American Journal of Psychiatry*, *169*(8), 790–804. https://doi.org/10.1176/appi.ajp.2012.11111616.

World Health Organization. (2017). *Depression and other common mental disorders: Global health estimates*. Available at Geneve: WHO. https://apps.who.int/iris/bitstream/handle/10665/254610/WHO-MSD-MER-2017.2-eng.pdf. accessed on December 5, 2019.

Zigmond, A. S., & Snaith, R. P. (1983). The hospital anxiety and depression scale. *Acta Psychiatrica Scandinavica*, *67*, 361–370. https://doi.org/10.1111/j.1600-0447.1983.tb09716.x.

Chapter 7

# Depression in disasters and traumatic events

**Mao-Sheng Ran and Man-Man Peng**

*Department of Social Work and Social Administration, Faculty of Social Sciences, University of Hong Kong, Hong Kong SAR, China*

## List of abbreviations

**BDI** the Beck Depression Inventory
**CES-D** the Center for Epidemiological Studies-Depression Scale
**GDS** the Geriatric Depression Scale
**HDS** the Hamilton Depression Rating Scale
**HSCL** the Hopkins Symptom Checklist
**MADRS** the Montgomery Asberg Depression Rating Scale
**PHQ-9** the Patient Health Questionnaire-9

## Introduction

Depression (major depressive disorder) is defined as a serious mental disorder that pose negative impacts on a person's thought, action, and daily function (American Psychiatric Association, 2013). Besides biochemical and genetic attributions, the determinants of depression include personality and environmental factors (American Psychiatric Association, 2013; Cadoret, O'Gorman, Heywood, & Troughton, 1985; Caspi & Moffitt, 2006). Natural disasters, such as earthquake, typhoon, hurricane, tsunami, and flood, often exert a devastating impact on the disaster area and its population (Guo et al., 2015). Given limited public resources, people in less developed areas were found to be affected by natural disasters in a greater degree than those in developed countries or regions (Wang et al., 2000). Exposure to natural disasters may result in mental health problems among the population in the disaster-affected area (Neria, Nandi, & Galea, 2008). Previous studies have indicated that depression is common after natural disasters (Guo et al., 2015). A previous meta-analysis study documented that the prevalence of depression after natural disasters ranged from 5.8% to 54.0% among the adult population, while the percentage ranged from 7.5% to 44.8% among children and adolescent (Tang, Liu, Liu, Xue, & Zhang, 2014). The prevalence of mental health issues and influencing factors of the mental health sequelae were varied over time (Guo, He, Qu, Wang, & Liu, 2017).

A traumatic event that a person witnesses or encounters is defined as "an actual or threatened death or serious injury, or a threat to the physical integrity of self or others" (American Psychiatric Association, 2000). Exposure to nonnatural traumatic events often also exert psychological effects among the people with experiences such as exposure to violence, terrorism, transportation disasters, civil unrest, and wars (Jacobson, Norman, Nguyen, & Brackbill, 2018; Nandi, Tracy, Beard, Vlahov, & Galea, 2009; Person, Tracy, & Galea, 2006). For instance, a previous study found a significant association of depression with exposure to an attack with a weapon in both males and females (Overstreet et al., 2016). University students experiencing traumatic events were also reported to be at a high risk of suffering from depression (Frazier et al., 2009).

The findings of the existing studies have indicated that more attention should be attached to the long-term consequences of natural disasters and traumatic events. This chapter aims to: (1) review the impacts of natural and nonnatural traumatic events on depression; (2) summarize clinical diagnosis and standardized assessment instruments for depression in previous studies; (3) review the consequences and risk factors of depression after natural disasters or traumatic events; and (4) provide coping strategies and suggestions for serving the people experiencing natural disasters or nonnatural traumatic events.

## Clinical diagnosis and standardized assessment instruments for depression

### Clinical diagnosis of depressive disorder

A face-to-face structured clinical interview is generally utilized to diagnose depressive disorders, which is usually conducted by trained professionals through asking the patients about a series of questions, including their perceived symptoms, duration of the depressive symptoms, potential causes of depression, preferred coping strategies, past and current medical history of the patients and their family, and the consequences 'of depression on daily functioning (Beidel & Frueh, 2018). Furthermore, in order to

The Neuroscience of Depression. https://doi.org/10.1016/B978-0-12-817933-8.00039-6

decrease the risk of misdiagnosis, mental health clinicians prefer to link an open-ended interview with a validated instrument or a semi-structured interview script (Beidel & Frueh, 2018), such as Hopkins Symptom Checklist (HSCL; Derogatis, Lipman, Rickels, Uhlenhuth, & Covi, 1974) and the Structured Clinical Interview for Diagnostic and Statistical Manual of Mental Disorders, Fourth Edition (SCID, DSM-IV) Axis I Disorders (First, Gibbon, Spitzer, Williams, & Benjamin, 1997).

### Standardized scales of depression

Various self-report instruments have been utilized to measure depression in nonpsychiatric settings among previous studies, such as the Beck Depression Inventory (BDI; Beck, Ward, Mendelson, Mock, & Erbaugh, 1961), the Geriatric Depression Scale (GDS; Brink et al., 1982; Yesavage et al., 1982), the Center for Epidemiological Studies-Depression Scale (CES-D; Radloff, 1977), the Hamilton Depression Rating Scale (HDS; Hamilton, 1960, 1967), the Montgomery Asberg Depression Rating Scale (MADRS-S; Montgomery & Åsberg, 1979), and the Patient Health Questionnaire-9 (PHQ-9). In particular, the CES-D has been widely used in the studies of depression after natural disasters (Cao et al., 2015; Guo et al., 2017; Qu et al., 2014). Although these assessments of depression have been developed over a long period of time, it remained controversial regarding the distinctions of culture, age, and particular setting for conducting the assessment (Beidel & Frueh, 2018).

## Depression after natural disasters and traumatic events

Table 1 shows the prevalence of depression after natural disasters and traumatic events in a few empirical studies. The prevalence of depression was reported to be 26.0% among adults from the earthquake-affected area (Cénat & Derivois, 2014). Among child and adolescent survivors after earthquakes, the prevalence of depression ranged from 13.6% to 76.0% in a western study (Şalcıoğlu & Başoğlu, 2008), while the prevalence of depression was revealed to be 24.5% at 6 months (Fan et al., 2011) and 29.4% at 18 months in previous eastern studies (Zhang et al., 2012). In another study focusing on depression after a natural disaster (Wang et al., 2012), it was found that the overall prevalence of depression after an earthquake was 19.5% among junior middle school students, with 24.0% among female adolescents, 14.7% among male adolescents.

## Consequences of depression after natural disasters and traumatic events

Previous literatures have documented the consequences of depression after natural disasters or nonnatural traumatic events in a diverse population (see Table 2). An emerging body of research found empirical evidences showing that exposure to traumatic events may yield negative psychological outcomes in the aftermath among both children

**TABLE 1** Examples of empirical studies on the prevalence of depression after natural disasters and traumatic events.

| Study | Sample | Study design | Measurement for depression | Key findings: prevalence of depression |
|---|---|---|---|---|
| Cénat and Derivois (2014) | Adults from the Republic of Haiti ($N$=1355) | Cross-sectional | Beck Depression Inventory (BDI) | 25.98% |
| Wang et al. (2012) | Students ($N$=2048) in Ningqiang county which lies approximately 327 km from the Wenchuan earthquake epicenter | Cross-sectional | Depression Self-Rating Scale for Children (DSRSC) | The overall prevalence was 19.5%, with 24.0% among females, 14.7% among males |
| Fan, Zhang, Yang, Mo, and Liu (2011) | Adolescents ($N$=2250) 6 months after the 2008 Wenchuan earthquake in China | Cross-sectional | Depression Self-rating Scale for Children | 24.5% in the 6 months after the earthquake |
| Pan et al. (2015) | Adolescents in Mao County which is 41 km from the epicenter ($N$=373) | Cross-sectional | Zung Self-rating Depression Scale | 44.8% |
| Zhang et al. (2012) | High school student survivors in a local boarding high school after the Wenchuan earthquake ($N$=548) | Longitudinal | Beck Depression Inventory (BDI) | 39.4% in the 6 months; 36.9% in the 12 months; and 29.4% in the 18 months after the earthquake |

**TABLE 2** Examples of empirical studies on consequences after natural disasters and traumatic events.

| Study | Sample | Study design | Measurement for depression/PTSD | Key findings: consequences after natural disasters and traumatic events |
|---|---|---|---|---|
| Zuromski et al., 2019 | Adolescent survivors after spring 2011 tornadoes in the areas surrounding Tuscaloosa, Alabama, and Joplin, Missouri ($N$=2000) | Cross-sectional | Diagnostic and Statistical Manual of Mental Disorders | suicidal ideation; suicidal behavior |
| Ran et al. (2015) | Student survivors in an affected high school ($N$=737) | Longitudinal | PTSD Checklist-Civilian Version (PCL-C) | The prevalence of suicidal ideation were 35.6%, 35.6%, and 30.7%, respectively, at 6-, 12-, and 18-month after the earthquake |
| Guo et al. (2017) | Adult survivors after the earthquake($N$=1369) | Cross-sectional | The Center for Epidemiologic Studies-Depression Scale (CES-D) | Worse quality of sleep<br>Negative affectivity<br>Unhealthy coping strategies |
| Miguel-Tobal et al. (2006) | Madrid residents after the March 11 train bombings ($N$=1589) | Cross-sectional | The SCID's Major Depressive Disorder (MDD) Interview | Mental health problem |
| North, Kawasaki, Spitznagel, and Hong (2004) | St. Louis area survivors ($N$=162) after the Great Midwestern Floods in South Australia | Cross-sectional | The DSM-III-R Diagnostic Interview Schedule/Disaster Supplement (DISIII-R) | Suicidal ideation |

and adult survivors, such as suicidal ideation and suicidal behavior (Meiser-Stedman, Dalgleish, Yule, & Smith, 2012; Zuromski et al., 2019). According to a longitudinal study, depression after an earthquake was an independent risk factor of suicidal ideation in adolescents (Ran et al., 2015). The prevalence of suicidal ideation was 35.6%, 35.6%, and 30.7%, respectively, among student survivors in an affected high school at 6-, 12-, and 18-month after the earthquake (Ran et al., 2015). Even for adult survivors, depression and suicidality were also found to be a long-term health issue of earthquake after a long period of time (Guo et al., 2017). People being affected by the earthquake with depressive symptoms had a higher possibility to have worse quality of sleep, negative affectivity, and unhealthy coping strategies, which made them more likely to contemplate suicide (Guo et al., 2017).

In the general population, the psychological effects of human-made disasters were suggested to be more far-reaching than that of natural disasters (Miguel-Tobal et al., 2006; Norris et al., 2002; Norris, Friedman, & Watson, 2002). Individuals with experience of human-made traumatic events tended to be at a high risk of being faced with a mental health problem (Miguel-Tobal et al., 2006). The associations between depression after traumatic events and suicidality have also been widely studied by scholars. For instance, a study conducted in South Australia revealed that exposure to traumatic events contributes to suicidal ideation in the general population (North et al., 2004).

## Causes and risk factors of depression

### Risk factors for depression after natural disasters among different population

Table 3 lists the possible risk factors for depression after natural disasters. Following the categories suggested by Tang et al. (2014), the risk factors for depression after natural disasters are classified into three types, which are demographic characteristics, trauma-related factors and other posttraumatic factors. Risk factors for depression after natural disasters vary among different population. Besides demographics, risk factors for depression among the general population in the earthquakes truck area included the loss of a family member, damage of residential place, subjective livelihood security, and not living in an urban area (Qu et al., 2014). Compared with posttraumatic stress disorder (PTSD), depression was found to be influenced by personal vulnerability and exposure to relevant stressors in a greater degree. For children and adolescents, previous studies have demonstrated multiple risk factors for depression after natural disasters. A meta-analysis study showed that children encountering or witnessing injury or death and

**TABLE 3** Examples of empirical studies on risk factors for depression after natural disasters.

| | | | | Key findings: risk factors for depression after natural disasters | | |
|---|---|---|---|---|---|---|
| **Study** | **Sample** | **Study design** | **Measurement for depression** | **Demographic characteristics** | **Trauma-related factors** | **Other factors** |
| Chen et al. (2017) | Children aged 7–15 years after the Wenchuan earthquake ($N$=20,749) | Cross-sectional | The Adolescent Depression Inventory | \ | \ | Introversion, psychoticism |
| Chui et al. (2017) | High school students after the Wenchuan earthquake ($N$=548) | Longitudinal | The Beck Depression Inventory (BDI) | Being female | Trauma-related self-injury, PTSD symptoms | Suicidal ideation |
| Guo et al. (2017) | People being severely affected by the earthquake ($N$=1369) | Cross-sectional | The Center for Epidemiologic Studies-Depression Scale (CES-D) | Female gender, low education | Traumatic experiences after the earthquake | Poor perceived health |
| Cao et al. (2015) | Survivors of Wenchuan Earthquake ($N$=1196) | Cross-sectional | The Center for Epidemiological Studies-Depression Scale (CES-D) | Older, lower educational level | Greater severity of earthquake-related exposure | \ |
| Ren et al. (2015) | Pregnant women in the epicenter area ($N$=128) | Cross-sectional | The Edinburgh Postnatal Depression Scale (EPDS) | Gestation age at the time of the earthquake, the number of children | Relatives wounded | Subjective support, and coping styles |
| Cénat and Derivois (2015) | Children and adolescent who survived the 2010 Haiti's earthquake ($N$=872) | Cross-sectional | Child Depression Inventory (CDI) | Female | The level of traumatic exposure | \ |
| Guo et al., (2015) | Survivors of Wenchuan earthquake in China ($N$=633) | Cross-sectional | The Children's Depression Inventory (CDI) | Women, living location | Households were severely damaged | Total social support, subjective support, support use and support from family members, neighbors and organizations |
| Qu et al. (2014) | Participants from all temporary camp communities near the epicenter ($N$=1514) | Cross-sectional | The Center for Epidemiologic Studies-Depression Scale (CES-D) | Female | Loss of a family member, residential house damage or collapse | Perceived livelihood security; not living in an urban area |

**TABLE 3** Examples of empirical studies on risk factors for depression after natural disasters—cont'd

| Study | Sample | Study design | Measurement for depression | Key findings: risk factors for depression after natural disasters | | |
|---|---|---|---|---|---|---|
| | | | | Demographic characteristics | Trauma-related factors | Other factors |
| Dong et al. (2013) | Pregnant women from an earthquake struck area ($N=253$) and a nonearthquakes truck area ($N=267$) | Cross-sectional | The Edinburgh Postnatal Depression Scale (EPDS) | \ | \ | Lack of partner support, perceived pressures of pregnancy |
| Lai, La Greca, Auslander, & Short, (2013) | Children from a district damaged by Hurricane ($N=277$) | Longitudinal | The Children's Depression Inventory (CDI) | \ | Initial loss/disruption Perceived life-threat | Recovery stressors (i.e., ongoing loss/disruption, major life events) |
| Sattler et al., (2006) | University students after El Salvador Earthquakes ($N=253$) | Cross-sectional | The Multiscore Depression Inventory | \ | Home damage | Condition resource loss |

having a lower degree of social support tended to experience depressive symptoms (Tang et al., 2014). Loss of parents was also reported as a significant risk factor of depression among adolescents with experience of a disaster (Goenjian et al., 2009). Another longitudinal study revealed that being female, trauma-related self-injury, PTSD symptoms, suicidal ideation at the baseline were found to predict depressive symptoms of high school students at 18 months after the Wenchuan earthquake in China (Chui et al., 2017). Compared with older adults, children encountering an earthquake might find it difficult to recover from the traumatic experience because of less knowledge or skill about seeking help from the social support system and external resources (Shenk, Mahon, Kalaw, Ramos, & Tufan, 2010). Another study investigating university students after El Salvador Earthquake found that home damage and condition resource loss were significant risk factors for their depression (Sattler et al., 2006). For pregnant women who were exposed to natural disasters, a greater level of perceived stress during pregnancy and the degree of social support derived from partners were also reported to predict depression (Dong et al., 2013). Depressive symptoms of pregnant women after the earthquake were found to be significantly related to the gestation age they were at, the number of children they had, whether their relatives were injured in the earthquake, and which kinds of coping style they preferred to utilize (Ren et al., 2015).

## Risk factors for depression among different trauma types

Table 4 lists the possible risk factors for depression after traumatic events. Discrepancies of risk factors exist across different trauma types. A study investigating adult residents of the New York City metropolitan area reported that being single, lower educational level, lower family income, and exposure to ongoing traumatic events increased the risk of depression among the participants after the September 11, 2011 terrorist attacks (Nandi et al., 2009). A study further revealed that low income and unemployment status were the demographic factors which increased the risk of depression among enrollees of the World Trade Center Health Registry (Jacobson et al., 2018). Besides the demographics, factors contributing to depressive symptoms after the 9/11 traumatic events included greater exposure to the traumatic event and panic attacks (Person et al., 2006), low social integration, low social support, chronic physical illness, and post-9/11 traumatic life events (Jacobson et al., 2018). Trauma-induced depression has also been researched in samples of journalists who had a traumatic experience during their work (Backholm & Björkqvist, 2012;

**TABLE 4** Examples of empirical studies on risk factors for depression after traumatic events.

| Study | Sample | Study design | Measurement for depression | Key findings: risk factors for depression after traumatic events | | |
|---|---|---|---|---|---|---|
| | | | | Demographic characteristics | Trauma factors | Other factors |
| Jacobson et al. (2018) | Enrollees of the World Trade Center Health Registry ($N$=21,258) | Longitudinal | Patient Health Questionnaire (PHQ)-8 | Low income, unemployment | Post-9/11 traumatic life events | Low social integration, low social support, chronic physical illness |
| Nandi et al. (2009) | Adult residents of the New York City metropolitan area ($N$=2282) | Longitudinal | Diagnostic and Statistical Manual of Mental Disorders, Fourth Edition (DSM-IV) | Single status, less educational attainment, lower household income | Exposure to ongoing traumatic events | \ |
| Person et al. (2006) | Representatives of the New York City metropolitan area ($N$=2700) | Cross-sectional | DSM-IV-TR (SCID) | \ | Greater exposure to the traumatic event | With panic attacks |
| Miguel-Tobal et al. (2006) | Madrid city residents ($N$=1589) After the Madrid March 11 Train Bombings | Cross-sectional | The SCID's Major Depressive Disorder (MDD) Interview | \ | Mass bereavement | Low social support |

Weidmann, Fehm, & Fydrich, 2008). In a study focusing on depression after the Madrid March 11, 2004 train bombings, mass bereavement and low social support were demonstrated as significant risk factors of depressive symptoms in the affected city residents (Miguel-Tobal et al., 2006).

## Biological factors and depression after disaster events

Table 5 lists the relationship between biological factors and depression after disaster events. Empirical evidences have indicated that depressive disorders are affected by genetic factors (Fan et al., 2017, 2018; Fu et al., 2007; Wang et al., 2015). For instance, the protective effect of *ACE* D allele on depression was observed among female adolescents without treatment at 6 and 12 months after the earthquake (Fan et al., 2018). A significant association between BDNF Val66Met and depression was found among Chinese adolescent survivors (Fan et al., 2018). In a longitudinal study, compared the female with GG homozygotes, female T allele carriers of *adiponectin* rs1501299 demonstrated a lower rate of prevalence of depression and showed a lower degree of depressive symptoms at 18 months after the earthquake (Wang, Zhu, Liu, Ran, & Fang, 2015). In an Italian study, a significant association was found among the gene 5-HTTLPR, childhood traumas, and the severity of depressive symptoms (Carli et al., 2011). By assessing the relationship of the monozygotic and dizygotic twin pairs with depressive symptoms, Fu et al. (2007) found that additive genetic factors were the significant risk factors for the development of major depression among male veterans in the United States.

## Implications and suggestions

Depression is common among the people with experiences of exposure to natural disasters or traumatic events, which has different features among different population within the diverse social contexts. As a crucial mental health sequelae, depression after natural disasters or nonnatural traumatic events should be attached more attention. In addition to the genetic factors (Fan et al., 2017, 2018; Fu et al., 2007; Wang et al., 2015), particular sociodemographic factors are

**TABLE 5 Examples of empirical studies on biological factors and depression after disaster events.**

| Study | Sample | Study design | Measurement for depression | Key findings: biological factors for depression after disaster events |
|---|---|---|---|---|
| Fan et al. (2018) | Adolescents after the 2008 Wenchuan earthquake ($N$=746) | Longitudinal | The Beck Depression Inventory (BDI) | Variants of *ACE I/D* |
| Fan et al. (2017) | Adolescents after the 2008 Wenchuan earthquake ($N$=746) | Longitudinal | The Beck Depression Inventory (BDI) | BDNF Val66Met |
| Wang, Zhu, Liu, Ran, and Fang (2015) | High school students at 6, 12, and 18 months after the earthquake ($N$=746) | Longitudinal | The Beck Depression Inventory (BDI) | Adiponectin rs1501299 |
| Carli et al. (2011) | Male prisoners detained in Italy ($N$=763) | Cross-sectional | Hamilton Depression Rating Scale (HDRS) | Gene 5-HTTLPR |
| Fu et al. (2007) | Middle-aged male–male monozygotic and dizygotic twins from the Vietnam Era Twin Registry ($N$=6744) | Cross-sectional | The Diagnostic Interview Schedule for the DSM-III-R | Additive genetic factor: monozygotic and dizygotic twin pairs |

also crucial when considering to implement a variety of intervention strategies, such as female (Cénat & Derivois, 2015; Guo et al., 2015, 2017; Qu et al., 2014), lower educational level (Cao et al., 2015; Guo et al., 2017), lower socioeconomic status ( Jacobson et al., 2018; Nandi et al., 2009). In addition, loss of property and residential house damage were widely found to be significant risk factors for depression after natural disasters in adult survivors (Guo et al., 2015; Qu et al., 2014), which indicated the need of providing social support to individuals in low socioeconomic status and with greater exposure to the disasters. Social support from the family, neighborhood, and social organizations were protective factors for depression in people encountering traumatic events (Guo et al., 2015). How to strengthen protective factors should also be important for specific interventions. Therefore, Engel's biopsychosocial model is still crucial when considering the coping strategies for mental health issues. Cultural-specific mental health preventions and interventions (e.g., mental health care services, peer support group) should be provided to help those at a high risk of being confronted with psychological issues in the affected communities after the disasters and traumatic events and to prevent negative disaster-related consequences such as suicidal behaviors.

## Key facts of depression

- Understanding the long-term psychological consequences (e.g., depression and suicidal behavior) of natural disasters and traumatic events is crucial for mental health research and clinical practices.
- Depression is common among the people with experiences of exposure to natural disasters or traumatic events, which has different features among different population within the diverse social contexts.
- Risk factors for depression after disasters and traumatic events include biological and psychosocial aspects.
- Depression has been found to be a long-term health issue of natural disasters and traumatic events after a long period of time.
- Long-term interventions should be developed for persons with depression after disasters and traumatic events.

## Summary points

- This chapter aims to review the impacts of natural and nonnatural traumatic events on depression.
- This chapter summarizes clinical diagnosis and standardized assessment instruments for depression in previous studies.
- This chapter reviews the consequences and risk factors of depression after natural disasters or traumatic events.
- Depression is common among the people with experiences of exposure to natural disasters or traumatic events, which has different features among different population within the diverse social contexts.
- This chapter provides coping strategies and suggestions for serving people with depression after natural disasters or nonnatural traumatic events.

## Mini-dictionary of terms

**Symptoms** A physical or mental feature which may indicate a condition of disease, particularly such a feature that is meaningful to the person with a disease.

**Depressive symptoms** A persistent feeling of sadness and loss of interest and can interfere with people's daily functioning.

**Trauma** A deeply distressing or disturbing experience, such as a disturbing experience of the earthquake, typhoon, hurricane, etc.

**Traumatic event** An incident that causes physical, emotional, spiritual, or psychological harm, such as earthquake, typhoon, hurricane, etc.

**Disaster** A serious disruption occurring over a short or long period of time that causes widespread human, material, economic or environmental loss. People who experienced disaster commonly need help from others. Developing countries suffer the greatest costs when a disaster hits than developed countries.

**Suicidal ideation** Thinking about suicide or wanting to take his/her own life, such as a person with depression thinks about suicide or wants to take his/her own life.

## References

American Psychiatric Association. (2000). *Diagnostic and statistical manual of mental disorders* (4th ed., text rev., p. 467). Washington, DC: Author.

American Psychiatric Association. (2013). Diagnostic and statistical manual of mental disorders. *BMC Medicine*, *17*, 133–137.

Backholm, K., & Björkqvist, K. (2012). The mediating effect of depression between exposure to potentially traumatic events and PTSD in news journalists. *European Journal of Psychotraumatology*, *3*(1), 18388.

Beck, A. T., Ward, C. H., Mendelson, M., Mock, J., & Erbaugh, J. (1961). An inventory for measuring depression. *Archives of General Psychiatry*, *4*(6), 561–571.

Beidel, D. C., & Frueh, B. C. (2018). *Adult psychopathology and diagnosis*. John Wiley & Sons.

Brink, T. L., Yesavage, J. A., Lum, O., Heersema, P. H., Adey, M., & Rose, T. L. (1982). Screening tests for geriatric depression. *Clinical Gerontologist*, *1*(1), 37–43.

Cadoret, R. J., O'Gorman, T. W., Heywood, E., & Troughton, E. (1985). Genetic and environmental factors in major depression. *Journal of Affective Disorders*, *9*(2), 155–164.

Cao, X., Wang, L., Cao, C., Zhang, J., Liu, P., Zhang, B., et al. (2015). Patterns of DSM-5 posttraumatic stress disorder and depression symptoms in an epidemiological sample of Chinese earthquake survivors: A latent profile analysis. *Journal of Affective Disorders*, *186*, 58–65.

Carli, V., Mandelli, L., Zaninotto, L., Roy, A., Recchia, L., Stoppia, L., et al. (2011). A protective genetic variant for adverse environments? The role of childhood traumas and serotonin transporter gene on resilience and depressive severity in a high-risk population. *European Psychiatry*, *26*(8), 471–478.

Caspi, A., & Moffitt, T. E. (2006). Gene–environment interactions in psychiatry: Joining forces with neuroscience. *Nature Reviews Neuroscience*, *7*(7), 583.

Cénat, J. M., & Derivois, D. (2014). Assessment of prevalence and determinants of posttraumatic stress disorder and depression symptoms in adults survivors of earthquake in Haiti after 30 months. *Journal of Affective Disorders*, *159*, 111–117.

Cénat, J. M., & Derivois, D. (2015). Long-term outcomes among child and adolescent survivors of the 2010 Haitian earthquake. *Depression & Anxiety*, *32*(1), 57–63.

Chen, X., Xu, J., Li, B., Li, N., Guo, W., Ran, M. S., et al. (2017). The role of personality and subjective exposure experiences in posttraumatic stress disorder and depression symptoms among children following Wenchuan earthquake. *Scientific Reports*, *7*(1), 17223.

Chui, C. H. K., Ran, M. S., Li, R. H., Fan, M., Zhang, Z., Li, Y. H., et al. (2017). Predictive factors of depression symptoms among adolescents in the 18-month follow-up after Wenchuan earthquake in China. *Journal of Mental Health*, *26*(1), 36–42.

Derogatis, L. R., Lipman, R. S., Rickels, K., Uhlenhuth, E. H., & Covi, L. (1974). The Hopkins symptom checklist (HSCL): A self-report symptom inventory. *Behavioral Science*, *19*(1), 1–15.

Dong, X., Qu, Z., Liu, F., Jiang, X., Wang, Y., Chui, C. H. K., et al. (2013). Depression and its risk factors among pregnant women in 2008 Sichuan earthquake area and non-earthquake struck area in China. *Journal of Affective Disorders*, *151*(2), 566–572.

Fan, M., Cao, T., Feng, Y., Su, M., Lin, J., Ran, M. S., et al. (2018). D allele of insertion/deletion polymorphism at angiotensin-converting enzyme gene is associated with reduced prevalence and severity of depression among Chinese adolescents at early stage after Wenchuan earthquake. *International Journal of Psychiatry in Clinical Practice*, *22*(2), 136–142.

Fan, M., Li, R. H., Hu, M. S., Xiao, L. Y., Zhou, X. D., Ran, M. S., et al. (2017). Association of Val66Met polymorphism at brain derived neurotrophic factor gene with depression among Chinese adolescents after Wenchuan earthquake: An 18 months longitudinal study. *Physiology & Behavior*, *179*, 16–22.

Fan, F., Zhang, Y., Yang, Y., Mo, L., & Liu, X. (2011). Symptoms of posttraumatic stress disorder, depression, and anxiety among adolescents following the 2008 Wenchuan earthquake in China. *Journal of Traumatic Stress*, *24*(1), 44–53.

First, M. B., Gibbon, M., Spitzer, R. L., Williams, J. B., & Benjamin, L. S. (1997). *Structured clinical interview for DSM-IV® Axis II personality disorders SCID-II*. American Psychiatric Pub.

Frazier, P., Anders, S., Perera, S., Tomich, P., Tennen, H., Park, C., et al. (2009). Traumatic events among undergraduate students: Prevalence and associated symptoms. *Journal of Counseling Psychology*, *56*(3), 450.

Fu, Q., Koenen, K. C., Miller, M. W., Heath, A. C., Bucholz, K. K., Lyons, M. J., et al. (2007). Differential etiology of posttraumatic stress disorder with conduct disorder and major depression in male veterans. *Biological Psychiatry*, *62*(10), 1088–1094.

Goenjian, A. K., Walling, D., Steinberg, A. M., Roussos, A., Goenjian, H. A., & Pynoos, R. S. (2009). Depression and PTSD symptoms among bereaved adolescents 6½ years after the 1988 Spitak earthquake. *Journal of Affective Disorders*, *112*(1–3), 81–84.

Guo, J., He, H., Qu, Z., Wang, X., & Liu, C. (2017). Post-traumatic stress disorder and depression among adult survivors 8 years after the 2008 Wenchuan earthquake in China. *Journal of Affective Disorders*, *210*, 27–34.

Guo, S., Tian, D., Wang, X., Xiao, Y., He, H., Qu, Z., et al. (2015). Protective effects of social support content and support source on depression and its prevalence 6 months after wenchuan earthquake. *Stress and Health*, *31*(5), 382–392.

Hamilton, M. (1960). A rating scale for depression. *Journal of Neurology, Neurosurgery, and Psychiatry*, *23*(1), 56.

Hamilton, M. A. X. (1967). Development of a rating scale for primary depressive illness. *British Journal of Social and Clinical Psychology*, *6*(4), 278–296.

Jacobson, M. H., Norman, C., Nguyen, A., & Brackbill, R. M. (2018). Longitudinal determinants of depression among world trade center health registry enrollees, 14–15 years after the 9/11 attacks. *Journal of Affective Disorders*, *229*, 483–490.

Lai, B. S., La Greca, A. M., Auslander, B. A., & Short, M. B. (2013). Children's symptoms of posttraumatic stress and depression after a natural disaster: Comorbidity and risk factors. *Journal of Affective Disorders*, *146*(1), 71–78.

Meiser-Stedman, R., Dalgleish, T., Yule, W., & Smith, P. (2012). Intrusive memories and depression following recent non-traumatic negative life events in adolescents. *Journal of Affective Disorders*, *137*(1–3), 70–78.

Miguel-Tobal, J. J., Cano-Vindel, A., Gonzalez-Ordi, H., Iruarrizaga, I., Rudenstine, S., Vlahov, D., et al. (2006). PTSD and depression after the Madrid March 11 train bombings. *Journal of Traumatic Stress*, *19*(1), 69–80.

Montgomery, S. A., & Åsberg, M. A. R. I. E. (1979). A new depression scale designed to be sensitive to change. *British Journal of Psychiatry*, *134*(4), 382–389.

Nandi, A., Tracy, M., Beard, J. R., Vlahov, D., & Galea, S. (2009). Patterns and predictors of trajectories of depression after an urban disaster. *Annals of Epidemiology*, *19*(11), 761–770.

Neria, Y., Nandi, A., & Galea, S. (2008). Post-traumatic stress disorder following disasters: A systematic review. *Psychological Medicine*, *38*(4), 467–480.

Norris, F. H., Friedman, M. J., & Watson, P. J. (2002). 60,000 disaster victims speak: Part II. Summary and implications of the disaster mental health research. *Psychiatry: Interpersonal and Biological Processes*, *65*(3), 240–260.

Norris, F. H., Friedman, M. J., Watson, P. J., Byrne, C. M., Diaz, E., & Kaniasty, K. (2002). 60,000 disaster victims speak: Part I. An empirical review of the empirical literature, 1981–2001. *Psychiatry: Interpersonal and Biological Processes*, *65*(3), 207–239.

North, C. S., Kawasaki, A., Spitznagel, E. L., & Hong, B. A. (2004). The course of PTSD, major depression, substance abuse, and somatization after a natural disaster. *The Journal of Nervous and Mental Disease*, *192*(12), 823–829.

Overstreet, C., Berenz, E. C., Sheerin, C., Amstadter, A. B., Canino, G., & Silberg, J. (2016). Potentially traumatic events, posttraumatic stress disorder, and depression among adults in Puerto Rico. *Frontiers in Psychology*, *7*, 469.

Pan, X., Liu, W., Deng, G., Liu, T., Yan, J., Tang, Y., et al. (2015). Symptoms of posttraumatic stress disorder, depression, and anxiety among junior high school students in worst-hit areas 3 years after the Wenchuan earthquake in China. *Asia Pacific Journal of Public Health*, *27*(2), NP1985–NP1994.

Person, C., Tracy, M., & Galea, S. (2006). Risk factors for depression after a disaster. *The Journal of Nervous and Mental Disease*, *194*(9), 659–666.

Qu, Z., Wang, C. W., Zhang, X., Ho, A. H., Wang, X., & Chan, C. L. (2014). Prevalence and determinants of depression among survivors 8 months after the Wenchuan earthquake. *The Journal of Nervous and Mental Disease*, *202*(4), 275–279.

Radloff, L. S. (1977). The CES-D scale: A self-report depression scale for research in the general population. *Applied Psychological Measurement*, *1*(3), 385–401.

Ran, M. S., Zhang, Z., Fan, M., Li, R. H., Li, Y. H., Ou, G. J., et al. (2015). Risk factorsof suicidal ideation among adolescents after Wenchuan earthquake in China. *Asian Journal of Psychiatry*, *13*, 66–71.

Ren, J., Jiang, X., Yao, J., Li, X., Liu, X., Pang, M., et al. (2015). Depression, social support, and coping styles among pregnant women after the lushan earthquake in ya'an, China. *PLoS One*, *10*(8), e0135809.

Şalcıoğlu, E., & Başoğlu, M. (2008). Psychological effects of earthquakes in children: Prospects for brief behavioral treatment. *World Journal of Pediatrics*, *4*(3), 165–172.

Sattler, D. N., De Alvarado, A. M. G., De Castro, N. B., Male, R. V., Zetino, A. M., & Vega, R. (2006). El Salvador earthquakes: Relationships among acute stress disorder symptoms, depression, traumatic event exposure, and resource loss. *Journal of Traumatic Stress: Official Publication of The International Society for Traumatic Stress Studies*, *19*(6), 879–893.

Shenk, D., Mahon, J., Kalaw, K. J., Ramos, B., & Tufan, I. (2010). Understanding the disaster experience of older adults by gender: The experience of survivors of the 2007 earthquake in Peru. *Health Care for Women International*, *31*(11), 965–980.

Tang, B., Liu, X., Liu, Y., Xue, C., & Zhang, L. (2014). A meta-analysis of risk factors for depression in adults and children after natural disasters. *BMC Public Health*, *14*(1), 623.

Wang, Q., Zhu, X. C., Liu, H., Ran, M. S., & Fang, D. Z. (2015). A longitudinal study of the association of adiponectin gene rs1501299 with depression in Chinese Han adolescents after Wenchuan earthquake. *Journal of Affective Disorders*, *175*, 86–91.

Wang, W., Fu, W., Wu, J., Ma, X. C., Sun, X. L., Huang, Y., et al. (2012). Prevalence of PTSD and depression among junior middle school students in a rural town far from the epicenter of the Wenchuan earthquake in China. *PLoS One*, *7*(7), e41665.

Wang, X., Gao, L., Shinfuku, N., Zhang, H., Zhao, C., & Shen, Y. (2000). Longitudinal study of earthquake-related PTSD in a randomly selected community sample in North China. *American Journal of Psychiatry*, *157*(8), 1260–1266.


Wang, Q., Zhu, X. C., Liu, H., Ran, M. S., & Fang, D. Z. (2015). A longitudinal study of the association of adiponectin gene rs1501299 with depression in Chinese Han adolescents after Wenchuan earthquake. *Journal of Affective Disorders*, *175*, 86–91.


Weidmann, A., Fehm, L., & Fydrich, T. (2008). Covering the tsunami disaster: Subsequent post-traumatic and depressive symptoms and associated social factors. *Stress and Health: Journal of the International Society for the Investigation of Stress*, *24*(2), 129–135.

Yesavage, J. A., Brink, T. L., Rose, T. L., Lum, O., Huang, V., Adey, M., et al. (1982). Development and validation of a geriatric depression screening scale: A preliminary report. *Journal of Psychiatric Research*, *17*(1), 37–49.

Zhang, Z., Ran, M. S., Li, Y. H., Ou, G. J., Gong, R. R., Li, R. H., et al. (2012). Prevalence and risk factors of posttraumatic stress disorder among adolescents after Wenchuan earthquake in China. *Psychological Medicine*, *42*, 1687–1693.

Zuromski, K. L., Resnick, H., Price, M., Galea, S., Kilpatrick, D. G., & Ruggiero, K. (2019). Suicidal ideation among adolescents following natural disaster: The role of prior interpersonal violence. *Psychological Trauma: Theory, Research, Practice, and Policy*, *11*(2), 184.

Chapter 8

# Depression and associated Alzheimer's disease

**Nikita Patil and Girdhari Lal Gupta**

*Department of Pharmacology, Shobhaben Pratapbhai Patel School of Pharmacy and Technology Management, Mumbai, India*

## Abbreviations

| | |
|---|---|
| **5HTP** | 5-hydroxytryptophan |
| **AADC** | aromatic L-amino acid decarboxylase |
| **ACTH** | adrenocorticotropic hormone |
| **BMAA** | β-*N*-methylamino-L-alanine |
| **BPSD** | behavioral and psychological symptoms of dementia |
| **CRP** | C-reactive protein |
| **IDO** | indoleamine-2,3-dioxygenase |
| **KYN** | kynurenine |
| **LTD** | long-term depression |
| **MDD** | major depressive disorder |
| **PFC** | prefrontal cortex |
| **QA** | quinolinic acid |
| **TGF-β1** | transforming growth factor-beta 1 |
| **TRP** | tryptophan |
| **TRPOH** | tryptophan hydroxylase |

## Introduction

Depression is a psychiatric syndrome and life-threatening form of mental illness. Depression mainly characterized by low mood, anhedonia, feelings of sadness along with somatic and cognitive disturbances (Marvel & Paradiso, 2004). In other words, depression is a heterogeneous illness and based on various factors, e.g., psychological, social, genetic, and biological factors (Chandan et al., 2019; Laird, Krause, Funes, & Lavretsky, 2019). According to, epidemiology study, depression affects more than 350 million people worldwide and is expected to be the second leading cause of disability after heart disease by the year 2020 (Mossie, Kindu, & Negash, 2016). There are various causes of depression but mainly depression is thought to be caused by an imbalance in neurotransmitters. Some of the common factors that may cause depression are genetics (hereditary), trauma, high levels of stress, mental illnesses such as schizophrenic disorder, individuals with low self-esteem, cancer, HIV, Rheumatoid arthritis, serious medical conditions such as heart disease, alcohol dependence, and drug abuse (Nerurkar, Siebert, McInnes, & Cavanagh, 2019; Wakefield et al., 2015). A number of genetic risk factors involved in the etiology of depression have been identified. Various genes responsible for the regulation of monoamines, corticotropin-releasing hormone (CRH) and brain-derived neurotrophic factor (BDNF) were found to be altered in depression (Shadrina, Bondarenko, & Slominsky, 2018). Prefrontal structures and their coordinated circuits have been engaged in the regulation, including the lateral orbital prefrontal cortex (PFC), dorsolateral PFC, ventromedial PFC, anterior cingulate cortex, amygdala, and the hippocampus. These regions regulate learning and contextual memory processes, executive function, emotion and reward, and have been implicated in depression and antidepressant action (Rao & Chen, 2008). Table 1 shows general features on Alzheimer's.

Alzheimer's disease (AD) is a neurodegenerative disorder and also established as a new clinical entity more than 100 years ago after an illustration of the famous patient Auguste D. by Alois Alzheimer. Currently, worldwide nearly 50 million people above the age of 60 years are diagnosed with AD and the projected figure is estimated to be 130 million in 2050 (Andrews et al., 2017). On the basis of genetics and various other factors, there are two forms of AD, a familial (early-onset) AD, and sporadic (late-onset) AD. It is reported that the prevalence of sporadic AD continues to rise and now it may become the sixth leading cause of death (Alzheimer's Association, 2019). These two forms of AD are accountable for severe cognitive decline in AD patients in their latest stage. Various primary etiological factors are responsible for the development of AD. Especially, inherited variation or mutation in the gene that encodes for APP, presenilin 1, presenilin 2, and other genes, act as predisposing factors for the initiation of AD. Currently, 95% of all AD cases belong to the sporadic form of AD and represent the most common cause of dementia in elderly patients (Mravec, Horvathova, & Padova, 2018).

The Neuroscience of Depression. https://doi.org/10.1016/B978-0-12-817933-8.00027-X

There are two main hallmarks of AD, amyloid-beta plaques and neurofibrillary tangles (NFTs). In AD, due to improper proteolytic cleavage of amyloid protein by β- and γ-secretases leads to the formation of amyloid-beta plaques. These plaques are responsible for neurodegeneration by accumulating on the surface of the cell. In addition, the hyperphosphorylation of microtubular tau protein also leads to the formation of intraneuronal NFTs (Lane, Hardy, & Schott, 2018). Mounting evidence also shows that endosomal trafficking can also be influenced by aggregation of abnormally enlarged endosomes in AD. Recent research also revealed that the production of reactive oxygen species (ROS) could encourage by metal intoxication which contributes to brain atrophy (Xu et al., 2017). There are several factors involved in AD pathology. Epigenetics and environmental are two related prominent factors in AD pathology. Epigenetics does not contribute to changes in the DNA sequence but associated with transmittable patterns of gene expression and genomic role. Many individual studies have recommended an attainable role for epigenetic changes in AD etiology (Mastroeni et al., 2010). The most common epigenetic consequence is DNA methylation, which modifies the genomic and chromosomal functioning. DNA methylation patterns are negatively affected by several environmental toxins, pollutants, and metals. For example, prenatal methylmercury exposure causes hypermethylation of BDNF gene and severe depression-like symptoms in the mouse hippocampus (Lardenoije et al., 2015). Exposure of air pollution especially damages the blood-brain barrier (BBB) which triggers pro-inflammatory and inflammatory response resulting in the neuroinflammation and AD pathogenesis. Some toxic agents from environments, such as β-*N*-methylamino-L-alanine (BMAA), cause misfolding and accumulation of proteins. It is reported that chronic exposure of BMAA in individuals, can initiate amyloid-beta plaques formation and later on production NFTs (Liu, Jiao, & Shen, 2018). Furthermore, self-aggregation of amyloid-beta in the presence of various metals like Zn, Fe, and Cu into amyloid fibrils, which recently brought a "metal hypothesis of AD" into focus. Numerous studies reported that particular metals and metalloids responsible for Aβ aggregation and tauopathy (Drew, 2017).

## Coexistence of depression and Alzheimer's disease

Various genetic factors are responsible for the development of depression-induced AD. Current evidence shows molecular mechanisms and cascade involved in the pathology of depression such as chronic inflammation, HPA axis deregulation, and low-level neurotropic factors. For more than 5 decades, a variety of reports and studies have shown a robust clinical linkage between chronic major depressive disorder (MDD) and AD (Caraci et al., 2018; Wuwongse, Chang, & Law, 2010). For many years, dementia and depression have been established as some well-defined clinical entities. The term "pseudodementia" recognized as a depression related to cognition impairment (Lange et al., 2019). The common relation and pathophysiological mechanisms between depression and AD have been identified. In several in vivo and in vitro studies, impairment of the serotonergic system has been observed in AD. There are some early behavioral and psychological symptoms of dementia (BPSD) observed in depressed patients such as recognition difficulty, change in mood, change in appetite, wake-sleep cycle alteration, confusion, and agitation (Elferink, van Tilborg, & Kessels, 2015).

## Dysfunction of the hypothalamic-pituitary-adrenal axis

Epidemiological studies suggested that stress can act as a predisposing factor for depression and AD (Bao, Meynen, & Swaab, 2008). Several studies clearly show that chronic exposure to stress may enhance the pathogenesis of depression. Chronic stress mainly causes dysfunction of the HPA axis. Furthermore, the corticotropin-releasing factor released through the hypothalamus in response to stressful stimuli, which further promotes the release of adrenocorticotropic hormone (ACTH) from the pituitary gland into the circulation. Eventually, under the influence of ACTH, glucocorticoids release from the adrenal cortex. Glucocorticoids can easily cross the BBB and enter into hippocampus region of the brain, exert a negative impact on the HPA axis, which further participates in the regulation of normal cellular biological processes and other central nervous system (CNS) functions, such as memory and learning (García-Pérez, Ferenczi, Kovács, & Milanés, 2019). In depression and AD, acute or chronic stress responsible for alteration in HPA axis regulation, which leads to a high cortisol level. Various study indicates that deregulation of the HPA axis is an early event responsible for AD pathogenesis. It is reported that high cortisol level is associated with more rapid cognitive decline. This high level of cortisol induces the retraction of hippocampal apical dendrites, reduces neurogenesis in the hippocampus region of the brain and causes neuronal loss (Melón, Hooper, Yang, Moss, & Maguire, 2018). Specifically, glucocorticoids promote neuronal death and also develop susceptibility of neurons to different toxins, such as excitotoxins and radical oxygen species. Especially, AD animal models show steady-state levels of β-APP cleaving enzyme can be enhanced by glucocorticoid administration. Eventually increases the β-amyloid formation and also encourages the production of NFTs, which strongly related to the severity of dementia. All these data established a

**TABLE 1** Other features associated with Alzheimer's disease.

| Alzheimer's disease features | Symptoms |
|---|---|
| Psychological problems | Depression, hallucination, or paranoia |
| Behavioral changes | Aggression, agitation, difficulty with self-care, irritability, meaningless repetition of own words, restlessness |
| Infection | Fever, cough, shortness of breath, pneumonia, and other respiratory infections |
| Loss of appetite, dehydration | Dry mouth, headaches, dry skin, sleepiness, irritability |
| Incontinence | Not being able to react to the sensation of needing the toilet, mobility issues |
| Dental problems | Sores, decayed teeth |

fundamental role in high levels of glucocorticoids and HPA axis dysfunction both in depression and AD (Canet, Chevallier, Zussy, Desrumaux, & Givalois, 2018; Green & Zhang, 2016).

## Chronic inflammation in depression and Alzheimer's disease

Chronic inflammation shows a critical role in the progression of AD and also involve in molecular events associated with AD pathogenesis. Increased plasma levels of IL-6 and interleukin-1β (IL-1β) in cerebrospinal fluid have been observed in AD patients. Aβ oligomers trigger inflammatory responses, which show a crucial role in the progression of AD. Different studies demonstrated that β-amyloid trigger activation of microglia which further release pro-inflammatory cytokines such as IL-1β, IL-6, and tumor necrosis factor-α (TNF-α) (Surendranathan et al., 2018). Cytokine production and β-amyloid-induced neurodegeneration have a direct correlation. Activation of pro-inflammatory cytokines imparts to neuronal death. It is reported that pro-inflammatory cytokines trigger the activation of several kinases but mainly cyclin-dependent kinase 5 (CDK5), one of the primary kinase, responsible for tau hyperphosphorylation thus ultimately promote the formation of NFTs (Calsolaro & Edison, 2016; Miller & Raison, 2016). Evidence indicates that inflammation plays a crucial part in depression and AD (Miller & Raison, 2016). Depressed patients show an increase in the level of pro-inflammatory and inflammatory markers, such as cytokines and chemokines (Haapakoski, Mathieu, Ebmeier, Alenius, & Kivimäki, 2015). Moreover, the elevated level of peripheral blood markers of inflammation increases the severity of inflammation and may increase the progression of depression. Recent data indicate that elevated levels of plasma cytokines correlate with increased CNS glutamate and decreased motor activity in patients with MDD, which associated with symptoms of anhedonia and motor slowing. Therefore, chronic inflammation can be considered as a common pathophysiological connection between AD and depression (Felger et al., 2020). In depressed patients, the existence of HPA dysfunction has been considered as an ensuing state of altered activation of the immune system and chronic inflammation. It has been hypothesized that HPA axis dysregulation, glucocorticoid receptor resistance, and high cortisol levels can be found in depressed patients. High levels of pro-inflammatory markers, cytokines, chemokines, an acute-phase protein, and cellular adhesion molecules have been observed in depression as well as in AD patients (Schiepers, Wichers, & Maes, 2005). Conversely, diminished levels of anti-inflammatory cytokines such as transforming growth factor β1 (TGF-β1), interleukin-10 (IL-10), and interleukin-4 (IL-4) have been found in the plasma of depressed patients. Different studies have demonstrated a positive correlation between the severity of depression symptoms and the increase in the inflammatory state. Pro-inflammatory cytokines interpose with various pathological processes of depression such as altering the metabolism of serotonin, alteration in synaptic plasticity, and decreased hippocampal neurogenesis. Furthermore, an increase in the level of pro-inflammatory cytokines and chronic inflammation act as predisposing factors for neurodegenerative processes both in AD and depression (Maes et al., 2009). TNF-α has been considered as a common potential target for both disease and a relevant neurobiological link between depression and AD. Various studies demonstrated that activation of microglia and abnormal TNF-α signaling can be enhanced by Aβ oligomer. This abnormal activation, alter the level of the monoamine and decreases serotonergic tone, which plays a cardinal role in Aβ-induced depressive-like behavior. Several studies concluded that abnormal TNF-α signaling and neuroinflammation could play a possible role in late-life depression and AD.

### Impairment of neurotrophin signaling in depression and Alzheimer's disease

Neurotrophins play a very essential role in synaptic plasticity and neuronal homeostasis. The neurotrophin, BDNF binds to tyrosine kinase receptor TrkB and supports neuronal survival, promotes connectivity and plasticity in the adult brain. Chronic unpredictable stress-induced animal models of depression show altered levels and activity of BDNF and TGF-β1. Antidepressants have been demonstrated to increase in BDNF-mediated signaling, whereas stress reduces BDNF-mediated signaling in the hippocampus (Duman & Li, 2012). In depression or AD, there is an alteration of BDNF in the specific region of the brain, e.g., chronic stress shows an elevated amount of BDNF in the nucleus accumbens, whereas the low level of BDNF has been observed in the substantia nigra region of the brain (Wook Koo et al., 2016). Altered BDNF signaling and treatment-resistant depression significantly impairs the release of BDNF (Cai et al., 2017; Goda et al., 2013). Synaptic dysfunction, loss of neurons, and finally cognitive deficits are three prominent outcomes associated with altered levels of BDNF and tropomyosin receptor kinase B (TrkB) receptor (Adlard, Perreau, & Cotman, 2005). Chronic stress can contribute to the altered BDNF level. Furthermore, other neurotrophic factors such as TGF-β1 also interact with BDNF and show neuroprotective function. TGF-β1 also regulates neuronal survival and maintain the proportion between T helper-1 and T helper-2 cytokines. Moreover, TGF-β1 also defends against excitotoxicity and most importantly β-amyloid (Diniz et al., 2014). TGF-β1 also promotes the expression of TrkB and BDNF. Improved synaptic plasticity in the presence of TGF-β1 has been observed, which further supports neurogenesis, improves connectivity and enhances synaptic plasticity in the brain (Park & Poo, 2013). Altered BDNF signaling due to the TrkB receptor is also related to the pathogenesis of mood disorders (Lindholm & Castrén, 2014).

## Alteration in monoamines level

The monoamine hypothesis of depression has a crucial role in the therapeutic effects of antidepressant drugs via alteration of the concentrations of dopamine, serotonin, and noradrenaline in the synaptic cleft. Therefore, most of the currently available antidepressants are based on the monoamine hypothesis and balance the level of the monoamine in the brain (Boku, Nakagawa, Toda, & Hishimoto, 2018). In AD, serotonergic 5-$HT_2$ receptors are majorly affected. Moreover, in the temporal cortex, decreased 5-$HT_{1A}$ receptor binding has also been shown to correlate with AD symptoms. Various in vitro studies suggested that diminution in 5-$HT_{1A}$ receptor binding in the hippocampus region correlated with cognitive decline in AD patients (Šimić et al., 2017). Due to the alteration of the serotonergic system and various other factors, postmortem AD brains show more amyloid-beta plaques and NFTs and reduced neuronal numbers as compared to healthy control brains. Specific behavioral changes can be found in AD patients due to loss in the serotonergic system. It is reported that a reduced level of 5-$HT_{1A}$ and 5-$HT_{2A}$ receptor binding sites have been observed in frontal, temporal, and parietal lobes, the amygdala and the hippocampus region of the AD brain (Theodore et al., 2012). A diminution in 5-HTT binding sites from frontal and temporal lobes to several subcortical structures has also been reported (Kepe et al., 2006; Yokokura et al., 2019). It is predicted that the altered level of 5-$HT_2$ receptors in the cortex has been associated with a decrease in the level of BDNF and NFT formation. AD patients show structural and functional alterations in the serotonergic system. The altered level of serotonin is directly related to the production of Aβ. Therefore, an increase in the level of serotonin promotes the activation of the non-amyloidogenic APP pathway and decreases the formation of neurotoxic Aβ (Nautiyal & Hen, 2017).

The common molecular pathway between depression and AD represented in Fig. 1. Each pathway represents a neurobiological link between depression and dementia-specific pathogenesis including (1) HPA axis dysregulation, (2) altered monoamines level, (3) chronic inflammation, (4) altered TGF-β1 signaling, and (5) deficits of nerve growth factors or neurotrophins.

## TGF signaling

TGF-β1 belongs to TGF-β superfamily, which comprises multifunctional signaling protein and plays a very important role in cell growth, differentiation, immune system suppression and cell repair (Caraci et al., 2018). TGF-β1 act as anti-inflammatory cytokine and exert a neuroprotective effect in a various animal model of the neurodegenerative disorder including AD. Moreover, due to its anti-inflammatory and neuroprotective effect, TGF-β1 also imparts a role in cognition. The transition from early to late long-term potentiation (LTP) mediated through decreased TGF-β1 signaling could increase the severity of depression associated with AD condition. TGF-β1 exerts its neuroprotective effect by encouraging the expression of CDK inhibitors (p21, p27) and reduces Aβ toxicity. This mechanism further inhibits GSK-3β and also prevents Aβ-induced tau hyperphosphorylation. Hence, specific impairment of Smad-dependent TGF-β1 could be assisted by Aβ-induced neuronal cell cycle reactivation. Hence, the multifunctional role of TGF-β1 can be considered as a potential target for the management of AD and depression (Herrup, 2010; Tichauer et al., 2014).

Moreover, perturbation of tryptophan metabolism is implicated in depression and AD pathogenesis. The Kynurenine (KYN) pathway dysfunction is associated with the depression and AD pathogenesis. Aberrant tryptophan metabolism triggers the KYN pathway in activated microglia and decreases the production of KA (kynurenic acid) in resting astrocytes, which negatively affect the TGF-β1 signaling and further increases Aβ accumulation in the brain and ultimately compromise the neuroprotective and anti-excitotoxic functions of TGF-β1. Increased KYN and diminished level of 5-HT, TGF-β1 is strongly related to

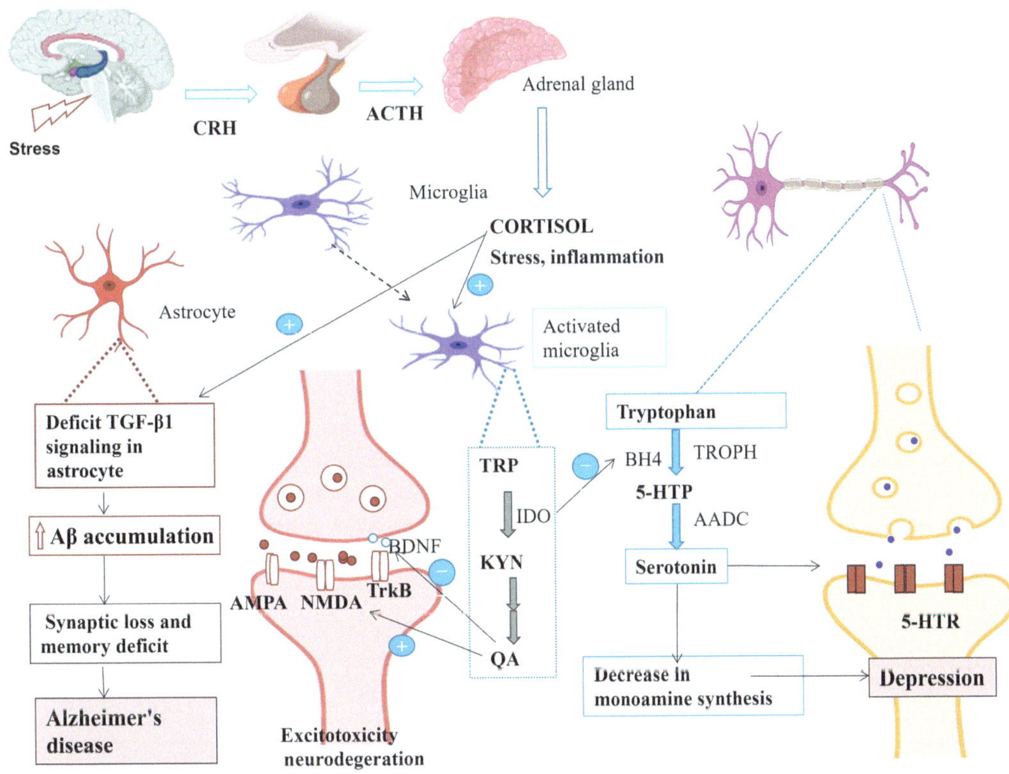


FIG. 1 Molecular links between depression and Alzheimer's disease pathophysiology. Depression has been associated with elevated levels of CRH, ACTH, and cortisol. A high level of cortisol increases inflammation which can lead to the stimulation of the indoleamine 2,3 dioxygenase (IDO) (tryptophan degrading enzyme). IDO also decreases the formation of serotonin via inhibition of tetrahydrobiopterin (BH4) levels. IDO diverts the tryptophan use toward the formation of neurotoxic 3-hydroxykynurenine (KYN) and quinolinic acid (QA) in the kynurenine pathway in the neuron as well as glial cell. Further, both neurotoxins activate glutamate receptors excitotoxicity and decrease in the BDNF level. Stress and inflammation are the key factors for the deficit of TGF-β1 signaling in astrocytes. Decreased level of TGF-β1, enhances the Aβ accumulation which further contributes to synaptic loss and cognition impairment. Due to inflammation activation of QA pathway, the level of BH4 decreases which hampers the synthesis of monoamines thus contribute to the depression and AD pathology.

increased excitotoxicity, lipid peroxidation, and neurodegeneration. These altered mechanistic pathways contribute to major depression and AD (Réus et al., 2015).

## Long-term potentiation alteration

Synaptic plasticity mainly concerned with the strength of synapse and can be defined as activity-dependent modifications in the efficacy and strength of synaptic transmission of preexisting synapses (Abraham & Williams, 2003). Synaptic plasticity alteration has two main forms long term depression (LTD) and LTP. Altered synaptic plasticity primarily, LTD and LTP are observed in neurodegenerative disorders including AD (Lacor et al., 2007). Disturbing levels of several cytokines and neuropeptides have been observed during AD. Therefore, any alteration of their functions could possibly play an imperative role in altered synaptic plasticity during AD. Altered protein expression and trafficking of neuropeptides have been observed in AD. Association of neuropeptides in the severity of AD symptoms has also been recognized. Somatostatin and Neuropeptide Y are the main neuropeptides which are mainly diminished during AD pathology. The majority of the excitatory synapses in the brain made up of glutamatergic synapses and connect many of the regions related to depression and stress, including the amygdala, hippocampus, and PFC. Altered synaptic plasticity, glutamatergic systems also play an essential role in the progression of depression (Vose & Stanton, 2017).

## Oxidative stress

Stress and dysfunction of the HPA axis can also accelerate oxidative damage and decrease antioxidant defenses (Huang, Zhang, & Chen, 2016). Commonly, in AD pathophysiology, the level of oxidative stress with an increase in inflammatory cytokines are frequent with aging and comorbid depression. Antioxidant and anti-inflammatory activities paradoxically decrease with aging and comorbid depression (Tönnies & Trushina, 2017). Oxidative stress occurs when the disbalance occurs between the oxygen free radicals and antioxidants. Stress and depression increase the production of ROS which ultimately decreases the effect of protective or reparative processes. Furthermore, increased oxidative stress intensifies the process of biological aging. Cellular oxidative damage is an essential element of the aging process. Hence, psychological stress could negatively affect the ratio of antioxidants (vitamin E) to serum oxidized lipids (F2-isoprostanes). Recurrent exposure to oxidative stress could intensify and encourage aging-linked comorbid diseases. It is reported that chronic psychological stress-induced depressed patients show increased plasma oxidative stress markers as compared to antioxidant compounds, such as vitamins C and E. Similarly, an increased level of oxidative stress and lower level of antioxidants is associated with lower BDNF activity (Mendiola-Precoma, Berumen, Padilla, & Garcia-Alcocer, 2016).

## Therapeutic management of AD

Currently, cholinesterase inhibitors and NMDA receptor antagonists are only FDA-approved medications available in the market for the management of AD. These include donepezil, rivastigmine, and galantamine belongs to the category of cholinesterase inhibitors whereas, memantine, is a noncompetitive NMDA receptor modulator. However, all these drugs produce asymptomatic effects. Unfortunately, it has been described that none of these drugs cures AD; however, they only have palliative measures. Since 2003 no new drug has been approved by the FDA for the treatment of AD. In various stages of the clinical trial, more than 20 compounds have completed large phase 3, randomized control trials, double-blind studies, but no drug showed any efficacy in improving brain functioning or slowing cognitive decline. Numerous phase 2 and phase 3 clinical trials have failed to demonstrate benefits. Hence currently, no disease-modifying treatment exists. At present, for the successful management and treatment of various chronic diseases, such as hypertension epilepsy, HIV, diabetes, congestive heart failure, and others, a combination treatment strategy show more clinical successes than monotherapy. By considering, underlying complexity and multifactorial nature of AD pathology, it seems irrational to use a single agent for the management of AD. Numerous studies show that the perspective of using a combination strategy targeting multiple pathways would be highly efficient in slowing disease progression as compared to monotherapy. It is reported that multiple medications spanning different drug classes can be effectively managed by many diseases. Worldwide, various researchers already implicated the combination strategy in the clinical trials, e.g., the Lilly trailblazer tested the combination of Aβ monoclonal antibody donanemab, with BACE1 inhibitor LY3202626. Unfortunately, in phase 2 clinical trial, LY3202626 showed a slight cognitive decline in AD patients, so the combination treatment arm was ultimately discontinued. However, we should continue to encourage various combination strategies in the clinical trials. Ideally, to increase the success of combination therapy in the clinical trials, it is necessary to implicate this type of treatment strategy at the preclinical stage (Long & Holtzman, 2019). The neurogenesis hypothesis of MDD suggested that adult hippocampal neurogenesis can be increased by antidepressants. Currently, several classes of antidepressants available in the market for the treatment of depression. Selective serotonin reuptake inhibitors (SSRIs) showed a lower risk than other antidepressants available in the market and show fewer chances of drug-drug interaction. Hence, SSRI frequently used for people with Alzheimer's and depression. Various studies demonstrated that various classes of antidepressants increase adult hippocampal neurogenesis. This recommends that antidepressants show positive effects on adult hippocampal neurogenesis. Over the past decade, clinical and animal studies have suggested that several neurotrophic growth factors play an essential role to improve the efficacy of antidepressants (Kajitani et al., 2012). Some studies reported that the treatment of depression in elderly patients improves memory, leading to improved cognitive performance, or execution and may reduce symptoms related to dementia. The treatment of AD (i.e., pharmacological, behavioral, or other modalities) can be used for the betterment of cognition result. A combination of pharmacological and non-pharmacological treatment strategies can be considered as a good approach to prolong cognitive decline and to improve cognition. (Byers & Yaffe, 2011).

## Conclusion

Epidemiological studies have shown a possible clinical association between depression and AD. Depressive symptoms can act as a potential risk factor for the progression of AD. Moreover, depressive symptoms in mild cognitive impairment as well as in normal aging are associated with cognitive performance diminution. Current evidence shows a common link between depression and AD pathogenesis. Furthermore, people with Alzheimer's can also develop mental health issues, especially depression or mood disorders. According to the discussed data, we propose that chronic inflammation, dysregulated HPA axis, altered neurotrophin signaling, diminished monoamines level, and increased oxidative stress can be recognized as a common neurobiological connection between AD and depression and also considered as a possible new target area for pharmacological treatment of these two disorders. Ultimately, more studies are required to help us understand the advanced neurobiological causes of the high comorbidity between depression and AD.

## Key facts

- Depression affects more than 350 million people worldwide and is expected to be the second leading cause of disability after heart disease by the year 2020.
- Currently, worldwide nearly 50 million people above the age of 60 years are diagnosed with AD and the projected figure is estimated to be 130 million in 2050.
- Many individual studies have recommended an attainable role for epigenetic changes in AD and depression etiology.
- There are some early BPSD observed in depressed patients.
- Since 2003, no new drug has been approved by the FDA for the treatment of AD. Hence currently, no disease-modifying treatment exists.
- Some studies reported that treatment of depression in elderly patients improves memory, leading to improved cognitive performance or execution and may reduce symptoms related to dementia.

## Summary points

- This chapter mainly focuses on the common molecular pathway and neurological link between AD and depression.
- A variety of reports and studies have shown a robust clinical linkage between chronic MDD and AD.
- Current evidence shows a common neurobiological connection between AD and depression such as chronic inflammation, dysregulated HPA axis, altered neurotrophin signaling, diminished monoamines level, and increased oxidative stress.
- A combination of pharmacological and non-pharmacological treatment strategy can be considered as a good approach to improve cognition and to reduce depressive behavior.
- Eventually, more studies are required to understand the advanced neurobiological causes of the high comorbidity between depression and AD.

## Mini-dictionary of terms

**Neurotoxic Aβ** Amyloid-beta oligomers or plaques accumulate between neurons and responsible for neuronal death.

**HPA axis dysregulation** Hypothalamic-pituitary-adrenal axis located in the brain and regulates various hormones. Any imbalance in secretion and regulation in the HPA axis is recognized as HPA axis dysregulation.

**Neuroprotective** Neuroprotective agents protect against neuronal injury and responsible for neuronal cell survival.

**Oxidative damage** Oxidative stress defined as a disturbance in the balance between the production of reactive oxygen species (free radicals) and antioxidants in the body which negatively affect cellular functions.

**Anterior cingulate cortex** In the human brain, the anterior cingulate cortex (ACC) is a frontal part of the corpus callosum and involved in complex cognitive functions.

**Endosomal trafficking** Endosomes are intracellular organelles and trafficking is the process by which proteins and other macromolecules are distributed throughout the cell.

**Long-term potentiation (LTP)** LTP related to synaptic strength and long-lasting strengthening of the response of a postsynaptic nerve cell.

## References


Abraham, W. C., & Williams, J. M. (2003). Properties and mechanisms of LTP maintenance. *The Neuroscientist*, *9*(6), 463–474. https://doi.org/10.1177/1073858403259119.

Adlard, P. A., Perreau, V. M., & Cotman, C. W. (2005). The exercise-induced expression of BDNF within the hippocampus varies across life-span. *Neurobiology of Aging*, *26*(4), 511–520. https://doi.org/10.1016/j.neurobiola
ging.2004.05.006.

Alzheimer's Association (2019). 2019 Alzheimer's disease facts and figures. *Alzheimer's & Dementia*, *15*(3), 321–387. https://doi.org/10.1016/J.JALZ.2019.01.010.

Andrews, S. J., Eramudugolla, R., Velez, J. I., Cherbuin, N., Easteal, S., & Anstey, K. J. (2017). Validating the role of the Australian National University Alzheimer's Disease Risk Index (ANU-ADRI) and a genetic risk score in progression to cognitive impairment in a population-based cohort of older adults followed for 12 years. *Alzheimer's Research & Therapy*, *9*(1), 16. https://doi.org/10.1186/s13195-017-0240-3.

Bao, A.-M., Meynen, G., & Swaab, D. F. (2008). The stress system in depression and neurodegeneration: Focus on the human hypothalamus. *Brain Research Reviews*, *57*(2), 531–553. https://doi.org/10.1016/j.brainresrev.2007.04.005.

Boku, S., Nakagawa, S., Toda, H., & Hishimoto, A. (2018). Neural basis of major depressive disorder: Beyond monoamine hypothesis. *Psychiatry and Clinical Neurosciences*, *72*(1), 3–12. https://doi.org/10.1111/pcn.12604.

Byers, A. L., & Yaffe, K. (2011). Depression and risk of developing dementia. *Nature Reviews Neurology*, *7*, 323–331. https://doi.org/10.1038/nrneurol.2011.60.

Cai, X., Shi, X., Zhang, X., Zhang, A., Zheng, M., & Fang, Y. (2017). The association between brain-derived neurotrophic factor gene polymorphism and migraine: A meta-analysis. *The Journal of Headache and Pain*, *18*(1), 13. https://doi.org/10.1186/s10194-017-0725-2.

Calsolaro, V., & Edison, P. (2016). Neuroinflammation in Alzheimer's disease: Current evidence and futuredirections. *Alzheimer's & Dementia*, *12*(6), 719–732. https://doi.org/10.1016/j.jalz.2016.02.010.

Canet, G., Chevallier, N., Zussy, C., Desrumaux, C., & Givalois, L. (2018). Central role of glucocorticoid receptors in Alzheimer's disease and depression. *Frontiers in Neuroscience*, *12*, 739. https://doi.org/10.3389/fnins.2018.00739.

Caraci, F., Spampinato, S. F., Morgese, M. G., Tascedda, F., Salluzzo, M. G., Giambirtone, M. C., et al. (2018). Neurobiological links between depression and AD: The role of TGF-β1 signaling as a new pharmacological target. *Pharmacological Research*, *130*, 374–384. https://doi.org/10.1016/j.phrs.2018.02.007.

Chandan, J. S., Thomas, T., Gokhale, K. M., Bandyopadhyay, S., Taylor, J., & Nirantharakumar, K. (2019). The burden of mental ill health associated with childhood maltreatment in the UK, using The Health Improvement Network database: A population-based retrospective cohort study. *The Lancet Psychiatry*, *6*(11), 926–934. https://doi.org/10.1016/S2215-0366(19)30369-4.

Diniz, B. S., Reynolds, C. F., Begley, A., Dew, M. A., Anderson, S. J., Lotrich, F., et al. (2014). Brain-derived neurotrophic factor levels in late-life depression and comorbid mild cognitive impairment: A longitudinal study. *Journal of Psychiatric Research*, *49*, 96–101. https://doi.org/10.1016/j.jpsychires.2013.11.004.

Drew, S. C. (2017). The case for abandoning therapeutic chelation of copper ions in Alzheimer's disease. *Frontiers in Neuroscience*, *11*, 317. https://doi.org/10.3389/fnins.2017.00317.

Duman, R. S., & Li, N. (2012). A neurotrophic hypothesis of depression: Role of synaptogenesis in the actions of NMDA receptor antagonists. *Philosophical Transactions of the Royal Society B: Biological Sciences*, *367*(1601), 2475–2484. https://doi.org/10.1098/rstb.2011.0357.

Elferink, M. W.-O., van Tilborg, I., & Kessels, R. P. C. (2015). Perception of emotions in mild cognitive impairment and Alzheimer's dementia: Does intensity matter? *Translational Neuroscience*, *6*(1), 139–149. https://doi.org/10.1515/tnsci-2015-0013.

Felger, J. C., Haroon, E., Patel, T. A., Goldsmith, D. R., Wommack, E. C., Woolwine, B. J., et al. (2020). What does plasma CRP tell us about peripheral and central inflammation in depression? *Molecular Psychiatry*, *25*(6), 1301–1311. https://doi.org/10.1038/s41380-018-0096-3.

García-Pérez, D., Ferenczi, S., Kovács, K. J., & Milanés, M. V. (2019). Distinct regulation pattern of Egr-1, BDNF and Arc during morphine-withdrawal conditioned place aversion paradigm: Role of glucocorticoids. *Behavioural Brain Research*, *360*, 244–254. https://doi.org/10.1016/j.bbr.2018.12.026.

Goda, A., Ohgi, S., Kinpara, K., Shigemori, K., Fukuda, K., & Schneider, E. B. (2013). Changes in serum BDNF levels associated with moderate-intensity exercise in healthy young Japanese men. *SpringerPlus*, *2*(1), 678. https://doi.org/10.1186/2193-1801-2-678.

Green, C., & Zhang, S. (2016). Predicting the progression of Alzheimer's disease dementia: A multidomain health policy model. *Alzheimer's & Dementia*, *12*(7), 776–785. https://doi.org/10.1016/J.JALZ.2016.01.011.

Haapakoski, R., Mathieu, J., Ebmeier, K. P., Alenius, H., & Kivimäki, M. (2015). Cumulative meta-analysis of interleukins 6 and 1β, tumour necrosis factor α and C-reactive protein in patients with major depressive disorder. *Brain, Behavior, and Immunity*, *49*, 206–215. https://doi.org/10.1016/j.bbi.2015.06.001.

Herrup, K. (2010). Reimagining Alzheimer's disease—An age-based hypothesis. *Journal of Neuroscience*, *30*(50), 16755–16762. https://doi.org/10.1523/JNEUROSCI.4521-10.2010.

Huang, W. J., Zhang, X., & Chen, W. W. (2016, May). Role of oxidative stress in Alzheimer's disease (review). *Biomedical Reports*, *4*, 519–522. https://doi.org/10.3892/br.2016.630.

Kajitani, N., Hisaoka-Nakashima, K., Morioka, N., Okada-Tsuchioka, M., Kaneko, M., Kasai, M., et al. (2012). Antidepressant acts on astrocytes leading to an increase in the expression of neurotrophic/growth factors: Differential regulation of FGF-2 by noradrenaline. *PLoS One*, *7*(12), e51197, https://doi.org/10.1371/journal.pone.0051197.

Kepe, V., Barrio, J. R., Huang, S.-C., Ercoli, L., Siddarth, P., Shoghi-Jadid, K., et al. (2006). Serotonin 1A receptors in the living brain of Alzheimer's disease patients. *Proceedings of the National Academy of Sciences*, *103*(3), 702–707. https://doi.org/10.1073/pnas.0510237103.

Lacor, P. N., Buniel, M. C., Furlow, P. W., Sanz Clemente, A., Velasco, P. T., Wood, M., et al. (2007). Aoligomer-induced aberrations in synapse composition, shape, and density provide a molecular basis for loss of connectivity in Alzheimer's disease. *Journal of Neuroscience*, *27*(4), 796–807. https://doi.org/10.1523/JNEUROSCI.3501-06.2007.

Laird, K. T., Krause, B., Funes, C., & Lavretsky, H. (2019). Psychobiological factors of resilience and depression in late life. *Translational Psychiatry*, *9*(1), 88. https://doi.org/10.1038/s41398-019-0424-7.

Lane, C. A., Hardy, J., & Schott, J. M. (2018). Alzheimer's disease. *European Journal of Neurology*, *25*(1), 59–70. https://doi.org/10.1111/ene.13439.

Lange, M., Castel, H., Le Fel, J., Tron, L., Maillet, D., Bernaudin, M., et al. (2019). How to assess and manage cognitive impairment induced by treatments of non-central nervous system cancer. *Neuroscience & Biobehavioral Reviews*, *107*, 602–614. https://doi.org/10.1016/j.neubiorev.2019.09.028.

Lardenoije, R., Iatrou, A., Kenis, G., Kompotis, K., Steinbusch, H. W. M., Mastroeni, D., et al. (2015). The epigenetics of aging and neurodegeneration. *Progress in Neurobiology*, *131*, 21–64. https://doi.org/10.1016/j.pneurobio.2015.05.002.

Lindholm, J. S. O., & Castrén, E. (2014). Mice with altered BDNF signaling as models for mood disorders and antidepressant effects. *Frontiers in Behavioral Neuroscience*, *8*, 143. https://doi.org/10.3389/fnbeh.2014.00143.

Liu, X., Jiao, B., & Shen, L. (2018). The epigenetics of Alzheimer's disease: Factors and therapeutic implications. *Frontiers in Genetics*, *9*, 579. https://doi.org/10.3389/fgene.2018.00579.

Long, J. M., & Holtzman, D. M. (2019). Alzheimer disease: An update on pathobiology and treatment strategies. *Cell*, *179*(2), 312–339. https://doi.org/10.1016/j.cell.2019.09.001.

Maes, M., Yirmyia, R., Noraberg, J., Brene, S., Hibbeln, J., Perini, G., et al. (2009). The inflammatory & neurodegenerative (I&ND) hypothesis of depression: Leads for future research and new drug developments in depression. *Metabolic Brain Disease*, *24*(1), 27–53. https://doi.org/10.1007/s11011-008-9118-1.

Marvel, C. L., & Paradiso, S. (2004). Cognitive and neurological impairment in mood disorders. *Psychiatric Clinics of North America*, *27*(1), 19–36. https://doi.org/10.1016/S0193-953X(03)00106-0.

Mastroeni, D., Grover, A., Delvaux, E., Whiteside, C., Coleman, P. D., & Rogers, J. (2010). Epigenetic changes in Alzheimer's disease: decrements in DNA methylation. *Neurobiology of Aging*, *31*(12), 2025–2037. https://doi.org/10.1016/j.neurobiolaging.2008.12.005.

Melón, L. C., Hooper, A., Yang, X., Moss, S. J., & Maguire, J. (2018). Inability to suppress the stress-induced activation of the HPA axis during the peripartum period engenders deficits in postpartum behaviors in mice. *Psychoneuroendocrinology*, *90*, 182–193. https://doi.org/10.1016/j.psyneuen.2017.12.003.

Mendiola-Precoma, J., Berumen, L. C., Padilla, K., & Garcia-Alcocer, G. (2016). Therapies for prevention and treatment of Alzheimer's disease. *BioMed Research International*, *2016*, 1–17. https://doi.org/10.1155/2016/2589276.

Miller, A. H., & Raison, C. L. (2016). The role of inflammation in depression: From evolutionary imperative to modern treatment target. *Nature Reviews Immunology*, *16*(1), 22–34. https://doi.org/10.1038/nri.2015.5.

Mossie, A., Kindu, D., & Negash, A. (2016). Prevalence and severity of depression and its association with substance use in jimma town, southwest Ethiopia. *Depression Research and Treatment*, *2016*, 1–7. https://doi.org/10.1155/2016/3460462.

Mravec, B., Horvathova, L., & Padova, A. (2018). Brain under stress and Alzheimer's disease. *Cellular and Molecular Neurobiology*, *38*(1), 73–84. https://doi.org/10.1007/s10571-017-0521-1.

Nautiyal, K. M., & Hen, R. (2017). Serotonin receptors in depression: From A to B. *F1000Research*, 6, 123. https://doi.org/10.12688/f1000research.9736.1.

Nerurkar, L., Siebert, S., McInnes, I. B., & Cavanagh, J. (2019). Rheumatoid arthritis and depression: An inflammatory perspective. *The Lancet Psychiatry*, *6*(2), 164–173. https://doi.org/10.1016/S2215-0366(18)30255-4.

Park, H., & Poo, M. (2013). Neurotrophin regulation of neural circuit development and function. *Nature Reviews Neuroscience*, *14*(1), 7–23. https://doi.org/10.1038/nrn3379.

Rao, U., & Chen, L.-A. (2008). Neurobiological and psychosocial processes associated with depressive and substance-related disorders in adolescents. *Current Drug Abuse Reviews*, 1(1), 68–80. Retrieved from http://www.ncbi.nlm.nih.gov/pubmed/19630707.

Réus, G. Z., Jansen, K., Titus, S., Carvalho, A. F., Gabbay, V., & Quevedo, J. (2015). Kynurenine pathway dysfunction in the pathophysiology and treatment of depression: Evidences from animal and human studies. *Journal of Psychiatric Research*, *68*, 316–328. https://doi.org/10.1016/j.jpsychires.2015.05.007.

Schiepers, O. J. G., Wichers, M. C., & Maes, M. (2005). Cytokines and major depression. *Progress in Neuro-Psychopharmacology and Biological Psychiatry*, *29*(2), 201–217. https://doi.org/10.1016/j.pnpbp.2004.11.003.

Shadrina, M., Bondarenko, E. A., & Slominsky, P. A. (2018). Genetics factors in major depression disease. *Frontiers in Psychiatry*, *9*, 334. https://doi.org/10.3389/fpsyt.2018.00334.

Šimić, G., Babić Leko, M., Wray, S., Harrington, C. R., Delalle, I., Jovanov-Milošević, N., et al. (2017). Monoaminergic neuropathology in Alzheimer's disease. *Progress in Neurobiology*, *151*, 101–138. https://doi.org/10.1016/j.pneurobio.2016.04.001.

Surendranathan, A., Su, L., Mak, E., Passamonti, L., Hong, Y. T., Arnold, R., et al. (2018). Early microglial activation and peripheral inflammation in dementia with Lewy bodies. *Brain*, *141*(12), 3415–3427. https://doi.org/10.1093/brain/awy265.

Theodore, W. H., Wiggs, E. A., Martinez, A. R., Dustin, I. H., Khan, O. I., Appel, S., et al. (2012). Serotonin 1A receptors, depression, and memory in temporal lobe epilepsy. *Epilepsia*, *53*(1), 129–133. https://doi.org/10.1111/j.1528-1167.2011.03309.x.

Tichauer, J. E., Flores, B., Soler, B., Eugenín-von Bernhardi, L., Ramírez, G., & von Bernhardi, R. (2014). Age-dependent changes on TGFβ1 Smad3 pathway modify the pattern of microglial cell activation. *Brain, Behavior, and Immunity*, *37*, 187–196. https://doi.org/10.1016/j.bbi.2013.12.018.

Tönnies, E., & Trushina, E. (2017). Oxidative stress, synaptic dysfunction, and Alzheimer's disease. *Journal of Alzheimer's Disease*, *57*(4), 1105–1121. https://doi.org/10.3233/JAD-161088.

Vose, L. R., & Stanton, P. K. (2017). Synaptic plasticity, metaplasticity and depression. *Current Neuropharmacology*, *15*(1), 71–86. https://doi.org/10.2174/1570159X14666160202121111.

Wakefield, C. E., Butow, P. N., Aaronson, N. A., Hack, T. F., Hulbert-Williams, N. J., Jacobsen, P. B., et al. (2015). Patient-reported depression measures in cancer: A meta-review. *The Lancet Psychiatry*, *2*(7), 635–647. https://doi.org/10.1016/S2215-0366(15)00168-6.

Wook Koo, J., Labonté, B., Engmann, O., Calipari, E. S., Juarez, B., Lorsch, Z., et al. (2016). Essential role of mesolimbic brain-derived neurotrophic factor in chronic social stress-induced depressive behaviors. *Biological Psychiatry*, *80*(6), 469–478. https://doi.org/10.1016/j.biopsych.2015.12.009.

Wuwongse, S., Chang, R. C.-C., & Law, A. C. K. (2010). The putative neurodegenerative links between depression and Alzheimer's disease. *Progress in Neurobiology*, *91*(4), 362–375. https://doi.org/10.1016/J.PNEUROBIO.2010.04.005.

Xu, J., Church, S. J., Patassini, S., Begley, P., Waldvogel, H. J., Curtis, M. A., et al. (2017). Evidence for widespread, severe brain copper deficiency in Alzheimer's dementia. *Metallomics*, *9*(8), 1106–1119. https://doi.org/10.1039/C7MT00074J.

Yokokura, M., Terada, T., Bunai, T., Nakaizumi, K., Kato, Y., Yoshikawa, E., et al. (2019). Alterations in serotonin transporter and body image-related cognition in anorexia nervosa. *NeuroImage: Clinical*, *23*, 101928. https://doi.org/10.1016/j.nicl.2019.101928.

Chapter 9

# Comorbidities of depression and Parkinson's disease


Tanvi Pingale and Girdhari Lal Gupta

*Department of Pharmacology, Shobhaben Pratapbhai Patel School of Pharmacy & Technology Management, SVKM'S NMIMS, Mumbai, India*


## List of abbreviations

| | |
|---|---|
| **AADC** | L-aromatic amino acid decarboxylase |
| **AMPA** | adenosine monophosphate |
| **BH4** | tetrahydrobiopterin |
| **Cldn5** | claudin-5 |
| **DA** | dopamine |
| **DAT** | dopamine transporter |
| **DOC** | DOPA decarboxylase |
| **IDO** | indoleamine-2,3-dioxygenase |
| **KAT-II** | kynurenine aminotransferase II |
| **KMO** | kynurenine monooxygenase |
| **KYNU** | kynureninase |
| **LC** | locus coeruleus |
| **NA** | noradrenaline |
| **NAergic** | noradrenergic |
| **NE** | norepinephrine |
| **NMDA** | *N*-methyl-D-aspartate |
| **PD** | Parkinson's disease |
| **PHA** | phenylalanine hydroxylase |
| **ROS** | reactive oxygen species |
| **SNpc** | substantia nigra pars compacta |
| **TH** | tyrosine hydroxylase |
| **VMAT** | vesicular monoamine transporter |
| **6OHDA** | 6-hydroxy dopamine |
| **5HT** | 5-hydroxytryptophan |
| **3HK** | h-hydroxy kynurenine |
| **3-HAA** | 3-hydroxyanthranilic acid |
| **3HAAD** | 3-hydroxyanthranilic acid dehydrogenase |

## Introduction

Around 6.2 million people in the world are affected by Parkinson's disease (PD) which has progressive devastating effects on dopamine neurons. The progressive loss of dopaminergic neurons of substantia nigra pars compacta (SNpc) region of midbrain results in decreased dopamine release. The deficit of striatal dopamine affects the basal nuclei motor activity which leads to tremor, rigidity, bradykinesia, and postural instability. Apart from the dopaminergic neurons of SNc and striatal dopamine, other areas such as the mesolimbic pathway, the raphe nuclei, and the locus coeruleus are also affected which are linked to PD nonmotor symptoms. Nonmotor symptoms are noticeable much before the motor symptoms in patients and affect the patient's life quality (Dickson, 2018). Nonmotor symptoms in PD can include sleep apnoea, pain, depression, anxiety, and fatigue. Patients also show cognitive dysfunction which can later evolve into dementia along with conceded thinking, memory, and language. Due to loss of nigrostriatal dopamine and mesocortical depletion, patients suffer from working memory and functional impairment. In PD patients, symptoms of depression and anxiety are allied with each other. These behavioral symptoms lead to poor quality of life in influencing their daily events and become additional predominant in their progressive stage (Fontoura et al., 2017; Reich & Savitt, 2019). Table 1 shows general features on Parkinson's.

## Depression in Parkinson's disease

The occurrence of depression is common in patients with Parkinson's disease. Depression occurrence and prevalence varies in Parkinson's patient due to difference in diagnostic criteria. In spite of high prevalence and influence on patients' life, no precise diagnosis method is available for depression associated with PD. The Diagnostic and Statistical Manual of Mental Disorders (DSM—V) described criteria for the diagnosis of major depression in PD patients (Habib, Gold, & Chrousos, 2001). Symptom overlapping must be addressed properly as it becomes difficult to diagnose depressed patients as symptoms such as sleep disturbance, depression and cognitive deficit in PD patients (Fontoura et al., 2017). Many studies reported a surge in depression incidence in Parkinson patients to occur before the commencement of motor indications, signifying that depression accompanying PD is the result of the degeneration of neurons in the brain. Long-term study showed that the progression of PD was elevated in a patient with


**The Neuroscience of Depression. https://doi.org/10.1016/B978-0-12-817933-8.00008-6**

**TABLE 1 Other features of Parkinson's disease.**

| Features | Description in short |
|---|---|
| Thinking difficulties | Experience cognitive problems (dementia) and thinking difficulties |
| Depression and emotional changes | Experience depression and mood disorders sometimes in the very early stages |
| Anxiety, panic attacks | Feeling of fear |
| Apathy | Lack of interest, enthusiasm, or concern |
| Chewing and eating problems | Late-stage Parkinson's disease affects the muscles in the mouth, making chewing difficult. This can lead to choking and poor nutrition |
| Swallowing problems/ sialorrhea | Saliva may accumulate in the mouth due to slowed swallowing, leading to drooling |
| Sleep problems and sleep disorders | Sleep problems including waking up frequently throughout the night, waking up early or falling asleep during the day |
| Constipation | Mainly due to a slower digestive tract |
| Bladder problems | Including being unable to control urine or having difficulty urinating |
| Sexual dysfunction | Decrease in sexual desire or performance |
| Blood pressure changes | Feel dizzy or lightheaded when you stand due to a sudden drop in blood pressure (orthostatic hypotension) |
| Fatigue | Lose energy and experience fatigue, especially later in the day |
| Smell dysfunction/ hyposmia | Difficulty identifying certain odors or the difference between odors |
| Pain and fatigue | Pain either in specific areas of bodies or throughout their bodies |
| Dysautonomia | Failure of the sympathetic or parasympathetic components of the ANS |
| Excessive sweating | Sweating too much (hyperhidrosis) due to ANS dysfunction |
| Seborrhea | Red, itchy rash, and white scales in scalp causes dandruff |
| Insomnia | REM behavior disorder |
| Restless legs syndrome | Periodic limb movements in sleep |
| Decreased visual motion perception | Decreased visual contrast and color discrimination |
| Tingling effects | Abnormal sensations, such as paresthesia |

depression which considered depression as a major risk factor for neurodegeneration in PD. Other factors of depression such as cognitive impairment and motor symptoms have occurred in Parkinson even before 40 years of age (Le, Sayana, & Jankovic, 2014). Discrete features have been seen in depression related to Parkinson which is not linked to PD. Various indications such as irritability, sorrow, sadness, and bitterness are more common while guiltiness, failure feeling and suicidal attempts are less common in PD patients with depression. Reports also showed that few percentages of Parkinson patients are affecting major depression and most of them are observed with light depressive symptoms. Such indications gathered less attention than motor instabilities and are not appropriately treated, which leads to the risk of morbidity, disability & decrement in patients' life quality (Dallé & Mabandla, 2018; Youdim & Bakhle, 2006).

## Pathophysiology of stress leading to depression

Stress comprises abrupt unpredictable physical, physiological, social, and environmental alterations felt by the human mind. Early stages of stress include psychiatric disorders such as generalized anxiety, depression and/or learning deficits. There are several hypotheses about stress leading to depression; most include a change in the concentration of neurotransmitters such as epinephrine, glutamate, dopamine, noradrenaline, serotonin, and γ-aminobutyric acid (GABA) (Vanle et al., 2018). Neurotransmitters transmit signals from the brain to different parts of the body with the help of nerve cells. Depending on the activation of the receptor most neurotransmitters can be classified either excitatory or inhibitory. Catecholamine modulatory neurotransmitters such as dopamine can be both inhibitory and excitatory which is mainly involved in PD. Another catecholaminergic neurotransmitter, noradrenaline/norepinephrine, participate in various activities associated with mood, motivation, cognitive functions, and emotion (Shah, Gilani, Sharma, & Vohora, 2005). The deficiency of norepinephrine is associated with depression. Another neurotransmitter serotonin is essential for the maintenance of a stable mood. Many studies have shown the connection of serotonin exhaustion with depression, so altered neurotransmitter release or level plays a pivotal role in the progression of a psychiatric disorder such as depression and Parkinson' disease (Dallé & Mabandla, 2018; Zheng et al., 2005).

## Biochemical scheme of depression

Many theories debated regarding the possible pathways that hint to depression associated with Parkinson's and their relation to other psychiatric symptoms like anxiety and apathy. Further acknowledged theories like Schildkraut's theories suggested that depression is related to the discrepancy in monoaminergic neurotransmitters. This theory is mostly based on the first- and second-generations' antidepressants mechanism of action, which blocks the norepinephrine and/or serotonin presynaptic reuptake and enhancing their transmission. Serotonin activates 5-HT receptors, a family which has 14 subtypes. Other studies indicate 5-$HT_{1A}$, 5-$HT_7$ are involved in depressive behavior. 5-$HT_{2C}$antagonists can also be used for the treatment of both PD and depression. Also, due to their role in the modification of dopamine release, they are deemed to be linked to the depressive behavior of patients (Loane et al., 2013). Along with serotonin and norepinephrine, dopamine may also play a role in depression progression in PD patients. Depressive behavior in PD patients can be efficiently improved using pramipexole, a dopamine agonist. Stimulation $D_3$ receptors located in the mesolimbic system are involved to improve such kind of situations and actively contributing to mood and behavior regulation (Fallon, Bor, Hampshire, Barker, & Owen, 2017).

## Coexistence of Parkinson' disease and depression

Parkinson's disease and depression both are CNS disorders, along with alteration in level of brain monoamines level, i.e., DA, 5HT, and NA, respectively, and formation of reactive oxygen species (ROS) (Marsh, 2013). In presynaptic nerve terminals from substantia nigra (SN), L-phenylalanine gets converted to tyrosine with the help of the enzyme phenylalanine hydroxylase. Tyrosine hydroxylase helps to convert tyrosine to L-DOPA and finally, L-DOPA is converted into dopamine (DA) by DOPA decarboxylase enzyme. After synthesis, DA is packaged into synaptic vesicles via vesicular monoamine transporter (VMAT) and dopamine transporter (DAT) and stored there until its release into the synapse during neurotransmission. In the synapse, this DA gets metabolized via monoamine oxidase (MAO) and carboxy-o-methyl transferase (COMT) into toxic metabolite 3,4-dihydroxy phenylacetic acid (DOPAC), 3-methoxy tyramine (3MT) and homovanillic acid (HVA) (Fig. 1).

In PD, oxidative stress increases as the level of toxic metabolite increases. It leads to mitochondrial dysfunction which inhibits complex I in ATP production, due to this more ROS gets generated and bind with protein α-synuclein present in the substantia nigra. Thereafter, the α helix structure of α-synuclein protein gets converted to β-plated sheet and form aggregation of an α-synuclein protein called Lewy bodies. This Lewy bodies interfere with DA synthesis and decrease TH level. It also activates microglia, ultimately leading to the production of pro-inflammatory cytokines like IL-1, IL-6, TNF-α, etc., and induce neuroinflammation (Mishra, Singh, & Shukla, 2018). In depression, peripherally pro-inflammatory cytokines increase the indoleamine 2,3 dioxygenase (IDO) (tryptophan degrading enzyme) level. Tryptophan is a serotonin (5-HT) precursor, transported to the brain via BBB with the help of L-type amino acid transporters. In astrocytes, L-kynurenine with the help of KAT-II is converted to kynurenic acid (KA), an NMDA receptor antagonist that act as an anti-excitotoxic and provides neuroprotection. In microglia, L-kynurenine is converted to quinolinic acid via 3HK and 3HAA using KMO and KYNU enzymes, respectively, and 3-HAAD helps to convert 3-HAA. In stress, quinolinic acid enhances ROS production and release of pro-inflammatory cytokines. These cytokines interfere with BH4, DAT, VMAT, and inhibit DA synthesis at presynaptic terminals (Bansal, Singh, Parhar, Kuhad, & Soga, 2019).

## 5HT receptors and Parkinson's disease

5-$HT_{1A}$ are the G-protein coupled receptors that cause inhibition of adenylate cyclase and prevent c-AMP formation, further diminish the activity of protein kinase A (PK-A). Besides, 5-$HT_{1A}$ receptors activated by the intracellular $K^+$ efflux initiated by G-protein and inhibit the firing of

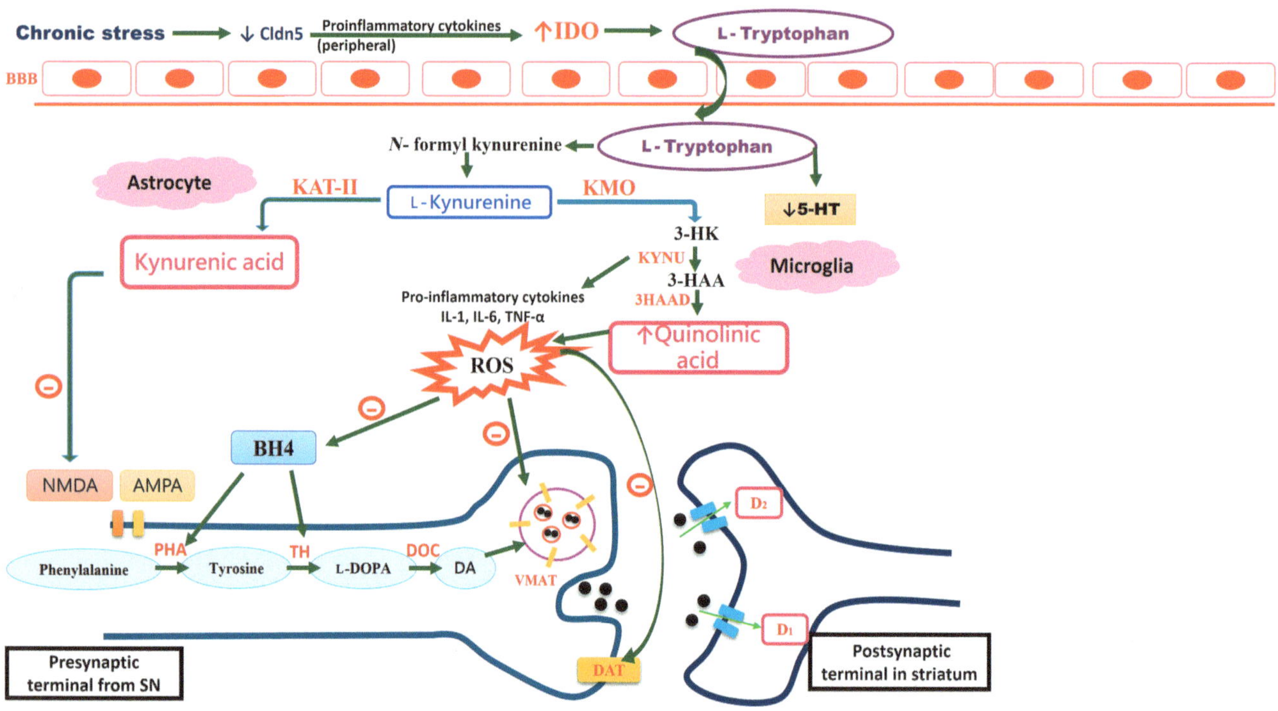


FIG. 1 Molecular links between depression and Parkinson's disease pathophysiology. In chronic stress, decrease Cldn5 level and peripherally proinflammatory cytokines increases the activity of indoleamine 2,3 dioxygenase (IDO) (tryptophan degrading enzyme). Tryptophan, the precursor of serotonin is transported to the brain via BBB with the help of nonspecific competitive L-type amino acid transporters. After crosses BBB tryptophan is metabolized into L-kynurenine via *N*-formyl kynurenine. In astrocytes with the help of KAT-II, L-kynurenine converted to kynurenic acid (KA). KA, NMDA receptor antagonist, acts as an anti-excitotoxic and provides neuroprotection. On the other hand, L-kynurenine is converted to 3HK and 3HAA using KMO and KYNU enzymes, respectively. This 3-HAAD helps to convert 3-HAA into quinolinic acid in microglia. Formation of ROS occurs, as an increase in the concentration of quinolinic acid and releases of pro-inflammatory cytokines. These cytokines interfere with BH4, DAT, VMAT and inhibit DA synthesis at presynaptic terminals in SN.

the neuron. These receptors are highly expressed in limbic, septum and raphe nuclei regions. Apart from these regions, moderate-to-low densities of 5-$HT_{1A}$ receptors are observed in thalamus, cerebral cortex, basal ganglia, and hypothalamus. 5-$HT_{1A}$ receptors work not only on presynaptic receptors found in the raphe nuclei but also in postsynaptic receptors present in other brain areas. These receptors negatively regulate stomatal 5-HT neuron activity and dendritic region of 5-HT neurons present in the raphe. Through this, 5-$HT_{1A}$ auto-receptors also regulate the general movement of the serotonergic system. 5-$HT_{1A}$ receptors are located on postsynaptic membranes of neurons or nerve terminals (heteroreceptors) innervated by 5-HT neurons (De Deurwaerdère & Di Giovanni, 2017). Loss of 5-HT neurons in raphe nuclei leads to a reduction in the density of 5-HT transporters in the forebrain region. The upregulation of postsynaptic 5-$HT_{1A}$ and 5-$HT_{2A}$ receptors occurs after the 5-HT receptors deficit. A clinical study demonstrated that 5-HT neuron dysfunction is more prominent in PD patients with depression than those without depression which indicates the close association of Parkinson and serotonergic system (Nicholson & Brotchie, 2002).

## Role of serotonin in depression associated with Parkinson's disease

Serotonergic cells in raphe nuclei divided into two groups: the rostral group which appears mainly in the forebrain, amygdala, the basal ganglia, the medial cerebral cortex and the caudal group which present in caudal brainstem, spinal cord. Serotonergic systems attributed several functions like including cognition, emotion, and motor behavior; thus, altered serotonergic neurotransmission commonly associated with PD and may contribute to the motor and nonmotor features (Weinstock, 2008). Evidence from preclinical and clinical studies has suggested that striatal serotonergic terminals release dopamine by nonphysiological mechanisms and contribute to the progression of levodopa-induced dyskinesias (LIDs). In PD, the degeneration of 5-HT transporters in raphe nuclei leads to a decrease in 5-HT transmission. However, dopamine produced from systemically administered levodopa can be stored and released via 5-HT neurons. For example, the 6-OHDA model comprises severe nigrostriatal dopaminergic neurodegeneration in rat, it has been revealed that 5-HT transporters have ability to reuptake of striatal levodopa-derived DA. Further, it has been observed that striatal dopamine reuptake and metabolism can be modified by monoamine transporter inhibitors like selective serotonin reuptake inhibitors (SSRIs) to improve motor symptoms of PD. Inhibition of 5-HT transporters provides a new choice of treatment which enhances the antiparkinson effect of drugs with the potential of an increase in extracellular DA level (Politis & Niccolini, 2015). The motor and nonmotor symptoms of Parkinson are the consequence of altered transmission of the serotonergic signal which is also due to weak modulation of DA. Emotional stress-induced in day-to-day lifestyle may further increase motor symptoms and damages the nigrostriatal pathway in PD patients. In early-onset PD, an increase in stress level in the brain exacerbates neuronal damage and availability of dopamine and 5-HT which can damage neurons synergistically or separately. The depression has become the indicator of Parkinson's and its timely treatment might help in the management of Parkinson's disease (Cai et al., 2015).

## Clinically available serotonergic drugs

Different anxiolytic agents such as buspirone, tandospirone, eltoprazine and fluoxetine have been evaluated for depression in Parkinson's disease. The antipsychotics and antidepressants like clozapine and quetiapine and mirtazapine also exhibited selective action. Each of the compounds comprises not only 5-$HT_{2A}$ antagonistic action but also activity at other different targets. The experiments performed to find treatment for depression in Parkinson patients displayed low efficacy but outcomes were positive. Therefore, further studies are required to find precise targets for the pharmacological action of these drugs in the depression of PD. The SSRI displayed their effect by indirect agonistic effect on 5-$HT_{1A}$ and/or 5-$HT_{1B/1D}$ receptors. Buspirone at the dose of 20 mg per day diminished LID deprived of deteriorating Parkinsonian debility in 7 out of 10 Parkinson patients. In PD, estimation of the role of 5-$HT_{1A}$ receptors is done through PET imaging study. Acute exposure of buspirone before L-DOPA treatment resulted in abridged LID through one-third without any effect on motor signs. The selected dose was bind to 5-$HT_{1A}$ rather than dopamine $D_2$ receptors. Thus, long-term assessment of buspirone requisite for its benefit determination (Fasipe, 2019; Mori, Murata, Matsuo, Takemoto, & Mine, 2014).

## Noradrenergic systems in the central nervous system

The Noradrenergic System (NAergic systems) comprises two important neurons associated with medulla oblongata (with A1/C1 nucleus) and rostral (with A2/C2 nucleus) situated in the pons. The caudal NAergic system not only corresponds to diffuse groups of cells in the lower brainstem but also the rostral NAergic system corresponds mainly to the locus coeruleus (LC, A6) located in the pons. Noradrenaline act as a strong neuromodulator & displays its effect on synaptic transmission, altering membrane of potential, the excitability of neurons and synaptic plasticity via adrenergic receptors (ARs). Two subtypes of ARs, viz., Alpha-ARs (namely α1 and α2) and beta-ARs (namely β1, β2, and β3) have been described till date which is located in regions such as striatum and substantial. Diverse subtypes of ARs are linked to different G proteins. By the mechanism of action, α1 and β show excitatory effects by postsynaptic ARs and α2 show inhibitory effects by presynaptic ARs (Arcos et al., 2003; Delaville, de Deurwaerdère, & Benazzouz, 2011).

## The locus coeruleus (LC)

The LC nucleus is located in all mammalian species and signifies the main source of NA for the central nervous system. The activity of LC neurons influenced by synaptic inputs. The Locus Coerulus noradrenergic system modulates cortical and subcortical areas by its projections. Various biochemical evaluations conducted are still unable to discriminate LC nucleus from its surroundings owing to its small size. Retrograde and anterograde amnesia have been linked to numerous afferent and efferent of LC neurons. The review of the document also displayed the orbitofrontal and anterior cingulate cortices become the origin of the glutamatergic afferent drive to the locus coer uleus (Benarroch, 2009). Diverse central functions modulate through LC by NA release into regions such as hippocampus, thalamus, substantia nigra, neocortex and subthalamic nucleus. L-DOPA plays a vital role in the synthesis of dopamine as it is the precursor of dopamine but due to its biochemistry, it also becomes a metabolic precursor of noradrenaline (NA), which indicates its relevance in PD. In a normal physiological process, amino acid tyrosine converted into L-Dopa catalyzed by tyrosine hydroxylase (TH). The conversion of synthetic L-DOPA to dopamine is catalyzed by L-aromatic amino acid decarboxylase (AADC) (Henry, Fox, Peggs, Crossman, & Brotchie, 1999). NA synthesizes from DA in the cytosol with the help of DA beta-hydroxylase and moved into vesicles by the vesicular monoamine transporter (VMAT) and degraded by monoamine oxidase A. Extracellular reuptake of NA is carried out by noradrenaline transporters and its release is controlled by activation of auto-inhibitory α2 ARs. Apart from that, extracellular DA is also transported inside the neuron by norepinephrine Transporter (NET) (Aston-Jones & Cohen, 2005). The LC NAergic system helps to perform various activities such as maintenance of synaptic plasticity, modulation of pain, and equilibrium of energy. LC neurons fire into both modes-tonic as well as phasic modes. Tonic activity with regular discharge patterns is high during wakefulness and reduces during sleep which plays a pivotal role in the sleep-waking cycle. NAergic receptors stimulation induces variation in vigilance. Environmental stimuli also increase tonic discharge which provokes behavioral excitement and exploratory behavior. LC neurons help to become more focused and performs the accurate task by reducing tonic firing and respond task associated with physical activity. Thus, it plays a crucial role in attention, stress and arousal (Aston-Jones & Cohen, 2005).

## Role of NA in PD symptoms

Although a variety of crucial biochemical variations is relatively common in Parkinson's patients, a unique role is performed by each neurotransmitter in the progression of the disease. Postmortem studies exhibited an explicit biochemical pattern that exists in a patient with PD. In PD the neocortical monoamine terminal loss (including NA), and to elucidate its potential role in the complex cognitive and affective impairment can be evaluated in vivo using PET imaging studies. This part is focused on exploring the part of NA exhaustion in specific motor and nonmotor symptoms linked to Parkinson's in human patients as well as animals.

(a) Motor symptoms: A combination of DA and NA scarcity affects PD patients more as motor symptoms become wider in action. L-Dopa's antiparkinsonian action also potentiates α2ARs agonism. It is also suggested that reserpine-induced tremor and rigidity are produced due to alpha-adrenergic mechanisms and in particular α2ARs. NAergic system modulates akathisia and tremor which are improved by drugs belong to class α2ARs agonist or a βARs antagonist.

Different studies stated the participation of the LC NA pathway in freezing, the "on–off" phenomenon, wearing-off effect, postural instability, and akinesia (Hill & Brotchie, 1999).

(b) Nonmotor symptoms: The major concern of any preclinical and clinical studies of Parkinson's disease is to improve motor symptoms by medication and less attention was given to nonmotor symptoms. In PD the nonmotor symptoms may appear many years before the onset of the motor complications. Not only dopaminergic but also noradrenergic role should be kept in mind for the treatment of Parkinson. The loss of dopaminergic and noradrenergic innervation of the limbic system is associated with neurobehavioral and cognitive difficulties, including depression, anxiety, dementia, and attention deficits (Wu, Liscic, Kim, Sorbi, & Yang, 2017). Degeneration of striatal dopaminergic neurons and alteration of noradrenergic neurons in the prefrontal cortex of the rat model may be caused due to emotional disturbance and anxiety. In naïve rats, these findings are constant as NAergic controls the sensitivity to external stimuli such as anxiety. In elderly patients, reports suggested that enhanced anxiety levels might be due to loss of NAergic and DAnergic neurons (Chaudhuri & Odin, 2010).

Depression is the most underrated nonmotor symptom of PD. The prevalence of depression associated with PD is about 40% found in some studies. The reason is still unknown but in some patients, pramipexole (dopamine agonist) improves the symptom which might be due to exhaustion of three monoamines. In a unilateral rat model of PD, DSP-4 was used to induce combined depletion of DA and NA while parachlorophenylalanine was used to deplete 5-HT which enhanced immobility time in Porsolt test (Tibar et al., 2018). The depletion of all three monoaminergic systems showed behavior with depressive symptoms. In humans, binding of [$^{11}$C]RTI-32 in LC and different limbic regions displayed an equal affinity for DAT and NET but a comparatively low affinity for 5-HT transporters. Thus, in depression associated with PD, the DAergic and NAergic innervations are involved and might explicitly play a crucial role in apathy, which is a major characteristic of depression. Histology of LC neurons exhibited that morphology is more affected in depression associated PD patients than other PD patients. Loss of cortical and limbic NAergic neurons due to LC mutilation leads to a decrease in NA count along with metabolites and associated enzymes in Parkinson. All these variations are prominent in dementia patients than those who lack dementia (Remy, Doder, Lees, Turjanski, & Brooks, 2005).

## Indication for the interaction between DA and NA in PD

Few studies establish DA role in noradrenergic PD modulation. Apart from the increased LC NA neuron firing rate, the density of β ARs in the different brain regions such as the cerebral cortex, hippocampus, thalamic nuclei are increased in 6-OHDA rats with a lesion which proves that diminishing DAergic activity enhances NAergic action (Wang, Zhang, Liu, Wu, & Wang, 2009). Anatomical supervision of DA revealed that inputs from the NAergic system advance transmission of the DAergic system. Ponzio, Hallman, and Jonsson (1981) proved that in the cerebral cortex and the striatum functional activity of the DAergic nerve terminals can be regulated by NAergic nerve terminals originating from the LC which further appears excitatory (Ponzio et al., 1981). Different pharmacological activities showed that blocking of α1ARs reduces the responsiveness of the mesolimbic system. Most of the antipsychotic drugs possess D2 receptor and α1ARs blocking capacity which also make them capable of improving psychosis condition by blocking postsynaptic and presynaptic

DA receptors with the help of their α1ARs antagonistic action. In the striatum DA extracellular levels elevated by stimulation of α2ARs and β1Ars. The β1ARs and α2Ars act in neostriatal neurons by regulation of DARPP-32 phosphorylation. Gi proteins activated by α2ARs inhibit D1 receptor-mediated Gs/PKA signaling and enhances D2 receptor/Gi signaling in striatopallidal neurons. However, a study conducted by Reches and Meiner in 1992 did not report any changes in the synthesis of DA and the dorsal NA bundle (DNEB) release. Reches and Meiner's study helped to understand the effect of LC on DA transmission in striatum not facilitated by the DNEB (Hara et al., 2010). LC NArgic neuron exerts its neuroprotective effect on SNc DAergic neuron which was derived from NA's role in the survival of DAergic neurons in MPTP studies of primates and mice. Results of the studies indicated that MPTP induced damage to DAergic neuron enhanced when rats were pretreated with DSP4. The obstruction of NET leads to an increase in extracellular NA and prevent damage of DA neurons from MPTP. The modification of the LC NAergic system involved in pathologies such as attention deficit, stress disorder, sleep disorder, hyperactivity disorder, etc. Due to its capability to modulate signals in various pathologies, it becomes a target for pharmacological treatment (Rommelfanger & Weinshenker, 2007).

## The locus coeruleus noradrenergic system in PD

In PD, the LC is often severely affected in neurodegenerative diseases and a rising amount of biochemical and anatomical information has strained. It is reported that in the PD brain as there is a drastic decrease in NAergic cell bodies, significant depletions (>80%) of NA concentration. Reports of positive emission tomography (PET) stated that levels of [$^{11}$C] RTI-32 have been abridged in numerous areas including areas with poorly innervated by DAergic fibers. Also, NAergic synapses can change the size and shape of presynaptic, postsynaptic components and morphological alterations of the mitochondria. An improvement in the density of α1 and β1 ARs in the 6-OHDA rat model and also in the prefrontal cortex of Parkinson's patients exhibited. Results obtained from animal and human studies implicated that LC NAergic impairment contributes to PD's motor and nonmotor symptoms(Aston-Jones & Cohen, 2005).

## Conclusion

Psychological indications often coexist with Parkinson's disease and due to poor diagnosis they are mostly remaining untreated. Studies showed that depression associated with Parkinson's get worse with time. The psychological symptoms like anxiety and depression are mostly observed with Parkinson's disease. Depression is the usual and devastating feature of Parkinson's disease mostly affecting 40%–50% of PD patients. Diagnosis of depression is not easy due to overlapping of clinical manifestation with Parkinson's disease. Depletion in brain catecholamines such as serotonin, norepinephrine, and dopamine in a patient with Parkinson's disease usually leads to depression. General strategies for the treatment of depression should be improved as they are not as useful in depression associated with Parkinson's disease. Evidence from various studies indicated that dopamine agonists might have some efficacy against mild depression associated with PD and antidepressants exerting its action of 5-HT and noradrenaline physiology can help to reduce depression in Parkinson's patients. Refining knowledge of depression associated with Parkinson's becomes very important as it has inoculated much-needed attention into neuropharmacological physiology.

## Key facts of comorbidities of depression and Parkinson's disease

- Around 6.2 million people in the world are affected by Parkinson's disease (PD) with the progressive loss of dopaminergic neurons of substantia nigra pars compacta (SNpc) region of the midbrain.
- The occurrence of depression is common in patients with Parkinson's disease.
- Clinical management of PD requires attention to both motors as well as nonmotor symptoms.
- In the clinical management of disease, depressive symptoms are unrecognized or even when recognized it remains untreated.
- Very few studies gathered attention to find a link between depression and Parkinson's disease.
- Understanding of biochemical pathways involved in depression associated with Parkinson's disease helps to find treatment strategy before leading to the advanced stage.

## Mini-dictionary of terms

**Bradykinesia** Slowness of movement and is one of the cardinal manifestations of Parkinson's disease.

**Levodopa-induced dyskinesia (LID)** Dyskinesia associated with levodopa which involves hyperkinetic movements includes chorea, dystonia and athetosis.

**Locus coeruleus (LC)** It is a nucleus present in the pons of the brainstem involved in responses in panic and stress condition.

**"on–off" phenomenon** It occurs in Parkinson's disease which refers to a switch between mobility and immobility. 'Off' time means reemerge of tremor, rigidity, and slow movement when levodopa is not working well.

**Wearing-off effect** Symptoms of Parkinson's disease reappear or deteriorate before the next dose of levodopa is due and recover after taking of next dose.

## Summary points

- Current chapters focus on biochemical pathways involved in the progression of depression in Parkinson's disease.
- Neurotransmitters such as epinephrine, dopamine, noradrenaline, serotonin and GABA are involved in the advancement of depression.
- Preclinical studies indicated that altered serotonin transmission in Parkinson's disease worsens the motor and nonmotor features.
- The formation of Lewy bodies interfere with dopamine synthesis and activates microglia to produce proinflammatory cytokines like IL-1, IL-6, TNF-α, etc. which increase the indoleamine 2,3 dioxygenase (IDO) (tryptophan degrading enzyme) level and leads to depression.

## References


Arcos, D., Sierra, A., Nuñez, A., Flores, G., Aceves, J., & Arias-Montaño, J. A. (2003). Noradrenaline increases the firing rate of a subpopulation of rat subthalamic neurones through the activation of α1-adrenoceptors. *Neuropharmacology*, *45*(8), 1070–1079. https://doi.org/10.1016/S0028-3908(03)00315-0.

Aston-Jones, G., & Cohen, J. D. (2005). Adaptive gain and the role of the locus coeruleus-norepinephrine system in optimal performance. *Journal of Comparative Neurology*, *493*(1), 99–110. https://doi.org/10.1002/cne.20723.

Bansal, Y., Singh, R., Parhar, I., Kuhad, A., & Soga, T. (2019). Quinolinic acid and nuclear factor erythroid 2-related factor 2 in depression: Role in neuroprogression. *Frontiers in Pharmacology*, *10*, . https://doi.org/10.3389/fphar.2019.00452.

Benarroch, E. E. (2009). The locus ceruleus norepinephrine system: Functional organization and potential clinical significance. *Neurology*, *73*(20), 1699–1704. https://doi.org/10.1212/WNL.0b013e3181c2937c.

Cai, L., Li, R., Tang, W. J., Meng, G., Hu, X. Y., & Wu, T. N. (2015). Antidepressant-like effect of geniposide on chronic unpredictable mild stress-induced depressive rats by regulating the hypothalamus-pituitary-adrenal axis. *European Neuropsychopharmacology*, *25*(8), 1332–1341. https://doi.org/10.1016/j.euroneuro.2015.04.009.

Chaudhuri, K. R., & Odin, P. (2010). The challenge of non-motor symptoms in Parkinson's disease. *Progress in Brain Research*, *184*, 325–341. https://doi.org/10.1016/S0079-6123(10)84017-8.

Dallé, E., & Mabandla, M. V. (2018). Early life stress, depression and Parkinson's disease: A new approach. *Molecular Brain*, *11*, 18. https://doi.org/10.1186/s13041-018-0356-9.

De Deurwaerdère, P., & Di Giovanni, G. (2017). Serotonergic modulation of the activity of mesencephalic dopaminergic systems: Therapeutic implications. *Progress in Neurobiology*, *151*, 175–236. https://doi.org/10.1016/j.pneurobio.2016.03.004.

Delaville, C., de Deurwaerdère, P., & Benazzouz, A. (2011). Noradrenaline and Parkinson's disease. *Frontiers in Systems Neuroscience*, *5*, 31. https://doi.org/10.3389/fnsys.2011.00031.

Dickson, D. W. (2018). Parkinsonism and related disorders neuropathology of Parkinson disease. *Parkinsonism and Related Disorders*, *46*, S30–S33. https://doi.org/10.1016/j.parkreldis.2017.07.033.

Fallon, S. J., Bor, D., Hampshire, A., Barker, R. A., & Owen, A. M. (2017). Spatial structure normalises working memory performance in Parkinson's disease. *Cortex*, *96*, 73–82. https://doi.org/10.1016/j.cortex.2017.08.023.

Fasipe, O. J. (2019). The emergence of new antidepressants for clinical use: Agomelatine paradox versus other novel agents. *IBRO Reports*, *6*, 95–110. https://doi.org/10.1016/j.ibror.2019.01.001.

Fontoura, J. L., Baptista, C., Pedroso, F. d. B., Pochapski, J. A., Miyoshi, E., & Ferro, M. M. (2017). Depression in Parkinson's disease: The contribution from animal studies. *Parkinson's Disease*, *2017*, 9124160. https://doi.org/10.1155/2017/9124160.

Habib, K. E., Gold, P. W., &Chrousos, G. P. (2001). Neuroendocrinology of stress.Endocrinology and Metabolism Clinics of North America, 30 (3), 695–728; vii–viii. Retrieved from http://www.ncbi.nlm.nih.gov/pubmed/11571937

Hara, M., Fukui, R., Hieda, E., Kuroiwa, M., Bateup, H. S., Kano, T. , et al. (2010). Role of adrenoceptors in the regulation of dopamine/DARPP-32 signaling in neostriatal neurons. *Journal of Neurochemistry*, *113*(4), 1046–1059. https://doi.org/10.1111/j.1471-4159.2010.06668.x.

Henry, B., Fox, S. H., Peggs, D., Crossman, A. R., &Brotchie, J. M. (1999). The α 2-adrenergic receptor antagonist idazoxan reduces dyskinesia and enhances anti-parkinsonian actions of L-dopa in the MPTP-lesioned primate model of Parkinson's disease. Movement Disorders, 14(5), 744–753. https://doi.org/10.1002/1531-8257(199909)14:5<744::AID-MDS1006>3.0.CO;2-7

Hill, M. P., & Brotchie, J. M. (1999). The adrenergic receptor agonist, clonidine, potentiates the anti-parkinsonian action of the selective κ-opioid receptor agonist, enadoline, in the monoamine-depleted rat. *British Journal of Pharmacology*, *128*(7), 1577–1585. https://doi.org/10.1038/sj.bjp.0702943.

Le, W., Sayana, P., & Jankovic, J. (2014). Animal models of Parkinson's disease: A gateway to therapeutics? *Neurotherapeutics*, *11*, 92–110. https://doi.org/10.1007/s13311-013-0234-1.

Loane, C., Wu, K., Bain, P., David, J., Piccini, P., & Politis, M. (2013). Serotonergic loss in motor circuitries correlates with severity of action-postural tremor in PD. *Neurology*, *80*(20), 1850–1855. https://doi.org/10.1212/WNL.0b013e318292a31d.

Marsh, L. (2013). Depression and parkinson's disease: Current knowledge topical collection on movement disorders. *Current Neurology and Neuroscience Reports*, *13*(12). https://doi.org/10.1007/s11910-013-0409-5.

Mishra, A., Singh, S., & Shukla, S. (2018). Physiological and functional basis of dopamine receptors and their role in neurogenesis: Possible implication for Parkinson's disease. *Journal of Experimental Neuroscience*, *12*, 1179069518779829. https://doi.org/10.1177/1179069518779829.

Mori, M., Murata, Y., Matsuo, A., Takemoto, T., & Mine, K. (2014). Chronic treatment with the 5-HT1A receptor partial agonist tandospirone increases hippocampal neurogenesis. *Neurology and Therapy*, *3*(1), 67–77. https://doi.org/10.1007/s40120-013-0015-0.

Nicholson, S. L., & Brotchie, J. M. (2002). 5-Hydroxytryptamine (5-HT, serotonin) and Parkinson's disease—Opportunities for novel therapeutics to reduce the problems of levodopa therapy. *European Journal of Neurology*, *9*, 1–6. https://doi.org/10.1046/j.1468-1331.2002.00272.x.

Politis, M., & Niccolini, F. (2015). Serotonin in Parkinson's disease. *Behavioural Brain Research*, *277*, 136–145. https://doi.org/10.1016/j.bbr.2014.07.037.

Ponzio, F., Hallman, H., &Jonsson, G. (1981). Noradrenaline and dopamine interaction in rat brain during development. Medical Biology, 59(3), 161–169. Retrieved from http://www.ncbi.nlm.nih.gov/pubmed/6796787

Reich, S. G., & Savitt, J. M. (2019). Parkinson's disease. *Medical Clinics of North America*, *103*, 337–350. https://doi.org/10.1016/j.mcna.2018.10.014.

Remy, P., Doder, M., Lees, A., Turjanski, N., & Brooks, D. (2005). Depression in Parkinson's disease: Loss of dopamine and noradrenaline innervation in the limbic system. *Brain*, *128*(6), 1314–1322. https://doi.org/10.1093/brain/awh445.

Rommelfanger, K. S., & Weinshenker, D. (2007). Norepinephrine: The redheaded stepchild of Parkinson's disease. *Biochemical Pharmacology*, *74*, 177–190. https://doi.org/10.1016/j.bcp.2007.01.036.

Shah, Z. A., Gilani, R. A., Sharma, P., & Vohora, S. B. (2005). Attenuation of stress-elicited brain catecholamines, serotonin and plasma corticosterone levels by calcined gold preparations used in Indian system of medicine. *Basic & Clinical Pharmacology & Toxicology*, *96*(6), 469–474. https://doi.org/10.1111/j.1742-7843.2005.pto_10.x.

Tibar, H., El Bayad, K., Bouhouche, A., Haddou, E. H. A. B., Benomar, A., Yahyaoui, M. , et al. (2018). Non-motor symptoms of Parkinson's disease and their impact on quality of life in a cohort of Moroccan patients. *Frontiers in Neurology*, *9*(Apr). https://doi.org/10.3389/fneur.2018.00170.

Vanle, B., Olcott, W., Jimenez, J., Bashmi, L., Danovitch, I., & Ishak, W. W. (2018). NMDA antagonists for treating the non-motor symptoms in Parkinson's disease. *Translational Psychiatry*, *8*, 117. https://doi.org/10.1038/s41398-018-0162-2.

Wang, T., Zhang, Q. J., Liu, J., Wu, Z. H., & Wang, S. (2009). Firing activity of locus coeruleus noradrenergic neurons increases in a rodent model of parkinsonism. *Neuroscience Bulletin*, *25*(1), 15–20. https://doi.org/10.1007/s12264-009-1023-z.

Weinstock, M. (2008). The long-term behavioural consequences of prenatal stress. *Neuroscience and Biobehavioral Reviews*, *32*, 1073–1086. https://doi.org/10.1016/j.neubiorev.2008.03.002.

Wu, S.-L., Liscic, R. M., Kim, S., Sorbi, S., & Yang, Y.-H. (2017). Non-motor symptoms of Parkinson's disease. *Parkinson's Disease*, *2017*, 4382518. https://doi.org/10.1155/2017/4382518.

Youdim, M. B. H., & Bakhle, Y. S. (2006). Monoamine oxidase: Isoforms and inhibitors in Parkinson's disease and depressive illness. *British Journal of Pharmacology*, *147*, S287–S296. https://doi.org/10.1038/sj.bjp.0706464.

Zheng, H., Gal, S., Weiner, L. M., Bar-Am, O., Warshawsky, A., Fridkin, M. , et al. (2005). Novel multifunctional neuroprotective iron chelator-monoamine oxidase inhibitor drugs for neurodegenerative diseases: In vitro studies on antioxidant activity, prevention of lipid peroxide formation and monoamine oxidase inhibition. *Journal of Neurochemistry*, *95*(1), 68–78. https://doi.org/10.1111/j.1471-4159.2005.03340.x.

Chapter 10

# Understanding the relationship between depression and alcohol among students

Daniel Teixeira dos Santos[a], Guilherme de Souza Paulo Filho[b], Marco Aurélio dos Santos Carvalho[b], and Vinícius Medeiros Henriques[b]

[a]*Department of Neurology, Hospital of Clinics of Porto Alegre, Porto Alegre, Rio Grande do Sul, Brazil,* [b]*School of Medicine, Universidade do Grande Rio, Rio de Janeiro, Brazil*

## Introduction

Depression and alcohol use disorders are common and often comorbidly associated with social, legal, academic, financial, and health burdens. Global estimates indicate that, in 2016, 4.4% of the population suffered from a major depressive disorder (World Health Organization, 2017). Furthermore, also in 2016, the prevalence of current drinking among individuals aged 15 or older was 43%, while heavy episodic drinkers amounted to 18.2%, harmful alcohol use was 2.5%, and alcohol dependence accounted for 2.6% (World Health Organization, 2019). The harmful use of alcohol resulted in 5.3% of all worldwide deaths in 2016, being greater than those of tuberculosis (2.3%), diseases of the gastrointestinal tract (4.5%), and road injuries (2.5%) (World Health Organization, 2019). Likewise, in 2015, depressive disorders led to over 50 million years lived with disability globally, being ranked as the single largest contributor to nonfatal health loss (World Health Organization, 2017). Table 1 shows the most common risk factors independently for depression and alcohol use disorders and Table 2 shows common risk factors for both conditions. As shown in Fig. 1, a variety of factors may modulate the relationship between depression and alcohol such as social support, suicidal ideation, anxiety, gender, social motives, coping motives, impulsivity and ruminative thinking. These correlations will be further reviewed in this chapter.

Depression and alcohol have a higher prevalence among students in comparison to the general population. A study observed that 26.6% of undergraduates have depressive symptoms (dos Santos, Nazário, Freitas, Henriques, & de Paiva, 2019), while those who fit alcohol abuse criteria ranged from 12.52% to 28.4% (Blanco et al., 2008; de Andrade et al., 2012; dos Santos et al., 2019). Meanwhile, among middle and high school students, estimates indicate that 7.3% are current drinkers (Guo et al., 2016) and that 17.7% have depressive symptoms (Kaur et al., 2014).

## Examining the association between alcohol and depression in students

Table 3 provides data regarding some of the most relevant studies addressing alcohol and depression in students.

### Alcohol-related variables and depression: Which ones do correlate and among who?

The relationship between alcohol and depression is different among student subpopulations and types of alcohol-related variables. The association between depression and alcohol use (frequency and/or intensity of alcohol consumption) is different between undergraduates or middle and high school students. Research encompassing the latter two groups found consistent results indicating a positive association between those conditions (Johannessen et al., 2017; Kaur et al., 2014; Rojas et al., 2012; Tomlinson & Brown, 2012; Torikka et al., 2001), whereas some studies with university students indicate a positive correlation (Pan et al., 2015), but the majority of other studies don't show any relationship (Armeli et al., 2014; El Ansari et al., 2014; Gonzalez et al., 2009; Grant et al., 2009; Heinz et al., 2009; Piumatti, 2017; Sebena et al., 2012; Weitzman, 2004). These findings indicate that alcohol use is more consistently associated with depression in middle and high school students in comparison to undergraduates.

Binge drinking is associated with depressive symptoms in middle and high school students (Johannessen et al., 2017; Tomlinson & Brown, 2012). Among university students, a positive relationship between those variables was found in a largely sampled cross-sectional study (Pedrelli et al., 2016). Also, a cohort investigation identified that depressive symptoms predicted binge drinking in women after a 1-week follow-up, although binge drinking did not

The Neuroscience of Depression. https://doi.org/10.1016/B978-0-12-817933-8.00004-9

**TABLE 1 Risk factors to alcohol use disorder and depression.**

| Depression | Alcohol use disorder |
|---|---|
| Family history | Graduated from high school |
| Female gender | Low income |
| Low income | Male gender |
| Native American race | Mood disorder |
| Not graduated from high school | Native American race |
| Serious medical illness | Other substance abuse |
| Single individuals | Single individuals |
| Stressful life events | Younger adults (18–30 years old) |
| Substance abuse | |
| Younger adults (18–30 years old) | |

predict those symptoms after the same follow-up period (Mushquash et al., 2013). However, other studies failed to find any cross-sectional relationship between those variables (El Ansari et al., 2014; Gonzalez et al., 2009; Weitzman, 2004), conflicting findings which may be explained by differences among studies' methodology, but, as shown by Mushquash et al. (2013), depressive symptoms seems to be the causal variable in the relationship between this variable and binge drinking when this correlation is present.

Both alcohol-related problems and alcohol-related medical diagnoses have been investigated among university students and show a positive correlation to depression, which is to be expected, as the former is one of the diagnostic criteria for the latter. Most studies, cross-sectional or longitudinal, agree that different levels of alcohol-related medical diagnoses (e.g., abuse, dependence, alcohol use

**TABLE 2 Risk factors in common between alcohol use disorder and depression.**

| |
|---|
| Native American race |
| Low income |
| Substance abuse |
| Younger adults (18–30 years old) |
| Alcohol use disorder (for depression) |
| Depression (for alcohol use disorder) |

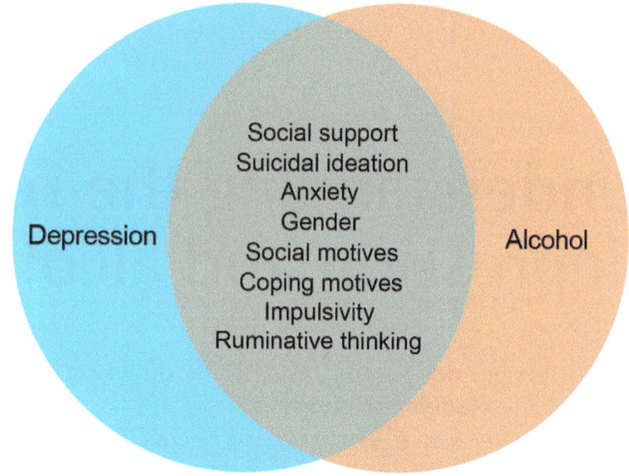


FIG. 1 Associated factors between alcohol and depression among students.

disorder) are correlated to depression or its severity (dos Santos et al., 2019; El Ansari et al., 2014; Huurre et al., 2010; Weitzman, 2004), while only one study found no correlation between alcohol abuse with depression (dos Santos et al., 2019), difference probably attributable to different cut-off points and instruments of measurement. Alcohol-related problems correlate with depression independently of the method of measurement (Armeli et al., 2014; El Ansari et al., 2014; Gonzalez et al., 2009; Grant et al., 2009; Heinz et al., 2009; Kenney, Abar, O'Brien, Clark, & LaBrie, 2016; Merrill et al., 2014; Sebena et al., 2012; Weitzman, 2004).

## Interventions regarding alcohol and depression among students

A brief motivational intervention had no effect in reducing alcohol-related problems after a 12-month follow-up among students with alcohol misuse, some effect on reduction in weekly drinks was observed among depressed men and on both weekly drinks and heavy frequency among nondepressed women. These results indicate that depressive status and gender moderate response to environmental interventions regarding alcohol use (Merrill et al., 2014). Likewise, brief web-based intervention regarding alcohol use, depression, or both did not reduce depressive symptoms, alcohol use, or alcohol-related problems in a 1-month follow-up among students with comorbid alcohol-use disorders and depression (Geisner et al., 2015).

## Exploring the causal relationships between alcohol and depression among students

The relationship between alcohol use and depression is complex as, sometimes, depression can be both a result

**TABLE 3** Association between depression and alcohol-related variables in students.

| Source | Country | *N* | Setting | Design | Outcomes |
|---|---|---|---|---|---|
| Torikka, Kaltiala-Heino, Rimpelä, Rimpelä, and Rantanen (2001) | Finland | 16.464 | High School | Cross-sectional | Positive correlation between alcohol use (1 question) and depression severity (13-Item BDI) |
| Gonzalez, Bradizza, and Lorraine Collins (2009) | United States | 91 | University | Cross-sectional | Positive correlation between drinking to cope with negative affect (DCS) and depression severity (BDI-II) |
| | | | | | Positive correlation between alcohol-related problems (YAACQ) and depression severity (BDI-II) |
| | | | | | No correlation between alcohol use (modified NIAAA) and depression severity (BDI-II) |
| | | | | | No correlation between binge drinking (1 question) and depression severity (BDI-II) |
| Heinz, Veilleux, and Kassel (2009) | United States | 182 | University | Cohort | Positive correlation between baseline alcohol-related problems (HAS) and baseline depression severity (IDD) |
| | | | | | Positive correlation between baseline drinking to cope with negative affect (DMQ) and baseline depression severity (IDD) |
| | | | | | No correlation between baseline alcohol use (2 questions) and baseline depression severity (IDD) |
| Grant, Stewart, and Mohr (2009) | Canada | 146 | University | Cohort | Positive correlation between baseline drinking to cope with negative affect (DMQ-R) and baseline anhedonic depression (MASQ) |
| | | | | | Positive correlation between baseline alcohol-related problems (RAPI) and baseline anhedonic depression (MASQ) |
| | | | | | No correlation between baseline alcohol use (1 question) and baseline anhedonic depression severity (MASQ) |
| El Ansari, Sebena, and Stock (2014) | England | 322 | University | Cross-sectional | Positive correlation between alcohol dependence (CAGE) and depression severity (modified BDI) |
| | | | | | Positive correlation between alcohol-related problems (CAGE) and depression severity (modified BDI) |
| | | | | | No correlation between alcohol use (1 question) and depression severity (modified BDI) |
| | | | | | No correlation between binge drinking (1 quuestion) and depression severity (modified BDI) |
| Johannessen, Andersson, Bjørngaard, and Pape (2017) | Norway | 6.238 | Middle School | Cross-sectional | Positive correlation between alcohol consumption before 15 years (1 question) and depression severity (DMI) |
| | | | | | Positive correlation between alcohol use (1 question) and depression severity (DMI) |
| | | | | | Positive correlation between binge drinking (1 question) and depression severity (DMI) |
| Pedrelli, Borsari, Lipson, Heinze, and Eisenberg (2016) | United States | 61.561 | University | Cross-sectional | Positive correlation between binge drinking (1 question) and depression (PHQ-9) |

*Continued*

**TABLE 3 Association between depression and alcohol-related variables in students—cont'd**

| Source | Country | *N* | Setting | Design | Outcomes |
|---|---|---|---|---|---|
| Merrill, Reid, Carey, and Carey (2014) | United States | 330 | University | Clinical Trial | Brief motivational intervention reduced weekly drinks (DDQ) and heavy frequency (DDQ) in non-depressed (CES-D) women after a 12 month follow-up, having no effect on depressed women |
| | | | | | Brief motivational intervention reduced weekly drinks (DDQ) in depressed (CES-D) men after a 12 month follow-up, having no effect on heavy frequency (DDQ) and in non-depressed men |
| | | | | | Brief motivational intervention did not reduce alcohol-related problems (RAPI) after a 12 month follow-up |
| | | | | | Positive correlation between alcohol-related problems (RAPI) after a 12 month follow-up and depression (CES-D) at baseline |
| Kenney, Jones, and Barnett (2015) | United States | 233 | University | Cohort | Positive correlation between baseline alcohol-related problems (RAPI) and baseline depression (CES-D) among women |
| | | | | | Positive correlation between alcohol-related problems at 1 month follow-up (RAPI) and baseline depression (CES-D) among women |
| | | | | | No correlation between alcohol-related problems (RAPI) at 6 month follow-up and baseline depression (CES-D) among women |
| dos Santos et al. (2019) | Brazil | 659 | University | Cross-sectional | Positive correlation between alcohol dependence (AUDIT) and depression (PHQ-9) |
| | | | | | No correlation between alcohol abuse (AUDIT) and depression (PHQ-9) |
| Piumatti (2017) | Italia | 247 | University | Cohort | No correlation between alcohol use (unstandardized question) and depression severity at baseline and 6 months follow-up (PHQ-2) |
| Pan et al. (2015) | China | 9.010 | University | Cross-sectional | Positive correlation between alcohol use (unstandardized question) and depression (BDI-II) |
| Geisner, Varvil-Weld, Mittmann, Mallett, and Turrisi (2015) | United States | 311 | University | Clinical Trial | Brief web-based intervention regarding alcohol use, depression or both did not reduce depressive symptoms (BDI) or alcohol use (AUDIT) in a 1 month follow-up among students with comorbid alcohol-use disorders (AUDIT) and depression (BDI) |
| | | | | | Brief web-based intervention regarding alcohol use, depression or both did not reduce alcohol-related problems (RAPI) in a 1 month follow-up among students with comorbid alcohol-use disorders (AUDIT) and depression (BDI) |
| Armeli et al. (2014) | United States | 844 | University | Cohort | Positive correlation between 3-month aggregate drinking to cope with negative affect (DMQ) and depression severity (BDI-II) |
| | | | | | Positive correlation between 3-month aggregate alcohol-related problems (YAACQ) and depression severity (BDI-II) |
| | | | | | No correlation between 3-month aggregate alcohol use (unstandardized questions) and depression severity (BDI-II) |

**TABLE 3** Association between depression and alcohol-related variables in students—cont'd

| Source | Country | *N* | Setting | Design | Outcomes |
|---|---|---|---|---|---|
| Mushquash et al. (2013) | Canada | 200 | University | Cohort | Depressive symptoms (CES-D) predicted binge drinking (modified NIAAA) over a 1 week follow-up among women |
| | | | | | Binge drinking (modified NIAAA) did not predict depressive symptoms (CES-D) over a 1 week follow-up among women |
| Tomlinson and Brown (2012) | United States | 400 | Middle School | Cross-sectional | Positive correlation between alcohol use and depressive symptoms (both: unstandardized questions) |
| | | | | | Positive correlation between binge drinking and depressive symptoms (both: unstandardized questions) |
| Huurre et al. (2010) | Finland | 1.471 | Middle School | Cohort | Positive correlation between adolescent's depressive symptoms (unstandardized questions) and alcohol use disorder or abuse (AUDIT) in males, but not females, after a 16 year follow-up |
| Weitzman (2004) | United States | 27.409 | University | Cross-sectional | Positive correlation between alcohol-related problems (unstandardized questions) and depression (modified SF-36) |
| | | | | | Positive correlation between alcohol abuse (unstandardized questions compared to DSM-IV criteria) and depression (modified SF-36) |
| | | | | | No correlation between binge drinking (unstandardized questions) and depression (modified SF-36) |
| | | | | | No correlation between alcohol use (unstandardized questions) and depression (modified SF-36) |
| Rojas et al. (2012) | Chile | 2.597 | High School | Cross-sectional | Positive correlation between alcohol use (7 questions), depression and depression severity (BDI-II) |
| Kaur et al. (2014) | Malaysia | 28.738 | Middle and High School | Cross-sectional | Positive correlation between alcohol use (1 question) and depression (DASS21) |
| Sebena, El Ansari, Stock, Orosova, and Mikolajczyk (2012) | European Countries | 2.529 | University | Cross-sectional | Positive correlation between alcohol-related problems (CAGE) and depression (BDI-II) |
| | | | | | No correlation between alcohol use (unstandardized questions) and depression (BDI-II) |

of and a reason for alcohol use and both could be the result of common social, environmental, and psychological predisposing factors. While depression as a withdrawal symptom and a consequence of drinking is indicated by studies involving clinical samples (Torikka et al., 2001) and a longitudinal cohort of 25 years of follow-up since birth (Tomlinson & Brown, 2012), relevant research regarding middle and high school and university students demonstrated that depression predicted short- and long-term alcohol use disorders or alcohol-related problems, and usually not the opposite (Huurre et al., 2010; Kenney et al., 2016; Merrill et al., 2014; Mushquash et al., 2013). Even though not a definitive conclusion due to conflicting results in the literature, these latter findings suggest that the relationship between alcohol and depression among students could be best explained by the self-medication hypothesis, a theory for substance use that hypothesizes that individuals may use alcohol to alleviate or cope with negative affect, which has environmental plausibility among students as this population often treats alcohol consumption as normative, thus establishing a proper scenario for its use as a coping motive.

## Influence of mental health on alcohol and depression

Table 4 provides data regarding studies associating mental health variables with alcohol and depression in students.

### Alcohol and depression: How do they relate to suicidal ideation?

The majority of studies addressing the relationship between alcohol use and alcohol-related problems, depression, and suicidal ideation among students are cross sectional and identified a consistent positive correlation between these variables among undergraduates (Dvorak et al., 2013; Lamis et al., 2016) and middle and high school Chinese students taken from a large multicenter study (Guo et al., 2016). One of those studies investigated the role of hopelessness and social support in the relationship between alcohol, depression, and suicidal ideation. While the former acted as a risk factor for the investigated conditions, the latter was found to be a protective factor, findings that corroborate the three-step theory, in which pain and hopelessness would generate suicidal ideation while connectedness, represented by social support, would act as a protective factor (Lamis et al., 2016). Also, middle and high school students with poor or average relationships with their teachers had higher levels of alcohol-related problems (Guo et al., 2016), which may be linked to Lamis et al. (2016) findings regarding the lack of social support, however, there is still need of longitudinal studies to clarify the nature of this relationship.

### The relationship between alcohol, depression, and anxiety

Anxiety symptoms relate to alcohol and depression in different ways. Reports suggest that social avoidant undergraduates were more likely to drink to cope with depression, which would explain the rise in alcohol problems at 12 months follow-up. However, other coping motives, including drinking to cope with anxiety, were not associated with increases in alcohol-related problems (Collins et al., 2018). In partial contradiction to this finding, another study with university students found that social anxiety was positively correlated to both drinking to cope with anxiety and depression and that students with high social anxiety had stronger urges to use alcohol immediately after a brief social interaction, followed by an exercise that led the student to ruminate about the aforementioned interaction (Potter et al., 2016).

## Influence of sociodemographic variables on alcohol and depression

Table 5 provides data regarding some of the relevant studies addressing the influence of socio-demographic factors on alcohol and depression in students.

### The role of gender in the relationship between alcohol and depression

Studies have found differences in the relationship between gender, alcohol, and depression. For instance, a Norwegian study with high school students found that the severity of depressive symptoms was tied to increase in alcohol consumption among both males and females ( Johannessen et al., 2017). Among college students, one study found that drinking to cope mediated the relationship between depressive symptoms and alcohol-related problems in both genders (Pedrelli et al., 2016), while another identified that drinking to cope was associated with alcohol-related problems for women and that social motives predicted those problems in males (Kenney et al., 2015). A cross-sectional study found a similar degree of association between major depressive disorder and frequency of heavy episodic drinking in both genders. The intensity differed though, with women having a higher intensity of heavy episodic drinking (Pedrelli et al., 2016).

A brief motivational intervention showed a reduction of alcohol consumption among college women regardless of the severity of depressive symptoms, having a positive outcome only among highly depressed college men, indicating that the relationship between depressive symptom's severity and alcohol consumption varies among genders (Merrill et al., 2014). Not only increased depression severity was correlated to higher alcohol consumption among men and the opposite has been described to women, but also depressive symptoms were correlated to compulsive drinking and daily alcohol consumption in females and not in males (Pedrelli et al., 2011).

The authoritativeness quality in the paternal relationship had different effects on genders. In men, it was found that an authoritative parental style was correlated with having a negative bond with their fathers. This correlation did not exist in women. A negative parental bond was shown to increase the likelihood of depressive symptoms in both genders which were associated with increases in alcohol-related problems. Interestingly, a negative bond with one's

**TABLE 4** Association between mental health variables on the association between alcohol and depression.

| Source | Country | *N* | Setting | Design | Outcomes |
|---|---|---|---|---|---|
| Lamis, Ballard, May, and Dvorak (2016) | United States | 2.034 | University | Cross-sectional | Positive correlations among depressive symptoms (BDI-II), hopelessness (BHS), alcohol problems (RAPI), and SI (MSSI) |
| | | | | | Depressive symptoms (BDI-II) and Hopelessness (BHS) were associated with high level of alcohol problems (RAPI) |
| | | | | | Social support inversely correlated with depressive symptoms (BDI-II), hopelessness (BHS), alcohol problems (RAPI), and SI (MSSI) |
| Dvorak, Lamis, and Malone (2013) | United States | 1.100 | University | Cross-sectional | Positive correlation between alcohol use (AUDIT) and suicide proneness (LAS-SF) - depressive symptoms (BDI-II) |
| | | | | | At high levels of alcohol use (AUDIT) the association between depressive symptoms (BDI-II) and suicide proneness was stronger. |
| Guo et al. (2016) | China | 105.752 | Middle and High School | Cross sectional | Students with suicide ideation were at higher risk of having alcohol- related problems (1 question) |
| | | | | | Students with higher scores on depressive symptoms (CES-D) were at higher risk of having alcohol-related problems (1 question) |
| | | | | | Students with average and poor relationship with teachers were positively associated with alcohol-related problems (1 question) |
| Collins, Thompson, Sherry, Glowacka, and Stewart (2018) | Canada | 219 | University | Cohort | Social avoidance (LSAS-S) predicted increases in coping with depression drinking motives (M DMQ-R) 6 months later, which in turn predicted increases in alcohol-related problems (RAPI) at 12 months |
| | | | | | Conformity motives and coping with anxiety motives (M DMQ-R) were not associated with increases in alcohol -related problems (RAPI) at 12 months |
| Potter, Galbraith, Jensen, Morrison, and Heimberg (2016) | United States | 83 | University | Cross-sectional | Positive association between social anxiety (SIAS) and drinking to cope with anxiety and depression motives (MDMQ-R) during the follow-up week |
| | | | | | Students with high social anxiety (SIAS) reported a stronger urge to use alcohol (Single question) immediately after the manipulation * than those with low social anxiety |

mother correlated negatively with depressive symptoms (Patock-Peckham & Morgan-Lopez, 2007).

## Other factors related to alcohol and depression

The context of drinking was also related to depressive symptoms. A cross-sectional study found a positive correlation between heavy drinking when alone and depressive symptoms. Social heavy drinkers were more akin to non-heavy drinkers than people who drank heavily while alone in regard to emotional coping and depressive symptoms (Christiansen et al., 2002). Another study reiterated the relationship between drinking alone and depressive symptoms, finding a positive correlation between them. Drinking at parties was a predictor for alcohol-related problems while no association was found when the subject drank with family and/or partner (Keough et al., 2015).

**TABLE 5** **Association between socio-environmental factors on the association between alcohol and depression.**

| Source | Country | *N* | Setting | Design | Outcomes |
|---|---|---|---|---|---|
| Johannessen et al. (2017) | Norway | 6.238 | High School | Cross-sectional | Positive correlation between levels of depressive symptoms (HSC-25) and early onset of alcohol consumption, frequent consumption and intoxication (unstandardized questions) |
| | | | | | The severity of depressive symptoms (HSC-25) was associated with an increase in early onset of alcohol consumption, frequent consumption and intoxication (unstandardized questions) |
| | | | | | Increasing severity of anxiety symptoms was only associated with an increase in early onset of alcohol consumption, frequent consumption and intoxication (unstandardized questions) among girls |
| Pedrelli et al. (2016) | United States | 163 | University | Cross-sectional | Depressive symptoms (CES-D) in males were positively correlated with greater coping motives (DMQ) and alcohol-related problems (B-YAACQ) |
| | | | | | Association between depressive symptoms (CES-D) and alcohol-related problems (BYAAC-Q) was mediated by coping motives (DMQ) in both males and females |
| | | | | | Frequency of heavy drinking (CDDR) did not mediate the association between coping motives (DCM) and alcohol-related problems (BYAAC-Q) |
| Kenney et al. (2016) | United States | 233 | University | Cohort | Depressed (CES-D) female students had bigger levels of alcohol-related negative consequences (RAPI) than non-depressed ones |
| | | | | | There was a faster decline in negative alcohol-related consequecences (RAPI) in the depressed group |
| | | | | | No difference was found between baseline weekly alcohol consumption (DDQ) and heavy episodic drinking (DDQ) between the depressed and non depressed (CES-D) |
| Kenney et al. (2015) | United States | 614 | University | Cohort | Depressive symptoms (CES-D) in men at college entry independently and directly predicted greater levels of alcohol consequences (YAAPST) |
| | | | | | College coping motives (DMQ) was associated with college alcohol consequences (YAAPST) for women, college social motives (DMQ) was associated with consequences for men |
| | | | | | Depressed mood at matriculation was not predictive of alcohol use (unstandardized question) or alcohol expectancies (B-CEOA) for men or women |
| Lambe, Mackinnon, and Stewart (2015) | Canada | 100 | University | Cohort | Drinking to cope with depression (M DMQ-R) mediated the association between dyadic conflict and alcohol-related problems (RAPI) |
| Keough, O'Connor, Sherry, and Stewart (2015) | Canada | 295 | University | Cross-sectional | Positive correlation found between depressive symptoms (MASQ) and alcohol-related problems (RAPI) but not alcohol use (unstandardized question) |
| | | | | | Positive correlation between depressive symptoms (MASQ) and drinking alone (unstandardized question) |
| | | | | | Drinking at parties was a positive predictor of alcohol use (unstandardized question) and alcohol-related problems (RAPI) |
| | | | | | Negative correlation between depressive symptoms (MASQ) and drinking with family and with a partner (unstandardized question) |

**TABLE 5** Association between socio-environmental factors on the association between alcohol and depression—cont'd

| Source | Country | *N* | Setting | Design | Outcomes |
|---|---|---|---|---|---|
| Merrill et al. (2014) | United States | 330 | University | Clinical Trial | Brief motivational intervention reduced weekly drinks (DDQ) and heavy frequency (DDQ) in non-depressed (CES-D) women after a 12 month follow-up, having no effect on depressed women |
| | | | | | Brief motivational intervention reduced weekly drinks (DDQ) in depressed (CES-D) men after a 12 month follow-up, having no effect on heavy frequency (DDQ) and in non-depressed men |
| Iwamoto, Liu, and Mccoy (2011) | United States | 167 | University | Cross-sectional | Positive correlation between depressive symptoms (CES-D) and alcohol use (MEDAS) |
| | | | | | Binge drinking was not correlated to depressive symptoms (CES-D). It was only positively correlated to perceived peer use |
| | | | | | Perceived peer use was a signficant predictor of alcohol use (MEDAS) |
| | | | | | Positive correlation between depressive symptoms (CES-D) and illicit drug use (MEDAS) |
| Pedrelli et al. (2011) | United States | 904 | University | Cross-sectional | As depressive symptoms increased (BDI), males reported drinking more while females reported less alcohol consumption (CHQ) |
| | | | | | The relationship between compulsive drinking (CHQ) and daily alcohol consumption (CHQ) was not moderated by depressive symptoms (BDI) in males |
| | | | | | Depressive symptoms (BDI) moderated the relationship between compulsive drinking (CHQ) and daily alcohol consumption (CHQ) in females |
| Patock-Peckham and Morgan-Lopez (2007) | United States | 441 | University | Cross-sectional | For men, father authoritativeness (PAQ) was negatively related to paternal negative bond (PBI) |
| | | | | | For women, father authoritativeness (PAQ) was positively related to paternal negative bond (PBI) |
| | | | | | For men and women, paternal negative bond (PBI) increased the likelihood of depressive symptoms (CES-D). Depressive symptoms (CES-D) were positively associated with alcohol-related problems (unstandardized question) |
| | | | | | The feeling of being rejected by one's father was positively correlated to depressive symptoms (CES-D) and drinking related problems (unstandardized question) while being rejected by one's mother was not |
| Christiansen, Vik, and Jarchow (2002) | United States | 424 | University | Cross-sectional | Positive correlation found between heavy drinking while alone (unstandardized question) and depressive symptoms (CES-D) |
| | | | | | Social heavy drinkers were found to be more similar to non-heavy drinkers than alone heavy drinkers (unstandardized question) in relation to emotional coping (AEQ) and depressive symptoms (CES-D) |

*Continued*

**TABLE 5** Association between socio-environmental factors on the association between alcohol and depression—cont'd

| Source | Country | *N* | Setting | Design | Outcomes |
|---|---|---|---|---|---|
| Pedrelli et al. (2016) | United States | 61.561 | University | Cross-sectional | College students with Major Depressive Disorder (PHQ-9) reported higher frequency and intensity of heavy episodic drinking (unstandardized question) |
| | | | | | Association between Major Depressive Disorder (PHQ-9) and high intensity for heavy episodic drinking (unstandardized question) was stronger in females |
| | | | | | Both men and women reported similar degree of association between major depressive disorder (PHQ-9) and frequency of heavy episodic drinking |

Other factors such as dyadic conflict (Lambe et al., 2015) were also correlated to depression and alcohol and their findings are displayed in Table 4.

## Influence of drinking motives on alcohol and depression

In order to understand the cause-and-effect relationship or even the correlation between depressive symptoms and alcohol use and its related problems, its crucial to discern the variety of factors regulating these processes, including the reasons why people drink. There are four classes of motives for alcohol consumption widely accepted: internally generated (drinking to enhance positive affect) or externally generated (drinking to obtain positive social rewards) positive reinforcement motives, and internally generated (drinking to reduce or regulate negative affect) or externally generated (drinking to avoid social rejection) negative reinforcement motives (Cooper, 1994). Among college students, studies have shown that all classes of motives are positively correlated with depressive symptoms and alcohol variables (Diulio et al., 2015).

Table 6 provides a comprehensive examination of noteworthy studies that correlate depressive symptoms, alcohol-related variables, and drinking motives in college students.

### Relationship between depressive symptoms, drinking to cope, and other variables

Drinking to cope is widely accepted as having a positive correlation between depressive symptoms and alcohol use (Armeli et al., 2015; Diulio et al., 2015; Gonzalez et al., 2011; Kenney et al., 2017), usually with a nonsocial drinking pattern: fewer social drinking, more nonsocial drinking, and less social companions (O'Hara et al., 2015). This is partially explained by student's belief that alcohol consumption leads to positive outcomes, which prompts great reliance on drinking as a coping mechanism (Gonzalez et al., 2011; Kenney et al., 2017). Thus, as positive expectancy theory proposes, our choices are guided by expectancies regarding the outcome of behavior as either positive or negative, which in turn, not only may increase the likelihood to initiate alcohol use but also may predict problematic drinking and alcohol-related problems. Recent studies suggest that these misleading expectations are due to a different perception of positive and negative consequences by depressed students (Kenney et al., 2017). However, these are short-term positive expectancies, as shown by the negative correlation between depressive symptoms, alcohol-related problems, and consideration for future consequences, indicating that depressed students tend to drink to maximize the immediate positive reinforcement disregarding negative long-term outcomes (Soltis et al., 2017). A 30-day longitudinal study found that drinking to cope motives were associated with higher levels of negative affect in the subsequent month, suggesting that these motivations not only precede negative affect but also allows its perpetuation (Armeli et al., 2015). A cohort analysis found not only that depressive symptoms and coping motives were positively correlated to higher negative alcohol consequences, but also that coping was a risk-enhancing factor among students with fewer depressive symptoms (Kenney et al., 2017).

Coping motives are also associated with theories of negative urgency, drinking excessively to achieve immediate desired objectives is often a response to intense negative emotions (Kenney et al., 2017). The latter is one of four core factors that make up for impulsivity traits along aside with lack of perseverance, lack of premeditation, and sensation seeking. Those four variables are often positively correlated with depressive symptoms and alcohol consumption, frequently leading to alcohol-related problems (King et al., 2011). Negative urgency can independently lead to rash actions and higher odds of experiencing alcohol-related problems. Premeditation can regulate the relation between

**TABLE 6 Influence of drinking motives on alcohol and depression.**

| Source | Country | *N* | Setting | Design | Outcomes |
|---|---|---|---|---|---|
| Diulio et al. (2015) | United States | 210 | University | Cross-sectional | Positive correlation between coping motives, enhancement motives, social motives and conformity motives (DMQ-R), alcohol use disorders (AUDIT) and depression (CES-D) |
| | | | | | Social motives (DMQ-R) were significantly more likely to be reported than other motives |
| Blevins and Stephens (2016) | United States | 170 | University | Clinical Trial | Providing feedback on using alcohol to cope (DMQ-R) reduced reports of using alcohol to cope with anxious and depressed affect in comparison to a traditional personalized feedback intervention that did not incorporate motives feedback |
| Kenney, Merrill, and Barnett (2017) | United States | 652 | University | Cohort | Negative and positive alcohol-related consequences (YAAPST) significantly decreased over the course of freshman and sophomore years |
| | | | | | Coping motives (DMR-Q) and depressive symptoms (CES-D) independently predicted higher levels of alcohol-related negative consequences during college |
| Armeli, Sullivan, and Tennen (2015) | United States | 521 | University | Cohort | Positive correlation between higher levels of drinking to cope motivation (DMQ-R) and higher levels of negative affect (RAPI) in the subsequent month |
| O'Hara, Armeli, and Tennen (2015) | United States | 722 | University | Cohort | Positive correlation between drinking for social reasons (DMQ-R & MDMQ-R) and higher quantities of social drinking (1 question), as well as heavy alcohol use (1 question) among social companions |
| | | | | | Negative correlation between enhancement motives (DMQ-R & MDMQ-R) and nonsocial drinking, but positive with perceived drinking quantity among social companions |
| | | | | | Positive correlation between coping–depression motives (DMQ-R & MDMQ-R) and nonsocial drinking, but negative with social drinking and number of social companions |
| Soltis, McDevitt-Murphy, and Murphy (2017) | United States | 393 | University | Cross-sectional | Positive correlation between depressive and stress symptoms (DASS-21) and craving (PACS) and alcohol-related problems (YAACQ), but negative with consideration of future consequences (CFC Scale) |
| | | | | | Depression, stress, craving and demand intensity significantly mediated alcohol-related problems |
| | | | | | Consideration of future consequences and craving significantly mediated the relation between depressive symptoms and alcohol-related problems |
| Gonzalez, Reynolds, and Skewes (2011) | United States | 143 | University | Cross-sectional | The relationship of depression (BDI-II) and alcohol problems (YAACQ) was mediated by drinking to cope (DMQ-R) and negative urgency (UPPS Impulsive Behavior Scale) |
| | | | | | The relationship of depression (BDI-II) and alcohol problems (YAACQ) was not mediated by lack of premeditation, lack of perseverance, sensation seeking, and delay discounting (UPPS Impulsive Behavior Scale) |

Continued

**TABLE 6 Influence of drinking motives on alcohol and depression—cont'd**

| Source | Country | *N* | Setting | Design | Outcomes |
|---|---|---|---|---|---|
| King, Karyadi, Luk, and Patock-Peckham (2011) | United States | 573 | University | Cross-sectional | Negative correlation between depressive symptoms (ASR) and alcohol-related problems (YAAPST) at high premeditation (UPPS Impulsive Behavioral Scale) |
| | | | | | Positive correlation between depressive symptoms (ASR) and alcohol-related problems (YAAPST) at low premeditation (UPPS Impulsive Behavioral Scale) |
| | | | | | Positive correlation between premeditation, negative urgency, sensation seeking and perseverance (UPPS Impulsive Behavioral Scale) and alcohol-related problems (YAAPST) |
| Villarosa, Messer, Madson, and Zeigler-Hill (2018) | United States | 566 | University | Cross-sectional | Positive correlation between depressive symptoms (BDI-II), coping with depression motives (MDMQ-R) and weekly alcohol consumption (DDQ), but negative with controlled comsuption |
| | | | | | Positive correlation between depressive symptoms (BDI-II), coping with depression motives (MDMQ-R) and hazardous drinking (AUDIT), but negative with controlled comsuption |
| | | | | | Positive correlation between depressive symptoms (BDI-II), conformity motives (MDMQ-R), and hazardous drinking (AUDIT), but negative with harm reduction strategies (PBSS-r) |
| Linden-Carmichael, Braitman, and Henson (2015) | United States | 260 | University | Cohort | Negative correlation between depressive symptoms (IPIP) and PBS use (PBSM), which was positively correlated with heavy alcohol consumption (DDQ) |
| | | | | | No correlation between depressive symptoms (IPIP), PBS use (PBSM) and negative alcohol-related consequences (RAPI) |
| Smith and Tran (2007) | United States | 233 | University | Cross-sectional | Positive correlation between depressive sympstoms (BDI), alcohol-related problems (AUDIT) and readiness to change (RTCQ) |
| | | | | | Brief web-based intervention regarding alcohol use, depression or both did not reduce alcohol-related problems (RAPI) in a 1 month follow-up among students with comorbid alcohol-use disorders (AUDIT) and depression (BDI) |
| Bravo et al. (2018) | United States, Argentina and Spain | 1864 | University | Cross-sectional | No ruminative facet - problem-focused thoughts, counterfactual thinking, repetitive thoughts, and anticipatory thoughts (RTSQ) mediated individually the relationship between depressive symptoms (CES-D) and drinking motives (DMQ-R) |
| | | | | | A second-order latent factor of ruminative thinking (RTSQ) did significantly mediate the associations between depressive symptoms (CES-D) and thre drinking motives - drinking to cope. enchancement motives and social motives (DMQ-R) |

**TABLE 6 Influence of drinking motives on alcohol and depression—cont'd**

| Source | Country | *N* | Setting | Design | Outcomes |
|---|---|---|---|---|---|
| Bravo, Pearson, and Henson (2017) | United States | 403 | University | Cross-sectional | Problem-focused toughts (RTSQ) and drinking to cope motives (DMQ-R) individually mediated the relationship between depressive symptoms (CESD-R) and alcohol-related problems (YAACQ) |
| | | | | | Positive correlation between depressive symptoms (CESD-R) and drinking to cope motives (DMQ-R) |
| | | | | | Positive correlation between drinking to cope (DMQ-R) motives alcohol-related problems (YAACQ) |

depressive symptoms and alcohol-related problems. High premeditation attenuated the correlation between these two variables, acting as a protective factor, while low premeditation enhanced the association, acting as a risk factor (King et al., 2011).

Another important construct that has been proposed as a mediator is ruminative thinking through its four main facets: problem-focused thoughts, counterfactual thinking, repetitive thoughts, and anticipatory thoughts. The response styles theory of depression, suggests that one's response to negative emotion determines whether that negative affect is reduced, maintained, or intensified, which proposes that rumination enhances negative thinking, impairs problem-solving, and diminish social support, thereby enhancing the correlation between depressive symptoms and alcohol-related problems. All four ruminative facets were positively correlated with depressive symptoms. Counterfactual thinking and anticipatory thoughts were positively correlated with coping motives (Bravo et al., 2017). Problem-focused thoughts individually mediated the relationship between depressive symptoms and drinking to cope with motives in one study (Bravo et al., 2018).

## Coping motives as a hindering factor for nonpharmacological interventions

Recent studies have shown that providing information about the risks of drinking to cope with negative affect and its associations with negative long-term consequences led to reduced reports of alcohol use to cope with depressive symptoms in comparison with interventions that did not cover motives for drinking (Blevins & Stephens, 2016). Protective behavioral strategies are a set of brief self-regulatory interventions used to lower harm while drinking through controlled consumption or harm reduction strategies. However, it has been observed that students with higher depressive symptoms and are drinking to cope with it, reported lower adherence to controlled consumption strategies, leading to more weekly alcohol consumption and more hazardous drinking. Not only engaging in self-harming behaviors as a means of coping could be a response to the negative view of self and others, highlighting the high comorbidity between depressive symptoms and alcohol use (Villarosa et al., 2018). By self-medicating their depressive symptoms with alcohol, students may also be less likely to engage in protective behavioral strategies due to the reduction of the desired effect from alcohol (Linden-Carmichael et al., 2015). Knowing that in order to reduce alcohol-related problems, it is important not only to choose the right intervention but also to know the state of mind of the target individuals. Readiness to change is modulated by the balance between costs and benefits of substance use, and when the former outweighs the latter, that individual starts the process of reduction of substance use. As discussed previously, negative affect may be a long-term consequence of drinking to cope associated with depression, and by assessing those motives, interventions may be more effective by identifying individuals with a higher readiness to change mindset (Smith & Tran, 2007).

## Other drinking motives and its relationship with depressive symptoms

Drinking for social reasons is often more likely to be reported than any other motivation (Diulio et al., 2015), being also positively correlated with higher quantities of social drinking consumption and heavy alcohol use among drinking companions (O'Hara et al., 2015). In some cases, social and coping motives may overlap, mainly due to students using social outcomes as a positive reinforcement stimulus by drinking to regulate negative emotions (Kenney et al., 2017). Enhancement motives were positively correlated with depressive symptoms and alcohol consumption (Diulio et al., 2015) and higher values of these motives were associated with less nonsocial drinking and higher perceived drinking quantity within social milieu

(O'Hara et al., 2015). Conformity motives were also positively correlated with depressive symptoms and alcohol consumption (Diulio et al., 2015; Villarosa et al., 2018).

## Key facts of alcohol and depression among students

- Students have a higher prevalence of alcohol use, alcohol use disorders, and depression in comparison to the general population.
- Depression is more prevalent among women, whereas alcohol misuse is often common among men.
- These conditions cause significant social, legal, academic, and health burdens.
- Alcohol and depression are often correlated in this subpopulation.
- Alcohol use disorders and depression are underrecognized and underdiagnosed among students.

## Summary points

- Depression, alcohol use, and alcohol use disorders have a higher prevalence among students in comparison to the general population and are often comorbid conditions that cause significant social, legal, academic, financial, and health burdens.
- Alcohol use and binge drinking are positively correlated to depression or its severity in middle and high school students, whereas conflicting associations between these variables are found in university students.
- Alcohol-related problems and alcohol-related medical diagnoses have been mostly studied amid undergraduates and show an important positive correlation to depression or its severity in this population.
- Depressive symptoms are the most common causal variable in the relationship between alcohol and depression among students by leading to more frequent and harmful alcohol consumption through a self-medication mechanism.
- The association between alcohol, depression, and suicidal ideation is consistent among university and middle and high school students, while anxiety had conflicting findings in regard to coping with anxiety motives to drink and their effect on alcohol.
- Alcohol problems are associated with social motives in men and coping motives in women, and depressed women have a higher intensity of heavy episodic drinking in comparison to depressed men.
- Heavy drinking while alone is correlated to depressive symptoms and drinking with a partner or family did not correlate to alcohol-related problems.
- Positive expectancies regarding alcohol use are associated with drinking to cope and predicts higher alcohol consumption among university students reporting depressive symptoms, not only correlating to short-term positive consequences but also to long-term negative consequences.
- Impulsivity traits and ruminative thinking are positively correlated to depressive symptoms, alcohol consumption, and alcohol-related problems.
- Nonpharmacologic interventions did not reduce alcohol-related problems and depressive symptoms, with little effect on alcohol use, and drinking to cope with depressive symptoms is associated with lower adherence to controlled consumption strategies not only for self-medicating reasons but also as a self-harming behavior as a means of coping.

## Mini-dictionary of terms

**Alcohol abuse** DSM-IV diagnosis involving a maladaptive pattern of drinking, leading to clinically significant impairment or distress, often involving alcohol-related problems.

**Alcohol dependence** Similar to DSM-IV's criteria as alcohol abuse but with symptoms of dependence, tolerance, or abstinence.

**Alcohol use disorder** DSM-V diagnosis in which alcohol abuse and dependence were combined in a single condition, having much of the criteria that both had.

**Binge drinking** The consumption of more than four doses of alcohol in a short period of time.

**Years lived with disability** A measurement of the burden of a disease calculated with the prevalence of a disorder and the number of years lost due to it.

## References

Armeli, S., Dranoff, E., Tennen, H., Austad, C. S., Fallahi, C. R., Raskin, S., et al. (2014). A longitudinal study of the effects of coping motives, negative affect and drinking level on drinking problems among college students. *Anxiety, Stress, and Coping*, *27*(5), 527–541.

Armeli, S., Sullivan, T. P., & Tennen, H. (2015). Drinking to cope motivation as a prospective predictor of negative affect. *Journal of Studies on Alcohol and Drugs*, *76*(4), 578–584.

Blanco, C., Okuda, M., Wright, C., Hasin, D. S., Grant, B. F., Liu, S.-M., et al. (2008). Mental health of college students and their non-college-attending peers: Results from the National Epidemiologic Study on Alcohol and Related Conditions. *Archives of General Psychiatry*, *65*(12), 1429–1437.

Blevins, C. E., & Stephens, R. S. (2016). The impact of motives-related feedback on drinking to cope among college students. *Addictive Behaviors*, *58*, 68–73.

Bravo, A. J., Pearson, M. R., & Henson, J. M. (2017). Drinking to cope with depressive symptoms and ruminative thinking: A multiple mediation model among college students. *Substance Use & Misuse*, *52*(1), 52–62.

Bravo, A. J., Pilatti, A., Pearson, M. R., Mezquita, L., Ibáñez, M. I., & Ortet, G. (2018). Depressive symptoms, ruminative thinking, drinking motives, and alcohol outcomes: A multiple mediation model among college students in three countries. *Addictive Behaviors*, *76*, 319–327.

Christiansen, M., Vik, P. W., & Jarchow, A. (2002). College student heavy drinking in social contexts versus alone. *Addictive Behaviors*, *27*(3), 393–404.

Collins, J.-L., Thompson, K., Sherry, S. B., Glowacka, M., & Stewart, S. H. (2018). Drinking to cope with depression mediates the relationship between social avoidance and alcohol problems: A 3-wave, 18-month longitudinal study. *Addictive Behaviors*, *76*, 182–187.

Cooper, M. L. (1994). Motivations for alcohol use among adolescents: Development and validation of a four-factor model. *Psychological Assessment*, *6*(2), 117–128.

de Andrade, A. G., Duarte, P.d. C. A. V., Barroso, L. P., Nishimura, R., Alberghini, D. G., & Oliveira, L. G.d. (2012). Use of alcohol and other drugs among Brazilian college students: Effects of gender and age. *Revista Brasileira de Psiquiatria (Sao Paulo, Brazil: 1999)*, *34*(3), 294–305.

Diulio, A. R., Dutta, N. M., Gauthier, J. M., Witte, T. K., Correia, C. J., & Angarano, D. (2015). Associations among depressive symptoms, drinking motives, and risk for alcohol-related problems in veterinary students. *Journal of Veterinary Medical Education*, *42*(1), 11–17.

dos Santos, D. T., Nazário, F. P., Freitas, R. A., Henriques, V. M., & de Paiva, I. S. (2019). Alcohol abuse and dependence among Brazilian medical students: Association to sociodemographic variables, anxiety and depression. *Journal of Substance Use*, *24*, 285–292. https://doi.org/10.1080/14659891.2018.1562574.

Dvorak, R. D., Lamis, D. A., & Malone, P. S. (2013). Alcohol use, depressive symptoms, and impulsivity as risk factors for suicide proneness among college students. *Journal of Affective Disorders*, *149*(1-3), 326–334.

El Ansari, W., Sebena, R., & Stock, C. (2014). Do importance of religious faith and healthy lifestyle modify the relationships between depressive symptoms and four indicators of alcohol consumption? A survey of students across seven universities in England, Wales, and Northern Ireland. *Substance Use & Misuse*, *49*(3), 211–220.

Geisner, I. M., Varvil-Weld, L., Mittmann, A. J., Mallett, K., & Turrisi, R. (2015). Brief web-based intervention for college students with comorbid risky alcohol use and depressed mood: Does it work and for whom? *Addictive Behaviors*, *42*, 36–43.

Gonzalez, V. M., Bradizza, C. M., & Lorraine Collins, R. (2009). Drinking to cope as a statistical mediator in the relationship between suicidal ideation and alcohol outcomes among underage college drinkers. *Psychology of Addictive Behaviors*, *23*, 443–451. https://doi.org/10.1037/a0015543.

Gonzalez, V. M., Reynolds, B., & Skewes, M. C. (2011). Role of impulsivity in the relationship between depression and alcohol problems among emerging adult college drinkers. *Experimental and Clinical Psychopharmacology*, *19*(4), 303–313.

Grant, V. V., Stewart, S. H., & Mohr, C. D. (2009). Coping-anxiety and coping depression motives predict different daily mood-drinking relationships. *Psychology of Addictive Behaviors: Journal of the Society of Psychologists in Addictive Behaviors*, *23*(2), 226–237.

Guo, L., Deng, J., He, Y., Deng, X., Huang, J., Huang, G., et al. (2016). Alcohol use and alcohol-related problems among adolescents in China: A large-scale cross-sectional study. *Medicine*, *95*(38), e4533.

Heinz, A. J., Veilleux, J. C., & Kassel, J. D. (2009). The role of cognitive structure in college student problem drinking. *Addictive Behaviors*, *34* (2), 212–218.

Huurre, T., Lintonen, T., Kaprio, J., Pelkonen, M., Marttunen, M., & Aro, H. (2010). Adolescent risk factors for excessive alcohol use at age 32 years. A 16-year prospective follow-up study. *Social Psychiatry and Psychiatric Epidemiology*, *45*, 125–134. https://doi.org/10.1007/s00127-009-0048-y.

Iwamoto, W., Liu, W. M., Mccoy, T. E., et al. (2011). An Exploratory Model of Substance Use Among Asian American Women: The Role of Depression, Coping, Peer Use and Asian Values. *Journal Ethnic Substance Abuse*, *10* (4), 295–315. https://doi.org/10.1080/15332640.2011.623494.

Johannessen, E. L., Andersson, H. W., Bjørngaard, J. H., & Pape, K. (2017). Anxiety and depression symptoms and alcohol use among adolescents—A cross sectional study of Norwegian secondary school students. *BMC Public Health*, *17*(1), 494.

Kaur, J., Cheong, S. M., Mahadir Naidu, B., Kaur, G., Manickam, M. A., Mat Noor, M., et al. (2014). Prevalence and correlates of depression among adolescents in Malaysia. *Asia-Pacific Journal of Public Health/Asia-Pacific Academic Consortium for Public Health*, *26*(5 Suppl), 53S–62S.

Kenney, S., Abar, C. C., O'Brien, K., Clark, G., & LaBrie, J. W. (2016). Trajectories of alcohol use and consequences in college women with and without depressed mood. *Addictive Behaviors*, *53*, 19–22.

Kenney, S., Jones, R. N., & Barnett, N. P. (2015). Gender differences in the effect of depressive symptoms on prospective alcohol expectancies, coping motives, and alcohol outcomes in the first year of college. *Journal of Youth and Adolescence*, *44*(10), 1884–1897.

Kenney, S. R., Merrill, J. E., & Barnett, N. P. (2017). Effects of depressive symptoms and coping motives on naturalistic trends in negative and positive alcohol-related consequences. *Addictive Behaviors*, *64*, 129–136.

Keough, M. T., O'Connor, R. M., Sherry, S. B., & Stewart, S. H. (2015). Context counts: Solitary drinking explains the association between depressive symptoms and alcohol-related problems in undergraduates. *Addictive Behaviors*, *42*, 216–221.

King, K. M., Karyadi, K. A., Luk, J. W., & Patock-Peckham, J. A. (2011). Dispositions to rash action moderate the associations between concurrent drinking, depressive symptoms, and alcohol problems during emerging adulthood. *Psychology of Addictive Behaviors: Journal of the Society of Psychologists in Addictive Behaviors*, *25*(3), 446–454.

Lambe, L., Mackinnon, S. P., & Stewart, S. H. (2015). Dyadic conflict, drinking to cope, and alcohol-related problems: A psychometric study and longitudinal actor-partner interdependence model. *Journal of Family Psychology: JFP: Journal of the Division of Family Psychology of the American Psychological Association*, *29* (5), 697–707.

Lamis, D. A., Ballard, E. D., May, A. M., & Dvorak, R. D. (2016). Depressive symptoms and suicidal ideation in college students: The mediating and moderating roles of hopelessness, alcohol problems, and social support. *Journal of Clinical Psychology*, *72*(9), 919–932.

Linden-Carmichael, A. N., Braitman, A. L., & Henson, J. M. (2015). Protective behavioral strategies as a mediator between depressive symptom fluctuations and alcohol consumption: A longitudinal examination among college students. *Journal of Studies on Alcohol and Drugs*, *76*(1), 80–88.

Merrill, J. E., Reid, A. E., Carey, M. P., & Carey, K. B. (2014). Gender and depression moderate response to brief motivational intervention for alcohol misuse among college students. *Journal of Consulting and Clinical Psychology*, *82*(6), 984–992.

Mushquash, A. R., Stewart, S. H., Sherry, S. B., Sherry, D. L., Mushquash, C. J., & Mackinnon, A. L. (2013). Depressive symptoms are a vulnerability factor for heavy episodic drinking: A short-term, four-wave longitudinal study of undergraduate women. *Addictive Behaviors*, *38*(5), 2180–2186.

O'Hara, R. E., Armeli, S., & Tennen, H. (2015). College students' drinking motives and social-contextual factors: Comparing associations across levels of analysis. *Psychology of Addictive Behaviors: Journal of the Society of Psychologists in Addictive Behaviors*, *29* (2), 420–429.

Pan, X.-F., Wen, Y., Zhao, Y., Hu, J.-M., Li, S.-Q., Zhang, S.-K., et al. (2015). Prevalence of depressive symptoms and its correlates among medical students in China: A national survey in 33 universities. *Psychology, Health & Medicine*, *21*, 882–889. https://doi.org/10.1080/13548506.2015.1127395.

Patock-Peckham, J. A., & Morgan-Lopez, A. A. (2007). College drinking behaviors: Mediational links between parenting styles, parental bonds, depression, and alcohol problems. *Psychology of Addictive Behaviors: Journal of the Society of Psychologists in Addictive Behaviors*, *21*(3), 297–306.

Pedrelli, P., Borsari, B., Lipson, S. K., Heinze, J. E., & Eisenberg, D. (2016). Gender differences in the relationships among major depressive disorder, heavy alcohol use, and mental health treatment engagement among college students. *Journal of Studies on Alcohol and Drugs*, *77*(4), 620–628.

Pedrelli, P., Farabaugh, A. H., Zisook, S., Tucker, D., Rooney, K., Katz, J., et al. (2011). Gender, depressive symptoms and patterns of alcohol use among college students. *Psychopathology*, *44*(1), 27–33.

Piumatti, G. (2017). Motivation, health-related lifestyles and depression among university students: A longitudinal analysis. *Psychiatry Research*, *260*, 412–417. https://doi.org/10.1016/j.psychres.2017.12.009.

Potter, C. M., Galbraith, T., Jensen, D., Morrison, A. S., & Heimberg, R. G. (2016). Social anxiety and vulnerability for problematic drinking in college students: The moderating role of post-event processing. *Cognitive Behaviour Therapy*, *45*(5), 380–396.

Rojas, G., Gaete, J., Guajardo, V., Martínez, V., Barroihlet, S., Meneses, J., et al. (2012). Consumo de sustancias y síntomas depresivos en escolares de liceos municipales de Santiago. *Revista Médica de Chile*, *140*, 184–191. https://doi.org/10.4067/s0034-98872012000200006.

Sebena, R., El Ansari, W., Stock, C., Orosova, O., & Mikolajczyk, R. T. (2012). Are perceived stress, depressive symptoms and religiosity associated with alcohol consumption? A survey of freshmen university students across five European countries. *Substance Abuse Treatment, Prevention, and Policy*, *7*, 21.

Smith, J. P., & Tran, G. Q. (2007). Does negative affect increase change readiness among college hazardous drinkers? *Addictive Behaviors*, *32*(10), 2281–2285.

Soltis, K. E., McDevitt-Murphy, M. E., & Murphy, J. G. (2017). Alcohol demand, future orientation, and craving mediate the relation between depressive and stress symptoms and alcohol problems. *Alcoholism, Clinical and Experimental Research*, *41*(6), 1191–1200.

Tomlinson, K. L., & Brown, S. A. (2012). Self-medication or social learning? A comparison of models to predict early adolescent drinking. *Addictive Behaviors*, *37*, 179–186. https://doi.org/10.1016/j.addbeh.2011.09.016.

Torikka, A., Kaltiala-Heino, R., Rimpelä, A., Rimpelä, M., & Rantanen, P. (2001). Depression, drinking, and substance use among 14- to 16-year-old Finnish adolescents. *Nordic Journal of Psychiatry*, *55*(5), 351–357.

Villarosa, M. C., Messer, M. A., Madson, M. B., & Zeigler-Hill, V. (2018). Depressive symptoms and drinking outcomes: The mediating role of drinking motives and protective behavioral strategies among college students. *Substance Use & Misuse*, *53*(1), 143–153.

Weitzman, E. R. (2004). Poor mental health, depression, and associations with alcohol consumption, harm, and abuse in a national sample of young adults in college. *The Journal of Nervous and Mental Disease*, *192*, 269–277. https://doi.org/10.1097/01.nmd.0000120885.17362.94.

World Health Organization. (2017). Depression and other common mental disorders. *Global Health Estimates*, *24*, 8–9.

World Health Organization. (2019). *Global status report on alcohol and health 2018*. World Health Organization.

Chapter 11

# Depression in obesity

**Ioannis D. Morres, Antonis Hatzigeorgiadis, and Yannis Theodorakis**
*Department of Physical Education and Sport Science, University of Thessaly, Trikala, Greece*

## List of abbreviations

**AMA** American Medical Association
**BMI** body mass index
**IOTS** International Obesity Task Force
**OECD** Organization for Economic Cooperation and Development
**OMA** Obesity Medicine Association
**PHQ-9** Patient Health Questionnaire-9

## Introduction

Depression is a disabling and life-threatening mood disorder, which affects large proportions of today's society. According to recent evidence from the Global Burden of Disease study, depression affects almost 264 million people worldwide ( James et al., 2018). This serious mental illness includes disturbed cognitive, affective, and physiological components. The main depressive disorders include dysthymia, bipolar depression, and major depressive disorder also referred to as a clinical depression. Major depressive disorder is the most common type of depression and is projected to become the top health problem by 2030. The diagnosis of major depressive disorder is typically based on the Diagnostic and Statistical Manual-5 (DSM-5) (American Psychiatric Association, 2013) or the International Classification of Disease (ICD-10) (World Health Organization [WHO], 1992). According to the DSM-5, major depressive disorder affects women twice as common as men and has the following diagnostic criteria. One of the two symptoms, low mood or anhedonia (lack of interest or pleasure) is essential, in addition to ≥4 of the following symptoms: (1) lack of concentration, (2) disturbed sleep, (3) disturbed appetite or weight, (4) fatigue or loss of energy, (5) psychomotor retardation or agitation, (6) feelings of worthlessness or guilt, and (7) persistent thoughts of death or suicide. The symptoms should be present for at least 2 weeks and cause impaired psychosocial functioning.

## Obesity

Obesity is defined as the body fat accumulation corresponding to a body mass index [BMI; weight in kilograms (kg) divided by the square of height in meters ($m^2$)] of $\geq 30\,kg/m^2$. The prevalence of obesity is increasing at an alarming rate. For example, the OECD reports that the prevalence of obesity in the United Kingdom has sharply risen from 14% in 1990 to 27% in 2016 (OECD, 2017). Obesity is linked to increased risks for physical health problems (e.g., cardiovascular disease, cancer, type 2 diabetes, metabolic syndrome, osteoarthritis, etc.) and to physical (multi)-morbidity, which may lead to psychosocial disability. Health and social care systems are thus, faced with escalating costs.

Obesity is typically attributed to an imbalance between energy consumption (e.g., caloric intake) and energy expenditure (e.g., metabolic rate or physical activity). Hence, clinicians, researchers, stakeholders, and the public may often consider obesity the outcome of poor health behaviors, e.g., energy overconsumption and physical inactivity. To this extent, obesity is often related to social stigma. However, obesity is increasingly recognized as a disease rather than as a simple fat accumulation due to poor health behaviors. The report "Obesity: Preventing and Managing the Global Epidemic" by WHO and IOTS in 1997 is a cross point of this notion. In this report, obesity was recognized as a complex, serious, chronic, and partly understood disease, which is part of a group of noncommunicable diseases demanding individual/social interventions for prevention and management (WHO, 1998). Ever since, an increasing number of associations (e.g., American Medical Association House of Delegates, 2013; Royal College of Physicians, 2019) and countries (e.g., Japan, Portugal, Italy) recognize obesity as a disease. Moreover, the OMA has recently stated that obesity includes impaired metabolic function with abnormal hunger, satiety, and fullness processing; this

The Neuroscience of Depression. https://doi.org/10.1016/B978-0-12-817933-8.00041-4

impairment is due to pathological consequences in bodily system imbalance, not due to fat accumulation from excessive calories (Bray, Kim, & Wilding; on behalf of the World Obesity Federation, 2017).

## Depression in obesity

A significant number of patients experience depression-obesity comorbidity. Data from the US National Health and Nutrition Examination Surveys 2005–2010 (Pratt & Brody, 2014) indicate that almost 55% of adult participants receiving drug treatment for moderate or severe depression were classified as obese; also, one-third of participants treated for mild or no depressive symptoms were classified as obese. A cross-sectional survey in general practice settings in Australia ($N = 3.361$) found that 23% of the patients classified in the obesity category were diagnosed as depressed (Carey et al., 2014). Similarly, a large study with obese patients in the United Kingdom ($N = 363.037$) reported that 25% of the sample was diagnosed with depression (Moussa et al., 2019). Moreover, obese people demonstrate a higher rate of depression than nonobese people (Fabricatore & Wadden, 2004). To this extent, epidemiological studies with cross-sectional or prospective design often with Mendelenian approaches have explored the depression-obesity combination. Systematic reviews and meta-analyses have also accumulated evidence from cross-sectional and/or prospective studies. Researchers appear to be recording relatively common findings suggesting the relationship of obesity with depression development, especially among women. Representative evidence is described below.

## Cross-sectional or prospective studies

Roberts, Deleger, Strawbridge, and Kaplan (2003) in a population cohort ($N = 1.886$) reported cross-sectional findings where obese vs. nonobese were twice likely to be depressed, whereas depressed vs. nondepressed were 1.8 times as likely to be obese. Prospective findings for 5 years later showed that obese vs nonobese had twice the risk of depression. Exploring if depression predicts obesity at the same follow-up, depressed vs. nondepressed had less than twice the risk of being obese adjusting for baseline obesity, whereas obesity predicted depression adjusting for baseline depression. No finding changed after controlling for social/health covariates. Conversely, depressed people were not at increased risk of developing obesity in 5 years' time. A gender-obesity interaction was nonsignificant, thus the obesity-depression association was not different for men and women.

A cross-sectional study in population cohort ($N = 9.125$) by Simon et al. (2006) has found that lifetime prevalence estimates for diagnosed major depression among obese people were higher than nonobese people. Similarly, the last 12-month prevalence estimate for major depression was higher among obese vs. nonobese people. No finding changed after adjusting for smoking, age, or gender. Hence, obesity was associated with significant increases in a lifetime and last 12-month diagnosis of major depression, especially among non-Hispanic whites and college graduates.

In the cross-sectional epidemiological study by Zhao et al. (2009) ($N = 177.047$), overweight/obese women showed a higher prevalence of current and lifetime depression in comparison to overweight/obese men. Also, after adjusting for demographics, obese women were significantly more likely to record current and lifetime depression vs. women with normal weight (BMI: 18–24.99). The same finding was found in men after adjusting for demographics and for a series of covariates including emotional support, leisure-time exercise, life satisfaction, and general health.

An epidemiological prospective study ($N = 4.363$) with repeated measures (four measures within 19 years) in a British population has found that common mental disorders including depression represent a risk factor for future obesity (Kivimäki et al., 2009). Also, a longitudinal cohort study by Herva et al. (2006) from Northern Finland included two follow-up measures (14 years and 31 years) in a large population sample ($N = 12.058$). Those authors provided evidence for the association of adolescent obesity with future depression in young adulthood. By measuring also, the waist-to-hip-ratio, Herva et al. (2006) found that male abdominal obesity was linked to concomitant depression. This study has also demonstrated that adolescent obesity and adulthood obesity emerged as risk factors for female depression.

Recent studies have used more sophisticated analyses and even larger population cohorts to examine the depression-obesity combination. Particularly, Speed, Jefsen, Børglum, Speed, and Østergaard (2019) employed a Mendelian randomization approach. Those authors tested the link between anthropometric measures of height (short stature), fat mass, nonfat mass to depression by employing genome-wide association study results from the UK Bionank ($N = 332.000$) and the Psychiatric Genomics Consortium ($N = 480.000$). The study showed that BMI is a causal risk factor for depression and supported the widely used approach to examine BMI rather than weight as a potential causal factor to depression. Both fat percentage and whole-body fat mass were found to be causal risk factors to depression, however, whole body nonfat mass did not show a causal relationship to depression, whereas the reduction of fat mass could lower the risk of depression. In addition, the study reported that depression was not a causal risk factor for any of the anthropometric measures.

Similarly, Tyrrell et al. (2019) employed a large sample ($N = 340.786$) to explore the causal contribution of BMI to depression using the Mendelian randomization method. In this study, higher BMI was represented by two genetic indicators, one of the two indicators is typically associated with metabolic side effects. Results showed that both indicators were linked to higher odds of depression in both sexes, and in women only. Meta-analytic findings provided additional evidence for the findings in all individuals. Collectively, BMI indicators were likely to bring about a causal effect on developing depression regardless of their association with metabolic side effects.

## Systematic reviews

Systematic reviews have also examined the obesity-depression combination. A systematic review from the late 2000s by Atlantis and Baker (2008) has found that cross-sectional studies report supportive evidence for the association between obesity and increased prevalence of depression but only in populations from the United States. Prospective studies, however, uniformly reported that obesity may increase the incidence of depression, although this evidence was derived from studies with increased risk of bias (Atlantis & Baker, 2008).

A systematic review and meta-analysis from the same period by Blaine (2008) has examined if depression represents a causal factor for weight gain. By including 16 longitudinal studies ($N = 33.000$), this review demonstrated that depressed people were almost two times more likely to become obese or gain weight vs. nondepressed people. The association of depression with obesity was relatively unaffected (and remained significant) after controlling for confounders including baseline BMI. Finally, no differences in the causal effects of depression on obesity were found after controlling for gender or age.

A recent systematic review and meta-analysis by Pereira-Miranda, Costa, Queiroz, Pereira-Santos, and Santana (2017) have reviewed eight cross-sectional and one cohort study ($N = 171.701$). Results showed that obese people were more likely (by 32%) to experience depression vs. nonobese people, as the prevalence ratio was 1.32 (CI 95%, 1.26–1.38). Gender analysis showed that this association remained significant only among women. Finally, Hyppönen, Mulugeta, Zhou, and Santhanakrishnan (2019) in their recent review explored evidence for the causal association of the BMI genetic risk score with various diseases including depression in a mega data set of a UK Biobank ($N = 337.536$). Based on 80 Mendelenian randomization studies, this study found that the BMI genetic risk score recorded a causal effect on depression; however, stringent analyses decreased confidence for the findings.

### Commit suicide attempts/suicides, depression, and obesity

One of the most critical issues in depression treatment deals with the potential to commit suicides. In obese people, obesity-related measures (e.g., BMI), disorders (e.g., body dimorphic disorder), and behaviors (e.g., weight control attempts including bariatric surgery) appear to be showing increased odds of suicidal ideation and suicides regardless the presence or mediation of depression. However, relatively controversial findings are also reported concerning the view on the association of increased risk of suicide attempts and suicides with obesity.

In particular, Ju et al. (2016) has investigated the association between weight control failure and suicidal ideation in a sample of overweight and obese people ($N = 6.621$) within the context of the Korea National Health and Nutrition Examination Survey. Those authors found no significant association between weight control failure and suicidal ideation among overweight participants or obese men. However, subgroup analysis for obese women showed a significant association between weight control failure and suicidal ideation. Also, it is noteworthy that depression was significantly correlated with suicidal ideation among all participants.

Carpenter, Hasin, Allison, and Faith (2000) recruited a large US population group ($N = 40.086$); after adjustment of covariates, these authors found that increases in BMI increased the risk of past 12-month suicide ideation and attempts by approximately 22% among women. The findings of this study are of special importance given that depression was assessed via the DSM criteria and not via psychometric lists for depression scores, which are widely used in large-scale studies. In contrast Bjerkeset, Romundstad, Evans, and Gunnell (2007) have demonstrated an inverse association between BMI and risk of suicide in a large Norwegian sample ($N = 44.396$).

Also, Gao, Juhaeri, Reshef, and Dai (2013) have examined if obesity is linked to the incidence rate of suicide attempts and suicide. Employing a large sample of British adults ($N = 849.434$), this study reported an inverse relationship between obesity (BMI) and the incidence of suicide attempt only in depressed men. Among depressed women, the study did not provide clear findings; however, the incidence of suicide attempts was higher in the under weight than in the normal weight subgroup.

## Conclusion

Depression-obesity comorbidity is a major health problem. Depression represents a mental health disorder and relevant treatments are available, but obesity is typically treated as an imbalance between energy consumption and energy expenditure and interventions focus on weight-loss lifestyle modifications. However, the scientific community and decision-making centers are gradually recognizing obesity as a complex health disorder instead. This is a valuable step forward because the depression-obesity comorbidity is prevalent and warrants a multidimensional therapeutic approach. Particularly, obesity indicators are causal factors of depression development regardless of their association with metabolic side effects, whereas depression contributes to obesity development and to increased risk of commit suicide ideation. Further, people with obesity are affected by depression more often than nonobese people, and people with extreme obesity ($BMI \geq 40\,kg/m^2$) are particularly vulnerable to commit suicide ideation. The recognition of obesity as a complex health disorder needs to lead to more effective treatments for the challenging health issue of depression-obesity comorbidity.

## Key facts of obese people in bariatric surgeries

- Bariatric surgery is the most effective intervention for people with extreme obesity ($BMI \geq 40\,kg/m^2$).
- Approximately 30% of obese people in bariatric surgeries are found with current depression.
- Most obese people report unfeasible and persistent post-surgery goals (e.g., weight loss of 50%), although preoperative feedback by health professionals indicates that surgery cannot bring about such effects.
- Obese people after bariatric surgery seem to show almost two times higher commit suicide rates vs obese people not-seeking bariatric surgery treatment; this trend suggests increased commit suicide rates among bariatric surgery patients.
- Preoperative mental health problems including depression are linked to the trend of increased commit suicide rates among obese people after bariatric surgery.
- Preoperative mental health therapies are essential to either ameliorate depression and/or enable obese people to set realistic weight-loss goals.

## Summary points

- Depression is a prevalent, severe, and life-threatening mental health disorder.
- The scientific community is currently redefining the phenomenon of obesity.
- Obesity is increasingly recognized as a complex, serious, and complex disorder rather than an imbalance between energy consumption and energy expenditure.
- The depression-obesity comorbidity is seen in a significant number of people receiving the widely used antidepressant medication.
- Obese people demonstrate a higher prevalence rate of depression compared to nonobese people.
- Obesity appears to be carrying causal effects on depression development.
- Commit suicide attempts and suicides appear to be related to increased obesity among patients diagnosed with major depression.
- Obesity-related behaviors including bariatric surgery or weight control attempts may often be linked to depression and increased suicide rates.

## Examples of mini-dictionary of terms

**Bariatric surgery** Types of interventions performed on the human stomach (e.g., gastric bypass, sleeve gastrectomy, adjustable gastric band, or biliopancreatic diversion with duodenal switch gastric bypass) with the aim to help people lose weight by restricting the amount of food the stomach can hold.

**Cross-sectional design** Data collection at a particular time by an observational study.

**Epidemiological studies** Large-scale studies exploring the association of certain conditions with diseases.

**Longitudinal design** Repeated data collection over a period of time by an observational study.

**Mendelian randomization** An increasingly used method by epidemiological studies, which explores the causal contribution of genetic factors to diseases outcomes.

## References


American Medical Association House of Delegates. (2013). Recognition of obesity as a disease. In *American Medical Association*. Retrieved from https://www.npr.org/documents/2013/jun/ama-resolution-obesity.pdf.

American Psychiatric Association. (2013). *Diagnostic and statistical manual of mental disorders* (5th ed.). Washington, DC: American Psychiatric Association.

Atlantis, E., & Baker, M. (2008). Obesity effects on depression: Systematic review of epidemiological studies. *International Journal of Obesity, 32*, 881. https://doi.org/10.1038/ijo.2008.54.

Bjerkeset, O., Romundstad, P., Evans, J., & Gunnell, D. (2007). Association of adult body mass index and height with anxiety, depression, and suicide in the general population: The HUNT study. *American Journal of Epidemiology, 167*(2), 193–202. https://doi.org/10.1093/aje/kwm280.

Blaine, B. (2008). Does depression cause obesity?: A meta-analysis of longitudinal studies of depression and weight control. *Journal of Health Psychology, 13*(8), 1190–1197. https://doi.org/10.1177/1359105308095977.

Bray, G. A., Kim, K. K., Wilding, J. P. H., & on behalf of the World Obesity Federation. (2017). Obesity: A chronic relapsing progressive disease process. A position statement of the World Obesity Federation. *Obesity Reviews, 18*(7), 715–723. https://doi.org/10.1111/obr.12551.

Carey, M., Small, H., Yoong, S. L., Boyes, A., Bisquera, A., & Sanson-Fisher, R. (2014). Prevalence of comorbid depression and obesity in general practice: A cross-sectional survey. *British Journal of General Practice, 64*(620), e122–e127. https://doi.org/10.3399/bjgp14X677482.

Carpenter, K. M., Hasin, D. S., Allison, D. B., & Faith, M. S. (2000). Relationships between obesity and DSM-IV major depressive disorder, suicide ideation, and suicide attempts: Results from a general population study. *American Journal of Public Health, 90*(2), 251–257. https://doi.org/10.2105/ajph.90.2.251.

Fabricatore, A. N., & Wadden, T. A. (2004). Psychological aspects of obesity. Clinics in Dermatology, 22(4), 332-337. https://doi.org/10.1016/j.clindermatol.2004.01.006.

Gao, S., Juhaeri, J., Reshef, S., & Dai, W. S. (2013). Association between body mass index and suicide, and suicide attempt among British adults: The health improvement network database. *Obesity*, *21*(3), E334–E342. https://doi.org/10.1002/oby.20143.

Herva, A., Laitinen, J., Miettunen, J., Veijola, J., Karvonen, J. T., Läksy, K., et al. (2006). Obesity and depression: Results from the longitudinal Northern Finland 1966 Birth Cohort Study. *International Journal of Obesity*, *30*(3), 520–527. https://doi.org/10.1038/sj.ijo.0803174.

Hyppönen, E., Mulugeta, A., Zhou, A., & Santhanakrishnan, V. K. (2019). A data-driven approach for studying the role of body mass in multiple diseases: A phenome-wide registry-based case-control study in the UK Biobank. The Lancet Digital Health, 1(3), e116-e126. https://doi.org/10.1016/S2589-7500(19)30028-7.

James, S. L., Abate, D., Abate, K. H., Abay, S. M., Abbafati, C., Abbasi, N., et al. (2018). Global, regional, and national incidence, prevalence, and years lived with disability for 354 diseases and injuries for 195 countries and territories, 1990–2017: A systematic analysis for the global burden of disease study 2017. *The Lancet*, *392*(10159), 1789–1858. https://doi.org/10.1016/S0140-6736(18)32279-7.

Ju, Y. J., Han, K.-T., Lee, T.-H., Kim, W., Park, J. H., & Park, E.-C. (2016). Association between weight control failure and suicidal ideation in overweight and obese adults: A cross-sectional study. *BMC Public Health*, *16*(1), 259. https://doi.org/10.1186/s12889-016-2940-1.

Kivimäki, M., Lawlor, D. A., Singh-Manoux, A., Batty, G. D., Ferrie, J. E., Shipley, M. J., et al. (2009). Common mental disorder and obesity: Insight from four repeat measures over 19 years: Prospective Whitehall II cohort study. *BMJ*, *339*, b3765. https://doi.org/10.1136/bmj.b3765.

Moussa, O. M., Ardissino, M., Kulatilake, P., Faraj, A., Muttoni, E., Darzi, A., et al. (2019). Effect of body mass index on depression in a UK cohort of 363 037 obese patients: A longitudinal analysis of transition. *Clinical Obesity*, *9*(3). https://doi.org/10.1111/cob.12305.

Organisation for Economic Cooperation and Development. (2017). *OECD. Health at a glance. OECD indicators. How does the United Kingdom compare?*. Retrieved from https://www.oecd.org/unitedkingdom/Health-at-a-Glance-2017-Key-Findings-UNITED-KINGDOM.pdf.

Pereira-Miranda, E., Costa, P. R. F., Queiroz, V. A. O., Pereira-Santos, M., & Santana, M. L. P. (2017). Overweight and obesity associated with higher depression prevalence in adults: A systematic review and meta-analysis. *Journal of the American College of Nutrition*, *36*(3), 223–233. https://doi.org/10.1080/07315724.2016.1261053.

Pratt, L., & Brody, D. (2014). *Depression and obesity in the US adult household population, 2005-2010. NCHS case brief, 167*. Hyattsville, MD: National Center for Health Statistics.

Roberts, R. E., Deleger, S., Strawbridge, W. J., & Kaplan, G. A. (2003). Prospective association between obesity and depression: Evidence from the Alameda County study. *International Journal of Obesity*, *27*(4), 514–521. https://doi.org/10.1038/sj.ijo.0802204.

Royal College of Physicians. (2019). *Obesity should be recognised as a disease*. Royal College of Physicians. Retrieved from https://www.rcplondon.ac.uk/news/rcp-calls-obesity-be-recognised-disease.

Simon, G. E., Von Korff, M., Saunders, K., Miglioretti, D. L., Crane, P. K., van Belle, G., et al. (2006). Association between obesity and psychiatric disorders in the US adult population. *Archives of General Psychiatry*, *63*(7), 824–830. https://doi.org/10.1001/archpsyc.63.7.824.

Speed, M. S., Jefsen, O. H., Børglum, A. D., Speed, D., & Østergaard, S. D. (2019). Investigating the association between body fat and depression via Mendelian randomization. *Translational Psychiatry*, *9*(1), 184. https://doi.org/10.1038/s41398-019-0516-4.

Tyrrell, J., Mulugeta, A., Wood, A. R., Zhou, A., Beaumont, R. N., Tuke, M. A., et al. (2019). Using genetics to understand the causal influence of higher BMI on depression. *International Journal of Epidemiology*, *48*(3), 834–848. https://doi.org/10.1093/ije/dyy223.

World Health Organization. (1992). *The ICD-10 classification of mental and behavioural disorders: Clinical descriptions and diagnostic guidelines*. Geneva: World Health Organization.

World Health Organization. (1998). Obesity: Preventing and managing the global epidemic. In *Report of a WHO consultation on obesity, Geneva, 3–5 June 1997*.

Zhao, G., Ford, E. S., Dhingra, S., Li, C., Strine, T. W., & Mokdad, A. H. (2009). Depression and anxiety among US adults: Associations with body mass index. *International Journal of Obesity*, *33*(2), 257–266. https://doi.org/10.1038/ijo.2008.268.

Chapter 12

# Heart rate variability and depression

**Renerio Fraguas and Bruno Pinatti Ferreira de Souza**
*Departamento e Instituto de Psiquiatria, Hospital das Clinicas, Faculdade de Medicina da Universidade de São Paulo, São Paulo, SP, Brazil*

## List of abbreviations

| | |
|---|---|
| **ACS** | acute coronary syndrome |
| **ANS** | autonomic nervous system |
| **BP** | blood pressure |
| **CAD** | coronary artery disease |
| **CHD** | coronary heart disease |
| **CHF** | chronic heart failure |
| **CVD** | cardiovascular disease |
| **DLPFC** | dorsolateral prefrontal cortex |
| **ECG** | electrocardiogram |
| **GDS** | geriatric depression scale |
| **HF-HRV** | High-frequency power heart rate variability |
| **HRV** | heart rate variability |
| **HRVB** | heart rate variability biofeedback |
| **IBIs** | interbeat intervals |
| **LF-HRV** | low-frequency power heart rate variability |
| **LF/HF ratio** | ratio of low frequency to high-frequency heart rate variability |
| **Ln** | natural logarithm |
| **LnHF** | natural log HF |
| **LnLF** | natural log LF |
| **MDD** | major depressive disorder |
| **MI** | myocardial infarction |
| **PFC** | prefrontal cortex |
| **PNS** | parasympathetic nervous system |
| **RMSSD** | root mean square of successive NN interval differences |
| **RSA** | respiratory sinus arrhythmia |
| **SDANN** | standard deviation of the averages of NN intervals |
| **SDNN** | standard deviation of NN intervals |
| **SDNNIDX** | mean of the standard deviations of all NN intervals |
| **sgACC** | subgenual anterior cingulate cortex |
| **SNS** | sympathetic nervous system |
| **SSRI** | serotonin selective reuptake inhibitor |
| **TCA** | tricyclic antidepressant |
| **ULF-HRV** | ultralow-frequency heart rate variability |
| **VLF-HRV** | very low-frequency heart rate variability |
| **VNS** | vagus nerve stimulation |

## Introduction

Heart rate variability (HRV) is defined as the changes in the time intervals between consecutive heartbeats, called interbeat intervals (IBIs) (Shaffer & Ginsberg, 2017). These changes are expected to occur and reflect the autonomic nervous system (ANS) adjusting the body to its physical and emotional states, and environmental stimulus. The ANS regulates heartbeat through its two branches, sympathetic (SNS) and parasympathetic nervous systems (PNS). The simplest way to interpret HVR is that high HRV reflects a healthy ANS response and low HRV reflects excessive sympathetic and/or inadequate parasympathetic modulation of heart rate. Even though the fundamental concepts of IBIs and sympathetic/parasympathetic modulation are simple, there are several forms of assessing HRV and several confounders that turn the interpretation of measurements not so simple. For example, the three main domains (i.e., time domain, frequency domain, and nonlinear metrics) used to measure HRV gather near 30 variables that are not interchangeable. Furthermore, HRV may be influenced by several factors, such as medications, diseases, gender, and age.

The clinical relevance of HRV was first reported in 1965 as a risk for fetal death (Hon & Lee, 1965). It has been largely studied in patients with cardiovascular disease (CVD) and is established as an important parameter of prognosis and mortality in this population. In patients with coronary heart disease (CHD), excessive sympathetic or reduced parasympathetic activity is associated with myocardial ischemia, ventricular tachycardia, ventricular fibrillation, and sudden cardiac death (Carney & Freedland, 2009). In patients following an acute myocardial infarction (MI), low HRV is a strong predictor of mortality and of arrhythmic events (TaskForce, 1996).

Depression is associated with higher incidence and worse prognosis of CVD. In patients hospitalized due to acute coronary syndrome (ACS) depression reaches a prevalence of up to 20% and is associated with higher all-cause mortality, cardiac mortality, and nonfatal events (Lichtman et al., 2014). Considering patients with chronic heart failure (CHF), depression prevalence varies according to the functional class and may reach 40% in patients with severe clinical impairment (Rutledge, Reis, Linke, Greenberg, & Mills, 2006). In this cardiac condition, depression is also

The Neuroscience of Depression. https://doi.org/10.1016/B978-0-12-817933-8.00038-4

associated with poor prognosis, at 1-year-follow-up patients with depression have a mortality odds ratio of 2.23 and rehospitalization odds ratio of 3.07 when compared to non-depressed patients ( Jiang et al., 2001). Changes in ANS function assessed by HRV analysis has been considered one of the possible pathophysiological links between depression and CVD (Carney et al., 1988).

In this chapter, we address the HRV measurement; the association of depression with HRV; HRV in depression and cardiovascular comorbidity; antidepressants effect on HRV; HRV and response to depression treatment; and treating depression by intervention on the autonomic system.

## HRV measurement

There are several ways to measure HRV and debates still exist regarding which method(s) should be used and how to interpret the physiological and clinical meaning of the various indices. Task forces have been setup in order to standardize terms, measures, and clinical relevance of the various methods of measuring HRV (Sassi et al., 2015; TaskForce, 1996). Currently, the three major measurements of HRV are time domain, frequency domain, and nonlinear metrics. It is important to mention that HRV indices are not interchangeable.

Time-domain indices of HRV quantify the amount of variability of IBIs. Table 1 describes the main indices of time domain and their frequency domain correlates.

Frequency-domain indices estimate the distribution of absolute or relative power in four frequency bands: ultra-low-frequency (ULF), very-low-frequency (VLF), low-frequency (LF), and high-frequency (HF) bands (Shaffer & Ginsberg, 2017). Table 2 illustrates the bands of frequency domain and their correlation with ANS. Some variables of these domains can be expressed in original units or as the natural logarithm (Ln) of original units to achieve a more normal distribution.

Nonlinear measurements require more complex analyses and quantify the unpredictability and complexity of a series of IBIs (Shaffer & Ginsberg, 2017).

Even though all HRV methods are based on electrocardiogram (ECG), the duration of the recordings may vary from a few minutes to more than 24 h and are usually classified as short-term and long-term recordings. This differentiation is important because the length of the recording may imply different physiological meanings.

## Association of depression and HRV

### Indices of HRV that are associated with depression

Depression has traditionally been associated with reduced HRV. Controversies have been found regarding the potential influence of antidepressants on this association and the real existence of a direct relationship between HRV and depression. However, meta-analysis studies have supported a specific association between depression and low HRV indices independently of antidepressants (Kemp et al., 2010; Koch, Wilhelm, Salzmann, Rief, & Euteneuer, 2019). A gradient effect has been found, HRV decreases as depression severity increase (Kemp et al., 2010). The most recent meta-analysis showed an association of depression with time domain and frequency domain measures, including low indices of HF-HRV, LF-HRV, SDNN, RMSSD, and high indices of LF/HF ratio. The significant reduction in LF-HRV was not found in a previous meta-analysis (Kemp et al., 2010), and some authors defend that depression may actually be associated with increased LF-HRV. The traditional concept that LF-HRV is an index of sympathetic tonus has been questioned and it has been proposed that it is influenced by both parasympathetic and sympathetic systems, what difficult the interpretation of its clinical meaning (Reyes del Paso, Langewitz, Mulder, van Roon, & Duschek, 2013). Both meta-analyses found increased levels of LF/HF ratio in depressed subjects, however consistent arguments have precluded to consider this ratio as an index of sympathetic/parasympathetic imbalance meaning (Reyes del Paso et al., 2013). Also, differently from previous meta-analysis, the most recent found small to moderate effect sizes suggesting that lower HRV explains a small part of the cardiovascular risk associated with depression. The findings of low indices of HF-HRV, SDNN, RMSSD indicate that depression is associated with low parasympathetic tonus.

### HRV and depression, cause and consequence

The meta-analyses mentioned above included cross-sectional studies, therefore they cannot support the conclusion of whether depression leads to an autonomic imbalance and to a negative effect on HRV, or an autonomic imbalance (assessed by HRV) leads to depression. Longitudinal studies have shown an association between changes in the severity of depression and changes in HRV indices. The study of de Guevara et al. found an inverse correlation between changes in depressive symptoms and changes in HF-HRV and LF-HRV among older adults over 6 months after acute coronary episode (de Guevara et al., 2004). However, such correlation also does not give a clue regarding which variable (i.e., depression or HRV indices) changed first (i.e., the cause) and which variable changed as the consequence.

Traditionally, it has been conceived that depression is the cause and changes in HRV a consequence. In this conception, depression is characterized by altered functional connections of brain structures (Ruiz Vargas, Soros, Shoemaker, & Hachinski, 2016) that are associated with control of sympathetic/parasympathetic tonus and

**TABLE 1** Characteristics of time domain measures of HRV.

| Time-domain indices | Unit | Description | Approximate frequency domain correlate |
|---|---|---|---|
| SDNN | Milliseconds | Standard deviation of all NN intervals | Total power |
| SDANN | Milliseconds | Standard deviation of the averages of NN intervals in all 5 min segments of the entire recording | Ultralow-frequency Heart rate variability |
| SDNNIDX | Milliseconds | Mean of the standard deviations of all NN intervals for all 5 min segments of the entire recording | Mean of 5-min total power |
| RMSSD | Milliseconds | Root mean square of successive NN interval differences | High-frequency heart rate variability |
| pNN50 | % | NN50 count divided by the total number of all NN intervals | High-frequency heart rate variability |

Based on Carney and Freedland (2009); Hartmann et al. (2018); TaskForce (1996).

**TABLE 2** Characteristics of frequency domain measures of HRV.

| Frequency-domain Indices | Unit | Description | Frequency range (Hz) | Autonomic nervous system correlate |
|---|---|---|---|---|
| Total power (24 h) | Milliseconds$^2$ | Variance of all NN intervals | Approximately $\leq 0.4$ | – |
| ULF | Milliseconds$^2$ | Power in the ultralow frequency range | $\leq 0.003$ | No consensus regarding ANS correlation |
| VLF | Milliseconds$^2$ | Power in the very low-frequency range | 0.003–0.04 | Parasympathetic nervous system seems to be the predominant ANS determinant |
| LF | Milliseconds$^2$ | Power in the low-frequency range | 0.04–0.15 | Seems to reflect both sympathetic and parasympathetic activity |
| HF | Milliseconds$^2$ | Power in the high-frequency range | 0.15–0.4 | Efferent vagal activity is the predominant determinant |
| LF/HF | Milliseconds$^2$/ Milliseconds$^2$ | Ratio LF/HF | | Higher values are considered to reflect predominant sympathetic activity and lower values reflect predominant vagal modulation activity (questioned) |

Based on Carney and Freedland (2009); Hartmann et al. (2018); Shaffer and Ginsberg (2017); TaskForce (1996). Column "Autonomic nervous system correlate" is restrict to ANS activity and does not address other factors that may influence the variables.

consequent reduction of HRV (see Fig. 1), indicating that depression has a negative impact on the cardiovascular system.

Recent data have supported that disturbances of cardiac autonomic status, particularly those indicating low vagal modulation, precede depressive symptoms. The Whitehall II study revealed that lower heart rate and higher SDNN, LF-HRV, HF-HRV, and RMSSD predicted lower incidence of depressive symptoms in those without depressive symptoms at baseline (Jandackova, Britton, Malik, & Steptoe, 2016). On the other hand, depressive symptoms did not predict subsequent lower HRV indices. In this matter, it has been proposed that that low vagal modulation, assessed by HRV, may be a risk factor for both depression and CHD (Jandackova et al., 2016) (see Fig. 2). Low HRV would be a mark of general vulnerability to psychopathology.

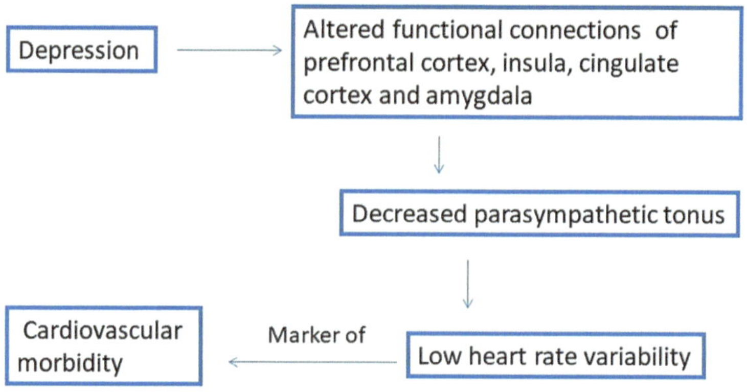


FIG. 1 Traditional concept of the link between depression and HRV.

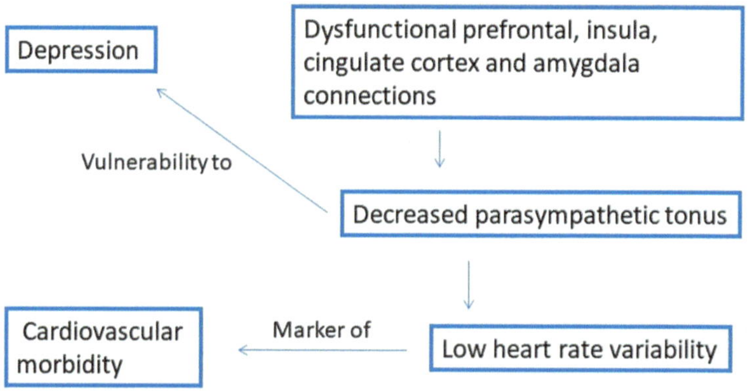


FIG. 2 Vagal vulnerability concept of the link between depression and HRV.

However, contradictory findings do exist for instance this relationship may differ among elderly patients. The recent study of An, Han et al. comprising 253 euthymic individuals, 65 years old or older, found that baseline HF-HRV was positively, and baseline LF-HRV and LF/HF ratio were negatively associated with the 5-year follow-up Geriatric Depression Scale (GDS) score suggesting that the onset of depressive symptoms may be preceded by high parasympathetic activity in old patients (An et al., 2020).

## Heart rate variability as a marker of depression

Heart rate variability has also been considered a potential marker for the diagnosis of depression. Kim et al. (2017) combining a biomarker panel of proteins and HRV indices achieved a model including RMSSD and four other parameters (apolipoprotein B, group-specific component, ceruloplasmin, and Sample entropy) presenting 80.1% of accuracy to distinguish MDD individuals from the control groups (Kim et al., 2017). Byun et al. used a protocol to assess autonomic responses to mental stress and recovery to differentiate patients with MDD from healthy controls. They found that the model including nonlinear and Poincaré plot HRV features showed better performance than that using linear indices, reaching 74.4% accuracy, 73% sensitivity, and 75.6% specificity in differentiating MDD and healthy controls (Byun et al., 2019). Using a paradigm including reactivity of HRV to the mental task and during paced respiration (i.e., 5-s inhalation and 5-s exhalation) Kobayashi, Guanghao, Shinba, Matsui, and Kirimoto (2017) could detect patients with MDD with a sensitivity of 86% under paced respiration and mental task conditions, higher than 68% found under mental task condition alone (Kobayashi et al., 2017).

The autonomic dysregulation of depression has been shown to be different from that of anxiety disorder. A study revealed that MDD patients had low HR-HRV and high LF/HF ratio during the initial-rest condition, and HF-HRV was less reactive to the task, while patients with generalized anxiety disorder showed high HF-HRV, and no impairment on autonomic reactivity. This high parasympathetic tone in anxious patients could reflect a parasympathetic response to anxiety while in depressed patients the results suggested the existence of an autonomic shift toward sympathetic activation and reduced reactivity to the task (Shinba, 2017). Also, in line with the specificity of HRV changes, measures of HF-HRV in melancholic, but not in non-melancholic patients have been found to be decreased compared to healthy controls (Kemp, Quintana, Quinn, Hopkinson, & Harris, 2014).

The association with lower indices of HRV (i.e., RMSSD, SDNN, Poincare Plot) compared to control subjects has been reported to occur in unipolar and bipolar depression (Bassett et al., 2016). It has been proposed that specificities in HRV could be used as a biomarker to distinguish bipolar from unipolar depression. Patients with bipolar II depression have been reported to have significantly lower mean R-R intervals variance (total HRV), LF-HRV, and HF-HRV, but higher LF/HF ratio than unipolar ones. Such results suggest that bipolar depression would have a cardiac sympathetic excitation and a vagal impairment, while unipolar depression would have an only vagal impairment (Chang, Chang, Kuo, & Huang, 2015). However, other studies have pointed to an absence of specificity of HRV changes. Actually, HRV changes have been associated with various psychopathologies including schizophrenia, attention problems, autism among others, and it has been proposed that HF-HRV marks the prefrontal cortex (PFC) function, and its impairment indicates a general vulnerability to psychopathology (Beauchaine & Thayer, 2015).

## Influence of gender and age in HRV of depressed and nondepressed subjects

It has been described that in nondepressed subjects, some HRV parameters of males significantly differ from those of females. Most studies have revealed lower LF-HRV and lower LF/HF ratio in females, suggesting that females have lower sympathetic modulation or higher parasympathetic modulation than males at rest (Koenig, Kemp, Beauchaine, Thayer, & Kaess, 2016). Some studies have reported higher HF-HRV in females compared to males, which could indicate higher parasympathetic tonus among females, but such findings have not been confirmed by other studies. Females have also been reported to have lower SDNN compared to males (Koenig et al., 2016). Such findings are intriguing considering that low SDNN may indicate low parasympathetic tonus. Incongruences regarding the difference in HRV between genders may in part be explained by the influence of age on HRV. A study investigating the influence of age and gender found that females have decreased absolute LF-HRV in the age range of 25–44 years and increased HF-HRV between 35 and 54 years compared to males. Such differences disappear at ages >44 years for LF-HRV and >54 years for HF-HRV (Voss, Schroeder, Heitmann, Peters, & Perz, 2015).

Data have supported that the effect of depression on HRV might differ according to gender. During the deep breathing, male patients with depression have been reported to have lower LF/HF than male controls, while significant differences were not observed in females (Kuang et al., 2019). Such findings suggest that the autonomic control of males and females could be differently altered by depression. It has been proposed that the imbalance of autonomic function in depressed women might explain the increased prevalence of CVD in these patients (Voss, Boettger, Schulz, Gross, & Bar, 2011).

Among depressed subjects, HRV time-domain indices (i.e., SDNN, SDANN1, RMSSD, and pNN50) have been inversely correlated to age for both genders, and old adults with clinical depression have been reported to have a reduction in LF-HRV (Brown et al., 2018).

## HRV in depression and cardiovascular comorbidity

Depression has been associated with changes in HRV in patients with coronary artery disease (CAD) and in patients with the acute coronary syndrome (ACS). During an ACS depression has been associated with changes in all three HRV domains. Time domain (SDNN, SDANN, and SDNNIDX) is significantly lower in patients with depression (Catipovic-Veselica et al., 2007; Wilkowska, Rynkiewicz, Wdowczyk, Landowski, & Cubala, 2019). Considering the frequency domain, depressed patients exhibit lower HF-HRV and higher LF/HF ratio, suggesting higher sympathetic activity (Guinjoan et al., 2004; Vigo et al., 2004).

In patients with CAD depression has been associated with lower SDNN, SDNNIDX, SDANN, pNN50, and RMSSD (Aydin Sunbul, Sunbul, & Gulec, 2017; Carney et al., 1995). Some studies have found a gradient relationship between depression severity and HRV. Higher severity of depression has been associated with lower pNN50 (Aydin Sunbul et al., 2017) and lower SDNN (Krittayaphong et al., 1997). Such a relationship may occur mostly for patients with moderately-to-severely depression compared to nondepressed individuals (Stein et al., 2000).

However, divergence does exist among studies, and in part, they may be explained by methodologic issues. For example, Gehi et al. assessed 873 patients with stable CHD, of which 195 were diagnosed with major depression (Gehi, Mangano, Pipkin, Browner, & Whooley, 2005) and found no association between depression and HRV. Later on, de Jonge et al. used the same sample, but compared somatic versus cognitive depression symptom dimensions instead of depressed and nondepressed, and found that somatic symptom scores were inversely associated with SDNN, SDANN, natural log LF (LnLF), and natural log HF (LnHF) (de Jonge, Mangano, & Whooley, 2007). However, after adjusting for medical comorbidities or lifestyle factors no association remained significant. The cognitive depression symptom dimension was not associated with HRV at any point in the analysis. These studies bring up important aspects to be considered in the association of HRV and depression. First, the definition of depression used today may not reflect the ideal construct to be pursued in the association with HRV, future studies assessing depressive symptoms and depression subtypes could bring relevant results. Second, since HRV is influenced by several factors, it is extremely important to perform studies with adequate sample size, assessment, and statistical analysis.

## The effect of antidepressants on HRV

Despite the amelioration of depression severity, in general, antidepressant treatment has not shown to reverse the decreased HRV in depressed patients. On the contrary, antidepressants have been consistently associated with negative cardiovascular impact. Actually, effect sizes of tricyclic (TCA) or noradrenergic antidepressants in reducing HRV are larger than those reported for depression (Licht, de Geus, van Dyck, & Penninx, 2010). The negative impact of TCA in HRV is probably associated with their anticholinergic and alpha 1-adrenergic properties. The antidepressant effect in HRV is reversible and systematically reverts with antidepressant discontinuation.

SSRIs, the first-line treatment for MDD, were not associated with an increase or decrease of HRV according to the first meta-analysis performed by Kemp et al. (2010). However, that does not exclude the existence of differences among new antidepressants regarding their effects on HRV. For example, compared to paroxetine, agomelatine treatment of MDD was associated with increases in LF-HRV and HF-HRV, while no changes on HRV indices were found among the paroxetine-treated group, suggesting that agomelatine has a more vagotonic effect (Yeh et al., 2016). Actually, differently from other SSRIs, paroxetine has an antimuscarinic (i.e., anticholinergic) effect, which may confer a different action in HRV. Such an effect is dose dependent, doses above 40mg/day approximate the effect of paroxetine in HRV to that of TCAs. The higher effect of paroxetine in decreasing HRV has been confirmed by a more recent meta-analysis either relative to nonusers, or users of other SSRI (Kemp et al., 2016). According to that review, among SSRI, fluoxetine was the only medication not to display significant alterations in HR or HRV.

## HRV and response to depression treatment

An intriguing question is that if once remission of depression is achieved its related changes in HRV could be reverted. Larger changes in vagal tone from before to after treatment have been associated with larger decreases in depressive severity even for nonpharmacological treatment (Chambers & Allen, 2002). Even for a short period interval of 2 weeks of antidepressant treatment the increase in HF-HRV associated with the decrease of severity of depression was enough to abolish a difference in HRV between depressive and controls subjects found at baseline (Hartmann, Schmidt, Sander, & Hegerl, 2018). Although improvement of depression has been associated with improvement of HRV, this improvement may be restricted to some but not HRV indices. In the study of Carney et al. (2000) treating depression with cognitive behavior therapy, only RMSSD improved to a level comparable to controls. These findings may indicate that some HRV changes in depression may not be restored. Actually, the study of Brunoni et al revealed that HRV indices did not change following treatment with either sertraline or transcranial direct current stimulation or even with the response to treatment. Findings that support the possibility that reduced HRV could be a trait marker of MDD, predisposing patients to negative health conditions (Brunoni et al., 2013). This matter still deserves clarification, one point to consider is that the most frequently adopted definitions of remission permit the presence of various depressive symptoms. In this matter, it is possible that some symptoms of depression and not all are responsible for the association with HRV changes. It has been found that symptoms such as "Guilt" and "loss of interest/pleasure in activities" and "psychomotor retardation" could be fundamental for the association of depression with changes in HRV (Borrione et al., 2018). In such a case, the reversal of HRV changes would require the full remission of these symptoms.

Another question that has been pursuing is if baseline HRV predicts response to antidepressants. For patients with anxious depression, higher HF-HRV has been found to predict good outcomes, whereas lower HF-HRV predicted poorer outcomes (Kircanski, Williams, & Gotlib, 2019). A study using escitalopram or quetiapine fumarate in two separate and consecutively run trials reveal that baseline respiratory sinus arrhythmia (RSA), and LF-HRV were significantly higher in responders than nonresponders (Hage, Sinacore, Heilman, Porges, & Halaris, 2017).

## Treating depression by intervention on the autonomic system

Considering that an autonomic dysregulation might be a putative underlying mechanism in depression, it is reasonable to hypothesize that treatments targeting this dysregulation could be well succeeded in ameliorating depression. Vagus nerve stimulation (VNS), originally applied to epilepsy, has also been reported to be effective for treating resistant depression, particularly in the long term (Rush et al., 2005). VNS was delivered originally by a generator surgically implanted in the left chest wall to deliver stimulus to the left vagus nerve. Lately, it has also been administered by a noninvasive method, transcutaneously delivering stimulus to vagal branches at the auricle of the ear. It modulates brain cortical-subcortical functional activity and cardiac function. VNS produces increases in HRV accessed by the RMSSD in depressed patients (Sperling et al., 2010). At the brain level, VNS would modulate the dorsolateral prefrontal cortex (DLPFC), subgenual anterior cingulate cortex (sgACC) networks (Iseger, van Bueren, Kenemans, Gevirtz, & Arns, 2020).

Heart rate variability biofeedback (HRVB) is a biobehavioral intervention targeting the enhancement of vagal control of heart rate. It enables the patient to obtain voluntary control of HRV. The patient views an instantaneous electronic display real-time information on HRV and tries to achieve its highest variability by controlling respiration. Heart rate variability is influenced by respiration. Each individual has a specific resonant frequency in which, as the individual inhales, HR rises, blood pressure (BP) falls, and the consequent baroreflex response produces a further increase in HR, with corresponding effects during exhalation. Achieving the resonance frequency, the individual produces large increases in HRV (Lehrer et al., 2003). Significant improvement of depression has been reported after four sessions of HRVB as add-on therapy in patients with MDD (Karavidas et al., 2007).

## Key facts

- The electrocardiogram was invented in the 19th century. This exam and its variations provide the basic data for HRV assessment.
- Clinical relevance of HRV was established in the 1960s and its research in depression started in the 1980s.
- A model using HRV variables reached 74.4% accuracy, 73% sensitivity, and 75.6% specificity in differentiating patients with the major depressive disorder from healthy controls.
- HRV alterations are associated with diseases other than mental ones such as diabetes, cirrhosis, and cancer.
- Other HRV utilities include assessment and guidance in sport performance.
- Tricyclic antidepressants and paroxetine above 40 mg/day seem to have the most negative impact in HRV, but this effect is reversible and systematically reverts with antidepressant discontinuation.

## Summary points

- HRV can be measured by time domain, frequency domain, and nonlinear methods and they are not interchangeable.
- The association of depression with high-frequency power HRV (HF-HRV), standard deviation of NN intervals (SDNN), and root mean square of successive NN interval differences (RMSSD) is well established.
- The LF/HF ratio has been found to be increased in depressed subjects, however, its role as an index of sympathetic/parasympathetic imbalance has been questioned.
- The association of depression with LF-HRV is controversial so is the interpretation of LF-HRV as an index of sympathetic tonus.
- The reduction of HRV is traditionally considered as one of the mechanisms behind the association of depression and cardiovascular disease (CVD).
- In contrast to the traditional view, data support that HRV changes may precede depression and could represent a mark of decreased parasympathetic tonus and vulnerability do depression.
- The amelioration of depression severity may be accompanied by an increase in HRV.
- Antidepressant treatment may decrease HRV precluding its possible improvement by amelioration of depression.

## Mini-dictionary of terms

**Acute coronary syndrome** Comprehend several conditions that course with a sudden reduction in heart blood flow, such as myocardial infarction and unstable angina.

**Cardiovascular disease** This term embraces several diseases that affect the heart and vascular systems, such as myocardial infarction, abnormal heart rhythms, stroke, and peripheral arteries diseases.

**Coronary heart disease** Characterized by impairment in the heart's arteries blood flow, usually due to arteriosclerosis.

**Heart rate variability (HRV)** Changes in the time intervals between consecutive heartbeats, called interbeat intervals.

**Heart rate variability (HRV) biofeedback** A biobehavioral intervention targeting the enhancement of vagal control of heart rate.

**Poincare plot** A method of nonlinear analysis of heart rate variability (HRV).

**Respiratory sinus arrhythmia (RSA)** An oscillation in the beat-to-beat heart rate pattern that occurs in the frequency of spontaneous breathing. It has been considered an estimate of cardiac vagal tone.

## References

An, H., Han, J. W., Jeong, H. G., Kim, T. H., Lee, J. J., Lee, S. B., et al. (2020). Parasympathetic predominance is a risk factor for future depression: A prospective cohort study. *Journal of Affective Disorders*, *260*, 232–237. https://doi.org/10.1016/j.jad.2019.09.015.

Aydin Sunbul, E., Sunbul, M., & Gulec, H. (2017). The impact of major depression on heart rate variability and endothelial dysfunction in patients with stable coronary artery disease. *General Hospital Psychiatry*, *44*, 4–9. https://doi.org/10.1016/j.genhosppsych.2016.10.006.

Bassett, D., Bear, N., Nutt, D., Hood, S., Bassett, S., & Hans, D. (2016). Reduced heart rate variability in remitted bipolar disorder and recurrent depression. *The Australian and New Zealand Journal of Psychiatry*, *50*(8), 793–804. https://doi.org/10.1177/0004867416652734.

Beauchaine, T. P., & Thayer, J. F. (2015). Heart rate variability as a transdiagnostic biomarker of psychopathology. *International Journal of Psychophysiology*, *98*(2 Pt. 2), 338–350. https://doi.org/10.1016/j.ijpsycho.2015.08.004.

Borrione, L., Brunoni, A. R., Sampaio-Junior, B., Aparicio, L. M., Kemp, A. H., Bensenor, I., et al. (2018). Associations between symptoms of depression and heart rate variability: An exploratory study. *Psychiatry Research*, *262*, 482–487. https://doi.org/10.1016/j.psychres.2017.09.028.

Brown, L., Karmakar, C., Gray, R., Jindal, R., Lim, T., & Bryant, C. (2018). Heart rate variability alterations in late life depression: A meta-analysis. *Journal of Affective Disorders*, *235*, 456–466. https://doi.org/10.1016/j.jad.2018.04.071.

Brunoni, A. R., Kemp, A. H., Dantas, E. M., Goulart, A. C., Nunes, M. A., Boggio, P. S., et al. (2013). Heart rate variability is a trait marker of major depressive disorder: Evidence from the sertraline vs. electric current therapy to treat depression clinical study. *The International Journal of Neuropsychopharmacology*, *16*(9), 1937–1949. https://doi.org/10.1017/S1461145713000497.

Byun, S., Kim, A. Y., Jang, E. H., Kim, S., Choi, K. W., Yu, H. Y., et al. (2019). Detection of major depressive disorder from linear and nonlinear heart rate variability features during mental task protocol. *Computers in Biology and Medicine*, *112*, 103381. https://doi.org/10.1016/j.compbiomed.2019.103381.

Carney, R. M., & Freedland, K. E. (2009). Depression and heart rate variability in patients with coronary heart disease. *Cleveland Clinic Journal of Medicine*, *76*(Suppl. 2), S13–S17. https://doi.org/10.3949/ccjm.76.s2.03.

Carney, R. M., Freedland, K. E., Stein, P. K., Skala, J. A., Hoffman, P., & Jaffe, A. S. (2000). Change in heart rate and heart rate variability during treatment for depression in patients with coronary heart disease. *Psychosomatic Medicine*, *62*(5), 639–647. https://doi.org/10.1097/00006842-200009000-00007.

Carney, R. M., Rich, M. W., teVelde, A., Saini, J., Clark, K., & Freedland, K. E. (1988). The relationship between heart rate, heart rate variability and depression in patients with coronary artery disease. *Journal of Psychosomatic Research*, *32*(2), 159–164. https://doi.org/10.1016/0022-3999(88)90050-5.

Carney, R. M., Saunders, R. D., Freedland, K. E., Stein, P., Rich, M. W., & Jaffe, A. S. (1995). Association of depression with reduced heart rate variability in coronary artery disease. *The American Journal of Cardiology*, *76*(8), 562–564. https://doi.org/10.1016/s0002-9149(99)80155-6.

Catipovic-Veselica, K., Galic, A., Jelic, K., Baraban-Glavas, V., Saric, S., Prlic, N., et al. (2007). Relation between major and minor depression and heart rate, heart-rate variability, and clinical characteristics of patients with acute coronary syndrome. *Psychological Reports*, *100* (3 Pt. 2), 1245–1254. https://doi.org/10.2466/pr0.100.4.1245-1254.

Chambers, A. S., & Allen, J. J. B. (2002). Vagal tone as an indicator of treatment response in major depression. *Psychophysiology*, *39*(6), 861–864. https://doi.org/10.1111/1469-8986.3960861.

Chang, H. A., Chang, C. C., Kuo, T. B., & Huang, S. Y. (2015). Distinguishing bipolar II depression from unipolar major depressive disorder: Differences in heart rate variability. *The World Journal of Biological Psychiatry*, *16*(5), 351–360. https://doi.org/10.3109/15622975.2015.1017606.

de Guevara, M. S., Schauffele, S. I., Nicola-Siri, L. C., Fahrer, R. D., Ortiz-Fragola, E., Martinez-Martinez, J. A., et al. (2004). Worsening of depressive symptoms 6 months after an acute coronary event in older adults is associated with impairment of cardiac autonomic function. *Journal of Affective Disorders*, *80*(2-3), 257–262. https://doi.org/10.1016/S0165-0327(03)00105-8.

de Jonge, P., Mangano, D., & Whooley, M. A. (2007). Differential association of cognitive and somatic depressive symptoms with heart rate variability in patients with stable coronary heart disease: Findings from the Heart and Soul Study. *Psychosomatic Medicine*, *69*(8), 735–739. https://doi.org/10.1097/PSY.0b013e31815743ca.

Gehi, A., Mangano, D., Pipkin, S., Browner, W. S., & Whooley, M. A. (2005). Depression and heart rate variability in patients with stable coronary heart disease: Findings from the Heart and Soul Study. *Archives of General Psychiatry*, *62*(6), 661–666. https://doi.org/10.1001/archpsyc.62.6.661.

Guinjoan, S. M., de Guevara, M. S., Correa, C., Schauffele, S. I., Nicola-Siri, L., Fahrer, R. D., et al. (2004). Cardiac parasympathetic dysfunction related to depression in older adults with acute coronary syndromes. *Journal of Psychosomatic Research*, *56*(1), 83–88. https://doi.org/10.1016/S0022-3999(03)00043-6.

Hage, B., Sinacore, J., Heilman, K., Porges, S. W., & Halaris, A. (2017). Heart rate variability predicts treatment outcome in major depression. *Journal of Psychiatry and Brain Science*, *2*(6).

Hartmann, R., Schmidt, F. M., Sander, C., & Hegerl, U. (2018). Heart rate variability as indicator of clinical state in depression. *Frontiers in Psychiatry*, *9*, 735. https://doi.org/10.3389/fpsyt.2018.00735.

Hon, E. H., & Lee, S. T. (1965). Electronic evaluations of the fetal heart rate patterns preceding fetal death, further observations. *American Journal of Obstetrics and Gynecology*, *87*, 814–826.

Iseger, T. A., van Bueren, N. E. R., Kenemans, J. L., Gevirtz, R., & Arns, M. (2020). A frontal-vagal network theory for major depressive disorder: Implications for optimizing neuromodulation techniques. *Brain Stimulation*, *13*(1), 1–9. https://doi.org/10.1016/j.brs.2019.10.006.

Jandackova, V. K., Britton, A., Malik, M., & Steptoe, A. (2016). Heart rate variability and depressive symptoms: A cross-lagged analysis over a 10-year period in the Whitehall II study. *Psychological Medicine*, *46*(10), 2121–2131. https://doi.org/10.1017/S003329171600060X.

Jiang, W., Alexander, J., Christopher, E., Kuchibhatla, M., Gaulden, L. H., Cuffe, M. S., et al. (2001). Relationship of depression to increased risk of mortality and rehospitalization in patients with congestive heart failure. *Archives of Internal Medicine*, *161*(15), 1849–1856. https://doi.org/10.1001/archinte.161.15.1849.

Karavidas, M. K., Lehrer, P. M., Vaschillo, E., Vaschillo, B., Marin, H., Buyske, S., et al. (2007). Preliminary results of an open label study of heart rate variability biofeedback for the treatment of major depression. *Applied Psychophysiology and Biofeedback*, *32*(1), 19–30. https://doi.org/10.1007/s10484-006-9029-z.

Kemp, A. H., Fráguas, R., Brunoni, A. R., Bittencourt, M. S., Nunes, M. A., Dantas, E. M., et al. (2016). Differential associations of specific selective serotonin reuptake inhibitors with resting-state heart rate and heart rate variability: Implications for health and well-being. *Psychosomatic Medicine*, *78*(7), 810–818. https://doi.org/10.1097/PSY.0000000000000336. 27219492.

Kemp, A. H., Quintana, D. S., Gray, M. A., Felmingham, K. L., Brown, K., & Gatt, J. M. (2010). Impact of depression and antidepressant treatment on heart rate variability: A review and meta-analysis. *Biological Psychiatry*, *67*(11), 1067–1074. https://doi.org/10.1016/j.biopsych.2009.12.012.

Kemp, A. H., Quintana, D. S., Quinn, C. R., Hopkinson, P., & Harris, A. W. (2014). Major depressive disorder with melancholia displays robust alterations in resting state heart rate and its variability: Implications for future morbidity and mortality. *Frontiers in Psychology*, *5*, 1387. https://doi.org/10.3389/fpsyg.2014.01387.

Kim, E. Y., Lee, M. Y., Kim, S. H., Ha, K., Kim, K. P., & Ahn, Y. M. (2017). Diagnosis of major depressive disorder by combining multimodal information from heart rate dynamics and serum proteomics using machine-learning algorithm. *Progress in Neuro-Psychopharmacology & Biological Psychiatry*, *76*, 65–71. https://doi.org/10.1016/j.pnpbp.2017.02.014.

Kircanski, K., Williams, L. M., & Gotlib, I. H. (2019). Heart rate variability as a biomarker of anxious depression response to antidepressant medication. *Depression and Anxiety*, *36*(1), 63–71. https://doi.org/10.1002/da.22843.

Kobayashi, M., Guanghao, S., Shinba, T., Matsui, T., & Kirimoto, T. (2017). Simple and objective screening of major depressive disorder by heart rate variability analysis during paced respiration and mental task conditions. *Conference Proceedings: Annual International Conference of the IEEE Engineering in Medicine and Biology Society*, *2017*, 1316–1319. https://doi.org/10.1109/EMBC.2017.8037074.

Koch, C., Wilhelm, M., Salzmann, S., Rief, W., & Euteneuer, F. (2019). A meta-analysis of heart rate variability in major depression. *Psychological Medicine*, *49*(12), 1948–1957. https://doi.org/10.1017/S0033291719001351.

Koenig, J., Kemp, A. H., Beauchaine, T. P., Thayer, J. F., & Kaess, M. (2016). Depression and resting state heart rate variability in children and adolescents—A systematic review and meta-analysis. *Clinical Psychology Review*, *46*, 136–150. https://doi.org/10.1016/j.cpr.2016.04.013.

Krittayaphong, R., Cascio, W. E., Light, K. C., Sheffield, D., Golden, R. N., Finkel, J. B., et al. (1997). Heart rate variability in patients with coronary artery disease: Differences in patients with higher and lower depression scores. *Psychosomatic Medicine*, *59*(3), 231–235. https://doi.org/10.1097/00006842-199705000-00004.

Kuang, D., Cui, L., Kuang, S., Yang, R., Chen, X., Zhang, L., et al. (2019). Effect of gender-related depression on heart rate variability during an autonomic nervous test. *Psychiatry Research*, *272*, 258–264. https://doi.org/10.1016/j.psychres.2018.12.099.

Lehrer, P. M., Vaschillo, E., Vaschillo, B., Lu, S. E., Eckberg, D. L., Edelberg, R., et al. (2003). Heart rate variability biofeedback increases baroreflex gain and peak expiratory flow. *Psychosomatic Medicine*, *65*(5), 796–805. https://doi.org/10.1097/01.psy.0000089200.81962.19.

Licht, C. M., de Geus, E. J., van Dyck, R., & Penninx, B. W. (2010). Longitudinal evidence for unfavorable effects of antidepressants on heart rate variability. *Biological Psychiatry*, *68*(9), 861–868. https://doi.org/10.1016/j.biopsych.2010.06.032.

Lichtman, J. H., Froelicher, E. S., Blumenthal, J. A., Carney, R. M., Doering, L. V., Frasure-Smith, N., et al. (2014). Depression as a risk factor for poor prognosis among patients with acute coronary syndrome: Systematic review and recommendations: A scientific statement from the American Heart Association. *Circulation*, *129*(12), 1350–1369. https://doi.org/10.1161/CIR.0000000000000019.

Reyes del Paso, G. A., Langewitz, W., Mulder, L. J., van Roon, A., & Duschek, S. (2013). The utility of low frequency heart rate variability as an index of sympathetic cardiac tone: A review with emphasis on a reanalysis of previous studies. *Psychophysiology*, *50*(5), 477–487. https://doi.org/10.1111/psyp.12027.

Ruiz Vargas, E., Soros, P., Shoemaker, J. K., & Hachinski, V. (2016). Human cerebral circuitry related to cardiac control: A neuroimaging meta-analysis. *Annals of Neurology*, *79*(5), 709–716. https://doi.org/10.1002/ana.24642.

Rush, A. J., Sackeim, H. A., Marangell, L. B., George, M. S., Brannan, S. K., Davis, S. M., et al. (2005). Effects of 12 months of vagus nerve stimulation in treatment-resistant depression: A naturalistic study. *Biological Psychiatry*, *58*(5), 355–363. https://doi.org/10.1016/j.biopsych.2005.05.024.

Rutledge, T., Reis, V. A., Linke, S. E., Greenberg, B. H., & Mills, P. J. (2006). Depression in heart failure a meta-analytic review of prevalence, intervention effects, and associations with clinical outcomes. *Journal of the American College of Cardiology*, *48*(8), 1527–1537. https://doi.org/10.1016/j.jacc.2006.06.055.

Sassi, R., Cerutti, S., Lombardi, F., Malik, M., Huikuri, H. V., Peng, C. K., et al. (2015). Advances in heart rate variability signal analysis: Joint position statement by the e-Cardiology ESC Working Group and the European Heart Rhythm Association co-endorsed by the Asia Pacific Heart Rhythm Society. *Europace*, *17*(9), 1341–1353. https://doi.org/10.1093/europace/euv015.

Shaffer, F., & Ginsberg, J. P. (2017). An overview of heart rate variability metrics and norms. *Frontiers in Public Health*, *5*, 258. https://doi.org/10.3389/fpubh.2017.00258.

Shinba, T. (2017). Major depressive disorder and generalized anxiety disorder show different autonomic dysregulations revealed by heart-rate variability analysis in first-onset drug-naive patients without comorbidity. *Psychiatry and Clinical Neurosciences*, *71*(2), 135–145. https://doi.org/10.1111/pcn.12494.

Sperling, W., Reulbach, U., Bleich, S., Padberg, F., Kornhuber, J., & Mueck-Weymann, M. (2010). Cardiac effects of vagus nerve stimulation in patients with major depression. *Pharmacopsychiatry*, *43*(1), 7–11. https://doi.org/10.1055/s-0029-1237374.

Stein, P. K., Carney, R. M., Freedland, K. E., Skala, J. A., Jaffe, A. S., Kleiger, R. E., et al. (2000). Severe depression is associated with markedly reduced heart rate variability in patients with stable coronary heart disease. *Journal of Psychosomatic Research*, *48*(4-5), 493–500. https://doi.org/10.1016/s0022-3999(99)00085-9.

TaskForce. (1996). Heart rate variability: Standards of measurement, physiological interpretation, and clinical use. Task Force of the European Society of Cardiology and the North American Society of Pacing and Electrophysiology. European Heart Journal, 17(3), 354–381. Retrieved from https://www.ncbi.nlm.nih.gov/pubmed/8737210.

Vigo, D. E., Nicola Siri, L., Ladrón De Guevara, M. S., Martínez-Martínez, J. A., Fahrer, R. D., Cardinali, D. P., et al. (2004). Relation of depression to heart rate nonlinear dynamics in patients > or = 60 years of age with recent unstable angina pectoris or acute myocardial infarction. *The American Journal of Cardiology*, *93*(6), 756–760. https://doi.org/10.1016/j.amjcard.2003.11.056.

Voss, A., Boettger, M. K., Schulz, S., Gross, K., & Bar, K. J. (2011). Gender-dependent impact of major depression on autonomic cardiovascular modulation. *Progress in Neuro-Psychopharmacology & Biological Psychiatry*, *35*(4), 1131–1138. https://doi.org/10.1016/j.pnpbp.2011.03.015.

Voss, A., Schroeder, R., Heitmann, A., Peters, A., & Perz, S. (2015). Short-term heart rate variability—Influence of gender and age in healthy subjects. *PLoS One*, *10*(3). https://doi.org/10.1371/journal.pone.0118308, e0118308.

Wilkowska, A., Rynkiewicz, A., Wdowczyk, J., Landowski, J., & Cubala, W. J. (2019). Heart rate variability and incidence of depression during the first six months following first myocardial infarction. *Neuropsychiatric Disease and Treatment*, *15*, 1951–1956. https://doi.org/10.2147/NDT.S212528.

Yeh, T. C., Kao, L. C., Tzeng, N. S., Kuo, T. B., Huang, S. Y., Chang, C. C., et al. (2016). Heart rate variability in major depressive disorder and after antidepressant treatment with agomelatine and paroxetine: Findings from the Taiwan Study of Depression and Anxiety (TAISDA). *Progress in Neuro-Psychopharmacology & Biological Psychiatry*, *64*, 60–67. https://doi.org/10.1016/j.pnpbp.2015.07.007.

Chapter 13

# Neuroinflammation and depression

B. García Bueno, K. MacDowell, J.L.M. Madrigal, and J.C. Leza
*Department of Pharmacology and Toxicology, Faculty of Medicine, University Complutense, Cibersam, IUIN, Imas12, Madrid, Spain*

## List of abbreviations

**AP-1** activator protein-1
**ARE** antioxidant response elements
**BBB** blood-brain barrier
**BDNF** brain-derived neurotrophic factor
**BH4** tetrahydrobiopterin cofactor
**CMS** chronic mild stress
**CNS** central nervous system
**COX2** inducible isoform of cyclooxygenase
**CRP** C-reactive protein
**CSF** cerebrospinal fluid
**DAMPs** damage-associated molecular patterns
**GFAP** glial fibrillary acidic protein
**GR** glucocorticoid receptor
**HMGB1** high mobility group box 1 protein
**HPA** hypothalamic pituitary adrenal axis
**IDO** indoleamine 2.3-dioxygenase
**IFNγ** interferon gamma
**IL *xx*** interleukins: IL1β, IL6, IL10, IL18
**IL1R KO** IL-1 receptor knockout
**iNOS** inducible isoform of nitric oxide synthase
**LPS** lipopolysaccharide
**MAPKs** mitogen-activated protein kinases
**MCP-1 or CCL2** monocyte chemoattractant protein 1
**MD** major depression
**NFκB** nuclear factor kappa B
**NLRP3** NOD-, LRR- and pyrin domain-containing protein 3
**NMDAR** *N*-methyl-D-aspartate receptor
**NOD-like receptors** nucleotide-binding oligomerization domain-like receptors
**NRF2** nuclear transcription factor (erythroid-derived 2)-like 2
**NSAID** non-steroidal anti-inflammatory drug
**O&NS** oxidative and nitrosative stress
**PAMPs** molecular patterns associated with pathogens
**PET** positron emission tomography
**PGE2** prostaglandin E2
**SNRI** serotonin–norepinephrine reuptake inhibitors
**SSRI** selective serotonin reuptake inhibitors
**TCA** tricyclic antidepressants
**TGFβ** transforming growth factor beta
**TLR** toll-like receptor
**TNFα** tumor necrosis factor α
**TNFR1** tumor necrosis factor receptor 1
**TSPO** translocator protein

## General aspects of neuroinflammation

The term inflammation describes the physiological response activated in the organisms in order to protect cells, tissues, and systems from harmful agents. These agents, in some cases, are microorganisms such as bacteria and viruses, but many other different kinds of stimuli can trigger an inflammatory response including mechanical injuries, toxic substances, cellular debris, and stress (physical or psychological).

The detection of these potential threats activates a series of actions leading to the displacement of specialized cells to the compromised area in order to eliminate the source of damage. The associated changes are more easily detected when the affected area is in outer parts of the body such as the skin: the increased blood flow facilitates the supply of immune cells that can leave the blood vessels and migrate to a location closer to the target. The vasodilatation and the extravasation lead to redness, swelling (edema), and heating of the affected area. These alterations combined with the sensation of pain and "loss of function" constitute the five classical signs of inflammation. However, given its characteristics, this kind of response does not occur in the central nervous system (CNS) the same way it does in other parts of the body. The mere swelling of the tissue, which may constitute a burden of relative severity in the periphery, can have catastrophic consequences when it affects the brain. In addition, the blood-brain barrier, which provides additional protection to the CNS by isolating it from potentially harmful agents present in the blood, also obstructs the entrance of immune cells.

Therefore, the regular inflammatory response does not take place in the CNS. But it possesses alternative mechanisms, which constitute what is known as *neuroinflammation*. The most relevant cell type responsible for the surveillance of the CNS is the microglia. In normal conditions, these cells are continuously analyzing the environment. When they detect a potential threat to the homeostasis of the

The Neuroscience of Depression. https://doi.org/10.1016/B978-0-12-817933-8.00001-3

tissue, microglia cells respond with a series of changes that include the retraction of their projections, an enlargement of their cell bodies that gives them an amoeboid shape and the production of cytokines and reactive oxygen and nitrogen species. Yet, this transition from the inactivated to the activated state is a complex process in which microglia can present different phenotypes and activities also included the release of anti-inflammatory mediators in order to control the whole process.

Astrocytes also seem to play a relevant role in the neuroinflammatory response. Like microglia, these cells also experience morphological changes when they respond to potentially noxious stimuli. The most characteristic ones include the enlargement of the cell body (hypertrophy) and their proliferation. However, the alteration more commonly used to detect the activation of these cells is their accumulation of the protein known as glial fibrillary acidic protein (GFAP).

Both microglia and astrocytes can detect phagocyte and eliminate noxious agents. In order to achieve this, these cells secrete diverse substances that are toxic to bacteria or other invading microorganisms. This response can be triggered in a relatively short time and help to prevent early this way the proliferation of dangerous cells. However, while it seems convenient to get rid quickly of the deleterious agent, this procedure is not very selective and is usually accompanied by the loss of healthy cells that are located near the core of inflammation, including neurons. In normal conditions, this loss of some healthy cells could be an acceptable price to pay in order to avoid the proliferation of dangerous organisms, however, in certain situations, the inflammatory response is conducted in an exaggerated way, producing large amounts of cytotoxic substances, or it is maintained for periods of time longer than necessary. The last seems to be the case in neurodegenerative diseases such as Alzheimer´s or Parkinson´s diseases and is the focus of numerous studies nowadays (Kinney et al., 2018; Krashia, Nobili, & D'Amelio, 2019). But other neurological and psychiatric pathologies, including depression, have also been confirmed to be associated with neuroinflammation.

## Evidence about neuroinflammation in depression

Depression has been considered as a psychiatric disease resulting from monoamine deficiency in some regions of the brain, but the complexity of this disease transcends beyond the monoaminergic hypothesis, leaving lots of unknowns, such as the longtime of latency in the response to traditional drug treatment and the resistance of some patients to it.

During the last decades, an extensive amount of evidence has emerged from animal models and clinical studies that point to a connection between inflammation and depression, mainly based on the association between the presence of some pro-inflammatory cytokines such as TNFα and IL6 and depressive symptoms. From the initial findings (rev in Dowlati et al., 2010), the inflammatory hypothesis of depression began to develop, being complementary with the monoamine-related pathophysiology, emphasizing the role of immune-inflammatory dysfunctions present in the disease.

Also, in the last two decades, many evidences have been presented focusing on stress exposure as one of the main sources of inflammation. Nowadays it is known that acute intense or long-lasting stressful stimuli can affect the organism in several aspects beyond the release of pro-inflammatory cytokines. This topic is further discussed in the Section "Neuroinflammation as commonplace for stress and depression."

### Preclinical research

A wide range of animals ranging from zebrafish to non-human primates are used for the study of depression, but rodents remain the gold standard for the study of this disease. Although the symptoms induced in animal models cannot be compared to the clinical depression suffered by patients, rodents can replicate some aspects of the disease. In this way, they give us an important tool to delve into the pathogenesis of the disease, to elucidate the possible mechanisms involved in the pathophysiology, and to search for new therapeutic strategies.

Within the rat and mouse animal models, those that cover a greater number of neuroanatomical and biochemical—including inflammation—alterations similar to those present in depression are based on chronic stress exposure (rev. in Gururajan, Reif, Cryan, & Slattery, 2019). Also, at the behavioral level, stress-based models of depression induce an adaptive depressive response known as "Sickness Behavior," which includes lethargy, anhedonia, decreased food intake, and social withdrawal, among others.

A large number of acute and chronic stress exposure studies have reported an increase in the peripheral levels of pro-inflammatory cytokines such as IL1β, IL6, and TNFα (Cheng, Jope, & Beurel, 2015; López-López et al., 2016; Wohleb, Franklin, Iwata, & Duman, 2016); it has also been observed that systemic injection of pro-inflammatory cytokines induce depressive behaviors in animal models (Dantzer, 2001) and even that treatment with antidepressants is able to normalize the levels of these cytokines (Lu et al., 2017). In addition, studies using genetically modified models for the IL1 receptor (IL1R KO) have shown that after a 5-week chronic mild stress (CMS) protocol the expected depressive-type behavior does not occur (Goshen et al., 2008). Similarly, KO mice for TNFαR1 present an antidepressant behavior (Kaster, Gadotti,

Calixto, Santos, & Rodrigues, 2012). On the other hand, it has been observed that anti-inflammatory cytokines such as IL10 and TGFβ, which regulate the production of pro-inflammatory cytokines, are decreased in the brain of CMS exposed animals (You et al., 2011).

This "cytokine environment" present in the brain in experimental models of depression is similar to that described in the periphery (Dantzer, O'Connor, Freund, Johnson, & Kelley, 2008) and could be a consequence of the intrinsic effect of stress, or due to a higher permeability of the brain-periphery communication, in which cytokines can pass the blood-brain barrier (BBB) (Cheng et al., 2018). Another contribution to the inflammatory environment is given by peripheral immune cells that migrate to the cerebral vasculature, the choroid plexus, and the meninges. These peripheral cells have a higher inflammatory potential than the microglia resident in the brain (Schwartz, Kipnis, Rivest, & Prat, 2013). Pro-inflammatory cytokines also activate the canonical NFkB pathway by binding to their specific receptors, which leads to the intracellular increase of inducible pro-inflammatory enzymes, iNOS, and COX2, which will induce the production of pro-inflammatory mediators such as PGE2 and oxidative/nitrosative species, which will contribute to the increase in lipid peroxidation and cellular damage present in the brain (Madrigal et al., 2002).

Another important source of danger signals that comes from the periphery and contributes to neuroimmune deregulation and inflammation, is the gut-brain axis. It has been observed that stress increases intestinal permeability, an effect also known as "leaky gut," that enables that structural components present in the microbiota (molecular patterns associated with pathogens, PAMPs), can go into circulation and reach the brain, where they will be recognized by specific innate immunity receptors. In animals subjected to CMS protocols, it has been found an increase in the expression of the innate immunity sentinel receptor TLR4 (specific LPS receptor, a component of the Gram-negative bacteria cellular wall) and activation of the pathway signaling through NFkB, increasing the expression of pro-inflammatory mediators (Gárate et al., 2011; Martín-Hernández et al., 2016). Also, mice exposed to social defeat have an upregulation of TLR2 and TLR4 (Wohleb et al., 2012). Downstream to TLR4, NLRP3, a component of the Nod-like receptors family, detects endogenous signals associated with damage such as HMGB1. In this vein, it has been observed that exposure to stress increases HMGB1 levels, inducing the secretion of pro-inflammatory cytokines by microglia (Weber, Frank, Tracey, Watkins, & Maier, 2015). In fact, blocking NLRP3 signaling in mice exposed to CMS causes a decrease in IL1β levels in the hippocampus and an improvement in depressive symptoms (Zhang et al., 2015). The intracellular consequences after experimental models of depressive behavior via TLR-4 and NLRP3 lead to the activation of MAPKs and the subsequent induction of AP-1, which coordinates the transcription of inflammatory genes (Martín-Hernández et al., 2016; Su et al., 2017).

Microglia are the principal immune cellular type of the brain and play an important role in neuroplasticity, neuroprotection, and maintaining the balance of cytokines, but it is also the main source of neuroinflammation, capable of exacerbating depressive symptoms during stress exposure (Singhal & Baune, 2017). Microglia can change between different roles and functions depending on their activation status and specific stimulus exposed. It has been found in animal models of depression a reduction in the density of neuroprotective microglia (M2 type) and an increment in the hyperactivity of pro-inflammatory microglia (M1 type) in the dentate gyrus, hippocampus, prefrontal cortex, and amygdala, although recent evidences suggest that the dichotomic M1/M2 profile could be too simplistic.

In addition to microglia, in the last years, it is emerging the study of the role of CNS-border associated non-parenchymal myeloid cells, such as perivascular (PVM), meningeal (MGM), and plexus choroideus (CPM) macrophages in the regulation of neuroinflammation in stress-related neuropathologies.

## Human subjects research

All the scientific evidence supporting the existence of systemic inflammation, in human major depression (MD) needs complementary studies focused on the characterization of neuroinflammation by imaging techniques. Before summarizing what is known about neuroinflammatory imaging in human depression it is worth to indicate that some authors have alerted about the need for more selective markers of microglial activation than the classic biomarker of neuroinflammation 18 kDa translocator protein (TSPO) (formerly known as the peripheral benzodiazepine receptor) (Notter, Coughlin, Sawa, & Meyer, 2018).

By means of TSPO positron emission tomography (PET) imaging as a biomarker of microgliosis, there is increasing scientific evidence supporting the existence of a "low-grade" neuroinflammatory response with relevance in the pathophysiology of psychiatric diseases (van der Doef, Doorduin, van Berckel, & Cervenka, 2015). In particular, for MD, some authors have reported an increase in this marker in brain regions such as the prefrontal cortex, anterior cingulate cortex and insula and, what it is more relevant, correlations with symptoms severity (Setiawan et al., 2015).

More recent studies have delved into this idea, exploring the possible correlations of TSPO imaging with other parameters, such as systemic inflammation, signs of neuroinflammation measured in postmortem brain tissue, symptomatology or current antidepressant treatments, increasing the utility of neuroimaging for the optimization of the diagnosis and the treatment of the disease. In this vein, a recent meta-analysis has reported a concomitant increase in the levels of pro-inflammatory cytokines in CSF and postmortem brain tissue and in the TSPO expression measured by PET in the anterior cingulate cortex and temporal cortex of patients with MD compared with controls (Enache, Pariante, & Mondelli, 2019). However, abnormalities in CSF and PET inflammatory markers were not correlated with those in peripheral blood (Enache et al., 2019). Other studies have found similar results, with TSPO levels only correlating with some central inflammatory markers (IL5 levels in CSF) but not with the peripheral ones. In this study, unmedicated patients diagnosed with MD had the highest level of TSPO binding, followed by medicated patients, who did not differ from control individuals (Richards et al., 2018). TSPO increased binding in the anterior cingulate cortex (ACC), prefrontal cortex, and insula has been related to suicidal thinking during a moderate to the severe major depressive episode (Holmes et al., 2018). In addition, TSPO total distribution volume is greater in patients with long-lasting major depressive disorder and less antidepressant treatment than in patients with short periods of no antidepressant treatment (Setiawan et al., 2018). Increased TSPO total distribution volume has also been related to cognitive deficits in major depressive disorders and reduced by cognitive-behavioral therapy (Li, Sagar, & Keri, 2018).

In summary, microglial activation, one of the hallmarks of neuroinflammation, measured by imaging techniques appears to be increased in the frontal cortex of MD patients, being directly associated with more severe symptomatology and inversely related to response to antidepressant treatment. There are still lots of knowledge gaps in this "hot topic" of psychiatry, but there are critical challenges on the use of TSPO in psychiatric disease (Guilarte, 2019) to consider and clinical trials to follow up (Yrondi et al., 2018).

## Mechanisms whereby neuroinflammation leads to alterations in brain structure/function in depression: Lessons from animal models

In stress-based models, different pathophysiological pathways have been identified that lead to specific alterations in the neurotransmitter systems, affecting the development of depressive symptomatology. When an organism is exposed to psychosocial stressors, the CNS first releases catecholamines into the bloodstream causing a myriad of systemic effects, including the synthesis and release of myeloid cells from the bone marrow. These circulating immune cells can recognize DAMPs and PAMPs through their toll-like or NOD-like receptors. The stimulation of TLR4 will activate the NFkB pathway that will induce the synthesis of pro-inflammatory mediators, including the synthesis of the immature forms of IL1β and IL18 among others. In turn, the activation of the inflammasome will induce the activation of caspase-1 that will transform proIL1 β and proIL18 in their mature forms. In addition, caspase 1 by cleavage of the glucocorticoid receptor will induce resistance to glucocorticoids, contributing to a maintained inflammation. The cytokines secreted into the bloodstream can access the brain through humoral and neural pathways, inducing the activation of resident immune cells and those present in the neurovasculature. On the other hand, stress can directly activate microglia toward a pro-inflammatory M1 phenotype, which will secrete CCL2 chemokine, which will attract more circulating myeloid cells to the brain, thus perpetuating the inflammatory response (Miller & Raison, 2016).

In the brain, pro-inflammatory cytokines will decrease the availability of monoamines in neurons by three mechanisms. First, by activating the p38 pathway (MAPK), the expression and functioning of presynaptic reuptake will be induced. Second, by decreasing the availability of the tetrahydrobiopterin cofactor (BH4) necessary for the synthesis of monoamines, since in an inflammatory environment, its use is directed to the synthesis of nitric oxide (NO) by iNOS. Third, pro-inflammatory cytokines will induce the activity of the enzyme indoleamine 2.3-dioxygenase (IDO), converting tryptophan into kynurenine, thus leaving no precursor for serotonin synthesis. In addition, in pro-inflammatory conditions, kynurenine can be transformed into quinolinic acid, a neurotoxic metabolite capable of binding to the glutamate receptor (NMDAR) and, together with the pro-inflammatory cytokine environment, will induce a decrease in the glutamate reuptake by astrocytes. These pro-inflammatory cytokines also will induce a greater release of glutamate by astrocytes, generating an excessive concentration of glutamate in the synaptic cleft, which will lead to excitotoxicity and decrease the BDNF neurotrophic factor. The environment of pro-inflammatory cytokines affects the processes of neural plasticity and neurogenesis in the hippocampus, on the one hand, by inhibiting the action of BDNF in neural stem cells and, on the other hand, NFkB inhibits the differentiation of these cells toward the neural progenitors, which affect learning processes and memory (Martín-Hernández et al., 2016).

Neuroimaging and animal studies have described a reduced volume of the prefrontal cortex and hippocampus in depression, indicating neural atrophy in layers II, III, and V of pyramidal neurons of PFC characterized by a decrease of apical dendrites, dendritic spine density, NMDA receptors, and synaptic proteins. In contrast, some changes observed in GABAergic medium spiny neurons from the *nucleus accumbens* show neuronal hypertrophy characterized for the increment of dendritic processes, spine density, synaptic plasticity, and BDNF levels (Price & Drevets, 2012; Wohleb et al., 2016). Similar findings were detected in pyramidal and stellate neurons in the basolateral complex of the amygdala on animals exposed to chronic stress (Vyas, Mitra, Shankaranarayana Rao, & Chattarji, 2002)

The neurochemistry inflammation-induced changes produced in basal ganglia, ventromedial prefrontal cortex, subgenual, and dorsal anterior cingulate cortex have been associated with some inhibitory aspects of reward motivation, increased sensitivity to aversive stimuli and anhedonia. Also, changes involving the amygdala, hippocampus, dorsal anterior cingulate cortex, and insula have been associated with increased activation of threat- and anxiety-related neurocircuitry (Wohleb et al., 2016).

## Peripheral inflammation and brain function in depression

A multisystemic impact in the form of a persistent low-grade inflammation from the prodromal phase has been recognized for most psychiatric diseases, depression included (Khandaker, Pearson, Zammit, Lewis, & Jones, 2014). In fact, some authors have noticed the relevant role of peripheral inflammation in frequent comorbidities with depression such as cardiovascular, inflammatory bowel, and metabolic diseases (Khandaker, Dantzer, & Jones, 2017). In addition, peripheral inflammation also affects brain function and the study of this interaction is a current priority to disentangle new and blood accessible biomarkers for early diagnosis, monitoring the evolution and to predict treatment response. In this vein, a recent meta-analysis has found that alterations in peripheral cytokine levels were associated with antidepressant treatment outcomes in MD (Liu et al., 2020).

On the other hand, there is increasing evidence supporting the use of peripheral immune cell count and ratios as diagnostic biomarkers for depression (Kayhan, Gunduz, Ersoy, Kandeger, & Annagur, 2017). In addition, several pro-inflammatory effects for infiltrated immune cells from the periphery to CNS have been proposed in situations where the function and structure of the BBB is compromised (Benatti et al., 2016), and also in the resolution of neuroinflammation (Dokalis & Prinz, 2019). Further research is needed to elucidate the precise role of these immune cells in the physiopathology of depression and other stress-related neuropathologies.

There are multiple neuro-immune pathways for which systemic inflammation reaches the CNS: (1) humoral pathway through the BBB (different inflammatory mediators such as cytokines and even inflammatory cells can use this route to passage into CNS); (2) signaling through structures in the brain that lack a normal BBB (circumventricular organs); (3) signaling through the blood-cerebrospinal fluid barrier formed by the choroid plexus and meningeal arachnoid membrane; (4) direct recruitment of immune stimulus-primed peripheral immune cells to the brain parenchyma; and (5) neural pathway by the activation of primary sensory afferent nerves such as the vagus nerve.

By means of these nonexclusive and sometimes synergic pathways, peripheral inflammation can affect brain function at different levels of complexity. Some examples of the myriad of inflammation-related processes with relevance in the physiopathology of depression are HPA stress axis activation, monoamine, and glutamate neurotransmission, sleep disturbances, anhedonia, anorexia, social apathy, chronic fatigue, pain, anxiety, irritability, impaired learning, and memory (Dantzer et al., 2008; Haase & Brown, 2015).

## Possible origins of increased neuroinflammation in depression

Although the nature of the relationship between inflammation and depression (cause, consequence, or concurrency) is not clear, one relevant topic to clarify is the possible origin/s of increased inflammation in depression. There are multiple non-excluding possibilities to explore:

1. Genetic predisposition to depression's risk, severity, and response to treatment. The most studied genetic variants include polymorphisms in the genes for pro-inflammatory cytokines, chemokines, enzymes, and acute phase proteins (Barnes, Mondelli, & Pariante, 2017). However, genetic predisposition is shared between major psychiatric diseases and further studies should explore the existence of discriminatory genetic factors for depression. It is also worthy to remark that the individual genotype depends on multiple environmental and epigenetic modulatory factors to conclude in a phenotype more susceptible or resilient to develop depression.
2. Activation of the immune system of the mother or fetus before/during birth. This activation can be caused by viral/bacterial/protozoan infection during the late first and early second trimester of pregnancy, or by exposition to psychical (obstetric complications), psychological stress (trauma) or nutritional deficits (Ronovsky, Berger, Molz, Berger, & Pollak, 2016). The common notion is that these factors "prime" an immature fetal immune system that will remain impaired for a lifetime.
3. Inflammation in particular episodes during the lifetime. Diverse environmental factors, such as episodes of psychosocial stress, infections (viral reactivations, poor hygiene), dietary deficiencies, alcohol, tobacco,

psychotropic drug abuse, medications. could maintain a "mild chronic inflammatory status" that could increase the susceptibility or to aggravate depression (Berk et al., 2013).

4. Onset or evolution of comorbid inflammation-related pathologies. The presence of other pathologies such as autoimmune/allergic disorders, obesity, metabolic syndrome, chronic pain, atherosclerosis, sleep alterations, and others could be a synergic and sometimes indistinguishable source of systemic inflammation in depression. Nowadays, the scientific effort in this issue is focused on two pathological conditions: gut disorders and periodontal disease. Increased gastrointestinal permeability (*leaky gut*) with an increased translocation of lipopolysaccharide (LPS) from Gram-negative bacteria may play a role in the pathophysiology of depression. Specifically, it has been suggested that increased LPS translocation may activate the pro-inflammatory pathway driven by the innate immune toll-like receptor 4 in some patients with major depression and may induce specific sickness behavior symptomatology (Maes, Kubera, & Leunis, 2008). Another source of increased and potentially deleterious levels of bacteria is periodontitis. Periodontitis is a chronic, multifactorial, polymicrobial disease-causing inflammation in the supporting structures of the teeth and at a systemic level. Depression is associated with less dental care and some authors have reported a direct correlation between the severity of periodontal disease and the severity of depressive symptoms. However, the precise molecular mechanisms involved in the relationship between both pathologies are still elusive although its nature is probably bidirectional (Araujo et al., 2016).

## Neuroinflammation-related mediators as biomarkers of depression

Multiple studies have been published demonstrating the alteration of neuroinflammation mediators in biological samples obtained from depressed patients. Because of this, some meta-analyses have been performed in order to help elucidate which mediators could have more potential as biomarkers. One of these meta-analyses (Liu et al., 2012) used 29 selected studies and confirmed that certain cytokines are elevated in the plasma from patients suffering MD. These include TNFα and IL-6. But also, the receptor for another cytokine such as IL-2 was detected to be increased in blood samples from MD patients. In agreement with these results, another meta-analysis comprehending 136 studies also concludes that TNFα and IL-6 are elevated in the blood from depressed subjects (Dowlati et al., 2010). This analysis also considered studies in which other cytokines such as IL-1β, IL-4, IL-2, IL-8, IL-10, and IFNγ were evaluated, but the results obtained indicate that the differences observed for these mediators are not statistically significant. Another meta-analysis (Howren, Lamkin, & Suls, 2009) also concludes the relevance of IL-6 as a biomarker of depression, but this study concludes that also IL-1β and C-reactive protein (CRP) are associated with depression. This was further confirmed by a systematic review and meta-analysis of longitudinal studies in which they also observed a significant association between increased CRP and depressive symptoms (Valkanova, Ebmeier, & Allan, 2013). Furthermore, elevated levels of IL-6 have been suggested as a risk factor for depression (Khandaker et al., 2014).

Chemokines, a particular group of cytokines play a relevant role in the recruitment of different cell types to the area in which the inflammatory response is taking place. These mediators are known as chemokines due to their ability to function as chemical signals that stimulate the movement of those cells expressing the specific chemokine receptors. Besides their ability to attract cells, chemokines have also been demonstrated to participate in many other physiological processes including the development of depression. A large number of studies focused on the involvement of chemokines in depression has allowed to also perform meta-analyses on this subject. This way, one meta-analysis including 15 studies concludes that the concentrations of monocyte chemotactic protein (MCP-1 or CCL2) are significantly higher in depressed patients compared with healthy ones (Eyre et al., 2016).

## Neuroinflammation as commonplace for stress and depression

Depression is paradigmatic stress-related neuropathology. Both chronic exposure or a high degree of an acute episode of psychological stress are relevant etiological factors to develop depression in a lifetime (McEwen, 1998). The deregulation of the hypothalamic-pituitary-adrenal axis of stress has been proposed as a biomarker for the diagnosis of depression and to predict treatment response. In addition, the most used experimental models for depression are based on the chronic/subchronic exposition to stressors of different natures (psychical, psychological, or mixed) and intensity (Willner, 2005). The diathesis-stress model for psychiatric diseases was proposed time ago (Zubin & Spring, 1977), and recently revisited (Colodro-Conde et al., 2018), to explain the relationship between stress and depression based on the personal predisposition or vulnerability to develop the disease. This vulnerability could be the result of several factors of genetic, psychological, biological, or situational nature. In this context, the concept of "stress-resilience" emerges as an attractive target to focus on research. Stress resilience is defined as the ability of some individuals to cope with stressful experiences without displaying the common signs of psychiatric neuropathology (Osorio, Probert, Jones, Young, & Robbins, 2017). One of the most studied signs of stress-related neuropathology is the activation of the neuroinflammatory response. In fact, stress is widely considered a neuroinflammatory condition both in preclinical and in clinical scenarios (García-Bueno, Caso, & Leza, 2008). However, the bi-directional interactions between the stress and the immune systems are complex and depend on multiple variables (McEwen, 1998). The classic view argues that stress exposure results in an impairment of the immune response. Two predominant mechanisms have been proposed: (1) direct pro-inflammatory actions of the main stress hormones glucocorticoids through the activation of their nuclear receptor GR. These GC actions can exacerbate injury-induced neuronal death depending on dose, timing and length of exposition (Sorrells, Caso, Munhoz, & Sapolsky, 2009) and (2) *"glucocorticoids resistance,"* a phenomenon based on a dysfunction of the GR that can lead to an impaired HPA axis negative feedback and a deficiency in the GR-dependent regulation of anti-inflammatory genes.

In the study of the relationship between stress and depression, we must keep in mind that stress is an inherent component of the physiopathology of every disease and it is considered a survival and generalized response, which makes its pharmacological/genetic manipulation difficult. An excellent example of the two faces of stress is found in the neuroinflammatory response elicited after its exposure. On one hand, different acute and chronic stress protocols show a pro-inflammatory response in the brain and other systems, mainly characterized by a complex release of several inflammatory mediators such as cytokines, prostanoids, free radicals, and transcription factors activation (García-Bueno et al., 2008) and, on the other, in response to acute/subchronic stress endogenous anti-inflammatory and antioxidant mechanisms are concomitantly regulated in the CNS. Two candidates are currently receiving special attention: (1) activation of the gamma isoform of peroxisome proliferator-activated nuclear receptors (PPAR$\gamma$), a transcription factor whose main effect is to mitigate inflammation (García-Bueno et al., 2005); (2) activation of the antioxidant pathway orchestrated by the nuclear transcription factor (erythroid-derived 2)-like 2 (NRF2) (Chen et al., 2019). In the presence of oxidative stress signals, NRF2 translocates into the nucleus where it binds to consensus sequences of antioxidant response elements (ARE). ARE encode a wide variety of antioxidant enzymes. In light of these findings, while the focus has traditionally been on antagonizing pro-inflammatory pathways, additional effort should be made to potentiate the anti-inflammatory/antioxidant side of inflammatory in chronic stress conditions (Leza et al., 2015).

Finally, the investigation on stress-based animal models of depression should continue in order to elucidate new mechanisms with a putative role in the immune system-related physiopathology of depression such as the polarization of microglia to the anti-inflammatory phenotype M2, the activation of inflammasomes, kynurenine pathway, endoplasmic reticulum stress/unfolded protein response, and mitophagy and autophagy.

## Anti-inflammatory effects of antidepressants

Even though promising results were obtained in vitro and in preclinical studies with antidepressants, limited clinical trials have been performed. The class that was mostly tested was the SSRI, and fewer tests were made with TCA and SNRI (Felger, 2017).

Interestingly, the mechanism of action of some antidepressant drugs could include the inhibition of certain cytokines. This way, a meta-analysis performed for different studies (Hannestad, DellaGioia, & Bloch, 2011), in which the effect of antidepressant drugs on the serum levels of the most relevant pro-inflammatory cytokines was analyzed, demonstrates that the levels of IL-1b are reduced by these treatments. This was also confirmed in a more recent meta-analysis (Wiedlocha et al., 2018). In addition, IL-6 also was reduced in most studies analyzed. Interestingly, SSRI seemed to have a clearer effect on IL-6 serum levels than other types of antidepressants. On the contrary, while some studies describe the ability of certain antidepressants (mainly SSRI) to reduce serum levels of TNF$\alpha$, none of these meta-analyses could confirm the existence of this effect.

This anti-inflammatory effect of antidepressants could be applied in the treatment of certain diseases in which neuroinflammation plays a relevant role. In fact, the selective noradrenaline reuptake inhibitor reboxetine has been recently demonstrated to reduce neurodegeneration in animal models of Parkinson´s (Wiedlocha et al., 2018) and Alzheimer´s disease (Gutiérrez et al., 2019).

## Antiinflammatory agents in depression

After the demonstration of the existence of pro-inflammatory mediators in patients, the potential use of anti-inflammatory agents for the treatment of depression has been analyzed in numerous studies. However, several discrepancies exist between them. A different meta-analysis recently performed concluded that, since cytokines have a causal role in depression, cytokine modulators could be considered as therapeutic agents to treat depression in chronically inflamed subjects (Husain, Strawbridge, Stokes, & Young, 2017; Kappelmann, Lewis, Dantzer, Jones, & Khandaker, 2018). Among anti-cytokine drugs, the anti-TNFα infliximab and others were one of the most relevant groups analyzed, and a significant antidepressant effect was observed. In some cases, the antidepressant action of these anti-inflammatory drugs was observed to be independent of the sex and age of the patients analyzed. Some other studies indicate improvement only in patients with preexisting inflammatory diseases (psoriasis, rheumatoid arthritis, hepatitis C, Crohn's disease, or systemic lupus erythematosus). In general, all available data on this are from a small number of clinical trials that provided a proof of concept for the antidepressant effects of cytokine blockers. Further large well-designed clinical trials assessing cytokine blockers in MD are merited.

Various clinical studies focused on the usage of a nonsteroidal anti-inflammatory drug (NSAID) and depression. The possible preventive effect of NSAIDS indicated a small beneficial effect in the case of aspirin (nonselective COX inhibitor) in individuals using it for 5—10 years, but the use of other nonselective COXs (naproxen) or selective COX-2 (celecoxib) inhibitors was not beneficial (Felger, 2017). Also, a series of studies have been conducted to evaluate the enhancement of anti-inflammatory strategies in combination with NSAIDs, such as the activation of the anti-inflammatory nuclear factor PPARγ with its agonist, the antidiabetic drug pioglitazone. The most promising agent for combination therapies will probably be celecoxib. More clinical studies with larger samples size and a better design should be conducted in order to have more substantial evidence for the additional efficacy of add-on NSAID treatment. Moreover, when assessing the adverse effects of add-on NSAID treatment in the reported trials, no relevant adverse effects were found. This might be attributable to the short treatment period that ranged from 6 to 8 weeks. This period might be too short to identify any adverse effects.

## Clinical implications and future research

Despite the evidence, there are still questions remaining on inflammation and depression, and we are still far from having ideally effective, rapid, and safe treatments to offer to our patients. There is therefore a need for a change in the drug discovery strategy, mainly based on a better understanding of pathophysiology. Because the low-grade systemic inflammation is observed only in half of the patients with depression (rev. in Dooley et al., 2018), it is still too soon to consider pro-inflammatory cytokines and/or their signaling pathways a possible unique strategy to treat depression. Current studies using new anti-inflammatory and antioxidant pharmacological approaches might include patients with different degrees of inflammation, taking into consideration that the inflammatory response is highly nonspecific, and it is activated in response to multiple endogenous and exogenous factors in the context of psychiatric diseases. All these parameters need to be controlled and considered before unequivocally implicating particular inflammatory and oxido/nitrosative stress (O&NS) pathways (Fig. 1). These confounding factors contribute to the heterogeneity of the available data are the effect of stress, as the main factor of epigenetic susceptibility, disease state/duration, body mass index, and comorbidities (obesity, diabetes, cardiovascular diseases). Similar to what occurs in psychosis, and despite all the basic and clinical evidence presented to date, three "hot" questions remain to be elucidated in the future (Leza et al., 2015):

1. The unresolved question of which came first, the chicken or the egg. Efforts should be made to determine if the symptomatic onset of the disease occurs in a vulnerable brain (immunologically/inflammatory primed months/years before symptoms manifest), or if a genetically prone subject develops symptoms that will increase the deleterious effects of inflammation and/or infection. Also, to be determined is the role of continuous inflammation and O&NS throughout the disease (poor hygiene, increase in BMI and/or obesity, toxic abuse, etc.) and especially the possible role of gut/mouth dysbiosis and microbiome.
2. Is it possible to find a unique golden marker for the disease? Great efforts to find biomarkers with broad platforms notwithstanding, a new trend is to study complete and robust pathways with all elements of the intracellular and intercellular pathways, including the elements of balancing mechanisms and their relationship with symptomatology. Inflammation and O&NS pathways are important, as are anti-inflammatory and anti-O&NS pathways.
3. How can the modest clinical effects reported with anti-inflammatory drugs be explained? First, one limitation of previous clinical trials of adjunctive NSAIDs in depression is the possible inclusion of patients with a mild inflammatory process. Stricter exclusion criteria should be used to avoid this possible confounding factor. Current studies focus on controlling inflammation via direct anti-inflammatory effects while forgetting the possibilities of pharmacological stimulation of anti-inflammatory pathways.

## Key facts

- Inflammatory markers are elevated in the plasma of many, but not all, patients with depression.
- The biomarkers most frequently elevated are the cytokines TNFa, IL6, and CPR.
- Experimental preclinical models (stress based) indicate a systemic low grade of inflammation.

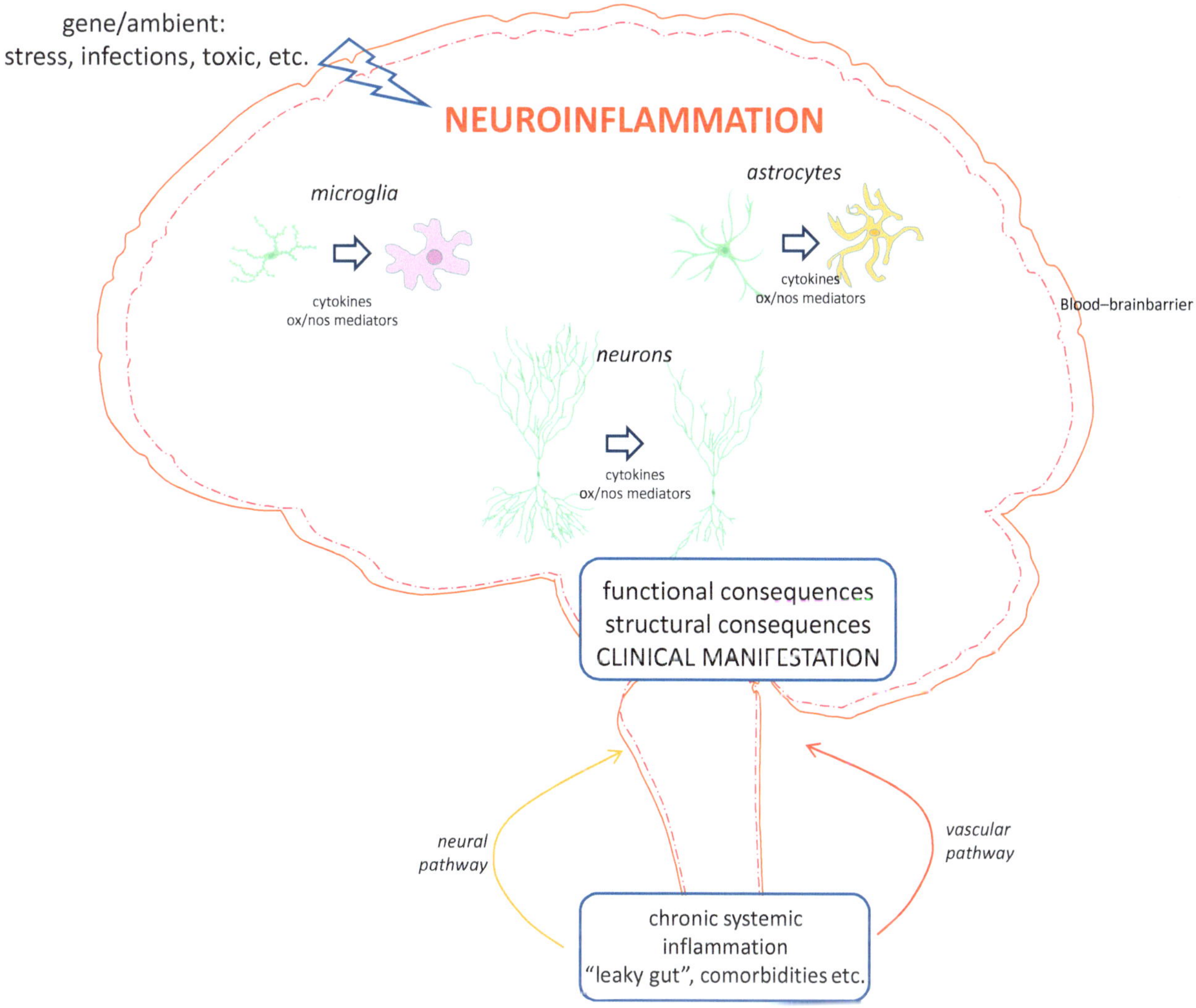


FIG. 1 Internal and external factors affecting brain neuroinflammation. Both external (gene/ambient, mainly stress) and internal factors (*leaky gut*, chronic inflammation related comorbidities) might influence brain physiology throughout sensory organs, or directly via vascular (blood) or neural. Neuroinflammation occurs after functional and structural changes in microglia (mainly), astrocytes and neurons, with intra- and intercellular pro-inflammatory and oxidative mediators (cytokines, transcription factors, inducible enzymes, etc.), leading to functional (neurotransmitter, receptor and circuitry mechanisms), structural (changes in volume in certain brain areas), and finally, clinical manifestations of depression.

## Summary points

- There is still no clear biomarker of inflammation in depressed patients.
- Some antidepressants have anti-inflammatory properties.
- Some anti-inflammatory agents (anti-cytokine and NSAIDs) are being studied as add-on treatments in major depression, but none has clear results.

## Mini-dictionary of terms

**Inflammation** This term describes the physiological response activated in the organisms in order to protect cells, tissues, and systems from harmful agents. In general, is a nonspecific response, with increased blood flow in order to facilitate the supply of immune cells that can leave the blood vessels and migrate to a location closer to the target. This leads to redness, swelling (edema), and heating of the affected area. These alterations combined with the sensation of pain and "loss of function" constitute the five classical signs of inflammation.

**Neuroinflammation** In the brain, inflammation does not occur exactly the same way it does in other parts of the body. Edema would have catastrophic consequences when it affects the brain. The blood-brain barrier, which provides additional protection to the CNS by isolating it from potentially harmful agents present in the blood, also obstructs the entrance of immune cells. The most relevant cell type responsible for the surveillance of the brain is the microglia.

## References


Araujo, M. M., Martins, C. C., Costa, L. C. M., Cota, L. O. M., Faria, R. L. A. M., Cunha, F. A., et al. (2016). Association between depression and periodontitis: A systematic review and meta-analysis. *Journal of Clinical Periodontology*, *43*(3), 216–228.

Barnes, J., Mondelli, V., & Pariante, C. M. (2017). Genetic contributions of inflammation to depression. *Neuropsychopharmacology*, *42*(1), 81–98.

Benatti, C., Blom, J. M., Rigillo, G., Alboni, S., Zizzi, F., Torta, R., et al. (2016). Disease-induced neuroinflammation and depression. *CNS Neurological Disorders Drug Targets*, *15*(4), 414–433.

Berk, M., Williams, L. J., Jacka, F. N., O'Neil, A., Pasco, J. A., Moylan, S., et al. (2013). So depression is an inflammatory disease, but where does the inflammation come from? *BMC Medicine*, *11*, 200.

Chen, H. C., Lee, J. K., Yip, T., Sernia, C., Lavidis, N. A., & Spiers, J. G. (2019). Sub-acute restraint stress progressively increases oxidative/nitrosative stress and inflammatory markers while transiently upregulating antioxidant gene expression in the rat hippocampus. *Free Radical Biology & Medicine*, *130*, 446–457.

Cheng, Y., Desse, S., Martinez, A., Worthen, R. J., Jope, R. S., & Beurel, E. (2018). TNFalpha disrupts blood brain barrier integrity to maintain prolonged depressive-like behavior in mice. *Brain, Behavior, and Immunity*, *69*, 556–567.

Cheng, Y., Jope, R. S., & Beurel, E. (2015). A pre-conditioning stress accelerates increases in mouse plasma inflammatory cytokines induced by stress. *BMC Neuroscience*, *16*, 31.

Colodro-Conde, L., Couvy-Duchesne, B., Zhu, G., Coventry, W. L., Byrne, E. M., Gordon, S., et al. (2018). A direct test of the diathesis-stress model for depression. *Molecular Psychiatry*, *23*(7), 1590–1596.

Dantzer, R. (2001). Cytokine-induced sickness behavior: Where do we stand? *Brain, Behavior, and Immunity*, *15*(1), 7–24.

Dantzer, R., O'Connor, J. C., Freund, G. G., Johnson, R. W., & Kelley, K. W. (2008). From inflammation to sickness and depression: When the immune system subjugates the brain. *Nature Reviews. Neuroscience*, *9*(1), 46–57.

Dokalis, N., & Prinz, M. (2019). Resolution of neuroinflammation: Mechanisms and potential therapeutic option. *Seminars in Immunopathology*, *41*, 699–709.

Dooley, L. N., Kuhlman, K. R., Robles, T. F., Eisenberger, N. I., Craske, M. G., & Bower, J. E. (2018). The role of inflammation in core features of depression: Insights from paradigms using exogenously-induced inflammation. *Neuroscience & Biobehavioral Reviews*, *94*, 219–237.

Dowlati, Y., Herrmann, N., Swardfager, W., Liu, H., Sham, L., Reim, E. K., et al. (2010). A meta-analysis of cytokines in major depression. *Biological Psychiatry*, *67*(5), 446–457.

Enache, D., Pariante, C. M., & Mondelli, V. (2019). Markers of central inflammation in major depressive disorder: A systematic review and meta-analysis of studies examining cerebrospinal fluid, positron emission tomography and post-mortem brain tissue. *Brain, Behavior, & Immunity*, *81*, 24–40.

Eyre, H. A., Air, T., Pradhan, A., Johnston, J., Lavretsky, H., Stuart, MJ., et al. (2016). A meta-analysis of chemokines in major depression. *Progress in Neuro-Psychopharmacology & Biological Psychiatry*, *68*, 1–8.

Felger, J. C. (2017). The role of dopamine in inflammation-associated depression: mechanisms and therapeutic implications. *Current Topics in Behavioral Neurosciences*, *31*, 199–219.

Gárate, I., García-Bueno, B., Madrigal, J. L., Bravo, L., Berrocoso, E., Caso, J. R., et al. (2011). Origin and consequences of brain Toll-like receptor 4 pathway stimulation in an experimental model of depression. *Journal of Neuroinflammation*, *8*, 151.

García-Bueno, B., Caso, J. R., & Leza, J. C. (2008). Stress as a neuroinflammatory condition in brain: Damaging and protective mechanisms. *Neuroscience and Biobehavioral Reviews*, *32*(6), 1136–1151.

García-Bueno, B., Madrigal, J. L., Lizasoain, I., Moro, M. A., Lorenzo, P., & Leza, J. C. (2005). Peroxisome proliferator-activated receptor gamma activation decreases neuroinflammation in brain after stress in rats. *Biological Psychiatry*, *57*(8), 885–894.

Goshen, I., Kreisel, T., Ben-Menachem-Zidon, O., Licht, T., Weidenfeld, J., Ben-Hur, T., et al. (2008). Brain interleukin-1 mediates chronic stress-induced depression in mice via adrenocortical activation and hippocampal neurogenesis suppression. *Molecular Psychiatry*, *13*(7), 717–728.

Guilarte, T. R. (2019). TSPO in diverse CNS pathologies and psychiatric disease: A critical review and a way forward. *Pharmacology & Therapeutics*, *194*, 44–58.

Gururajan, A., Reif, A., Cryan, J. F., & Slattery, D. A. (2019). The future of rodent models in depression research. *Nature Reviews. Neuroscience*, *20*(11), 686–701.

Gutiérrez, I. L., González-Prieto, M., Caso, J. R., Garcia-Bueno, B., Leza, J. C., & Madrigal, J. L. M. (2019). Reboxetine treatment reduces neuroinflammation and neurodegeneration in the 5xFAD mouse model of Alzheimer's disease: role of CCL2. *Molecular Neurobiology*, *56*(12), 8628–8642.

Haase, J., & Brown, E. (2015). Integrating the monoamine, neurotrophin and cytokine hypotheses of depression- -a central role for the serotonin transporter? *Pharmacology & Therapeutics*, *147*, 1–11.

Hannestad, J., DellaGioia, N., & Bloch, M. (2011). The effect of antidepressant medication treatment on serum levels of inflammatory cytokines: A meta-analysis. *Neuropsychopharmacology*, *36*(12), 2452–2459.

Holmes, S. E., Hinz, R., Conen, S., Gregory, C. J., Matthews, J. C., Anton-Rodriguez, J. M., et al. (2018). Elevated translocator protein in anterior cingulate in major depression and a role for inflammation in suicidal thinking: A positron emission tomography study. *Biological Psychiatry*, *83*(1), 61–69.

Howren, M. B., Lamkin, D. M., & Suls, J. (2009). Associations of depression with C-reactive protein, IL-1, and IL-6: A meta-analysis. *Psychosomatic Medicine*, *71*(2), 171–186.

Husain, M. I., Strawbridge, R., Stokes, P. R., & Young, A. H. (2017). Anti-inflammatory treatments for mood disorders: Systematic review and meta-analysis. *Journal of Psychopharmacology*, *31*(9), 1137–1148.

Kappelmann, N., Lewis, G., Dantzer, R., Jones, P. B., & Khandaker, G. M. (2018). Antidepressant activity of anti-cytokine treatment: A systematic review and meta-analysis of clinical trials of chronic inflammatory conditions. *Molecular Psychiatry*, *23*(2), 335–343.

Kaster, M. P., Gadotti, V. M., Calixto, J. B., Santos, A. R., & Rodrigues, A. L. (2012). Depressive-like behavior induced by tumor necrosis factor-alpha in mice. *Neuropharmacology*, *62*(1), 419–426.

Kayhan, F., Gunduz, S., Ersoy, S. A., Kandeger, A., & Annagur, B. B. (2017). Relationships of neutrophil-lymphocyte and platelet-lymphocyte ratios with the severity of major depression. *Psychiatry Research*, *247*, 332–335.

Khandaker, G. M., Dantzer, R., & Jones, P. B. (2017). Immunopsychiatry: Important facts. *Psychological Medicine*, *47*(13), 2229–2237.

Khandaker, G. M., Pearson, R. M., Zammit, S., Lewis, G., & Jones, P. B. (2014). Association of serum interleukin 6 and C-reactive protein in childhood with depression and psychosis in young adult life a population-based longitudinal study. *JAMA Psychiatry*, *71*(10), 1121–1128.

Kinney, J. W., Bemiller, S. M., Murtishaw, A. S., Leisgang, A. M., Salazar, A. M., & Lamb, B. T. (2018). Inflammation as a central mechanism in Alzheimer's disease. *Alzheimer's and Dementia: Translational Research and Clinical Interventions*, *4*, 575–590.

Krashia, P., Nobili, A., & D'Amelio, M. (2019). Unifying hypothesis of dopamine neuron loss in neurodegenerative diseases: Focusing on Alzheimer's disease. *Frontiers in Molecular Neuroscience*, *17*(12), 123.

Leza, J. C., García-Bueno, B., Bioque, M., Arango, C., Parellada, M., Do, K., et al. (2015). Inflammation in schizophrenia: A question of balance. *Neuroscience and Biobehavioral Reviews*, *55*, 612–626.

Li, H., Sagar, A. P., & Keri, S. (2018). Translocator protein (18 kDa TSPO) binding, a marker of microglia, is reduced in major depression during cognitive-behavioral therapy. *Progress in Neuropsychopharmacology & Biological Psychiatry*, *83*, 1–7.

Liu, F., Guo, W., Yu, D., Gao, Q., Gao, K., Xue, Z., et al. (2012). Classification of different therapeutic responses of major depressive disorder with multivariate pattern analysis method based on structural MR scans. *PLoS One*, *7*(7), e40968.

Liu, J. J., Wei, Y. B., Strawbridge, R., Bao, Y., Chang, S., Shi, L., et al. (2020). Peripheral cytokine levels and response to antidepressant treatment in depression: A systematic review and meta-analysis. *Molecular Psychiatry*, *25*(2), 339–350.

López-López, A. L., Jaime, H. B., Escobar Villanueva, M. D. C., Padilla, M. B., Palacios, G. V., & Aguilar, F. J. A. (2016). Chronic unpredictable mild stress generates oxidative stress and systemic inflammation in rats. *Physiology & Behavior*, *161*, 15–23.

Lu, Y., Ho, C. S., Liu, X., Chua, A. N., Wang, W., McIntyre, R. S., et al. (2017). Chronic administration of fluoxetine and pro-inflammatory cytokine change in a rat model of depression. *PLoS One*, *12*(10), e0186700.

Madrigal, J. L., Hurtado, O., Moro, M. A., Lizasoain, I., Lorenzo, P., Castrillo, A., et al. (2002). The increase in TNF-alpha levels is implicated in NF-kappaB activation and inducible nitric oxide synthase expression in brain cortex after immobilization stress. *Neuropsychopharmacology*, *26*(2), 155–163.

Maes, M., Kubera, M., & Leunis, J. C. (2008). The gut-brain barrier in major depression: Intestinal mucosal dysfunction with an increased translocation of LPS from gram negative enterobacteria (leaky gut) plays a role in the inflammatory pathophysiology of depression. *Neuroendocrinology Letters*, *29*(1), 117–124.

Martín-Hernández, D., Caso, J. R., Bris, Á. G., Maus, S. R., Madrigal, J. L. M., García-Bueno, B., et al. (2016). Bacterial translocation affects intracellular neuroinflammatory pathways in a depression-like model in rats. *Neuropharmacology*, *103*, 122–133.

McEwen, B. S. (1998). Protective and damaging effects of stress mediators. *The New England Journal of Medicine*, *338*(3), 171–179.

Miller, A. H., & Raison, C. L. (2016). The role of inflammation in depression: From evolutionary imperative to modern treatment target. *Nature Reviews. Immunology*, *16*(1), 22–34.

Notter, T., Coughlin, J. M., Sawa, A., & Meyer, U. (2018). Reconceptualization of translocator protein as a biomarker of neuroinflammation in psychiatry. *Molecular Psychiatry*, *23*(1), 36–47.

Osorio, C., Probert, T., Jones, E., Young, A. H., & Robbins, I. (2017). Adapting to stress: Understanding the neurobiology of resilience. *Behavioral Medicine*, *43*(4), 307–322.

Price, J. L., & Drevets, W. C. (2012). Neural circuits underlying the pathophysiology of mood disorders. *Trends in Cognitive Sciences*, *16*(1), 61–71.

Richards, E. M., Zanotti-Fregonara, P., Fujita, M., Newman, L., Farmer, C., Ballard, E. D., et al. (2018). PET radioligand binding to translocator protein (TSPO) is increased in unmedicated depressed subjects. *EJNMMI Research*, *8*(1), 57.

Ronovsky, M., Berger, S., Molz, B., Berger, A., & Pollak, D. D. (2016). Animal models of maternal immune activation in depression research. *Current Neuropharmacology*, *14*(7), 688–704.

Schwartz, M., Kipnis, J., Rivest, S., & Prat, A. (2013). How do immune cells support and shape the brain in health, disease, and aging? *The Journal of Neuroscience*, *33*(45), 17587–17596.

Setiawan, E., Attwells, S., Wilson, A. A., Mizrahi, R., Rusjan, P. M., Miler, L., et al. (2018). Association of translocator protein total distribution volume with duration of untreated major depressive disorder: A cross-sectional study. *Lancet Psychiatry*, *5*(4), 339–347.

Setiawan, E., Wilson, A. A., Mizrahi, R., Rusjan, P. M., Miler, L., Rajkowska, G., et al. (2015). Role of translocator protein density, a marker of neuroinflammation, in the brain during major depressive episodes. *JAMA Psychiatry*, *72*(3), 268–275.

Singhal, G., & Baune, B. T. (2017). Microglia: An interface between the loss of neuroplasticity and depression. *Frontiers in Cellular Neuroscience*, *11*, 270.

Sorrells, S. F., Caso, J. R., Munhoz, C. D., & Sapolsky, R. M. (2009). The stressed CNS: When glucocorticoids aggravate inflammation. *Neuron*, *64*(1), 33–39.

Su, W. J., Zhang, Y., Chen, Y., Gong, H., Lian, Y. J., Peng, W., et al. (2017). NLRP3 gene knockout blocks NF-kappaB and MAPK signaling pathway in CUMS-induced depression mouse model. *Behavioral Brain Research*, *322*(Pt. A), 1–8.

Valkanova, V., Ebmeier, K. P., & Allan, C. L. (2013). CRP, IL-6 and depression: A systematic review and meta-analysis of longitudinal studies. *Journal of Affective Disorders*, *150*(3), 736–744.

van der Doef, T. F., Doorduin, J., van Berckel, B. N. M., & Cervenka, S. (2015). Assessing brain immune activation in psychiatric disorders: Clinical and preclinical PET imaging studies of the 18-kDa translocator protein. *Clinical and Translational Imaging*, *3*(6), 449–460.

Vyas, A., Mitra, R., Shankaranarayana Rao, B. S., & Chattarji, S. (2002). Chronic stress induces contrasting patterns of dendritic remodeling in hippocampal and amygdaloid neurons. *Journal of Neuroscience*, *22*(15), 6810–6818.

Weber, M. D., Frank, M. G., Tracey, K. J., Watkins, L. R., & Maier, S. F. (2015). Stress induces the danger-associated molecular pattern HMGB-1 in the hippocampus of male Sprague Dawley rats: A priming stimulus of microglia and the NLRP3 inflammasome. *Journal of Neuroscience*, *35*(1), 316–324.

Wiedlocha, M., Marcinowicz, P., Krupa, R., Janoska-Jazdzik, M., Janus, M., Debowska, W., et al. (2018). Effect of antidepressant treatment on peripheral inflammation markers—A meta-analysis. *Progress in Neuropsychopharmacology & Biological Psychiatry*, *80*(Pt. C), 217–226.

Willner, P. (2005). Chronic mild stress (CMS) revisited: Consistency and behavioural-neurobiological concordance in the effects of CMS. *Neuropsychobiology*, *52*(2), 90–110.

Wohleb, E. S., Fenn, A. M., Pacenta, A. M., Powell, N. D., Sheridan, J. F., & Godbout, J. P. (2012). Peripheral innate immune challenge exaggerated microglia activation, increased the number of inflammatory CNS macrophages, and prolonged social withdrawal in socially defeated mice. *Psychoneuroendocrinology*, *37*(9), 1491–1505.

Wohleb, E. S., Franklin, T., Iwata, M., & Duman, R. S. (2016). Integrating neuroimmune systems in the neurobiology of depression. *Nature Reviews. Neuroscience*, *17*(8), 497–511.

You, Z., Luo, C., Zhang, W., Chen, Y., He, J., Zhao, Q., et al. (2011). Pro- and anti-inflammatory cytokines expression in rat's brain and spleen exposed to chronic mild stress: Involvement in depression. *Behavioral Brain Research*, *225*(1), 135–141.

Yrondi, A., Aouizerate, B., El-Hage, W., Moliere, F., Thalamas, C., Delcourt, N., et al. (2018). Assessment of translocator protein density, as marker of neuroinflammation, in major depressive disorder: A pilot, multicenter, comparative, controlled, brain PET study (INFLADEP Study). *Frontiers in Psychiatry*, *9*, 326.

Zhang, Y., Liu, L., Liu, Y. Z., Shen, X. L., Wu, T. Y., Zhang, T., et al. (2015). NLRP3 inflammasome mediates chronic mild stress-induced depression in mice via neuroinflammation. *The International Journal of Neuropsychopharmacology*, *18*(8), pyv006.

Zubin, J., & Spring, B. (1977). Vulnerability—New view of Schizophrenia. *Journal of Abnormal Psychology*, *86*(2), 103–126.

Chapter 14

# Interlinking antidepressants and the immune system

Katarzyna A. Lisowska, Krzysztof Pietruczuk, and Łukasz P. Szałach
*Department of Pathophysiology, Medical University of Gdańsk, Gdańsk, Poland*

## List of abbreviations

| | |
|---|---|
| **IL** | interleukin |
| **INF** | interferon |
| **MAOIs** | monoamine oxidase inhibitors |
| **NaSSAs** | noradrenergic and specific serotonergic antidepressants |
| **NDRIs** | norephineprine-dopamine reuptake inhibitors |
| **SARIs** | serotonin antagonist and reuptake inhibitors |
| **SNRIs** | serotonin norepinephrine reuptake inhibitors |
| **SPARIs** | serotonin partial agonist-reuptake inhibitors |
| **SSRIs** | selective serotonin reuptake inhibitors |
| **TCAs** | tricyclicantidepressants |
| **TNF** | tumor necrosis factor |

## Introduction

Since the beginning of the 20th century, clinical observations have been showing that the mental state, in particular, chronic stress, an essential factor in the development of depression, significantly affects the functioning of the immune system. In the 1980s and 1990s, studies were published showing that lymphocytes taken from widowed women or the husbands of women with advanced breast carcinoma have significantly reduced response to stimulation with mitogens (Bartrop, Luckhurst, Lazarus, Kiloh, & Penny, 1977; Schleifer, Keller, Camerino, Thornton, & Stein, 1983). The above observations are related to the well-known relationships between the immune, nervous, and endocrine systems. These relations are of interest to psychoneuroimmunology, which was established by the study of Ader and Cohen (1975). In the depression, the functioning of the immune system is disturbed, which can be both an element of the pathogenesis of changes in the central nervous system and its final effect.

This chapter summarizes knowledge about immune disorders in the course of depression and presents the results of studies showing the influence of antidepressants from different groups on selected immunologic parameters. Attention will also be paid to potential immunological indicators that may help to predict patient's response to therapy.

## The immune system in the depression

### Innate immunity in the depression

There are many studies documenting disorders of innate responses in the course of depression. Most of the research focuses primarily on the assessment of the level of cytokines that play an essential role in the chronic inflammatory process, which may be a factor initiating the depressive episode or a condition accompanying depression.

Various meta-analyses confirm that patients with depression have an increased serum levels of two proinflammatory cytokines, IL-6 and TNF-α (Dowlati et al., 2010; Liu, Ho, & Mak, 2012; Strawbridge, Arnone, Danese, & Papadopoulos, 2015). Both cytokines are secreted mainly by macrophages and are responsible for initiating acute phase response and stimulation of hypothalamic-pituitary-adrenal (HPA) axis. T lymphocytes activated during the depressive episode could also be the other source of TNF-α and IL-6 (Duggal, Upton, Phillips, Hampson, & Lord, 2014). Other studies also showed an increased level of IL-1β (Alcocer-Gómez et al., 2017; Dahl et al., 2014; Pérez-Sánchez et al., 2018). Moreover, the increased concentration of IL-1 in depressed patients appears to be associated with increased activity of inflammasome. The study of Alcocer-Gómez et al. (2017) demonstrated that increased expression of inflammasome NLRP3 is responsible for the increased serum concentration of IL-1β and IL-18 in patients suffering from a major depressive disorder.

The studies also showed that the number of white blood cells is significantly increased in depressed patients, mainly because of the increase of neutrophils (Demir et al., 2015; Kronfol & House, 1989). As neutrophils not only are the most abundant group of granulocytes but also they are active participants in the acute inflammatory process, their

The Neuroscience of Depression. https://doi.org/10.1016/B978-0-12-817933-8.00050-5

increase could be an expression of the intensification of the inflammatory process in the course of depression. A more recent study of Aydin Sunbul et al. (2016) demonstrated that the severity of depression is significantly associated with increased neutrophil/lymphocyte ratio and could, according to authors, serve an independent predictor of severe or very severe depression.

Depressed patients have decreased number (Başterzi et al., 2010; Ivanova et al., 2007; Schleifer, Keller, & Bartlett, 1999) as well as activity (Başterzi et al., 2010; Kook, Mizruchin, Odnopozov, Gershon, & Segev, 1995; Mizruchin et al., 1999) of natural killers (NK) cells, which are responsible for natural cytotoxicity against viruses and cancer cells. However, according to Hernandez et al. (2010) depressed patients have a higher percentage of NK cells in blood samples compared with healthy people.

## Adaptive immunity in the depression

Numerous studies confirm that patients with depression have a reduced percentage of lymphocytes (Demir et al., 2015; Grosse et al., 2016; Kronfol & House, 1989). An essential feature of lymphocytes from depressed patients is the reduced mitochondrial potential (Pietruczuk, Jakuszkowiak, Landowski, & Witkowski, 2005), which translates into reduced ATP production. The final result is not only an increased lymphocyte susceptibility to apoptosis (Ivanova et al., 2007; Szuster-Ciesielska et al., 2008), but also reduced proliferative responses to mitogens, e.g., phytohaemagglutinin (PHA) or concanavalin A (ConA) (Darko et al., 1989; Kronfol & House, 1989; Kronfol, House, Silva, Greden, & Carroll, 1986), which may translate into a reduced percentage of lymphocytes in the blood of depressed patients. At the same time, the concentration of IL-2, a cytokine essential for differentiation of T lymphocytes into effector cells, is increased in depressed patients (Dahl et al., 2014; Pérez-Sánchez et al., 2018; Schmidt et al., 2014). However, its stimulating effect is suppressed by the presence of soluble IL-2R in the blood, where it can bind IL-2 (Liu et al., 2012).

For specific groups of T lymphocytes, the obtained results are less obvious. Kronfol and House (1989) showed that while the percentage of total lymphocytes is lower in depressed patients compared with healthy people, the absolute number of lymphocytes is the same. Müller, Hofschuster, Ackenheil, Mempel, and Eckstein (1993) demonstrated that $CD3^+$ T lymphocytes, especially $CD4^+$ (helper) T lymphocytes, increase in the depression, while $CD8^+$ (cytotoxic) T lymphocytes decrease. The same result could be found in the article by Maes et al. (1992a). A meta-analysis of Zorrilla et al. (2001) reported an increase of $CD4^+/CD8^+$ ratio in patients with depression. However, the most recent studies show that there is no evidence that the percentage of T lymphocytes or their two central populations: $CD4^+$ or $CD8^+$ cells changes in the course of depression (Başterzi et al., 2010; Grosse et al., 2016; Patas et al., 2018). T lymphocytes of depressed patients also present significantly reduced expression of chemokine receptors CXCR3 and CCR6 (Patas et al., 2018), which play an important role in the recruitment of leukocytes to places where the inflammatory process is present. Also, cells have increased expression of some activation markers: CD95, which is also involved in the process of apoptosis (Ivanova et al., 2007; Szuster-Ciesielska et al., 2008), CD69, an early activation marker, (Duggal et al., 2014; Pietruczuk et al., 2005) and CD25, which is an alpha chain of the IL-2 receptor (Maes et al., 1991; Pietruczuk et al., 2005).

Recent studies highlight the abnormalities of $CD4^+$ T lymphocyte subsets. We currently know many different subsets of $CD4^+$ cells: Th1 cells that are responsible for cell-mediated immunity, Th2 cells promoting humoral (antibody-dependent) responses, Th3 cells, and Tregs (regulatory T lymphocytes) that regulate the immune response, and Th17 cells responsible for developing inflammatory processes. Depression seems to be dominated primarily by Th1 and Th17 cells, which can activate macrophages and exaggerate the inflammatory process in the course of depression. Depressed patients present high concentrations of IL-12 (Lee & Kim, 2006; Schmidt et al., 2014) responsible for promoting the Th1-type immune responses as well as for stimulating the production of high levels of IFN-$\gamma$ (Myint, Leonard, Steinbusch, & Kim, 2005; Schmidt et al., 2014). The second cytokine is mainly produced by Th1 cells and can activate macrophages to produce proinflammatory cytokines and promote cell-mediated immunity. Recently, attention is also being paid to Th17 cells, a subset of proinflammatory T helper cells, which also increases in the course of depression (Chen et al., 2011; Davami et al., 2016). Their increase is most likely the result of high IL-6 levels and may contribute to further intensification of inflammatory processes in depressed patients. Depressed patients also are characterized by a decreased percentage of $CD4^+CD25^+FoxP3^+$ cells, which are regulatory T lymphocytes responsible for maintaining tolerance against self-antigens, thus preventing autoimmune reactions (Chen et al., 2011; Grosse et al., 2016; Li et al., 2010). The percentage of Th3 cells that exert their regulatory function primarily by secreting transforming growth factor $\beta$ (TGF-$\beta$) is also significantly reduced in depressed patients when compared with healthy people (Lee & Kim, 2006; Myint et al., 2005).

Little we know about the disorders in the humoral response for which B lymphocytes are responsible. Older research suggested that depressed patients are characterized by an increase in the percentage of B lymphocytes (Maes et al., 1992b) and an increase in some classes of immunoglobulins (Maes et al., 2013), but more recent studies do not

**TABLE 1** Disturbances of innate and adaptive immunity in the depression.

| | Innate immunity | Adaptive immunity |
|---|---|---|
| Cytokines | ↑ IL-1β, IL-6, TNF-α | ↑ IL-2, sIL-2R, IL-12, IFN-γ |
| Cells | ↑ Neutrophils<br>↓ NK cells | ↓ Lymphocytes<br>↓ $CD4^+/CD8^+$<br>↓ Tregs, Th3<br>↑ Th1/Th2, Th17 |

confirm the existence of a relationship between the occurrence of depressive symptoms and the humoral response (Dubois, Zdanowicz, Reynaert, & Jacques, 2016; Hernandez et al., 2010).

Changes in the immunological parameters of patients with depression are summarized in Table 1.

## Influence of antidepressants on the immune system

In general, antidepressants work by inhibiting the reuptake of specific neurotransmitters, e.g., serotonin, hence increasing their levels around the nerves within the brain, which improves the mood of the depressed patient. There are several groups of antidepressant drugs, the most important of which are selective serotonin reuptake inhibitors (SSRIs), serotonin-norepinephrine reuptake inhibitors (SNRIs), and tricyclic antidepressants (TCAs). SSRIs and SNRIs are potent inhibitors of the reuptake of serotonin and norepinephrine. Both neurotransmitters play an important role in mood. TCAs act primarily as SNRIs by blocking the serotonin transporter (SERT or 5-HTT) and the norepinephrine transporter (NET). However, studies show that their effects may have a greater range (Table 2).

According to research data, antidepressants can modulate melatonin synthesis. Melatonin, a hormone produced primarily in the pineal gland, regulates the sleep-wake cycle, the process which is strongly disturbed in the depression. Carvalho, Gorenstein, Moreno, Pariante, and Markus (2009) demonstrated that melatonin production increases after treatment with fluoxetine or duloxetine by measuring its urinary metabolite, 6-sulphatoxymelatonin (aMT6s). Also, Miller, Ekstrom, Mason, Lydiard, and Golden (2001) showed that changes in urinary excretion of aMT6s could distinguish antidepressant responders from nonresponders.

A known problem in depression is the development of glucocorticoid resistance and disruption of the HPA axis. Glucocorticoid resistance in depressed patients is associated with decreased expression of glucocorticoid receptors (GR). Studies showed that SSRIs and TCAs promote GR nuclear translocation in the cell, enhance their function, and in this way, normalize hyperactivity of the HPA axis (Pariante & Miller, 2001).

Another problem in depression is an increased risk of cardiovascular disease (CVD) development, associated with, among others, reduced heart rate variability (HRV) due to alterations in the autonomic nervous system. Although individual studies have suggested that antidepressants may affect HRV in depressed patients, the meta-analysis by Kemp et al. (2010) did not explicitly confirm this effect. According to one of the latest studies, however, the main three groups of antidepressants can be associated with lower measures of HRV (O'Regan, Kenny, Cronin, Finucane, & Kearney, 2015).

**TABLE 2** The effect of antidepressants on the nervous and endocrine systems.

| | Other effects identified |
|---|---|
| SSRIs | ↑ melatonin production<br>↑ GR expression and function<br>↓ HRV |
| SNRIs | ↑ melatonin production<br>↓ HRV |
| TCAs | ↑ melatonin production<br>↑ GR expression and function<br>↓ HRV |

Nevertheless, most reports of nonantidepressant activity of antidepressants are found in the field of immunology. It is associated with a significant relationship between the nervous, endocrine, and immune systems.

Since proinflammatory cytokines not only have a confirmed effect on the HPA axis but may also contribute to neurodegeneration and impede transmission in the nervous system, most authors present the impact of antidepressants on the level of cytokines in the blood of depressed patients and correlate it with the response to treatment.

A meta-analysis by Strawbridge et al. (2015) based on 35 clinical trials performed in over 20 years revealed that in patients suffering from unipolar depression, who have responded to widely understood antidepressant treatment, serum level of TNF-α and IL-6 significantly decreased even though there was no significant change in CRP concentration before and after treatment. Moreover, patients who did not respond to therapy tended to have higher baseline inflammation, while a decrease in TNF-α level was associated with response to treatment. Other meta-analysis involving 22 studies showed that the level of IL-1β in patients' serum significantly decreased regardless patients received SSRIs, SNRIs, or TCAs, while IL-6 level decreased slightly only in patients receiving SSRIs (Hannestad, Dellagioia, & Bloch, 2011). The authors also demonstrated that serum TNF-α levels did not change significantly regardless of the type of antidepressant even though treatment reduced depressive symptoms. The latest study by Köhler et al. (2018) confirmed the observation of Strawbridge and colleagues that treatment with SSRIs and SNRIs decreases the peripheral concentration of IL-6 and TNF-α. Moreover, Alcocer-Gómez et al. (2017) demonstrated that the decrease in IL-1β level is most probably associated with a reduction of inflammasome activity in mononuclear cells isolated from the blood of patients treated with SSRIs, SNRIs, TCAs, or NaSSAs (noradrenergic and specific serotonergic antidepressants).

Since the parameters of acquired immunity also change in the course of depression, the following are reports on the impact of different groups of antidepressants on selected immunologic parameters. Many of the studies presented below point out that in addition to the therapeutic effect of the antidepressant used, their anti-inflammatory and immunosuppressive effects may be beneficial to the patient.

## Selective serotonin reuptake inhibitors

Because SSRIs are the most commonly used antidepressants, our knowledge of their immunomodulatory effects is also the greatest. Studies showed that almost all SSRIs (paroxetine, fluvoxamine, fluoxetine, and escitalopram) can increase the number (Başterzi et al., 2010) as well as activity of NK cells (Kook et al., 1995; Mizruchin et al., 1999; Park, Lee, Jeong, Han, & Jeon, 2015). It should be emphasized that the increase in the percentage (Başterzi et al., 2010) and activity (Kook et al., 1995) of NK cells was observed only in patients responding to therapy and, therefore, could serve as potential marker of response to antidepressant treatment. Similar observations were published in the late 1990s when Frank, Hendricks, Johnson, Wieseler, and Burke (1999) demonstrated in their study that both fluoxetine and paroxetine treatment was associated with augmented NK cell activity in a subgroup of depressed patients exhibiting the low activity of NK cells before treatment both in vivo and in vitro.

It seems that using SSRIs may aggravate existing disorders in the functioning of T lymphocytes. Several studies demonstrated that different drugs (paroxetine, sertraline, and fluoxetine) not only decrease the viability of mitogen-stimulated T lymphocytes (Taler et al., 2007) but also inhibit their proliferation capacity (Edgar, Sterin-Borda, Cremaschi, & Genaro, 1999). However, according to Edgar et al. (1999), the influence of fluoxetine depends on mitogen concentrations and has a dual effect on T-cell proliferation in vitro—fluoxetine stimulated proliferation only when suboptimal dose of mitogen was used, but with optimal mitogen dose it reduced proliferation capacity of lymphocytes. This observation could explain why some authors saw no changes in the percentage of T lymphocytes, $CD4^{+}$ or $CD8^{+}$ cells in the blood of patients during therapy (Başterzi et al., 2010), while others demonstrated that it declines in some cases, like during escitalopram treatment (Canan & Ataoglu, 2009). Moreover, these results suggest that disturbances in adaptive immunity could be independent of the development of depression and therefore, may not necessarily improve with treatment, despite the patient's well-being. In addition, Taler et al. (2007) demonstrated that reduced proliferative activity of T lymphocytes under the influence of SSRIs is also accompanied by a reduced production of TNF-α, which may translate into a decrease in the concentration of TNF-α in patients' serum during treatment. Also, research by Eller, Vasar, Shlik, and Maron (2008) showed that the persistence of high serum levels of TNF-α during escitalopram therapy predicts lack of clinical response to treatment.

Some studies showed that therapy with different SSRIs may contribute to a decrease of serum concentration of IL-12 (Lee & Kim, 2006) and at the same time increase levels of TGF-β (Lee & Kim, 2006; Myint et al., 2005) as well as IL-10, another anti-inflammatory cytokine (Ho, Yeh, Huang, & Liang, 2015), which may reduce the intensity of cell-mediated immunity and inflammation. Maes et al. (1999) and Kubera et al. (2001) suggested that changes in the cytokine levels are most likely the result of changes in whole blood cells due to treatment. They stimulated whole blood cells taken from patients and healthy people with PHA or lipopolysaccharide (LPS), endotoxin which is a potent stimulator of monocytes, in the presence of antidepressants and they proved that fluoxetine (Kubera et al., 2001) and sertraline (Maes et al., 1999) decrease production of IFN-γ and increase IL-10. The positive effect of these changes is a decline of the Th1/Th2 ratio, so the immune response is redirected toward humoral activity, and cytotoxic responses offset.

At this point, it is worth mentioning that individual reports suggested that SSRIs such as escitalopram or citalopram may, however, increase the production of cytokines, e.g., IL-1β, IL-6, TNF-α, or IL-17 in vitro (Munzer et al., 2013). Different results may result from differences in the doses of drugs used or different time of cell incubation with drugs in vitro. Also, when comparing the results obtained in vivo and in vitro, it must be taken into account that the in vitro results only show the effect of the drugs themselves on blood cells. Meanwhile, in vivo, the activity of lymphocytes and monocytes can also be modified by changes in serotonin levels because of the presence of SERT on their surface, whose density may change in the course of depression (Barkan, Gurwitz, Levy, Weizman, & Rehavi, 2004).

## Serotonin norepinephrine reuptake inhibitors

Since SNRIs are the second important group of antidepressants, many authors who investigated the immunomodulatory effects of SSRIs also analyzed the influence of SNRIs. And so, for example, Başterzi et al. (2010) demonstrated that by the 6th week, responders of treatment with venlafaxine showed significantly higher NK cell numbers than non-responders. Grosse et al. (2016) reported that non-responders are characterized by a decreased percentage of natural killer (NK) cells before starting treatment compared with responders and therefore suggested it could serve as a predictive of antidepressant therapy. However, according to their results, antidepressant treatment with venlafaxine, whether successful or not, did not influence levels of NK cells. In the same study, the authors also demonstrated that treatment with venlafaxine increases percentages of Tregs regardless of the treatment outcome. Furthermore, patients with the highest rates of $CD8^+$ T lymphocytes before treatment with SNRIs were at the highest risk of not-responding to antidepressant, but, as with NK cells, treatment didn't change their percentage.

Lee and Kim (2006) showed that treatment with venlafaxine, like fluoxetine or paroxetine, causes a reduction in serum concentration of IL-12 and, at the same time, increases the level of TGF-β. Kubera et al. (2001), who demonstrated that in fluoxetine has an anti-inflammatory effect in vitro, obtained similar results for venlafaxine—it increased the production of IL-10 in stimulated whole blood cells thus lowering IFN-γ/IL-10 ratio. As with SSRIs, studies showed that persistent high serum TNF levels in depressed patients are associated with a lack of response to venlafaxine (Li et al., 2013). A recent study by Chen et al. (2018) showed that the immunomodulatory effect of venlafaxine is stronger than that of paroxetine. In patients treated with venlafaxine, serum levels of not only TNF-α, IFN-γ, IL-1β or IL-8, but also IL-4, IL-5, IL-10 associated with Th2 responses, decline during therapy. Furthermore, the authors showed that paroxetine paradoxically appeared to increase the concentration of some of these cytokines, IFN-γ and IL-6 in particular, in the blood of patients, which is in opposition to many other scientific reports.

In turn, the recent study of Avuloglu Yilmaz, Unal, and Yuzbasioglu (2017) demonstrated that milnacipran decreases the mitogenic index of human lymphocytes in vitro and has a genotoxic effect in a dose-dependent manner.

## Tricyclicantidepressants

In vitro studies showed that different TCAs (clomipramine, imipramine) not only exhibit anti-inflammatory properties by reducing the production of IL-1β and TNF-α by monocytes but they also decrease the release of IFN-γ and IL-2 from stimulated lymphocytes (Xia, DePierre, & Nässberger, 1996). Study of Kubera et al. (2001) revealed that whole blood cells of depressed patients stimulated with PHA or LPS in the presence of imipramine produce more anti-inflammatory IL-10, while Maes et al. (1999) showed that another drug from this group of antidepressants, clomipramine, reduces the secretion of IFN-γ.

Grosse et al. (2016) reported that treatment with imipramine, similar to venlafaxine, increased percentages of Tregs without any influence on other populations of T lymphocytes.

Interestingly, according to Schleifer et al. (1999), the only authors who demonstrated that in patients with depression T cell response to mitogenic stimulation is significantly increased compared to healthy people, in patients receiving TCAs lymphocyte activity in response to PHA or ConA significantly decreases.

Table 3 compares the effect of antidepressants from different groups on selected immune parameters.

## Noradrenergic and specific serotonergic antidepressants and others

At the moment, we know the least about the immunomodulatory effects of drugs from other groups: NaSSAs, SARIs (serotonin antagonist and reuptake inhibitors), SPARIs (serotonin partial agonist-reuptake inhibitors), NDRIs (norepinephrine-dopamine reuptake inhibitors), or MAOIs (monoamine oxidase inhibitors). The following are the results of individual reports that appear in the literature.

Mirtazapine has been shown to increase the expression of SERT on the surface of T lymphocytes, which may

TABLE 3 Influence of three main groups of antidepressants on the immune parameters.

| | Cytokines | Lymphocytes |
|---|---|---|
| SSRIs | ↓ IL-1β, IL-6, TNF-α, IL-12, IFN-γ<br>↑ IL-10, TGF-β<br>(both in vivo and in vitro) | ↑ Tregs (in vivo)<br>↑ NK cells (in vivo)<br>↑ The activity of NK cells (in vitro)<br>↓ Lymphocyte proliferation (in vitro) |
| SNRIs | ↓ IL-1β (both in vivo and in vitro)<br>↓ IFN-γ, TNF-α, IL-12 (in vivo)<br>↑ IL-10 (in vitro)<br>↑ TGF-β (in vivo) | ↑ Tregs (in vivo)<br>↑ NK cells (in vivo)<br>↑ The activity of NK cells (in vitro)<br>↓ Lymphocyte proliferation (in vitro) |
| TCAs | ↓ IL-1β, TNF-α (both in vivo and in vitro)<br>↓ IFN-γ, IL-2 (in vitro)<br>↑ IL-10 (in vitro) | ↑ Tregs (in vivo)<br>↓ Lymphocyte proliferation (in vitro) |

improve the reuptake of serotonin thus lowering its concentration in the environment of lymphocytes and, consequently, reducing their ability to proliferate (Peña, Baccichet, Urbina, Carreira, & Lima, 2005). Meanwhile, Norizadeh Tazehkand and Topaktas (2015) demonstrated that mirtazapine has a cytotoxic effect on human lymphocytes and reduces their proliferation capacity in vitro.

Maes et al. (1999) showed that trazodone, like clomipramine or sertraline, inhibits the production of IFN-γ in whole blood cells of depressive patients stimulated by PHA or LPS. Moreover, according to the recent study of Avuloglu Yilmaz et al. (2017) trazodone has a cytotoxic effect on human lymphocytes, but mainly in high concentrations. Current data also show that buspirone also demonstrates cytotoxic potential—it induces oxidative stress in human lymphocytes through the formation of reactive oxygen species, which leads to mitochondrial and lysosomal damages (Salimi, Razianm, & Pourahmad, 2018).

Landmann, Schaub, Link, and Wacker (1997) in the late 1990s demonstrated that moclobemide does not affect inflammatory parameters in depressed patients. However, Lin et al. (2000) several years later published a brief report, in which he proved that moclobemide suppresses the production of TNF-α and IL-8, and significantly enhances the production of IL-10 in monocytes of healthy people.

## Conclusions

The finding discussed above clearly show that commonly used antidepressants, SSRIs, SNRIs, and TCAs cause a decrease in the concentration of proinflammatory cytokines but also the mobilization of regulatory cells. These changes are beneficial for the patient in whom we observe the intensification of inflammatory processes in the course of depression. Moreover, some of the studies indicate that changes in selected immunological parameters before and during therapy may predict the patient's response to treatment, which can help in recognizing treatment-resistant depression.

## Key facts

### Key facts of the immune system in the depression

- Depressed patients have increased levels of proinflammatory cytokines (IL-1β, IL-6, and TNF-α).
- An increased IL-1 concentration is associated with increased activity of inflammasome.
- The rate of NK cells is significantly reduced in depressed patients.
- The percentage of Th1 and Th17 cells is increased compared with healthy people.
- The proportions of regulatory T lymphocytes (Tregs and Th3 cells) are significantly decreased in depressed patients.

### Key facts of the influence of antidepressants on the immune system

- SSRIs, SNRIs, and TCAs decrease the level of IL-1β, IL-6, and TNF-α, and increase the concentration of IL-10, TGF-β.
- Treatment with antidepressants increases the percentages of NK cells and Tregs.
- The increase of NK cell number or activity is associated with a good response to SSRIs and SNRIs.
- Antidepressants significantly reduce the proliferation capacity of lymphocytes.
- NaSSAs, SARIs, and SPARIs have been shown to have cytotoxic effects on human T lymphocytes.

## Summary points

- This chapter focuses on the immune disorders in the course of depression and shows how antidepressants influence selected immunologic parameters.
- Depressed patients not only have increased levels of proinflammatory cytokines but also present increased percentages of Th1 and Th17 cells, which attenuate pro-inflammatory processes.
- Depression is accompanied by reduced rates of regulatory T lymphocytes that in healthy people dampen inflammatory processes.
- Commonly used antidepressants, SSRIs, SNRIs as well as TCAs, decrease the concentration of proinflammatory cytokines and increase the level of antiinflammatory cytokines.
- Treatment with antidepressants also increases the percentages of NK cells and regulatory T lymphocytes.
- Changes in NK cell number or activity are associated with response to antidepressant treatment.
- The in vitro studies show that antidepressants significantly reduce the proliferation capacity of lymphocytes.
- In general, the influence of commonly used antidepressants is beneficial for patients because it reduces inflammation.

## Mini-dictionary of terms

**Inflammasome** A multioligomer expressed in myeloid cells that is responsible for the activation of inflammatory responses, mainly by inducing the production of IL-1β and IL-18.

**Apoptosis** Programmed and controlled cell death in the multicellular organism. The process leads to removal old or damaged cells.

**Mitogen** A chemical compound, e.g., protein, that induces mitosis of cells, in particular, lymphocytes.

## References


Ader, R., & Cohen, N. (1975). Behaviorally conditioned immunosuppression. *Psychosomatic Medicine*, *37*, 333–340.

Alcocer-Gómez, E., Casas-Barquero, N., Williams, M. R., Romero-Guillena, S. L., Cañadas-Lozano, D., Bullón, P., et al. (2017). Antidepressants induce autophagy dependent-NLRP3-inflammasome inhibition in major depressive disorder. *Pharmacological Research*, *121*, 114–121.

Avuloglu Yilmaz, E., Unal, F., & Yuzbasioglu, D. (2017). Evaluation of cytogenetic and DNA damage induced by the antidepressant drug-active ingredients, trazodone and milnacipran, in vitro. *Drug and Chemical Toxicology*, *40*, 57–66.

Aydin Sunbul, E., Sunbul, M., Yanartas, O., Cengiz, F., Bozbay, M., Sari, I., et al. (2016). Increased neutrophil/lymphocyte ratio in patients with depression is correlated with the severity of depression and cardiovascular risk factors. *Psychiatry Investigation*, *13*, 121–126.

Barkan, T., Gurwitz, D., Levy, G., Weizman, A., & Rehavi, M. (2004). Biochemical and pharmacological characterization of the serotonin transporter in human peripheral blood lymphocytes. *European Neuropsychopharmacology*, *14*, 237–243.

Bartrop, R. W., Luckhurst, E., Lazarus, L., Kiloh, L. G., & Penny, R. (1977). Depressed lymphocyte function after bereavement. *Lancet*, *1*, 834–836.

Başterzi, A. D., Yazici, K., Buturak, V., Cimen, B., Yazici, A., Eskandari, G., et al. (2010). Effects of venlafaxine and fluoxetine on lymphocyte subsets in patients with major depressive disorder: A flow cytometric analysis. *Progress in Neuro-Psychopharmacology and Biological Psychiatry*, *34*, 70–75.

Canan, F., & Ataoglu, A. (2009). Effect of escitalopram on white blood cells in patients with major depression. *Journal of Clinical Medicine Research*, *1*, 290–291.

Carvalho, L. A., Gorenstein, C., Moreno, R., Pariante, C., & Markus, R. P. (2009). Effect of antidepressants on melatonin metabolite in depressed patients. *Journal of Psychopharmacology*, *23*, 315–321.

Chen, Y., Jiang, T., Chen, P., Ouyang, J., Xu, G., Zeng, Z., et al. (2011). Emerging tendency towards autoimmune process in major depressive patients: A novel insight from Th17 cells. *Psychiatry Research*, *188*, 224–230.

Chen, C. Y., Yeh, Y. W., Kuo, S. C., Liang, C. S., Ho, P. S., Huang, C. C., et al. (2018). Differences in immunomodulatory properties between venlafaxine and paroxetine in patients with major depressive disorder. *Psychoneuroendocrinology*, *87*, 108–118.

Dahl, J., Ormstad, H., Aass, H. C., Malt, U. F., Bendz, L. T., Sandvik, L., et al. (2014). The plasma levels of various cytokines are increased during ongoing depression and are reduced to normal levels after recovery. *Psychoneuroendocrinology*, *45*, 77–86.

Darko, D. F., Gillin, J. C., Risch, S. C., Bulloch, K., Golshan, S., Tasevskam, Z., et al. (1989). Mitogen-stimulated lymphocyte proliferation and pituitary hormones in major depression. *Biological Psychiatry*, *26*, 145–155.

Davami, M. H., Baharlou, R., Ahmadi Vasmehjani, A., Ghanizadeh, A., Keshtkar, M., Dezhkam, I., et al. (2016). Elevated IL-17 and TGF-β serum levels: A positive correlation between T-helper 17 cell-related pro-inflammatory responses with major depressive disorder. *Basic and Clinical Neuroscience*, *7*, 137–142.

Demir, S., Atli, A., Bulut, M., İbiloğlu, A. O., Güneş, M., Kaya, M. C., et al. (2015). Neutrophil–lymphocyte ratio in patients with major depressive disorder undergoing no pharmacological therapy. *Neuropsychiatric Disease and Treatment*, *11*, 2253–2258.

Dowlati, Y., Herrmann, N., Swardfager, W., Liu, H., Sham, L., Reim, E. K., et al. (2010). A meta-analysis of cytokines in major depression. *Biological Psychiatry*, *67*, 446–457.

Dubois, T., Zdanowicz, N., Reynaert, C., & Jacques, D. (2016). Depression, family and immunity: Influence of humoral immunity on family relationships and on the severity of depression. *Psychiatria Danubina*, *28*, 170–174.

Duggal, N. A., Upton, J., Phillips, A. C., Hampson, P., & Lord, J. M. (2014). Depressive symptoms post hip fracture in older adults are associated with phenotypic and functional alterations in T cells. *Immunity & Ageing*, *11*, 25.

Edgar, V. A., Sterin-Borda, L., Cremaschi, G. A., & Genaro, A. M. (1999). Role of protein kinase C and cAMP in fluoxetine effects on human T-cell proliferation. *European Journal of Pharmacology*, *372*, 65–73.

Eller, T., Vasar, V., Shlik, J., & Maron, E. (2008). Pro-inflammatory cytokines and treatment response to escitalopram in major depressive disorder. *Progress in Neuro-Psychopharmacol & Biological Psychiatry*, *32*, 445–450.

Frank, M. G., Hendricks, S. E., Johnson, D. R., Wieseler, J. L., & Burke, W. J. (1999). Antidepressants augment natural killer cell activity: In vivo and in vitro. *Neuropsychobiology*, *39*, 18–24.

Grosse, L., Carvalho, L. A., Birkenhager, T. K., Hoogendijk, W. J., Kushner, S. A., Drexhage, H. A., et al. (2016). Circulating cytotoxic T cells and natural killer cells as potential predictors for antidepressant response in melancholic depression. Restoration of T regulatory cell populations after antidepressant therapy. *Psychopharmacology*, *233*, 1679–1688.

Hannestad, J., Dellagioia, N., & Bloch, M. (2011). The effect of antidepressant medication treatment on serum levels of inflammatory cytokines: A meta-analysis. *Neuropsychopharmacology*, *36*, 2452–2459.

Hernandez, M. E., Martinez-Fong, D., Perez-Tapia, M., Estrada-Garcia, I., Estrada-Parra, S., & Pavón, L. (2010). Evaluation of the effect of selective serotonin-reuptake inhibitors on lymphocyte subsets in patients with a major depressive disorder. *European Neuropsychopharmacology*, *20*, 88–95.

Ho, P. S., Yeh, Y. W., Huang, S. Y., & Liang, C. S. (2015). A shift toward T helper 2 responses and an increase in modulators of innate immunity in depressed patients treated with escitalopram. *Psychoneuroendocrinology*, *53*, 246–255.

Ivanova, S. A., Semke, V. Y., Vetlugina, T. P., Rakitina, N. M., Kudyakova, T. A., & Simutkin, G. G. (2007). Signs of apoptosis of immunocompetent cells in patients with depression. *Neuroscience and Behavioral Physiology*, *37*, 527–530.

Kemp, A. H., Quintana, D. S., Gray, M. A., Felmingham, K. L., Brown, K., & Gatt, J. M. (2010). Impact of depression and antidepressant treatment on heart rate variability: A review and meta-analysis. *Biological Psychiatry*, *67*, 1067–1074.

Köhler, C. A., Freitas, T. H., Stubbs, B., Maes, M., Solmi, M., Veronese, N., et al. (2018). Peripheral alterations in cytokine and chemokine levels after antidepressant drug treatment for major depressive disorder: Systematic review and meta-analysis. *Molecular Neurobiology*, *55*, 4195–4206.

Kook, A. I., Mizruchin, A., Odnopozov, N., Gershon, H., & Segev, Y. (1995). Depression and immunity: The biochemical interrelationship between the central nervous system and the immune system. *Biological Psychiatry*, *37*, 817–819.

Kronfol, Z., & House, J. D. (1989). Lymphocyte mitogenesis, immunoglobulin and complement levels in depressed patients and normal controls. *Acta Psychiatrica Scandinavica*, *80*, 142–147.

Kronfol, Z., House, J. D., Silva, J., Greden, J., & Carroll, B. J. (1986). Depression, urinary free cortisol excretion and lymphocyte function. *The British Journal of Psychiatry*, *148*, 70–73.

Kubera, M., Lin, A. H., Kenis, G., Bosmans, E., van Bockstaele, D., & Maes, M. (2001). Anti-inflammatory effects of antidepressants through suppression of the interferon-gamma/interleukin-10 production ratio. *Journal of Clinical Psychopharmacology*, *21*, 199–206.

Landmann, R., Schaub, B., Link, S., & Wacker, H. R. (1997). Unaltered monocyte function in patients with major depression before and after three months of antidepressive therapy. *Biological Psychiatry*, *41*(6), 675–681.

Lee, K. M., & Kim, Y. K. (2006). The role of IL-12 and TGF-beta1 in the pathophysiology of major depressive disorder. *International Immunopharmacology*, *6*, 1298–1304.

Li, Z., Qi, D., Chen, J., Zhang, C., Yi, Z., Yuan, C., et al. (2013). Venlafaxine inhibits the upregulation of plasma tumor necrosis factor-alpha (TNF-α) in the Chinese patients with major depressive disorder: A prospective longitudinal study. *Psychoneuroendocrinology*, *38*, 107–114.

Li, Y., Xiao, B., Qiu, W., Yang, L., Hu, B., Tian, X., et al. (2010). Altered expression of CD4(+)CD25(+) regulatory T cells and its 5-HT(1a) receptor in patients with major depression disorder. *Journal of Affective Disorders*, *124*, 68–75.

Lin, A., Song, C., Kenis, G., Bosmans, E., De Jongh, R., Scharpé, S., et al. (2000). The in vitro immunosuppressive effects of moclobemide in healthy volunteers. *Journal of Affective Disorders*, *58*, 69–74.

Liu, Y., Ho, R. C., & Mak, A. (2012). Interleukin (IL)-6, tumour necrosis factor alpha (TNF-α) and soluble interleukin-2 receptors (sIL-2R) are elevated in patients with major depressive disorder: A meta-analysis and meta-regression. *Journal of Affective Disorders*, *139*, 230–239.

Maes, M., Bosmans, E., Suy, E., Vandervorst, C., De Jonckheere, C., & Raus, J. (1991). Immune disturbances during major depression: Upregulated expression of interleukin-2 receptors. *Neuropsychobiology*, *24*, 115–120.

Maes, M., Kubera, M., Mihaylova, I., Geffard, M., Galecki, P., Leunis, J. C., et al. (2013). Increased autoimmune responses against autoepitopes modified by oxidative and nitrosative damage in depression: Implications for the pathways to chronic depression and neuroprogression. *Journal of Affective Disorders*, *149*, 23–29.

Maes, M., Song, C., Lin, A. H., Bonaccorso, S., Kenis, G., De Jongh, R., et al. (1999). Negative immunoregulatory effects of antidepressants: Inhibition of interferon-gamma and stimulation of interleukin-10 secretion. *Neuropsychopharmacology*, *20*, 370–379.

Maes, M., Stevens, W., DeClerck, L., Bridts, C., Peeters, D., Schotte, C., et al. (1992a). Immune disorders in depression: Higher T helper/T suppressor-cytotoxic cell ratio. *Acta Psychiatrica Scandinavica*, *86*, 423–431.

Maes, M., Stevens, W. J., DeClerck, L. S., Bridts, C. H., Peeters, D., Schotte, C., et al. (1992b). A significantly increased number and percentage of B cells in depressed subjects: Results of flow cytometric measurements. *Journal of Affective Disorders*, *24*, 127–134.

Miller, H. L., Ekstrom, R. D., Mason, G. A., Lydiard, R. B., & Golden, R. N. (2001). Noradrenergic function and clinical outcome in antidepressant pharmacotherapy. *Neuropsychopharmacology*, *24*, 617–623.

Mizruchin, A., Gold, I., Krasnov, I., Livshitz, G., Shahin, R., & Kook, A. I. (1999). Comparison of the effects of dopaminergic and serotonergic activity in the CNS on the activity of the immune system. *Journal of Neuroimmunology*, *101*, 201–214.

Müller, N., Hofschuster, E., Ackenheil, M., Mempel, W., & Eckstein, R. (1993). Investigations of the cellular immunity during depression and the free interval: Evidence for an immune activation in affective psychosis. *Progress in Neuro-Psychopharmacology & Biological Psychiatry*, *17*, 713–730.

Munzer, A., Sack, U., Mergl, R., Schönherr, J., Petersein, C., Bartsch, S., et al. (2013). Impact of antidepressants on cytokine production of depressed patients in vitro. *Toxins (Basel)*, *5*, 2227–2240.

Myint, A. M., Leonard, B. E., Steinbusch, H. W., & Kim, Y. K. (2005). Th1, Th2, and Th3 cytokine alterations in major depression. *Journal of Affective Disorders*, *88*, 167–173.

Norizadeh Tazehkand, M., & Topaktas, M. (2015). The in vitro genotoxic and cytotoxic effects of remeron on human peripheral blood lymphocytes. *Drug and Chemical Toxicology*, *38*, 266–271.

O'Regan, C., Kenny, R. A., Cronin, H., Finucane, C., & Kearney, P. M. (2015). Antidepressants strongly influence the relationship between depression and heart rate variability: Findings from The Irish Longitudinal Study on Ageing (TILDA). *Psychological Medicine*, *45*, 623–636.

Pariante, C. M., & Miller, A. H. (2001). Glucocorticoid receptors in major depression: Relevance to pathophysiology and treatment. *Biological Psychiatry*, *49*, 391–404.

Park, E. J., Lee, J. H., Jeong, D. C., Han, S. I., & Jeon, Y. W. (2015). Natural killer cell activity in patients with major depressive disorder treated with escitalopram. *International Immunopharmacology*, *28*, 409–413.

Patas, K., Willing, A., Demiralay, C., Engler, J. B., Lupu, A., Ramien, C., et al. (2018). T cell phenotype and T cell receptor repertoire in patients with major depressive disorder. *Fronters in Immunology*, *9*, 291.

Peña, S., Baccichet, E., Urbina, M., Carreira, I., & Lima, L. (2005). Effect of mirtazapine treatment on serotonin transporter in blood peripheral lymphocytes of major depression patients. *International Immunopharmacology*, *5*, 1069–1076.

Pérez-Sánchez, G., Becerril-Villanueva, E., Arreola, R., Martínez-Levy, G., Hernández-Gutiérrez, M. E., Velasco-Velásquez, M. A., et al. (2018). Inflammatory profiles in depressed adolescents treated with fluoxetine: An 8-week follow-up open study. *Mediators of Inflammation*, *2018*, 4074051.

Pietruczuk, K., Jakuszkowiak, K., Landowski, J., & Witkowski, J. M. (2005). Comparison of chosen immune system parameters in patients suffering from depression and healthy donors. *Psychiatria*, *2*, 210–216.

Salimi, A., Razianm, M., & Pourahmad, J. (2018). Analysis of toxicity effects of buspirone, cetirizine and olanzapine on human blood lymphocytes: In vitro model. *Current Clinical Pharmacology*, *13*, 120–127.

Schleifer, S. J., Keller, S. E., & Bartlett, J. A. (1999). Depression and immunity: Clinical factors and therapeutic course. *Psychiatry Research*, *85*, 63–69.

Schleifer, S. J., Keller, S. E., Camerino, M., Thornton, J. C., & Stein, M. (1983). Suppression of lymphocyte stimulation following bereavement. *JAMA*, *250*, 374–377.

Schmidt, F. M., Lichtblau, N., Minkwitz, J., Chittka, T., Thormann, J., Kirkby, K. C., et al. (2014). Cytokine levels in depressed and non-depressed subjects, and masking effects of obesity. *Journal of Psychiatric Research*, *55*, 29–34.

Strawbridge, R., Arnone, D., Danese, A., & Papadopoulos, A. (2015). Inflammation and clinical response to treatment in depression : A meta-analysis. *European Neuropsychopharmacology*, *25*, 1532–1543.

Szuster-Ciesielska, A., Słotwińska, M., Stachura, A., Marmurowska-Michałowska, H., Dubas-Slemp, H., Bojarska-Junak, A., et al. (2008). Accelerated apoptosis of blood leukocytes and oxidative stress in blood of patients with major depression. *Progress in Neuro-Psychopharmacology & Biological Psychiatry*, *32*, 686–694.

Taler, M., Gil-Ad, I., Lomnitski, L., Korov, I., Baharav, E., Bar, M., et al. (2007). Immunomodulatory effect of selective serotonin reuptake inhibitors (SSRIs) on human T lymphocyte function and gene expression. *European Neuropsychopharmacology*, *17*, 774–780.

Xia, Z., DePierre, J. W., & Nässberger, L. (1996). Tricyclic antidepressants inhibit IL-6, IL-1 beta and TNF-alpha release in human blood monocytes and IL-2 and interferon-gamma in T cells. *Immunopharmacology*, *34*, 27–37.

Zorrilla, E. P., Luborsky, L., McKay, J. R., Rosenthal, R., Houldin, A., Tax, A., et al. (2001). The relationship of depression and stressors to immunological assays: A meta-analytic review. *Brain, Behavior, and Immunity*, *15*, 199–226.

Part II

# Biomarkers and diagnosis

Chapter 15

# Assessment scoring tools of depression

Clarice Gorenstein[a,b], Elaine Henna[a,c], and Yuan-Pang Wang[a]
[a]*Department and Institute of Psychiatry, LIM-23, Hospital das Clínicas, University of São Paulo Medical School, São Paulo, SP, Brazil,* [b]*Department of Pharmacology, Institute of Biomedical Sciences, University of São Paulo, São Paulo, Brazil,* [c]*Department of Medicine, Discipline of Psychiatry, College of Medical and Health Sciences, Pontifícial Catholic University of Sao Paulo, Sorocaba, Brazil*

## List of abbreviations

**BDI** Beck Depression Inventory
**CDI** the Children's depression inventory
**CESD** the Center for Epidemiologic Studies Depression Scale
**CIDI** Composite International Diagnostic Interview
**CIS-R** Clinical Interview Schedule-Revised
**CSDD** Cornell Scale for Depression in Dementia
**EPSD** Edinburgh Postnatal Depression Scale
**GDS** the Geriatric Depression Scale
**HADS** the Hospital Anxiety and Depression Scale
**HAM-D** Hamilton Rating Scale for Depression
**K-SADS** Schedule for Affective Disorders and Schizophrenia for School-Age children
**MADRS** Montgomery-Åsberg Depression Rating Scale
**MINI** Mini International Neuropsychiatric Interview
**PDSS** Postpartum Depression Screening Scale
**PHQ-9** Patient Health Questionnaire
**SADS** Schedule for Affective Disorders and Schizophrenia
**SCAN** Schedule for Clinical Assessment in Neuropsychiatry
**SCID** Structured Clinical Interview for DSM-5 Disorders

## Introduction

Traditionally, the assessment of psychopathology should precede treatment recommendations. The diagnosis of most medical diseases can be supported by ancillary methods, such as laboratory tests, physiological measures, and image exams. Objective indicators of body temperature, weight, and blood pressure can be easily measured through simple devices. In psychiatry, however, there are no objective biomarkers to define the boundary between normal and abnormal states. This limit is often based on arbitrary personal judgment, making it hard to establish an unquestionable diagnosis in mental health settings.

There are several psychometric instruments designed to measure the subjective experience, e.g., mood disorders and symptoms. The prototypes of modern assessment tools were tailored since the 19th century and refined after the discovery of antidepressants in the 20th century. In this chapter, we focus on the assessment tools designed to screen, diagnose, and measure the intensity of depressive symptoms, respectively brief screening tools, interviews, and rating scales. We selected validated instruments to assess the psychopathology of depression.

## Depression as a public health issue

Major depressive disorder (MDD) comprises a constellation of symptoms and signs that alter dimensions of mood, cognition, and behavior. MDD is pervasive and interferes negatively in daily life functioning throughout the lifespan. In extreme conditions, a depressed person might commit suicide. The sex ratio is around 2:1 for women and men, which suggests the role of biological and psychological sex differences determining its manifestation (Lim et al., 2018). MDD is responsible for an economic burden of $210.5 billion/year and accounts for 10% of medical appointments in the United States (Greenberg, Fournier, Sisitsky, Pike, & Kessler, 2015). Therefore, the epidemic of depression is a global matter of public health concern.

MDD encompasses the presence of negative affect (sadness, guilt), absence of positive affect (anhedonia), and several physical symptoms (diurnal variation, fatigue, alterations in appetite and sleep, and diminished concentration). The manifestations of depression vary according to age, sex, genetics, and environmental factors. The core source for the development of the measurement tools of depressive symptoms comes from a large amount of information about patients suffering from depression, either in community or in clinical settings.

In the past five decades, a great joint effort has been dedicated to building assessment tools for screening depression in large community studies. Similarly, several structured interviews were proposed to improve the reliability of the diagnosis of psychiatric disorders. Currently, the diagnosis of depression is established thought a polythetic construct (Lim et al., 2018). This means the presence of at least five symptoms, including at least sad mood and/or anhedonia, lasting 15 days or over, and has to cause significant distress

The Neuroscience of Depression. https://doi.org/10.1016/B978-0-12-817933-8.00051-7

**BOX 1 DSM-5 criteria for major depressive disorder**

A. Presence of five (or more) of the following symptoms in the same 2-week period, which represents a change from previous functioning; at least one of the symptoms is either (1) depressed mood or (2) loss of interest or pleasure.

*Note:* symptoms that are clearly attributable to a medical condition should not be included.

1. Depressed mood (sadness, emptiness, hopelessness) most of the day
2. Diminished interest or pleasure in almost all activities
3. Significant weight loss (when not dieting) or weight gain or decrease or increase in appetite
4. Insomnia or hypersomnia
5. Psychomotor agitation or retardation
6. Fatigue or loss of energy
7. Feelings of worthlessness or excessive guilt
8. Diminished ability to think or concentrate; indecisiveness
9. Recurrent thoughts of death, recurrent suicidal ideation without a specific plan, or a suicide attempt or specific plan for committing suicide.

B. The symptoms cause significant distress or impairment in social, occupational, or other important areas of functioning.

C. The symptoms are not attributable to the physiological effects of a substance (e.g., a drug of abuse, a medication) or another medical condition.

D. The disturbance is not better explained by a persistent schizoaffective disorder, schizophrenia, delusional disorder, or other specified or unspecified schizophrenia spectrum and other psychotic disorders.

E. There has never been a manic episode or a hypomanic episode.

*Source:* American Psychiatric Association (2013).

*Note:* Criteria A–C represent a major depressive episode.

or impairment in normal functioning (Box 1). Also, several tools rate depressive symptoms in pharmacological trials.

## Overview

Assessment tools refer to instruments to screen, diagnose, and monitor health outcomes. In general, screening measures are brief self-reported instruments aiming to identify possible cases of depression in the community. Conversely, diagnostic instruments are based in an interview conducted by a clinician. Lastly, rating scales are tools designed to evaluate the intensity of psychopathology and monitor symptomatic changes.

The scientific field dedicated to measuring psychopathological symptoms is known as psychometrics. Its main task is to establish the performance of assessment tools in objective indicators, such as reliability and validity.

Reliability refers to the accuracy of an instrument, i.e., the ability to detect a true, reproducible score. However, the error-free condition is impossible to achieve with a psychological instrument. Sources of errors could be generated by respondents (providing incorrect data or endorsing systematically high or low scores), and interviewers, whose background bias the assessment of psychopathology. The lower the errors are, the higher is the reliability. Examples of indicators of reliability are internal consistency (Cronbach alpha coefficient) and test-retest coefficient (intraclass coefficient), both ranging from 0 to 1. If the conditions remain the same across the time, the scores should be similar. The retest framework can indicate the presence or absence of the effect of an intervention.

Validity refers to the ability of an instrument to measure what it proposes to measure, encompassing the target construct. For example, the construct of depression comprises several subdimensions, such as affective, cognitive, anxious, somatic, and psychotic symptoms. The scale content coverage is not perfect, that is, they can measure at best some of depressive symptoms. The types of validity are the face, criterion, structure, and construct. Each type requires specific analytical strategies, for example, indicators of sensibility and specificity of a given scale can establish the best cutoff point for discriminating case/noncase and be obtained through receiver-operating characteristics (ROC) analysis.

Assessment tools might be self-reported or applied by an external assessor. While the self-rating is more subjective, the observer instruments require clinical judgment to rate the presence and/or intensity of symptoms. The self-report format is recommended for screening instruments and some rating scales, whereas diagnostic instruments are applied by a trained interviewer.

The self-assessment instruments are simple to administer, cost-effective, do not require time-consuming training programs and suffer little interference from interviewers' expectations. However, they depend on the respondent's cooperation, and their ability to understand the questions. In general, self-assessment scales are less suitable for measuring observable behaviors (e.g., motor retardation or agitation) and self-perception (e.g., hypochondriac concern or depressive delusions). Also, schooling is an issue, particularly in low-income countries. A critique of the self-report format is the liability of omitting or exaggerating symptoms, either intentionally or unintentionally. Frequently, the respondent found difficulties to score a subjective experience into a numerical representation, either due to inability or nonfamiliarity. On the other hand, this strategy of scale self-administration favors trustworthy responses to embarrassing questions as unusual sexual behavior, suicidal ideation, and the use of psychotropic substances.

Differently, observer instruments require training and familiarity with the theoretical concepts of the underlying construct covered by the instrument. Thus, the agreement between different observers is fundamental to ensure the accuracy of the assessment. During the interview, the respondent might answer the questions in a positive light (social desirability bias), and the interviewer might be influenced by his previous experience and expectation. These instruments are onerous to be implemented in large studies.

## Screening depression

From the perspective of public health, screening depression in community and specific populations is of paramount importance. Depression should be routinely screened in primary care to provide early support for the general population. The US Preventive Services Task Force recommends screening depression in adult, pregnant, and postpartum patients in primary care (Siu, and USPSTF, 2016). Generally, screening measurements are self-reported. No screening measure can surrogate the diagnosis of depression.

Recommended screening instruments of depression are the Center for Epidemiological Studies-Depression (CES-D) and the Patient Health Questionnaire (PHQ-9). For special populations, the main instruments are the Hospital Anxiety and Depression Scales (HADS), the Children's Depression Inventory (CDI), the Elderly Geriatric Depression Scale (GDS), the Edinburgh Postnatal Depression Scale (EPDS), and Postpartum Depression Screening Scale (PDSS) (Tables 1 and 2).

The 20-item Center for Epidemiologic Studies Depression Scale (CES-D; Radloff, 1977) covers depressive mood, behavior, and self-perception. Each item is rated according to how often the respondent has felt during the past week. Responses range from 0 to 3, with a total score ranging from 0 to 60. The recommended cutoff is 16, but better indicators were reported for 20 (Vilagut, Forero, Barbaglia, & Alonso, 2016). The internal consistency of the scale ranges from 0.81 to 0.92 (Cosco, Prina, Stubbs, & Wu, 2017). The CES-D is a public domain instrument.

The Patient Health Questionnaire (PHQ-9; Kroenke et al., 2001) is one of the most used screening tools for depression (Kroenke, Spitzer, Williams, & Lowe, 2010; Smarr & Keefer, 2011). It contains nine DSM-items plus an item about impairment. Respondents should rate the items on a four-point ordinal scale indicating how often the symptom was present over the past 2 weeks. Total scores range from 0 to 27. In primary care, a recommended score of 10 is the cutoff for probable cases (Tomitaka et al., 2018). For additional details, please refer to Chapter 18.

The Hospital Anxiety and Depression Scale (HADS, Zigmond & Snaith, 1983) was built to screen anxiety and depression in medical outpatient clinics. This scale excludes somatic complaints that mimic depression-related symptoms, e.g., fatigue, abnormal appetite, and sleep. It comprises 14 items, seven for anxiety and seven for depression, that are rated from 0 to 3, according to how often the respondent has felt during the past week. Anxiety and depression are scored separately with good psychometric properties (Bjelland, Dahl, Haug, & Neckelmann, 2002),

**TABLE 1** Screening tools for depression.

| Instrument | Target population | No. of items |
|---|---|---|
| CES-D | Adults<br>Adolescents | 20 |
| PHQ-9 | Adults | 9 |
| HADS | Adults | 14 |
| CDI | Children<br>Adolescents | 27 |
| GDS | Elderly | 30 |
| EPDS | Pregnancy<br>Postnatal | 10 |
| PDSS | Postnatal | 35 |

*CDI*, The Children's Depression Inventory; *CES-D*, The Center for Epidemiologic Studies Depression Scale; *EPSD*, Edinburgh Postnatal Depression Scale; *GDS*, The Geriatric Depression Scale; *PHQ-9*, Patient Health Questionnaire; *HADS*, The Hospital Anxiety and Depression Scale; *PDSS*, Postpartum Depression Screening Scale.

**TABLE 2 Psychometric properties of screening tools for depression.**

| Instrument | Cutoff | Psychometric indexes | References |
|---|---|---|---|
| CES-D | ≥16 | Alpha: 0.90<br>Sensitivity: 74.6%<br>Specificity: 73.4% | Radloff (1977) |
| PHQ-9 | ≥10 | Alpha: 0.86<br>Sensitivity: 88%<br>Specificity: 88% | Kroenke, Spitzer, and Williams (2001) |
| HADS | ≥8 | Alpha: 0.83<br>Sensitivity: 80%<br>Specificity: 80% | Zigmond and Snaith (1983) |
| CDI | ≥16 | Alpha: 0.80 | Kovacs (1992) |
| GDS | ≥11 | Alpha: 0.91<br>Sensitivity: 80%<br>Specificity: 75% | Yesavage et al. (1983) |
| EPDS | 9–13 | Alpha: 0.84<br>Sensitivity: 86%<br>Specificity: 78% | Cox, Holden, and Sagovsky (1987) |
| PDSS | ≥80 | Alpha: 0.83–0.94<br>Sensitivity: 94%<br>Specificity: 98% | Beck and Gable (2000) |

*CDI*, The Children´s Depression Inventory; *CES-D*, The Center for Epidemiologic Studies Depression Scale; *EPSD*, Edinburgh Postnatal Depression Scale; *GDS*, The Geriatric Depression Scale; *HADS*, The Hospital Anxiety and Depression Scale; *PHQ-9*, Patient Health Questionnaire; *PDSS*, Postpartum Depression Screening Scale.

Various cutoff points have been established (Herrmann, 1997), varying from 8 to 11 (Zigmond & Snaith, 1983), being 8/9 regarded as an optimal cutoff point (Bjelland et al., 2002). The HADS can be purchased from GL Assessment (https://www.gl-assessment.co.uk/).

Some instruments are designed for accounting age-related differences in the expression of depression. The diagnostic criteria for depression in youth are the same as adults, except that irritability goes along with depressed mood and anhedonia. Melancholia occurs more often in adolescents, whereas hypersomnia in children (Weiss & Garber, 2003). Usually, the own adolescent can answer a screening instrument.

The Children's Depression Inventory (CDI; Kovacs, 1992) is a 27-item self-rated screening scale of depression for respondents aged between 7 and 16 years. The CDI covers five domains during the past two weeks: negative mood, interpersonal problems, ineffectiveness, anhedonia, and negative self-esteem. Responses are scored from 0 to 2. The cutoff score of 16 indicates an optimal trade-off between sensitivity and specificity (Timbremont, Braet, & Dreessen, 2004). Probable cases should be referred to interviews such as the Kiddie Schedule for Affective Disorders and Schizophrenia (K-SADS) to confirm the diagnosis. The copyright of CDI belongs to the WPS Publisher.

The Geriatric Depression Scale (GDS; Yesavage et al., 1983) is designed to screen depression in the elderly with mild cognitive impairment. Aging-related cognitive problems can compromise the accuracy of self-reports, just as depression can impair cognitive skills. Some symptoms were excluded, such as slowness, insomnia, hyposexuality, and dementia. The GDS has 30 yes/no items that assess the presence or absence of mood, cognition, anxiety, and behaviors in the past week. Scores vary from 0 to 10 and a cutoff of 10/11 is evidence of probable depression. A shorter 15-item version is also available (Sheikh & Yesavage, 1986), mitigating the burden for the elderly who easily become fatigued or distracted (Balsamo, Cataldi, Carlucci, Padulo, & Fairfield, 2018).

The two most used scales to screening depression in perinatal periods are the Edinburgh Postnatal Depression Scale (EPDS) and the Postpartum Depression Screening Scale (PDSS).

The EPDS (Cox et al., 1987) was designed to screen depression in the postnatal period and during pregnancy. It contains 10 questions focusing on the cognitive and affective symptoms of depression that occur 7 days before the interview. Each symptom is rated from 0 to 3, totaling a maximum score of 30. For further details about EPDS, please go to Chapter 20.

The PDSS (Beck & Gable, 2000) is a 35-item self-report questionnaire to screen how the mother feels after the labor. The answers are rated on a 5-point Likert scale, ranging from 1 (strongly disagree) to 5 (strongly agree). The total score varies from 35 to 175, where score ≥80 indicates a probable case of MDD (Beck & Gable, 2000). The scale should be purchased with the WPS publisher.

## The diagnosis of depression

Considerable subjectivity exists in the psychiatric assessment because there are no objective measures for emotional symptoms. Patients might speak out about symptoms they believe relevant to the interviewer and hide uncomfortable or embarrassing symptoms such as death thoughts, abnormal sexual or eating behavior, etc. Psychiatrists have to examine subjective complaints reported by patients to set a hypothetical diagnosis. The interpretation of these psychopathologies might differ according to the background and the training of the interviewer. For example, while psychotherapists would understand symptoms of sadness and anhedonia as the outcome of a grief reaction, biological psychiatrists would explain pessimistic feelings and motor retardation as results of an imbalanced neurotransmitter system in the brain.

Regarding the type of interview, there are unstructured or structured formats. The unstructured interview is the usual type in clinical settings and depends on the interviewer's experience. A clinical interviewer might explore unrestrainedly the patient's complaint in this strategy. Unreliable diagnoses yielded from unstructured interviews can be partially fixed through operational criteria homogenized by classification systems, such as the DSM and ICD. However, some novice interviewers might avoid investigating distressing topics, compromising the coverage completeness of investigations.

The structured interview is anchored in a uniform schedule to inquire about respondents' symptoms. Usually, diagnosis can be achieved following a classification system. In general, the coverage of an interview is a predefined set of psychopathological areas. Standardized training for their use is mandatory and the familiarity with diagnostic criteria improves the between-rater agreement. Examples of popular standardized structured interview are the Structured Clinical Interview for DSM Disorders (SCID), the Composite International Diagnostic Interview (CIDI), the Mini International Neuropsychiatric Interview (MINI), the Schedule for Affective Disorders and Schizophrenia (SADS), the Schedule for Clinical Assessment in Neuropsychiatry (SCAN), the Clinical Interview Schedule-Revised (CIS-R), etc. (Tables 3 and 4). Currently, the most popular instruments are SCID and CIDI, although alternative interviews with evidence of reliability can be adopted per local preference.

The Structured Clinical Interview for DSM Disorders (SCID) is a semistructured diagnostic interview (First, Williams, Karg, & Spitzer, 2015), elaborated under the auspices of the American Psychiatry Association. The SCID takes into account the clinical judgment to establish a top-down diagnosis following the DSM system. In other words, the interviewer should formulate a hypothetical

**TABLE 3** Structured interviews for diagnosis of depression.

| Interview | Target population | Format |
|---|---|---|
| SCID | Adults<br>Children | Semistructured interview<br>12 modules |
| CIDI | Adults | Structured interview<br>41 modules |
| MINI | Adults | Structured interview<br>19 modules |
| SADS<br>K-SADS | Adults<br>Children (6–17 year) | Semistructured interview<br>20 modules |
| SCAN | Adults | Structured interview<br>27 sections |
| CIS-R | Adults<br>Adolescents | Structured interview (ICD-10)<br>14 symptom groups |

*CIDI*, Composite International Diagnostic Interview; *CIS-R*, Clinical Interview Schedule-Revised; *K-SADS*, Schedule for Affective Disorders and Schizophrenia for School-Age Children-Present and Lifetime Version; *MINI*, Mini International Neuropsychiatric Interview; *SCAN*, The Schedules for Clinical Assessment in Neuropsychiatry; *SCID*, Structured Clinical Interview for Diagnosis.

**TABLE 4 Psychometric properties of diagnostic interviews.**

| Interview | Psychometric indexes | References |
|---|---|---|
| SCID | Interrater reliability:<br>Kappa: 0.61 (current diagnosis)<br>Kappa: 0.68 (lifetime diagnosis) | Williams et al. (1992) |
| CIDI | Interrater reliability:<br>Kappa: 0.94<br>Test-retest: 0.71 | Robins et al. (1988) |
| MINI | Interrater reliability:<br>Kappa: 0.85 | Sheehan et al. (1998) |
| SADS | Alpha: >0.60<br>Interrater reliability:<br>Kappa: >0.82<br>Test-retest: 0.70 | Endicott and Spitzer (1978) |
| SCAN | Interrater reliability:<br>Kappa: 0.53<br>Sensitivity: 59%<br>Specificity: 90% | Wing et al. (1990) |
| CIS-R | Interrater reliability:<br>Kappa: 0.23<br>Sensitivity: 43%<br>Specificity: 80%–90% | Brugha et al. (1999) |

*CIDI*, Composite International Diagnostic Interview; *CIS-R*, Clinical Interview Schedule-Revised; *K-SADS*, Schedule for Affective Disorders and Schizophrenia for School-Age Children-Present and Lifetime Version; *MINI*, Mini International Neuropsychiatric Interview; *SCAN*, The Schedules for Clinical Assessment in Neuropsychiatry; *SCID*, Structured Clinical Interview for Diagnosis.

diagnosis at the end of the initial assessment, and then all suspected mental disorders would be scrutinized one-by-one, through the operational criteria of the DSM system. It covers mood, psychotic, anxiety, alcohol and substance abuse, somatoform, and eating disorders.

There are versions for research, clinical trials, and primary care (SCID-R, SCID-CT, and SCID-CV, respectively), as well as further versions for children (KIDS-SCID) and personality disorders (SCID-5-PD). Previous versions of SCID have indicated moderate-to-excellent inter-rater reliability for mood disorders (Lobbestael, Leurgans, & Arntz, 2011; Zanarini & Frankenburg, 2001). Its application takes about 90min to be completed. The SCID can be purchased at http://www.appi.org.

The Composite International Diagnostic Interview (CIDI; Robins et al., 1988) is widely applied in epidemiological studies (Kessler, Aguilar-Gaxiola, et al., 2009; Kessler, Avenevoli, et al., 2009). This is a fully structured interview and can be administered by lay interviewers. The CIDI allows for generating comparable datasets across cultures and languages. Its application takes around one hour and over to complete. The diagnosis is classified through an algorithm as present, subliminal or absent, either in DSM or ICD system. Additional information about the CIDI can be found on the website https://www.hcp.med.harvard.edu/wmhcidi/about-the-who-wmh-cidi/.

## Rating scales

Rating scales are invaluable complementary sources of the psychopathological assessment after a proper diagnosis. This group of assessment tools aims to measure the intensity of the clinical conditions, being applicable in different phases of clinical treatment or pharmacological trial. The baseline severity of depression might be adopted as an inclusion criterion for eligibility to enter or be excluded in a clinical trial. The objectivity and the same content coverage of a scale allow selecting participants above or below a threshold, which can ensure a more homogenous group of patients in the trial. The interpretation of the severity score might help the choice of suitable therapeutic modality, either psychotherapy or pharmacotherapy.

Repeated measures of the same scale provide a consistent indicator of the clinical state over time. A comparison between baseline and posttreatment scores estimates the magnitude of the observed change after the intervention. Currently, the inclusion of rating scales in clinical trials is largely adopted worldwide, some of them presenting well-established patterns of outcome for the response, remission, and refractoriness of a treatment.

The most used scales of depression are the Hamilton Rating Scale for Depression (HAM-D), the Montgomery-Åsberg Depression Rating Scale (MADRS), the Beck

**TABLE 5** Rating scales for depression.

| Rating scale | Psychological measure | Target population | Format |
|---|---|---|---|
| HAM-D | Severity | Adults | Clinician-rated<br>21 items<br>17 items |
| MADRS | Severity<br>Treatment improvement | Adults | Self-report (SR) or<br>Clinician-rated (CR)<br>10 items |
| BDI | Severity | Adults<br>≥13 year | Self-report or<br>Clinician-rated<br>21 questions |
| CSDD | Severity | Patients with dementia | Clinician-rated<br>19 items and 2 parts (for patient and caregivers) |

*CSDD*, Cornell Scale for Depression in Dementia; *BDI*, Beck Depression Inventory; *HAM-D*, Hamilton Rating Scale for Depression; *MADRS*, Montgomery-Åsberg Depression Rating Scale.

**TABLE 6** Psychometric properties of rating scales for depression.

| Rating scale | Cutoff | Psychometric indexes | References |
|---|---|---|---|
| HAM-D | ≥9 | Alpha: 0.46–0.81<br>Interrater reliability:<br>Kappa: 0.82–0.98<br>Test-retest: 0.81–0.98 | Bagby et al. (2004) |
| MADRS | ≥17 | Interrater reliability:<br>Kappa: 0.62–0.96 (item-by-item) | Montgomery and Åsberg (1979) |
| BDI-II | ≥11 | Alpha: 0.92–0.93<br>Test-retest: 0.73–0.96<br>Sensitivity: ≥70% | Beck et al. (1996) |
| CSDD | ≥6 | Alpha: 0.84<br>Interrater reliability:<br>Kappa: 0.74<br>Sensitivity: 81% | Alexopoulos et al. (1988) |

*BDI*, Beck Depression Inventory; *CSDD*, Cornell Scale for Depression in Dementia; *HAM-D*, Hamilton Rating Scale for Depression; *MADRS*, Montgomery-Åsberg Depression Rating Scale.

Depression Inventory (BDI), and the Cornell Scale for Depression in Dementia (CSDD) (Table 5 and 6).

The HAM-D (Hamilton, 1967) was developed to evaluate the severity of depression in inpatients, with greater emphasis on somatic symptoms. Cognitive and somatic symptoms cover around half of the HAM-D items, which make it sensitive to detect changes during treatment. Its first version composed of 21 items, but nowadays the 17-item version is the most adopted. A Structured Interview Guide for the Hamilton Depression Rating Scale (SIGH-D) was developed as a schedule to standardize its interview-based administration (Williams, 1988). Although HAM-D has been widely criticized for its psychometric and conceptual flaws (Bagby, Ryder, Schuller, & Marshall, 2004), this scale is still considered one of the most adopted tools for measuring the severity of depression in applied researches (Santor, Gregus, & Welch, 2006). For additional details go to Chapter 17.

The MADRS (Montgomery & Åsberg, 1979) consists of 10 items covering core depressive symptoms, mainly mood (sadness, tension, lassitude, pessimism, and suicidal thoughts). Nine items are scored based on patient report and one based on the interviewer observation (apparent sadness). Each item is scored from 0 to 6 in an increasing scale of severity, resulting in a total score varying from 0 to 60. The higher is the total score, the more severe is the depression. The MADRS was designed to be sensitive to detect changes in treatment according to the following ranges: 0–8: remission; 9–17: mild depression; 18–34: moderate depression; ≥35: severe depression (Müller-Thomsen, Arlt, Mann, Mass, & Ganzer, 2005).

Besides several versions of the MADRS, a structured interview guide (SIGMA; Williams & Kobak, 2008) was developed to help standardize its application. The self-report version (MADRS-S, Svanborg & Åsberg, 1994) is recommended to be used concurrently to the MADRS-SIGMA in clinical trials. A recent parent form (MADRS-P; Torres et al., 2017) was developed for screening of MDD in adolescents.

The Beck Depression Inventory (BDI; Beck, Ward, Mendelson, Mock, & Erbaugh, 1961) is a 21-item self-report questionnaire on how respondents feel recently. The scale is rated in an increasing ordinal scale of severity, ranging from 0 to 3. The total score varies from 0 to 63. The most recent BDI version, BDI-II (Beck, Steer, & Brown, 1996), covers psychological and somatic manifestations of a 2-week major depressive episode, as operationalized in the DSM-IV (APA, 2013). The BDI has been also used as a screening instrument considering scores over 11 as probable depression. For additional details go to Chapter 16.

Likewise screening instruments, some rating scales are also developed for special populations whose clinical expression of depression might differ greatly.

The Cornell Scale for Depression in Dementia (CSDD; Alexopoulos, Abrams, Young, & Shamoian, 1988) aims to quantify the severity of depression in patients with dementia. Its administration takes around 30 min and requires two steps. First, the clinician asks the patient's caregiver on each of the 19 items of the scale and then briefly interviews the patient. The CSDD is rated on a severity scale ranging from 0 to 2, summing a maximum score of 38. The time frame is the past 7 days.

## Comments

Each type of assessment tools has well-defined objectives. The identification of depressive conditions with simple and fast screening tools could beneficiate several patients, referring them to proper diagnosis and treatment. Researchers using standardized interviews could greatly reduce the unreliable diagnosis of depression to emerge. In the era of evidence-based clinical practice, the systematic use of psychological measures is increasingly regarded as an objective demonstration of change of mental states, or evidence of intervention efficacy in clinical trials. From the perspective of public health, a wise selection of assessment tools of depression might mitigate underrecognition, undertreatment, and the associated societal burden of depression.

## Key facts

- Psychometric tools may quantify the magnitude of different aspects of depression, but their coverage is usually imperfect.
- Psychiatric diagnosis relies on personal judgment, which reliability is debatable.
- Assessment tools with evidence of reliability and validity can screen, diagnose, and follow-up depressive states.
- Instruments to screen or rate depressive symptomatology are just complementary psychometric tools, they should not replace the clinical interview.
- Despite the availability of assessment tools for measuring depression, this condition remains underrecognized, misdiagnosed, and undertreated.

## Summary points

- This chapter highlights instruments developed to screen, diagnose, and evaluate treatment efficacy.
- Tools to screen depression should be short, fast, easily understandable, and reliable. Examples of the most used ones are the CES-D, the PHQ-9, and the HADS.
- The diagnosis of major depressive disorders relies on clinical judgment, preferably through structured interviews like the SCID and CIDI.
- Interviewers of structured interviews should be trained to apply the tool and be acquainted with clinical psychopathology.
- Rating scales measure the intensity of clinical conditions in different phases of pharmacological trials and investigations. The most used scales are the HAMD, the MADRS, and the BDI-II.

## Mini-dictionary of terms

**Assessment** In healthcare settings, an assessment is an operation aimed to identify needs of the patient and how those needs should be addressed, by gathering comprehensive information about a patient's physiological, psychological, sociological, and spiritual status. The assessment can be performed through observation or interview by a healthcare professional or self-reported by patients. Usually, the answers are registered as scores in rating scales or categorized as a diagnosis.

**Anhedonia** This term refers to aninability to experience pleasure in situations and people before enjoyable and persistently low motivation. Even though anhedonia represents a core symptom of depression, this symptom also occurs in other mental disorders such as schizophrenia.

**Positive affect** Positive affect reflects the one's capacity of experiencing positive and pleasurable feelings and expressions. Manifestations of positive affect are closely related to changes in the mood. Examples of positive affectivity are pride, enthusiasm, joy, cheerfulness, energy, and interest.

**Negative affect** Negative affect is related to the extent to which a subject can experience emotional distress and unpleasant emotions, such as fear, anger, disgust, irritability, sadness, shame, and guilt.

**Receiver-operating characteristic (ROC) curve** The ROC curve is a graphical plot to illustrate the performance of a binary assessment system, the true positive rate against the false positive rate. Several threshold values are calculated until establishing the more accurate cut-off value against the gold-standard.

## References


Alexopoulos, G. A., Abrams, R. C., Young, R. C., & Shamoian, C. A. (1988). Cornell scale for depression in dementia. *Biological Psychiatry*, *23*(3), 271–284. https://doi.org/10.1016/0006-3223(88)90038-8.

American Psychiatric Association. (2013). *Diagnostic and statistical manual of mental disorders* (5th ed.). Washington, DC: APA. https://doi.org/10.1176/appi.books.9780890425596.

Bagby, R. M., Ryder, A. G., Schuller, D. R., & Marshall, M. B. (2004). The Hamilton Depression Rating Scale: has the gold standard become a lead weight? *American Journal of Psychiatry*, *161*(12), 2163–2177. https://doi.org/10.1176/appi.ajp.161.12.2163.

Balsamo, M., Cataldi, F., Carlucci, L., Padulo, C., & Fairfield, B. (2018). Assessment of late-life depression via self-report measures: A review. *Clinical Interventions in Aging*, *13*, 2021–2044. https://doi.org/10.2147/CIA.S178943. eCollection 2018.

Beck, C. T., & Gable, R. K. (2000). Postpartum depression screening scale: Development and psychometric testing. *Nursing Research*, *49*(5), 272–282. https://doi.org/10.1097/00006199-200009000-00006.

Beck, A. T., Steer, R. A., & Brown, G. K. (1996). *Manual for the Beck Depression Inventory-II*. San Antonio, TX: Psychological Corporation.

Beck, A. T., Ward, C., Mendelson, M., Mock, J., & Erbaugh, J. (1961). An inventory for measuring depression. *Achieves of General Psychiatry*, *4* (6), 561–571. https://doi.org/10.1001/archpsyc.1961.01710120031004.

Bjelland, I., Dahl, A. A., Haug, T. T., & Neckelmann, D. (2002). The validity of the Hospital Anxiety and Depression Scale. An updated literature review. *Journal of Psychosomatic Research*, *52*(2), 69–77. https://doi.org/10.1016/s0022-3999(01)00296-3.

Brugha, T. S., Bebbington, P. E., Jenkins, R., Meltzer, H., Taub, N. A., Janas, M., et al. (1999). Cross validation of a general population survey diagnostic interview: a comparison of CIS-R with SCAN ICD-10 diagnostic categories. *Psychological Medicine*, *24*, 16–27. https://doi.org/10.1017/s0033291799008892.

Cosco, T. B., Prina, M., Stubbs, B., & Wu, Y.-T. (2017). Reliability and validity of the Center for Epidemiologic Studies Depression Scale in a population-based cohort of middle-aged U.S. adults. *Journal of Nursing Measurement*, *25*(3), 476–486. https://doi.org/10.1891/1061-3749.25.3.476.

Cox, J. L., Holden, J. M., & Sagovsky, R. (1987). Detection of postnatal depression: Development of the 10-item Edinburgh Postnatal Depression Scale. *British Journal of Psychiatry*, *150*, 782–786. https://doi.org/10.1192/bjp.150.6.782.

Endicott, J., & Spitzer, R. L. (1978). A diagnostic interview: Schedule for affective disorders and schizophrenia. *Archives of General Psychiatry*, *35*, 837–853.

First, M. B., Williams, J. B. W., Karg, R. S., & Spitzer, R. L. (2015). *Structured clinical interview for DSM-5: Research version*. Arlington, VA: American Psychiatric Association.

Greenberg, P. E., Fournier, A. A., Sisitsky, T., Pike, C. T., & Kessler, R. C. (2015). The economic burden of adults with major depressive disorder in the United States (2005 and 2010). *Journal of Clinical Psychiatry*, *76*(2), 155–162. https://doi.org/10.4088/JCP.14m09298.

Hamilton, M. (1967). Development of a rating scale for primary depressive illness. *The British Journal of Social and Clinical Psychology*, *6*(4), 278–296. https://doi.org/10.1111/j.2044-8260.1967.tb00530.x.

Herrmann, C. (1997). International experiences with the Hospital Anxiety and Depression Scale—A review of validation data and clinical results. *Journal of Psychosomatic Research*, *42*(1), 17–41. https://doi.org/10.1016/s0022-3999(96)00216-4.

Kessler, R. C., Aguilar-Gaxiola, S., Alonso, J., Chatterji, S., Lee, S., & Ustün, T. B. (2009). The WHO World Mental Health (WMH) surveys. *Psychiatrie (Stuttgart, Germany)*, *6*(1), 5–9.

Kessler, R. C., Avenevoli, S., Green, J., Gruber, M. J., Guyer, M., He, Y., et al. (2009). National comorbidity survey replication adolescent supplement (NCS-A): III. Concordance of DSM-IV/CIDI diagnoses with clinical reassessments. *Journal of the American Academy of Child and Adolescent Psychiatry*, *48*(4), 386–399. https://doi.org/10.1097/CHI.0b013e31819a1cbc.

Kovacs, M. (1992). *Children's depression inventory*. North Tonawanda, NY: Multi-Health Systems, Inc.

Kroenke, K., Spitzer, R. L., & Williams, J. B. (2001). The PHQ-9: validity of a brief depression severity measure. *Journal of General Internal Medicine*, *16*(9), 606–613. https://doi.org/10.1046/j.1525-1497.2001.016009606.x.

Kroenke, K., Spitzer, R. L., Williams, J. B. W., & Lowe, B. (2010). The patient health questionnaire somatic, anxiety, and depressive symptom scales: A systematic review (PDF). *General Hospital Psychiatry*, *32* (4), 345–359. https://doi.org/10.1016/j.genhosppsych.2010.03.006.

Lim, G. Y., Tam, W. W., Lu, Y., Ho, C. S., Zhang, M. W., & Ho, R. C. (2018). Prevalence of depression in the community from 30 countries between 1994 and 2014. *Scientific reports*, *8*(1), 2861. https://doi.org/10.1038/s41598-018-21243-x.

Lobbestael, J., Leurgans, M., & Arntz, A. (2011). Inter-rater reliability of the structured clinical interview for DSM-IV Axis I disorders (SCID I) and Axis II disorders (SCID II). *Clinical Psychology & Psychotherapy*, *18*(1), 75–79. https://doi.org/10.1002/cpp.693.

Montgomery, S. A., & Åsberg, M. (1979). A new depression scale designed to be sensitive to change. *British Journal of Psychiatry*, *134*, 382–389. https://doi.org/10.1192/bjp.134.4.382.

Müller-Thomsen, T., Arlt, S., Mann, U., Mass, R., & Ganzer, S. (2005). Detecting depression in Alzheimer's disease: Evaluation of four different scales. *Archives of Clinical Neuropsychology*, *20*(2), 271–276. https://doi.org/10.1016/j.acn.2004.03.010.

Radloff, L. S. (1977). The CES-D scale: A self-report depression scale for research in the general population. *Applied Psychological Measurement*, *1*(3), 385–401. https://doi.org/10.1177/014662167700100306.

Robins, L. N., Wing, J., Wittchen, H. U., Helzer, J. E., Babor, T. F., Burke, J., et al. (1988). The Composite International Diagnostic Interview. An epidemiologic instrument suitable for use in conjunction with different diagnostic systems and in different cultures. *Archives of General Psychiatry*, *45*(12), 1069–1077. https://doi.org/10.1001/archpsyc.1988.01800360017003.

Santor, D. A., Gregus, M., & Welch, A. (2006). Eight decades of measurement in depression. *Measurement: Interdisciplinary Research and Perspectives*, *4*(3), 135–155. https://doi.org/10.1207/s15366359mea0403_1.

Sheehan, D. V., Lecrubier, Y., Sheehan, K. H., Amorim, P., Janavs, J., Weiller, E., et al. (1998). The Mini-International Neuropsychiatric Interview (M.I.N.I): The development and validation of a structured diagnostic psychiatric interview for DSM-IV and ICD-10. *The Journal of Clinical Psychiatry*, *59*(20), 22–33.

Sheikh, J. I., & Yesavage, J. A. (1986). Geriatric Depression Scale (GDS): Recent evidence and development of a shorter version. *Clinical Gerontologist*, *5*, 165–172. https://doi.org/10.1300/j018v05n01_09.

Siu, A. L., & The US Preventive Services Task Force (USPSTF). (2016). Screening for depression in adults: US preventive services task force recommendation statement. *The Journal of the American Medical Association*, *315*(4), 380–387. https://doi.org/10.1001/jama.2015.18392.

Smarr, K. L., & Keefer, A. L. (2011). Measures of depression and depressive symptoms: Beck Depression Inventory-II (BDI-II), Center for Epidemiologic Studies Depression Scale (CES-D), Geriatric Depression Scale (GDS), Hospital Anxiety and Depression Scale (HADS), and Patient Health Questionnaire-9 (PHQ-9). *Arthritis Care & Research*, *63*(Suppl. 11), S454–S466. https://doi.org/10.1002/acr.20556.

Svanborg, P., & Åsberg, M. (1994). A new self-rating scale for depression and anxiety states based on the Comprehensive Psychopathological Rating Scale. *Acta Psychiatrica Scandinavica*, *89*(1), 21–28. https://doi.org/10.1111/j.1600-0447.1994.tb01480.x.

Timbremont, B., Braet, C., & Dreessen, L. (2004). Assessing depression in youth: Relation between the children's depression inventory and a structured interview. *Journal of Clinical Child & Adolescent Psychology*, *33* (1), 149–157. https://doi.org/10.1207/S15374424JCCP3301_14.

Tomitaka, S., Kawasaki, Y., Ide, K., Akutagawa, M., Ono, Y., & Furukawa, T. A. (2018). Stability of the distribution of patient health questionnaire-9 scores against age in the general population: Data from the National Health and Nutrition Examination Survey. *Frontiers in Psychiatry*, *9*, 390. https://doi.org/10.3389/fpsyt.2018.00390.

Torres, S. C., Olofsdotter, S., Vadlin, S., Ramklint, M., Nilsson, K. W., & Sonnby, K. (2017). Diagnostic accuracy of the Montgomery-Åsberg Depression Rating Scale parent report among adolescent psychiatric outpatients. *Nordic Journal of Psychiatry*, *72*(3), 184–190. https://doi.org/10.1080/08039488.2017.1414873.

Vilagut, G., Forero, C. G., Barbaglia, G., & Alonso, J. (2016). Screening for depression in the general population with the Center for Epidemiologic Studies Depression (CES-D): A systematic review with meta-analysis. *PLoS One*, *11*(5). https://doi.org/10.1371/journal.pone.0155431, e0155431.

Weiss, B., & Garber, J. (2003). Developmental differences in the phenomenology of depression. *Developmental Psychopathology*, *15*, 403–430. https://doi.org/10.1017/S0954579403000221.

Williams, J. B. (1988). A structured interview guide for the Hamilton Depression Rating Scale. *Archives of General Psychiatry*, *45*(8), 742–747. https://doi.org/10.1001/archpsyc.1988.01800320058007.

Williams, J. B. W., Gibbon, M., First, M. B., Spitzer, R. L., Davies, M., Borus, J., et al. (1992). The structured clinical interview for DSM-III-R (SCID): II. Multisite test-retest reliability. *Archives of General Psychiatry*, *49*(8), 630–636. https://doi.org/10.1001/archpsyc.1992.01820080038006.

Williams, J. B., & Kobak, K. A. (2008). Development and reliability of a structured interview guide for the Montgomery Asberg Depression Rating Scale (SIGMA). *British Journal of Psychiatry*, *192*(1), 52–58. https://doi.org/10.1192/bjp.bp.106.032532.

Wing, J. K., Babor, T., Brugha, T., Burke, J., Cooper, J. E., Giel, R., et al. (1990). SCAN. Schedules for clinical assessment in neuropsychiatry. *Archives of General Psychiatry*, *47*(6), 589–593. https://doi.org/10.1001/archpsyc.1990.01810180089012.

Yesavage, J., Brink, T. L., Rose, T. L., Lum, O., Huang, V., Adey, M., et al. (1983). Development and validation of a geriatric depression screening scale: A preliminary report. *Journal of Psychiatry Research*, *17*(1), 37–49. https://doi.org/10.1016/0022-3956(82)90033-4.

Zanarini, M. C., & Frankenburg, F. R. (2001). Attainment and maintenance of reliability of Axis I and Axis II disorders over the course of a longitudinal study. *Comprehensive Psychiatry*, *42*, 369–374. https://doi.org/10.1053/comp.2001.24556.

Zigmond, A. S., & Snaith, R. P. (1983). The hospital anxiety and depression scale. *Acta Psychiatrica Scandinavica*, *67*, 361–370. https://doi.org/10.1111/j.1600-0447.1983.tb09716.x.

Chapter 16

# The Beck depression inventory: Uses and applications

Yuan-Pang Wang[a] and Clarice Gorenstein[a,b]

[a]*Department and Institute of Psychiatry, LIM-23, Hospital das Clínicas, University of São Paulo Medical School, São Paulo, SP, Brazil;* [b]*Department of Pharmacology, Institute of Biomedical Sciences, University of São Paulo, São Paulo, Brazil*

## List of abbreviations

| | |
|---|---|
| **AUDIT** | alcohol use disorders identification test |
| **BAI** | Beck anxiety inventory |
| **BDI** | Beck depression inventory |
| **BDI-FS** | BDI-fast screen for medical patients (7 items) |
| **BDI-$I_A$** | Beck depression inventory, version $I_A$ |
| **BDI-II** | Beck depression inventory, version II |
| **BDI-PC** | BDI-for primary care (7 items) |
| **BDI-SF** | BDI-short form (13 items) |
| **BHS** | Beck helpless scale |
| **CES-D** | Center for Epidemiologic Studies of Depression |
| **DAST** | drug abuse screening test |
| **DSM** | diagnostic and statistical manual |
| **HAM-A** | Hamilton anxiety rating scale |
| **HAM-D** | Hamilton depression rating scale |
| **IRT** | item response theory |
| **K10** | Kessler's psychological distress scale |
| **MADRS** | Montgomery-Åsberg depression rating scale |
| **NICE** | National Institute for Health and Clinical Excellence |
| **ROC** | receiver operating characteristics |
| **SCID-I** | structured clinical interview for DSM axis I diagnosis |
| **SRQ-20** | self-reporting questionnaire |
| **SSI** | scale for suicidal ideation |
| **STAI** | Spielberger's state–trait anxiety inventory |

## Introduction

Depression is among the most common psychiatric disorders in the community and its prevalence is forecast as increasing worldwide (Lim et al., 2018). Variation of its manifestation might be the joint effect of biological factors (like genetic, physiologic, and endocrine determinants) and social or demographic factors (such as gender-specific roles and coping strategies). Some demographic determinants, like gender and age, have been strongly associated with depression. The prevalence of depression is higher among women when compared to men, with a sex ratio of around 2:1 of diagnosed depression after adolescence (Kessler & Bromet, 2013). The peak age of onset of a first depressive episode is around early 20', although depression occurring after labor and in elderly (Salk, Hyde, & Abramson, 2017) also requires close care. Accurate identification of depression through comparable assessment tools might improve its undertreatment and mitigate its burden in the community.

Among existing tools tailored for measuring depressive symptoms, the Beck depression inventory (BDI) is one of the most accepted self-report instruments for the detection and documentation of psychopathological severity of depressive conditions (Beck, Steer, & Garbin, 1988; Richter, Werner, Heerlein, Kraus, & Sauer, 1998). The 21-item BDI is among one of three instruments (BDI-II, Hospital Anxiety and Depression Scale, Patient Health Questionnaire-9) endorsed by the National Institute for Health and Clinical Excellence (NICE) for use in primary care in measuring baseline depression severity and responsiveness to treatment. The BDI was developed by Beck, Ward, Mendelson, Mock, and Erbaugh (1961) for quantitative assessment of depressive symptoms. This instrument is projected for recording the presence and intensity of depression. Experts (Furukawa, 2010; McDowell, 2006) regard the current version, the BDI-II, as one of the most popular instruments for the identification of depression and one of the greatest screening measures of depression. Although this version has been refurbished to reflect the core symptoms of DSM-IV operational criteria for the diagnosis of a major depressive episode (American Psychiatric Association, 1994), its application must be accompanied by a clinical evaluation for interpretation and to avoid the risk of false-positive tests.

What makes the BDI-II so successful relies on its easy applicability, to most individuals of different age range and patients in different clinical settings. Also, this tool is viewed as a paramount ground for plenty of psychometric studies. Since its original publication, we have found over

The Neuroscience of Depression. https://doi.org/10.1016/B978-0-12-817933-8.00020-7

14 thousand articles dedicated to reporting its use in several languages and settings in the MedLine database by the end of 2019. Possibly, some millions of people around the world have already responded to this instrument since its launch.

The present chapter aims to discuss the uses and applications of this tool. After describing existing versions of the BDI, some psychometric properties are underscored along with recommendations of application and interpretation.

## Versions

The 21-item BDI was built by Beck and colleagues in 1961 at the Center for Cognitive Therapy (Beck et al., 1961) for use in cognitive psychotherapy to rate the presence of depression-related cognitive distortion. The original version has received several reformulations to improve clinical and research needs. The first revision occurred in 1979, also popularized as BDI-$I_A$, differs from the original version regarding the reference time of the assessment ("last week" instead of "today") and several changes in the wording of some items (Beck, Rush, Shaw, & Emery, 1979). The latest version, the BDI-II, was updated in 1996 (Beck, Steer, & Brown, 1996) with significant changes (Table 1). The most important modifications are:

(A) the items about "changes in appetite" and "sleep pattern" also have an option to increase, what account further atypical features of depression; and
(B) the respondent must score the presence of typical symptoms of depression in the last 15 days—depressed mood and/or loss of pleasure or hedonic ability—with accessory symptoms, such as behavioral symptoms and depressive cognitions.

The original English version of the BDI-II has been translated into more than 20 languages, including several Anglo-Saxon, Latin, and Asian languages (Wang & Gorenstein, 2013a). There are several shorter versions: the abbreviated form, with 13 items (Short Form, BDI-SF; Beck & Steer, 1993), for use in clinical patients, and the version with 7 items, for Primary Care (BDI-PC), also called "BDI-FS" (Fast Screen for medical patients; Beck, Steer, & Brown, 2000). The BDI – FS, formerly known as the BDI – PC, excludes some somatic items and includes cognitive and affective items from the BDI – II to assess individuals with medical or substance abuse problems (Beck et al., 2000). The timeframe on the BDI – FS is the same as the BDI – II.

## Content description

The BDI – II includes items measuring cognitive, affective, somatic, and vegetative symptoms of depression. It contains 21 items, as following: sadness, pessimism, failure, loss of pleasure, guilt, punishment, self-esteem, self-criticism, suicidal ideas, crying, agitation, loss of interest, indecision, devaluation, lack of energy, changes without a sleep pattern, irritability, changes in appetite, difficulty concentrating, tiredness, and loss of interest in sex. Each item consists of four options, displayed as an ordinal scale of 0–3, with the highest scores representing the maximum symptom intensity (Table 2, *Sample question #1*). The total score refers to the sum of all individual items and ranges from 0 to 63. There are no recommended subscales of the BDI-II.

The content validity of the BDI-II appears to be adequate but narrower than that of its former version. While the BDI-I included six of the nine criteria for DSM-based depression (Richter et al., 1998), the BDI-II reflected an amended specificity to indicate DSM-based depression. As a consequence, the sensitivity of BDI-II to detect a broader concept of depression has been increased.

The qualitative representation of the theoretical trait to be measured is critical to test validity. It is true that the content coverage of this DSM-based BDI-II allows reliable comparisons in an array of settings and facilitates tailoring therapeutic interventions, but this reform might not represent the very same construct of depression (Maj, 2012).

## Target population

The BDI-II was validated using college students, adult psychiatric outpatients, and adolescent psychiatric outpatients (Beck et al., 1996). Therefore, this tool can be

**TABLE 1** Difference between items of the Beck Depression Inventory version $I_A$ and II.

| Eliminated from BDI-$I_A$ in BDI-II | Included new items to BDI-II | Items of BDI-$I_A$ amended in BDI-II |
|---|---|---|
| Weight loss | Agitation | Changes in appetite (either increase and decrease) |
| Distortion of body image | Worthlessness | Changes in sleeping pattern (either increase and decrease) |
| Somatic preoccupation | Concentration | Loss of interest (interpersonal relationship and activities) |
| Work inhibition | Loss of energy | |

*BDI-$I_A$*, Beck Depression Inventory, version $I_A$; *BDI-II*, Beck Depression Inventory, version II.

**TABLE 2** Sample questions[a] of the Beck Depression Inventory.

| |
|---|
| *Question #1*<br>0—I don't get strained more than usual.<br>1—I get strained more easily than I used to.<br>2—I get strained from doing almost everything.<br>3—I am too strained to do anything. |
| *Question #2*<br>0—I don't get tired more than usual.<br>1a—I get tired more easily than I used to.<br>1b—I get tired less easily than I used to.<br>2a—I get tired from doing almost everything.<br>2b—I get less tired from doing almost everything.<br>3a—I am too tired to do anything.<br>3b—I am less tired to do anything. |

[a]*These are hypothetical statements for each item of the BDI-II, which are protected under copyright laws.*

recommended for use in different populations samples, including individuals in the community, psychiatric and clinical patients, and underage groups since the age of 13. The BDI-FS was validated using general medical inpatients referred for psychiatric consultation and outpatients seen by family practice, pediatrics, and internal medicine (Beck et al., 2000).

The former BDI $-$ I$_A$ version was developed and validated using psychiatric and healthy populations (Beck & Steer, 1987). Similarly, the BDI $-$ II was validated using college students, adult psychiatric outpatients, and adolescent psychiatric outpatients (Beck et al., 1996). The BDI $-$ FS was validated using general medical inpatients referred for psychiatric consultation and outpatients seen by family practice, pediatrics, and internal medicine (Beck et al., 2000).

## Application

Paper and pencil self-application is the standardized way of instrument administration, either in the group or individual format. Besides, it is permissible for the applicator to read the instructions to the respondent in situations where reading ability is compromised (e.g., low visual acuity, low educational level, concentration problems). For the original English version of BDI-II, the recommended schooling level is at least 5th or 6th grade. In some countries, local organizations recommend their clinical application and interpretation made by psychologists who are familiar with psychological educational tests. There is no recommendation for specific training in its administration. The burden of application of the BDI-II is low; it takes around 5–10min and can reach 15min in conditions of an oral application. Elderly and patients with severe depression or obsessive-compulsive disorders often take longer than the average stipulated time for the normative population. In general, the BDI-II has good acceptance by respondents.

## Guidelines

It is convenient to examine the responses provided by the patient for completeness and duplicate item endorsement. When a respondent indicates the same type of response for each of the 21 items of the BDI-II, it is recommended that the applicator explains to him that rarely all symptoms are felt with the same degree of intensity, suggesting the review and reconsideration of some answers. Special attention should pay to the correct score of changes in sleep patterns (item 16) and appetite (item 18). These two items contain seven ordered options (0, 1a, 1b, 2a, 2b, 3a, and 3b) to differentiate between increased and reduced behavior (Table 2, *Sample question #2*). Nevertheless, the possible scores of these two items also vary between 0 and 3.

## Interpretation of scores

Although the BDI-II can be easily applied and scored without extensive training, determining the severity and establishing the diagnosis of the major depressive episode are actions that require additional evaluation by a clinician. Ideally, the diagnosis of major depression must be established by clinical evaluation, because BDI-II's score just reflects the degree of depression. The interpretation of findings should be made by a professional with clinical experience to indicate an appropriate therapeutic intervention. For example, the responses of item 2 (pessimism) and item 9 (suicidal thoughts or desires) require careful assessment, because positive responses might indicate a risk of suicide.

In general, the BDI-II is a highly reliable instrument. Its internal consistency (Cronbach's alpha coefficient) is as high as 0.93 for university students and 0.92 for psychiatric patients (Beck et al., 1996). The internal consistency of the BDI-II was good to excellent for the versions translated into different languages, with alpha coefficients ranging from 0.83 to 0.95. The average item-total correlation of the BDI-II was 0.59, considered appropriate for 17 of the 21 items of the English version and 15 items of the Portuguese version (Gorenstein, Wang, Argimon, & Werlang, 2011). The content validity is higher than the previous BDI-I$_A$, where few items were found adequate to reflect the DSM-IV's criteria.

Previous literature indicates that the BDI-II items are relatively homogeneous (Wang & Gorenstein 2013a). The stability of the instrument over time, or the reliability test-retest, was monitored by the mean of the intraclass coefficient (Pearson's $r$) as 0.93 in university students.

The application of the instrument in other international studies showed similar values of test-retest reliability, ranging from adequate to excellent, between 0.73 and 0.96. Conversely, there is limited information for this type of reliability in psychiatric or medical samples. Therefore, practitioners should be watchful when interpreting BDI-II scores, which might be related to the heterogeneous characteristics of depressive conditions and sample representativeness.

The BDI-II is sensitive to change in depression in cross-cultural studies: a 5-point difference corresponded to a minimally important clinical difference, 10–19 points to a moderate difference, and ≥20 points to a large difference (Smarr & Keefer, 2011).

## Validity

### Criterion validity

The criterion validity of the original BDI-II is investigated in diagnosed patients through standardized clinical interviews. To validate the BDI-II as a screening instrument to detect major depression, some cutoff points obtained by analyzing the receiver operating characteristics (ROC) curves are established to classify the intensity of depression in psychiatric patients. For the BDI-II, the ranges are 0–13, for minimal depression; 14–19, for mild depression; 20–28, for moderate depression; and 29–63, for severe depression (Beck et al., 1996). For the 7-item BDI-FS, the guidelines are: minimal = 0–3, mild depression = 4–8, moderate depression = 9–12, and severe depression = 13–21 (Table 3). The mean score of BDI-II ranged from 5.1 to 38.4 and relied on the type of participants (Wang & Gorenstein, 2013b). In general, nonclinical samples presented the lowest mean scores, medical samples intermediate, and psychiatric samples the highest. Because sample standardization might not reflect demographically representation of the population and little evidence has been provided regarding the gender and culture fairness of the items and total score, the original authors recommended the development of local norms.

Considering the Structured Clinical Interview for DSM-IV Axis-I Disorders (SCID-I) as the gold standard, the performance of the BDI-II was demonstrated from the indicators of sensitivity and specificity in different samples. In a sample of adults from the São Paulo community (Gomes-Oliveira, Gorenstein, Lotufo-Neto, Andrade, & Wang, 2012), the best cutoff point for detecting possible cases of depression was 10/11, which showed a sensitivity of 70% of cases and a specificity of 84.4%. The area under the ROC curve indicated that the instrument can adequately identify 82.1% of depression cases. This overall performance shows the good ability of BDI-II to detect probable cases of depression in adults in the community. Further analysis of the discriminating function of the instrument, calculated using the canonical discriminating functions method, also has shown its adequacy to discriminate against varying degrees of intensity of depressive symptoms (Gomes-Oliveira et al., 2012).

The ROC analysis for different nonclinical and patients showed a wide range of optimal thresholds (Fig. 1). For nonclinical samples, the best cutoff fell between 10 and 16; for psychiatric patients, between 18 and 25; and for medical patients, between 7 and 20 (Wang & Gorenstein, 2013a). This is in line with a recent meta-analysis on the optimal cutoff point for the BDI-II (von Glischinski, von Brachel, & Hirschfeld, 2019). Most studies fail to acknowledge that reported results are only specific estimates and subject to random fluctuations. Considerable conflicting recommendations for practitioners are found. In summary, researchers should use different or higher cut points to screen for depression in primary care and healthy populations (a score of 13 and higher) and psychiatric settings (a score of 19 and higher).

### Construct validity

Evidence of the convergent and divergent validity of the BDI-II was demonstrated through Pearson's correlation coefficients (*r*) between the BDI-II scores and those of other validated psychometric tests applied simultaneously. In general, the BDI-II presents an adequate convergent

**TABLE 3** Recommended range for score interpretation of the BDI-II and BDI-FS.

| Score range | BDI-II | BDI-FS |
|---|---|---|
| No or minimal depression | 0–13 | 0–3 |
| Mild depression | 14–19 | 4–8 |
| Moderate depression | 20–28 | 9–12 |
| Severe depression | 29–63 | 13–21 |

*BDI-II*, Beck Depression Inventory, version II; *BDI-FS*, Beck Depression Inventory, Fast Screen in Medical Patients.
Source based on Beck, A. T., Steer, R. A., & Brown, G. K. (1996). *BDI-II: Beck Depression Inventory Manual.* (2nd. ed.). San Antonio: Psychological Corporation.

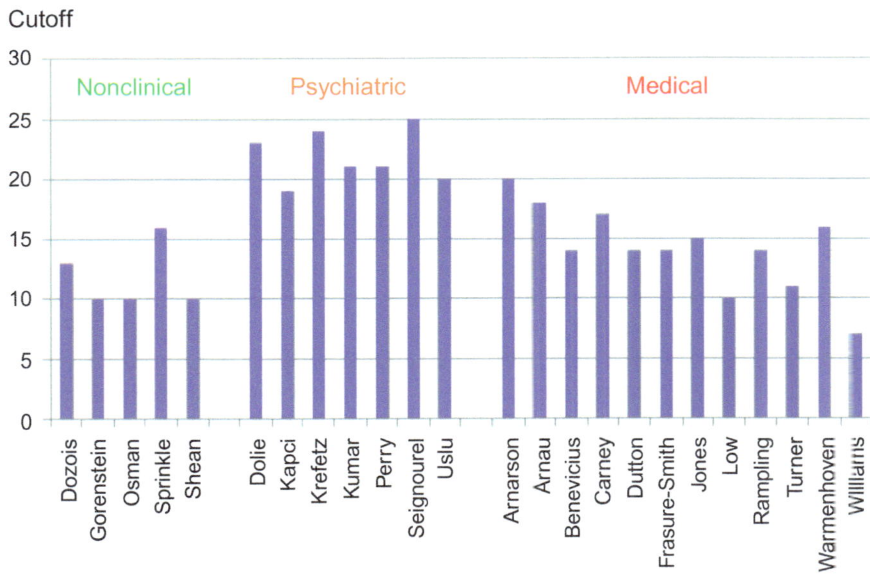


**FIG. 1** Cutoff point for distinguishing depressed patients and nondepressed individuals in nonclinical, psychiatric, and medical settings. All individual studies can be retrieved in the reference section of the article. *Adapted from Wang, Y. P., & Gorenstein, C. (2013a). Psychometric properties of the Beck Depression Inventory-II: A comprehensive review.* Brazilian Journal of Psychiatry, *35(4), 416–431. doi: 10.1590/1516-4446-2012-1048.*

validity with measures of depression, anxiety, and common mental disorders. The convergent validity between BDI-I and BDI-II was high ($r$ ranging from 0.82 to 0.94). For scales of depression, there is high construct convergence (Beck et al., 1996; Wang & Gorenstein, 2013a), such as the Hamilton Depression Rating Scale (HAM-D, $r=0.66$–$0.75$), the Montgomery-Åsberg Depression Scale (MADRS, $r=0.68$–$0.75$), the Center of Epidemiological Studies for Depression (CES-D, $r=0.66$–$0.86$), the Beck Hopelessness Scale (BHS, $r=0.55$–$0.69$), the Scale for Suicidal Ideation (SSI, $r=0.37$), etc.

Likewise, considerable convergence is found with scales of anxiety, such as the Hamilton Rating Scale of Anxiety (HAM-A, $r=0.47$–$0.67$), the Beck Anxiety Inventory (BAI, $r=0.56$–$0.69$), the Spielberger's State–Trait Anxiety Inventory (STAI, $r=0.37$–$0.83$), among others.

Regarding common mental disorders, the correlation of the BDI-II with instruments for screening varied between 0.67 and 0.89 for the Self-Reporting Questionnaire (SRQ-20) and between 0.63 and 0.93 for the Psychological Distress Scale (K10), for example.

Regarding the divergent validity, a poor correlation ($r<0.3$) was found for Alcohol Use Disorder Identification Test (AUDIT) and Drug Abuse Screening Test (DAST), indicating that the construct assessed by BDI-II does not or poorly overlaps with those for scales of alcohol and drug use.

## Structural validity

Although no formal assignment of items to subscales has been made for the BDI-I, Beck et al. (1988) concluded that reviews of factor analyses have identified three common factors: negative attitudes toward self, performance impairment, and somatic disturbances. A similar factorial structure was later described by Shafer (2006) in his meta-analytical review of existing studies.

Because the BDI-II was built on nontheoretical assumptions, investigators often choose factor analysis to account for variance in test performance and determine which psychological events make up test performance. The current trend in psychometrics is the adoption of the technique of confirmatory factor analysis to demonstrate the instrument's construct validity.

Beck et al. (1996) reported a structure of two oblique factors for the BDI-II, represented by the cognitive-affective and somatic-vegetative dimensions (between-factor correlation $r=0.62$ and 0.66, respectively for student and outpatient samples). In general, the construct covered by the BDI-II describes a bidimensional structure, either for psychiatric samples or medical patients (Wang & Gorenstein, 2013a).

Although most international studies have found a similar bidimensional structure, some of them have observed heterogeneous factorial structures, in terms of number, importance of factors, and item distribution. The accessory dimension of performance difficulty was described in different investigations. High between-factor correlation ($>0.50$, range 0.49–0.87) accounts for a large amount of common data variance, also pointing out the possibility of a general factor of depression. All these conflicting findings posit the existence of alternative structural models. As such, the controversial factorial structure of the BDI should be examined in terms of structural equivalence and integrated into a quantitative synthesis in cross-cultural comparisons. Sophisticated alternative structural analysis of the BDI-II, such as the hierarchical model and the bifactor model, can strengthen future investigations.

## Item response theory

The item analysis of a psychometric tool can be conducted through either classical test theory (approach of cumulative item score) or item response theory (IRT, the approach of the latent trait of scale items) (de Sá Junior, de Andrade, Andrade, Gorenstein, & Wang, 2018). The IRT is a family of statistical techniques to examine the item performance of a scale, in terms of item-wise discrimination, difficulty, and the effect of item guessing. In other words, the IRT-based model aims to calculate the probability of getting an expected response to an item along a continuum of the latent trait (e.g., level of ability or depression severity along a latent dimension theta [$\theta$]). Unlike classical test theory, the estimated latent trait is sample invariant and represents an accurate measure of the true construct free of errors (Hambleton & Jones, 1993). A comprehensive assessment of the functioning of each BDI-II item might refine a psychometric tool, through the identification of malfunctioning items and those highly discriminant items that can inform about the severity of depression. This type of analysis can select efficient items to improve the detection of depressive conditions and in different samples.

For a large sample of college students (de Sá Junior et al., 2018, 2019), all BDI-II items present good discrimination of the underlying construct of depression. Regarding the IRT model to assess the item difficulty of the BDI-II, the results showed that the scale is comprised of moderate-to-high difficult items, which best indicate its applicability in clinically depressed patients.

The most endorsed or easiest items along the depression continuum $\theta$ in this student sample are "changes in sleep," "self-criticalness," "tiredness or fatigue," "loss of energy," "changes in appetite." The following items, such as "concentration difficulty," "irritability," "indecisiveness," "guilty feeling," "agitation," and "loss of interest," express a moderate severity of the underlying construct. Finally, the items indicating a high level of severity of depression (difficult endorsement) are: "sadness," "pessimism," "crying," "self-dislike," "worthlessness," "past failure," "punishment feelings," "loss of interest in sex," and "suicide thoughts." We can find in Fig. 2, the graphical comparison (item characteristic curves by IRT analysis) of item "changes in sleep," "loss of interest," and "suicide thought," representing respectively an item of low, moderate, and high depressive severity. Furthermore, respondents' gender and age might influence the response pattern of depressive symptoms, but the measures of self-reported symptoms did not inflate severity scores (de Sá Junior et al., 2019). Two items "crying" and "loss of interest in sex" are flagged for differential item functioning, respectively for gender and age, but the global weight of these items on the total score was negligible.

## BDI-II in medical settings

Depressive symptoms are often present in medical conditions such as heart disease, neurological diseases, stroke, cancer, diabetes, and HIV infection (Katon, Lin, & Kroenke, 2007). The BDI-II is one of the most used instruments to assess depressive states in health-care services to provide proper recognition and early treatment of depression, aiming for a faster recovery and shorter hospitalization.

A systematic review (Wang & Gorenstein, 2013a) found validation studies with BDI-II in several clinical settings, from primary care to highly specialized outpatient environments. In general, the BDI-II performed well in medical patients, showing high reliability (Cronbach's alpha around 0.9) and a good correlation with measures of depression and anxiety.

As expected, medical patients tend to score higher on somatic symptoms, such as "change in sleep pattern" and "change in appetite." Tiredness or fatigue is a symptom that may have a special clinical significance in patients with chronic fatigue syndrome (Brown, Kaplan, & Jason, 2012) or coronary heart disease (Bunevicius, Staniute, Brozaitiene, & Bunevicius, 2012). Interestingly, a study showed that postacute myocardial infarction patients did not have higher somatic symptom scores than psychiatric outpatients who were matched on similar cognitive/affective scores (Thombs, Ziegelstein, Beck, & Pilote, 2008). Compared with undergraduate students, somatic scores of cardiac patients' were only approximately one point higher, indicating that the variation in somatic symptoms is not necessarily related to depression in medical and nonmedical respondents.

When the purpose of BDI-II is to screen likely cases of major depression in medical settings, the sensitivity is the most important indicator for minimizing the chance of false-negative cases. Because clinical patients tend to score more in items that deal with physical complaints, the BDI-II may overestimate the frequency of depression in these patients.

According to the sample (Wang & Gorenstein, 2013b), medical studies reported good BDI-II performance with high sensitivity (72%–100%). Occasionally, the researcher may wish to improve specificity to select a pure sample of depressed patients, which implies the need to adjust the cutoffs according to the type of patient. Based on sample characteristics the best cutoff point may vary widely, ranging from 7 to 22.

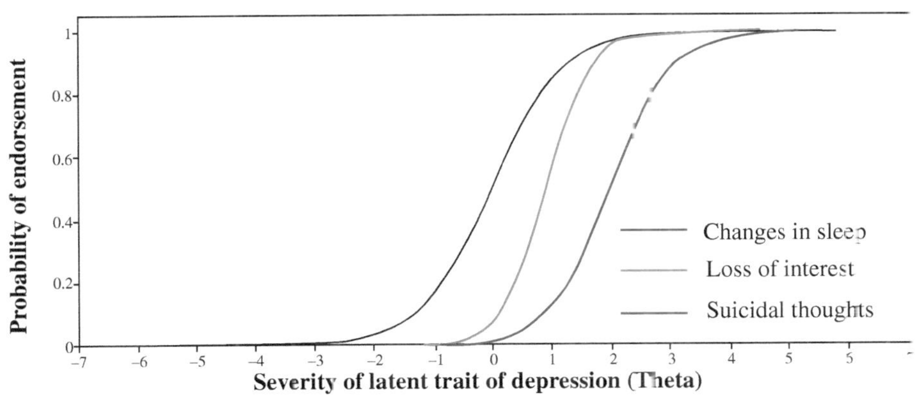


FIG. 2 Item Characteristics Curve for selected BDI-II items (changes in sleep, loss of interest, and suicidal thoughts). *From de Sá Junior, A. R., de Andrade, A. G., Andrade, L. H., Gorenstein, C., & Wang, Y. P. (2018). Response pattern of depressive symptoms among college students: What lies behind items of the Beck Depression Inventory-II?* Journal of Affective Disorders, 234, *124–130. doi:10.1016/j.jad.2018.02.064, with reprint permission.*

In summary, BDI-II can be easily adapted under most clinical conditions to detect major depression and recommend appropriate intervention. Although this scale represents a solid way to detect depression in medical conditions, the clinician should search for further evidence of score interpretability before using BDI-II to make clinical decisions.

## Factors that affect the score

Several respondents' features can influence the point use of a psychometric instrument of depression. The gender effect is consistent between studies: the average scores of female respondents are significantly higher than those of males. Even though this consistency might reflect a higher prevalence of depression in women, a gender-specific response style to the self-report questionnaire should be considered. In some cultures, a significantly higher total score was found among women and a higher proportion of symptomatic women to the proportion of symptomatic men (Delisle, Beck, Dobson, Dozois, & Thombs, 2012). Also, the effect of age and socioeconomic level was likewise observed. Future psychometric research should demonstrate the factors that may affect the score, such as gender, age, education, socioeconomic level, and the presence of concomitant clinical and psychiatric diseases.

Cultural factors are mostly derived from the linguistic translation of the scale. For example, large differential item functioning values were found for 12 BDI-II items between Turkish and US students with the same level of depression (Canel-Çinarbas, Cui, & Lauridsen, 2011), what suggest an equivalence problem with the Turkish version. Participants from different cultural backgrounds might respond in a different way to a local language version of the instrument. In Europe, the cross-cultural equivalence of the BDI was analyzed for five countries (Nuevo et al., 2009). The measurement invariance could not be assumed across the five countries, with the Spanish sample accounting for most of the item malfunctioning and variance difference. Consequently, further powerful analyses like IRT models should be conducted to evaluate sources of this dissimilarity.

## Limitations

BDI-II presents the same problems as other self-applicable inventories, in which the score can be easily exaggerated, minimized or even falsified by the respondents. The environment of application of the instrument (e.g., filling in front of other people and the clinical environment) might generate a distinct final result. Respondents with a concomitant medical disease(s) used to over-report physical complaints, such as fatigue and sleep changes, which can increase scores of the BDI even in the absence of depressive symptoms. Therefore, researchers should consider the type of study of a population in the interpretation of the instrument's score. The authors recommend the alternative short version of BDI (BDI-PC or BDI-FS) for medical patients, with seven items on depressive cognitions. Sometimes, BDI is inadvertently used by health professionals to "quickly" diagnose cases of depression. It is never too late to reiterate that this instrument was designed to be used to detect depressive symptoms, but the diagnosis of depression remains to be confirmed later. We find in Table 4 further strengths and weaknesses of the BDI-II.

## Comments

Assessment tools need to measure effectively the heterogeneity of depression for several purposes. First, dissimilar presentations of depression may respond differently to various interventions, or have a diverse course. Second, there is a potential of relating depressive disorder subtypes to specific genetic vulnerabilities. Third, the study of associations between clinical phenomena and the neural

**TABLE 4 Strengths and weaknesses of the Beck Depression Inventory-II.**

| Strengths | Weaknesses |
|---|---|
| ▪ Concise and user friendly<br>▪ Broad content coverage of depressive symptoms<br>▪ Good reliability across languages<br>▪ Positive correlation with other psychological tests and health outcomes<br>▪ Allows easy symptom screening and reassessment<br>▪ Flexible application in different settings | ▪ Overidentification in women and underidentification in men<br>▪ Easy to fake responses<br>▪ Low discrimination of score options<br>▪ Aligns with DSM-IV, but not DSM-5<br>▪ Nonrepresentativeness of most samples<br>▪ Does not take into account major life events<br>▪ Needs purchase license for use |

substrates and circuitry of mood and emotion might be improved with advanced techniques to study neurophysiological processes. Therefore, sophisticated measurement is a requirement for studies that examine such associations, link genetic diatheses to particular forms of depressive disorder, or guide the development of future treatments.

Among existing tools tailored for measuring depression, the BDI is one of the most accepted self-report instruments for the detection and psychopathological documentation of depressive conditions. Its robust psychometric properties confirm the versatility of this instrument in a variety of samples and health care settings (Beck et al., 1988; Wesley, Gatchel, Garofalo, & Polatin, 1999).

In general, experts consider that the clinical utility of BDI has been improved in version II. The flexible cutoff points of the BDI-II are a good alternative for the screening of probable cases of depression when followed by a clinical interview or a diagnostic instrument. Other advantages of the BDI-II include its easy application (self-application), simple wording, the straightforward scoring method, and good acceptance by users. The reliability and validity of BDI-II were extensively investigated, using sophisticated techniques, in different populations and countries. Summarizing, the usability of the BDI-II in both clinical and research contexts is high. There is factual evidence of its utility in identifying depression in and nonclinical populations. However, its ability to monitor changes in the depressive state over time, providing a quantitative measure of the improvement and effectiveness of therapeutic methods, has not yet been sufficiently established. The BDI-II is progressively consolidating itself as one of the most investigated and adopted instruments for the assessment of depression.

## How to obtain

The use of the BDI-II is protected under local copyright laws. For purchasing a license of the BDI-II and BDI-FS (manual and instrument), you should contact the Pearson Assessments (www.pearsonassessments.com). The BDI-I is no longer sold to the public. Computer software is available from Pearson Assessments for onscreen administration or the input of data from a desktop scanner. The computer program may be used to administer a single questionnaire or to integrate the results of sequential administrations.

## Key facts

- The current version of the BDI, the BDI-II (Beck et al., 1996), is one of the most recommended self-report tools for depression.
- This BDI-II was amended to convey the 2-week DSM-IV criteria of the major depressive episode.
- The BDI-II can be administered to respondents over 13 years, in different communities and health-care settings (either psychiatric or medical).
- The paper-and-pencil application of BDI-II is straightforward and user friendly.
- The inventory has 21 items, which are written in concise statements.
- It is scored from 0 to 3, whose total score ranges from 0 to 63. Threshold score over 13 indicates "possible" depression.
- This tool presents robust psychometric properties, in terms of reliability and validity.
- Translations of the BDI-II to main modern languages are available for use in different cultures and countries.
- The BDI-II is a useful measurement tool for easy detection of depressive psychopathologies, but its scores do not replace a careful clinical diagnosis of major depression.

## Summary points

- This chapter highlights the applicability and utility of the BDI-II for measuring depressive symptoms.
- The content coverage of the BDI-II is aligned with the DSM-IV criteria of major depressive episode.
- The reliability of the BDI-II is good to excellent, across samples and languages.

- The BDI-II presents a good correlation with established measures of depression (e.g., CES-D, MADRS, and HADS), anxiety, and general distress.
- The performance of the BDI-II against gold-standard clinical interviews for diagnosis of major depressive disorders is diverse, being reliant on the sample's characteristics.
- In general, the BDI-II presents a two-dimensional factorial structure, cognitive-affective and somatic-vegetative ones. Alternative models are described.
- Item analysis of the BDI-II has shown a heterogeneous ability to detect depression in a continuum of severity.
- The interpretation of scores should rely on clinical expertise because BDI-II items hold the differential capacity to indicate depression and might be misrepresented.
- Robust psychometric properties, flexible use in different settings, and good acceptance by users.

## Mini-dictionary of terms

**Classical test theory** Also known as the *true score model* is the mathematics behind creating and answering tests and measurement scales. The goal of classical test theory is to improve tests, particularly the reliability and validity of tests.

**Item response theory** Also known as *latent trait* theory, *strong true score* theory, or *modern mental test* theory, it is a paradigm for the design, analysis, and scoring of tests, questionnaires, and similar instruments measuring abilities, attitudes, or other variables. This is a theory of testing based on the relationship between individuals' performances on a test item and the test takers' levels of performance on an overall measure of the ability that item was designed to measure.

**Reliability** In psychometrics, reliability is the overall consistency of a measure. This is the characteristic of a set of test scores that relates to the amount of random error from the measurement process that might be embedded in the scores. A measure is said to have high reliability if it produces similar results under consistent conditions. Highly reliable scores are accurate, reproducible, and consistent from one testing occasion to another.

**Validity** In psychometrics, validity or test validity indicates "the degree to which evidence and theory support the interpretations of test scores." In a broader sense, validity means the extent to which a concept, construct, conclusion, or measurement is well founded and likely corresponds accurately to the real world. The main types of validity are face, content, criterion, construct, structural, or factorial validity. The demonstration of the validity of a scoring tool is vital to confirm its quality and applicability in different groups of respondents.

## References


American Psychiatric Association (1994). *Diagnostic and Statistical Manual of Mental Disorders (DSM-IV)* (4th ed.). Washington, DC: American Psychiatric Association.

Beck, A. T., Rush, A. J., Shaw, B. F., & Emery, G. (1979). *Cognitive therapy of depression*. New York: Guilford.

Beck, A. T., & Steer, R. A. (1987). *Manual for the Beck Depression Inventory*. San Antonio, TX: The Psychological Corporation.

Beck, A. T., & Steer, R. A. (1993). *Beck Depression Inventory Manual*. San Antonio: Psychological Corporation.

Beck, A. T., Steer, R. A., & Brown, G. K. (1996). *BDI-II: Beck Depression Inventory Manual* (2nd. ed.). San Antonio: Psychological Corporation.

Beck, A. T., Steer, R. A., & Brown, G. K. (2000). *BDI: Fast screen for medical patients manual*. San Antonio, TX: The Psychological Corporation.

Beck, A. T., Steer, R. A., & Garbin, M. G. (1988). Psychometric properties of the Beck Depression Inventory: Twenty-five years of evaluation. *Clinical Psychology Review*, *8*(1), 77–100. https://doi.org/10.1016/0272-7358(88)90050-5.

Beck, A. T., Ward, C. H., Mendelson, M., Mock, J., & Erbaugh, J. (1961). An inventory for measuring depression. *Archives of General Psychiatry*, *4*, 561–571. https://doi.org/10.1001/archpsyc.1961.01710120031004.

Brown, M., Kaplan, C., & Jason, L. (2012). Factor analysis of the Beck Depression Inventory-II with patients with chronic fatigue syndrome. *Journal of Health Psychology*, *17*(6), 799–808. https://doi.org/10.1177/1359105311424470.

Bunevicius, A., Staniute, M., Brozaitiene, J., & Bunevicius, R. (2012). Diagnostic accuracy of self-rating scales for screening of depression in coronary artery disease patients. *Journal of Psychosomatic Research*, *72*, 22–25. https://doi.org/10.1016/j.jpsychores.2011.10.006.

Canel-Çinarbas, D. Cui, Y., & Lauridsen, E. (2011). Cross-cultural validation of the Beck Depression Inventory-II across US and Turkish samples. *Measurement and Evaluation in Counseling and Development*, *44*, 77–91.

de Sá Junior, A. R., de Andrade, A. G., Andrade, L. H., Gorenstein, C., & Wang, Y. P. (2018). Response pattern of depressive symptoms among college students: What lies behind items of the Beck Depression Inventory-II? *Journal of Affective Disorders*, *234*, 124–130. https://doi.org/10.1016/j.jad.2018.02.064.

de Sá Junior, A. R., Liebel, G., de Andrade, A. G., Andrade, L. H., Gorenstein, C., & Wang, Y. P. (2019). Can gender and age impact on response pattern of depressive symptoms among college students? A differential item functioning analysis. *Frontiers in Psychiatry*, *10*, 50. https://doi.org/10.3389/fpsyt.2019.00050.

Delisle, V. C., Beck, A. T., Dobson, K. S., Dozois, D. J., & Thombs, B. D. (2012). Revisiting gender differences in somatic symptoms of depression: Much ado about nothing? *PLoS One*, *7*(2). https://doi.org/10.1371/journal.pone.0032490, e32490.

Furukawa, T. A. (2010). Assessment of mood: Guides for clinicians. *Journal of Psychosomatic Research*, *68*(6), 581–589. https://doi.org/10.1016/j.jpsychores.2009.05.003.

Gomes-Oliveira, M. H., Gorenstein, C., Lotufo-Neto, F., Andrade, L. H., & Wang, Y. P. (2012). Validation of the Brazilian Portuguese version of the Beck Depression Inventory-II in a community sample. *Brazilian Journal of Psychiatry*, *34*(4), 389–394. https://doi.org/10.1016/j.rbp.2012.03.005.

Gorenstein, C., Wang, Y. P., Argimon, I. L., & Werlang, B. S. G. (2011) *Manual do Inventário de Depressão de Beck—BDI-II: adaptação brasileira*. São Paulo: Casa do Psicólogo

Hambleton, R. K., & Jones, R. W. (1993). Comparison of classical test theory and item response theory and their applications to test development. *Educational Measurement: Issues and Practice*, *12*(3), 38–47.

Katon, W., Lin, E. H. B., & Kroenke, K. (2007). The association of depression and anxiety with medical symptom burden in patients with chronic medical illness. *General Hospital Psychiatry*, *29*, 147–155. https://doi.org/10.1016/j.genhosppsych.2006.11.005.

Kessler, R. C., & Bromet, E. J. (2013). The epidemiology of depression across cultures. *Annual Review of Public Health*, *34*, 119–138. https://doi.org/10.1146/annurev-publhealth-031912-114409.

Lim, G. Y., Tam, W. W., Lu, Y., Ho, C. S., Zhang, M. W., & Ho, R. C. (2018). Prevalence of depression in the community from 30 countries between 1994 and 2014. *Scientific Reports*, *8*(1), 2861. https://doi.org/10.1038/s41598-018-21243-x.

Maj, M. (2012). Development and validation of the current concept of major depression. *Psychopathology*, *45*, 135–146.

McDowell, I. (2006). *Measuring health: A guide to rating scales and questionnaires* (3rd ed.). New York: Oxford University.

Nuevo, R., Dunn, G., Dowrick, C., Vázquez-Barquero, J. L., Casey, P., Dalgard, O. S., et al. (2009). Cross-cultural equivalence of the Beck Depression Inventory: A five-country analysis from the ODIN study. *Journal of Affective Disorders*, *114*(1–3), 156–162. https://doi.org/10.1016/j.jad.2008.06.021.

Richter, P., Werner, J., Heerlein, A., Kraus, A., & Sauer, H. (1998). On the validity of the Beck Depression Inventory. A review. *Psychopathology*, *31*(3), 160–168. https://doi.org/10.1159/000066239.

Salk, R. H., Hyde, J. S., & Abramson, L. Y. (2017). Gender differences in depression in representative national samples: Meta-analyses of diagnoses and symptoms. *Psychological Bulletin*, *143*(8), 783–822. https://doi.org/10.1037/bul0000102.

Shafer, A. B. (2006). Meta-analysis of the factor structures of four depression questionnaires: Beck, CES-D, Hamilton, and Zung. *Journal of Clinical Psychology*, *62*(1), 123–146. https://doi.org/10.1002/jclp.20213.

Smarr, K. L., & Keefer, A. L. (2011). Measures of depression and depressive symptoms: Beck Depression Inventory-II (BDI-II), Center for Epidemiologic Studies Depression Scale (CES-D), Geriatric Depression Scale (GDS), Hospital Anxiety and Depression Scale (HADS), and Patient Health Questionnaire-9 (PHQ-9). *Arthritis Care & Research*, *63*(Suppl. 11), S454–S466. https://doi.org/10.1002/acr.20556.

Thombs, B. D., Ziegelstein, R. C., Beck, C. A., & Pilote, L. (2008). A general factor model for the Beck Depression Inventory-II: Validation in a sample of patients hospitalized with acute myocardial infarction. *Journal of Psychosomatic Research*, *65*, 115–121. https://doi.org/10.1016/j.jpsychores.2008.02.027.

von Glischinski, M., von Brachel, R., & Hirschfeld, G. (2019). How depressed is "depressed"? A systematic review and diagnostic meta-analysis of optimal cut points for the Beck Depression Inventory revised (BDI-II). *Quality of Life Research*, *28*(5), 1111–1118. https://doi.org/10.1007/s11136-018-2050-x.

Wang, Y. P., & Gorenstein, C. (2013a). Psychometric properties of the Beck Depression Inventory-II: A comprehensive review. *Brazilian Journal of Psychiatry*, *35*(4), 416–431. https://doi.org/10.1590/1516-4446-2012-1048.

Wang, Y. P., & Gorenstein, C. (2013b). Assessment of depression in medical patients: Systematic review of the utility of the Beck Depression Inventory-II. *Clinics*, *68*(9), 1274–1287. https://doi.org/10.6061/clinics/2013(09)15.

Wesley, A. L., Gatchel, R. J., Garofalo, J. P., & Polatin, P. B. (1999). Toward more accurate use of the Beck Depression Inventory with chronic back pain patients. *Clinical Journal of Pain*, *15*(2), 117–121. https://doi.org/10.1097/00002508-199906000-00008.

Chapter 17

# Hamilton depression rating scale: Uses and applications

**Lubova Renemane and Jelena Vrublevska**
*Department of Psychiatry and Narcology, Riga Stradins University, Riga, Latvia*

## List of abbreviations

| | |
|---|---|
| **HDRS, HRSD, HAM-D** | Hamilton depression rating scale |
| **SIGH-D** | structured guide for the HDRS |
| **TCAs** | tricyclic antidepressants |
| **SSRIs** | selective reuptake inhibitors |

## Introduction

The Hamilton Depression Rating Scale, also known as the Hamilton Rating Scale for Depression (often abbreviated to HDRS, HRSD, or HAM-D), was originally designed to measure the severity of depression and to evaluate the performance of the first group of antidepressants.

In 1960, Max Hamilton, a psychiatrist from the department of psychiatry at Leeds University, presented a rating scale for depression to the psychiatric society (Hamilton, 1960). This scale quickly gained great notability and became widely used within the medical community to assess both the severity of depression in dynamics and the effectiveness of treatment of patients with depressive disorder (Braund, Tillman, Palmer, & Harris, 2020; Cerou, Peigne, Comets, & Chenel, 2020; Liu et al., 2020). It was subsequently revised in 1966 (Hamilton, 1966), 1967 (Hamilton, 1967), 1969 (Hamilton, 1969), and 1980 (Hamilton, 1980). Over the past 50 years, the HAM-D has been modified into 11 versions and applied in various patient populations in psychiatric, medical, and other scientific settings (Rohan et al., 2016; Williams, 2001). Still, the HDRS is widely available in the public domain and is not protected by copyright.

Attempts were made to modify the scale, both by including more signs of depression and by increasing the degree of gradation of each of item, a more detailed assessment of their intensity or frequency in different revisions of scale. For more than 40 years it has been the gold standard for the assessment of depression and serves as a point of reference for more recently developed scales (Table 1). However, criticism of the instrument has been increasing (Bagby, Ryder, Schuller, & Marshall, 2005).

## Administration and uses

The HDRS is an interviewer-rated scale for adults with already identified depression. The HADS scale should not be used as a diagnostic instrument. It requires a trained rater with knowledge of the tool and symptoms of the depressive illness. The HADS is designed to be used by a health care professional during a clinical interview and to measure the outcome of treatment, especially drugs. Hamilton indicated that the value of the standard HADS scale depends on the skills of the interviewer (Hamilton, 1960).

In 1988, the Structured Guide for the HDRS (SI-HDRS) was developed to standardize the manner of the administration of the scale (Williams, 1988). A study conducted on a series of psychiatric inpatients demonstrated a better level of agreement for most of the HDRS items. Not only experienced clinicians can apply the SI-HDRS items, but also trained interviewers who lack psychiatric background. In addition, some studies have pointed out that a structured interview does not differ between face-to-face and telephone administration groups. Thus, this instrument can be used economically in large-scale, community-based research projects (Potts, Daniels, Burnam, & Wells, 1990).

Through repeated and consistent use of the scale, a clinician can document the results of treatment. The assessment typically takes 15–30 min and can be used both for inpatients and for patients outside a psychiatric clinical setting.

## Scoring and interpretation

The original version contained 17 items, but during the last 5 decades, the researchers have modified it for certain

The Neuroscience of Depression. https://doi.org/10.1016/B978-0-12-817933-8.00019-0

**TABLE 1** Rating scales for depression.

| Scale | Acronym | Number of items | Administration time | Indication |
|---|---|---|---|---|
| *Scales competed by professionals* | | | | |
| Composite International Diagnostic Interview[a] | CIDI | 41 sections of items | 20–30 min | Diagnostic tool based on DSM-IV |
| Hamilton Depression Rating Scale[b] | HDRS<br>HAM-D<br>HRSD | 17–21 | 15–30 min | Severity of depression<br>Treatment outcomes |
| Montgomery-Asberg Depression Rating Scale[c] | MADRS | 10 | 20–40 min | Severity of depression<br>Treatment outcomes |
| Structured Clinical Interview for DSM[d] | SCID-I<br>SCID-II<br>SCID-5 | Depend on version | 15 min–2 h | Diagnostic tool based on DSM-IV or DSM-5 |
| Symptom Checklist 90-Revision[e] | SCL-90-R | 90 | 12–15 min | Severity of symptoms<br>Treatment outcomes |
| The Mini International Neuropsychiatric Interview[f] | M.I.N.I. | 17 modules of items | 15 min | Diagnostic tool based on DSM-III-R, DSM-IV, DSM-5 and ICD-10 |
| *Scales competed by patient, self-rating scales* | | | | |
| Beck Depression Inventory[g] | BDI | 21 | 5–10 min | Screening tool<br>Severity of DS<br>Treatment outcomes |
| Center of Epidemiologic Studies—depression[h] | CES-D | 20 | 15 min | Screening tool |
| Geriatric Depression Scale[i] | GDS-4<br>GDS-15<br>GDS-30 | 4<br>15<br>30 | 3 min<br>5–7 min<br>10–20 min | Screening tool<br>Severity of DS<br>Treatment outcomes |
| Hospital Anxiety and Depression scale[j] | HADS | 14 | 2–5 min | Screening tool<br>Severity of symptoms<br>Treatment outcomes |
| Inventory of depression and Anxiety Symptoms[k] | IDAS | 99 | 20–30 min | Measures symptoms of depression and anxiety |
| Patient Health Questionnaire for depression—9[l] | PHQ-9 | 9 | 3 min | Screening tool<br>Severity of DS<br>Treatment outcomes |
| Self-Report Questionnaire—20[m] | SRQ-20 | 20 | 10–15 min | Screening tool<br>Treatment outcomes |
| WHO Well-Being Index[n] | WHO-5 | 5 | 3 min | Screening tool<br>Treatment outcomes |

**TABLE 1** Rating scales for depression—cont'd

| Scale | Acronym | Number of items | Administration time | Indication |
|---|---|---|---|---|
| Zung's Self-Rating Depression Scale[o] | SDS<br>ZSDS | 20 | 10–20 min | Screening tool<br>Severity of DS<br>Treatment outcomes |
| *Scales completed by professionals and patients* | | | | |
| The Primary Care Evaluation of Mental Disorders[p] | PRIME-MD | 27 and clinician interview | 10 min | Screening tool<br>Diagnostic tool |

[a]*Kessler and Ustun (2004).*
[b]*Hamilton (1960, 1966, 1967, 1969, 1980).*
[c]*Montgomery and Åsberg (1979).*
[d]*Shankman et al. (2018).*
[e]*Derogatis (1977).*
[f]*Sheehan et al. (1998).*
[g]*Beck, Ward, Mendelson, Mock, and Erbaugh (1961).*
[h]*Radloff (1977).*
[i]*Yesavage et al. (1982).*
[j]*Zigmond and Snaith (1983).*
[k]*Watson et al. (2007).*
[l]*Spitzer, Kroenke, and Williams (1999).*
[m]*Beusenberg et al. (1994).*
[n]*Topp, Østergaard, Søndergaard, and Bech (2015).*
[o]*Zung (1965).*
[p]*Spitzer et al. (1994).*
Rating scales of depression are comprised of three groups: scales competed by professionals, completed by patients (self-rating scales) and by both. The administration time varies from 3 min to 2 h.
*DS*, depressive symptoms; *DSM-IV*, the Diagnostic and Statistical Manual of Mental Disorders, 4th Edition; *DSM-V*, 5th Edition; *DSM-III-R*, 3rd Edition; *ICD-10*, International Classification of Diseases 10th Edition.
Data are from

needs. Subsequently, attempts were made to expand or shorter the scale, both by cutting-off or including more items, ranging from 6 to 36, or transforming the assessment into a more detailed one concerning changes in intensity or frequency (McIntyre, Kennedy, Bagby, & Bakish, 2002; Willaims, 2001).

For example, a shortened form with seven items that are identified and designated as the Toronto HAM-D7 with cutoff scores for full remission offers a useful tool that physicians can readily employ in clinical practice (McIntyre et al., 2002). Longer versions were developed to cover reverse neurovegetative, atypical symptoms of depression (Willaims, 2001). It should be noted that special studies showed little success in expanding the list of features included in the HDRS since this did not enhance the reliability and validity of this scale.

However, the most commonly used versions in the studies are either a 17- or a 21-item scale.

The first 17 items measure the severity of depressive symptoms experienced over the past week and include depressed mood, guilt, suicide, initial insomnia, middle insomnia, delayed insomnia, work and interests, retardation, agitation, psychic anxiety, somatic anxiety, gastrointestinal symptoms, general somatic symptoms, genital symptoms, hypochondriasis, loss of insight and weight (Table 2).

A 21-item version (HDRS-21) includes additional 4 items (18–21 items) intended to identify a subtype of depression and to measure factors that might be related to depression, but are not thought to be measures of severity, such as paranoia, obsessional symptoms, depersonalization and derealization, and diurnal variation (Table 2). These points are not taken into account when calculating the total score of the Hamilton scale, which determines the severity of depressive disorder, thus decreasing the internal consistency as a whole.

**TABLE 2 Hamilton depression rating scale: Symptoms and score range.**

| Item no. | Symptom | Score range |
|---|---|---|
| 1. | Depressed mood | 0–4 |
| 2. | Guilt | 0–4 |
| 3. | Suicide | 0–4 |
| 4. | Initial insomnia | 0–2 |
| 5. | Middle insomnia | 0–2 |
| 6. | Delayed insomnia | 0–2 |
| 7. | Work and interests | 0–4 |
| 8. | Retardation | 0–4 |
| 9. | Agitation | 0–2 |
| 10. | Psychic anxiety | 0–4 |
| 11. | Somatic anxiety | 0–4 |
| 12. | Gastrointestinal symptoms | 0–2 |
| 13. | General somatic symptoms | 0–2 |
| 14. | Genital symptoms | 0–2 |
| 15. | Hypochondriasis | 0–4 |
| 16. | Loss of insight | 0–2 |
| 17. | Loss of weight | 0–2 |
| 18. | Diurnal variation | 0–2 |
| 19. | Depersonalization | 0–4 |
| 20. | Paranoid symptoms | 0–4 |
| 21. | Obsessional symptoms | 0–2 |

Data are from Hamilton, M. (1960). A rating scale for depression. *Journal of Neurology, Neurosurgery & Psychiatry, 23*(1), 56. https://doi.org/10.1136/jnnp.23.1.56.
The symptoms and score range are shown for the 21-item scale.

**TABLE 3** Hamilton depression rating scale: Grading of scores.

| | Severity of symptom |
|---|---|
| *Score range 0–4* | |
| 0. | Absent |
| 1. | Mild or trivial |
| 2. | Moderate |
| 3. | |
| 4. | Severe |
| *Score range 0–2* | |
| 0. | Absent |
| 1. | Slight or doubtful |
| 2. | Clear present |

Data are from Hamilton, M. (1960). A rating scale for depression. *Journal of Neurology, Neurosurgery & Psychiatry, 23*(1), 56. https://doi.org/10.1136/jnnp.23.1.56.
The symptoms severity are evaluated on a five-point scale (0–4) and on a three-point scale (0–2).

**TABLE 4** Hamilton depression rating scale 17-item: Interpretation.

| Score range | Severity level |
|---|---|
| 7 or < | Absence of depression or being normal |
| 8–16 | Mild depression |
| 17–23 | Moderate depression |
| 24 or > | Severe depression |
| The total score range from 0 to the maximum score 52 | |

Data are from Zimmerman, M., Martinez, J. H., Young, D., Chelminski, I., & Dalrymple, K. (2013). Severity classification on the Hamilton depression rating scale. *Journal of Affective Disorders, 150*(2), 384–388. https://doi.org/10.1016/j.jad.2013.04.028.
The scores less than 7 or 7 are indicative of the absence of depression or being normal, 8–16 consider mild depression, 17–23 suggest moderate depression and scores 24 or over are indicative of severe depression.

**TABLE 5** Hamilton depression rating scale 17-item: Response to treatment.

- A decrease of 50% or more in the HDRS score is often considered as a positive treatment response
- A score of 7 or less considered equivalent to a remission

Data are from Zimmerman, M., Chelminski, I., & Posternak, M. (2004). A review of studies of the Hamilton depression rating scale in healthy controls: Implications for the definition of remission in treatment studies of depression. *The Journal of Nervous and Mental Disease, 192*, 595–601.
[a] *Zimmerman, M., Chelminski, I. & Posternak, M. 2004. A Review of Studies of the Hamilton Depression Rating Scale in Healthy Controls: Implications for the Definition of Remission in Treatment Studies of Depression. The Journal of Nervous and Mental Disease, 192, 595–601*
The interpretation of changes in the scores during the treatment is shown in Table 5.

Depending on the item, it is scored between 0 (not present) and 4 (severe) points using either a three-point or a five-point scale and summed up to obtain the total score. The HDRS comprises 17 items, of which 9 are evaluated on a five-point scale (0–4) and 8—on a three-point scale (0–2). The total score can range from 0 to the maximum score 52 on a 17-point scale, with higher scores indicating more serious depression (Tables 2 and 3).

For the HDRS-17, scores less than 7 are indicative of the absence of depression or being normal, 8–16 consider mild depression, 17–23 suggest moderate depression and scores over 24 are indicative of severe depression (Zimmerman, Chelminski, & Posternak, 2004) (Table 4).

A decrease of 50% or more in the HDRS score is often considered as a positive treatment response; on the other hand, a score of 7 or less considered equivalent to a remission (Zimmerman et al., 2004) (Table 5).

## Indication for the use of the HDRS

In most clinical trials, the HDRS is used as a standard to determine the severity of depression and response to the treatment. The HDRS is the most common tool for evaluating the effectiveness of antidepressants in clinical trials (Leader, O'Connell, & VandenBerg, 2019). According to the criteria of the National Institute for Clinical Excellence, the effect of an antidepressant compared to placebo is clinically significant, if the difference between the effect of antidepressant and placebo is only three points or more on the Hamilton scale (Andrews, Thomson, Amstadter, & Neale, 2012). In other studies, evaluating the level of depression, the HDRS is used to assess the effectiveness of psychotropic drugs of different classes, but not only antidepressants (An et al., 2019; Fayed et al., 2019; Kimball et al., 2019).

In addition, there are many studies evaluating the effectiveness of non-pharmacological interventions on depression. It was used to measure the effectiveness of psychotherapy, neurostimulation, complementary, and alternative treatment of depression (chronotherapy, light therapy, physical exercises, and others) (An et al., 2019; Arslanoglou, Banerjee, Pantelides, Evans, & Kiosses, 2019; Elboga, Karayagmurlu, Kocamer Sahin, & Altindag, 2019; Erbay, Zayman, Erbay, & Unal, 2019; Wang et al., 2019; Yrondi et al., 2019). In several studies, the HDRS was used as a tool to measure suicidality (Shen et al., 2019; Williams et al., 2019).

Although Hamilton recommended to use the scale only with patients already diagnosed affective disorder of the depressive type, many researchers use a scale to measure the level of depression for patients suffering from psychotic disorders, alcohol and drug addictions, neurotic disorders and other mental disorders (Kimball et al., 2019; Saglam, Aslan, & Hursitoglu, 2020; Tonna et al., 2019).

Scientists used this scale to measure the level of depression in patients suffering from general medical conditions, neurological or neurodegenerative diseases (Che et al., 2019; Chen et al., 2019; Fayed et al., 2019; Hassaan et al., 2019; Shen et al., 2019).

The HDRS can be used in place of self-report scales, when a patient is unable to read or in case when there are concerns about the accuracy of the patient's self-report (Halfaker, Akeson, Hathcock, Mattson, & Wunderlich, 2011).

This scale can be useful for patients with cognitive impairment who experience difficulty managing self-reporting tools; and the scale has demonstrated reliability, validity, and efficiency in older adults above 60 years (Arslanoglou et al., 2019; Tulaci & Ekinci, 2020) (Table 6).

**TABLE 6 Hamilton depression rating scale: Indication for the use.**

| |
|---|
| • To determine the severity of depression and response to treatment[a] |
| • For evaluating the effectiveness of antidepressants[b] and others psychotropic medications[c,d,e] |
| • For evaluating the effectiveness of nonpharmacological interventions: psychotherapy, neurostimulation, complementary and alternative treatment of depression (chronotherapy, light therapy, physical exercises and others)[d,f,g,h,i,j] |
| • To measure suicidality[k,l] |
| • To measure the level of depression for patients suffering from psychotic disorders, alcohol and drug addictions, neurotic disorders and other mental disorders[m,n,o] |
| • To measure the level of depression in patients suffer from general medical conditions, neurological or neurodegenerative diseases[c,e,l,m,n,o,p] |
| • Can be used in place of self-reports scales (if patient unable to read or patients has cognitive impairment)[c,q] |

Data are from
[a]*Hamilton (1960).*
[b]*Leader et al. (2019).*
[c]*Fayed et al. (2019).*
[d]*An et al. (2019).*
[e]*Kimball et al. (2019).*
[f]*Arslanoglou et al. (2019).*
[g]*Elboga et al. (2019).*
[h]*Wang et al. (2019).*
[i]*Yrondi et al. (2019).*
[j]*Erbay et al. (2019).*
[k]*Williams (1988).*
[l]*Shen et al. (2019).*
[m]*Saglam et al. (2020).*
[n]*Tonna et al. (2019), Chen et al. (2019).*
[o]*Che et al. (2019).*
[p]*Hassaan et al. (2019).*
[q]*Tulaci and Ekinci (2020).*
The general indication for the use of scale is shown in Table 6.

## Validity and reliability

Since 1960, validation of the scale has been performed in a wide array of studies. The researchers also evaluated consistency of the results obtained in different studies. The reviews of recent years demonstrate that the scale has generally satisfactory psychometric characteristics.

Major psychometric properties of the HDRS were examined by Bagby et al. (2005). In the current review of 70 studies, the authors confirmed that the internal, interrater and retest reliability estimates are mostly good for the overall rating score. However, the internal reliability coefficients show that some items are problematic, and internal and retest coefficient are weak for many individual items. The data of the review suggest that the convergent, discriminant, and predictive validity meet established criteria (Bagby et al., 2005).

However, the validity of the HAMD has often been criticized in recent years. Some items of the scale are distinguished by constructive inadequacy, which influences the accuracy of assessing the severity of depression. It is noted that threshold criteria that distinguish a mild degree of depression from remission do not meet criteria of validity. Some of the items exhibit weak construct validity (Romera, Pérez, Menchón, Polavieja, & Gilaberte, 2011; Sawamura, Ishigooka, & Nishimura, 2018).

## Limitations

The HDRS was developed before the publication of the Diagnostic and Statistical Manual of Mental Disorders, 4th or 5th Edition and does not evaluate all domains of the diagnostic criteria for Major Depressive Disorder such as feelings of worthlessness and anhedonia (Santor & Coyne, 2001).

The HRSD is criticized for being used in clinical practice as it places more emphasis on insomnia rather than on feelings of hopelessness, self-destructive thoughts, suicidal cognitions and actions (Brown et al., 2000); it also favors somatic signs and symptoms and can miss atypical symptoms, such as overeating and oversleeping. A meta-analysis data suggests that after therapeutic treatments, the HRSD is more 'sensitive to change' on retesting in comparison to the Beck Depression Inventory (self-rating instrument) and this is probably why it has been so widely used in clinical trials (Brown et al., 2000). In addition, from a critical point of view, in case when suicidal thoughts increase and sleep is improved, an antidepressant may show statistical efficacy in a clinical trial. Some studies show that the HDRS has priority sedative tricyclic antidepressants (TCAs) above selective reuptake inhibitors (SSRIs) (Möller, 2001; Santor and Coyne, 2001; Worboys, 2012). The side effect of TCAs could influence the sleep and somatic items; and such side effects as insomnia, agitation, gastrointestinal and sexual side effect of SSRIs could be rated on a scale and may be registered as being less effective

**TABLE 7** **Hamilton depression rating scale (HRSD): Limitations.**

| |
|---|
| • It does not evaluate all domains of the diagnostic criteria for Major Depressive Disorder in DSM-IV or DSM-V such as feelings of worthlessness and anhedonia[a] |
| • It places more emphasis on insomnia rather than on feelings of hopelessness, self-destructive thoughts, suicidal cognitions and actions[b] |
| • It favors somatic signs and symptoms and can miss atypical symptoms, such as overeating and oversleeping[b] |
| • After therapeutic treatments, the HRSD is more "sensitive to change" on retesting in comparison to the Beck Depression Inventory[b] |
| • In case when suicidal thoughts increase and sleep is improved, an antidepressant may show statistical efficacy |
| • The side effect of TCAs could influence the sleep and somatic items[a,c,d] |
| • The side effects of SSRIs as insomnia, agitation, gastrointestinal and sexual side effect of SSRIs could be rated on scale[c,e,f] |
| • Atypical symptoms of depression such as hypersomnia and hyperphagia are not assessed |

Data are from
[a]*Santor and Coyne (2001).*
[b]*Brown et al. (2000).*
[c]*Möller (2001).*
[d]*Worboys (2012).*
[e]*Bagby et al. (2005).*
[f]*Linden et al. (1995).*
The limitation of the scale are mentioned in Table 7.
*DSM-IV*, the Diagnostic and Statistical Manual of Mental Disorders, 4th Edition; *DSM-V*: the Diagnostic and Statistical Manual of Mental Disorders, 5th Edition; *TCAs*, tricyclic antidepressants; *SSRIs*, selective reuptake inhibitors.

in treating the depression than it actually is (Bagby et al., 2005; Linden, Borchelt, Barnow, & Geiselmann, 1995; Möller, 2001). Another limitation of the HDRS-17 is that atypical symptoms of depression such as hypersomnia and hyperphagia are not assessed (Table 7).

## Key facts of the Hamilton depression rating scale

- The Hamilton Depression Rating Scale (HDRS) was originally designed to measure the severity of depression symptoms and response to treatment.
- The HDRS is an interviewer-rated scale for adults with already identified depression.
- The HADS scale should not be used as a diagnostic instrument.
- In 1988, the Structured Guide for the HDRS was developed to standardize the manner of the administration of the scale.
- The most commonly used versions of the HDRS in the studies are either a 17- or a 21-item scale.
- The first 17 items measure the severity of depressive symptoms experienced over the past week.
- A 21-item version includes additional four items (18–21 items) intended to identify a subtype of depression. These points are not taken into account when calculating the total score of the Hamilton scale.
- The HDRS does not evaluate all domains of the diagnostic criteria for Major Depressive Disorder. The side effects of antidepressants could influence the score of items.

## Summary points

- This chapter focuses on the Hamilton Depression Rating Scale which is a widely used measure instrument to assess the severity of depression in dynamics and the effectiveness of treatment of patients with already identified depressive disorder.
- A trained rater with knowledge of the tool and symptoms of the depressive illness should administer it.
- The most commonly used versions in the studies are either a 17- or a 21-item scale.
- The scoring of the severity of the depressive symptoms is based on the first 17 items. It is scored between 0 (not present) and 4 (severe) points using either a three-point or a five-point scale and summed up to obtain the total score.
- The assessment generally takes 15–30 min.
- The scale does not evaluate all domains of the diagnostic criteria for Major Depressive Disorder and the side effects of antidepressants could influence the score of items.

## Mini-dictionary of terms

**Depression rating scale** A measuring instrument intended for a comparative assessment of the severity of depressive symptoms in dynamics in cases where clinical diagnosis is already established.

**Structured guide for rating scale** Standardized way of interviewing with a series of the same questions in the same order, designed to assess the severity of symptoms of the scale.

**Treatment response.** A rating scale score changes during a continuous measure to evaluate the treatment outcome (response, remission, nonresponse, partial response, relapse, recurrence, recovery, and depressive breakthrough).

**Positive treatment response** A decrease of 50% or more in the initial rating scale scores.

**Nonresponse to treatment** Changes below 50% in the initial rating scale scores.

**Remission** A score of the Hamilton Depression Rating Scale 7 or less.

## References


An, J. H., Jang, E. H., Kim, A. Y., Fava, M., Mischoulon, D., Papakostas, G. I., et al. (2019). Ratio of plasma BDNF to leptin levels are associated with treatment response in major depressive disorder but not in panic disorder: A 12-week follow-up study. *Journal of Affective Disorders*, *259*, 349–354. https://doi.org/10.1016/j.jad.2019.08.021.

Andrews, P., Thomson, J. A., Amstadter, A., & Neale, M. (2012). Primum non Nocere: An evolutionary analysis of whether antidepressants do more harm than good. *Frontiers in Psychology*, *3*(117). https://doi.org/10.3389/fpsyg.2012.00117.

Arslanoglou, E., Banerjee, S., Pantelides, J., Evans, L., & Kiosses, D. N. (2019). Negative emotions and the course of depression during psychotherapy in suicidal older adults with depression and cognitive impairment. *American Journal of Geriatric Psychiatry*, *27*(12), 1287–1295. https://doi.org/10.1016/j.jagp.2019.08.018.

Bagby, R., Ryder, A., Schuller, D., & Marshall, M. (2005). The Hamilton depression rating scale: Has the gold standard become a lead weight? *The American Journal of Psychiatry*, *161*, 2163–2177. https://doi.org/10.1176/appi.ajp.161.12.2163.

Beck, A. T., Ward, C. H., Mendelson, M., Mock, J., & Erbaugh, J. (1961). An inventory for measuring depression. *Archives of General Psychiatry*, *4*(6), 561–571.

Beusenberg, M., Orley, J. H., & World Health Organization. Division of Mental Health. (1994). *A User's guide to the self reporting questionnaire* (SRQ/compiled by M. Beusenberg and J. Orley). Geneva: World Health Organization.

Braund, T. A., Tillman, G., Palmer, D. M., & Harris, A. W. F. (2020). Verbal memory predicts treatment outcome in syndromal anxious depression: An iSPOT-D report. *Journal of Affective Disorders*, *260*, 245–253. https://doi.org/10.1016/j.jad.2019.09.028.

Brown, G., Beck, A., Conwell, Y., Goldston, D., Jobes, D., Linehan, M., et al. (2000). *A review of suicide assessment measures for intervention research with adults and older adults*.

Cerou, M., Peigne, S., Comets, E., & Chenel, M. (2020). Application of item response theory to model disease progression and agomelatine effect in patients with major depressive disorder. *AAPS Journal*, *22* (1), 13. https://doi.org/10.1208/s12248-019-0379-x.

Che, B. Z., Zhong, C. K., Ge, J. Z., Li, R. Y., Zhu, Z. B., Bu, X. Q., et al. (2019). Serum matrix metalloproteinase-9 is associated with depression after acute ischemic stroke. *Circulation Journal*, *83*(11). https://doi.org/10.1253/circj.CJ-19-0376.

Chen, Y. L., Huang, H. Y., Ning, P. P., Zhao, Q. Z., Wang, H., Shen, Q. Y., et al. (2019). Frequency and factors related to drooling in Chinese patients with multiple system atrophy: A cross-sectional study. *Clinical Autonomic Research*, *29*(6), 595–601. https://doi.org/10.1007/s10286-019-00602-2.

Derogatis, L. R. (1977). *The SCL-90 Manual I: Scoring, administration and procedures for the SCL-90*. Baltimore, MD: Clinical Psychometric Research.

Elboga, G., Karayagmurlu, E., Kocamer Sahin, S., & Altindag, A. (2019). Analysis of elderly patients treated with electroconvulsive therapy and complication rates: A single center experience. *Anadolu Psikiyatri Dergisi-Anatolian Journal of Psychiatry*, *20*(6), 613–618. https://doi.org/10.5455/apd.27764.

Erbay, M. F., Zayman, E. P., Erbay, L. G., & Unal, S. (2019). Evaluation of transcranial magnetic stimulation efficiency in major depressive disorder patients: A magnetic resonance spectroscopy study. *Psychiatry Investigation*, *16*(10), 745–750. https://doi.org/10.30773/pi.2019.07.17.3.

Fayed, N., Oliven, B., del Hoyo, Y. L., Andres, E., Perez-Yus, M. C., Fayed, A., et al. (2019). Changes in metabolites in the brain of patients with fibromyalgia after treatment with an NMDA receptor antagonist. *The Neuroradiology Journal*, *32*(6), 408–419. https://doi.org/10.1177/1971400919857544.

Halfaker, D. A., Akeson, S. T., Hathcock, D. R., Mattson, C., & Wunderlich, T. L. (2011). Psychological aspects of pain. In T. A. Lennard, S. Walkowski, A. K. Singla, & D. G. Vivian (Eds.), *Pain procedures in clinical practice* (3rd ed., pp. 13–22). Saint Louis: Hanley & Belfus.

Hamilton, M. (1960). A rating scale for depression. *Journal of Neurology, Neurosurgery & Psychiatry*, *23*(1), 56. https://doi.org/10.1136/jnnp.23.1.56.

Hamilton, M. (1966). Assessment of change in psychiatric state by means of rating scales. *Proceedings of the Royal Society of Medicine*, *59* (Suppl. 1), 10–13.

Hamilton, M. (1967). Development of a rating scale for primary depressive illness. *British Journal of Social and Clinical Psychology*, *6*(4), 278–296. https://doi.org/10.1111/j.2044-8260.1967.tb00530.x.

Hamilton, M. (1969). Standardised assessment and recording of depressive symptoms. *Psychiatria, Neurologia, Neurochirurgia*, *72*(2), 201–205.

Hamilton, M. (1980). Rating depressive patients. *Journal of Clinical Psychiatry*, *41*(12), 21–24.

Hassaan, S. H., Darwish, A. M., Khalifa, H., Ramadan, H. K. A., Hassany, S. M., Ahmed, G. K., et al. (2019). Assessment of cognitive functions and psychiatric symptoms in hepatitis C patients receiving pegylated interferon alpha and ribavirin: A prospective cohort study. *International Journal of Psychiatry in Medicine*, *54*(6), 424–440. https://doi.org/10.1177/0091217419858277.

Kessler, R. C., & Ustun, T. B. (2004). The World Mental Health (WMH) survey initiative version of the World Health Organization (WHO) composite international diagnostic interview (CIDI). *International Journal of Methods in Psychiatric Research*, *13*(2), 93–121.

Kimball, A., Schorr, M., Meenaghan, E., Bachmann, K. N., Eddy, K. T., Misra, M., et al. (2019). A randomized placebo-controlled trial of low-dose testosterone therapy in women with anorexia nervosa. *Journal of Clinical Endocrinology & Metabolism*, *104*(10), 4347–4355. https://doi.org/10.1210/jc.2019-00828.

Leader, L. D., O'Connell, M., & VandenBerg, A. (2019). Brexanolone for postpartum depression: Clinical evidence and practical considerations. *Pharmacotherapy*, *39*(11), 1105–1112. https://doi.org/10.1002/phar.2331.

Linden, M., Borchelt, M., Barnow, S., & Geiselmann, B. (1995). The impact of somatic morbidity on the Hamilton depression rating scale in the very old. *Acta Psychiatrica Scandinavica*, *92*(2), 150–154. https://doi.org/10.1111/j.1600-0447.1995.tb09559.x.

Liu, X. Y., Song, D. H., Yin, Y. Y., Xie, C. M., Zhang, H. S., Zhang, H. X., et al. (2020). Altered Brain Entropy as a predictor of antidepressant response in major depressive disorder. *Journal of Affective Disorders*, *260*, 716–721. https://doi.org/10.1016/j.jad.2019.09.067.

McIntyre, R., Kennedy, S., Bagby, R. M., & Bakish, D. (2002). Assessing full remission. *Journal of Psychiatry & Neuroscience: JPN*, *27*(4), 235–239.

Möller, H. (2001). Methodological aspects in the assessment of severity of depression by the Hamilton depression scale. *European Archives of Psychiatry and Clinical Neuroscience*, *251*(Suppl. 2), II13–20. https://doi.org/10.1007/BF03035121.

Montgomery, S. A., & Åsberg, M. (1979). A new depression scale designed to be sensitive to change. *British Journal of Psychiatry*, *134*(4), 382–389.

Potts, M. K., Daniels, M., Burnam, M. A., & Wells, K. B. (1990). A structured interview version of the Hamilton depression rating scale: Evidence of reliability and versatility of administration. *Journal of Psychiatric Research*, *24*(4), 335–350. https://doi.org/10.1016/0022-3956(90)90005-B.

Radloff, L. S. (1977). The CES-D scale: A self-report depression scale for research in the general population. *Applied Psychological Measurement*, *1*(3), 385–401.

Rohan, K. J., Rough, J. N., Evans, M., Ho, S.-Y., Meyerhoff, J., Roberts, L. M., et al. (2016). A protocol for the Hamilton rating scale for depression: Item scoring rules, rater training, and outcome accuracy with data on its application in a clinical trial. *Journal of Affective Disorders*, *200*, 111–118. https://doi.org/10.1016/j.jad.2016.01.051.

Romera, I., Pérez, V., Menchón, J. M., Polavieja, P., & Gilaberte, I. (2011). Optimal cutoff point of the Hamilton rating scale for depression according to normal levels of social and occupational functioning. *Psychiatry Research*, *186*(1), 133–137. https://doi.org/10.1016/j.psychres.2010.06.023.

Saglam, F., Aslan, E., & Hursitoglu, O. (2020). Emotion regulation difficulties of patients with bipolar disorder and first-degree relatives. *Anadolu Psikiyatri Dergisi-Anatolian Journal of Psychiatry*, *21*(1), 30–36. https://doi.org/10.5455/apd.36087.

Santor, D. A., & Coyne, J. C. (2001). Examining symptom expression as a function of symptom severity: Item performance on the Hamilton rating scale for depression. *Psychological Assessment*, *13*(1), 127–139. https://doi.org/10.1037/1040-3590.13.1.127.

Sawamura, J., Ishigooka, J., & Nishimura, K. (2018). Re-evaluation of the definition of remission on the 17-item Hamilton depression rating scale based on recovery in health-related quality of life in an observational post-marketing study. *Health and Quality of Life Outcomes*, *16*(1), 14. https://doi.org/10.1186/s12955-018-0838-6.

Shankman, S. A., Funkhouser, C. J., Klein, D. N., Davila, J., Lerner, D., & Hee, D. (2018). Reliability and validity of severity dimensions of psychopathology assessed using the Structured Clinical Interview for DSM-5 (SCID). *International Journal of Methods in Psychiatric Research*, *27*(1).

Sheehan, D. V., Lecrubier, Y., Sheehan, K. H., Amorim, P., Janavs, J., Weiller, E., et al. (1998). The Mini-International Neuropsychiatric Interview (M.I.N.I.): the development and validation of a structured diagnostic psychiatric interview for DSM-IV and ICD-10. *The Journal of Clinical Psychiatry*, *59*(Suppl. 20), 22–33. quiz 34-57.

Shen, Y. M., Wu, F. C., Zhou, Y. J., Ma, Y. J., Huang, X. B., Ning, Y. P., et al. (2019). Association of thyroid dysfunction with suicide attempts in first-episode and drug naive patients with major depressive disorder. *Journal of Affective Disorders*, *259*, 180–185. https://doi.org/10.1016/j.jad.2019.08.067.

Spitzer, R. L., Kroenke, K., & Williams, J. B. W. (1999). Validation and utility of a self-report version of PRIME-MD the PHQ primary care study. *JAMA*, *282*(18), 1737–1744.

Spitzer, R. L., Williams, J. B., Kroenke, K., Linzer, M., deGruy, F. V., 3rd, Hahn, S. R., et al. (1994). Utility of a new procedure for diagnosing mental disorders in primary care. The PRIME-MD 1000 study. *JAMA*, *272*(22), 1749–1756.

Tonna, M., Ossola, P., Marchesi, C., Bettini, E., Lasalvia, A., Bonetto, C., et al. (2019). Dimensional structure of first episode psychosis. *Early Intervention in Psychiatry*, *13*(6), 1431–1438. https://doi.org/10.1111/eip.12789.

Topp, C., Østergaard, S., Søndergaard, S., & Bech, P. (2015). The WHO-5 well-being index: A systematic review of the literature. *Psychotherapy and Psychosomatics*, *84*, 167–176.

Tulaci, R. G., & Ekinci, O. (2020). Quality of life, disability, and residual depressive symptoms in older remitted and partly remitted patients with major depressive disorder. *Anadolu Psikiyatri Dergisi-Anatolian Journal of Psychiatry*, *21*(1), 45–52. https://doi.org/10.5455/apd.47873.

Wang, Z. Q., Li, S. S., Wu, L. Y., Qi, Q., Liu, H. R., Jin, X. M., et al. (2019). Effect of acupuncture on lung cancer-related fatigue: Study protocol for a multi-center randomized controlled trial. *Trials*, *20*(1), 9. https://doi.org/10.1186/s13063-019-3701-0.

Watson, D., O'Hara, M. W., Simms, L. J., Kotov, R., Chmielewski, M., McDade-Montez, E. A., et al. (2007). Development and validation of the Inventory of Depression and Anxiety Symptoms (IDAS). *Psychol Assess*, *19*(3), 253–268.

Willaims, J. B. W. (2001). Standardizing the Hamilton depression rating scale: Past, present, and future. *European Archives of Psychiatry and Clinical Neuroscience*, *251*(2). https://doi.org/10.1007/bf03035120. 115–112.

Williams, J. B. W. (1988). A structured interview guide for the Hamilton depression rating scale. *Archives of General Psychiatry*, *45*(8), 742–747. https://doi.org/10.1001/archpsyc.1988.01800320058007.

Williams, J. B. (2001). Standardizing the Hamilton depression rating scale: Past, present, and future. *European Archives of Psychiatry and Clinical Neuroscience*, *251*(Suppl. 2), Ii6–12. https://doi.org/10.1007/bf03035120.

Williams, N. R., Heifets, B. D., Bentzley, B. S., Blasey, C., Sudheimer, K. D., Hawkins, J., et al. (2019). Attenuation of antidepressant and antisuicidal effects of ketamine by opioid receptor antagonism. *Molecular Psychiatry*, *24*(12), 1779–1786. https://doi.org/10.1038/s41380-019-0503-4.

Worboys, M. (2013). The Hamilton rating scale for depression: The making of a "gold standard" and the unmaking of a chronic illness, 1960–1980. *Chronic Illness*, *9*(3), 202–219. https://doi.org/10.1177/1742395312467658.

Yesavage, J. A., Brink, T. L., Rose, T. L., Lum, O., Huang, V., Adey, M., et al. (1982). Development and validation of a geriatric depression screening scale: a preliminary report. *Journal of Psychiatric Research*, *17*(1), 37–49.

Yrondi, A., Nemmi, F., Billoux, S., Giron, A., Sporer, M., Taib, S., et al. (2019). Grey Matter changes in treatment-resistant depression during electroconvulsive therapy. *Journal of Affective Disorders*, *258*, 42–49. https://doi.org/10.1016/j.jad.2019.07.075.

Zigmond, A. S., & Snaith, R. P. (1983). The hospital anxiety and depression scale. *Acta Psychiatr Scand*, *67*(6), 361–370.

Zimmerman, M., Chelminski, I., & Posternak, M. (2004). A review of studies of the Hamilton depression rating scale in healthy controls: Implications for the definition of remission in treatment studies of depression. *The Journal of Nervous and Mental Disease*, *192*(9), 595–601. https://doi.org/10.1097/01.nmd.0000138226.22761.39.

Zung, W. W. K. (1965). A self-rating depression scale. *Archives of General Psychiatry*, *12*(1), 63–70.

Chapter 18

# The patient health questionnaire (PHQ)

Maria Iglesias-González and Crisanto Diez-Quevedo
*Department of Psychiatry and Legal Medicine, School of Medicine, Universitat Autònoma de Barcelona, Hospital Universitari Germans Trias i Pujol, Badalona, Spain*

## List of abbreviations

| | |
|---|---|
| **AUC** | area under the curve |
| **BDI-II** | Beck depression inventory |
| **CES-D** | Center for Epidemiological Studies Scale |
| **COPD** | chronic obstructive pulmonary disease |
| **DOR** | diagnostic odd ratio |
| **DSM-III-R** | diagnostic and statistical manual of mental disorders, third edition revised |
| **GDS-15** | geriatric depression scale |
| **GHQ-12** | general health questionnaire |
| **HADS** | hospital anxiety and depression scale |
| **HCV** | hepatitis C virus |
| **HDRS** | Hamilton depression rating scale |
| **MINI** | mini international neuropsychiatric interview |
| **PHQ** | patient health questionnaire |
| **PHQ-2** | two-items version of the depression module of the PHQ |
| **PHQ-8** | eight-items version of the depression module of the PHQ |
| **PHQ-9** | nine-items version (full version) of the depression module of the PHQ |
| **PHQ-A** | 14-items version of the PHQ for adolescents |
| **PRIME-MD** | primary care evaluation of mental disorders |
| **PROMIS** | patient-reported outcomes measurement information system |
| **SCID-I** | structured clinical interview for DSM-III-R axis I disorders |
| **SCL-90** | symptom checklist |

## Introduction

Since many cases of depression are treated by general practitioners and are usually comorbid with different medical entities, it is often difficult to accurately diagnose depressive disorders and it can lead to either an underestimation or an overdiagnosis, with the consequent aftermaths on therapeutic management. The routine usage of screening instruments can be useful to address this problem (see Table 1 for a list of most used tools in depression). Two are the main objectives of the application of instruments: screening itself, that is, to rule out those individuals without a disorder; and case finding, that is, the confirmation of those who do present the disorder.

With these goals in mind, the Primary Care Evaluation of Mental Disorders (PRIME-MD) was an instrument developed and validated in the early 1990s to simplify the screening and diagnosis in primary care of the five most common mental disorders: depressive, anxiety, and somatoform disorders, alcohol use, and bulimia /binge eating disorder (Spitzer et al., 1994).

PRIME-MD included a first self-administered instrument that served as a screening tool, followed by a clinician-administered structured interview adapted from the Structured Clinical Interview for the Diagnostic and Statistical Manual of Mental Disorders, Third Edition Revised (DSM-III-R) (SCID) (Spitzer et al., 1992). Subsequently, from two large studies with 6000 patients, a fully self-administered version, the Patient Health Questionnaire (PHQ), was developed and validated (Spitzer et al., 1999; Spitzer, Williams, Kroenke, Hornyak, & McMurray, 2000). Afterward, the depression module of the PHQ, the PHQ-9, was used separately (Kroenke, Spitzer, & Williams, 2001), and since then it has become the most widely used instrument in research and clinical practice both for screening and diagnosis.

## Content and scoring

The PHQ-9 includes nine items that replicate the nine diagnostic criteria for the major depressive episode from the DSM-III-R, along with an additional 10th item of impairment in work, daily living, or interpersonal relationships areas due to major depression. The patient is asked to respond to each item regarding his/her condition during the 2 weeks prior to assessment. The answers are scored on a four-point scale that ranges from 0 to 3 depending on the frequency with which symptoms appear (0 = Not at all, 1 = Several days, 2 = More than half of the days, 3 = Nearly every day) (Kroenke et al., 2001). Methods of administration include pencil and paper self-report, computer touch screen, direct interview, or telephone interview. The PHQ-9 behaves the same regardless of the type of administration (Fann et al., 2009). The usual administration time is about

The Neuroscience of Depression. https://doi.org/10.1016/B978-0-12-817933-8.00029-3

**TABLE 1 Most common tools used in depression screening and assessment.**

*Structured interviews for diagnosis of depression*

- SCID-I (Structured Clinical Interview for DSM-III-R) (Spitzer, Williams, Gibbon, & First, 1992)
- MINI (Mini International Neuropsychiatric Interview) (Sheehan et al., 1998)
- PRIME-MD (Primary Care Evaluation of Mental Disorders) (Spitzer et al., 1994)
- CIDI (Composite International Diagnostic Interview) (Kessler & Ustun, 2004)
- DIS (Diagnostic Interview Schedule) (Robins, Helzer, Croughan, & Ratcliff, 1981)
- SDDS-PC (Symptom Driven Diagnostic System for Primary Care) (Olfson et al., 1995)
- SADS (Schedule for Affective Disorders and Schizophrenia) (Endicott & Spitzer, 1978)
- SCAN (Schedules for Clinical Assessment in Neuropsychiatry) (Wing et al., 1990)
- SPIFA-A (Structured Psychiatric Interview for General Practice) (Dahl et al., 2009)

*Case-finding instruments*

- PHQ (Patient Health Questionnaire) (Spitzer, Kroenke, & Williams, 1999)
- BDI-II (Beck Depression Inventory) (Beck, Ward, Mendelson, Mock, & Erbaugh, 1961)
- Depression subscale of the HADS (Hospital Anxiety and Depression Scale) (Zigmond & Snaith, 1983)
- CES-D (Center for Epidemiologic Studies of Depression Scale) (Radloff, 1977)
- GHQ-12 (General Health Questionnaire – 12 items) (Goldberg, 1972)
- CDSS (Calgary Depression Scale for Schizophrenia) (Addington, Addington, Maticka-Tyndale, & Joyce, 1992)
- MDI (Major Depression Inventory) (Bech, Rasmussen, Olsen, Noerholm, & Abildgaard, 2001)
- EPDS (Edinburgh postnatal depression scale) (Matijasevich, Munhoz, Tavares, et al., 2014)
- WHO-5 (World Health Organization - Five Well-Being Index) (WHO, 1998)
- 4DSQ (Four-Dimensional Symptom Questionnaire) (Terluin et al., 2006)

*Instruments for severity measurement*

- PHQ (Patient Health Questionnaire) (Spitzer et al., 1999)
- BDI-II (Beck Depression Inventory) (Beck et al., 1961)
- MADRS (Montgomery Asberg Depression Rating Scale) (Montgomery & Asberg, 1979)
- Hamilton Depression Rating Scale-17 items (HDRS-17) (Hamilton, 1960)
- IDS (Inventory of Depressive Symptomatology) (Rush et al., 1986)
- ZDS (Zung Self-Rated Depression Scale) (Zung, 1965)
- CDSS (Calgary Depression Scale for Schizophrenia) (Addington et al., 1992)
- QIDS (16-Item Quick Inventory of Depressive Symptomatology) (Rush et al., 2003)

3 min and there is no need for professional training either for administration or scoring. However, it is true that it will be essential to be able to offer a therapeutic approach, either pharmacological or psychotherapeutic, to all those patients who can be diagnosed with depression with this instrument. Otherwise, its use would not be adequate.

Scoring the PHQ-9 can be done through both a diagnostic algorithm and a continuous measure of the sum of scores of all nine items. In the first case, to make a diagnosis of major depression it is required that (1) at least five of the nine items score positive (i.e., score "2" or "3" in all cases except for item 9, about suicidal thoughts and self-harm, in which the score "1" is considered also positive), (2) at least one of the first two items score positive (sad mood or loss of interest), and (3) there is a functional impairment according to the response to item 10. Another diagnosis of "minor depression" or "other depressive disorder" can be done following the same criteria but requiring the presence of at least two positive items. In this case, it is recommended to complete clinical evaluation to finish fitting the diagnosis.

In the second case, total scores range between 0 and 27. The authors proposed the following cut-off points to measure the severity of depression: 1–4 no depression, 5–9 mild depression, 10–14 moderate depression, 15–19 moderately severe depression, and 20–27 severe depression. Scores $\geq 10$ are considered indicative of diagnosing major depression (Kroenke et al., 2001). This method of quantitative correction with the sum of scores of all items has demonstrated better psychometric features of sensitivity and specificity than the algorithmic method (Manea, Gilbody, & McMillan, 2015; Pettersson, Boström, Gustavsson, & Ekselius, 2015).

PHQ-9 has been extensively translated and adapted to numerous languages, including, but not limited to Spanish (Diez-Quevedo, Rangil, Sanchez-Planell, Kroenke, & Spitzer, 2001), Afrikaans, Arabic, Chinese, French,

German, Dutch, Greek, Hebrew, Italian, Japanese, Persian, Thai, and Swahili. Most of these translations are available at https://www.phqscreeners.com/

## Abbreviated versions

Two shortened versions of the PHQ-9 have been developed, one of 2 items (PHQ-2) and the other of 8 (PHQ-8). The two-item version includes the first two items of PHQ-9: depressive mood and loss of interest. It has demonstrated good psychometric properties, suggesting its potential benefit as a tool for depression screening (Kroenke, Spitzer, & Williams, 2003; Löwe, Kroenke, & Gräfe, 2005). The scoring ranges between 0 and 6, and the cut-off point of ≥3 has been considered an indicator of clinically significant depression. In this case, assessment should be completed by the PHQ-9 or a clinical interview. Arroll et al. (2010) reported the largest validation study of the PHQ-2, compared with a reference standard interview, undertaken in an exclusively primary care population. They suggested that a PHQ-2 score of ≥2 rather than ≥3 resulted in more depressed patients being correctly identified. The PHQ-2 also seems sensitive to monitoring therapeutic response (Löwe et al., 2005).

The PHQ-8 omits the ninth item that asks about ideas of death or self-harm. It is applicable to epidemiological studies in which it would be an uncommon response or where depression is a secondary measure. The results of the algorithmic diagnosis are good and significantly correlate with the full version (Kroenke et al., 2009). Either its diagnostic algorithm or a cut-off point ≥10 can be used for defining current depression.

## Psychometric characteristics

The diagnostic validity of the PHQ was first established in two studies involving 3000 patients in eight primary care centers and 3000 patients in seven obstetrics-gynecology clinics. In the first study (Spitzer et al., 1999) the PHQ-9 demonstrated diagnostic validity comparable to that of the original PRIME-MD, with even higher sensitivity of the PHQ for major depressive disorder (0.73 vs 0.57). In the second study (Spitzer et al., 2000), the authors did also demonstrate that the PHQ had a comparable construct validity to the original PRIME-MD interview and was more efficient to use. Construct validity was established by the association between PHQ diagnoses and indices of functional impairment, showing that patients with PHQ diagnoses had more functional impairment, disability days, and health care use than did patients without PHQ diagnoses. The PHQ-9 demonstrated a high level of internal consistency, with a Cronbach's alpha of 0.89 and 0.86 (Kroenke et al., 2001; Spitzer et al., 2000). The test-retest reliability was also high, with correlations between self-administration and a telephone revaluation in the first 48h of 0.84–0.95 (Kroenke et al., 2001; Pinto-Meza, Serrano-Blanco, Peñarrubia, Blanco, & Haro, 2005) and in 7 days of 0.81–0.96 (Löwe, Unützer, Callahan, Perkins, & Kroenke, 2004). Criterion validity was demonstrated by an agreement with an independent mental health professional interview. A year later, Kroenke et al. (2001) examined the validity of the PHQ-9 as a new measure of depression severity. They used the same sample of patients to evaluate construct validity using the 20-item Short-Form General Health Survey, self-reported sick days and clinic visits, and symptom-related difficulties. Criterion validity was assessed against and independent mental health professional. As PHQ-9 depression severity increased, there was a substantial decrease in functional status, and symptom-related difficulties, sick days, and health care utilization increased. Using the mental health professional reinterview as the criterion standard, a PHQ-9 score ≥ 10 had a sensitivity of 0.88 and a specificity of 0.88 for major depression. Results were similar in primary care and obstetrics-gynecology samples.

In the following years, Löwe, Unützer, et al. (2004) validated the PHQ-9 as an outcome measure for depression and as a tool for measuring psychopharmacological treatment response. They first investigated sensitivity to change the PHQ-9 in three groups of patients whose depression status either improved, remained unchanged or deteriorated over time. PHQ-9 change scores differed significantly between the three depression outcome groups. They later developed a prospective, non-interventional, observational study including depressed patients treated with sertraline. The PHQ-9 change scores were considerably greater in therapy responders than in nonresponders (Löwe, Schenkel, Carney-Doebbeling, & Göbel, 2006).

The PHQ-9 was also validated as a screening instrument for depression in the general population. Martin, Rief, Klaiberg, and Braehler (2006) found a strong positive association between depression and disability and a strong negative association between depression and functional health status in patients with both major depressive disorders and other depressive disorders including minor depression, highlighting its ability to identify not only major depression, but also subthreshold depressive symptoms in the general population.

And regarding the measurement of suicide risk, Simon et al. (2013), in a study with 84,418 patients and 207,625 administrations of the questionnaire, found that the cumulative risk of suicidal behaviors during a year increased from 0.4% among those who had answered "Not at all" to item 9 of the PHQ-9 to 4% among those who had answered "Nearly every day," while that of suicide death increased from 0.03% to 0.3%, respectively. The score on item 9 of the PHQ-9 was a powerful predictor of suicide even after adjusting for age, sex, treatments, and severity of depression.

## Screening and case-finding properties

In 2007, two systematic reviews and meta-analysis were published on the diagnostic accuracy of the self-report version of the PHQ-9. In the first one, Wittkampf, Naeije, Schene, Huyser, and van Weert (2007) identified 12 validation studies of PHQ-9 with a total of 6394 participants. The PHQ-9 showed a high specificity (0.94) but quite low sensitivity (0.77) when using the algorithm method, indicating that the PHQ-9 is a valid instrument to detect depression, but further assessment should be developed in order to properly diagnose the major depressive disorder, especially in settings with a low prevalence of depression. Similar results were obtained in the second systematic review by Gilbody, Richards, Brealey, and Hewitt (2007), who identified 14 validation studies of PHQ-9 about the diagnosis of major depression with a total of 5026 participants. There was a high specificity (0.92) and a moderate sensitivity (0.80). In this study, no differences in diagnostic properties of the PHQ-9 were found when compared to the cut-off score of ≥10 or the diagnostic algorithm method. However, some other cut-off points were explored, providing similar levels of sensitivity and specificity, and thus suggesting that the choice of the cut-off point may vary according to the clinical population.

Manea, Gilbody, and McMillan (2012) published a meta-analysis of 18 validation studies with a total of 7180 participants summarizing the psychometric properties of the PHQ-9 across a range of studies and cut-off scores to select the optimal cut-off for detecting depression. They found that PHQ-9 has acceptable diagnostic properties at a range of cut-off scores between ≥8 and ≥11, without significant differences in sensitivity or specificity at a cut-off score of ≥10 compared with other scores within this interval. Although there is still limited evidence on the diagnostic accuracy for each of the cut-off scores, it could be useful to be able to apply different cut-off points depending on the characteristics of the different populations to which they are applied. A cut-off score of ≥10 may result in many false negatives in hospital settings, while more false positives in primary care.

Seven years after the first published reviews, the number of validation studies for the PHQ-9 doubled, and more extended systematic reviews were performed. All of them placed on emphasis high levels of heterogeneity between studies supporting the hypotheses that similar cut-off points might not be appropriate in all settings and the selection of the most appropriate cut-off point should consider also the prevalence of the major depressive disorder in specific populations. However, mainly based on the selective reporting bias, it is very difficult to explore in which way different cut-off points perform differently in different settings.

Consistent with the previous meta-analysis, Moriarty, Gilbody, McMillan, and Manea (2015) included 36 studies with 21,292 participants and found that the sensitivity of the PHQ-9 at cut-off point ≥10 was lower than that reported in the original validation study (0.78), whereas the specificity was similar (0.87). This may have consequences for the use of the PHQ-9 as a screening instrument, as high sensitivity is required in order to ensure that few people with depression are missed. At this cut-off, the PHQ-9 was found to be a better screening tool in primary care than secondary care settings.

Manea et al. (2015) conducted a systematic review of the PHQ-9 validation studies which used the algorithmic scoring method. They found 27 validation studies conducted in various settings (primary care, out- or inpatients from different medical specialties, or community samples), with 15,615 participants. This meta-analysis underlined that this method of scoring may lead to problematically low sensitivity in both primary care and hospital settings (0.53). In both settings, this method would miss many patients with major depressive disorder. The summed-item scoring method at cut-off point ≥10 had better sensitivity (0.77) and maintained good specificity (0.85).

Mitchell, Yadegarfar, Gill, and Stubbs (2016) performed a meta-analysis on 26 publications which included 14,760 adults. Sensitivity and specificity were 0.81 and 0.85 for the PHQ-9 sum score, 0.57 and 0.93 for the algorithmic PHQ-9, and 0.89 and 0.76 for the PHQ-2. For case finding, none of the methods was suitable, but for screening all methods showed good clinical utility, although an adjustment of the cut-off points should be considered (≥14 on the PHQ-9 and ≥6 on the PHQ-2). A recent individual participant data meta-analysis with more than 9000 patients clearly suggested that original cut-off overestimated the prevalence of depression (Levis, Benedetti, et al., 2020). Otherwise, their use can be improved by a two-step procedure using first PHQ-2 and then PHQ-9 only following a positive initial PHQ-2 assessment, thus improving specificity without altering sensitivity (Levis, Sun, et al., 2020).

Some authors have also suggested possible adjustments of the categorical algorithm in order to improve sensitivity. Zuithoff et al. (2010) defined an adjusted categorical algorithm where all responses other than "Not at all" accounted for symptom presence. This adjustment resulted in a sensitivity of 0.84 and specificity of 0.81 compared to the 0.28 and 0.98, respectively, in the original algorithm; and to a cut-off score of ≥6.

An important aspect raised by Manea, Boehnke, Gilbody, Moriarty, and McMillan (2017) was the possibility of an authoring effect so that studies carried out by the original or associated PHQ-9 developers could show better results than those published by other authors, the so-called *allegiance* phenomenon. They collected a total of seven studies from the original authors and 20 independent studies. Pooled diagnostic odds ratio (DOR) for the allegiant group was 64.40 compared to 15.05 for the independent group, allegiance status being a significant predictor of the variation of DOR. In the case of the cut-off studies, DOR was 49.31 vs 24.96. In allegiant studies, sensitivity was 0.77 and specificity 0.94. In the independent studies, figures were 0.48 and 0.94. Heterogeneity was high ($I^2 = 68.1\%$). Allegiance status explained 51.5% of the heterogeneity, although there were other variables that could explain differences, for example, the use of more appropriate translations, or a shorter period between the test and the reference assessment.

## Comparison with other psychometric instruments

In 38 studies with more than 32,000 primary care patients, the characteristics of the PHQ-9 were equal or superior to those of other instruments for the evaluation of depression (Williams, Pignone, Ramirez, & Perez Stellato, 2002). This data were replicated in two subsequent studies (Henkel et al., 2004; Löwe et al., 2004).

In 2015, (Pettersson et al. (2015)) reviewed a total of 20 instruments for the diagnosis of depression. Among them, only two structured interviews, the Structured Clinical Interview for DSM-IV-Axis-I Disorders (SCID-I), with a sensitivity of 0.85 and a specificity of 0.92, and the Mini International Neuropsychiatric Interview (MINI), with figures of 0.95 and 0.84 respectively, and only one self-report questionnaire, the PHQ-9 with a cut-off point of 10, with respective figures of 0.88 and 0.78, met the minimum criteria of specificity of $\geq$0.80 and a specificity of $\geq$0.70. The use of the rest of the scales did not seem to be supported by enough evidence. The Hospital Anxiety and Depression Scale (HADS), with a cut-off point of 7, had too low sensitivity to be useful in clinical practice, and the Beck Depression Inventory (BDI-II), with a cut-off point of 14, had too low specificity.

Ren, Yang, Browning, Thomas, and Liu (2015) conducted a systematic review of eight studies with different instruments in patients with coronary heart disease. Regarding screening, in which sensitivity must be maximized, the best instrument was the Center for Epidemiological Studies Scale (CES-D) (cut-off $\geq$16) and the Symptom Checklist (SCL-90) (cut-off $\geq$25), with sensitivities of 0.93 and 0.96 respectively, and a negative predictive value of 0.98 and 0.99. However, they had a very low positive predictive value, 0.55 and 0.32. In the case of PHQ-9 (cut-off $\geq$10), the sensitivity was low (0.54) but the specificity was high (0.91) and also the positive predictive value (0.58 and 0.60 in two studies), being therefore much more suitable for diagnosis.

Another systematic review of screening instruments in primary care included 66 publications with 48,234 participants and 55 instruments (El-Den, Chen, Gan, Wong, & O'Reilly, 2018). Figures for sensitivity and specificity were, respectively, 0.28–0.95 and 0.61–0.93 for PHQ-9, 0.42–0.95 and 0.61–0.95 for PHQ-2, 0.61–0.88 and 0.72–0.85 for CES-D, and 0.72–0.87 and 0.63–0.75 for General Health Questionnaire (GHQ-12).

## Special populations

The PHQ-9 behaves also similar in different populations regarding sex, age, ethnic groups or medical morbidities.

Several studies have been conducted in elderly populations. Although the sample was low ($n = 71$), Phelan et al. (2010) studied the accuracy of the questionnaire in adults over 65 years. For major depression, the PHQ-9 showed an area under the curve (AUC) of 0.87, while for the PHQ-2 it was 0.81 and for the Geriatric Depression Scale (GDS-15) it was 0.81, with no significant differences between them. Similar figures were found when studying an elderly population ($n = 713$) with chronic medical diseases (diabetes mellitus and chronic obstructive pulmonary disease—COPD), favoring the quantitative correction method (Lamers et al., 2008).

Versions for adolescents have also been developed, such as one of 14 items with dichotomous yes/no answers, the so-called PHQ-A (Johnson, Harris, Spitzer, & Williams, 2002; Zuckerbrot & Jensen, 2006). The validity of PHQ-2 among 444 adolescents between 13 and 17 years has also been investigated. With a cut-off point of $\geq$3, the sensitivity was 0.74 and the specificity was 0.75 for major depression, with an AUC of 0.84 (Richardson et al., 2010).

In patients with neurological disorders, PHQ-9 has been studied at least in Parkinson's disease, stroke, epilepsy, multiple sclerosis, spinal cord injury, and cognitive impairment. Williams et al. (2012) compared the properties of nine psychometric measures in 229 patients with Parkinson's disease and found that all scales were similarly effective for the detection of depression. In the case of PHQ-9, applying a cut-off point of $\geq$6, the sensitivity was 0.66, specificity 0.80, positive predictive value 0.69, and negative predictive value 0.77, with an AUC of 0.81. Meader, Moe-Byrne, Llewellyn, and Mitchell (2014) reviewed the properties of several instruments in 24 studies, finding that both CES-D, as well as Hamilton Depression Rating Scale (HDRS) and PHQ-9 were optimal instruments for screening for poststroke depression, although the usefulness for case finding was weak. In the case of PHQ-9, the sensitivity figures found were 0.86 and specificity 0.79. In the case of multiple sclerosis, systematic evidence reveals that the PHQ-9 at cut-off point of 11 was a suitable tool to screen for major depressive disorder, with a sensitivity of 0.95 and a specificity of 0.88. However, its use as a diagnostic tool cannot currently be recommended (Patrick & Connick, 2019). Amtmann et al. (2014) studied also various instruments in a sample of 455 individuals with multiple sclerosis, finding that PHQ-9, as well as CES-D and PROMIS Depression Short Form (PROMIS-D-8) were appropriate instruments. The PHQ-9 has also demonstrated high reliability and validity as a screening tool for the detection of current major depression in patients with epilepsy, with an AUC of 0.89 and a specificity of 0.84 and sensitivity of 0.83 at cut-off score > 6 (Xia et al., 2019).

McManus, Pipkin, and Whooley (2005) compared the characteristics of four instruments for screening depression in 1024 patients with coronary heart disease, including PHQ-9 and PHQ-2. The AUC were similar for all instruments, between 0.84 and 0.87. The sensibility of the PHQ-9 with a cut-off point of ≥10 was 0.54 and the specificity of 0.90, with a positive likelihood ratio of 5.4. Similarly, Stafford, Berk, and Jackson (2007), in a small sample of 193 patients, found no difference between PHQ-9 and HADS to detect major or minor depression, although the algorithmic method of PHQ-9 had a higher positive likelihood ratio. It also seemed necessary to reduce the cut-off point of the PHQ-9 to ≥5/6 in order to increase the sensitivity, while the specificity was maintained. Recently, Yuan et al. (2019) compared PHQ-9 and HADS-D for the screening of major depression in 782 Chinese patients with the acute coronary syndrome. The diagnostic accuracy of both scales was similar, with AUC of 0.84 and 0.81, respectively. In this study, the optimal cut-off point was the usual one of ≥10. The specificity of both scales was similar, 0.85 vs 0.86, while the sensitivity was higher for the PHQ-9, 0.87 vs 0.76.

Monahan et al. (2009) studied PHQ-9 and PHQ-2 in a population of 347 individuals living with HIV in Kenya. Factor analysis revealed acceptably high factor loadings on a major core depressive factor and adequate item discrimination values, supporting construct validity. Test-retest reliability of 0.59 for PHQ-9 total score was moderate but acceptable in relation to the setting and context of the study. The PHQ-2 also demonstrated very good operating characteristics. The best cut-off point for sensitivity and specificity (≥3) provided very high specificity for any depressive disorder. The PHQ proved reliable, valid, and useful as a screening instrument to detect depression, generalized anxiety, and panic disorders in 500 consecutive HCV patients attending a liver unit in a general hospital. The study presented a good agreement between PHQ and SCID-I diagnoses showing high sensitivity of 0.84 and even higher specificity 0.97 (Navinés et al., 2012).

Van Dijk et al. (2018) carried out a systematic review of different instruments applied to the measurement of depressive symptoms in populations with diabetes mellitus, type 1 or type 2. They collected 21 studies evaluating nine different questionnaires. For the PHQ-9, there was strong evidence of good internal consistency and criterion validity. Construct validity (hypothesis testing) was rated "good" with a moderate level of evidence. The evidence for structural validity was inconclusive since two studies of at least good quality found different factor structures.

Grapp, Terhoeven, Nikendei, Friederich, and Maatouk (2019) compared the usefulness of somatic symptoms vs cognitive-emotional symptoms of PHQ-9 in a sample of 4705 cancer patients. Nonsomatic symptoms played a leading role in the diagnosis of major depression, while somatic symptoms gained relevance in the diagnosis of other depressive disorders (including milder forms of depression). Thus, nonsomatic symptoms are less discriminatory, but ignoring them could lead to an underestimation of depressive syndromes.

The validity of the PHQ-9 has also been tested in rheumatological disorders (Rosemann et al., 2007), dermatological disorders (Picardi et al., 2004), ear-nose-throat disorders (Persoons, Luyckx, Desloovere, Vandenberghe, & Fischler, 2003), chronic pain (Dobscha et al., 2009; Kroenke et al., 2007), pulmonary diseases (von Siemens et al., 2019), and perinatal depression (Gjerdingen, Crow, McGovern, Miner, & Center, 2009; Hanusa, Scholle, Haskett, Spadaro, & Wisner, 2008; Yonkers et al., 2009).

## Conclusions

The method of quantitative correction of the PHQ-9 with the sum of scores of all items has demonstrated better psychometric properties of sensitivity and specificity than the algorithmic method. The two-item version, the PHQ-2, has also demonstrated good psychometric properties. The PHQ-9 is also adequate for measuring depression outcome.

Several systematic reviews and meta-analysis usually found high specificity but moderate sensitivity, with controversy about which cut-off score should be applied depending on the characteristics of the population and the settings. A cut-off score of ≥10 may result in many false negatives in hospital settings, while more false positives in primary care.

The use of these screening tools can be improved by a two-step procedure using first PHQ-2 and then PHQ-9 only following a positive initial PHQ-2 assessment.

The PHQ-9 behaves similarly in different populations regarding sex, age, ethnic groups, or medical morbidities.

## Key facts of PHQ

- PHQ is a fully self-administered version of the PRIME-MD.
- PHQ-9 is the depression module of the PHQ.
- PHQ-9 includes nine items that replicate the nine diagnostic criteria for major depressive episode from the DSM-III-R, along with an additional 10th item of

impairment in work, daily living, or interpersonal relationships areas due to major depression.

- Scoring the PHQ-9 can be done through both a diagnostic algorithm or a continuous measure of the sum of scores of all nine items.
- PHQ-9 has been extensively translated and adapted to numerous languages.
- Two shortened versions of the PHQ-9 have been developed, PHQ-2 and PHQ-8.
- PHQ-9 is a good screening and case-finding tool, with moderate to good psychometric properties, and a good instrument to monitor depression outcomes.
- The use of these screening tools can be improved by a two-step procedure using first PHQ-2 and then PHQ-9 only following a positive initial PHQ-2 assessment.
- The PHQ-9 behaves similarly in different populations regarding sex, age, ethnic groups, or medical morbidities.

## Summary points

- The routine usage of screening instruments can be useful to assist in the diagnosis of depression in nonpsychiatric clinical settings.
- PHQ-9 shows high specificity but moderate sensitivity, with controversy about which cut-off score should be applied depending on the characteristics of the population and the settings.
- The use of these screening tools can be improved by a two-step procedure using first PHQ-2 and then PHQ-9 only following a positive initial PHQ-2 assessment.
- The PHQ-9 behaves similarly in different populations regarding sex, age, ethnic groups, or medical morbidities.
- When using PHQ-9 it will be essential to be able to offer thereafter a therapeutic approach to all those patients who can be diagnosed with depression with this instrument. Otherwise, its use would not be adequate.

## Mini-dictionary of terms

**Screening instruments** Questionnaires used to rule out noncases of a given entity in a population.

**Case-finding instruments** Questionnaires used to diagnose cases of a given entity in a population.

**Sensitivity** The probability of correctly classifying a sick individual. Sensitivity is, therefore, the ability of the test to detect the disease. Sensitivity is the percentage of true positives or the probability that the test is positive if the disease is present.

**Specificity** The probability that a healthy subject has a negative test result. Specificity is the percentage of true negatives or the probability that the test is negative if the disease is not present.

**Algorithmic correction** Correction through a set of instructions and rules defined and nonambiguous, orderly and finite that typically allows to offer a diagnosis.

## References


Addington, D., Addington, J., Maticka-Tyndale, E., & Joyce, J. (1992). Reliability and validity of a depression rating scale for schizophrenics. *Schizophrenia Research*, *6*, 201–208.

Amtmann, D., Kim, J., Chung, H., Bamer, A. M., Askew, R. L., Wu, S., et al. (2014). Comparing CESD-10, PHQ-9, and PROMIS depression instruments in individuals with multiple sclerosis. *Rehabilitation Psychology*, *59*, 220–229.

Arroll, B., Goodyear-Smith, F., Crengle, S., Gunn, J., Kerse, N., Fishman, T., et al. (2010). Validation of PHQ-2 and PHQ-9 to screen for major depression in the primary care population. *Annals of Family Medicine*, *8*, 348–353.

Bech, P., Rasmussen, N. A., Olsen, L. R., Noerholm, V., & Abildgaard, W. (2001). The sensitivity and specificity of the major depression inventory, using the present state examination as the index of diagnostic validity. *Journal of Affective Disorders*, *66*, 159–164.

Beck, A. T., Ward, C. H., Mendelson, M., Mock, J., & Erbaugh, J. (1961). An inventory for measuring depression. *Archives of General Psychiatry*, *4*, 561–571.

Dahl, A. A., Kruger, M. B., Dahl, N. H., Karlson, H., Knorring, L. V., & Stordal, E. (2009). SPIFA-A presentation of the structured psychiatric interview for general practice. *Nordic Journal of Psychiatry*, 1–11.

Diez-Quevedo, C., Rangil, T., Sanchez-Planell, L., Kroenke, K., & Spitzer, R. L. (2001). Validation and utility of the patient health questionnaire in diagnosing mental disorders in 1003 general hospital Spanish inpatients. *Psychosomatic Medicine*, *63*, 679–686.

Dobscha, S. K., Corson, K., Perrin, N. A., Hanson, G. C. Leibowitz R. Q., Doak, M. N., et al. (2009). Collaborative care for chronic pain in primary care: A cluster randomized trial. *JAMA*, *301*, 1242–1252.

El-Den, S., Chen, T. F., Gan, Y. L., Wong, E., & O'Reilly, C. L. (2018). The psychometric properties of depression screening tools in primary healthcare settings: A systematic review. *Journal of Affective Disorders*, *225*, 503–522.

Endicott, J., & Spitzer, R. L. (1978). A diagnostic interview: The schedule for affective disorders and schizophrenia. *Arch General Psychiatry*, *35*, 837–844.

Fann, J. R., Berry, D. L., Wolpin, S., Austin-Seymour, M., Bush, N., Halpenny, B., et al. (2009). Depression screening using the Patient Health Questionnaire-9 administered on a touch screen computer. *Psychooncology*, *18*, 14–22.

Gilbody, S., Richards, D., Brealey, S., & Hewitt, C. (2007). Screening for depression in medical settings with the Patient Health Questionnaire (PHQ): A diagnostic meta-analysis. *Journal of General Internal Medicine*, *22*, 1596–1602.

Gjerdingen, D., Crow, S., McGovern, P., Miner, M., & Center, B. (2009). Postpartum depression screening at well-child visits: Validity of a 2-question screen and the PHQ-9. *Annals of Family Medicine*, *7*(1), 63–70.

Goldberg, D. P. (1972). *The detection of psychiatric illness by questionnaire (Maudsley monograph no. 21)*. Oxford, UK: Oxford University Press.

Grapp, M., Terhoeven, V., Nikendei, C., Friederich, H. C., & Maatouk, I. (2019). Screening for depression in cancer patients using the PHQ-9: The accuracy of somatic compared to non-somatic items. *Journal of Affective Disorders*, *254*, 74–81.

Hamilton, M. (1960). A rating scale for depression. *Journal of Neurology, Neurosurgery, and Psychiatry*, *23*, 56–62.

Hanusa, B. H., Scholle, S. H., Haskett, R. F., Spadaro, K., & Wisner, K. L. (2008). Screening for depression in the postpartum period: A comparison of three instruments. *Journal of Women's Health*, *17*(4), 585–596.

Henkel, V., Mergl, R., Kohnen, R., Allgaier, A. K., Möller, H. J., & Hegerl, U. (2004). Use of brief depression screening tools in primary care: Consideration of heterogeneity in performance in different patient groups. *General Hospital Psychiatry*, *26*, 190–198.

Johnson, J. G., Harris, E. S., Spitzer, R. L., & Williams, J. B. (2002). The patient health questionnaire for adolescents: Validation of an instrument for the assessment of mental disorders among adolescent primary care patients. *The Journal of Adolescent Health*, *30*, 196–204.

Kessler, R. C., & Ustun, T. B. (2004). The World Mental Health (WMH) survey initiative version of the World Health Organization (WHO) Composite International Diagnostic Interview (CIDI). *International Journal of Methods in Psychiatric Research*, *13*, 93–121.

Kroenke, K., Bair, M., Damush, T., Hoke, S., Nicholas, G., Kempf, C., et al. (2007). Stepped Care for Affective Disorders and Musculoskeletal Pain (SCAMP) study: Design and practical implications of an intervention for comorbid pain and depression. *General Hospital Psychiatry*, *29*, 506–517.

Kroenke, K., Spitzer, R. L., & Williams, J. B. (2001). The PHQ-9: Validity of a brief depression severity measure. *Journal of General Internal Medicine*, *16*, 606–613.

Kroenke, K., Spitzer, R. L., & Williams, J. B. (2003). The patient health Questionnaire-2: Validity of a two-item depression screener. *Medical Care*, *41*, 1284–1292.

Kroenke, K., Strine, T. W., Spitzer, R. L., Williams, J. B., Berry, J. T., & Mokdad, A. H. (2009). The PHQ-8 as a measure of current depression in the general population. *Journal of Affective Disorders*, *114*, 163–173.

Lamers, F., Jonkers, C. C., Bosma, H., Penninx, B. W., Knottnerus, J. A., & van Eijk, J. T. (2008). Summed score of the Patient Health Questionnaire-9 was a reliable and valid method for depression screening in chronically ill elderly patients. *Journal of Clinical Epidemiology*, *61*, 679–687.

Levis, B., Benedetti, A., Ioannidis, J. P. A., Sun, Y., Negeri, Z., He, C., et al. (2020). Patient Health Questionnaire-9 scores do not accurately estimate depression prevalence: Individual participant data meta-analysis. *Journal of Clinical Epidemiology*, *122*, 115–128.

Levis, B., Sun, Y., He, C., Wu, Y., Krishnan, A., Bhandari, P. M., et al. (2020). Accuracy of the PHQ-2 alone and in combination with the PHQ-9 for screening to detect major depression: Systematic review and meta-analysis. *JAMA*, *323*(22), 2290–2300.

Löwe, B., Kroenke, K., & Gräfe, K. (2005). Detecting and monitoring depression with a two-item questionnaire (PHQ-2). *Journal of Psychosomatic Research*, *58*, 163–171.

Löwe, B., Schenkel, I., Carney-Doebbeling, C., & Göbel, C. (2006). Responsiveness of the PHQ-9 to psychopharmacological depression treatment. *Psychosomatics*, *47*, 62–67.

Löwe, B., Spitzer, R. L., Gräfe, K., Kroenke, K., Quenter, A., Zipfel, S., et al. (2004). Comparative validity of three screening questionnaires for DSM-IV depressive disorders and physicians' diagnoses. *Journal of Affective Disorders*, *78*, 131–140.

Löwe, B., Unützer, J., Callahan, C. M., Perkins, A. J., & Kroenke, K. (2004). Monitoring depression treatment outcomes with the patient health questionnaire-9. *Medical Care*, *42*, 1194–1201.

Manea, L., Boehnke, J. R., Gilbody, S., Moriarty, A. S., & McMillan, D. (2017). Are there researcher allegiance effects in diagnostic validation studies of the PHQ-9? A systematic review and meta-analysis. *BMJ Open*, *7*, e015247.

Manea, L., Gilbody, S., & McMillan, D. (2012). Optimal cut-off score for diagnosing depression with the Patient Health Questionnaire (PHQ-9): A meta-analysis. *CMAJ*, *184*, E191–E196.

Manea, L., Gilbody, S., & McMillan, D. (2015). A diagnostic meta-analysis of the Patient Health Questionnaire-9 (PHQ-9) algorithm scoring method as a screen for depression. *General Hospital Psychiatry*, *37*, 67–75.

Martin, A., Rief, W., Klaiberg, A., & Braehler, E. (2006). Validity of the brief patient health questionnaire mood scale (PHQ-9) in the general population. *General Hospital Psychiatry*, *28*, 71–77.

Matijasevich, A., Munhoz, T. N., Tavares, B. F., et al. (2014). Validation of the Edinburgh postnatal depression scale (EPDS) for screening of major depressive episode among adults from the general population. *BMC Psychiatry*, *14*, 284.

McManus, D., Pipkin, S. S., & Whooley, M. A. (2005). Screening for depression in patients with coronary heart disease (data from the Heart and Soul Study). *The American Journal of Cardiology*, *96*, 1076–1081.

Meader, N., Moe-Byrne, T., Llewellyn, A., & Mitchell, A. J. (2014). Screening for poststroke major depression: A meta-analysis of diagnostic validity studies. *Journal of Neurology, Neurosurgery, and Psychiatry*, *85*, 198–206.

Mitchell, A. J., Yadegarfar, M., Gill, J., & Stubbs, B. (2016). Case finding and screening clinical utility of the Patient Health Questionnaire (PHQ-9 and PHQ-2) for depression in primary care: A diagnostic metaanalysis of 40 studies. *BJPsych Open*, *2*, 127–138.

Monahan, P. O., Shacham, E., Reece, M., Kroenke, K., Ong'or, W. O., Omollo, O., et al. (2009). Validity/reliability of PHQ-9 and PHQ-2 depression scales among adults living with HIV/AIDS in western Kenya. *Journal of General Internal Medicine*, *24*, 189–197.

Montgomery, S. A., & Asberg, M. (1979). A new depression scale designed to be sensitive to change. *The British Journal of Psychiatry*, *134*, 382–389.

Moriarty, A. S., Gilbody, S., McMillan, D., & Manea, L. (2015). Screening and case finding for major depressive disorder using the Patient Health Questionnaire (PHQ-9): A meta-analysis. *General Hospital Psychiatry*, *37*, 567–576.

Navinés, R., Castellví, P., Moreno-España, J., Gimenez, D., Udina, M., Cañizares, S., et al. (2012). Depressive and anxiety disorders in chronic hepatitis C patients: Reliability and validity of the Patient Health Questionnaire. *Journal of Affective Disorders*, *138*, 343–351.

Olfson, M., Leon, A. C., Broadhead, W. E., Weissman, M. M., Barrett, J. E., Blacklow, R. S., et al. (1995). The SDDS-PC: A diagnostic aid for multiple mental disorders in primary care. *Psychopharmacology Bulletin*, *31*, 415–420.

Patrick, S., & Connick, P. (2019). Psychometric properties of the PHQ-9 depression scale in people with multiple sclerosis: A systematic review. *PLoS One*, *14*, e0197943.

Persoons, P., Luyckx, K., Desloovere, C., Vandenberghe, J., & Fischler, B. (2003). Anxiety and mood disorders in otorhinolaryngology outpatients presenting with dizziness: Validation of the self-administered PRIME-MD Patient Health Questionnaire and epidemiology. *General Hospital Psychiatry*, *25*, 316–323.

Pettersson, A., Boström, K. B., Gustavsson, P., & Ekselius, L. (2015). Which instruments to support diagnosis of depression have sufficient accuracy? A systematic review. *Nordic Journal of Psychiatry*, *69*, 497–508.

Phelan, E., Williams, B., Meeker, K., Bonn, K., Frederick, J., Logerfo, J., et al. (2010). A study of the diagnostic accuracy of the PHQ-9 in primary care elderly. *BMC Family Practice*, *11*, 63.

Picardi, A., Amerio, P., Baliva, G., Barbieri, C., Teofoli, P., Bolli, S., et al. (2004). Recognition of depressive and anxiety disorders in dermatological outpatients. *Acta Dermato-Venereologica*, *84*, 213–217.

Pinto-Meza, A., Serrano-Blanco, A., Peñarrubia, M. T., Blanco, E., & Haro, J. M. (2005). Assessing depression in primary care with the PHQ-9: Can it be carried out over the telephone? *Journal of General Internal Medicine*, *20*, 738–742.

Radloff, L. S. (1977). The CES-D scale: A self-report depression scale for research in the general population. *Applied Psychological Measurement*, *1*, 385–401.

Ren, Y., Yang, H., Browning, C., Thomas, S., & Liu, M. (2015). Performance of screening tools in detecting major depressive disorder among patients with coronary heart disease: A systematic review. *Medical Science Monitor*, *21*, 646–653.

Richardson, L. P., Rockhill, C., Russo, J. E., Grossman, D. C., Richards, J., McCarty, C., et al. (2010). Evaluation of the PHQ-2 as a brief screen for detecting major depression among adolescents. *Pediatrics*, *125*, e1097–e1103.

Robins, L. N., Helzer, J. E., Croughan, J., & Ratcliff, K. S. (1981). National Institute of Mental Health Diagnostic Interview Schedule. Its history, characteristics, and validity. *Archives of General Psychiatry*, *38*, 381–389.

Rosemann, T., Backenstrass, M., Joest, K., Rosemann, A., Szecsenyi, J., & Laux, G. (2007). Predictors of depression in a sample of 1,021 primary care patients with osteoarthritis. *Arthritis and Rheumatism*, *57*, 415–422.

Rush, A. J., Giles, D. E., Schlesser, M. A., Fulton, C. L., Weissenburger, J., & Burns, C. (1986). The Inventory for Depressive Symptomatology (IDS): Preliminary findings. *Psychiatry Research*, *18*, 65–87.

Rush, A. J., Trivedi, M. H., Ibrahim, H. M., Carmody, T. J., Arnow, B., Klein, D. N., et al. (2003). The 16-Item Quick Inventory of Depressive Symptomatology (QIDS), clinician rating (QIDS-C), and self-report (QIDS-SR): A psychometric evaluation in patients with chronic major depression. *Biological Psychiatry*, *54*, 573–583.

Sheehan, D. V., Lecrubier, Y., Sheehan, K. H., Amorim, P., Janavs, J., Weiller, E., et al. (1998). The Mini-International Neuropsychiatric Interview (M.I.N.I.): The development and validation of a structured diagnostic psychiatric interview for DSM-IV and ICD-10. *The Journal of Clinical Psychiatry*, *59*(Suppl. 20), 22–33.

Simon, G. E., Rutter, C. M., Peterson, D., Oliver, M., Whiteside, U., Operskalski, B., et al. (2013). Do PHQ depression questionnaires completed during outpatient visits predict subsequent suicide attempt or suicide death? *Psychiatric Services*, *64*, 1195–1202.

Spitzer, R. L., Kroenke, K., & Williams, J. B. (1999). Validation and utility of a self-report version of PRIME-MD: The PHQ primary care study. Primary Care Evaluation of Mental Disorders. Patient Health Questionnaire. *JAMA*, *282*, 1737–1744.

Spitzer, R. L., Williams, J. B., Gibbon, M., & First, M. B. (1992). The Structured Clinical Interview for DSM-III-R (SCID). I: History, rationale, and description. *Archives of General Psychiatry*, *49*, 624–629.

Spitzer, R. L., Williams, J. B., Kroenke, K., Hornyak, R., & McMurray, J. (2000). Validity and utility of the PRIME-MD patient health questionnaire in assessment of 3000 obstetric-gynecologic patients: The PRIME-MD Patient Health Questionnaire Obstetrics-Gynecology Study. *American Journal of Obstetrics and Gynecology*, *183*, 759–769.

Spitzer, R. L., Williams, J. B., Kroenke, K., Linzer, M., de Gruy, F. V., Hahn, S. R., et al. (1994). Utility of a new procedure for diagnosing mental disorders in primary care. The PRIME-MD 1000 study. *JAMA*, *272*, 1749–1756.

Stafford, L., Berk, M., & Jackson, H. J. (2007). Validity of the hospital anxiety and depression scale and patient health Questionnaire-9 to screen for depression in patients with coronary artery disease. *General Hospital Psychiatry*, *29*, 417–424.

Terluin, B., van Marwijk, H. W. J., Adèr, H. J., de Vet, H. C. W., Penninx, B. W. J. H., Hermens, M. L. M., et al. (2006). The four-dimensional symptom questionnaire (4DSQ): A validation study of a multidimensional self-report questionnaire to assess distress, depression, anxiety and somatization. *BMC Psychiatry*, *6*, 34.

van Dijk, S. E. M., Adriaanse, M. C., van der Zwaan, L., Bosmans, J. E., van Marwijk, H. W. J., van Tulder, M. W., et al. (2018). Measurement properties of depression questionnaires in patients with diabetes: A systematic review. *Quality of Life Research*, *27*, 1415–1430.

von Siemens, S. M., Jörres, R. A., Behr, J., Alter, P., Lutter, J., Lucke, T., et al. (2019). Effect of COPD severity and comorbidities on the result of the PHQ-9 tool for the diagnosis of depression: results from the COSYCONET cohort study. *Respiratory Research*, *20*(1), 30.

WHO. (1998). *Wellbeing measures in primary health care/the Depcare project*. Copenhagen: WHO Regional Office for Europe.

Williams, J. R., Hirsch, E. S., Anderson, K., Bush, A. L., Goldstein, S. R., Grill, S., et al. (2012). A comparison of nine scales to detect depression in Parkinson disease: Which scale to use? *Neurology*, *78*, 998–1006.

Williams, J. W., Jr., Pignone, M., Ramirez, G., & Perez Stellato, C. (2002). Identifying depression in primary care: A literature synthesis of case-finding instruments. *General Hospital Psychiatry*, *24*, 225–237.

Wing, J. K., Babor, T., Brugha, T., Burke, J., Cooper, J. E., Giel, R., et al. (1990). SCAN. Schedules for clinical assessment in neuropsychiatry. *Archives of General Psychiatry*, *47*, 589–593.

Wittkampf, K. A., Naeije, L., Schene, A. H., Huyser, J., & van Weert, H. C. (2007). Diagnostic accuracy of the mood module of the patient health questionnaire: A systematic review. *General Hospital Psychiatry*, *29*, 388–395.

Xia, N. G., Lin, J. H., Ding, S. Q., Dong, F. R., Shen, J. Z., Du, Y. R., et al. (2019). Reliability and validity of the Chinese version of the Patient Health Questionnaire 9 (C-PHQ-9) in patients with epilepsy. *Epilepsy & Behavior*, *95*, 65–69.

Yonkers, K. A., Smith, M. V., Lin, H., Howell, H. B., Shao, L., & Rosenheck, R. A. (2009). Depression screening of perinatal women: An evaluation of the healthy start depression initiative. *Psychiatric Services*, *60*(3), 322–328.

Yuan, J., Ding, R., Wang, L., Sheng, L., Li, J., & Hu, D. (2019). Screening for depression in acute coronary syndrome patients: A comparison of patient health questionnaire-9 versus hospital anxiety and depression scale-depression. *Journal of Psychosomatic Research*, *121*, 24–28.

Zigmond, A. S., & Snaith, R. P. (1983). The hospital anxiety and depression scale. *Acta Psychiatrica Scandinavica*, *67*, 361–370.

Zuckerbrot, R. A., & Jensen, P. S. (2006). Improving recognition of adolescent depression in primary care. *Archives of Pediatrics & Adolescent Medicine*, *160*, 694–704.

Zuithoff, N. P., Vergouwe, Y., King, M., Nazareth, I., van Wezep, M. J., Moons, K. G., et al. (2010). The patient health Questionnaire-9 for detection of major depressive disorder in primary care: Consequences of current thresholds in a crosssectional study. *BMC Family Practice*, *11*, 98.

Zung, W. W. (1965). A self-rating depression scale. *Archives of General Psychiatry*, *12*, 63–70.

Chapter 19

# Screening for antenatal depression (AND) using self-report questionnaires: Conceptual issues and measurement limitations

Colin R. Martin[a] and Caroline J. Hollins Martin[b]

[a]*Institute for Clinical and Applied Health Research, University of Hull, Hull, United Kingdom;* [b]*School of Health and Social Care, Edinburgh Napier University, Edinburgh, United Kingdom*

## List of abbreviations

**AND** antenatal depression
**EPDS** Edinburgh postnatal depression scale
**HADS** Hospital Anxiety and Depression Scale
**PDD** perinatal depression disorders
**PHQ-9** Patient Health Questionnaire-9 item
**PND** postnatal depression

## Introduction

Antenatal depression (AND) represents a comparatively under-researched area of mood disorder associated with childbearing women, with postnatal depression (PND) generally receiving far more attention. This inequity in research focus between AND and PND is surprising for four reasons. First, rates of AND and PND are similar, with some evidence that AND incidence rates may be slightly higher than those for PND (Gavin et al., 2005). Second, there is convincing evidence that AND may be a significant predictor of PND (Beck, 2001). Third, the provision of research-informed etiology and interventions are limited for AND, with a general focus more upon PND (Chojenta, Lucke, Forder, & Loxton, 2016; Dennis & Dowswell, 2013a, 2013b; Lee et al., 2007; Morrell et al., 2016). Fourth, the historical focus about screening for PND, its clinical significance and presentation, has to date occluded possibility that its detection may be strongly linked with AND (Jomeen & Martin, 2008).

There is a dynamic landscape of depression rates over the antenatal and postnatal period (Jomeen & Martin, 2008; Martin & Redshaw, 2018). Hence to gain more understanding, emphasis has to be placed to contextualize the various types of depression that arise over the childbearing spectrum. Within this chapter, we will call these perinatal depression disorders (PDD), which classically includes PND and the more recently labeled AND. This embodiment has etiological importance in terms of understanding whether or not PND is a specific type of depression with a unique etiological pathway, or straightforwardly a continuation of AND, which may or may not have been diagnosed during pregnancy. A conceptual comparison of postnatal-specific PND versus the continuity model of PND being an extension of AND is shown in Fig. 1. This chapter intends to focus specifically on conceptual issues that impact upon use and accuracy of self-report measures that screen for AND.

## The etiological paradox of PDD

Recent insights into PND that develops in fathers and partners (Cox, 2005; Psouni, Agebjorn, & Linder, 2017; Roubinov, Luecken, Crnic, & Gonzales, 2014; Shaheen et al., 2019) emphasizes need for an integrated approach being applied to understanding perinatal depression that challenges the already established and accepted etiological models. For example, there exists convincing evidence to support that a distinct biological component underpins the development of PND, which encompasses an etiological model that describes physiological effects mostly at a hormonal level (Barry et al., 2015; Mah, 2016; Rogers, Hughes, Tomlinson, & Blissett, 2016). However, since the observation was made that fathers can also experience PND, which incidentally is not labeled as classic depression, but instead a specific type of partner PND. The point made here is that fathers can also develop PND, which is a viewpoint that challenges and undermines

The Neuroscience of Depression. https://doi.org/10.1016/B978-0-12-817933-8.00090-6

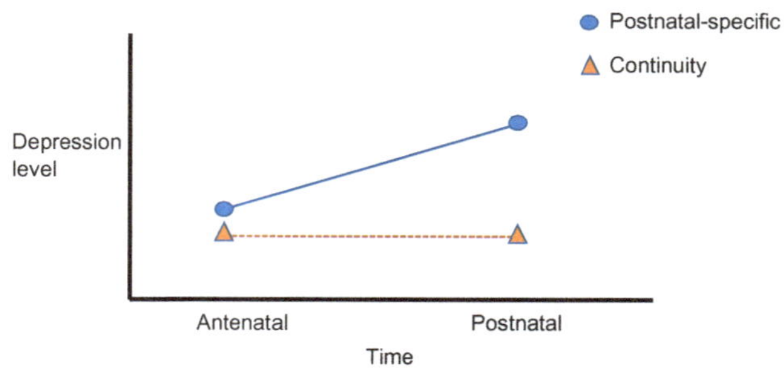


FIG. 1 "Conceptual comparison of postnatal-specific vs. continuity models of depression." The anticipated trajectory of depression symptoms over time, in particular between the antenatal and postnatal period gives insights into the debate over whether perinatal depression is representative of a postnatal depression-specific etiology (*circle*) or a continuity model of undiagnosed depression pre-existing antenatally (*triangle*).

the majority of etiological biological models that discuss causal factors of PND. The idea that fathers/partners can also experience PND is antagonistic to models that support hormonal causes and the premise that pharmacological interventions are useful for treating PND (e.g., selective serotonin reuptake inhibitors) (Ishikawa & Shiga, 2017; Milgrom et al., 2015; Molyneaux et al., 2018). Hence, conflicting etiological models about causes of PND have manifested an incomplete and ambiguous clinical picture of the cause, effect, and treatment of PDD more generally. Given that the main screening focus is upon PND, it is recommended that an equal attention and focus be paid to diagnosis and treatment of AND. This realization of the significant existence of AND is now essentially beginning to penetrate contemporary clinical guidelines (Hayes, 2010; National Collaborating Centre for Mental Health, 2018). In truth, the process of screening and intervention to manage AND PND in terms of protocol could effectively be the same in both the antenatal and postnatal period (Fig. 2). With this idea in mind, the focus will now move to explore possible screening approaches towards specific identification of AND. This exploration includes

- An examination of measurement characteristics of self-report measures that can be used to initially indicate that a childbearing woman has AND.
- A discussion about how self-report measures can be used to indicate, monitor, treat, and monitor improvement/deterioration of AND.
- A debate about possible advantages and limitations of using self-report measures to understand more about AND.

## Screening for AND: Measurement issues

To date, clinical research has focused primarily upon PDD and in particular upon the variety of self-report measures that are used to diagnose, monitor, gauge improvement/deterioration, and potential extension of AND into the postnatal period. However, at present, there is some confusion over which measuring tools are the most appropriate to use for exploring AND. This conundrum about what screening measures to use is important, simply because the majority of self-report measures have been specifically designed and validated for use post-childbirth. For example, the Edinburgh Postnatal Depression Scale (EPDS) (Cox, Holden, & Sagovsky, 1987) is the most widely used measure and is widely perceived to be the "gold standard" screening measure for initial detection of PND. Since its development, the EPDS has also been validated to detect AND (Murray & Cox, 1990), with a higher threshold for case classification (Murray & Cox, 1990). In many respects, the EPDS has a number of potential advantages compared with other measures used to diagnose the variety of

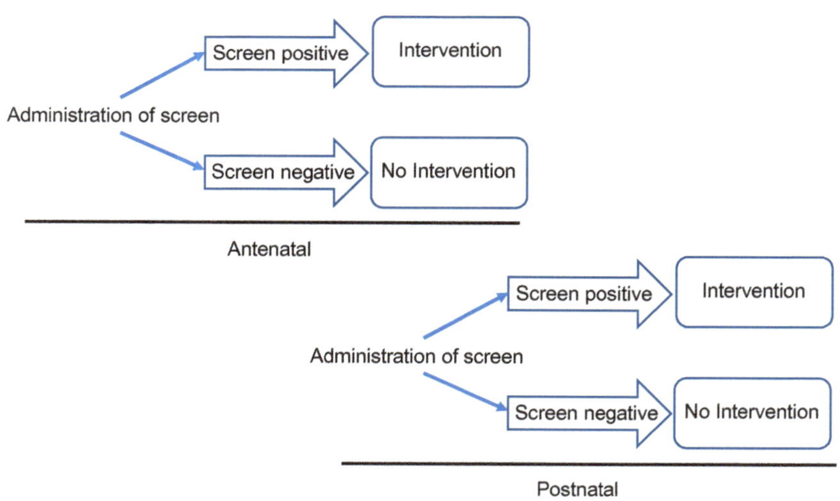


FIG. 2 "Screening pathways for perinatal depression differentiated by antenatal screen and postnatal screen." The screening process describes the outcome of a positive compared to a negative screen result at the antenatal and postnatal period. It is clear that the process is identical and only differentiated by the time of screening.

PDD. However, the use of different threshold scores between the antenatal and postnatal period is potentially problematic. First, there is no universal or accepted agreement about one single EPDS threshold screening score for indicating a diagnosis of AND or PND, with this lack of agreement about cut-off scores causing differences in detection rates between the antenatal and postnatal period. These threshold score irregularities may profoundly influence differences in logged occurrences of AND PND (Fig. 3).

## Measuring continuity of PDD across the reproductive spectrum

Using the same self-report measure in both the antenatal and postnatal period has several advantages. First, there is the advantage of providing continuity in terms of the same questions being asked at two observation points (antenatal/postnatal). This dual scoring approach offers an opportunity for total and subscale scores to be meaningfully compared between the two timepoints using the same metric. Second, this dual scoring approach offers a theoretically grounded and contextually relevant approach to reflect upon the idea that both AND PND represent different types of PDD. Using robust validated psychometric self-report measures allows clinicians and researchers to record a baseline in the antenatal period, against which a second observation point in the postnatal period can be compared. To expand the debate, a recorded significant consistency in scores between the antenatal and postnatal observation points supports the etiological argument that PND may simply be a continuation of an ongoing AND. This side of the debate conceptualizes that PND is not a type-specific form of depression that occurs post-childbirth. Third, inherent fidelity and robustness of measurement are more reliable when the same self-report measure is used both before and after the woman gives birth. The rationale for using the same self-report measure is that the robustness of the statistical psychometric characteristics creates consistency in scores across observational timepoints.

## Psychometric characteristics

The EPDS has an extensive legacy of psychometric evaluation in terms of measurement characteristics of validity and reliability (Coates, Ayers, & de Visser, 2017; Gollan et al., 2017; Reichenheim, Moraes, Oliveira, & Lobato, 2011). The factor structure of the EPDS has been explored extensively across the childbearing spectrum (Adouard, Glangeaud-Freudenthal, & Golse, 2005; Guedeney & Fermanian, 1998; Kozinszky, Toreki, Hompoth, Dudas, & Nemeth, 2017; Teissedre & Chabrol, 2004). During validation, instead of being the unidimensional scale, the EPDS was originally designed to be, it has proved itself to be multidimensional. To illustrate this, in contrast to its prima facie specified unidimensional measurement model (Fig. 4), the EPDS has been shown to consist of two discrete domains of both depression and anxiety (Matthey, 2008) (Fig. 5). To add, a recent study speculates that the EPDS is comprised of three subscales (Reichenheim et al., 2011). Irrespective of the number of domains the EPDS is reported to have, an important observation is the consistency between studies in reporting multidimensionality (Martin & Redshaw, 2018). In addition, these reported psychometric observations are argued in relation to theoretically-based and clinically-relevant conceptualizations of depression (Kozinszky et al., 2017). Further studies that have examined the underlying factor structure of the EPDS longitudinally across observation points have found the tool to be reliable across time (Kubota et al., 2018; Martin & Redshaw, 2018). These psychometric reports allow clinicians and

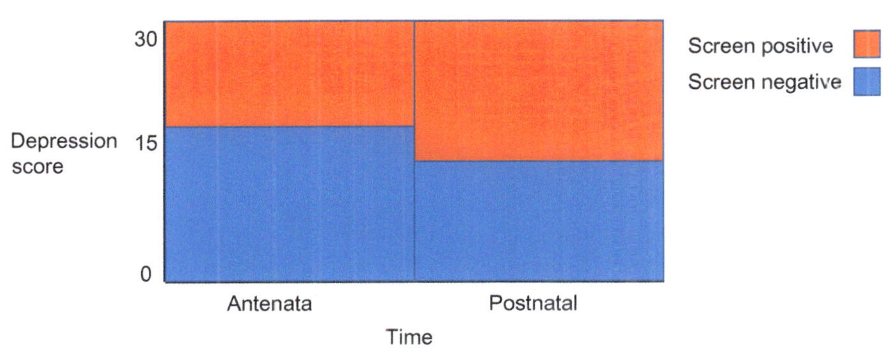


FIG. 3 "Impact of different cut-score threshold screening scores between antenatal and postnatal periods on detection rates." The use of the same tool to screen antenatally and postnatally for depression but with suggested different antenatal and postnatal threshold scores potentially produces different detection rates within the same tool. This is not essentially a problem if the threshold scores are consistent (consistently different in absolute terms) but they are not.

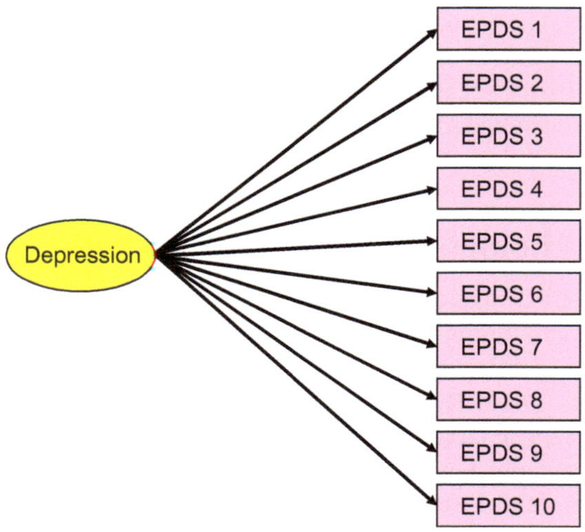


FIG. 4 "The original conceptual model of the EPDS specified as a measurement model." The original conceptual model of the EPDS specifies that all items measure a single domain of depression, thus each individual EPDS item loads on to the latent variable or domain of "depression." This model is central to the use and operationalization of the single total score of EPDS items in order to establish a "depression" score which can then be assessed against threshold cut-offs for caseness.

researchers to be confident that the EPDS is measuring the same constructs *consistently* across time, even when scores and screening threshold categorizations change across the reproductive cycle.

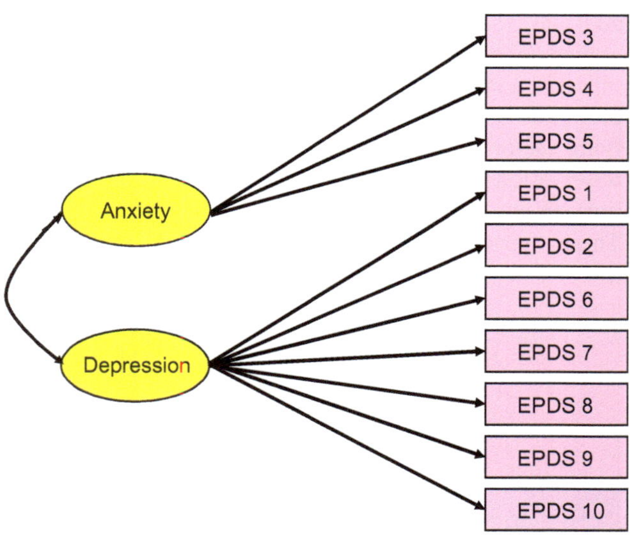


FIG. 5 "Multidimensional model of the EPDS specified as a measurement model." An example of a multidimensional model of the EPDS based on factor analysis. There are many multidimensional models of the EPDS that have been described using factor analysis. This model shows two correlated factors of depression and anxiety and the specific items which load on the relevant factor. This circumscribes the EPDS beyond its original use and conceptual alignment but may have added clinical benefit, for example the use of the three "Anxiety" items as a separate subscale for screening for anxiety disorder.

## Screening measure item content and overlaps

Turning to the actual content of the many self-report depression measures, one assumption that can be made when screening for depression symptomology is that each one has been intended to measure the same thing. Consequently, and in terms of purpose, these tools are therefore potentially interchangeable (Fried, 2017). Accepting this premise of exchangeability, a variety of robust psychometric self-report measures are available for use. Assuming the idea of interchangeability of self-report measures is correct, the depression subscale of the *Hospital Anxiety and Depression Scale* (HADS) (Zigmond & Snaith, 1983) and the *Hamilton Depression Rating Scale* (HDRS) (Hamilton, 1967) should produce extremely similar relative scores when screening for non-depressed/depressed categorizations. However, contradicting this assumption of tool exchangeability, a highly illuminative study by Fried (2017) compared seven commonly used robust self-report measures for depression (not including the EPDS or HADS) and found the degree of overlap between the measures to be modest at best (Fried, 2017). A representation of this issue of overlap in relation to two hypothetical depression measures can be viewed in Fig. 6.

Consequently, this lack of exchangeability of tools, which is highlighted in the Fried (2017) study, presents a conundrum for clinical practitioners about what tools they recommend for measuring depression. For example, and in relation to measuring AND, some clinical practitioners may prefer to use

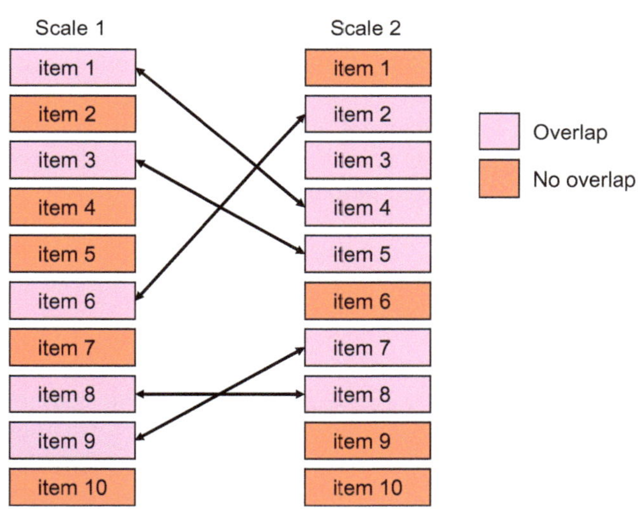


FIG. 6 "Item overlap between two hypothetical measures of depression." The issue of item overlap is shown in relation to two hypothetical 10-item depression scales. Items which are equivalent across scales are indicated with the double-headed arrows. Brief screening measures comprise relatively few items thus non-equivalent items may comprise a large proportion of items between scales. Evidence of a proportion of items being non-equivalent would indicate that scales should not be used interchangeably. This aspect of the use of measures is seldom investigated.

alternative measures for screening, such as the 9-item version of the *Patient Health Questionnaire* (PHQ) (Sidebottom, Harrison, Godecker, & Kim, 2012; Spitzer et al., 1994).

It is important to note that the Fried (2017) study was focused upon detecting general depression, as opposed to screening specifically for PDD. Also, there is no reason to believe that the same issues of lack of content overlap also apply to perinatal specific screening measures. Given the Fried (2017) findings and the importance of clinical staff place in using evidence-based perinatal mental health screening guidelines, more studies are required to evaluate the effectiveness of evaluation tools and their accuracy at indicating AND PND. It is also important to identify the optimum receiver operating characteristics of how to use specific depression measures within the perinatal context. In view of this dearth of information, an empirical content analysis that uses the same methodology as Fried (2017) is recommended as a matter of urgency.

## Screening for AND: Which measures perform best?

The prior section has highlighted a number of key issues that may effect the measurement characteristics of commonly used measures that screen for depression, both generally and over the childbearing spectrum. It is incumbent upon the reader to decide whether the issues raised are inherent limitations, which are important to consider when choosing and using depression screening measures. Given that screening is undeniably the first step towards detecting depression, valid and reliable self-report measures are a cost-effective method of doing this.

The EPDS has already been highlighted as a validated psychometric self-report measure that is fit for use in antenatal populations. Moreover, the EPDS is the instrument that has been most researched in terms of measurement properties and screening veracity during the childbearing spectrum. It should be noted, however, that the majority of these studies have focused on the use of the EPDS during the postnatal period, with significantly fewer studies conducted in the antenatal period. Consequently, compared with the postnatal period, the antenatal period holds a comparatively impoverished evidence-base in terms of using the EPDS to identify and manage AND. One issue that has emerged in relation to antenatal use of the EPDS concerns the cut-off-scores for case classification. Murray and Cox (1990) have recommended a higher threshold for EPDS case classification when identifying AND, compared with detecting PND. In addition, Bergink et al. (2011) suggest that overall *lower* cut-scores should be used. Noting that most studies that have attempted to evaluate perinatal depression screening measures, including the EPDS, are cross sectional. This lack of testing across multiple observation sites is an inherent limitation of most of these studies. Taking just one study as an example that has taken an alternative approach, Martin and Redshaw (2018) explored the performance of the EPDS across two postpartum observation points (3 and 6 months). The Martin and Redshaw (2018) study found that although scores changed over the two time periods, the underlying structure of the EPDS remained consistent. A further recommendation is made that more studies be conducted to examine the characteristics of tool performance and its validity across timepoints.

## The nine-item Patient Health Questionnaire (PHQ-9)

PHQ-9 (Spitzer et al., 1994) has been used to detect the depressive disorder in pregnancy and in many respects has a number of attributes to commend its use. The most notable attribute of the PHQ-9 is its consistent cut-off-score for detection of depression (9/10). However, it should be noted that the PHQ-9 has both simple and complex scoring approaches, and it is important to be specific about which scoring approach should be used in any screening protocol. In terms of screening acuity, the PHQ-9 has similar sensitivity and specificity performance as the EPDS. This comparable performance suggests that both these tools are suitable for screening for antenatal depression (Zhong et al., 2014). An added advantage of the EPDS is that, in addition to depression, it may also be suitable for screening for the presence of anxiety disorder, with this ability an area of screening practice is beyond the original specification of the measure. It should be noted that the caveat regarding the content overlap of items with other self-report depression measures still applies. Yet and despite similar screening performance, the PHQ-9 and the EPDS should not be assumed to be interchangeable instruments.

## The "Whooley questions"

The two "Whooley questions" (Whooley, Avins, Miranda, & Browner, 1997) are valid and reliable brief screener questions used in the UK to detect prior and present mental health issues. In relation to pregnant women, they are asked by the midwife at the woman's first antenatal "booking visit." The "Whooley questions" have been found to have adequate utility in terms of sensitivity, specificity, and suitability for an initial depression screen:

- During the past month have you often been bothered by feeling down, depressed, or hopeless?
- During the last month have you often been bothered by having little interest or pleasure in doing things?

The two "Whooley questions" are derived from the PHQ-9, with one modification of the woman providing a simple dichotomous "yes" or "no" response. In the event of a positive screen, a third follow-up question is asked:

- Is this something with which you would like to help?

Arroll, Goodyear-Smith, Kerse, Fishman, and Gunn (2005) provide further confirmation (or otherwise) of the initial positive two "Whooley questions" screen.

Clinical guidelines published by the National Collaborating Centre for Mental Health (2018) suggest the practice of an initial screen with the "Whooley questions." In the event of a positive screen, the woman is given a self-report measure to complete (e.g., EPDS or PHQ-9). If AND or PND is indicated, a post-interview diagnosis will be made by an appropriately qualified mental health professional. The first and second steps in this process are fundamentally brief, with the 10-item EPDS taking around 10 min to complete and score (National Collaborating Centre for Mental Health, 2018).

Given that substantially more is known about the screening utility of the EPDS, there is a convincing rationale for its use in undertaking a one-step screen. However, this suggestion is based upon the established psychometric characteristics of the EPDS. As such, it is important to note that within the context of clinical guidelines, the psychometric characteristics of the EPDS (e.g., factor structure) are assumed to be robust. Also, the clinical validity of the EPDS is largely based upon sensitivity and specificity analysis, which has been compared against "gold standards," such as clinical diagnosis according to the National Collaborating Centre for Mental Health (2018).

## Beyond guidelines: Other questionnaires for the detection of depression

Conciseness is key when considering the practical clinical application of using self-report depression screening measures across the childbearing spectrum. During the process of selecting appropriate instruments for inclusion in maternity care providers clinical guidelines, often arbitrary thresholds are chosen to determine whether an instrument might be suitable for use as a screener. To provide an example, according to the National Collaborating Centre for Mental Health (2018), the main criteria to consider when selecting a self-report depression measure, was that the instrument should consist of less than 12 items. This restriction in the number of questions is in some respects unfortunate, simply because some potentially very useful instruments that have credible utility in measuring depression are excluded. In terms of practicality, these disqualified self-report depression measures may take a similar time to complete. One example of an alternative quality measure for measuring depression is the *Hospital Anxiety Depression Scale* (HADS), which has had copious amounts of explorations into its measurement characteristics (Christensen et al., 2020; Martin & Thompson, 2000; Norton, Cosco, Doyle, Done, & Sacker, 2013), which include the context of pregnancy and the postnatal period ( Jomeen & Martin, 2004; Karimova & Martin, 2003; Waqas, Aedma, Tariq, Meraj, & Naveed, 2019). When considering potential utility of the 14 item HADS and its use within an antenatal context, it is useful to know that from the outset the scale was designed to comprise of two separate subscales (1) anxiety and (2) depression. Hence, accepting that the HADS has four additional questions to the EPDS (i.e., 14 vs 10, respectively), the advantage that the HADS has is that it screens for both depression and anxiety. This additional screening for anxiety is particularly useful, given that anxiety has been reported to co-exist with depression during the childbearing spectrum (National Collaborating Centre for Mental Health, 2018). In addition to depression, the merits of detecting anxiety with little extra effort in terms of cost and time should not be underemphasized. In general, the HADS is the most commonly used screening measure for detecting anxiety and depression, and it has a large evidence-base to support its use across a broad range of clinical groups. In addition, and consistent with the EPDS, the HADS includes no somatic body-related items that are likely to be influenced by physiological changes. For example, changes in physiology that occur during pregnancy can affect sleep or appetite.

## Conclusion

Self-report measures that screen for depression are useful for identifying women who may have developed depression during the antenatal or postnatal period. Use of psychometrically robust screening measures to detect depression will help facilitate timely and appropriate treatment interventions, at the same time as being economical with time and resources. Measures used in a perinatal context, which are incorporated into clinical guidelines are generally assumed to be psychometrically valid and reliable. However, this assumption is based on a very narrow definition of validity in terms of sensitivity and specificity. Broader concepts of validity, such as assessing factor structure are generally not considered important when incorporating self-report measures into guidelines and protocols. This oversight presents several challenges. First, the measurement models underlying the measures domains may have been found to be statistically different from the original dimensional specification of the instrument. As such, this infers potential for generating implicit measurement errors. In terms of sensitivity and specificity characteristics, measures which appear to perform well and are thus deemed suitable for screening for PDD, are the EPDS and the PHQ-9. The "Whooley questions" are also valid to use for a quick initial mental health screen. Post assimilating a confirmatory response to the "Whooley questions," clinical guidelines recommend follow-up with either the EPDS or PHQ-9. However, the choice of whether to use either the EPDS or PHQ-9 is arbitrary and rests with both availabilities of the measure and/or clinician preference. This quandary about what self-report measure to use, in itself, presents a clinical and measurement problem. This quandary is based upon the premise that both measures assess exactly the same phenomena (no depression/depression), which makes them essentially interchangeable. In relation to this concept of exchangeability of self-report depression measures, recent reports support that there is often little overlap in the content of commonly used screening scales. Literature advocates that further research in the area of antenatal screening for depression is both required and encouraged. One relevant example would be to conduct studies that are intended to improve clinical pathways, which advocate the most appropriate evidence-based self-report measure to use.

## Key facts

- Antenatal depression (AND) rates are similar to postnatal depression (PND) rates, yet the focus on screening has traditionally been on PND.
- Recognizing the inequity in focus between AND and PND, contemporary research has begun to focus on the occurrence of depression in the antenatal period.
- AND can be successfully and economically identified through the use of self-report screening questionnaires.
- The use of self-report questionnaires can be enhanced by an awareness of the measurement characteristics of the instruments themselves, including an appreciation of the underlying measurement models of the tools, both conceptually and in practice.
- Evidence from self-report depression screening measures that are used more generally suggest that there is limited overlap in the depression-specific items between measures. This notion suggests that measures are not interchangeable and that using different measures during the antenatal and postnatal period may be an important source of measurement error.

## Summary points

- Screening for antenatal depression (AND) is as important as screening for postnatal depression (PND).
- A two-stage process using the (1) "Whooley questions" and if affirmative (2) a valid and reliable depression inventory, is suggested for initial case detection.
- Self-report questionnaires are a valid and reliable means of screening for AND.
- Clinicians need to be more aware of the underlying measurement characteristics of screening questionnaires that initially detect for AND.
- Further research is required to examine the degree of overlap of questionnaire items between measures, to determine whether they have an acceptable commonality in assessing a consistent conceptual domain of depression.

## Mini-dictionary of terms

**Antenatal depression (AND)** Depression occurring during pregnancy.

**Factor analysis** The mathematical process of identifying the underlying measurement model of a questionnaire and associated factors.

**Factor structure** The underlying measurement model or inherent domains embedded within a questionnaire measure.

**Interchangeability** The assumption that different measures (in this context measures of depression) measure the same phenomena and are thus interchangeable.

**Measure continuity** The use of the same measure over two or more occasions, generally but not exclusively in this context related to an antenatal and postnatal observation.

**Overlap** The degree of commonality of items across self-report questionnaire measures (in this context self-report depression items).

**Psychometrics** The measurement characteristics of a measure including the main indices of validity, reliability, and factor structure.

**Screening by questionnaire** The process of accurate detection of a case of depression by a valid, reliable, accurate and self-report questionnaire.

**Sensitivity** The proportion of true positives of all cases diagnosed with a disorder (antenatal depression in this context) in the population.

**Specificity** The proportion of true negatives of all cases that are not diagnosed with a disorder (antenatal depression in this context) in the population.

## References


Adouard, F., Glangeaud-Freudenthal, N. M., & Golse, B. (2005). Validation of the Edinburgh postnatal depression scale (EPDS) in a sample of women with high-risk pregnancies in France. *Archives of Women's Mental Health*, *8*(2), 89–95. https://doi.org/10.1007/s00737-005-0077-9.

Arroll, B., Goodyear-Smith, F., Kerse, N., Fishman, T., & Gunn, J. (2005). Effect of the addition of a 'help' question to two screening questions on specificity for diagnosis of depression in general practice: Diagnostic validity study. *British Medical Journal*, *331*, 884.

Barry, T. J., Murray, L., Fearon, R. M. Moutsiana, C., Cooper, P., Goodyer, I. M., et al. (2015). Maternal postnatal depression predicts altered offspring biological stress reactivity in adulthood. *Psychoneuroendocrinology*, *52*, 251–260. https://doi.org/10.1016/j.psyneuen.2014.12.003.

Beck, C. T. (2001) Predictors of postpartum depression: An update. *Nursing Research*, *50*(5), 275–285. https://doi.org/10.1097/00006199-200109000-00004.

Bergink, V., Kooistra, L., Lambregtse-van den Berg, M. P., Wijnen, H., Bunevicius, R., van Baar, A., et al. (2011). Validation of the Edinburgh depression scale during pregnancy. *Journal of Psychosomatic Research*, *70*(4), 385–389. https://doi.org/10.1016/j.jpsychores.2010.07.008.

Chojenta, C. L., Lucke, J. C., Forder, P. M., & Loxton, D. J. (2016). Maternal health factors as risks for postnatal depression: A prospective longitudinal study. *PLoS One*, *11*(1). https://doi.org/10.1371/journal.pone.0147246.

Christensen, A. V., Dixon, J. K., Juel, K., Ekholm, O., Rasmussen, T. B., Borregaard, B., et al. (2020). Psychometric properties of the Danish Hospital Anxiety and Depression Scale in patients with cardiac disease: results from the DenHeart survey. *Health and Quality of Life Outcomes*, *18*(1). https://doi.org/10.1186/s12955-019-1264-0.

Coates, R., Ayers, S., & de Visser, R. (2017). Factor structure of the Edinburgh Postnatal Depression Scale in a population-based sample. *Psychological Assessment*, *29*(8), 1016–1027. https://doi.org/10.1037/pas0000397.

Cox, J. (2005). Postnatal depression in fathers. *Lancet*, *366*(9490), 982. https://doi.org/10.1016/S0140-6736(05)67372-2.

Cox, J. L., Holden, J. M., & Sagovsky, R. (1987). Detection of postnatal depression. development of the 10-item Edinburgh postnatal depression scale. *British Journal of Psychiatry*, *150*, 782–786. https://doi.org/10.1192/bjp.150.6.782.

Dennis, C. L., & Dowswell, T. (2013a). Interventions (other than pharmacological, psychosocial or psychological) for treating antenatal depression. *Cochrane Database of Systematic Reviews*, *7*. https://doi.org/10.1002/14651858.CD006795.pub3.

Dennis, C. L., & Dowswell, T. (2013b). Psychosocial and psychological interventions for preventing postpartum depression. *Cochrane Database of Systematic Reviews*, *2*. https://doi.org/10.1002/14651858.CD001134.pub3.

Fried, E. I. (2017). The 52 symptoms of major depression: Lack of content overlap among seven common depression scales. *Journal of Affective Disorders*, *208*, 191–197. https://doi.org/10.1016/j.jad.2016.10.019.

Gavin, N. I., Gaynes, B. N., Lohr, K. N., Meltzer-Brody, S., Gartlehner, G., & Swinson, T. (2005). Perinatal depression: A systematic review of prevalence and incidence. *Obstetrics and Gynecology*, *106*(5 Pt. 1), 1071–1083. https://doi.org/10.1097/01.AOG.0000183597.31630.db.

Gollan, J. K., Wisniewski, S. R., Luther, J. F., Eng, H. F., Dills, J. L., Sit, D., et al. (2017). Generating an efficient version of the Edinburgh postnatal depression scale in an urban obstetrical population. *Journal of Affective Disorders*, *208*, 615–620. https://doi.org/10.1016/j.jad.2016.10.013.

Guedeney, N., & Fermanian, J. (1998). Validation study of the French version of the Edinburgh Postnatal Depression Scale (EPDS): New results about use and psychometric properties. *European Psychiatry*, *13*(2), 83–89. https://doi.org/10.1016/S0924-9338(98)80023-0.

Hamilton, M. (1967). Development of a rating scale for primarydepressive illness. *The British Journal of Social and Clinical Psychology*, *6*, 278–296.

Hayes, B. A. (2010). From 'postnatal depression' to 'perinatal anxiety and depression': Key points of the National Perinatal Depression Plan for nurses and midwives in Australian primary health care settings. *Contemporary Nurse*, *35*(1), 58–67. https://doi.org/10.5172/conu.2010.35.1.058.

Ishikawa, C., & Shiga, T. (2017). The postnatal 5-HT1A receptor regulates adult anxiety and depression differently via multiple molecules. *Progress in Neuro-Psychopharmacology and Biological Psychiatry*, *78*, 66–74. https://doi.org/10.1016/j.pnpbp.2017.04.014.

Jomeen, J., & Martin, C. R. (2004). Is the hospital anxiety and depression scale (HADS) a reliable screening tool in early pregnancy? *Psychology & Health*, *19*(6), 787–800. https://doi.org/10.1080/0887044042000272895.

Jomeen, J., & Martin, C. R. (2008). Reflections on the notion of post-natal depression following examination of the scoring pattern of women on the EPDS during pregnancy and in the post-natal period. *Journal of Psychiatric and Mental Health Nursing*, *15*(8), 645–648. https://doi.org/10.1111/j.1365-2850.2008.01282.x.

Karimova, G., & Martin, C. (2003). A psychometric evaluation of the hospital anxiety and depression scale during pregnancy. *Psychology, Health & Medicine*, *8*(1), 89–103. https://doi.org/10.1080/1354850021000059296.

Kozinszky, Z., Toreki, A., Hompoth, E. A., Dudas, R. B., & Nemeth, G. (2017). A more rational, theory-driven approach to analysing the factor structure of the Edinburgh Postnatal Depression Scale. *Psychiatry Research*, *250*, 234–243. https://doi.org/10.1016/j.psychres.2017.01.059.

Kubota, C., Inada, T., Nakamura, Y., Shiino, T., Ando, M., Aleksic, B., et al. (2018). Stable factor structure of the Edinburgh Postnatal Depression Scale during the whole peripartum period: Results from a Japanese prospective cohort study. *Scientific Reports*, *8*(1). https://doi.org/10.1038/s41598-018-36101-z.

Lee, A. M., Lam, S. K., Sze Mun Lau, S. M., Chong, C. S., Chui, H. W., & Fong, D. Y. (2007). Prevalence, course, and risk factors for antenatal anxiety and depression. *Obstetrics and Gynecology*, *110*(5), 1102–1112. https://doi.org/10.1097/01.AOG.0000287065.59491.70.

Mah, B. L. (2016). Oxytocin, postnatal depression, and parenting: A systematic review. *Harvard Review of Psychiatry*, *24*(1), 1–13. https://doi.org/10.1097/HRP.0000000000000093.

Martin, C. R., & Redshaw, M. (2018). Establishing a coherent and replicable measurement model of the Edinburgh postnatal depression scale. *Psychiatry Research*, *264*, 182–191. https://doi.org/10.1016/j.psychres.2018.03.062.

Martin, C. R., & Thompson, D. R. (2000). A psychometric evaluation of the Hospital Anxiety and Depression Scale in coronary care patients following acute myocardial infarction. *Psychology, Health & Medicine*, *5*(2), 193–201. https://doi.org/10.1080/713690189.

Matthey, S. (2008). Using the Edinburgh postnatal depression scale to screen for anxiety disorders. *Depression and Anxiety*, *25*(11), 926–931. https://doi.org/10.1002/da.20415.

Milgrom, J., Gemmill, A. W., Ericksen, J., Burrows, G., Buist, A., & Reece, J. (2015). Treatment of postnatal depression with cognitive behavioural therapy, sertraline and combination therapy: a randomised controlled trial. *Australian and New Zealand Journal of Psychiatry*, *49* (3), 236–245. https://doi.org/10.1177/0004867414565474.

Molyneaux, E., Telesia, L. A., Henshaw, C., Boath, E., Bradley, E., & Howard, L. M. (2018). Antidepressants for preventing postnatal depression. *Cochrane Database of Systematic Reviews*, (4). https://doi.org/10.1002/14651858.CD004363.pub3.

Morrell, C. J., Sutcliffe, P., Booth, A., Stevens, J., Scope, A., Stevenson, M., et al. (2016). A systematic review, evidence synthesis and meta-analysis of quantitative and qualitative studies evaluating the clinical effectiveness, the cost-effectiveness, safety and acceptability of interventions to prevent postnatal depression. *Health Technology Assessment*, *20*(37), 1–414. https://doi.org/10.3310/hta20370.

Murray, D., & Cox, J. L. (1990). Screening for depression during pregnancy with the Edinburgh Depression Scale (EDDS). *Journal of Reproductive and Infant Psychology*, *8*, 99–107.

National Collaborating Centre for Mental Health. (2018). Antenatal and postnatal mental health. Clinical management and service guidance. Updated edition. In *National clinical guideline number 192*. Retrieved from CHECK LEICESTER OR LONDON??.

Norton, S., Cosco, T., Doyle, F., Done, J., & Sacker, A. (2013). The hospital anxiety and depression scale: A meta confirmatory factor analysis. *Journal of Psychosomatic Research*, *74*(1), 74–81. https://doi.org/10.1016/j.jpsychores.2012.10.010.

Psouni, E., Agebjorn, J., & Linder, H. (2017). Symptoms of depression in Swedish fathers in the postnatal period and development of a screening tool. *Scandinavian Journal of Psychology*, *58*(6), 485–496. https://doi.org/10.1111/sjop.12396.

Reichenheim, M. E., Moraes, C. L., Oliveira, A. S., & Lobato, G. (2011). Revisiting the dimensional structure of the Edinburgh Postnatal Depression Scale (EPDS): Empirical evidence for a general factor. *BMC Medical Research Methodology*, *11*. https://doi.org/10.1186/1471-2288-11-93.

Rogers, S. L., Hughes, B. A., Tomlinson, J. W., & Blissett, J. (2016). Cortisol metabolism, postnatal depression and weight changes in the first 12 months postpartum. *Clinical Endocrinology*, *85*(6), 881–890. https://doi.org/10.1111/cen.13150.

Roubinov, D. S., Luecken, L. J., Crnic, K. A., & Gonzales, N. A. (2014). Postnatal depression in Mexican American fathers: Demographic, cultural, and familial predictors. *Journal of Affective Disorders*, *152–154*, 360–368. https://doi.org/10.1016/j.jad.2013.09.038

Shaheen, N. A., AlAtiq, Y., Thomas, A., Alanazi, H. A., AlZahrani, Z. E., Younis, S. A. R., et al. (2019). Paternal postnatal depression among fathers of newborn in Saudi Arabia. *American Journal of Men's Health*, *13*(1). https://doi.org/10.1177/1557988319831219.

Sidebottom, A. C., Harrison, P. A., Godecker, A., & Kim, H. (2012). Validation of the Patient Health Questionnaire (PHQ)-9 for prenatal depression screening. *Archives of Women's Mental Health*, *15*(5), 367–374. https://doi.org/10.1007/s00737-012-0295-x.

Spitzer, R. L., Williams, J. B., Kroenke, K., Linzer, M., deGruy, F. V., 3rd, Hahn, S. R., et al. (1994). Utility of a new procedure for diagnosing mental disorders in primary care. The PRIME-MD 1000 study. *JAMA*, *272*(22), 1749–1756. Retrieved from https://www.ncbi.nlm.nih.gov/pubmed/7966923.

Teissedre, F., & Chabrol, H. (2004). A study of the edinburgh postnatal depression scale (EPDS) on 859 mothers: Detection of mothers at risk for postpartum depression. *Encephale 30*(4), 376–381. https://doi.org/10.1016/s0013-7006(04)95451-6.

Waqas, A., Aedma, K. K., Tariq, M., Meraj, H., & Naveed, S. (2019). Validity and reliability of the Urdu version of the hospital anxiety & depression scale for assessing antenatal anxiety and depression in Pakistan. *Asian Journal of Psychiatry*, *45*, 20–25. https://doi.org/10.1016/j.ajp.2019.08.008.

Whooley, M. A., Avins, A. L., Miranda, J., & Browner, W. S. (1997). Case-finding instruments for depression. Two questions are as good as many. *Journal of General Internal Medicine*, *12*(7), 439–445. https://doi.org/10.1046/j.1525-1497.1997.00076.x.

Zhong, Q., Gelaye, B., Rondon, M., Sanchez, S. E., Garcia, P. J., Sanchez, E., et al. (2014). Comparative performance of patient health questionnaire-9 and Edinburgh postnatal depression scale for screening antepartum depression. *Journal of Affective Disorders*, *162*, 1–7. https://doi.org/10.1016/j.jad.2014.03.028.

Zigmond, A. S., & Snaith, R. P. (1983). The hospital anxiety and depression scale. *Acta Psychiatrica Scandinavica*, *67*(6), 361–370. https://doi.org/10.1111/j.1600-0447.1983.tb09716.x.

Chapter 20

# Edinburgh postnatal depression scale: Description and applications

Jacqueline K. Gollan[a], Gabrielle A. Mesches[a], and Isabel A. Gortner[b]

[a]*Department of Psychiatry and Behavioral Sciences, Department of Obstetrics and Gynecology, Feinberg School of Medicine, Northwestern University, Chicago, IL, United States;* [b]*Latin School of Chicago, Chicago, IL, United States*

## Abbreviations

**EPDS** Edinburgh postnatal depression scale
**PPD** postpartum depression

## Introduction

Depression with peripartum onset (PPD) is defined as a depressive episode that occurs during pregnancy or within 4 weeks (5th ed., DSM-5, American Psychiatric Association, 2013) or 6 weeks of delivery (ICD-10, World Health Organization, 1990), though research and clinical reports suggest that the phase of onset should be extended to 1-year postdelivery (Wisner et al., 2013). A depressive episode is identified when a person reports a pervasive depressed mood and/or the lack of interest or pleasure, along with functional impairment and/or distress in at least four of the following areas of function: appetite or weight, sleep quality or quantity, motor behavior, energy level, cognition (attention and decision-making), guilt and self-regard, and will to live. These symptoms should be present most days for a minimum of 2 consecutive weeks (American Psychiatric Association, 2013).

Given that PPD has a period prevalence of 21.9% (Wisner et al., 2013), the use of an efficient screening tool to identify new cases of PPD could reduce maternal morbidity and mortality. Women who report PPD symptoms experience a higher likelihood of recurrent depressive episodes as well as increased risk of onset during the postpartum phase with subsequent pregnancies (Josefsson & Sydsjo, 2007; Nylen et al., 2010; Rasmussen, Strom, Wohlfahrt, Videbech, & Melbye, 2017). Also, PPD exerts adverse effects on maternal physical and emotional health (Pereira et al., 2014), maternal responsiveness (Jarde et al., 2016), long-term infant bonding (Moehler, Brunner, Wiebel, Reck, & Resch, 2006), and infant development (McLearn, Minkovitz, Strobino, Marks, & Hou, 2006; Taveras et al., 2003). With the high probability of recurrence of PPD (Cohen et al., 2006) and its adverse impact on maternal and infant health and function (Kendig et al., 2017; Mitchell & Goodman, 2018; Stein et al., 2014), reducing PPD is a key public health objective (American College of Obstetricians and Gynecologists' Committee on Obstetric Practice, 2005).

Using an evidence-based, standardized screening approach, the Edinburgh postnatal depression scale (EPDS; Cox, Holden, & Sagovsky, 1987; Evins Theofrastous, & Galvin, 2000) can improve health outcomes for mothers and their families. The purpose of this chapter is to provide clinicians and researchers with a resource on the EPDS. This chapter is organized into two sections. The first section includes an overview of the EPDS format, purpose, administration, scoring, interpretation, psychometric properties, and cultural considerations. The second section includes an overview of various ways in which the EPDS can be applied toward identifying PPD symptoms early in the perinatal phase, evaluating the efficacy of treatments and management of PPD, furthering neuroscience research, and monitoring relapse.

## Edinburgh postnatal depression scale

The EPDS is the most commonly used screening tool worldwide (Cox et al., 1987). This measure is designed to provide information to clinicians who work in the fields of primary care, pediatrics, and obstetrics and gynecology to identify mothers who are experiencing postpartum depression symptoms. The EPDS was developed in the United Kingdom to screen for symptoms of depression among women seeking perinatal medical services (Cox et al., 1987). The EPDS can be obtained from the authors and from the web.

The EPDS is a self-report instrument with 10 statements pertaining to mood and anxiety. The purpose of this

The Neuroscience of Depression. https://doi.org/10.1016/B978-0-12-817933-8.00002-5

measure is to assess the presence and severity of depression symptoms among women who are within the 6- to 8-week postpartum phase, though it has been validated for use in late postpartum (Cox, Chapman, Murray, & Jones, 1996). Positive responses to the questions potentially reflect symptoms or other problems and require further clinical inquiry. The EPDS can be used along with a clinical interview or by itself to quickly screen PPD symptoms. It is not able to detect other mental health conditions, specifically bipolar disorder, panic or generalized anxiety disorders, and phobias or trauma symptoms.

The 10 items on the EPDS are reflected as statements on the person's enjoyment level (items #1 and #2), self-blame (item #3), anxiety and panic (items #4 and #5), coping (item #6), unhappiness (items #7, #8, #9), and suicide (item #10). The person is asked to underline the response with which they most identify with using their data from the past 7 days. The items for each response are based on a Likert scale, and thus, each statement can be scored 0, 1, 2, and 3 to reflect the increasing severity/intensity of the symptom. Items 1, 2, and 4 are ranked from "0" (i.e., "*as much as I always could*") to "3" ("*not at all*"). The remaining seven items (items 1, 5, 6, 7, 8, 9, and 10) are reverse scored from "3" ("*Yes, most of the time*") to "0" (i.e., "*No, never*"). The scores of each item are summed to generate a total score.

Numerous studies suggest that the EPDS is an effective tool in identifying depression beyond clinical judgment (e.g., Cox et al., 1996; Gollan et al., 2017). The scale developers suggest that a cutoff score of $\geq$12 reflects depression (Cox et al., 1987), with a sensitivity of 86% and specificity of 78% (Evins et al., 2000). Additional studies suggest that a total score of 12 or 13 may be an alternative, more sensitive threshold with a sensitivity of 100% and specificity of 95% (Boyce, Stubbs, & Todd, 1993; Gaynes et al., 2005). Finally, other studies suggest that a cutoff score of 10 holds the best sensitivity and specificity for women from low socioeconomic urban context (Morris-Ruch, Freda, & Bernstein, 2003).

The measurement structure of the EPDS is viewed as a unitary test of depression, though studies suggest that a two- or three-dimensional factor structure is more representative. For instance, three of the EPDS items fail to contribute to the depression factor (Astbury, Brown, Lumley, & Small, 1994; Des Rivieres-Pigeon et al., 2000; Gollan et al., 2017; Reichenheim, Moraes, Oliveira, & Lobato, 2011). In comparison, when all items are considered, a two-factor structure can be identified among samples of women during pregnancy (Adouard, Glangeaud-Freudenthal, & Golse, 2005) as well as adolescent (Logsdon, Usui, & Nering, 2009) and adult women in postpartum (Astbury et al., 1994, Matthey, Henshaw, Elliott, & Barnett, 2006, Pallant, Miler, & Tennant, 2006). Finally, a three-factor structure is evident among samples of women during postpartum (Brouwers, van Baar, & Pop, 2001; Chabrol & Teissedre, 2004; Jomeen & Martin, 2007; Ross, Gilbert Evans, Sellers, & Romach, 2003). Despite the variation in results, the psychometric properties of the EPDS are well established and the EPDS can be recommended for clinical and research applications. As the EPDS can be distilled into a seven-item, one-factor scale of depression (Gollan et al., 2017), the seven-item version can serve as an efficient measure of depression in clinical and research activities. If used in its current format, the EPDS can offer information about the depressive and anxious features of PPD.

The EPDS can obtain clear data from geographically diverse, non-English speaking patient populations (Banerjee, Banerjee, Kriplani, Saxena, & Banerjee, 2000; Fisch, Tadmor, Dankner, & Diamant, 1997). The EPDS has been translated into 18 languages and validated on community and clinical populations globally (Wickberg & Hwang, 1996). Many of these translated copies are available on-line (Western Australia Department of Health, retrieved online November 20, 2019). A comprehensive review of the psychometrics of translated EPDS assessments suggests that the EPDS continues to hold high sensitivity and specificity for women across the globe (Vega-Dienstmaier, Mazzotti Suarez, & Campos Sanchez, 2002). However, the recommended cutoff score of the EPDS can differ due to varied diagnostic systems. For instance, highest sensitivity and specificity is achieved with a cutoff score of 10 or 11 on the Spanish EPDS (Garcia-Esteve, Ascaso, Ojuel, & Navarro, 2003), 11 or 12 on both the Swedish and Danish EPDS (Bagedahl-Strindlund & Monsen Borjesson, 1998; Smith-Nielsen, Matthey, Lange, & Vaever, 2018), and a higher cutoff score of 14 or 15 on the Vietnamese EPDS (Matthey, Barnett, & Elliott, 1997).

Cultural issues in the assessment of symptoms using translated versions of EPDS may affect the psychometric properties and the recommended cutoff score. Women from primary health-care clinics in Turkey showed less understanding of items #3, #5, #6 on the Turkish translation of the EPDS (Aydin, Inandi, Yigit, & Hodoglugil, 2004). In comparison, Vietnamese women showed reluctance in discussing their symptoms (Matthey & Barnett, 1997), and consequently, the EPDS achieved modest positive predictive values (45.5%; Matthey, Barnett, & Minas, 1996).

The English version of the EPDS has been adapted for diverse patient populations. Fisher, Kopeman, and O'Hara (2012) assessed a partner version of the assessment (EPDS-P) with maternal reports of paternal depression. The EPDS-P was proven to be a reliable and valid measure of paternal depression. The EPDS was developed at a third grade reading level (Logsdon & Hutti, 2006) so that individuals at varying reading comprehension levels may complete the self-report. The EPDS has not yet been assessed for individuals with intellectual or learning disabilities, though concerns have been reported about using the EPDS for women with psychomotor depressive symptoms (Guedeney, Fermanian, Guelfi, & Kumar, 2000). Women presenting with these symptoms may take a substantially longer amount of time to complete the survey than most, which lead to false-negative reports of depression in the study. Despite a need for further adaptation for special-needs populations, the EPDS has been found to be a valid and reliable measure for detecting postpartum depression (McBride, Wiens, McDonald, Cox, & Chan, 2014).

## Applications

### *Neuroscience research*

The EPDS can be used in research studies to characterize clinical severity and to quantify change of clinical status. The EPDS has been used to quantify the effect of PPD on maternal emotional recognition capacity of infant emotion (Parsons et al., 2019) as well as define clinical response to repetitive transcranial magnetic stimulation (Ganho-Avila, Poleszczyk, Mohamed, & Osorio, 2019). Moreover, the EPDS can be used to characterize group assignment to differentiate differences in white matter integrity during postpartum (Silver et al., 2018). Finally, the EPDS as a research tool is its direct application to the clinical setting therein increasing the generalizability of research results to clinical settings.

## Case detection

The EPDS can identify new and repeated cases of PPD. Rates of new cases of PPD hover at 14.5% within the first 90days of childbirth (Gavin et al., 2005). Moreover, between 9% and 16% of women globally report PPD within the first year of childbirth (Schmeid et al., 2013; Underwood, Waldie, D'Souza, Peterson, & Morton, 2016; Woody, Ferrari, Siskind, Whiteford, & Harris, 2017). The EPDS can also track women with a history of depression as they face a 1.52 increased risk of depression onset during their first year of postpartum (Davidson & Robertson, 1985; Kumar & Robson, 1984; Moses-Kolko & Roth, 2004). For example, 50%–62% of women with prior PPD onset with depression relative to a much smaller percentage (e.g., 2%–5%) of women with no history of depression. This is consistent with report that 33% of women with a history of perinatal depression will have depression during or after their next pregnancy (Wisner et al., 2013). Clinically, asking patients complete the EPDS twice during pregnancy (e.g., in their second and third trimester) and at the first postpartum visit (usually 6–8 weeks after delivery) can increase case detection and inform incidence and prevalence data.

## Clinical characterization

The EPDS can enhance a clinician's capacity to detect mood disorder. For instance, with a sample of 10,000 women during postpartum who screened positive on the EPDS with a score $\geq$ 10, 68.5% had a diagnosis of current major depression and 22.6% of bipolar disorder. Moreover, data from item #10 (e.g., thoughts of self-harm) on the EPDS indicates severity of illness. Of the sample described above, 19.3% had thoughts of self-harm (Wisner et al., 2013). However, the EPDS does not measure bipolar symptoms (Wisner et al., 2013) which can impede effective treatment (Dimidjian et al., 2017; Lenze & Potts, 2017 Spinelli & Endicott, 2003; Wisner & Wheeler, 1994) Though the EPDS cannot identify anxiety, sleep, or bipolar disorders, positive responses on the EPDS support the rationale for issuing a clinical referral to a patient for follow-up care.

## Case formulation and treatment implementation

Developing an informed description about the patient's symptoms using the EPDS can support the clinician's ability to generate a clear case conceptualization. With a case formulation, the clinician can tailor the treatment program to modify the underlying neurobiological or psychologic substrates related to PPD. Also, knowledge of specific symptoms can drive treatment plans. For instance, positive responses on the EPDS item #10 should elicit a psychiatric case review, a timely appointment with a mental health provider, and consideration of psychotropic medications and/or evidence-based psychotherapy. Also, the EPDS can be used in the context of clinical services including outpatient OB/GYN services and OB/GYN-Psychiatry Collaborative Care programs, wherein treatment teams from different clinical disciplines can track the patient and coordinate care. Finally, the EPDS offers an efficient method for collecting data about the patient's clinical status over time. A longitudinal prospective follow-up of the patient's symptoms can determine treatment response/nonresponse. The EPDS can be used to track patient status after the treatment ends toward identifying early symptom relapse.

In summary, the EPDS is one of the most extensively researched self-report measures designed to screen for PPD. The simplicity of the EPDS permits patients to report on the severity of their mood and anxiety symptoms within the past week. Its quick scoring approach and intuitive interpretation of results are simple and inform clinicians toward their case formulations and treatment plans. Further, the EPDS has a wealth of psychometric data to support its use in diverse clinical and research settings. Based on this foundation, the EPDS can be used to detect PPD symptoms during the perinatal phase, assess the efficacy of treatments, further the neuroscience of depression research, and set benchmarks for clinicians to coordinate patient care.

## Key facts

- Postpartum depression (PPD) is the presentation of a depressive episode detected within the first two to three months after childbirth.
- Between 9% and 16% of women globally report PPD within the first year of childbirth.
- The EPDS is an efficient tool to identify new cases of PPD and to guide treatment strategies can decrease maternal disability and infant exposure to maternal depression.

## Summary points

- One of the most extensively researched self-report tools to screen for PPD is the Edinburgh postnatal depression scale (EPDS; Cox et al., 1987).
- The EPDS is a 10-item measure that asks women to select the statements that best describe the severity of their mood and anxiety symptoms in the past 7 days, therein generating a total score that reflects PPD severity.
- This measure is designed to provide information to clinicians who work in primary care and obstetrics, pediatrics, and obstetrics and gynecology to identify mothers who are experiencing postpartum depression symptoms.
- Clinicians may use the EPDS to quantify and track symptoms over time in both clinical and research settings.

## Mini-dictionary of terms

**Postpartum depression** Postpartum depression (PPD) describes a depressive episode occurring within two to three months after childbirth.

**EPDS** The Edinburgh postnatal depression scale includes a 10-item Likert scale to assess symptoms and severity of depression.

## References

Adouard, F., Glangeaud-Freudenthal, N. M., & Golse, B. (2005). Validation of the Edinburgh postnatal depression scale (EPDS) in a sample of women with high-risk pregnancies in France. *Archives of Women's Mental Health*, *8*(2), 89–95.

American College of Obstetricians and Gynecologists' Committee on Obstetric Practice. (2005). The American college of obstetricians and gynecologists committee opinion no. 630. Screening for perinatal depression. *Journal of Obstetric, Gynecologic & Neonatal Nursing*, *125*, 1268–1271.

American Psychiatric Association. (2013). *Diagnostic and statistical manual of mental disorders* (5th ed.). Washington, DC: American Psychiatric Association.

Astbury, J., Brown, S., Lumley, J., & Small, R. (1994). Birth events, birth experiences and social differences in postnatal depression. *Australian Journal of Public Health*, *18*(2), 176–184.

Aydin, N., Inandi, T., Yigit, A., & Hodoglugil, N. N. S. (2004). Validation of the Turkish version of the Edinburgh postnatal depression scale among women within their first postpartum year. *Social Psychiatry and Psychiatric Epidemiology*, *39*, 483–484.

Bagedahl-Strindlund, M., & Monsen Borjesson, K. (1998). Postnatal depression: A hidden illness. *Acta Psychiatrica Scandinavica*, *98*, 272–275.

Banerjee, N., Banerjee, A., Kriplani, A., Saxena, S., & Banerjee, A. (2000). Evaluation of postpartum depression using the Edinburgh postnatal depression scale in evaluation of postpartum depression in a rural community in India. *International Journal of Social Psychiatry*, *46*, 74–75.

Boyce, P., Stubbs, J., & Todd, A. (1993). The Edinburgh postnatal depression scale: Validation for an Australian sample. *Australian and New Zealand Journal of Psychiatry*, *27*(3), 472–476.

Brouwers, E. P. M., van Baar, A. L., & Pop, V. J. M. (2001). Does the Edinburgh Postnata depression scale measure anxiety? *Journal of Psychosomatic Research*, *51*, 659–663.

Chabrol, H., & Teissedre, F. (2004). Relation between Edinburgh postnatal depression scale scores at 2-3 days and 4-6 weeks postpartum. *Journal of Reproductive and Infant Psychology*, *22*(1), 33–39.

Cohen, L. S., Altshuler, L. L., Harlow, B. L., Nonacs, R., Newport, D. J., Viguera, A. C., et al. (2006). Relapse of major depression during pregnancy in women who maintain or discontinue antidepressant treatment. *JAMA*, *295*(5), 499–507.

Cox, J. L., Chapman, G., Murray, D., & Jones, P. (1996). Validation of the Edinburgh postnatal depression scale (EPDS) in non-postnatal women. *Journal of Affective Disorders*, *39*, 185–189.

Cox, J. L., Holden, J. M., & Sagovsky, R. (1987). Detection of postnatal depression. Development of the 10-item Edinburgh postnatal depression scale. *The British Journal of Psychiatry*, *150*(6), 282–286.

Davidson, J., & Robertson, E. (1985). A follow-up study of postpartum illness, 1946-1978. *Acta Psychiatrica Scandinavica*, *71*, 451–457.

Des Rivieres-Pigeon, C., Seguin, L., Brodeur, J. M., Perreault, M., Boyer, G., Colin, C., et al. (2000). The Edinburgh postnatal depression scale: The validity of its Quebec version for a population low socioeconomic status mothers. *Canadian Journal of Community Mental Health = Revue Canadienne de Sante Mentale Communautaire*, *19*(1), 201–214.

Dimidjian, S., Goodman, S. H., Sherwood, N. E., Simone, G. E., Ludman, E., Gallop, R., et al. (2017). A pragmatic randomized clinical trial of behavioral activation for depressed pregnant women. *Journal of Consulting and Clinical Psychology*, *85*(1), 26–36.

Evins, G. G., Theofrastous, J. P., & Galvin, S. L. (2000). Postpartum depression: A comparison of screening and routine clinical evaluation. *American Journal of Obstetrics and Gynecology*, *182*, 1080–1082.

Fisch, R. Z., Tadmor, O. P., Dankner, R., & Diamant, Y. Z. (1997). Postnatal depression: A prospective study of its prevalence, incidence and psychosocial determinants in an Israeli sample. *Journal of Obstetrics and Gynaecology*, *23*(6), 547–554.

Fisher, S., Kopeman, R., & O'Hara, M. W. (2012). Partner report of paternal depression using the Edinburgh postnatal depression scale-partner. *Archives of Women's Mental Health, 15*(4), 283–288.

Ganho-Avila, A., Poleszczyk, A., Mohamed, M. M. A., & Osorio, A. (2019). Efficacy of rTMS in decreasing postnatal depression symptoms: A systematic review. *Psychiatry Review, 279*, 315–322.

Garcia-Esteve, L., Ascaso, C., Ojuel, J., & Navarro, P. (2003). Validation of the Edinburgh postnatal depression scale (EPDS) in Spanish mothers. *Journal of Affective Disorders, 75*(1), 71–76.

Gavin, N. I., Gaynes, B. N., Lohr, K., Meltzer-Brody, W., Gartlehner, G., & Swinson, T. (2005). Perinatal depression: A systematic review of prevalence and incidence. *Obstetrics and Gynecology, 106*(5), 1071–1083.

Gaynes, B. N., Gavin, N., Meltzer-Brody, S., Lohr, K. N., Swinson, T., Gartlehner, G., et al. (2005). Perinatal depression: Prevalence, screening accuracy, and screening outcomes: Summary. In *AHRQ evidence report summaries*. Maryland: Agency for Healthcare Research and Quality (US).

Gollan, J. K., Wisniewski, S. R., Luther, J. F., Eng, H. F., Dills, J. L., Sit, D., et al. (2017). Generating an efficient version of the Edinburgh postnatal depression scale in an urban obstetrical population. *Journal of Affective Disorders, 208*, 615–620.

Guedeney, N., Fermanian, J., Guelfi, J. D., & Kumar, R. C. (2000). The Edinburgh postnatal depression scale (EPDS) and the detection of major depressive disorder in early postpartum: Some concerns about false negatives. *Journal of Affective Disorders, 61*(1–2), 107–112.

Jarde, A., Morais, M., Kingston, D., Giallo, R., MacQueen, G. M., Giglia, L., et al. (2016). Neonatal outcomes in women with untreated antenatal depression compared with women without depression: A systematic review and meta-analysis. *JAMA Psychiatry, 73*(8), 826–837.

Jomeen, J., & Martin, C. R. (2007). Replicability and stability of the multidimensional model of the Edinburgh postnatal depression scale in late pregnancy. *Journal of Psychiatric and Mental Health Nursing, 14*, 319–324.

Josefsson, A., & Sydsjo, G. (2007). A follow-up study of postpartum depressed women: Recurrent maternal depressive symptoms and child behaviour after four years. *Archives of Women's Mental Health, 10*(4), 141–145.

Kendig, S., Keats, J. P., Hoffman, M. C., Kay, L. B., Miller, E. S., Simas, T. A., et al. (2017). Consensus bundle on maternal mental health: Perinatal depression and anxiety. *Journal of Obstetric, Gynecologic & Neonatal Nursing, 46*(2), 272–281.

Kumar, R., & Robson, K. M. (1984). A prospective national study of emotional disorders in childbearing women. *British Journal of Psychiatry, 144*, 35–47.

Lenze, S. N., & Potts, M. A. (2017). Brief interpersonal psychotherapy for depression during pregnancy in a low-income population: A randomized controlled trial. *Journal of Affective Disorders, 210*, 151–157.

Logsdon, M. C., & Hutti, M. H. (2006). Readability: An important issue impacting healthcare for women with postpartum depression. *The American Journal of Maternal/Child Nursing, 31*(6), 350–355.

Logsdon, M. C., Usui, W. M., & Nering, M. (2009). Validation of Edinburgh postnatal depression scale for adolescent mothers. *Archives of Women's Mental Health, 12*(6), 433–440.

Matthey, S., & Barnett, B. (1997). Translation and validation of the Edinburgh postnatal depression scale into Vietnamese and Arabic. In B. Ferguson, & D. Barnes (Eds.), *Perspectives on transcultural mental health* (pp. 77–82). Sydney: Transcultural Mental Health Centre.

Matthey, S., Barnett, B. E. W., & Elliott, A. (1997). Vietnamese and Arabic women's responses to the diagnostic interview schedule (depression) and self-report questionnaires: Cause for concern. *Australian and New Zealand Journal of Psychiatry, 31*, 360–369.

Matthey, S., Barnett, B., & Minas, I. H. (1996). Postnatal depression and social support in Vietnamese and Arabic women in South West Sydney. In I. H. Minas (Ed.), *Recent developments in mental health* (pp. 164–170). Melbourne: Centre for Cultural Studies in Health, University of Melbourne.

Matthey, S., Henshaw, C., Elliott, S., & Barnett, B. (2006). Variability in use of cut-off scores and formats on the Edinburgh postnatal depression scale: Implications for clinical and research practice. *Archives of Women's Mental Health, 9*(6), 309–315.

McBride, H. L., Wiens, R. M., McDonald, M. J., Cox, D. W., & Chan, E. K. H. (2014). The Edinburgh postnatal depression scale (EPDS): A review of the reported validity evidence. In B. Zumbo, & E. Chan (Eds.), *Validity and validation in social, behavioral, and health sciences* (pp. 157–174). Switzerland: Springer, Cham.

McLearn, K. T., Minkovitz, C. S., Strobino, D. M., Marks, E., & Hou, W. (2006). Maternal depressive symptoms at 2 to 4 months post partum and early parenting practices. *Archives of Pediatrics & Adolescent Medicine, 160*(3), 279–284.

Mitchell, J., & Goodman, J. (2018). Comparative effects of antidepressant medications and untreated major depression on pregnancy outcomes: A systematic review. *Archives of Women's Mental Health, 21*(5), 505–516.

Moehler, E., Brunner, R., Wiebel, A., Reck, C., & Resch, F. (2006). Maternal depressive symptoms in the postnatal period are associated with long-term impairment of mother-child bonding. *Archives of Women's Mental Health, 9*(5), 273–278.

Morris-Ruch, J. K., Freda, M. C., & Bernstein, P. S. (2003). Screening for postpartum depression in an inner city population. *American Journal of Obstetrics and Gynecology, 188*(5), 1217–1219.

Moses-Kolko, E. L., & Roth, E. K. (2004). Antepartum and postpartum depression: Healthy mom, healthy baby. *Journal of the American Medical Women's Association, 59*(3), 181–191.

Nylen, K. J., O'Hara, M. W., Brock, R., Moel, J., Gorman, L., & Stuart, S. (2010). Predictors of the longitudinal course of postpartum depression following interpersonal psychotherapy. *Journal of Consulting and Clinical Psychology, 78*(5), 757–763.

Pallant, J. F., Miller R. L., & Tennant, A. (2006). Evaluation of the Edinburgh post Natal depression scale using rasch analysis. *BMC Psychiatry, 6*(1), 28.

Parsons, C. E., Nummenmaa, L., Sinerva, E., Korja, R., Kajanctja, J., Young, K. S. et al. (2019). Investigating the effects of perinatal status and gender on adults' responses to infant and adult facial emotion. *Emotion*, 2019. online first publication November 21.

Pereira, A. T., Marques, M., Soares, M. J., Maia, B. R., Bos, S., Valente, J. et al. (2014). Profile of depressive symptoms in women in the perinatal and outside the perinatal period: Similar or not? *Journal of Affective Disorders, 166*, 71–78.

Rasmussen, M. H., Strom, M., Wohlfahrt, J., Videbech, P., & Melbye M. (2017). Risk, treatment duration, and recurrence risk of postpartum affective disorder in women with no prior psychiatric history: A population-based cohort study. *PLoS Medicine, 14*(9), e1002392.

Reichenheim, M. E., Moraes, C. L., Oliveira, A. S., & Lobato, G. (2011). Revisiting the dimensional structure of the Edinburgh postnatal depression scale (EPDS): Empirical evidence for a general factor. *BMC Medical Research Methodology*, *11*, 93.

Ross, L. E., Gilbert Evans, S. E., Sellers, E. M., & Romach, M. K. (2003). Measurement issues in postpartum depression part 1: Anxiety as a feature of postpartum depression. *Archives of Women's Mental Health*, *6*, 51–57.

Schmeid, V., Johnson, M., Naidoo, N., Austin, M. P., Matthey, S., Kemp, L., et al. (2013). Maternal mental health in Australia and New Zealand: A review of longitudinal studies. *Women and Birth*, *26*(3), 167–178.

Silver, M., Moore, C. M., Villamarin, V., Jaitly, N., Hall, J. E., Rothschild, A. J., et al. (2018). *Neuropsychopharmacology*, *43*(7), 1573–1580.

Smith-Nielsen, J., Matthey, S., Lange, T., & Vaever, M. S. (2018). Validation of the Edinburgh postnatal depression scale against both the DSM-5 and ICD-10 diagnostic criteria for depression. *BMC Psychiatry*, *18*(1), 393.

Spinelli, M. G., & Endicott, J. (2003). Controlled clinical trial of interpersonal psychotherapy versus parenting education program for depressed pregnant women. *American Journal of Psychiatry*, *160* (3), 555–562.

Stein, A., Pearson, R. M., Goodman, S. H., Rapa, E., Rahman, A., McCallum, M., et al. (2014). Effects of perinatal mental health disorders on the fetus and child. *Lancet*, *384*(9956), 1800–1819.

Taveras, E. M., Capra, A. M., Braveman, P. A., Jensvold, N. G., Escobar, G. J., & Lieu, T. A. (2003). Clinician support and psychosocial risk factors associated with breastfeeding discontinuation. *Pediatrics*, *112*(1), 108–115.

Underwood, L., Waldie, K., D'Souza, S., Peterson, E. R., & Morton, S. (2016). A review of longitudinal studies on antenatal and postnatal depression. *Archives of Women's Mental Health*, *19*(5), 711–720.

Vega-Dienstmaier, J. M., Mazzotti Suarez, G., & Campos Sanchez, M. (2002). Validation of a Spanish version of the Edinburgh postnatal depression scale. *Actas Españolas de Psiquiatría*, *30*(2), 106–111.

Wickberg, B., & Hwang, C. P. (1996). The Edinburgh postnatal depression scale: Validation on a Swedish community sample. *Acta Psychiatrica Scandinavica*, *94*, 181–184.

Wisner, K. L., Sit, D. K., McShea, M. C., Rizzo, D. M., Zoretich, R. A., Hughes, C. L., et al. (2013). Onset timing, thoughts of self-harm, and diagnoses in postpartum women with screen-positive depression findings. *JAMA Psychiatry*, *70*(5), 490–498.

Wisner, K. L., & Wheeler, S. B. (1994). Prevention of recurrent postpartum major depression. *Psychiatric Services*, *45*(12), 1191–1196.

Woody, C. A., Ferrari, A. J., Siskind, D. J., Whiteford, H. A., & Harris, M. G. (2017). A systematic review and meta-regression of the prevalence and incidence of perinatal depression. *Journal of Affective Disorders*, *219*, 86–92.

World Health Organization. (1990). International statistical classification and related health problems, tenth ed. In *Geneva*.

Chapter 21

# The death depression scale: Description and applications

David Lester[a] and Mahboubeh Dadfar[b]

[a]Psychology Program, Stockton University, Galloway, NJ, United States, [b]Department of Master for Public Health, School of Behavioral Sciences and Mental Health-Tehran Institute of Psychiatry, International Campus, School of Public Health, Student Committee of Education and Development Center (EDC), Iran University of Medical Sciences, Tehran, Iran

## List of abbreviations

| | |
|---|---|
| **CLFDS** | Collett-Lester fear of death scale |
| **CMDS** | Chinese death metaphor scale |
| **DAS** | death anxiety scale |
| **DCS** | death concern scale |
| **DDS** | death depression scale |
| **DDS-R** | death depression scale-revised |
| **DOS** | death obsession scale |
| **RFDS** | reason for fearing death scale |
| **WDS** | wish to be dead scale |

The prospect of death results in many emotions, attitudes, and thoughts, among which psychologists have focused on three components: death anxiety, death depression, and death obsession (Limonero, 1996), and these three components were labeled as death distress by Abdel-Khalek (2012). The concept of death depression was introduced by Templer, Lavoie, Chalgujian, and Thomas-Dobson (1990) because depression appeared to be present in many people as they anticipated death and were in the process of dying. Templer et al. defined death depression as feelings of despair, loneliness, dread, and sadness that may be aroused by contemplation of one's own death or experiencing the death of others.

Depression is the fourth stage in the dying process proposed by Kübler-Ross (1969): denial, anger, bargaining, depression, and acceptance. In the fourth stage, people become depressed over their impending death and may become silent, refuse visitors, and spend much of the time sullen and in mourning (Afonso & Minayo, 2013; Kübler-Ross & Kessler, 2014). Related to this, Erikson (1959) proposed that depression and despair are common in the last stage of life when people, who lack a coherent sense of self, view themselves as failures. Depression and depressive symptoms in general are associated with, or exacerbated by, existential despair and a lack of meaning in life (Havens & Chaemi, 2005). Depressed individuals respond to reminders of death with a defense of their cultural worldview (Simon, Greenberg, Harmon-Jones, Solomon, & Pyzzczynsid, 1996), and depression may be associated with less buffering against death concerns. The concept of death depression is also central to some systems of psychotherapy. For example, Nassar (2010) noted its relevance for logotherapy, an existential psychotherapy based on Victor Frankl's ideas (Frankl, 2006).

In one of the first efforts to assess reactions to death, Templer (1970) developed a death anxiety scale (DAS). However, Lester (1990) judged the content of Templer's scale to be very heterogeneous, and so he developed the Collett-Lester Fear of Death Scale (CLFDS), which has four subscales that differentiate between this fear for one's own death vs the death of others and also between the dying process and the state of being dead. Lester and Blustein (1980) developed a separate scale to measure attitudes toward funerals. Later, other scales were developed to measure death depression (DDS; Templer et al., 1990), death concern (DCS; Dickstein, 1972), death obsession (DOS; Abdel-Khalek, 1998b), and even a desire for death (WDS; Lester, 2013).[a] This chapter focuses on the description and application of Templer's death depression scale (DDS).

## The death depression scale (DDS)

The DDS devised by Templer et al. (1990) is a 17-item self-report questionnaire and consists of six elements: death despair (items 8, 11, and 16), death loneliness (items 4, 9, 10, and 13), death dread/fear (items 14, 15, and 16), death sadness (items 2 and 3), death depression (items 2 and 12), and death finality/end (items 6 and 7). The DDS has two

a. There are others relevant scales that, for example, assess the reasons for fearing death (RFDS; Abdel-Khalek, 2002b) and metaphors for death (DMD; Yin, Fang, Zhou, Shen, & He, 2017).

The Neuroscience of Depression. https://doi.org/10.1016/B978-0-12-817933-8.00034-7

different formats: a true/false or yes/no format, and a five-point Likert scale. Templer preferred the true/false format. In a small sample of undergraduates, scores using a true/false correlated strongly with scores using a Likert-type format (Pearson $r = 0.77$). Perusal of the content of these items reveals problems. Items 8 and 16, for example, seem to tap philosophical issues, while items 14 and 15 seem more relevant to the fear of death of self and others. It would be expected that item analyses and factor analyses would identify several clusters of items, only some of which are pertinent to death depression. Two items (items 11 and 12) are reversed scored in order to control for an acquiescence response set.[b]

There are two important issues for understanding the value of the DDS: (1) How heterogeneous are the items, a question typically answered by factor analysis, and (2) What are correlates of DDS scores?

## The DDS items

The DDS has 17 items (chosen from a pool of 50 items) which judges thoughts that were relevant to death depression and whose item-total correlations were greater than 0.30. Typical items are: *I get depressed when I think about death* and *I feel sad when I dream of death*, and also *Death means terrible loneliness*.

Several studies have demonstrated that the DDS has good (coefficients >0.80) internal consistency (e.g., Cronbach alpha coefficients), split-half reliability, Spearman-Brown coefficients, and test-retest reliability (e.g., Abdel-Khalek, 1997, 1998a, 2004a, 2004b; Aghazadeh, Mohammadzadeh, & Rezaie, 2014; Almostadi, 2012; Dadfar & Lester, 2017; Dadfar, Lester, Asgharnejad Farid, Atef Vahid, & Birashk, 2014; Mohammadzadeh, Rezaei, & Aghazadeh, 2016; Rajabi, Begdeli, & Naderi, 2015; Sharif Nia et al., 2017; Templer et al., 1990, 2002; Tomás-Sabádo & Gómez-Benito, 2003, 2005, 2007).

Mohammadzadeh et al. (2016) found that the reliability of the DDS was better using a Likert format than using a true/false format. However, when the DDS is scored for the subscales identified by factor analysis, the reliability coefficients are only moderate [for example, 0.34–0.62 (Aghazadeh et al., 2014)]. Both Aghazadeh et al. (2014) and Mohammadzadeh et al. (2016) found that the scores for the different subscales of the DDS correlated strongly with the DDS total score.

b. Lester and Abdel-Khalek (2003) found that one- or two reverse-scored items impaired the reliability of the CLFDS, and the revised version of that scale removed the reverse-scored item. To properly control for an acquiescence response set, half of the items should be reversed scored.

## Death depression, death anxiety, and death obsession

Templer et al. (1990) reported correlations of DDS scores with scores for death anxiety ($r = 0.67$) and weaker correlations with scores for general depression and anxiety. In a study of Spanish nurses, Tomás-Sabádo and Gómez-Benito (2003) found that DDS scores were strongly associated with death anxiety scores and weakly associated with general depression and general anxiety, replicating the Templer et al. results. Maltby and Day (2000) also reported high correlations between DDS scores and scores for death anxiety and death obsession (0.68 and 0.62, respectively), as well as with general depression (0.25) and general anxiety (0.28). Interestingly, Abdel-Khalek (1997) found that death anxiety scores, but not DDS scores, were associated with a measure of general depression.

The moderate to strong associations between DSS, DAS, and DOS scores and with scores on other death attitude scales have been reported in many studies: with American university students (Alvarado, Templer, Bresler, & Dobson-Thomas, 1993; Lester, 2003), nurses in Kuwait (Ayyad, 2013), Iranian university students (Mohammadzadeh & Najafi, 2020), Iranian hospital staff (Dadfar & Lester, 2017), and many other studies.[c] Triplett et al. (1995) suggesting using compound scores based on the scales of death depression and death anxiety, such as the DDS + DAS score and DDS − DAS score, but it is not clear how this improves the validity of the DDS scale.

It would seem important in future research, therefore, to find ways of studying death depression independently of death anxiety, death obsession, and other aspects of death thoughts and emotions. This could be done by controlling death anxiety, for example, using multiple regression or by devising a revised DDS scale that has less heterogeneity.

## Heterogeneity of DDS items

Templer et al. (1990) identified six factors which he labeled death despair, death loneliness, death dread/fear, death sadness, death depression, and death finality. The death depression factor had only two items (items 1 and 12[d]). Tomás-Sabádo and Gómez-Benito (2003) examined a Spanish version of the DDS with a sample of nurses. Their factor analysis identified four factors (labeled as death sadness, death finality, meaningless of life, and feelings

c. Abdel-Khalek (1997, 1998b, 2000, 2002a, 2004a, 2012), Almostadi (2012), Chibnall, Videen, Duckro, and Miller (2002), Dadfar and Lester (2016, 2017, 2018), Dadfar, Abdel-Khalek, and Lester (2017a, 2018), Dadfar, Lester, & Abdel-Khalek, 2020, Dadfar et al. (2014), Mohammadzadeh, Ashouri, Vahedi, and Asgharipour (2018), Tomás-Sabádo and Gómez-Benito (2005).

d. Templer misidentified the items in his report.

of loss), and items concerned with death depression and death sadness loaded on the same factor.

Using a true/false format for the DDS in a sample of Iranian university students, Aghazadeh et al. (2014) obtained four factors which they labeled: death despair, death finality, death loneliness, and death acceptance. Using a Likert format of the DDS in Iranian university students, Mohammadzadeh et al. (2016) obtained three factors labeled: death despair/finality, death loneliness, and death acceptance. In a sample of Iranian hospital staff using a true/false format, Dadfar and Lester (2017) identified four factors for the DDS labeled: death finality/end, death dread/fear, death despair/depression, and death loneliness. To complicate the situation, Abdel-Khalek (1997) found that the factor structure of the items of the DDS differed for men and for women in Egyptian college students.

Alvarado et al. (1993) factor analyzed all the items from DDS and DAS together and obtained 10 factors but reported only five factors in their paper. No one factor was heavily loaded with items from the DDS. The best factor was factor 2 which had three items from the DDS loading highly (>0.50) out of 17 items and no death anxiety items, and this study again indicates the heterogeneity of the items in the DDS. In contrast to these results, Mohammadzadeh et al. (2018) had Iranian university students complete the DDS, DAS, and DOS and factor analyzed the responses to all items. They identified three factors that matched (with a few minor exceptions) the three scales. However, these studies administered the scales separately, rather than mixing up the items from all of the scales examined. Mixing up the items is more methodologically sound since it would disguise the fact that separate scales are being administered and also would eliminate fatigue and other time-related effects.

On the whole, therefore, factor analyses indicate great heterogeneity in the items of the DDS. Future research should explore increasing the number of items that focus on death depression and eliminating items that focus on other aspects of death, such as death anxiety and death loneliness.

## Correlates of DDS scores

### *Religiosity*

In a mixed sample of undergraduates and hospital employees and spouses, Alvarado, Templer, Bresler, and Dobson-Thomas (1995) reported that low DDS scores were associated with a greater strength of religious conviction and a belief in a life after death. Harville, Stokes, Templer, and Rienzi (2004) found that more certainty about life after death was associated with lower DDS scores. Afterlife beliefs played a greater role in lowering DDS scores than a belief in God. A study among Iranian patients with cancer showed that the DDS scores were significantly associated with religious coping (Pahlevan Sharif et al., 2018). DDS scores were more strongly correlated with negative religious coping in the earlier stages of cancer than in the later stages.

In samples of Iranian students, Roshdieh, Templer, Cannon, and Canfield (1999) found that the DDS score were positively associated with weaker religious beliefs, while Mohammadzadeh and Ashouri (2017) found that engaging in the practices of religious beliefs (such as attending services) was not associated with DDS scores. However, negative religious coping mechanisms were positively associated with DDS scores, as were both anxious and avoidant attachments to God. Mohammadzadeh and Najafi (2020) classified Iranian university students as having positive or negative religious coping styles and found that those with a negative coping style had higher DDS scores.

### *Age*

Sisco, Reimer, Yanovsky, Thomas-Dobson, and Templer (1992) examined four measures: death depression, death anxiety, death distress (adding the *z*-scores for both death depression and death anxiety), and death discomfort differential (death depression *z*-score minus death anxiety *z*-score). Older subjects had lower DDS scores and less death distress, and those living with a significant other had lower DDS scores and less death distress.

In a sample of undergraduates and graduate students (with, therefore, a limited range of ages), Robak, Griffin, Lacomb, and Quint (2000) found no association between DDS scores and knowledge and attitudes about aging. Templer et al. (1990) reported that older undergraduates had lower DDS scores.

### *Sex*

Templer et al. (1990) reported a slight tendency for female undergraduates to score higher on the DDS than their male counterparts. Sisco et al. (1992) reported that women had higher DDS scores and higher death anxiety scores, greater death distress, and a lower death discomfort differential. In a study of nurses (Kuwaiti and non-Kuwaiti) working in Kuwait, Ayyad (2013) found a significant sex difference in DDS scores with men having higher scores. In contrast, in a study of Iranian university students, Roshdieh et al. (1999) found that women had higher DDS scores than men, while Bahrami, Dadfar, Lester, and Abdel-Khalek (2014) found no differences by sex in a sample of Iranian older adults. Abdel-Khalek (1997, 2000-2001) reported higher DDS scores in female Kuwaiti undergraduates than in men. On the whole, therefore, the majority of studies report that women score higher on the DDS than do men.

### *Other correlates*

Hintze, Templer, Cappelletty, and Frederick (1993) studied gay men who were HIV positive and found that those men whose illness was more severe and whose disability was greater had higher DDS and death anxiety scores. (In this sample, DDS scores and death anxiety scores were strongly associated [$r=0.91$].)

Reimer and Templer (1996) studied a sample of high school students and their parents in the United States and the Philippines. The correlation between DDS scores of mothers with the DDS scores of fathers was strong and statistically significant in both countries, as were the correlations between each parent and each child. In the Philippine sample, father-son, mother-son, father-daughter, and mother-daughter correlations were similar. In the United States, however, the mother-daughter correlation was stronger than the father-daughter correlation, and the father-son correlation was higher than the mother-son correlation, although all were statistically significant. In the United States sample, Catholics had higher DDS scores than did Protestants, and Hispanics had higher DDS scores than Euro-Americans.

In a sample of Iranian students, Roshdieh et al. (1999) found that DDS scores were positively associated with injury to friends and relatives during the 1980–88 war with Iraq, death of friends and relatives during the war, and not believing in life after death. DDS scores were higher in those who had experienced electrical outages during the war and shortages of food and water.

Dadfar et al. (2014) found that Iranian female nurses had higher DDS scores than did female hospital staff for the death despair, death loneliness, and death finality/end components of the DDS.

### Interventions

Effective psychosocial interventions for death depression have been reported in Iranian samples (Dadfar & Lester, 2016, 2017, 2020; Dadfar, Abdel-Khalek, & Lester, 2018; Dadfar et al., 2016). İnci and Öz (2009) reported that death education for nurses significantly decreased both death depression and death anxiety, while Dadfar, Asgharnejad Farid, Lester, Atef Vahid, and Birashk (2016) reported also that death education programs reduced death depression and death obsession in Iranian nurses. In samples of Iranian elderly, Abdollahzadeh and Khabbazi (2017) demonstrated that integrative reminiscence reduced death depressed mood, while Kabiri Nasab and Abdollahzadeh (2017) found that acceptance and commitment therapy (ACT) reduced death depression. Using the 8A model death education program,[e] Dadfar and Lester (2020) reported that there was a significant difference between pretest and posttest scores for DDS scores in Iranian nurses.

## The death depression scale-revised (DDS-R)

Templer et al. (2002) decided to revise the DDS in an attempt to make DDS scores less strongly associated with death anxiety scores. The DDS-R has 21 items, scored on a five-point Likert scale. The Cronbach alpha of the DDS-R was 0.92. A factor analysis identified four factors, labeled death sadness, anergia (loss of energy), existential vacuum (a sense of meaningless in life), and anhedonia. DDS-R scores were associated with DAS scores ($r=0.50$) and less strongly with scores for general depression and general anxiety.

### Reliability

As with the original DDS, the DDS-R scale has a high Cronbach alpha and test-retest reliability in a variety of subjects and countries (Abdel-Khalek, 2005; Al-Sabwah & Abdel-Khalek, 2006a; Goudarzian et al., 2019; Harville et al., 2004; Rajabi & Naderi Nobandegani, 2017; Sharif Nia et al., 2017; Tomás-Sabádo, Limonero, Templer, & Gómez-Benito, 2005).

### Heterogeneity of items

Most factor analyses of the DDS-R identify four factors, as did Templer in his original report. For example, in a sample of Spanish students, Tomás-Sabádo et al. (2005) identified four factors labeled: anergia and vacuum, death sadness, others' death, and anhedonia. Harville et al. (2004) also identified four factors, labeled death sadness, reduced activity (anergia), death hopelessness/existential vacuum, and death of someone close. Using a reduced 15-item version of the DDS-R, Rajabi et al. (2015) identified only three factors. Of particular interest, Sharif Nia, Pahlevan Sharif, Lehto, Boyle, et al. (2017) found only one factor and in a sample of Iranian patients with advanced cancer. This suggests that those close to death do not make complex judgments about their death distress.

e. The 8A model is a community-wide death education project called Empowerment Network for Adjustment to Bereavement and Loss in End-of-life (ENABLE) that was developed in Hong Kong (Chan, Tin, Chan, Chan, & Tang, 2010). The 8A model consists of: alienation, avoidance, access, acknowledgment, action, acceptance, appreciation, and actualization.

In a sample of Spanish nurses, Tomás-Sabádo and Gómez-Benito (2005) factor analyzed the items from the DDS-R and from their own DAS and identified four factors. The first factor contained only death anxiety items, the second factor only DDS-R items, while the third and fourth factors had items from both scales. In a sample of undergraduates, Tomás-Sabádo and Limonero (2007) factor analyzed responses to the items of the DDS-R and the death obsession scale and found that responses to the two scales loaded, on the whole, on two distinct factors, suggesting that the two constructs were different from each other (while a third factor tapped death anergia and anhedonia). Tomás-Sabádo and Gómez-Benito (2003) argued that death anxiety and death depression are distinct constructs but, again, it would make more sense for researchers to use, for example, only the items loading strongly on a depression factor in future research.

## Associations with death anxiety and death obsession

In a sample of Kuwaiti undergraduates, Abdel-Khalek (2004b) found that scores from the DDS-R, DAS, and DOS loaded on a single factor, and he suggested that there was a general construct of death distress. In a similar sample, Abdel-Khalek (2005) also reported that DDS-R scores were strongly associated with DAS and DOS scores. Rajabi et al. (2015) explored a Farsi version of a 15-item DDS-R scale and found that DDS-R scores correlated strongly with scores for death anxiety and death obsession (0.75 and 0.70), and less strongly with scores for general depression (0.52). Abdel-Khalek (2004c) found that DDS-R scores were strongly associated with scores on his Arabic DAS, both for men and women, as did Al-Sabwah and Abdel-Khalek (2006b).

Rajabi et al. (2015) reported significant correlations between DDS-R, DAS, and DOS scores and also with depression symptoms in Iranian nurses. In a sample of Iranian married nurses, Rajabi and Naderi Nobandegani (2017) reported that death obsession, general depression, and death anxiety predicted DDS-R scores (0.62% of the total variance of DDS-R scores).

## Correlates of DDS-R scores

Abdel-Khalek and Al-Sabwah (2005) found that Egyptian nursing students with higher DDS-R scores had lower satisfaction with life scores, but the same was true for death anxiety. Al-Sabwah and Abdel-Khalek (2006b) found that religiosity was negatively associated with DDS-R scores (and also with death anxiety scores).

In a sample of Kuwaiti undergraduates, Abdel-Khalek and Lester (2006) found no association between DDS-R scores and attitudes toward physician-assisted suicide. In Egyptian nursing students, Al-Sabwah and Abdel-Khalek (2006a) found no differences in DDS-R scores between students from the 4 years of study. In a sample of Kuwaiti female college students, Abdel-Khalek (2007) found a negative correlation between DDS-R scores and scores on the love of life scale. Abdel-Khalek (2005) found that women had significantly higher DDS-R scores than did men. DDS-R scores were negatively correlated with scores on a happiness scale for women, but not significantly for men.

In a sample of American undergraduates and graduate students, Harville et al. (2004) found that scores on the DDS-R were significantly associated with scores on all nine subscales of a life attitude scale with, for example, a negative association with *purpose in life* and a positive association with *existential vacuum*. On a religious attitude scale, certainty of belief in God and in a life after death were both associated with low DDS-R scores. DDS-R scores were higher in Asian Americans and lower in African Americans and those from the Middle East. DDS-R scores were also higher in Catholics and Buddhists and nonbelievers and lower in Muslims. DDS-R scores were lower in older, more educated subjects and those living with a significant other.

Sharif Nia, Pahlevan Sharif, Lehto, Boyle, et al. (2017) found that cancer stage, economic status, and religious coping behaviors predicted death depression in Iranian patients with cancer, while Goudarzian et al. (2019) found that cancer stage, income, and an absence of history of drug use predicted DDS-R scores in Iranian patients with cancer. Postolică, Enea, Dafinoiu, Pretrov, and Azoicăi (2019), in a study of cancer patients, found that DDS-R scores were negatively associated with a measure of a sense of coherence, while death anxiety scores were not significantly associated with a sense of coherence scores. DDS-R scores were not associated with scores on a measure of supernatural beliefs (including a belief in God and in the devil). However, there was evidence of a curvilinear relationship (an inverted-U shaped curve), with DDS-R scores lower in those with the weakest and the strongest supernatural beliefs. Agnostics, with median scores for supernatural beliefs, had the highest DDS-R scores. Incidentally, younger cancer patients (<50 years of age) had higher DDS-R scores than older cancer patients.

## Comment

Even though the DDS-R does seem to be a marginally improved measure of death depression psychometrically, the DDS-R does not improve the distinctiveness of death depression from death anxiety and death obsession. It is puzzling why researchers do not use the seven items comprising the death sadness factor of the DDS-R as a separate scale for measuring death depression.

## Discussion

The concept of depression aroused by the prospect of death would seem to be a useful construct in the research on depression. Unfortunately, the DDS and DDS-R scales devised by Templer and his colleagues appear to be heterogeneous in content, and scores on the scales are strongly associated with scores on measures of death anxiety and death obsession. If a DDS cannot be devised that is more focused on depression rather than anxiety and other emotions, then more complex statistical techniques should be used in research to tease out the impact of death depression as distinct from death anxiety.

As is typical in research, much of the research has used samples that are convenient, mainly undergraduate students and nursing students. It is disappointing that more research has not been conducted on other populations, such as psychiatric patients and those who have severe illnesses. It was interesting in research that has been conducted on patients with cancer that they seemed to make less complex judgments about their death distress than those who are not seriously ill (Sharif Nia, Pahlevan Sharif, Lehto, Boyle, et al., 2017).

The concept of death depression suggests that research into depression might consider other sources of depression other than the dying process and death. Death is an inevitable event *in the future*, but depression can be caused by one's past experiences and current circumstances, as well as other factors. Research that distinguishes between depression caused by past events, present circumstances, and future anticipated events (as well as the type of events and circumstances) would extend our understanding of the phenomenology of depression beyond a simple determination of "Are you depressed" and how severe the depression is.

## Key facts of the death depression scale (DDS)

- There are two versions of the DDS
- Both DDS versions have acceptable reliability
- Both DDS versions have heterogeneous item content and merit more homogeneity of content
- Few correlates of death depression have been identified to date

## Mini-dictionary of terms

**Death depression** The state of being depressed by the prospect of one's own death.

## References


Abdel-Khalek, A. M. (1997). Death, anxiety, and depression. *Omega*, *35*, 219–229.

Abdel-Khalek, A. M. (1998a). Death, anxiety, and depression in Lebanese undergraduates. *Omega*, *37*, 289–302.

Abdel-Khalek, A. M. (1998b). The structure and measurement of death obsession. *Personality & Individual Differences*, *24*, 159–165.

Abdel-Khalek, A. M. (2000-2001). Death, anxiety and depression in Kuwaiti undergraduates. *Omega*, *42*, 309–320.

Abdel-Khalek, A. M. (2002a). Death, anxiety, and depression. *Omega*, *45*, 277–287.

Abdel-Khalek, A. M. (2002b). Why do we fear death? *Death Studies*, *26*, 669–680.

Abdel-Khalek, A. M. (2004a). Death anxiety, death depression, and death obsession. *Psychological Reports*, *94*, 1212–1214.

Abdel-Khalek, A. M. (2004b). The Arabic scale of death anxiety (ASDA). *Death Studies*, *28*, 435–457.

Abdel-Khalek, A. M. (2005). Happiness and death distress. *Death Studies*, *29*, 949–958.

Abdel-Khalek, A. M. (2007). Love of life and death distress. *Omega*, *55*, 267–278.

Abdel-Khalek, A. M. (2012). The death distress construct and scale. *Omega*, *64*, 171–184.

Abdel-Khalek, A. M., & Al-Sabwah, M. N. (2005). Satisfaction with life and death distress. *Psychological Reports*, *97*, 790–792.

Abdel-Khalek, A. M., & Lester, D. (2006). Correlations of attitudes toward physician-assisted suicide, death depression, death obsession, and trait anxiety. *Psychological Reports*, *98*, 734.

Abdollahzadeh, H., & Khabbazi, M. (2017). Effectiveness of integrative reminiscence on health anxiety, death depression and perceived stress in elderly. *Health Psychology*, *6*(2), 101–114.

Afonso, S. B., & Minayo, M. C. (2013). A reappraisal of the works of Elisabeth Kubler-Ross. *Ciência & Saúde Coletiva*, *18*, 2729–2732.

Aghazadeh, S. E., Mohammadzadeh, A., & Rezaie, A. (2014). Validation of death depression scale (DDS) in university students in 2012. *Journal of Research & Behavioral Sciences*, *12*, 433–442.

Almostadi, D. A. (2012). *The relationship between death depression and death anxiety among cancer patients in Saudi Arabia*. Graduate dissertation USA: University of South Florida.

Al-Sabwah, M. N., & Abdel-Khalek, A. M. (2006a). Four-year cross-sectional comparison of death distress among nursing college students. *Omega*, *52*, 237–248.

Al-Sabwah, M. N., & Abdel-Khalek, A. M. (2006b). Religiosity and death distress in Arabic college students. *Death Studies*, *30*, 365–375.

Alvarado, K. A., Templer, D. I., Bresler, C., & Dobson-Thomas, S. (1993). Are death anxiety and death depression distinct entities?. *Omega*, *26*, 113–118.

Alvarado, K. A., Templer, D. I., Bresler, C., & Dobson-Thomas, S. (1995). The relationship of religious variables to death depression and death anxiety. *Journal of Clinical Psychology*, *51*, 202–204.

Ayyad, F. (2013). Death distress among two samples of lower and higher stress in health care professionals. *Psychological Reports*, *113*, 318–327.

Bahrami, F., Dadfar, M., Lester, D., & Abdel-Khalek, A. M. (2014). Death distress in Iranian older adult. *Advances in Environmental Biology*, *8* (12), 56–62.

Chan, W. C. H., Tin, A. F., Chan, C. H. Y., Chan, C. L. W., & Tang, A. C. W. (2010). Introducing the 8A model in death education training: Promoting planning for end-of-life care for Hong Kong Chinese. *Illness, Crisis & Loss*, *18*, 49–62.

Chibnall, J. T., Videen, S. D., Duckro, P. N., & Miller, D. K. (2002). Psychosocial-spiritual correlates of death distress in patients with life-threatening medical conditions. *Palliative Medicine*, *16*, 331–338.

Dadfar, M., Abdel-Khalek, A. M., & Lester, D. (2017). Psychometric characteristics of the reasons for death fear scale with Iranian nurses. *International Journal of Nursing Sciences*, *4*(4), 384–388.

Dadfar, M., Abdel-Khalek, A. M., & Lester, D. (2018). Validation of the Farsi version of death obsession scale with nurses. *International Journal of Nursing Sciences*, *5*(2), 186–192.

Dadfar, M., Asgharnejad Farid, A. A., Lester, D., Atef Vahid, M. K., & Birashk, B. (2016). Effectiveness of death education program by methods of didactic, experimental, and 8A model on the reduction of death distress among nurses. *International Journal of Medical Research & Health Sciences*, *5*(7), 60–71.

Dadfar, M., & Lester, D. (2016). The reliability, validity, and factorial structure of the Collett-Lester fear of death scale in a sample of Iranian nurses. *International Journal of Medical Research & Health Sciences*, *5*(7S), 306–317.

Dadfar, M., & Lester, D. (2017). Cronbach's α reliability, concurrent validity, and factorial structure of the death depression scale in an Iranian hospital staff sample. *International Journal of Nursing Sciences*, *4*, 135–141.

Dadfar, M., & Lester, D. (2018). The Farsi translation, reliability and validity of the death concern scale. *Trends in Psychiatry & Psychotherapy*, *40*, 114–125.

Dadfar, M., & Lester, D. (2020). The effectiveness of 8A model death education on the reduction of death depression. *Nursing Open*, *7*, 294–298.

Dadfar, M., Lester, D., & Abdel-Khalek, A. M. (2020). Validity and reliability of the Farsi version of the death anxiety scale with nurses. *Illness, Crisis & Loss*. In press.

Dadfar, M., Lester, D., Asgharnejad Farid, A. A., Atef Vahid, M. K., & Birashk, B. (2014). Death depression in Iranian nurses. *Advances in Environmental Biology*, *8*(13), 218–222.

Dickstein, L. S. (1972). Death concern: Measurement and correlates. *Psychological Reports*, *30*, 563–571.

Erikson, E. H. (1959). *Identity and the life cycle*. New York: International Universities Press.

Frankl, V. E. (2006). *Man's search for meaning*. Boston, MA: Beacon Press.

Goudarzian, A., Sharif Nia, H., Sheikh Raberi, A., Tahmasbi, M., Farokhmanesh, K., Soleimani, M., et al. (2019). Determinant factors of death depression in Iranian patients with cancer. *Iranian Journal of Cancer Care*, *1*(1), 8–14.

Harville, M., Stokes, S. J., Templer, D. I., & Rienzi, B. (2004). Relation of existential and religious variables to the death depression scale-revised. *Omega*, *48*, 165–184.

Havens, L. L., & Chaemi, S. N. (2005). Existential despair and bipolar disorder. *American Journal of Psychotherapy*, *59*, 137–147.

Hintze, J., Templer, D. I., Cappelletty, G. G., & Frederick, W. (1993). Death depression and death anxiety in HIV infected males. *Death Studies*, *17*, 333–341.

İnci, F., & Öz, F. (2009). Effects of death education on nurses' death anxiety, depression regarding death, and attitudes towards the dying patient. *Anatolian Journal of Psychiatry*, *10*(4), 253–260

Kabiri Nasab, Y., & Abdollahzadeh, H. (2017). Effect of Acceptance and *Commitment Therapy* (ACT) on quality of life, death depression, and inner shame of the elderly. In *Paper presented at the 6th congress of Iranian psychology association, Tehran, Iran*.

Kübler-Ross, E. (1969). *On death and dying*. New York: Routledge.

Kübler-Ross, E., & Kessler, D. (2014). *On grief & grieving*. New York: Scribner.

Lester, D. (1990). The Collett-Lester fear of death scale. *Death Studies*, *14*, 451–468.

Lester, D. (2003). Death anxiety, death depression, and death obsession. *Psychological Reports*, *93*, 695–696.

Lester, D. (2013). A scale to measure the desire to be dead. *Omega*, *67*, 323–327.

Lester, D., & Abdel-Khalek, A. (2003). The Collett-Lester fear of death scale. *Death Studies*, *27*, 81–85.

Lester, D., & Blustein, J. (1980). Attitudes toward funerals. *Psychological Reports*, *46*, 1074.

Limonero, J. T. (1996). The phenomenon of death as an example of research on emotions. *Revista de Psicología General y Aplicada*, *49*, 249–265.

Maltby, J., & Day, L. (2000). The reliability and validity of the death obsession scale among English university and adult samples. *Personality & Individual Differences*, *28*, 695–700.

Mohammadzadeh, A., & Ashouri, A. (2017). The study of some religious correlates of death depression among university students. *Iranian Journal of Psychiatry & Clinical Psychology*, *23*(1), 68–77.

Mohammadzadeh, A., Ashouri, A., Vahedi, M., & Asgharpour, N. (2018). Death distress dimensions: Death anxiety, death depression, and death obsession. *Journal of Fundamentals of Mental Health*, *20*(6), 395–404.

Mohammadzadeh, A., & Najafi, M. (2020). The comparison of death anxiety, obsession, and depression between Muslim population with positive and negative religious coping. *Journal of Religion and Health*, *59*, 1055–1064.

Mohammadzadeh, A., Rezaei, A., & Aghazadeh, S. E. (2016). Validation of Likert form death depression scale in a university student sample. *Journal of Ilam University of Medical Sciences*, *24*(1), 89–97.

Nassar, S. L. (2010). A measure of interest to logotherapy researchers. *The International Forum for Logotherapy*, *33*, 56–57.

Pahlevan Sharif, S., Lehto, R. H., Sharif Nia, H., Goudarzian, A. H., Haghdoost, A. A., Yaghoobzadeh, A., et al. (2018). Religious coping and death depression in Iranian patients with cancer: Relationships to disease stage. *Supportive Care in Cancer*, *26*, 2571–2579.

Postolică, R., Enea, V., Dafinoiu, I., Petrov, I., & Azoicăi, D. (2019). Association of sense of coherence and supernatural beliefs with death anxiety and death depression among Romanian cancer patients. *Death Studies*, *43*, 9–19.

Rajabi, G. R., Begdeli, Z., & Naderi, Z. (2015). Psychometric properties of the Persian version of the death depress scale among nurses. *Death Studies*, *39*, 342–346.

Rajabi, G. R., & Naceri Nobandegani, Z. (2017). Death obsession, death anxiety, and depression as predictors of death depression in nurses. *Jentashapir Journal of Health Research*, *8*(6), e62835.

Reimer, W., & Templer, D. I. (1996). Death anxiety, death depression, death distress, and death discomfort differential. *Omega*, *32*, 319–330.

Robak, R. W., Griffin, P. W., Lacomb, M., & Quint, W. (2000). Perceptions of aging and their relationship with death depression and sex. *Perceptual & Motor Skills*, *90*, 1179–1183.

Roshdieh, S., Templer, D. I., Cannon, W. G., & Canfield, M. (1999). The relationship of death anxiety and death depression to religion and civilian war-related experiences in Iranians. *Omega*, *39*, 201–210.

Sharif Nia, H., Pahlevan Sharif, S., Lehto, R. H., Allen, K. A., Goudarzian, A. H., Yaghoobzadeh, A., et al. (2017). Psychometric evaluation of Persian version of the death depression scale in Iranian patients with acute myocardial infraction. *Iranian Journal of Psychiatry*, *12*(3), 172–181.

Sharif Nia, H., Pahlevan Sharif, S., Lehto, R. H., Boyle, C., Yaghoobzadeh, A., Kaveh, O., et al. (2017). Development and psychometric evaluation of a Persian version of the death depression scale-revised. *Japanese Journal of Clinical Oncology*, *47*, 713–719.

Simon, L., Greenberg, J., Harmon-Jones, E., Solomon, S., & Pyzzczynsid, T. (1996). Mild depression, mortality salience and defense of the worldview evidence of intensified terror management in the mildly depressed. *Personality and Social Psychology Bulletin*, *22*, 81–90.

Sisco, K., Reimer, W., Yanovsky, A., Thomas-Dobson, S., & Templer, D. I. (1992). Death depression versus death anxiety. *Psychological Reports*, *71*, 1191–1194.

Templer, D. I. (1970). The construction and validation of a death anxiety scale. *Journal of General Psychology*, *82*, 165–177.

Templer, D. I., Harville, M., Hutton, S., Underwood, R., Tomeo, M., Russell, M., et al. (2002). Death depression scale-revised. *Omega*, *44*, 105–112.

Templer, D. I., Lavoie, M., Chalgujian, H., & Thomas-Dobson, S. (1990). The measurement of death depression. *Journal of Clinical Psychology*, *46*, 834–839.

Tomás-Sabádo, J., & Gómez-Benito, J. (2003). The Spanish for of the death depression scale. *Perceptual & Motor Skills*, *96*, 49–53.

Tomás-Sabádo, J., & Gómez-Benito, J. (2005). Death anxiety and death depression in Spanish nurses. *Psychological Reports*, *97*, 21–24.

Tomás-Sabádo, J., & Limonero, J. T. (2007). Death depression and death obsession. *Psychological Reports*, *100*, 755–758.

Tomás-Sabádo, J., Limonero, J. T., Templer, D. I., & Gómez-Benito, J. (2005). The death depression scale revised. *Omega*, *50*, 43–52.

Triplett, G., Roshdieh, S., Cohen, D., Stanczak, E. M., Reimer, W., Siscoe, K., et al. (1995). Death discomfort differential. *Omega*, *31*, 295–304.

Yin, F., Fang, J., Zhou, N., Shen, H., & He, Y. (2017). Development and validation of the Chinese death metaphors scale-revised. *Death Studies*, *41*, 87–93.

## Further reading

Sharif Nia, H., Pahlevan Sharif, S., Esmaeili, R., Goudarzian, A. H., Tahmasbi, B., Yaghoobzadeh, A., et al. (2017). Factors influencing the level of death depression in patients with cancer: A path analysis. *Journal of Mazandaran University Medical Sciences*, *26*(145), 318–331.

Chapter 22

# The Depression Anxiety Stress Scale: Features and applications

Jennifer C.P. Gillies and David J.A. Dozois
*Department of Psychology, The University of Western Ontario, London, ON, Canada*

## Introduction

The Depression Anxiety Stress Scales (DASS; Lovibond & Lovibond, 1995b) and its short form were designed to assess the constructs of depression, anxiety, and stress while maximizing discrimination among them. The DASS is now used widely in clinical and research settings with diverse populations (e.g., Nieuwenhuijsen, de Boer, Verbeek, Blonk, & van Dijk, 2003; Norton, 2007; Persons, Koerner, Eidelman, Thomas, & Liu, 2016). The widespread use of the DASS, in combination with research suggesting strong psychometric properties of both versions (e.g., Crawford & Henry, 2003; Lee, Lee, & Moon, 2019), has seemingly led researchers to extend its use to additional populations and alternative forms of test administration. For example, it has been translated into at least 50 languages, used with child and adolescent samples (e.g., Burckhardt, Manicavasagar, Batterham, Hadzi-Pavolic, & Shand, 2017; Ruble, George, Gallicchio, & Gamaldo, 2015), and administered via a computer (e.g., Lee, 2019; Sinclair et al., 2012; Wardenaar, Wanders, Jeronimus, & de Jonge, 2018).

This chapter offers a (1) description of the DASS and its development, (2) summary and an evaluation of the research on the psychometric properties of both the DASS and DASS-21, as well as the research examining the scale's use with extended populations and alternative forms of administration, (3) review of the research and clinical applications of the DASS, and (4) summary of its strengths, limitations, and recommended future directions.

## Test description

The DASS is a 42-item self-report state measure that assesses depression, anxiety, and stress (Lovibond & Lovibond, 1995b). It was developed to facilitate the advancement of how these constructs are defined, understood, and measured, in part by maximizing discrimination among them. The Depression subscale measures sadness, hopelessness, devaluation of life, worthlessness, lack of interest, anhedonia, and inertia. The Anxiety subscale measures autonomic arousal, situational anxiety, and the subjective experience of anxious affect. The Stress subscale assesses chronic and nonspecific arousal, including nervous arousal, irritability/overreactivity, impatience, difficulty relaxing, and the tendency to get upset/agitated easily. Sample items include "I felt that I had nothing to look forward to" (Depression), "I felt scared without any good reason" (Anxiety), and "I found it difficult to relax" (Stress).

The DASS-21, a half-length version of the scale, was derived from the 42-item DASS. According to the "Frequently Asked Questions" (FAQ) page on the DASS website (http://www2.psy.unsw.edu.au/dass//DASSFAQ.htm), seven items from each scale were selected, ensuring three criteria were met: (1) each item had "good" factor loadings,[a] (2) all "subscales" within each scale were represented,[b] and (3) converted DASS-21 scale scores were equivalent to full-scale DASS scores. Accumulating evidence suggests similar psychometric properties for the full-scale DASS and the DASS-21 (e.g., Page, Hooke, & Morrison, 2007).

Using a four-point Likert-type scale ranging from 0 (*Did not apply to me*) to 3 (*Applied to me very much, or most of the time*), respondents are instructed to indicate the extent to which each statement applied to them over the past week. Copies of the full-length DASS, DASS-21, and the scoring template can be downloaded at no cost from the DASS website (www.psy.unsw.edu.au/dass/). While these resources are freely available, details of the original psychometric properties, normative data, and clinical cutoff scores/severity labels can be found only in the manual, which is proprietary and must be purchased.

a. The website explanation does not define what was considered a "good" factor loading.

b. Though the authors use the term "subscales," the term is best understood as describing the groups of items deemed to measure the specific facets of the three scales (e.g., dysphoria, within the Depression scale), rather than as subscales that are scored and interpreted.

The Neuroscience of Depression. https://doi.org/10.1016/B978-0-12-817933-8.00025-6

Total scores for the Depression, Anxiety, and Stress scales are obtained by summing the scores for the relevant items. Interpretations based on general population normative information may be made by using the table of severity ratings included in the manual. The table presents five descriptive severity labels (normal, mild, moderate, severe, and extremely severe) for each of the three scales, alongside clinical cutoff scores in the form of Z scores, percentiles, and raw total scale scores.

The manual indicates that DASS-21 total scale scores may be converted (by multiplying scale scores by 2) and then compared to DASS norms. Although the researchers themselves provide no empirical evidence to support the comparability of converted DASS-21 scores with DASS scores, at least two studies have since offered supportive evidence (Davies, Caputi, Skarvelis, & Ronan, 2015; Henry & Crawford, 2005).

The FAQ page on the DASS website indicates that a total or average score of the three scale scores may also be calculated to produce a measure of negative emotional symptoms. Instructions on how to interpret overall scores using severity labels from the normative data are correspondingly included on the FAQ webpage. The few studies that have examined the psychometric properties of composite scores generally indicate excellent internal consistency (Crawford & Henry, 2003; Henry & Crawford, 2005; Page et al., 2007).

### Theoretical basis

In the manual, the authors cite empirical evidence demonstrating the considerable overlap between measurement and clinical manifestation of anxiety and depression. In spite of this overlap, Lovibond and Lovibond (1995b) maintained their assertion that depression and anxiety are conceptually distinct. They posit that the existing measures of depression and anxiety were limiting the definition and understanding of depression and anxiety, which was simultaneously hindering discrimination of the two constructs. The authors therefore aimed to construct psychometrically strong self-report scales measuring depression and anxiety that captured "the full range of [their] core symptoms" (p. 6) and that were also maximally discriminant.

Additionally, Lovibond and Lovibond (1995b) conceptualized depression and anxiety as dimensional, rather than categorical. That is, they argued that the experiences of depression, anxiety, and stress in clinical vs nonclinical populations are not qualitatively different, but instead reflect gradations of severity.

### Test development

The developers' original intention to construct a measure that accurately assessed, and maximally differentiated, depression and anxiety led to the development of the Self-Analysis Questionnaire. The Self-Analysis Questionnaire was later refined to form the 42-item DASS. Early in the development process, a third factor, which represented a dimension of stress and/or tension, emerged. This factor was ultimately labeled "Stress."

The authors' perspectives that the existing measures of depression and anxiety were limiting their discrimination and that the two constructs were dimensional influenced the development procedures of the scale. For example, in the first stage of development, scale items were generated by a group of clinical researchers rather than being selected from a pool of items taken from the existing anxiety and depression scales. Additionally, diagnostic inclusion criteria were not required at most stages of development. That is, once the original items had been evaluated using a clinical sample, nonclinical samples were also used. Validity checks were then completed, wherein the DASS was administered to participant samples drawn from a variety of specific populations (e.g., individuals experiencing insomnia, depression, and/or anxiety; Lovibond & Lovibond, 1995b).

During the final stage of DASS development, psychometric evaluation of data from a sample of nonclinical participants confirmed the three-factor structure of the scale and indicated satisfactory internal consistency and convergent and divergent validity (Lovibond & Lovibond, 1995b).

## Psychometric properties

### Reliability

Many studies employing community and/or clinical samples have reported excellent internal consistency estimates for the full-length DASS scales. Cronbach's alpha values[c] from eight of these studies are presented in Table 1 ($M_{Depression}=0.95$; $M_{Anxiety}=0.89$; $M_{Stress}=0.93$; range = 0.84 to 0.97). Similarly, a systematic review of the measurement properties of the DASS-21 determined that the internal consistency of the DASS-21 was well supported in community and/or clinical samples (Lee et al., 2019). Coefficient alphas from 10 studies employing the - DASS-21 are presented in Table 2 ($M_{Depression}=0.88$; $M_{Anxiety}=0.82$; $M_{Stress}=0.87$; range = 0.73 to 0.94).

In contrast, there is a dearth of empirical evaluation of the test–retest reliability of the DASS. Although one study (Brown, Chorpita, Korotitsch, & Barlow, 1997) found support for the 2-week test–retest reliability of the full-length DASS, several relevant and inherent challenges to measuring its test–retest reliability should be noted. For

c. Studies that assessed internal consistency using coefficient omega within the context of a bifactor model have also supported strong internal consistency (e.g., Osman et al., 2012).

**TABLE 1** **Internal consistency estimates for the full-length Depression Anxiety Stress Scale (DASS).**

| | | Coefficient alpha | | | |
|---|---|---|---|---|---|
| **Reference** | **Sample** | **D** | **A** | **S** | **Tot.** |
| Antony et al. (1998) | 258 adult outpatients with affective disorders and 49 nonclinical adults (Canada) | 0.97 | 0.92 | 0.95 | – |
| Brown et al. (1997) | 437 adult patients with mood and anxiety disorders (US) | 0.96 | 0.89 | 0.93 | – |
| Clara, Cox, and Enns (2001) | 439 adult outpatients with mood disorders (Canada) | 0.96 | 0.89 | 0.94 | – |
| Crawford, Cayley, Lovibond, Wilson, and Hartley (2011) | 497 adults (Australia) | 0.95 | 0.88 | 0.94 | – |
| Crawford and Henry (2003) | 1771 adults (UK) | 0.95 | 0.90 | 0.93 | 0.97 |
| Lovibond and Lovibond (1995b) | 717 university students (Australia) | 0.91 | 0.84 | 0.90 | – |
| Page et al. (2007; Study 1) | 124 adult inpatients and day patients with depression (Australia) | 0.96 | 0.92 | 0.95 | 0.97 |
| Page et al. (2007; Study 2) | 816 inpatients with depression, ages 14–83 (Australia) | 0.96 | 0.91 | 0.93 | 0.96 |

*Note.* Tot., total DASS score (i.e., sum of Depression, Anxiety, and Stress scale scores).

**TABLE 2** **Internal consistency estimates for the 21-item Depression Anxiety Stress Scale (DASS-21).**

| | | Coefficient alpha | | |
|---|---|---|---|---|
| **Reference** | **Sample** | **D** | **A** | **S** |
| Antony et al. (1998)[a] | 258 outpatients and 49 nonclinical adults (Canada) | 0.94 | 0.87 | 0.91 |
| Clara et al. (2001)[a] | 439 adult outpatients (Canada) | 0.92 | 0.81 | 0.88 |
| Crawford et al. (2011)[b] | 497 adults (Australia) | 0.90 | 0.79 | 0.89 |
| Gomez, Summers, Summers, Wolf, and Summers (2014)[b] | 687 adults (Australia) | 0.89 | 0.92 | 0.86 |
| Henry and Crawford (2005)[a] | 1794 adults (UK) | 0.88 | 0.82 | 0.90 |
| Lee et al. (2019)[b] | 255 university students (91 US and 164 South Korean) | 0.90 | 0.82 | 0.87 |
| Lovibond and Lovibond (1995b)[c] | 717 university students (Australia) | 0.81 | 0.73 | 0.81 |
| Norton (2007)[b] | 895 university students (US) | 0.83 | 0.78 | 0.87 |
| Osman et al. (2012)[b] | 887 university students (US) | 0.85 | 0.81 | 0.88 |
| Sinclair et al. (2012)[b] | 499 adults (US) | 0.91 | 0.80 | 0.84 |

[a]*Full-length DASS presented to participants, and DASS-21 items extracted to calculate Cronbach's alpha.*
[b]*DASS-21 version presented to participants.*
[c]*Not reported whether DASS-21 items presented as DASS-21 or extracted from full-length DASS.*

example, too much time between administrations could allow for valid changes in symptomatology, which would lead to underestimations of retest reliability. On the other hand, the reliability of the DASS may be overestimated if the scale is readministered too soon and responses are biased by respondents' recollections of their original responses. Furthermore, research (Longwell & Truax, 2005) has suggested that other symptom measures are prone to retest effects, whereby the repeated administration of the measure leads to an artificial decrease in scores that reflects measurement artifacts rather than actual changes in symptom severity.

## Validity

The construct validity of the full-length DASS and its short form has been well supported by studies utilizing both clinical (e.g., Antony, Bieling, Cox, Enns, & Swinson, 1998; Brown et al., 1997) and nonclinical samples (e.g., Crawford & Henry, 2003; Lovibond & Lovibond, 1995a; Sinclair et al., 2012). For example, correlations between each DASS scale and other measures tapping the same constructs tend to be significantly higher than the correlations between each DASS scale and measures that are not proposed to measure the same construct. Specifically, correlations between each of the DASS scales and existing measures of depression have been significantly higher for the Depression scale than for the Anxiety or Stress scales (Brown et al., 1997; Gloster et al., 2008; Lovibond & Lovibond, 1995a; Weiss, Aderka, Lee, Beard, & Björgvinsson, 2015).

Similar patterns of correlations support convergent validity for the Anxiety and Stress scales (e.g., Brown et al., 1997; Crawford & Henry, 2003; Henry & Crawford, 2005). For instance, measures of anxiety and fearfulness have correlated more strongly with the DASS-Anxiety scale than with the Depression or Stress scales.

There is also empirical evidence to support the concurrent validity of the DASS and DASS-21. Antony et al. (1998) found that participants diagnosed with depression or anxiety reported significantly higher DASS scale scores than did nonclinical participants. These results were replicated in an analysis of DASS-21 items (Antony et al., 1998). Visual inspection of mean scale scores from a variety of studies evinces a similar pattern (see Tables 3 and 4). That is, mean scale scores in nonclinical samples tend to be lower than those in clinical samples. Mean scores are also consistently higher for inpatient than for outpatient samples. Moreover, reported mean scale scores tend to correspond to the severity categories listed in the manual (Lovibond & Lovibond, 1995b, p. 26).

Study findings have also generally supported the original three-factor structure for the DASS and DASS-21 (Antony et al., 1998; Brown et al., 1997; Crawford & Henry, 2003; Lee et al., 2019; Lovibond & Lovibond, 1995b; Page et al., 2007). It should be noted, however, that the authors of a recent systematic review of the DASS-21's measurement properties (Lee et al., 2019) concluded that there was also sufficient high-quality support for a bifactor solution. That is, a number of studies have found that the best-fitting model included a common negative affectivity factor in addition to the three constructs of depression, anxiety, and stress.

## Validity of extending the DASS to additional populations and administration formats

The validity of the DASS and DASS-21 is strongly supported for assessing the constructs of depression, anxiety, and stress in the English-speaking adult populations from Western

**TABLE 3** Mean scores and standard deviations for each Depression, Anxiety, Stress Scales (DASS) scale and total scores reported by studies utilizing the full-length DASS.

| Reference | Sample | Depression *M (SD)* | Anxiety *M (SD)* | Stress *M (SD)* | Total *M (SD)* |
|---|---|---|---|---|---|
| *Clinical* | | | | | |
| Brown et al. (1997)[a] | Outpatients | 13.37 (9.74) | 12.39 (8.00) | 18.31 (9.16) | – |
| Davies et al. (2015) | Outpatients | 20.49 (13.33) | 16.48 (11.57) | 22.67 (12.21) | 54.74 (34.43) |
| Page et al. (2007); Study 1 | Inpatients and day patients | 24.15 (11.97) | 17.85 (10.00) | 23.07 (11.15) | – |
| Page et al. (2007); Study 2 | Inpatients | 30.2 (10.9) | 19.9 (10.1) | 27.2 (10.0) | – |
| *Nonclinical* | | | | | |
| Antony et al. (1998) | Nonclinical adults | 2.18 (2.83) | 1.43 (1.86) | 4.12 (3.81) | – |
| Lovibond and Lovibond (1995b) | Normative sample | 6.34 (6.97) | 4.70 (4.91) | 10.11 (7.91) | – |
| Henry and Crawford (2005) | Nonclinical adults | 5.55 (7.48) | 3.56 (5.39) | 9.27 (8.04) | 18.38 (18.82) |

[a]*Calculated by weighted averaging of mean scores and standard deviations grouped under individual clinical diagnoses.*

**TABLE 4** Mean scores and standard deviations for each Depression, Anxiety Stress Scales-21 (DASS-21) scale and total scores reported by studies utilizing the short version of the DASS.

| Reference | Sample | Depression *M* (*SD*) | Anxiety *M* (*SD*) | Stress *M* (*SD*) | Total *M* (*SD*) |
|---|---|---|---|---|---|
| *Clinical* | | | | | |
| Davies et al. (2015) | Outpatients | 21.02 (13.51) | 17.06 (11.64) | 23.17 (12.35) | 61.78 (34.25) |
| Weiss et al. (2015) | Outpatients | 18.64 (13.14) | 9.58 (7.84) | 14.53 (9.67) | – |
| Ronk et al. (2013) | Outpatients | 13.32 (11.10) | 9.09 (8.82) | 15.01 (10.00) | – |
| | Inpatients | 32.15 (11.41) | 23.45 (11.76) | 30.40 (10.44) | – |
| | Inpatients | 28.27 (11.84) | 20.53 (10.85) | 29.96 (10.24) | – |
| Ng et al. (2007) | Inpatients | 29.01 (–) | 26.67 (–) | 20.32 (–) | – |
| Ronk, Hooke, and Page (2012) | Inpatients | 35.05 (8.35) | – | – | – |
| *Nonclinical* | | | | | |
| Antony et al. (1998) | Nonclinical adults | 2.12 (3.64) | 1.22 (1.77) | 3.51 (3.78) | – |
| Henry and Crawford (2005) | Nonclinical adults | 5.66 (7.74) | 3.76 (5.90) | 9.46 (8.40) | 18.86 (19.32) |
| Sinclair et al. (2012) | Nonclinical adults | 5.70 (8.20) | 3.99 (6.27) | 8.12 (7.62) | 17.80 (20.18) |
| Osman et al. (2012) | Undergraduate students | 8.36 (7.20) | 5.86 (6.76) | 10.58 (9.14) | 25.22 (20.46) |
| Norton (2007) | Undergraduate students | 8.54 (8.64) | 7.70 (7.58) | 12.22 (8.76) | – |

*Note.* Values have been doubled to be comparable to DASS-42, as advised in the manual.

countries via the paper-and-pencil administration format. The DASS has, however, also been used in a variety of additional populations and administered via new formats. This section therefore highlights the use of the DASS and its short form (a) with children and young adolescents, (b) in translated formats, and (c) when administered via computer.

## Age of administration

The manual and website provide ambiguous and slightly contradictory information regarding the recommended age of administration. In the manual, the developers suggest that the scale may be used in children 12 years or older, but also mention that no participants younger than 17 years were included in the normative sample (Lovibond & Lovibond, 1995b, p. 3). In contrast, the FAQ webpage states that the DASS "has been used with adolescents," but that there are "little data from that age range to confirm the validity of the scales." Although the authors do not explicitly state that they recommend using the scale with adolescents, the ambiguity of their statements may lead test users to infer that the authors support its use with this population.

Making such an inference would be problematic. The authors of a recent systematic review of the measurement properties of the DASS-21 concluded that the validity of using the DASS-21 with child or young adolescent samples could not be confirmed, owing to a dearth of empirical data (Lee et al., 2019). Furthermore, an examination of the few studies that have investigated the validity of the DASS-21 in child and young adolescent samples has revealed inconsistent findings—results from only six studies (Duffy, Cunningham, & Moore, 2005; Mellor et al., 2014; Moore, Dowdy, & Furlong, 2017; Patrick, Dyck, & Bramston, 2010; Shaw, Campbell, Runions, & Zubrick, 2017; Szabó, 2010) provided support for four different factor structures.

Given this lack of empirical clarity, the DASS is not currently recommended for use with children or young adolescents. Further research that elucidates the validity of using the DASS and DASS-21 with children and young adolescents is certainly required prior to employing this measure with younger groups.

## Translations

The DASS (and/or its short form) has been translated into at least 50 different languages. A caution at the top of the translation page of the DASS website warns that the validity of the translated versions of the DASS cannot be guaranteed. Some of the translation webpages include a link to a peer-reviewed article or unpublished manuscript explaining the translation process and/or validation of the translated version.

When evaluating a translation/test adaptation, it is important to consider that bias can be introduced when translating a scale from the original language to another (van de Vijver & Tanzer, 2004). An assessment scale that was created for use in a particular language in a specific region will not necessarily measure the same construct in the same way when translated. Indeed, nonequivalence of translated assessment scales is not uncommon (e.g., Byrne & Watkins, 2003).

To avoid introducing bias that results in nonequivalence, Byrne (2016) recommends a rigorous four-stage process for assessment scale adaptation. Notably, most of the translated versions of the DASS would not be considered to have met these rigorous standards. More concerning is that few of the research groups have even examined the psychometric properties of the translated questionnaires (Scholten, Velten, Bieda, Zhang, & Margraf, 2017).

Readers interested in using a translated version in their own research or who would like to interpret/compare findings from studies that employed a translated version are therefore strongly encouraged to gather all existing information on the adaptation process for that translation. Once the translation process is deemed appropriate, interested readers should conduct a comprehensive literature review of the empirical evidence evaluating the measurement properties of the translated version of interest. This will allow readers to critically evaluate the equivalence of that translated version (see Byrne, 2016; International Test Commission [ITC], 2005) and to draw an informed conclusion regarding the validity of the scale for the intended use.

## Computerized administration

The ITC's (2005) guidelines on computer-based and internet-delivered testing clearly state that the equivalence of a paper-and-pencil test and its computerized counterpart cannot be assumed. The FAQ page on the DASS website indicates that the DASS "can be administered by a computer." It should be noted, however, that the scale developers do not cite empirical evidence supporting the equivalence of the computerized and the paper-and-pencil DASS (e.g., that the psychometric properties are comparable; that norms for the computerized version are equivalent). To our knowledge, no studies have specifically evaluated the equivalence of these two administration formats. Such research is thus required before computer/internet administration of the DASS can be recommended.

# Applications

## Research applications

Although the DASS was specifically developed to provide the maximum meaningful discrimination between depression and anxiety, the correlations observed among the three DASS scales in the development samples and in numerous other studies were still relatively high (see Table 5). However, Lovibond and Lovibond (1995b) point out that, in their normative sample, none of the DASS items loaded meaningfully (>0.25) on another scale. They consequently suggest that the high correlations among the DASS scales are the result of common causes of depression, anxiety, and stress, rather than a reflection of the scales measuring overlapping constructs. The general support for a three- or bi-factor model of the DASS and its short form also provides support for the assertion that each of the scales measures a related, yet distinct construct. A measure that can be used to discriminately assess levels of depression, anxiety, and stress in both clinical and nonclinical populations is a highly useful tool for researchers investigating the etiology and course of various forms of psychopathology.

## Clinical applications

### *Screening*

The manual states that the DASS is suitable "for screening normal adolescents and adults" (p. 3). There is very little empirical evidence explicitly examining the use of the paper-and-pencil administration of the DASS to screen for depression or anxiety disorders in a general population. Although a handful of studies recruiting from diverse populations found support for the use of the DASS as a screening instrument in specific populations (e.g., Guest, Tran, Gopinath, Cameron, & Craig, 2018), these findings are inherently limited in their generalizability.

Despite the manual and website's recommendation for using the DASS as a screening tool, there is a dearth of research evaluating its sensitivity, specificity, and positive and negative predictive power to screen for depression and anxiety disorders in the general population. Consequently, further research evaluating the validity of the DASS as a

**TABLE 5** Scale-level correlations among the Depression Anxiety Stress Scale (DASS) and DASS-21.

| Reference | Sample | Version | D-A | D-S | A-S |
|---|---|---|---|---|---|
| *Clinical* | | | | | |
| Antony et al. (1998) | Outpatients | Full | 0.44 | 0.63 | 0.74 |
| | | Short | 0.46 | 0.57 | 0.72 |
| Brown et al. (1997) | Outpatients | Full | 0.51 | 0.64 | 0.65 |
| Brown et al. (1997) | Outpatients | Full | 0.45 | 0.66 | 0.66 |
| Davies et al. (2015) | Outpatients | Full | 0.75 | 0.80 | 0.82 |
| | | Short | 0.71 | 0.78 | 0.77 |
| Ng et al. (2007) | Inpatients | Short | 0.55 | 0.62 | 0.72 |
| Ronk et al. (2016) | Inpatients | Short | 0.52 | 0.57 | 0.69 |
| *Nonclinical* | | | | | |
| Crawford and Henry (2003) | Nonclinical adults | Full | 0.70 | 0.72 | 0.71 |
| Lovibond and Lovibond (1995b) | Normative sample | Full | 0.54 | 0.56 | 0.65 |
| Sinclair et al. (2012) | Nonclinical adults | Short | 0.73 | 0.72 | 0.63 |

screening tool must be conducted with diverse samples across a variety of settings before it can be recommended for this application.

## *Measuring treatment outcomes*

Several studies collectively provide support for using the DASS and its short form as a routine clinical treatment outcome measure. For example, Ng et al. (2007) observed statistically significant decreases in adult inpatients' DASS-21 scores from admission to discharge that paralleled significant improvements observed on other outcome measures. These findings have been replicated in studies using both the DASS-21 (Ronk, Hooke, & Page, 2016) and the full-length DASS (Page et al., 2007).

Furthermore, Ronk, Korman, Hooke, and Page (2013) found support for using the DASS-21 to assess the *clinically* significant change in both outpatient and inpatient samples. Ronk et al. calculated cutoff points separating the normal range from the outpatient range and the outpatient range from the inpatient range. These cutoff values (see Ronk et al., 2013) are consistent with the mean DASS scores reported in the normal, outpatient, and inpatient samples summarized in Tables 3 and 4, and correspond closely to the severity rating cutoff scores in the manual (i.e., normal vs mild; moderate vs severe). The concordance of Ronk et al.'s cutoff values, the severity labels in the manual, and reported mean DASS scores across normal, outpatient, and inpatient samples provide converging evidence that DASS scores can yield meaningful information about an individual's clinical severity. Ronk et al. also calculated minimum reliable change scores within each of the population distributions as well as between them.

Taken together, these findings provide empirical support for the use of the DASS and its short form as a treatment outcome measure in both inpatient and outpatient populations. Additionally, Ronk and her colleagues (2013) provide users with useful DASS-21 cutoff scores that can be used to evaluate whether patients have demonstrated reliable change and to differentiate among patients who are considered recovered, recovering, improved, unchanged, or deteriorated.

## *Progress monitoring*

Although the DASS developers do not list progress monitoring as a specific use of the DASS, it has been used as such (e.g., Persons et al., 2016). In comparison to the accumulating research evaluating the validity of using the DASS and DASS-21 to measure treatment *outcome*, there is a paucity of research examining the validity of routinely administering either version of the DASS to monitor treatment *progress*. Notably, ceiling effects have been observed for the Depression and Stress scales, particularly when used with clinically severe populations. For example, when both the DASS and the Beck Depression Inventory (BDI) were administered to a sample of inpatients and day patients diagnosed with a depressive disorder, 8.9% of patients scored the maximum Depression scale score (Page et al., 2007). In contrast, no participants scored the maximum BDI score. Analogous results were found in a follow-up study of inpatients diagnosed with a mood disorder (Page et al., 2007).

Ceiling effects can significantly reduce a measure's sensitivity to change. The observation of such effects for the Depression and Stress scales in samples of severely depressed individuals thus warrants caution in their use for progress monitoring in this population. Moreover, there are no currently available empirically supported recommendations on how frequently the DASS should be administered to patients, nor on how to operationalize progress (or lack thereof). Further investigations of the utility of using the DASS and DASS-21 as a routinely administered measure to monitor treatment progress are therefore required.

## Summary

### Strengths

The DASS evidences a number of strengths that support its use across research and clinical settings. First, numerous studies have demonstrated excellent internal consistency and strong factorial and construct validity for both versions of the DASS. It is important to note, however, that these investigations have employed the paper-and-pencil administration using samples comprising English-speaking individuals living in Western countries.

Second, the DASS's suitability for use with both clinical and nonclinical samples supports the integration of psychological research and clinical practice. Additionally, because the DASS distinguishes between the constructs of depression, anxiety, and stress, its use may be particularly advantageous in research aiming to advance understanding of the etiology, presentation, and maintenance of complexly related internalizing disorders. The fact that the DASS is free to administer is also an advantage for researchers who may be collecting data from large samples.

Third, the DASS has been demonstrated to be a useful measure for assessing treatment outcomes. Additionally, it possesses the qualities associated with valuable treatment outcome measures (Hickie, Andrews, & Davenport, 2002): The DASS is inexpensive, quick to administer, completed by the patient, broad in its coverage of symptoms, and responsive to change in clinical status. Clinically meaningful change indices provided by Ronk et al. (2013) render the DASS even more useful as an outcome measure.

### Limitations and future directions

There are also a number of limitations of the DASS that researchers and clinicians must keep in mind. For instance, the validity of the original DASS cannot simply be assumed to be the same as the validity of the DASS in different formats or populations. Notwithstanding the recommendations made in the manual and on the website, there is insufficient empirical evidence that the DASS can be (1) employed in child and adolescent populations, (2) administered via computer or the internet, or (3) used as a screening or progress-monitoring tool. Relatedly, most of the translated versions of the DASS would not meet ITC guidelines for test adaptation. Moreover, the psychometric properties of many of the translations have not been examined. Before using a translation (or interpreting the results of a study that employed a translated version), readers are encouraged to evaluate its translation process and to conduct a literature review of psychometric properties of the adaptation. Further research elucidating the extension of the DASS to these new formats and populations is needed.

Another limitation is the potential for ceiling effects for the Depression and Stress scales in samples of severely depressed individuals. Clinicians using the DASS with such individuals should remain cognizant of this possibility, as ceiling effects can significantly reduce a measure's sensitivity to change. The use of the DASS to monitor treatment progress is also precluded by the lack of any empirically supported recommendations on how frequently it should be administered to patients or on how to interpret changes in their scores.

Research supporting the use of the DASS as a screening measure is also limited. Before it can be recommended for this use, its sensitivity, specificity, and positive and negative predictive power must be evaluated. Investigations on the validity of using the DASS as a screening tool and to monitor treatment progress would thus be valuable.

## Conclusion

In sum, the English version of the DASS and its short form can be used to assess the constructs of depression, anxiety, and stress in clinical and nonclinical populations. Numerous studies have demonstrated excellent internal consistency and strong factorial and construct validity for the paper-and-pencil administration of the DASS (and DASS-21) in samples of English-speaking individuals living in Western countries. Its use as a treatment outcome measure in this population is also well supported by empirical evidence. The fact that the DASS is inexpensive, quick and easy to administer, broad in its coverage of symptoms, and responsive to change in clinical status makes it a practical choice for outcome measurement.

## Key facts of translations/adaptations of tests or measures

- An assessment scale that was created for use in a particular language in a specific region will not necessarily measure the same construct in the same way when translated into a different language and/or used in a different region.

- It is not uncommon for translated or adapted assessment scales to demonstrate nonequivalence both within and across national borders.
- The language of an original measure is often referred to as the "source language."
- The language into which an existing measure has been translated is often referred to as the "target language."
- Construct bias, method bias, and/or item bias can be introduced when translating a scale from a source language to a target language.

## Summary points

- The Depression Anxiety Stress Scale (DASS) and its short form (DASS-21) were designed to assess the constructs of depression, anxiety, and stress while maximizing discrimination among them.
- The DASS is used widely in clinical and research settings with diverse populations.
- The validity of the paper-and-pencil English version of the DASS and DASS-21 is strongly supported for assessing the constructs of depression, anxiety, and stress in clinical and nonclinical adult populations.
- There is insufficient empirical evidence that the DASS or its short form can be (1) employed in child and adolescent populations, (2) administered via computer or the internet, or (3) used as a screening or progress-monitoring tool.
- The use of the DASS as a treatment outcome measure in this population is well supported by empirical evidence. The fact that it is inexpensive, quick and easy to administer, broad in its coverage of symptoms, and responsive to change in clinical status makes the DASS a practical choice for outcome measurement.

## Mini-dictionary of terms

**Internalizing symptoms** Cognitive, affective, and physiological expressions of disorders characterized by withdrawal and inward distress.

**Negative affectivity** A disposition or trait reflecting variations in the degree to which individuals tend to exhibit negative emotionality and a negative self-concept. Considered distinct from positive affectivity.

**Positive affectivity** A disposition or trait reflecting variations in the degree to which individuals tend to exhibit positive emotionality. Considered distinct from negative affectivity.

**Sensitivity** A method for understanding how accurately a test or measure predicts who does and who does not have a disorder (i.e., the percentage of individuals with a disorder correctly classified by a test as having the disorder).

**Specificity** A method for understanding how accurately a test or measure predicts who does and who does not have a disorder (i.e., refers to the percentage of individuals without a disorder correctly classified by a test as not having the disorder).

## Acknowledgments

During the preparation of this chapter, David J.A. Dozois was supported by an Insight Grant from the Social Sciences and Humanities Research Council of Canada. This support is gratefully acknowledged.

## References

Antony, M. M., Bieling, P. J., Cox, B. J., Enns, M. W., & Swinson, R. P. (1998). Psychometric properties of the 42-item and 21-item versions of the depression anxiety stress scales in clinical groups and a community sample. *Psychological Assessment*, *10*, 176–181.

Brown, T. A., Chorpita, B. F., Korotitsch, W., & Barlow, D. H. (1997). Psychometric properties of the depression anxiety stress scales (DASS) in clinical samples. *Behaviour Research and Therapy*, *35*, 79–89.

Burckhardt, R., Manicavasagar, V., Batterham, P. J., Hadzi-Pavolic, D., & Shand, F. (2017). Acceptance and commitment therapy universal prevention program for adolescents: A feasibility study. *Child and Adolescent Psychiatry and Mental Health*, *11*, 27.

Byrne, B. M. (2015). Adaptation of assessment scales in cross-national research: Issues, guidelines, and caveats. *International Perspectives in Psychology: Research, Practice, Consultation*, *5*, 51–65.

Byrne, B. M., & Watkins, D. (2003). The issue of measurement invariance revisited. *Journal of Cross-Cultural Psychology*, *34*, 155–175.

Clara, I. P., Cox, B. J., & Enns, M. W. (2001). Confirmatory factor analysis of the Depression-Anxiety-Stress Scales in depressed and anxious patients. *Journal of Psychopathology and Behavioral Assessment*, *23*, 61–67.

Crawford, J., Cayley, C., Lovibond, P. F., Wilson, P. H., & Hartley, C. (2011). Percentile norms and accompanying interval estimates from an Australian general adult population sample for self-report mood scales (BAI, BDI, CRSD, CES-D, DASS, DASS-21, STAI-X, STAI-Y, SRDS, and SRAS). *Australian Psychologist*, *46*, 3–14.

Crawford, J. R., & Henry, J. D. (2003). The depression anxiety stress scales (DASS): Normative data and latent structure in a large non-clinical sample. *British Journal of Clinical Psychology*, *42*, 111–131.

Davies, G., Caputi, P., Skarvelis, M., & Ronan, N. (2015). The depression anxiety and stress scales: Reference data from a large psychiatric outpatient sample. *Australian Journal of Psychology*, *67*, 97–104.

Duffy, C. J., Cunningham, E. G., & Moore, S. M. (2005). Brief report: The factor structure of mood states in an early adolescent sample. *Journal of Adolescence*, *28*, 677–680.

Gloster, A. T., Rhoades, H. M., Novy, D., Klotsche, J., Senior, A., Kunik, M., et al. (2008). Psychometric properties of the depression anxiety stress Scale-21 in older primary care patients. *Journal of Affective Disorders*, *110*, 248–259.

Gomez, R., Summers, M., Summers, A., Wolf, A., & Summers, J. (2014). Depression anxiety stress Scales-21 measurement and structural invariance across ratings of men and women. *Assessment*, *21*, 418–426.

Guest, R., Tran, Y., Gopinath, B., Cameron, I. D., & Craig, A. (2018). Prevalence and psychometric screening for the detection of major depressive disorder and post-traumatic stress disorder in adults injured in a motor vehicle crash who are engaged in compensation. *BMC Psychology*, *6*(1), 4.

Henry, J. D., & Crawford, J. R. (2005). The short-form version of the depression anxiety stress scales (DASS-21): Construct validity and normative data in a large non-clinical sample. *British Journal of Clinical Psychology*, *44*, 227–239.

Hickie, I. B., Andrews, G., & Davenport, T. A. (2002). Measuring outcomes in patients with depression or anxiety: An essential part of clinical practice. *The Medical Journal of Australia*, *177*, 205–207.

International Test Commission (2005). International Guidelines on Computer-Based and Internet Delivered Testing. [www.intestcom.org].

Lee, D. (2019). The convergent, discriminant, and nomological validity of the depression anxiety stress scales-21 (DASS-21). *Journal of Affective Disorders*, *259*, 136–142.

Lee, J., Lee, E.-H., & Moon, S. H. (2019). Systematic review of the measurement properties of the depression anxiety stress scales–21 by applying updated COSMIN methodology. *Quality of Life Research*, *28*, 2325–2339.

Longwell, B. T., & Truax, P. (2005). The differential effects of weekly, monthly, and bimonthly administrations of the Beck depression inventory-II: Psychometric properties and clinical implications. *Behavior Therapy*, *36*, 265–275.

Lovibond, P. F., & Lovibond, S. H. (1995a). The structure of negative emotional states: Comparison of the depression anxiety stress scales (DASS) with Beck depression and anxiety inventories. *Behaviour Research and Therapy*, *33*, 335–343.

Lovibond, S. H., & Lovibond, P. F. (1995b). *Manual for the depression anxiety stress scales*. Sydney, Australia: Psychology Foundation of Australia.

Mellor, D., Vinet, E. V., Xu, X., Hidayah Bt Mamat, N., Richardson, & Romàn, F. (2014). Factorial invariance of the DASS-21 among adolescents in four countries. *European Journal of Psychological Assessment*, *31*, 138–142.

Moore, S. A., Dowdy, E., & Furlong, M. J. (2017). Using the depression anxiety stress scales–21 with U.S. adolescents: An alternate models analysis. *Journal of Psychoeducational Assessment*, *35*, 581–598.

Ng, F., Trauer, T., Dodd, S., Callaly, T., Campbell, S., & Berk, M. (2007). The validity of the 21-item version of the depression anxiety stress scales as a routine clinical outcome measure. *Acta Neuropsychiatrica*, *19*, 304–310.

Nieuwenhuijsen, K., de Boer, A. G. E. M., Verbeek, J. H. A. M., Blonk, J. H. A. M., & van Dijk, F. J. H. (2003). The depression anxiety stress scales (DASS): Detecting anxiety disorder and depression in employees absent from work because of mental health problems. *Occupational and Environmental Medicine*, *60*, 77–82.

Norton, P. J. (2007). Depression anxiety stress scales (DASS-21): Psychometric analysis across four racial groups. *Anxiety, Stress, and Coping*, *20*, 253–265.

Osman, A., Wong, J. L., Bagge, C. L., Freedenthal, S., Gutierrez, P. M., & Lozano, G. (2012). The depression anxiety stress scales—21 (DASS-21): Further examination of dimensions, scale reliability, and correlates. *Journal of Clinical Psychology*, *68*, 1322–1338.

Page, A. C., Hooke, G. R., & Morrison, D. L. (2007). Psychometric properties of the depression anxiety stress scales (DASS) in depressed clinical samples. *British Journal of Clinical Psychology*, *46*, 283–297.

Patrick, J., Dyck, M., & Bramston, P. (2010). Depression anxiety stress scale: Is it valid for children and adolescents? *Journal of Clinical Psychology*, *66*, 996–1007.

Persons, J. B., Koerner, K., Eidelman, P., Thomas, C., & Liu, H. (2016). Increasing psychotherapists' adoption and implementation of the evidence-based practice of progess monitoring. *Behaviour Research and Therapy*, *76*, 24–31.

Ronk, F. R., Hooke, G. R., & Page, A. C. (2012). How consistent are clinical significance classifications when calculation methods and outcome measures differ? *Clinical Psychology: Science and Practice*, *19*, 167–179.

Ronk, F. R., Hooke, G. R., & Page, A. C. (2016). Validity of clinically significant change classifications yielded by Jacobson-Truax and Hageman-Arrindell methods. *BMC Psychiatry*, *16*, 187.

Ronk, G. R., Korman, J. R., Hooke, G. R., & Page, A. C. (2013). Assessing clinical significance of treatment outcomes using the DASS-21. *Psychological Assessment*, *25*, 1103–1110.

Ruble, K., George, A., Gallicchio, L., & Gamaldo, C. (2015). Sleep disordered breathing risk in childhood cancer survivors: An exploratory study. *Pediatric Blood & Cancer*, *62*, 693–697.

Scholten, S., Velten, J., Bieda, A., Zhang, X. C., & Margraf, J. (2017). Testing measurement invariance of the depression, anxiety, and stress scales (DASS-21) across four countries. *Psychological Assessment*, *29*, 1376–1390.

Shaw, T., Campbell, M. A., Runions, K. C., & Zubrick, S. R. (2017). Properties of the DASS-21 in an Australian community adolescent population. *Journal of Clinical Psychology*, *73*, 879–892.

Sinclair, S. J., Siefert, C. J., Slavin-Mulford, J. M., Stein, M. B., Renna, M., & Blais, M. A. (2012). Psychometric evaluation and normative data for the depression anxiety stress Scales-21 (DASS-21) in a nonclinical sample of U.S. adults. *Evaluation & the Health Professions*, *35*, 259–279.

Szabó, M. (2010). The short version of the depression anxiety stress scales (DASS–21): Factor structure in a young adolescent sample. *Journal of Adolescence*, *33*, 1–8.

van de Vijver, F. J. R., & Tanzer, N. K. (2004). Bias and equivalence in cross-cultural assessment: An overview. *European Review of Applied Psychology/Revue Européenne de Psychologie Appliquée*, *47*, 263–279.

Wardenaar, K. J., Wanders, R. B. K., Jeronimus, B. R., & de Jonge, P. (2018). The psychometric properties of an internet-administered version of the depression anxiety and stress scales (DASS) in a sample of Dutch adults. *Journal of Psychopathology and Behavioral Assessment*, *40*, 318–333.

Weiss, R. B., Aderka, I. M., Lee, J., Beard, C., & Björgvinsson, T. (2015). A comparison of three brief depression measures in an acute psychiatric population: CES-D-10, QIDS-SR, and DASS-21-DEP. *Journal of Psychopathology and Behavioral Assessment*, *37*, 217–230.


## Further reading


Psychology Foundation of Australia. (2018), DASS translations. Retrieved from http://www2.psy.unsw.edu.au/dass/translations.htm.

Chapter 23

# Arabic version of the two-question quick inventory of depression (QID-2-AR): Description and applications

Amani Ahmed[a] and Muaweah Ahmad Alsaleh[b,c,d,e]

[a]Caen University Hospital, Emergency Department, Caen, France, [b]Psychological Studies, Neuropsychologist, Psychotherapist, Psychology Researcher, Behavioural Science, Psychotherapy Practice and Clinical Psychopathology, Psychology Biostatistics, CERReV Laboratory; Public Health-Health Ethics, INSERM & Adapted Physical Activities and Health, COMETE & Pain Management, University of Caen Normandy, Caen, & New Navarre Hospital, & University of Lorraine, Interpsy Laboratory Nancy, France, [c]Independent Psychotherapist, France Asylum Land & Reception Service for Unaccompanied Foreign Minors, Caen, France, [d]Faculty of Education, Counseling Psychology University of Aleppo, Aleppo, Syria, [e]Psychological Courses Developer, University of the People, Pasadena, CA, United States

## List of abbreviations

| | |
|---|---|
| **LMHPCS** | Life, Mental Health and Primary Care Settings |
| **MMHPCHSW** | Medical, Mental Health, Primary Care, Hospital Settings, Wartime |
| **PPPCN** | Practitioners, Psychotherapist, Physicians, Caretakers, and/or Nurses |
| **QID-2-Ar** | Two-question Quick Inventory of Depression-Arabic |
| **RCPPPPPCH** | Routine Consultations, Prenatal, Postnatal, Pediatric, Primary, Psychiatric Care, Hospitalizations |

## Introduction

In the Arab world, depression affects many persons. Studies in psychology, medicine, psychiatric, and epidemiology on the health status of people (children, teens, university students, patients, adults, refugees, etc.) indicate that depression in these populations is high, most prevalent, and frequent (Alsaleh & Kubitary, 2017; Kubitary, Alomer, & Alsaleh, 2017; Kubitary & Alsaleh, 2017, 2018). Depressive disorders are a major public health problem in everyday life, in mental health settings, and in primary care settings (LMHPCS) with far-reaching consequences. Depression is often treatable, yet is often overlooked in hospital departments and clinics (busy neurology clinics, busy outpatient clinic, etc.). In addition, if it is detected, more often than not it is inadequately treated. In the world, almost all patients/persons are under direct psychological pressure because of diseases, society, war, and/or enviro-psychological factors. Therefore, the rapid screening for depression can help to improve treatment effectiveness in all patients/persons with and/or without the disease, in students, elderly, postpartum, refugees, children, teenagers, adults, etc.

## Human plague

"*Depression is a major human blight*" (Smith, 2014). "*If the extent of human suffering were used to decide which diseases deserve the most medical attention, then depression would be near the top of the list*" (Ledford, 2014). Depression is a serious psychic manifestation and one of the most common disorders in the world (Kubitary & Alsaleh, 2018; Mann, 2005; Nature, 2014). Depression is up sharply in the last decade around the world (Alsaleh, 2016a, 2016b; Alsaleh et al., 2019; Kubitary & Alsaleh, 2018). Depression is the fourth largest burden of disease in the world, according to a widely cited WHO study and could reach the second-highest in 2030 worldwide, not so much because of an increasing the number of cases of depression or their severity, but because the first three causes of morbidity (perinatal problems, lower respiratory tract infections, HIV) will be better detected, diagnosed and treated (Alsaleh, 2016a, 2016b; Smith, 2014).

## Negative impacts of depression

People with depression are, compared to nondepressive ones, about half as likely to follow their studies, their medications, etc. Depression in the population has negative impacts on the person's social situation, the person's future

The Neuroscience of Depression. https://doi.org/10.1016/B978-0-12-817933-8.00055-4

well-being, and quality of life. Depression reduces the social and professional life of the person (work, studies, academic achievement social relationships, deterioration in relationships, daily tasks, marital problems, and affect future employment, etc.). In severe cases, life can lose all interest. Depression can also have negative consequences on the evolution of the mental state and leads to the risk of the suicidal act (Alsaleh, 2016a, 2016b; Kubitary & Alsaleh, 2018).

The negative effects of depression (see Table 1) have been demonstrated in the clinical and nonclinical populations (Alsaleh, 2016a, 2016b; Alsaleh et al., 2019; Alsaleh & Lebreuilly, 2017; Alsaleh, Lebreuilly, & Defer, 2015; Keller et al., 2000; Kubitary & Alsaleh, 2018; Mohr, Hart, Julian, & Tasch, 2007), with one of them being poor drug adherence (Alsaleh, 2014; Alsaleh et al., 2015, 2019; Kubitary & Alsaleh, 2018). Depression in persons has negative repercussions on their social and relational situations. It affects feel, think, and handle daily activities, such as sleeping, eating, or working. It is associated with more marked impairments in psychosocial function and work performance, which significantly influences their quality of life. It leads people to exaggerate their cognitive difficulties, fatigue, pain, more frequent suicide attempts, increased health-care utilization and hospitalization, and poor medication adherence. Poor drug adherence or lack of medication adherence is worsened by depression (Alsaleh et al., 2019; Fragoso et al., 2014; Kubitary et al., 2017; Kubitary & Alsaleh, 2018; Tarrants, Oleen-Burkey, Castelli-Haley, & Lage, 2011). During pregnancy, depression has harmful impact on maternal health. In addition, increasing evidence indicate that depression also harms fetuses and infants (Stewart, 2005).

The presence of depression produces complications in the treatment of both diseases and neurological lesions. Depression can lead to poor compliance with treatment resulting in worsening of the situation. It puts the patients at a higher risk of suicide and may produce a desire for hastened death. It leads to a decline in patient satisfaction with medical care and predicts disease progression and impairs quality of life not only of the patients but also of their caregivers.

**TABLE 1 Impacts of depression disorders.**

1. Reduces the social and professional life of the person
2. Impairment in psychosocial function and work performance
3. Affects the feel, think, and handle daily activities, such as sleeping, eating, or working
4. Leads to the risk of the suicidal act
5. Increased health-care utilization, hospitalization, and poor medication adherence
6. Harms fetuses during pregnancy
7. Disease progression and low therapeutic efficacy

## Depression undetected and undiagnosed

Depression is often underdiagnosed, underdetected, and undertreated (Alsaleh, 2014, 2016a, 2016b; Alsaleh et al., 2019; Fragoso et al., 2014; Kubitary & Alsaleh, 2018; Marrie et al., 2009; Skokou, Soubasi, & Gourzis, 2012). Depression is a frequent disease and a frequently observed morbidity, yet is undetected, undiagnosed, and untreated, which may delay or render management inadequate.

"*Many patients are treated in primary care settings, where diagnosis and treatment of depression need improvement*" (Mojtabai, 2014). "*There is no simple and perfect way to screen for depression*" (Fragoso et al., 2014). Depression undetected, undiagnosed, and undertreated stay often lifelong.

Prompt, effective, and systematic screening improve early diagnosis and treatment, which mitigates adverse effects of depression on life, education, career, relationships, and improve patients' well-being and their satisfaction with the quality of life and of medical care. Effective treatment of depression and therapeutic control of a depressive state can improve personal life (e.g., university life, primary care, quality of life, success). It reduces mortality, improves the outcome after acute myocardial infarction, or stroke, neurological lesion, and lowers the risk of suicide. It improves also the mental and physical health of people and patients.

"*Busy clinicians must balance their focus on the presenting disease of a patient with a duty to detect other comorbidities that the patient might be experiencing*" (Rabins, 2016). Because depression touches so much of persons with significant pathology, and/or without pathology, busy clinicians should take specific and effective steps to detect depression and then refer or evaluate those who have symptoms for further assessment and treatment (Kubitary & Alsaleh, 2018). "*Several questions need to be addressed before focusing on the timing and benefit of early intervention including developing a universally validated screening tool, developing a definitive definition, and establishing acceptable treatment recommendations*" (Dwyer Hollender, 2014). It is necessary to develop a universally validated screening tool for prompt, effective, and systematic screening, early diagnosis for depression in life, in any age, any sex, and any condition by the simpler and faster, reliable, and effective method.

The main objective of this chapter is to make available to practitioners, and busy clinicians, as early as possible, a universally validated screening tool that is the simpler and faster method to evaluate the depression. Thus, the aim is to provide, in the hands of the clinicians and practitioners, the description and applications the QID-2 (two-question Quick Inventory of Depression) in any conditions which the much simpler approach for evaluation of depression in people enduring the diseases or no and the wartime in all world, especially Arab world.

## Burden time and effort to screening of depression

Many PPPCN believe that the screening process requires too much time and effort and are a burden for both patients and clinicians (Alsaleh, 2016a, 2016b; Alsaleh et al., 2019; Kubitary & Alsaleh, 2018). This makes that depression in the population is largely underdiagnosed and is often neither known nor treated.

## Difficulty of detecting depression

The difficulty of detecting depression and its overlap with the physical symptoms of diseases presents a challenge for the clinician to differentiate whether certain symptoms are related to depression or physical illness. The most frequently cited examples of confounding factors are fatigue and age. Other potential confounding factors are insomnia, impaired appetite, and impaired memory and concentration. Thus, difficulties in detecting and/or diagnosing depression can lead to false diagnoses of depression (Alsaleh, 2016a, 2016b; Alsaleh et al., 2019; Feinstein, 2011; Hackett & Jardine, 2017; Kubitary et al., 2017; Kubitary & Alsaleh, 2018).

The question here is: How can we best identify the depression in busy clinical practices, in other settings, and in wartime for good treatment? A brief, valid, reliable instrument with good sensitivity, specificity, and a low false-positive rate is needed. The gap between the demand and delivery of mental health services in the hospital world, busy clinics, during diseases and/or wartime can be reduced by validating freely available and psychometrically sound psychological tools.

## Depressed mood and anhedonia

Depression is much related to depressed mood and anhedonia and is manifested by sadness. Health-care providers and clinicians have to recognize and differentiate it. A state of sadness or "depressed" does not constitute depression. To establish the correct diagnosis, it must be ensured that the person has one of two symptoms (depressed mood and anhedonia), but also different symptoms listed in DSM-V. These symptoms occur on a daily basis with some intensity for at least 2 weeks. Depression causes a change in functioning in peoples' professional, social, or family life and generates real distress. Finally, in the woman who has just given birth, it is necessary to differentiate the "baby blues," emotional, and physical backlash of the childbirth that does not last of the depression of the "postpartum." The QID-2 solves these problems.

## Effective screening

Establishing the effective screening and diagnosis of depression is warranted. Prior to diagnosis, prompt, effective, and systematic screening for early identification of depressive symptoms in sick people or non-sick is essential. Depressed mood and anhedonia (decreased interest or pleasure in activities) are the two main depressive symptoms listed in DSM-V, which must be present most of the day nearly every day for 2 weeks (American Psychiatric Association, 2013; Kubitary et al., 2017; Kubitary & Alsaleh, 2018). Some psychiatric disorders are related with depressed mood and anhedonia (Alsaleh et al., 2019; Chapman, Chapman, & Raulin, 1976; Der-Avakian & Markou, 2012; Kubitary & Alsaleh, 2017, 2018; Olivares & Berrios, 1998). This screening is performed using two simple questions (two-question Quick Inventory of Depression "QID-2"). These two questions assess depressed mood and anhedonia (Alsaleh, 2016a, 2016b; Alsaleh et al., 2019; Baillon, Dennis, Lo, & Lindesay, 2014; Kroenke, Spitzer, & Williams, 2003; Kubitary & Alsaleh, 2018; Lebrun & Cohen, 2009; Spitzer, Kroenke, & Williams, 1999). Depressive mood and anhedonia are the two main depressive symptoms listed in DSM-V (Alsaleh et al., 2019; Kubitary & Alsaleh, 2018). The UK National Institute for Health and Care Excellence (NICE) and studies (Alsaleh et al., 2019; Kubitary & Alsaleh, 2018) recommended regular screening of depression for patients and the populations and suggested the use of two screening questions. In addition, psychometric evaluation of the clinical course of depression in clinics, after, before, and during treatment is essential. This allows the health-care providers (PPPCN) to determine the severity of the depressive state as well as the effectiveness of medication and/or psychotherapeutic treatments. This estimate thus guides their decisions, in particular to adjust the dosage or to change the antidepressant and the psychotherapy technique.

## Multiple cultures recommended of QID-2

What has led to increased interest in using QID-2 (see Table 2) is that there is very much a need for an effective, efficient method/way to improve compliance with screening (Alsaleh et al., 2019; Kubitary & Alsaleh, 2018). QID-2, two simple questions, is validated on many populations and many cultures and is recommended (see Table 2 and Fig. 1) to screen for depression in chronic diseases, general population, in normal and abnormal conditions, and wartime (Alsaleh et al., 2019; Gelaye et al., 2016; Kubitary & Alsaleh, 2017, 2018; Lebrun & Cohen, 2009; Li, Friedman, Conwell, & Fiscella, 2007; Löwe, Kroenke, & Gräfe, 2005; Mohr et al., 2007;

**TABLE 2** QID-2-En.

| | | |
|---|---|---|
| | Yes | No |
| 1. During the past 2 weeks, have you often been bothered by feeling down, depressed, or hopeless? | | |
| 2. During the past 2 weeks, have you often been bothered by little interest or pleasure in doing things? | | |
| **Arabic version-QID-2-Ar (النسخة العربية)** | | |
| | نعم | لا |
| 1. خلال الأسبوعين الماضيين، هل شعرت مسبقاً بالإحباط، أو الكآبة أو اليأس؟ | | |
| 2. خلال الأسبوعين الماضيين، هل شعرت مسبقاًبقلة الاهتمام أو الاستمتاع في حياتك اليومية؟ | | |
| **French version-QID-2-Fr** | | |
| | Oui | Non |
| 1. Durant les deux dernières semaines, vous êtes-vous déjà senti faible, déprimé ou sans espoir ? | | |
| 2. Durant les deux dernières semaines, avez-vous déjà ressenti peu d'intérêt ou de plaisir dans votre vie quotidienne? | | |

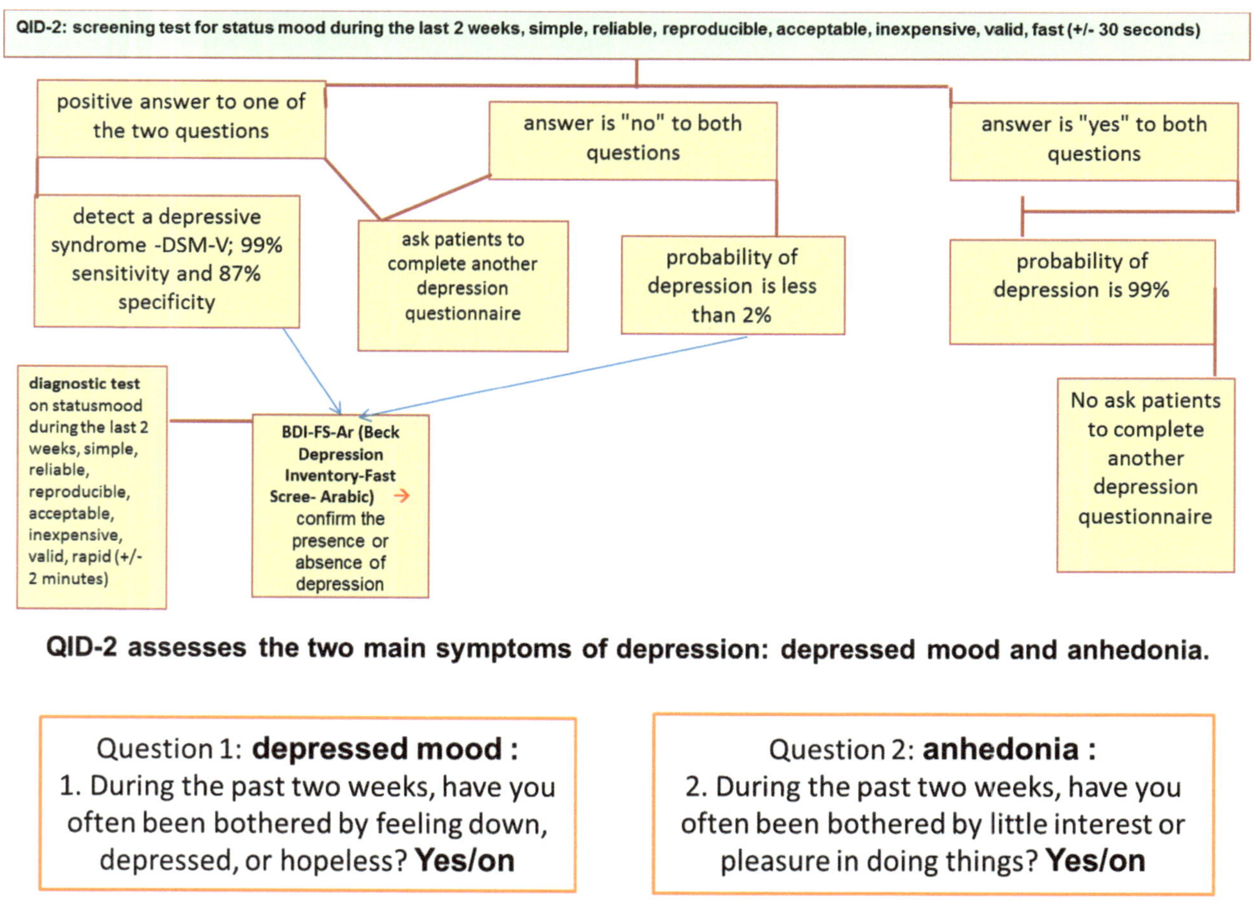


**FIG. 1** Recommendations on the diagnosis of depression with QID-2.

Sheeran et al., 2010; Spitzer et al., 1999). These two probe questions (QID-2) are most promising and recommended by the United States and Canadian task forces on preventive health care and common to many of the instruments.

In clinics (as busy neurology clinics, busy outpatient clinic), in routine consultations, in hospital departments, hospitalizations, before/after treatment, QID-2 can identify the state of depression among patients/persons in nonclinical/clinical conditions and in wartime conditions. These QID-2 are a reasonable alternative screening measure, given its brevity and potential for administration during the clinical interview. The Arabic and French version of the QID-2 also has highly favorable psychometric properties (Alsaleh et al., 2019; Kubitary & Alsaleh, 2018). The Arabic and French versions of the QID-2 are available from the author.

The QID-2 is an effective tool that may also improve compliance with recommended or mandated screening and may even possibly be applied to other types of patients as well as in any age. The QID-2 can also alert the PPPCN to cases requiring more extensive evaluation among patients enduring the diseases and enduring the abnormal adverse conditions of war (Alsaleh et al., 2019; Kubitary & Alsaleh, 2017, 2018). Taken together, findings in the studies support the cross-cultural reliability and validity of the QID-2 in the world (Alsaleh et al., 2019; Gelaye et al., 2016; Kubitary & Alsaleh, 2017, 2018; Lebrun & Cohen, 2009; Li et al., 2007; Löwe et al., 2005; Spitzer et al., 1999). Hence, QID-2 was found to be valid and reliable for use in cross-cultural research settings with little time constraints, during routine consultations, hospitalizations, busy clinics, and wartime.

## QID-2 test alternative of scales

These two questions are an alternative to the depression scales to assess the required criterion for the diagnosis of depression as a screening question (Alsaleh et al., 2019; Kubitary & Alsaleh, 2018). However, based on the results of the studies (Alsaleh et al., 2019; Baillon et al., 2014; Chae, Chae, Tyndall, Ramirez, & Winter, 2012; Kubitary & Alsaleh, 2017, 2018; Li et al., 2007; Mohr et al., 2007), the QID-2 is a useful tool for screening for depression in patients/persons and refugees. Therefore, it is recommended that PPPCN add the QID-2 to their routine clinical consultations of patients as in the pre-treatment setting. During diseases and/or wartime, such an evaluation is even more necessary and should be routine during every consultation.

These two items are almost as effective as longer tools (Alsaleh et al., 2019; Canadian Task Force on Preventive Health Care, 2003; Kubitary & Alsaleh, 2018; U.S. Preventive Services Task Force, 2002). The NICE guidelines suggested a two-stage screening process in which the two questions are the initial screen, followed by a second screening for those who respond positively to either or both questions to reduce the number of false positives being subjected to further assessments for depression (Alsaleh et al., 2019; Kubitary & Alsaleh, 2018; National Collaborating Centre for Mental Health (UK), 2010).

## Utility of QID-2

The QID-2 has been used with mothers attending well-child visits at three rural pediatric practices (see Fig. 1). The brief depression screening (QID-2) was well accepted by mothers and usually did not prolong visits. In postpartum depression, QID-2 is an effective tool in screening. During the clinical interview, QID-2 is a reasonable alternative screening test given its brevity, manageability, and ease of management (Chae et al., 2012; Seehusen, Baldwin, Runkle, & Clark, 2005).

The QID-2 has been validated in patients, students, primary care, busy outpatient practice, mental health services, obstetrics-gynecology practices, cancer, lateral epicondylitis, women during pregnancy and the postpartum period, multiple sclerosis, chronic kidney disease, and wartime (Alsaleh, 2016a, 2016b Alsaleh et al., 2019; Arroll et al., 2010; Arroll, Khin, & Kerse, 2003; Bennett et al., 2008; Chae et al., 2012; Kroenke et al., 2003; Kubitary & Alsaleh, 2018; Li et al., 2007; Mohr et al., 2007; Randall, Voth, Burnett, Bazhenova, & Bardwell, 2013). The QID-2 has found to be as simpler and effective as other depression screens or scales. The QID-2 has been validated as a reliable, simpler, and faster depression screening tool. In hospitals, clinicians can now use a simple two-item questionnaire (QID-2) to quickly screen for depression during routine consultations, in prenatal care, postnatal care, pediatric care, primary care, psychiatric care, hospitalizations (RCPPPPPCH), MMHPCHSW, LMHPCS, and visiting to hospital departments such as nephrology, oncology, and neurology. Including these two questions either under the usual or abnormal (wartime) conditions as part of the clinical routine for every consultation could alert PPPCN to cases requiring more extensive evaluation.

## Description of QID-2

The QID-2 consists of two self-assessment items and is a good screening test for detecting depression (Alsaleh et al., 2019; Kubitary & Alsaleh, 2017, 2018). The QID-2 consists of two items that align with DSM-V criteria for depression and assesses both depressive mood and anhedonia. When used in a dichotomous yes/no fashion, a positive response to either item constitutes a positive result (Alsaleh et al., 2019; Kubitary & Alsaleh, 2018; Spitzer et al., 1999; Whooley, Avins, Miranda, & Browner, 1997).

## Description of diagnostic cutoff value for the QID-2

The two items evaluate mood status, including depressive mood and loss of pleasure (anhedonia), over the past 2 weeks, giving a maximum total score of 2 and a minimum of 0. The QID-2 only provides three data points: 0, 1, or 2. This step requires approximately 30 s (Alsaleh et al., 2019; Kubitary & Alsaleh, 2018).

The QID-2 fulfilled all the positive parameters of a screening test, as shown in other studies. It appears to be a useful tool for identifying patients with depression with a simple "yes" response to one of two questions (cutoff value of 1). The diagnostic cutoff value for the QID-2 is $\geq 1$. Using a threshold score of $\geq 1$ rather than 2 resulted in more depressed patients identified (Alsaleh et al., 2019; Kubitary & Alsaleh, 2018). It has been said that a screening instrument is satisfactory if sensitivity and specificity are balanced and close to 80%. In the studies, a good balance between sensitivity and specificity was reached with a QID-2 score $\geq 1$, with values between 0.70 and 1.00 for sensitivity, specificity, and positive predictive value, which can be considered high. Studies (Alsaleh et al., 2019; Kubitary & Alsaleh, 2018; Lebrun & Cohen, 2009) have said that a positive response to one of the two questions can detect a depressive syndrome, according to the DSM-V criteria, with a sensitivity of 99% and specificity of 87%. PPPCN have to ask all persons the QID-2. If the response of a person is "no" to both questions, the probability of depression in this person is less than 2%. If the response is "yes" to both, then probability is 99% (Alsaleh et al., 2019; Fragoso et al., 2014; Kubitary & Alsaleh, 2018).

## Recommendations the threshold score of QID-2

The threshold score of "2" for the QID-2 is recommended (see Fig. 1) to be more certain that all those with depression are detected without having to use a complete depression questionnaire such as the Beck Depression Inventory (BDI)-II or other depression questionnaires. However, patients may be asked to complete such a questionnaire if their scores are "0" or "1," but not if their score is "2," which, therefore, reduces the case-finding load. A threshold score of 2 has clinical advantages over a score of "1" because patients with depression being correct will be detected (Alsaleh et al., 2019; Kubitary & Alsaleh, 2017, 2018). QID-2 is a useful tool for researchers and practitioners.

## Applications of QID-2 and recommendations

Early identification of depression is the first step for effective treatment, I recommend that patient/person should be systematically screened in the RCPPPPPCH, MMHPCHSW, and LMHPCS using QID-2 that is brief, highly acceptable to patients and staff and reliably detects the presence of depression.

The psychometric properties of the QID-2 (reliability, accuracy, and validity) are highly satisfactory. International studies demonstrate that the QID-2 has adequate psychometric characteristics to allow their use for subjects with depressive disorders. In terms of applicability, the execution takes less time ($\pm 30$ s for QID-2). Therefore, this scale is not a burden for patients or personnel to assess and is very easy (see Fig. 1).

Depression is treatable and is a very important factor in the quality of life in patients. Therefore, clinicians must provide correct screening and diagnosis of depression while avoiding overlap with disease symptoms. However, they have few opportunities to meet this challenge (avoiding false-positive diagnoses and false-negative diagnoses) during their routine consultation. To minimize false diagnoses, the two questions (QID-2) are useful for screening, diagnosis, and clinical trials. Concerning PPPCN, they use the QID-2, to detect depression and alert physicians before confirming the diagnosis with the other scales as the Beck Depression Inventory-Fast Screen (BDI-FS) and the presence of five of the symptoms listed in DSM-V.

Finally, given the prevalence of depression in the population and the impact of the disease on the individual and social level, objective tools for measuring depression are important for therapeutic management and optimal clinical monitoring. This chapter is the first that makes available to PPPCN, as early as possible, a simpler and faster method to evaluate depression in RCPPPPPCH, MMHPCHSW, and LMHPCS to screen depression as soon as possible in all people in any conditions to put in place an effective and adapted treatment.

## Discussion

Now, the epidemic of depression spreads like wildfire in all persons, any age, sex, and time in the world. Depression is a major contributor to poor quality of life, suicide, sustained incapacity, deterioration of emotional and psychological status, lack of sexual desire, and disability. However, there are difficulties to assess the depression in patients due to overlapping symptoms between these conditions; for example, in children and adolescents due to their dysfunctions; in students due to independent living and academic challenges; and in patients due to their physical symptoms.

The need for a careful screening and diagnostic of depressive symptoms and their proper management is primary in MMHPCHSW. Based on studies and literature data, QID-2 is particularly suitable for screening patients and other people because it allows reducing the number of false diagnoses (false-positive diagnoses and false-negative diagnoses) and recognizes subthreshold mood symptoms with minimal contamination by somatic conditions.

Early recognition improves the treatment of depression and their symptoms improving persons' well-being, perception, and satisfaction about the quality of medical care. QID-2 allows assessing the depression in both patients and persons as well as primary caregivers (PCs). QID-2 allows PPPCN to judge the proposed psychotherapy treatment efficacy and the likely course of a disease.

## Implications of QID-2 in care for good clinical practice

The QID-2 is economical in terms of research effort, time, and professional expertise required, under conditions of wartime and in RCPPPPPCH and LMHPCS. QID-2 is more widely used because of its simplicity compared with multiple assessments that can drain limited resources. The QID-2 could also be of value where symptoms need to be measured on a daily basis, such as in the assessment of early antidepressant treatment effects, especially on the two main symptoms of depressive mood and anhedonia. The QID-2 might also be useful as a patient's self-reporting tool or for combined reports by clinicians. The screen has the practical advantages of rapid administration, thereby reducing the burden in both patients and clinicians. In busy outpatient clinics and during wartime, self-assessment scales such as the BDI are time consuming and, therefore, not the most appropriate tools, especially considering that screening for depression can be accurately done through simpler techniques that are easy to implement and less time consuming. Thus, QID-2 may be recommended as a replacement for the BDI and/or other scales because of limited resources particularly during wartime or in busy clinics, the use of other questionnaires to confirm the presence or the absence of depression may be suggested as BDI-FS. In addition, it may not be necessary for patients/persons to complete multiple questionnaires, although referral to a psychiatrist or psychotherapist for interview and treatment is recommended, depending on availability.

## QID-2, depression, clinicians, patients, busy clinics, and wartime

Clinicians should be aware that, at wartime and in busy clinics, patients have a greater risk of depression. They should evaluate depression at every consultation, as it may have negative effects on cognitive function and adherence to treatment.

Patients who already have depression may further develop other, more common, symptoms because of war or long waiting times in busy clinics, such as anxiety and stress or equally distressing psychiatric symptoms. Under normal conditions, clinicians should be alert to the risk of depression and suicide in patients/persons and assess it. During the war, RCPPPPPCH, and LMHPCS, such an evaluation is even more necessary and should be routine for every consultation. For these reasons and more, international studies have validated two questions (QID-2), which provide a much simpler approach.

## Conclusion

QID-2 helps clinicians to correctly screen and diagnose the depression in medical and nonmedical populations of any age in routine consultations, a busy clinic, and wartime for early treatment. QID-2, the simplest, fastest, most valid, most reliable, and most effective, is developed for use by PPPCN in RCPPPPPCH, MMHPCHSW, and reducing burden in patient-clinician.

## Key facts of depression

- According to the World Health Organization, more than 264 million people of all ages are affected by depression.
- The DSM-5 outlines the following criteria to make a diagnosis of depression. The individual must be experiencing five or more symptoms during the same 2-week period and at least one of the symptoms should be either (1) depressed mood or (2) loss of interest or pleasure (anhedonia).
- Studies suggest that depression is associated with a global burden and is an important cause of suicide, burden of disability, dissatisfaction, loss of social function, morbidity and mortality worldwide, reduction in the quality of life, and increase in mortality and desire for hastened death.
- Studies suggest that depression is widely undiagnosed and untreated.
- Depression affects the treatment and prognosis of the disease.

## Key facts of QID-2

- QID-2 (two-question quick inventory of depression) is a scale that is simple and fast, reliable, and effective and has cross-cultural and international reliability, accuracy, and validity, for example, the Arab world
- QID-2 helps PPPCN facilitate, its effective and systematic screening, and diagnosis.
- QID-2 reduces the time and effort required for the screening process as well as the burden on both patients and clinicians, especially in wartime.

## Summary points

- This chapter focuses on the efficiency of the two simple questions assessing depressed mood and anhedonia (two-question quick inventory of depression "QID-2"), which were reformulated according to the DSM-V criteria for major depression, and Arabic culture and society.
- This chapter shows that how depression, which is the most common mood disorder in the world, is widely undiagnosed and untreated.
- Depression is not detected because the screening and diagnosis need improvement.
- The lack of scale simple, lack of developing a universally validated screening tool, and the lack of time in practitioners, psychotherapist, physicians, caretakers, and/or nurses (PPPCN) are the principal facts that do not diagnose depression.
- QID-2 is a simpler and faster method for evaluating the depression, which is to make available for PPPCN in medical settings, in mental health settings, in primary care, hospital settings, and in wartime (MMHPCHSW).
- QID-2 is the most efficient, most reliable, and safest in the world.
- The QID-2 can easily be implemented into the daily clinical practice of mental health professionals, routine consultations, hospitalizations, and in care facilitating the depression diagnostic process, especially in case of comorbid depression.
- In addition, this chapter supported the cross-cultural reliability, accuracy, and validity of the QID-2 among all people in the world, especially in the Arab world and wartime.

## Mini-dictionary of terms

**Depressed mood** A mood involving feeling down, depressed, hopeless, or sad.

**Anhedonia** A condition in which one is unable to experience pleasure. It is the loss of pleasure, little interest, or pleasure in doing things as much as one used to or one used to enjoy.

**QID-2 (screening instrument)** It is not only a valid screening and diagnostic instrument for depression in university life, primary care but also in mental health settings and wartime.

**Dichotomous** It is divided into two parts or classifications, for example, dichotomous yes/no fashion.

**Psychometric** It is the science that helps us to validate and develop the scales in psychology, for example, QID-2.

**Insomnia** A condition in which one has difficult in falling asleep (onset) or staying asleep (maintenance), even when one has the chance to do so.

**Pretreatment** Any treatment received beforehand to make other processes more effective.

## Disclosure of potential conflicts of interest

The authors declare that they have no competing interest.

## References


Alsaleh, M. (2014). Dépistage des états dépressifs: Cas de la personne âgée et de la sclérose en plaques. *11e Journées Européennes de la Dépression.*

Alsaleh, M. (2016a). *Analyse psychosociale et cognitive de la santé mentale chez les étudiants de première année: Validation du questionnaire des pensées positives et négatives et du questionnaire de la dépression de Beck: Effet des pensées positives et des facteurs psychosociaux.* Consulté à l'adresse http://www.theses.fr/2016CAEN1031.

Alsaleh, M. (2016b). *Dépression et Sclérose En Plaques Dépistage, diagnostic, échelles d'évaluation et traitement de la dépression dans la SEP.* Consulté à l'adresse http://nbn-resolving.de/urn:nbn:de:101:1-201610222044.

Alsaleh, M., & Kubitary, A. (2017). *Verdicts of water DropsTRPPT on brain and life new frontiers in neurocognitive positive psychotherapy in war and peace.* Normal-Abnormal-Medical-non-Medical Conditions. Consulté à l'adresse http://nbn-resolving.de/urn:nbn:de:101:1-201702112098.

Alsaleh, M., & Lebreuilly, R. (2017). Validation de la traduction française d'un questionnaire court de dépression de Beck (BDI-FS-Fr). *Annales Médico-psychologiques, revue psychiatrique, 175*(7), 608–616. https://doi.org/10.1016/j.amp.2016.06.015.

Alsaleh, M., Lebreuilly, J., & Defer, G. (2015). Screening and evaluation of the frequency of depressive states in the first years of multiple sclerosis: What impact on the treatment of patients? In *International Mental Health Congress 'Mental health for all: Connecting people and sharing experience'*.

Alsaleh, M., Videloup, L., Lobbedez, T., Lebreuilly, J., Morello, R., & Thuillier Lecouf, A. (2019). Improved detection and evaluation of depression in patients with chronic kidney disease: validity and reliability of screening (PHQ-2) and diagnostic (BDI-FS-Fr) tests of depression in chronic kidney disease. *Kidney Diseases*, 1–11. https://doi.org/10.1159/000497352.

American Psychiatric Association. (2013). *Diagnostic and statistical manual of mental disorders: Dsm-5* (5th revised edition). Washington, DC: American Psychiatric Publishing.

Arroll, B., Goodyear-Smith, F., Crengle, S., Gunn, J., Kerse, N., Fishman, T., et al. (2010). Validation of PHQ-2 and PHQ-9 to screen for major depression in the primary care population. *Annals of Family Medicine, 8*(4), 348–353. https://doi.org/10.1370/afm.1139.

Arroll, B., Khin, N., & Kerse, N. (2003). Screening for depression in primary care with two verbally asked questions: Cross sectional study. *BMJ (Clinical Research Ed.), 327*(7424), 1144–1146. https://doi.org/10.1136/bmj.327.7424.1144.

Baillon, S., Dennis, M., Lo, N., & Lindesay, J. (2014). Screening for depression in Parkinson's disease: The performance of two screening questions. *Age and Ageing, 43*(2), 200–205. https://doi.org/10.1093/ageing/aft152.

Bennett, I. M., Coco, A., Coyne, J. C., Mitchell, A. J., Nicholson, J., Johnson, E., et al. (2008). Efficiency of a two-item pre-screen to reduce the burden of depression screening in pregnancy and postpartum: An IMPLICIT network study. *Journal of the American Board of Family Medicine: JABFM, 21*(4), 317–325. https://doi.org/10.3122/jabfm.2008.04.080048.

Canadian Task Force on Preventive Health Care. (2003). New grades for recommendations from the Canadian Task Force on Preventive Health Care. *Canadian Medical Association Journal, 169*(3), 207–208.

Chae, S. Y., Chae, M. H., Tyndall, A., Ramirez, M. R., & Winter, R. O. (2012). Can we effectively use the two-item PHQ-2 to screen for postpartum depression? *Family Medicine*, *44*(10), 698–703.

Chapman, L. J., Chapman, J. P., & Raulin, M. L. (1976). Scales for physical and social anhedonia. *Journal of Abnormal Psychology*, *85*(4), 374–382.

Der-Avakian, A., & Markou, A. (2012). The neurobiology of anhedonia and other reward-related deficits. *Trends in Neurosciences*, *35*(1), 68–77. https://doi.org/10.1016/j.tins.2011.11.005.

Dwyer Hollender, K. (2014). Screening, diagnosis, and treatment of post-stroke depression. *The Journal of Neuroscience Nursing: Journal of the American Association of Neuroscience Nurses*, *46*(3), 135–141. https://doi.org/10.1097/JNN.0000000000000047.

Feinstein, A. (2011). Multiple sclerosis and depression. *Multiple Sclerosis (Houndmills, Basingstoke, England)*, *17*(11), 1276–1281. https://doi.org/10.1177/1352458511417835.

Fragoso, Y. D., Adoni, T., Anacleto, A., da Gama, P. D., Gonçalves, M. V. M., da Matta, A. P. C., et al. (2014). Recommendations on diagnosis and treatment of depression in patients with multiple sclerosis. *Practical Neurology*, *14*(4), 206–209. https://doi.org/10.1136/practneurol-2013-000735.

Gelaye, B., Wilson, I., Berhane, H. Y., Deyessa, N., Bahretibeb, Y., Wondimagegn, D., et al. (2016). Diagnostic validity of the Patient Health Questionnaire-2 (PHQ-2) among Ethiopian adults. *Comprehensive Psychiatry*, *70*, 216–221. https://doi.org/10.1016/j.comppsych.2016.07.011.

Hackett, M. L., & Jardine, M. J. (2017). We need to talk about depression and dialysis: But what questions should we ask, and does anyone know the answers? *Clinical Journal of the American Society of Nephrology*, *12*(2), 222–224. https://doi.org/10.2215/CJN.13031216.

Keller, M. B., McCullough, J. P., Klein, D. N., Arnow, B., Dunner, D. L., Gelenberg, A. J., et al. (2000). A comparison of nefazodone, the cognitive behavioral-analysis system of psychotherapy, and their combination for the treatment of chronic depression. *The New England Journal of Medicine*, *342*(20), 1462–1470. https://doi.org/10.1056/NEJM200005183422001.

Kroenke, K., Spitzer, R. L., & Williams, J. B. W. (2003). The Patient Health Questionnaire-2: Validity of a two-item depression screener. *Medical Care*, *41*(11), 1284–1292. https://doi.org/10.1097/01.MLR.0000093487.78664.3C.

Kubitary, Amani, Alomer, M., & Alsaleh, M. (2017). Mental Health: Services, assessment and perspectives: Validity of the arabic version of beck depression inventory-fast screen (BDI-FS-Ar) in War conditions: The Psychometric Properties in Medical and Non-Medical Syrian Populations. Consulté 17 octobre 2017, à l'adresse https://www.novapublishers.com/catalog/product_info.php?products_id=62650.

Kubitary, A., & Alsaleh, M. (2017). Validation and reliability studies of instruments: syria-arabic newer self-rating scales for PsyWar syndrome: QID-2-Ar (Quick Inventory of Depression-Two questions-Arabic), BDI-FS-Ar (Beck Depression Inventory-Fast Screen-Arabic), and MADRS-S-Ar (Montgomery-Asberg Depression Rating Scale Self-assessment-Arabic): Clinical assessments of depression and depressive symptoms in Syria-War conditions. *Psychology and Behavioral Science International Journal*, *4*, 555642.

Kubitary, A., & Alsaleh, M. A. (2018). Validity of Arabic version of the two-question Quick Inventory of Depression (QID-2-Ar): Screening for multiple sclerosis in an Arab environment and during the Syrian war. *Revue Neurologique*, *174*(3), 137–144. https://doi.org/10.1016/j.neurol.2017.06.026.

Lebrun, C., & Cohen, M. (2009). Dépression et sclérose en plaques. *Revue Neurologique*, *165*(Suppl. 4), S156–S162. https://doi.org/10.1016/S0035-3787(09)72128-9.

Ledford, H. (2014). Medical research: If depression were cancer. *Nature News*, *515*(7526), 182. https://doi.org/10.1038/515182a.

Li, C., Friedman, B., Conwell, Y., & Fiscella, K. (2007). Validity of the Patient Health Questionnaire 2 (PHQ-2) in identifying major depression in older people. *Journal of the American Geriatrics Society*, *55*(4), 596–602. https://doi.org/10.1111/j.1532-5415.2007.01103.x.

Löwe, B., Kroenke, K., & Gräfe, K. (2005). Detecting and monitoring depression with a two-item questionnaire (PHQ-2). *Journal of Psychosomatic Research*, *58*(2), 163–171. https://doi.org/10.1016/j.jpsychores.2004.09.006.

Mann, J. J. (2005). The medical management of depression. *The New England Journal of Medicine*, *353*(17), 1819–1834. https://doi.org/10.1056/NEJMra050730.

Marrie, R. A., Horwitz, R., Cutter, G., Tyry, T., Campagnolo, D., & Vollmer, T. (2009). The burden of mental comorbidity in multiple sclerosis: Frequent, underdiagnosed, and undertreated. *Multiple Sclerosis (Houndmills, Basingstoke, England)*, *15*(3), 385–392. https://doi.org/10.1177/1352458508099477.

Mohr, D. C., Hart, S. L., Julian, L., & Tasch, E. S. (2007). Screening for depression among patients with multiple sclerosis: Two questions may be enough. *Multiple Sclerosis (Houndmills, Basingstoke, England)*, *13*(2), 215–219. https://doi.org/10.1177/1352458506070926.

Mojtabai, R. (2014). Diagnosing depression in older adults in primary care. *New England Journal of Medicine*, *370*(13), 1180–1182. https://doi.org/10.1056/NEJMp1311047.

National Collaborating Centre for Mental Health (UK). (2010). *Depression: The treatment and management of depression in adults* (Updated Edition). Consulté à l'adresse http://www.ncbi.nlm.nih.gov/books/NBK63748/.

Nature. (2014). The burden of depression. *Nature*, *515*(7526), 163. https://doi.org/10.1038/515163a.

Olivares, J. M., & Berrios, G. E. (1998). The anhedonias: Clinical and neurobiological aspects. *International Journal of Psychiatry in Clinical Practice*, *2*(3), 157–171. https://doi.org/10.3109/13651509809115352.

Rabins, P. V. (2016). Depressive symptoms in ophthalmology patients. *JAMA Ophthalmology*, *134*(9), 1015. https://doi.org/10.1001/jamaophthalmol.2016.2367.

Randall, J. M., Voth, R., Burnett, E., Bazhenova, L., & Bardwell, W. A. (2013). Clinic-based depression screening in lung cancer patients using the PHQ-2 and PHQ-9 depression questionnaires: A pilot study. *Supportive Care in Cancer*, *21*(5), 1503–1507. https://doi.org/10.1007/s00520-012-1712-4.

Seehusen, D. A., Baldwin, L.-M., Runkle, G. P., & Clark, G. (2005). Are family physicians appropriately screening for postpartum depression? *The Journal of the American Board of Family Practice*, *18*(2), 104–112.

Sheeran, T., Reilly, C. F., Raue, P. J., Weinberger, M. I., Pomerantz, J., & Bruce, M. L. (2010). The PHQ-2 on OASIS-C. *Home Healthcare Nurse*, *28*(2), 92–104. https://doi.org/10.1097/NHH.0b013e3181cb560f.

Skokou, M., Soubasi, E., & Gourzis, P. (2012). Depression in multiple sclerosis: A review of assessment and treatment approaches in adult and pediatric populations. *ISRN Neurology*, *2012*, 427102. https://doi.org/10.5402/2012/427102.

Smith, K. (2014). Mental health: A world of depression. *Nature News*, *515*(7526), 180. https://doi.org/10.1038/515180a.

Spitzer, R. L., Kroenke, K., & Williams, J. B. (1999). Validation and utility of a self-report version of PRIME-MD: The PHQ primary care study. Primary Care Evaluation of Mental Disorders. Patient Health Questionnaire. *JAMA*, *282*(18), 1737–1744.

Stewart, D. (2005). Depression during pregnancy. *Canadian Family Physician*, *51*(8), 1061–1063.

Tarrants, M., Oleen-Burkey, M., Castelli-Haley, J., & Lage, M. J. (2011). The impact of comorbid depression on adherence to therapy for multiple sclerosis [research article]. *Multiple Sclerosis International*, *2011*. https://doi.org/10.1155/2011/271321.

U.S. Preventive Services Task Force. (2002). Screening for depression: Recommendations and rationale. *Annals of Internal Medicine*, *136* (10), 760–764.

Whooley, M. A., Avins, A. L., Miranda, J., & Browner, W. S. (1997). Case-finding instruments for depression. *Journal of General Internal Medicine*, *12*(7), 439–445. https://doi.org/10.1046/j.1525-1497.1997.00076.x.

Chapter 24

# Depression and biomarkers of cardiovascular disease

Allison J. Carroll[a] and Olivia E. Bogucki[b]

[a]Department of Psychiatry and Behavioral Sciences, Northwestern University Feinberg School of Medicine, Chicago, IL, United States; [b]Department of Psychiatry and Psychology, Mayo Clinic, Rochester, MN, United States

## List of abbreviations

**ANS** autonomic nervous system
**BNP** brain-type natriuretic peptide
**CAC** coronary artery calcification
**cIMT** carotid intima-media thickness
**CRP** C-reactive protein
**CVD** cardiovascular disease
**HPA** hypothalamic-pituitary-adrenal axis
**HRV** heart rate variability
**IL-1** interleukin-1
**IL-6** interleukin-6
**MetS** metabolic syndrome
**PNS** parasympathetic nervous system
**RSA** respiratory sinus arrythmia
**SNS** sympathetic nervous system

## Depression and cardiovascular disease

Since the 1990s, research has demonstrated a bidirectional relationship between depression and cardiovascular disease (CVD). While preliminary research described the role of depression in mortality and poor outcomes among patients with prevalent CVD (Lichtman et al., 2014), additional studies have given us a greater understanding of *how* depression contributes to the development of CVD (Baune et al., 2012; Parissis et al., 2007). Several biological mechanisms have been implicated in the relationship between depression and CVD, including genetic, metabolic, immune-inflammatory, autonomic, and hypothalamic-pituitary-adrenal (HPA) axis dysregulations (Table 1; Penninx, 2017).

## What is a biomarker?

The term biomarker is an amalgamation of two terms: "biological" and "marker." The National Institutes of Health Biomarkers Definitions Working Group (2001) define a biomarker as, "a characteristic that is objectively measured and evaluated as an indicator of normal biological processes, pathogenic processes, or pharmacologic responses to a therapeutic intervention." The Biomarkers Working Group further identified the various ways in which biomarkers may be used, including screening, diagnosis, classification, prognosis, treatment response, or as a surrogate end point in research to evaluate the safety and efficacy of novel treatments.

As outlined by Singh and Rose (2009), biomarkers have been a trending topic in psychological research since the early 2000s. Multiple goals underlie this line of research. First, the use of biomarkers in diagnostic decision-making has been proposed to increase diagnostic precision. Biological variables are often thought to be more objective than subjective, self-reported symptoms frequently associated with mental health assessments. Second, biomarkers may be useful for screening adults who are at increased risk for psychopathology. Elucidating biologically-based predisposing or precipitating factors for mental health conditions may improve primary prevention efforts in vulnerable populations. Finally, biomarkers may be informative for treatment selection. In line with precision medicine, biological information may help to determine specific psychological or pharmacological treatments that would be most effective for a particular patient.

## Associations between depression and biomarkers of cardiovascular disease

A multitude of biomarkers may indicate that a person is at increased risk for CVD (Vasan, 2006). While a comprehensive review of CVD biomarkers is beyond the scope of this chapter, it is important to note that many of these CVD biomarkers have also been implicated in the development, maintenance, and treatment of depression. Theoretically, depression incites a complex cascade of

The Neuroscience of Depression. https://doi.org/10.1016/B978-0-12-817933-8.00018-9

**TABLE 1 Summary of potential mechanisms linking depression to increased cardiovascular risk.**

| | |
|---|---|
| ***Causal mediating mechanisms*** | |
| Pathophysiology | Metabolic dysregulations |
| | Immuno-inflammatory dysregulations |
| | Automatic dysregulations |
| | HPA axis dysregulations |
| Unhealthy lifestyle | Smoking |
| | Excessive alcohol use |
| | Physical inactivity/sedentary behavior |
| | Unhealthy diet |
| | Poor adherence to and engagement in medical care |
| *Alternative mechanisms* | |
| Residual confounding | Depression picks up on or is prodromal of not yet discovered or not yet measured (sub)clinical conditions |
| Third underlying factors | Childhood stressors |
| | Personality |
| | Genetic pleiotropy |

Table adapted from Penninx, B.W. (2017). Depression and cardiovascular disease: Epidemiological evidence on their linking mechanisms. *Neuroscience and Biobehavioral Reviews, 74*, 277–286, table 1, used with permission from Elsevier.

biological processes, which are often interrelated and interdependent, that ultimately impacts cardiovascular health (Fig. 1; Parissis et al., 2007). Therefore, it stands to reason that the biomarkers implicated in these conditions overlap.

In this chapter, the biomarkers of interest have been classified into three categories: functional, circulating, and structural biomarkers (Table 2). The content herein should be evaluated within the context of the following: (1) research regarding relationships between depression and biomarkers is mixed, with several studies failing to find support for these relationships (see section "Limitations of measuring biomarkers"); (2) it is unlikely that the pathophysiological mechanisms represented by these biomarkers fully explain the relationship between depression and CVD (see section "Psychosocial factors impacting the depression-biomarkers relationship"); (3) this chapter emphasizes the pathway from depression to CVD pathophysiology, though these relationships are likely bidirectional; and (4) this discussion is focused on depression causing initial CVD, rather than depression as it relates to poor prognosis among CVD patients, as the disease process of CVD itself may have deleterious effects on pathophysiological, behavioral, and psychological functioning.

## Functional biomarkers

Functional biomarkers are measured using physiological recordings to evaluate how efficiently organs (e.g., heart) or regulatory systems (e.g., nervous system) are working. They are relatively easy to assess with the correct equipment. Functional biomarkers include measures of autonomic, metabolic, and endothelial dysfunction.

### Autonomic dysfunction

Autonomic dysfunction occurs when there is an imbalance between the two branches of the autonomic nervous system (ANS): the sympathetic (SNS) and parasympathetic (PNS) nervous systems. More specifically, autonomic dysfunction is characterized by increased SNS activation (i.e., the "fight or flight" response) and decreased PNS activation (i.e., the "rest and digest" response). Multiple areas of the brain influence control over the ANS and are implicated in autonomic functioning (Fig. 2; Thayer & Lane, 2009). The ANS is responsible for regulating the activity of organs and therefore has widespread effects throughout the body. Autonomic dysfunction is a broad, overarching term that encompasses multiple biologically based measures. Two measures of autonomic dysfunction are discussed below: cardiac vagal tone and heart rate variability (HRV).

#### *Cardiac vagal tone*

The vagus nerve is the main component of the PNS. One of the 12 cranial nerves that connects the brain and organs, the vagus nerve exerts control over the heart and determines the rate at which the heart beats. Cardiac vagal tone has been defined as the degree of control the vagus nerve exerts on the heart (Laborde, Mosley, & Thayer, 2017), and is commonly represented using respiratory sinus arrythmia (RSA). RSA is defined as the beat-to-beat variability in heart rate specifically during the respiration cycle. RSA approximates the amount of control that the PNS exerts over the heart, with higher values (i.e., greater variability) viewed as more adaptive (Berntson et al., 1997). While adults with and without depression show similar RSA at rest, studies have indicated that adults with depression exhibited lower RSA in response to physical, cognitive, and emotional stress compared to healthy adults (Ehrenthal, Herrmann-Lingen, Fey, & Schauenburg, 2010; Nugent, Bain, Thayer, Sollers 3rd, & Drevets, 2011). In addition, adults with depression who exhibited lower RSA during sad films reported greater depressive symptoms and were less likely to experience

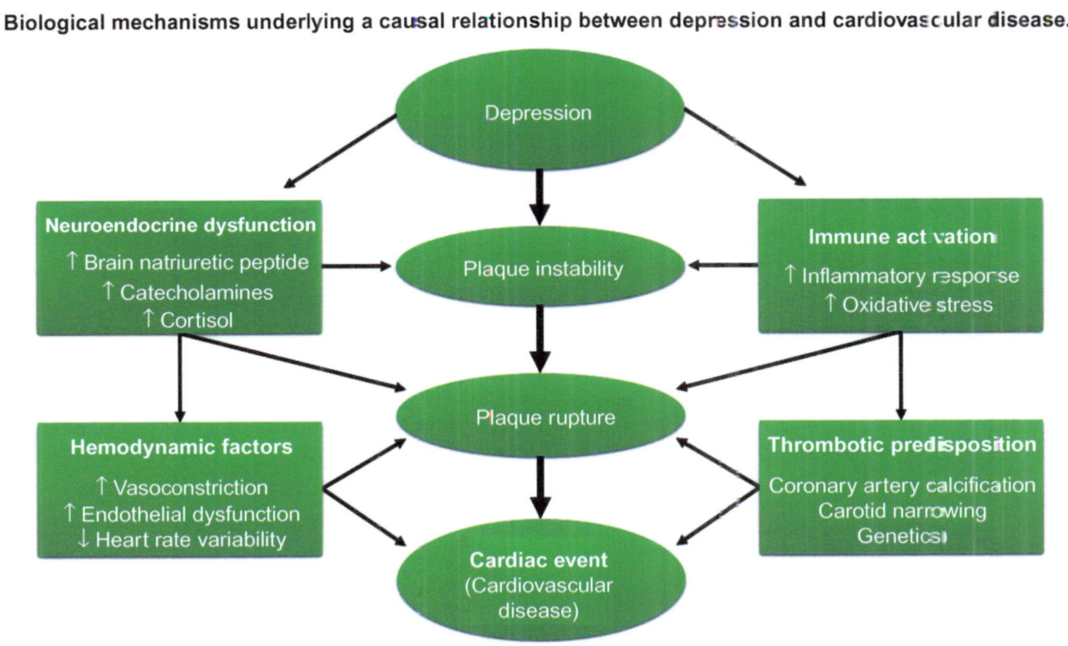


**FIG. 1** This figure demonstrates the complex, intersecting pathways by which depression can lead to cardiovascular disease. *Figure adapted from Parissis, J. T., Fountoulaki, K., Filippatos, G., Adamopoulos, S., Paraskevaidis, I., & Kremastinos, D. (2007). Depression in coronary artery disease: Novel pathophysiologic mechanisms and therapeutic implications.* International Journal of Cardiology, 116, *153–160, fig. 1, used with permission from Elsevier.*

**TABLE 2** Summary of biomarkers of the pathophysiological processes by which depression is associated with increased risk for CVD.

| Category | Biological process | Biomarker |
|---|---|---|
| Functional | Autonomic dysfunction | Cardiac vagal tone<br>Respiratory sinus arrythmia (RSA)<br>Heart rate variability (HRV) |
| | Metabolic dysfunction | Cluster of central obesity, dyslipidemia, high blood pressure, and blood sugar |
| | Endothelial dysfunction | Flow-mediated dilation<br>Cell adhesion molecules<br>P-selection |
| Circulating | Inflammation | C-reactive protein (CRP)<br>Fibrinogen<br>Interleukin 1 (IL-1)<br>Interleukin 6 (IL-6)<br>Tumor necrosis factor<br>Myeloperoxidase |
| | Oxidative stress | 8-OHdG<br>Isoprostanes<br>Malondialdehyde<br>Hexanoyl-lysine<br>Protein carbonyls<br>Antioxidants<br>Antioxidant enzymes |
| | Cardiac stress | Brain-type natriuretic peptide (BNP) |
| | HPA axis stress response | Cortisol |
| | Autonomic dysfunction | Catecholamines |
| Structural | Atherosclerosis | Coronary artery calcification (CAC)<br>Carotid intima-media thickness (cIMT) |

The biomarkers, which represent distinct but interrelated and interdependent biological processes, have been categorized into *functional measures*, assessed by physiological recordings; *circulating measures*, assessed by blood, saliva, urine, or cerebrospinal fluid; and *structural measures*, assessed by imaging techniques.

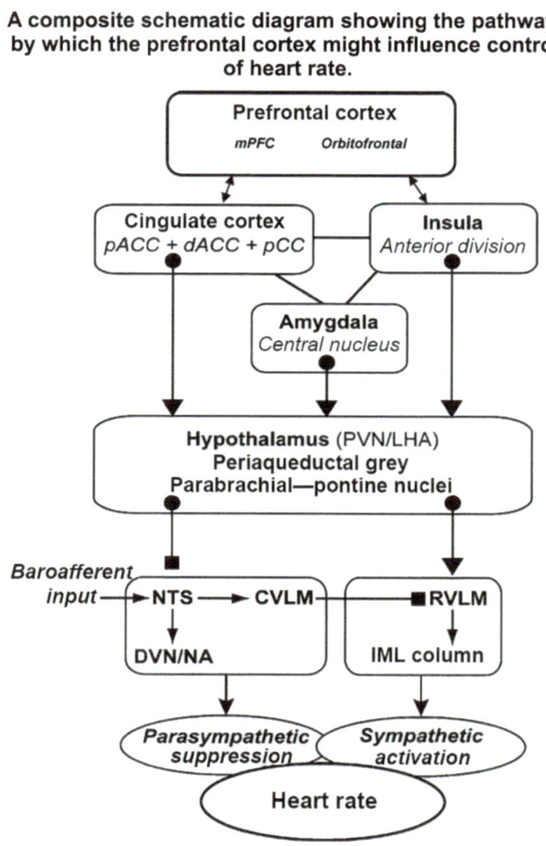


FIG. 2 The complex cascade of biological processes that influences heart rate begins when the prefrontal cortex initiates activation in different brain regions to inhibit the parasympathetic nervous system and activate the sympathetic nervous system, which results in increased heart rate. Abbreviations: *mPFC*, medial prefrontal cortex; *pACC*, perigenual anterior cingulate; *dACC*, dorsal anterior cingulate; *pCC*, posterior cingulate cortex; *PVN*, paraventricular nucleus; *LHA*, lateral hypothalamus; *NTS*, nucleus of the solitary tract; *CVLM*, caudal ventrolateral medullary; *RVLM*, rostral ventrolateral medullary; *DVN/NA*, dorsal vagal motor nucleus/nucleus accumbens; *IML*, intermediolateral nucleus. *Figure from Thayer, J.F., & Lane, R.D. (2009). Claude Bernard and the heart-brain connection: Further elaboration of a model of neurovisceral integration.* Neuroscience and Biobehavioral Reviews, 33, *81–88, fig. 1, used with permission from Elsevier.*

remission of depression over time (Panaite et al., 2016; Rottenberg, Salomon, Gross, & Gotlib, 2005).

### *Heart rate variability*

HRV is the beat-to-beat variability in heart rate. HRV can be calculated using multiple methods, which are thought to represent different aspects of autonomic functioning. HRV is typically thought to index the balance of SNS and PNS control over the heart (Task Force of the European Society of Cardiology and the North American Society of Pacing and Electrophysiology, 1996). Much like RSA, higher values represent more adaptive functioning. A meta-analysis demonstrated that adults with depression exhibited lower HRV at rest compared to healthy adults (Kemp et al., 2010). While preliminary studies have failed to find differences between adults with and without depression experiencing sadness (Bogucki, 2019), research has indicated that adults with depression exhibited lower HRV in response to physical and cognitive stress compared to healthy adults (Kemp et al., 2010; Nugent et al., 2011).

### Metabolic dysfunction

Metabolism is the series of complex biochemical process through which our bodies break down food to be used as energy. Metabolic dysfunction occurs when there are disruptions in these biochemical processes, which lead to disorder and disease. Metabolic syndrome (MetS) is defined by the co-occurrence of abdominal obesity, dyslipidemia, high blood pressure, and hyperglycemia. MetS is composed of independent risk factors for CVD and is itself associated with increased risk for CVD (Alberti et al., 2009). The conditions that comprise MetS are often used as indicators of metabolic dysfunction and have been implicated in depression. A meta-analysis of 39 studies found a significant association between depression and MetS; a subset of 14 prospective studies established a bidirectional relationship between the 2 conditions, with depression inflating the risk for MetS and MetS increasing the likelihood of developing depression (Pan et al., 2012). A follow-up meta-analysis of 18 studies replicated and expanded these results, estimating the prevalence of MetS in depression across studies as 30% (Vancampfort et al., 2014).

### Endothelial dysfunction

Endothelial dysfunction occurs when endothelial cells release lower levels of nitrous oxide, which results in reduced blood vessel dilation and contributes to the development of atherosclerosis. While endothelial dysfunction is implicated in the development of CVD, it also appears to be present in adults with depression who have not yet developed CVD. A meta-analysis of 12 studies found that adults with depression or depressive symptoms exhibited lower flow-mediated dilation compared to healthy adults. A greater effect size was found for studies including adults with CVD or CVD risk factors, both of which are linked to endothelial dysfunction (Cooper et al., 2011). Studies using circulating measures to evaluate endothelial dysfunction have found similar results—for example, adults with depression exhibited elevated levels of some, but not all, cell adhesion molecules (van Dooren et al., 2016).

## Circulating biomarkers

Circulating biomarkers are measured using blood, saliva, urine, or cerebrospinal fluid samples. They are commonly used in research studies as they are relatively easy and inexpensive to obtain. Some examples of circulating biomarkers include markers of inflammation, markers of oxidative stress, brain natriuretic peptide (BNP), cortisol, and catecholamines.

## Inflammation

Inflammation has been identified as a primary driver of the development of atherosclerosis leading to CVD morbidity and mortality (Ross, 1999). Depression is also highly correlated with the inflammatory response and some have gone so far as to consider depression to be an inflammatory disease (Berk et al., 2013), though there is significant debate regarding this hypothesis (e.g., Miller & Raison, 2016). Regardless, via an acute or chronic stress response, depression causes a release of inflammatory biomarkers, thereby inciting a cascade of physiological effects (Fig. 3; Miller et al., 2009).

Levels of circulating inflammatory markers are consistently higher among adults with depression compared to healthy adults. A meta-analysis found small but significant differences between adults with and without depression on measures of C-reactive protein (CRP), interleukin-1 (IL-1), and interleukin-6 (IL-6) (Howren, Lamkin, & Suls, 2009); moreover, there was a dose-dependent relationship between depression and inflammation wherein greater exposure to depression (e.g. greater symptom severity or longer-lasting symptoms) was associated with higher levels of inflammatory biomarkers. Some research has further demonstrated that antidepressant pharmacotherapies are associated with lower levels of inflammatory biomarkers (Lindqvist et al. 2017), indicating that improved depression is associated with a corresponding decrease in the inflammatory response. Several studies have also examined the relationships between depression, inflammatory biomarkers, and cardiac outcomes. In a large sample of women with suspected coronary disease, Vaccarino et al. (2007) found that baseline depression was associated with more than 2.5 times greater risk of having a cardiac event

Stress-induced activation of the inflammatory response.

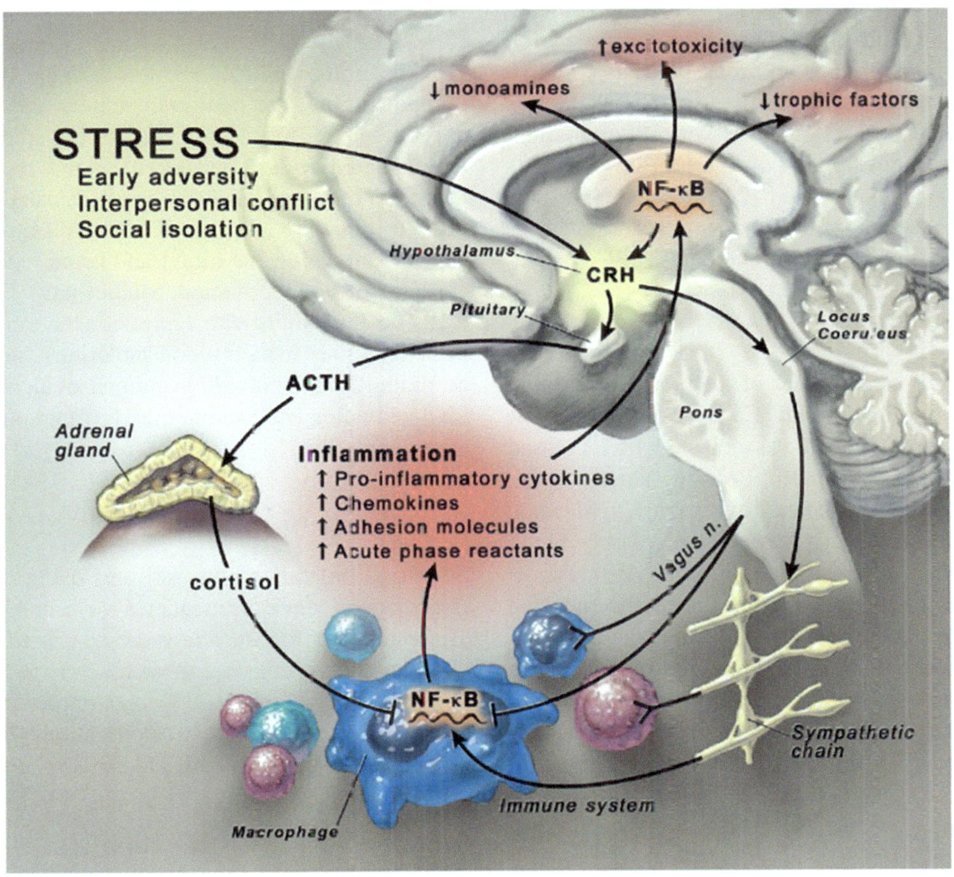


FIG. 3 Psychosocial stressors activate the central nervous system stress circuitry. Catecholamines are released from sympathetic nerve endings, work via the immune system, and result in the release of inflammatory mediators that promote inflammation. In turn, inflammatory biomarkers can affect the brain to induce further inflammatory pathways that affect cardiovascular health. Typically, there is an inhibitory pathway that eventually decreases the inflammatory response. However, in the context of chronic stress (e.g., prolonged depression), activation of inflammatory pathways may become less sensitive to the inhibitory effects, leading to higher, persistent levels of circulating inflammatory biomarkers. Abbreviations: *ACTH*, adrenocorticotropic hormone; *CRH*, corticotropin-releasing hormone; *NF-κB*, nuclear factor kappa B. *Figure from Miller, A. H., Maletic, V., & Raison, C. L. (2009). Inflammation and its discontents: The role of cytokines in the pathophysiology of major depression.* Biological Psychiatry, 65, *732–741, fig. 2, used with permission from Elsevier.*

overtime. However, once the inflammatory biomarkers were accounted for, the association between depression and a cardiac event was attenuated by 13% with CRP and 4% with IL-6, indicating that cardiac risk can be partially attributed to the inflammatory process associated with depression.

### Oxidative stress

Oxidative stress is defined by an abundance of free radicals in the body that overwhelms the antioxidant defense systems, leading to atherosclerosis (Sack, Fyhrquist, Saijonmaa, Fuster, & Kovacic, 2017). It is a multifaceted process that can be captured via multiple biomarkers such as those that index damage to DNA (e.g., 8-OHdG), damage to lipids and proteins (e.g., isoprostanes), or markers of antioxidants or antioxidant enzymes (e.g., superoxide dismutase). Across these processes, studies have consistently shown elevated biomarkers of oxidative damage and depleted levels of antioxidants among adults with depression compared to healthy adults (Black, Bot, Scheffer, Cuijpers, & Penninx, 2015; Palta, Samuel, Miller 3rd, & Szanton, 2014). Importantly, the association between depressive symptoms and measures of oxidative stress is independent of other cardiovascular risk factors (Hirose et al., 2016). Interestingly, as with inflammation, antidepressant treatment appears to have positive effects on oxidative stress (Lindqvist et al., 2017). No studies have yet evaluated the association between depression and oxidative stress in relation to cardiac outcomes.

### Brain natriuretic peptide

BNP is a commonly used marker of diagnosis, prognosis, and treatment of patients with CVD, especially heart failure, and has been shown to be associated with increased risk for incident CVD (Natriuretic Peptides Studies et al., 2016). BNP is released by the heart in response to elevated pressure and strain and plays a role in regulating blood pressure, blood volume, and sodium volume. At this time, few studies have assessed associations between depression and BNP among healthy adults who have not yet developed CVD. One study found depressive symptoms and BNP to be independently, but not synergistically, associated with incident heart failure (van den Broek et al., 2011). In a secondary analysis of a large sample of at-risk primary care patients, depression was associated with elevated BNP levels, even after adjusting for risk factors (Kabutoya, Hoshide, & Kario, 2016).

### Cortisol

Stress, generally speaking, is a well-known contributor to CVD (American Heart Association, 2014). Cortisol is the primary hormone that is released, via the HPA axis, in response to a stressor. Adults with depression, compared to those without, appear to have a dysregulated stress response, as evidenced by impaired recovery following a stressor such that cortisol levels remain elevated for a longer period of time in adults with depression (Burke, Davis, Otte, & Mohr, 2005). Others have shown that the stress response is hypoactive among adults with depression compared to healthy adults (Taylor et al., 2006). However, few studies have directly connected depression with HPA axis reactivity or cortisol levels and subsequent CVD-related outcomes. This lack of evidence may be because stress is such an amorphous term—it functions under both acute and chronic conditions, it occurs across a variety of contexts (e.g., occupational stress, social stress, anticipatory stress, and post-traumatic stress)—and it is implicated more broadly in poor health beyond cardiac function. Therefore, understanding how cortisol is specifically and distinctly associated with depression-related risk for CVD is challenging.

### Catecholamines

Catecholamines, which include dopamine, epinephrine, and norepinephrine, are neurotransmitters and hormones produced in response to stress by the sympathetic-adrenomedullary axis and adrenal gland, respectively. After the stress response has been initiated, catecholamines continue to circulate (Brown, 2006). Consequently, levels of catecholamines in the blood and urine have been used as an index of SNS activation. Studies have shown that medically healthy adults with depression have elevated levels of catecholamines in their blood and urine (Carney, Freedland, & Veith, 2005), providing indirect evidence of increased SNS activation (i.e., autonomic dysfunction) in individuals with depression.

## Structural biomarkers

Structural biomarkers are assessed using radiation, radiotracer, sound wave, or magnetic wave imaging techniques. These markers are currently the best predictors of future CVD, and some consider these markers to actually be subclinical measures of CVD. Two examples of structural biomarkers are coronary artery calcification (CAC) and carotid intima-media thickness (cIMT).

### Coronary artery calcification

CAC, assessed by noninvasive electron-beam computed tomography, is a measure of plaque buildup in the arteries that supply blood directly to the heart. CAC is an independent and consistent predictor of future cardiac events among otherwise healthy adults, where the presence of any CAC is associated with significantly increased risk for CVD (Greenland, LaBree, Azen, Doherty, & Detrano, 2004). Depression has been associated with CAC, with larger effects when depression is assessed as a clinical diagnosis (as opposed to depressive symptoms) or when it is present at multiple exams (Lin, Zhang, & Ma, 2018, e.g., Fig. 4). As is the case with CVD, there are race- and sex-specific associations between depression and CAC. In a large study of women's lifetime health, black women demonstrated a stronger association between depression and CAC than white women (Lewis et al., 2009). Finally, evidence supports a mediated or moderated association between depression and CAC by other CVD risk factors, such as cigarette smoking (Carroll et al., 2017). While this is an exciting avenue for research, CAC is primarily used clinically to stratify at-risk individuals and is not indicated for multiple measurements.

## Carotid intima-media thickness

cIMT measured via ultrasound assesses the extent of carotid atherosclerotic vascular disease within the carotid artery, which transports blood from the heart to the brain. Depression has been associated with the presence of cIMT, especially among older adults, as well as the progression of cIMT when assessed longitudinally (Faramawi et al., 2007; Pizzi et al., 2014). However, there is compelling evidence that this association is mediated by other risk factors, such as inflammation (Jorge et al., 2017). In addition, a reverse causation relationship is prominent between vascular disease or events (e.g., stroke) and depression, suggesting that the directionality of this association may be primarily that vascular disease, or progression of disease as is being assessed by cIMT, was present first and subsequently caused depression (Prugger et al., 2015).

## Psychosocial factors impacting the depression-biomarkers relationship

Myriad behavioral, psychological, and social factors have been implicated in the relationship between depression and CVD, including these pathophysiological mechanisms (Hare, Toukhsati, Johansson, & Jaarsma, 2014). First, many researchers believe that most, if not all, of the association between depression and CVD is related to behavioral factors such as cigarette smoking, alcohol and substance use, physical inactivity, poor diet, and medical nonadherence, all of which are more prevalent among adults with depression (Penninx, 2017). In addition, these poor health behaviors tend to cluster together and thus interdependently increase risk for CVD (Spring, Moller, & Coons, 2012).

Second, psychological factors often present together. This includes comorbid psychiatric diagnoses: an estimated 50%–90% of adults presenting with depression also present with anxiety, and vice versa (Gorman, 1996). Accumulation of life stress via exposure to acute and chronic stressors is related to cardiac functioning and negatively impacts cardiovascular health, potentially via stress-induced depression as well as independent pathways (Slavich, 2016). Finally, various personality traits and characteristics have demonstrated associations with cardiovascular health and disease onset, including both risk (e.g., type D personality) and protective factors (e.g., optimism) (Sahoo, Padhy, Padhee, Singla & Sarkar, 2018).

Third, social, cultural, and contextual factors, often referred to as social determinants of health, can influence biomarkers and thus the relationship between depression and CVD. For example, exposure to violence or other traumas, discrimination, food insecurity, and level of acculturation, among many other factors, can negatively impact cardiovascular functioning (Everson-Rose & Lewis, 2004; Flores-Torres et al., 2017). On the other hand, social factors,

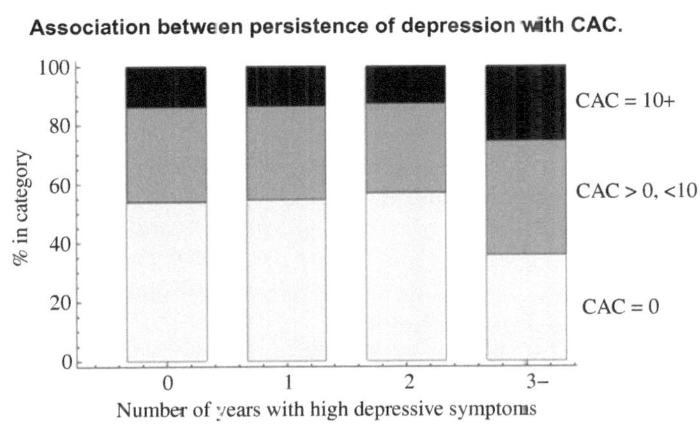


FIG. 4 Level of CAC score by number of years with elevated depressive symptoms (assessed by Center for Epidemiologic Studies Depression (CES-D) scale ≥16) among women in the Study of Women's Health Across the Nation (SWAN) Heart study. Women with more persistent depression had greater prevalence of CAC. Abbreviations: *CAC*, coronary artery calcification. *Figure from Janssen, I., Powell, L. H., Matthews, K. A., Jasielec, M. S., Hollenberg, S. M., Bromberger, J. T., et al. (2016). Relation of persistent depressive symptoms to coronary artery calcification in women aged 46 to 59 years.* American Journal of Cardiology, 117, *1884–1889, fig. 1, used with permission from Elsevier.*

such as (perceived) social support, can be cardioprotective (Everson-Rose & Lewis, 2004). Though difficult to implement, these factors should be considered in the study of depression and CVD.

## Limitations of measuring biomarkers

While biomarkers may offer a more objective measure of a disease or disorder, there are limitations to their use. First, CVD biomarkers currently have limited reliability. As detailed in this chapter, research has identified many CVD biomarkers that have been associated with depression. However, consistency is limited across the literature, with some studies failing to replicate results. While this is a common phenomenon throughout the literature, the cause of contradictory evidence with regard to biomarkers in particular may be due to natural variability within clinical populations (Strimbu & Tavel, 2010). That is to say, depression is a highly heterogeneous disorder characterized by diverse etiology (e.g., vascular), presentation (e.g., atypical depression), and course (e.g., treatment-resistance) across individuals that may result in distinct biomarker profiles.

Second, these biomarkers currently have limited validity. To be applied in clinical practice, biomarkers must accurately capture the construct of depression as it relates to CVD. The available biomarkers lack sufficient sensitivity and specificity, or the ability to correctly categorize patients, when used alone. Thus, multiple biomarkers and measures may be needed for diagnostic, prognostic, and treatment evaluation purposes (Schmidt, Shelton, & Duman, 2011).

Finally, there is a significant financial cost associated with measuring biomarkers. Some biomarker categories (e.g., structural) are expensive to collect and analyze. While other biomarker categories (e.g., circulating) are relatively inexpensive on an individual level, the costs associated with these tests can be in the multimillion or billion dollar range when conducted at a population level (Ioannidis & Tzoulaki, 2012). In addition, several of these biomarkers have limited clinical utility and thus are less likely to be available for low-cost observational research. Consequently, the expenses associated with biomarkers could place a significant burden on the health-care system.

## Future directions

Despite the variety of biomarkers of cardiovascular health that have been investigated in relation to depression, it remains a newer area of research and thus is open for multiple avenues for ongoing and new exploration. For example, it is becoming increasingly important to cross-tabulate biomarkers that represent similar and distinct pathophysiology to begin to limit the focus of research on the most relevant mechanistic pathways. Likewise, given the abundance of evidence indicating that pathophysiological mechanisms are at least partially responsible for the connection between depression and CVD, it is important to begin including and examining these measures in clinical treatment trials to move the research from bench to bedside. This includes both behavioral and pharmacological treatments, such as the application of anti-inflammatory drugs to treat depression (Maes et al., 2012). Finally, as the science of genomics advances, this is an area of exploration to identify at-risk individuals and targeted treatment.

## Summary and conclusions

Depression is associated with CVD, but the mechanisms of action have yet to be fully elucidated. In this chapter, we outlined research supporting relationships between depression and biomarkers of various pathophysiological mechanisms related to CVD. Specifically, we explored functional, circulating, and structural biomarkers and evaluated the evidence by which they provide evidence for depression as an independent risk factor for CVD. Despite this evidence, pathophysiological mechanisms are likely influenced by numerous biological and psychosocial factors and, therefore, their study via biomarkers is significantly limited. Finally, biomarkers present with several methodological and economic concerns. Ongoing and future research will pursue various avenues to identify options to effectively treat depression and prevent CVD.

## Key facts

Key facts of depression and cardiovascular disease:

- Depression is associated with increased risk of future CVD.
- Among adults with CVD, depression predicts poor cardiac outcomes and mortality.
- Adults with CVD have 2–4 times higher rates of depression than healthy adults.
- Severity of CVD is unrelated to severity of depression, except among adults with heart failure.
- Tailored behavioral treatments for depression among adults with CVD effectively reduce depressive symptoms and improve quality of life.
- Treating depression among adults with CVD has not been shown to reduce risk of mortality.

## Summary points

- Depression is associated with functional, circulating, and structural biomarkers of CVD.
- A biomarker is an objective measure of a pathophysiological process.

- Functional biomarkers measure autonomic, metabolic, and endothelial processes.
- Circulating biomarkers have bidirectional associations with depression.
- Structural biomarkers are the best predictors of future CVD, but have limited clinical utility.
- Biomarker research is limited by reliability, validity, and economic implications.
- Future directions in this research include narrowing the focus of the pathophysiology and expanding treatment options for depression and CVD prevention.

## Mini-dictionary of terms

**Autonomic dysfunction** Imbalance between the sympathetic and parasympathetic nervous systems

**Brain natriuretic peptide** Hormone secreted by the heart in response to pressure and strain

**Cardiac vagal tone** Amount of control the vagus nerve exerts on the heart

**Carotid intima-media thickness** Amount of narrowing of the carotid artery

**Catecholamines** Neurotransmitters and hormones secreted by the sympathetic-adrenomedullary axis and adrenal glands in response to stress

**Coronary artery calcification** Amount of calcium buildup in the arteries

**Cortisol** Stress hormone released by the adrenal glands

**Endothelial dysfunction** Disruptions in the release of nitrous oxide by endothelial cells

**Heart rate variability** Beat-to-beat variability in heart rate

**Inflammation** Immune system response to remove harmful stimuli and promote healing

**Metabolic dysfunction** Disruptions in the biochemical process to break down food into energy

**Oxidative stress** Imbalance between free radicals and antioxidants

**Respiratory sinus arrhythmia** Beat-to-beat variability in heart rate specifically during the respiration cycle

## References


Alberti, K. G., Eckel, R. H., Grundy, S. M., Zimmet, P. Z., Cleeman, J. I., Donato, K. A., et al. (2009). Harmonizing the metabolic syndrome: A joint interim statement of the International Diabetes Federation Task Force on Epidemiology and Prevention; National Heart, Lung, and Blood Institute; American Heart Association; World Heart Federation; International Atherosclerosis Society; and International Association for the Study of Obesity. *Circulation*, *120*, 1640–1645.

American Heart Association. (2014). *Stress and heart health*. Retrieved 12/1/2019 from https://www.heart.org/en/healthy-living/healthy-lifestyle/stress-management/stress-and-heart-health.

Baune, B. T., Stuart, M., Gilmour, A., Wersching, H., Heindel, W., Arolt, V., et al. (2012). The relationship between subtypes of depression and cardiovascular disease: A systematic review of biological models. *Translational Psychiatry*, *2*, e92.

Berk, M., Williams, L. J., Jacka, F. N., O'Neil, A., Pasco, J. A., Moylan, S. et al. (2013). So depression is an inflammatory disease, but where does the inflammation come from? *BMC Medicine*, *11*, 200.

Berntson, G. G., Bigger, J. T., Jr., Eckberg, D. L., Grossman, P., Kaufmann P. G., Malik, M., et al. (1997). Heart rate variability: Origins, methods and interpretive caveats. *Psychophysiology*, *34*, 623–648.

Biomarkers Definitions Working Group, Atkinson, A. J., Jr., Colburn, W A., DeGruttola, V. G., DeMets, D. L., Downing, G. J., et al. (2001). Biomarkers and surrogate endpoints: Preferred definitions and conceptual framework. *Clinical Pharmacology and Therapeutics*, *69*, 89–95.

Black, C. N., Bot, M., Scheffer, P. G., Cuijpers, P., & Penninx, B. W. (2015). Is depression associated with increased oxidative stress? A systematic review and meta-analysis. *Psychoneuroendocrinology*, *51*, 164–175.

Bogucki, O. E. (2019). *Cognitive, mood, and cardiovascular reactivity and recovery in response to sadness in remitted major depressive disorder*. Doctoral dissertation. Retrieved 9/7/2020 from https://digitalcommons.library.umaine.edu/etd/3047/.

Brown, D. E. (2006). Measuring hormonal variation in the sympathetic nervous system: Catecholamines. In G. D. James, & G. E. Ice (Eds.), *Measuring stress in humans: A practical guide for the field* (pp. 94–121). Cambridge: Cambridge University Press.

Burke, H. M., Davis, M. C., Otte, C., & Mohr D. C. (2005). Depression and cortisol responses to psychological stress: A meta-analysis. *Psychoneuroendocrinology*, *30*, 846–856.

Carney, R. M., Freedland, K. E., & Veith, R. C. (2005). Depression, the autonomic nervous system, and coronary heart disease. *Psychosomatic Medicine*, *67*(Suppl. 1), S29–S33.

Carroll, A. J., Carnethon, M. R., Liu, K., Jacobs, D. R., Colangelo, L. A., Stewart, J. C., et al. (2017). Interaction between smoking and depressive symptoms with subclinical heart disease in the Coronary Artery Risk Development in Young Adults (CARDIA) study. *Health Psychology*, *36*, 101–111.

Cooper, D. C., Tomfohr, L. M., Milic, M. S., Natarajan, L., Bardwell, W. A., Ziegler, M. G., et al. (2011). Depressed mood and flow-mediated dilation: A systematic review and meta-analysis. *Psychosomatic Medicine*, *73*, 360–369.

Ehrenthal, J. C., Herrmann-Lingen, C., Fey, M., & Schauenburg, H. (2010). Altered cardiovascular adaptability in depressed patients without heart disease. *The World Journal of Biological Psychiatry*, *11*, 586–593.

Everson-Rose, S. A., & Lewis, T. T. (2004). Psychosocial factors and cardiovascular diseases. *Annual Review of Public Health*, *26*, 469–500.

Faramawi, M. F., Gustat, J., Wildman, R. P., Rice, J., Johnson, E., & Sherwin, R. (2007). Relation between depressive symptoms and common carotid artery atherosclerosis in American persons > or = 65 years of age. *The American Journal of Cardiology*. *99*, 1510–1613.

Flores-Torres, M. H., Lynch, R., Lopez-Ridaura, R., Yunes, E., Monge, A., Ortiz-Panozo, E., et al. (2017). Exposure to violence and carotid artery intima-media thickness in Mexican women. *Journal of the American Heart Association*, *6*, e006249.

Gorman, J. M. (1996). Comorbid depression and anxiety spectrum disorders. *Depression and Anxiety*, *4*, 160–168.

Greenland, P., LaBree, L., Azen, S. P., Doherty, T. M., & Detrano, R. C. (2004). Coronary artery calcium score combined with Framingham score for risk prediction in asymptomatic individuals. *JAMA*, *291*, 210–215.

Hare, D. L., Toukhsati, S. R., Johansson, P., & Jaarsma, T. (2014). Depression and cardiovascular disease: A clinical review. *European Heart Journal*. *35*, 1365–1372.

Hirose, A., Terauchi, M., Akiyoshi, M., Owa, Y., Kato, K., & Kubota, T. (2016). Depressive symptoms are associated with oxidative stress in middle-aged women: A cross-sectional study. *BioPsychoSocial Medicine*, *10*, 12.

Howren, M. B., Lamkin, D. M., & Suls, J. (2009). Associations of depression with C-reactive protein, IL-1, and IL-6: A meta-analysis. *Psychosomatic Medicine*, *71*, 171–186.

Ioannidis, J. P., & Tzoulaki, I. (2012). Minimal and null predictive effects for the most popular blood biomarkers of cardiovascular disease. *Circulation Research*, *110*, 658–662.

Jorge, A., Lertratanakul, A., Lee, J., Pearce, W., McPherson, D., Thompson, T., et al. (2017). Depression and progression of subclinical cardiovascular disease in systemic lupus erythematosus. *Arthritis Care & Research (Hoboken)*, *69*, 5–11.

Kabutoya, T., Hoshide, S., & Kario, K. (2016). Abstract 17718: Depression was associated with brain natriuretic peptide and home morning systolic blood pressure in male patients at risk for cardiovascular disease. *Circulation*, *134*, A17718.

Kemp, A. H., Quintana, D. S., Gray, M. A., Felmingham, K. L., Brown, K., & Gatt, J. M. (2010). Impact of depression and antidepressant treatment on heart rate variability: A review and meta-analysis. *Biological Psychiatry*, *67*, 1067–1074.

Laborde, S., Mosley, E., & Thayer, J. F. (2017). Heart rate variability and cardiac vagal tone in psychophysiological research—Recommendations for experiment planning, data analysis, and data reporting. *Frontiers in Psychology*, *8*, 213.

Lewis, T. T., Everson-Rose, S. A., Colvin, A., Matthews, K., Bromberger, J. T., & Sutton-Tyrrell, K. (2009). Interactive effects of race and depressive symptoms on calcification in African American and white women. *Psychosomatic Medicine*, *71*, 163–170.

Lichtman, J. H., Froelicher, E. S., Blumenthal, J. A., Carney, R. M., Doering, L. V., Frasure-Smith, N., et al. (2014). Depression as a risk factor for poor prognosis among patients with acute coronary syndrome: Systematic review and recommendations: A scientific statement from the American Heart Association. *Circulation*, *129*, 1350–1369.

Lin, S., Zhang, H., & Ma, A. (2018). The association between depression and coronary artery calcification: A meta-analysis of observational studies. *Journal of Affective Disorders*, *232*, 276–282.

Lindqvist, D., Dhabhar, F. S., James, S. J., Hough, C. M., Jain, F. A., Bersani, F. S., et al. (2017). Oxidative stress, inflammation and treatment response in major depression. *Psychoneuroendocrinology*, *76*, 197–205.

Maes, M., Fisar, Z., Medina, M., Scapagnini, G., Nowak, G., & Berk, M. (2012). New drug targets in depression: Inflammatory, cell-mediated immune, oxidative and nitrosative stress, mitochondrial, antioxidant, and neuroprogressive pathways. And new drug candidates—Nrf2 activators and GSK-3 inhibitors. *Inflammopharmacology*, *20*, 127–150.

Miller, A. H., Maletic, V., & Raison, C. L. (2009). Inflammation and its discontents: The role of cytokines in the pathophysiology of major depression. *Biological Psychiatry*, *65*, 732–741.

Miller, A. H., & Raison, C. L. (2016). The role of inflammation in depression: From evolutionary imperative to modern treatment target. *Nature Reviews. Immunology*, *16*, 22–34.

Natriuretic Peptides Studies Collaboration, Willeit, P., Kaptoge, S., Welsh, P., Butterworth, A. S., Chowdhury, R., et al. (2016). Natriuretic peptides and integrated risk assessment for cardiovascular disease: An individual-participant-data meta-analysis. *The Lancet Diabetes and Endocrinology*, *4*, 840–849.

Nugent, A. C., Bain, E. E., Thayer, J. F., Sollers, J. J., 3rd, & Drevets, W. C. (2011). Heart rate variability during motor and cognitive tasks in females with major depressive disorder. *Psychiatry Research*, *191*, 1–8.

Palta, P., Samuel, L. J., Miller, E. R., 3rd, & Szanton, S. L. (2014). Depression and oxidative stress: Results from a meta-analysis of observational studies. *Psychosomatic Medicine*, *76*, 12–19.

Pan, A., Keum, N., Okereke, O. I., Sun, Q., Kivimaki, M., Rubin, R. R., et al. (2012). Bidirectional association between depression and metabolic syndrome: A systematic review and meta-analysis of epidemiological studies. *Diabetes Care*, *35*, 1171–1180.

Panaite, V., Hindash, A. C., Bylsma, L. M., Small, B. J., Salomon, K., & Rottenberg, J. (2016). Respiratory sinus arrhythmia reactivity to a sad film predicts depression symptom improvement and symptomatic trajectory. *International Journal of Psychophysiology*, *99*, 108–113.

Parissis, J. T., Fountoulaki, K., Filippatos, G., Adamopoulos, S., Paraskevaidis, I., & Kremastinos, D. (2007). Depression in coronary artery disease: Novel pathophysiologic mechanisms and therapeutic implications. *International Journal of Cardiology*, *116*, 153–160.

Penninx, B. W. (2017). Depression and cardiovascular disease: Epidemiological evidence on their linking mechanisms. *Neuroscience and Biobehavioral Reviews*, *74*, 277–286.

Pizzi, C., Costa, G. M., Santarella, L., Flacco, M. E., Capasso, L., Bert, F., et al. (2014). Depression symptoms and the progression of carotid intima-media thickness: A 5-year follow-up study. *Atherosclerosis*, *233*, 530–536.

Prugger, C., Godin, O., Perier, M. C., Ritchie, K., Helmer, C., Empana, J. P., et al. (2015). Longitudinal association of carotid plaque presence and intima-media thickness with depressive symptoms in the elderly: The three-city study. *Arteriosclerosis, Thrombosis, and Vascular Biology*, *35*, 1279–1283.

Ross, R. (1999). Atherosclerosis—An inflammatory disease. *The New England Journal of Medicine*, *340*, 115–126.

Rottenberg, J., Salomon, K., Gross, J. J., & Gotlib, I. H. (2005). Vagal withdrawal to a sad film predicts subsequent recovery from depression. *Psychophysiology*, *42*, 277–281.

Sack, M. N., Fyhrquist, F. Y., Saijonmaa, O. J., Fuster, V., & Kovacic, J. C. (2017). Basic biology of oxidative stress and the cardiovascular system: Part 1 of a 3-part series. *Journal of the American College of Cardiology*, *70*, 196–211.

Sahoo, S., Padhy, S. K., Padhee, B., Singla, N., & Sarkar, S. (2018). Role of personality in cardiovascular diseases: An issue that needs to be focused too! *Indian Heart Journal*, *70*(Suppl. 3), S471–s77.

Schmidt, H. D., Shelton, R. C., & Duman, R. S. (2011). Functional biomarkers of depression: Diagnosis, treatment, and pathophysiology. *Neuropsychopharmacology*, *36*, 2375–2394.

Singh, I., & Rose, N. (2009). Biomarkers in psychiatry. *Nature*, *460*, 202–207.

Slavich, G. M. (2016). Life stress and health: A review of conceptual issues and recent findings. *Teaching of Psychology*, *43*, 346–355.

Spring, B., Moller, A. C., & Coons, M. J. (2012). Multiple health behaviours: Overview and implications. *Journal of Public Health (Oxford, England)*, *34*(Suppl. 1), i3–i10.

Strimbu, K., & Tavel, J. A. (2010). What are biomarkers? *Current Opinion in HIV and AIDS*, *5*(6), 463–466.

Task Force of the European Society of Cardiology and the North American Society of Pacing and Electrophysiology. (1996). Heart rate variability: Standards of measurement, physiological interpretation and clinical use. *Circulation*, *93*, 1043–1065.

Taylor, C. B., Conrad, A., Wilhelm, F. H., Neri, E., DeLorenzo, A., Kramer, M. A., et al. (2006). Psychophysiological and cortisol responses to psychological stress in depressed and nondepressed older men and women with elevated cardiovascular disease risk. *Psychosomatic Medicine*, *68*, 538–546.

Thayer, J. F., & Lane, R. D. (2009). Claude Bernard and the heart-brain connection: Further elaboration of a model of neurovisceral integration. *Neuroscience and Biobehavioral Reviews*, *33*, 81–88.

Vaccarino, V., Johnson, B. D., Sheps, D. S., Reis, S. E., Kelsey, S. F., Bittner, V., et al. (2007). Depression, inflammation, and incident cardiovascular disease in women with suspected coronary ischemia: The National Heart, Lung, and Blood Institute-sponsored WISE study. *Journal of the American College of Cardiology*, *50*, 2044–2050.

van den Broek, K. C., Defilippi, C. R., Christenson, R. H., Seliger, S. L., Gottdiener, J. S., & Kop, W. J. (2011). Predictive value of depressive symptoms and B-type natriuretic peptide for new-onset heart failure and mortality. *The American Journal of Cardiology*, *107*, 723–729.

van Dooren, F. E., Schram, M. T., Schalkwijk, C. G., Stehouwer, C. D., Henry, R. M , Dagnelie, P. C., et al. (2016). Associations of low grade inflammation and endothelial dysfunction with depression—The Maastricht Study. *Brain, Behavior, and Immunity*, *56*, 390–396.

Vancampfort, D., Correll, C. U., Wampers M., Sienaert, P., Mitchell, A. J., De Herdt, A., et al. (2014). Metabolic syndrome and metabolic abnormalities in patients with major depressive disorder: A meta-analysis of prevalences and moderating variables *Psychological Medicine*, *44*, 2017–2028.

Vasan, R. S. (2006) Biomarkers of cardiovascular disease: Molecular basis and practical considerations. *Circulation*, *113*, 2335–2362.

Chapter 25

# Thioredoxin as an antioxidant protein as a marker in depression

Efruz Pirdogan Aydin[a] and Ece Turkyilmaz Uyar[b]
[a]*Department of Psychiatry, University of Health Sciences, Sisli Etfal Teaching and Research Hospital, Istanbul, Turkey;* [b]*Department of Psychiatry, Okmeydani Teaching and Research Hospital, Istanbul, Turkey*

## Abbreviations

| | |
|---|---|
| **8-OHdG** | 8-hydroxy-2′-deoxyguanosine |
| **ADHD** | attention deficit hyperactivity disorder |
| **AP-1** | activator protein 1 |
| **ASK1** | apoptosis signal-regulating kinase 1 |
| **CAT** | catalase |
| **CNS** | central nervous system |
| **GSH** | glutathione |
| **Gpx** | glutathione peroxidase |
| **HIF-1α** | hypoxia-inducible factor 1 |
| **HPA** | hypothalamic-pituitary-adrenal |
| **IDO** | indoleamine 2,3-dioxygenase |
| **IFN-α** | interferon alpha |
| **IL** | interleukin |
| **JNK** | c-Jun-N-terminal kinase |
| **MCI** | mild cognitive impairment |
| **MDA** | malondialdehyde |
| **NF-κB** | nuclear factor kappa B |
| **NLRP3** | Nod-like receptor protein-3 |
| **Nrf-2** | nuclear factor erythroid 2-related factor-2 |
| **OCD** | obsessive-compulsive disorder |
| **PTSD** | post-traumatic stress disorder |
| **ROS** | reactive oxygen species |
| **RNR** | ribonucleotide reductase |
| **RNS** | reactive nitrogen species |
| **SOD** | superoxide dimutase |
| **Sp-Trx** | spermatid-specific thioredoxin |
| **SSRIs** | selective serotonin reuptake inhibitors |
| **TAOC** | total antioxidant capacity |
| **TNF** | tumor necrosis factor |
| **Trx** | thioredoxin |
| **TrxR** | thioredoxine reductase |
| **Txip** | thioredoxin-interacting protein |

## Introduction

Depressive disorder is a multifactorial disease, which is accompanied by varying degrees of neurovegetative, psychological, and cognitive symptoms. Depressive disorder is usually observed comorbid with age-related somatic diseases such as cardiovascular disease, diabetes, arthritis (Read, Sharpe, Modini, & Dear, 2017) and cancer (Krebber et al., 2014) as well as neurodegenerative diseases such as multiple sclerosis (Morris et al., 2018), Alzheimer's, Parkinson's, and Huntington's disease (Galts et al., 2019). A considerable number of patients with depression present with a neuroprogressive course with relapses despite treatment, unresponsiveness to treatment, cognitive impairment following the episodes, and an increased risk of age-related somatic diseases (Bakunina, Pariante, & Zunszain, 2015; Criolo, Grande, Martin-Santos, Vieta, & Carvalho, 2018). Although the ethiopathogenesis of depression is based on the monoamine deficiency hypothesis due to the treatment effect of antidepressants, this hypothesis is insufficient today to explain the aforementioned points. Neuroinflammation and oxidative stress are believed to play an important role in the development of depression as in the neurodegenerative diseases (Bakunina et al., 2015; Maes et al., 2009; Maurya et al., 2016). Long-term IFN-α treatment in chronic hepatitis C patients leads to depressive-like symptoms and a tumor necrosis factor-alpha antagonist, etanercept used in psoriasis has antidepressant effect. These two observations above show that systemic and central nervous system inflammation and oxidative stress have an important role in depression (Dantzer, O'Connor, Lawson, & Kelley, 2011).

People with traumatic childhood experiences and stressful life events may be predisposed to developing depression. Chronic and repetitive psychological and physical stress leads to hyperactivation of hypothalamic-pituitary-adrenal (HPA) axis and induces inflammatory response (Dean & Keshavan, 2017; Maes et al., 2009). Increased pro-inflammatory cytokines stimulate indoleamine 2,3-dioxygenase (IDO) in tryptophan mechanism and activate kynurenine pathway, which in turn decreases serotonin levels and causes accumulation of neurotoxic metabolites such as quinolinic acid (Dantzer et al., 2011).

The Neuroscience of Depression. https://doi.org/10.1016/B978-0-12-817933-8.00024-4

Due to their effects of increasing oxidative stress proinflammatory cytokines, hypercortisolemia and neurotoxins are responsible for neurodegenerative processes in depression (Dean & Keshavan, 2017; Maes et al., 2009). The dominance of such neurodegenerative factors over neuroprotective factors leads to a decrease in neurogenesis, neuronal apoptosis, and decreased neuronal plasticity at relevant regions. The relationship between inflammation and oxidative stress, which are responsible for ethiopathogenesis of depression, is shown in Fig. 1. In depression, these neurodegenerative processes lead to structural changes in prefrontal cortex, hippocampus, and amygdala, which are brain areas responsible for emotional regulation, response to stress, and cognitive functions (Schmaal et al., 2016).

Reactive oxygen species (ROS) and reactive nitrogen species (RNS) are free radicals formed as a result of mitochondrial metabolic processes or inflammation. Under normal conditions, ~~formed~~ ROS and RNS are buffered by coenzyme Q10, α-tocopherol (vitamin E), vitamin C and some antioxidant proteins such as thioredoxin (Trx) and glutathione (GSH) as well as enzymes like superoxide dismutase (SOD), glutathione peroxidase (Gpx), catalase (CAT), and thioredoxine reductase (TrxR) (Nordberg & Arnér, 2001). GSH and Trx systems account for the majority of cellular redox balance (Sen & Chakraborty, 2011) and play an important role in neurodegenerative diseases (Conrad, Schick, & Angeli, 2013; Ren et al., 2017; Silva-Adaya, Gonsebatt, & Guevara, 2014). Under pathological conditions, imbalance between these systems increases oxidative stress, which damages biomolecules like cellular proteins, lipids, and DNA (Czarny, Wigner, Galecki, & Sliwinski, 2018; Halliwell, 2006; Maes et al., 2009).

Accumulated oxidative stress causes accelerated cellular aging and organ malformations by damaging these macromolecules and shortening telomere lengths (Wolkowitz et al., 2011). Lifelong oxidative damages result in aging, cognitive decline, age-related disorders, and depression (Dröge & Schipper, 2007; Liguori et al., 2018).

## Role of oxidative stress in depressive disorder

There is growing evidence that inflammation and oxidative stress play a major role in depression (Bakunina et al., 2015; Black, Bot, Scheffer, Cuijpers, & Penninx, 2015; Czarny et al., 2018; Liu et al., 2015; Maes et al., 2009). In postmortem studies of patients with depression, it is reported that there is increased inflammation and apoptosis in samples taken from prefrontal cortex (Shelton et al., 2011) and RNA damage is increased in hippocampal cells as a result of oxidative stress (Che, Wang, Shao, & Young, 2010).

A meta-analysis study reported that 8-hydroxy-2′-deoxyguanosine (8-OHdG) indicating DNA damage and F2-isoprostanes indicating lipid damage are increased in patients with depression as oxidative stress markers (Black et al., 2015). Moreover, in other meta-analyses, increased levels of lipid peroxidation (Mazereeuw, Herrmann, Andreazza, Khan, & Lanctôt, 2015) and decreased levels of antioxidants (Liu et al., 2015) are found. In addition, many studies reported low levels of antioxidants such as vitamin E (Owen, Batterham, Probst, Grenyer, & Tapsell, 2005) and CoQ10 (Maes et al., 2009); alteration in levels of antioxidant enzymes such as SOD (Bilici et al., 2001; Herken et al., 2007; Liu et al., 2015), CAT (Camkurt, Fındıklı, İzci, Kurutaş, & Tuman,

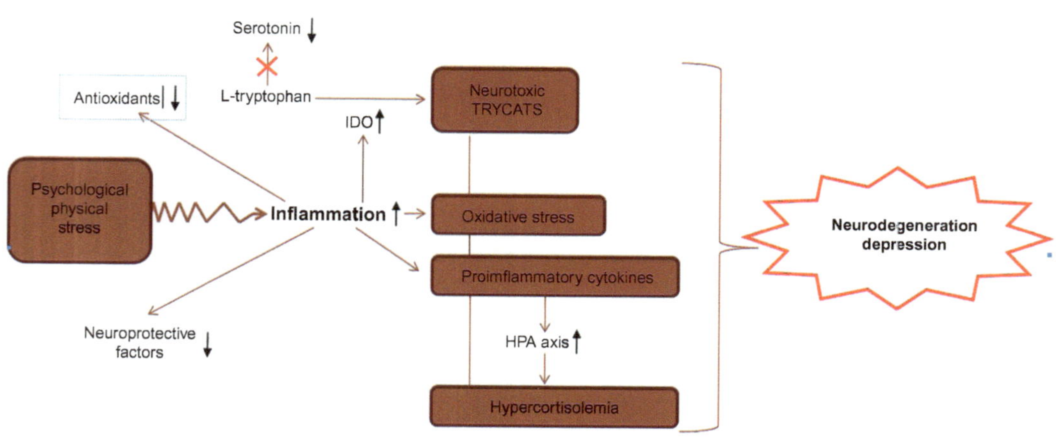


FIG. 1 The role of inflammation and oxidative stress in depression ethiopathogenesis. Inflammation is induced by psychological and physical stress in depression. The inflammatory response increases the pro-inflammatory cytokines, oxidative stress while decrease antioxidants and neuroprotective factors. The pro-inflammatory cytokines cause to increase indoleamine-2,3-dioxygenase (IDO) levels with inactivate L-tryptophan-serotonin pathways and consequently accumulation of neurotoxic tryptophan catabolites (TRYCATs) in brain. And, the pro-inflammatory cytokines stimulate hypothalamic-pituitary-adrenal (HPA) axis and then result in hypercortisolemia. Finally, neurotoxic TRYCATS, oxidative stress, pro-inflammatory cytokines, and hypercortisolemia lead to neurodegeneration in depression.

2016; Gałecki, Szemraj, Bieńkiewicz, Zboralski, & Gałecka, 2009; Tsai & Huang, 2016), GPx (Gałecki et al., 2009; Sarandol et al., 2007; Stefanescu & Ciobica, 2012); and decreased total antioxidant capacity (TAOC) (Cumurcu, Ozyurt, Etikan, Demir, & Karlidag, 2009; Liu et al., 2015; Sarandol et al., 2007). Furthermore, decreased levels of oxidative stress and inflammation were reported in patients whose depressive symptoms improved after treatment with selective serotonin reuptake inhibitors (SSRIs), while no decrease was detected in levels of baseline pro-inflammatory cytokines and oxidative stress in patients with treatment-resistant depression (Cumurcu et al., 2009; Lindqvist et al., 2017). Some biomarkers indicating oxidative stress and inflammation in depression are summarized in Table 1. In addition to these, thioredoxin (Trx) was reported to be a good biomarker indicating oxidative stress in many disorders (Lillig & Holmgren, 2007). However, there are only a few studies conducted

**TABLE 1 A list of biomarkers for depression monitoring**

| Mechanisms | Biomarkers | Results | Consistency in multiple studies | References[a] |
|---|---|---|---|---|
| **Inflammatory biomarkers** | | | | |
| Acute phase protein | CRP | ↑ | Y | Howren, Lamkin, and Suls (2009), Haapakoski, Mathieu, Ebmeier, Alenius, and Kivimäki (2015), and Valkanova, Ebmeier, and Allan (2013) |
| Cytokines | IL-1β | ↑/No change | N | Dowlati et al. (2010), Haapakoski et al. (2015), Howren et al. (2009), and Köhler et al. (2017) |
| | IL-6 | ↑ | Y | Dowlati et al. (2010), Haapakoski et al. (2015), Liu, Ho, and Mak (2012), and Köhler et al. (2017) |
| | TNF-α | ↑/No change | N | Dowlati et al. (2010), Haapakoski et al. (2015), Liu et al. (2013), and Köhler et al. (2017) |
| | sIL-2R | ↑ | Y | Liu et al. (2013) and Köhler et al. (2017) |
| | IL-1RA, IL-12, IL-13, IL-18, sTNFR2 | ↑ | – | Köhler et al. (2017) |
| | IFN-γ | ↓/No change | N | Dowlati et al. (2010) and Köhler et al. (2017) |
| | IL-10 | ↑/No change | N | Dowlati et al. (2010) and Köhler et al. (2017) |
| Chemokines | CCL2, CXCL8 | ↑ | Y | Eyre et al. (2016), Köhler et al. (2017), and Leighton et al. (2018) |
| | CXCL4, CXCL7 | ↑ | – | Leighton et al. (2018) |
| | CCL4 | ↓ | – | Leighton et al. (2018) |
| Neopterin | | ↑ | – | *Maes, Mihaylova, Kubera, and Ringel (2012)* |
| Lipid preoxidation | F2-isoprostenes | ↑ | Y | Black et al. (2015) and Liu et al. (2015) |
| | MDA | ↑ | Y | Mazereeuw et al. (2015) and Liu et al. (2015) |
| DNA damage | 8-OHdG | ↑ | – | Black et al. (2015) |
| Enzymatic antioxidants | CAT | ↑/No change | N | *Camkurt et al. (2016), Gałecki et al. (2009), and Tsai and Huang (2016)* |
| | GPx | ↓/↑ No change | N | *Gałecki et al. (2009), Sarandol et al. (2007), and Stefanescu and Ciobica (2012)* |
| | SOD | ↑/No change | N | *Bilici et al. (2001), Herken et al. (2007), and Liu et al. (2015)* |
| | PON | ↓ | – | Liu et al. (2015) |

*Continued*

**TABLE 1** A list of biomarkers for depression monitoring—cont'd

| Mechanisms | Biomarkers | Results | Consistency in multiple studies | References |
|---|---|---|---|---|
| **Inflammatory biomarkers** | | | | |
| Nonenzymatic antioxidants | HDL | ↓ | – | Liu et al. (2015) |
| | coQ10 | ↓ | – | *Maes et al. (2009)* |
| | Vitamin C | ↓ | – | Liu et al. (2015) |
| | Vitamin E | ↓ | – | *Maes et al. (2000) and Owen et al. (2005)* |
| | Zinc | ↓ | Y | Liu et al. (2015) and Swardfager et al. (2013) |
| Others | TAC | ↓ | – | Liu et al. (2015) |

[a]*Meta-analytic studies are shown in normal font and others in* italic *font.*
*CAT,* catalase; *coQ10,* coenzyme Q10; *CRP,* C-reactive protein; *Gpx,* glutathione peroxidase; *MDA,* malondialdehyde; *PON,* paraoxonase; *SOD,* superoxide dismutase; *TAC,* total antioxidant capacity; *8-OHdG,* 8-hydroxy-2′-deoxyguanosine.

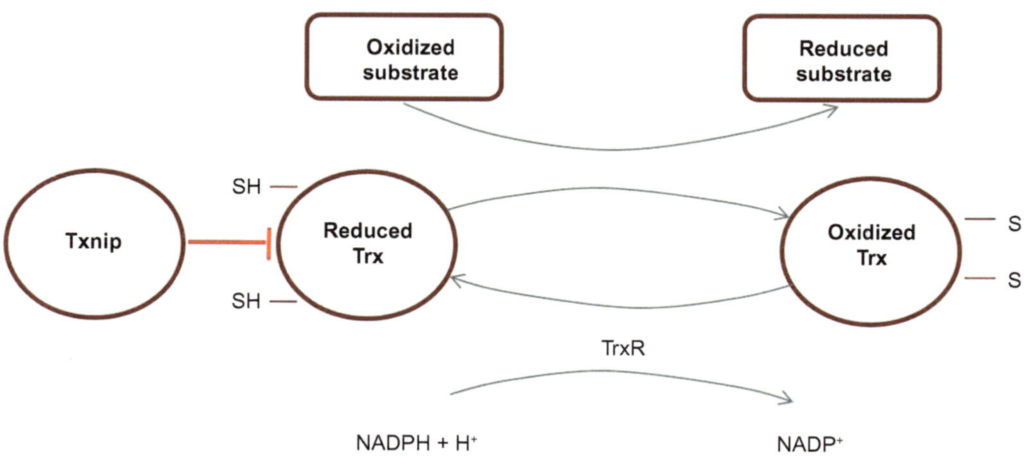


FIG. 2 Thioredoxin antioxidant system. Trx system contains Trx protein, thioredoxin reductase (TrxR), NADPH coenzyme, and thioredoxin-interacting protein (Txip). Trx can find in reduced form (Trx-$(SH)_2$) and oxidized form (Trx-$S_2$) at redox reactions. Oxidized Trx is converted to active and reduced form by receiving electrons from NADPH via TrxR. Active Trx can reduce oxidized substrate to reduced substrate. Txnip can bind reduced Trx, thus inhibits Trx.

with Trx in depression. Unfortunately, data from these studies are yet limited and the results are conflictive (Aydın et al., 2018; Bharti, Tan, Deol, Wu, & Wang, 2019; Wang et al., 2015; Zhou, Tan, Letourneau, & Wang, 2019).

## Thioredoxin antioxidant system

Trx system consists of Trx protein, thioredoxin reductase (TrxR), NADPH coenzyme, and thioredoxin-interacting protein (Txip) (Matsuzawa, 2017) (Fig. 2). Trx protein is a part of the antioxidant system. Trx family of proteins includes cytosolic Trx (Trx1), mitochondrial Trx (Trx2), and spermatid-specific isoform Trx (Sp-Trx). The most extensively studied member of the family is Trx1. Active catalytic fragment of Trx1 consists of two thiol moieties within redox-active cysteine residues (Cys32 and Cys35 are active sites for Trx1) (Nordberg & Arnér, 2001). Several important cellular functions such as maintenance of redox balance, signal transduction, gene transcription, enzymatic reaction, metal binding, and structural stabilization are regulated by transformation of dithiol (–SH–SH) to disulfide (–S–S) by these protein cysteine residues (Lu & Holmgren, 2012). Trx can reverse protein oxidative modifications such as sulfenylation and nitrosylation and reduces oxidized peroxiredoxins by acting as an electron transmitter for oxidized peroxiredoxins, and thus, facilitates the clearance of peroxides (Zhou et al., 2019). Oxidized Trx is converted to active and reduced form by receiving electrons from NADPH via TrxR (Matsuzawa, 2017; Nordberg & Arnér, 2001). While ROS and RNS are disrupting factors in maintenance of integrity and stability of DNA, Trx is involved in DNA replication and repair by reducing disulfide bonds of ribonucleotide reductase (RNR) (Nordberg & Arnér, 2001). Oxidative stress and TNF-α activate apoptosis signal-regulating kinase 1 (ASK1). Consequently, when ASK1-c-Jun-N-terminal kinase (JNK)/p38 signaling pathway is activated, cellular apoptosis and inflammation process begins. When the cell is in resting state, Trx1 binds to ASK1 and ASK1 remains as an inactive complex (Silva-Adaya et al., 2014; Spindel, World, & Berk, 2012). Thus, Trx antioxidant system has a major role in inhibition of inflammation and cellular apoptosis (Lu & Holmgren, 2012). In addition, Trx regulates gene expression by reducing cysteine residues and leads to bind target DNA sites of several transcription factors such as hypoxia-inducible factor 1 (HIF-1α) associated with hypoxia; activator protein 1 (AP-1) associated with cellular growth, differentiation, and adhesion; nuclear factor erythroid 2-related factor-2 (Nrf-2) associated with antioxidant defenses; p53 associated with apoptosis; nuclear factor kappa B (NF-κB) associated with inflammation mediators; and estrogen and glucocorticoid receptors (Lu & Holmgren, 2012; Matsuzawa, 2017; Nordberg & Arnér, 2001; Silva-Adaya et al., 2014). Roles of Trx described above are shown in Fig. 3.

Txnip inhibits Trx by interacting with active site of Trx and consequently, causes oxidative stress by increasing nitrosylation and sulfenylation in proteins as well as inflammation and apoptosis by activation of ASK1/JNK-p38 pathway (Spindel et al., 2012; Zhou et al., 2019). Txnip also binds to nod-like receptor protein-3 (NLRP3), and thus, facilitates the formation of NLRP3 inflammasome complex form. This causes conversion from pro-caspase-1 to caspase-1 and activation of caspase-1 results in formation of mature interleukin (IL)-1β and IL-18, contributing to inflammation process (Zhou et al., 2019) (Fig. 4). Collectively, these data suggest that Trx system also plays a role in various biological functions such as regulation of apoptosis, redox regulation of transcription factors, cellular defense against oxidative stress, and immunomodulation.

Increased oxidative stress is thought to cause different pathophysiological conditions in different tissues or organs. An increase in Trx levels has observed due to oxidative stress in age-related diseases such as cardiovascular disorder (Nagarajan, Oka, & Sadoshima, 2017), lung disorders (Xu, Li, Wu, & Xu, 2012), diabetes (Vencjärvi et al., 2014), and cancer (Arner & Holmgren, 2006). These studies demonstrated that Trx is a novel biomarker of oxidative stress

## Trx system in neurodegenerative diseases

High levels of glucose metabolism and oxygen consumption in brain cause overproduction of ROS. In addition, brain becomes even more sensitive to oxidative stress because of the reasons like reduced activities of some antioxidant enzymes in the brain such as SOD, CAT and Gpx; high levels of metal ions resulting in the production of free radicals; vulnerability to ROS attacks due to the richness in polyunsaturated fatty acids; increased NO

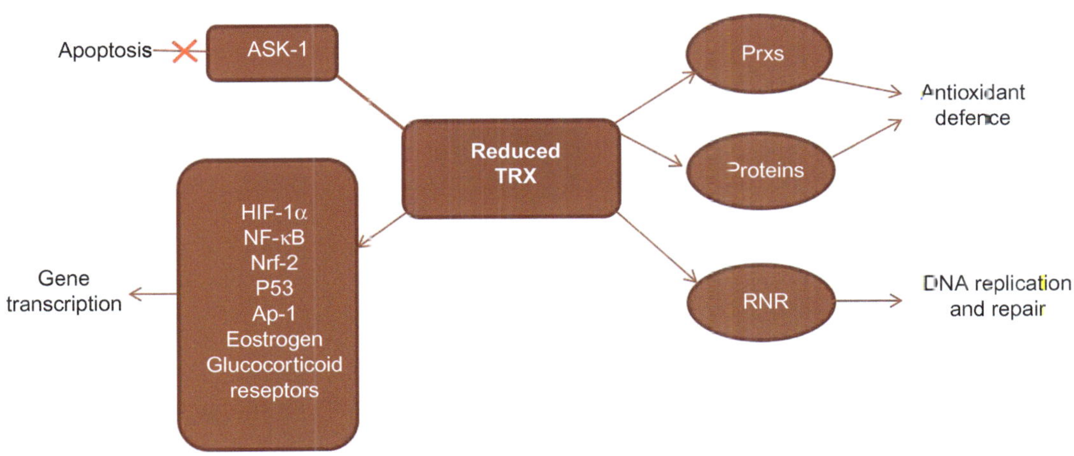


FIG. 3 Functions of thioredoxin. Thioredoxin (Trx) can reduce oxidized proteins and peroxiredoxins by transformation disulfide to dithiol bonds. Trx involves in DNA replication and repair by reducing disulfide bonds of ribonucleotide reductase (RNR) and regulate gene transcription by reducing hypoxia-inducible factor 1 (HIF-1α), activator protein 1 (AP-1), nuclear factor erythroid 2-related factor-2 (Nrf-2), p53, nuclear factor kappa B (NF-κB), estrogen, and glucocorticoid receptors. Reduced Trx prevents apoptosis via binding to apoptosis signal-regulating kinase 1 (ASK-1).

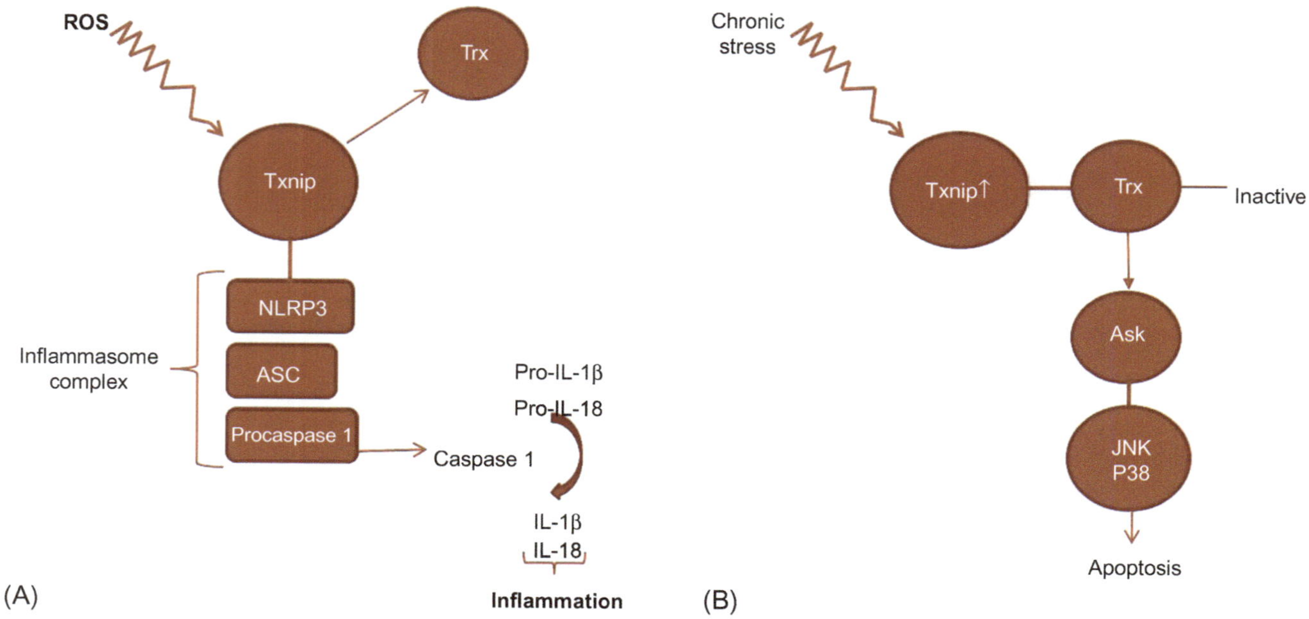


FIG. 4 (A) The role of thioredoxin-interacting protein. When reactive oxygen species (ROS) increase, thioredoxin (Trx) dissociates from thioredoxin-interacting protein (Txnip). Txnip can bind nod-like receptor protein-3 (NLRP3)-inflammasome complex. This leading from pro-caspase-1 to caspase-1, subsequently mature IL-1β and IL-18. ASC, apoptosis-associated speck-like. (B) Txnip levels are increased by chronic stress. Txnip binding to Trx and Trx transforms inactive, in this way apoptosis signal-regulating kinase 1 (ASK1)/c-Jun-N-terminal kinase (JNK)-p38 pathways activate.

production as a result of activated microglial cells as an immune system response; and glutamate receptor-mediated excitotoxicity leading to neuronal death (Halliwell, 2006). Aging is a major risk factor in neurodegenerative diseases in association with increased ROS production. Increased apoptosis and accumulation of proteins, DNA and membrane damage cause ischemia/reperfusion injuries and neurodegenerative diseases of the brain. Impairments in several cellular functions such as elimination of ROS, mitochondrial functions, protein degradation, and neuroinflammatory response contribute to these pathological processes (Halliwell, 2006; Ren et al., 2017; Silva-Adaya et al., 2014). As two major antioxidant systems working in parallel with each other, Trx and GSH have an important role in redox signaling of brain (Conrad et al., 2013; Ren et al., 2017). Trx proteins are widely expressed in different sites of mammary CNS (Silva-Adaya et al., 2014). It has demonstrated that by aging cellular redox balance is disrupted due to increased Txnip expression and decreased Trx activation (Oberacker et al., 2018). Several studies showed increases in oxidative stress and changes in Trx levels in many disorders from mild cognitive impairment (MCI) to neurodegenerative diseases such as Alzheimer's and Parkinson's disease (Ren et al., 2017; Silva-Adaya et al., 2014). In addition, rat studies reported that Trx1 is involved in neuronal development (Silva-Adaya et al., 2014), plays a neuroprotective role in optic nerve against neurodegenerative effects of cytokines (Kitaoka et al., 2016), increases neurogenesis in hippocampus, and ensures cognitive improvement in cerebral ischemia (Tian et al., 2014).

However, there are limited studies on Trx system in psychiatric disorders. Studies in schizophrenia, which is among the neurodegenerative diseases, reported an increase in Trx levels in first-episode patients (Bas et al., 2017; Owe-Larsson et al., 2011; Zhang et al., 2009) and chronic patients (Owe-Larsson et al., 2011; Zhang et al., 2013). Decreased levels of Trx were observed in patients with bipolar disorder during manic episodes compared to healthy subjects (Genc et al., 2015). High levels of Trx were found in neurodevelopmental disorders such as autism (Zhang, Gao, & Zhao, 2015) and attention deficit hyperactivity disorder (ADHD) (Guney et al., 2019) in which Trx was suggested to be diagnostic biomarker. On the other hand, no changes were reported in Trx levels of patients with obsessive-compulsive disorder (OCD) (Bilgiç, Çolak Sivri, & Kılınç, 2018). Studies on Trx in psychiatric disorders are shown in Table 2.

Relationship between oxidative stress and impairment of cognitive functions has been extensively studied within the context of neurodegenerative diseases and aging. Several studies showed a strong relationship between cognitive functions and Trx levels (Aydın et al., 2018; Bas et al., 2017; Yang, Liu, Xue, & Chen, 2012; Zhang et al., 2013). Hyperbaric oxygen therapy in post-traumatic stress disorder (PTSD)-induced rats is reported to increase TrxR

**TABLE 2 Studies on thioredoxin in psychiatric disorders**

| Psychiatric disorders | Trx levels | Cognitive functions—Trx correlations | References |
|---|---|---|---|
| Schizoprenia—first episode | ↑ | Psychmotor speed, attention, visual screening | Bas et al. (2017) |
| | ↑ | – | Owe-Larsson et al. (2011) |
| | ↑ | – | Zhang et al. (2009) |
| Schizoprenia—chronic | ↑ | – | Owe-Larsson et al. (2011) |
| | ↔ | Attention | Zhang et al. (2013) |
| Bipolar disorder—manic episode | ↓ | – | Genc et al. (2015) |
| MDD—treatment-resistant | ↔ | Verbal fluency | Aydın et al. (2018) |
| Autism | ↑ | – | Zhang et al. (2015) |
| ADHD | ↑ | Not correlate to executive functions | Guney et al. (2019) |
| OCD | ↔ | – | Bilgic et al. (2013) |

*ADHD*, attention deficit hyperactivity disorder; *MDD*, major depressive disorder; *OCD*, obsessive-compulsive disorder; *Trx*, thioredoxin; ↑: increase; ↓: decrease; ↔: no change.

expression in hippocampus and thus, reduce anxiety-like symptoms and eliminate learning and memory impairments (Peng et al., 2010). An association is reported between Trx levels and cognitive areas such as attention (Bas et al., 2017; Zhang et al., 2013), psychomotor speed, and visual screening (Bas et al., 2017).

## Trx system in depression

In literature, there are limited studies investigating Trx in depression and chronic stress models. In a study, rats were exposed to intermittent cold for 14 days and it is observed that the resulting chronic stress causes increase in Trx levels in hippocampus (Wang et al., 2015).

In another study, rats were exposed to chronic unpredictable stress that resulted in depression-like behaviors. This study showed no changes in Trx levels of rats but reported increases in protein cysteine sulfenylation and nitrosylation; ASK1 and Txnip levels in hippocampus and frontal cortex (Zhou et al., 2019). Same study suggested that increased Txnip levels in hippocampus and frontal cortex initiated oxidative stress process at relevant sites by reducing activation of Trx and activating ASK1 signaling pathway and NLRP3 inflammation pathway and thus, led to depressive-like condition. In another study, HT22 rat hippocampal cells were incubated with fluoxetine or venlafaxine for 5 days. Levels of Trx and TrxR were observed to increase after treatment independent of increased ROS in hippocampal cells. It was emphasized that antidepressants may confer a protective effect against oxidative stress by increasing Trx levels and Trx can be a potential therapeutic target in treatment of depression (Bharti et al., 2019). However, in our study, Trx levels of treatment-resistant depression patients were not statistically significantly different than those of healthy controls. The fact that these treatment-resistant patients were on long-term antidepressant therapy might have influenced Trx levels. As no other study has yet been conducted in patients with first-episode depression or without treatment, it is difficult to interpret the results of Trx levels in depressive patients with available limited data.

It is also hard to comment on the activation of Trx since Txnip and TrxR levels were not evaluated concomitantly in our study. By the way, we found that depressive patients with high levels of Trx had better verbal fluency (Aydın et al., 2018). Impairments are observed in verbal fluency, memory, and executive functions in patients with depression and these impairments can be persistent even in the remission periods (Knight & Baune, 2018). Repetitive and untreated depression and late-life depression can result in cognitive impairments and declines or even, can progress to dementia (Diniz, Butters, Albert, Dew, & Reynolds, 2013; Panza et al., 2010). Neuroinflammatory changes and oxidative damage in brain are thought to have an important role in this process, which is associated with increasing cognitive impairments from depression to dementia (Herbert & Lucassen, 2016). Therefore, our study supports that Trx is protective against oxidative stress and positively affects cognitive functions in patients with depression (Aydın et al., 2018).

In conclusion, all of the above studies indicate that the data on Trx are yet limited. However it is supposed that Trx system may have an important place in depression, neurodegenerative, and age-related diseases because of its close relationship with oxidative stress, apoptosis, and inflammation pathways and it is a potential therapeutic target in treatment of depression. Thus, there is a need for further studies to be conducted with Trx in the future.

## Key facts of thioredoxin

- Thioredoxin protein has two cysteine residues in its active site that is a part of antioxidant system.
- Thioredoxin system consists of Trx protein, thioredoxin reductase, NADPH coenzyme, and thioredoxin-interacting protein.
- Trx system also plays a role in various biological functions such as regulation of apoptosis, redox regulation of transcription factors, cellular defense against oxidative stress, and immunomodulation.
- Thioredoxin is a good marker for showing oxidative stress.
- Thioredoxin proteins are widely expressed in mammary central nervous system.
- Thioredoxin involves neuronal development, plays neuroprotective role, and increases neurogenesis.

## Summary points

- Neuroinflammation and oxidative stress are believed to play an important role in development of depression.
- Oxidative stress, pro-inflammatory cytokines, hypercortisolemia, and neurotoxins are responsible for neurodegenerative process in depression.
- Thioredoxin is reported to be a good biomarker indicating oxidative stress in many disorders.
- Thioredoxin level is altered in many neurodegenerative disease and is indicated to be associated with cognitive function.
- Thioredoxin may be a potential therapeutic target in treatment of depression though data on thioredoxin are yet limited.

## Mini-dictionary of terms

**Neurodegeneration** Is the progressive loss of structure or function of neurons, including apoptosis, decreased neurogenesis, and neuronal plasticity.

**Neuroprotective factor** That factors are serving to protect neurons from injury and degeneration that comprises brain-derived neurotrophic factor (BDNF), fibroblast growth factor (FGF), neural cell adhesion molecule (NCAM), omega-3 fatty acids, etc.

**Reactive oxygen and nitrogen species** Are free radicals such as superoxide radical ($O_2\bullet^-$), hydrogen peroxide ($H_2O_2$), hydroxyl radical (•OH), nitric oxide (NO), and peroxynitrite ($ONOO^-$) that produced by all aerobic organisms in result of mitochondrial metabolic process or inflammation.

**Oxidative stress** Is an imbalance between the production of free radicals and antioxidant defenses in body.

**Redox balance** It is an equilibrium of "reduction" and "oxidation" reactions in cell.

**Treatment-resistant depression** Inadequate response to antidepressant treatments of enough doses and duration.

## References

Arner, E. S., & Holmgren, A. (2006, December). The thioredoxin system in cancer. *Seminars in Cancer Biology*, *16*(6), 420–426. Academic Press.

Aydın, E. P., Genç, A., Dalkıran, M., Uyar, E. T., Deniz, I., Özer, Ö. A., et al. (2018). Thioredoxin is not a marker for treatment-resistance depression but associated with cognitive function: An rTMS study. *Progress in Neuro-Psychopharmacology and Biological Psychiatry*, *80*, 322–328.

Bakunina, N., Pariante, C. M., & Zunszain, P. A. (2015). Immune mechanisms linked to depression via oxidative stress and neuroprogression. *Immunology*, *144*(3), 365–373.

Bas, A., Gultekin, G., Incir, S., Bas, T. O., Emul, M., & Duran, A. (2017). Level of serum thioredoxin and correlation with neurocognitive functions in patients with schizophrenia using clozapine and other atypical antipsychotics. *Psychiatry Research*, *247*, 84–89.

Bharti, V., Tan, H., Deol, J., Wu, Z., & Wang, J. F. (2019). Upregulation of antioxidant thioredoxin by antidepressants fluoxetine and venlafaxine. *Psychopharmacology*, 1–10.

Bilgiç, A., Çolak Sivri, R., & Kılınç, İ. (2018). 8-F2-isoprostane, thioredoxin and thioredoxin reductase levels in children with obsessive–compulsive disorder. *Nordic Journal of Psychiatry*, *72*(7), 484–488.

Bilici, M., Efe, H., Köroğlu, M. A., Uydu, H. A., Bekaroğlu, M., & Değer, O. (2001). Antioxidative enzyme activities and lipid peroxidation in major depression: Alterations by antidepressant treatments. *Journal of Affective Disorders*, *64*(1), 43–51.

Black, C. N., Bot, M., Scheffer, P. G., Cuijpers, P., & Penninx, B. W. (2015). Is depression associated with increased oxidative stress? A systematic review and meta-analysis. *Psychoneuroendocrinology*, *51*, 164–175.

Camkurt, M. A., Fındıklı, E., İzci, F., Kurutaş, E. B., & Tuman, T. C. (2016). Evaluation of malondialdehyde, superoxide dismutase and catalase activity and their diagnostic value in drug naïve, first episode, non-smoker major depression patients and healthy controls. *Psychiatry Research*, *238*, 81–85.

Che, Y., Wang, J. F., Shao, L., & Young, L. T. (2010). Oxidative damage to RNA but not DNA in the hippocampus of patients with major mental illness. *Journal of Psychiatry & Neuroscience*, *35*(5), 296–302.

Conrad, M., Schick, J., & Angeli, J. P. F. (2013). Glutathione and thioredoxin dependent systems in neurodegenerative disease: What can be learned from reverse genetics in mice. *Neurochemistry International*, *62*(5), 738–749.

Cumurcu, B. E., Ozyurt, H., Etikan, I., Demir, S., & Karlidag, R. (2009). Total antioxidant capacity and total oxidant status in patients with major depression: Impact of antidepressant treatment. *Psychiatry and Clinical Neurosciences*, *63*(5), 639–645.

Czarny, P., Wigner, P., Galecki, P., & Sliwinski, T. (2018). The interplay between inflammation, oxidative stress, DNA damage, DNA repair and mitochondrial dysfunction in depression. *Progress in Neuro-Psychopharmacology and Biological Psychiatry*, *80*, 309–321.

Dantzer, R., O'Connor, J. C., Lawson, M. A., & Kelley, K. W. (2011). Inflammation-associated depression: From serotonin to kynurenine. *Psychoneuroendocrinology*, *36*(3), 426–436.

Dean, J., & Keshavan, M. (2017). The neurobiology of depression: An integrated view. *Asian Journal of Psychiatry*, *27*, 101–111.

Diniz, B. S., Butters, M. A., Albert, S. M., Dew, M. A., & Reynolds, C. F. (2013). Late-life depression and risk of vascular dementia and Alzheimer's disease: Systematic review and meta-analysis of community-based cohort studies. *The British Journal of Psychiatry*, *202*(5), 329–335.

Dowlati, Y., Herrmann, N., Swardfager, W., Liu, H., Sham, L., Reim, E. K., et al. (2010). A meta-analysis of cytokines in major depression. *Biological Psychiatry*, *67*(5), 446–457.

Dröge, W., & Schipper, H. M. (2007). Oxidative stress and aberrant signaling in aging and cognitive decline. *Aging Cell*, *6*(3), 361–370.

Eyre, H. A., Air, T., Pradhan, A., Johnston, J., Lavretsky, H., Stuart, M. J., et al. (2016). A meta-analysis of chemokines in major depression. *Progress in Neuro-Psychopharmacology and Biological Psychiatry*, *68*, 1–8.

Gałecki, P., Szemraj, J., Bieńkiewicz, M., Zboralski, K., & Gałecka, E. (2009). Oxidative stress parameters after combined fluoxetine and acetylsalicylic acid therapy in depressive patients. *Human Psychopharmacology: Clinical and Experimental*, *24*(4), 277–286.

Galts, C. P., Bettio, L. E., Jewett, D. C., Yang, C. C., Brocardo, P. S., Rodrigues, A. L. S., et al. (2019). Depression in neurodegenerative diseases: Common mechanisms and current treatment options. *Neuroscience & Biobehavioral Reviews*, *102*, 56–84.

Genc, A., Kalelioglu, T., Karamustafalioglu, N., Tasdemir, A., Gungor, F. C., Genc, E. S., et al. (2015). Level of plasma thioredoxin in male patients with manic episode at initial and post-electroconvulsive or antipsychotic treatment. *Psychiatry and Clinical Neurosciences*, *69* (6), 344–350.

Guney, E., Buyuktaskin, D., Torun, Y. T., Arslan, B., Gulbahar, O., Ozaslan, A., et al. (2019). Increased serum thioredoxin levels are not correlated with executive functions in children with attention deficit hyperactivity disorder. *Neuroscience Letters*, *705*, 118–123.

Haapakoski, R., Mathieu, J., Ebmeier, K. P., Alenius, H., & Kivimäki, M. (2015). Cumulative meta-analysis of interleukins 6 and 1β, tumour necrosis factor α and C-reactive protein in patients with major depressive disorder. *Brain, Behavior, and Immunity*, *49*, 206–215.

Halliwell, B. (2006). Oxidative stress and neurodegeneration: Where are we now? *Journal of Neurochemistry*, *97*(6), 1634–1658.

Herbert, J., & Lucassen, P. J. (2016). Depression as a risk factor for Alzheimer's disease: Genes, steroids, cytokines and neurogenesis—What do we need to know? *Frontiers in Neuroendocrinology*, *41*, 153–171.

Herken, H., Gurel, A., Selek, S., Armutcu, F., Ozen, M. E., Bulut, M., et al. (2007). Adenosine deaminase, nitric oxide, superoxide dismutase, and xanthine oxidase in patients with major depression: Impact of antidepressant treatment. *Archives of Medical Research*, *38*(2), 247–252.

Howren, M. B., Lamkin, D. M., & Suls, J. (2009). Associations of depression with C-reactive protein, IL-1, and IL-6: A meta-analysis. *Psychosomatic Medicine*, *71*(2), 171–186.

Kitaoka, Y., Tanito, M., Kojima, K., Sase, K., Kaidzu, S., Munemasa, Y., et al. (2016). Axonal protection by thioredoxin-1 with inhibition of interleukin-1β in TNF-induced optic nerve degeneration. *Experimental Eye Research*, *152*, 71–76.

Knight, M. J., & Baune, B. T. (2018). Cognitive dysfunction in major depressive disorder. *Current Opinion in Psychiatry*, *31*(1), 26–31.

Köhler, C. A., Freitas, T. H., Maes, M. D., De Andrade, N. Q., Liu, C. S., Fernandes, B. S., et al. (2017). Peripheral cytokine and chemokine alterations in depression: A meta-analysis of 82 studies. *Acta Psychiatrica Scandinavica*, *135*(5), 373–387.

Krebber, A. M. H., Buffart, L. M., Kleijn, G., Riepma, I. C., De Bree, R. Leemans, C. R., et al. (2014). Prevalence of depression in cancer patients: A meta-analysis of diagnostic interviews and self-report instruments. *Psycho-Oncology*, *23*(2), 121–130.

Leighton, S. P., Nerurkar, L., Krishnadas, R., Johnman, C., Graham, G. J. & Cavanagh, J. (2018). Chemokines in depression in health and in inflammatory illness: A systematic review and meta-analysis. *Molecular Psychiatry*, *23*(1), 48–58.

Liguori, I., Russo, G., Curcio, F., Bulli, G., Aran, L., Della-Morte, D., et al. (2018). Oxidative stress, aging, and diseases. *Clinical Interventions in Aging*, *13*, 757–772.

Lillig, C. H., & Holmgren, A. (2007). Thioredoxin and related molecules—From biology to health and disease. *Antioxidants & Redox Signaling*, *9* (1), 25–47.

Lindqvist, D., Dhabhar, F. S., James, S. J., Hough, C. M., Jain, F. A., Bersani, F. S., et al. (2017). Oxidative stress, inflammation and treatment response in major depression. *Psychoneuroendocrinology*, *76*, 197–205.

Liu, Y., Ho, R. C. M., & Mak, A. (2012). Interleukin (IL)-6, tumour necrosis factor alpha (TNF-α) and soluble interleukin-2 receptors (sIL-2R) are elevated in patients with major depressive disorder: A meta-analysis and meta-regression. *Journal of Affective Disorders*, *139*(3), 230–239.

Liu, T., Zhong, S., Liao, X., Chen, J., He, T., Lai, S., et al. (2015). A meta-analysis of oxidative stress markers in depression. *PLoS One*, *10*(10), e0138904.

Lu, J., & Holmgren, A. (2012). Thioredoxin system in cell death progression. *Antioxidants & Redox Signaling*, *17*(12), 1738–1747.

Maes, M., De Vos, N., Pioli, R., Demedts, P., Wauters, A., Neels, H., et al. (2000). Lower serum vitamin E concentrations in major depression: Another marker of lowered antioxidant defenses in that illness. *Journal of Affective Disorders*, *58*(3), 241–246.

Maes, M., Mihaylova, I., Kubera, M., & Ringel, K. (2012). Activation of cell-mediated immunity in depression: Association with inflammation, melancholia, clinical staging and the fatigue and somatic symptom cluster of depression. *Progress in Neuro-Psychopharmacology and Biological Psychiatry*, *36*(1), 169–175.

Maes, M., Yirmyia, R., Noraberg, J., Brene, S., Hibbeln, J., Perini, G., et al. (2009). The inflammatory & neurodegenerative (I&ND) hypothesis of depression: Leads for future research and new drug developments in depression. *Metabolic Brain Disease*, *24*(1), 27–53.

Matsuzawa, A. (2017). Thioredoxin and redox signaling: Roles of the thioredoxin system in control of cell fate. *Archives of Biochemistry and Biophysics*, *617*, 101–105.

Maurya, P. K., Noto, C., Rizzo, L. B., Rios, A. C., Nunes, S. O., Barbosa, D. S., et al. (2016). The role of oxidative and nitrosative stress in accelerated aging and major depressive disorder. *Progress in Neuro-Psychopharmacology and Biological Psychiatry*, *65*, 134–144.

Mazereeuw, G., Herrmann, N., Andreazza, A. C., Khan, M. M., & Lanctôt, K. L. (2015). A meta-analysis of lipid peroxidation markers in major depression. *Neuropsychiatric Disease and Treatment*, *11*, 2479–2491.

Morris, G., Reiche, E. M. V., Murru, A., Carvalho, A. F., Maes, M., Berk, M., et al. (2018). Multiple immune-inflammatory and oxidative and nitrosative stress pathways explain the frequent presence of depression in multiple sclerosis. *Molecular Neurobiology*, *55*(8), 6282–6306.

Nagarajan, N., Oka, S., & Sadoshima, J. (2017). Modulation of signaling mechanisms in the heart by thioredoxin 1. *Free Radical Biology and Medicine*, *109*, 125–131.

Nordberg, J., & Arnér, E. S. (2001). Reactive oxygen species, antioxidants, and the mammalian thioredoxin system. *Free Radical Biology and Medicine*, *31*(11), 1287–1312.

Oberacker, T., Bajorat, J., Ziola, S., Schroeder, A., Röth, D., Kastl, L., et al. (2018). Enhanced expression of thioredoxin-interacting-protein regulates oxidative DNA damage and aging. *FEBS Letters*, *592*(13), 2297–2307.

Oriolo, G., Grande, I., Martin-Santos, R., Vieta, E., & Carvalho, A. F. (2018). Pathways driving Neuroprogression in depression: The role of immune activation. In *Inflammation and Immunity in Depression* (pp. 173–198). Academic Press.

Owe-Larsson, B., Ekdahl, K., Edbom, T., Ösby, U., Karlsson, H., Lundberg, C., et al. (2011). Increased plasma levels of thioredoxin-1 in patients with first episode psychosis and long-term schizophrenia. *Progress in Neuro-Psychopharmacology and Biological Psychiatry*, *35*(4), 1117–1121.

Owen, A. J., Batterham, M. J., Probst, Y. C., Grenyer, B. F., & Tapsell, L. C. (2005). Low plasma vitamin E levels in major depression: Diet or disease? *European Journal of Clinical Nutrition*, *59*(2), 304–306.

Panza, F., Frisardi, V., Capurso, C., D'Introno, A., Colacicco, A. M., Imbimbo, B. P., et al. (2010). Late-life depression, mild cognitive impairment, and dementia: possible continuum? *The American Journal of Geriatric Psychiatry*, *18*(2), 98–116.

Peng, Y., Feng, S. F., Wang, Q., Wang, H. N., Hou, W. G., Xiong, L., et al. (2010). Hyperbaric oxygen preconditioning ameliorates anxiety-like behavior and cognitive impairments via upregulation of thioredoxin reductases in stressed rats. *Progress in Neuro-Psychopharmacology and Biological Psychiatry*, *34*(6), 1018–1025.

Read, J. R., Sharpe, L., Modini, M., & Dear, B. F. (2017). Multimorbidity and depression: A systematic review and meta-analysis. *Journal of Affective Disorders*, *221*, 36–46.

Ren, X., Zou, L., Zhang, X., Branco, V., Wang, J., Carvalho, C., et al. (2017). Redox signaling mediated by thioredoxin and glutathione systems in the central nervous system. *Antioxidants & Redox Signaling*, *27*(13), 989–1010.

Sarandol, A., Sarandol, E., Eker, S. S., Erdinc, S., Vatansever, E., & Kirli, S. (2007). Major depressive disorder is accompanied with oxidative stress: Short-term antidepressant treatment does not alter oxidative–antioxidative systems. *Human Psychopharmacology: Clinical and Experimental*, *22*(2), 67–73.

Schmaal, L., Veltman, D. J., van Erp, T. G., Sämann, P. G., Frodl, T., Jahanshad, N., et al. (2016). Subcortical brain alterations in major depressive disorder: Findings from the ENIGMA major depressive disorder working group. *Molecular Psychiatry*, *21*(6), 806–812.

Sen, S., & Chakraborty, R. (2011). The role of antioxidants in human health. *Oxidative stress: Diagnostics, prevention, and therapy* (pp. 1–37). American Chemical Society.

Shelton, R. C., Claiborne, J., Sidoryk-Wegrzynowicz, M., Reddy, R., Aschner, M., Lewis, et al. (2011). Altered expression of genes involved in inflammation and apoptosis in frontal cortex in major depression. *Molecular Psychiatry*, *16*(7), 751–762.

Silva-Adaya, D., Gonsebatt, M. E., & Guevara, J. (2014). Thioredoxin system regulation in the central nervous system: Experimental models and clinical evidence. *Oxidative Medicine and Cellular Longevity*, *2014*, 1–13.

Spindel, O. N., World, C., & Berk, B. C. (2012). Thioredoxin interacting protein: Redox dependent and independent regulatory mechanisms. *Antioxidants & Redox Signaling*, *16*(6), 587–596.

Stefanescu, C., & Ciobica, A. (2012). The relevance of oxidative stress status in first episode and recurrent depression. *Journal of Affective Disorders*, *143*(1–3), 34–38.

Swardfager, W., Herrmann, N., Mazereeuw, G., Goldberger, K., Harimoto, T., & Lanctôt, K. L. (2013). Zinc in depression: A meta-analysis. *Biological Psychiatry*, *74*(12), 872–878.

Tian, L., Nie, H., Zhang, Y., Chen, Y., Peng, Z., Cai, M., et al. (2014). Recombinant human thioredoxin-1 promotes neurogenesis and facilitates cognitive recovery following cerebral ischemia in mice. *Neuropharmacology*, *77*, 453–464.

Tsai, M. C., & Huang, T. L. (2016). Increased activities of both superoxide dismutase and catalase were indicators of acute depressive episodes in patients with major depressive disorder. *Psychiatry Research*, *235*, 38–42.

Valkanova, V., Ebmeier, K. P., & Allan, C. L. (2013). CRP, IL-6 and depression: A systematic review and meta-analysis of longitudinal studies. *Journal of Affective Disorders*, *150*(3), 736–744.

Venojärvi, M., Korkmaz, A., Aunola, S., Hällsten, K., Virtanen, K., Marniemi, J., et al. (2014). Decreased thioredoxin-1 and increased HSP90 expression in skeletal muscle in subjects with type 2 diabetes or impaired glucose tolerance. *BioMed Research International*, *2014*, 1–6.

Wang, X., Che, H., Zhang, W., Wang, J., Ke, T., Cao, R., et al. (2015). Effects of mild chronic intermittent cold exposure on rat organs. *International Journal of Biological Sciences*, *11*(10), 1171–1180.

Wolkowitz, O. M., Mellon, S. H., Epel, E. S., Lin, J., Dhabhar, F. S., Su, Y., et al. (2011). Leukocyte telomere length in major depression: Correlations with chronicity, inflammation and oxidative stress-preliminary findings. *PLoS One*, *6*(3), e17837.

Xu, J., Li, T., Wu, H., & Xu, T. (2012). Role of thioredoxin in lung disease. *Pulmonary Pharmacology & Therapeutics*, *25*(2), 154–162.

Yang, X. H., Liu, H. G., Xue, L. I. U., & Chen, J. N. (2012). Thioredoxin and impaired spatial learning and memory in the rats exposed to intermittent hypoxia. *Chinese Medical Journal*, *125*(17), 3074–3080.

Zhang, Q. B., Gao, S. J., & Zhao, H. X. (2015). Thioredoxin: A novel, independent diagnosis marker in children with autism. *International Journal of Developmental Neuroscience*, *40*, 92–96.

Zhang, X. Y., Xiu, M. H., De Yang, F., Tan, Y. L., He, S., Kosten, T. A., et al. (2013). Thioredoxin, a novel oxidative stress marker and cognitive performance in chronic and medicated schizophrenia versus healthy controls. *Schizophrenia Research*, *143*(2–3), 301–306.

Zhang, X. Y., Xiu, M. H., Wang, F., Qi, L. Y., Sun, H. Q., Chen, S., et al. (2009). The novel oxidative stress marker thioredoxin is increased in first-episode schizophrenic patients. *Schizophrenia Research*, *113*(2–3), 151–157.

Zhou, H., Tan, H., Letourneau, L., & Wang, J. F. (2019). Increased thioredoxin-interacting protein in brain of mice exposed to chronic stress. *Progress in Neuro-Psychopharmacology and Biological Psychiatry*, *88*, 320–326.

Chapter 26

# Methods of neuroimaging in depression: Applications to resting-state functional connectivity

Moon-Soo Lee

*Department of Psychiatry, Korea University, College of Medicine, Seoul, Republic of Korea*

## List of abbreviations

| | |
|---|---|
| **ACC** | anterior cingulate cortex |
| **ALFF** | amplitude of low-frequency fluctuations |
| **BOLD** | blood-oxygen-level-dependent |
| **CEN** | central executive network |
| **DMN** | default mode network |
| **dmPFC** | dorsomedial prefrontal cortex |
| **HC** | healthy control |
| **ICA** | independent component analysis |
| **iFC** | intrinsic functional connectivity |
| **LLD** | late-life depression |
| **LLD** | late-life depression |
| **MDD** | major depressive disorder |
| **MPFC** | medial prefrontal cortex |
| **MTG** | middle temporal gyrus |
| **PCA** | principal component analysis |
| **PCC** | posterior cingulate cortex |
| **ROI** | region of interest |
| **RSFC** | resting-state functional connectivity |
| **rs-fMRI** | resting-state functional magnetic resonance imaging |
| **RSN** | resting-state network |
| **SCA** | seed-based correlational analysis |
| **SN** | salience network |

## Introduction

In this review, we will present a comprehensive appraisal of existing neuroimaging research using rs-fMRI in depression. For comparison purposes, childhood is defined as the period from infancy to prepubescence, adolescence as peri- and post-pubescence up to the age of 18, adulthood from ages 19 to 65, and late life as over age 65.

The brain is a network comprised of a collection of spatially distributed brain regions—nodes—showing correlated and synchronized activity patterns. BOLD signals are often used as a surrogate measure for neuronal activity. Initially, fMRI analysis techniques were used to assess BOLD signal fMRI data during model-driven, task-evoked activation. The techniques used to explore task-related fMRI are also useful for resting-state research exploring the spontaneous, or default neural activity of the brain at baseline. Biswal et al. first observed that spontaneous, low-level synchronous fluctuations (<0.1 Hz) in the BOLD signal occurred independently of task stimulation (Biswal, Yetkin, Haughton, & Hyde, 1995). Following this, several studies observing resting-state activity have consistently reported functionally linked resting-state networks during rest. It is now possible to identify large-scale brain networks that show consistent, large-scale spatial patterns of coherent signal, as well as human neural functional architecture, by using fMRI to examine changes in spontaneous or resting-state fluctuations in the BOLD signal. These brain networks show patterns of connectivity between multiple ROIs within spatially distributed, large-scale networks; these patterns have been termed RSNs (Beckmann & Smith, 2005). RSNs can be reliably identified through independent studies (Cole, Smith, & Beckmann, 2010), although these studies use different subjects, methods, and types of magnetic resonance acquisition protocol parameters. RSNs are localized to gray matter. These networks can be detected during the "resting state" when a subject is not engaged in any specific task. As resting-state conditions require minimal instructions and low demands on attention, data can be gathered from various populations with relative ease. The absence of specific tasks also benefits the generalizability, reproducibility, and comparability of independent studies. By observing synchronous, spontaneous activity in different regions of the brain, researchers can investigate how different nodes and networks communicate. This communication is known as RSFC. RSFC measures the temporal coherence of the BOLD signal within or between nodes

The Neuroscience of Depression. https://doi.org/10.1016/B978-0-12-817933-8.00091-8

or networks during rest. Accordingly, rs-fMRI has been used to explore the pathophysiological mechanisms that underlie specific psychiatric disorders, including depressive disorders. There are two major methods of studying resting-state fMRI data: model-dependent methods (SCA) and mode-free methods (PCA, ICA, etc.) (Van den Heuvel & Hulshoff Pol, 2010). Only two research methods—SCA and ICA—will be explained here. In SCA, researchers use a ROI based on a hypothesis and calculate correlations between the BOLD signal and all other voxels. Increases in functional connectivity indicate increased synchronization between two regions. On the contrary, ICA does not require any prior hypothesis or ROI. Instead, ICA is an exploratory model-free method, which locates temporally coherent, spatially independent components of interest using a multivariate voxel-based approach (Beckmann, DeLuca, Devlin, & Smith, 2005). The results from ICA can be split into subnetworks. The dimensionality of the ICA model order shows the number of components that the data are decomposed into, where a higher model order indicates the existence of subnetworks within larger networks (Cole et al., 2010; Mulders, van Eijndhoven, Schene, Beckmann, & Tendolkar, 2015). An increase in functional connectivity in ICA demonstrates the degree to which a voxel's signal is correlated with a specific network or component overtime.

Recent studies have identified many functional brain networks by monitoring brain activity during specific tasks or during resting state, including the default mode network (DMN)—the first RSN identified—the salience network (SN), and the central executive network (CEN) (Fig. 1). The existence of these large-scale brain networks suggests that there is also functional connection or connectivity of specific voxels or individual voxel regions of interest (ROI). Three of these identified networks (DMN, SN, and CEN) have been considered as the 'core' neurocognitive networks implicated in depressive disorders (Mulders et al., 2015). These networks represent brain functioning during rest (DMN), cognition (CEN), and emotional processing (SN). These networks are also implicated in pathophysiological changes in depression. In MDD, most rs-fMRI findings are associated with one of these three major neural networks (Mulders et al., 2015). Now, I will go more deeply into each of these important networks. Each of these important networks will now be discussed in further detail.

## Default mode network

The default mode corresponds to decreased signals during task performance. The DMN ("task-negative" network) exhibits synchronized deactivation during tasks and prominent activation during passive states like rest. The DMN is often divided into two subnetworks: the anterior subnetwork (including mainly the MPFC) and the posterior subnetwork (including the posterior CC and the precuneus cortex). The DMN extends into the subgenual ACC (Connolly et al., 2013). Both the PCC and the precuneus (associated with the DMN) form the midline core with dorsomedial prefrontal cortex, which is also associated with the CEN. These cortical structures are involved in spontaneous cognition, self-referential processing, and affective decision-making (Gabbay et al., 2013). In addition to these core regions, the inferior parietal lobe, which is involved in spontaneous or self-generated cognition, is also closely linked to DMN structures. The default network was identified by its consistent activation during passive task states. The discovery of the brain's default network was entirely accidental when researchers measured brain activity during undirected mental states (Buckner, Andrews-Hanna, & Schacter, 2008).

## Central executive network (CEN)

The CEN is also known as the cognitive control network. In contrast, the CEN, composed of the dorsolateral and dorsomedial prefrontal cortices and the posterior parietal cortex, is engaged in cognitively demanding tasks, such as decision-making and working memory (Cao et al., 2016; Mulders et al., 2015).

## Salience network

The SN includes the frontoinsular cortex, dorsal ACC, amygdala, and temporal poles. The insula is thought to mediate interpretation of sensory information from the body (Craig, 2002). The ACC can be divided anatomically based on cognitive (caudal or dorsal) and emotional (rostral or ventral) components (Bush, Luu, & Posner, 2000). The ACC can be also divided into pregenual and subgenual regions (Stevens, Hurley, & Taber, 2011). The SN also possesses extended connections with other subcortical nodes consisting of the putamen, dorsomedial thalamus, ventral striatum, substantia nigra, and ventral tegmental area.

As a whole, frontoinsular the cortico-striato-thalamo-cortical loop contributes to the SN, and is associated with detecting events salient to homeostasis, as well as playing a central role in emotional control (Peters, Dunlop, & Downar, 2016). This network is involved in evaluating the relevance of internal and external stimuli in order to generate appropriate responses (Pannekoek et al., 2014).

Specific pathophysiological changes reported in depressive disorder across age groups—childhood and adolescence, adulthood including late life—will now be discussed in detail.

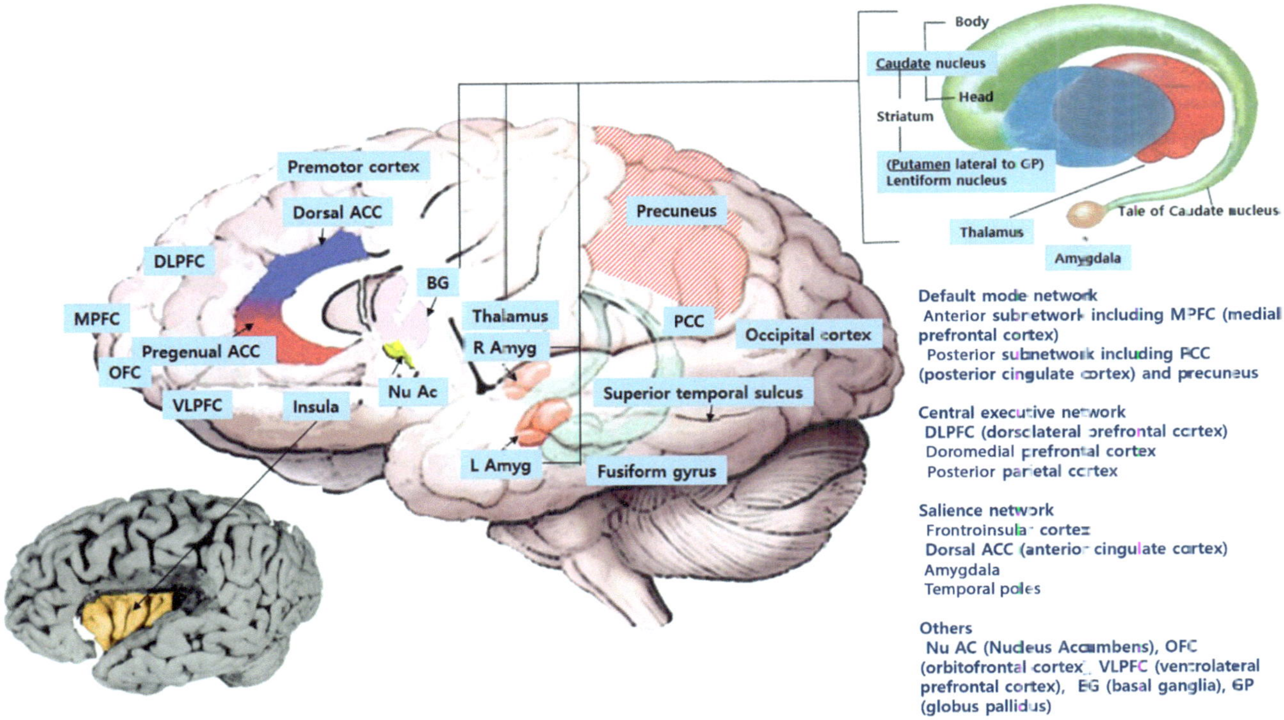


**FIG. 1** Orientation to key functional region and three major networks deployed in depressive disorder from the resting-state functional connectivity studies.

## Childhood and adolescence

Recent neuroimaging studies of MDD in patients under 18 have used task-based and resting-state fMRIs. There are very few RSFC studies in children, as most studies have been conducted with adolescents (Gaffrey, Luby, Botteron, Repovs, & Barch, 2012). Adolescence is a transitional period between childhood and adulthood. There are some studies demonstrating altered functional connectivity in adolescents with MDD, although these are few in number compared to studies with adults (Connolly et al., 2013; Cullen et al., 2009, 2014; Davey, Harrison, Yucel, & Allen, 2012; Gabbay et al., 2013; Jiao et al., 2011; Jin et al., 2011; Lee et al., 2019; Pannekoek et al., 2014) (Table 1). Pediatric depression must be considered within the context of physiologic brain development and psychopathologic changes in the brain. Relevant network-specific study findings will now be explored in depth.

### Default mode network

DMN activity is often investigated using seeds in the PCC and precuneus. When the PCC was used as a seed, medication-naïve school-age children with preschool onset of clinically significant depression showed decreased connectivity between the PCC and regions within the middle temporal gyrus and inferior parietal lobe—regions associated with the DMN. One interpretation is the cortical area (for example, the inferior parietal lobe) involved in facilitating the downregulation or disengagement of the DMN is less strongly connected in children with a history of depression (Gaffrey et al., 2012).

Connolly et al. reported increased connectivity between the subgenual ACC and insula, and between the subgenual ACC and amygdala in adolescents with MDD compared to controls. Both the insula and the amygdala are components of the SN. Subgenual hyperactivation has been observed in other studies of adolescents with depression, although this finding was not directly from a resting-state fMRI study (Yang et al., 2009). In another study by Davey et al., patients with depression (aged 15–24 years) showed increased connectivity between the subgenual ACC and dorsomedial prefrontal cortex. This connectivity was positively correlated with illness severity (Davey et al., 2012). In addition, the subgenual ACC and the amygdala are anatomically interconnected by white matter fibers in the uncinate fasciculus (Kier, Staib, Davis, & Bronen, 2004). On the contrary, decreased connectivity between the subgenual ACC and the precuneus, which constitutes the DMN was also demonstrated (Connolly et al., 2013). Similarly, Cullen et al. showed that adolescents with MDD exhibit decreased functional connectivity in the subgenual ACC-based network, including the supragenual ACC,

**TABLE 1 Resting-state functional connectivity studies performed in children and adolescents.**

| Study | Participants characteristics | Research results |
|---|---|---|
| Default mode network connectivity in children with a history of preschool onset depression (Gaffrey et al., 2012) | 21 children with a history of preschool depression and 18 healthy controls | Decreased connectivity between the PCC and regions within the MTG, inferior parietal lobule, and cerebellum in children with known depression during the preschool period. Increased connectivity between the PCC and regions within the subgenual and anterior cingulate cortices and anterior MTG bilaterally was also found in these children. A clinically relevant "brain-behavior" relationship between atypical functional connectivity of the PCC and disruptions in emotion regulation was identified |
| Resting-state functional connectivity of subgenual anterior cingulate cortex in depressed adolescents (Connolly et al., 2013) | 45 healthy control subjects and 30 with MDD | Elevated connectivity between the sgACC and the insula and between the sgACC and the amygdala in the MDD group compared with the control subjects. Decreased connectivity between the sgACC and the precuneus in the MDD group relative to the control subjects. Within the MDD group, higher levels of depression significantly correlated with decreased connectivity between the sgACC and left precuneus. Increased rumination was significantly associated with reduced connectivity between sgACC and the middle and inferior frontal gyri in the MDD group |
| A preliminary study of functional connectivity in comorbid adolescent depression (Cullen et al., 2009) | Adolescents with MDD ($n=12$) and healthy adolescents ($n=14$) | Adolescents with MDD have decreased functional connectivity in a subgenual ACC-based neural network that includes the supragenual ACC (BA 32), the right medial frontal cortex (BA 10), the left inferior (BA 47) and superior frontal cortex (BA 22), superior temporal gyrus (BA 22), and the insular cortex (BA 13) |
| Abnormal amygdala resting-state functional connectivity in adolescent depression (Cullen et al., 2014) | 41 adolescents and young adults aged 12–19 years with MDD and 29 healthy adolescents | Adolescents with MDD showed lower positive RSFC between the amygdala and hippocampus, parahippocampus, and brain stem; this connectivity was inversely correlated with general depression, dysphoria, and lassitude and was positively correlated with well-being. Patients also demonstrated greater amygdala-precuneus RSFC in contrast to negative amygdala-precuneus RSFC in the adolescents serving as controls |
| Regionally specific alterations in functional connectivity of the anterior cingulate cortex in major depressive disorder (Davey et al., 2012) | 18 patients with MDD, aged 15–24 years, matched with 20 healthy controls | The depressed participants showed altered patterns of connectivity with ventral cingulate subregions. They showed increased connectivity between subgenual ACC and dorsomedial frontal cortex, with connectivity strength showing positive correlation with illness severity. Depressed participants also showed increased connectivity between pregenual ACC and left dorsolateral frontal cortex, and decreased connectivity between pregenual ACC and the caudate nucleus bilaterally |
| Striatum-based circuitry of adolescent depression and anhedonia (Gabbay et al., 2013) | 21 psychotropic drug-naïve adolescents with MDD and 21 healthy controls | Adolescents with MDD showed increased iFC between all striatal regions bilaterally and the dorsomedial prefrontal cortex, as well as |

**TABLE 1 Resting-state functional connectivity studies performed in children and adolescents—cont'd**

| Study | Participants characteristics | Research results |
|---|---|---|
| | | between the right ventral caudate and the ACC. MDD severity was associated with iFC between the striatum and midline structures including the precuneus, posterior cingulate cortex, and dmPFC. Distinct striatal iFC patterns involving the pregenual ACC, subgenual ACC, supplementary motor area, and supramarginal gyrus were associated with anhedonia severity |
| Increased activity imbalance in fronto-subcortical circuits in adolescents with major depression (Jiao et al., 2011) | 18 MDD and 18 healthy adolescents | Resting brain activity levels measured by ALFF was higher in the frontal cortex than that in the subcortical system involving mainly (para) limbic-striatal regions in both healthy controls and MDD adolescents. The difference of ALFF values between frontal and subcortical systems was increased in MDD adolescents as compared with the controls |
| A preliminary study of the dysregulation of the resting networks in first-episode medication-naïve adolescent depression (Jin et al., 2011) | 16 first-episode and unmedicated adolescents with MDD and 16 healthy controls | The topological properties of depressed adolescents' networks were significantly disrupted compared with HC. Dysregulation of brain regions was found in the anterior cingulate cortex, dorsolateral, medial and inferior prefrontal cortex, insula, amygdala, and the temporal cortices. The connectivity degree of amygdala-related functional connection was positively correlated with the duration of depression |
| Resting-state functional connectivity in medication-naïve adolescents with major depressive disorder (Lee et al., 2019) | 23 drug-naïve adolescents diagnosed with first-episode MDD and 27 healthy participants | The adolescents with MDD exhibited the following characteristics: a lower RSFC between the right amygdala and right superior frontal gyrus; a lower RSFC between the right hippocampus and clusters including the right insula and right middle frontal gyrus; a higher RSFC between the left insula and clusters including the bilateral middle frontal gyrus, right superior frontal gyrus, and right frontal pole; and a higher RSFC between the left dorsal anterior cingulate cortex and a cluster including the left insula |
| Aberrant resting-state functional connectivity in limbic and salience networks in treatment-naïve clinically depressed adolescents (Pannekoek et al., 2014) | 26-treatment-naïve, clinically depressed adolescents and 26 matched HCs | Compared to HCs, clinically depressed adolescents showed increased RSFC of the left amygdala with right parietal cortical areas and decreased right amygdala RSFC with left frontal cortical areas including the ACC, as well as with right occipito-parietal areas. The bilateral dACC showed decreased RSFC with the right middle frontal gyrus, frontal pole, and inferior frontal gyrus in clinically depressed adolescents |

portions of the frontal cortex, superior temporal gyrus, and insula from the SCA. When considering the insula has a central role in interoception, it is possible that decreased functional connectivity in the subgenual ACC-based network might be related to the somatic complaints and negative perception of interpersonal feedback commonly seen in depression (Cullen et al., 2009). From these research findings, the subgenual ACC-based network can be considered as an important structure in the pathophysiology of depression.

### Central executive network

Few papers have used ICA and SCA to investigate the CEN. More research is required to facilitate a detailed discussion.

### Salience network

The most widely investigated seed region associated with the SN is the amygdala. The amygdala is important emotion processing and related physiologic responses (LeDoux, 2000). The amygdala is closely connected to the hippocampus and parahippocampus. Accordingly, amygdala-hippocampal connections may facilitate the modulation of emotional memories. Positive RSFC between these regions has been shown in healthy adults (Roy et al., 2009) and is also implicated in depression. Cullen et al. performed a seed-based whole-brain approach to examine RSFC involving the amygdala in adolescents with depression (Cullen et al., 2014). The results suggested impaired RSFC between the amygdala and the hippocampus and parahippocampus, which was also inversely correlated with depressive symptom scores, such that patients with lower positive RSFC scored more highly on the depression scale. This suggests that patients with MDD exhibit an under-connected amygdala-hippocampus circuit during rest. This may be a compensatory process to offset the hyperconnectivity during the processing of negative emotional memories and the general hyperactivity of the amygdala in depression (Cullen et al., 2014). The aberrant connections of the amygdala to other brain regions have been reported in Pannekoek's studies (Pannekoek et al., 2014). Adolescents with clinical depression also showed decreased RSFC between the right amygdala and left frontal cortical areas, including the ACC. The connection between the amygdala and the pregenual ACC is associated with the regulation of affective processing. The ACC is associated with emotion regulation through its inhibitory control on amygdala function. Accordingly, it can be assumed there is decreased ACC regulation of limbic activation in depression. In another study, adolescents with depression also showed decreased RSFC between the right amygdala and the right superior frontal gyrus when compared to healthy controls (Lee et al., 2019). Taken together, these findings demonstrate the importance of amygdala RSFC in adolescent depression.

The striatum also contributes to the SN. Gabbay et al. reported that adolescents with MDD showed increased intrinsic functional connectivity between the striatum and midline structures including the PCC, precuneus, and dorsomedial prefrontal cortex—structures that also constitute the DMN. As these cortical midline structures are involved in spontaneous cognition, self-referential processing, and affective decision-making, they may also mediate sense and feeling of self (Gabbay et al., 2013).

### Other approaches

Jin et al. used graph theory analysis to investigate complex brain networks ( Jin et al., 2011). They demonstrated that adolescents with depression exhibited significant disruption to the topological properties of networks involving the ACC, dorsolateral, medial and inferior prefrontal cortex, insula, amygdala, and the temporal lobe. These areas were previously identified in the abovementioned studies, although different methodological approaches were used. In addition, amygdala connectivity was positively correlated with the duration of depression. Jiao et al. measured resting brain activity using ALFF in adolescents with MDD and identified an increased imbalance between the frontal control system and the (para) limbic-striatal emotional processing system ( Jiao et al., 2011).

## Adulthood

There is also evidence to demonstrating altered functional connectivity in adults with MDD. Many adult studies show consistent findings with the youth literature (Kerestes, Davey, Stephanou, Whittle, & Harrison, 2014). However, due to limited studies and small sample sizes, it is still unclear whether adolescents with MDD exhibit similar dysfunction (Zhang, 2017).

There are currently very few meta-analyses evaluating resting state in MDD (Kaiser, Andrews-Hanna, Wager, & Pizzagalli, 2015; Yan et al., 2019). Recently, the Enhancing Neuroimaging Genetics through Meta-Analysis (ENIGMA) MDD Consortium has published its findings. The researchers collated data from 45 MDD study cohorts from 14 countries across six continents to address issues such as poor replication, unreliable results, and overestimation of effect sizes (Schmaal et al., 2019).

LLD is known to have an average prevalence of 13.5% (Beekman, Copeland, & Prince, 1999). In terms of connectivity, LLD is known to emerge from circuit dysfunctions within different brain networks, rather than localized brain lesions (Tadayonnejad & Ajilore, 2014). There is limited research on functional connectivity in LLD (Alexopoulos et al., 2012; Sexton et al., 2012). A summary of this research will now be provided and compared to previous findings.

### Default mode network

In Yan et al.'s meta-analysis, MDD patients showed significantly lower DMN within-network functional connectivity than healthy controls (Yan et al., 2019). The subgroup analysis revealed that first-episode drug-naïve participants did not differ significantly from healthy controls. However, patients with recurrent MDD showed significantly decreased functional connectivity compared to healthy controls. There was a significant association between decreased

DMN functional connectivity and greater symptom severity (as measured by the Hamilton Depression Rating Scale) in patients with recurrent MDD, though this relationship was not significant in first-episode drug-naïve patients.

However, these results need to be interpreted carefully, as there are several studies that demonstrate increased, as opposed to decreased, DMN functional connectivity in MDD. These ICA studies show an increase in connectivity within the nodes of the anterior DMN relative to healthy controls. Greicius et al. reported significantly greater resting-state DMN functional connectivity between the subgenual cingulate and thalamus in subjects with depression. In addition, the depression group showed a positive correlation between the length of the current depressive episode and functional connectivity in the subgenual cingulate (Greicius et al., 2007). In another study, both of the spatially identified DMN subnetworks—anterior and posterior—showed increased functional connectivity in pretreatment MDD patients, and differences in the posterior subnetwork were normalized after 12 weeks of antidepressant treatment (Li et al., 2013). Similarly, the results of a meta-analysis implementing heterogeneous analysis strategies and the use of homogeneous Chinese participants, rather than Caucasian samples, also demonstrated increased DMN functional connectivity (Yan et al., 2019).

In the case of SCA, the subgenual ACC and PCC have been commonly used as seeds, although the subgenual ACC is not always considered as part of the DMN. Kaiser et al.'s meta-analysis comprised 27 seed-based voxel-wise RSFC results from 25 publications. Patients with MDD exhibited hyperconnectivity between the DMN seeds and regions of the hippocampus, extending to the middle temporal gyrus and MPFC. The MPFC is hyperconnected with other DMN systems and plays a role in self-referential thinking and autobiographical memory retrieval (Kim, 2012). These regions are hyperactive in depressed patients and can bias attention toward self-focused thinking. In addition, hypoconnectivity between affective network seeds and regions of the MPFC which belong to the DMN was also observed. The MPFC is known to play a role in upregulation and downregulation of negative emotions (Ochsner et al., 2004). When considering that the MPFC mediates emotional regulation, and that limbic regions in the affective network are involved in affective responses, then the hypoconnectivity between the MPFC and affective network in depressed patients can manifest as both blunted positive communication (between the MPFC and nucleus accumbens, which is known to play a role in the processing of positive emotion; Badgaiyan, 2010) and excessive negative communication (between the MPFC and amygdala), i.e., impaired emotional regulation (Kaiser et al., 2015).

## Central executive network

The literature on the CEN in depression is limited. A meta-analysis by Kaiser et al. reported hypoconnectivity between CEN seeds and the bilateral posterior parietal cortex (Kaiser et al., 2015). The CEN is involved in cognitive control, and hypoconnectivity within the CEN can manifest as impaired rule-based problem solving and decision-making in the context of goal-directed behavior. In SCA studies, the dorsolateral prefrontal cortex has been used as a seed region, with the majority finding hypoconnectivity between the dorsolateral prefrontal cortex and other regions of the CEN. Hypoconnectivity within the CEN has also been observed in late-life depression (Alexopoulos et al., 2012) and first-episode patients with MDD (Ye et al., 2012). However, when treated with transcranial magnetic stimulation over 5 weeks, patients with depression exhibited increased functional connectivity within the DMN and decreased connectivity within the CEN. Five weeks after treatment, transcranial magnetic stimulation normalized depression-related subgenual hyperconnectivity in the DMN, but did not change connectivity within the CEN. Hypoconnectivity within the CEN persisted even after 5-week course of treatment (Liston et al., 2014). From these research findings, it can be assumed that hypoconnectivity within the CEN might predict a lower remission rate and the persistence of depressive symptoms and signs post-treatment (Alexopoulos et al., 2012).

## Salience network

Networks involved in emotion regulation have been investigated using both ICA and SCA. Manoliu et al. used high model order ICA to analyze rs-fMRI data and identify the SN. Patients with MDD showed decreased intra-intrinsic functional connectivity within the SN's right anterior insula, which was inversely correlated with the severity of depressive symptoms (as measured by both the Hamilton Depression Rating Scale and Beck Depression Inventory). These findings suggest aberrant connectivity within the right anterior insula (Manoliu et al., 2014). In addition, the most widely investigated seed regions in the SN used for SCA are the insula and the amygdala. Considering the extensive structural and functional connectivity between the insula and other brain structures, the interoceptive deficits in MDD and the insula's role in interoception, it can be inferred that it may also play a role in MDD. SCA revealed increased connectivity between the insula and the pregenual ACC belonging to the anterior DMN (Horn et al., 2010). On the contrary, there was typically decreased connectivity between the amygdala (another important node in the SN) and various brain regions (Yue et al., 2013).

## Conclusions

Many functional brain networks have been identified by monitoring brain activity during specific tasks or during resting state. Three of these networks (the DMN, SN, and CEN) are considered the 'core' neurocognitive networks implicated in depressive disorder. These networks are also associated with pathophysiological changes in depression. More specifically, altered functional connectivity in several brain regions implicated in processing, regulation, emotional memory; cognitive processes related to rumination; impaired concentration; and physiological arousal in patients with depression.

## Summary points

- There are two major methods of studying resting-state fMRI data: model-dependent methods (SCA) and mode-free methods.
- Three of identified large-scale brain networks (DMN, SN, and CEN) have been considered as the "core" neurocognitive networks in depressive disorders.
- The subgenual ACC-based network can be considered an important structure in the pathophysiology of depression.
- Adolescents with depression show changed RSFC between the amygdala and other areas.
- There are several studies that demonstrate mixed (increased or decreased) within-network DMN functional connectivity in adult patients with MDD.
- Hypoconnectivity within the CEN might predict a lower remission rate and the persistence of depressive symptoms in adult patients with depressive disorders.

## Acknowledgments

This work was supported by the Basic Science Research Program through the National Research Foundation of Korea (NRF) and funded by the Ministry of Education (grant numbers NRF-2014R1A1A2054904 and NRF-2017R1D1A1B03028672).

### Disclosure of conflicts of interest

None.

## References

Alexopoulos, G. S., Hoptman, M. J., Kanellopoulos, D., Murphy, C. F., Lim, K. O., & Gunning, F. M. (2012). Functional connectivity in the cognitive control network and the default mode network in late-life depression. *Journal of Affective Disorders*, *139*(1), 56–65. https://doi.org/10.1016/j.jad.2011.12.002.

Badgaiyan, R. D. (2010). Dopamine is released in the striatum during human emotional processing. *Neuroreport*, *21*(18), 1172–1176. https://doi.org/10.1097/WNR.0b013e3283410955.

Beckmann, C. F., DeLuca, M., Devlin, J. T., & Smith, S. M. (2005). Investigations into resting-state connectivity using independent component analysis. *Philosophical Transactions of the Royal Society, B: Biological Sciences*, *360*(1457), 1001–1013. https://doi.org/10.1098/rstb.2005.1634.

Beckmann, C. F., & Smith, S. M. (2005). Tensorial extensions of independent component analysis for multisubject FMRI analysis. *NeuroImage*, *25*(1), 294–311. https://doi.org/10.1016/j.neuroimage.2004.10.043.

Beekman, A. T., Copeland, J. R., & Prince, M. J. (1999). Review of community prevalence of depression in later life. *The British Journal of Psychiatry*, *174*, 307–311. https://doi.org/10.1192/bjp.174.4.307.

Biswal, B., Yetkin, F. Z., Haughton, V. M., & Hyde, J. S. (1995). Functional connectivity in the motor cortex of resting human brain using echo-planar MRI. *Magnetic Resonance in Medicine*, *34*(4), 537–541. https://doi.org/10.1002/mrm.1910340409.

Buckner, R. L., Andrews-Hanna, J. R., & Schacter, D. L. (2008). The brain's default network: Anatomy, function, and relevance to disease. *Annals of the New York Academy of Sciences*, *1124*, 1–38. https://doi.org/10.1196/annals.1440.011.

Bush, G., Luu, P., & Posner, M. I. (2000). Cognitive and emotional influences in anterior cingulate cortex. *Trends in Cognitive Sciences*, *4*(6), 215–222. https://doi.org/10.1016/s1364-6613(00)01483-2.

Cao, W., Cao, X., Hou, C., Li, T., Cheng, Y., Jiang, L., et al. (2016). Effects of cognitive training on resting-state functional connectivity of default mode, salience, and central executive networks. *Frontiers in Aging Neuroscience*, *8*, 70. https://doi.org/10.3389/fnagi.2016.00070.

Cole, D. M., Smith, S. M., & Beckmann, C. F. (2010). Advances and pitfalls in the analysis and interpretation of resting-state FMRI data. *Frontiers in Systems Neuroscience*, *4*, 8. https://doi.org/10.3389/fnsys.2010.00008.

Connolly, C. G., Wu, J., Ho, T. C., Hoeft, F., Wolkowitz, O., Eisendrath, S., et al. (2013). Resting-state functional connectivity of subgenual anterior cingulate cortex in depressed adolescents. *Biological Psychiatry*, *74*(12), 898–907. https://doi.org/10.1016/j.biopsych.2013.05.036.

Craig, A. D. (2002). How do you feel? Interoception: The sense of the physiological condition of the body. *Nature Reviews. Neuroscience*, *3*(8), 655–666. https://doi.org/10.1038/nrn894.

Cullen, K. R., Gee, D. G., Klimes-Dougan, B., Gabbay, V., Hulvershorn, L., Mueller, B. A., et al. (2009). A preliminary study of functional connectivity in comorbid adolescent depression. *Neuroscience Letters*, *460*(3), 227–231. https://doi.org/10.1016/j.neulet.2009.05.022.

Cullen, K. R., Westlund, M. K., Klimes-Dougan, B., Mueller, B. A., Houri, A., Eberly, L. E., et al. (2014). Abnormal amygdala resting-state functional connectivity in adolescent depression. *JAMA Psychiatry*, *71*(10), 1138–1147. https://doi.org/10.1001/jamapsychiatry.2014.1087.

Davey, C. G., Harrison, B. J., Yucel, M., & Allen, N. B. (2012). Regionally specific alterations in functional connectivity of the anterior cingulate cortex in major depressive disorder. *Psychological Medicine*, *42*(10), 2071–2081. https://doi.org/10.1017/S0033291712000323.

Gabbay, V., Ely, B. A., Li, Q., Bangaru, S. D., Panzer, A. M., Alonso, C. M., et al. (2013). Striatum-based circuitry of adolescent depression and anhedonia. *Journal of the American Academy of Child and Adolescent Psychiatry*, *52*(6), 628–641. e613 https://doi.org/10.1016/j.jaac.2013.04.003.

Gaffrey, M. S., Luby, J. L., Botteron, K., Repovs, G., & Barch, D. M. (2012). Default mode network connectivity in children with a history of preschool onset depression. *Journal of Child Psychology and Psychiatry*, *53*(9), 964–972. https://doi.org/10.1111/j.1469-7610.2012.02552.x.

Greicius, M. D., Flores, B. H., Menon, V., Glover, G. H., Solvason, H. B., Kenna, H., et al. (2007). Resting-state functional connectivity in major depression: Abnormally increased contributions from subgenual cingulate cortex and thalamus. *Biological Psychiatry*, *62*(5), 429–437. https://doi.org/10.1016/j.biopsych.2006.09.020.

Horn, D. I., Yu, C., Steiner, J., Buchmann, J., Kaufmann, J., Osoba, A., et al. (2010). Glutamatergic and resting-state functional connectivity correlates of severity in major depression—The role of pregenual anterior cingulate cortex and anterior insula. *Frontiers in Systems Neuroscience*, *4*, 33. https://doi.org/10.3389/fnsys.2010.00033.

Jiao, Q., Ding, J., Lu, G., Su, L., Zhang, Z., Wang, Z., et al. (2011). Increased activity imbalance in fronto-subcortical circuits in adolescents with major depression. *PLoS One*, *6*(9). https://doi.org/10.1371/journal.pone.0025159.

Jin, C., Gao, C., Chen, C., Ma, S., Netra, R., Wang, Y., et al. (2011). A preliminary study of the dysregulation of the resting networks in first-episode medication-naive adolescent depression. *Neuroscience Letters*, *503*(2), 105–109. https://doi.org/10.1016/j.neulet.2011.08.017.

Kaiser, R. H., Andrews-Hanna, J. R., Wager, T. D., & Pizzagalli, D. A. (2015). Large-scale network dysfunction in major depressive disorder: A meta-analysis of resting-state functional connectivity. *JAMA Psychiatry*, *72*(6), 603–611. https://doi.org/10.1001/jamapsychiatry.2015.0071.

Kerestes, R., Davey, C. G., Stephanou, K., Whittle, S., & Harrison, B. J. (2014). Functional brain imaging studies of youth depression: A systematic review. *Neuroimage: Clinical*, *4*, 209–231. https://doi.org/10.1016/j.nicl.2013.11.009.

Kier, E. L., Staib, L. H., Davis, L. M., & Bronen, R. A. (2004). MR imaging of the temporal stem: Anatomic dissection tractography of the uncinate fasciculus, inferior occipitofrontal fasciculus, and Meyer's loop of the optic radiation. *American Journal of Neuroradiology*, *25*(5), 677–691. Retrieved from https://www.ncbi.nlm.nih.gov/pubmed/15140705.

Kim, H. (2012). A dual-subsystem model of the brain's default network: Self-referential processing, memory retrieval processes, and autobiographical memory retrieval. *NeuroImage*, *61*(4), 966–977. https://doi.org/10.1016/j.neuroimage.2012.03.025.

LeDoux, J. E. (2000). Emotion circuits in the brain. *Annual Review of Neuroscience*, *23*, 155–184. https://doi.org/10.1146/annurev.neuro.23.1.155.

Lee, J., Pavuluri, M. N., Kim, J. H., Suh, S., Kim, I., & Lee, M. S. (2019). Resting-state functional connectivity in medication-naive adolescents with major depressive disorder. *Psychiatry Research: Neuroimaging*, *288*, 37–43. https://doi.org/10.1016/j.pscychresns.2019.04.008.

Li, B., Liu, L., Friston, K. J., Shen, H., Wang, L., Zeng, L. L., et al. (2013). A treatment-resistant default mode subnetwork in major depression. *Biological Psychiatry*, *74*(1), 48–54. https://doi.org/10.1016/j.biopsych.2012.11.007.

Liston, C., Chen, A. C., Zebley, B. D., Drysdale, A. T., Gordon, R., Leuchter, B., et al. (2014). Default mode network mechanisms of transcranial magnetic stimulation in depression. *Biological Psychiatry*, *76*(7), 517–526. https://doi.org/10.1016/j.biopsych.2014.01.023.

Manoliu, A., Meng, C., Brandl, F., Doll, A., Tahmasian, M., Scherr, M., et al. (2014). Insular dysfunction within the salience network is associated with severity of symptoms and aberrant inter-network connectivity in major depressive disorder. *Frontiers in Human Neuroscience*, *7*, 930. https://doi.org/10.3389/fnhum.2013.00930.

Mulders, P. C., van Eijndhoven, P. F., Schene, A. H., Beckmann, C. F., & Tendolkar, I. (2015). Resting-state functional connectivity in major depressive disorder: A review. *Neuroscience and Biobehavioral Reviews*, *56*, 330–344. https://doi.org/10.1016/j.neubiorev.2015.07.014.

Ochsner, K. N., Ray, R. D., Cooper, J. C., Robertson, E. R., Chopra, S., Gabrieli, J. D., et al. (2004). For better or for worse: Neural systems supporting the cognitive down- and up-regulation of negative emotion. *NeuroImage*, *23*(2), 483–499. https://doi.org/10.1016/j.neuroimage.2004.06.030

Pannekoek, J. N., van der Werff, S. J., Meens, P. H., van den Bulk, B. G., Jolles, D. D., Veer, I. M., et al. (2014). Aberrant resting-state functional connectivity in limbic and salience networks in treatment—Naive clinically depressed adolescents. *Journal of Child Psychology and Psychiatry*, *55*(12), 1317–1327. https://doi.org/10.1111/jcpp.12266.

Peters, S. K., Dunlop, K., & Downar, J. (2016). Cortico-striatal-thalamic loop circuits of the salience network: A central pathway in psychiatric disease and treatment. *Frontiers in Systems Neuroscience*, *10*, 104. https://doi.org/10.3389/fnsys.2016.00104.

Roy, A. K., Shehzad, Z., Margulies, D. S., Kelly, A. M., Uddin, L. Q., Gotimer, K., et al. (2009). Functional connectivity of the human amygdala using resting state fMRI. *NeuroImage*, *45*(2), 614–626. https://doi.org/10.1016/j.neuroimage.2008.11.030.

Schmaal, L., Pozzi, E., Ho, T., Velzen, L.v., Veer, I., Opel, N., et al. (2019). *ENIGMA MDD: Seven years of global neuroimaging studies of major depression through worldwide data sharing*. Retrieved from PsyArXiv Preprints.

Sexton, C. E., Allan, C. L., Le Masurier, M., McDermott, L. M., Kalu, U. G., Herrmann, L. L., et al. (2012). Magnetic resonance imaging in late-life depression: Multimodal examination of network disruption. *Archives of General Psychiatry*, *69*(7), 680–689. https://doi.org/10.1001/archgenpsychiatry.2011.1862.

Stevens, F. L., Hurley, R. A., & Taber, K. H. (2011). Anterior cingulate cortex: Unique role in cognition and emotion. *The Journal of Neuropsychiatry and Clinical Neurosciences*, *23*(2), 121–125. https://doi.org/10.1176/appi.neuropsych.23.2.121.

Tadayonnejad, R., & Ajilore, O. (2014). Brain network dysfunction in late-life depression: A literature review. *Journal of Geriatric Psychiatry and Neurology*, *27*(1), 5–12. https://doi.org/10.1177/0891988713516539.

Van den Heuvel, M. P., & Hulshoff Pol, H. E. (2010). Exploring the brain network: A review on resting-state fMRI functional connectivity. *European Neuropsychopharmacology*, *20*(8), 519–534. https://doi.org/10.1016/j.euroneuro.2010.03.008.

Yan, C. G., Chen, X., Li, L., Castellanos, F. X., Bai, T. J., Bo, Q. J., et al. (2019). Reduced default mode network functional connectivity in patients with recurrent major depressive disorder. *Proceedings of the National Academy of Sciences of the United States of America*, *116*(18), 9078–9083. https://doi.org/10.1073/pnas.1900390116

Yang, T. T., Simmons, A. N., Matthews, S. C., Tapert, S. F., Frank, G. K., Bischoff-Grethe, A., et al. (2009). Depressed adolescents demonstrate greater subgenual anterior cingulate activity. *Neuroreport*, *20*(4), 440–444. https://doi.org/10.1097/WNR.0b013e3283262e10.

Ye, T., Peng, J., Nie, B., Gao, J., Liu, J., Li, Y., et al. (2012). Altered functional connectivity of the dorsolateral prefrontal cortex in first-episode patients with major depressive disorder. *European Journal of Radiology*, *81*(12), 4035–4040. https://doi.org/10.1016/j.ejrad.2011.04.058.

Yue, Y., Yuan, Y., Hou, Z., Jiang, W., Bai, F., & Zhang, Z. (2013). Abnormal functional connectivity of amygdala in late-onset depression was associated with cognitive deficits. *PLoS One*, *8*(9). https://doi.org/10.1371/journal.pone.0075058.

Zhang, F. (2017). Resting-state functional connectivity abnormalities in adolescent depression. *eBioMedicine*, *17*, 20–21. https://doi.org/10.1016/j.ebiom.2017.02.028.

Chapter 27

# Neural markers of depression in MRI

Chien-Han Lai[a,b]

[a]*Institute of Biophotonics, National Yang-Ming University, Taipei, Taiwan,* [b]*PhD Psychiatry & Neuroscience Clinic, Taoyuan, Taiwan*

## Abbreviations

| | |
|---|---|
| **ACC** | anterior cingulate cortex |
| **ATR** | anterior thalamic radiation |
| **CC** | corpus callosum |
| **DTI** | diffusion tensor imaging |
| **FA** | fractional anisotropy |
| **FMRI** | functional magnetic resonance imaging |
| **GM** | gray matter |
| **MRI** | magnetic resonance imaging |
| **Rs-FMRI** | resting-state functional magnetic resonance imaging |
| **SLF** | superior longitudinal fasciculus |
| **UF** | uncinate fasciculus |
| **VBM** | voxel-based morphometry |
| **WM** | white matter |

## Introduction

The burden of depression has gradually increased in these years, which is associated with economic impacts, influences of life quality, mental health, and physical health. Recent study results showed that socioeconomic development and time-related progress would not really control of the disease burden and progress (Greenberg, Fournier, Sisitsky, Pike, & Kessler, 2015; Murray et al., 2012). The most typical representation of depression is major depressive disorder (MDD), which causes a lot of losses on the economics and productivities due to its significant social and occupational functional impairments (Souery et al., 2007). Since the impacts of MDD on the mental health and society are so significant, the understanding of pathophysiology model of MDD will be an emergent topic. Among the potential targets of pathophysiology in MDD, the brain absolutely plays an important role in this perspective. The clinical symptoms of depression are usually associated with the cognition and affect regulation impairments, which are related to the underlying etiology of brain structure and function. For the researchers to understand the brain, the neuroimaging method will be a crucial tool. There are different dimensions of spatial and temporal resolutions in the different kinds of neuroimaging methods. For example, the positron emission tomography has a great temporal resolution. However, this method suffers from the limitation of spatial resolution. For the assistance and compensation of the limitation, magnetic resonance imaging (MRI) is usually applied simultaneously to localize the significant clusters in the positron emission tomography results. Therefore, the relative balanced profile of spatial resolution and temporal resolution in MRI makes this neuroimaging method more acceptable and popular in the translational medicine study to understand the structure and function of brain. MRI can survey the brain structure, such as the gray matter (GM), white matter (WM), magnetic resonance angiography, cortical thickness, total brain volume, cerebrospinal fluid volume, and other brain structural targets. In addition, MRI can explore the brain function in many kinds of perspectives, such as the functional MRI (FMRI), magnetic resonance spectroscopy, arterial spin labeling, and other methods. In this chapter, focus is on the neural markers revealed by MRI in depression. For the structural MRI, the GM and WM-related studies will be the major components of literature review. For FMRI, the task FMRI and resting-state FMRI (Rs-FMRI) will be the crucial studies in this perspective (Fig. 1).

## Introduction and concept of MRI and FMRI

MRI provides the powerful tool in the field of translational medicine, which can be applied in understanding the brain structure and function. The hydrogen ion fluctuations, which are derived from the directional magnetic field, will produce the MRI contrast. The formation of images relies on the MRI contrast. In addition to the traditional and clinical application, more important perspectives of MRI in study field will depend on the branching of MRI technology, which is FMRI. The FMRI is crucial for the translational medicine and neuroscience study. The changes in the local vessel blood flow and subsequent oxygenation of blood hemoglobin while neural activities will cause the changes in the oxyhemoglobin and deoxyhemoglobin

The Neuroscience of Depression. https://doi.org/10.1016/B978-0-12-817933-8.00037-2

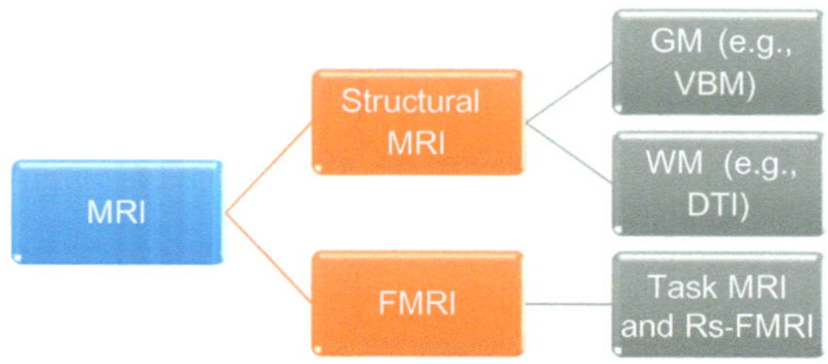


FIG. 1 Structural MRI and FMRI gross categories. The structural MRI includes GM and WM studies. The representative methods of GM and WM studies are VBM and DTI studies. The FMRI includes task FMRI and Rs-FMRI.

percentage in the active areas. The changes in oxyhemoglobin and deoxyhemoglobin percentage will cause the fluctuations of brain magnetic effects, which can be measured as the blood-oxygen-level-dependent signals. The signal is the important base of FMRI. This tool can detect the brain function by different tasks, such as emotion, cognitive, and language tasks. This type of FMRI can be named as task MRI (Chen, Liu, Zhu, & Zhang, 2009). In addition to task MRI, recently another type of FMRI, which is called Rs-FMRI, is popular in study of the brain activities at resting state (Andellini, Cannata, Gazzellini, Bernardi, & Napolitano, 2015). The resting-state activities represent the default brain activities. In addition to the basis of blood oxygen-level-dependent signals, the default brain activities are associated with the coupling between astrocytes and neurons. The Rs-FMRI can measure many kinds of parameters of default brain activities, such as regional homogeneity, low-frequency fluctuations and functional connectivity network, etc. The functional connectivity network can be explained by the social network calculation, which is named as "small-world" network. In the theory of small-world network, a high-efficiency connectivity will be associated with better ability of adaptation. The degree distribution and connection level of brain networks are also crucial for the default network functioning (Guye, Bettus, Bartolomei, & Cozzone, 2010).

The methods of structural MRI in GM and WM include voxel- and surface-based segmentation, voxel-based morphometry (VBM), and cortical thickness (Helms, 2016). The VBM can obtain the GM from the segmentation and perform a nonparametric independent 2-sample test to identify the significantly altered structural regions. In addition to the VBM analysis, the structural brain analysis can measure the volume, thickness, and shape of brain (Bigler, 2015). For WM tracts, diffusion MRI-based techniques can obtain the neuroanatomical structures of white matter tract by analyzing the water diffusion property and water molecule free movement within the fasciculus under the basis of Brownian motion theory (Bastiani & Roebroeck, 2015). Among the diffusion MRI-based techniques, the diffusion tensor imaging (DTI) MRI and tractography technique are the crucial tools (Richardson & Price, 2009) to investigate the micro-integrity architectures of tracts, such as the macro-, meso-, and microscopic organization of brain connectivity and anatomy via the parameters of fractional anisotropy (FA), medial diffusivity, radial diffusivity, and axial diffusivity (Fig. 2; Table 1).

## Hypothetic model in depression

According to the stress theory in depression, the stress-related influences, such as elevated cortisol level, excessive glutamate, and decreased neurotrophin factors, would be related to structural and functional alterations of brain in MDD. The hypothetic model for MDD pathophysiology focuses on the "limbic-cortical-striatal-pallidal-thalamic tract" regions (Sheline, 2000). The proposed limbic regions mostly include the hippocampus and amygdala. The limbic regions have intense and wide connections with cortical-striatal-pallidal-thalamic regions. In the recent years, the complex term of "limbic-cortical-striatal-pallidal-thalamic tract" model (Sheline, 2000) has been replaced by fronto-limbic model due to that the hippocampus, amygdala, thalamus, and caudate nucleus can be categorized into the limbic system regions (Fig. 3). In addition, several structural MRI studies also introduce the concept of fronto-limbic model to address the pathophysiology of MDD (de Kwaasteniet et al., 2013; Lai, Hsu, & Wu, 2010; van Tol et al., 2010). The FMRI studies also supported the concept of fronto-limbic model of MDD (Ubl et al., 2015). In addition to the fronto-limbic model, the default mode network of brain is related to the self-referential thoughts and negative ruminations. The default mode network, such as the ventromedial prefrontal cortex, anterior cingulate, and lateral parietal cortex, can be revealed by Rs-FMRI technique (Sheline et al., 2009). It also suggests that

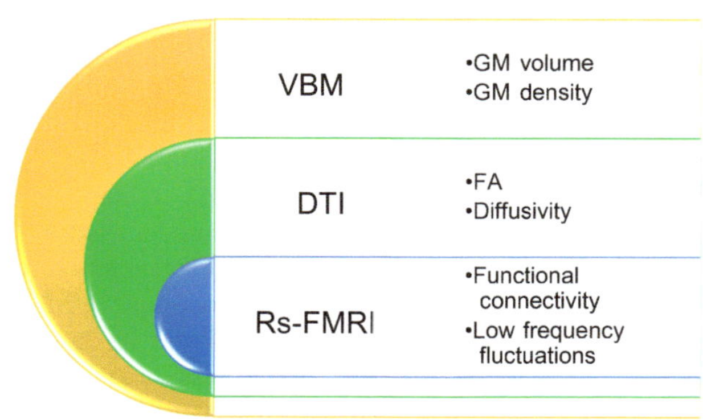


FIG. 2 Introduction of VBM, DTI, and Rs-FMRI parameters. VBM parameters include GM volume and density. DTI can measure WM micro-integrity via FA and different kinds of diffusivity parameters. The Rs-FMRI parameters include the functional connectivity, low-frequency fluctuations, etc.

**TABLE 1** A list of MRI markers and techniques for depression analysis.

| Techniques | Parameters |
|---|---|
| VBM | GM volume and density |
| DTI | FA, mean diffusivity, radial diffusivity, and axial diffusivity |
| Task FMRI | Task functional connectivity and activity |
| Rs-FMRI | Resting functional connectivity and low-frequency fluctuations |

VBM, DTI, task FMRI, and Rs-FMRI are major MRI techniques for depression analysis. The parameters of these techniques are also listed within this table.

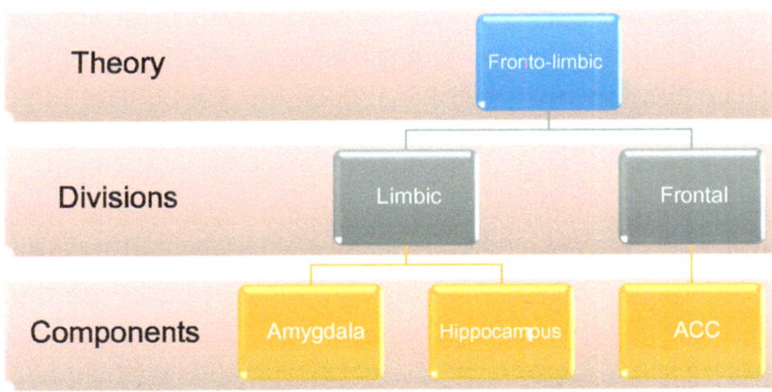


FIG. 3 Fronto-limbic model. This theory is composed of two major divisions. Frontal division includes the frontal regions and ACC. Limbic division includes amygdala, hippocampus, etc.

fronto-limbic model might mainly constitute the modern model to explain the pathophysiology of MDD. However, the other regions revealed by the default mode network, such as parietal regions, still might play a significant role in the pathophysiology of depression. In addition, the traditional concept of which type of model would be specific to which kind of mental illness should be reevaluated. A possible share of common alterations in the same regions might exist in different kinds of mental illnesses. A possible pattern of severity dimension and spreading degrees of altered regions in different kinds of mental illnesses might be the future trend. Therefore, even the current studies mostly support the fronto-limbic model in the MDD pathophysiology. However, the fronto-limbic model may be discovered in other mental illnesses. The advanced analysis, such as pattern recognition, can be used to identify the cumulative severities and spreading regions of MDD when compared to other mental illnesses. This type of sophisticated model would be more precise to identify what level of severities of structural and functional alterations in MDD (Lai, 2019). Therefore, state-of-the-art hypothetic model of MDD might be divided to two dimensions, mainly involving regions of fronto-limbic model and quantitative pattern recognition analysis for the alterations in these regions to find a more specific "identity" of structural and functional alterations in MDD.

## Structural neural marker in MDD: GM aspect

The previous VBM reports show that GM alterations in the fronto-limbic regions of MDD (Lai et al., 2010; van Tol et al., 2010; Yuan et al., 2008). The clinical symptoms and cognitive impairments in the MDD might be associated with GM alterations in the frontal regions (van Tol et al., 2010). In addition to the cognition influences, the depressive severity and illness duration might be associated with GM alterations in the frontal regions, such as right medial frontal cortex (Vasic, Walter, Hose, & Wolf, 2008), superior frontal gyrus, and medial frontal gyrus (Serra-Blasco et al., 2013). Several meta-analytic reports summarize the GM alterations in the anterior cingulate cortex (ACC), frontal cortex, orbitofrontal cortex, hippocampus, and striatum, which are compatible with fronto-limbic model in MDD (Arnone, McIntosh, Ebmeier, Munafo, & Anderson, 2012; Bora, Fornito, Pantelis, & Yucel, 2012; Lai, 2013; Wise, Cleare, Herane, Young, & Arnone, 2014).

The recent biomarker study using GM volume in MDD also replicates the importance of fronto-limbic regions, such as the ACC, temporal regions, and prefrontal cortex (Hellewell et al., 2019). The specific genotype of NCAN rs1064395 is also associated with GM volume alterations in the fronto-limbic regions, such as the prefrontal cortex, hippocampus, and amygdala, which are crucial for emotional processing and regulation in MDD (Dannlowski et al., 2015). The cumulative illness severity and hospitalization history are also inversely correlated with GM volume of bilateral hippocampus in MDD, which also suggest the possible predictor values of hippocampus (Zaremba et al., 2018). Another study of serotonin 5-HT1A autoreceptor binding and GM volume also shows the inverse correlation between 5-HT1A binding and GM volume in the ACC and the positive correlation between 5-HT1A binding and GM volume in the temporal lobe of MDD patients (Zanderigo et al., 2018). The explorative comparison study between MDD and panic disorder shows MDD-related GM volume reductions in fronto-limbic regions, such as bilateral medial frontal cortex, right superior frontal gyrus, and right superior temporal gyrus. However, MDD is also associated with GM volume alterations in the additional regions beyond the fronto-limbic model, such as bilateral cerebellums (Lai & Wu, 2015). A more recent meta-analysis of VBM studies find smaller GM volumes in the fronto-limbic regions and beyond fronto-limbic regions, such as the temporal, parietal, and cerebellar regions, which suggest a possible amendment of current fronto-limbic model (Wise et al., 2017). Another meta-analysis also finds the similar GM volume alterations in the fronto-limbic regions, such as the ventrolateral prefrontal cortex, hippocamps, and thalamus. In addition, GM alterations of MDD are found in the beyond fronto-limbic regions, such as parietal regions and cerebellums (Arnone et al., 2016).

From the above literature, it seems that GM alterations in MDD involve more extensive than fronto-limbic model. A possible beyond fronto-limbic model in the perspective of GM alterations in MDD might exist. The beyond fronto-limbic model might include the temporal, parietal, and cerebellar regions. The researchers should consider exploring and focusing on the beyond fronto-limbic model regions, such as the temporo-parieto-cerebellar regions (Fig. 4).

## Structural neural marker in MDD: WM aspect

The study of WM micro-integrity alterations in depression usually focuses on the FA values. The WM tracts in the fronto-limbic model are usually observed with altered FA values, such as the superior longitudinal fasciculus (SLF), anterior thalamic radiation (ATR), and corpus callosum (CC). The WM alterations in MDD seem more significant in the late-onset subgroup due to the degenerative characteristics of aging. The higher cortical functions of SLF include the language and cognitive functions. The SLF connects the language areas, frontal areas, and limbic regions to form a circuit for higher cortical function. In addition, the SLF provides a connection between cortical and subcortical structures (Bernal & Altman, 2010). Lower FA values in the left SLF and internal capsule of patients with late-onset MDD also suggest the role of fronto-limbic circuit in the etiology of MDD. In addition, the combined WM alterations of SLF and CC might disturb the control of cognitive and affective functions in MDD. The decreased FA values in the SLF are a relatively stable trait of WM micro-integrity alterations in MDD (Murphy & Frodl, 2011). Apart from SLF, ATR is another region that has a lower FA in the etiology of MDD. The ATR connects the frontal regions, thalamus, and hippocampus regions via the medial forebrain bundle to converge in the anterior internal capsule and medial prefrontal cortex, which is also crucial for the fronto-limbic model for MDD. The alterations in the WM micro-integrity of ATR are found in the MDD and earlier age-onset MDD is prone to have reduced hippocampal volume, which suggests the ATR connection between frontal and hippocampus will be influenced in MDD. The WM alterations of ATR might also support the theories of vascular and glucocorticoid cascades in depression (Sexton et al., 2012). A more recent meta-analytic study also confirmed the role of SLF in the pathophysiology of MDD. In addition, the cerebellar WM tracts and body of CC have robust reductions of FA values. The lower FA values in the genu of CC and ATR seem specific for the medicine-naïve MDD patients ( Jiang et al., 2017). The CC is another important structure for cognition and emotion, which might be also crucial for the pathophysiology of MDD. The combined alterations of FA values in the CC and SLF also reconfirm the crucial roles of these two regions. In addition, the negative correlation between FA values and symptom severity in the CC and SLF also suggests core regions of WM alterations (Cole et al., 2012). The disturbances in interhemispheric connections of CC might be altered in the meta-analytic study in MDD (Liao et al., 2013). Another meta-analytic study of MDD also confirms the stable characteristics of WM micro-integrity alterations in the genu of CC, which connects the prefrontal cortex of bilateral hemispheres, to influence the ability of mood regulation (Wise et al., 2016). Recently, the ENIGMA consortium also confirms the subtle and widespread alterations of FA values in the brain and with largest differences in the CC (van Velzen et al., 2020) (Fig. 5).

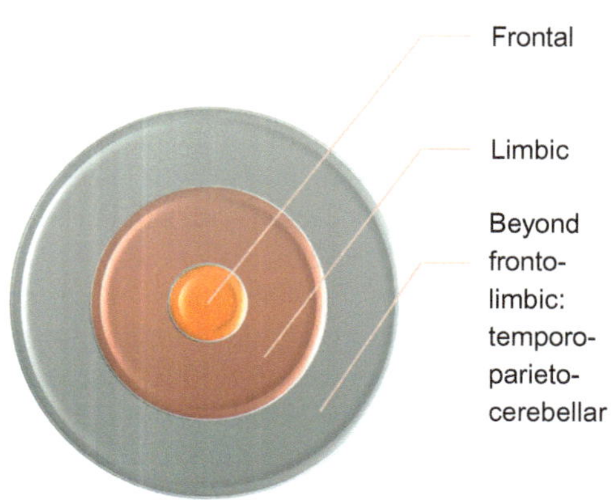


FIG. 4 MRI GM neural markers in MDD. VBM studies prove the importance of fronto-limbic model. However, the beyond fronto-limbic regions should not be ignored.

## Functional neural marker in MDD: Task FMRI aspect

MDD has different silhouettes under different kinds of tasks. The emotion regulation task FMRI still supports

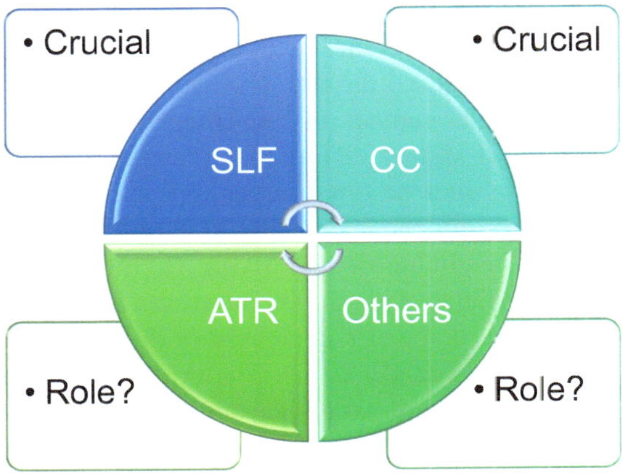


FIG. 5 MRI WM neural markers in MDD. SLF and CC might be the core structures of micro-integrity alterations in MDD. ATR and other WM tracts might also play a role.

the importance of ACC and greater depression symptom severity seems associated with less engagement of ACC (Fitzgerald, Klumpp, Langenecker, & Phan, 2019). The emotional memory formation task FMRI is related to hippocampus-amygdala activation, especially for the symptom severity (Ai et al., 2020). The reward task FMRI shows that MDD patients have alterations of reward prediction errors, which are blunted in the striatum and ventral tegmental area. The impaired reward learning seems associated with dopamine dysregulation in depression (Kumar et al., 2018). The MDD patients might need the compensatory mechanism to activate the frontal regions, such as dorsolateral prefrontal cortex and ACC, to a more extent to handle the working memory loading. More mental efforts will be needed to successfully perform the working memory task (Gartner et al., 2018). The self-referential processing task FMRI reveals that MDD patients are incapable to activate rostral ACC during the negative self-referential stimuli, which suggests the lower ability to drift attention away from the negative ruminations inside the mind (Wagner, Schachtzabel, Peikert, & Bar, 2015). The imaging stress task FMRI shows that remitted and current depression subjects have reduced activation in the frontoparietal regions, such as ventromedial prefrontal cortex and precuneus. In addition, the stress-related changes in the activities of ventromedial prefrontal cortex are negatively with cortisol response. The striatum is also responsible for stress reaction in remitted MDD patients (Ming et al., 2017). The stress task can correspond to the stress theory in the formation of depression symptoms. Therefore, the stress-related cortisol response and stress response still suggest the crucial role of fronto-limbic regions in the stress model of MDD. The theory of fronto-limbic model is still supported by the task FMRI studies. The ACC and frontal regions are responsible for the top-down regulation activities, such as the emotion regulation, modulation of self-referential ruminations, stress modulation, cognitive ability, and executive function, to control the exaggerated responses from the limbic regions, such as amygdala, striatum, and hippocampus. The limbic regions can show their abilities of bottom-up feedback, such as the primitive emotional responses, stress reaction, emotional memory formation, and reward learning, to communicate with the frontal regions about the impacts and strengths of top-down regulations. The balanced top-down and bottom-up mechanisms will lead to the possible solution for many kinds of mental illnesses, especially for depression. However, the complexities and inconsistences of task FMRI results will interrupt the researchers to come to the real solution of brain alterations in depression. A recent meta-analysis shows no significant results across cognitive and emotional processing tasks. In addition, no convergence is obtained across positive, negative affect, and memory. The inconsistence might be related to different task designs and contrasts, uncorrected inference calculations with liberal statistical threshold, varying inclusion and exclusion criteria, and heterogeneity in clinical population (Muller et al., 2017). Therefore, the researchers still need to spend more efforts to find a more consistent answer of the task FMRI-related brain pathophysiology (Fig. 6).

FIG. 6 Task FMRI neural markers in MDD. Different kinds of tasks will determine the results. The fronto-limbic model is confirmed in the emotion and cognitive tasks. The reward task marker points to striatum.

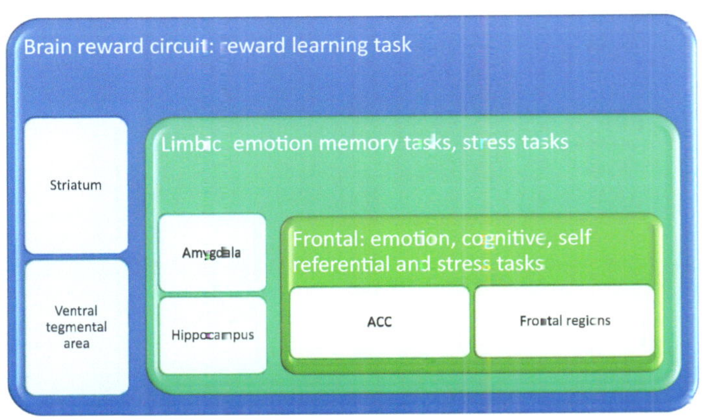

## Functional neural marker in MDD: Rs-FMRI aspect

There are many kinds of parameters in Rs-FMRI, such as the functional connectivity, low-frequency fluctuations and regional homogeneity, etc. Therefore, the inconsistence of Rs-FMRI findings in MDD seems more significant than task FMRI. Therefore, the meta-analyses of Rs-FMRI studies also reveal such inconsistent findings. A meta-analysis of seed-based resting-state functional connectivity studies using multilevel kernel density shows the hypo-connectivity of frontal and parietal regions, which are important for cognitive control of emotion regulation and attention. The parietal regions of dorsal attention system are hypo-connected with frontoparietal network. In addition, the hyper-connectivity within default mode network is associated with impairments in the self-referential thoughts and ruminations, which suggest the imbalance of self-evaluation of internal world in MDD. The altered self-ruminations will influence the attention to the external world and cause depressive biases, which might be associated with alterations in cognitive control, salience, and emotion processing (Kaiser, Andrews-Hanna, Wager, & Pizzagalli, 2015). However, the seed-based functional connectivity is still influenced by the selection of seeds and underlying theory to choose these seeds. Therefore, the alterations of frontoparietal and default mode network can't represent the whole perspective of functional connectivity studies in MDD. Another meta-analysis of first-episode medicine-naïve MDD using activation likelihood estimation shows alterations of brain activities in the fronto-limbic regions and default mode network, such as the dorsolateral prefrontal cortex, putamen, superior temporal gyrus, precuneus, and posterior cingulate cortex (Zhong, Pu, & Yao, 2016). From this point, it is obvious that the selection of studies also determines the meta-analysis results. In addition, the meta-analytic methods also influence the results (Fig. 7).

Another meta-analysis of low-frequency fluctuations in MDD using signed differential mapping shows that increased amplitude of low-frequency fluctuations in the ACC. In addition, the increasing phenomenon is more prominent in medicine-treated MDD. The increased activation is associated with illness duration. The medicine-naïve MDD has a specific decreased brain activity in the cerebellum (Zhou et al., 2017). The crucial role of ACC is emphasized in the meta-analysis of low-frequency fluctuations in MDD. The meta-analysis of regional homogeneity in first-episode medicine-naïve MDD shows increased regional homogeneity in left hippocampus and decreased regional homogeneity in left orbitofrontal cortex, which also emphasizes the possible architecture of fronto-limbic model from the viewpoint of Rs-FMRI studies (Hao, Chen, Mao, Zhong, & Dai, 2019). The alterations of regional homogeneity in the hippocampus and orbitofrontal cortex might suggest the impairments of cognitive control, emotional memory, and emotion regulation. Another meta-analysis in MDD shows hyperactivity in the fronto-limbic regions, such as middle frontal gyrus, amygdala, and hippocampus. The hypoactivity is significant in the fronto-occipital and cerebellar regions. In addition, the illness duration is positively correlated with the hyperactivity in the hippocampus and hypoactivity in the cerebellum (Ma et al., 2019). The importance of fronto-limbic model can be confirmed in several kinds of parameters, such as the low-frequency fluctuations and regional homogeneity. However, the selection of different kinds of Rs-FMRI parameters in different kinds of meta-analytic methods still determines the results in a crucial way.

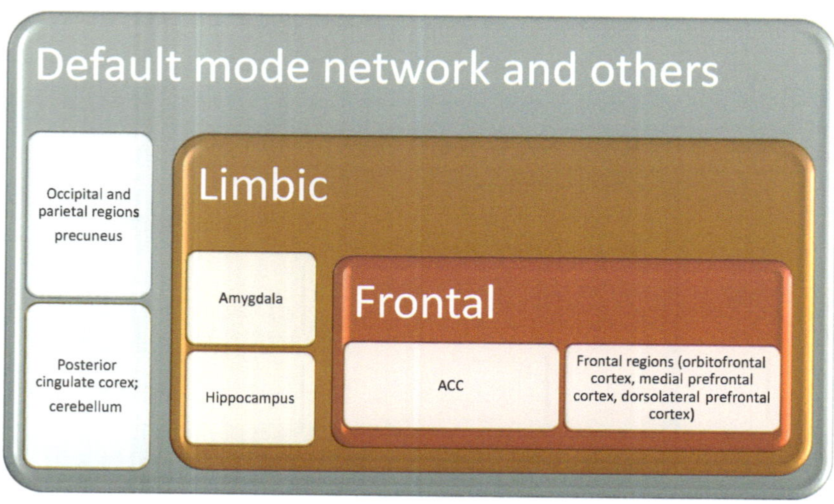


FIG. 7 Rs-FMRI neural markers in MDD. The role of frontal regions is still crucial. In addition, the limbic regions play a role. The default mode network, parietal, occipital, and cerebellar regions might be the additional components.

## Key facts of neural MRI markers in MDD

- The GM alterations in MDD are not limited to the fronto-limbic regions. More extensive involvements of GM alterations in the beyond fronto-limbic regions, such as temporal, parietal, and cerebellar regions.
- The WM alterations in MDD might be localized in the SLF, ATR, and CC. However, the diversities of WM micro-integrity seem more obvious than GM studies due to the nature characteristics of WM tract and analytic methods.
- The findings of task FMRI studies in MDD still not reach the consistent state till now. The different kinds of tasks, study designs, and statistical settings might lead to such inconsistences. More efforts will be warranted in this perspective.
- The findings of Rs-FMRI seem more inconsistent than task FMRI. The different kinds of Rs-FMRI parameters, meta-analytic methods, and MDD characteristics might determine the results in a crucial way.
- Beyond fronto-limbic regions are still emerging in the recent MRI studies of MDD.

## Summary points

- The perspective of GM alterations in MDD should include the fronto-limbic model and beyond fronto-limbic model, which might extend to the temporal, parietal, and cerebellar regions.
- The WM micro-integrity alterations, especially for FA, mostly will include the fronto-limbic model-related WM tracts, such as the SLF, ATR, and CC. However, the diverse and inconsistent results should be taken into consideration.
- The findings of task FMRI depends on the task characteristics. The emotion-related tasks might detect the alterations in the fronto-limbic regions, such as the ACC, hippocampus, and amygdala. The reward-related tasks mostly reveal the findings in the reward regions, such as the striatum. In spite of the inconsistences, the importance of fronto-limbic regions should not be ignored in task FMRI study.
- The fronto-limbic model seems more crucial in the regional homogeneity and low-frequency fluctuations of MDD. The ACC might be not so significant. However, the frontal regions seem still crucial in most Rs-FMRI studies. In addition, the default mode network still plays a role in Rs-FMRI studies in MDD.
- Fronto-limbic model might be the backbone of the MRI neural markers in MDD.

## Mini-dictionary of terms

**FMRI** A MRI method to detect the changes in magnetic effects of deoxyhemoglobin due to blood flow changes coupling with neural activities during task or resting-state status.

**Rs-FMRI** This method can detect the resting-state brain activities and find the default mode network of brain, which can constitute the small-world network of brain.

**VBM** A structural MRI method to quantify the GM in the voxel-wise volume or density forms. Then, the nonparametric 2-sample test was performed to identify the significant regions between two groups, such as controls and patients.

**DTI** A FMRI method to identify the WM tract micro-integrity alterations using the diffusion parameters, such as FA and diffusivity.

**Task FMRI** The type of FMRI which is modulated by the different tasks of emotion, cognition and reward perspectives, etc. The task FMRI can detect the specific-activated regions induced by different kinds of tasks.

## References


Ai, H., Opmeer, E. M., Marsman, J. C., Veltman, D. J., van der Wee, N. J. A., Aleman, A., et al. (2020). Longitudinal brain changes in MDD during emotional encoding: Effects of presence and persistence of symptomatology. *Psychological Medicine*, *50*(8), 1316–1326.

Andellini, M., Cannata, V., Gazzellini, S., Bernardi, B., & Napolitano, A. (2015). Test-retest reliability of graph metrics of resting state MRI functional brain networks: A review. *Journal of Neuroscience Methods*, *253*, 183–192.

Arnone, D., Job, D., Selvaraj, S., Abe, O., Amico, F., Cheng, Y., et al. (2016). Computational meta-analysis of statistical parametric maps in major depression. *Human Brain Mapping*, *37*, 1393–1404.

Arnone, D., McIntosh, A. M., Ebmeier, K. P., Munafo, M. R., & Anderson, I. M. (2012). Magnetic resonance imaging studies in unipolar depression: Systematic review and meta-regression analyses. *European Neuropsychopharmacology*, *22*, 1–16.

Bastiani, M., & Roebroeck, A. (2015). Unraveling the multiscale structural organization and connectivity of the human brain: the role of diffusion MRI. *Frontiers in Neuroanatomy*, *9*, 77.

Bernal, B., & Altman, N. (2010). The connectivity of the superior longitudinal fasciculus: A tractography DTI study. *Magnetic Resonance Imaging*, *28*, 217–225.

Bigler, E. D. (2015). Structural image analysis of the brain in neuropsychology using magnetic resonance imaging (MRI) techniques. *Neuropsychology Review*, *25*, 224–249.

Bora, E., Fornito, A., Pantelis, C., & Yucel, M. (2012). Gray matter abnormalities in Major Depressive Disorder: A meta-analysis of voxel based morphometry studies. *Journal of Affective Disorders*, *138*, 9–18

Chen, W., Liu, X., Zhu, X. H., & Zhang, N. (2009). Functional MRI study of brain function under resting and activated states. *Conference Proceedings: Annual International Conference of the IEEE Engineering in Medicine and Biology Society*, *2009*, 4061–4063.

Cole, J., Chaddock, C. A., Farmer, A. E., Aitchison, K. J., Simmons, A., McGuffin, P., et al. (2012). White matter abnormalities and illness severity in major depressive disorder. *The British Journal of Psychiatry*, *201*, 33–39.

Dannlowski, U., Kugel, H., Grotegerd, D., Redlich, R., Suchy, J., Opel, N., et al. (2015). NCAN cross-disorder risk variant is associated with limbic gray matter deficits in healthy subjects and major depression. *Neuropsychopharmacology*, *40*, 2510–2516.

de Kwaasteniet, B., Ruhe, E., Caan, M., Rive, M., Olabarriaga, S., Groefsema, M., et al. (2013). Relation between structural and functional connectivity in major depressive disorder. *Biological Psychiatry*, *74*, 40–47.

Fitzgerald, J. M., Klumpp, H., Langenecker, S., & Phan, K. L. (2019). Transdiagnostic neural correlates of volitional emotion regulation in anxiety and depression. *Depression and Anxiety*, *36*, 453–464.

Gartner, M., Ghisu, M. E., Scheidegger, M., Bonke, L., Fan, Y., Stippl, A., et al. (2018). Aberrant working memory processing in major depression: Evidence from multivoxel pattern classification. *Neuropsychopharmacology*, *43*, 1972–1979.

Greenberg, P. E., Fournier, A. A., Sisitsky, T., Pike, C. T., & Kessler, R. C. (2015). The economic burden of adults with major depressive disorder in the United States (2005 and 2010). *The Journal of Clinical Psychiatry*, *76*, 155–162.

Guye, M., Bettus, G., Bartolomei, F., & Cozzone, P. J. (2010). Graph theoretical analysis of structural and functional connectivity MRI in normal and pathological brain networks. *Magma*, *23*, 409–421.

Hao, H., Chen, C., Mao, W., Zhong, J., & Dai, Z. (2019). Aberrant brain regional homogeneity in first-episode drug-naive patients with major depressive disorder: A voxel-wise meta-analysis. *Journal of Affective Disorders*, *245*, 63–71.

Hellewell, S. C., Welton, T., Maller, J. J., Lyon, M., Korgaonkar, M. S., Koslow, S. H., et al. (2019). Profound and reproducible patterns of reduced regional gray matter characterize major depressive disorder. *Translational Psychiatry*, *9*, 176.

Helms, G. (2016). Segmentation of human brain using structural MRI. *Magma*, *29*, 111–124.

Jiang, J., Zhao, Y. J., Hu, X. Y., Du, M. Y., Chen, Z. Q., Wu, M., et al. (2017). Microstructural brain abnormalities in medication-free patients with major depressive disorder: A systematic review and meta-analysis of diffusion tensor imaging. *Journal of Psychiatry & Neuroscience*, *42*, 150–163.

Kaiser, R. H., Andrews-Hanna, J. R., Wager, T. D., & Pizzagalli, D. A. (2015). Large-scale network dysfunction in major depressive disorder: A meta-analysis of resting-state functional connectivity. *JAMA Psychiatry*, *72*, 603–611.

Kumar, P., Goer, F., Murray, L., Dillon, D. G., Beltzer, M. L., Cohen, A. L., et al. (2018). Impaired reward prediction error encoding and striatal-midbrain connectivity in depression. *Neuropsychopharmacology*, *43*, 1581–1588.

Lai, C. H. (2013). Gray matter volume in major depressive disorder: A meta-analysis of voxel-based morphometry studies. *Psychiatry Research*, *211*, 37–46.

Lai, C. H. (2019). The neural markers of MRI to differentiate depression and panic disorder. *Progress in Neuro-Psychopharmacology & Biological Psychiatry*, *91*, 72–78.

Lai, C. H., Hsu, Y. Y., & Wu, Y. T. (2010). First episode drug-naive major depressive disorder with panic disorder: Gray matter deficits in limbic and default network structures. *European Neuropsychopharmacology*, *20*, 676–682.

Lai, C. H., & Wu, Y. T. (2015). The gray matter alterations in major depressive disorder and panic disorder: Putative differences in the pathogenesis. *Journal of Affective Disorders*, *186*, 1–6.

Liao, Y., Huang, X., Wu, Q., Yang, C., Kuang, W., Du, M., et al. (2013). Is depression a disconnection syndrome? Meta-analysis of diffusion tensor imaging studies in patients with MDD. *Journal of Psychiatry & Neuroscience*, *38*, 49–56.

Ma, X., Liu, J., Liu, T., Ma, L., Wang, W., Shi, S., et al. (2019). Altered resting-state functional activity in medication-naive patients with first-episode major depression disorder vs. healthy control: A quantitative meta-analysis. *Frontiers in Behavioral Neuroscience*, *13*, 89.

Ming, Q., Zhong, X., Zhang, X., Pu, W., Dong, D., Jiang, Y., et al. (2017). State-independent and dependent neural responses to psychosocial stress in current and remitted depression. *The American Journal of Psychiatry*, *174*, 971–979.

Muller, V. I., Cieslik, E. C., Serbanescu, I., Laird, A. R., Fox, P. T., & Eickhoff, S. B. (2017). Altered brain activity in unipolar depression revisited: Meta-analyses of neuroimaging studies. *JAMA Psychiatry*, *74*, 47–55.

Murphy, M. L., & Frodl, T. (2011). Meta-analysis of diffusion tensor imaging studies shows altered fractional anisotropy occurring in distinct brain areas in association with depression. *Biol Mood Anxiety Disord*, *1*, 3.

Murray, C. J., Vos, T., Lozano, R., Naghavi, M., Flaxman, A. D., Michaud, C., et al. (2012). Disability-adjusted life years (DALYs) for 291 diseases and injuries in 21 regions, 1990-2010: A systematic analysis for the Global Burden of Disease Study 2010. *Lancet*, *380*, 2197–2223.

Richardson, F. M., & Price, C. J. (2009). Structural MRI studies of language function in the undamaged brain. *Brain Structure & Function*, *213*, 511–523.

Serra-Blasco, M., Portella, M. J., Gomez-Anson, B., de Diego-Adelino, J., Vives-Gilabert, Y., Puigdemont, D., et al. (2013). Effects of illness duration and treatment resistance on grey matter abnormalities in major depression. *The British Journal of Psychiatry*, *202*, 434–440.

Sexton, C. E., Le Masurier, M., Allan, C. L., Jenkinson, M., McDermott, L., Kalu, U. G., et al. (2012). Magnetic resonance imaging in late-life depression: Vascular and glucocorticoid cascade hypotheses. *The British Journal of Psychiatry*, *201*, 46–51.

Sheline, Y. I. (2000). 3D MRI studies of neuroanatomic changes in unipolar major depression: The role of stress and medical comorbidity. *Biological Psychiatry*, *48*, 791–800.

Sheline, Y. I., Barch, D. M., Price, J. L., Rundle, M. M., Vaishnavi, S. N., Snyder, A. Z., et al. (2009). The default mode network and self-referential processes in depression. *Proceedings of the National Academy of Sciences of the United States of America*, *106*, 1942–1947.

Souery, D., Oswald, P., Massat, I., Bailer, U., Bollen, J., Demyttenaere, K., et al. (2007). Clinical factors associated with treatment resistance in major depressive disorder: Results from a European multicenter study. *The Journal of Clinical Psychiatry*, *68*, 1062–1070.

Ubl, B., Kuehner, C., Kirsch, P., Ruttorf, M., Flor, H., & Diener, C. (2015). Neural reward processing in individuals remitted from major depression. *Psychological Medicine*, *45*, 3549–3558.

van Tol, M. J., van der Wee, N. J., van den Heuvel, O. A., Nielen, M. M., Demenescu, L. R., Aleman, A., et al. (2010). Regional brain volume in depression and anxiety disorders. *Archives of General Psychiatry*, *67*, 1002–1011.

van Velzen, L. S., Kelly, S., Isaev, D., Aleman, A., Aftanas, L. I., Bauer, J., et al. (2020). White matter disturbances in major depressive disorder: A coordinated analysis across 20 international cohorts in the ENIGMA MDD working group. *Molecular Psychiatry*, *25*(7), 1511–1525.

Vasic, N., Walter, H., Hose, A., & Wolf, R. C. (2008). Gray matter reduction associated with psychopathology and cognitive dysfunction in unipolar depression: a voxel-based morphometry study. *Journal of Affective Disorders*, *109*, 107–116.

Wagner, G., Schachtzabel, C., Peikert, G., & Bar, K. J. (2015). The neural basis of the abnormal self-referential processing and its impact on cognitive control in depressed patients. *Human Brain Mapping*, *36*, 2781–2794.

Wise, T., Cleare, A. J., Herane, A., Young, A. H., & Arnone, D. (2014). Diagnostic and therapeutic utility of neuroimaging in depression: An overview. *Neuropsychiatric Disease and Treatment*, *10*, 1509–1522.

Wise, T., Radua, J., Nortje, G., Cleare, A. J., Young, A. H., & Arnone, D. (2016). Voxel-based meta-analytical evidence of structural disconnectivity in major depression and bipolar disorder. *Biological Psychiatry*, *79*, 293–302.

Wise, T., Radua, J., Via, E., Cardoner, N., Abe, O., Adams, T. M., et al. (2017). Common and distinct patterns of grey-matter volume alteration in major depression and bipolar disorder: Evidence from voxel-based meta-analysis. *Molecular Psychiatry*, *22*, 1455–1463.

Yuan, Y., Zhu, W., Zhang, Z., Bai, F., Yu, H., Shi, Y., et al. (2008). Regional gray matter changes are associated with cognitive deficits in remitted geriatric depression: An optimized voxel-based morphometry study. *Biological Psychiatry*, *64*, 541–544.

Zanderigo, F., Pantazatos, S., Rubin-Falcone, H., Ogden, R. T., Chhetry, B. T., Sullivan, G., et al. (2018). In vivo relationship between serotonin 1A receptor binding and gray matter volume in the healthy brain and in major depressive disorder. *Brain Structure & Function*, *223*, 2609–2625.

Zaremba, D., Enneking, V., Meinert, S., Forster, K., Burger, C., Dohm, K., et al. (2018). Effects of cumulative illness severity on hippocampal gray matter volume in major depression: A voxel-based morphometry study. *Psychological Medicine*, *48*, 2391–2398.

Zhong, X., Pu, W., & Yao, S. (2016). Functional alterations of fronto-limbic circuit and default mode network systems in first-episode, drug-naive patients with major depressive disorder: A meta-analysis of resting-state fMRI data. *Journal of Affective Disorders*, *206*, 280–286.

Zhou, M., Hu, X., Lu, L., Zhang, L., Chen, L., Gong, Q., et al. (2017). Intrinsic cerebral activity at resting state in adults with major depressive disorder: A meta-analysis. *Progress in Neuro-Psychopharmacology & Biological Psychiatry*, *75*, 157–164.

Part III

# Pharmacological treatments for depression

Chapter 28

# Angiotensin receptor 1 blockade as an antidepression strategy

Lilla Lenart[a,b] and Andrea Fekete[a,b]
[a] 1st Department of Pediatrics, Semmelweis University, Budapest, Hungary; [b] SE Diabetes Research Group, Semmelweis University, Budapest, Hungary

## Abbreviations

| | |
|---|---|
| ACE | angiotensin-converting enzyme |
| ANG | angiotensin |
| ARB | angiotensin receptor blocker |
| ATR1 | angiotensin receptor 1 |
| BDNF | brain-derived neurotrophic factor |
| DM | diabetes mellitus |
| IRAP | insulin receptor aminopeptidase |
| MAS1 | MAS proto-oncogene G protein-coupled receptor |
| mBDNF | mature brain-derived neurotrophic factor |
| MMP | matrixmetalloprotease |
| p75NTR | p75 neurotrophin receptor |
| proBDNF | precursor BDNF |
| RAAS | renin-angiotensin-aldosterone system |
| RAASi | renin-angiotensin-aldosterone system inhibitors |
| SPECT-MRI | single-photon emission computed tomography-magnetic resonance imaging |
| TrkB | tropomyosin receptor kinase B |

## Introduction

Depression is a major health concern, which possesses a huge social and economic burden in Euro-Atlantic societies. According to the World Health Organization, more than 300 million people, 4.4% of the world's population suffer from depression (WHO, 2017). The pathomechanism of depression is very complex and not fully understood; moreover, most therapeutic interventions focus on the monoamine system. Medications are partially effective, since remission rate is around 30%–50% following first-line treatment regardless of the type of antidepressant (Entsuah, Huang, & Thase, 2001). (Table 1).

Neuroinflammation appears to be important in the pathology of brain diseases such as depression. The inflammatory hypothesis suggests that inflammation contributes to the development of depression. This theory is based on the fact that increased immune pathway activity leads to inflammation and nitro-oxidative stress, which alter monoamine levels (Galecki & Talarowska, 2018). Indeed, depressed patients have elevated proinflammatory cytokine levels and normalized levels of inflammatory markers are connected to the remission rate of depression (Hannestad, DellaGioia, & Bloch, 2011). Furthermore, elevated proinflammatory mediator levels are linked to nonresponse to antidepressant treatment.

It is well known that angiotensin (ANG) II is a proinflammatory factor in the periphery; anti-inflammatory effects of ANGII receptor blockers (ARBs) are well documented in hypertension or diabetes (Majewski & Bakris, 2016). However, a growing body of evidence reveals that ANGII is also an important player in the central nervous system. Extensive cerebral inflammation via angiotensin receptor 1 (ATR1) leads to neuronal damage.

Renin-angiotensin-aldosterone system inhibitors (RAASi) are a group of drugs, which act by blocking RAAS elements and are commonly used in the treatment of hypertension and certain types of chronic kidney disease. Clinical observational studies have shown that interpersonal sensitivity and depression scores are improved and less antidepressant drugs are needed in patients taking RAASi (Ahola, Harjutsalo, Forsblom, & Groop, 2014; Pavlatou et al., 2008). Thus, inhibition of RAAS may be a novel therapeutic option in depression.

## Renin-angiotensin-aldosterone system in the brain

Classic systemic RAAS was discovered in the 19th century and is crucial for regulating blood pressure, water and electrolyte balance, and cardiovascular homeostasis. In the classic RAAS, renin is synthetized in the kidney and is responsible for the formation of ANGI by cleaving angiotensinogen. Angiotensin-converting enzyme (ACE) or chymase converts ANGI into ANGII, which is the main effector through ATR1 and ATR2.

It was surprising when—almost 100 years later than the classic RAAS was identified—a separate RAAS was

The Neuroscience of Depression. https://doi.org/10.1016/B978-0-12-817933-8.00003-7

**TABLE 1 Experimental and clinical evidence of the protective effects of ARBs.**

| Study design | Findings | Studies |
|---|---|---|
| *Experimental data* | | |
| Drug: losartan (0.1, 1.0, and 5 mg/kg, i.p.)<br>Species: Swiss albino mice<br>Model: healthy controls<br>Test: forced swim test | Antidepressant effect | Vijayapandi and Nagappa (2005) |
| Drug: losartan (10 mg/kg, i.p.)<br>Species: Swiss albino mice, Wistar rat<br>Model: healthy controls<br>Test: tail suspension test, forced swim test, plus maze | Antidepressant and anti-anxiolytic effect | Nayak and Patil (2008) |
| Drug: losartan (15 mg/kg, p.o)<br>Model: healthy controls in repeated forced to swim (for 15 days)<br>Species: Sprague-Dawley rat<br>Test: forced swim test | Antidepressant effect | Pedreañez et al. (2011) |
| Drug: valsartan (40 mg/kg, p.o)<br>Species: C57BL/6J mice<br>Model: unpredictable chronic mild stress<br>Test: forced swimming test, tail suspension test, open field test, sucrose preference test | Antidepressant and antianxiolytic effect | Ping et al. (2014) |
| Drug: telmisartan (0.05 mg/kg, p.o.)<br>Species: Wistar rat<br>Model: diabetes-induced depression<br>Test: forced swimming test, open field test | Antidepressant effect, reduced proinflammatory mediators | Aswar et al. (2017) |
| Drug: irbesartan (40 mg/kg, p.o.)<br>Species: Swiss albino mice<br>Model: unpredictable chronic mild stress<br>Test: forced swimming test, tail suspension test, open field test | Antidepressant effect, reduced oxidative stress, elevated brain 5-HT levels | Ayyub et al. (2017) |
| Drug: losartan (10 and 45 mg/kg, i.p.)<br>Species: C57BL/6J mice, Wistar rat<br>Model: healthy controls<br>Test: forced swimming test | Antidepressant effect | Diniz et al. (2018) |
| Drug: losartan (20 mg/kg, p.o.)<br>Species: Wistar rat<br>Model: diabetes-induced depression<br>Test: forced swimming test, open field test | Antidepressant effect, reduced proinflammatory mediators | Lenart et al. (2019), Lenart, Lakat, Hodrea, and Fekete (2019) |
| *Clinical data* | | |
| ARBs in hypertensive patients | ARBs improved quality of life compared with standard therapies | Weber, 2005 |
| Open-label study of candesartan in diabetic patients (*n*:17) | Improved intrapersonal sensitivity and depression scores | Pavlatou et al. (2008) |
| 378 patients in a primary care clinic | ARB treatment is associated with lower antidepressant usage | Nasr et al. (2011) |
| Cross-sectional data from 1705 type 1 diabetic participants in the Finnish Diabetic Nephropathy (FinnDiane) Study | ARB was associated with lower frequency of antidepressant usage | Ahola, Harjutsalo, Forsblom, and Groop (2014) |
| 5-year cohort of 144,660 patients treated with ACEi/ARB | 53% reduced risk of depression | Boal et al. (2016) |

Representative studies of antidepressant effect of angiotensin-receptor blockers from different preclinical animal models and clinical observations. *ACEi*, angiotensin-converting enzyme inhibitor, *ARB*, angiotensin receptor blocker.

detected in the brain (Ganten et al., 1971). Around the third and fourth ventricles, there are circumventricular organs containing fenestrated capillaries that allow access for peripheral RAAS components. However, the majority of brain regions are impermeable for RAAS since elements cannot cross through the blood-brain barrier.

Brain RAAS is partly independent of the peripheral system, but expresses all elements of the classic RAAS. Both angiotensinogen and renin are synthetized by neurons and astrocytes. In astrocytes, these peptides are constitutively secreted for conversion, while in neurons both can be secreted or remain intracellular (Jackson, Eldahshan, Fagan, & Ergul, 2018). The main effector, ANGII is mainly produced in the hypothalamus, hypophysis, cortex, amygdala, cerebellum, and hippocampus (Phillips & Stenstrom, 1985). There are two different ways of ANGII production: (1) *extracellular:* ANGII is generated from angiotensinogen extracellularly and acts as a neurohormone and (2) *intracellular*: neurons convert angiotensinogen to ANGII within the cells then secreted into the extracellular space and act as co- or neurotransmitter (Paul, Bader, Steckelings, Voigtlander, & Ganten, 1993).

RAAS in the brain regulates blood pressure, body temperature, and locomotor activity as well as memory, behavior, and learning. ANGII induces vasoconstriction, increases the release of catecholamines and aldosterone synthesis. ANGII is increased during acute or chronic stress; it induces inflammatory processes and activates the HPA axis, which could lead to depressive-like behavior. Both animal and human studies confirmed that increased local RAAS activity in the brain is associated with diseases of the nervous system, such as Alzheimer's (Wang et al., 2007), Parkinson's (Reardon, Mendelsohn, Chai, & Horne, 2000), or depression (Isingrini, Desmidt, Belzung, & Camus, 2009).

In the brain besides classical RAAS components, nonclassical elements are synthetized as well, such as ACE2, ANG 1–7, ANGIV peptides; MAS1; and insulin receptor aminopeptidase (IRAP). These newly discovered components act via two pathways: the ACE2/ANG1–7/MAS1 and the ANGIV/IRAP axis. Both axes have been shown to be vasoprotective and to improve neurological deficits through anti-oxidative and anti-inflammatory effects, which act opposite to the classical ANGII/ATR1 axis mentioned above (McKinley et al., 2003).

## Experimental data and clinical studies

A body of preclinical evidence suggests that drugs targeting RAAS have antidepressant effect. In angiotensinogen deficient mice, depressive-like behavior was reduced as shown by forced swim test (Okuyama, Sakagawa, Sugiyama, Fukamizu, & Murakami, 1999). Pharmacologically blocking ANGII production by inhibiting ACE also showed antidepressant-like actions (Giardina & Ebert, 1989). Moreover, inhibition of ATR1 by its specific blocker losartan (Nayak & Patil, 2008), telmisartan (Aswar, Chepurwar, Shintre, & Aswar, 2017), valsartan (Ping, Qian, Song, & Zhaochun, 2014) or irbesartan (Ayyub, Najmi, & Akhtar, 2017) also minimized depression-like symptoms in chronic stress or diabetes-induced depression models.

These preclinical data have a strong translational value, since several clinical observational studies have also established a protective antidepressant role of RAASi. The first evidence that RAASi have antidepressant properties in humans was observed in hypertensive patients who demonstrated reduced depressive symptoms after receiving ACE inhibitors (Germain & Chouinard, 1988, 1989). Clinical evidence is predominantly derived from patients taking ACE inhibitors, but similar findings were reported after ARB treatment, which not only improved quality of life (Weber, 2005) but also ameliorated interpersonal sensitivity and depression scores (Pavlatou et al., 2008). Furthermore, the recent papers point out that taking ARBs is associated with significantly lower rates of antidepressant usage in depressed hypertensive (Nasr, Crayton, Agarwal, Wendt, & Kora, 2011) or type 1 diabetic patients (Ahola, Harjutsalo, Forsblom, & Groop, 2014) (Table 2). Even though these clinical observations support the clinical efficacy of RAASi—especially ARBs—in the treatment of depression, future clinical studies (e.g., randomized controlled trials) targeting RAAS are needed to confirm the utility of these drugs in the treatment of depression.

Mechanisms of the antidepressant effects of ARBs are not entirely clarified yet. While ARBs have a pivotal role in the regulation of cerebral circulation by blocking ATR1s in cerebrovascular endothelial and vascular smooth muscle cells, studies indicate that their neuroprotective effects go beyond blood pressure reduction.

The first studies supporting this hypothesis showed that ARBs in very low, nondepressor dosage (telmisartan or candesartan) improved neurological function, cognitive deficit, and significantly attenuated posttraumatic inflammation in normotensive rats (Timaru-Kast et al., 2012; Tsukuda et al., 2009). Furthermore, in spontaneously hypertensive rats, ARBs have superior neuroprotective effects when compared to calcium channel blockers (amlodipine) (Ito et al., 2002). In vitro studies also highlighted the direct neuroprotective and antiinflammatory effects of ARBs using neurons, which are not connected to cardiovascular regulation.

**TABLE 2** Drug groups used in depression.

| Drug subclasses | Actions of drugs | Examples of the most frequently used drugs |
|---|---|---|
| *Typical antidepressant* | | |
| Monoamine oxidase inhibitors | Elevating the levels of neurotransmitters by inhibiting of monoamine oxidase | Phenelzine<br>Tranylcypromine<br>Isocarboxazid |
| Tricyclic antidepressants | Increasing levels of neurotransmitters by blocking of serotonin transporter and norepinephrine transporter | Amitriptyline<br>Clomipramine<br>Desipramine<br>Doxepin<br>Imipramine<br>Nortriptyline<br>Protriptyline |
| Selective serotonin reuptake inhibitors | Increasing serotonine activity by inhibition of reuptake of serotonin | Citalopram<br>Escitalopram<br>Fluoxetine<br>Fluvoxamine<br>Paroxetine<br>Sertraline |
| Serotonin-norepinephrine reuptake inhibitors | Increasing the levels of serotonin and norepinephrine by inhibition of reuptake of both serotonin and norepinephrine | Duloxetine<br>Levomilnacipran<br>Milnacipran<br>Venlafaxine |
| *Atypical antidepressant* | | |
| Tetracyclic antidepressants | Inhibiting the reuptake of norepinephrine and serotonine receptor | Amoxapine<br>Maprotiline<br>Mianserin<br>Mirtazapine<br>Setiptiline |
| Norepinephrine-dopamine reuptake inhibitors | Blocking the norepinephrine transporter and dopamine transporter | Bupropion<br>Diphenylprolinol |
| Norepinephrine reuptake inhibitors | Increasing the level of norepinephrin and epinephrin by blocking of norepinephrine transporter | Atomoxetine<br>Reboxetine Teniloxazine<br>Viloxazine |
| Serotonin antagonists | Antagonizing serotonin receptors | Nefazodone<br>Trazodone |

## Brain-derived neurotrophic factor

Brain-derived neurotrophic factor (BDNF) belongs to the neurotrophin subfamily and it is mainly expressed in the central and peripheral nervous system. The neurotrophin subfamily contains four neurotrophins; BDNF, neural growth factor, neurotrophins 3 and 4, which are mainly involved in neuronal growth, differentiation, survival, and synaptic plasticity. BDNF is initially synthesized as pre-pro BDNF in the endoplasmic reticulum after which it is either directly secreted to the extracellular space or is cleaved by endopeptidases or furin to produce the mature form. The transformation to mature BDNF (mBDNF) can also be completed extracellularly by plasmin or matrix metalloproteases (MMP). Both proBDNF and mBDNF are biologically active, but elicit opposing effects (Lu, Pang, & Woo, 2005). mBDNF binds to tropomyosin receptor kinase (TrkB) and subsequently activates several signaling cascades including the Ras-mitogen-activated protein kinase (MAPK), the phosphatidylinositol-3-kinase (PI3K), and the phospholipase Cγ (PLC-γ) pathway. These pathways contribute to axon and dendrite development, neurotransmitter synthesis, and elevation of presynaptic and postsynaptic plasticity (Cunha, Brambilla, & Thomas, 2010). It has recently been discovered that uncleaved proBDNF is also active and facilitates long-term depression and induces proapoptotic processes via p75 neurotrophin receptor ($p75^{NTR}$)-mediated pathways (Lee, Kermani, Teng, & Hempstead, 2001).

Among neurotrophins, the role of BDNF is the most investigated in the development of depression. Multiple lines of evidence highlight the role of BDNF in the pathomechanism of neuropsychiatric disorders such as depression. In healthy subjects, elevated BDNF and TrkB levels were detected in the hippocampus and prefrontal cortex. Studies reported decreased brain and serum BDNF levels in depressed patients (Cunha et al., 2006) and BDNF increment in response to antidepressant treatment (paroxetine and milnacipran) (Yoshimura et al., 2007). Furthermore, in postmortem studies, BDNF expression increased in the dentate gyrus, hilus, and supragranular regions in patients treated with antidepressant medications at the time of death (Chen, Dowlatshahi, MacQueen, Wang, & Young, 2001). Similarly, experimental data show that decreased BDNF levels correlate with the development of depression-like behavior. Increased level of BDNF after local infusion into the brain has also been reported to have beneficial effect in animal models of depression (Russo-Neustadt, Ha, Ramirez, & Kesslak, 2001). Interestingly enough, though BDNF has an antidepressant effect in the hippocampus and prefrontal cortex, it may have opposing impact in other brain regions. Injection of BDNF into the ventral segmental area nucleus accumbens or amygdala facilitates depression-like or anxiety-like symptoms (Yu & Chen, 2011).

Recently, a variant in the human BDNF gene was found, leading to a valine to methionine change in the pro-domain of the BDNF protein (position 66). This variant decreases activity-dependent BDNF secretion (Chen et al., 2004; Egan et al., 2003) and is associated with increased susceptibility to neuropsychiatric disorders including depression (Lang et al., 2005; Post, 2007). The role of this polymorphism in the development of depression was confirmed by animal studies. BDNF$^{Met/Met}$ mice showed increased anxiety-related behavior, which could not be reversed by the antidepressant fluoxetine (Chen et al., 2006).

## Diabetes and comorbid depression

Diabetes mellitus (DM) and associated complications cause an enormous burden both to the patients and society. Cognitive decline and depression incidence are twice as high in patients with DM, and most of the cases remain underdiagnosed. Concurring occurrence of the diseases impairs quality of life, decreases life expectancy, and increases the risk of serious diabetic complications (Lin et al., 2010).

A common pathophysiologic background of DM and depression has been proposed, but the exact pathomechanism is not known to date. Current research aims to clarify the relation of the diseases on a molecular level. Common immunological processes, similar reaction to oxidative stress as well as common endocrinologic and neurobiological factors are suspected to be involved.

The fact that activated RAAS is a typical feature of both diabetes and depression make RAASi great candidates for the treatment of both diseases (Fig. 1).

Our group showed that diabetes induces behavioral despair in streptozotocin-induced experimental diabetes, which is a sign of depression-like behavior in rodents (Lenart et al., 2016). ARB losartan reduced depression-like behavior suggesting a strong antidepressant effect (Lenart, Balogh, et al. 2019). Our SPECT-MRI measurements revealed decreased cerebral perfusion in every segmented brain region of diabetic rats. Of note, ARBs neither alter mean arterial pressure and nor improved cerebral blood flow. Thus, these findings also verified that the protective antidepressant effect of losartan is independent of its antihypertensive properties.

The causal role of proinflammatory mediators has been widely proposed in the pathophysiology of both depression and diabetes (Felger & Lotrich, 2013; Tsalamandris et al., 2019). Microglia, astrocyte activation, and elevated proinflammatory mediators—as a sign of neuroinflammation—were associated with the development of depression. In our study, diabetes induced neuroinflammation via activating microglia and astrocytes and via inducing pro-inflammatory mediator production. ARB treatment led to suppressed inflammatory response and decreased glial activation.

Our findings on the neuroprotective capacity of ARBs are supported by the recent studies using lipopolysaccharide-induced inflammation and hypertensive models in which glial activation and inflammatory mediators were inhibited by ARBs (Bhat, Goel, Shukla, & Hanif, 2016; Bhat, Goel, Shukla, Shukla, & Hanif, 2018).

Recent studies have confirmed that diabetes and depression-associated inflammation reduces the expression of neurotrophins, especially BDNF in the brain (Calabrese et al., 2014; Guan & Fang, 2006). Lower level of BDNF is associated with the development of diabetes comorbid depression (Lenart et al., 2016). Based on the specific structural abnormalities (reduced hippocampal and prefrontal cortex volume), the role of BDNF has emerged in the development of comorbidity. We identified that in diabetes BDNF is mostly produced in astrocytes and the dentate gyrus. Our group proved that proapoptotic proBDNF-p75NTR signaling remains unchanged, while the prosurvival mBDNF-TrkB pathway is substantially impaired in diabetes-associated depression. ARBs promote fast conversion of proBDNF to mBDNF by increasing the level of both intra- (furin) and extracellular (MMP3) cleavage enzymes. Consequently, proBDNF is immediately converted to mBDNF instead of activating p75NTR and proapoptotic signaling. ARB losartan increases mBDNF, TrkB, and CREB levels, indicating that this pathway is activated and induces neuronal survival (Fig. 2). Since several studies indicate that other RAASi (such as ACE inhibitors) have

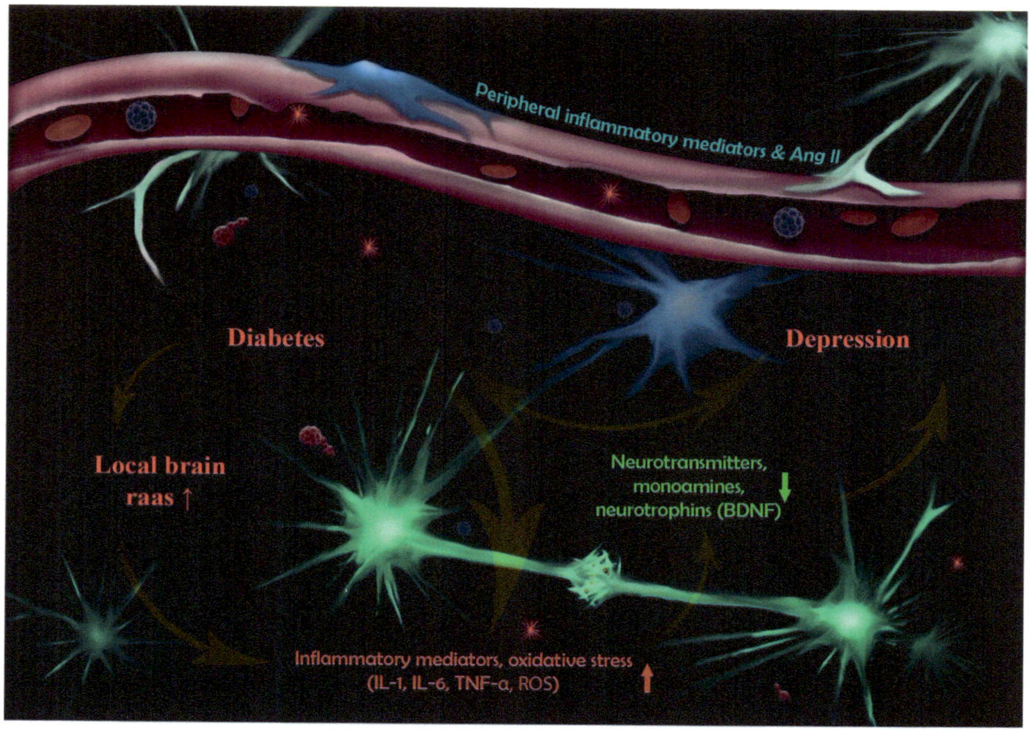


FIG. 1 Development of diabetes comorbid depression. In diabetic patients, elevated blood glucose levels and possible obesity increase the systemic levels of certain inflammatory cytokines, which induce inflammatory processes in the central nervous system. Furthermore, oxidative stress increases the amount of reactive oxygen species (ROS). As a result, levels of monoamines and neurotrophins, including brain-derived neurotrophic factor (BDNF) are reduced, leading to depression-like symptoms. Subclinical inflammation associated with depression contributes to reduced beta-cell function and hence to insulin resistance and type 2 diabetes. Furthermore, renin-angiotensin-aldosterone system (RAAS) inhibitors have been proved to improve depressive symptoms and cognitive function, confirming the role of local brain RAAS in the development of neuroprotection and the pathomechanism of depression. *From Lenart, L., Lakat, T., Hodrea, J., & Fekete, A. (2019). The role of renin-angiotensin-aldosterone system in diabetes associated depression.* Diabetologia Hungarica, 27*(N3), 181–187. doi:10.24121/dh.2019.14 with permission.*

neuroprotective and antidepressant properties, we suggest that these actions are universal for all types of RAASi. All these results strengthen the concept that inhibition of ANGII by ARBs or (other RAASi) may be of considerable significance in the therapy of diabetes-associated depression.

## Key facts of role of RAAS in mood disorders

- Depression is a major health concern affecting more than 300 million people worldwide.
- The prevalence of depression is higher in diseases with overactivated renin-angiotensin-aldosterone system (e.g., hypertension or diabetes).
- Pathomechanism of depression is very complex and not fully clear yet, the medications are only partially effective.
- Clinical observational studies revealed that depression is improved in patients taking angiotensin receptor blockers (ARBs) or angiotensin-converting enzyme inhibitors.
- Anti-inflammatory, anti-oxidative, and antiapoptotic actions of ARBs are contributed to their neuroprotective and antidepressant effects.
- Protective antidepressant effect of ARBs is beyond their blood pressure lowering property.

## Summary points

- Depression possesses huge public health problem affecting millions of people; it demands a large economic burden on the individual and national health-care systems.
- Neuroinflammation—which is characterized by activation of glial cells and elevated production of pro-inflammatory mediators—is a key process in the pathology of depression.
- Activation of local renin-angiotensin-aldosterone system (RAAS) in the brain is associated with diseases of the nervous system, such as depression.
- Currently, RAAS inhibitors (RAASi) represent the first-line therapy for hypertension and kidney disease in diabetic patients; however, experimental data and clinical

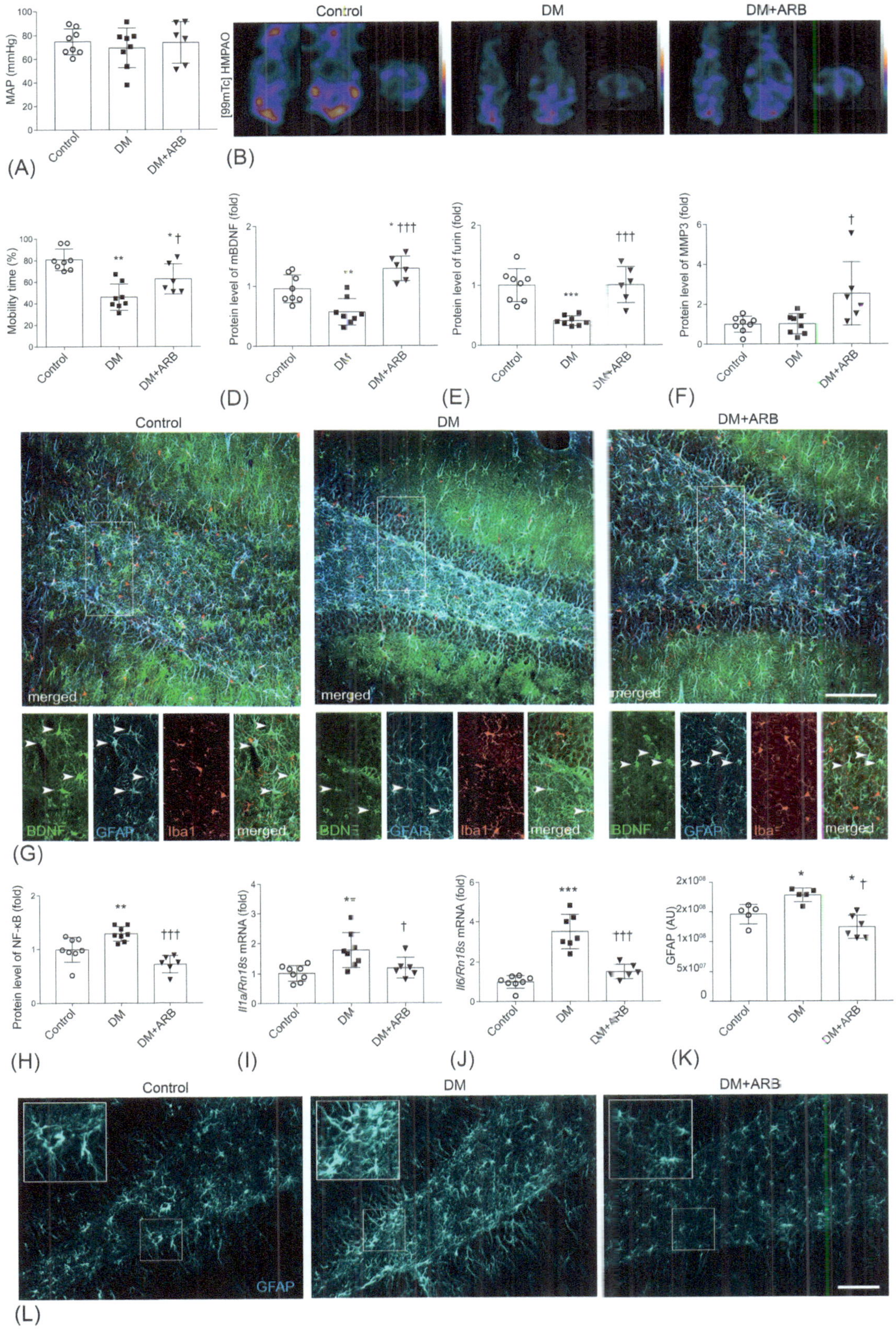


**FIG. 2** Summary of the protective antidepressant and antiinflammatory effects of angiotensin receptor blocker in the experimental model of diabetes-associated depression. (A–C) Unaltered mean arterial pressure and $^{99mTc}$HMPAO uptake in SPECT coregistration suggest that ARB's protective effects are independent of blood pressure lowering effect. (C) ARB attenuated depressive-like behavior in diabetic rats. (D–G) ARB normalized BDNF production and facilitated BDNF maturation in the hippocampi. Triple immunofluorescent staining showing BDNF, GFAP-positive astrocytes, and Iba1-positive microglia in the brain. Arrowheads on inserts indicate astrocytes with BDNF immunopositivity. Scale bar: 200 μm. (H–I) ARB mitigated proinflammatory cytokine responses and reduced astrocyte activation in diabetic brain. *Arrowheads* showing P2Y12-positive microglia with thicker cell bodies and a partial withdrawal of processes (indicative of higher level of microglial activity (Scale bar: 50 μm). *$P < 0.05$ vs control; **$P < 0.01$ vs control; ***$P < 0.001$ vs control; †$p < 0.05$ vs DM; ††$p < 0.01$ vs DM; †††$p < 0.001$ vs DM. *ARB*, angiotensin receptor blocker; *BDNF*, brain-derived neurotrophic factor; *DM*, diabetes mellitus; *MAP*, mean arterial pressure; *MMP*, matrix metalloproteinase; *NF-κB*, nuclear factor kappa-light-chain-enhancer of activated B cells; *Il1a*, interleukin 1 alpha; *Il6*, interleukin 6. *From Lenart, L., Balogh, B. D., Lenart, N., Barczi, A., Hosszu, A., Farkas, T., et al. (2019). Novel therapeutic potential of angiotensin receptor 1 blockade in a rat model of diabetes-associated depression parallels altered BDNF signalling.* Diabetologia, *62(8): 1501–1513 (in press) with permission.*

studies suggest that these drugs may have additional neuroprotective and antidepressant effects as well.

- Lower level of brain-derived neurotrophic factor (BDNF) contributes to the development of depression.
- Depression incidence is two times higher in patients with diabetes and concurring occurrence of the diseases impairs quality of life, decreases life expectancy, and increases the risk of serious diabetic complications.
- Blockade of angiotensin II by RAASi has antiinflammatory effect in the brain; moreover facilitates the production and signaling of BDNF, contributing to their neuroprotective property.

## Mini-dictionary of terms

**Angiotensin receptor blocker** Type of drugs, which are mainly applied as antihypertensive agents. Mechanism of action of ARBs is based on selective inhibition of angiotensin II by competitive antagonism of the angiotensin II receptors.

**Brain-derived neurotrophic factor** A protein, which is expressed in the central and peripheral nervous system. Its main function is to regulate neuronal growth, differentiation, survival, and synaptic plasticity.

**Depression** It is a type of psychiatric disorder, which is characterized by sadness loss of interest, hopelessness and, sometimes, suicidal thoughts.

**Diabetes mellitus** Diabetes is a complex metabolic disorder, which is characterized by high blood sugar level and lack of insulin.

**Renin-angiotensin-aldosterone system** A hormone system, which regulates blood pressure, water and electrolyte balance, and cardiovascular homeostasis.

**Neuroinflammation** Type of inflammation process in the central nervous system, which is linked to sustained activation of glial cells and elevated production of pro-inflammatory mediators.

**Funding** Funding for this study was provided by National Research, Development and Innovation Office (FK-124491, NN-114607, 2017-1.3.1-VKE-2017-00006), and Higher Education Institutional Excellence Program of Semmelweis University (FIKP), TKP-2019, OTKA-PD 131637.

## References

Ahola, A. J., Harjutsalo, V., Forsblom, C., & Groop, P. H. (2014). Renin-angiotensin-aldosterone-blockade is associated with decreased use of antidepressant therapy in patients with type 1 diabetes and diabetic nephropathy. *Acta Diabetologica*, *51*(4), 529–533. https://doi.org/10.1007/s00592-013-0547-x.

Aswar, U., Chepurwar, S., Shintre, S., & Aswar, M. (2017). Telmisartan attenuates diabetes induced depression in rats. *Pharmacological Reports*, *69*(2), 358–364. https://doi.org/10.1016/j.pharep.2016.12.004.

Ayyub, M., Najmi, A. K., & Akhtar, M. (2017). Protective effect of irbesartan an angiotensin (AT1) receptor antagonist in unpredictable chronic mild stress induced depression in mice. *Drug Research*, *67* (1), 59–64. https://doi.org/10.1055/s-0042-118172.

Bhat, S. A., Goel, R., Shukla, R., & Hanif, K. (2016). Angiotensin receptor blockade modulates NFkappaB and STAT3 signaling and inhibits glial activation and neuroinflammation better than angiotensin-converting enzyme inhibition. *Molecular Neurobiology*, *53*(10), 6950–6967. https://doi.org/10.1007/s12035-015-9584-5.

Bhat, S. A., Goel, R., Shukla, S., Shukla, R., & Hanif, K. (2018). Angiotensin receptor blockade by inhibiting glial activation promotes hippocampal neurogenesis via activation of Wnt/beta-catenin signaling in hypertension. *Molecular Neurobiology*, *55*(6), 5282–5298. https://doi.org/10.1007/s12035-017-0754-5.

Boal, A. H., Smith, D. J., McCallum, L., Muir, S., Touyz, R. M., Dominiczak, A. F, & Padmanabhan, S. (2016). Monotherapy with major antihypertensive drug classes and risk of hospital admissions for mood disorders. *Hypertension*, *68*(5), 1132–1138. https://doi.org/10.1161/HYPERTENSIONAHA.116.08188.

Calabrese, F., Rossetti, A. C., Racagni, G., Gass, P., Riva, M. A., & Molteni, R. (2014). Brain-derived neurotrophic factor: A bridge between inflammation and neuroplasticity. *Frontiers in Cellular Neuroscience*, *8*, 430. https://doi.org/10.3389/fncel.2014.00430.

Chen, B., Dowlatshahi, D., MacQueen, G. M., Wang, J. F., & Young, L. T. (2001). Increased hippocampal BDNF immunoreactivity in subjects treated with antidepressant medication. *Biological Psychiatry*, *50*(4), 260–265. https://doi.org/10.1016/s0006-3223(01)01083-6.

Chen, Z. Y., Jing, D., Bath, K. G., Ieraci, A., Khan, T., Siao, C. J., et al. (2006). Genetic variant BDNF (Val66Met) polymorphism alters anxiety-related behavior. *Science*, *314*(5796), 140–143. https://doi.org/10.1126/science.1129663.

Chen, Z. Y., Patel, P. D., Sant, G., Meng, C. X., Teng, K. K., Hempstead, B. L., et al. (2004). Variant brain-derived neurotrophic factor (BDNF) (Met66) alters the intracellular trafficking and activity-dependent secretion of wild-type BDNF in neurosecretory cells and cortical neurons. *The Journal of Neuroscience*, *24*(18), 4401–4411. https://doi.org/10.1523/JNEUROSCI.0348-04.2004.

Cunha, C., Brambilla, R., & Thomas, K. L. (2010). A simple role for BDNF in learning and memory? *Frontiers in Molecular Neuroscience*, *3*, 1. https://doi.org/10.3389/neuro.02.001.2010.

Cunha, A. B., Frey, B. N., Andreazza, A. C., Goi, J. D., Rosa, A. R., Gonçalves, C. A., et al. (2006). Serum brain-derived neurotrophic factor is decreased in bipolar disorder during depressive and manic episodes. *Neuroscience Letters*, *398*(3), 215–219. https://doi.org/10.1016/j.neulet.2005.12.085.

Diniz, C. R. A. F., Casarotto, P. C., Fred, S. M., Biojone, C., Castrén, E., & Joca, S. R. L. (2018). Antidepressant-like effect of losartan involves TRKB transactivation from angiotensin receptor type 2 (AGTR2) and recruitment of FYN. *Neuropharmacology*, *135*, 163–171. https://doi.org/10.1016/j.neuropharm.2018.03.011.

Egan, M. F., Kojima, M., Callicott, J. H., Goldberg, T. E., Kolachana, B. S., Bertolino, A., et al. (2003). The BDNF val66met polymorphism affects activity-dependent secretion of BDNF and human memory and hippocampal function. *Cell*, *112*(2), 257–269.

Entsuah, A. R., Huang, H., & Thase, M. E. (2001). Response and remission rates in different subpopulations with major depressive disorder administered venlafaxine, selective serotonin reuptake inhibitors, or placebo. *The Journal of Clinical Psychiatry*, *62*(11), 869–877. https://doi.org/10.4088/jcp.v62n1106.

Felger, J. C., & Lotrich, F. E. (2013). Inflammatory cytokines in depression: Neurobiological mechanisms and therapeutic implications. *Neuroscience*, *246*, 199–229. https://doi.org/10.1016/j.neuroscience.2013.04.060.

Galecki, P., & Talarowska, M. (2018). Inflammatory theory of depression. *Psychiatria Polska*, *52*(3), 437–447. https://doi.org/10.12740/PP/76863.

Ganten, D., Marquez-Julio, A., Granger, P., Hayduk, K., Karsunky, K. P., Boucher, R., et al. (1971). Renin in dog brain. *The American Journal of Physiology*, *221*(6), 1733–1737.

Germain, L., & Chouinard, G. (1988). Treatment of recurrent unipolar major depression with captopril. *Biological Psychiatry*, *23*(6), 637–641.

Germain, L., & Chouinard, G. (1989). Captopril treatment of major depression with serial measurements of blood cortisol concentrations. *Biological Psychiatry*, *25*(4), 489–493.

Giardina, W. J., & Ebert, D. M. (1989). Positive effects of captopril in the behavioral despair swim test. *Biological Psychiatry*, *25*(6), 697–702. https://doi.org/10.1016/0006-3223(89)90240-0.

Guan, Z., & Fang, J. (2006). Peripheral immune activation by lipopolysaccharide decreases neurotrophins in the cortex and hippocampus in rats. *Brain, Behavior, and Immunity*, *20*(1), 64–71. https://doi.org/10.1016/j.bbi.2005.04.005.

Hannestad, J., DellaGioia, N., & Bloch, M. (2011). The effect of antidepressant medication treatment on serum levels of inflammatory cytokines: A meta-analysis. *Neuropsychopharmacology*, *36*(12), 2452–2459. https://doi.org/10.1038/npp.2011.132.

Isingrini, E., Desmidt, T., Belzung, C., & Camus, V. (2009) Endothelial dysfunction: A potential therapeutic target for geriatric depression and brain amyloid deposition in Alzheimer's disease? *Current Opinion in Investigational Drugs*, *10*(1), 46–55.

Ito, T., Yamakawa, H., Bregonzio, C., Terron, J. A., Falcon-Neri, A., & Saavedra, J. M. (2002). Protection against ischemia and improvement of cerebral blood flow in genetically hypertensive rats by chronic pretreatment with an angiotensin II AT1 antagonist. *Stroke*, *33*(9), 2297–2303. https://doi.org/10.1161/01.str.0000027274.03779.f3.

Jackson, L., Eldahshan, W., Fagan, S. C., & Ergul, A. (2018). Within the brain: The renin angiotensin system. *International Journal of Molecular Sciences*, *19*(3), 876. https://doi.org/10.3390/ijms19030876.

Lang, U. E., Hellweg, R., Kalus, P., Bajbouj, M., Lenzen, K. P., Sander, T., et al. (2005). Association of a functional BDNF polymorphism and anxiety-related personality traits. *Psychopharmacology*, *180*(1), 95–99. https://doi.org/10.1007/s00213-004-2137-7.

Lee, R., Kermani, P., Teng, K. K., & Hempstead, B. L. (2001). Regulation of cell survival by secreted proneurotrophins. *Science*, *294*(5548), 1945–1948. https://doi.org/10.1126/science.1065057.

Lenart, L., Balogh, B. D., Lenart, N., Barczi, A., Hosszu, A., Farkas, T., et al. (2019). Novel therapeutic potential of angiotensin receptor 1 blockade in a rat model of diabetes-associated depression parallels altered BDNF signalling. *Diabetologia*, *62*(8), 1501–1513. https://doi.org/10.1007/s00125-019-4888-z.

Lenart, L., Hodrea, J., Hosszu, A., Koszegi, S., Zelena, D., Balogh, D., et al. (2016). The role of sigma-1 receptor and brain-derived neurotrophic factor in the development of diabetes and comorbid depression in streptozotocin-induced diabetic rats. *Psychopharmacology*, *233*(7), 1269–1278. https://doi.org/10.1007/s00213-016-4209-x.

Lenart, L., Lakat, T., Hodrea, J., & Fekete, A. (2019). The role of renin-angiotensin-aldosterone system in diabetes associated depression. *Diabetologia Hungarica*, *27*(N3), 181–187. https://doi.org/10.24121/dh.2019.14.

Lin, E. H., Rutter, C. M., Katon, W., Heckbert, S. R., Ciechanowski, P. Oliver, M. M., et al. (2010). Depression and advanced complications of diabetes: A prospective cohort study. *Diabetes Care*, *33*(2), 264–269. https://doi.org/10.2337/dc09-1068

Lu, B., Pang, P. T., & Woo, N. H. (2005). The yin and yang of neurotrophin action. *Nature Reviews. Neuroscience*, *6*(8), 603–614. https://doi.org/10.1038/nrn1726.

Majewski, C., & Bakris, G. L. (2016). Has RAAS blockade reached its limits in the treatment of diabetic nephropathy? *Current Diabetes Reports*, *16*(4), 24. https://doi.org/10.1007/s11892-016-0713-y.

McKinley, M. J., Albiston, A. L., Allen, A. M., Mathai, M. L., May, C. N., McAllen, R. M., et al. (2003). The brain renin-angiotensin system: Location and physiological roles. *The International Journal of Biochemistry & Cell Biology*, *35*(6), 901–918.

Nasr, S. J., Crayton, J. W., Agarwal, B., Wendt, B., & Kora, R. (2011). Lower frequency of antidepressant use in patients on renin-angiotensin-aldosterone system modifying medications. *Cellular and Molecular Neurobiology*, *31*(4), 615–618. https://doi.org/10.1007/s10571-011-9656-7.

Nayak, V., & Patil, P. A. (2008). Antidepressant activity of fosinopril, ramipril and losartan, but not of lisinopril in depressive paradigms of albino rats and mice. *Indian Journal of Experimental Biology*, *46*(3), 180–184.

Okuyama, S., Sakagawa, T., Sugiyama, F., Fukamizu, A., & Murakami, K. (1999). Reduction of depressive-like behavior in mice lacking angiotensinogen. *Neuroscience Letters*, *261*(3), 167–170. https://doi.org/10.1016/s0304-3940(99)00002-6.

Paul, M., Bader, M., Steckelings, U. M., Voigtlander, T., & Ganten, D. (1993). The renin-angiotensin system in the brain. Localization and functional significance. *Arzneimittel-Forschung*, *43*(2A), 207–213.

Pavlatou, M. G., Mastorakos, G., Lekakis, I., Liatis, S., Vamvakou, G., Zoumakis, E., et al. (2008). Chronic administration of an angiotensin II receptor antagonist resets the hypothalamic-pituitary-adrenal (HPA) axis and improves the affect of patients with diabetes mellitus type 2: Preliminary results. *Stress*, *11*(1), 62–72. https://doi.org/10.1080/10253890701475621.

Pedreañez, A., Arcaya, J. L., Carrizo, E., Rincón, J. Viera, N., Peña, C., … Mosquera, J. (2011). Experimental depression induces renal oxidative stress in rats. *Physiology & Behavior*, *104*(5), 1002–1009. https://doi.org/10.1016/j.physbeh.2011.06.021.

Phillips, M. I., & Stenstrom, B. (1985). Angiotensin II in rat brain comigrates with authentic angiotensin II in high pressure liquid chromatography. *Circulation Research*, *56*(2), 212–219.

Ping, G., Qian, W., Song, G., & Zhaochun, S. (2014). Valsartan reverses depressive/anxiety-like behavior and induces hippocampal neurogenesis and expression of BDNF protein in unpredictable chronic mild stress mice. *Pharmacology, Biochemistry, and Behavior*, *124*, 5–12. https://doi.org/10.1016/j.pbb.2014.05.006.

Post, R. M. (2007). Role of BDNF in bipolar and unipolar disorder: Clinical and theoretical implications. *Journal of Psychiatric Research*, *41*(12), 979–990. https://doi.org/10.1016/j.jpsychires.2006.09.009.

Reardon, K. A., Mendelsohn, F. A., Chai, S. Y., & Horne, M. K. (2000). The angiotensin converting enzyme (ACE) inhibitor, perindopril, modifies the clinical features of Parkinson's disease. *Australian and New Zealand Journal of Medicine*, *30*(1), 48–53.

Russo-Neustadt, A., Ha, T., Ramirez, R., & Kesslak, J. P. (2001). Physical activity-antidepressant treatment combination: Impact on brain-derived neurotrophic factor and behavior in an animal model. *Behavioural Brain Research*, *120*(1), 87–95.

Timaru-Kast, R., Wyschkon, S., Luh, C., Schaible, E. V., Lehmann, F., Merk, P., et al. (2012). Delayed inhibition of angiotensin II receptor type 1 reduces secondary brain damage and improves functional recovery after experimental brain trauma. *Critical Care Medicine*, *40*(3), 935–944. https://doi.org/10.1097/CCM.0b013e31822f08b9.

Tsalamandris, S., Antonopoulos, A. S., Oikonomou, E., Papamikroulis, G. A., Vogiatzi, G., Papaioannou, S., et al. (2019). The role of inflammation in diabetes: Current concepts and future perspectives. *European Cardiology*, *14*(1), 50–59. https://doi.org/10.15420/ecr.2018.33.1.

Tsukuda, K., Mogi, M., Iwanami, J., Min, L. J., Sakata, A., Jing, F., et al. (2009). Cognitive deficit in amyloid-beta-injected mice was improved by pretreatment with a low dose of telmisartan partly because of peroxisome proliferator-activated receptor-gamma activation. *Hypertension*, *54*(4), 782–787. https://doi.org/10.1161/HYPERTENSIONAHA.109.136879.

Vijayapandi, P., & Nagappa, A. N. (2005). Biphasic effects of losartan potassium on immobility in mice. *Yakugaku Zasshi*, *125*(8), 653–657. https://doi.org/10.1248/yakushi.125.653.

Wang, J., Ho, L., Chen, L., Zhao, Z., Zhao, W., Qian, X., et al. (2007). Valsartan lowers brain beta-amyloid protein levels and improves spatial learning in a mouse model of Alzheimer disease. *The Journal of Clinical Investigation*, *117*(11), 3393–3402. https://doi.org/10.1172/JCI31547.

Weber, M. A. (2005). Angiotensin-II receptor blockers for hypertension and heart failure: Quality of life and outcomes. *Managed Care Interface*, *18*(2), 47–54.

WHO. (2017). Depression and other common mental disorders: Global health estimates. In *Licence: CC BY-NC-SA 3.0 IGO*.

Yoshimura, R., Mitoma, M., Sugita, A., Hori, H., Okamoto, T., Umene, W., et al. (2007). Effects of paroxetine or milnacipran on serum brain-derived neurotrophic factor in depressed patients. *Progress in Neuro-Psychopharmacology & Biological Psychiatry*, *31*(5), 1034–1037. https://doi.org/10.1016/j.pnpbp.2007.03.001.

Yu, H., & Chen, Z. Y. (2011). The role of BDNF in depression on the basis of its location in the neural circuitry. *Acta Pharmacologica Sinica*, *32*(1), 3–11. https://doi.org/10.1038/aps.2010.184.

Chapter 29

# The link between cannabinoids and depression

Mohaddeseh Ebrahimi-Ghiri[a] and Fatemeh Khakpai[b]

[a]Department of Biology, Faculty of Sciences, University of Zanjan, University Blvd., Zanjan, Iran, [b]Cognitive and Neuroscience Research Center (CNRC), Islamic Azad University of Medical Sciences, Tehran, Iran

## List of abbreviations

| | |
|---|---|
| **$\Delta^9$-THC** | delta-9-tetrahydrocannabinol |
| **2-AG** | 2-arachidonoylglycerol |
| **AEA** | anandamide |
| **Akt** | phosphoinositide 3-kinase (PI3K)/protein kinase B |
| **BDNF** | brain-derived neurotrophic factor |
| **cAMP** | cyclic adenosine monophosphate |
| **CB1R** | CB1 cannabinoid receptor |
| **CB2R** | CB2 cannabinoid receptor |
| **FAAH** | fatty acid amide hydrolase |
| **FST** | forced swim test |
| **GABA** | gamma-aminobutyric acid |
| **GPCRs** | G-protein-coupled receptors |
| **HPA** | hypothalamus-pituitary-adrenal gland |
| **MAGL** | monoacylglyceride lipase |
| **MAPKs** | mitogen-activated protein kinases |
| **NADA** | N-arachidonyl dopamine |
| **PKA** | protein kinase A |
| **TRPV1** | transient receptor potential vanilloid 1 |
| **TST** | tail suspension test |

## Introduction

The plant *Cannabis sativa*, better known as marijuana, is known for its recreational and medicinal properties (Ligresti, De Petrocellis, & Di Marzo, 2016). This plant comprises over 400 different compounds, of which 104 are known as cannabinoids (Lafaye, Karila, Blecha, & Benyamina, 2017). Cannabis exerts therapeutic potential in a wide range of medical conditions, including neurological and psychiatric disorders and the addictive and harmful effects, including cardiovascular and respiratory diseases, cognitive alterations, psychosis, schizophrenia, and mood disorders (Cohen, Weizman, & Weinstein, 2019). The increase in the efficacy of cannabis and a change in the ratio between the two primary components, delta-9-tetrahydrocannabinol ($\Delta^9$-THC) and cannabidiol, are responsible for the enhanced complications of cannabis use (Lafaye et al., 2017). Subsequent investigation led to the discovery of the endocannabinoid system, which consists of CB1 and CB2 cannabinoid receptors (CB1R and CB2R), endogenous ligands, and enzymes responsible for endocannabinoid biosynthesis and degradation (Poleszak et al., 2018). Endocannabinoids belong to the derivatives of the omega-6-polyunsaturated fatty acids and are produced from precursor components within the cellular membrane (Mangieri & Piomelli, 2007). Due to the lipophilic nature, endocannabinoids are not able to dissolve in water and store in the synaptic vesicles, but they are metabolized immediately. After the stimulus, these endogenous substances are synthesized and released by the postsynaptic neurons "on-demand." Then, they are diffused to the synaptic cleft via the exocytosis process, where their receptors are located (Poleszak et al., 2018). Endocannabinoids act as so-called reversed neurotransmitters, preventing the release of other neurotransmitters, such as glutamic acid, gamma-aminobutyric acid (GABA), noradrenaline, and serotonin (Komorowski & Stepien, 2007) (Fig. 1). Five endocannabinoids have been recognized so far: anandamide (AEA), 2-arachidonoylglycerol (2-AG), arachidonyl-glycerol ether (noladine ether), virodhamine (arachidonic acid ester with ethanolamine), and N-arachidonyl dopamine (endovanilloid, NADA). AEA and 2-AG are the most important (Poleszak et al., 2018). AEA is a partial agonist at both CB1Rs and CB2Rs, displaying a higher affinity for the CB1R (Howlett et al., 2002). 2-AG is a full agonist at both CB1Rs and CB2Rs but shows less affinity than AEA for both receptors. Endocannabinoids are degraded through two specific enzymes: the fatty acid amide hydrolase (FAAH) and the monoacylglyceride lipase (MAGL) enzymes. FAAH degrades AEA, whereas the MAGL degrades 2-AG (Cravatt et al., 1996; Dinh et al., 2002).

The Neuroscience of Depression. https://doi.org/10.1016/B978-0-12-817933-8.00013-X

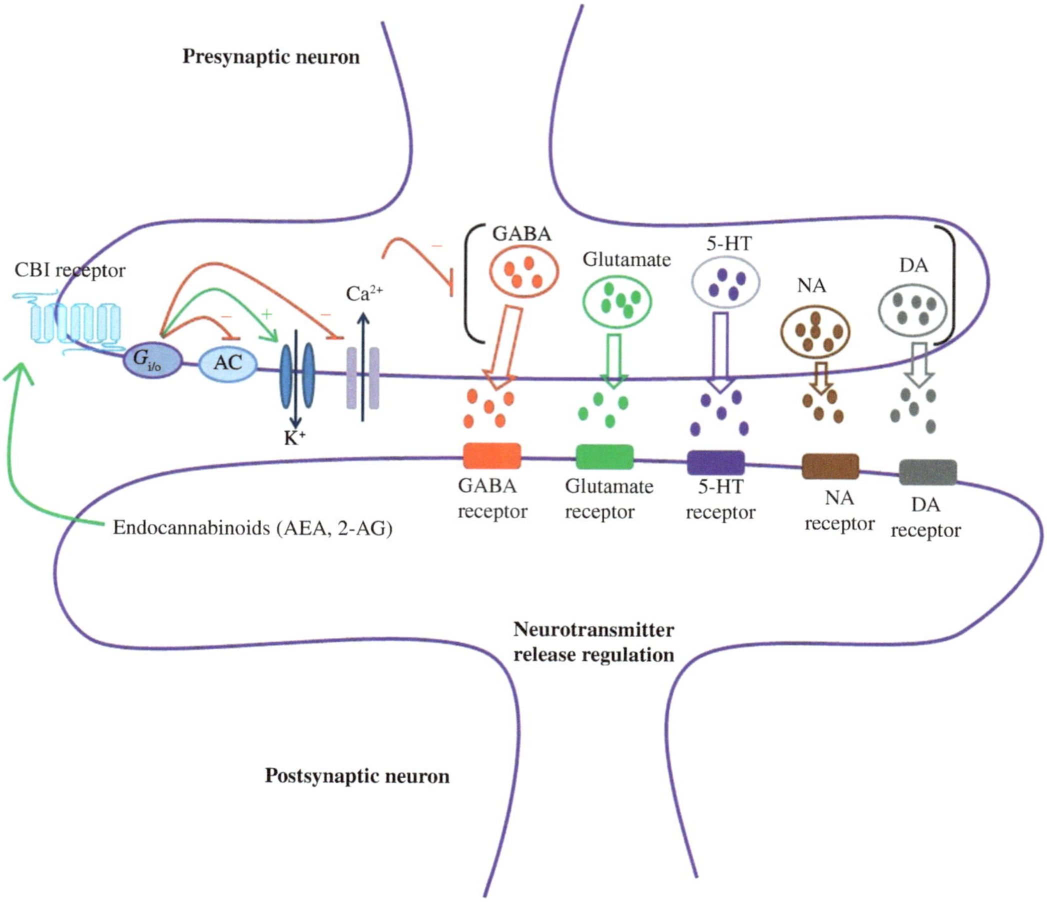


FIG. 1 Schematic figure of modulation the release of neurotransmitter via CB1 receptors. The endocannabinoids are synthesized and released through the postsynaptic neurons "on-demand." Then, they are diffused to the synaptic cleft through the exocytosis process, where their receptors are located. Therefore, they regulate the release of other neurotransmitters, for example, glutamate, GABA, serotonin (5-HT), noradrenaline (NA), and dopamine (DA).

## Distribution of cannabinoid receptors in the central nervous system

Two well-known cannabinoid receptors, CB1R and CB2R, mediate the biological effects induced by cannabinoids. These receptors are inhibitory G-protein-coupled receptors (GPCRs), which are characterized by seven transmembrane helices. The CB1Rs represent the most abundant class of GPCRs in the brain with a widespread expression throughout the brain suggestive of a neuromodulatory role for the receptor in both humans and rodents (Howlett & Abood, 2017). In humans, the brain regions with the highest level of CB1R expression include olfactory bulb, hippocampus, basal ganglia, and cerebellum, whereas CB1R expression is moderate in the cerebral cortex, septum, amygdala, hypothalamus, and parts of the brain stem and the dorsal horn of the spinal cord. The thalamus and the ventral horn of spinal cord have low CB1R density (Mackie, 2005). In rodents, CB1Rs are widely expressed in neocortex, hippocampus, striatum, substantia nigra, and the cerebellum. In contrast, they have a moderate distribution in the cingulate, entorhinal and piriform cortical areas, olfactory bulbs, amygdala, and nucleus accumbens. CB1R density is much lower in the diencephalon (thalamus and hypothalamus) and midbrain, and virtually null in the medulla (Herkenham et al., 1990; Tsou, Brown, Sanudo-Pena, Mackie, & Walker, 1998).

The cannabinoid CB2Rs are mainly distributed in immune tissues and inflammatory cells, although they are recently detected in several brain structures such as the cerebral cortex, hippocampus, amygdala, hypothalamus, and cerebellum (Gong et al., 2006). The presence of CB2Rs in brain regions responsible for stress response proposes their contribution in the regulation of mood (Onaivi et al., 2008).

## Cannabinoid receptor signaling pathways

Typically, both the CB1Rs and CB2Rs are coupled to pertussis toxin-sensitive $G_{i/o}$ proteins. Activation of CB1Rs and CB2Rs suppresses the adenylyl cyclase activity, leading to a following reduction in the cyclic adenosine monophosphate (cAMP) cascade and the activity of protein kinase A (PKA) (Howlett et al., 2002). Endocannabinoids also bind to other receptors such as transient receptor potential vanilloid 1 (TRPV1), peroxisome proliferator-activated receptors, GPR55, and GPR119 and this non-CB1/2R activity of endocannabinoids accounts for the differential effects of specific cannabinoid agonists and pharmacological modulators of endocannabinoid tone (Ross, 2003; Ryberg et al., 2007). Under certain circumstances, the CB1R but not the CB2R can switch its coupling of G protein from Gi/o to Gs or Gq (Bonhaus, Chang, Kwan, & Martin, 1998; Lauckner, Hille, & Mackie, 2005). Moreover, the CB1R modulates the activity of several types of ion channels. The CB1R can suppress calcium influx via voltage-gated calcium channels (Brown, Safo, & Regehr, 2004) or augment potassium currents via inward-rectifying potassium channels (Guo & Ikeda, 2004). Several mitogen-activated protein kinases (MAPKs), including ERK1/2, p38, and JNK, are activated by the CB1R. The phosphoinositide 3-kinase (PI3K)/protein kinase B (Akt) pathway is activated by CB1R as well (Fig. 1) (Zou & Kumar, 2018). Furthermore, CB1Rs are found on presynaptic axon terminals of glutamatergic neurons as well as on GABAergic neurons (Freund, Katona, & Piomelli, 2003); thus, CB1Rs are ideally positioned to modulate the balance of excitation and inhibition within a given neural circuit. Available evidence demonstrates that CB1Rs are present on serotonergic (Haring, Marsicano, Lutz, & Monory, 2007) and noradrenergic (Oropeza, Mackie, & Van Bockstaele, 2007) neurons, indicating that these receptors also directly modulate monoamine efflux as well (Fig. 2).

## Cannabinoids and depression disorder: Clinical evidence

The idea that the endocannabinoid system may be engaged in affective disorders stems from cannabis users reporting mood improvement (Bambico & Gobbi, 2008). The existing evidence shows that cannabis can exert a bidirectional effect on depression. Some evidence points to the antidepressant activity of cannabinoids (Denson & Earleywine, 2006), while others report an increase in depressive symptoms in cannabis users (Lev-Ran et al., 2014). The effect of cannabinoids on depression can be evaluated in levels of cannabinoid receptors endogenous ligands, and cannabinoid degradative enzymes. Depression was observed as side effects of the treatment of obese patients with rimonabant (SR141716A), a selective CB1 blocker (Despres, Golay, & Sjostrom, 2005). Regarding this observation, it is hypothesized that blockade of CB1Rs induce depression in humans. Other studies have reported a hyperfunctionality of the endocannabinoid system, including upregulation of CB1Rs, and increased CB1R-

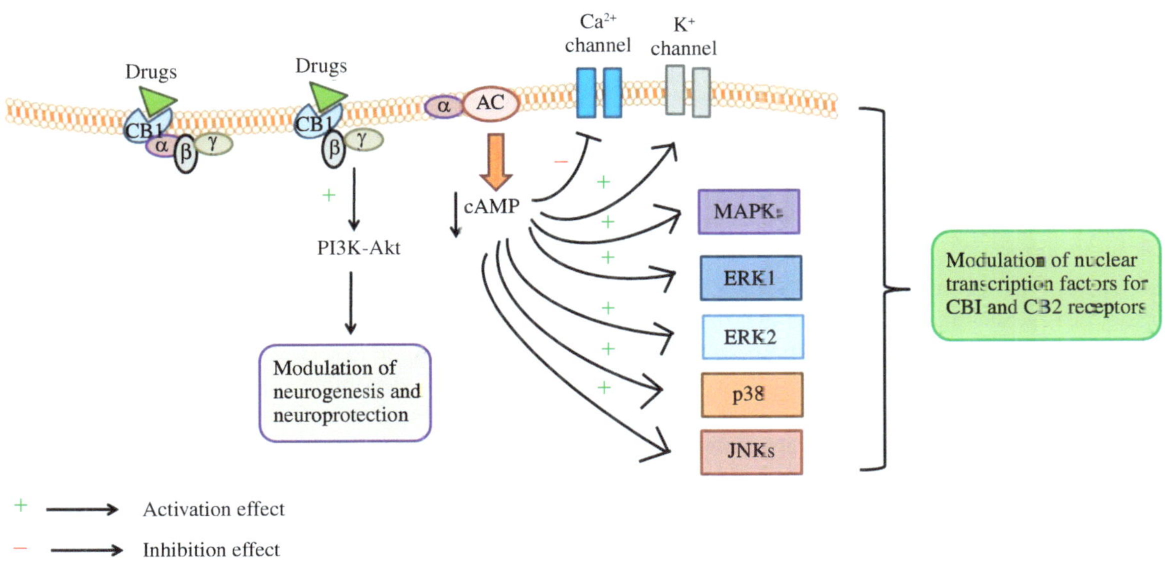


FIG. 2 Schematic diagram of signal transduction pathways of CB1 receptors. The activation of CB1 receptors prevents cAMP production through its coupling Gi/Go protein. This causes to the inhibition of calcium channel activity, and activation of inwardly rectifying potassium currents as well as stimulation of MAPKs, ERK1, ERK2, p38, and JNKs. These various MAPKs participate in downstream signal transduction and modulation of nuclear transcription factors for both CB1 and CB2 receptors. Also, the CB1 receptor activates the PI3K-Akt pathway, which is possibly essential for cannabinoid-mediated effects of neurogenesis and neuroprotection.

mediated signaling in the prefrontal cortex of depressed or alcoholic suicides (Vinod et al., 2005). By contrast, a downregulation of the endocannabinoid activity (Koethe et al., 2007) and lower levels of 2-AG have been observed in major depression, while minor depression seems to be associated with high levels of serum AEA (Hill, Miller, Carrier, Gorzalka, & Hillard, 2009). A postmortem study demonstrated that FAAH activity in the ventral striatum of alcohol-dependent suicides is higher than alcohol-dependent nonsuicides indicating a crucial role of increased FAAH activity for depression, and suicide in depressed patients (Vinod et al., 2010). According to some authors (Gold & Chrousos, 2002), suppression of the endocannabinoid degradation could be a better antidepressant option than cannabis use, particularly in patients with melancholic depression, insomnia, hypophagia, anhedonia, and anxiety.

Few studies have explored the role of CB2Rs in depressed humans. A recent study reported that CB2R and GPR55 gene expressions were significantly lower in the dorsolateral prefrontal cortex of suicide victims (Garcia-Gutierrez et al., 2018). Another study showed the presence of an association between bipolar disorder and a polymorphism in the CB2R gene (Minocci et al., 2011), supporting the hypothesis that the CB2R may be engaged in the pathogenesis of this affective disorder.

## Cannabinoids and depression disorder: Preclinical evidence

Preclinical studies have evidenced that activation of CB1Rs through direct administration of agonists (i.e., oleamide, HU210, or AEA) or by inhibition of the endocannabinoid uptake (by AM404, MAGL, or URB597) induces antidepressant-like activity (Hill & Gorzalka, 2005; Umathe, Manna, & Jain, 2011). It has been shown that CB1R knockout mice present increased depressive-like responses in the both forced swim test (FST) and tail suspension test (TST) (Aso et al., 2008). They also exhibit changes in the mRNA level of some of genes related to the development of depression (Aso, Ozaita, Serra, & Maldonado, 2011). Shearman et al. noted that the blockade of CB1Rs by AM251 results in an antidepressant-like effect in animal models of depression in a dose-dependent manner (Shearman et al., 2003). The levels of endocannabinoid degradative enzyme, FAAH, increase in the hippocampus of rats exposed to chronic mild stress and Wistar-Kyoto rats as animal models of depression (Vinod et al., 2012). In agreement with this finding, FAAH knockout mice showed antidepressive behavior (Bambico et al., 2010). Another supporting argument about the contribution of the endocannabinoid system in depression is the fact that the available antidepressant drugs, depending on the class of drugs and on the targeted brain regions, affect the endocannabinoid signaling through CB1Rs (Smaga et al., 2017). It is evident that the cannabinoid effects are exerted through the endocannabinoid system that modulates the effects of neurotransmitters and other cellular mechanisms in ways that are not yet fully understood (Stoner, 2017). Moreira and Lutz suggested a "set-point hypothesis" for the biphasic effects observed by the cannabinoid system in different behaviors. Based on this hypothesis, the endocannabinoid system works as a balance by modulation of excitatory and inhibitory neurotransmission. Therefore, either ablation of signaling (due to blockage of the receptors) or overactivation of the system (due to high doses of agonists) will shift the balance toward depressive states. Thus, pharmacological intervention strategies modulating CB1 receptor signaling should aim at establishing an optimal dosage window, i.e., to restore the "set point" (Moreira & Lutz, 2008).

Animal models of depression have shown different changes in the CB2R density. Mice of chronic stress model have reduced the expression of CB2Rs in the hippocampus (Garcia-Gutierrez, Gutierrez-Adan, & Manzanares, 2010), while rats of maternal deprivation model have increased the CB2R levels in the hippocampus and frontal cortex (Marco et al., 2013). Meanwhile, there is no evidence demonstrating the alteration in CB2R expression induced by antidepressant drugs.

## Possible mechanisms in the effects of cannabinoids on depression

One of the possible mechanisms of cannabinoid effects on depression is through the modulation of other neurotransmitters release. The endocannabinoid system as a neuromodulatory system is involved in the regulation of excitatory, inhibitory, and monoaminergic networks in the brain. There is an argument that deficits in corticolimbic endocannabinoid signaling, either through downregulation of endogenous ligands and/or alterations in CB1R functionality, may have implications for the pathological development of depression (Mclaughlin & Bambico, 2011). It has been shown that $\Delta^9$-THC enhances extracellular dopamine and glutamate level and decreases GABA levels in the prefrontal cortex of rats (Pistis et al., 2002) and blocks the reuptake of dopamine, norepinephrine, and serotonin (Steffens & Feuerstein, 2004). By contrast, CB1R blockage intensifies the efflux of norepinephrine, serotonin, and dopamine in brain regions associated with depression (Tzavara et al., 2003). Animal studies show the contribution of 5-HTA1 receptors in the antidepressant-like effects of cannabidiol (Sartim, Guimarães, & Joca, 2016; Zanelati, Biojone, Moreira, Guimarães, & Joca, 2010). Moreover, interactions between endocannabinoid and cholinergic system (Kruk-slomka, Michalak, & Biala, 2015), glutamatergic system (Ebrahimi-Ghiri, Shahini, Khakpai, &

Zarrindast, 2019; Khakpai, Ebrahimi-Ghiri, Alijanpour, & Zarrindast, 2019), and β-carbolines (Ebrahimi-Ghiri, Nasehi, & Zarrindast, 2019) in depression-related behaviors have been proposed. It is also worth noting that antidepressant drugs increase the density of the CB1R in the hippocampus, hypothalamus, and amygdala of rodents (Hill, Ho, Hillard, & Gorzalka, 2008).

Another possible mechanism of cannabinoid effects on depression-related behaviors is mediated via modulation of the hypothalamus-pituitary-adrenal gland (HPA) axis activity. Clear evidence shows that CB1R activation inhibits HPA axis activity, while impairment in the CB1R signaling increases HPA axis activity under both basal conditions and following stress exposure (Riebe & Wotjak, 2011).

Available evidence suggests that endocannabinoid signaling assists neurotrophin expression and hippocampal neurogenesis. Stimulation of CB1R encourages brain-derived neurotrophic factor (BDNF) expression and promotes neurogenesis, while FAAH-deficient mice enhance cell proliferation in the hippocampus. Accordingly, disruption of endocannabinoid tone results in impaired cell proliferation, neurogenesis, and BDNF levels. Therefore, CB1R facilitation mimics the effects of conventional antidepressants on neurotrophic signaling and neurogenesis processes in the hippocampus, proposing that this system may prevent hippocampal neurodegeneration in depressed patients (Fig. 3) (Smaga, Bystrowska, Gawli, & Przegali, 2014).

## Conclusion

There is evidence that the endocannabinoid system is involved in the physiopathology of depression. They are: (1) CB1Rs and their endogenous ligands are distributed mainly in neuroanatomical structures related to depression, (2) activation of CB1Rs initiates multiple signaling pathways that produce antidepressant effects, (3) preclinical and clinical findings report a disruption of endocannabinoid signaling in depressed patients and animal models of depression, and (4) the antidepressant exposures provoke some modulations within the endocannabinoid system through CBRs. Preclinical data report the inconsistent results from the role of the endocannabinoid in the pathophysiology of depression. Differences in the animal model of depression, the duration of endocannabinoids injection, or the route of injection may be useful in these inconsistent results. To determine the precise role of the endocannabinoids in the pathophysiology of depression, further studies are needed.

## Key facts

- According to the World Health Organization (WHO), about 147 million people, or 2.5% of the global population, use cannabis (marijuana), making it the world's most widely cultivated, trafficked, and abused illicit substance.
- The endocannabinoid system is a neuromodulatory system involved in a wide range of physiological processes in the central nervous system and the periphery.
- This system typically functions "on-demand," depending on environmental stimuli and the emotional state of the organism.
- It has a wide neuroanatomical distribution, modulating brain regions with different functions in responses to aversive stimuli.
- Endocannabinoids modulate cellular processes through the regulation of other neurotransmitters, neurogenesis, and HPA axis activity.
- Recent evidence has shown that the endocannabinoid system participates in the pathogenesis of the depression; however, its role in this disease is still poorly understood.

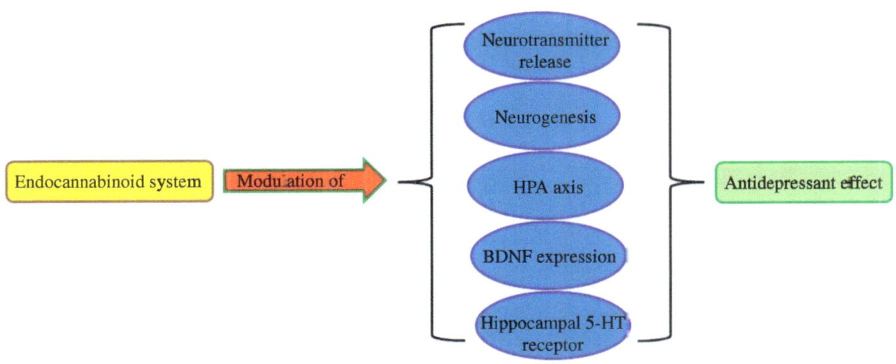


FIG. 3 The possible action mechanisms of the endocannabinoid system in the induction of antidepressive-like behaviors. The possible mechanisms of cannabinoid effects on depression mediated by the modulation of neurotransmitters release, neurogenesis, HPA axis activity, and BDNF expression as well as hippocampal 5-HT receptors.

## Summary points

- The endocannabinoid system is comprised of two GPCRs (CB1R and CB2R), their endogenous small lipid ligands (AEA and 2-AG), and the proteins for endocannabinoid biosynthesis and deactivation (FAAH and MAGL).
- The distribution of endocannabinoids and CB1Rs in depression-associated neuroanatomical structures and neural circuits supports a secure link between the endocannabinoid system, and depressive disorders.
- It has been proposed that impairments in endocannabinoid tone engage to the development of depressive illness.
- Endocannabinoids via multiple pathways (increased cell proliferation and neurogenesis, dampened HPA axis activation, reduced monoamine metabolism and uptake, increased 5-HT neuron firing, and increased neurotrophin production) can produce behavioral and biochemical effects as the conventional antidepressant treatment.
- A deficiency in the endocannabinoid tone has been found in animal models of depression and depressed patients.

## Mini-dictionary of terms

**FST** Forced swim test (FST) is the most widely used paradigm to assess antidepressant-like behavior. In this test, rodents are forced to swim in an inescapable cylinder filled with water and eventually adopt a characteristic immobile posture, which is interpreted as a passive stress-coping strategy or depressive-like behavior (behavioral despair).

**TST** In the tail suspension test (TST), mice are suspended by their tails for a defined period of time, and several antidepressants decrease their immobility.

## References


Aso, E., Ozaita, A., Serra, M.-A., & Maldonado, R. (2011). Genes differentially expressed in CB1 knockout mice: Involvement in the depressive-like phenotype. *European Neuropsychopharmacology : The Journal of the European College of Neuropsychopharmacology*, *21*(1), 11–22. https://doi.org/10.1016/j.euroneuro.2010.06.007.

Aso, E., Ozaita, A., Valdizan, E. M., Ledent, C., Pazos, A., Maldonado, R., et al. (2008). BDNF impairment in the hippocampus is related to enhanced despair behavior in CB1 knockout mice. *Journal of Neurochemistry*, *105*(2), 565–572. https://doi.org/10.1111/j.1471-4159.2007.05149.x.

Bambico, F. R., Cassano, T., Dominguez-Lopez, S., Katz, N., Walker, C. D., Piomelli, D., et al. (2010). Genetic deletion of fatty acid amide hydrolase alters emotional behavior and serotonergic transmission in the dorsal raphe, prefrontal cortex, and hippocampus. *Neuropsychopharmacology : Official Publication of the American College of Neuropsychopharmacology*, *35*(10), 2083–2100. https://doi.org/10.1038/npp.2010.80.

Bambico, F. R., & Gobbi, G. (2008). The cannabinoid CB1 receptor and the endocannabinoid anandamide: Possible antidepressant targets. *Expert Opinion on Therapeutic Targets*, *12*(11), 1347–1366. https://doi.org/10.1517/14728222.12.11.1347.

Bonhaus, D. W., Chang, L. K., Kwan, J., & Martin, G. R. (1998). Dual activation and inhibition of adenylyl cyclase by cannabinoid receptor agonists: Evidence for agonist-specific trafficking of intracellular responses. *The Journal of Pharmacology and Experimental Therapeutics*, *287*(3), 884–888.

Brown, S. P., Safo, P. K., & Regehr, W. G. (2004). Endocannabinoids inhibit transmission at granule cell to Purkinje cell synapses by modulating three types of presynaptic calcium channels. *The Journal of Neuroscience : The Official Journal of the Society for Neuroscience*, *24*(24), 5623–5631. https://doi.org/10.1523/JNEUROSCI.0918-04.2004.

Cohen, K., Weizman, A., & Weinstein, A. (2019). Positive and negative effects of Cannabis and cannabinoids on health. *Clinical Pharmacology and Therapeutics*, *105*(5), 1139–1147. https://doi.org/10.1002/cpt.1381.

Cravatt, B. F., Giang, D. K., Mayfield, S. P., Boger, D. L., Lerner, R. A., & Gilula, N. B. (1996). Molecular characterization of an enzyme that degrades neuromodulatory fatty-acid amides. *Nature*, *384*(6604), 83–87. https://doi.org/10.1038/384083a0.

Denson, T. F., & Earleywine, M. (2006). Decreased depression in marijuana users. *Addictive Behaviors*, *31*(4), 738–742. https://doi.org/10.1016/j.addbeh.2005.05.052.

Despres, J.-P., Golay, A., & Sjostrom, L. (2005). Effects of rimonabant on metabolic risk factors in overweight patients with dyslipidemia. *The New England Journal of Medicine*, *353*(20), 2121–2134. https://doi.org/10.1056/NEJMoa044537.

Dinh, T. P., Carpenter, D., Leslie, F. M., Freund, T. F., Katona, I., Sensi, S. L., et al. (2002). Brain monoglyceride lipase participating in endocannabinoid inactivation. *Proceedings of the National Academy of Sciences of the United States of America*, *99*(16), 10819–10824. https://doi.org/10.1073/pnas.152334899.

Ebrahimi-Ghiri, M., Nasehi, M., & Zarrindast, M.-R. (2019). Anxiolytic and antidepressant effects of ACPA and harmaline co-treatment. *Behavioural Brain Research*, *364*, 296–302. https://doi.org/10.1016/j.bbr.2019.02.034.

Ebrahimi-Ghiri, M., Shahini, F., Khakpai, F., & Zarrindast, M.-R. (2019). Antinociceptive and antidepressive efficacies of the combined ineffective doses of S-ketamine and URB597. *Naunyn-Schmiedeberg's Archives of Pharmacology*, *392*(11), 1393–1400. https://doi.org/10.1007/s00210-019-01676-5.

Freund, T. F., Katona, I., & Piomelli, D. (2003). Role of endogenous cannabinoids in synaptic signaling. *Physiological Reviews*, *83*(3), 1017–1066. https://doi.org/10.1152/physrev.00004.2003.

Garcia-Gutierrez, M. S., Perez-Ortiz, J. M., Gutierrez-Adan, A., & Manzanares, J. (2010). Depression-resistant endophenotype in mice overexpressing cannabinoid CB(2) receptors. *British Journal of Pharmacology*, *160*(7), 1773–1784. https://doi.org/10.1111/j.1476-5381.2010.00819.x.

Garcia-Gutierrez, M. S., Navarrete, F., Navarro, G., Reyes-Resina, I., Franco, R., Lanciego, J. L., et al. (2018). Alterations in gene and protein expression of cannabinoid CB2 and GPR55 receptors in the dorsolateral prefrontal cortex of suicide victims. *Neurotherapeutics: The Journal of the American Society for Experimental NeuroTherapeutics*, *15*(3), 796–806. https://doi.org/10.1007/s13311-018-0610-y.

Gold, P. W., & Chrousos, G. P. (2002). Organization of the stress system and its dysregulation in melancholic and atypical depression: High vs low CRH/NE states. *Molecular Psychiatry*, *7*(3), 254–275. https://doi.org/10.1038/sj.mp.4001032.

Gong, J.-P., Onaivi, E. S., Ishiguro, H., Liu, Q.-R., Tagliaferro, P. A., Brusco, A., et al. (2006). Cannabinoid CB2 receptors: Immunohistochemical localization in rat brain. *Brain Research*, *1071*(1), 10–23. https://doi.org/10.1016/j.brainres.2005.11.035.

Guo, J., & Ikeda, S. R. (2004). Endocannabinoids modulate N-type calcium channels and G-protein-coupled inwardly rectifying potassium channels via CB1 cannabinoid receptors heterologously expressed in mammalian neurons. *Molecular Pharmacology*, *65*(3), 665–674. https://doi.org/10.1124/mol.65.3.665.

Haring, M., Marsicano, G., Lutz, B., & Monory, K. (2007). Identification of the cannabinoid receptor type 1 in serotonergic cells of raphe nuclei in mice. *Neuroscience*, *146*(3), 1212–1219. https://doi.org/10.1016/j.neuroscience.2007.02.021.

Herkenham, M., Lynn, A. B., Little, M. D., Johnson, M. R., Melvin, L. S., de Costa, B. R., et al. (1990). Cannabinoid receptor localization in brain. *Proceedings of the National Academy of Sciences of the United States of America*, *87*(5), 1932–1936. https://doi.org/10.1073/pnas.87.5.1932.

Hill, M. N., & Gorzalka, B. B. (2005). Pharmacological enhancement of cannabinoid CB1 receptor activity elicits an antidepressant-like response in the rat forced swim test. *European Neuropsychopharmacology : The Journal of the European College of Neuropsychopharmacology*, *15*(6), 593–599. https://doi.org/10.1016/j.euroneuro.2005.03.003.

Hill, M. N., Ho, W.-S. V., Hillard, C. J., & Gorzalka, B. B. (2008). Differential effects of the antidepressants tranylcypromine and fluoxetine on limbic cannabinoid receptor binding and endocannabinoid contents. *Journal of Neural Transmission (Vienna, Austria : 1996)*, *115*(12), 1673–1679. https://doi.org/10.1007/s00702-008-0131-7.

Hill, M. N., Miller, G. E., Carrier, E. J., Gorzalka, B. B., & Hillard, C. J. (2009). Circulating endocannabinoids and N-acyl ethanolamines are differentially regulated in major depression and following exposure to social stress. *Psychoneuroendocrinology*, *34*(8), 1257–1262. https://doi.org/10.1016/j.psyneuen.2009.03.013.

Howlett, A. C., & Abood, M. E. (2017). CB1 and CB2 receptor pharmacology. *Advances in Pharmacology (San Diego, Calif.)*, *80*, 169–206. https://doi.org/10.1016/bs.apha.2017.03.007.

Howlett, A. C., Barth, F., Bonner, T. I., Cabral, G., Casellas, P., Devane, W. A., et al. (2002). International Union of Pharmacology. XXVII. Classification of cannabinoid receptors. *Pharmacological Reviews*, *54*(2), 161–202.

Khakpai, F., Ebrahimi-Ghiri, M., Alijanpour, S., & Zarrindast, M.-R. (2019). Ketamine-induced antidepressant like effects in mice: A possible involvement of cannabinoid system. *Biomedicine & Pharmacotherapy = Biomédecine & Pharmacothérapie*, *112*, 108717. https://doi.org/10.1016/j.biopha.2019.108717.

Koethe, D., Llenos, I. C., Dulay, J. R., Hoyer, C., Torrey, E. F., Leweke, F. M., et al. (2007). Expression of CB1 cannabinoid receptor in the anterior cingulate cortex in schizophrenia, bipolar disorder, and major depression. *Journal of Neural Transmission (Vienna, Austria : 1996)*, *114*(8), 1055–1063. https://doi.org/10.1007/s00702-007-0660-5.

Komorowski, J., & Stepien, H. (2007). The role of the endocannabinoid system in the regulation of endocrine function and in the control of energy balance in humans. *Postepy higieny i medycyny doswiadczalnej (Online)*, *61*, 99–105.

Kruk-slomka, M., Michalak, A., & Biala, G. (2015). Antidepressant-like effects of the cannabinoid receptor ligands in the forced swimming test in mice : Mechanism of action and possible interactions with cholinergic system. *Behavioural Brain Research*, *284*, 24–36. https://doi.org/10.1016/j.bbr.2015.01.051.

Lafaye, G., Karila, L., Blecha, L., & Benyamina, A. (2017). Cannabis, cannabinoids, and health. *Dialogues in Clinical Neuroscience*, *19*(3), 309–316.

Lauckner, J. E., Hille, B., & Mackie, K. (2005). The cannabinoid agonist WIN55,212-2 increases intracellular calcium via CB1 receptor coupling to Gq/11 G proteins. *Proceedings of the National Academy of Sciences of the United States of America*, *102*(52), 19144–19149. https://doi.org/10.1073/pnas.0509588102.

Lev-Ran, S., Roerecke, M., Le Foll, B., George, T. P., McKenzie, K., & Rehm, J. (2014). The association between cannabis use and depression: A systematic review and meta-analysis of longitudinal studies. *Psychological Medicine*, *44*(4), 797–810. https://doi.org/10.1017/S0033291713001438.

Ligresti, A., De Petrocellis, L., & Di Marzo, V. (2016). From phytocannabinoids to cannabinoid receptors and endocannabinoids: Pleiotropic physiological and pathological roles through complex pharmacology. *Physiological Reviews*, *96*(4), 1593–1659. https://doi.org/10.1152/physrev.00002.2016.

Mackie, K. (2005). Distribution of cannabinoid receptors in the central and peripheral nervous system. *Handbook of Experimental Pharmacology*, *168*, 299–325.

Mangieri, R. A., & Piomelli, D. (2007). Enhancement of endocannabinoid signaling and the pharmacotherapy of depression. *Pharmacological Research*, *56*(5), 360–366. https://doi.org/10.1016/j.phrs.2007.09.003.

Marco, E. M., Valero, M., de la Serna, O., Aisa, B., Borcel, E., Ramirez, M. J., et al. (2013). Maternal deprivation effects on brain plasticity and recognition memory in adolescent male and female rats. *Neuropharmacology*, *68*, 223–231. https://doi.org/10.1016/j.neuropharm.2012.08.014.

Mclaughlin, R. J., & Bambico, F. R. (2016). Role of the endocannabinoid system in etiopathology of major depression. *Frontiers in Neuroscience: Neurobiology of Depression*, 203–222.

Minocci, D., Massei, J., Martino, A., Milianti, M., Piz, L., Di Bello, D., et al. (2011). Genetic association between bipolar disorder and 524A >C (Leu133Ile) polymorphism of CNR2 gene, encoding for CB2 cannabinoid receptor. *Journal of Affective Disorders*, *134*(1–3), 427–430. https://doi.org/10.1016/j.jad.2011.05.023.

Moreira, F. A., & Lutz, B. (2008). The endocannabinoid system: Emotion, learning and addiction. *Addiction Biology*, *13*(2), 195–212. https://doi.org/10.1111/j.1369-1600.2008.00104.x.

Onaivi, E. S., Ishiguro, H., Gong, J.-P., Patel, S., Meozzi, P. A., Myers, L., et al. (2008). Functional expression of brain neuronal CB2 cannabinoid receptors are involved in the effects of drugs of abuse and in depression. *Annals of the New York Academy of Sciences*, *1139*, 434–449. https://doi.org/10.1196/annals.1432.036.

Oropeza, V. C., Mackie, K., & Van Bockstaele, E. J. (2007). Cannabinoid receptors are localized to noradrenergic axon terminals in the rat frontal cortex. *Brain Research*, *1127*(1), 36–44. https://doi.org/10.1016/j.brainres.2006.09.110.

Pistis, M., Ferraro, L., Pira, L., Flore, G., Tanganelli, S., Gessa, G. L., et al. (2002). Delta(9)-tetrahydrocannabinol decreases extracellular GABA and increases extracellular glutamate and dopamine levels in the rat prefrontal cortex: An in vivo microdialysis study. *Brain Research*, *948*(1–2), 155–158. https://doi.org/10.1016/s0006-8993(02)03055-x.

Poleszak, E., Wosko, S., Slawinska, K., Szopa, A., Wrobel, A., & Serefko, A. (2018). Cannabinoids in depressive disorders. *Life Sciences*, *213*, 18–24. https://doi.org/10.1016/j.lfs.2018.09.058.

Riebe, C. J., & Wotjak, C. T. (2011). Endocannabinoids and stress. *Stress (Amsterdam, Netherlands)*, *14*(4), 384–397. https://doi.org/10.3109/10253890.2011.586753.

Ross, R. A. (2003). Anandamide and vanilloid TRPV1 receptors. *British Journal of Pharmacology*, *140*(5), 790–801. https://doi.org/10.1038/sj.bjp.0705467.

Ryberg, E., Larsson, N., Sjogren, S., Hjorth, S., Hermansson, N.-O., Leonova, J., et al. (2007). The orphan receptor GPR55 is a novel cannabinoid receptor. *British Journal of Pharmacology*, *152*(7), 1092–1101. https://doi.org/10.1038/sj.bjp.0707460.

Sartim, A. G., Guimarães, F. S., & Joca, S. R. L. (2016). Antidepressant-like effect of cannabidiol injection into the ventral medial prefrontal cortex—Possible involvement of 5-HT1A and CB1 receptors. *Behavioural Brain Research*, *303*, 218–227. https://doi.org/10.1016/j.bbr.2016.01.033.

Shearman, L. P., Rosko, K. M., Fleischer, R., Wang, J., Xu, S., Tong, X. S., et al. (2003). Antidepressant-like and anorectic effects of the cannabinoid CB1 receptor inverse agonist AM251 in mice. *Behavioural Pharmacology*, *14*(8), 573–582. https://doi.org/10.1097/01.fbp.0000104880.69384.38.

Smaga, I., Bystrowska, B., Gawli, D., & Przegali, E. (2014). The endocannabinoid/endovanilloid system and depression. *Current Neuropharmacology*, *12*(5), 462–474.

Smaga, I., Zaniewska, M., Gawlinski, D., Faron-Gorecka, A., Szafranski, P., Cegla, M., et al. (2017). Changes in the cannabinoids receptors in rats following treatment with antidepressants. *Neurotoxicology*, *63*, 13–20. https://doi.org/10.1016/j.neuro.2017.08.012.

Steffens, M., & Feuerstein, T. J. (2004). Receptor-independent depression of DA and 5-HT uptake by cannabinoids in rat neocortex—Involvement of Na(+)/K(+)-ATPase. *Neurochemistry International*, *44*(7), 529–538. https://doi.org/10.1016/j.neuint.2003.08.009.

Stoner, S. A. (2017). *Effects of marijuana on mental health: Depression* (pp. 1–6). University of Washington: Alcohol & Drug Abuse Institute.

Tsou, K., Brown, S., Sanudo-Pena, M. C., Mackie, K., & Walker, J. M. (1998). Immunohistochemical distribution of cannabinoid CB1 receptors in the rat central nervous system. *Neuroscience*, *83*(2), 393–411. https://doi.org/10.1016/s0306-4522(97)00436-3.

Tzavara, E. T., Davis, R. J., Perry, K. W., Li, X., Salhoff, C., Bymaster, F. P., et al. (2003). The CB1 receptor antagonist SR141716A selectively increases monoaminergic neurotransmission in the medial prefrontal cortex: Implications for therapeutic actions. *British Journal of Pharmacology*, *138*(4), 544–553. https://doi.org/10.1038/sj.bjp.0705100.

Umathe, S. N., Manna, S. S. S., & Jain, N. S. (2011). Involvement of endocannabinoids in antidepressant and anti-compulsive effect of fluoxetine in mice. *Behavioural Brain Research*, *223*(1), 125–134. https://doi.org/10.1016/j.bbr.2011.04.031.

Vinod, K. Y., Arango, V., Xie, S., Kassir, S. A., Mann, J. J., Cooper, T. B., et al. (2005). Elevated levels of endocannabinoids and CB1 receptor-mediated G-protein signaling in the prefrontal cortex of alcoholic suicide victims. *Biological Psychiatry*, *57*(5), 480–486. https://doi.org/10.1016/j.biopsych.2004.11.033.

Vinod, K. Y., Kassir, S. A., Hungund, B. L., Cooper, T. B., Mann, J. J., & Arango, V. (2010). Selective alterations of the CB1 receptors and the fatty acid amide hydrolase in the ventral striatum of alcoholics and suicides. *Journal of Psychiatric Research*, *44*(9), 591–597. https://doi.org/10.1016/j.jpsychires.2009.11.013.

Vinod, K. Y., Xie, S., Psychoyos, D., Hungund, B. L., Cooper, T. B., & Tejani-Butt, S. M. (2012). Dysfunction in fatty acid amide hydrolase is associated with depressive-like behavior in Wistar Kyoto rats. *PLoS One*, *7*(5), e36743. https://doi.org/10.1371/journal.pone.0036743.

Zanelati, T. V., Biojone, C., Moreira, F. A., Guimarães, F. S., & Joca, S. R. (2010). Antidepressant-like effects of cannabidiol in mice : Possible involvement of 5-HT 1A receptors. *British Journal of Pharmacology*, *159*(1), 122–128. https://doi.org/10.1111/j.1476-5381.2009.00521.x.

Zou, S., & Kumar, U. (2018). Cannabinoid receptors and the endocannabinoid system : Signaling and function in the central nervous system. *International Journal of Molecular Sciences*, *19*(3), 833. https://doi.org/10.3390/ijms19030833.

Chapter 30

# Agomelatine: Profile and applications to depression

Trevor R. Norman

*Department of Psychiatry, Austin Hospital, University of Melbourne, Heidelberg, VIC, Australia*

## List of abbreviations

| | |
|---|---|
| **5HT** | serotonin |
| **5HT2B** | serotonin type two B receptor |
| **5HT2C** | serotonin type two C receptor |
| **ARCI** | Addiction Research Center Inventory |
| **AUSPAR** | Australian Public Assessment Report |
| **BPD** | bipolar disorder |
| **CHMP** | Committee for Medicinal Products for Human Use |
| **CYP1A2/CYP2C9/ CYP2C19** | subtypes of cytochrome P450 family of liver metabolizing enzymes |
| **DSPS** | delayed sleep phase syndrome |
| **EMEA** | European Evaluation Medicines Agency |
| **GABA** | gamma aminobutyric acid |
| **HAM-D 17** | Hamilton depression rating scale, 17 items |
| **MDD** | major depressive disorder |
| **MT** | melatonin |
| **MT1** | melatonin type 1 receptor |
| **MT2** | melatonin type 2 receptor |
| **PSG** | polysomnography |
| **S32006** | 5HT2C antagonist from Servier Laboratories |
| **SMD** | standard mean difference |

## Introduction

Medications used successfully for the pharmacotherapy of depressive disorders have encompassed a range of compounds with diverse chemical structures yet with a remarkable uniformity of putative mechanisms of action. These agents have affected, at least acutely, the transporter molecules for the reuptake of neurotransmitters by presynaptic nerve endings (e.g., tricyclic antidepressants and selective serotonin reuptake inhibitors) or the prevention of the catabolism of monoamines (e.g., monoamine oxidase inhibitors). While such pharmacological actions are unlikely to fully explain the alleviation of depressive symptomatology, they have provided the impetus for the development of most new antidepressants. Against this background a molecule with primary pharmacological actions independent of effects on transporters or monoamine catabolism is rare. The introduction of agomelatine into clinical practice represented a radical departure from these "traditional" pathways.

Agomelatine was developed based on the circadian hypothesis of depression: that depression is accompanied by alterations in the 24-h rhythms of sleep, hormone secretion, and basal body temperature (de Bodinat et al., 2010). It is unclear how such desynchrony arises, but it has been speculated that the link to sleep disturbances in depression are pivotal (Boivin, 2000). Normalization of rhythms should result in the alleviation of the symptoms of the disorder.

## Pharmacology

The pharmacological profile of agomelatine is unique in that it exhibits agonist actions at the melatonin receptors (MT1 and MT2) as well as antagonist effects at 5HT2 receptors (Guardiola-Lemaitre et al., 2014). It has virtually no affinity for a range of other neurotransmitter receptors including GABA, dopamine, adenosine, adrenoceptors, muscarinic, nicotinic, histamine, excitatory amino acid, benzodiazepine, and sigma receptors, as well as sodium, potassium, and calcium channels. It also has no affinity for most serotonin receptors other than 5HT2C and 5HT2B, where it acts as an antagonist. It does not interact with 5HT2A receptors. Agonist actions at melatonin receptors and serotonin antagonist effects appear to be necessary for antidepressant activity.

## Pharmacodynamics

In view of its melatonin agonist effects, agomelatine has demonstrated resynchronization of circadian rhythms in animal models of circadian rhythm disruption (Popoli, 2009). For example, rodents kept in constant darkness establish a free-running circadian rhythm, which can be reentrained to the timing of agomelatine administration

The Neuroscience of Depression. https://doi.org/10.1016/B978-0-12-817933-8.00032-3

(Martinet, Guardiola-Lemaitre, & Mocaer, 1996). In a model of delayed sleep phase syndrome (DSPS) agomelatine dose-dependently shifted the phase toward the onset of darkness (Armstrong, McNulty, Guardiola-Lemaitre, & Redman, 1993). The effect was comparable to that of melatonin.

An important insight into the mechanism of action of agomelatine was demonstrated in microdialysis studies designed to assess the effect of the compound on central concentrations of monoamine neurotransmitters. A dose-dependent increase in dopamine concentrations in the frontal cortex, but not nucleus accumbens or striatum, was demonstrated (Millan et al., 2003). Similarly, noradrenaline concentrations were increased in frontal cortex. Both of these effects were attributed to the 5HT2C antagonist effects of agomelatine. Serotonin concentrations were not affected by agomelatine.

The importance of dual actions as a melatonin agonist and 5HT2C antagonist was demonstrated in studies using various animal models predictive of antidepressant activity. Thus, in the forced swim test agomelatine was active after repeated administration whereas melatonin was not (Bourin, Mocaër, & Porsolt, 2004). A selective 5HT2C antagonist, S32006, has been used in animal models to examine the role of this receptor subtype in the mode of action of agomelatine. Neither S32006 nor melatonin demonstrated full activity in the tree shrew chronic psychosocial stress model whereas agomelatine was fully active (Schmelting et al., 2014). Similarly, although S32006 and melatonin showed some partial activity in the olfactory bulbectomized rat, their effects were not maximal compared to agomelatine (Norman et al., 2012). Neurochemical studies examining the increase in hippocampal BDNF production in rodents following chronic agomelatine administration also demonstrated that both pharmacological actions are necessary for the antidepressant-like activity (Racagni et al., 2011). Synergism between the two systems is implied by these results and suggests that clinical antidepressant effects are likely to arise from both actions.

## Pharmacokinetics, metabolism, and drug interactions

Agomelatine, although it is rapidly and extensively absorbed (>80%) after oral administration, has a low estimated bioavailability (~3%–4%) which demonstrates substantial interindividual (160%) and intraindividual (104%) variability (CHMP, 2008). Much of this high variability can be accounted for by the interindividual differences of CYP1A2 activity (PI, 2010). Peak plasma concentrations are reduced ~20%–30% when agomelatine is administered with food, but overall absorption and bioavailability are not markedly affected. Orally administered agomelatine is extensively metabolized in the liver by cytochromes CYP1A2 (90%) and CYP2C9/CYP2C19 (10%). Several metabolites are generated but the major metabolites are hydroxylated and demethylated agomelatine, neither of which is pharmacologically active. Both metabolites are rapidly conjugated and eliminated in the urine. Mean plasma elimination half-life of agomelatine is 1–2h. The pharmacokinetics is essentially linear in the therapeutic dose range. Repeated administration does not alter the pharmacokinetics. Neither gender nor age has significant effects on the pharmacokinetic profile of the drug. While altered renal function does not significantly affect the single-dose pharmacokinetic profile, there are marked alterations due to impaired hepatic function. Agomelatine is not recommended for patients with significant hepatic impairment (PI, 2010).

Given the known enzymes involved in the metabolism of agomelatine it can be anticipated that potent CYP1A2 inhibitors (fluvoxamine, ciprofloxacin, and propranolol) and inducers (cigarette smoking and rifampicin) would be expected to affect the kinetics of the drug. To a lesser extent, CYP2C9/CYP2C19 inducers and inhibitors might also be expected to interact with agomelatine.

## Clinical efficacy in depressive episodes

### Major depression

Evaluation of the efficacy of agomelatine for the short-term treatment of major depressive disorder was undertaken using what has become established practice for potential new antidepressant medications: 6- to 8-week trials against a placebo or comparative studies (with or without a placebo arm) against established medications. Most of these studies were conducted using doses of agomelatine between 25 and 50mg/day administered at night. Extensive evaluation of the dose-response effect for agomelatine has not been undertaken. However, studies conducted with 1 and 5mg/day suggested the drug was not effective, while toxicity issues (see later) are likely to curtail the use of higher doses (Norman & Olver, 2019). Prescribing information recommends doses between 25 and 50mg/day, at night. Clinical trials have been the subject of a number of systematic or meta-analysis reviews of the data. One meta-analysis evaluating the efficacy vs placebo identified 12 studies including some unpublished trials and concluded that agomelatine was superior to placebo (Taylor, Sparshatt, Varma, & Olofinjana, 2014). The measure of treatment effect was the standardized mean difference (SMD), i.e., the difference in mean final values between agomelatine and placebo accounting for the standard deviation. Based on this measure the effect size was small, 0.24, but demonstrated that agomelatine was superior to placebo based on response rates. The difference between agomelatine and placebo was not significant when the remission rate was used as the outcome measure. This later finding is perhaps not surprising given that remission from depression is unlikely to be achieved in the time frame of the clinical trials included in the analysis.

A later network meta-analysis compared 21 commonly used antidepressant medications for the short-term treatment of major depressive disorder and included both published and unpublished, double-blind, randomized controlled trials (Cipriani et al., 2018). All antidepressants were shown to be more effective than placebo using the primary outcome variable of response rate, but the effect sizes were relatively modest. The response rate was defined as a reduction of ≥50% of the baseline score of an observer-rated depression rating scale. Among the antidepressants included in the analysis, amitriptyline had the greatest effect size for efficacy and reboxetine the least, while agomelatine was in the middle rank. On the other hand, patient acceptability of treatment, using the surrogate measure of treatment dropout rate compared to placebo, ranked agomelatine highest.

Taken together these two meta-analyses would suggest that agomelatine is an effective short-term treatment for major depressive disorders with at least equivalent efficacy to other commonly used antidepressants and more effective than placebo. In everyday clinical practice, antidepressants are more likely to be prescribed over longer time frames than a few weeks. Long-term efficacy/prevention of relapse of agomelatine was not addressed in either meta-analysis. This gap in the knowledge of the use of agomelatine was highlighted in the EMEA assessment of the medication along with limited data concerning use in elderly depressed patients (CHMP, 2008). Since that evaluation, some studies have attempted to address these perceived short comings.

The efficacy of agomelatine in the prevention of relapse of depression has been evaluated in one published (Goodwin et al., 2009) and two unpublished evaluations (CAGO178A2304, 2010; CHMP, 2008). Following open treatment with 25 or 50 mg/day agomelatine for 8–10 weeks, responders (HAM-D17 < 10; CGI ≤ 2) were randomized to continue agomelatine or to placebo in a double-blind manner and followed for up to 24 weeks (Goodwin et al., 2009). Time to relapse (HAM-D 17 ≥ 16, or withdrawal from the study for lack of efficacy, suicide, and suicide attempt) was compared between the two treatments using Kaplan-Meier survival analysis. A statistically significant difference in relapse rates favored agomelatine over placebo ($P < 0.0001$). At the end of the study, the cumulative relapse rate was 21.7% for agomelatine and 46.6% for placebo. A post hoc analysis in patients who had more severe depression (HAM-D17 > 25) at the study baseline, suggested a better response in patients remaining on agomelatine (relapse rate 21.9%) than on those assigned to placebo (45.1%).

A multicentre study used an identical design to the previous study except that all patients received 8-week open treatment with 25 mg/day agomelatine (CHMP, 2008). Responders to open-label agomelatine were randomly and double-blind, assigned to continue agomelatine or placebo. The main outcome criterion was the time to relapse during the continuation phase as judged by the HAM-D total score. The relapse rates were 25.9% (48/185) in the agomelatine vs 23.5% (42/179) in the placebo group during 8- to 34-week period. This difference was not statistically significant. Kaplan-Meier survival analysis showed no difference between the groups. A further unpublished study found no advantage for agomelatine over placebo in the prevention of relapse (CAGO178A2304, 2010). Patients with major depressive disorder (HAM-D17 ≥ 22 at baseline) entered an open-label treatment period of 4–12 weeks during which they received agomelatine 25 or 50 mg/day depending on response. Patients demonstrating a response to treatment entered a further 12-week stabilization phase. Responders were then randomized to either placebo or to continue agomelatine and followed for up to 52 weeks. Relapse was experienced by 23.0% of agomelatine and 26.2% of placebo-treated patients in the continuation phase, while clinical remission was achieved by 51.4% of agomelatine and 56.4% of placebo-treated patients. The failure to find significant differences in this study may reflect the patient population studied: highly censored study population in that all had received 16- to 24-week open-label treatment with medication prior to double-blind allocation.

The efficacy of agomelatine compared to three SSRIs (fluoxetine, sertraline, and escitalopram) was assessed in a meta-analysis of the outcomes after 24-week treatment (Demyttenaere et al., 2013). The HAM-D 17 ratings were lower at 24 weeks in agomelatine-treated patients compared to SSRI-treated patients but the overall difference, although statistically significant, was clinically minor (mean of 1.04 points). Response (~75%–80%) and remission (40%–50%) rates were not much different between treatments. The analysis suggests that agomelatine is at least as effective as SSRIs in maintaining response over 6 months.

The efficacy of agomelatine for the treatment of depression in the elderly is not well established. An early open evaluation conducted in 23 elderly patients (>65 years of age) with severe major depression demonstrated significant improvement following treatment with 25–50 mg/day agomelatine for 12 weeks (Lužný, 2012). Two double-blind, placebo-controlled evaluations of agomelatine reached opposite conclusions. On the one hand, no significant difference between treatments was observed in a 6 week, fixed dose 25 mg/day study in patients aged 60 years or more (Goodwin et al., 2017). Using a more flexible dose of 25–50 mg/day over 8 weeks, agomelatine was superior to placebo in patients aged >65 years (Heun et al., 2013). A network meta-analysis of the efficacy of antidepressants in specific elderly populations (>65 years) did not enthusiastically endorse agomelatine as an effective treatment (Krause et al., 2019). The role of agomelatine in treating elderly patients requires further evaluation particular in the "old-old" subgroup where evidence-based approaches are lacking.

## Bipolar depression

The switch into mania/hypomania or mixed states in bipolar disorder patients treated with antidepressant medications is a significant consideration in the treatment of depressive episodes. All classes of antidepressant medications have been implicated with the switching rate estimated at between 20% and 40% of bipolar patients (Goldberg & Truman, 2003). The propensity for agomelatine to induce "switches" has been an incidental finding in studies designed to address the antidepressant efficacy of the agent in major depressive episodes in patients with bipolar disorder. In open-label evaluations agomelatine added to existing mood stabilizing medications has been reported to improve depression scores in bipolar I (Calabrese, Guelfi, & Perdrizet-Chevallier, 2007) and bipolar II patients (Fornaro et al., 2013). Both studies used 25 mg/day of agomelatine and evaluated responses at the end of 6-week treatment. Extension of treatment for up to 1 year suggested that the drug maintained clinical gains observed in the initial period. However, interpretation of the longer-term outcome findings were challenged, at least for bipolar I patients, where responses were suggested to be lower than those reported (Eppel, 2008). Both studies were associated with some patients who "switched" into mania/hypomania or demonstrated agitation. The later may represent subthreshold hypomania or a "mixed" state.

A more rigorous evaluation of agomelatine as an adjunctive treatment to mood stabilizing agents was performed in a randomized, double-blind, placebo-controlled study in 344 bipolar I patients (Yatham et al., 2016). The study evaluated the response of depressive symptoms to 8-week treatment with 25–50 mg/day (adjusted according to the clinical response) with the possibility of an extension of treatment for up to 52 weeks. Both agomelatine and placebo improved depression rating scores equally from baseline to week 8. Analysis of the LOCF population from baseline to the end of the extension phase was not different between agomelatine and placebo for responders (61.9% vs 55.0%) or remitters (57.7% vs 52.6%). The rates of emergent of mania/hypomania in this study were not specifically reported other than to state that the combination:

> *was not associated with a high level of emergence of mania or hypomania symptoms over the study period.*

## Depression in medical comorbidities

Depression arising in the context of medical illness is common, with almost a quarter of patients with one or more physical complaints having a depressive disorder (Moussavi et al., 2007). The use of antidepressant medications is frequently indicated. Clearly proven treatments with a low potential for drug-drug interactions are a primary consideration. Agomelatine has been evaluated in depression arising in medical comorbidities: Parkinson's disease, Type II diabetes, and cardiovascular disease (Norman & Olver, 2019). These investigations have for the most part been case series or open evaluations. Indications from these studies have been that agomelatine at doses similar to those used in major depressive disorder (i.e., 25 or 50 mg/day) provides relief from the symptoms of depression without affecting the underlying disease.

## Side effect profile

The recommended dose of agomelatine is 25 mg/day administered at night and increased to 50 mg/day if there is no response after 2 weeks. The side effects associated with the use of the medication at these doses are well characterized from short-term clinical registration studies. Slightly more than half (52.8%) of patients receiving agomelatine at recommended doses over 6–8 weeks will experience an emergent adverse event (AUSPAR, 2010). This rate of emergent effects was similar to that reported in placebo and fluoxetine (20 mg/day) treated patients, but less than those treated with paroxetine (20 mg/day). Events which were considered to be related to drug administration occurred in 32.5% of agomelatine recipients. Common events in agomelatine patients (incidence ≥2%) were headache (14.1%), nausea (7.7%), dizziness (5.5%), dry mouth (3.5%), diarrhea (3.1%), and somnolence (2.9%). Events more common for agomelatine than placebo were dizziness, somnolence, diarrhea, fatigue, and upper abdominal pain. Side effects attributable to medication were dose dependent with more events reported in those receiving 50 mg (40.5%) than 25 mg (30.4%). With repeated dosing, drug-related side effects tended to diminish.

For patients receiving medication for 6–24 weeks, emergent events were reported in 14.1% of agomelatine recipients, 13.8% of placebo recipients. These events were also dose-dependent being reported in 13.3% of patients who took agomelatine 25 mg and 16.7% of patients who took agomelatine 50 mg.

Considerations in the use of antidepressants recently have been given to the effects on sexual function and sleep disturbance. Most antidepressants have a deleterious effect on sexual function in both men and women (Baldwin, 2004). Agomelatine produced fewer sexual side effects in healthy volunteers receiving 25 or 50 mg/day for 8 weeks than paroxetine (20 mg/day) (Montejo et al., 2010). Similarly, in depressed patients receiving 25 or 50 mg/day of agomelatine fewer sexual side effects were reported than in patients receiving sertraline, paroxetine, or venlafaxine, as assessed with a specific rating scale, the Arizona Sexual Experience Scale (Montejo, Majadas, Rizvi, & Kennedy, 2011).

A review of the effects of agomelatine (25 or 50 mg/day) on sleep parameters in major depressive disorder, measured both objectively with polysomnography or subjectively with standard rating scales, concluded that the drug improved the sleep-wake cycle (Quera-Salva, Lemoine, & Guilleminault, 2010). Improvements occurred from the first week of treatment. Based on PSG there were improvements in sleep efficiency and slow-wave sleep. Slow-wave sleep was resynchronized with the first sleep cycle of the night. Coupled with its known circadian resynchronizing abilities (vide supra), agomelatine may prove useful in depressed patients presenting with a delayed phase sleep disturbance.

## Serious adverse events

### Liver function abnormalities

All antidepressant medications are associated with a risk of hepatotoxicity, with agomelatine rated as one of the more notable culprits (Voican, Corruble, Naveau, & Perlemuter, 2014). Indeed, most authorities require adherence to a strict liver function monitoring protocol when prescribing agomelatine. The guidelines recommend liver function tests before initiating treatment then at 3, 6, 12, and 24 weeks of treatment or if the dose is increased to 50 mg/day or when clinically indicated (PI, 2010). The guidelines correspond with a retrospective pooled analysis of transaminases measured in 49 clinical studies of the drug compared to placebo (Perlemuter et al., 2016). Transaminases were increased to >3 times ULN in 1.3% (25 mg/day) and 2.5% (50 mg/day) of patients treated with agomelatine compared to 0.5% for placebo. In most patients (64%) the onset occurred before 12 weeks. These findings are in broad agreement with an earlier systematic review which reported hepatotoxicity in 4.6% of agomelatine and 2.1% of placebo-treated patients (Freiesleben & Furczyk, 2015).

### Withdrawal syndrome

There is little evidence that agomelatine is associated with a specific withdrawal syndrome or that it has abuse potential. However, empirical evidence for these conclusions is sparse. After 12 weeks of 25 mg/day agomelatine treatment, half of the patients continued medication while half were switched to placebo for 2 weeks in a double-blind trial (Montgomery, Kennedy, Burrows, Lejoyeux, & Hindmarch, 2004). In those who switched to placebo, symptoms were not different from those who remained on agomelatine. This was taken as evidence for a lack of a discontinuation syndrome. In relapse prevention studies (vide supra) patients abruptly switched to placebo experienced few early relapses, which was attributed by the authors to a lack of discontinuation syndrome (Goodwin et al., 2009).

In healthy volunteers, agomelatine did not demonstrate abuse potential as measured with the Addiction Research Center Inventory (ARCI) check-list (PI, 2010).

## Safety on overdose

A retrospective study analyzed outcomes and symptoms following acute overdoses with agomelatine (Rauber-Lüthy et al., 2014). A total of 42 patients ingested a mean overdose of 678 mg (25–2450 mg) with no severe cases and no fatalities. Symptoms observed were drowsiness, dizziness, somnolence, psychomotor slowing, nausea, abdominal pain, vomiting, headache, tachycardia 100–120 bpm, QTc-prolongation ≤500 ms, rhabdomyolysis, tremor, confusion, and dry mouth. The intensity of the symptoms was reported as mild to moderate. A retrospective review of calls to a poisons center identified 16 cases of agomelatine overdose taken alone (Wong, Lee, & Lee, 2018). The majority of cases ($n = 9$; 56%) were reported to be asymptomatic. Remaining overdose patients developed drowsiness, dizziness, or nausea within 4 h of drug ingestion. The median overdose was 250 mg.

## Conclusions

Agomelatine remains the sole exemplar in a potential class of antidepressant agents exhibiting melatonin agonist and 5HT2C antagonist properties. While clinical efficacy in the short-term treatment of major depressive disorder appears to be well established, efficacy in the longer term has not been as readily demonstrated. Furthermore, for some population subgroups, efficacy is also less firmly established. Thus, in the elderly, particularly the "old-old," there remain questions over the usefulness of agomelatine as either a short- or long-term alternative to other antidepressants. In patients with bipolar disorder, agomelatine treatment for a depressive episode, as an adjunct to mood stabilizing agents, requires further rigorous evaluation to fully assess its potential role. In general, the medication has a relatively benign side effect profile compared to other antidepressants. The lack of sexual side effects and the ability to consolidate sleep confer clinical advantages in certain patient populations. This comparative tolerability advantage over other antidepressants was revealed by the network meta-analysis. Agomelatine, for the time being, is likely to be an important alternative treatment of major depression, one used as a second-line approach rather than as the first-line approach.

## Key facts of agomelatine

- Agomelatine was developed at Servier Laboratories (France) in the early 1990s and given the code number S20098.
- Chemically agomelatine is one of a series of naphthalenic derivatives related to melatonin.
- Agomelatine is a 5HT2C antagonist and MT1 and MT2 agonist.
- Agomelatine has demonstrated resynchronization of circadian rhythms.
- Oral administration results in rapid and extensive absorption with low bioavailability.
- Mean plasma elimination half-life of agomelatine is 1–2 h.
- Agomelatine is metabolized in the liver by CYP1A2 and CYP2C9/CYP2C19.
- Metabolites of agomelatine are not pharmacologically active.
- Antidepressant activity of agomelatine superior to placebo and equivalent to other antidepressants has been demonstrated in short-term clinical trials.
- Long-term efficacy and responses in the elderly are not as well established.
- A network meta-analysis suggests that agomelatine has equivalent efficacy and is better tolerated than most antidepressants.
- Efficacy in bipolar depression is not well established although switching to hypomania/mania appears to be low.
- In depression in medical illness agomelatine may be useful although there is limited evidence.
- Agomelatine is generally well tolerated with side effect profile similar to placebo.
- Elevated liver function tests are associated with agomelatine administration necessitating ongoing monitoring.
- Agomelatine does not appear to be associated with a withdrawal syndrome on abrupt discontinuation.
- Agomelatine was approved for use in Europe in 2009 and Australia in 2010. It is *not* approved for use in the United States.

## Mini-dictionary of terms

**Circadian rhythm** Refers to the approximate 24-h endogenous rhythm of certain hormones and behavioral patterns in humans, other animals, plants, and even bacteria. The term is derived from the Latin words *circa* meaning about and *diem* meaning day. Rhythms are generated endogenously but can be influenced by external factors (or *zeitgebers* from the German "time giver"). One of the more powerful zeitgebers is exposure to sunlight, which helps to entrain human endogenous rhythms to the 24-h clock. One hypothesis of the etiology of depressive disorders is that circadian rhythms are awry resulting in the manifestation of the symptoms of the disorder.

**Melatonin** *N*-acetyl-5-methoxy tryptamine is a neurohormone synthesized primarily in the pineal gland at night from serotonin by a series of enzymatically catalyzed steps.[a] Concentrations of the hormone typically rise with the onset of darkness and begin to decline with the onset of light. Melatonin secretion thus has a circadian pattern which is regulated by signals from the suprachiasmatic nucleus. Measurement of plasma (or salivary) concentrations of melatonin can provide a reliable measure of an individual's circadian rhythm.

**Melatonin receptors** Three subtypes of melatonin receptor have been cloned. All are G-protein-coupled receptors which bind melatonin. MT1 and MT2 receptors are mammalian receptors while MT3 receptors are present in amphibians and birds. Binding of melatonin to the MT1 receptor appears to be most closely associated with REM sleep and its phase-shifting properties, while binding to MT2 receptors may be involved in modulation of NREM sleep. Activation of MT2 receptors in the retina inhibits the release of dopamine.

**5HT2C receptor** Is a G-protein-coupled receptor that mediates excitatory transmission. Activation of the receptor suppresses the release of dopamine and noradrenaline in certain brain areas. It is believed to be involved in mood appetite and sexual behavior. Unlike other serotonin receptors it is subject to RNA editing which may result in reduced affinity for G-proteins.

**Rhythm resynchronization** Refers to the resetting of the circadian rhythm to the timing of the 24-h clock. This can be accomplished with either bright light or melatonin. In delayed phase sleep syndrome, for example, giving melatonin in the evening will shift the rhythm toward the time of the melatonin administration, or alternatively, the same phase shift can be accomplished by a shining bright light on the eyes in the mornings.

**Network meta-analysis** Also called a mixed treatments comparison or multiple treatments comparison meta-analysis, compares three or more treatments by either direct comparison of randomized controlled trials or indirect by comparison to a common comparator.

## References


Armstrong, S. M., McNulty, O. M., Guardiola-Lemaitre, B., & Redman, J. R. (1993). Successful use of S20098 and melatonin in an animal model of delayed sleep-phase syndrome (DSPS). *Pharmacology, Biochemistry, and Behavior*, *46*, 45–49.

AUSPAR. (2010). *Australian public assessment report for agomelatine*. https://www.tga.gov.au/sites/default/files/auspar-valdoxan.pdf.

Baldwin, D. S. (2004). Sexual dysfunction associated with antidepressant drugs. *Expert Opinion on Drug Safety*, *3*, 457–470.

Boivin, D. B. (2000). Influence of sleep–wake and circadian rhythm disturbance in psychiatric disorders. *Journal of Psychiatry and Neuroscience*, *25*, 446–458.


a. For a detail description of the biological control of melatonin synthesis, see Reiter (1991).

Bourin, M., Mocaër, E., & Porsolt, R. (2004). Antidepressant-like activity of S 20098 (agomelatine) in the forced swimming test in rodents: Involvement of melatonin and serotonin receptors. *Journal of Psychiatry & Neuroscience*, *29*, 126–133.

CAGO178A2304. (2010). *Novartis. Clinical trial results database [online]*. Available from URL: *https://www.novctrd.com/CtrdWeb/displaypdf.nov?trialresultid=3420*.

Calabrese, J. R., Guelfi, J. D., & Perdrizet-Chevallier, C. (2007). Agomelatine adjunctive therapy for acute bipolar depression: Preliminary open data. *Bipolar Disorders*, *9*, 628–635.

CHMP. (2008). *Assessment report for valdoxan*. https://www.ema.europa.eu/en/documents/assessment-report/valdoxan-epar-public-assessment-report_en.pdf.

Cipriani, A., Furukawa, T. A., Salanti, G., Chaimani, A., Atkinson, L. Z., Ogawa, Y., et al. (2018). Comparative efficacy and acceptability of 21 antidepressant drugs for the acute treatment of adults with major depressive disorder: A systematic review and network meta-analysis. *Lancet*, *391*, 1357–1366.

de Bodinat, C., Guardiola-Lemaitre, B., Mocaer, E., Renard, P., Munoz, C., & Millan, M. J. (2010). Agomelatine, the first melatonergic antidepressant: Discovery, characterization and development. *Nature Reviews. Drug Discovery*, *9*, 628–642.

Demyttenaere, K., Corruble, E., Hale, A., Quera-Salva, M. A., Picarel-Blanchot, F., & Kasper, S. (2013). A pooled analysis of six-month comparative efficacy and tolerability in four randomized clinical trials: Agomelatine versus escitalopram, fluoxetine, and sertraline. *CNS Spectrums*, *18*, 163–170.

Eppel, A. B. (2008). Agomelatine adjunctive therapy for acute bipolar depression: Preliminary open data. *Bipolar Disorders*, *10*, 749–750.

Fornaro, M., McCarthy, M. J., De Berardis, D., De Pasquale, C., Tabaton, M., Martino, M., et al. (2013). Adjunctive agomelatine therapy in the treatment of acute bipolar II depression: A preliminary open label study. *Neuropsychiatric Disease and Treatment*, *9*, 243–251.

Freiesleben, S. D., & Furczyk, K. (2015). A systematic review of agomelatine-induced liver injury. *Journal of Molecular Psychiatry*, *3*, 4–12.

Goldberg, J. F., & Truman, C. J. (2003). Antidepressant-induced mania: An overview of current controversies. *Bipolar Disorders*, *5*, 407–420.

Goodwin, G. M., Emsley, R., Rembry, S., Rouillon, F., & Agomelatine Study Group. (2009). Agomelatine prevents relapse in patients with major depressive disorder without evidence of a discontinuation syndrome: A 24-week randomized, double-blind, placebo-controlled trial. *Journal of Clinical Psychiatry*, *70*, 1128–1137.

Goodwin, G. M., Thomas, P., Heun, R., Boyer, P., Picarel-Blanchot, F., & de Bodinat, C. (2017). Antidepressant effect in older depressed patients: The lessons of two agomelatine trials. *International Clinical Psychopharmacology*, *32*, 184–194.

Guardiola-Lemaitre, B., De Bodinat, C., Delagrange, P., Millan, M. J., Munoz, C., & Mocaër, E. (2014). Agomelatine: Mechanism of action and pharmacological profile in relation to antidepressant properties. *British Journal of Pharmacology*, *171*, 3604–3619.

Heun, R., Ahokas, A., Boyer, P., Gimenez-Montesinos, N., Pontes-Soares, F., & Olivier, V. (2013). The efficacy of agomelatine in elderly patients with recurrent major depressive disorder: A placebo-controlled study. *Journal of Clinical Psychiatry*, *74*, 587–594.

Krause, M., Gutsmiedl, K., Bighelli, I., Schneider-Thoma, J., Chaimani, A., & Leucht, S. (2019). Efficacy and tolerability of pharmacological and non-pharmacological interventions in older patients with major depressive disorder: A systematic review, pairwise and network meta-analysis. *European Neuropsychopharmacology*, *29*, 1003–1022.

Lužný, J. (2012). Agomelatine in elderly—Finally a patient friendly antidepressant in psychogeriatry? *Actas Españolas de Psiquiatría*, *40*, 304–307.

Martinet, L., Guardiola-Lemaitre, B., & Mocaer, E. (1996). Entrainment of circadian rhythms by S-20098, a melatonin agonist, is dose and plasma concentration dependent. *Pharmacology, Biochemistry, and Behavior*, *54*, 713–718.

Millan, M. J., Gobert, A., Lejeune, F., Dekeyne, A., Newman-Tancredi, A., Pasteau, V., et al. (2003). The novel melatonin agonist agomelatine (S20098) is an antagonist at 5-hydroxytryptamine2C receptors, blockade of which enhances the activity of frontocortical dopaminergic and adrenergic pathways. *The Journal of Pharmacology and Experimental Therapeutics*, *306*, 954–964.

Montejo, A., Majadas, S., Rizvi, S. J., & Kennedy, S. H. (2011). The effects of agomelatine on sexual function in depressed patients and healthy volunteers. *Human Psychopharmacology*, *26*, 537–542.

Montejo, A. L., Prieto, N., Terleira, A., Matias, J., Alonso, S., Paniagua, G., et al. (2010). Better sexual acceptability of agomelatine (25 and 50 mg) compared with paroxetine (20 mg) in healthy male volunteers. An 8-week, placebo-controlled study using the PRSEXDQ-SALSEX scale. *Journal of Psychopharmacology*, *24*, 111–120.

Montgomery, S. A., Kennedy, S. H., Burrows, G. D., Lejoyeux, M., & Hindmarch, I. (2004). Absence of discontinuation symptoms with agomelatine and occurrence of discontinuation symptoms with paroxetine: A randomized, double-blind, placebo-controlled discontinuation study. *International Clinical Psychopharmacology*, *19*, 271–280.

Moussavi, S., Chatterji, S., Verdes, E., Tandon, A., Patel, V., & Ustun, B. (2007). Depression, chronic diseases, and decrements in health: Results from the world health surveys. *Lancet*, *370*, 851–858.

Norman, T. R., Cranston, I., Irons, J. A., Gabriel, C., Dekeyne, A., Millan, M. J., et al. (2012). Agomelatine suppresses locomotor hyperactivity in olfactory bulbectomised rats: A comparison to melatonin and to the 5-HT2C antagonist, S32006. *European Journal of Pharmacology*, *674*, 27–32.

Norman, T. R., & Olver, J. S. (2019). Agomelatine for depression: Expanding the horizons? *Expert Opinion on Pharmacotherapy*, *20*, 647–656.

Perlemuter, G., Cacoub, P., Valla, D., Guyader, D., Saba, B., Batailler, C., et al. (2016). Characterisation of agomelatine-induced increase in liver enzymes: Frequency and risk factors determined from a pooled analysis of 7605 treated patients. *CNS Drugs*, *30*, 877–888.

Popoli, M. (2009). Agomelatine: Innovative pharmacological approach in depression. *CNS Drugs*, *23*(Suppl. 2), 27–34.

Product Information. (2010). https://www.ebs.tga.gov.au/ebs/picmi/picmirepository.nsf/pdf?OpenAgent&id=CP-2010-PI-07273-3.

Quera-Salva, M. A., Lemoine, P., & Guilleminault, C. (2010). Impact of the novel antidepressant agomelatine on disturbed sleep-wake cycles in depressed patients. *Human Psychopharmacology*, *25*, 222–229.

Racagni, G., Riva, M. A., Molteni, R., Musazzi, L., Calabrese, F., Popoli, M., et al. (2011). Mode of action of agomelatine: Synergy between melatonergic and 5-HT2C receptors. *World Journal of Biological Psychiatry*, *12*, 574–587.

Rauber-Lüthy, C., Prasa, D., Heistermann, E., Seidel, C., Stedtler, U., Gross, S., et al. (2014). Favorable acute toxicity profile of agomelatine. *Clinical Toxicology*, *52*, 767–768.

Reiter, R. J. (1991). Pineal melatonin: Cell biology of its synthesis and of its physiological interactions. *Endocrine Reviews*, *12*, 151–180.

Schmelting, B., Corbach-Sohle, S., Kohlhause, S., Schlumbohm, C., Flugge, G., & Fuchs, E. (2014). Agomelatine in the tree shrew model of depression: Effects on stress-induced nocturnal hyperthermia and hormonal status. *European Neuropsychopharmacology*, *24*, 437–447.

Taylor, D., Sparshatt, A., Varma, S., & Olofinjana, O. (2014). Antidepressant efficacy of agomelatine: Meta-analysis of published and unpublished studies. *British Medical Journal*, *348*, g1888.

Voican, C. S., Corruble, E., Naveau, S., & Perlemuter, G. (2014). Antidepressant-induced liver injury: A review for clinicians. *American Journal of Psychiatry*, *171*, 404–415.

Wong, A., Lee, C., & Lee, J. (2018). Agomelatine overdose and related toxicity. *Toxicology Communications*, *2*, 62–65.

Yatham, L. N., Vieta, E., Goodwin, G. M., Bourin, M., de Bodinat, C., Laredo, J., et al. (2016). Agomelatine or placebo asadjunctive therapy to a mood stabiliser in bipolar I depression: Randomised double-blind placebo-controlled trial. *British Journal of Psychiatry*, *208*, 78–86.

Chapter 31

# Bumetanide and use in depressive states

M. Tessier[a], A. Rezzag[a], C. Pellegrino[a], and C. Rivera[a,b]
[a]Inmed, INSERM, Aix-Marseille University, Marseille, France, [b]Neuroscience Center, University of Helsinki, Helsinki, Finland

## List of abbreviations

| | |
|---|---|
| APTT | partial thromboplastin time |
| BDNF | brain-derived neurotrophic factor |
| $Cl^-$ | chloride ions |
| ECT | electroconvulsive therapy |
| GABA | gamma amino butyric acid |
| GABAA | ionotropic receptor of GABA |
| GABAB | metabotropic receptor of GABA |
| GAD67 | glutamate decarboxylase 67 |
| HPA axis | hypothalamic-pituitary-adrenal axis |
| LTD | long-term depression |
| LTP | long-term potentiation |
| MAO | monoamine oxidase inhibitors |
| SSRIs | selective serotonin reuptake inhibitors |

## Summary points

- GABAergic neurotransmission controls adult hippocampal neurogenesis.
- GABAergic neurotransmission is involved in depression.
- KCC2 and NKCC1 are the two main transporters responsible for chloride homeostasis regulation, which determine the effect of GABAergic transmission.
- Changes in the expression and the functionality of these transporters after brain insults modify brain physiology.
- TBI induces alteration of adult hippocampal neurogenesis and changes in GABAergic transmission that are involved in the etiology of posttraumatic depression.
- Depression is a common feature of many neurological disorders such as schizophrenia, autism, and epilepsy.
- Hippocampal formation is one of the main structure involved in depression.
- Bumetanide is a powerful diuretic also known to be an antagonist of the main chloride ion importer NKCC1.
- Bumetanide have been shown to be effective in many disorders such as autism and traumatic brain injury.
- Bumetanide also has a variety of positive effects on different brain trauma by acting on secondary neurogenesis and cell death.
- Bumetanide is effective in treating depressive-like behaviors and anxiety.

## Key facts

- Bumetanide is a potent molecule to prevent depression
- Chloride homeostasis needs to be maintained to prevent depression
- Hippocampal structure coherence is important to fight against depression
- Neurogenesis is a key point in the setting up of depression
- Neuronal death is involved in depression

## Mini-dictionary of terms

**Depression** Pathological state marked by sadness with moral pain, loss of self-esteem, psychomotor retardation.

**Mental disorder** Any clinically significant behavioral or psychological syndrome characterized by distressing symptoms, significant impairment of functioning, or significantly increased risk of death, pain, or other disability. Mental disorders are assumed to result from some behavioral, psychological, or biological dysfunction in the individual. The concept does not include deviant behavior, disturbances that are essentially conflicts between the individual and society.

**Homeostasis** Regulatory process by which the body maintains the various constants of the body within the limits of normal values.

## Chloride homeostasis and depression

### Hippocampal plasticity

The brain possesses remarkable plasticity; it is capable of rapidly creating and eliminating synapses as well as modify functional circuits, particularly during learning processes. One of the most studied molecular factors necessary for this neuroplasticity is the brain-derived neurotrophic factor (BDNF). BDNF is a neurotrophin that promotes the survival of existing neurons and encourages the growth and differentiation of new neurons and synapses (Acheson et al., 1995). Serum BDNF levels have been shown to be reduced in patients diagnosed with major depressive disorder (Monteleone, Serritella, Martiadis, & Maj, 2008), indicating a possible role for BDNF in the pathophysiology of depression.

The Neuroscience of Depression. https://doi.org/10.1016/B978-0-12-817933-8.00053-0

The hippocampus is the brain region most often studied in depression research. Structurally, the hippocampus is part of the limbic system and is connected to emotionally related brain regions such as the prefrontal cortex and the amygdala. The synaptic plasticity of the hippocampus, as modeled by long-term potentiation (LTP) and long-term depression (LTD), is an important mechanism for memory (Malenka & Bear, 2004). Later chronic stress has been shown to impair hippocampal-dependent explicit memory in animal models of depression (Anacker & Hen, 2017; Pittenger & Duman, 2008).

Indeed, significant stress can alter the LTP and reinforce the LTD in the rodent hippocampus (Xu, Anwyl, & Rowan, 1997). It has also been shown that a negative stimuli can also decrease the branching and plasticity of dendrites in the hippocampus (Son et al., 2012) as well as sensory deprivation (Popova & Naumenko, 2019), changes in neurohormones (Xu, Zheng, An, Wang, & Zhang, 2016) and changes in hippocampal volume as one of the final consequences of depression (Darren et al., 2018). Abnormalities of the hypothalamic-pituitary-adrenal (HPA) axis exist in patients with major depressive disorders, which may have an impact on glucocorticoid release. In particular, several components of the HPA axis have been implicated in the development of major depressive disorders, including the hippocampus, which contains elevated levels of glucocorticoid and glutamate receptors, making it more susceptible to stress and depression (Pittenger & Duman, 2008; Ulrich-Lai & Herman, 2009). In addition to that GABAergic neurotransmission has also been involved, the interaction between major depressive disorders (MDDs) and GABAergic neurotransmission has been suggested in a genetic mice model of GABA(B)-R *knock-out* (Mombereau et al., 2005) and in studies showing an antidepressant effect of potent and selective blockage of GABA(A) transmission (Rudolph & Knoflach, 2011) at both the hippocampus (Boldrini et al., 2013) and mesolimbic system (Kandratavicius, Hallak, Carlotti, Assirati, & Leite, 2014).

Finally, one of the most widespread conclusions is that the volume of the hippocampus decreases in patients with major depressive disorders, and this has been confirmed by several magnetic resonance imaging (MRI) studies (McKinnon, Yucel, Nazarov, & MacQueen, 2009; Videbech & Ravnkilde, 2004). Several mechanisms have been suggested as possible causes of this reduction in hippocampal volume such as synaptic plasticity, deficit in secondary neurogenesis, and neuronal death (Czéh & Lucassen, 2007; Sapolsky, 2000). These last two elements are detailed in the following paragraphs.

## Hippocampal neurogenesis

Neurogenesis in adult individuals involves the generation of new neurons and neuronal connections in the dentate gyrus of the hippocampus and the subventricular area of the lateral ventricles (Curtis, Low, & Faull, 2012). Neurogenesis of the hippocampus occurs only in the dentate gyrus, with, nearly 9000 new cells being born every day in rodents. The rate of neurogenesis in the hippocampus decreases slightly with age. Compared to the millions of cells in the granular layer of the hippocampus, the newly formed neurons are few in number but maybe sufficient to achieve functional significance (Spalding et al., 2013).

In humans, the subject is somewhat more controversial. Indeed, despite the observation of new neurons in the human adult brain as early as 1988, two articles revived the controversy a few years ago. One, published in Cell by Boldrini et al. (2018) in April 2018, showed that there is neurogenesis throughout life even in adulthood in humans, even in elderly subjects. The second, published in Nature in March 2018 by Sorrells et al. (2018) showed, on the contrary, an absence of neurogenesis in adults, with a halt in the production of new cells in the brain from the end of adolescence. A third article put an end to the controversy 1 year later, with rigorous methodological conditions, the authors showed without ambiguity the presence of adult neurogenesis in humans, even in very old subjects (Moreno-Jiménez et al., 2019). In the dentate gyrus of the hippocampus (GD), newly formed neurons are produced in the subgranular region (Duan, Kang, Liu, Ming, & Song, 2008; Encinas & Sierra, 2012). This area is located at the edge of the granular cell layer and the hilus. They then migrate into the granular layer where they will undergo a process of maturation and integration into the existing neural network. They represent about 4% of the total granular cell population (Cameron & McKay, 2001) and these cells are more easily excitable and respond to external stimuli (Ge et al., 2006).

Although the rates of neuronal regeneration are comparable in middle-aged humans and mice, the neurogenesis patterns of the adult hippocampus are significantly different. In humans, about one-third of the neurons in the hippocampus are renewed. In contrast, this proportion is 10% in mice (Imayoshi et al., 2008). In rodents, proliferative cells migrate from a subgranular area to the granular layer, whereas in primates, the subgranular area is less easily defined and proliferative cells appear in the polymorphic and granular layers and in the hilum (Boldrini et al., 2013; Lavenex, Lavenex, Bennett, & Amaral, 2009).

Adult hippocampal neurogenesis is a sequential process in which proliferating cells exit the cell cycle and undergo different stages until they become fully mature neurons. This maturation sequence is said to be similar to that observed during embryonic development (Espósito et al., 2005; Laplagne et al., 2007) and is divided into four different phases organized temporally, from (i) maintenance of the stem cell pool; (ii) proliferation of progenitors; (iii) migration of immature newly formed cells, and finally

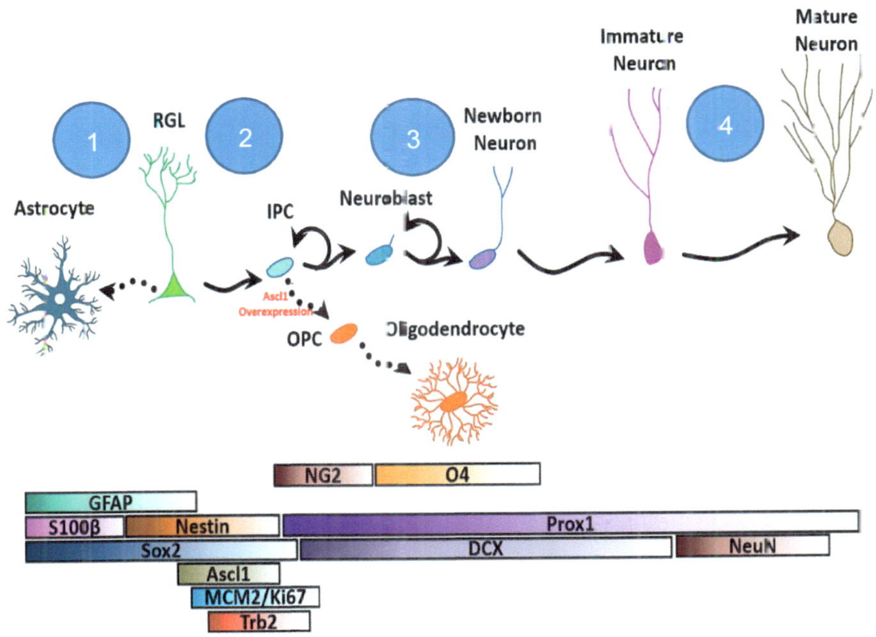


FIG. 1 The four phases of neurogenesis in the dentate gyrus of the hippocampus: (1) maintenance of the progenitor pool, (2) proliferation of progenitors, (3) survival and migration, and (4) maturation and differentiation. Different markers used in immunohistochemistry. *(Adapted from Duan, X., Kang, E., Liu, C. Y., Ming, G.-L., & Song, H. (2008). Development of neural stem cell in the adult brain.* Current Opinion in Neurobiology, *18(1), 108–115. 10.1016/j.conb.2008.04.001; Encinas, J. M., & Sierra, A. (2012). Neural stem cell deforestation as the main force driving the age-related decline in adult hippocampal neurogenesis.* Behavioural Brain Research, *227(2), 433–439. 10.1016/j.bbr.2011.10.010.)*

(iv) Maturation and functional integration of newly formed neurons (Fig. 1). The hypothesis that altered neurogenesis in the adult hippocampus leads to depression is increasingly being studied (Goubert et al., 2019; Jacobs, van Praag, & Gage, 2000; Petrik, Lagace, & Eisch, 2012; Schinder & Gage, 2004). It has even been shown that alterations in adult neurogenesis in the hippocampus and depression can be reciprocally causal (Miller & Hen, 2015).

In humans, the total number of granular cells in the dentate gyrus and the size of the latter in depressed patients on drug therapy are higher than in untreated patients based on postmortem studies (Boldrini et al., 2013). This is one of the more valid hypotheses to explain the efficacy of antidepressant treatments known to modulate the production of new neurons in GD. Indeed, many antidepressant treatments (e.g., SSRIs, MAO inhibitors, APTTs, ECTs, and mood stabilizers) have been shown to facilitate neurogenesis and, as such, improve the treatment of depression (Anacker et al., 2011). Gulbins and his collaborators demonstrated one of the mechanisms linking antidepressants and neurogenesis. They showed that antidepressants such as imipramine significantly decreased levels of ceramides (a sphingolipid that blocks the maturation of brain cells), which increased neurogenesis (Gulbins et al., 2015).

It should be noted, however, that research findings on the neurogenesis and pathophysiology of depression are contradictory. For example, a defect in neurogenesis in some animal models does not always produce depressive-like symptoms and a number of effects of antidepressants are independent of neurogenesis. Yet most researchers in this field agree that neurogenesis remains an important factor in understanding the pathophysiology and treatment of depression.

## Hippocampal apoptosis during depression

Proliferation, differentiation, and apoptosis are the principal steps of the life of neurons in the hippocampus. However, numerous studies have shown that depression and stress may induce greater apoptosis of hippocampal neurons in rodents and mammals (Lucassen et al., 2006). Studies have shown an increase in apoptosis in the dentate gyrus of female rats that have undergone repeated separation of their offspring and altered memory capacity with behavioral changes similar to depression (Sung et al., 2010). In these maternal separation rat models, tadalafil, a phosphodiesterase type 5 inhibitor, exerts antidepressant effects by suppressing maternal separation-induced apoptosis and increasing cell proliferation in the dentate gyrus. However, there are differences in the effects of chronic and acute depression on apoptosis. In animal models and human studies, chronic depression appears to induce greater apoptosis and much more spread over time than acute depression. In addition to tadalafil (mentioned above), several types of drugs may have antidepressant effects by acting on apoptosis in the hippocampus. For example, venlafaxine, a serotonin/norepinephrine dual reuptake inhibitor, suppresses apoptosis of the hippocampus by regulating BDNF (Huang et al., 2014).

## GABAergic neurotransmission impairment

There is abundant evidence that major depressive disorder is associated with various deficits in GABAergic transmission. Indeed, studies in depressed patients show reduced levels of GABA in the brain (Sanacora et al., 2004, 1999), plasma (Petty & Schlesser, 1981; Petty & Sherman, 1984), and cerebrospinal fluid (Hare et al., 1982). These results are supported by studies showing a reduction in the protein and mRNA coding for glutamic acid decarboxylase GAD67, a key enzyme for the synthesis of GABA, in the prefrontal cortex and tonsil of depressed patients (Guilloux et al., 2012; Karolewicz et al., 2010). These reduced levels of GAD67 and GABA are consistent with studies showing selective loss of GABAergic interneurons in the cortex of depressed patients (Rajkowska, O'Dwyer, Teleki, Stockmeier, & Miguel-Hidalgo, 2007). Reduced expression of GABAA receptors has also been observed in depressed patients (Klumpers et al., 2010) as well as changes in their subunits. By comparing the prefrontal cortex of depressed suicide victims with that of nondepressed persons who died from other causes, Merali et al. found a reduction in mRNA expression in the subunits α1, 3, 4, and δ (Merali et al., 2004). In animals, genetically modified mice rendered heterozygous for the subunit γ2 or deficient for the subunit α2 present the behavioral, cognitive, and cellular alterations expected in an animal model of depression (Duveau et al., 2011; Shen et al., 2010). Finally, studies show the antidepressant effect of GABAA receptor agonists (Boldrini et al., 2013; Rudolph & Knoflach, 2011), and GABAA receptor antagonists that allow modulation of anxiety and depressive behavior (Mombereau et al., 2005).

It is interesting to note that these genetic animal models of depression show a deficit in the survival and maturation of newly formed neurons in the hippocampus, in other words, have deficits in secondary neurogenesis (Earnheart et al., 2007) and also synaptogenesis (Ren et al., 2015). Indeed, secondary neurogenesis is controlled by local GABAergic interneurons and GABAA receptors. In particular, the release of GABA by parvalbumin interneurons controls the quiescence output of RGL (radial glia-like cell) in the subgranular cell layer of the dentate gyrus (Song et al., 2012). This mechanism is controlled by the α4βγ2 subunits of GABAA receptors and early progenitors (Duveau et al., 2011; Song et al., 2012). The maturation of dendrites and spines of progenitors is also dependent on the α2βγ2 receptors but is independent of the receptors containing the δ subunit, which themselves appear to act as mediators of GABAAergic inhibition of mature granular cells (Duveau et al., 2011; Ren et al., 2015). Therefore, it is readily conceivable that alterations in GABAergic transmission induced by excessive stress and excessive glutamate release contribute to long-lasting adverse effects on hippocampal function such as the development of depressive-type disorders.

## Depression and chloride homeostasis hypothesis

Chloride ions play an important role in controlling the excitability of the main neurons of the central nervous system. At inhibitory synapses, GABA activates GABAA receptors that are permeable to chloride ions. The direction of the resulting chloride flow depends on the intracellular chloride concentration ($[Cl^-]i$) of the postsynaptic neuron. Regulation of this intraneuronal chloride concentration by cationic chloride cotransporters is therefore essential for neuronal activity. The GABA neurotransmission is the main inhibitory system of the central nervous system. GABA is synthetized and released by the interneurons at the synaptic cleft where then it can modulate inhibition by acting either on ionotropic or metabotropic receptors. The strength of inhibition is tuned by the fine control of intracellular chloride concentration (Kaila, Ruusuvuori, Seja, Voipio, & Puskarjov, 2013; Medina et al., 2014).

There is a large correlation between chloride-dependent changes in GABAergic transmission and the events highly associated with the development of depression such as pathologic insults including traumatic brain injury. First of all, it is important to understand from which pathway the chloride may enter the cell. There are three possible mechanisms: (i) passive diffusion acting in the direction of the chloride gradient, (ii) primary active ATP-dependent chloride pumps acting against the gradient, and finally (iii) secondary active chloride transport that is also an ATP-dependent mechanism acting against the gradient but able to transport two ions in the same or a different direction.

This last mechanism of chloride ions entry can be electroneutral or not and thus able to modify the membrane potential (Duran, Thompson, Xiao, & Hartzell, 2010). Also, there are channels permeable to chloride. Apart from the transporters, there are ionotropic and metabotropic GABA channels activated by GABA or glycine depending on the site (Avila, Nguyen, & Rigo, 2013; Demarque et al., 2002). In addition, there are volume-dependent channels known to be located on astrocytes and responsible for cell death during ischemic episodes in the hippocampus (Zhang, Cao, Kimelberg, & Zhou, 2011). Another type of chloride channel are the calcium-gated chloride channels, those channels regulate action potential duration and are important at developmental stages (Huang et al., 2012). Some chloride channels may also present cotransporters properties such as ClC-0 ( Jentsch, 2008) that can be both voltage-gated channel or proton exchangers, acting both in pH and cell volume regulation (Medina et al., 2014; Zifarelli & Pusch, 2007). Finally the most popular of chloride transporters is the CCC family for cation-chloride cotransporters. In mammals, there are the main regulatory proteins for chloride homeostasis (Blaesse, Airaksinen, Rivera, & Kaila, 2009). Each member is defined by its expression profile during development and its subcellular localization (Kaila, 1994). Under normal conditions, [Cl-]i is low, so activation of the GABAA receptor results in increased membrane conductance and an influx of chloride ions into the neuron, causing hyperpolarization. This decreases the probability that the postsynaptic neuron will trigger an action potential that is the inhibitory effect of GABAA signaling. In contrast to mature neurons, [Cl-]i is kept elevated in developing neurons, resulting in an electrochemical gradient of chloride ions in the opposite direction. In immature neurons, activation of GABAA receptors causes an influx of chloride ions, resulting in depolarization (Ben-Ari, Cherubini, Corradetti, & Gaiarsa, 1989; Sulis Sato et al., 2017). The depolarizing effect of GABAergic signaling in immature neurons promotes trophic growth and is essential for the early establishment of neuronal circuits (Ben-Ari, Gaiarsa, Tyzio, & Khazipov, 2007; Wang & Kriegstein, 2011). However, under certain pathophysiological conditions, the control of [Cl-]i is disrupted and therefore affects the inhibition mediated by GABAergic transmission (Goubert et al., 2019; Pallud et al., 2014). There is growing evidence that altered homeostasis of chloride ions is implicated in a wide variety of neurological and psychiatric conditions (Bragin, Sanderson, Peterson, Connor, & Müller, 2009; Kourdougli et al., 2017; Medina et al., 2014).

The main chloride exporter in neurons is the $K^+$/Cl type 2 cotransporter (KCC2) encoded by the *slc12a5* gene, which force chloride ions to exit the neuron against its concentration gradient (Rivera et al., 1999) creating the driving force leading to an inward flux of Cl-ions through GABAA receptor channels producing postsynaptic hyperpolarization upon GABA release (Blaesse et al., 2009; Chamma et al., 2013; Payne, Rivera, Voipio, & Kaila, 2003). On the other hand, the $Na^+/K^+/Cl^-$ type 1 cotransporter (NKCC1) encoded by the *slc12a2* gene located on chromosome 5 (Delpire, Rauchman, Beier, Hebert, & Gullans, 1994; Yamada et al., 2004) is considered the main "importer" of chloride ions. It transports chloride, potassium, and sodium ions in the neuron, using the electrochemical gradient of $Na^+$. Together, KCC2, and NKCC1 are the two main transporters responsible for the regulation of [Cl-]i, their activity controls the level of intraneuronal chloride ions, which in turn determines the postsynaptic effect of GABAergic transmission. Altered expression and/or activity of either of these cotransporters has been associated with a wide variety of brain disorders. The maintenance of intra- and extracellular $Cl^-$ concentrations by the specific transporters, NKCC1 and KCC2, has also been shown to play a major role in the proper functioning of brain inhibition and the prevention of, e.g., trauma-induced neuronal death (Kourdougli, Varpula, Chazal, & Rivera, 2015; Pellegrino et al., 2011; Rivera et al., 1999; Shulga & Rivera, 2013; Shulga et al., 2008) and in preventing the development of depressive-type behavior (Goubert et al., 2019; Xu et al., 1997).

## Perspective in the use of bumetanide as a therapeutic agent

Interestingly, bumetanide, a NKCC1 transporter antagonist, has been shown to alleviate many disorders such as autism spectrum disorders, Parkinson's disease, and schizophrenia. This underscores the therapeutic potential of restoring low levels of (Cl)i and effective GABAergic inhibition (Ben-Ari, 2017; Damier, Hammond, & Ben-Ari, 2016; Lemonnier et al., 2012). It has also been shown that bumetanide may have a variety of positive effects on brain insults such as TBI, stroke or epilepsy (Goubert et al., 2019; Hui et al., 2016; Zhang, Liu, & Xu, 2008) and may also act on secondary neurogenesis in stroke and brain trauma (Hu et al., 2017). Finally, bumetanide has been shown to be effective in treating depressive-type behaviors and induced anxiety in maternal separation models (Hu et al., 2017; Sung et al., 2010) as well as in brain trauma models where depression is one of the most prevalent long-term sequelae (Goubert et al., 2019). Normalizing pathological changes in chloride homeostasis maybe a promising therapeutical avenue to treat term consequences of brain dysfunction such as depression.

### Traumatic brain injury (TBI)-induced depression

TBI affects millions of people worldwide each year. People who undergo TBI may have significant long-term sequelae (Bondi et al., 2015; Singh, Mason, Lecky, & Dawson, 2018), including a decrease in cognitive performance with, in particular the appearance of a depressive or depressive-like behaviors (Perry et al., 2015). Alteration in adult hippocampal neurogenesis (Alvarez et al., 2016), as well as changes in GABAergic transmission (Goubert et al., 2019), maybe involved in the etiology of these comorbidities.

Binding of GABA to GABAA receptors triggers depolarization of damaged neurons due to brain trauma, unlike normal mature neurons where GABA has a hyperpolarizing effect (Medina et al., 2014). Many studies showed that this change in the functioning of the GABA profile results from changes in the expression and functionality of chloride ion cotransporters NKCC1 and KCC2, as detailed earlier (Kourdougli et al., 2017; Rivera et al., 1999) So the depolarization caused by activation of the GABAA receptor is directly linked to the reduction of the extrusion of $Cl^-$ by KCC2 and the maintenance or accumulation of $Cl^-$ by NKCC1(Kaila, 1994). Blocking NKCC1 restores hyperpolarization following this abnormal activation of NKCC1 and blocks the dependence on trophic factors of damaged neurons (Shulga et al., 2008; Shulga et al., 2012). Remarkably blocking this depolarization with the specific NKCC1 antagonist, bumetanide, at early phase after TBI leads to changes in newborn neurons production (Ibrahim et al., 2016) and reduces inflammation (Anderson, Ao, & Sofroniew, 2014; Burda, Bernstein, & Sofroniew, 2015) in the dentate gyrus, a hippocampal structure often associated with depressive-like behavior. It has been shown that modification in KCC2 expression is linked to the survival mechanism of the injured principal neuron and could change the excitability of the network (Hsieh, Gruber, Jenkins, & Ranganath, 2014; Pellegrino et al., 2011). In several brain pathologies as TBI, principal neurons are not the only one to be impacted. It has been proposed as well that parvalbumin-positive interneurons are a very sensitive cell population to hyperexcitability that could trigger them to death (Bao et al., 2017; Hsieh et al., 2016; Huusko, Römer, Ndode-Ekane, Lukasiuk, & Pitkänen, 2015). Some studies showed that bumetanide application could avoid trauma-induced interneurons death at least in the DG by blocking posttraumatic GABAA-induced depolarization caused by KCC2 downregulation (Goubert et al., 2019; Shulga & Rivera, 2013). In their study, Goubert et al. showed the prophylactic action of bumetanide on cognitive effects and established that the mechanism was linked to neurogenesis of the DG.

## What is bumetanide?

Bumetanide or3-(butylamino)-4-phenoxy-5-sulfamoyl-benzoic acid, with the formula $C_{17}H_{20}N_2O_5S$ and a molecular weight of about 364.4 g/mol is a potent sulfamyl derivatives, member of the benzoic acid family (Fig. 2) and has powerful diuretic action on Henle's loop of the kidney. Bumetanide antagonizes sodium-potassium-chloride transport, has an estimated half-life of 60–90 min and 80% is absorbed by the body. The excretion rate through the urine is around 80% and only 2% by the biliary. In case of overdose, it might cause an intense loss of water, accompanied by loss of volumes, dehydration, decreased in blood volume, and at worst a circulatory collapse followed by thrombosis and vascular embolism. Electrolyte depletion may manifest itself as weakness, dizziness, mental confusion, anorexia, lethargy, vomiting, and cramps. Treatment consists of replacing fluid and electrolyte loss with careful monitoring of urine and electrolyte output and serum electrolyte levels.

## Way of action

Bumetanide is known to interact with several proteins (Table 1), it is also mainly used for the treatment of edema associated with congestive heart failure, liver, and kidney diseases (Kharod et al., 2019). Bumetanide specifically inhibits Na-K-Cl transport. In the kidneys this action blocks the active reabsorption of chloride ions and possibly sodium in the Henle ascending loop, thereby altering the transfer of electrolytes into the proximal tubule. This results in excretion of sodium, chloride ions, and water and, consequently, diuresis. It is often used in patients in whom high doses of furosemide are ineffective.

**TABLE 1** Known interactors of bumetanide molecule.

| Genes | Proteins | Type of action |
|---|---|---|
| *slc12a1* | NKCC2 | Antagonist |
| *slc12a2* | NKCC1 | Antagonist |
| *slc12a5* | KCC2 | Antagonist if concentration > 100 μM |
| *GPR35* | G35 coupled receptors | Agonist |

In the brain, bumetanide can prevent epileptic seizures in newborns or in pathological situations in adults by blocking the chloride cotransporter NKCC1 (Blauwblomme et al., 2018; Goubert et al., 2019; Lemonnier & Ben-Ari, 2010). This inhibits the absorption of chloride ions into the neurons, thereby decreasing the internal concentration of chloride ions, which blocks the excitatory effect of GABA.

## Analogs

- Furosemide (Fig. 3)

Furosemide or 4-chloro-2-(furan-2-ylmethylamino)-5-sulfamoylbenzoic acid, with the chemical formula $C_{12}H_{11}ClN_2O_5S$ (molecular weight: 330.74 g/mol) is a derivative of sulfamyl acid, a member of chlorobenzoic acids, it has a primary role as loop diuretic. It is a potent inhibitor of the sodium-potassium-chloride NKCC cotransporters. Furosemide has an estimated half-life of 120 min, is absorbed by about 60% and excreted more than 80% in the urine. Furosemide is indicated for the treatment of edema associated with congestive heart failure, cirrhosis of the liver and renal disease, including nephrotic syndrome, in adults and children. Oral furosemide is indicated alone for the treatment of mild to moderate hypertension or severe hypertension in combination with other antihypertensive drugs. Intravenous furosemide is indicated as adjunctive therapy in acute pulmonary edema when rapid diuresis is desired. Although in vivo and in vitro studies have demonstrated an anticonvulsant effect of furosemide, a loop diuretic, the precise mechanism of this effect is still debated.

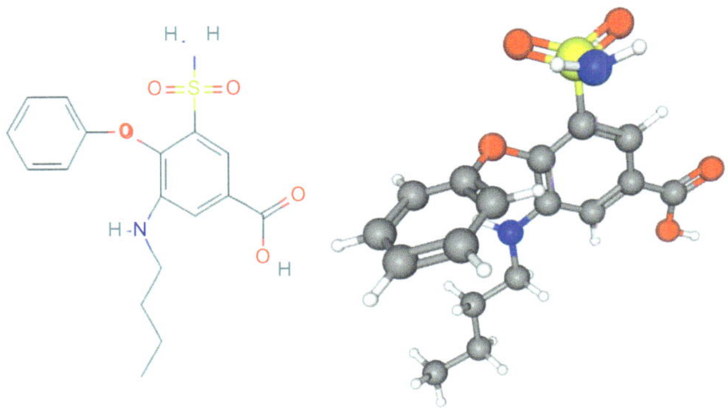

**FIG. 2** 2D and 3D chemical structure of bumetanide. (*Source: https://pubchem.ncbi.nlm.nih.gov/compound/2471.*)

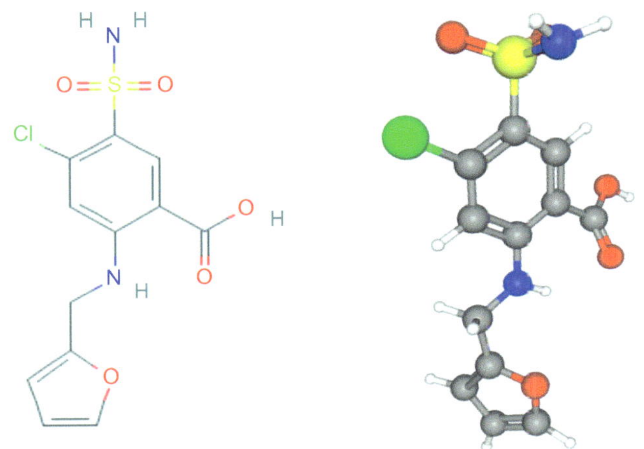

**FIG. 3** 2D and 3D chemical structure of furosemide. (*Source: https://pubchem.ncbi.nlm.nih.gov/compound/3440.*)

- Azosemide (Fig. 4)

Azosemide or 2-chloro-5-(2H-tetrazol-5-yl)-4-(thiophen-2 ylmethylamino)-benzene sulfonamide with the chemical formula $C_{12}H_{11}ClN_6O_2S_2$ is a monosulfamyl belonging to the class of loop diuretics. This compound with a molecular weight of 370.8 g/mol inhibits the reabsorption of sodium and chloride in the thick ascending branch of Henle's loop. The oral bioavailability of this drug is estimated to be 20.4%, has an estimated half-life of 2–3 h and is excreted more than 80% in the urine. The exact mechanism of action is unclear. However, it acts primarily on the Henle loop, both in the medullary and cortical segments of the thick ascending limb.

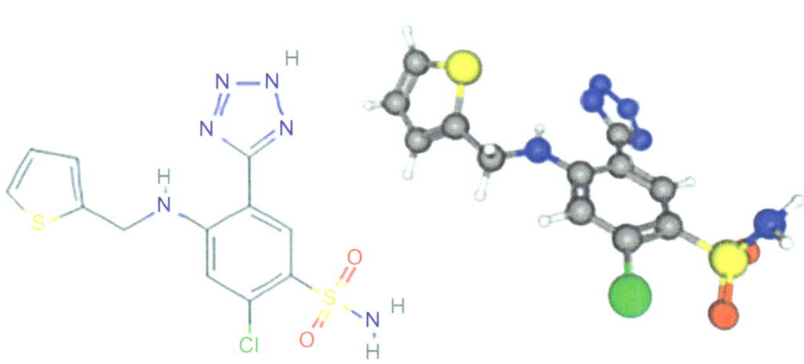

**FIG. 4** 2D and 3D chemical structure of azosemide. (*Source: https://pubchem.ncbi.nlm.nih.gov/compound/2273.*)

- Piretanide (Fig. 5)

Piretanide or 4-phenoxy-3-(pyrrolidin-1-yl)-5-sulfamoylbenzoic acid, of the chemical formula $C_{17}H_{18}N_2O_5S$ is a sulfamoylbenzoic acid belonging to the class of loop diuretics. This compound with a molecular weight of 362.4 g/mol is an antagonist of the sodium-potassium-chloride cotransporters. Piretanide is structurally related to furosemide and bumetanide.

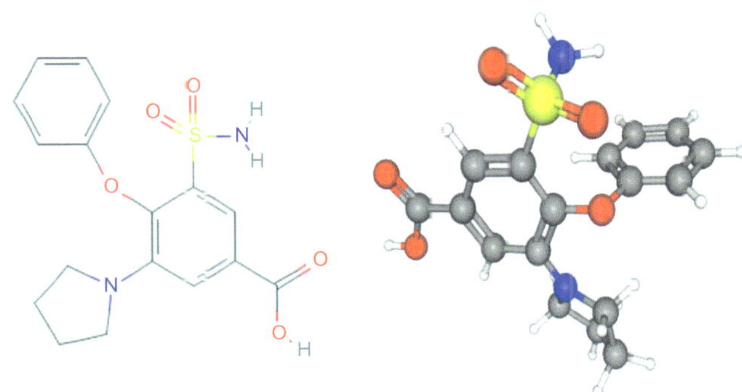

**FIG. 5** 2D and 3D chemical structure of piretanide. (*Source: https://pubchem.ncbi.nlm.nih.gov/compound/4849.*)

- Torsemide (Fig. 6)

Torsemide or torasemide, 1-[4-(3-methylanilino)pyridin-3-yl]sulfonyl-3-propan-2-ylurea, of the chemical formula $C_{16}H_{20}N_4O_3S$ is a sulfonylurea anilinopyridine belonging to the class of loop diuretics. This compound with a molecular weight of 348.4 g/mol is an antagonist of the sodium-potassium-chloride cotransporters and is also an antihypertensive agent. Torsemide is the diuretic with the highest oral bioavailability, even in the advanced stages

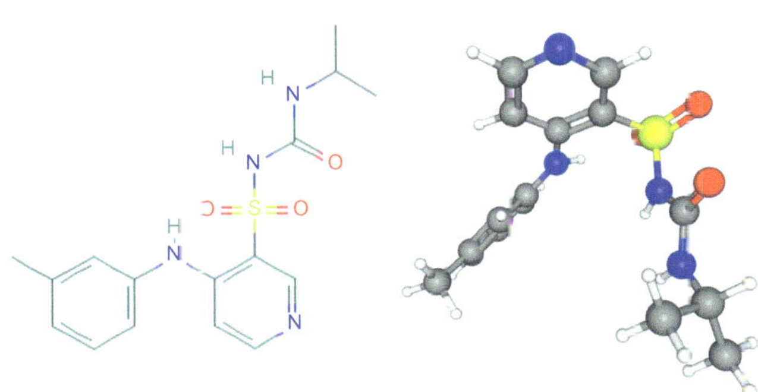

**FIG. 6** 2D and 3D chemical structure of torsemide. (*Source: https://pubchem.ncbi.nlm.nih.gov/compound/41781.*)

of chronic renal failure. This bioavailability tends to be greater than 80%, regardless of the patient's condition. Torsemide is mainly treated by the hepatic route and excreted in the feces, from which around 70%–80% of the administered dose is excreted by this route. On the other hand, approximately 20%–30% of the administered dose is excreted in the urine. It has a prolonged duration of action compared to other loop diuretics with an estimated half-life of 3.5 h.

Torsemide is indicated for the treatment of edema associated with congestive heart failure, kidney, or liver disease. It is also approved for use as an antihypertensive agent, alone or in combination with other antihypertensive agents. In addition, there have been reports that torsemide may reduce myocardial fibrosis by decreasing collagen accumulation.

## Postischemic depression and bumetanide

According to the World Health Organization, stroke remains the leading cause of motor disability over the world (Goldstein et al., 2011). About 14 million people are affected every year. It is the first cause of death for women and the third for men. In France, it is estimated that more than 140,000 individuals have a stroke per year. It represents 1 stroke every 4 min. The majority of stroke survivors preserve more or less severe neurological deficits (60%), including, among others, sensorimotor, cognitive, metabolic, visual, and language disorders. Only one patient on three can return to professional activity. The risk to develop dementia increases twofold following a stroke. Hence, it is not surprising that stroke represents more than half of the hospitalizations for neurological reasons, resulting in a very high economic cost per year (≈27 billiard in Europe). Interestingly, ischemia survivors will ultimately develop depression (Kronenberg, Katchanov, & Endres, 2006). Clinical studies exposed evidence of the association between lesion characteristics and appearance of depression (Kronenberg et al., 2012). Many studies confirmed that after a cerebral ischemia NKCC1 expression increases and KCC2 expression reduces together with GABA depolarization, the chloride channel on the cell membrane became wide open allowing a massive influx of chloride ions, thus causing membrane depolarization after GABAA activation. This phenomenon as previously described here mediates neuronal excitability that could lead to neuronal death (Blaesse et al., 2009; Jaenisch, Witte, & Frahm, 2010; Nabekura et al., 2002). As joint phenomenon, membrane depolarization allows a large calcium influx leading to an intracellular calcium overcharge which can be harmful to cell survival (Moubarak et al., 2007; Nabekura et al., 2002; Olmos & Lladó, 2014; Shulga et al., 2012). The plasticity of the nervous system after cerebral ischemic-stroke is crucial to undergo nerve regeneration and dendritic development in the brain but also in the spinal cord (Canty et al., 2013; Damjanac et al., 2007; Drevets, Price, & Furey, 2008). But a large number of newborn neurons do not survive for a long period (Belousov et al., 2012; Kaushal & Schlichter, 2008) and the local microenvironment is not leading to self-repair (Fritschy & Sarter, 2012). Thus, the need to seek new drugs that could exert a potent neuroprotective effect in the brain such as shown by riluzole (Chen & Manji, 2006; Molinaro et al., 2008) is of major importance.

Xu et al. (2016) discussed the effects of using bumetanide on neurogenesis and neuroprotection after brain ischemia, they demonstrated that bumetanide not only promotes neuronal precursor cell regeneration and dendrite development but also mediates the recovery of cognitive function and protects brain tissue. However, they showed that the utilization of bumetanide had no effect on nerve regeneration under normal conditions. Meaning that bumetanide application can set up a microenvironment specifically under pathological conditions that can help the regeneration and survival of nerve and creating a neuroprotective effect in the brain. Yet, the effect of bumetanide on neurogenesis and depressive behavioral improvement after cerebral ischemia needs more investigation.

## Bumetanide in epilepsy

Almost one-third of epilepsy cases suffer from depression and anxiety, making depression the most common neurological disorder to affect people with epilepsy (Huberfeld et al., 2007). Brain tissue resected from patients with epilepsy showed alteration in KCC2 and NKCC1 expression in neurons (Huberfeld et al., 2011; Pallud et al., 2014). Considering the upregulated expression of NKCC1 relative to KCC2 seems to be responsible of the depolarizing effect of GABA in the nervous system (Alvarado-Rojas et al., 2014; Fukuda & Watanabe, 2018), it is obvious to conclude that blocking NKCC1 could effectively and in a potent manner inhibits GABA depolarizing effect. The application of bumetanide has produced conflicting results in epilepsy. Neonatal seizures have been the most studied in this regard. Short-term prevention of epileptiform activity has been established in vitro and in neonatal rats (Dzhala et al., 2005; Nardou, Ben-Ari, & Khalilov, 2009). Bumetanide has also been shown to reduce the development of recurrent glutamatergic mossy fiber sprouting in the dentate gyrus of pilocarpine-treated rats (Kourdougli et al., 2017), which is implicated in the generation of seizures in both patients with temporal lobe epilepsy and in animal models (Epsztein, Represa, Jorquera, Ben-Ari, & Crépel, 2005; Golarai, Greenwood, Feeney, & Connor, 2001). Dzhala and colleagues in 2005 showed the capacity of bumetanide to shift the chloride gradient and abolish the epileptiform activity in rats hippocampus (Dzhala et al., 2005). In other works, they demonstrated that bumetanide application can boost the effectiveness of phenobarbital in neonatal seizures model (Dzhala, Brumback, & Staley, 2008). Using Phenobarbital alone can suppress not more than 30% of seizures, but when bumetanide is added, recurrent seizures are abolished around 70%. In the remaining cases, the coadministration of bumetanide and phenobarbital was able to diminish the frequency, duration, and power of seizures (Dzhala et al., 2008). On the other hand, work by the group of Antoine Depaulis has shown that in the unilateral hippocampal kinate infusion model of adult temporal loop epilepsy this mechanism is more complicated (Stamboulian-Platel et al., 2016). Despite typical changes in the expression of chloride transporters activation of GABAA receptors was equally effective in reducing the severity of seizures. The many positive effects of bumetanide in different trauma models suggest that application of bumetanide as a therapy protocols in epileptic subjects may meliorate comorbidities such as depression.

## Bumetanide and autism

Autism is a neurodevelopmental disorder described by impairments in verbal and nonverbal communications and social interactions as well as a limited interest in the surrounding environment combined with repetitive patterns of behavior (Åsberg Johnels et al., 2016). GABA excitatory/inhibitory dysregulation, in selective neuronal circuits, can lead to disturbance in developing brain which may cause some neurodevelopmental disorders such as Autism (Cellot & Cherubini, 2014). Many studies have highlighted the role of GABAergic neurotransmission as a key element in the excitatory/inhibitory shift during development (Ben-Ari et al., 2007) that is caused by large changes in chloride homeostasis (Rivera et al., 1999). In Autism, there is an abnormal development of inhibitory GABA caused by the deficiency of maturation of GABAergic neurons (Feng, Li, Wang, Shan, & Jia, 2020; Tyzio et al., 2014). Remarkably blocking NKCC1 function with bumetanide shifts GABA from excitatory to inhibitory (Nardou et al., 2011) resulting in drastic changes in the pathology as reported by the latest studies on autistic children (Hadjikhani et al., 2015; Lemonnier et al., 2012). The mechanism is believed to restore low intracellular chloride concentration in neurons by restoring chloride homeostasis and GABAA receptor-mediated inhibition (Hadjikhani et al., 2015; Lemonnier & Ben-Ari, 2010). The exact molecular mechanism for the effect of bumetanide in this pathology though is not known.

In the last decade, diverse antipsychiatrics and antidepressants were used to treat related behavioral problem in autism, like depression, anxiety, and obsessive-compulsive behavior. To this day there is no efficient cure for autism but many approaches were used such as risperidone and aripiprazole (Wink, Erickson, & McDougle, 2010), but both of them failed to treat symptoms of autism and even cause negative side effects like sedation, extrapyramidal symptoms, drowsiness, drooling, and body weight gain (Lemonnier et al., 2017). Bumetanide was originally designed to treat chronic disorders like hypertension and nephritic syndrome. The repurposing of bumetanide was successfully used in children with autistic syndromes, their results suggested that bumetanide has improved clinical symptoms with minimal side effects (Lemonnier & Ben-Ari, 2010).

Following this first study, a second was carried on autistic adolescents treated with bumetanide for 10 months (Hadjikhani et al., 2015). This showed an improvement of facial emotion recognition and activation of brain regions as the inferior occipital cortex, and an increase of activity in areas responsible for rewards, motivation, and emotion. In a more recent study involving 88 children with autism, confirmed a remarkable therapeutic benefit with only very low sides effects as hypokalemia clinically manageable with potassium supplementation (Lemonnier et al., 2017).

Altogether, these results have shown that bumetanide, is a safe, available, and inexpensive promising treatment strategy to ameliorate autistic symptoms without serious side effects, making it interesting for more clinical trials to confirm its efficacy in autism disorders. It is to note that a phase 3 clinical trial is ongoing at European scale.

## Parkinson's disease and bumetanide

Parkinson's disease (PD) is a neurological disorder associated mainly with loss of dopaminergic neurons in substantia nigra (Kalia & Lang, 2015). The treatment of PD is almost entirely based on the use of dopamine replacement drugs to correct dopamine deficiency in the brain. However, these treatments failed to slow the neurodegenerative process and cause several negative effects like dyskinesia and psychiatric disorders (Lees, Hardy, & Revesz, 2009). Finding alternative drugs that act on other targets is the new challenge in PD. As the deep brain stimulation has brought an effective approach to treat PD symptoms but is supported by an invasive surgery (Kalia, Sankar, & Lozano, 2013). Some studies have shown that GABAergic medium spiny neurons provoke abnormal Giant GABAergic Currents in animal models of PD (Dehorter et al., 2009). In a genetic model of PD, these giant GABAergic currents could be used as marker of dopaminergic neurons depletion as observed in the pathology (Dehorter et al., 2012). Considering these observations, it appears that the striatal GABAergic system is altered in PD even before the first motor signs of the pathology. Taken together this pinpoints the use of drugs able to target those giant GABAergic currents and thus could provide new perspectives to treat PD. In such context, Damier and colleagues reported results of four PD patients treated with bumetanide for 2 months and showed that bumetanide was tolerated and demonstrated a clear improvement of PD motor symptoms in all patients (Damier et al., 2016). However, bumetanide is not labeled to treat PD patients, and these encouraging observations deserve larger investigations.

## Bumetanide and schizophrenia

Clinical observations of schizophrenia include depressive mood, mood swing, apathy, lack of interest, anhedonia, lack of attention and focus disorders, sleep disorders, social behavior deterioration, compulsive disorders, speech anomalies, and impulsive-aggressive behavior. Experimental studies on animal models of schizophrenia showed a major role of the prefrontal cortex (PFC) and its maturation in the etiology of the pathology (Kesner & Churchwell, 2011). Interestingly the maturation profile of the PFC is slightly delayed on time compared to the other brain regions such as the hippocampus or the lateral cortex (Bouamrane et al., 2016). A genetic model of schizophrenia has shown that layer 5/6 of the prefrontal cortex displays abnormal development of GABAergic maturation and has highlighted a change in the excitatory/inhibitory imbalance during development underlying the changes in the network properties (Iafrati et al., 2014). As the PFC is known to play a major role in the higher cognitive and executive tasks (Fuster, 1991; Goldman-Rakic, 1995) it would be interesting to test the potent action of bumetanide during development and in model of Schizophrenia. The literature shows that patients with schizophrenia have been treated with bumetanide faced opposite effects. One study showed a positive effect of long-term treatment on hallucinations in a patient with schizophrenia (Lemonnier, Lazartigues, & Ben-Ari, 2016) while a second showed no effect (Rahmanzadeh et al., 2017). These controversial effects need more investigation to conclude about the potential therapeutic value of bumetanide for this disease.

## References


Acheson, A., Conover, J. C., Fandl, J. P., DeChiara, T. M., Russell, M., Thadani, A., et al. (1995). A BDNF autocrine loop in adult sensory neurons prevents cell death. *Nature*, *374*(6521), 450–453. https://doi.org/10.1038/374450a0.

Alvarado-Rojas, C., Huberfeld, G., Baulac, M., Clemenceau, S., Charpier, S., Miles, R., et al. (2014). Different mechanisms of ripple-like oscillations in the human epileptic subiculum. *Annals of Neurology*, https://doi.org/10.1002/ana.24324.

Alvarez, D. D., Giacomini, D., Yang, S. M., Trinchero, M. F., Temprana, S. G., Büttner, K. A., et al. (2016). A disynaptic feedback network activated by experience promotes the integration of new granule cells. *Science (New York, N.Y.)*, *354*(6311), 459–465. https://doi.org/10.1126/science.aaf2156.

Anacker, C., & Hen, R. (2017). *Adult hippocampal neurogenesis and cognitive flexibility—Linking memory and mood* (pp. 1–12).

Anacker, C., Zunszain, P. A., Cattaneo, A., Carvalho, L. A., Garabedian, M. J., Thuret, S., et al. (2011). Antidepressants increase human hippocampal neurogenesis by activating the glucocorticoid receptor. *Molecular Psychiatry*, *16*(7), 738–750. https://doi.org/10.1038/mp.2011.26.

Anderson, M. A., Ao, Y., & Sofroniew, M. V. (2014). Heterogeneity of reactive astrocytes. *Neuroscience Letters*, *565*, 23–29. https://doi.org/10.1016/j.neulet.2013.12.030.

Åsberg Johnels, J., Hovey, D., Zürcher, N., Hippolyte, L., Lemonnier, E., Gillberg, C., et al. (2016). Autism and emotional face-viewing. *Autism Research: Official Journal of the International Society for Autism Research*, https://doi.org/10.1002/aur.1730.

Avila, A., Nguyen, L., & Rigo, J.-M. (2013). Glycine receptors and brain development. *Frontiers in Cellular Neuroscience*, *7*, 184. https://doi.org/10.3389/fncel.2013.00184.

Bao, H., Asrican, B., Li, W., Gu, B., Wen, Z., Lim, S.-A., et al. (2017). Long-range GABAergic inputs regulate neural stem cell quiescence and control adult hippocampal neurogenesis. *Cell Stem Cell*, *21*(5), 604–617.e5. https://doi.org/10.1016/j.stem.2017.10.003.

Belousov, A. B., Wang, Y., Song, J.-H., Denisova, J. V., Berman, N. E., & Fontes, J. D. (2012). Neuronal gap junctions play a role in the secondary neuronal death following controlled cortical impact. *Neuroscience Letters*, https://doi.org/10.1016/j.neulet.2012.06.065.

Ben-Ari, Y. (2017). NKCC1 chloride importer antagonists attenuate many neurological and psychiatric disorders. *Trends in Neurosciences*, *40*(9), 536–554. https://doi.org/10.1016/j.tins.2017.07.001.

Ben-Ari, Y., Cherubini, E., Corradetti, R., & Gaiarsa, J. L. (1989). Giant synaptic potentials in immature rat CA3 hippocampal neurones. *The Journal of Physiology*, *416*(1), 303–325. https://doi.org/10.1113/jphysiol.1989.sp017762.

Ben-Ari, Y., Gaiarsa, J.-L., Tyzio, R., & Khazipov, R. (2007). GABA: A pioneer transmitter that excites immature neurons and generates primitive oscillations. *Physiological Reviews*, *87*(4), 1215–1284. https://doi.org/10.1152/physrev.00017.2006.

Blaesse, P., Airaksinen, M. S., Rivera, C., & Kaila, K. (2009). Cation-chloride cotransporters and neuronal function. *Neuron*, *61*(6), 820–838. https://doi.org/10.1016/j.neuron.2009.03.003.

Blauwblomme, T., Dossi, E., Pellegrino, C., Goubert, E., Gal Iglesias, B., Sainte-Rose, C., et al. (2018). GABAergic transmission underlies interictal epileptogenicity in pediatric FCD. *Annals of Neurology*, https://doi.org/10.1002/ana.25403.

Boldrini, M., Fulmore, C. A., Tartt, A. N., Simeon, L. R., Pavlova, I., Poposka, V., et al. (2018). Human hippocampal neurogenesis persists throughout aging. *Cell Stem Cell*, *22*(4), 589–599.e5. https://doi.org/10.1016/j.stem.2018.03.015.

Boldrini, M., Santiago, A. N., Hen, R., Dwork, A. J., Rosoklija, G. B., Tamir, H., et al. (2013). Hippocampal granule neuron number and dentate gyrus volume in antidepressant-treated and untreated major depression. *Neuropsychopharmacology: Official Publication of the American College of Neuropsychopharmacology*, *38*(6), 1068–1077. https://doi.org/10.1038/npp.2013.5.

Bondi, C. O., Semple, B. D., Noble-Haeusslein, L. J., Osier, N. D., Carlson, S. W., Dixon, C. E., et al. (2015). Found in translation: Understanding the biology and behavior of experimental traumatic brain injury. *Neuroscience and Biobehavioral Reviews*, *58*, 123–146. https://doi.org/10.1016/j.neubiorev.2014.12.004.

Bouamrane, L., Scheyer, A. F., Lassalle, O., Iafrati, J., Thomazeau, A., & Chavis, P. (2016). Reelin-haploinsufficiency disrupts the developmental trajectory of the E/I balance in the prefrontal cortex. *Frontiers in Cellular Neuroscience*, *10*(1805), 308. https://doi.org/10.3389/fncel.2016.00308.

Bragin, D. E., Sanderson, J. L., Peterson, S., Connor, J. A., & Müller, W. S. (2009). Development of epileptiform excitability in the deep entorhinal cortex after status epilepticus. *European Journal of Neuroscience*, *30*(4), 611–624. https://doi.org/10.1111/j.1460-9568.2009.06863.x.

Burda, J. E., Bernstein, A. M., & Sofroniew, M. V. (2015). Astrocyte roles in traumatic brain injury. *Experimental Neurology*, https://doi.org/10.1016/j.expneurol.2015.03.020.

Cameron, H. A., & McKay, R. D. (2001). Adult neurogenesis produces a large pool of new granule cells in the dentate gyrus. *The Journal of Comparative Neurology*, *435*(4), 406–417. https://doi.org/10.1002/cne.1040.

Canty, A. J., Teles-Grilo Ruivo, L. M., Nesarajah, C., Song, S., Jackson, J. S., Little, G. E., et al. (2013). Synaptic elimination and protection after minimal injury depend on cell type and their prelesion structural dynamics in the adult cerebral cortex. *The Journal of Neuroscience: The official journal of the Society for Neuroscience*, *33*(25), 10374–10383. https://doi.org/10.1523/JNEUROSCI.0254-13.2013.

Cellot, G., & Cherubini, E. (2014). GABAergic signaling as therapeutic target for autism spectrum disorders. *Frontiers in Pediatrics*, *2*(16), 70. https://doi.org/10.3389/fped.2014.00070.

Chamma, I., Heubl, M., Chevy, Q., Renner, M., Moutkine, I., Eugène, E., et al. (2013). Activity-dependent regulation of the K/Cl transporter KCC2 membrane diffusion, clustering, and function in hippocampal neurons. *The Journal of Neuroscience: The Official Journal of the Society for Neuroscience*, *33*(39), 15488–15503. https://doi.org/10.1523/JNEUROSCI.5889-12.2013.

Chen, G., & Manji, H. K. (2006). The extracellular signal-regulated kinase pathway: An emerging promising target for mood stabilizers. *Current Opinion in Psychiatry*, *19*(3), 313–323. https://doi.org/10.1097/01.yco.0000218604.63463.cd.

Curtis, M. A., Low, V. F., & Faull, R. L. M. (2012). Neurogenesis and progenitor cells in the adult human brain: A comparison between hippocampal and subventricular progenitor proliferation. *Developmental Neurobiology*, *72*(7), 990–1005. https://doi.org/10.1002/dneu.22028.

Czéh, B., & Lucassen, P. J. (2007). What causes the hippocampal volume decrease in depression? Are neurogenesis, glial changes and apoptosis implicated? *European Archives of Psychiatry and Clinical Neuroscience*, *257*, 250–260. https://doi.org/10.1007/s00406-007-0728-0. 5.

Damier, P., Hammond, C., & Ben-Ari, Y. (2016). Bumetanide to treat Parkinson disease: A report of 4 cases. *Clinical Neuropharmacology*, *39* (1), 57–59. https://doi.org/10.1097/WNF.0000000000000114.

Damjanac, M., Rioux Bilan, A., Barrier, L., Pontcharraud, R., Anne, C., Hugon, J., et al. (2007). Fluoro-Jade B staining as useful tool to identify activated microglia and astrocytes in a mouse transgenic model of Alzheimer's disease. *Brain Research*, *1128*(1), 40–49. https://doi.org/10.1016/j.brainres.2006.05.050.

Darren, W., Roddy, M. D., Chloe Farrell, P., Kelly Doolin, P., Roman, E., Leonardo Tozzi, M. P., et al. (2018). The hippocampus in depression: More than the sum of its parts? advanced hippocampal substructure segmentation in depression. *Biological Psychiatry*, 1–54. https://doi.org/10.1016/j.biopsych.2018.08.021.

Dehorter, N., Guigoni, C., Lopez, C., Hirsch, J., Eusebio, A., Ben-Ari, Y., et al. (2009). Dopamine-deprived striatal GABAergic interneurons burst and generate repetitive gigantic IPSCs in medium spiny neurons. *The Journal of Neuroscience: The Official Journal Of The Society for Neuroscience*, *29*(24), 7776–7787. https://doi.org/10.1523/JNEUROSCI.1527-09.2009.

Dehorter, N., Lozovaya, N., Mdzomba, B. J., Michel, F. J., Lopez, C., Tsintsadze, V., et al. (2012). Subthalamic lesion or levodopa treatment rescues giant GABAergic currents of PINK1-deficient striatum. *The Journal of Neuroscience: The Official Journal of the Society for Neuroscience*, *32*(50), 18047–18053. https://doi.org/10.1523/JNEUROSCI.2474-12.2012.

Delpire, E., Rauchman, M. I., Beier, D. R., Hebert, S. C., & Gullans, S. R (1994). Molecular cloning and chromosome localization of a putative basolateral Na(+)-K(+)-2Cl-cotransporter from mouse inner medullary collecting duct (mIMCD-3) cells. *The Journal of Biological Chemistry*, *269*(41), 25677–25683.

Demarque, M., Represa, A., Becq, H., Khalilov, I., Ben-Ari, Y., & Aniksztejn, L. (2002). Paracrine intercellular communication by a Ca2+- and SNARE-independent release of GABA and glutamate prior to synapse formation. *Neuron*, *36*(6), 1051–1061.

Drevets, W. C., Price, J. L., & Furey, M. L. (2008). Brain structural and functional abnormalities in mood disorders: Implications for neurocircuitry models of depression. *Brain Structure & Function*, *213*(1–2) 93–118. https://doi.org/10.1007/s00429-008-0189-x.

Duan, X., Kang, E., Liu, C. Y., Ming, G.-L., & Song, H. (2008). Development of neural stem cell in the adult brain. *Current Opinion in Neurobiology* *18*(1), 108–115. https://doi.org/10.1016/j.conb.2008. 04.001.

Duran, C., Thompson, C. H., Xiao, Q., & Hartzell, H. C. (2010). Chloride channels: Often enigmatic, rarely predictable. *Annual Review of Physiology*, *72*, 95–121. https://doi.org/10.1146/annurev-physiol-021909-135811.

Duveau, V., Laustela, S., Barth, L., Gianolini, F., Vogt, K. E., Keist, R., et al. (2011). Spatiotemporal specificity of GABAA receptor-mediated regulation of adult hippocampal neurogenesis. *The European Journal of Neuroscience*, *34*(3), 362–373. https://doi.org/10.1111/j.1460-9568.2011.07782.x.

Dzhala, V. I., Brumback, A. C., & Staley, K. J. (2008). Bumetanide enhances phenobarbital efficacy in a neonatal seizure model. *Annals of Neurology*, *63*(2), 222–235. https://doi.org/10.1002/ana.21229.

Dzhala, V. I., Talos, D. M., Sdrulla, D. A., Brumback, A. C., Mathews, G. C., Benke, T. A., et al. (2005). NKCC1 transporter facilitates seizures in the developing brain. *Nature Medicine*, *11*(11), 1205–1213. https://doi.org/10.1038/nm1301.

Earnheart, J. C., Schweizer, C., Crestani, F., Iwasato, T., Itohara, S., Mohler, H., et al. (2007). GABAergic control of adult hippocampal neurogenesis in relation to behavior indicative of trait anxiety and depression states. *The Journal of Neuroscience: The Official Journal Of the Society for Neuroscience*, *27*(14), 3845–3854. https://doi.org/10.1523/JNEUROSCI.3609-06.2007.

Encinas, J. M., & Sierra, A. (2012). Neural stem cell deforestation as the main force driving the age-related decline in adult hippocampal neurogenesis. *Behavioural Brain Research*, *227*(2), 433–439. https://doi.org/10.1016/j.bbr.2011.10.010.

Epsztein, J., Represa, A., Jorquera, I., Ben-Ari, Y., & Crépel, V. (2005). Recurrent mossy fibers establish aberrant kainate receptor-operated synapses on granule cells from epileptic rats. *The Journal of Neuroscience: The Official Journal Of the Society for Neuroscience*, *25* (36), 8229–8239. https://doi.org/10.1523/JNEUROSCI.1469-05.2005.

Espósito, M. S., Piatti, V. C., Laplagne, D. A., Morgenstern, N. A., Ferrari, C. C., Pitossi, F. J., et al. (2005). Neuronal differentiation in the adult hippocampus recapitulates embryonic development. *The Journal of Neuroscience: The Official Journal Of the Society for Neuroscience*, *25*(44), 10074–10086. https://doi.org/10.1523/JNEUROSCI.3114-05.2005.

Feng, J.-Y., Li, H.-H., Wang, B., Shan, L., & Jia, F.-Y. (2020). Successive clinical application of vitamin D and bumetanide in children with autism spectrum disorder: A case report. *Medicine*, *99*(2). https://doi.org/10.1097/MD.0000000000018661.

Fritschy, J.-M., & Sarter, M. (2012). Early brain repair and protection. *The European Journal of Neuroscience*, *35*(12), 1810. https://doi.org/10.1111/j.1460-9568.2012.08173.x.

Fukuda, A., & Watanabe, M. (2018). Pathogenic potential of human SLC12A5 variants causing KCC2 dysfunction. *Brain Research*, https://doi.org/10.1016/j.brainres.2018.12.025.

Fuster, J. M. (1991). The prefrontal cortex and its relation to behavior. *Progress in Brain Research*, *87*, 201–211. https://doi.org/10.1016/s0079-6123(08)63053-8.

Ge, S., Goh, E. L. K., Sailor, K. A., Kitabatake, Y., Ming, G.-L., & Song, H. (2006). GABA regulates synaptic integration of newly generated neurons in the adult brain. *Nature*, *439*(7076), 589–593. https://doi.org/10.1038/nature04404.

Golarai, G., Greenwood, A. C., Feeney, D. M., & Connor, J. A. (2001). Physiological and structural evidence for hippocampal involvement in persistent seizure susceptibility after traumatic brain injury. *The Journal of Neuroscience: The Official Journal of the Society for Neuroscience*, *21*(21), 8523–8537.

Goldman-Rakic, P. S. (1995). Architecture of the prefrontal cortex and the central executive. *Annals of the New York Academy of Sciences*, *769*, 71–83. https://doi.org/10.1111/j.1749-6632.1995.tb38132.x. 1 Structure and.

Goldstein, L. B., Bushnell, C. D., Adams, R. J., Appel, L. J., Braun, L. T., Chaturvedi, S., et al. (2011). Guidelines for the primary prevention of stroke: A guideline for healthcare professionals from the American Heart Association/American Stroke Association. *Stroke: A journal of cerebral circulation*, https://doi.org/10.1161/STR.0b013e3181fcb238. Lippincott Williams & Wilkins, Hagerstown, MD.

Goubert, E., Altvater, M., Rovira, M.-N., Khalilov, I., Mazzarino, M., Sebastiani, A., et al. (2019). Bumetanide prevents brain trauma-induced depressive-like behavior. *Frontiers in Molecular Neuroscience*, *12*, 12. https://doi.org/10.3389/fnmol.2019.00012.

Guilloux, J.-P., Douillard-Guilloux, G., Kota, R., Wang, X., Gardier, A. M., Martinowich, K., et al. (2012). Molecular evidence for BDNF- and GABA-related dysfunctions in the amygdala of female subjects with major depression. *Molecular Psychiatry*, *17*(11), 1130–1142. https://doi.org/10.1038/mp.2011.113.

Gulbins, E., Walter, S., Becker, K. A., Halmer, R., Liu, Y., Reichel, M., et al. (2015). A central role for the acid sphingomyelinase/ceramide system in neurogenesis and major depression. *Journal of Neurochemistry*, *134*(2), 183–192. https://doi.org/10.1111/jnc.13145.

Hadjikhani, N., Zürcher, N. R., Rogier, O., Ruest, T., Hippolyte, L., Ben-Ari, Y., et al. (2015). Improving emotional face perception in autism with diuretic bumetanide: A proof-of-concept behavioral and functional brain imaging pilot study. *Autism: The International Journal of Research and Practice*, *19*(2), 149–157. https://doi.org/10.1177/1362361313514141.

Hare, T. A., Wood, J. H., Manyam, B. V., Gerner, R. H., Ballenger, J. C., & Post, R. M. (1982). Central nervous system gamma-aminobutyric acid activity in man. Relationship to age and sex as reflected in CSF. *Archives of Neurology*, *39*(4), 247–249. https://doi.org/10.1001/archneur.1982.00510160053012.

Hsieh, L.-T., Gruber, M. J., Jenkins, L. J., & Ranganath, C. (2014). Hippocampal activity patterns carry information about objects in temporal context. *Neuron*, *81*(5), 1165–1178. https://doi.org/10.1016/j.neuron.2014.01.015.

Hsieh, T.-H., Lee, H. H. C., Hameed, M. Q., Pascual-Leone, A., Hensch, T. K., & Rotenberg, A. (2016). Trajectory of parvalbumin cell impairment and loss of cortical inhibition in traumatic brain injury. *Cerebral Cortex*, https://doi.org/10.1093/cercor/bhw318.

Hu, J.-J., Yang, X.-L., Luo, W.-D., Han, S., Yin, J., Liu, W.-H., et al. (2017). Bumetanide reduce the seizure susceptibility induced by pentylenetetrazol via inhibition of aberrant hippocampal neurogenesis in neonatal rats after hypoxia-ischemia. *Brain Research Bulletin*, *130*, 188–199. https://doi.org/10.1016/j.brainresbull.2017.01.022.

Huang, Z.-L., Liu, R., Bai, X.-Y., Zhao, G., Song, J.-K., Wu, S., et al. (2014). Protective effects of the novel adenosine derivative WS0701 in a mouse model of posttraumatic stress disorder. *Acta Pharmacologica Sinica*, *35*(1), 24–32. https://doi.org/10.1038/aps.2013.143.

Huang, W. C., Xiao, S., Huang, F., Harfe, B. D., Jan, Y. N., & Jan, L. Y. (2012). Calcium-activated chloride channels (CaCCs) regulate action potential and synaptic response in hippocampal neurons. *Neuron*, *74*(1), 179–192. https://doi.org/10.1016/j.neuron.2012.01.033.

Huberfeld, G., Menendez de la Prida, L., Pallud, J., Cohen, I., Le Van Quyen, M., Adam, C., et al. (2011). Glutamatergic pre-ictal discharges emerge at the transition to seizure in human epilepsy. *Nature Neuroscience*, *14*(5), 627–634. https://doi.org/10.1038/nn.2790.

Huberfeld, G., Wittner, L., Clemenceau, S., Baulac, M., Kaila, K., Miles, R., et al. (2007). Perturbed chloride homeostasis and GABAergic signaling in human temporal lobe epilepsy. *The Journal of Neuroscience: The Official Journal of the Society for Neuroscience*, *27*(37), 9866–9873. https://doi.org/10.1523/JNEUROSCI.2761-07.2007.

Hui, H., Rao, W., Zhang, L., Xie, Z., Peng, C., Su, N., et al. (2016). Inhibition of Na+-K+-2Cl-cotransporter-1 attenuates traumatic brain injury-induced neuronal apoptosis via regulation of Erk signaling. *Neurochemistry International*, https://doi.org/10.1016/j.neuint.2016.02.002.

Huusko, N., Römer, C., Ndode-Ekane, X. E., Lukasiuk, K., & Pitkänen, A. (2015). Loss of hippocampal interneurons and epileptogenesis: A comparison of two animal models of acquired epilepsy. *Brain Structure & Function*, *220*(1), 153–191. https://doi.org/10.1007/s00429-013-0644-1.

Iafrati, J., Orejarena, M. J., Lassalle, O., Bouamrane, L., Gonzalez-Campo, C., & Chavis, P. (2014). Reelin, an extracellular matrix protein linked to early onset psychiatric diseases, drives postnatal development of the prefrontal cortex via GluN2B-NMDARs and the mTOR pathway. *Molecular Psychiatry*, *19*(4), 417–426. https://doi.org/10.1038/mp.2013.66.

Ibrahim, S., Hu, W., Wang, X., Gao, X., He, C., & Chen, J. (2016). Traumatic brain injury causes aberrant migration of adult-born neurons in the hippocampus. *Scientific Reports*, *6*(1), 21793. https://doi.org/10.1038/srep21793.

Imayoshi, I., Sakamoto, M., Ohtsuka, T., Takao, K., Miyakawa, T., Yamaguchi, M., et al. (2008). Roles of continuous neurogenesis in the structural and functional integrity of the adult forebrain. *Nature Neuroscience*, *11*(10), 1153–1161. https://doi.org/10.1038/nn.2185.

Jacobs, B. L., van Praag, H., & Gage, F. H. (2000). Adult brain neurogenesis and psychiatry: A novel theory of depression. *Molecular Psychiatry*, *5*(3), 262–269. https://doi.org/10.1038/sj.mp.4000712.

Jaenisch, N., Witte, O. W., & Frahm, C. (2010). Downregulation of potassium chloride cotransporter KCC2 after transient focal cerebral ischemia. *Stroke: A Journal of Cerebral Circulation*, *41*(3), e151–e159. https://doi.org/10.1161/STROKEAHA.109.570424.

Jentsch, T. J. (2008). CLC chloride channels and transporters: From genes to protein structure, pathology and physiology. *Critical Reviews in Biochemistry and Molecular Biology*, *43*(1), 3–36. https://doi.org/10.1080/10409230701829110.

Kaila, K. (1994). Ionic basis of GABAA receptor channel function in the nervous system. *Progress in Neurobiology*, *42*(4), 489–537.

Kaila, K., Ruusuvuori, E., Seja, P., Voipio, J., & Puskarjov, M. (2013). GABA actions and ionic plasticity in epilepsy. *Current Opinion in Neurobiology*, *26C*, 34–41. https://doi.org/10.1016/j.conb.2013. 11.004.

Kalia, L. V., & Lang, A. E. (2015). Parkinson's disease. *Lancet (London, England)*, *386*(9996), 896–912. https://doi.org/10.1016/S0140-6736(14)61393-3.

Kalia, S. K., Sankar, T., & Lozano, A. M. (2013). Deep brain stimulation for Parkinson's disease and other movement disorders. *Current Opinion in Neurology*, *26*(4), 374–380. https://doi.org/10.1097/WCO.0b013e3283632d08.

Kandratavicius, L., Hallak, J. E., Carlotti, C. G., Assirati, J. A., & Leite, J. P. (2014). Neurotrophin receptors expression in mesial temporal lobe epilepsy with and without psychiatric comorbidities and their relation with seizure type and surgical outcome. *Acta Neuropathologica Communications*, *2*, 81. https://doi.org/10.1186/s40478-014-0081-2.

Karolewicz, B., Maciag, D., O'Dwyer, G., Stockmeier, C. A., Feyissa, A. M., & Rajkowska, G. (2010). Reduced level of glutamic acid decarboxylase-67 kDa in the prefrontal cortex in major depression. *The International Journal of Neuropsychopharmacology*, *13*(4), 411–420. https://doi.org/10.1017/S1461145709990587.

Kaushal, V., & Schlichter, L. C. (2008). Mechanisms of microglia-mediated neurotoxicity in a new model of the stroke penumbra. *The Journal of Neuroscience: The Official Journal Of the Society for Neuroscience*, *28*(9), 2221–2230. https://doi.org/10.1523/JNEUROSCI.5643-07.2008.

Kesner, R. P., & Churchwell, J. C. (2011). An analysis of rat prefrontal cortex in mediating executive function. *Neurobiology of Learning and Memory*, *96*(3), 417–431. https://doi.org/10.1016/j.nlm.2011.07.002.

Kharod, S. C., et al. (2019). Off-label use of bumetanide for brain disorders: An overview. *Frontiers in Neuroscience*, *13*, 310–330. https://doi.org/10.3389/fnins.2019.00310.

Klumpers, U. M. H., Veltman, D. J., Drent, M. L., Boellaard, R., Comans, E. F. I., Meynen, G., et al. (2010). Reduced parahippocampal and lateral temporal GABAA-[11C]flumazenil binding in major depression: Preliminary results. *European Journal of Nuclear Medicine and Molecular Imaging*, *37*(3), 565–574. https://doi.org/10.1007/s00259-009-1292-9.

Kourdougli, N., Pellegrino, C., Renko, J.-M., Khirug, S., Chazal, G., Kukko-Lukjanov, T.-K., et al. (2017). Depolarizing GABA contributes to glutamatergic network rewiring in epilepsy. *Annals of Neurology*, https://doi.org/10.1002/ana.24870.

Kourdougli, N., Varpula, S., Chazal, G., & Rivera, C. (2015). Detrimental effect of post status epilepticus treatment with ROCK inhibitor Y-27632 in a pilocarpine model of temporal lobe epilepsy. *Frontiers in Cellular Neuroscience*, *9*, 413. https://doi.org/10.3389/fncel.2015.00413.

Kronenberg, G., Balkaya, M., Prinz, V., Gertz, K., Ji, S., Kirste, I., et al. (2012). Exofocal dopaminergic degeneration as antidepressant target in mouse model of poststroke depression. *Biological Psychiatry*, *72*(4), 273–281. https://doi.org/10.1016/j.biopsych.2012.02.026.

Kronenberg, G., Katchanov, J., & Endres, M. (2006). Post-stroke depression: Clinical aspects, epidemiology, therapy, and pathophysiology. *Der Nervenarzt*, *77*(10), 1176, 1179–82–1184–5 https://doi.org/10.1007/s00115-006-2130-y.

Laplagne, D. A., Kamienkowski, J. E., Espósito, M. S., Piatti, V. C., Zhao, C., Gage, F. H., et al. (2007). Similar GABAergic inputs in dentate granule cells born during embryonic and adult neurogenesis. *European Journal of Neuroscience*, *25*(10), 2973–2981. https://doi.org/10.1111/j.1460-9568.2007.05549.x.

Lavenex, P., Lavenex, P. B., Bennett, J. L., & Amaral, D. G. (2009). Postmortem changes in the neuroanatomical characteristics of the primate brain: Hippocampal formation. *The Journal of Comparative Neurology*, *512*(1), 27–51. https://doi.org/10.1002/cne.21906.

Lees, A. J., Hardy, J., & Revesz, T. (2009). Parkinson's disease. *Lancet (London, England)*, *373*(9680), 2055–2066. https://doi.org/10.1016/S0140-6736(09)60492-X.

Lemonnier, E., & Ben-Ari, Y. (2010). The diuretic bumetanide decreases autistic behaviour in five infants treated during 3 months with no side effects. *Acta Paediatrica (Oslo, Norway: 1992)*, *99*(12), 1885–1888. https://doi.org/10.1111/j.1651-2227.2010.01933.x.

Lemonnier, E., Degrez, C., Phelep, M., Tyzio, R., Josse, F., Grandgeorge, M., et al. (2012). A randomised controlled trial of bumetanide in the treatment of autism in children. *Translational Psychiatry* 2. https://doi.org/10.1038/tp.2012.124.

Lemonnier, E., Lazartigues, A., & Ben-Ari, Y. (2016). Treating schizophrenia with the diuretic bumetanide: A case report. *Clinical Neuropharmacology*, *39*(2), 115–117. https://doi.org/10.1097/WNF.0000000000000136

Lemonnier, E., Villeneuve, N., Sonie, S., Serret, S., Rosier, A., Roue, M., et al. (2017). Effects of bumetanide on neurobehavioral function in children and adolescents with autism spectrum disorders. *Translational Psychiatry*, *7*(3). https://doi.org/10.1038/tp.2017.10.

Lucassen, P. J., Heine, V. M., Muller, M. B., van der Beek, E. M., Wiegant, V. M., De Kloet, E. R., et al. (2006). Stress, depression and hippocampal apoptosis. *CNS & Neurological Disorders Drug Targets*, *5*(5), 531–546. https://doi.org/10.2174/187152706778559273.

Malenka, R. C., & Bear, M. F. (2004). LTP and LTD: An embarrassment of riches. *Neuron*, *44*(1), 5–21. https://doi.org/10.1016/j.neuron.2004.09.012.

McKinnon, M. C., Yucel, K., Nazarov, A., & MacQueen, G. M. (2009). A meta-analysis examining clinical predictors of hippocampal volume in patients with major depressive disorder. *Journal of Psychiatry & Neuroscience*, *34*(1) 41–54.

Medina, I., Friedel, P., Rivera, C., Kahle, K. T., Kourdougli, N., Uvarov, P., et al. (2014). Current view on the functional regulation of the neuronal K(+)-Cl(-) cotransporter KCC2. *Frontiers in Cellular Neuroscience*, *8*, 27. https://doi.org/10.3389/fncel.2014.00027.

Merali, Z., Du, L., Hrdina, P., Palkovits, M., Faludi, G., Poulter, M. O., et al. (2004). Dysregulation in the suicide brain: mRNA expression of corticotropin-releasing hormone receptors and GABA(A) receptor subunits in frontal cortical brain region. *The Journal of Neuroscience: The Official Journal of the Society for Neuroscience*, *24*(6), 1478–1485. https://doi.org/10.1523/JNEUROSCI.4734-03.2004.

Miller, B. R., & Hen, R. (2015). The current state of the neurogenic theory of depression and anxiety. *Current Opinion in Neurobiology*, *30*, 51–58. https://doi.org/10.1016/j.conb.2014.08.012.

Molinaro, G., Battaglia, G., Riozzi, B., Storto, M., Fucile, S., Eusebi, F., et al. (2008). GABAergic drugs become neurotoxic in cortical neurons pre-exposed to brain-derived neurotrophic factor. *Molecular and Cellular Neurosciences*, *37*(2), 312–322. https://doi.org/10.1016/j.mcn.2007.10.009.

Mombereau, C., Kaupmann, K., Gassmann, M., Bettler, B., van der Putten, H., & Cryan, J. F. (2005). Altered anxiety and depression-related behaviour in mice lacking GABAB(2) receptor subunits. *Neuroreport*, *16*(3), 307–310.

Monteleone, P., Serritella, C., Martiadis, V., & Maj, M. (2008). Decreased levels of serum brain-derived neurotrophic factor in both depressed and euthymic patients with unipolar depression and in euthymic patients with bipolar I and II disorders. *Bipolar Disorders*, *10*(1), 95–100. https://doi.org/10.1111/j.1399-5618.2008.00459.x.

Moreno-Jiménez, E. P., Flor-García, M., Terreros-Roncal, J., Rábano, A., Cafini, F., Pallas-Bazarra, N., et al. (2019). Adult hippocampal neurogenesis is abundant in neurologically healthy subjects and drops sharply in patients with Alzheimer's disease. *Nature Medicine*, *25*(4), 554–560. https://doi.org/10.1038/s41591-019-0375-9.

Moubarak, R. S., Yuste, V. J., Artus, C., Bouharrour, A., Greer, P. A., Menissier-de Murcia, J., et al. (2007). Sequential activation of poly (ADP-ribose) polymerase 1, calpains, and Bax is essential in apoptosis-inducing factor-mediated programmed necrosis. *Molecular and Cellular Biology*, *27*(13), 4844–4862. https://doi.org/10.1128/MCB.02141-06.

Nabekura, J., Ueno, T., Okabe, A., Furuta, A., Iwaki, T., Shimizu-Okabe, C., et al. (2002). Reduction of KCC2 expression and GABAA receptor-mediated excitation after in vivo axonal injury. *The Journal of Neuroscience: The Official Journal of the Society for Neuroscience*, *22*(11), 4412–4417.

Nardou, R., Ben-Ari, Y., & Khalilov, I. (2009). Bumetanide, an NKCC1 antagonist, does not prevent formation of epileptogenic focus but blocks epileptic focus seizures in immature rat hippocampus. *Journal of Neurophysiology*, *101*(6), 2878–2888. https://doi.org/10.1152/jn.90761.2008.

Nardou, R., Yamamoto, S., Chazal, G., Bhar, A., Ferrand, N., Dulac, O., et al. (2011). Neuronal chloride accumulation and excitatory GABA underlie aggravation of neonatal epileptiform activities by phenobarbital. *Brain: A Journal of Neurology*, *134*(Pt. 4), 987–1002. https://doi.org/10.1093/brain/awr041.

Olmos, G., & Lladó, J. (2014). Tumor necrosis factor alpha: A link between neuroinflammation and excitotoxicity. *Mediators of Inflammation*, *2014*, 861231. https://doi.org/10.1155/2014/861231.

Pallud, J., Le Van Quyen, M., Bielle, F., Pellegrino, C., Varlet, P., Labussiere, M., et al. (2014). Cortical GABAergic excitation contributes to epileptic activities around human glioma. *Science Translational Medicine*, *6*(244), 244ra89. https://doi.org/10.1126/scitranslmed.3008065.

Payne, J. A., Rivera, C., Voipio, J., & Kaila, K. (2003). Cation-chloride co-transporters in neuronal communication, development and trauma. *Trends in Neurosciences*, *26*(4), 199–206.

Pellegrino, C., Gubkina, O., Schaefer, M., Becq, H., Ludwig, A., Mukhtarov, M., et al. (2011). Knocking down of the KCC2 in rat hippocampal neurons increases intracellular chloride concentration and compromises neuronal survival. *The Journal of Physiology*, *589*(Pt. 10), 2475–2496. https://doi.org/10.1113/jphysiol.2010.203703.

Perry, D. C., Sturm, V. E., Peterson, M. J., Pieper, C. F., Bullock, T., Boeve, B. F., et al. (2015). Association of traumatic brain injury with subsequent neurological and psychiatric disease: A meta-analysis. *Journal of Neurosurgery*, 1–16. https://doi.org/10.3171/2015.2.JNS14503.

Petrik, D., Lagace, D. C., & Eisch, A. J. (2012). The neurogenesis hypothesis of affective and anxiety disorders: Are we mistaking the scaffolding for the building? *Neuropharmacology*, *62*(1), 21–34. https://doi.org/10.1016/j.neuropharm.2011.09.003.

Petty, F., & Schlesser, M. A. (1981). Plasma GABA in affective illness. A preliminary investigation. *Journal of Affective Disorders*, *3*(4), 339–343. https://doi.org/10.1016/0165-0327(81)90003-3.

Petty, F., & Sherman, A. D. (1984). Plasma GABA levels in psychiatric illness. *Journal of Affective Disorders*, *6*(2), 131–138. https://doi.org/10.1016/0165-0327(84)90018-1.

Pittenger, C., & Duman, R. S. (2008). Stress, depression, and neuroplasticity: A convergence of mechanisms. *Neuropsychopharmacology: Official Publication of the American College of Neuropsychopharmacology*, *33*(1), 88–109. https://doi.org/10.1038/sj.npp.1301574.

Popova, N. K., & Naumenko, V. S. (2019). Neuronal and behavioral plasticity: The role of serotonin and BDNF systems tandem. *Expert Opinion on Therapeutic Targets*, *14728222*(2019), 1572747. https://doi.org/10.1080/14728222.2019.1572747.

Rahmanzadeh, R., Shahbazi, A., Ardakani, M.-R. K., Mehrabi, S., Rahmanzade, R., & Joghataei, M. T. (2017). Lack of the effect of bumetanide, a selective NKCC1 inhibitor, in patients with schizophrenia: A double-blind randomized trial. *Psychiatry and Clinical Neurosciences*, *71*(1), 72–73. https://doi.org/10.1111/pcn.12475.

Rajkowska, G., O'Dwyer, G., Teleki, Z., Stockmeier, C. A., & Miguel-Hidalgo, J. J. (2007). GABAergic neurons immunoreactive for calcium binding proteins are reduced in the prefrontal cortex in major depression. *Neuropsychopharmacology: Official Publication of the American College of Neuropsychopharmacology*, *32*(2), 471–482. https://doi.org/10.1038/sj.npp.1301234.

Ren, Z., Sahir, N., Murakami, S., Luellen, B. A., Earnheart, J. C., Lal, R., et al. (2015). Defects in dendrite and spine maturation and synaptogenesis associated with an anxious-depressive-like phenotype of GABAA receptor-deficient mice. *Neuropharmacology*, *88*, 171–179. https://doi.org/10.1016/j.neuropharm.2014.07.019.

Rivera, C., Voipio, J., Payne, J. A., Ruusuvuori, E., Lahtinen, H., Lamsa, K., et al. (1999). The K+/Cl- co-transporter KCC2 renders GABA hyperpolarizing during neuronal maturation. *Nature*, *397*(6716), 251–255. https://doi.org/10.1038/16697.

Rudolph, U., & Knoflach, F. (2011). Beyond classical benzodiazepines: Novel therapeutic potential of GABAA receptor subtypes. *Nature Reviews. Drug Discovery*, *10*(9), 685–697. https://doi.org/10.1038/nrd3502.

Sanacora, G., Gueorguieva, R., Epperson, C. N., Wu, Y.-T., Appel, M., Rothman, D. L., et al. (2004). Subtype-specific alterations of gamma-aminobutyric acid and glutamate in patients with major depression. *Archives of General Psychiatry*, *61*(7), 705–713. https://doi.org/10.1001/archpsyc.61.7.705.

Sanacora, G., Mason, G. F., Rothman, D. L., Behar, K. L., Hyder, F., Petroff, O. A., et al. (1999). Reduced cortical gamma-aminobutyric acid levels in depressed patients determined by proton magnetic resonance spectroscopy. *Archives of General Psychiatry*, *56*(11), 1043–1047. https://doi.org/10.1001/archpsyc.56.11.1043.

Sapolsky, R. M. (2000). Glucocorticoids and hippocampal atrophy in neuropsychiatric disorders. *Archives of General Psychiatry*, *57*(10), 925–935. https://doi.org/10.1001/archpsyc.57.10.925.

Schinder, A. F., & Gage, F. H. (2004). A hypothesis about the role of adult neurogenesis in hippocampal function. *Physiology (Bethesda, Md.)*, *19* (5), 253–261. https://doi.org/10.1152/physiol.00012.2004.

Shen, H., Sabaliauskas, N., Sherpa, A., Fenton, A. A., Stelzer, A., Aoki, C., et al. (2010). A critical role for alpha4betadelta GABAA receptors in shaping learning deficits at puberty in mice. *Science (New York, N.Y.)*, *327*(5972), 1515–1518. https://doi.org/10.1126/science.1184245.

Shulga, A., Magalhães, A. C., Autio, H., Plantman, S., di Lieto, A., Nykjær, A., et al. (2012). The loop diuretic bumetanide blocks posttraumatic p75NTR upregulation and rescues injured neurons. *The Journal of Neuroscience: The Official Journal of the Society for Neuroscience*, *32*(5), 1757–1770. https://doi.org/10.1523/JNEUROSCI.3282-11.2012.

Shulga, A., & Rivera, C. (2013). Interplay between thyroxin, BDNF and GABA in injured neurons. *Neuroscience*, *239*, 241–252. https://doi.org/10.1016/j.neuroscience.2012.12.007.

Shulga, A., Thomas-Crusells, J., Sigl, T., Blaesse, A., Mestres, P., Meyer, M., et al. (2008). Posttraumatic GABA(A)-mediated [Ca2+]i increase is essential for the induction of brain-derived neurotrophic factor-dependent survival of mature central neurons. *The Journal of Neuroscience: The Official Journal of the Society for Neuroscience*, *28*(27), 6996–7005. https://doi.org/10.1523/JNEUROSCI.5268-07.2008.

Singh, R., Mason, S., Lecky, F., & Dawson, J. (2018). Prevalence of depression after TBI in a prospective cohort: The SHEFBIT study. *Brain Injury*, *32*(1), 84–90. https://doi.org/10.1080/02699052.2017.1376756.

Son, H., Banasr, M., Choi, M., Chae, S. Y., Licznerski, P., Lee, B., et al. (2012). Neuritin produces antidepressant actions and blocks the neuronal and behavioral deficits caused by chronic stress. *Proceedings of the National Academy of Sciences*, *109*(28), 11378–11383. https://doi.org/10.1073/pnas.1201191109.

Song, J., Zhong, C., Bonaguidi, M. A., Sun, G. J., Hsu, D., Gu, Y., et al. (2012). Neuronal circuitry mechanism regulating adult quiescent neural stem-cell fate decision. *Nature*, *489*(7414), 150–154. https://doi.org/10.1038/nature11306.

Sorrells, S. F., Paredes, M. F., Cebrian-Silla, A., Sandoval, K., Qi, D., Kelley, K. W., et al. (2018). Human hippocampal neurogenesis drops sharply in children to undetectable levels in adults. *Nature*, *555*(7696), 377–381. https://doi.org/10.1038/nature25975.

Spalding, K. L., Bergmann, O., Alkass, K., Bernard, S., Salehpour, M., Huttner, H. B., et al. (2013). Dynamics of hippocampal neurogenesis in adult humans. *Cell*, *153*(6), 1219–1227. https://doi.org/10.1016/j.cell.2013.05.002.

Stamboulian-Platel, S., Legendre, A., Chabrol, T., Platel, J.-C., Pernot, F., Duveau, V., et al. (2016). Activation of GABAA receptors controls mesiotemporal lobe epilepsy despite changes in chloride transporters expression: In vivo and in silico approach. *Experimental Neurology*, *284*(Pt A), 11–28. https://doi.org/10.1016/j.expneurol.2016.07.009.

Sulis Sato, S., Artoni, P., Landi, S., Cozzolino, O., Parra, R., Pracucci, E., et al. (2017). Simultaneous two-photon imaging of intracellular chloride concentration and pH in mouse pyramidal neurons in vivo. *Proceedings of the National Academy of Sciences of the United States of America*, *114* (41), E8770–E8779. https://doi.org/10.1073/pnas.1702861114.

Sung, Y.-H., Shin, M.-S., Cho, S., Baik, H.-H., Jin, B.-K., Chang, H.-K., et al. (2010). Depression-like state in maternal rats induced by repeated separation of pups is accompanied by a decrease of cell proliferation and an increase of apoptosis in the hippocampus. *Neuroscience Letters*, *470*(1), 86–90. https://doi.org/10.1016/j.neulet.2009.12.063.

Tyzio, R., Nardou, R., Ferrari, D. C., Tsintsadze, T., Shahrokhi, A., Eftekhari, S., et al. (2014). Oxytocin-mediated GABA inhibition during delivery attenuates autism pathogenesis in rodent offspring. *Science (New York, N.Y.)*, *343*(6171), 675–679. https://doi.org/10.1126/science.1247190.

Ulrich-Lai, Y. M., & Herman, J. P. (2009). Neural regulation of endocrine and autonomic stress responses. *Nature Reviews Neuroscience*, *10*(6), 397–409. https://doi.org/10.1038/nrn2647.

Videbech, P., & Ravnkilde, B. (2004). Hippocampal volume and depression: A meta-analysis of MRI studies. *The American Journal of Psychiatry*, *161*(11), 1957–1966. https://doi.org/10.1176/appi.ajp.161.11.1957.

Wang, D. D., & Kriegstein, A. R. (2011). Blocking early GABA depolarization with bumetanide results in permanent alterations in cortical circuits and sensorimotor gating deficits. *Cerebral Cortex*, *21*(3), 574–587. https://doi.org/10.1093/cercor/bhq124.

Wink, L. K., Erickson, C. A., & McDougle, C. J. (2010). Pharmacologic treatment of behavioral symptoms associated with autism and other pervasive developmental disorders. *Current Treatment Options in Neurology*, *12*(6), 529–538. https://doi.org/10.1007/s11940-010-0091-8.

Xu, L., Anwyl, R., & Rowan, M. J. (1997). Behavioural stress facilitates the induction of long-term depression in the hippocampus. *Nature*, *387* (6632), 497–500. https://doi.org/10.1038/387497a0.

Xu, X., Zheng, C., An, L., Wang, R., & Zhang, T. (2016). Effects of dopamine and serotonin systems on modulating neural oscillations in Hippocampus-prefrontal cortex pathway in rats. *Brain Topography*, https://doi.org/10.1007/s10548-016-0485-3.

Yamada, J., Okabe, A., Toyoda, H., Kilb, W., Luhmann, H. J., & Fukuda, A. (2004). Cl⁻ uptake promoting depolarizing GABA actions in immature rat neocortical neurones is mediated by NKCC1. *The Journal of Physiology*, *557*(Pt. 3), 829–841. https://doi.org/10.1113/jphysiol.2004.062471.

Zhang, H., Cao, H. J., Kimelberg, H. K., & Zhou, M. (2011). Volume regulated anion channel currents of rat hippocampal neurons and their contribution to oxygen-and-glucose deprivation induced neuronal death. *PLoS One*, *6*(2). https://doi.org/10.1371/journal.pone.0016803.

Zhang, W., Liu, L.-Y., & Xu, T.-L. (2008). Reduced potassium-chloride co-transporter expression in spinal cord dorsal horn neurons contributes to inflammatory pain hypersensitivity in rats. *Neuroscience*, *152*(2), 502–510. https://doi.org/10.1016/j.neuroscience.2007.12.037.

Zifarelli, G., & Pusch, M. (2007). CLC chloride channels and transporters: A biophysical and physiological perspective. *Reviews of Physiology, Biochemistry and Pharmacology*, *158*, 23–76.

Chapter 32

# Linking citalopram, serotonin reuptake inhibitors, and depressed pregnant women

Marta Weinstock

*Institute of Drug Research, School of Pharmacy, The Hebrew University Medical Centre, Ein Kerem, Jerusalem, Israel*

## List of abbreviations

| | |
|---|---|
| **ASD** | autism spectrum disorder |
| **ADHD** | attention deficit hyperactivity disorder |
| **DLB** | depressive-like behavior |
| **DRN** | dorsal raphe nuclei |
| **E** | embryonic day |
| **GABA** | gamma aminobutyric acid |
| **MDD** | major depressive disorder |
| **P** | postnatal day |
| **PFC** | prefrontal cortex |
| **SSRI** | selective serotonin reuptake inhibitor |
| **SERT** | serotonin transporter |
| **TP** | Tryptophan |
| **TPH2** | tryptophan hydroxylase 2 |

## Introduction

Depression and anxiety disorders are a major cause of morbidity, affecting more than 16% of adults during their lifetime. The drugs most frequently prescribed for treatment of these conditions are SSRIs, inhibitors of the serotonin transporter (SERT) that terminates the action of serotonin by recapturing it into the nerve terminal. Of these, citalopram is the most selective inhibitor for the uptake of serotonin. The clinical use of SSRIs stems from the serotonin (5-HT) hypothesis, which postulated that depression results from a deficit in serotoninergic transmission (Owens & Nemeroff, 1994). It was originally thought that by inhibiting SERT, SSRIs increase extracellular 5-HT enabling more to interact with its receptors. However, this hypothesis did not account for the time lag of 3–4 weeks until symptoms of depression are ameliorated, or for the fact that the increase in extracellular 5-HT is relatively small, and in normal animals, actually lowers impulse conduction in 5-HT neurons.

The efficacy of SSRIs for treating depression has recently been disputed. Four out of six antidepressant trials did not find that the drugs were more effective than placebo (Blackburn, 2019). Another mega analysis that included all the known antidepressants in 522 placebo-controlled or drug comparison trials, concluded that SSRIs have only a modestly better effect than placebo in subjects with major depressive disorder (MDD), but are not more effective in its less severe forms (Cipriani et al., 2018).

SSRIs are also most frequently prescribed for the treatment of anxiety and depression in pregnant women, of whom, 10%–16% fulfill the criteria for MDD (Yonkers, Blackwell, Glover, & Forray, 2014). Their use in pregnancy has been questioned because of reports of adverse effects on the children. However, these were based on the outcome of drug treatment in mothers with MDD that was compared to that of normal mothers. The reports did not account for the significant proportion of children of depressed mothers that present with similar problems in development and behavior to those ascribed to the drugs.

SSRIs can cross the placenta and reach the fetal brain (Hendrick et al., 2003). By inhibiting SERT in the brain and other organs in which 5-HT is found, they can adversely affect development and behavior which are critically dependent on optimal levels and activity of 5-HT in the fetus. Differences among SSRIs in pharmacokinetics, time of administration, and actions other than SERT inhibition could also affect the outcome in the offspring.

Although animal models (usually rats or mice) of psychiatric conditions can allow a better control of some variables encountered in the human studies, they cannot sufficiently recapture the human condition and its complexities (Andersen & Thompson, 2011).

## Development of serotonergic systems

The serotonergic system reaches its highest functional status before birth and gradually declines during childhood and adulthood. Serotonergic neurons are present in the human brain by 5–7 weeks of gestation. Organization of 5-HT cell bodies in a complex of raphe nuclei begins in the 7th week and is complete by the 20th week. Axons spread from the raphe nuclei to their target areas during the following

The Neuroscience of Depression. https://doi.org/10.1016/B978-0-12-817933-8.00092-X

weeks of gestation and undergo differentiation into more complex forms (Kinney, Belliveau, Trachtenberg, Rava, & Paterson, 2007). Brain 5-HT levels continue to increase throughout the first 2 years of life before declining to adult levels by 5 years of age.

Much more is known about the stages of development of serotonergic system in rodents. 5-HT synthesis begins from ≈ embryonic day (E)11 in raphe nuclei (named B1-B9) with the appearance of tryptophan hydroxylase 2 (TPH2) and the amino acid TP, which is supplied to the fetal rat brain from the placenta (Bonnin & Levitt, 2011). TP is converted to 5-hydroxytryptophan by TPH2. 5-hydroxytryptophan is decarboxylated to 5-HT by L-aromatic acid decarboxylase. Development of the rodent serotonergic system occurs in three stages; axon elongation from E13-E16, differentiation of specific pathways from E15-E19, innervation of terminal fields in the hypothalamus, basal ganglia, hippocampus and other parts of the limbic system and their elaboration, from E19 until 21 days of age (Lauder, 1990). Beginning on E16, axons originating in nuclei B1-B5 innervate the spinal cord (Tork, 1990) from which they can regulate autonomic responses, pain, and motor activity. 5-HT also plays a crucial role in cardiac morphogenesis.

The appearance of SERT in the rat closely parallels that of TPH2. SERT is expressed in the raphe nuclei on E11 and facilitates the organization of neuronal connectivity by maintaining 5-HT near its target receptors. In rodents and humans, SERT is also transiently expressed ectopically during development and also in sensory, thalamic, hippocampal, and hypothalamic neurons, and in the prefrontal cortex (PFC), throughout development. In 5-HT neurons, SERT expression is highest during fetal life and decreases during adolescence and adulthood.

## Role of 5-HT in neural development and behavior

5-HT neurons are highly branched and do not all make synaptic contacts, but release 5-HT in a paracrine manner, enabling it to play an important role in guiding axons to their targets. 5-HT also promotes cell division, differentiation, migration, myelination, synaptogenesis, spinogenesis, and dendritic pruning (Bonnin, Torii, Wang, Rakic, & Levitt, 2007) through its actions on a large family of 14 receptors, which are divided into seven subfamilies ($5\text{-HT}_1$–$5\text{-HT}_7$) (Glennon & Dukat, 1991). $5\text{-HT}_{1A}$R and $5\text{-HT}_{2A}$R subtypes are present from mid-gestation before the serotonin fibers reach their destination.

5-HT regulates the function of the hypothalamic–pituitary–adrenal and sympathetic nervous systems that control the response to stress (McEwen, 2005). The serotonergic activity also influences complex social behaviors, circadian rhythms, thermoregulation, mood, reward, and memory. Therefore, stress or drugs that alter 5-HT levels can affect both brain development and these behaviors (Deneris & Gaspar, 2018).

Unlike glutamate or GABA, the main excitatory and inhibitory neurotransmitters, 5-HT acts as a neuromodulator, increasing or decreasing neuronal activity according to the identity of the 5-HT receptors activated. $5\text{-HT}_{1A}$Rs mediate the inhibitory actions of 5-HT in the soma and dendrites of 5-HT neurons and serve as autoreceptors in the median and dorsal raphe nuclei (DRN). Their activation inhibits the formation of cyclic AMP, thereby reducing neuronal firing and synaptic release of 5-HT (Raymond, Mukhin, Gettys, & Garnovskaya, 1999). $5\text{-HT}_{1A}$Rs are also expressed postsynaptically on dendrites of excitatory glutamatergic pyramidal cells and on inhibitory GABAergic interneurons in brain regions involved in the regulation of stress and emotion. Activation of postsynaptic $5\text{-HT}_{2A}$Rs increases cellular levels of inositol triphosphate and diacylglycerol and mediates the excitatory actions of serotonin on target neurons. High densities of $5\text{-HT}_{2A}$Rs are found throughout the cortex, and lower densities in subcortical structures (Beliveau et al., 2017).

In the brains of subjects with MDD, the concentrations of 5-HT and its main metabolite, 5-hydroxyindoleacetic acid, are lower than that in controls. Binding of 5-HT to $5\text{-HT}_{1A}$R is higher in the rostral part of the DRN (Boldrini, Underwood, Mann, & Arango, 2008). Since activation of $5\text{-HT}_{1A}$R inhibits firing in axons arising from the DRN, it can explain the reduction of 5-HT signaling in areas of the limbic system in MDD. It has been postulated that SSRIs relieve depression by increasing 5-HT in the vicinity of the $5\text{-HT}_{1A}$R autoreceptors in the DRN, thereby causing their downregulation. This brings about a gradual restoration of 5-HT release in target areas (Stahl, 1998).

## Pharmacokinetics of SSRIs

The six SSRI medications currently in clinical use are fluoxetine, fluvoxamine, paroxetine, sertraline, citalopram, and escitalopram. They are racemates, except escitalopram and paroxetine, which are S-enantiomers. Their clinical recommended doses and pharmacokinetic properties are summarized in Table 1. Fluoxetine has a 10-fold higher affinity for the SERT than for the noradrenaline transporter. It is rapidly demethylated in the liver to an active metabolite, norfluoxetine. Because of its relatively long half-life of 1–4 days, fluoxetine takes much longer than the other SSRIs to reach steady-state levels.

All the other SSRIs show a higher selectivity than fluoxetine for SERT than for the noradrenaline transporter. Fluvoxamine is rapidly metabolized to at least nine metabolites (Overmars, Scherpenisse, & Post, 1983), which do not appear to inhibit SERT. Paroxetine is metabolized by CYP2D6 and methylated by catechol-o-methyl transferase.

**TABLE 1** Pharmacokinetic parameters of SSRIs.

| Drug | Daily dose (mg) | Half-life (h) (range) | Metabolizing enzymes | % Change in plasma levels in pregnancy | | Concentration fetal/maternal blood | Concentration milk/maternal blood |
|---|---|---|---|---|---|---|---|
| | | | | 2nd trimester | 3rd trimester | Ratio (range) | Ratio (range) |
| Fluoxetine | 60–100 | 45 (24–144) | CYP2C9 = CYP2D6 ↑ | ↓8 | ↓13 | 0.65 (0.33–1.06) | 0.2 (0–1.1) |
| Fluvoxamine | 50–100 | 15 (9–28) | CYP2D6 ↑ CYP1A2 ↑ | ↓38* | ↓56* | – | 0.9 (0.3–1.4) |
| Paroxetine | 20–60 | 18 (7–65) | CYP2D6 ↑ CYP3A4 = | ↓34 | ↓51** | 0.51 (0.05–0.91) | 0.6 (0–2.4) |
| Sertraline | 50–200 | 26 (22–36) | CYP2C19 ↑ and others | ↑36 | ↑68** | 0.29 (0.1–0.66) | 0.9 (0–5.2) |
| Citalopram | 20–60 | 33 (23–45) | CYP2D6 ↑ | ↓15 | ↓24* | 0.85 (0.42–1.42) | 1.7 (0.9–2.8) |
| Escitalopram | 10–20 | 30 (27–33) | CYP2C19 ↓ | ↑4 | ↑7 | – | – |

Direction of activity change in pregnancy ↑ increase; ↓ decrease; = unchanged.
Significant change *$P < 0.01$; **$P < 0.001$.

Sertraline also inhibits the dopamine transporter. Less is known about its metabolism than that of the other SSRIs, but it appears to be demethylated by CYP2D6. Citalopram is the most selective inhibitor of SERT relative to that of the noradrenaline transporter. Its pharmacological activity resides in the S-isomer, escitalopram, and both are metabolized by several CYP450 enzymes (Table 1).

A meta-analysis of data from more than 290 pregnant women receiving SSRIs found that blood levels of some but not fluoxetine and escitalopram were significantly changed during pregnancy, compared to those in nonpregnant women of comparable ages (Table 1). The increase in blood levels of paroxetine in the study of Westin, Brekke, Molden, Skogvoll, and Spigset (2017) only occurred in subjects that were poor metabolizers.

The fetal brain is particularly vulnerable to maternal SSRI exposure because of the low levels of drug-metabolizing liver enzymes. All SSRIs are able to penetrate cord blood and reach the fetus but the amounts of the individual SSRIs present in its blood vary, with ratios of their concentrations to those in maternal blood ranging from 0.29 for sertraline to 0.85 for citalopram (Table 1). All the SSRIs are secreted into breast milk. Ratios of their concentrations in milk to those in maternal plasma are also highest for citalopram (1.7) and lowest for fluoxetine (0.2) (Weissman et al., 2004). This may explain some of the differences in the actions of the drugs that are seen in the offspring after maternal treatment.

## Prenatal exposure to SSRIs in humans

### Deficits in early development

Exposure of pregnant women to SSRIs during the first semester increases the number of infants with low birth weight, cardiac defects, craniosynostosis, musculoskeletal system defects and persistent pulmonary hypertension (Alwan, Friedman, & Chambers, 2016). These conditions could occur as a result of higher synaptic levels of 5-HT in the heart, lungs, and skeletal muscle. Lower birth weight is still seen when the outcome is compared to that in untreated mothers with MDD. The incidence of cardiac defects is reduced, but still present, (Eke, Saccone, & Berghella, 2016). Treatment with citalopram, escitalopram, or sertraline significantly increases limb malformations, while paroxetine increases atrio-septal and respiratory system defects in the children (Gao et al., 2017). Citalopram causes the highest incidence of craniosynostosis and defects in the musculoskeletal system in the infants (Berard, Zhao, & Sheehy, 2017), in comparison with those of untreated mothers, possibly because higher concentrations of the drug reach the fetus from the maternal blood and milk than those of the other drugs (Hendrick et al., 2003; Rampono et al., 2009).

### Behavior

A significant number of children of depressed or chronically stressed mothers have emotional problems, poorer intellectual and language abilities and suffer from attention deficit hyperactivity disorder (ADHD) or autism spectrum disorder (ASD) (Weinstock, 2008). The early studies on the influence of prenatal SSRI administration on child development and behavior did not include a comparison group of such mothers. Therefore, they were unable to determine whether any abnormalities in the children resulted from drug treatment or from the maternal condition. Most of the information was obtained retrospectively from large medical databases and many, without patient interviews or records of the dose of the drug and when it was taken. This omission is significant because the consumption of SSRIs varies during pregnancy. Most women take drugs in the first trimester before they know they are pregnant. Drug intake decreases in the second trimester and increases again close to the time of birth, especially in mothers with more severe depression (Yonkers et al., 2014).

It was reported that children of women with untreated MDD, have higher scores of anxiety, depression, internalizing and externalizing behavior than controls at the age of 3–6 years. They also have deficits in executive function, but not in working memory (Table 2). The appearance of these behaviors appears to be related to the severity and duration of maternal depression (Hermansen et al., 2016; Lupattelli et al., 2018). Treatment with SSRIs reduces externalizing behavior (Hanley et al., 2015), does not affect executive function, internalizing and anxious behavior, but impairs working memory (El Marroun et al., 2017).

ADHD is characterized by inattention, hyperactivity, disorganization and/or impulsivity and occurs in a larger proportion of children of mothers with MDD than in the general population. Treatment of such mothers with SSRIs during all trimesters increases the incidence of ADHD in the children in comparison to those of control subjects, but not to those of untreated depressed mothers (Boukhris et al., 2017; Man et al., 2017) (Table 2).

Autism consists of a spectrum of behavioral disorders characterized by deficits in social communication and social reciprocity, together with repetitive and stereotyped behaviors that occur before 3 years of age. It is found in approximately 1% in the general population, including children of mothers with MDD, and is four times more prevalent in boys than in girls. Involvement of 5-HT signaling has been suggested because infants with ASD lack the peak levels of brain 5-HT at 2 years of age, while young adults show a decrease in 5-HT binding to 5-$HT_{1A}$R and 5-$HT_{2A}$R in select areas of the limbic system (Oblak, Gibbs, & Blatt, 2013). Citalopram treatment of children with ASD increases motor activity, impulsiveness, and sleep disturbance, without producing any evident therapeutic benefit (Volkmar, 2009).

**TABLE 2** Effect of treatment of maternal MDD with SSRIs on behavior in humans.

| Behavior problems | | | Compared to healthy controls | | Compared to untreated MDD | | |
|---|---|---|---|---|---|---|---|
| **All SSRIs** | **Age of children (years)** | **Trimester#** | **No. of subjects** | ***P*** | **No. subjects** | ***P*** | **Ref.** |
| Internalizing | 3 | All | | | 110 | 0.037 | 1 |
| Externalizing | 3 | All | | | | 0.635 | |
| Anxious/depressed | 3 | All | | | | 0.003 | |
| Internalizing | 6 | All | | | | 0.037 | |
| Anxious/depressed | 5 | All | | | | 0.127 | |
| Internalizing | 5–6 | All | 1382 | <0.01 | 624 | 0.056 | 2 |
| Externalizing | 5–6 | All | | <0.05 | | NS | |
| Anxious/depressed | 5 | 3 | 3775 | | 290 | 0.04 | 3 |
| | | **Data analyzed according to relative risk** | | | | | |
| | **Age of children (years)** | **Trimester#** | **No. of subjects** | **RR** | **No. of subjects** | **RR** | **Ref.** |
| Executive function | 4 | | 3906 | 3.28** | 44 | 5.58* | 4 |
| Working memory | | | | 1.29 | | 2.29** | |
| ADHD meta analyses | 3–9 | All | 189,002 | 2.23** | 425 | 0.70 | 5 |
| | | 1 | | 2.32** | | | |
| | | 2 | | 2.22** | | | |
| | | 3 | | 2.25** | | | |
| ADHD | >5 | All | | | 1547 | | 6 |
| | | 1 | | | | 1.0 | |
| | | 2 + 3 | | | | 1.2 | |
| | **Age of children (years)** | **Trimester#** | **No. of studies#** | **RR** | **No. of studies#** | **RR** | **Ref.** |
| ASD meta-analyses | | All | 5 | 1.60* | | | 7 |
| | | 1 | 4 | 1.88* | | | |
| | | 2 | 4 | 1.73* | | | |
| | | 3 | 4 | 1.59 | | | |
| | | All | 5 | 1.7* | 4 | 1.4 | 8 |
| | | 1 | 5 | 2.0* | 3 | 1.8* | |
| **Individual SSRIs** | | | **No of subjects** | **RR** | **No of subjects** | **RR** | **Ref.** |
| Citalopram/escitalopram | | All | 1097 | 1.71** | 639 | 1.47* | 9 |
| Sertraline | | All | 1016 | 1.35 | 672 | 1.17 | |
| Fluoxetine | | All | 458 | 1.04 | 327 | 1.08 | |
| Paroxetine | | All | 173 | 1.40 | 108 | 1.21 | |

RR, relative risk. #when the mothers were taking the SSRIs. cf. respective control. *$P < 0.05$; **$P < 0.01$; NS, not significant.
[1]Hanley, Brain, and Oberlander (2015); [2]Hermansen, Roysamb, Augusti, and Melinder (2016); [3]Lupattelli et al. (2018); [4]El Marroun et al. (2017); [5]Boukhris, Sheehy, and Berard (2017); [6]Man et al. (2017); [7]Halvorsen, Hesel, Ostergaard, and Danielsen (2019); [8]Brown, Hussain-Shamsy, Lunsky, Dennis, and Vigod (2017); [9]Viktorin, Uher, Reichenberg, Levine, and Sandin (2017).

Treatment with SSRIs as a group increases the number of children with ASD in comparison with those of control mothers (Halvorsen et al., 2019). This effect is greatest when the drugs are taken in the first or second semesters (Brown et al., 2017; Halvorsen et al., 2019). Among studies with individual SSRIs, citalopram, and escitalopram significantly increase the incidence of ASD, while fluoxetine has the least effect (Viktorin et al., 2017) (Table 2). The effect of citalopram is reduced, but still present, when compared to children of untreated, depressed women (Brown et al., 2017). It is not known whether SSRIs can alter 5-HT levels or receptor binding in the brain of children with ASD.

These studies indicate that SSRIs do not reduce internalizing behavior, anxiety and depression in young children of mothers with MDD and may even adversely affect working memory. There are indications that SSRIs may reduce the incidence of externalizing problems and ADHD, but their effect on ASD is not yet clear and may differ among individual drugs.

## Effects of prenatal SSRI administration in normal pregnant rodents

### Early development

All the SSRIs tested, at doses of 5–12 mg/kg, reduced the body weight of rat pups at some time up to 14 days of age (Table 3). Paroxetine also increased mortality in Fischer rats.

**TABLE 3** Perinatal treatment of control rats with antidepressants: Effect on body weight.

| Drug (dose mg/kg) | Species strain | Time admin | Route of admin. | Pup age when tested | Body weight | Ref. |
|---|---|---|---|---|---|---|
| Fluoxetine (5) | Rat S-D | E6–E22 | sc b.i.d. | 5 | ↓ | 10 |
| Citalopram (5) | Rat S-D | E6–E22 | sc b.i.d. | 11 | ↓ | |
| Citalopram (10) | Rat S-D | E6-E22 | sc b.i.d. | 14–17 | ↓ | 11 |
| Fluoxetine (12) | Rat Wistar | E12–P1 | Oral q.d. | 7 | ↓ | 12 |
| Paroxetine (10) | Rat Fischer 344 | E14-E21 | Oral DF | 1 | ↓ ↑ Mortality | 13 |
| Fluvoxamine (10) | Rat Outbred strain | P1-P14 | sc q.d. | 4–14 | ↓ | 14 |
| Fluoxetine (10) | Rat NIH Norway | P2-P7 | Oral q.d. | 7 | ↓ | 15 |
| Sertraline (5) | Mouse C57BL/6 | E1-P14 | ip | 3–14, 21 | ↑ [3–14] = [21] | 16 |

S-D, Sprague–Dawley; sc, subcutaneous injection; ip, intraperitoneal injection; q.d., once daily; DF, drinking fluid; [] refers to pup age, days after birth.
[10]Sprowles, J. L. N., Hufgard, J. R., Gutierrez, A., Bailey, R. A., Jablonski, S. A., Williams, M. T., et al. (2017). Differential effects of perinatal exposure to antidepressants on learning and memory, acoustic startle, anxiety, and open-field activity in Sprague–Dawley rats. *International Journal of Developmental Neuroscience, 61*, 92–111; [11]Sprowles, J. L., Hufgard, J. R., Gutierrez, A., Bailey, R. A., Jablonski, S. A., Williams, M. T., et al. (2016). Perinatal exposure to the selective serotonin reuptake inhibitor citalopram alters spatial learning and memory, anxiety, depression, and startle in Sprague–Dawley rats. *International Journal of Developmental Neuroscience, 54*, 39–52; [12]Olivier, J. D., Valles, A., van Heesch, F., Afrasiab-Middelman, A., Roelofs, J. J., Jonkers, M., et al. (2011). Fluoxetine administration to pregnant rats increases anxiety-related behavior in the offspring. *Psychopharmacology (Berl), 217*, 419–432; [13]Van den Hove, D. L., Blanco, C. E., Scheepens, A., Desbonnet, L., Myint, A. M., Leonard, B. E., et al. (2008). Prenatal maternal paroxetine treatment and neonatal mortality in the rat: A preliminary study. *Neonatology, 93*, 52–55; [14]Glazova, N. Y., Merchieva, S. A., Volodina, M. A., Sebentsova, E. A., Manchenko, D. M., Kudrun, V. S., et al. (2014). Effects of neonatal fluvoxamine administration on the physical development and activity of the serotoninergic system in white rats. *Acta Naturae, 6*, 98–105; [15]Smit-Rigter, L. A., Noorlander, C. W., von Oerthel, L., Chameau, P., Smidt, M. P., & van Hooft, J. A. (2012). Prenatal fluoxetine exposure induces life-long serotonin 5-HT(3) receptor-dependent cortical abnormalities and anxiety-like behavior. *Neuropharmacology, 62*, 865–870; [16]Meyer, L. R., Dexter, B., Lo, C., Kenkel, E., Hirai, T., Roghair, R. D., et al. (2018). Perinatal SSRI exposure permanently alters cerebral serotonin receptor mRNA in mice but does not impact adult behaviors. *The Journal of Maternal-Fetal & Neonatal Medicine, 31*, 1393–1401.

It is not known if there was any cardiac damage. In contrast, sertraline increased body weight in mice.

## Behavior

The effects of SSRIs on the behavior of the offspring of pregnant rats or mice are summarized in Table 4. In most, but not all studies, the authors claimed that the administered dose produced blood levels that were similar to those in humans. Prenatal administration of fluoxetine or citalopram (5 mg/kg b.i.d.) and citalopram (10 mg/kg b.i.d.), given by the same group in another study, impaired spatial learning in adult rat offspring of both sexes and the drugs increased activity in the marble burying test. This is indicative of compulsive, obsessive behavior and attention deficits (Taylor, Lerch, & Chourbaji, 2017), also seen in children of mothers with MDD. Fluoxetine impaired social activity of juvenile rats. When given prenatally and until weaning, fluoxetine did not cause depressive-like activity (DLB) in adult male rats or mice, but heightened their anxiety when injected directly into neonates, aged 2–21 days. Citalopram (10 mg/kg) induced DLB and anxiety in adult males, but not in females, while a higher dose of escitalopram (12.2 mg/kg) induced DLB in female rats. Together with those in the preceding section, the data show that fluoxetine, paroxetine, citalopram, and escitalopram can cause adverse effects on early development and behavior when given to normal rats or mice. There is insufficient data to enable one to conclude whether, or not, sertraline is safer than the other drugs.

## 5-HT signaling

Serotonergic activity in target regions like the hippocampus, amygdala, and PFC is largely dependent on impulse transmission in neurons arising in the DRN (Fig.1). This is decreased by activation of 5-$HT_{1A}$R in the DRN and increased by activation of 5-$HT_{1A}$R on GABAergic neurons in the lateral portions of the DRN and in the medial PFC. Activity in 5-HT neurons arising from the medial raphe and DRN is also tightly controlled by afferent pathways from the PFC (Celada, Puig, Amargos-Bosch, Adell, & Artigas, 2004).

Anxiety and/or DLB after neonatal administration of fluoxetine results from increased activation by 5-HT of $5HT_{1A}$R autoreceptors and postsynaptic 5-$HT_{2A/C}$Rs since both behaviors are prevented by concomitant administration of a $5HT_{2A/C}$, but not by a $5HT_{1A}$R antagonist. While the latter, given alone, induces anxiety in naïve rats, a 5-$HT_{2A}$R agonist induces both anxiety and DLB (Sarkar et al., 2014). The data suggest that SSRIs can cause anxiety and/or DLB in the offspring depending on the ultimate balance they induce between activation or downregulation of 5-$HT_{2A}$Rs and $5HT_{1A}$Rs, respectively.

Maternal treatment with citalopram decreases the density of 5-$HT_{1A}$R in GABAergic interneurons in the medial PFC of males. This would further lessen inhibition by GABA of pyramidal glutamatergic neurons, an effect akin to activation by 5-HT of 5-$HT_{2A}$R. Citalopram also decreased 5-$HT_{1A}$R expression in the DRN, thereby acting like the 5-$HT_{1A}$R antagonist in the experiments of Sarkar et al. (2014). Thus, treatment with citalopram, fluoxetine, and paroxetine causes similar pro-depressive and anxiogenic behavior to that found in mice with SERT deletion (Gingrich et al., 2017). This may result from an excess in 5-HT levels in the rodent cortex.

## SSRI administration to stressed mothers: Effect on offspring behavior

While SSRI administration to control pregnant rats clearly has adverse effects on their offspring, this may differ in those of chronically stressed depressed rats, if 5-HT signaling is decreased. This was examined in only five studies (Table 5). Chronic stress was shown to increase anxiety (Zohar et al., 2016) and DLB (Salari et al., 2016) in mothers. Both were prevented by drug treatment. Chronic treatment with citalopram during adolescence or adulthood reversed the behavioral abnormalities in prenatally stressed offspring, but adversely affected that of controls (Zohar, Dosoretz-Abittan, Shoham, & Weinstock, 2015).

In contrast to its effect in offspring of control rats, fluoxetine prevented the decrease in social interaction in adolescence and DLB and heightened anxiety in juvenile and adult offspring. Conversely, citalopram, did not alleviate anxiety or DLB in adult males, but prevented DLB, but not anxiety in females. The effect of fluoxetine on behavior in males and that of citalopram on DLB in females was associated with the restoration of 5-$HT_{1A}$R signaling in the PFC. While citalopram also restored 5-$HT_{1A}$R expression in GABAergic interneurons in the DRN in females, it had no effect on 5-$HT_{1A}$Rs in either the PFC or DRN in males (Fig. 1). 5-$HT_{2A}$R expression was not measured in these studies.

Taken together, the findings show that treatment with fluoxetine appears to able to overcome the changes caused by prenatal stress on $5HT_{1A}$R expression and their resulting effects on neuronal activity and behavior. On the other hand, citalopram does not prevent the appearance of anxiety, DLB, or correct 5-$HT_{1A}$R expression in males but can prevent DLB in females. Studies are lacking in both sexes and/or with other SSRIs, in rodents and human subjects, to enable any further conclusions about potential differences between the drugs on offspring behavior.

**TABLE 4** Perinatal treatment of control rats with antidepressants: Effect on behavior and expression of $5HT_{1A}R$.

| Drug (dose mg/kg) Species-strain | Time in gestation | Route of admin. | Age tested sex | Spatial memory [a] | Social behavior | Anxiety | Depression (FST) | $5HT_{1A}R$ in PFC | Ref. |
|---|---|---|---|---|---|---|---|---|---|
| Escitalopram (12.2) Rat S-D | -3*-E21 | sc MP | 28, 90 F | NT | ↓ [28] =[90] | = [28,90] | NT | NT | 17 |
| Paroxetine (10) Rat bHR/bLR S-D | -5*-P21 | Oral DF | 75 M | NT | NT | = | ↑ | NT | 18 |
| Citalopram (10) Rat Wistar | E1-P21 | Oral DF | 60 M, F | NT | NT | ↑ M=F | ↑ M=F | ↓ M=F | 19 |
| Fluoxetine (5) Rat S-D | E6–E22 | sc b.i.d. | 60 M, F | ↓M,F | = | ↑ MB | = | NT | 10 |
| Citalopram (5) Rat S-D | E6–E22 | sc b.i.d. | 60 M, F | ↓ | = | ↑ MB | = | NT | |
| Citalopram (10) Rat S-D | E6–E22 | sc b.i.d. | 60+ M, F | ↓ | = | ↑ MB | = | NT | 11 |
| Fluoxetine (12) Rat Wistar | E12–P1 | Oral q.d. | 28, 104 M | NT | ↓ [28]= [104] | = [104] | = [104] | NT | 12 |
| Fluoxetine (5/day) Rat S-D | P1-P21 | MP | 32–39 M, F | NT | NT | OF= | = | NT | 20 |
| Fluoxetine (50 μg) Rat S-D | P2-P21 | sc | 28, 72 | NT | NT | OF ↑ | NT | = | 21 |
| Fluoxetine (10) Rat S-D | P2-P21 | Oral | Adults | NT | NT | ↑ OF ↑ EPM | ↑ | NT | 22 |
| Escitalopram (10) Mouse Swiss CD1 | P5-P14 | sc | 60 M, F | NT | NT | = | ↑ | NT | 23 |
| Sertraline (5) Mouse C57BL/6 + (1.5) to pups | E1-P14 | ip | 147 M, F | = | = | = | NT | Cortex ↑ mRNA | 16 |
| Fluoxetine (8/day) Mouse NMRI | E10-E21 P0-P20 | DF | 70+ M | NT | NT | = | = | NT | 24 |

Legend as in Table 3. M, male; F, female; NT, not tested; MP, mini pumps; *before mating; [a] tested either in Morris water maze or in Barnes maze. EPM, elevated plus maze (time in open arms); OF, open field (time in center); MB, marble burying; FST, forced swim test (time immobile).

[17]Ehrlich, D. E., Neigh, G. N., Bourke, C. H., Nemeth, C. L., Hazra, R., Ryan, S. J., et al. (2015). Prenatal stress, regardless of concurrent escitalopram treatment, alters behavior and amygdala gene expression of adolescent female rats. *Neuropharmacology, 97*, 251–258; [18]Glover, M. E., Pugh, P. C., Jackson, N. L., Cohen, J. L., Fant, A. D., Akil, H., et al. (2015). Early-life exposure to the SSRI paroxetine exacerbates depression-like behavior in anxiety/depression-prone rats. *Neuroscience, 284*, 775–797; [19]Zohar, Shoham, and Weinstock (2016); [20]Rayen, I., van den Hove, D. L., Prickaerts, J., Steinbusch, H. W., & Pawluski, J. L. (2011). Fluoxetine during development reverses the effects of prenatal stress on depressive-like behavior and hippocampal neurogenesis in adolescence. *PLoS One, 6*, e24003; [21]Nagano, M., Liu, M., Inagaki, H., Kawada, T., & Suzuki, H. (2012). Early intervention with fluoxetine reverses abnormalities in the serotonergic system and behavior of rats exposed prenatally to dexamethasone. *Neuropharmacology, 63*, 292–300; [22]Sarkar, Chachra, and Vaidya (2014); [23]Popa, D., Lena, C., Alexandre, C., & Adrien, J. (2008). Lasting syndrome of depression produced by reduction in serotonin uptake during postnatal development: Evidence from sleep, stress, and behavior. *The Journal of Neuroscience, 28*, 3546–3554; [24]Salari, Fatehi-Gharehlar, Motayagheni, and Homberg (2016).

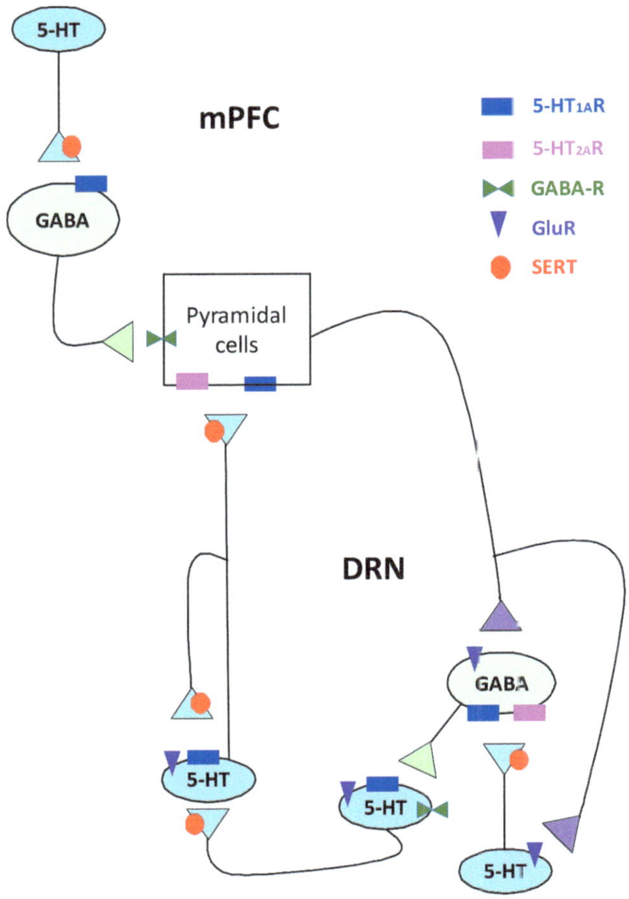


**FIG.1** Circuit of medial prefrontal cortex and dorsal raphe. Serotonergic activity in target regions like the PFC is largely dependent on impulse transmission in neurons arising in the DRN. This is decreased by activation of 5-$HT_{1A}$R on the 5-HT cells in the DRN and medial PFC and increased by activation of 5-$HT_{1A}$R on GABAergic neurons in the lateral portions of the DRN and in the PFC. Impulse transmission in 5-HT neurons from the DRN is also tightly controlled by afferents arising in the medial PFC and other brain areas. Activation of 5$HT_{2A}$R increases neuronal firing. SERT inhibition by SSRIs increases 5-HT in the synapse. In the healthy brain, this results in a decrease in 5-HT in target areas. In MDD or rat models of depression in which binding of 5-HT to 5-$HT_{1A}$R is increased, SSRIs gradually restore 5-HT release in target areas by downregulating these receptors. GABA-R, GABA receptor; GluR, glutamate receptor.

## Key facts

- Selective inhibitors of the transporter (SERT) that recaptures serotonin into the neuron are most commonly used to treat depression and anxiety in pregnant women.
- It is critical to maintain correct levels of serotonin during fetal development of the brain and other systems. Therefore, excess amounts resulting from SERT inhibition may cause adverse effects in the children.
- Citalopram is the most selective of 6 SERT inhibitors (SSRIs) and most frequently prescribed, after fluoxetine.
- Citalopram reaches relatively higher concentrations in the fetus from the placenta and breast milk than the other SSRIs.
- All SSRIs can lower birth weight and paroxetine can cause atrio-septal defects. Citalopram is more likely than the other drugs to cause limb malformations, musculoskeletal system defects, and craniosynostosis.
- Abnormal behaviors seen in the children of depressed mothers include internalizing behavior, anxiety, depression, attention deficit hyperactivity disorder (ADHD), and autism.
- Treatment with SSRIs may reduce the symptoms of ADHD, but not the other abnormal behaviors. SSRIs may adversely affect working memory.

## Summary points

- Selective inhibitors (SSRIs) of the protein (SERT) that transports 5-HT from the synapse into the neuron are most frequently used to treat depression in pregnancy.
- 5-HT plays a crucial role in the development and guidance of its own, and other neurons during embryonic development and can influence a variety of behaviors.
- There is a higher incidence of anxiety, depression, attention deficit hyperactivity disorder, and autism in children of mothers with major depression.
- Treatment with SSRIs as a group neither increases nor decreases the incidence of these behaviors.
- There is not enough information about the effects of individual SSRIs, but fluoxetine may be superior to citalopram in attenuating early developmental defects and behavioral abnormalities in children of mothers with depression

## Mini-dictionary of terms

**Autism spectrum disorder** Condition encompassing a range of symptoms that include impaired communication and social interaction.

**Autoreceptors** Receptors located in the cell body and presynaptic region of nerve cells.

**Craniosynostosis** Premature fusing of joints between the bone plates of an infant's skull.

**Externalizing behavior** Encompasses aggression, impulsivity.

**Internalizing behavior** Encompasses social withdrawal, poor communication.

**Paracrine** The action of a hormone on cells other than those that release it.

**Racemate** Contains equal amounts of left (S) and right (*R*)-handed enantiomers of a chiral molecule.

**Receptors** Proteins on the cell membrane that bind neurotransmitters and initiate a cellular response.

**Serotonin transporter** Membrane protein that transport serotonin (5-HT) from the synaptic space into the neuron.

**Spinogenesis** Development of dendritic spines in neurons.

**TABLE 5** Perinatal antidepressant treatment of stressed rats: Effect on behavior and expression of $5HT_{1A}R$.

| Drug, (dose mg/kg) Stress and its timing | Species strain | Time of drug admin. | Route | Pup age sex | Social behavior | Anxiety (EPM: OF) | Depression (FST: SP) | $5HT_{1A}R$ in PFC | Ref. |
|---|---|---|---|---|---|---|---|---|---|
| Fluoxetine (5) | Rat S-D | E1-P21 | Oral | 27 M, F | PS ↓ | NT | NT | NT | 25 |
| CUS 0–21 | | | | | Drug ↑ | | | | |
| Citalopram (10/day) | Rat | E1-P21 | Oral | 60 M, F | NT | | | PFC[a] | 19 |
| CVS 14–21 | | | | | | PS ↑ M=F | PS ↑ M, F | PS ↓ M=F | |
| | Wistar | | DF | | | Drug=M, F | Drug=M, F | Drug=M ↑ F | |
| Fluoxetine (5) | Rat | P1-P21 | MP | 32 M, F | NT | NT | PS ↑ | NT | 20 |
| 3× daily restr.15–21 | S-D | | | | | | Drug ↓ [32] | | |
| Fluoxetine (17.2 mg/kg/day) | Rat | P2-P21 | DF | 77–84 M | NT | PS ↑ | NT | PS ↓ [28,84] | 21 |
| (Dex 50 μ/kg) 16–21 | S-D | | | | | Drug ↓ [84] | | Drug ↑ [28,84] | |
| Fluoxetine (8/day) | Mouse | E10-E20 | DF | 68+ M | NT | PS ↑ | PS ↑ | NT | 24 |
| 3× daily restr. 5–19 | NMRI | P0-P20 | | | | Drug ↓ | Drug ↓ | | |

[a]*On GABAergic interneurons.*
Legend as in Tables 3 and 4.
CUS, chronic unpredictable stress; CVS, chronic varied stress; restr, restraint; Dex, dexamethasone; SP, sucrose preference.
[25]Gemmel, M., Hazlett, M., Bogi, E., De Lacalle, S., Hill, L. A., Kokras, N., et al. (2017). Perinatal fluoxetine effects on social play, the HPA system, and hippocampal plasticity in preadolescent male and female rats: Interactions with pregestational maternal stress. Psychoneuroendocrinology, 84, *159–171*.

## References


Alwan, S., Friedman, J. M., & Chambers, C. (2016). Safety of selective serotonin reuptake inhibitors in pregnancy: A review of current evidence. *CNS Drugs*, *30*, 499–515.

Andersen, S. L., & Thompson, B. S. (2011). Reply to Abramowitz et al.: Animal models of OCD. *Biological Psychiatry*, *69*(9), e31–e32.

Beliveau, V., Ganz, M., Feng, L., Ozenne, B., Hojgaard, L., Fisher, P. M., et al. (2017). A high-resolution in vivo atlas of the human Brain's serotonin system. *The Journal of Neuroscience*, *37*, 120–128.

Berard, A., Zhao, J. P., & Sheehy, O. (2017). Antidepressant use during pregnancy and the risk of major congenital malformations in a cohort of depressed pregnant women: An updated analysis of the Quebec pregnancy cohort. *BMJ Open*, *7*, e013372.

Blackburn, T. P. (2019). Depressive disorders: Treatment failures and poor prognosis over the last 50 years. *Pharmacology Research & Perspectives*, *7*, e00472.

Boldrini, M., Underwood, M. D., Mann, J. J., & Arango, V. (2008). Serotonin-1A autoreceptor binding in the dorsal raphe nucleus of depressed suicides. *Journal of Psychiatric Research*, *42*, 433–442.

Bonnin, A., & Levitt, P. (2011). Fetal, maternal, and placental sources of serotonin and new implications for developmental programming of the brain. *Neuroscience*, *197*, 1–7.

Bonnin, A., Torii, M., Wang, L., Rakic, P., & Levitt, P. (2007). Serotonin modulates the response of embryonic thalamocortical axons to netrin-1. *Nature Neuroscience*, *10*, 588–597.

Boukhris, T., Sheehy, O., & Berard, A. (2017). Antidepressant use in pregnancy and the risk of attention deficit with or without hyperactivity disorder in children. *Paediatric and Perinatal Epidemiology*, *31*, 363–373.

Brown, H. K., Hussain-Shamsy, N., Lunsky, Y., Dennis, C. E., & Vigod, S. N. (2017). The association between antenatal exposure to selective serotonin reuptake inhibitors and autism: A systematic review and meta-analysis. *The Journal of Clinical Psychiatry*, *78*, e48–e58.

Celada, P., Puig, M., Amargos-Bosch, M., Adell, A., & Artigas, F. (2004). The therapeutic role of 5-HT1A and 5-HT2A receptors in depression. *Journal of Psychiatry & Neuroscience*, *29*, 252–265.

Cipriani, A., Furukawa, T. A., Salanti, G., Chaimani, A., Atkinson, L. Z., Ogawa, Y., et al. (2018). Comparative efficacy and acceptability of 21 antidepressant drugs for the acute treatment of adults with major depressive disorder: A systematic review and network meta-analysis. *Lancet*, *391*, 1357–1366.

Deneris, E., & Gaspar, P. (2018). Serotonin neuron development: Shaping molecular and structural identities. *Wiley Interdisciplinary Reviews: Developmental Biology*, *7*(1), 1–38.

Eke, A. C., Saccone, G., & Berghella, V. (2016). Selective serotonin reuptake inhibitor (SSRI) use during pregnancy and risk of preterm birth: A systematic review and meta-analysis. *BJOG*, *123*, 1900–1907.

El Marroun, H., White, T. J., Fernandez, G., Jaddoe, V. W., Verhulst, F. C., Stricker, B. H., et al. (2017). Prenatal exposure to selective serotonin reuptake inhibitors and non-verbal cognitive functioning in childhood. *Journal of Psychopharmacology*, *31*, 346–355.

Gao, S. Y., Wu, Q. J., Zhang, T. N., Shen, Z. Q., Liu, C. X., Xu, X., et al. (2017). Fluoxetine and congenital malformations: A systematic review and meta-analysis of cohort studies. *British Journal of Clinical Pharmacology*, *83*, 2134–2147.

Gingrich, J. A., Malm, H., Ansorge, M. S., Brown, A., Sourander, A., Suri, D., et al. (2017). New insights into how serotonin selective reuptake inhibitors shape the developing Brain. *Birth Defects Research*, *109*, 924–932.

Glennon, R. A., & Dukat, M. (1991). Serotonin receptors and their ligands: A lack of selective agents. *Pharmacology Biochemistry, and Behavior*, *40*, 1009–1017.

Halvorsen, A., Hesel, B., Ostergaard, S. D. & Danielsen, A. A. (2019). In utero exposure to selective serotonin reuptake inhibitors and development of mental disorders: A systematic review and meta-analysis. *Acta Psychiatrica Scandinavica*, *139*, 493–507.

Hanley, G. E., Brain, U., & Oberlander, T. F. (2015). Prenatal exposure to serotonin reuptake inhibitor antidepressants and childhood behavior. *Pediatric Research*, *78*, 174–180.

Hendrick, V., Stowe, Z. N., Altshuler, L. L., Hwang, S., Lee, E., & Haynes, D. (2003). Placental passage of antidepressant medications. *The American Journal of Psychiatry*, *160*, 993–996.

Hermansen, T. K., Roysamb, E., Augusti, E. M., & Melinder, A. (2016). Behavior and inhibitory control in children with prenatal exposure to antidepressants and medically untreated depression. *Psychopharmacology*, *233*, 1523–1535.

Kinney, H. C., Belliveau, R. A., Trachtenberg, F. L., Rava, L. A., & Paterson, D. S. (2007). The development of the medullary serotonergic system in early human life. *Autonomic Neuroscience*, *132*, 81–102.

Lauder, J. M. (1990). Ontogeny of the serotonergic system in the rat: Serotonin as a developmental signal. *Annals of the New York Academy of Sciences*, *600*, 297–313. discussion 314

Lupattelli, A., Wood, M., Ystrom, E., Skurtveit, S., Handal, M., & Nordeng, H. (2018). Effect of time-dependent selective serotonin reuptake inhibitor antidepressants during pregnancy on behavioral, emotional, and social development in preschool-aged children. *Journal of the American Academy of Child and Adolescent Psychiatry*, *57*, 200–208.

Man, K. K. C., Chan, E. W., Ip, P., Coghill, D., Simonoff, E., Chan, P. K. L., et al. (2017). Prenatal antidepressant use and risk of attention-deficit/hyperactivity disorder in offspring: Population based cohort study. *BMJ*, *357*, j2350.

McEwen, B. S. (2005). Glucocorticoids, depression, and mood disorders: Structural remodeling in the brain. *Metabolism*, *54*(5 Suppl 1), 20–23.

Oblak, A., Gibbs, T. T., & Blatt, G. J. (2013). Reduced serotonin receptor subtypes in a limbic and a neocortical region in autism. *Autism Research*, *6*, 571–583.

Overmars, H., Scherpenisse, P. M., & Post, L. C. (1983). Fluvoxamine maleate: metabolism in man. *European Journal of Drug Metabolism and Pharmacokinetics*, *8*, 269–280.

Owens, M. J., & Nemeroff, C. B. (1994). Role of serotonin in the pathophysiology of depression: Focus on the serotonin transporter. *Clinical Chemistry*, *40*, 288–295.

Rampono, J., Simmer, K., Ilett, K. F., Hackett, L. P., Doherty, D. A., Elliot, R., et al. (2009). Placental transfer of SSRI and SNRI antidepressants and effects on the neonate. *Pharmacopsychiatry*, *42*, 95–100.

Raymond, J. R., Mukhin, Y. V., Gettys, T. W., & Garnovskaya, M. N. (1999). The recombinant 5HT1A receptor: G protein coupling and signalling pathways. *British Journal of Pharmacology*, *127*, 1751–1764.

Salari, A. A., Fatehi-Gharehlar, L., Motayagheni, N., & Homberg, J. R. (2016). Fluoxetine normalizes the effects of prenatal maternal stress on depression- and anxiety-like behaviors in mouse dams and male offspring. *Behavioural Brain Research*, *311*, 354–367.

Sarkar, A., Chachra, P., & Vaidya, V. A. (2014). Postnatal fluoxetine-evoked anxiety is prevented by concomitant 5-HT2A/C receptor blockade and mimicked by postnatal 5-HT2A/C receptor stimulation. *Biological Psychiatry*, *76*, 858–868.

Stahl, S. M. (1998). Mechanism of action of serotonin selective reuptake inhibitors. Serotonin receptors and pathways mediate therapeutic effects and side effects. *Journal of Affective Disorders*, *51*, 215–235.

Taylor, G. T., Lerch, S., & Chourbaji, S. (2017). Marble burying as compulsive behaviors in male and female mice. *Acta Neurobiologiae Experimentalis (Wars)*, *77*, 254–260.

Tork, I. (1990). Anatomy of the serotonergic system. *Annals of the New York Academy of Sciences*, *600*, 9–34 (discussion 34-35).

Viktorin, A., Uher, R., Reichenberg, A., Levine, S. Z., & Sandin, S. (2017). Autism risk following antidepressant medication during pregnancy. *Psychological Medicine*, *47*, 2787–2796.

Volkmar, F. R. (2009). Citalopram treatment in children with autism spectrum disorders and high levels of repetitive behavior. *Archives of General Psychiatry*, *66*(6), 581–582.

Weinstock, M. (2008). The long-term consequences of prenatal stress. *Neuroscience & Biobehavioral Reviews*, *32*, 1073–1086.

Weissman, A. M., Levy, B. T., Hartz, A. J., Bentler, S., Donohue, M., Ellingrod, V. L., et al. (2004). Pooled analysis of antidepressant levels in lactating mothers, breast milk, and nursing infants. *The American Journal of Psychiatry*, *161*, 1066–1078.

Westin, A. A., Brekke, M., Molden, E., Skogvoll, E., & Spigset, O. (2017). Selective serotonin reuptake inhibitors and venlafaxine in pregnancy: Changes in drug disposition. *PLoS One*, *12*, e0181082.

Yonkers, K. A., Blackwell, K. A., Glover, J., & Forray, A. (2014). Antidepressant use in pregnant and postpartum women. *Annual Review of Clinical Psychology*, *10*, 369–392.

Zohar, I., Dosoretz-Abittan, L., Shoham, S., & Weinstock, M. (2015). Sex dependent reduction by prenatal stress of the expression of 5HT1A receptors in the prefrontal cortex and CRF type 2 receptors in the raphe nucleus in rats: Reversal by citalopram. *Psychopharmacology*, *232*, 1643–1653.

Zohar, I., Shoham, S., & Weinstock, M. (2016). Perinatal citalopram does not prevent the effect of prenatal stress on anxiety, depressive-like behavior and serotonergic transmission in adult rat offspring. *The European Journal of Neuroscience*, *43*, 590–600.

Chapter 33

# Citalopram and its use in sleep-deprivation-induced depression

Afzal Misrani[a,b] and Cheng Long[a,c]
[a]School of Life Sciences, South China Normal University, Guangzhou, PR China; [b]School of Life Sciences, Guangzhou University, Guangzhou, PR China; [c]South China Normal University-Panyu Central Hospital Joint Laboratory of Translational Medical Research, Panyu Central Hospital, Guangzhou, PR China

## List of abbreviations

| | |
|---|---|
| **5-HT** | 5-hydroxytryptamine |
| **AD** | Alzheimer's disease |
| **BDNF** | brain-derived neurotrophic factor |
| **CaMKII** | calcium/calmodulin-dependent protein kinase II |
| **CNS** | central nervous system |
| **CREB** | cAMP response element-binding protein |
| **CTM** | citalopram |
| **FDA** | Food and Drug Administration |
| **LTP** | long-term potentiation |
| **MDD** | major depressive disorder |
| **NHIS** | National Health Interview Survey |
| **NSF** | National Sleep Foundation |
| **PFC** | prefrontal cortex |
| **REM** | rapid eye movement |
| **SD** | sleep deprivation |
| **SNRI** | serotonin-norepinephrine reuptake inhibitor |
| **SSRI** | selective serotonin reuptake inhibitor |
| **SWS** | slow-wave sleep |

## Introduction

A poorly understood but important problem in our 24/7 society is the increasing prevalence of sleep deprivation (SD), which negatively affects life quality and is becoming a major health concern (Rajaratnam & Arendt, 2001; Shochat, Cohen-Zion, & Tzischinsky, 2014). In the modern world, SD is caused by suboptimal work schedules, around-the-clock lifestyles, and psychosocial stress, which severely affect our physical and mental health, placing a huge financial burden on our economy (Hillman & Lack, 2013). Therefore, an understanding of the cellular and molecular pathways affected by SD is clearly of social and clinical importance.

Because sleep plays an important role in metabolic homeostasis, energy restoration, learning-related synapse formation and neuronal reactivation—all of which are necessary for normal brain function—insufficient sleep may have detrimental effects on brain function. The National Sleep Foundation (NSF) suggests that 7–8 h sleep per night is necessary to maintain good health. Several epidemiological studies suggest that chronically disrupted sleep leads to pathological anxiety and depression in humans (Roberts & Duong, 2014). Moreover, recent reports argue that chronic sleep disruption may slowly induce changes within the brain that are very similar to those reported in depressed patients (Buysse et al., 2008; Meerlo, Havekes, & Steiger, 2015). Recently, a relationship between SD and depression has been established using a large-scale community-based cohort study (Roberts & Duong, 2014), suggesting that a reduced amount of sleep increases the risk of depression, which in turn increases the risk of sleep disturbance. The authors provide evidence that adolescents with disrupted sleep report an increase in depression and anxiety, as well as poor academic performance. This study also confirmed that sleep loss has prolonged negative effects on the brain, which may contribute to depression. More than 90% of depression subjects complain about impairment in sleep quality (Chang, Chang, Lin, Wu, & Yen, 2017).

During the last decade, the number of patients who consult primary care physicians or psychiatrists for symptoms of depression has doubled (Horwitz, 2010; Kasper & Muller-Spahn, 2002). The majority of depressed patients are prescribed antidepressants. Selective serotonin reuptake inhibitors (SSRIs) are extensively prescribed for the treatment of several sleep disorders and depression. Repeated administration of SSRIs seems to exert antidepressant effects in rodents. Citalopram (CTM) hydrobromide, an SSRI, has been available as a Food and Drug Administration (FDA)-approved drug since the 1980s in the United States—and subsequently in other countries of the world—for the treatment of depression (Cipriani et al., 2012). Inhibition of the neuronal transporter for

The Neuroscience of Depression. https://doi.org/10.1016/B978-0-12-817933-8.00048-7

5-hydroxytryptamine (5-HT; serotonin) has long been established as an important mechanism of action of several antidepressants. CTM, a dicyclic phthalide derivative, is typical in this regard in that it causes inhibition of the reuptake of 5-HT in the brain. CTM is a highly selective and potent SSRI with relatively limited effects on the neuronal reuptake of norepinephrine and dopamine. There is an acute increase in extracellular 5-HT levels in animals following treatment with 10 mg/kg CTM twice-daily, which in turn results in activation of somato-dendritic 5-$HT_{1A}$ auto-receptors. The result of this acute activation is a feedback inhibition in the raphe nucleus, with downregulation of auto-receptors and a subsequent increase in serotonergic neurotransmission.

CTM is the most-frequently used antidepressant in clinical practice for the treatment of depression. Disrupted sleep is prevalent in depressed patients, but CTM treatment can improve sleep quality in patients with depression and Alzheimer's disease (AD) (Newell et al., 2016; Shahsavand-Ananloo et al., 2013). A recent study on humans showed that CTM at a dose of 20 mg/kg/day improves sleep latency, sleep duration, and the global Pittsburgh Sleep Quality Index (Wu, Chang, & Zu, 2015). Importantly, the acute administration of CTM facilitates memory consolidation in healthy humans and depression models (Harmer, Bhagwagar, Cowen, & Goodwin, 2002). Therefore, it seems that regulation of sleep might be important for understanding the pathophysiology and treatment of depression. Moreover, a clinical trial (Shahsavand-Ananloo et al., 2013) designed to compare the effects of CTM and fluoxetine on the sleep quality of patients with major depressive disorder (MDD) revealed that both CTM and fluoxetine improve sleep quality after 8 weeks of treatment at a dose of 20–40 mg/kg/day. In another trial, CTM (20 mg/kg/day) and the serotonin-norepinephrine reuptake inhibitor (SNRI) venlafaxine (75 mg/kg/day) both reduced sleep disturbance in postmenopausal women (Davari-Tanha et al., 2016). Efforts have also been made to investigate whether CTM and doxepin can alleviate insomnia and comorbid anxiety (Wu et al., 2015). Overall, a plethora of recent studies suggests a neuroprotective role for CTM against depression and other neuropsychiatric disorders where sleep is compromised.

## Prevalence of SD

An adequate amount of sleep is essential for a healthy life. The optimal amount of sleep for a normal adult is 8 h in a 24-h period. Studies of sleep in humans have shown the adverse effects of SD on human health according to metabolic, physical, psychological, and behavioral parameters (Adamantidis and de Lecea, 2008). These adverse effects can then lead to greater risks of severe disease and may affect learning and memory processes and mood because the restorative properties of sleep are not fully experienced during SD. Poor sleep negatively affects the quality of life and has long-term consequences on the brain function. Deficits in cognitive function because of insufficient sleep are experienced universally in modern civilizations. Moreover, insufficient sleep or sleep loss is a hallmark of modern society, contributing to major health issues and conferring a huge economic burden. In American adults and teenagers, sleep time has declined by 1.5–2 h over the last 50 years (Van Cauter, Spiegel, Tasali, & Leproult, 2008). A subjective sleep duration of <6 h was reported by >30% of 30–64-year-old Americans. A National Health Interview Survey (NHIS) suggests that chronic sleep disturbance and sleep disorders affect 70 million Americans (Foley, Ancoli-Israel, Britz, & Walsh, 2004). Various other surveys report decreased sleep duration and its association with mental illness in American adults (Krueger & Friedman, 2009; Seixas et al., 2015; Shockey & Wheaton, 2017). A recent study conducted on the Chinese population reported that 51.0% of adolescents do not achieve optimal sleep duration (<8 h per day) on weekdays, although this is less severe (9.8%) at the weekend (Chen et al., 2014). A large scale epidemiological study on Chinese high school students revealed that 18.0% experienced poor sleep (Tang et al., 2016). Moreover, this poor sleep quality increases with age (Wang et al., 2017).

## Sleep loss as a risk factor for depression

Sleep occupies about one-third of the human lifetime and plays a crucial role in energy restoration, neuronal recovery, and brain plasticity. These processes are not only associated with daytime brain function and body homeostasis but are also important for growth, mood regulation, and psychological and cognitive development. Major depression, also known as depressive disorder, is a typical mood disorder in which the individual shows a persistently depressed mood state. The morbidity of depression in the United States is 16.2%, and a gradually increasing trend has been observed. The prevalence of MDD ranges between 5 and 6% in teenagers (Birmaher et al., 2007), with a lifetime prevalence rate of 20% in adults. People with poor sleep quality are more likely to develop depression than individuals who sleep well (Szklo-Coxe, Young, Peppard, Finn, & Benca, 2010). Moreover, sleep disruption is now known to be a possible contributor to the development of depression. Indeed, insomnia is the first sign of depression to appear and the last sign to disappear. Polysomnographic studies have shown disturbed sleep architecture and continuity in depressed patients. These disturbances include decreased slow-wave sleep (SWS), intermittent awakening, prolonged sleep latency, shortened rapid eye movement (REM) latency, and elevated REM density, particularly during the first REM period (Baglioni et al., 2015; Palagini, Baglioni, Ciapparelli, Gemignani, & Riemann, 2013).

Many people regularly do not obtain sufficient sleep because of our modern around-the-clock psychosocial lifestyle, high work pressure, and psychosocial stress. Indeed, sleep complaints and restricted sleep have been identified as risk factors for several psychiatric disorders; in the case of depression, the association with sleep disturbance is particularly strong. Sleep disturbance is not only a symptom of depression, but it may contribute to the development of depression, because it acts as a causal factor that sensitizes individuals, contributes to the development of depression, worsens the severity of symptoms, and decreases the efficacy of various antidepressants (Baglioni et al., 2011). SD negatively affects various brain regions, but the main areas that govern mood and memory, i.e., the hippocampus and prefrontal cortex (PFC), are severely affected by SD, resulting in symptoms of depression and cognitive dysfunction. Sleep serves a fundamental role in regulating memory processes, neuronal plasticity, and synaptic strength, while disrupted sleep may lead to depression through alterations in these neuronal plasticity processes, resulting in depressive symptoms, memory impairments, and altered brain connectivity. The mechanism of such impairments is still unknown, but in general, SD causes alterations in the molecular pathway of neural plasticity in the hippocampus and PFC. In this chapter, we discuss one molecular pathway that is sensitive to SD-induced depression.

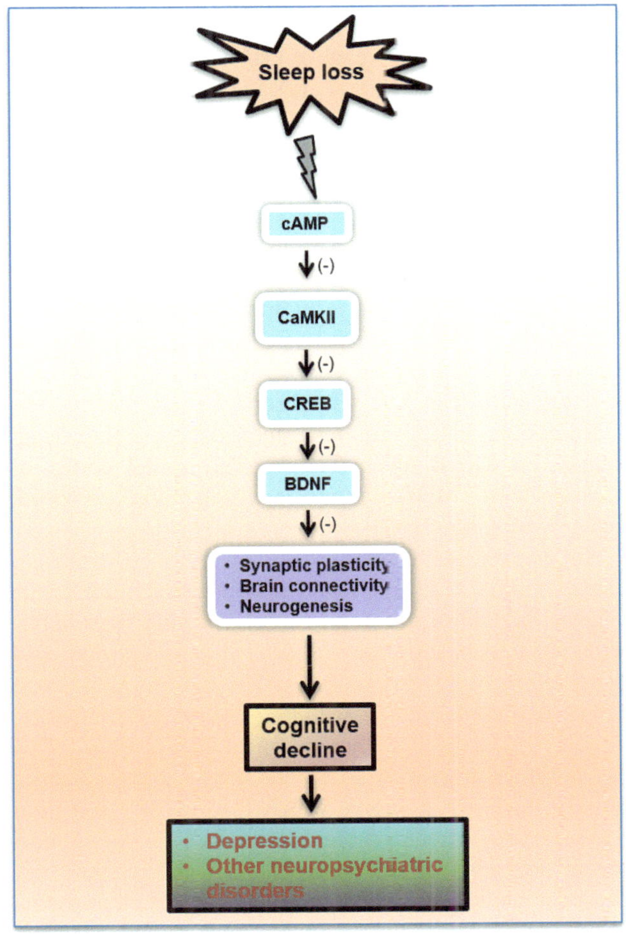


FIG. 1 Potential molecular mechanisms of deleterious effects of sleep deprivation, which contribute to the progression of depression and other neuropsychiatric disorders. Sleep deprivation negatively affects cAMP signaling pathway, which results in diminished CaMKII, CREB, and BDNF function, leading to decreased neurogenesis, disrupted LTP and brain connectivity between brain regions, which in turn cause memory impairment and mood disorders. If this molecular pathway remains disturbed over a long time frame, it may lead to major depression and various other neuropsychiatric disorders.

## Sleep deprivation-induced molecular deficits—Effects of antidepressants

In the central nervous system (CNS), action potentials activate the molecular pathway by means of $Ca^{2+}$ influx. The major components of this pathway are calcium/calmodulin kinase II (CaMKII), cyclic AMP-responsive element-binding protein (CREB), brain-derived neurotrophic factor (BDNF). The involvement of CaMKII-CREB-BDNF in mood disorders is well documented and SD is known to negatively affect this signaling pathway (Fig. 1). In this section, we discuss in detail the impact of SD and antidepressants on this signaling.

### Calcium/calmodulin kinase II (CaMKII)

CaMKII is a $Ca^{2+}$-activated kinase that is highly abundant in the brain and plays a vital role in the expression of synaptic plasticity. This kinase is present in large amounts at synapses and is the main protein of the so-called postsynaptic density, the electron-dense thickening underlying the postsynaptic membrane at excitatory synapses (Lisman, Schulman, & Cline, 2002). CaMKII is involved in sleep–wake regulation in various brain regions, such as the pedunculopontine tegmentum, pedunculopontine nucleus, and dorsal raphe nucleus in the brainstem, as well as the cortex and hippocampus (Cui et al., 2016; Datta, O'malley, & Patterson, 2011; Vyazovskiy, Cirelli, Pfister-Genskow, Faraguna, & Tononi, 2008). Furthermore, a CaMKIIα knockout mouse model exhibits modified behavior, while alterations in CaMKII signaling contribute to the pathophysiology of depression. Various antidepressants, when given to treat depression, generally increase CaMKII activity as a part of their mechanism of action. Moreover, several antidepressants including fluoxetine, desipramine, and reboxetine, markedly increase the enzyme activity of CaMKII and CaMKIV in rats (Vinet et al., 2004; Tiraboschi, Tardito, et al., 2004).

## cAMP response element-binding protein (CREB)

CREB functions downstream of CaMKII as a regulator of gene expression and is associated with many neuroprotective processes. Thus, CREB phosphorylation leads to transcriptional activation of genes, many of which are involved in mood disorders. Accumulating evidence indicates that sleep enhances CREB phosphorylation, which is consistent with a close relationship between CREB signaling and sleep regulation (Luo, Phan, Yang, Garelick, & Storm, 2013). CREB plays an important role in the pathophysiology of neuropsychiatric disorders, as well as in response to antidepressants (Tardito et al., 2009). A reduction in CREB levels is observed in animal models of depression, and postmortem studies show a significant decrease in CREB levels in the orbitofrontal cortex of patients with MDD (Yamada, Yamamoto, Ozawa, Riederer, & Saito, 2003).

Antidepressants are prescribed for up to 2–6 weeks to achieve optimal clinical efficacy in depression models; this suggests that these drugs might need a long-term application to activate downstream pathways. Regarding this phenomenon, a novel hypothesis is emerging that could change our views of the pathophysiology of depression and the mechanism of action of antidepressants. This hypothesis assumes that antidepressants trigger not only second messenger systems, which lead to the activation of transcription effector molecules such as CREB but also activate neurotrophic signaling pathways and enhance hippocampal neurogenesis. Moreover, most antidepressants activate and increase the levels of cAMP in the cell, possibly through activation of adrenoceptors or serotonin (5-HT) receptors. The ability of antidepressants to trigger protein kinases in the cell and the potential for these kinases to phosphorylate CREB suggests that antidepressants can activate CREB as part of their mechanism of action. In fact, fluoxetine administration in rats increases CREB levels in several brain areas, including the amygdala, cortex, hippocampus, and hypothalamus (Carlini et al., 2012; Qi et al., 2008). Interestingly, desipramine only increases levels of CREB in the dentate gyrus area of the hippocampus (Thome et al., 2000). Diminished cAMP signaling in the brain is supposed to occur in depression, while antidepressants increase levels of the critical downstream signaling target of cAMP signaling, i.e., CREB. Antidepressants enhance signaling of 5-HT, possibly by inhibiting reuptake to presynaptic terminals, inhibiting catabolism, binding to 5-HT or noradrenergic receptors, or a combination of such effects. These cellular events take place shortly after drug administration; however, the drugs require several weeks to achieve optimal clinical antidepressant effects. In our study, CTM treatment for 5 days increased pCREB levels in the hippocampus and PFC of an SD mouse model (Misrani et al., 2019, 2020). Reduction in CREB function has been associated with the etiology of depression. Various animal studies suggest that activation of CREB is linked with antidepressant efficacy (Blendy, 2006). Importantly, postmortem studies provide the strongest evidence in support of this hypothesis. Patients who were on antidepressant medication at the time of death showed a higher level of CREB (Dowlatshahi, Macqueen, Wang, & Young, 1998) than deceased patients who had not been taking antidepressants. Decreased levels of tCREB and pCREB were also reported in the orbitofrontal cortex of antidepressant-free patients with MDD (Yamada et al., 2003). Accordingly, CREB represents a molecular state marker for both depression and the response to antidepressant treatment.

## Brain-derived neurotrophic factor (BDNF)

BDNF has emerged as a key molecule for the control of neuronal differentiation and survival, synapse formation, and the regulation of activity-dependent changes in synapse structure and function (Leal, Afonso, Salazar, & Duarte, 2015). In recent years, increasing interest has focused on the molecular mechanisms involved in neuroplasticity, which may be relevant to the treatment of major depression (Krishnan & Nestler, 2008). In particular, SSRI antidepressants trigger a complex biochemical cascade that mediates the actions of the neurotrophin BDNF and ultimately promotes neuronal plasticity (Martinowich & Lu, 2008). Moreover, postmortem studies show decreased levels of BDNF in depression patients (Banerjee, Ghosh, Ghosh, Bhattacharyya, & Mondal, 2013). Disrupted BDNF signaling is thought to be a consequence of sleep loss (Schmitt, Holsboer-Trachsler, & Eckert, 2016) and various neuropsychiatric models are characterized by decreased levels of BDNF (Cunha et al., 2006).

BDNF is abundant in various brain regions implicated in mood disorders, including the PFC, hippocampus and amygdala, and shows a remarkable activity-dependent regulation of expression and secretion. Thus, its neurotrophic functions are connected to various physiological functions in the brain particularly, relevant to neuroplasticity, memory and sleep (Duman, 2014; Duman, Heninger, & Nestler, 1997; Faraguna, Vyazovskiy, Nelson, Tononi, & Cirelli, 2008), suggesting it has a biological role over the entire life span. Given the complexity of the BDNF system in terms of transcriptional, translational and posttranslational regulatory mechanisms, a disruption in BDNF signaling may lead to defects in neuronal survival and regeneration, as well as structural brain abnormalities and reduced neuronal plasticity, which in turn may impair learning and memory, and normal brain function. Our finding that CTM treatment maintains BDNF levels supports the hypothesis that BDNF mediates the action of antidepressants (Misrani et al., 2020). Moreover, these findings may have implications for our understanding of how SSRIs negate the deleterious effects of SD on the CaMKII-CREB-BDNF pathway. Consistent with this, our previous and various other studies suggest that antidepressants, including CTM maintain CaMKII-CREB-BDNF signaling (Chung et al., 2014; Misrani et al., 2019; Sairanen, O'leary, Knuuttila, & Castren, 2007; Tiraboschi, Giambelli, et al., 2004). The ability of antidepressants to trigger protein kinases in the cell and the potential for these kinases to phosphorylate CREB and BDNF suggest that antidepressants can activate this signaling pathway as part of their mechanism of action.

## Synaptic plasticity and SD

Synaptic plasticity is an activity-dependent change in neuronal plasticity that results in a persistent enhancement of synaptic transmission. In other words, synaptic plasticity can be defined as a long-lasting change in the strength of synaptic connections through the involvement of several cellular and molecular signaling cascades, together with protein synthesis. SD may disrupt the molecular events underlying synaptic plasticity and learning and memory (CaMKII-CREB-BDNF) and lead to depression. Disrupted synaptic plasticity is abundant in depression patients, suggesting a crucial role for synaptic mechanisms and the respective molecular pathways in the etiology of depression. For example, a molecular study demonstrated that acute SD causes an alteration in long-term potentiation (LTP) through disruption of cAMP signaling (Havekes et al., 2016), which might be reversed by increasing cAMP signaling (Havekes et al., 2014), suggesting a strong association between disrupted LTP and depression.

## Effects of CTM on SD-induced depression

Impaired sleep is very common in patients with depression and a range of other neuropsychiatric disorders. CTM can induce daytime sedation and improve the sleep quality of depressed patients and is currently one of the most-frequently used antidepressants in clinical practice for the treatment of depressive symptoms and sleep disturbances. Furthermore, CTM is associated with improved sleep quality in patients with depression and AD (Newell et al., 2016; Shahsavand-Ananloo et al., 2013). Recently, CTM has attracted much attention as a potential treatment for the cognitive deficits associated with AD. One study in an APP/PS1 mouse model showed that CTM at a dose of 10 mg/kg for 1 month could attenuate memory and synaptic plasticity deficits and decrease levels of Aβ (Zhang et al., 2018). A similar study suggested that CTM could prevent spatial memory and synaptic plasticity deficits and decrease the levels of insoluble $A\beta_{40}$ in 3xTgAD mice at a dose of 10 mg/kg (Wei et al., 2017). Moreover, we and others have established that CTM increases levels of CaMKII, CREB and BDNF, and improves learning and memory by promoting synaptic plasticity and hippocampal neurogenesis in SD mice and socially isolated rats (Gong et al., 2017; Misrani et al., 2019, 2020).

## Concluding remarks

Globally, a considerable proportion of the population does not obtain sufficient sleep. If sleep loss accumulates and remains untreated over a long-time frame, the resulting sleep-deprived and depressed individuals would represent an enormous social and economic burden. Drugs that enhance neuronal plasticity and modulate the associated genes provide hope that this situation might be alleviated. We now know that acute SD causes learning and memory deficits and alters the mood and synaptic plasticity. Prolonged SD may disrupt learning and memory-related signaling pathways, suppress neurogenesis, and even lead to depression and morphological changes in the brain, including hippocampal shrinkage. Because insufficient sleep is a hallmark of modern society that contributes to major health issues, any procedures that improve sleep may have both immediate and long-term benefits for cognitive function, including mood regulation, learning and memory, and normal brain function. Where professional commitments demand regular sleep sacrifice, CTM could offer a means of neuroprotection against the deleterious molecular, synaptic, and behavioral effects of SD (Fig. 2). In summary, CTM treatment represents a promising means of counteracting SD-induced alterations, thus, further research is warranted to verify the effects of CTM on mood disorders associated with poor sleep in humans.

## Key facts about sleep deprivation

- The NSF suggests that 7–8 h of sleep per 24-h period is necessary for the maintenance of good health.
- More than 90% of depression patients complain about impaired sleep quality.
- In American adults and teenagers, sleep time has declined by 1.5–2 h over the last 50 years.
- The NHIS suggests that chronic sleep disturbance and sleep disorders affect 70 million Americans.
- A recent study conducted on the Chinese population reported that 51.0% of adolescents do not achieve optimal sleep.

## Summary points

- Sleep deprivation is a hallmark of modern society.
- Sleep deprivation is known to be a possible contributor to the development of depression.

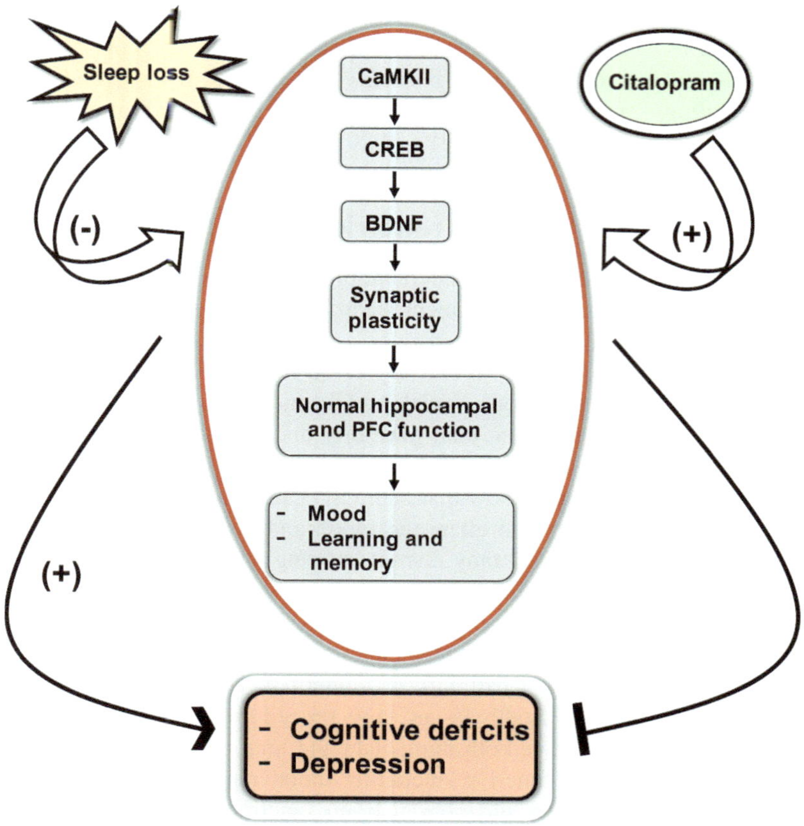


**FIG. 2** Underlying molecular pathways involved in the neuroprotective activity of citalopram against SD-induced depression. Sleep deprivation negatively affects the molecular pathway underpinning LTP (CaMKII, CREB, and BDNF), leading to memory impairment and depressive-like behavior. CTM maintains CaMKII, CREB, and BDNF activity, thereby preventing LTP deficits, memory impairment, and depression symptoms. *Modified from figure 8 of Misrani, A., Tabassum, S., Chen, X., Tan, S.Y., Wang, J.C., Yang, L., et al. (2019). Differential effects of citalopram on sleep-deprivation-induced depressive-like behavior and memory impairments in mice.* Progress in Neuro-Psychopharmacology & Biological Psychiatry, 88, *102–111.*

- Sleep deprivation causes disruption in the CaMKII-CREB-BDNF pathway and impairs synaptic plasticity.
- Citalopram treatment prevents sleep-deprivation-induced alteration in CaMKII-CREB-BDNF signaling and synaptic plasticity.
- Citalopram treatment restores sleep-deprivation-associated depression.

## Mini-dictionary of terms

**BDNF** Brain-derived neurotrophic factor (BDNF) is a neurotrophic growth factor involved in neuroprotection.

**CaMKII** CaMKII is a $Ca^{2+}$-activated kinase that is highly abundant in the brain and plays a vital role in the expression of synaptic plasticity; it is involved in the etiology of depression.

**Citalopram** Citalopram (CTM), an SSRI, is the most-frequently used antidepressant in clinical practice for the treatment of depression.

**CREB** cAMP response element-binding protein (CREB) is a cellular transcription factor. The activity of CREB is decreased in sleep deprivation and depression.

**Sleep deprivation** Sleep deprivation is a condition of curtailment of sleep in modern society.

## Acknowledgments

The present study was supported by grants from the National Natural Science Foundation of China [31871170], the Guangdong Grant "Key Technologies for Treatment of Brain Disorders" [2018B030332001] to CL.

## References

Adamantidis, A., & De Lecea, L. (2008). Sleep and metabolism: Shared circuits, new connections. *Trends in Endocrinology and Metabolism*, *19*, 362–370.

Baglioni, C., Battagliese, G., Feige, B., Spiegelhalder, K., Nissen, C., Voderholzer, U., et al. (2011). Insomnia as a predictor of depression: A meta-analytic evaluation of longitudinal epidemiological studies. *Journal of Affective Disorders*, *135*, 10–19.

Baglioni, C., Nanovska, S., Regen, W., Spiegelhalder, K., Feige, B., Nissen, C., et al. (2016). Sleep and mental disorders: A meta-analysis of polysomnographic research. *Psychological Bulletin*, *142*, 969–990.

Banerjee, R., Ghosh, A. K., Ghosh, B., Bhattacharyya, S., & Mondal, A. C. (2013). Decreased mRNA and protein expression of BDNF, NGF, and their receptors in the hippocampus from suicide: An analysis in human postmortem brain. *Clinical Medicine Insights Pathology*, *6*, 1–11.

Birmaher, B., Brent, D., Issues, A. W. G. O. Q., Bernet, W., Bukstein, O., Walter, H., et al. (2007). Practice parameter for the assessment and treatment of children and adolescents with depressive disorders. *Journal of the American Academy of Child and Adolescent Psychiatry*, *46*, 1503–1526.

Blendy, J. A. (2006). The role of CREB in depression and antidepressant treatment. *Biological Psychiatry*, *59*, 1144–1150.

Buysse, D. J., Angst, J., Gamma, A., Ajdacic, V., Eich, D., & Rossler, W. (2008). Prevalence, course, and comorbidity of insomnia and depression in young adults. *Sleep*, *31*, 473–480.

Carlini, V. P., Poretti, M. B., Rask-Andersen, M., Chavan, R. A., Ponzio, M. F., Sawant, R. S., et al. (2012). Differential effects of fluoxetine and venlafaxine on memory recognition: Possible mechanisms of action. *Progress in Neuro-Psychopharmacology & Biological Psychiatry*, *38*, 159–167.

Chang, L. Y., Chang, H. Y., Lin, L. N., Wu, C. C., & Yen, L. L. (2017). Disentangling the effects of depression on trajectories of sleep problems from adolescence through young adulthood. *Journal of Affective Disorders*, *217*, 48–54.

Chen, T., Wu, Z., Shen, Z., Zhang, J., Shen, X., & Li, S. (2014). Sleep duration in Chinese adolescents: Biological, environmental, and behavioral predictors. *Sleep Medicine*, *15*, 1345–1353.

Chung, S., Kim, H. J., Kim, H. J., Choi, S. H., Cho, J. H., Cho, Y. H., et al. (2014). Desipramine and citalopram attenuate pretest swim-induced increases in prodynorphin immunoreactivity in the dorsal bed nucleus of the stria terminalis and the lateral division of the central nucleus of the amygdala in the forced swimming test. *Neuropeptides*, *48*, 273–280.

Cipriani, A., Purgato, M., Furukawa, T. A., Trespidi, C., Imperadore, G., Signoretti, A., et al. (2012). Citalopram versus other anti-depressive agents for depression. *The Cochrane Database of Systematic Reviews*, *7*, CD006534.

Cui, S. Y., Li, S. J., Cui, X. Y., Zhang, X. Q., Yu, B., Sheng, Z. F., et al. (2016). Phosphorylation of CaMKII in the rat dorsal raphe nucleus plays an important role in sleep-wake regulation. *Journal of Neurochemistry*, *136*, 609–619.

Cunha, A. B., Frey, B. N., Andreazza, A. C., Goi, J. D., Rosa, A. R., Goncalves, C. A., et al. (2006). Serum brain-derived neurotrophic factor is decreased in bipolar disorder during depressive and manic episodes. *Neuroscience Letters*, *398*, 215–219.

Datta, S., O'malley, M. W., & Patterson, E. H. (2011). Calcium/calmodulin kinase II in the pedunculopontine tegmental nucleus modulates the initiation and maintenance of wakefulness. *The Journal of Neuroscience*, *31*, 17007–17016.

Davari-Tanha, F., Soleymani-Farsani, M., Asadi, M., Shariat, M., Shirazi, M., & Hadizadeh, H. (2016). Comparison of citalopram and venlafaxine's role in treating sleep disturbances in menopausal women, a randomized, double-blind, placebo-controlled trial. *Archives of Gynecology and Obstetrics*, *293*, 1007–1013.

Dowlatshahi, D., Macqueen, G. M., Wang, J. F., & Young L. T. (1998). Increased temporal cortex CREB concentrations and antidepressant treatment in major depression. *Lancet*, *352*, 1754–1755.

Duman, R. S. (2014). Neurobiology of stress, depression, and rapid acting antidepressants: Remodeling synaptic connections. *Depression and Anxiety*, *31*, 291–296.

Duman, R. S., Heninger, G. R., & Nestler, E. J. (1997). A molecular and cellular theory of depression. *Archives of General Psychiatry*, *54*, 597–606.

Faraguna, U., Vyazovskiy, V. V., Nelson, A. B., Tononi, G., & Cirelli, C. (2008). A causal role for brain-derived neurotrophic factor in the homeostatic regulation of sleep. *The Journal of Neuroscience*, *28*, 4088–4095.

Foley, D., Ancoli-Israel, S., Britz, P., & Walsh, J. (2004) Sleep disturbances and chronic disease in older adults: Results of the 2003 National Sleep Foundation Sleep in America Survey. *Journal of Psychosomatic Research*, *56*, 497–502.

Gong, W. G., Wang, Y. J., Zhou, H., Li, X. L., Bai, F., Ren, Q. G., et al. (2017). Citalopram ameliorates synaptic plasticity deficits in different cognition-associated brain regions induced by social isolation in middle-aged rats. *Molecular Neurobiology*, *54*, 1927–1938.

Harmer, C. J., Bhagwagar, Z., Cowen, P. J., & Goodwin, G. M. (2002). Acute administration of citalopram facilitates memory consolidation in healthy volunteers. *Psychopharmacology*, *163*, 106–110.

Havekes, R., Bruinenberg, V. M., Tudor, J. C., Ferri, S. L., Baumann, A., Meerlo, P., et al. (2014). Transiently increasing cAMP levels selectively in hippocampal excitatory neurons during sleep deprivation prevents memory deficits caused by sleep loss. *The Journal of Neuroscience*, *34*, 15715–15721.

Havekes, R., Park, A. J., Tolentino, R. E., Bruinenberg, V. M., Tudor, J. C., Lee, Y., et al. (2016). Compartmentalized PDE4A5 signaling impairs hippocampal synaptic plasticity and long-term memory. *The Journal of Neuroscience*, *36*, 8936–8946.

Hillman, D. R., & Lack, L. C. (2013). Public health implications of sleep loss: The community burden. *The Medical Journal of Australia*, *199*, S7–10.

Horwitz, A. V. (2010). How an age of anxiety became an age of depression. *The Milbank Quarterly*, *88*, 112–138.

Kasper, S., & Muller-Spahn, F. (2002). Intravenous antidepressant treatment: Focus on citalopram. *European Archives of Psychiatry and Clinical Neuroscience*, *252*, 105–109.

Krishnan, V., & Nestler, E. J. (2008). The molecular neurobiology of depression. *Nature*, *455*, 894–902.

Krueger, P. M., & Friedman, E. M. (2009). Sleep duration in the United States: A cross-sectional population-based study. *American Journal of Epidemiology*, *169*, 1052–1063.

Leal, G., Afonso, P. M., Salazar, I. L., & Duarte, C. B. (2015). Regulation of hippocampal synaptic plasticity by BDNF. *Brain Research*, *1621*, 82–101.

Lisman, J., Schulman, H., & Cline, H. (2002). The molecular basis of CaMKII function in synaptic and behavioural memory. *Nature Reviews. Neuroscience*, *3*, 175–190.

Luo, J., Phan, T. X., Yang, Y., Garelick, M. G., & Storm, D. R. (2013). Increases in cAMP, MAPK activity, and CREB phosphorylation during REM sleep: Implications for REM sleep and memory consolidation. *The Journal of Neuroscience*, *33*, 6460–6468.

Martinowich, K., & Lu, B. (2008). Interaction between BDNF and serotonin: Role in mood disorders. *Neuropsychopharmacology*, *33*, 73–83.

Meerlo, P., Havekes, R., & Steiger, A. (2015). Chronically restricted or disrupted sleep as a causal factor in the development of depression. *Current Topics in Behavioral Neurosciences*, *25*, 459–481.

Misrani, A., Tabassum, S., Chen, X., Tan, S. Y., Wang, J. C., Yang, L., et al. (2019). Differential effects of citalopram on sleep-deprivation-induced depressive-like behavior and memory impairments in mice. *Progress in Neuro-Psychopharmacology & Biological Psychiatry*, *88*, 102–111.

Misrani, A., Tabassum, S., Wang, M., Chen, J., Yang, L., & Long, C. (2020). Citalopram prevents sleep-deprivation-induced reduction in CaMKII-CREB-BDNF signaling in mouse prefrontal cortex. *Brain Research Bulletin*, *155*, 11–18.

Newell, J., Yesavage J. A., Taylor, J. L., Kraemer, H. C., Munro, C. A., Friedman, L., et al. (2016). Sedation mediates part of Citalopram's effect on agitation in Alzheimer's disease. *Journal of Psychiatric Research*, *74*, 17–21.

Palagini, L., Baglioni, C., Ciapparelli, A., Gemignani, A., & Riemann, D. (2013). REM sleep dysregulation in depression: State of the art. *Sleep Medicine Reviews*, *17*, 377–390.

Qi, X., Lin, W., Li, J., Li, H., Wang, W., Wang, D., et al. (2008). Fluoxetine increases the activity of the ERK-CREB signal system and alleviates the depressive-like behavior in rats exposed to chronic forced swim stress. *Neurobiology of Disease*, *31*, 278–285.

Rajaratnam, S. M., & Arendt, J. (2001). Health in a 24-h society. *Lancet*, *358*, 999–1005.

Roberts, R. E., & Duong, H. T. (2014). The prospective association between sleep deprivation and depression among adolescents. *Sleep*, *37*, 239–244.

Sairanen, M., O'leary, O. F., Knuuttila, J. E., & Castren, E. (2007). Chronic antidepressant treatment selectively increases expression of plasticity-related proteins in the hippocampus and medial prefrontal cortex of the rat. *Neuroscience*, *144*, 368–374.

Schmitt, K., Holsboer-Trachsler, E., & Eckert, A. (2016). BDNF in sleep, insomnia, and sleep deprivation. *Annals of Medicine*, *48*, 42–51.

Seixas, A. A., Nunes, J. V., Airhihenbuwa, C. O., Williams, N. J., Pandi-Perumal, S. R., James, C. C., et al. (2015). Linking emotional distress to unhealthy sleep duration: Analysis of the 2009 National Health Interview Survey. *Neuropsychiatric Disease and Treatment*, *11*, 2425–2430.

Shahsavand-Ananloo, E., Berenji, F., Sadeghniiat, K., Alimadadi, A., Zahiroddin, A. R., Tabatabaee, M., et al. (2013). Comparing effects of citalopram with fluoxetine on sleep quality in patients with major depressive disorder. *European Review for Medical and Pharmacological Sciences*, *17*, 1155–1161.

Shochat, T., Cohen-Zion, M., & Tzischinsky, O. (2014). Functional consequences of inadequate sleep in adolescents: A systematic review. *Sleep Medicine Reviews*, *18*, 75–87.

Shockey, T. M., & Wheaton, A. G. (2017). Short sleep duration by occupation group—29 states, 2013–2014. *MMWR. Morbidity and Mortality Weekly Report*, *66*, 207–213.

Szklo-Coxe, M., Young, T., Peppard, P. E., Finn, L. A., & Benca, R. M. (2010). Prospective associations of insomnia markers and symptoms with depression. *American Journal of Epidemiology*, *171*, 709–720.

Tang, D., Li, P., Guo, L., Xu, Y., Gao, X., Deng, J., et al. (2016). The prevalences of and association between nonmedical prescription opioid use and poor sleep among Chinese high school students. *Scientific Reports*, *6*, 30411.

Tardito, D., Musazzi, L., Tiraboschi, E., Mallei, A., Racagni, G., & Popoli, M. (2009). Early induction of CREB activation and CREB-regulating signalling by antidepressants. *The International Journal of Neuropsychopharmacology*, *12*, 1367–1381.

Thome, J., Sakai, N., Shin, K., Steffen, C., Zhang, Y. J., Impey, S., et al. (2000). cAMP response element-mediated gene transcription is upregulated by chronic antidepressant treatment. *The Journal of Neuroscience*, *20*, 4030–4036.

Tiraboschi, E., Giambelli, R., D'urso, G., Galietta, A., Barbon, A., De Bartolomeis, A., et al. (2004). Antidepressants activate CaMKII in neuron cell body by Thr286 phosphorylation. *Neuroreport*, *15*, 2393–2396.

Tiraboschi, E., Tardito, D., Kasahara, J., Moraschi, S., Pruneri, P., Gennarelli, M., et al. (2004). Selective phosphorylation of nuclear CREB by fluoxetine is linked to activation of CaM kinase IV and MAP kinase cascades. *Neuropsychopharmacology*, *29*, 1831–1840.

Van Cauter, E., Spiegel, K., Tasali, E., & Leproult, R. (2008). Metabolic consequences of sleep and sleep loss. *Sleep Medicine*, *9*(Suppl. 1), S23–S28.

Vinet, J., Carra, S., Blom, J. M., Brunello, N., Barden, N., & Tascedda, F. (2004). Chronic treatment with desipramine and fluoxetine modulate BDNF, CaMKKalpha and CaMKKbeta mRNA levels in the hippocampus of transgenic mice expressing antisense RNA against the glucocorticoid receptor. *Neuropharmacology*, *47*, 1062–1069.

Vyazovskiy, V. V., Cirelli, C., Pfister-Genskow, M., Faraguna, U., & Tononi, G. (2008). Molecular and electrophysiological evidence for net synaptic potentiation in wake and depression in sleep. *Nature Neuroscience*, *11*, 200–208.

Wang, S., Wu, Y., Ungvari, G. S., Ng, C. H., Forester, B. P., Gatchel, J. R., et al. (2017). Sleep duration and its association with demographics, lifestyle factors, poor mental health and chronic diseases in older Chinese adults. *Psychiatry Research*, *257*, 212–218.

Wei, Z., Junhong, G., Xiaoyuan, N., Jie, W., Zhaojun, W., Meina, W., et al. (2017). Citalopram ameliorates impairments in spatial memory and synaptic plasticity in female 3xTgAD mice. *BioMed Research International*, *2017*, 1238687.

Wu, J., Chang, F., & Zu, H. (2015). Efficacy and safety evaluation of citalopram and doxepin on sleep quality in comorbid insomnia and anxiety disorders. *Experimental and Therapeutic Medicine*, *10*, 1303–1308.

Yamada, S., Yamamoto, M., Ozawa, H., Riederer, P., & Saito, T. (2003). Reduced phosphorylation of cyclic AMP-responsive element binding protein in the postmortem orbitofrontal cortex of patients with major depressive disorder. *Journal of Neural Transmission (Vienna)*, *110*, 671–680.

Zhang, Q., Yang, C., Liu, T., Liu, L., Li, F., Cai, Y., et al. (2018). Citalopram restores short-term memory deficit and non-cognitive behaviors in APP/PS1 mice while halting the advance of Alzheimer's disease-like pathology. *Neuropharmacology*, *131*, 475–486.

Chapter 34

# Monoaminergic system and antidepressants

David Martín-Hernández[a], Cristina Ulecia-Morón[a], Álvaro G. Bris[a], Marta P. Pereira[b], and Javier R. Caso[a]

[a]*Department of Pharmacology and Toxicology, School of Medicine, CIBERSAM-ISCIII, Universidad Complutense de Madrid, Madrid, Spain;* [b]*Department of Molecular Biology and Center of Molecular Biology Severo Ochoa, Universidad Autonoma de Madrid, Madrid, Spain*

## List of abbreviations

**5-HT** serotonin
**CNS** central nervous system
**DA** dopamine
**FC** frontal cortex
**MAOI** monoamine oxidase inhibitor
**MDD** major depressive disorder
**NA** noradrenaline
**NaSSA** Noradrenergic and Specific Serotoninergic Antidepressant
**NDRI** noradrenaline and dopamine reuptake inhibitor
**NE** norepinephrine
**NET** norepinephrine transporter
**NRI** noradrenaline reuptake inhibitor
**SERT** serotonin transporter
**SNRI** serotonin–norepinephrine reuptake inhibitor
**SPARI** serotonin partial agonist reuptake inhibitor
**SSRI** selective serotonin reuptake inhibitor
**TCA** tricyclic antidepressant
**TRI** triple reuptake inhibitor

## Introduction

The etiology and pathophysiology of the major depressive disorder (MDD) are still unknown, although different hypotheses (not exclusive but complementary) have been proposed, trying to shed light on its mechanisms. However, the vast majority of available antidepressants are still based on the classic monoamine hypothesis of mood disorders namely, MDD is the result of a deficit in monoaminergic neurotransmission, especially involving serotonin (5-HT) and noradrenaline (-NA- also known as norepinephrine), but also dopamine (DA). This theory has ruled our knowledge of both the pathophysiology of MDD and the mechanism of action of pharmacological treatments till this day, and it has guided to the creation of more than a few generations of antidepressant drugs.

Consequently, current treatments try to enhance serotonergic and noradrenergic neurotransmission through actions on degradation enzymes (monoamine oxidase inhibitors—MAOIs), reuptake transporters (tricyclic antidepressants—TCAs—and selective reuptake inhibitors), receptors (multi-receptor agonists or antagonists) or on these last two elements simultaneously (serotonin partial agonist reuptake inhibitors—SPARIs) (Ferrari & Villa, 2017; Otte et al., 2016) (Table 1). Importantly, there is no conclusive evidence that any of these drugs are better substantially than any other from the perspective of antidepressant efficacy; the main factor for the selection of the antidepressant drug are the side effects and the toxicity profiles.

Thus, although there are significant limitations to the existing monoamine theory, including hypothalamic–pituitary–adrenal (HPA) axis dysfunctions, as well as neurotrophic and inflammatory alterations, this theory is the foundation of current pharmacological treatments. As a consequence, the aim of this chapter is to review the actions of antidepressants affecting the monoaminergic system. Particularly, it will examine the main monoamines involved in the MDD (i.e., 5-HT, NA, and DA) but also some of the less-studied monoamines in the MDD realm (e.g., histamine, melatonin, and tryptamine).

## Serotonin

Serotonin (5-hydroxitriptamine, -5-HT-) is the neurotransmitter traditionally linked to MDD. In the brain, serotoninergic neurons are organized in several groups in the *raphe nuclei*, projecting the more caudal to the spinal cord and the periphery, and the dorsal and medial ones to different brain regions, including frontal cortex (FC) and hippocampus.

5-HT influences a wide variety of processes such as cognitive control, emotional regulation, autonomic response, neuronal plasticity, and modulation of other neurotransmitter networks like the glutamatergic one (Lesch & Waider, 2012). Consequently, 5-HT is critically involved in psychiatric diseases, particularly in MDD.

The Neuroscience of Depression. https://doi.org/10.1016/B978-0-12-817933-8.00093-1

**TABLE 1** Antidepressants.

| Mechanism of action | Pharmacological target | Antidepressant |
|---|---|---|
| Prevention of NT breakdown | MAO | • *Nonselective inhibitors (MAOIs):* tranylcypromine, hydrazine<br>• *Selective and reversible inhibitors:* moclobemide |
| Inhibition of NT reuptake | Monoamine transport | *Tricyclic antidepressants (TCAs):* imipramine, chloripramine, desipramine, amitriptyline, nortryptiline<br>*Tetrayclic antidepresants (TeCAs):* maprotiline, mirtazapine, mianserin, setiptiline |
| | Serotonin transport | • *Selective reuptake inhibitor (SSRIs): citalopram, escitalopram, fluoxetine, fluvoxamine, paroxetine, sertraline*<br>• *Selective reuptake inhibitor and receptor modulator (SARIs, SPARIs): trazodone, vortioxetine, vilazodone* |
| | Norepinephrine transport | • *Selective reuptake inhibitor (NRIs):* Reboxetine, atomoxetine<br>• *Norepinephrine-dopamine inhibitor:* bupropion |
| | Serotonin and norepinephrine transport | • *Selective (SNRIs):* Venlafaxine, desvenlafaxine, duloxetine |

Current antidepressants, including their mechanisms of action and pharmacological targets.

In the brain, monoamine oxidases are in charge of 5-HT degradation and the serotonin transporter (SERT) reuptakes 5-HT from the synaptic cleft. Moreover, 5-HT exerts its biological actions through several specific receptors, and within these receptors are included the main pharmacological targets of current serotonin-based drugs for MDD (Table 2).

## Serotonin and noradrenaline transporters and antidepressants

In particular, the blockade of SERT to increase 5-HT neurotransmission represents the main mechanism of currently available antidepressants. Significantly, selective serotonin reuptake inhibitors (SSRIs) are the most prescribed group of antidepressants due to their proven efficacy and safety. Some of them are *citalopram*, *escitalopram*, *fluoxetine*, *fluvoxamine*, *paroxetine*, and *sertraline*.

SSRIs result from the evolution of TCAs, in an effort to reduce side effects and enhance the therapeutic outcome. TCAs share a common chemical structure which comprises three rings of atoms, have different affinities for monoaminergic reuptake transporters, and bind to receptors of other neurotransmission systems (Gillman, 2007). In this sense, some have a more powerful inhibition of SERT (e.g., *imipramine*), others preferentially inhibit the norepinephrine

**TABLE 2** Serotonin receptors as antidepressants.

| Family | Receptor | Antidepressant effect |
|---|---|---|
| 5-HT$_1$ | 5-HT$_{1A}$ | • Antagonism (Artigas, Romero, de Montigny, & Blier, 1996)<br>• Agonism (Depoortere et al., 2010) |
| | 5-HT$_{1B}$ | • Antagonism (Dawson et al., 2006)<br>• Agonism (Hwang & Chung, 2014) |
| 5-HT$_2$ | 5-HT$_{2A}$ | • Agonism (Sato, Skelin, & Diksic, 2010)<br>• Antagonism (Artigas, 2013) |
| | 5-HT$_{2B}$ | • Agonism (Diaz, Narboux-Neme, Boutourlinsky, Doly, & Maroteaux, 2016) |
| | 5-HT$_{2C}$ | • Agonism (Cryan & Lucki, 2000)<br>• Antagonism (Cremers et al., 2004) |
| 5-HT$_3$ | 5-HT$_3$ | • Antagonism (Gupta, Prabhakar, & Radhakrishnan, 2016) |
| 5-HT$_7$ | 5-HT$_7$ | • Antagonism (Canale et al., 2016) |

Families of serotonin receptors and the resulting antidepressant effect by their pharmacological modulation.

transporter (NET) over SERT (e.g., *desipramine*) and others have a more balanced inhibition of SERT and NET (e.g., *amitriptyline*). In clinical practice, SSRIs have practically ousted TCAs and the last ones are reserved only for special situations in which SSRIs do not reach the desired effect.

## Monoamine oxidase and antidepressants

MAOIs use, despite their importance in the formulation of monoaminergic hypothesis, is restricted in MDD to the most severe cases of nonresponder patients, due to interactions with other drugs or diet, and dangerous side effects such as hypertensive crisis. MAOIs can be selective for each isoform of MAO (A or B) or nonselective. *Moclobemide* (A-type), *selegiline* (B-type), and *iproniazid* (nonselective) stand out within this group of antidepressants.

## Serotonin receptors and antidepressants

5-HT signals through specific receptors. To date, 14 5-HT receptors grouped in 7 families have been described, mediating both excitatory and inhibitory transmissions through G-coupled proteins, with the exception of the 5-HT$_3$ ionic channel activated by ligand (Fig. 1).

There is a growing body of evidence about the antidepressant properties of each receptor modulation, leading not only to better knowledge about the implication of the complex serotonin system in MDD pathophysiology but also to new multi-target drug design. Even more, some SSRIs modulate the expression of 5-HT receptors, indicating other mechanisms of action apart from the blockade of SERT (Martín-Hernández et al., 2019). Considering this, new SSRIs with actions on 5-HT receptors, the Serotonin Partial Agonist/Reuptake Inhibitors (SPARIs), have been developed. That is the case of *trazodone* (5-HT$_{2A}$ antagonist), *vilazodone* (5-HT$_{1A}$ agonist) and *vortioxetine* (5-HT$_{1A}$ agonist, 5-HT$_{1B}$ partial agonist, 5-HT$_{1D}$ antagonist, 5-HT$_3$ antagonist, and 5-HT$_7$ antagonist).

Recently commercialized antidepressants deserve to be outlined for their combined effects on 5-HT and other neurotransmitter systems, such as *mianserin* (antagonist or inverse agonist of 5-HT$_{1D}$, 5-HT$_{1F}$, 5-HT$_{2A}$, 5-HT$_{2B}$, 5-HT$_{2C}$, 5-HT$_3$, 5-HT$_6$, 5-HT$_7$, $\alpha_1$, $\alpha_2$, histamine H$_1$, and NET inhibitor) and *mirtazapine* (antagonist or inverse agonist of 5-HT$_{2A}$, 5-HT$_{2C}$, $\alpha_1$,$\alpha_{2A}$, $\alpha_{2B}$, $\alpha_{2C}$, muscarinic, and histamine H$_1$).

Regarding 5-HT$_1$, research has focused on 5-HT$_{1A}$ and 5-HT$_{1B}$ due to the difficulties in studying 5-HT$_{1D}$, 5-HT$_{1E}$ and 5-HT$_{1F}$. Both inhibit open the potassium channels exerting a presynaptic inhibitory potential as autoreceptors and postsynaptic as heteroreceptors. As autoreceptors, they are critical for the negative feedback on the presynaptic neuron, but their distribution in the brain marks important differences. 5-HT$_{1A}$ are located in the *raphe nuclei* and their activation induces a decrease in the whole-brain 5-HT levels due to multiple projections (Sprouse & Aghajanian, 1987). Inhibition or desensitization of these receptors exhibits antidepressant effects (Bortolozzi et al., 2012; Hughes et al., 2007) and could be responsible for the delay in the effectivity of SSRIs (Artigas et al., 1996). The 5-HT$_{1A}$ heteroreceptors have an important presence in the hippocampus, septum, amygdala, and

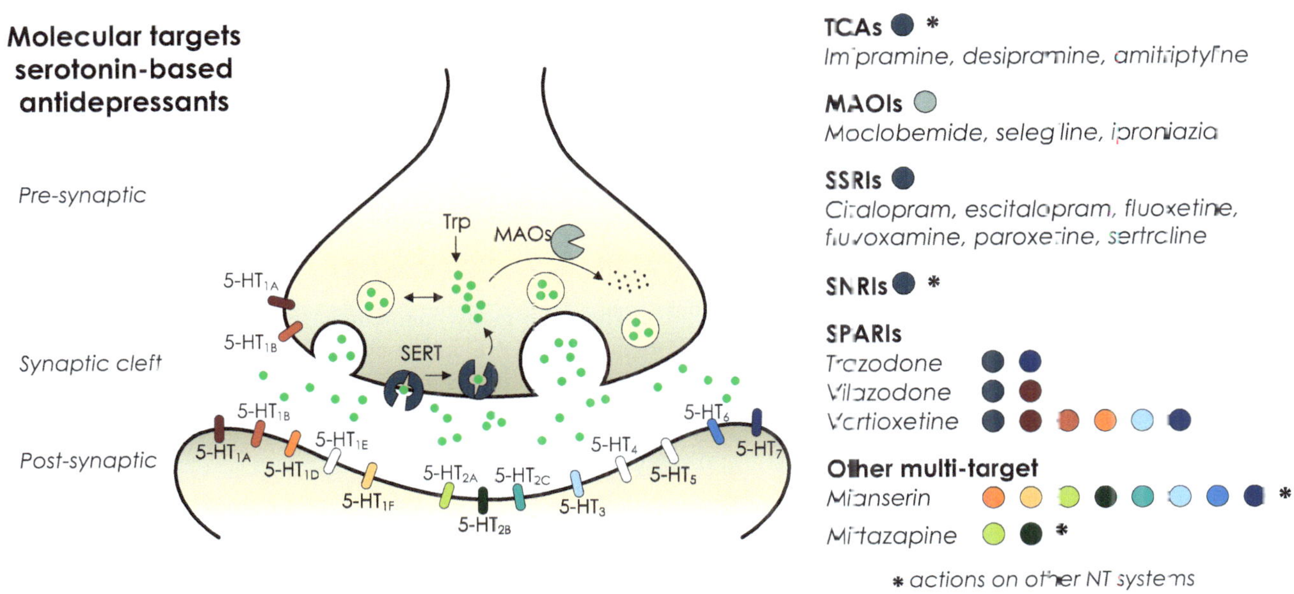


FIG. 1 Molecular targets serotonin-based antidepressants. Serotonergic synapsis including its receptors and the reuptake transporter (SERT) and the molecular targets for antidepressants.

corticolimbic regions. They stimulate neurogenesis in the hippocampal CA3 region potentially evoking antidepressant effects. In FC exert an indirect suppression of the 5-HT cell firing in the *dorsal raphe nucleus* (Altieri, Garcia-Garcia, Leonardo, & Andrews, 2013) playing also a role in the antidepressant response (Blier, Bergeron, & de Montigny, 1997). Thus, partial agonism of $5\text{-HT}_{1A}$ receptors has shown antidepressant properties and represents one of the actions of *vortioxetine*, recently approved for the treatment of depressive disorders (Mork et al., 2012).

$5\text{-HT}_{1B}$ autoreceptors are axonal and, therefore, have less influence on the global serotoninergic tone (Sprouse & Aghajanian, 1987) and $5\text{-HT}_{1B}$ heteroreceptors modulate other neurotransmitter networks (Sari, 2004). Hence, both agonism and antagonism of these receptors have shown antidepressant effects (Dawson et al., 2006; Hwang & Chung, 2014).

The $5\text{-HT}_2$ family, with a postsynaptic localization, is typically linked to antidepressant actions. For $5\text{-HT}_{2A}$, both agonism (Sato et al., 2010) and antagonism (Artigas, 2013) have demonstrated their ability to improve depressive-like behaviors. Due to their higher peripheral expression, $5\text{-HT}_{2B}$ has not been so much under the scope of MDD research but they seem to be necessary for the antidepressant efficacy through their activity in the dorsal raphe nuclei (Diaz, et al., 2016). Nevertheless, its potentiality as a pharmacological target is limited because of possible pulmonary and cardiovascular side effects. The $5\text{-HT}_{2C}$ is currently undergoing analysis to evaluate if its agonism, partial agonism, or antagonism could be relevant to new drug design (Cremers et al., 2004; Cryan & Lucki, 2000) without reaching clear conclusions at the moment.

The antidepressant potential of $5\text{-HT}_3$ antagonism has been proved and included in new drugs, although the exact mechanism which drives it is not utterly understood. Plausible explanations include the control of 5-HT release, reduction of HPA axis hyperactivity, attenuation of oxidative stress damage and neuroplasticity reestablishment (Gupta et al., 2016).

Lastly, the $5\text{-HT}_7$ seems significant for depressive disorders considering its involvement in learning, memory, circadian rhythms, sleeping, and mood (Guseva, Wirth, & Ponimaskin, 2014). The blockade of $5\text{-HT}_7$ signaling has antidepressant properties (Canale et al., 2016) and, furthermore, the ability of antipsychotics to regulate this receptor is under research as an enhancer of conventional antidepressant activity (Berman et al., 2009).

## Noradrenaline and adrenaline

Noradrenaline (NA), also called norepinephrine (NE), is a member of the catecholamine group that can act both as a neurotransmitter and as a hormone. The subsequent addition of a methyl group to nitrogen allows the formation of adrenaline (A), also known as epinephrine (E). In this way, both catecholamines derive from the amino acid tyrosine, being the main orchestrators of the sympathetic nervous system. Specifically, they are involved in the fight-or-flight physiological response, which is pivotal for survival. To achieve this, they need to exert their effects through their binding to α- and β-adrenergic receptors located in the cell surface. It is NE the main neurotransmitter of this system, rather than E.

Peripheral NE is responsible for the increase in blood pressure and heart rate. In the CNS, NE participates in the formation and consolidation of memories, cognition, vigilance, fear and arousal. It is synthetized and stored in the *locus coeruleus*, which is the cerebral structure implicated in these particular tasks.

The presence of depressive symptoms has been traditionally linked to NE deficiency. However, current experts in the area emphasize now the importance of its synaptic balance as one of the main contributors to the depression onset (Galecki, Mossakowska-Wojcik, & Talarowska, 2018). NE is largely present at the synapse of noradrenergic neurons when it is released after the action potential. Its removal from the synaptic cleft is very quick: it binds to presynaptic receptors to be again introduced in the cellular cytosol, where it could be either degraded or encapsulated in vesicles for consecutive liberations.

Although the pharmacological modulation of NE has demonstrated to be efficient in the management of depressive symptoms, the overall effect is partly due to its interactions with other monoaminergic pathways. Among them, the monoamines 5-HT and DA are the most relevant ones. For instance, it is known that classical serotoninergic neurons have in their membranes receptors for NE, and the other way around (Gorea & Adrien, 1988). Taken all together, the intersection at different levels of each of their pathways will affect the prospect outcome of the antidepressant treatment.

### NE and antidepressants

There are many compounds seeking to increase the total amount of NE in the synaptic cleft, which has shown to counteract the symptomatology of patients with MDD. Some of them have been already mentioned in the previous subheading of this chapter, as they also affect other monoamine neurotransmissions (Fig. 2).

#### *Tricyclics antidepressants (TCA)*

Desipramine and nortriptyline are the two antidepressants that cause the most significant impact on NE, at the same

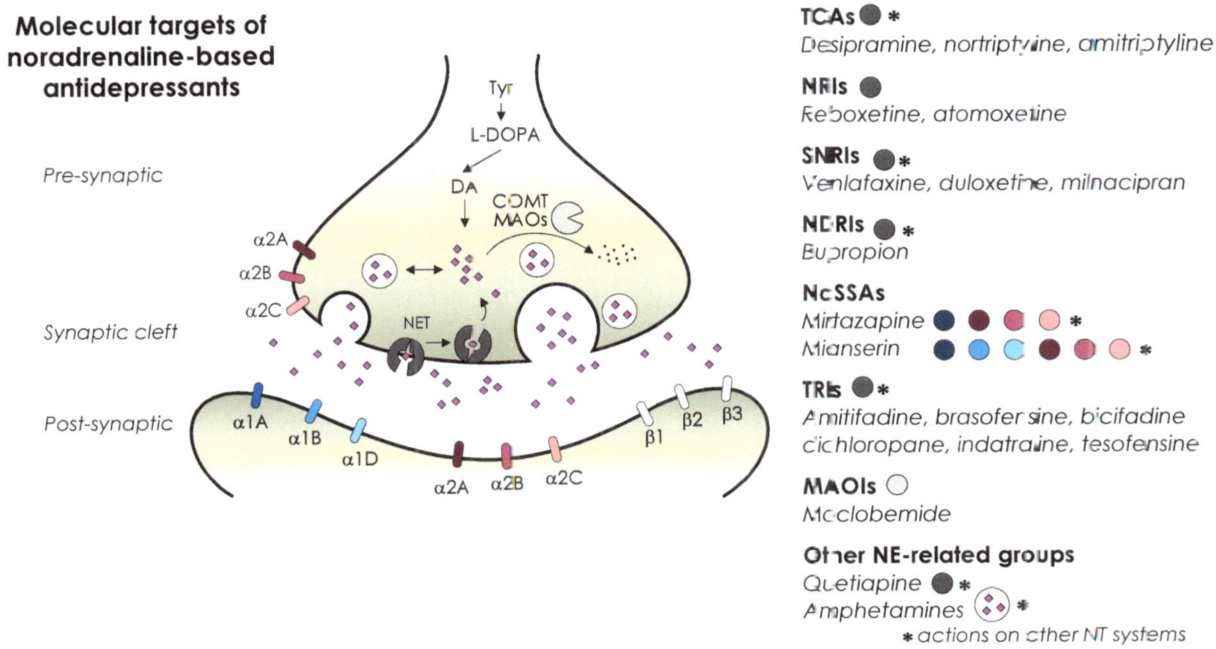


FIG. 2 Molecular targets noradrenaline-based antidepressants. Noradrenergic synapsis including its receptors and the reuptake transporter (NET) and the molecular targets for antidepressants.

time as they give rise to fewer side effects. They are nonselective monoamine reuptake inhibitors and their use has been reduced due to their many side effects.

- *Desipramine*
- *Nortriptyline*
- *Amitriptyline*

## *NRI (noradrenaline reuptake inhibitors)*

Taken the role of NE in the response to fear and cognition, this group of drugs are very useful for those depressive disorders coursing with anxiety episodes.

- *Reboxetine*: it targets the NE transporter (NET), increasing the amount of NE in the synaptic cleft. It has a mean half-life of 12h and it is metabolized mainly in the liver. Because its effectiveness is reduced when compared to other antidepressants, it is not currently used as selection treatment. However, it is particularly prescribed for seasonal depression and attention deficit hyperactivity disorder (ADHD). Under some circumstances, reboxetine is combined with SSRIs to increase the efficacy of the treatment.
- *Atomoxetine*: it was not initially synthetized as an antidepressant, but it exerts its function in the same way than reboxetine. It is predominantly employed in ADHD therapy since it did not show prominent improvements in patients with major depressive disorder (MDD). Contrary to reboxetine, atomoxetine has a short mean half-life, so it requires a twice-daily administration.

## *SNRI (serotonin-noradrenalin reuptake inhibitors)*

This group is known as dual antidepressants. They selectively increase the amount of both NE and 5-HT in the synaptic cleft by inhibiting their reuptake.

- *Venlafaxine*: it is a strong inhibitor of the reuptake of both neurotransmitters, especially the NE, having also some action in the dopamine reuptake, but in a much lower rate. Through metabolism, it derives to *O*-desmethylvenlafaxine (*desvenlafaxine*), whose effect is similar to venlafaxine on NE and 5-HT. In contrast to venlafaxine, this last derivative has the advantage of minor interactions, since its metabolism is carried out through the liver. Both are recommended for patients with MDD.
- *Duloxetine*: it highly inhibits NE and 5-HT reuptake, with little effect on the dopamine reuptake. It is well tolerated and prescribed for MDD, among other conditions, such as neuropathic pain. Venlafaxine and duloxetine have proven similar results in the treatment of depression.
- *Milnacipran*: it is particularly convenient for those patients taking different kinds of medication since it presents a very low rate of drug interactions. However,

because of its renal excretion, attention must be drawn to those patients suffering from renal failure. It is more frequently used for anxiety and fibromyalgia, considering that it is not the drug of choice in the treatment of MDD.

### NDRI (noradrenaline and dopamine reuptake inhibitors)

This group included many drugs whose use has been restricted due to their considerable side effects (e.g., amineptine and nomifensine). Nevertheless, the most important drug of this group is *bupropion*, which selectively inhibits the reuptake of DA in the synaptic cleft. Interestingly, bupropion's efficacy rests on the NE reuptake, which is mediated thanks to the metabolite hydroxybupropion. This occurs even though bupropion action is independent of NE reuptake (Hewett et al., 2010). Bupropion is metabolized in the liver, the same as hydroxybupropion, in addition to being also partially excreted via renal. Although its main indication is for smoke cessation, it is also convenient for the treatment of melancholic and seasonal depression. Due to the role of DA in the reward brain system, it can cause manic episodes in bipolar disorders.

### NaSSA (noradrenergic and specific serotoninergic antidepressants)

- *Mirtazapine*: it works by antagonizing the $\alpha_2$-adrenergic receptors and autoreceptors. This way of action is slightly particular: it acts blocking de autoreceptors in the presynaptic neurons, ultimately avoiding the NE reuptake. This process ends modifying the serotonin reuptake too by 5-$HT_2$ and 5-$HT_3$ receptors.
- *Mianserin*: it is a tetracyclic antidepressant, and its use is normally associated with depression coursing with anxiety, agitation, and insomnia.

### Triple reuptake inhibitors (TRI)

TRIs selectively reuptake the three main monoamines: 5-HT, NE, and DA. They are currently under study, given the controversial effects and applicability of the majority of them. It is currently thought that they would be suited for patients with MDD coursing with significant anhedonia. Among the most important ones, the following can be highlighted:

- *Amitifadine*
- *Brasofensine*
- *Dichloropane*
- *Bicifadine*
- *Indatraline*
- *Tesofensine*

The great advantage that TRIs add up to the treatment of psychiatric disorders is that the inclusion of DA reuptake may lead to the decrease of many common side effects (Subbaiah, 2018).

### MAO-Is

They work by inhibiting the isoform I of the monoamine oxidase enzyme, so that the blockade of the NE breakdown permits the increase of its concentration, together with that of 5-HT and DA. Their main problem is the lack of selectivity since other pathways are also altered. Consequently, there are a number of side effects coming from these drugs. They have been commonly prescribed to treat MDD coursing with atypical symptoms and anxiety.

- *Moclobemide*: it can be prescribed for cases of resistant depression and certain episodes of anxiety.

### Other NE-related drugs having antidepressant actions

- *Quetiapine*: drug of selection for the treatment of bipolar depression, it seems to carry out its effect on the NE metabolism through its metabolite norquetiapine, which tends to bind to the NET.
- *Amphetamines*: they also provoke the increase of NE through the release of the neurotransmitter at their synaptic cleft.

## Dopamine and antidepressants

Although in a smaller manner than other monoamines, the dopaminergic system has also a significant impact on the basis of depressive symptomatology being a target for treatments for MDD. In this sense, anhedonia, a core symptom of MDD, involves downregulation of the DA system.

Most DA neurons are spread in different places through the brainstem and most of the projections of DA neurons implicate the Ventral Tegmental Area (VTA) connecting with *nucleus accumbens* and *striatum* being part of the reward system. DA receptors have been also described as a potential target for antidepressants in other CNS areas like the hippocampus and typical depressive symptoms (i.e., anhedonia and lack of motivation), are highly linked to DA impairments (Grace, 2016) (Fig. 3).

### DA and first-line antidepressants

First-line antidepressants can exert effects at the dopaminergic neurotransmission, which is mainly downregulated in patients with MDD. Actually, SSRIs can increase D1/D2 receptors in the CNS from animal models and, in patients with MDD, some similar results have been shown

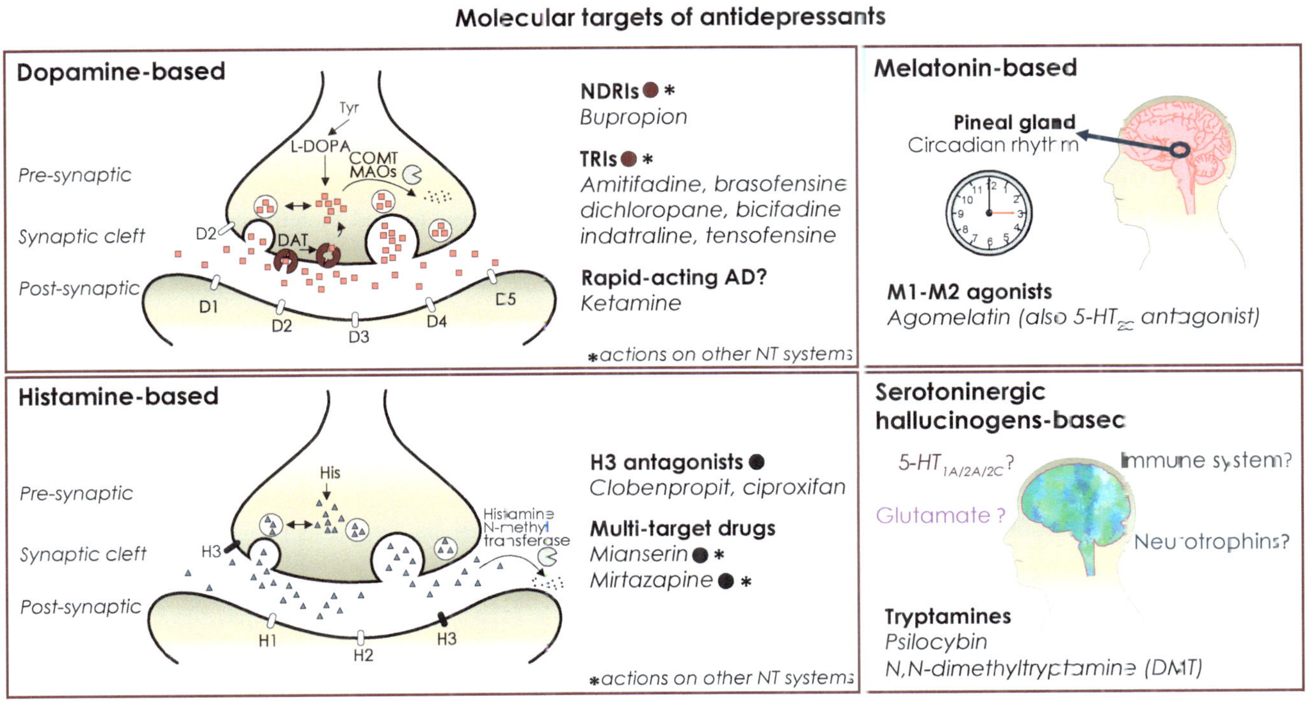


FIG. 3 Molecular targets DA-, histamine-, melatonin-, and serotonergic hallucinogens-based antidepressants. Dopaminergic and histaminergic synapses, and melatonin and serotoninergic hallucinogen molecular targets for antidepressants.

after fluoxetine treatment. MAOIs or atypical TCAs can also enhance DA neurotransmission (Penttila et al., 2004).

## DA and norepinephrine and dopamine reuptake inhibitors (NDRIs)

NDRIs are antidepressants that exert their action by blocking the transport of NA and DA and stopping their reuptake, being the *bupropion* the approved member of the NDRIs family. Bupropion side effects are numerous, being one of the most worrisome is the seizures. However, the risk of seizures at therapeutic dosage is minimum. Apart from seizures, other serious side effects of bupropion include insomnia, nausea, agitation, anxiety, abdominal pain, tremor, tachycardia, and hypertension. Bupropion can also provoke some dependence and the development of withdrawal symptoms.

Bupropion mechanism of action has not been completely determined. It affects NA and DA neurotransmission, the DA being more predominant than the NA uptake. Usually, it is used as second- or third-line antidepressant, when SSRIs or SNRIs cannot be administered. Interestingly, research suggests that bupropion may be a reasonable treatment option for depressed pregnant women who require pharmacotherapy (Hendrick, Suri, Gitlin, & Ortiz-Portillo, 2017).

Other NDRIs, such as nomifensine, amineptine, and medifoxamine, have been withdrawn from the market due to different complications with other diseases, their possible side effects, and/or other safety reasons. In any case, the antidepressant properties of these NDRIs support the potential role of the DA in the pathophysiology of the MDD.

## Dopamine and triple reuptake inhibitors (TRIs)

TRIs are other kinds of antidepressants that can be related to the DA neurotransmission, which again support the potential role of the DA in the pathophysiology of the MDD.

Due to its capacity to exert actions in multiple monoaminergic systems, they can be used to treat different impairments comorbid with MDD and other CNS disorders.

The main TRIs have been already mentioned in the NE and depression subheading of this chapter. Here, it is important to remember that the advantage that TRIs add up to the treatment of psychiatric disorders is that the inclusion of DA reuptake may lead to the decrease of many common side effects.

## DA and rapid-acting antidepressants

It has been shown that ketamine, a rapid-acting antidepressant, can also affect DA neurotransmission, restoring regular levels of DA after its decrease in depression-like models in rodents (Belujon & Grace, 2017) once again sustaining the view that DA might play an important role in the pathophysiology of the MDD.

## Other monoamines (histamine, melatonin, and tryptamine hallucinogens)

There are more monoamine neurotransmitters that serotonin, noradrenaline, and dopamine. In particular, research has also focused on histamine and melatonin neurotransmission. Thus, in the next section of this chapter, the possible role of these other monoamines will be analyzed (see Fig. 3).

### Histaminergic system and antidepressants

Histamine is a monoamine mainly produced by mast cells. In the CNS, histamine is also produced by the histaminergic neurons for regulated release that are pinpointed in the hypothalamus from where they project to most of the brain areas. The neuronal histaminergic system is implicated in a number of functions, including energy and endocrine homeostasis, cognition and attention, all of which have a tendency to be critically affected in neuropsychiatric disorders. Hence, some of the unwanted side effects of the antihistamine medications can be closely related to an undesirable mood status.

Experimental studies have shown that brain histamine may be involved in the actions of acute and chronic stress protocols as inductors of depressive-like behaviors and that increased brain histamine levels are associated with antidepressant-like effects in the Porsolt's forced swim test (Shan, Bao, & Swaab, 2015).

Data from patients and postmortem studies show discrepancies with the experimental models about the importance of the histaminergic system on the MDD. Interestingly, as is indicated by the pharmacodynamics of current drug treatments, many of these antidepressants have important effects on the histamine receptors producing undesirable side effects. Besides, there are studies showing that patients with MDD have decreased brain histamine receptor binding that can be correlated with the severity of the symptomatology. It has been therefore suggested that the histaminergic system modulation may be a suitable tool in the treatment of MDD.

In summary, the use of the histaminergic system as a therapeutic target is being mainly suggested by experimental models, usually involving histamine H3 receptor antagonists (e.g., *clobenpropit* and *ciproxifan*), but there is a lack of trials examining the role of antagonists or agonists of the histamine receptors on the MDD symptomatology.

### Melatonergic system and antidepressants

Melatonin, as well as 5-HT, is a monoamine from the subgroup of indolamines. The precursor of both is tryptophan, in a process involving several phases. As a result, levels of 5-HT influence levels of melatonin, and a 5-HT deficit may also end in diminished levels of melatonin. Melatonin is principally a hormone and therefore a molecule released into the bloodstream but, although the majority of the melatonin is synthesized outside of the CNS, it appears to be capable to cross the blood–brain barrier and act both centrally and peripherally.

As a neuromodulator, melatonin mainly acts through the two G-protein coupled receptors MT1 and MT2, regulating numerous brain functions, including circadian rhythms, sleep, and mood.

Different preclinical and clinical studies have not provided uniform results, but present data hint that activating the melatonergic system may lead to anxiolytic and antidepressant effects.

In this sense, *agomelatine* is an antidepressant acting as an antagonist for 5-HT2C receptors but also as an agonist of both MT1 and MT2 receptors, that it has called some attention. Thus, recent studies have proposed a potential therapeutic role of the MT receptors in patients with MDD (Comai et al., 2019).

The lack of selective MT1 or MT2 ligands has not yet allowed for the testing of the potential therapeutic actions on preclinical models of mood disorders. Furthermore, powered randomized controlled clinical trials are required to make clear a possible melatonin effect on mood disorders as well as to get long-term data from these trials.

Thus, the possible role of the melatonergic system as a therapeutic target for the treatment of patients with MDD is still under study as the obtained results to date are not conclusive.

### Serotonergic hallucinogens and antidepressants

In the search for treatments, bearing in mind the sometimes elevated number of treatment-resistant patients, some attention has been paid to the serotonergic hallucinogens (i.e., tryptamine hallucinogens) such as the tryptamines *N,N-dimethyltryptamine* (*DMT*) and *psilocybin* (4-phosphoryloxy-N,N-DMT). In fact, some clinical trials have been performed and they suggest favorable effects for treatment-resistant depression, anxiety, and depression associated with life-threatening diseases. However, conducted studies had small sample sizes, a high degree of heterogeneity and in some cases, important deficiencies in their design. Furthermore, these studies had an added limitation, namely the difficulty in unraveling placebo effects (Dos Santos et al., 2016). Thus, more profound and better design studies, including more patients, are needed to replicate these preliminary findings.

The mechanisms of action accountable for the favorable effects produced by DMT and psilocybin are not entirely comprehended. However, given that DMT and psilocybin are agonists of serotonin 5-HT1A/2A/2C receptors, involvement of these receptors in the therapeutic effects of these hallucinogenic drugs appears plausible. Furthermore, they have other possible mechanisms of therapeutic action, such as the modulation of glutamatergic transmission, increment of neurotrophic factors, and modulation of some immune system pathways (Baumeister, Barnes, Giaroli, & Tracy, 2014).

Thus, some data indicate that, when safely administered in controlled and organized settings, DMT and psilocybin might have antidepressant effects. However, further studies are very much warranted to advance the understanding of the actions of tryptamine hallucinogens in neuropsychiatric disorders.

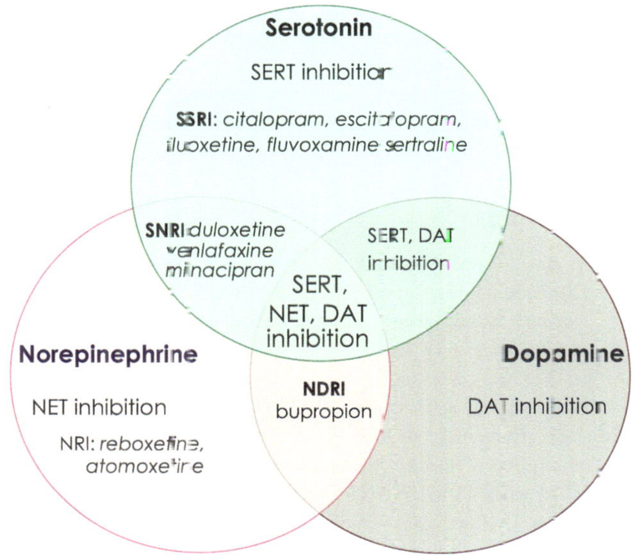


FIG. 4 Principal monoamines and antidepressants. Current antidepressants and their main actions on the major elements of the monoaminergic neurotransmission.

## Conclusion

The majority of antidepressant treatments pursue to heighten monoaminergic, in general, and serotonergic and noradrenergic, in particular, neurotransmissions. Thus, the inhibition of the monoamine reuptake still epitomizes the first pharmacological approach to treat mood disorders.

There is no definite proof that current monoaminergic antidepressants are superior substantively to any other from the perspective of antidepressant efficacy. Thus, the selection of the antidepressant is frequently based on the side effects and the toxicity profiles.

Monoaminergic antidepressants have several important pitfalls involving their efficacy, the generation of withdrawal symptoms at discontinuation or lessening of the treatment, unfavorable long-term outcomes and paradoxical effects, and worrying or persistent side effects. However, despite the rising of new therapeutic approaches involving other pathophysiological theories (e.g., esketamine and the glutamatergic neurotransmission), antidepressants acting on the monoaminergic system are still the standard pharmacological treatment for the MDD and they are responsible for many therapeutic successes as well as used widely in clinical practice (Fig. 4).

## Key facts of antidepressants

- The majority of antidepressant treatments pursue to heighten the monoaminergic neurotransmission.
- The selection of the antidepressant is frequently based on the side effects and the toxicity profiles.
- Monoaminergic antidepressants are still the standard pharmacological treatment for the MDD.
- Monoaminergic antidepressants have important pitfalls involving their efficacy and side effects.
- Most experts agree that an SSRI should be the first-line choice for patients with MDD.

## Summary points

- The vast majority of antidepressant drugs are based on the monoamine hypothesis of depression.
- The inhibition of the monoamine reuptake still epitomizes the first pharmacological approach to treat mood disorders.
- Selective serotonin reuptake inhibitors are the most prescribed group of antidepressants due to their proven efficacy and safety.
- There are significant number of patients with MDD who are resistant to treatment.
- Even though the important pitfalls of the monoaminergic antidepressants, they have contributed to ameliorate the quality of life of many patients around the world.

## Mini-dictionary of terms

**Monoamine** Molecule containing one amino group connected to an aromatic ring by a two-carbon chain that can act as a neurotransmitter or neuromodulator.

**Monoaminergic neurotransmission** Neurotransmission (transmission of a signal across a chemical synapse) involving monoamines.

**Monoaminergic theory of depression** Theory about the etiological origin on the MDD postulating that it is caused by a deficit in the monoaminergic neurotransmission.

**Monoaminergic antidepressant** Drug with antidepressive properties through its actions on the strengthening of the monoaminergic neurotransmission.

**Monoamine transporter** Protein that reuptakes monoamines (i.e., 5-HT, NE, or DA) from the synaptic cleft back to the presynaptic neuron.

## References

Altieri, S. C., Garcia-Garcia, A. L., Leonardo, E. D., & Andrews, A. M. (2013). Rethinking 5-HT1A receptors: Emerging modes of inhibitory feedback of relevance to emotion-related behavior. *ACS Chemical Neuroscience*, *4*(1), 72–83.

Artigas, F. (2013). Serotonin receptors involved in antidepressant effects. *Pharmacology & Therapeutics*, *137*(1), 119–131.

Artigas, F., Romero, L., de Montigny, C., & Blier, P. (1996). Acceleration of the effect of selected antidepressant drugs in major depression by 5-HT1A antagonists. *Trends in Neurosciences*, *19*(9), 378–383.

Baumeister, D., Barnes, G., Giaroli, G., & Tracy, D. (2014). Classical hallucinogens as antidepressants? A review of pharmacodynamics and putative clinical roles. *Therapeutic Advances in Psychopharmacology*, *4*(4), 156–169.

Belujon, P., & Grace, A. A. (2017). Dopamine system dysregulation in major depressive disorders. *The International Journal of Neuropsychopharmacology*, *20*(12), 1036–1046.

Berman, R. M., Fava, M., Thase, M. E., Trivedi, M. H., Swanink, R., McQuade, R. D., et al. (2009). Aripiprazole augmentation in major depressive disorder: A double-blind, placebo-controlled study in patients with inadequate response to antidepressants. *CNS Spectrums*, *14*(4), 197–206.

Blier, P., Bergeron R., & de Montigny, C. (1997). Selective activation of postsynaptic 5-HT1A receptors induces rapid antidepressant response. *Neuropsychopharmacology*, *16*(5), 333–338.

Bortolozzi, A., Castañé, A., Semakova, J., Santana, N., Alvarado, G., Cortés, R., et al. (2012). Selective siRNA-mediated suppression of 5-HT1A autoreceptors evokes strong anti-depressant-like effects. *Molecular Psychiatry*, *17*, 612–623.

Canale, V., Kurczab, R., Partyka, A., Satala, G., Lenda, T., Jastrzebska-Wiesek, M., et al. (2016). Towards new 5-HT7 antagonists among arylsulfonamide derivatives of (aryloxy)ethyl-alkyl amines: Multiobjective based design, synthesis, and antidepressant and anxiolytic properties. *European Journal of Medicinal Chemistry*, *108*, 334–346.

Comai, S., Lopez-Canul, M., De Gregorio, D., Posner, A., Ettaoussi, M., Guarnieri, F. C., et al. (2019). Melatonin MT1 receptor as a novel target in neuropsychopharmacology: MT1 ligands, pathophysiological and therapeutic implications, and perspectives. *Pharmacological Research*, *144* 343–356.

Cremers, T. I., Giorgetti, M., Bosker, F. J., Hogg, S., Arnt, J., Mork, A., et al. (2004). Inactivation of 5-HT(2C) receptors potentiates consequences of serotonin reuptake blockade. *Neuropsychopharmacology*, *29*(10), 1782–1789.

Cryan, J. F., & Lucki, I. (2000). Antidepressant-like behavioral effects mediated by 5-hydroxytryptamine (2C) receptors. *Journal of Pharmacology and Experimental Therapeutics*, *295*, 1120–1126.

Dawson, L. A., Hughes, Z. A., Starr, K. R., Storey, J. D., Bettelini, L., Bacchi, F., et al. (2006). Characterisation of the selective 5-HT1B receptor antagonist SB-616234-a (1-[6-(cis-3,5-dimethylpiperazin-1-yl)-2,3-dihydro-5-methoxyindol-1-yl]-1-[2′-methyl-4′-(5-methyl-1,2,4-oxadiazol-3-yl)biphenyl-4-yl]methanone hydrochloride): In vivo neurochemical and behavioural evidence of anxiolytic/antidepressant activity. *Neuropharmacology*, *50*(8), 975–983.

Depoortere, R., Auclair, A. L., Bardin, L., Colpaert, F. C., Vacher, B., & Newman-Tancredi, A. (2010). F15599, a preferential post-synaptic 5-HT1A receptor agonist: Activity in models of cognition in comparison with reference 5-HT1A receptor agonists. *European Neuropsychopharmacology*, *20*, 641–654.

Diaz, S. L., Narboux-Neme, N., Boutourlinsky, K., Doly, S., & Maroteaux, L. (2016). Mice lacking the serotonin 5-HT2B receptor as an animal model of resistance to selective serotonin reuptake inhibitors antidepressants. *European Neuropsychopharmacology*, *26*(2), 265–279.

Dos Santos, R. G., Osorio, F. L., Crippa, J. A., Riba, J., Zuardi, A. W., & Hallak, J. E. (2016). Antidepressive, anxiolytic, and antiaddictive effects of ayahuasca, psilocybin and lysergic acid diethylamide (LSD): A systematic review of clinical trials published in the last 25 years. *Therapeutic Advances in Psychopharmacology*, *6*(3), 193–213.

Ferrari, F., & Villa, R. F. (2017). The neurobiology of depression: An integrated overview from biological theories to clinical evidence. *Molecular Neurobiology*, *54*(7), 4847–4865.

Galecki, P., Mossakowska-Wojcik, J., & Talarowska, M. (2018). The anti-inflammatory mechanism of antidepressants—SSRIs, SNRIs. *Progress in Neuropsychopharmacology and Biological Psychiatry*, *80* (Pt. C), 291–294.

Gillman, P. K. (2007). Tricyclic antidepressant pharmacology and therapeutic drug interactions updated. *British Journal of Pharmacology*, *151*(6), 737–748.

Gorea, E., & Adrien, J. (1988). Serotonergic regulation of noradrenergic coerulean neurons: Electrophysiological evidence for the involvement of 5-HT2 receptors. *European Journal of Pharmacology*, *154*(3), 285–291.

Grace, A. A. (2016). Dysregulation of the dopamine system in the pathophysiology of schizophrenia and depression. *Nature Reviews. Neuroscience*, *17*(8), 524–532.

Gupta, D., Prabhakar, V., & Radhakrishnan, M. (2016). 5HT3 receptors: Target for new antidepressant drugs. *Neuroscience and Biobehavioral Reviews*, *64*, 311–325.

Guseva, D., Wirth, A., & Ponimaskin, E. (2014). Cellular mechanisms of the 5-HT7 receptor-mediated signaling. *Frontiers in Behavioral Neuroscience*, *8*, 306.

Hendrick, V., Suri, R., Gitlin, M. J., & Ortiz-Portillo, E. (2017). Bupropion use during pregnancy: A systematic review. *The Primary Care Companion for CNS Disorders*, *19*(5), 17r02160.

Hewett, K., Gee, M. D., Krishen, A., Wunderlich, H. P., Le Clus, A., Evoniuk, G., et al. (2010). Double-blind, placebo-controlled comparison of the antidepressant efficacy and tolerability of bupropion XR and venlafaxine XR. *Journal of Psychopharmacology*, *24*(8), 1209–1216.

Hughes, Z. A., Starr, K. R., Scott, C. M., Newson, M. J., Sharp, T., Watson, J. M., et al. (2007). Simultaneous blockade of 5-HT1A/B receptors and 5-HT transporters results in acute increases in extracellular 5-HT in both rats and guinea pigs: In vivo characterization of the novel 5-HT1A/B receptor antagonist/5-HT transport inhibitor SB-649915-B. *Psychopharmacology*, *192*, 121–133.

Hwang, E. K., & Chung, J. M. (2014). 5HT(1B) receptor-mediated pre-synaptic depression of excitatory inputs to the rat lateral habenula. *Neuropharmacology*, *81*, 153–165.

Lesch, K. P., & Waider, J. (2012). Serotonin in the modulation of neural plasticity and networks: Implications for neurodevelopmental disorders. *Neuron*, *76*(1), 175–191.

Martín-Hernández, D., Pereira, M. P., Tendilla-Beltran, H., Madrigal, J. L. M., Garcia-Bueno, B., Leza, J. C., et al. (2019). Modulation of Monoaminergic systems by antidepressants in the frontal cortex of rats after chronic mild stress exposure. *Molecular Neurobiology*, *56*(11), 7522–7533.

Mork, A., Pehrson, A., Brennum, L. T., Nielsen, S. M., Zhong, H., Lassen, A. B., et al. (2012). Pharmacological effects of Lu AA21004: A novel multimodal compound for the treatment of major depressive disorder. *The Journal of Pharmacology and Experimental Therapeutics*, *340*(3), 666–675.

Otte, C., Gold, S. M., Penninx, B. W., Pariante, C. M., Etkin, A., Fava, M., et al. (2016). Major depressive disorder. *Nature Reviews. Disease Primers*, *2*, 16065.

Penttila, J., Kajander, J., Aalto, S., Hirvonen, J., Nagren, K., Ilonen, T., et al. (2004). Effects of fluoxetine on dopamine D2 receptors in the human brain: A positron emission tomography study with [11C]raclopride. *The International Journal of Neuropsychopharmacology*, *7*(4), 431–439.

Sari, Y. (2004). Serotonin1B receptors: From protein to physiological function and behavior. *Neuroscience and Biobehavioral Reviews*, *28*(6), 565–582.

Sato, H., Skelin, I., & Diksic, M. (2010). Chronic buspirone treatment decreases 5-HT1B receptor densities and the serotonin transporter but increases the density of 5-HT2A receptors in the bulbectomized rat model of depression: An autoradiographic study. *Brain Research*, *1345*, 28–44.

Shan, L., Bao, A. M., & Swaab, D. F. (2015). The human histaminergic system in neuropsychiatric disorders. *Trends in Neurosciences*, *38*(3), 167–177.

Sprouse, J. S., & Aghajanian, G. K. (1987). Electrophysiological responses of serotoninergic dorsal raphe neurons to 5-HT1A and 5-HT1B agonists. *Synapse*, *1*(1), 3–9.

Subbaiah, M. A. M. (2018). Triple reuptake inhibitors as potential therapeutics for depression and other disorders: Design paradigm and developmental challenges. *Journal of Medicinal Chemistry*, *61*(6), 2133–2165.

Chapter 35

# Duloxetine usage in depression

Bing Hu
*Department of Neurology, University of California, San Francisco, San Francisco, CA, United States*

## List of abbreviations

| | |
|---|---|
| **5-HT** | serotonin |
| **CL/F** | apparent clearance |
| **IBS** | irritable bowel syndrome |
| **MDD** | major depressive disorder |
| **NE** | norepinephrine |
| **OBGYN** | obstetrics and gynecology |
| **PD** | Parkinson's disease |
| **PMDD** | premenstrual dysphoric disorder |
| **SIADH** | syndrome of inappropriate antidiuretic hormone secretion |
| **SNRI** | serotonin (5-HT) and norepinephrine (NE) reuptake inhibitor |
| **SNRIs** | serotonin (5-HT) and norepinephrine (NE) reuptake inhibitors |
| **SSRI** | selective serotonin reuptake inhibitor |
| **SSRIs** | selective serotonin reuptake inhibitors |
| **TEAEs** | treatment-emergent adverse events |
| **V d/F** | apparent volume of distribution |

## Introduction

Depression is considered one of the most prevailing and disabling psychiatric diseases and the treatment for this disorder remains challenging (Li et al., 2013). The treatment for depression includes pharmacotherapy and non-pharmacological intervention, such as tanscutaneous vagus nerve stimulation (Rong et al., 2016) and repetitive transcranial magnetic stimulation (Phillips, Burr, & Dunner, 2018).

To date, pharmacological treatment has been the main therapeutic approach (Feighner, 1999;Wagner et al., 2018). During the past decades, the neuropharmacology of depression developed quickly, and tricyclic antidepressants and monoamine oxidase blockers are primary stage antidepressants (Feighner, 1999). Due to their poor selectivity and undesirable and even dangerous side effects, they had limited use (Feighner, 1999). Then, selective serotonin reuptake inhibitors (SSRIs) emerged (Feighner, 1999). However, SSRIs only selectively increase the serotonin function (Feighner, 1999), which limits their antidepressive potency and spectrum. Therefore, the second-generation antidepressants, selective serotonin (5-HT), and norepinephrine (NE) reuptake inhibitors (SNRIs) were developed (Bymaster, Lee, Knadler, Detke, & Iyengar, 2005; Monteleone, Caputo, Tecce, & Capasso, 2011).

Duloxetine (Cymbalta(®)) is a potent SNRI to treat depression, especially major depressive disorder (MDD) (Dhillon, 2013) and it has been approved by the U.S. Food and Drug Administration to treat MDD (Kirwin & Gören, 2005). Duloxetine evidently improved depressive symptoms in young, adult, and elderly patients with depression (Girardi et al., 2009; Meighen, 2007; Rovera et al., 2016) and its antidepressive efficacy was superior to SSRI in certain conditions (Beyer & Johnson, 2018). Duloxetine is generally safe and well tolerable (Bitter, Filipovits, & Czobor, 2011) as duloxetine is more selective than the first-generation antidepressants (Monteleone et al., 2011). Moreover, duloxetine shows therapeutic superiority because it can extensively treat not only depression and its related symptoms, but also various comorbidities of depression (Dunner, Goldstein, Mallinckrodt, Lu, & Detke, 2003; Hu, Doods, Treede, & Ceci, 2016; Lewis-Fernandez et al., 2016; Papakostas, 2015; Volpe, 2008).

Therefore, duloxetine is considered a well-recognized, versatile first-line antidepressant (Beyer & Johnson, 2018;Lewis-Fernandez et al., 2016; Volpe, 2008). Moreover, the effort to improve the efficacy of duloxetine is ongoing. To enhance the distribution of duloxetine in the brain, duloxetine loaded nanostructured lipid carriers a new formulation of duloxetine that can be administrated intranasally and directly delivered to the brain. was developed (Alam et al., 2014). The primary study determined that intranasal administration of this formulation can deliver eightfold greater amount of duloxetine to the brain (Alam et al., 2014). Furthermore, to increase the oral bioavailability of duloxetine, a novel microemulsion formulation of duloxetine was formulated recently (Sindhu, Kumar, Iqbal, Ali, & Baboota, 2018). The preclinical study demonstrated that depressive behavior of the rats which were

The Neuroscience of Depression. https://doi.org/10.1016/B978-0-12-817933-8.00094-3

administrated with duloxetine microemulsion was remarkably ameliorated compared with duloxetine suspension (Sindhu et al., 2018). These formulation strategies could bring brighter prospects for augmentation of the efficacy of duloxetine.

This article highlights the effects of duloxetine on symptoms of depression, treatment-resistant depression, OBGYN-based depression (gynecology and obstetrics), and comorbidities of depression. In addition, the pharmacokinetics and side effects of duloxetine are also described. These shed light on the functions and prospects of duloxetine for clinicians and researchers.

## Pharmacokinetic profile

Considering duloxetine is unstable in acid, the slow-release formulation of duloxetine was developed to avoid to be dissolved by gastric acid (Knadler, Lobo, Chappell, & Bergstrom, 2011). This formulation of duloxetine can be released and absorbed promptly at the pH of intestinum tenue (Knadler et al., 2011).

According to the pharmacokinetic outcomes of patients, the rate of duloxetine absorption in the intestine is fast (absorption rate constant 0.168 h-1) (Lobo, Quinlan, O'Brien, Knadler, & Heathman, 2009). Following the oral single dose of 60 mg, the absolute oral bioavailability was in the range from 30% to 80%, and the average was 50% (Bymaster et al., 2005). Duloxetine reaches maximum plasma concentration around 6 h following oral administration (Knadler et al., 2011). As the elimination half-life is relatively long, which is about 10–12 h, a steady state can be maintained within 3 days (Knadler et al., 2011).

Duloxetine has a great average apparent steady-state volume of distribution, ranging from 1620 to 1800 L, which suggests that duloxetine is widely distributed all over the body (Sharma, Goldberg, & Cerimele, 2000). Duloxetine highly binds to proteins in human plasma since in vitro research showed that more than 90% of duloxetine binds to plasma proteins, which is supposed to have influence on distribution of duloxetine (Bymaster et al., 2005). There is evidence that duloxetine can distribute into the brain, as a research using positron emission tomography on healthy male volunteers determined the 5-HT transporter occupancy in the thalamus (Takano et al., 2006). Intriguingly, a study, which evaluated duloxetine concentrations of cerebrospinal fluid in patients administrated with 30–120 mg of duloxetine daily, displayed quite low concentrations of duloxetine in the cerebrospinal fluid (Paulzen et al., 2016). This may indicate that the drug is difficult to pass blood-cerebrospinal fluid barrier (Paulzen et al., 2016). Furthermore, this may also attribute to the active delivery mechanisms which deliver substance out of cerebrospinal fluid (Paulzen et al., 2016). Moreover, other mechanisms may be involved, which need to be clarified by future study.

Duloxetine is metabolized in the liver and becomes a variety of complex inactive metabolites after extensive metabolism (Gupta, Nihalani, & Masand, 2007; Knadler et al., 2011). The glucuronide conjugate of 4-hydroxy duloxetine, as well as the sulfate conjugate of 5-hydroxy-6-methoxy duloxetine, are the main metabolites of duloxetine (Lantz et al., 2003). It is worth noting that metabolites of duloxetine are inactive (Gupta et al., 2007). The study demonstrated that following multiple dosing from 20 to 40 mg, the average duloxetine clearance (CL/F) was 114 L/h in healthy men (Sharma et al., 2000). Duloxetine is primarily excreted in urine since around 72% of the radioactive substance was discharged from urine mainly as the substance of glucuronide and/or sulfate conjugates of the oxidative duloxetine metabolites, after administration of duloxetine labeled with $^{14}C$ (Knadler et al., 2011), although a comparatively small amount of radioactive substance was detected in stool as duloxetine (<5% of the dose) 96 h following administration (Knadler et al., 2011).

The clinical pharmacokinetics of duloxetine was analogous between children and adolescent patients suffering from major depressive disorder (MDD) (Lobo, Quinlan, & Prakash, 2014). Evaluations of V d/F as well as CL/F were supposed to be higher in the pediatric patients than those in adult populations; consequently, the mean steady-state concentration of duloxetine was less in the pediatric patients than that in adult populations (Lobo et al., 2014).

## Duloxetine for depression and its associated symptoms

Duloxetine, a selective 5-HT and NE reuptake blocker, is a robust and wide spectrum antidepressant (Bymaster et al., 2005; Dunner et al., 2003). Duloxetine has extensive effects on both unipolar and bipolar depression (Serafini et al., 2010), and can cure patients with mild, moderate, and major depression (Shelton et al., 2007).

As patients experiencing bipolar disorder may suffer from mood instability during duloxetine administration (Wilke, Gota, & Muzina, 2010), close attention should be paid to these patients. Moreover, the double-blind, randomized controlled clinical trial elucidated that duloxetine substantially ameliorated passive symptoms of depression, such as psychomotor retardation as well as sexual problems (Mowla, Dastgheib, & Razeghian Jahromi, 2016).

In general, the side events of duloxetine were from mild to moderate and showed no increase in the rate of suicide, and were well tolerable, safe, and even suitable for older patients (Bitter et al., 2011; Raskin et al., 2007). Therefore, duloxetine is commonly prescribed to elderly patients with MDD (Raskin et al., 2007).

An open-label clinical trial in elderly patients with MDD between 65 and 87 years of age was conducted and the duloxetine treatment obviously alleviated depressive symptoms (Rovera et al., 2016). Another clinical study revealed that duloxetine can act as rescue pharmacological treatment for depression in elderly patients who were refractory to the management of SSRI escitalopram (Karp et al., 2008). This case indicated that duloxetine can have good tolerability for elderly depressive populations even at the dosage of 120 mg/day (Karp et al., 2008). Moreover, there is a study reported that duloxetine showed latent favorable efficacy for young patients with depression (Xu et al., 2016).

Duloxetine shows superiority for treatment of depression as it can relieve not only depressed mood but also crucial symptoms of depression, such as pain, anxiety, and cognitive malfunction (Dunner et al., 2003; Gaynor et al., 2011; Papakostas, 2015) (Table 1).

Physical symptoms of pain frequently occur in patients experiencing MDD, and the presence of pain in MDD treatments indicates poorer therapeutic results (Fava, Mallinckrodt, Detke, Watkin, & Wohlreich, 2004; Gaynor et al., 2011). Duloxetine administration remarkably alleviated pain symptoms and enhanced the remission rate of depression in the double-blind randomized controlled trial (Gaynor et al., 2011).

Anxiety is a prevalent symptom in patients with MDD (Fawcett & Kravitz, 1983; Stein, Kirk, Prabhu, Grott, & Terepa, 1995). Duloxetine rapidly alleviated anxiety symptoms occurring in patients with depression and this therapeutic efficacy continued during the whole period of acute treatment (Dunner et al., 2003).

Cognitive malfunction is a frequent symptom in patients with MDD and it was implicated that cognitive malfunction is a pivotal factor affecting disability in the workplace (McIntyre & Lee, 2016). A clinical study revealed that duloxetine can considerably improve immediate and delayed memory in individuals with MDD, suggesting that duloxetine can promote the psychosocial recovery of individuals suffering from MDD (Papakostas, 2015).

More functions of duloxetine could be revealed by clinicians and researchers in future studies.

## Effects of duloxetine on depression in gynecology and obstetrics

Duloxetine has an extensive antidepressive spectrum which covers OBGYN-based depression.

Approximately 50%–80% of women suffer from mild-to-severe premenstrual symptoms, which include irascibility, depression, swelling, mastodynia, as well as cephalalgia during their reproductive age (Halbreich, Borenstein, Pearlstein, & Kahn, 2003), and an estimated 3%–8% of women with premenstrual symptoms develop the premenstrual dysphoric disorder (PMDD) (Cohen et al., 2002; Halbreich et al., 2003; Wittchen, Becker, Lieb, & Krause, 2002).

Clinical trials tested the efficacy of duloxetine on PMDD. Duloxetine (60 mg/d) therapy was efficacious, which resulted in 50% or more attenuation in baseline premenstrual symptoms of 65% of the study subjects following duloxetine treatment for three menstrual cycles (Ramos, Hara, & Rocha, 2009), and 50% reduction of daily symptom

**TABLE 1** Effective spectrum of duloxetine.

| Symptoms of depression | Depression in comorbidities | Special comorbidities |
|---|---|---|
| Depressed mood | Psychotic disorders | Irritable bowel syndrome |
| Painful physical symptoms | Parkinson's disease | Anxiety, chronic pain |
| Anxiety symptoms | Breast cancer | Trigeminal neuralgia |
| Cognitive dysfunction | Mitochondrial encephalomyopathy with lactic acidosis and stroke-like episodes | Binge eating disorder |

The first column illustrates the critical symptoms of depression, which can be improved by duloxetine. The second column represents comorbidities of depression and the depressive symptoms in these comorbidities can be attenuated by duloxetine. The third column represents special comorbidities and duloxetine can cure both depression and these comorbidities.
The table is created according to the references of (Biglia et al., 2018; Bonuccelli et al., 2012 Cozart, Diaz Vera, Denham, & Whiting, 2018; Dunner et al., 2003; Englisch et al., 2009; Gaynor et al., 2011; Guerdjikova et al., 2012;Hsu, Chang, Peng, & Liang, 2014 Lewis-Fernandez et al., 2016; Mancini, Perna, Rossi, & Petralia, 2010; Papakostas, 2015; Shelton et al., 2007; Wohlreich et al., 2009).

scores was obtained in 78% patients following duloxetine treatment for two menstrual cycles (Mazza, Harnic, Catalano, Janiri, & Bria, 2008), respectively. In general, duloxetine shows good tolerability in patients experiencing PMDD (Mazza et al., 2008; Ramos et al., 2009).

According to clinical reports, duloxetine treatment in the period of pregnancy enhanced the risk of spontaneous abortion (Hoog, Cheng, Elpers, & Dowsett, 2013; Kjaersgaard et al., 2013), and duloxetine exposure in late gestation may be related to poor neonatal adaptation syndrome (Eyal & Yaeger, 2008). However, no increase in major fetal malformations was observed in patients receiving pharmacotherapy of duloxetine (Einarson et al., 2012; Hoog et al., 2013). It was reported that a woman administrated duloxetine to manage her depression during uncomplicated pregnancy and breastfeeding (Briggs et al., 2009). No evidence of developmental disorder and other toxicity was found in the baby exposed to duloxetine in the posterior half of pregnancy as well as in breast milk of the first 32 days following delivery (Briggs et al., 2009), and according to clinical pharmacokinetic test, duloxetine can exist in the milk of breastfeeding women who received duloxetine, and the steady-state concentrations of duloxetine in human milk are approximately one-fourth of those in plasma of lactating women (Lobo et al., 2008). Further study demonstrated that duloxetine may be acceptable for breastfeeding woman (Sachs & Committee on Drugs, 2013).

Numerous women experience MDD as well as hot flashes during the menopausal transition (Avis, Brambilla, McKinlay, & Vass, 1994; Cohen, Soares, Vitonis, Otto, & Harlow, 2006; Joffe et al., 2002), and depression and hot flashes frequently coexist in the menopause transition as women suffering from hot flashes are at high risk for MDD (Joffe et al., 2002). Duloxetine was given to patients with MDD, anxiety, and hot flashes in the menopausal transition, and it was proved to remarkably attenuate depressive symptoms, anxiety, and frequency as well as the severity of their hot flashes (Freeman et al., 2013).

Moreover, duloxetine treatment considerably ameliorated both depression and concurrent symptoms of anxiety, pain, sleep disturbance, and vasomotor symptoms in postmenopausal women (Joffe et al., 2007).

## Effects of duloxetine on treatment-resistant depression

As treatment for treatment-resistant depression is challenging, multiple strategies of treatment with duloxetine were developed (Delgado, Saldana, Barzman, & Coffey, 2007; Papakostas et al., 2006; Rosso, Rigardetto, Bogetto, & Maina, 2012; Seguí et al., 2010).

Across-class switch of antidepressant is supposed to be one of the best choices to manage the treatment of resistant depression (Rosso et al., 2012). When patients whose depressive symptoms did not get relief following adequate SSRI administration were treated with SNRI antidepressant duloxetine, the depressive symptoms significantly improved (Rosso et al., 2012). This suggests that switching ineffective antidepressant to different type of responsive antidepressant is an efficacious strategy (Rosso et al., 2012). Moreover, a clinical case report determined that duloxetine monotherapy of across-class switching can also effectively cure medication-resistant MDD in an adolescent patient who poorly responded to previous combination therapy with other antidepressants (Delgado et al., 2007).

It is estimated that about one-third of patients suffering from MDD failed to attain remission with monotherapy of antidepressant (Rapaport et al., 2006). Therefore, a combined regimen was developed to manage refractory depression (Papakostas et al., 2006;Seguí et al., 2010).

In a preliminary clinical trial, the symptoms of patients experiencing treatment-resistant MDD, who did not get complete relief by either duloxetine or bupropion administration alone, were significantly reduced after combination treatment of duloxetine and bupropion (Papakostas et al., 2006). This result elucidated the feasibility of the combination of duloxetine and bupropion to cure treatment-resistant MDD (Papakostas et al., 2006).

Furthermore, for depression patients resistant to monotherapy of duloxetine, the adjunctive combination with other types of antidepressants were developed (Seguí et al., 2010). To treat duloxetine-resistant depressive patients, who have no response or partial response to duloxetine, reboxetine, another antidepressant with different pharmacological mechanism, was added to the treatment regimen as an adjunct (Seguí et al., 2010). The depressive symptoms were remarkably relieved and although common side effects were observed, they were tolerable (Seguí et al., 2010).

Duloxetine combined with multiple approaches can be an effective solution for medication-resistant major depression co-occurring with comorbidity (Hanretta & Malek-Ahmadi, 2006). It was reported that a female patient with medication-resistant MDD as well as post-traumatic stress disorder received treatment of duloxetine combined with olanzapine and electroconvulsive therapy (Hanretta & Malek-Ahmadi, 2006). This treatment completely ameliorated symptoms of major depression without any complications (Hanretta & Malek-Ahmadi, 2006).

As a broad-spectrum antidepressant, duloxetine has been prescribed for off-label aims sometimes and suprathreshold dosages may be critical for special cases (Safer & Arnow, 2012). The suprathreshold dosage duloxetine (initial dose may reach 180mg) was applied to a treatment-resistant major depression patient concurrent with anorexia nervosa binge-purging type and obsessive-compulsive disorder, and this special dosage of duloxetine effectively ameliorated the mood of the patient (Safer & Arnow, 2012). This suggests the feasibility of suprathreshold dosage of duloxetine for management of special refractory cases.

Duloxetine could be involved in more and more novel regimens to manage refractory depression in future treatments.

## Effects of duloxetine on depression with comorbidities

As a highly prevalent neuropsychiatric disorder, depression extensively coexists with various diseases. Duloxetine can treat not only depressive symptoms in comorbidities but also its comorbidities in certain cases (Table 1).

It is demonstrated that anxiety frequently coexists with depression, and duloxetine was effective to cure comorbid anxiety and MDD simultaneously (Mancini et al., 2010).

Moreover, MDD often coexists with irritable bowel syndrome (IBS), which is difficult to cure (Lewis-Fernandez et al., 2016). An open-label clinical trial determined duloxetine considerably ameliorated both the alimentary canal and depressive symptoms gradually (Lewis-Fernandez et al., 2016).

There is a high incidence of depression associated with chronic pain, and both preclinical and clinical studies determined that chronic pain can cause depression (Hu, Doods, Treede, & Ceci, 2009; Hudson & Pope Jr., 1996; Wolfe et al., 1990). Duloxetine displayed extensive effects on diverse pain disorders (Table 2).

Importantly, duloxetine showed superior effects to fluoxetine, another widely used antidepressant, since duloxetine can evidently relieve both depressive behavior in forced swimming test and pain response in the model of chronic neuropathic pain comorbid with depression, whereas fluoxetine failed (Hu et al., 2016) (Figs.1 and 2).

Duloxetine was effective and well tolerated in pediatric chronic pain patients with concomitant MDD at minimal dosages, as duloxetine rapidly relieved pain and depressed mood of two adolescent female patients with minimum to no detrimental effects (Meighen, 2007). Another adolescent patient suffering from femoral neuralgia with 'reactive depression' received duloxetine in the multistrategy management regimen, and her mental symptoms were also evidently relieved after treatment (Kachko, Ben Ami, Liberman, Birk, & Kronenberg, 2011).

In addition, for the arthritis and fibromyalgia patients comorbid with MDD, duloxetine also significantly relieved not only MDD severity but also the pain of arthritis and fibromyalgia (Leombruni et al., 2015; Wohlreich et al., 2009).

Chronic headache and MDD are closely related (Volpe, 2008). Clinical studies demonstrated that duloxetine evidently reduced both headache and depressive symptoms

**TABLE 2** Effects of duloxetine on both chronic pain and comorbid depression.

| Report | Type of pain | Subject | Dosage | Relief of pain and depression | |
|---|---|---|---|---|---|
| Meighen (2007) | Lower back and hip pain | Adolescent | 20–40mg/day | + | + |
| Kachko, Ben Ami, Liberman, Birk, & Kronenberg, (2011) | Femoral neuropathy | Adolescent | 20–60mg/day | + | + |
| Leombruni et al., (2015) | Fibromyalgia | Adult | 60mg/day | + | + |
| Volpe (2008) | Chronic headache | Adult | 60mg/day | + | + |
| Hsu et al. (2014) | Trigeminal neuralgia | Adult | 30–60mg/day | + | + |
| Curone et al. (2013) | Medication overuse | Adult | 30–60mg/day | + | + |
| Wohlreich et al. (2009) | Arthritis | Elderly | 60mg/day | + | + |

Summary of types of pain disorder comorbid with depression, and duloxetine is efficacious on these chronic pains and their concurrent depression. "+" means efficacious.
The table is created according to the references of (Curone et al., 2013; Hsu et al., 2014; Kachko et al., 2011; Leombruni et al., 2015; Meighen, 2007; Volpe, 2008; Wohlreich et al., 2009).

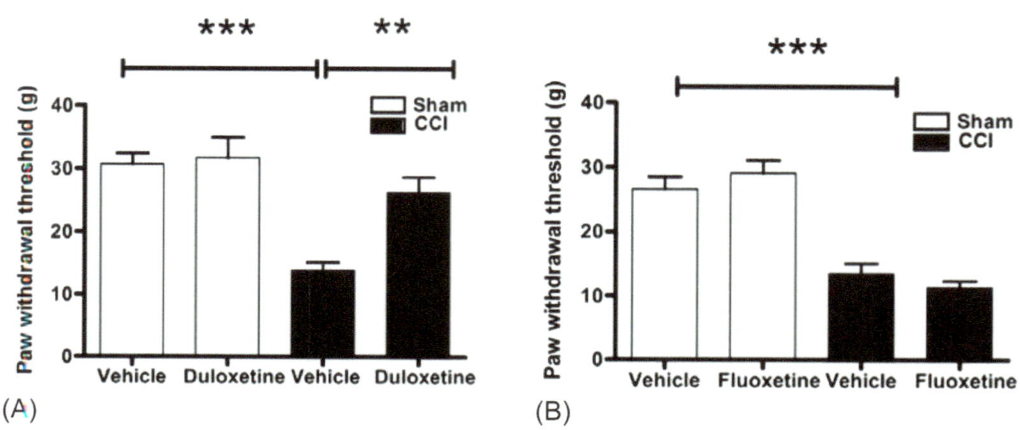


FIG. 1 Effect of duloxetine and fluoxetine on pain response in chronic neuropathic pain in animals with comorbid depression. Effect of duloxetine (20mg/kg) and fluoxetine (20mg/kg) on paw withdrawal thresholds (pain response) following chronic constriction injury of the sciatic nerve. Data are expressed as means ± SEM. $^{**}P < 0.01$, $^{***}P < 0.001$ indicate the values are significantly different. *Data are from Hu, B., Doods, H., Treede, R. D., & Ceci, A. (2016). Duloxetine and 8-OH-DPAT, but not fluoxetine, reduce depression-like behaviour in an animal model of chronic neuropathic pain.* Neuroscience Letters, 619, *162–167, with permission from the publishers.*

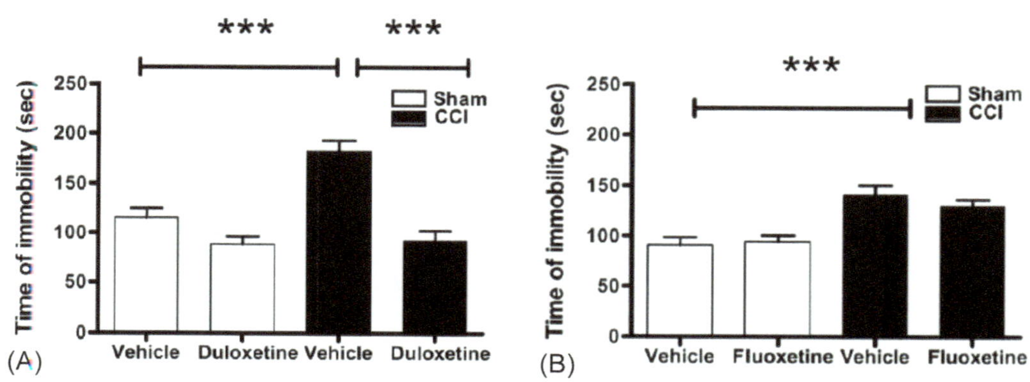


FIG. 2 Effect of duloxetine and fluoxetine on depressive behavior induced by chronic neuropathic pain. Effect of duloxetine (20mg/kg) and fluoxetine (20mg/kg) on time of immobility at 30cm water depth following chronic constriction injury of the sciatic nerve. Data are expressed as mean ± SEM. $^{***}P < 0.001$ indicates the values are significantly different. *Data are from Hu, B., Doods, H., Treede, R. D., & Ceci, A. (2016). Duloxetine and 8-OH-DPAT, but not fluoxetine, reduce depression-like behaviour in an animal model of chronic neuropathic pain.* Neuroscience Letters, 619, *162–167., with permission from the publishers.*

in medication overuse derived from persistent migraine and chronic headache patients concurrent with MDD (Curone et al., 2013; Volpe, 2008).

Similarly, there is a case reported of a woman suffering from trigeminal neuralgia who developed MDD (Hsu et al., 2014). After duloxetine administration, not only was her trigeminal neuralgia surprisingly and rapidly relieved but she also later experienced remission of her depressive symptoms (Hsu et al., 2014).

Moreover, duloxetine can treat depressive symptoms in less frequent or rare comorbidities of depression. Parkinson's disease (PD) and psychotic disorders are sometimes concomitant with depression, and duloxetine was efficacious in improving depression concurrent with PD and psychotic disorders without obvious adverse effects (Bonuccelli et al., 2012;Englisch et al., 2009).

Interestingly, a clinical study showed duloxetine not only evidently improved depression coexisting with breast cancer or binge eating disorder, but also considerably alleviated the onset rate of hot flushes in breast cancer survivors (Biglia et al., 2018) and symptoms and signs of binge eating patients (Guerdjikova et al., 2012). Moreover, duloxetine can also treat concomitant depression in case of a rare disease, Mitochondrial encephalomyopathy with lactic acidosis and stroke-like episodes (Cozart et al., 2018).

In conclusion, numerous studies showed the extensive effectiveness of duloxetine in diverse comorbidities of depression. Intriguingly, duloxetine provides additional benefit since it can treat not only depressive symptoms but also comorbidities (Guerdjikova et al., 2012; Hu et al., 2016; Lewis-Fernandez et al., 2016; Mancini et al., 2010; Volpe, 2008). Further effects of duloxetine on more depression associated comorbidities could be explored by future study.

## Side effects

Duloxetine has various side effects (Table 3). The most common treatment-emergent detrimental effects were observed in the digestive system (28.3%), psychiatric and nervous system (25.8%, 28.0%), such as sicchasia, diarrhea, drowsiness as well as dizziness, while the severity was mainly mild to moderate (Bitter et al., 2011; Pangallo et al., 2010) (Fig. 3).

Sexual dysfunction is a symptom of depression and also a common adverse effect of antidepressants (Werneke, Northey, & Bhugra, 2006). Clinical trials demonstrated that the incidence rate of acute treatment-emergent sexual malfunction in patients treated with duloxetine was higher than that in the placebo group (Delgado et al., 2005), but lower than those in the escitalopram and paroxetine groups (SSRI) (Clayton, Kornstein, Prakash, Mallinckrodt, & Wohlreich, 2007; Delgado et al., 2005).

Following the extensive usage of duloxetine in diverse patients for more than a decade, rare adverse effects emerged.

Mania occurs in patients treated with any class of antidepressant, which is a rare adverse effect of antidepressants (Peritogiannis et al., 2009; Viktrup, Perahia, & Tylee, 2004). It was reported that hypomania was caused in a non-bipolar patient by the administration of duloxetine (Peritogiannis et al., 2009). When duloxetine was ceased, the hypomanic symptoms started to diminish (Peritogiannis et al., 2009). This reminds us that although it is rare, the possibility of mania or hypomanic caused by duloxetine should not be ignored (Peritogiannis et al., 2009).

A case report showed that panic attacks occurred in a depressive patient who was treated with duloxetine, and when the treatment was ceased, these panic attacks discontinued (Sabljić et al., 2011).

Furthermore a large-scale study determined that patients lost weight following short-term duloxetine therapy, while patients modestly gained body weight following long-term therapy (Wise, Perahia, Pangallo, Losin, & Wiltse, 2006). The size of body weight changes indicates that duloxetine has only a minor influence on weight for most patients (Wise et al., 2006).

In addition, the syndrome of inappropriate antidiuretic hormone secretion (SIADH), which leads to approximately 60% of hyponatremias, is a well-recognized detrimental effect of SSRIs, but SNRIs seldom induces SIADH-associated hyponatremia (Amoako, Brown, & Riley, 2015). There was a report that hyponatremia occurred in an elderly woman who was treated with duloxetine, and her sodium recovered after the cessation of duloxetine and appropriate treatment (Amoako et al., 2015).

Impulse control disorders are an undesired effect of activation of dopamine (Miller & Gallagher, 2015). Interestingly, duloxetine was supposed to suppress dopamine transporter at higher dosages (Kihara & Ikeda, 1995). A case report indicated a young female with depression developed novel Self-Limited Kleptomania Symptoms (stealing behaviors) when her duloxetine dosage was raised from 60 to 90 mg (Miller & Gallagher, 2016). This could be due to the fact that dopamine transporter was suppressed by duloxetine, and the brain was unable to control the impulse (Miller & Gallagher, 2016). Another case was reported that

**TABLE 3** Side effects of duloxetine.

| | Common side effects | Unusual side effects |
|---|---|---|
| Digestive system | Nausea, dry mouth, decreased appetite, vomiting, constipation, diarrhea | — |
| Neuropsychiatric system | Somnolence, insomnia, dizziness, headache, agitation | Mania or hypomania. Kleptomania symptoms, retrobulbar optic neuritis, restless legs syndrome, panic attacks, parkinsonism symptoms, dysosmia, and dysgeusia |
| Other systems | Fatigue, asthenia, sexual dysfunction, hyperhidrosis, or increased sweating | Urinary retention, hyponatremia, body weight changes |

Summary of side effects of duloxetine including common, uncommon, and rare adverse effects.
The table is created according to the references of (Amoako, Brown, & Riley, 2015; Bayrak, Cetin, Meteris, & Kesebir, 2015; Bicer, Kosker, Celikay, & Gurdal, 2016; Bitter et al., 2011; Delgado et al., 2005; Englisch, Fritzinger, & Zink, 2008; Miller & Gallagher, 2016; Pangallo et al., 2010;Peritogiannis, Antoniou, Mouka, Mavreas, & Hyphantis, 2009; Rottach et al., 2008; Sabljić, Rakun, Ružić, & Grahovac, 2011; Wise, Perahia, Pangallo, Losin, & Wiltse, 2006; Yoshida, Fukuchi, & Sugawara, 2017).

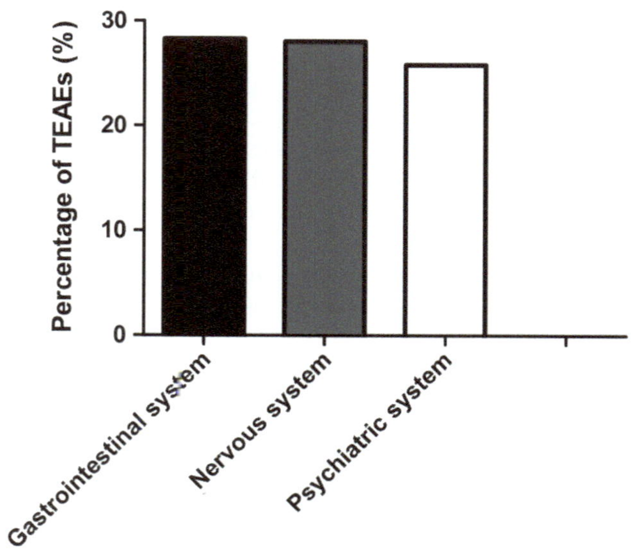


FIG. 3 The incidence rates of treatment-emergent adverse events in different System Organ Class. The incidence rates of treatment-emergent adverse events (TEAEs) in gastrointestinal, psychiatric and nervous systems, which are expressed as percentage. *The figure is created according to the reference of Pangallo, B.A., Zhang, Q., Desaiah, D., Perahia, D.G., Detke, M.J., & Kennedy, S.H.(2010).Long-term safety of duloxetine during open-label compassionate use treatment of patients who completed previous duloxetine clinical trials.* Current Medical Research and Opinion, 26*(11), 2643–2651.*

a patient developed parkinsonism symptoms following treatment with duloxetine for MDD (Bayrak et al., 2015). It may be prudent to advise both clinicians and patients that antidepressant related extrapyramidal system can happen and it is critical to monitor this adverse effect (Bayrak et al., 2015).

Recently, rare smelling and visual impairments possibly related to duloxetine treatment were reported, respectively (Bicer et al., 2016; Yoshida et al., 2017). An old Japanese woman got dysosmia and dysgeusia following duloxetine administration (Yoshida et al., 2017). Another case is a woman treated with duloxetine displayed vision impairment caused by retrobulbar neuritis (Bicer et al., 2016). Occasionally, restless legs syndrome can occur in patients treated with duloxetine (Rottach et al., 2008).

Side effects are typically undesirable. However, according to patient's condition, the side effect and therapeutic effect are dialectically related. For example, duloxetine can cause urinary retention, which is considered an adverse effect, but can also cure stress urinary incontinence (Englisch et al., 2008; Viktrup et al., 2004).

These studies provide the knowledge for appropriate usage of duloxetine in the clinic. Furthermore, the balance between therapeutic efficacy and side effects based on the condition of each patient should be considered.

## Key facts of duloxetine

- Duloxetine has superiority because it can treat not only depression but also certain comorbidities.
- Duloxetine evidently reduces depression and co-occurring anxiety.
- Duloxetine remarkably ameliorates depression and its various concurrent pain disorders, such as arthritis, trigeminal neuralgia, and chronic headache.
- MDD often coexists with IBS, and duloxetine can substantially relieve both gastrointestinal and depressive symptoms gradually.
- Depression can occur in breast cancer survivors, and duloxetine can alleviate not only depressed mood but also the onset rate of hot flushes of breast cancer survivors.
- Binge eating disorder sometimes is concomitant with depression and duloxetine can intriguingly relieve both depression and symptoms of binge eating disorder.

## Summary points

- This chapter focuses on the effects of versatile first-line antidepressant duloxetine on depression and its associated symptoms and comorbidities.
- Duloxetine is a potent dual serotonin (5-HT) and norepinephrine (NE) reuptake inhibitor.
- Duloxetine relieves both depressed mood and its related symptoms, such as anxiety, painful physical symptoms, and cognitive dysfunction.
- Various strategies with duloxetine were developed to treat resistant depression.
- Duloxetine effectively attenuates depression linked to OBGYN.
- Duloxetine can effectively treat not only depression but also its comorbidities, such as IBS and pain.
- Side effects of duloxetine are generally mild to moderate and are relatively safe and well tolerated.

## Mini-dictionary of terms

**Cerebrospinal fluid** A colorless and transparent liquid which surrounds and protects the brain and spinal cord. The substance can transfer between cerebrospinal fluid and brain.

**IBS** An intestinal disease, the main symptoms of which include abdominal pain or discomfort, excessive gas, diarrhea, and constipation.

**Menopausal transition** A period when women experience a series of physiological changes and is close to reproductive senility.

**Comorbidity** A condition that is characterized by multiple diseases or pathologies.

**Forced swimming test** A classical and widely used preclinical behavior test to detect depressive behavior of animals and antidepressive efficacy of drugs.

**Binge eating disorder** It is a grave eating disorder that is characterized by frequently eating a lot of food with the feeling of being unable to control that behavior.

**Hyponatremia** It is considered hyponatremia when the serum sodium concentration is less than 135 mmoL/L, which is the most prevalent electrolyte unbalance in clinic.

## Acknowledgment

The author acknowledges Russell Fitzgerald for the critical language checking.

Declare: When the author expresses an opinion (statements without citations), to the best of the author's knowledge, the author affirms that no literature was found which fitted that expression; therefore, no paper can be cited. If there is the same opinion published in the science journal before the author's article which the author did not know, please let the author know and the author will cite this article.

## References


Alam, M. I., Baboota, S., Ahuja, A., Ali, M., Ali, J., Sahni, J. K., et al. (2014). Pharmacoscintigraphic evaluation of potential of lipid nanocarriers for nose-to-brain delivery of antidepressant drug. *International Journal of Pharmaceutics*, *470*(1–2), 99–106.

Amoako, A. O., Brown, C., & Riley, T. (2015). Syndrome of inappropriate antidiuretic hormone secretion: A story of duloxetine-induced hyponatraemia. *BMJ Case Reports*, *2015*, 1–3.

Avis, N., Brambilla, D., McKinlay, S., & Vass, K. (1994). A longitudinal analysis of the association between menopause and depression: Results from the Massachusetts Women's Health Study. *Annals of Epidemiology*, *4*(3), 214–220.

Bayrak, A., Cetin, B., Meteris, H., & Kesebir, S. (2015). Parkinsonism secondary to duloxetine use: A case report. *Nothern Clinics of Istanbul*, *2* (3), 243–246.

Beyer, J. L., & Johnson, K. G. (2018). Advances in pharmacotherapy of late-life depression. *Current Psychiatry Reports*, *20*(5), 34.

Bicer, T., Kosker, M., Celikay, O., & Gurdal, C. (2016). A case of retrobulbar optic neuritis caused by duloxetine. *Cutaneous and Ocular Toxicology*, *35*(3), 251–253.

Biglia, N., Bounous, V. E., Susini, T., Pecchio, S., Sgro, L. G., Tuninetti, V., et al. (2018). Duloxetine and escitalopram for hot flushes: Efficacy and compliance in breast cancer survivors. *European Journal of Cancer Care*, *27*(1), 1–9.

Bitter, I., Filipovits, D., & Czobor, P. (2011). Adverse reactions to duloxetine in depression. *Expert Opinion on Drug Safety*, *10*(6), 839–850.

Bonuccelli, U., Meco, G., Fabbrini, G., Tessitore, A., Pierantozzi, M., Stocchi, F., et al. (2012). A non-comparative assessment of tolerability and efficacy of duloxetine in the treatment of depressed patients with Parkinson's disease. *Expert Opinion on Pharmacotherapy*, *13*(16), 2269–2280.

Briggs, G. G., Ambrose, P. J., Ilett, K. F., Hackett, L. P., Nageotte, M. P., & Padilla, G. (2009). Use of duloxetine in pregnancy and lactation. *The Annals of Pharmacotherapy*, *43*(11), 1898–1902.

Bymaster, F. P., Lee, T. C., Knadler, M. P., Detke, M. J., & Iyengar, S. (2005). The dual transporter inhibitor duloxetine: A review of its preclinical pharmacology, pharmacokinetic profile, and clinical results in depression. *Current Pharmaceutical Design*, *11*(12), 1475–1493.

Clayton, A., Kornstein, S., Prakash, A., Mallinckrodt, C., & Wohlreich, M. (2007). Changes in sexual functioning associated with duloxetine, escitalopram, and placebo in the treatment of patients with major depressive disorder. *The Journal of Sexual Medicine*, *4*(4 Pt 1), 917–929.

Cohen, L. S., Soares, C. N., Otto, M. W., Sweeney, B. H., Liberman, R. F., & Harlow, B. L. (2002). Prevalence and predictors of premenstrual dysphoric disorder (PMDD) in older premenopausal women. The Harvard study of moods and cycles. *Journal of Affective Disorders*, *70*(2), 125–132.

Cohen, L. S., Soares, C. N., Vitonis, A. F., Otto, M. W., & Harlow, B. L. (2006). Risk for new onset of depression during the menopausal transition: The Harvard study of moods and cycles. *Archives of General Psychiatry*, *63*(4), 385–390.

Cozart, B., Diaz Vera, J., Denham, J. D. & Whiting, W. L. (2018). Treatment of depression with duloxetine in mitochondrial encephalopathy, lactic acidosis, and stroke-like episodes. *Clinical Neuropharmacology*, *41*(3), 103–105.

Curone, M., Tullo, V., Savino, M., Proietti-Cecchini, A., Bussone, G., & D'Amico, D. (2013). Outcome of patients with chronic migraine with medication overuse and depression after duloxetine: Influence of coexisting obsessive compulsive disorder. *Neurological Sciences*, *34* (Suppl. 1), S175–S177.

Delgado, P. L., Brannan, S. K., Mallinckrodt, C. H. Tran, P. V., McNamara, R. K., Wang, F., et al. (2005). Sexual functioning assessed in 4 double-blind placebo- and paroxetine-controlled trials of duloxetine for major depressive disorder. *The Journal of Clinical Psychiatry*, *66*(6), 686–692.

Delgado, S. V., Saldana, S. N., Barzman, D. H., & Coffey, E. (2007). Response to duloxetine in a depressed, treatment-resistant adolescent female. *Journal of Child and Adolescent Psychopharmacology*, *17*(6), 889–894.

Dhillon, S. (2013). Duloxetine: A review of its use in the management of major depressive disorder in older adults. *Drugs & Aging*, *30*(1), 59–79.

Dunner, D. L., Goldstein, D. J., Mallinckrodt, C., Lu, Y., & Detke, M. J. (2003). Duloxetine in treatment of anxiety symptoms associated with depression. *Depression and Anxiety*, *18*(2), 53–61.

Einarson, A., Smart, K., Vial, T., Diav-Citrin, O., Yates, L., Stephens S., et al. (2012). Rates of major malformations in infants following exposure to duloxetine during pregnancy: A preliminary report. *The Journal of Clinical Psychiatry*, *73*(11), 1471.

Englisch, S., Fritzinger, M., & Zink, M. (2008). Urinary retention during combined treatment of postpsychotic depression with duloxetine and olanzapine. *Clinical Neuropharmacology*, *31*(5), 307–309.

Englisch, S., Knopf, U., Scharnholz, B., Kuwilsky, A., Deuschle, M., & Zink, M. (2009). Duloxetine for major depressive episodes in the course of psychotic disorders: An observational clinical trial. *Journal of Psychopharmacology*, *23*(8), 875–882.

Eyal, R., & Yaeger, D. (2008). Poor neonatal adaptation after in utero exposure to duloxetine. *The American Journal of Psychiatry*, *165* (5), 651.

Fava, M., Mallinckrodt, C. H., Detke, M. J., Watkin, J. G., & Wohlreich, M. M. (2004). The effect of duloxetine on painful physical symptoms in depressed patients: Do improvements in these symptoms result in higher remission rates? *The Journal of Clinical Psychiatry*, *65*(4), 521–530.

Fawcett, J., & Kravitz, H. M. (1983). Anxiety syndromes and their relationship to depressive illness. *The Journal of Clinical Psychiatry*, *44* (8 Pt. 2), 8–11.

Feighner, J. P. (1999). Mechanism of action of antidepressant medications. *The Journal of Clinical Psychiatry*, *60*(Suppl. 4), 4–11.

Freeman, M. P., Hirschberg, A. M., Wang, B., Petrillo, L. F., Connors, S., Regan, S., et al. (2013). Duloxetine for major depressive disorder and daytime and nighttime hot flashes associated with the menopausal transition. *Maturitas*, *75*(2), 170–174.

Gaynor, P. J., Gopal, M., Zheng, W., Martinez, J. M., Robinson, M. J., & Marangell, L. B. (2011). A randomized placebo-controlled trial of duloxetine in patients with major depressive disorder and associated painful physical symptoms. *Current Medical Research and Opinion*, *27*(10), 1849–1858.

Girardi, P., Pompili, M., Innamorati, M., Mancini, M., Serafini, G., Mazzarini, L., et al. (2009). Duloxetine in acute major depression: Review of comparisons to placebo and standard antidepressants using dissimilar methods. *Human Psychopharmacology*, *24*(3), 177–190.

Guerdjikova, A. I., McElroy, S. L., Winstanley, E. L., Nelson, E. B., Mori, N., McCoy, J., et al. (2012). Duloxetine in the treatment of binge eating disorder with depressive disorders: A placebo-controlled trial. *The International Journal of Eating Disorders*, *45*(2), 281–289.

Gupta, S., Nihalani, N., & Masand, P. (2007). Duloxetine: Review of its pharmacology, and therapeutic use in depression and other psychiatric disorders. *Annals of Clinical Psychiatry*, *19*(2), 125–132.

Halbreich, U., Borenstein, J., Pearlstein, T., & Kahn, L. S. (2003). The prevalence, impairment, impact, and burden of premenstrual dysphoric disorder (PMS/PMDD). *Psychoneuroendocrinology*, *28* (Suppl.3), 1–23.

Hanretta, A. T., & Malek-Ahmadi, P. (2006). Combined use of ECT with duloxetine and olanzapine: A case report. *The Journal of ECT*, *22*(2), 139–141.

Hoog, S. L., Cheng, Y., Elpers, J., & Dowsett, S. A. (2013). Duloxetine and pregnancy outcomes: Safety surveillance findings. *International Journal of Medical Sciences*, *10*(4), 413–419.

Hsu, C. C., Chang, C. W., Peng, C. H., & Liang, C. S. (2014). Rapid management of trigeminal neuralgia and comorbid major depressive disorder with duloxetine. *The Annals of Pharmacotherapy*, *48*(8), 1090–1092.

Hu, B., Doods, H., Treede, R. D., & Ceci, A. (2009). Depression-like behaviour in rats with mononeuropathy is reduced by the CB2-selective agonist GW405833. *Pain*, *143*(3), 206–212.

Hu, B., Doods, H., Treede, R. D., & Ceci, A. (2016). Duloxetine and 8-OH-DPAT, but not fluoxetine, reduce depression-like behaviour in an animal model of chronic neuropathic pain. *Neuroscience Letters*, *619*, 162–167.

Hudson, J. I., & Pope, H. G., Jr. (1996). The relationship between fibromyalgia and major depressive disorder. *Rheumatic Diseases Clinics of North America*, *22*(2), 285–303.

Joffe, H., Hall, J., Soares, C., Hennen, J., Reilly, C. J., Carlson, K., et al. (2002). Vasomotor symptoms are associated with depression in perimenopausal women seeking primary care. *Menopause*, *9*(6), 392–398.

Joffe, H., Soares, C. N., Petrillo, L. F., Viguera, A. C., Somley, B. L., Koch, J. K., et al. (2007). Treatment of depression and menopause-related symptoms with the serotonin-norepinephrine reuptake inhibitor duloxetine. *The Journal of Clinical Psychiatry*, *68*(6), 943–950.

Kachko, L., Ben Ami, S., Liberman, A., Birk, E., & Kronenberg, S. (2011). Duloxetine contributing to a successful multimodal treatment program for peripheral femoral neuropathy and comorbid "reactive depression" in an adolescent. *Pain Research & Management*, *16*(6), 457–459.

Karp, J. F., Whyte, E. M., Lenze, E. J., Dew, M. A., Begley, A., Miller, M. D., et al. (2008). Rescue pharmacotherapy with duloxetine for selective serotonin reuptake inhibitor nonresponders in late-life depression: Outcome and tolerability. *The Journal of Clinical Psychiatry*, *69*(3) 457–463.

Kihara, T., & Ikeda, M. (1995). Effects of duloxetine, a new serotonin and norepinephrine uptake inhibitor, on extracellular monoamine levels in rat frontal cortex. *The Journal of Pharmacology and Experimental Therapeutics*, *272*(1), 177–183.

Kirwin, J. L., & Gören, J. L. (2005). Duloxetine: A dual serotonin-norepinephrine reuptake inhibitor for treatment of major depressive disorder. *Pharmacotherapy*, *25*(3), 396–410.

Kjaersgaard, M. I., Parner, E. T., Vestergaard, M., Sørensen, M. J., Olsen, J., Christensen, J., et al. (2013). Prenatal antidepressant exposure and risk of spontaneous abortion—A population-based study. *PLoS One*, *8* (8), e72095.

Knadler, M. P., Lobo, E., Chappell, J., & Bergstrom, R. (2011). Duloxetine: Clinical pharmacokinetics and drug interactions. *Clinical Pharmacokinetics*, *50*(5), 281–294.

Lantz, R. J., Gillespie, T. A., Rash, T. J., Kuo, F., Skinner, M., Kuan, H. Y., et al. (2003). Metabolism, excretion, and pharmacokinetics of duloxetine in healthy human subjects. *Drug Metabolism and Disposition*, *31* (9), 1142–1150.

Leombruni, P., Miniotti, M., Colonna, F., Sica, C., Castelli, L., Bruzzone, M., et al. (2015). A randomised controlled trial comparing duloxetine and acetyl L-carnitine in fibromyalgic patients: Preliminary data. *Clinical and Experimental Rheumatology*, *33*(1 Suppl. 88), S82–S85.

Lewis-Fernandez, R., Lam, P., Lucak, S., Galfalvy, H., Jackson, E., Fried, J., et al. (2016). An open-label pilot study of duloxetine in patients with irritable bowel syndrome and comorbid major depressive disorder. *Journal of Clinical Psychopharmacology*, *36*(6), 710–715.

Li, K., Zhou, T., Liao, L., Yang, Z., Wong, C., Henn, F., et al. (2013). βCaMKII in lateral habenula mediates core symptoms of depression. *Science*, *341*(6149), 1016–1020.

Lobo, E. D., Loghin, C., Knadler, M. P., Quinlan, T., Zhang, L., Chappell, J., et al. (2008). Pharmacokinetics of duloxetine in breast milk and plasma of healthy postpartum women. *Clinical Pharmacokinetics*, *47*(2), 103–109.

Lobo, E. D., Quinlan, T., O'Brien, L., Knadler, M. P., & Heathman, M. (2009). Population pharmacokinetics of orally administered duloxetine in patients: Implications for dosing recommendation. *Clinical Pharmacokinetics*, *48*(3), 189–197.

Lobo, E. D., Quinlan, T., & Prakash, A. (2014). Pharmacokinetics of orally administered duloxetine in children and adolescents with major depressive disorder. *Clinical Pharmacokinetics*, *53*(8), 731–740.

Mancini, M., Perna, G., Rossi, A., & Petralia, A. (2010). Use of duloxetine in patients with an anxiety disorder, or with comorbid anxiety and major depressive disorder: A review of the literature. *Expert Opinion on Pharmacotherapy*, *11*(7), 1167–1181.

Mazza, M., Harnic, D., Catalano, V., Janiri, L., & Bria, P. (2008). Duloxetine for premenstrual dysphoric disorder: A pilot study. *Expert Opinion on Pharmacotherapy*, *9*(4), 517–521.

McIntyre, R. S., & Lee, Y. (2016). Cognition in major depressive disorder: A "systemically important functional index" (SIFI). *Current Opinion in Psychiatry*, *29*(1), 48–55.

Meighen, K. G. (2007). Duloxetine treatment of pediatric chronic pain and co-morbid major depressive disorder. *Journal of Child and Adolescent Psychopharmacology*, *17*(1), 121–127.

Miller, C. W., & Gallagher, K. E. (2016). Self-limited kleptomania symptoms as a side effect of duloxetine. *Case Reports in Psychiatry*, *2016*, 5472492.

Monteleone, F., Caputo, M., Tecce, M. F., & Capasso, A. (2011). Duloxetinein the treatment ofdepression: An overview. *Central Nervous System Agents in Medicinal Chemistry*, *11*(3), 174–183.

Mowla, A., Dastgheib, S. A., & Razeghian Jahromi, L. (2015). Comparing the effects of sertraline with duloxetine for depression severity and symptoms: A double-blind, randomized controlled trial. *Clinical Drug Investigation*, *36*(7), 539–543.

Pangallo, B. A., Zhang, Q., Desaiah, D., Perahia, D. G., Detke, M. J., & Kennedy, S. H. (2010). Long-term safety of duloxetine during open-label compassionate use treatment of patients who completed previous duloxetine clinical trials. *Current Medical Research and Opinion*, *26* (11), 2643–2651.

Papakostas, G. I. (2015). Antidepressants and their effect on cognition in major depressive disorder. *The Journal of Clinical Psychiatry*, *76*(8), e1046.

Papakostas, G. I., Worthington, J. J., 3rd, Iosifescu, D. V., Kinrys, G., Burns, A. M., Fisher, L. B., et al. (2006). The combination of duloxetine and bupropion for treatment-resistant major depressive disorder. *Depression and Anxiety*, *23*(3), 178–181.

Paulzen, M., Gründer, G., Veselinovic, T., Wolf, B., Hiemke, C., & Lammertz, S. E. (2016). Duloxetine enters the brain—But why is it not found in the cerebrospinal fluid. *Journal of Affective Disorders*, *189*, 159–163.

Peritogiannis, V., Antoniou, K., Mouka, V., Mavreas, V., & Hyphantis, T. N. (2009). Duloxetine-induced hypomania: Case report and brief review of the literature on SNRIs-induced mood switching. *Journal of Psychopharmacology*, *23*(5), 592–596.

Phillips, A. L., Burr, R. L., & Dunner, D. L. (2018). rTMS effects in patients with co-morbid somatic pain and depressive mood disorders. *Journal of Affective Disorders*, *241*, 411–416.

Ramos, M. G., Hara, C., & Rocha, F. L. (2009). Duloxetine treatment for women with premenstrual dysphoric disorder: A single-blind trial. *The International Journal of Neuropsychopharmacology*, *12*(8), 1081–1088.

Rapaport, M. H., Gharabawi, G. M., Canuso, C. M., Mahmoud, R. A., Keller, M. B., Bossie, C. A., et al. (2006). Effects of risperidone augmentation in patients with treatment-resistant depression: Results of open-label treatment followed by double-blind continuation. *Neuropsychopharmacology*, *31*(11), 2505–2513.

Raskin, J., Wiltse, C. G., Siegal, A., Sheikh, J., Xu, J., Dinkel, J. J., et al. (2007). Efficacy of duloxetine on cognition, depression, and pain in elderly patients with major depressive disorder: An 8-week, double-blind, placebo-controlled trial. *The American Journal of Psychiatry*, *164*(6), 900–909.

Rong, P., Liu, J., Wang, L., Liu, R., Fang, J., Zhao, J., et al. (2016). Effect of transcutaneous auricular vagus nerve stimulation on major depressive disorder: A nonrandomized controlled pilot study. *Journal of Affective Disorders*, *195*, 172–179.

Rosso, G., Rigardetto, S., Bogetto, F., & Maina, G. (2012). A randomized, single-blind, comparison of duloxetine with bupropion in the treatment of SSRI-resistant major depression. *Journal of Affective Disorders*, *136*(1–2), 172–176.

Rottach, K. G., Schaner, B. M., Kirch, M. H., Zivotofsky, A. Z., Teufel, L. M., Gallwitz, T., et al. (2008). Restless legs syndrome as side effect of second generation antidepressants. *Journal of Psychiatric Research*, *43*(1), 70–75.

Rovera, C., Mauri, M. C., Bertin, E., Di Pace, C., Paletta, S., Reggiori, A., et al. (2016). Duloxetine in elderly major depression disorder: Effectiveness and drug plasma level evaluation. *Human Psychopharmacology*, *31*(5), 349–355.

Sabljić, V., Rakun, R. Ružić, K., & Grahovac, T. (2011). Duloxetine-related panic attacks. *Psychiatria Danubina*, *23*(1), 114–116.

Sachs, H. C., & Committee on Drugs. (2013). The transfer of drugs and therapeutics into human breast milk: An update on selected topics. *Pediatrics*, *132*(3), e796–e809.

Safer, D. L., & Arnow, K. D. (2012). Suprathreshold duloxetine for treatment-resistant depression, anorexia nervosa binge-purging type, and obsessive-compulsive disorder: A case report. *Innovations in Clinical Neuroscience*, *9*(3), 13–16.

Seguí, J., López-Muñoz, F., Alamo, C., Camarasa, X., García-García, P., & Pardo, A. (2010). Effects of adjunctive reboxetine in patients with duloxetine-resistant depression: A 12-week prospective study. *Journal of Psychopharmacology*, *24*(8), 1201–1207.

Serafini, G., Pompili, M., Del Casale, A., Mancini, M., Innamorati, M., Lester, D., et al. (2010). Duloxetine versus venlafaxine in the treatment of unipolar and bipolar depression. *Clinica Terapeutica*, *161*(4), 321–327.

Sharma, A., Goldberg, M. J., & Cerimele, B. J. (2000). Pharmacokinetics and safety of duloxetine, a dual-serotonin and norepinephrine reuptake inhibitor. *Journal of Clinical Pharmacology*, *40*(2), 161–167.

Shelton, R. C., Prakash, A., Mallinckrodt, C. H., Wohlreich, M. M., Raskin, J., Robinson, M. J., et al. (2007). Patterns of depressive symptom response in duloxetine-treated outpatients with mild, moderate or more severe depression. *International Journal of Clinical Practice*, *61*(8), 1337–1348.

Sindhu, P., Kumar, S., Iqbal, B., Ali, J., & Baboota, S. (2018). Duloxetine loaded-microemulsion system to improve behavioral activities by upregulating serotonin and norepinephrine in brain for the treatment of depression. *Journal of Psychiatric Research*, *99*, 83–95.

Stein, M. B., Kirk, P., Prabhu, V., Grott, M., & Terepa, M. (1995). Mixed anxiety-depression in a primary-care clinic. *Journal of Affective Disorders*, *34*(2), 79–84.

Takano, A., Suzuki, K., Kosaka, J., Ota, M., Nozaki, S., Ikoma, Y., et al. (2006). A dose-finding study of duloxetine based on serotonin transporter occupancy. *Psychopharmacology*, *185*(3), 395–399.

Viktrup, L., Perahia, D. G., & Tylee, A. (2004). Duloxetine treatment of stress urinary incontinence in women does not induce mania or hypomania. *The Primary Care Companion to the Journal of Clinical Psychiatry*, *6*(6), 239–243.

Volpe, F. M. (2008). An 8-week, open-label trial of duloxetine for comorbid major depressive disorder and chronic headache. *The Journal of Clinical Psychiatry*, *69*(9), 1449–1454.

Wagner, G., Schultes, M. T., Titscher, V., Teufer, B., Klerings, I. & Gartlehner, G. (2018). Efficacy and safety of levomilnacipran, vilazodone and vortioxetine compared with other second-generation antidepressants for major depressive disorder in adults: A systematic review and network meta-analysis. *Journal of Affective Disorders*, *228*, 1–12.

Werneke, U., Northey, S., & Bhugra, D. (2006). Antidepressants and sexual dysfunction. *Acta Psychiatrica Scandinavica*, *114*(6), 384–397.

Wilke, W. S., Gota, C. E., & Muzina, D. J. (2010). Fibromyalgia and bipolar disorder: A potential problem? *Bipolar Disorders*, *12*(5), 514–520.

Wise, T. N., Perahia, D. G., Pangallo, B. A., Losin, W. G., & Wiltse, C. G. (2006). Effects of the antidepressant duloxetine on body weight: Analyses of 10 clinical studies. *The Primary Care Companion to the Journal of Clinical Psychiatry*, *8*(5), 269–278.

Wittchen, H. U., Becker, E., Lieb, R., & Krause, P. (2002). Prevalence, incidence and stability of premenstrual dysphoric disorder in the community. *Psychological Medicine*, *32*(1), 119–132.

Wohlreich, M. M., Sullivan, M. D., Mallinckrodt, C. H., Chappell, A. S., Oakes, T. M., Watkin, J. G., et al. (2009). Duloxetine for the treatment of recurrent major depressive disorder in elderly patients: Treatment outcomes in patients with comorbid arthritis. *Psychosomatics*, *50*(4), 402–412.

Wolfe, F., Smythe, H. A., Yunus, M. B., Bennett, R. M., Bombardier, C., Goldenberg, D. L., et al. (1990). The American College of Rheumatology 1990 Criteria for the Classification of Fibromyalgia. Report of the Multicenter Criteria Committee. *Arthritis and Rheumatism*, *33*(2), 160–172.

Xu, Y., Bai, S. J., Lan, X. H., Qin, B., Huang, T., & Xie, P. (2016). Randomized controlled trials of serotonin-norepinephrine reuptake inhibitor in treating major depressive disorder in children and adolescents: A meta-analysis of efficacy and acceptability. *Brazilian Journal of Medical and Biological Research*, *49*(6), 1–8.

Yoshida, K., Fukuchi, T., & Sugawara, H. (2017). Dysosmia and dysgeusia associated with duloxetine. *BMJ Case Reports*, *2017*, 1–3.

Chapter 36

# Escitalopram and blonanserin as antidepressant agents linking in neurotrophic mechanisms

**Wataru Ukai, Kenta Deriha, Eri Hashimoto, and Chiaki Kawanishi**
*Department of Neuropsychiatry, Graduate School of Medicine, Sapporo Medical University, Sapporo, Japan*

## List of abbreviations

| | |
|---|---|
| **ACC** | anterior cingulate cortex |
| **BDNF** | brain-derived neurotrophic factor |
| **CBT** | cognitive behavioral therapy |
| **CORT** | corticosterone |
| **CREB** | cyclic AMP-responsive element-binding protein |
| **EPS** | extrapyramidal side effects |
| **FST** | forced swim test |
| **GABA** | γ-aminobutyric acid |
| **GR** | glucocorticoid receptor |
| **HPA** | hypothalamic–pituitary–adrenal axis |
| **HSP** | heat-shock protein |
| **MB** | mammillary body primordia |
| **MDD** | major depressive disorder |
| **NAc** | nucleus accumbens |
| **NMDA** | *N*-methyl-D-aspartate |
| **PANSS** | positive and negative syndrome scale |
| **PFC** | prefrontal cortex |
| **PSD95** | postsynaptic density protein 95 |
| **PTSD** | posttraumatic stress disorder |
| **PV** | parvalbumin |
| **QOL** | quality of daily living |
| **SVZ** | subventricular zone |
| **TLR** | toll-like receptor |
| **TrkB** | tropomyosin-related kinase B |
| **TST** | tail suspension test |
| **VTA** | ventral tegmental area |

## Introduction: Modern molecular theory of depression

Depression is the disease of the modern health problem of serious concern. The lack of response to the clinically available antidepressants, developed on the basis of the monoamine hypothesis of depression, has led scientists to think about the mechanisms explaining the new pathogenesis of depression. In this context, the neurotrophic mechanism and the inflammatory theory have emerged to clarify many aspects of depression that the previous theories have failed to explain. The *N*-methyl-D-aspartate (NMDA) receptor, and γ-aminobutyric acid type A ($GABA_A$) receptor activation, and furthermore, dopamine $D_3$ receptor and Toll-like receptor (TLR) modulation have opened up new prospects for understanding pathophysiology of refractory, atypical depression. Treatment of these agents is suggested to ameliorate depressive symptoms and hypothalamic–pituitary–adrenal axis (HPA) hyperactivity, alongside reducing levels of hippocampal, amygdala, and prefrontal cortex (PFC) inflammatory cytokines, restoring glutamate/GABA balance, and enhancing hippocampal, amygdal, and PFC neurogenesis, by increasing brain-derived neurotrophic factor (BDNF) gene and protein expression in the animal models and patients of depression.

## BDNF as a leading player in depression

Brain-derived neurotrophic factor (BDNF), a member of the neurotrophin family, is a basic homodimeric protein that promotes neuronal survival and regulates the proliferation and differentiation of nerve cells in the peripheral and nervous systems (Lindsay, Wiegand, Altar, & DiStefano, 1994). BDNF has attracted attention among candidate downstream effectors involved in antidepressant effects. Multiple neurobiological mechanisms appear to be involved in mediating the therapeutic effects of antidepressant therapy. Some of these mechanisms may play a role in neuroprotection and neuro/synaptogenesis, neuroinflammation, and glutamate/GABA balance modulation.

Chronic treatments with different classes of antidepressants (e.g., SSRIs, SNRIs, and NassA) have been associated with increases in BDNF expression and neurogenesis in the hippocampus (Duman & Monteggia, 2006) and subventricular zone (SVZ) (Hitoshi et al., 2007). Circulating BDNF is found in human serum and plasma and a large amount is

The Neuroscience of Depression. https://doi.org/10.1016/B978-0-12-817933-8.00035-9

stored in platelets (Fujimura et al., 2002). Since the BDNF protein is reported to cross the blood–brain barrier in both directions (Pan, Banks, Fasold, Bluth, & Kastin, 1998) and positive correlation between blood BDNF levels and brain-tissue BDNF levels is demonstrated (Klein et al., 2011), peripheral BDNF is a promising candidate biomarker associated with depression and antidepressant responses.

## Role of BDNF in corticosterone hormone stress-induced depression model

Corticosteroids are thought to be the primary humoral mediators of stress. Under stressed conditions, excessive corticosteroids can trigger negative feedback regulation in different brain regions, and the high levels of circulating corticosteroids are reduced via the inhibition of HPA activity. Corticosterone (CORT) hormone stress can activate the HPA axis, and repeat HPA axis activation promotes physiological damage and impairs normal brain function (Kalynchuk, Gregus, Boudreau, & Perrot-Sinal, 2004), thus leading to serious changes in affective behaviors that are indicative of depression-like symptoms in rodents (Lee, Shim, Lee, Yang, & Hahm, 2009).

CORT is also known to decrease hippocampal neurogenic capacity (Brummelte & Galea, 2010), a deficit that may be associated with behavioral alterations, including depressive-like behaviors (Hanson, Owens, & Nemeroff, 2011). Chronic CORT exposure suggested to reduce sucrose consumption and the frequency of grooming in the splash test (Freitas et al., 2016), while increase latency to feed in the novelty suppressed feeding test (Gourley, Kiraly, Howell, Olausson, & Taylor, 2008), immobility times in the forced swim test (FST) (Zhao et al., 2008) and tail suspension test (TST) (Pazini et al., 2016). On the other hand, transgenic mice with enhanced adult neurogenesis were resistant to the depressive-like behavior elicited by the chronic administration of CORT (Hill, Sahay, & Hen, 2015), and a chronic treatment with an antidepressant abolished the behavioral abnormality (Conrad et al., 2007). Similarly, depressed patients with elevated cortisol levels reported to display anhedonia, cognitive dysfunction, and personal neglect (Morgan, Mitchell, & Jablensky, 2005) in human studies.

Although the precise role of BDNF in the CORT stress hormone-induced physiological brain damage mechanism remains unclear, the recent evidence suggests that glucocorticoid receptor (GR), one of the main receptors of CORT, downregulates BDNF gene expression through direct binding to BDNF regulatory sequences in mouse neuronal primary cultures and in newly characterized mouse hippocampal BZ cell line (Chen, Lombès, & Le Menuet, 2017). In this report, a short region within exon IV promoter responsible for GR-mediated BDNF repression was defined in the transient transfection assays, and GR recruitment onto this fragment, through unidentified transcription factor tethering demonstrated in the chromatin immunoprecipitation. Consistent with these findings, it was indicated that the long-term treatment of escitalopram could ameliorate depressive-like behaviors in the FST, and reversed BDNF levels in serum and the hippocampus, which were decreased by chronic CORT exposures (Furuse et al., 2019).

## BDNF signaling activation and underlying mechanisms

Many antidepressants upregulate the expression of BDNF via cyclic AMP-responsive element-binding protein (CREB) pathway (Nibuya, Nestler, & Duman, 1996). Decreased BDNF signaling may be associated with the pathophysiology of depression, and increased BDNF signaling is associated with the mechanisms of actions of antidepressant drugs (Castrén, 2014). Moreover, it is indicated that reduced serum BDNF levels may be attributed to decreased BDNF release from platelets (Watanabe et al., 2010). Deceased serum BDNF levels in patients with major depressive disorder (MDD) are corrected in those that respond to antidepressant treatments (Bocchio-Chiavetto et al., 2010; Kurita, Nishino, Kato, Numata, & Sato, 2012). The regional difference of BDNF level change has been observed in the brain of various depression models.

In case using chronic CORT administration-induced depression model, long-term escitalopram treatment demonstrated to increase BDNF in the brain, and the effect of escitalopram was specific to the hippocampus and not the anterior cingulate cortex (ACC) or the nucleus accumbens (NAc) (Furuse et al., 2019). Since BDNF is mainly stored in glutamatergic neurons (Lessmann & Brigadski, 2009) and glutamate is one of the main neurotransmitters inducing BDNF expression (Carvalho, Caldeira, Santos, & Duarte, 2008), different serotonergic innervations and glutamatergic activation (Taylor et al., 2010) may be responsible for the specific responses of the frontal cortex, including the hippocampus and NAc, to antidepressants with respect to other brain areas.

Studies are currently being conducted to clarify the structural as well as behavioral actions of antidepressants via increased BDNF signaling. It has been demonstrated that structural synaptic plasticity is mediated by activity-dependent release of BDNF and subsequent autocrine stimulation of tropomyosin-related kinase B (TrkB) signaling. Multiple downstream pathways of BDNF–TrkB signaling are suggested for antidepressant behavioral actions including mTORC1, Rho GTPases (Rac 1, RohA) signaling, they are critical regulators of the actin cytoskeleton and neuronal morphogenesis, enhancing BDNF-dependent spine-synapse plasticity and actin polymerization (Hedrick et al., 2016).

## Trophic mechanism in chronic, recurrent depression—Early life adversity

In regard to the trophic mechanisms of depression, several clinical meta-analyses have found that hippocampal and plasma/serum BDNF levels are low in unmedicated depressive patients and those were increased after antidepressant treatment (Kishi, Yoshimura, Ikuta, & Iwata, 2018). However, each study involves inconsistent result, not all antidepressant treatments exert increasing effect of serum BDNF levels (Molendijk et al., 2011), and increases in BDNF levels are indicated more prominent in responders than in nonresponders (Kurita et al., 2012; Lee & Kim, 2008). Moreover, since mature-BDNF levels and the mature-BDNF/pro-BDNF ratio may be significantly higher in bipolar disorder patients than in controls (Södersten et al., 2014), the predictive and pathological roles of BDNF in various types of depression remain elusive.

Repeated exposure to environmental adversities during development may be a risk factor for more frequent episodes, greater comorbidity, suicidality, and poorer treatment outcomes in depression (Spiwak et al., 2011). Therefore, the study to examine the questions about depression and antidepressant effects on serum BDNF levels for predicting drug response is highly demanded, especially in those refractory, atypical depression. In one study, examining for this point of view, revealed a severe behavioral abnormalities accompanied with synaptic plasticity molecule and neural circuit modifications in serum and the NAc at the level of BDNF protein expression changes using combined stress model of fetal ethanol and adolescent CORT treatment (Furuse et al., 2019).

In that experiment, longer immobility time was significantly attenuated by escitalopram, but not by another SSRI, sertraline, or newly developed atypical antipsychotic, blonanserin; these were administered during adolescent stress exposure. Interestingly, the serum BDNF levels, which were decreased in the CORT treatment-induced depression model, remained the same in combined stress-induced refractory depression model, and furthermore, they were decreased by the treatment of escitalopram or blonanserin (Fig. 1).

While, in the brain, sertraline and escitalopram, but not blonanserin indicated reverse effect on slightly decreased BDNF levels in the hippocampus. Neither sertraline, escitalopram, nor blonanserin recovered decreased BDNF levels in the ACC. In addition, it is worthy of notice that escitalopram indicated reduction of BDNF levels in the NAc, those were not affected by neither sertraline nor blonanserin treatment (Fig. 2). Although studies suggesting parallel findings between recovery effects in depressive-like behavior and BDNF level changes in serum or brain regions are limited, the current evidence may have profound implications for treatment against treatment-resistant depression (Samuels et al., 2011). Together, it must be possible that not only increase in hippocampus but also decrease in NAc BDNF levels is necessity for antidepressant medication responsible for refractory type of depression.

The evidence showing decrease in blood BDNF levels with effective antidepressant effect is controversial, and the mechanisms for the reductions observed in this neurotrophic factor remain unclear. However, as mentioned earlier, the recent studies demonstrating importance of glutamate/GABA balance modulation of antidepressant effect in the brain area responsible for processing emotions. Since the embryonic GABAergic system is vulnerable to various stressors (Cuzon, Yeh, Yanagawa, Obata, & Yeh, 2008), prenatal ethanol exposure may cause gross morphological changes in GABAergic cells during development. A report

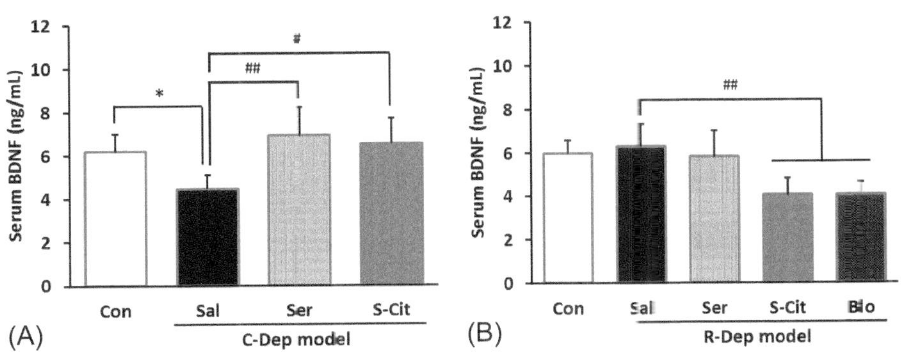


FIG. 1 Effects of antidepressants and antipsychotic on serum BDNF level change in two types of depression model rats. (A) Serum BDNF levels were significantly lower in corticosterone-induced depression (C-Dep) model rats than in control rats, and treatment with antidepressant sertraline (Ser) or escitalopram (S-Cit) significantly reversed the decrease to control levels. (B) Treatment with S-Cit or antipsychotic blonanserin (Blo) significantly reduced the serum BDNF levels than those in combined stress-induced refractory depression (R-Dep) model rats. *Data are from Furuse, K., Ukai, W., Hashimoto, E., Hashiguchi, H., Kigawa, Y., Ishii, T., et al. (2019). Antidepressant activities of escitalopram and blonanserin on prenatal and adolescent combined stress-induced depression model: Possible role of neurotrophic mechanism change in serum and nucleus accumbens.* Journal of Affective Disorders, 247, *97–104. https://doi.org/10.1016/j.jad.2019.01.007, with permission from the Publishers.*

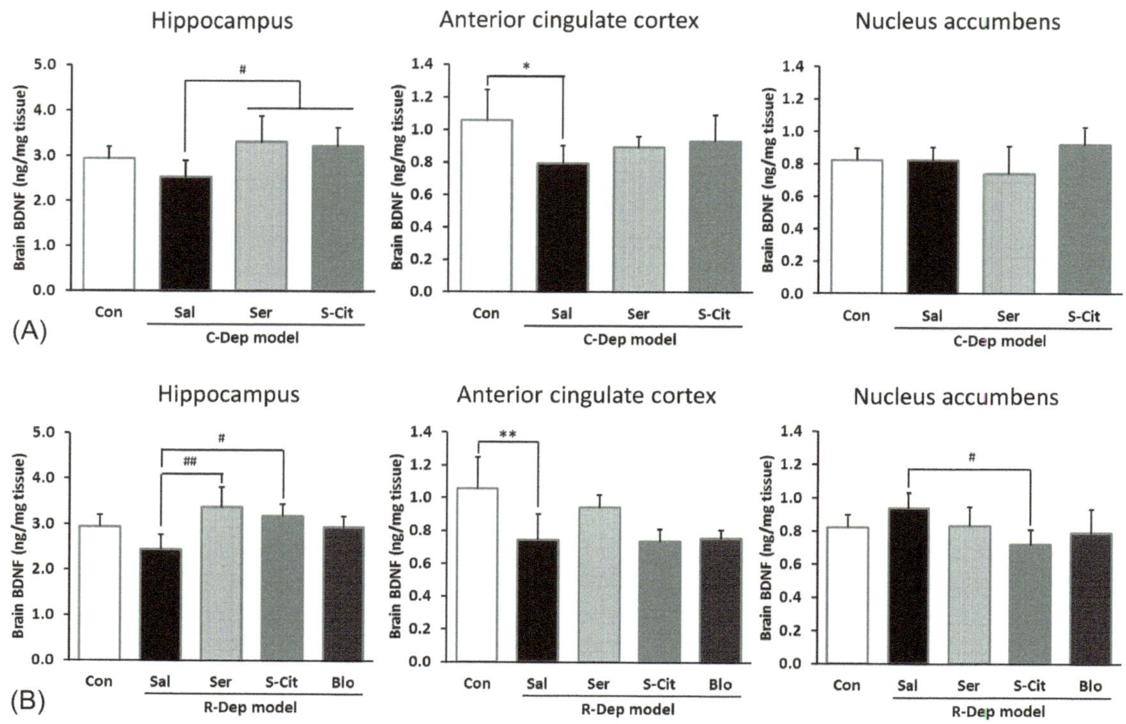


FIG. 2 Effects of antidepressants and antipsychotic on brain regional BDNF level change in two types of depression model rats. (A) In corticosterone-induced depression (C-Dep) model rats, antidepressant Ser or S-Cit treatment significantly increased BDNF levels in hippocampus. In anterior cingulate cortex (ACC), BDNF levels were significantly lower than those in control rats. (B) In combined stress-induced refractory depression (R-Dep) model rats, Ser or S-Cit significantly increased the BDNF levels in hippocampus. In AAC, BDNF levels were significantly lower than those in control rats. In nucleus accumbens (NAc), BDNF levels were significantly decreased by the treatment with S-Cit. *Data are from Furuse, K., Ukai, W., Hashimoto, E., Hashiguchi, H., Kigawa, Y., Ishii, T., et al. (2019). Antidepressant activities of escitalopram and blonanserin on prenatal and adolescent combined stress-induced depression model: Possible role of neurotrophic mechanism change in serum and nucleus accumbens.* Journal of Affective Disorders, *247, 97–104. https://doi.org/10.1016/j.jad.2019.01.007, with permission from the Publishers.*

indicated that the number of parvalbumin (PV)-positive GABAergic interneurons was reduced in the anterior cingulate cortex, amygdala, and hippocampus of prenatal ethanol and adolescent CORT exposure-induced refractory depression model rats (Kigawa et al., 2014).

In that study, it was indicated that a combined treatment with sertraline and intravenous neural stem cell administration restored reductions in GABAergic interneurons in these areas, and this was associated with the recovery of reductions in postsynaptic density protein 95 (PSD95) levels in refractory depression model rats and potential therapeutic efficacy against depressive-like behavior. Although the role of GABAergic interneurons in the pathophysiology of depression and actions of antidepressants remain unclear, it is plausible that GABAergic activation downregulates the glutamatergic functions contributing to the storage, production, and release of BDNF (Carvalho et al., 2008). This, in turn, may be followed by enhanced synaptogenesis, which may be mirrored by the neuroplasticity marker BDNF (Duman & Li, 2012).

One of the possible explanations for the lower serum BDNF levels and recovery responses of escitalopram was hinted by the BDNF function change on the mesolimbic dopamine pathway. In a clinical study longitudinally evaluated the relationship between serum BDNF level change and treatment responses of escitalopram in patients with posttraumatic stress disorder (PTSD), it was demonstrated that escitalopram monotherapy strongly ameliorated PTSD and comorbid depressive symptoms over the course, with no significant changes being observed in serum BDNF levels throughout the treatment. Nevertheless, lower mean serum BDNF levels over the course of the trial were a strong predictor of good PTSD responses to escitalopram (Berger et al., 2010) (Fig. 3). In contrast to the role of BDNF in the hippocampus, high BDNF levels in the mesolimbic dopamine pathway may play a crucial role in the development and maintenance of the pathophysiology of PTSD.

The mesolimbic dopamine pathway is composed of dopaminergic neurons in the midbrain ventral tegmental area (VTA) and their projections to the NAc (Berton et al., 2006). Accordingly, the inhibition of BDNF activity in the VTA-NAc pathway has indicated to exert antidepressant-like effects in rodent models and also

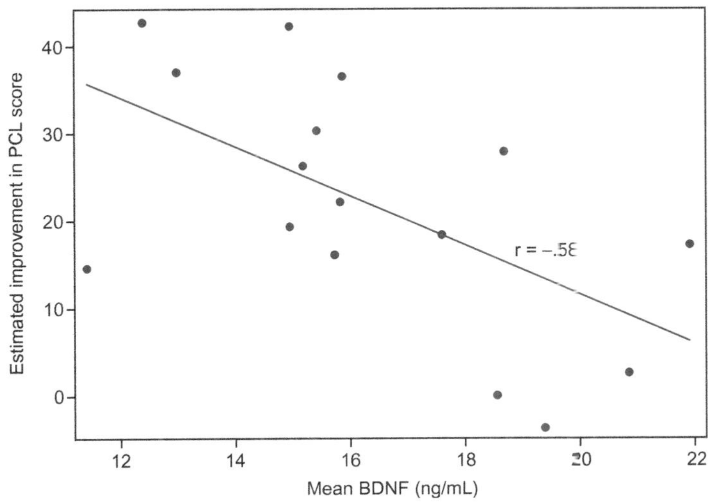


**FIG. 3** Relationship between PTSD symptom response to escitalopram and mean BDNF levels. Lower mean serum BDNF levels were associated with a greater decrease in PTSD symptoms measured by PTSD Checklist (PCL) over the course of 12-week open clinical trial. *Data are from Berger, W., Mehra, A., Lenoci, M., Metzler, T. J., Otte, C., Tarasovsky, G., et al. (2010). Serum brain-derived neurotrophic factor predicts responses to escitalopram in chronic posttraumatic stress disorder.* Progress in Neuro-Psychopharmacology & Biological Psychiatry, 34*(7), 1279–1284. https://doi.org/10.1016/j.pnpbp.2010.07.008, with permission from the Publishers.*

reversed the social avoidance behavior caused by a defeat stress paradigm (Berton et al., 2006; Eisch et al., 2003). Therefore, strong BDNF activity in the VTA-NAc pathway may be a source of treatment resistance. Thus, if serum BDNF is thought to reflect the activity in the VTA-NAc, their higher levels may be a predictor of poor responses to antidepressant treatment. The evidence of escitalopram-induced reduction of BDNF in NAc corresponded well to this hypothesis.

## Trophic mechanism in chronic, recurrent depression—Adolescent

Adolescent depression has been suggested to indicate a high risk for suicidal behaviors and impairments in social and educational functioning (Bridge, Tina, & David, 2006; Fergusson & Woodward, 2002), and frequently becoming a gateway to a chronic or recurrent depressive disorder and heightened risk for suicide and suicidal behavior that persists into adulthood (Weissman et al., 1999). Although when various antidepressant treatments were put to the test, even with the combination of cognitive behavioral therapy (CBT), their remission rates of adolescents with depression reported less than 60% (Kennard et al., 2009). In addition, the risk of suicidal events has been a common concern in those who do not respond to antidepressant treatment (Vitiello et al., 2009).

In a clinical study, examined the relationship between antidepressant effect and serum BDNF level change in depressed adolescents, it was indicated that there were no significant changes of BDNF levels in patients with 8-weeks escitalopram treatment. However, in more precise analysis has found that, the patients who responded to escitalopram show greater reduction of BDNF levels at first 2 weeks as compared with patients who did not respond to escitalopram (Lee et al., 2020) (Fig. 4).

This evidence seems at first sight quite opposite to the neurotrophic hypothesis. However, the recent studies have revealed an importance of valance mechanism of two different BDNF signalings of BDNF–TrkB and BDNF-p75 as a critical mediator for maintaining activity-dependent neuronal plasticity, both in the developing and adult central nervous system (Deppmann et al., 2008). In regard to the affective brain circuit, adolescence is corresponding to a critical period in development underlying hormonal and neural correlates including the HPA axis, amygdala, and PFC (Tottenham & Galván, 2016).

Therefore, the effect of antidepressant could be reflected in BDNF level change with therapeutic response in depressed adolescents, in different style from adults, who might be experiencing the critical period with dynamic changes in the affective brain circuit, and possessing different features from those in depressed adults, whose critical periods have already terminated to a completed network.

## Possible antidepressant activity as adjunctive agent

Antidepressant treatment has been shown to be effective for only one-third of patients with depression, partially useful for another third, and useless for the remaining third. If their outcomes are not optimal, patients often drop from treatment and come across detrimental risks such as

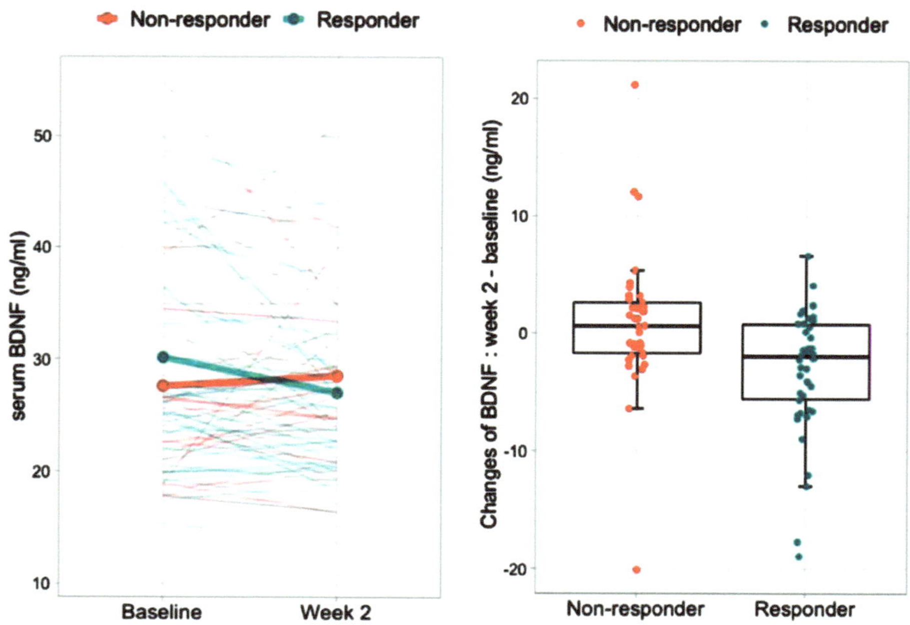


FIG. 4 Early decrease of serum BDNF levels in escitalopram responders in adolescent patients with depression. (A) Serum BDNF levels significantly decreased in responders but not in nonresponders to escitalopram treatment (8 weeks totally) from baseline to week 2. (B) The values of serum BDNF level change from baseline to week 2 represented in boxplot diagram. *Data are from Lee, J., Lee, K. H., Kim, S. H., Han, J. Y., Hong, S. B., Cho, S. C., et al. (2020). Early changes of serum BDNF and SSRI response in adolescents with major depressive disorder.* Journal of Affective Disorders, 265, *325–332. https://doi.org/10.1016/j.jad.2020.01.045, with permission from the Publishers.*

disability and suicide. Alternative antidepressants or augmentation therapies are common strategies when the first antidepressant treatment did not work well. For treatment-resistant depression, the main augmentation therapies include combination of two different antidepressants or combining antidepressant with an anticonvulsant agent (Bauer et al., 2007; Kennedy, Lam, Parikh, Patten, & Ravindran, 2009).

Recently, many clinical trials demonstrated that antidepressant augmented with second and third-generation antipsychotics, including brexpiprazole, aripiprazole, olanzapine, quetiapine, and risperidone, resulted in better treatment response or higher rates of remission than antidepressant monotherapy. These clinical studies investigated antipsychotic augmentations focusing on the clinical symptoms of depression, with an aim to improve treatment-resistant depression (Komossa, Depping, Gaudchau, Kissling, & Leucht, 2010). However, few studies have been done comparing the individual efficacy of these antipsychotics using as option for each patient.

Blonanserin is a novel antipsychotic that preferentially interacts with dopamine D2 and serotonin 2A receptors. In a randomized clinical study, it was reported that the significant difference was not observed in positive and negative syndrome scale (PANSS) total score and 5-factor score (negative symptoms, excitation, cognitive disorders, positive symptoms anxiety/depression), between blonanserin and risperidone treatment groups (Li et al., 2015), while the superior improvement of depressive symptoms and quality of daily living (QOL), verbal fluency, executive function, and work skills to risperidone was suggested in another study (Hori et al., 2014).

## Potential mechanism of antidepressant action induced by blonanserin

The evidence about potential antidepressant properties of blonanserin has just began to be found, however, various interesting mechanisms emerged from these work, and it might be worth investigation for contributing to understanding of new insight into the pathological mechanisms of refractory depression. In an animal model study, it was reported that blonanserin could exert an adjuvant anti-anxiety and antidepressant activity at very low doses treatments, evaluated by using elevated plus maze test and forced swim test (Limaye & Patil, 2016). In another study, interesting observations were found in a refractory depression model experiment, using fetal and adolescent period combined stress exposures, antidepressant escitalopram but not another antidepressant sertraline showed recovery effect of depression-like behaviors, and altered BDNF levels in serum, hippocampus, and NAc. On the other hand, the treatment with blonanserin did not affect depression-like behaviors, but reversed abnormal social interactions and altered BDNF levels in serum. Precisely, blonanserin treatment reduced serum BDNF levels, similar to the escitalopram treatment, but did not induce the antidepressant effects as exerted by escitalopram treatment (Furuse et al., 2019).

These findings indicate that blonanserin may exert therapeutic effects on social behavioral abnormalities in the early life combined stress-induced depression, which is related to the biological mechanism responsible for alterations in serum BDNF levels change in a manner similar to the escitalopram. Further studies are needed to identify specific brain regions in which blonanserin alters neuronal activity and serum BDNF levels. Recent studies have found that blonanserin and cariprazine exert their therapeutic efficacy through dopamine D3 receptor antagonism on cognitive and social dysfunctions in schizophrenia and bipolar mania (Caccia, Invernizzi, Nobili, & Pasina, 2013). D3 receptors are reported to be highly distributed in the olfactory tubercle area. Since the olfactory tubercle region is known to play an important role in emotional and reward-related behaviors, it may be one of the candidate brain areas related to the effect of blonanserin on abnormal social interaction and decreased serum BDNF levels.

## Possible BDNF/GABA activation by blonanserin through D3 receptors

The D3 receptor is also reported to be highly represented both at pre- and postsynapses in the ventral striatum (NAc and Calleja island) (Bouthenet et al., 1991; Murray, Ryoo, Gurevich, & Joyce, 1994). The report documented that dopamine acting through the D3 receptor can modulate the expression of innate anxiety-like behaviors involving a long-lasting, downregulation of GABAergic activity in lateral/basolateral amygdala (Diaz et al., 2000), NAc (Chen, Kittler, Moss, & Yan, 2006), and hippocampus neurons (Swant, Stramiello, & Wagner, 2008). It is also reported that D3 receptor expression could be regulated by BDNF (Guillin et al., 2001), which is involved in the control of $GABA_A$ receptor response in NAc (Koo et al., 2014).

Together, from these findings, selective expression of D3 receptor attracts great attention. Important evidences have been derived from an in situ hybridization study, which has revealed specific D3 receptor mRNA expression in proliferation zone of rat brain during development. It was demonstrated that at birth, D3 receptor mRNA expression appeared in differentiating neuronal fields of the NAc and medial mammillary body primordia (MB), and on P5, reached at a distribution similar to that found in adult, which D3 receptor gene expression continuing in the striatal proliferative subventricular zone (SVZ) (Diaz et al., 1997).

These evidences strongly suggest the neurogenetic modulatory role of D3 receptor in the adult brain as well as developing brain. There is an interesting report to examine the GABAergic interneuron genesis effects of antipsychotics in impaired differentiation induced by exposure to the NMDA receptor antagonist MK-801 in the adult SVZ-derived progenitor cells expressing proteoglycan NG2 cultures. The report demonstrated that second-generation antipsychotics including blonanserin but not first-generation antipsychotic haloperidol promoted or restored changes in the GABAergic interneuron differentiation of NG2(+) cells induced by exposure to MK-801, which was considered to be one of the drug-induced schizophrenia model (Kaneta et al., 2017) (Fig. 5).

As one of the candidate molecular mechanisms for proliferating action of antipsychotic, heat-shock protein (HSP) expression change was analyzed, and restoration of HSP productions by antipsychotics was also revealed in NG2 (+) cells with differentiation impairment. It can be inferred that concerning the mechanism of the driving potential of the proliferation of progenitor cells in order to promote SVZ neurogenesis would be keeping great attention as one of the new candidate antidepressant effect-related functions via D3 receptor modulation.

These current findings suggest alterations in serum and brain regional BDNF levels as a prototypical example of how various complicated stress-induced refractory depression and effective antidepressant/antipsychotic treatments produce neuronal activity and behavioral changes. Although many questions remain regarding their neural mechanisms of GABA activity modulation involving GABA interneuron genesis through D3 receptor, these findings might provide insights into the neurobiological mechanisms underlying the therapeutic potential in pharmaceutical interventions for treatment-resistant refractory depression.

## Key facts

- Many antidepressants upregulate the expression of BDNF via cyclic AMP-responsive element-binding protein (CREB) pathway.
- Decreased BDNF signaling might be associated with the pathophysiology of depression, and increased BDNF signaling is responsible for the mechanisms of antidepressant drugs.
- Decreased serum BDNF levels in patients with depression reported to be corrected in those that respond to antidepressant treatments.
- Repeated exposure to environmental adversities during development may be a risk factor for poor treatment outcomes in depression.
- In a clinical study with PTSD, lower serum BDNF levels over the course of the trial suggested as a strong predictor of treatment response to escitalopram.

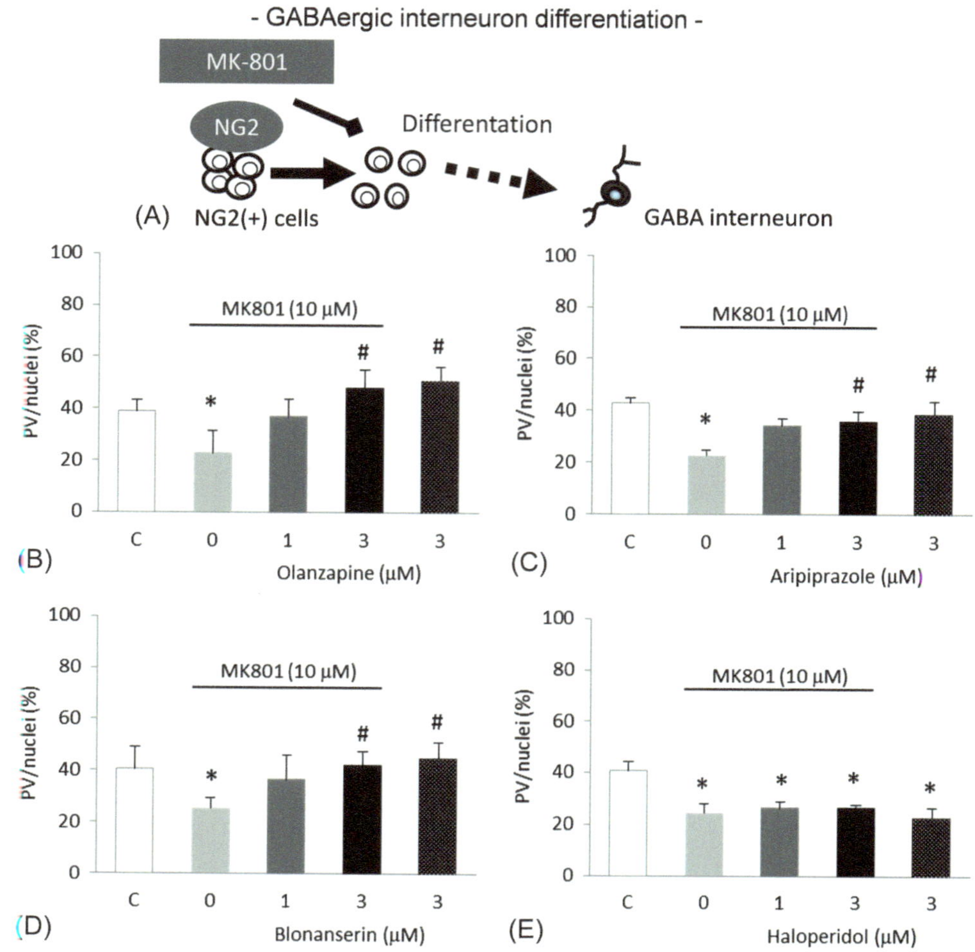


FIG. 5 Ameliorative effects of antipsychotics on MK801-induced GABAergic interneuron genesis impairment from NG2 (+) cells. (A) GABAergic interneuron differentiation was evaluated by the proportion of immunofluorescently stained cells with anti-parvalbumin (PV) antibody relative to the total number of cells revealed with nucleic DAPI counterstaining. Co-treatment with second-generation antipsychotic olanzapine (B), aripiprazole (C), or blonanserin (D), but not first-generation antipsychotic haloperidol (E) significantly recovered the NMDA receptor antagonist MK801-induced impairment of PV-positive GABAergic interneuron differentiation from adult subventricular zone (SVZ)-derived progenitor cells expressing proteoglycan NG2. *Data are from Kaneta, H., Ukai, W., Tsujino, H., Furuse, K., Kigawa, Y., Tayama, M., Ishii, T., Hashimoto, E., & Kawanishi, C. (2017). Antipsychotics promote GABAergic interneuron genesis in the adult rat brain: Role of heat-shock protein production. Journal of Psychiatric Research, 92, 108–118. https://doi.org/10.1016/j.jpsychires.2017.03.008, with permission from the Publishers.*

- Antidepressant augmented with second or third-generation antipsychotics, including blonanserin demonstrated in better treatment response or higher rates of remission than antidepressant monotherapy.
- Studies suggested blonanserin exerts its therapeutic efficacy through dopamine D3 receptor antagonism on cognitive and social dysfunctions in schizophrenia and bipolar mania.

## Summary points

- This chapter focuses on neurotrophic mechanisms of escitalopram and blonanserin on refractory depression caused by the repeated exposure to environmental adversities during development.
- There is a study demonstrated that escitalopram can attenuated depression-like behavior in fetal and adolescent period combined stress-induced refractory depression model.
- In the refractory depression model, serum BDNF levels indicated no decrease, however, escitalopram or blonanserin induced reductions of its levels.
- Escitalopram also decreased BDNF in NAc, where neuron activity reflected closely to the BDNF levels in serum.
- It can be hypothesized that not only increase in hippocampus but also decrease in NAc BDNF levels is

necessity for antidepressant medication responsible for refractory depression.

- Blonanserin exerted recovery effect of social behavioral abnormality in early life combined stress-induced refractory depression model.
- D3 receptors highly presented regions those include olfactory tubercle and NAc suggested to be pivotal brain areas responsible for abnormal social behavior and serum BDNF level reduction.

## Mini-dictionary of terms

**Refractory depression** Depression which does not achieve or keep sufficient treatment response to adequate treatments.

**Glutamate/GABA balance** Presynaptic functions of glutamate and GABA release controlled by synthesis and degradation enzyme activities.

**Corticosteroids** Primary humoral stress mediators which can activate hypothalamic–pituitary–adrenal (HPA) axis with physiological damage.

**BDNF signaling** The release of brain-derived neurotrophic factor (BDNF) stimulate tropomyosin-related kinase B (TrkB) and its signaling activates multiple downstream pathways important for antidepressant behavior and drug response.

**GABA interneuron** Inhibitory interneuron that uses GABA as neurotransmitter, integrate and filter incoming information, and dictate postsynaptic neuronal spike timing.

## References


Bauer, M., Bschor, T., Pfennig, A., Whybrow, P. C., Angst, J., Versiani, M., et al. (2007). World federation of societies of biological psychiatry (WFSBP) guidelines for biological treatment of unipolar depressive disorders in primary care. *World Journal of Biological Psychiatry*, *8*(2), 67–104. https://doi.org/10.1080/15622970701227829.

Berger, W., Mehra, A., Lenoci, M., Metzler, T. J., Otte, C., Tarasovsky, G., et al. (2010). Serum brain-derived neurotrophic factor predicts responses to escitalopram in chronic posttraumatic stress disorder. *Progress in Neuro-Psychopharmacology & Biological Psychiatry*, *34*(7), 1279–1284. https://doi.org/10.1016/j.pnpbp.2010.07.008.

Berton, O., McClung, C. A., Dileone, R. J., Krishnan, V., Renthal, W., Russo, S. J., et al. (2006). Essential role of BDNF in the mesolimbic dopamine pathway in social defeat stress. *Science (New York, N.Y.)*, *311*(5762), 864–868. https://doi.org/10.1126/science.1120972.

Bocchio-Chiavetto, L., Bagnardi, V., Zanardini, R., Molteni, R., Nielsen, M. G., Placentino, A., et al. (2010). Serum and plasma BDNF levels in major depression: A replication study and meta-analyses. *The World Journal of Biological Psychiatry: The Official Journal of the World Federation of Societies of Biological Psychiatry*, *11*(6), 763–773. https://doi.org/10.3109/15622971003611319.

Bouthenet, M. L., Souil, E., Martres, M. P., Sokoloff, P., Giros, B., & Schwartz, J. C. (1991). Localization of dopamine D3 receptor mRNA in the rat brain using in situ hybridization histochemistry: Comparison with dopamine D2 receptor mRNA. *Brain Research*, *564*(2), 203–219. https://doi.org/10.1016/0006-8993(91)91456-B.

Bridge, J., Tina, G., & David, B. (2006). Adolescent suicide and suicidal behavior. *Journal of Child Psychiatry*, *47*(3–4), 372–394. https://doi.org/10.1111/j.1469-7610.2006.01615.x. 16492264.

Brummelte, S., & Galea, L. A. M. (2010). Chronic high corticosterone reduces neurogenesis in the dentate gyrus of adult male and female rats. *Neuroscience*, *168*(3), 680–690. https://doi.org/10.1016/j.neuroscience.2010.04.023.

Caccia, S., Invernizzi, R. W., Nobili, A., & Pasina, L. (2013). A new generation of antipsychotics: Pharmacology and clinical utility of cariprazine in schizophrenia. *Therapeutics and Clinical Risk Management*, *9*(1), 319–328. https://doi.org/10.2147/TCRM.S35137.

Carvalho, A. L., Caldeira, M. V., Santos, S. D., & Duarte, C. B. (2008). Role of the brain-derived neurotrophic factor at glutamatergic synapses. *British Journal of Pharmacology*. https://doi.org/10.1038/sj.bjp.0707509.

Castrén, E. (2014). Neurotrophins and psychiatric disorders. *Handbook of Experimental Pharmacology*, *220*, 461–479. https://doi.org/10.1007/978-3-642-45106-5_17.

Chen, G., Kittler, J. T., Moss, S. J., & Yan, Z. (2006). Dopamine D3 receptors regulate GABAA receptor function through a phospho-dependent endocytosis mechanism in nucleus accumbens. *Journal of Neuroscience*, *26*(9), 2513–2521. https://doi.org/10.1523/JNEUROSCI.4712-05.2006.

Chen, H., Lombès, M., & Le Menuet, D. (2017). Glucocorticoid receptor represses brain-derived neurotrophic factor expression in neuron-like cells. *Molecular Brain*, *10*(1), 12. https://doi.org/10.1186/s13041-017-0295-x.

Conrad, C. D., McLaughlin, K. J., Harman, J. S., Foltz, C., Wieczorek, L., Lightner, E., et al. (2007). Chronic glucocorticoids increase hippocampal vulnerability to neurotoxicity under conditions that produce CA3 dendritic retraction but fail to impair spatial recognition memory. *The Journal of Neuroscience: The Official Journal of the Society for Neuroscience*, *27*(31), 8278–8285. https://doi.org/10.1523/JNEUROSCI.2121-07.2007.

Cuzon, V. C., Yeh, P. W. L., Yanagawa, Y., Obata, K., & Yeh, H. H. (2008). Ethanol consumption during early pregnancy alters the disposition of tangentially migrating GABAergic interneurons in the fetal cortex. *The Journal of Neuroscience: The Official Journal of the Society for Neuroscience*, *28*(8), 1854–1864. https://doi.org/10.1523/JNEUROSCI.5110-07.2008.

Deppmann, C. D., Mihalas, S., Sharma, N., Lonze, B. E., Niebur, E., & Ginty, D. D. (2008). A model for neuronal competition during development. *Science*, *320*(5874), 369–373. https://doi.org/10.1126/science.1152677.

Diaz, J., Pilon, C., Le Foll, B., Gros, C., Triller, A., Schwartz, J. C., et al. (2000). Dopamine D3 receptors expressed by all mesencephalic dopamine neurons. *Journal of Neuroscience*, *20*(23), 8677–8684. https://doi.org/10.1523/jneurosci.20-23-08677.2000.

Diaz, J., Ridray, S., Mignon, V., Griffon, N., Schwartz, J. C., & Sokoloff, P. (1997). Selective expression of dopamine D3 receptor mRNA in proliferative zones during embryonic development of the rat brain. *Journal of Neuroscience*, *17*(11), 4282–4292. https://doi.org/10.1523/jneurosci.17-11-04282.1997.

Duman, R. S., & Li, N. (2012). A neurotrophic hypothesis of depression: Role of synaptogenesis in the actions of NMDA receptor antagonists. *Philosophical Transactions of the Royal Society of London. Series B, Biological Sciences*, *367*(1601), 2475–2484. https://doi.org/10.1098/rstb.2011.0357.

Duman, R. S., & Monteggia, L. M. (2006). A neurotrophic model for stress-related mood disorders. *Biological Psychiatry*, *59*(12), 1116–1127. https://doi.org/10.1016/j.biopsych.2006.02.013.

Eisch, A. J., Bolaños, C. A., de Wit, J., Simonak, R. D., Pudiak, C. M., Barrot, M., et al. (2003). Brain-derived neurotrophic factor in the ventral midbrain-nucleus accumbens pathway: A role in depression. *Biological Psychiatry*, *54*(10), 994–1005. http://www.ncbi.nlm.nih.gov/pubmed/14625141.

Fergusson, D. M., & Woodward, L. J. (2002). Mental health, educational, and social role outcomes of adolescents with depression. *Archives of General Psychiatry*, *59*(3), 225–231. https://doi.org/10.1001/archpsyc.59.3.225.

Freitas, A. E., Egea, J., Buendia, I., Gómez-Rangel, V., Parada, E., Navarro, E., et al. (2016). Agmatine, by improving neuroplasticity markers and inducing Nrf2, prevents corticosterone-induced depressive-like behavior in mice. *Molecular Neurobiology*, *53*(5), 3030–3045. https://doi.org/10.1007/s12035-015-9182-6.

Fujimura, H., Altar, C. A., Chen, R., Nakamura, T., Nakahashi, T., Kambayashi, J., et al. (2002). Brain-derived neurotrophic factor is stored in human platelets and released by agonist stimulation. *Thrombosis and Haemostasis*, *87*(4), 728–734. http://www.ncbi.nlm.nih.gov/pubmed/12008958.

Furuse, K., Ukai, W., Hashimoto, E., Hashiguchi, H., Kigawa, Y., Ishii, T., et al. (2019). Antidepressant activities of escitalopram and blonanserin on prenatal and adolescent combined stress-induced depression model: Possible role of neurotrophic mechanism change in serum and nucleus accumbens. *Journal of Affective Disorders*, *247*, 97–104. https://doi.org/10.1016/j.jad.2019.01.007.

Gourley, S. L., Kiraly, D. D., Howell, J. L., Olausson, P., & Taylor, J. R. (2008). Acute hippocampal brain-derived neurotrophic factor restores motivational and forced swim performance after corticosterone. *Biological Psychiatry*, *64*(10), 884–890. https://doi.org/10.1016/j.biopsych.2008.06.016.

Guillin, O., Diaz, J., Carroll, P., Griffon, N., Schwartz, J. C., & Sokoloff, P. (2001). BDNF controls dopamine D3 receptor expression and triggers behavioural sensitization. *Nature*, *411*(6833), 86–89. https://doi.org/10.1038/35075076.

Hanson, N. D., Owens, M. J., & Nemeroff, C. B. (2011). Depression, antidepressants, and neurogenesis: a critical reappraisal. *Neuropsychopharmacology: Official Publication of the American College of Neuropsychopharmacology*, *36*(13), 2589–2602. https://doi.org/10.1038/npp.2011.220.

Hedrick, N. G., Harward, S. C., Hall, C. E., Murakoshi, H., McNamara, J. O., & Yasuda, R. (2016). Rho GTPase complementation underlies BDNF-dependent homo- and heterosynaptic plasticity. *Nature*, *538*(7623), 104–108. https://doi.org/10.1038/nature19784.

Hill, A. S., Sahay, A., & Hen, R. (2015). Increasing adult hippocampal neurogenesis is sufficient to reduce anxiety and depression-like behaviors. *Neuropsychopharmacology: Official Publication of the American College of Neuropsychopharmacology*, *40*(10), 2368–2378. https://doi.org/10.1038/npp.2015.85.

Hitoshi, S., Maruta, N., Higashi, M., Kumar, A., Kato, N., & Ikenaka, K. (2007). Antidepressant drugs reverse the loss of adult neural stem cells following chronic stress. *Journal of Neuroscience Research*, *85*(16), 3574–3585. https://doi.org/10.1002/jnr.21455.

Hori, H., Yamada, K., Kamada, D., Shibata, Y., Katsuki, A., Yoshimura, R., et al. (2014). Effect of blonanserin on cognitive and social function in acute phase Japanese schizophrenia compared with risperidone. *Neuropsychiatric Disease and Treatment*, *10*, 527–533. https://doi.org/10.2147/NDT.S59861.

Kalynchuk, L. E., Gregus, A., Boudreau, D., & Perrot-Sinal, T. S. (2004). Corticosterone increases depression-like behavior, with some effects on predator odor-induced defensive behavior, in male and female rats. *Behavioral Neuroscience*, *118*(6), 1365–1377. https://doi.org/10.1037/0735-7044.118.6.1365.

Kaneta, H., Ukai, W., Tsujino, H., Furuse, K., Kigawa, Y., Tayama, M., et al. (2017). Antipsychotics promote GABAergic interneuron genesis in the adult rat brain: Role of heat-shock protein production. *Journal of Psychiatric Research*, *92*, 108–118. https://doi.org/10.1016/j.jpsychires.2017.03.008.

Kennard, B. D., Silva, S. G., Tonev, S., Rohde, P., Hughes, J. L., Vitiello, B., et al. (2009). Remission and recovery in the treatment for adolescents with depression study (TADS): Acute and long-term outcomes. *Journal of the American Academy of Child and Adolescent Psychiatry*, *48*(2), 186–195. https://doi.org/10.1097/CHI.0b013e31819176f9.

Kennedy, S. H., Lam, R. W., Parikh, S. V., Patten, S. B., & Ravindran, A. V. (2009). Canadian network for mood and anxiety treatments (CANMAT) clinical guidelines for the management of major depressive disorder in adults. *Journal of Affective Disorders*, *117* (Suppl. 1). https://doi.org/10.1016/j.jad.2009.06.043.

Kigawa, Y., Hashimoto, E., Ukai, W., Ishii, T., Furuse, K., Tsujino, H., et al. (2014). Stem cell therapy: A new approach to the treatment of refractory depression. *Journal of Neural Transmission (Vienna, Austria : 1996)*, *121*(10), 1221–1232. https://doi.org/10.1007/s00702-014-1194-2.

Kishi, T., Yoshimura, R., Ikuta, T., & Iwata, N. (2018). Brain-derived neurotrophic factor and major depressive disorder: Evidence from meta-analyses. *Frontiers in Psychiatry*, *8*(Jan). https://doi.org/10.3389/fpsyt.2017.00308. Frontiers Media S.A.

Klein, A. B., Williamson, R., Santini, M. A., Clemmensen, C., Ettrup, A., Rios, M., et al. (2011). Blood BDNF concentrations reflect brain-tissue BDNF levels across species. *The International Journal of Neuropsychopharmacology*, *14*(3), 347–353. https://doi.org/10.1017/S1461145710000738.

Komossa, K., Depping, A. M., Gaudchau, A., Kissling, W., & Leucht, S. (2010). Second-generation antipsychotics for major depressive disorder and dysthymia. *The Cochrane Database of Systematic Reviews*, *12*. https://doi.org/10.1002/14651858.CD008121.pub2, CD008121.

Koo, J. W., Lobo, M. K., Chaudhury, D., Labonté, B., Friedman, A., Heller, E., et al. (2014). Loss of BDNF signaling in D1R-expressing NAc neurons enhances morphine reward by reducing GABA inhibition. *Neuropsychopharmacology: Official Publication of the American College of Neuropsychopharmacology*, *39*(11), 2646–2653. https://doi.org/10.1038/npp.2014.118.

Kurita, M., Nishino, S., Kato, M., Numata, Y., & Sato, T. (2012). Plasma brain-derived neurotrophic factor levels predict the clinical outcome of depression treatment in a naturalistic study. *PLoS One*, *7*(6). https://doi.org/10.1371/journal.pone.0039212, e39212.

Lee, H.-Y., & Kim, Y.-K. (2008). Plasma brain-derived neurotrophic factor as a peripheral marker for the action mechanism of antidepressants. *Neuropsychobiology*, *57*(4), 194–199. https://doi.org/10.1159/000149817.

Lee, J., Lee, K. H., Kim, S. H., Han, J. Y., Hong, S. B., Cho, S. C., et al. (2020). Early changes of serum BDNF and SSRI response in adolescents with major depressive disorder. *Journal of Affective Disorders*, *265*, 325–332. https://doi.org/10.1016/j.jad.2020.01.045.

Lee, B., Shim, I., Lee, H.-J., Yang, Y., & Hahm, D.-H. (2009). Effects of acupuncture on chronic corticosterone-induced depression-like behavior and expression of neuropeptide Y in the rats. *Neuroscience Letters*, *453*(3), 151–156. https://doi.org/10.1016/j.neulet.2009.01.076.

Lessmann, V., & Brigadski, T. (2009). Mechanisms, locations, and kinetics of synaptic BDNF secretion: An update. *Neuroscience Research*, *65* (1), 11–22. https://doi.org/10.1016/j.neures.2009.06.004.

Li, H., Yao, C., Shi, J., Yang, F., Qi, S., Wang, L., et al. (2015). Comparative study of the efficacy and safety between blonanserin and risperidone for the treatment of schizophrenia in Chinese patients: A double-blind, parallel-group multicenter randomized trial. *Journal of Psychiatric Research*, *69*, 102–109. https://doi.org/10.1016/j.jpsychires.2015.07.015.

Limaye, R. P., & Patil, A. N. (2016). Blonanserin—A novel antianxiety and antidepressant drug? An experimental study. *Journal of Clinical and Diagnostic Research: JCDR*, *10*(9), FC17–FC21. https://doi.org/10.7860/JCDR/2016/19347.8530.

Lindsay, R. M., Wiegand, S. J., Altar, C. A., & DiStefano, P. S. (1994). Neurotrophic factors: From molecule to man. *Trends in Neurosciences*, *17*(5), 182–190. http://www.ncbi.nlm.nih.gov/pubmed/7520198.

Molendijk, M. L., Bus, B. A. A., Spinhoven, P., Penninx, B. W. J. H., Kenis, G., Prickaerts, J., et al. (2011). Serum levels of brain-derived neurotrophic factor in major depressive disorder: State-trait issues, clinical features and pharmacological treatment. *Molecular Psychiatry*, *16*(11), 1088–1095. https://doi.org/10.1038/mp.2010.98.

Morgan, V. A., Mitchell, P. B., & Jablensky, A. V. (2005) The epidemiology of bipolar disorder: Sociodemographic, disability and service utilization data from the Australian National Study of low prevalence (psychotic) disorders. *Bipolar Disorders*, *7*(4), 326–337. https://doi.org/10.1111/j.1399-5618.2005.00229.x.

Murray, A. M., Ryoo, H. L., Gurevich, E., & Joyce, J. N. (1994). Localization of dopamine D3 receptors to mesolimbic and D2 receptors to mesostriatal regions of human forebrain. *Proceedings of the National Academy of Sciences of the United States of America*, *91*(23), 11271–11275. https://doi.org/10.1073/pnas.91.23.11271.

Nibuya, M., Nestler, E. J., & Duman, R. S. (1996). Chronic antidepressant administration increases the expression of cAMP response element binding protein (CREB) in rat hippocampus. *The Journal of Neuroscience: The Official Journal of the Society for Neuroscience*, *16*(7), 2365–2372. http://www.ncbi.nlm.nih.gov/pubmed/8601816.

Pan, W., Banks, W. A., Fasold, M. B., Bluth, J., & Kastin, A. J. (1998). Transport of brain-derived neurotrophic factor across the blood-brain barrier. *Neuropharmacology*, *37*(12), 1553–1561. http://www.ncbi.nlm.nih.gov/pubmed/9886678.

Pazini, F. L., Cunha, M. P., Rosa, J. M., Colla, A. R. S., Lieberknecht, V., Oliveira, Á., et al. (2016). Creatine, similar to ketamine, counteracts depressive-like behavior induced by corticosterone via PI3K/Akt/mTOR pathway. *Molecular Neurobiology*, *53*(10), 6818–6834. https://doi.org/10.1007/s12035-015-9580-9.

Samuels, B. A., Leonardo, E. D., Gadient, R., Williams, A., Zhou, J., David, D. J., et al. (2011). Modeling treatment-resistant depression. *Neuropharmacology*, *61*(3), 408–413. https://doi.org/10.1016/j.neuropharm.2011.02.017.

Södersten, K., Pålsson, E., Ishima, T., Funa, K., Landén, M., Hashimoto, K., et al. (2014). Abnormality in serum levels of mature brain-derived neurotrophic factor (BDNF) and its precursor proBDNF in mood-stabilized patients with bipolar disorder: A study of two independent cohorts. *Journal of Affective Disorders*, *160*, 1–9. https://doi.org/10.1016/j.jad.2014.01.009.

Spiwak, R., Pagura, J., Bolton, J. M., Elias, B., Beesdo-Baum, K., Lieb, R., et al. (2011). Childhood exposure to caregiver suicidal behavior and risk for adult suicide attempts: Findings from a National Survey. *Archives of Suicide Research*, *15*(4), 313–326. https://doi.org/10.1080/13811118.2011.615694.

Swant, J., Stramiello, M., & Wagner, J. J. (2008). Postsynaptic dopamine D3 receptor modulation of evoked IPSCs via GABA(A) receptor endocytosis in rat hippocampus. *Hippocampus*, *18*(5), 492–502. https://doi.org/10.1002/hipo.20408.

Taylor, M. J., Norbury, R., Murphy, S., Rudebeck, S., Jezzard, P., & Cowen, P. J. (2010). Lack of effect of citalopram on magnetic resonance spectroscopy measures of glutamate and glutamine in frontal cortex of healthy volunteers. *Journal of Psychopharmacology (Oxford, England)*, *24*(8), 1217–1221. https://doi.org/10.1177/0269881109105679.

Tottenham, N., & Galván, A. (2016). Stress and the adolescent brain: Amygdala-prefrontal cortex circuitry and ventral striatum as developmental targets. *Neuroscience and Biobehavioral Reviews*, *70*, 217–227. Elsevier Ltd https://doi.org/10.1016/j.neubiorev.2016.07.030.

Vitiello, B., Silva, S. G., Rohde, P., Kratochvil, C. J., Kennard, B. D., Reinecke, M. A., et al. (2009). Suicidal events in the treatment for adolescents with depression study (TADS). *Journal of Clinical Psychiatry*, *70*(5), 741–747. https://doi.org/10.4088/JCP.08m04607.

Watanabe, K., Hashimoto, E., Ukai, W., Ishii, T., Yoshinaga, T., Ono, T., et al. (2010). Effect of antidepressants on brain-derived neurotrophic factor (BDNF) release from platelets in the rats. *Progress in Neuro-Psychopharmacology & Biological Psychiatry*, *34*(8), 1450–1454. https://doi.org/10.1016/j.pnpbp.2010.07.036.

Weissman, M. M., Wolk, S., Goldstein, R. B., Moreau, D., Adams, P., Greenwald, S., et al. (1999). Depressed adolescents grown up. *Journal of the American Medical Association*, *281*(18), 1707–1713. https://doi.org/10.1001/jama.281.18.1707.

Zhao, Y., Ma, R., Shen, J., Su, H., Xing, D., & Du, L. (2008). A mouse model of depression induced by repeated corticosterone injections. *European Journal of Pharmacology*, *581*(1–2), 113–120. https://doi.org/10.1016/j.ejphar.2007.12.005.

Chapter 37

# (2*R*,6*R*)-Hydroxynorketamine as a novel antidepressant and its role in the antidepressant actions of (*R*,*S*)-ketamine

Shigeyuki Chaki and Jun-ichi Yamaguchi
*Research Headquarters, Taisho Pharmaceutical Co., Ltd. Saitama, Japan*

## List of abbreviations

| | |
|---|---|
| **BDNF** | brain-derived neurotrophic factor |
| **CSF** | cerebrospinal fluid |
| **HNK** | hydroxynorketamine |
| **i.c.v.** | intracerebroventricular |
| **MDD** | major depressive disorder |
| **mPFC** | medial prefrontal cortex |
| **mTORC1** | mechanistic target of rapamycin complex 1 |
| **NMDA** | *N*-methyl-D-aspartate |
| **PD** | pharmacodynamics |
| **PK** | pharmacokinetic |
| **TRD** | treatment-resistant depression |

## Introduction

The discovery of the rapid and sustained antidepressant effect of ketamine, a noncompetitive antagonist of the *N*-methyl-D-aspartate (NMDA) receptor, in patients with major depressive disorder (MDD), including patients with treatment-resistant depression (TRD) (Berman et al., 2000; Zarate et al., 2006) is regarded as the most outstanding finding in depression research in 60 years. Based on the studies of the antidepressant effect of ketamine, (*S*)-ketamine (esketamine), an enantiomer of ketamine, was approved for the treatment of TRD in the form of a nasal spray application (Spravato) in March 2019. However, since both ketamine and (*S*)-ketamine have given rise to a number of adverse events, including psychotomimetic/dissociative symptoms, abuse potential, and neurotoxicity, routine use of (*S*)-ketamine nasal spray in daily practice is greatly limited, and drug discovery to develop agents with the same antidepressant profile as ketamine but a better safety profile is underway. In order to develop such agents, it is important to fully understand the mechanisms underlying the antidepressant effect of ketamine and identifying the active substance(s) after administration of ketamine is the first step in determining its precise mechanisms.

Ketamine is a racemic mixture of (*S*)-ketamine and (*R*)-ketamine ((*R*,*S*)-ketamine), and is rapidly and stereoselectively metabolized into several metabolites. Interestingly, the recent studies have indicated that (*R*)-ketamine, which has lower affinity for the NMDA receptor than (*S*)-ketamine, is more potent and persistent in exerting the antidepressant effect in rodents. Hashimoto's group was the first to report that (*R*)-ketamine (arketamine) is more potent (Yang et al., 2015) than (*S*)-ketamine. Importantly, (*R*)-ketamine does not induce the abnormal behavior in rodents that is related to the adverse events associated with ketamine and (*S*)-ketamine (Yang et al., 2015). The results showing that (*R*)-ketamine is more potent than (*S*)-ketamine have been replicated at other institutions (Fukumoto et al., 2017; Zanos et al., 2016). (*R*)-ketamine may, therefore, be the more active stereoisomer of racemic ketamine in terms of an antidepressant effect, and it may provide a safer approach to treatment than racemic ketamine or (*S*)-ketamine. It is noteworthy that (*R*)-ketamine was recently found to exert a rapid antidepressant effect within 24 h in TRD patients without appearance of dissociative symptom in an open-label pilot study (Leal et al., 2020). Investigating the mechanisms underlying the antidepressant effect of (*R*)-ketamine can make it possible to distinguish between mechanisms of ketamine related to its antidepressant effect and its side effects, and identifying the active substance(s) following administration of (*R*)-ketamine is the first step in achieving this goal. It has recently been proposed that one of the metabolites of (*R*)-ketamine, not (*R*)-ketamine itself, is responsible for the antidepressant effect of (*R*)-ketamine. Zanos et al. (2016) found that (1) (*R*)-ketamine exerts a more potent antidepressant effect than (*S*)-ketamine, (2) partial blockade of (*R*,*S*)-ketamine

The Neuroscience of Depression. https://doi.org/10.1016/B978-0-12-817933-8.00011-6

metabolism to (2*S*,6*S*;2*R*,6*R*)-hydroxynorketamine (HNK) abolishes the antidepressant effect of (*R*,*S*)-ketamine, and (3) (2*R*,6*R*)-HNK exerts much more potent antidepressant effects than (2*S*,6*S*)-HNK in some animal models. Based on all of these findings taken together, authors proposed that (2*R*,6*R*)-HNK is responsible for the antidepressant effect of (*R*)-ketamine, and possibly of (*R*,*S*)-ketamine. An antidepressant effect of (2*R*,6*R*)-HNK was reported by several research groups after this initial report. By contrast, there are reports which oppose the proposal that (2*R*,6*R*)-HNK is essential for the antidepressant effects of (*R*)-ketamine. In this chapter, we summarize evidence for an antidepressant effect of (2*R*,6*R*)-HNK and its underlying mechanisms reported to date, and we discuss: (1) whether (2*R*,6*R*)-HNK has a ketamine-like antidepressant effect and (2) whether (2*R*,6*R*)-HNK is essential for the antidepressant effect of (*R*)-ketamine and (*R*,*S*)-ketamine to occur. We also consider the validity of the pharmacological data from a pharmacokinetic standpoint, but do not discuss the roles of other ketamine metabolites in depth, because they are not within the scope of this chapter.

## Metabolism of (*R*,*S*)-ketamine

(*R*,*S*)-Ketamine is used clinically in the form of the hydrochloride, mainly via non-oral routes, i.e., intravenously, intramuscularly, and intranasally, because of its low oral bioavailability (17%) as a result of its extensive first-pass metabolism (Clements, Nimmo, & Grant, 1982). (*R*,*S*)-Ketamine has a relatively low plasma protein binding rate (47%) and is rapidly and extensively distributed in the brain and in highly perfused tissues. It has a Vdss (distribution volume at steady state) of nearly 3 L/kg, and is rapidly eliminated with a terminal half-life of approximately 3 h (Clements & Nimmo, 1981; Dayton, Stiller, Cook, & Perel, 1983). The clearance mechanism of (*R*,*S*)-ketamine is metabolism and subsequent urinary excretion (Chang, Savory, Albin, Goulet, & Glazko, 1970).

(*R*,*S*)-Ketamine is extensively and stereoselectively metabolized into a variety of oxidative metabolites by cytochrome P450 (CYP) enzymes (Desta et al., 2012; Portmann et al., 2010). Fig. 1 shows the metabolic pathways of (*R*,*S*)-ketamine in humans. (*R*,*S*)-Ketamine undergoes predominantly *N*-demethylation, mainly by CYP2B6 and CYP3A4, to form norketamine ((*R*,*S*)-NK). (*R*,*S*)-NK is further subjected to oxidation at the C4-, C5-, or C6-position of the cyclohexanone moiety, which results in the formation of five diastereomeric HNK metabolites. The C6-hydroxylation occurs preferentially, mainly by CYP2B6 and CYP2A6, to form (2*S*,6*S*;2*R*,6*R*)-HNK. (*R*,*S*)-NK also undergoes dehydrogenation mainly by CYP2B6 to 5,6-dehydronorketamine ((*R*,*S*)-DHNK). Some of the (*R*,*S*)-

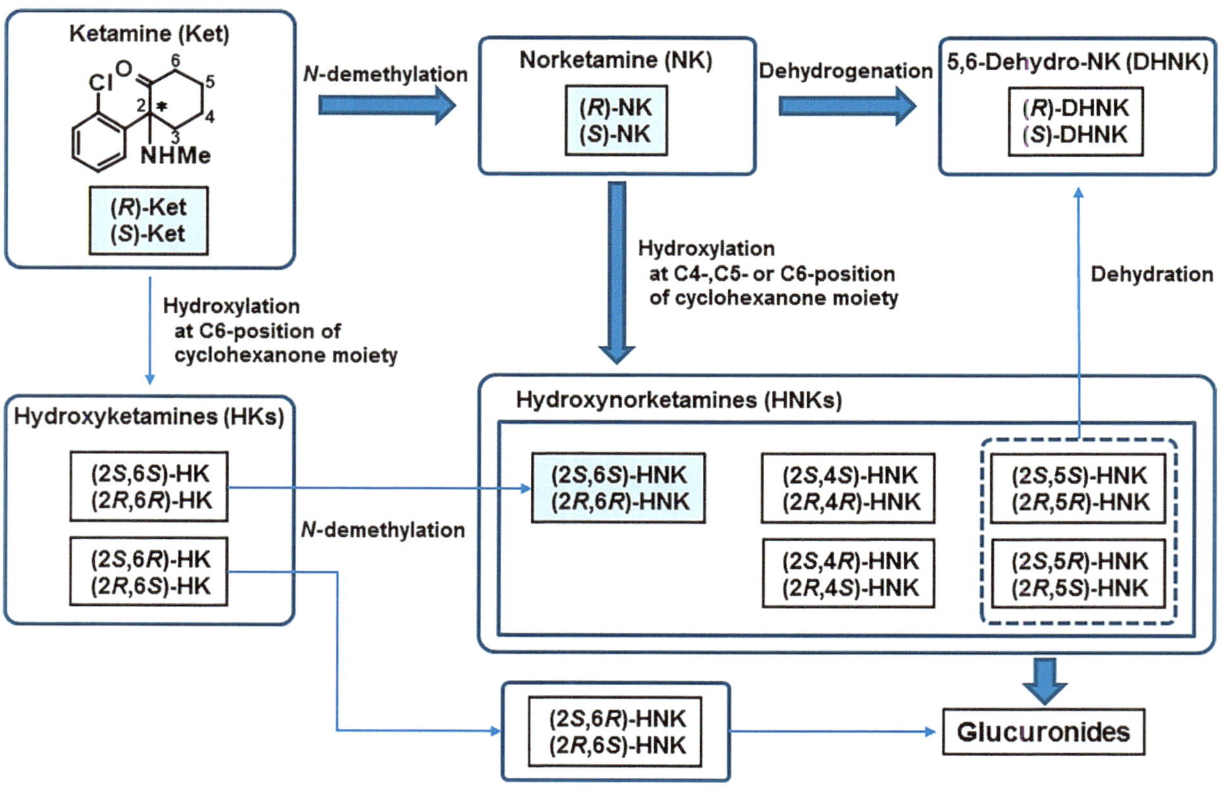


FIG. 1 Metabolic pathway of ketamine in humans.

DHNK has been reported to be formed via 5-HNKs. In addition, (*R*,*S*)-ketamine undergoes direct oxidation at the C6-position of the cyclohexanone moiety to form two diastereomeric hydroxylketamine (HK) metabolites, which subsequently undergo *N*-demethylation to form (2*S*,6*S*;2*R*,6*R*)-HNK and (2*S*,6*R*;2*R*,6*S*)-HNK. The several HNK metabolites are finally biotransformed into glucuronides, which are then excreted as major metabolites (Moaddel et al., 2010). (*R*,*S*)-Ketamine has been reported to be biotransformed into phenolic metabolites, in addition to being metabolized via the pathways shown in Fig. 1 (Turfus et al., 2009).

(*R*,*S*)-Ketamine, (*R*,*S*)-NK, (*R*,*S*)-DHNK, four of six diastereomeric HNKs, and HK were detected as circulating drug-related components of bipolar depression (BD) patients and MDD patients during the first 230 min after a 40-min (*R*,*S*)-ketamine infusion (0.5 mg/kg) (Zarate et al., 2012). (*R*,*S*)-DHNK was the major metabolite in four of the nine BD patients, (*R*,*S*)-NK in 3/9, and (2*S*,6*S*;2*R*,6*R*)-HNK in 2/9 (Zhao et al., 2012). In patients with complex regional pain syndrome, (2*S*,6*S*;2*R*,6*R*)-HNK and (2*S*,6*R*;2*R*,6*S*)-HNK were detected as the major circulating metabolites on day 3 of a 5-day continuous infusion of 40 mg/h of (*R*,*S*)-ketamine (Moaddel et al., 2010). In rodents, the unchanged drug, (*R*,*S*)-NK, and (2*S*,6*S*;2*R*,6*R*)-HNK, which easily penetrate into the brain, were detected as the predominant circulating drug components after intraperitoneal (i.p.) administration of (*R*,*S*)-ketamine (Moaddel et al., 2016; Yamaguchi et al., 2018; Zanos et al., 2016). As described below, one of the major metabolites of (*R*,*S*)-ketamine, (2*R*,6*R*)-HNK, which is formed from the (*R*)-ketamine enantiomer alone, has recently gained much attention as a novel antidepressant.

## Antidepressant effects of (2*R*,6*R*)-HNK

Antidepressant effects of (2*R*,6*R*)-HNK were first reported by Dr Gould's group in Nature in 2016 (Zanos et al., 2016). Like (*R*,*S*)-ketamine, the antidepressant effects of (2*R*,6*R*)-HNK were sustained for at least 3 days in the forced swimming test. It was noteworthy that (2*R*,6*R*)-HNK exerted these effects at a dose as low as 5 mg/kg or 10 mg/kg, i.p., much lower than the doses of (2*S*,6*S*)-HNK (15 times lower in the learned helplessness test). Comparisons between the antidepressant potency of (*R*,*S*)-ketamine and (2*R*,6*R*)-HNK were impossible in this study, because the effects of (*R*,*S*)-ketamine were tested at only 10 mg/kg, i.p. However, the results of the learned helplessness test showed that the antidepressant effects of (*R*,*S*)-ketamine (10 mg/kg, i.p.) were slightly more potent than those of (2*R*,6*R*)-HNK at the same dose, whereas the results of the novelty-suppressed feeding test showed that the potency of both compounds (10 mg/kg, i.p.) was similar. The antidepressant effects of (2*R*,6*R*)-HNK were further confirmed in the forced swimming test (24 h after administration) and novelty-suppressed feeding test (3 days after administration) in naive mice by Duman's group (Fukumoto et al., 2019). By contrast, (2*R*,6*R*)-HNK had no effect up to 30 mg/kg in the female urine sniffing test (5 days after administration), whereas (*R*,*S*)-ketamine (10 mg/kg) increased sniffing time, suggesting that (*R*,*S*)-ketamine was more potent than (2*R*,6*R*)-HNK in this test. Thus, the antidepressant profile of (2*R*,6*R*)-HNK is different from that of (*R*,*S*)-ketamine. Antidepressant effects of (2*R*,6*R*)-HNK have also been reported by Pham et al (2018) who showed that (2*R*,6*R*)-HNK (10 mg/kg, i.p.) decreased immobility in the forced swimming test 24 h after administration. Moreover, (2*R*,6*R*)-HNK (10 mg/kg, i.p.) has been reported to rapidly (<1 h) and persistently (>21 days) reverse depressive-like behaviors in a modified learned helplessness model in which rats were exposed to inescapable and unpredictable shocks for 6 days (Chou et al., 2018). Imipramine had no antidepressant effects under the same experimental conditions, whereas scopolamine showed rapid and sustained antidepressant effects, suggesting that the modified learned helplessness paradigm can be regarded as a treatment-resistant model. Chou et al. (2018) also found that (2*R*,6*R*)-HNK (10 mg/kg, i.p.) reversed depressive-like behaviors of rats induced by chronic restraint stress or spinal nerve ligation.

By contrast, some studies have reported observing a lack of antidepressant effects of (2*R*,6*R*)-HNK. In experiments on mice, Hashimoto's group found that (2*R*,6*R*)-HNK, at 10 mg/kg, i.p., did not reverse depressive-like behaviors induced by lipopolysaccharide (3 h after administration) or chronic social defeat stress (1–7 days after administration), whereas the same dose of (*R*)-ketamine had significant effects (Yang et al., 2017). In addition, Shirayama and Hashimoto (2018) found that neither (2*R*,6*R*)-HNK (20 and 40 mg/kg, i.p.) nor (*R*)-NK (20 mg/kg, i.p.) had any effects in the rat learned helplessness test, but that (*R*)-ketamine (20 mg/kg, i.p.) exerted antidepressant effects. Moreover, administration of (2*R*,6*R*)-HNK (10 mg/kg, i.p. or 20 mg/kg, i.p.) to mice did not reverse the increase in immobility in the forced swimming test or impairment of female preference in the female encounter test, both of which were induced by repeated corticosterone doses (Yokoyama et al., 2020). Interestingly, in their study, the same dose of (2*S*,6*S*)-HNK (20 mg/kg, i.p.) exhibited antidepressant effects, in contrast to the results reported by Zanos et al. (2016).

As described above, there have been discrepancies between the antidepressant effects of reported in studies conducted on (2*R*,6*R*)-HNK, although (2*R*,6*R*)-HNK may exert the antidepressant effects in rodents, at least at high exposure levels (pharmacokinetic (PK)/pharmacodynamics

(PD) relationship will be discussed in another section). There may be many reasons for the discrepancies, including differences in experimental conditions. However, (*R*,*S*)-ketamine or (*R*)-ketamine has been demonstrated to exhibit efficacy even in models in which (2*R*,6*R*)-HNK has no effects. The evidence clearly indicates that the antidepressant profile of (2*R*,6*R*)-HNK differs from those of (*R*,*S*)-ketamine and (*R*)-ketamine.

Identification of the regions in the brain responsible for the antidepressant-like effects of (2*R*,6*R*)-HNK and (*R*,*S*)-ketamine has been attempted. Intracerebroventricular (i.c.v.) injection of (2*R*,6*R*)-HNK was investigated in a mouse learned helplessness paradigm (Zanos et al., 2019). In helplessness experiments in rats, (2*R*,6*R*)-HNK (3 and 10nmol) reversed escape failures (1 day after injection), and (2*R*,6*R*)-HNK (1, 3, and 10nmol) reduced sucrose preference (2 days after injection). Notably, a 30nmol i.c.v. injection of (2*R*,6*R*)-HNK had much weaker antidepressant effects, indicating antidepressant effects of (2*R*,6*R*)-HNK describes an inverted U dose–response curve, the same as the effects of (*R*,*S*)-ketamine (Chowdhury et al., 2017). Fukumoto et al. (2019) observed that intra-mPFC injection of (2*R*,6*R*)-HNK (10ng/side) exerted antidepressant effects in a forced swimming test (1 day after administration) and in a novelty-suppressed feeding test (3 days after administration), without affecting locomotor activity. Similarly, Pham et al. (2018) found that intra-mPFC injection of (2*R*,6*R*)-HNK (1 nmol/side) reduced immobility in a forced swimming test (24h after injection) in naïve mice, the same as (*R*,*S*)-ketamine (1nmol/side) did. Chou et al. (2018) showed that intra-periaqueductal gray injection of (2*R*,6*R*)-HNK (1, 3, and 10ng) exerted dose-dependent antidepressant effects in both helplessness rats and naïve rats.

Hashimoto's group observed no antidepressant effects in a chronic social defeat stress model when (2*R*,6*R*)-HNK (20μg) was injected i.c.v., but rapid and sustained antidepressant effects were observed after i.c.v. injection of the same dose of (*R*)-ketamine (Zhang, Fujita, & Hashimoto, 2018).

It should be pointed out that i.c.v. injection or intra-mPFC injection of (*R*,*S*)-ketamine (Fukumoto, Iijima, & Chaki, 2016) or (*R*)-ketamine (Shirayama & Hashimoto, 2017) itself has been found to exert antidepressant effects in several animal models, and thus, it is likely that (*R*,*S*)-ketamine and (*R*)-ketamine themselves can elicit antidepressant effects without being converted to their metabolites.

The antidepressant-like effects of (2*R*,6*R*)-HNK in several animal models after systemic administration and local injection are summarized in Tables 1 and 2. Of note, Table 3 shows drugs or drug groups used in depression relative to ketamine.

## Mechanisms underlying the antidepressant effects of (2*R*,6*R*)-HNK

Blockade of NMDA receptors (presumably on GABA interneurons) has been proposed to be involved in the antidepressant effect of (*R*,*S*)-ketamine (Autry et al., 2011; Gerhard et al., 2020), but there are opposing opinions regarding this matter (Hashimoto, 2016). (2*R*,6*R*)-HNK has much lower affinity than (*R*,*S*)-ketamine for NMDA receptors. Indeed, Lumsden et al. (2019) reported that antidepressant-relevant dose of (2*R*,6*R*)-HNK is not sufficient inhibit NMDA-induced lethality, and that inhibitory effects of (2*R*,6*R*)-HNK on NMDA-induced responses in cells or tissues occurred at much higher concentrations than extracellular maximal concentration in the hippocampus (7.57μM) obtained by i.p. administration of 10mg/kg (2*R*,6*R*)-HNK (antidepressant-relevant dose). These observations clearly indicate that (2*R*,6*R*)-HNK does not functionally block NMDA receptors at the doses (or exposure levels) necessary to exert its antidepressant effects, and, thus, NMDA receptor blockade may not be responsible for the antidepressant effects of (2*R*,6*R*)-HNK. By contrast, Monteggia's group observed that (2*R*,6*R*)-HNK (50μM but not 10μM) significantly inhibited synaptic NMDA receptors in a similar manner to its parent compounds, and that it subsequently elicited the same signal transduction pathway (decrease in phosphorylation of eukaryotic elongation factor) associated with the NMDA receptor inhibition exerted by ketamine (Suzuki, Nosyreva, Hunt, Kavalali, & Monteggia, 2017). However, these effects occur at much higher concentration than at pharmacological doses, and the significance of NMDA receptor blockade in the actions of (2*R*,6*R*)-HNK is questionable.

(2*R*,6*R*)-HNK has been reported to exert its antidepressant effects via stimulation of AMPA receptor signaling, both by acute stimulation of the AMPA receptor and by an increase in AMPA receptor expression (Zanos et al., 2016). The antidepressant effects of (2*R*,6*R*)-HNK have been found to be attenuated by an AMPA receptor antagonist both before administration of (2*R*,6*R*)-HNK (approximately 24h prior to the forced swimming test) and 23.5h after (2*R*,6*R*)-HNK administration (0.5h prior to the forced swimming test). It was noteworthy that (2*R*,6*R*)-HNK has been shown to enhance excitatory synaptic transmission in the hippocampus through an increase in glutamate release in the absence of any direct actions on AMPA receptor function (Riggs et al., 2020), and that (2*R*,6*R*)-HNK, the same as (*R*,*S*)-ketamine, increased GluR1 and GluR2 in the hippocampus 24h after administration, but not 1h after administration (Zanos et al., 2016). Thus, the rapid action of (2*R*,6*R*)-HNK on the AMPA receptor may be exerted via an increase in glutamate release, while its sustained action (24h later) may be

**TABLE 1** **Effects of (2*R*,6*R*)-HNK in animal models (systemic administration).**

| Authors/ references | Model/test | Species/ strain | Dose (mg/ kg, i. p.) | Time after dosing | Results | Positive control |
|---|---|---|---|---|---|---|
| Zanos et al. (2016) | FST | CD-1 mouse | *1*, **5**, **25**, **75**, **125** | 1 and 24 h | Decrease immobility | (*R*,*S*)-ketamine 10 mg/kg |
| | LH | CD-1 mouse | *1*, **5**, **10**, **25**, **75** | 24 h | Decrease escape failure | (*R*,*S*)-ketamine 10 mg/kg |
| | NSFT | CD-1 mouse | *3*, **10** | 0.5 or 1 h | Decrease animals not eating | (*R*,*S*)-ketamine 10 mg/kg |
| | CORT/SPT, FUST | CD-1 mouse | **10** | 24 h | Reverse decreased sucrose preference (SPT) and female urine sniffing preference (FUST) | |
| | CSDS/SI | C57BL/ 6J mouse | **20** | 24 h | Reverse decreased social interaction | |
| Fukumoto et al. (2019) | FST | C57BL/ 6J mouse | *3*, **10**, **30** | 24 h | Decrease immobility | (*R*,*S*)-ketamine 10 mg/kg |
| | NSFT | C57BL/ 6J mouse | *3*, **10**, **30** | 72 h | Decrease latency to feed | (*R*,*S*)-ketamine 10 mg/kg |
| | FUST | C57BL/ 6J mouse | **30** | 24 h | Increase sniffing time | (*R*,*S*)-ketamine 10 mg/kg |
| | FUST | C57BL/ 6J mouse | *3*, *10*, *30* | 96 h | No effect | (*R*,*S*)-ketamine 10 mg/kg |
| Pham et al. (2018) | FST | BALB/cj mouse | **10** | 24 h | Decrease immobility | (*R*)-ketamine 10 mg/kg |
| Chou et al. (2018) | Modified LH/FST, SPT, LH | SD rat | **10** | 1 h–21 days | Reverse increased immobility (FST), decreased sucrose preference (SPT), and increased escape failure (LH) | |
| | Chronic restraint stress/FST, SPT | SD rat | **10** | | Reverse increased immobility (FST) and decreased sucrose preference (SPT) | |
| | Spinal nerve ligation/FST, SPT | SD rat | **10** | | Reverse increased immobility (FST) and decreased sucrose preference (SPT) | |
| Yang et al. (2017) | Lipopolysaccharide/ FST, TST | C57BL/6 mouse | *10* | 3 h, 1 day, 2 days | No effect | (*R*)-ketamine 10 mg/kg |
| | CSDS/TST, FST, SPT | C57BL/6 mouse | *10* | 3, 7 day | No effect | (*R*)-ketamine 10 mg/kg |

*Continued*

**TABLE 1 Effects of (2*R*,6*R*)-HNK in animal models (systemic administration)—cont'd**

| Authors/ references | Model/test | Species/ strain | Dose (mg/ kg, i. p.) | Time after dosing | Results | Positive control |
|---|---|---|---|---|---|---|
| Shirayama and Hashimoto (2018) | LH | SD rat | *20* | 1, 5 days | No effect | (*R*)-ketamine 10 mg/kg |
| Yokoyama et al. (2020) | CORT/FST,FET | C57BL/ 6J mouse | *10, 20* | 30 min (FST) 1 day (FET) | No effect | |

*Bold*, effective dose(s); *italic*, ineffective dose(s); *CORT*, repeated corticosterone treated model; *CSDS*, chronic social defeat stress model; *FET*, female encounter test; *FST*, forced swimming test; *FUST*, female urine sniffing test; *LH*, learned helplessness paradigm; *NSFT*, novelty-suppressed feeding test; *SL*, social interaction test; *SPT*, sucrose preference test; *TST*, tail suspension test.

**TABLE 2 Effects of (2*R*,6*R*)-HNK in animal models (local injection).**

| Authors/ references | Model/test | Species/ strain | Dose | Route | Time after dosing | Results |
|---|---|---|---|---|---|---|
| Zanos et al. (2019) | LH/SPT | CD-1 mouse | *1*, **3**, **10**, *30* nmol | i.c.v. | 24 h | Reverse increased immobility |
| | LH/SPT | CD-1 mouse | **1**, **3**, **10**, *30* nmol | i.c.v. | 48 h | Reverse decreased sucrose preference |
| | | | **3**, **10**, *30* nmol | i.c.v. | 5 days | Reverse decreased sucrose preference |
| Zhang et al. (2018) | CSDS/TST, FST, SPT | C57BL/6 mouse | *20* μg | i.c.v. | 4 h, 1, 2, 7 days | No effect |
| Fukumoto et al. (2019) | FST | C57BL/ 6J mouse | **10 ng/site** | intra-mPFC | 24 h | Decrease immobility |
| | NSFT | C57BL/ 6J mouse | **10 ng/site** | intra-mPFC | 72 h | Decrease latency to feed |
| Pham et al. (2018) | FST | BALB/cj mouse | **1 nmol/ site** | intra-mPFC | 24 h | Decrease immobility |
| Chou et al. (2018) | Modified LH/ FST, SPT | SD rat | *0.1, 0.3,* **1, 3, 10 ng** | intra-PAG | 1 h–21 days | Reverse increased immobility (FST) and decreased sucrose preference (SPT) |
| | Chronic restraint stress/ FST, SPT | SD rat | **1 ng** | intra-PAG | | Reverse increased immobility (FST) and decreased sucrose preference (SPT) |
| | Spinal nerve ligation/FST, SPT | SD rat | **1 ng** | intra-PAG | | Reverse increased immobility (FST) and decreased sucrose preference (SPT) |

*Bold*, effective dose(s); *italic*, ineffective dose(s); *CSDS*, chronic social defeat stress model; *FST*, forced swimming test; *LH*, learned helplessness paradigm; *mPFC*, medial prefrontal cortex; *NSFT*, novelty-suppressed feeding test; *PAG*, periaqueductal gray; *SPT*, sucrose preference test; *TST*, tail suspension test.

**TABLE 3 Drug groups used for depression.**

| Class/drug | Mode of action | Representative drug of the class | Remark |
|---|---|---|---|
| RIMA (reversible inhibitors of monoamine oxidase A) | Reversible inhibition of monoamine oxidase A | Moclobemide, etc. | |
| TCA (tricyclic antidepressant)/TeCA (tetracyclic antidepressant) | Inhibition of serotonin transporter and/or norepinephrine transporter | Imipramine, Mianserin, etc. | |
| SSRI (selective serotonin reuptake inhibitor) | Selective inhibition of serotonin transporter | Fluoxetine, Escitalopram, etc. | |
| SNRI (serotonin and norepinephrine reuptake inhibitor) | Inhibition of serotonin transporter and norepinephrine transporter | Venlafaxine, Duloxetine, etc. | |
| NDRI (norepinephrine and dopamine reuptake inhibitor) | Inhibition of norepinephrine transporter and dopamine transporter | Bupropion, etc. | |
| NaSSA (noradrenergic and specific serotonergic antidepressant) | Blockade of $\alpha_{2A}$, 5-$HT_{2A}$, 5-$HT_{2C}$, 5-$HT_3$, and $H_1$ receptors | Mirtazapine, etc. | |
| Vortioxetine | Inhibition of serotonin transporter and actions on 5-HT receptors | | |
| Agomelatine | Stimulation of $MT_{1/2}$ and blockade of 5-$HT_{2C}$ receptor | | |
| Brexanolone (allopregnanolone) | Positive allosteric modulation of $GABA_A$ receptor | | For postpartum depression |
| Esketamine | Blockade of NMDA receptor (?) | | For treatment-resistant depression |

mediated by an increase in AMPA receptor levels. On the other hand, since Pham et al. (2018) observed that (2*R*,6*R*)-HNK increased extracellular basal glutamate levels even at 24h after administration, direct stimulation of AMPA receptors by increased glutamate cannot be completely ruled out as the mechanism responsible for sustained antidepressant effect (effect at 24h after administration) of (2*R*,6*R*)-HNK. Moreover, Ho et al. (2018) found that (*R*,*S*)-ketamine, (2*R*,6*R*)-HNK, and (2*S*,6*S*)-HNK increased the level of AMPA receptor subunit expression in U2510MG cells in an estrogen-receptorα (ERα)-dependent manner at concentrations of 200–400nM. However, since (2*R*,6*R*)-HNK and (2*S*,6*S*)-HNK exhibited the same affinity for ERα, the significance of this mechanism in the regulation of AMPA receptor transmission needs to be further clarified.

A role of brain-derived neurotrophic factor (BDNF) signaling in the antidepressant effect of (2*R*,6*R*)-HNK has been suggested. Fukumoto et al. (2019) clearly demonstrated the involvement of BDNF signaling by showing that the antidepressant effect of (2*R*,6*R*)-HNK is attenuated in mice with a knock-in of the BDNF Val66Met allele (which blocks the processing and activity-dependent release of mature BDNF) and by intra-mPFC microinjection of an anti-BDNF neutralizing antibody or a TrkB inhibitor. In addition, (2*R*,6*R*)-HNK (0.5–10nM) was found to increase BDNF release through activation of the AMPA receptor in rat primary cortical cultures. Fukumoto et al. (2019) further demonstrated that (2*R*,6*R*)-HNK activates the mechanistic target of rapamycin complex 1 (mTORC1) signaling in the mPFC and that the mTORC1 signaling in the mPFC is involved in the antidepressant effect of (2*R*,6*R*)-HNK, presumably by increasing synaptic function and spine density. However, the effect of (2*R*,6*R*)-HNK on BDNF/TrkB/mTORC1 signaling was tested at a dose of 30mg/kg, i.p., and its effect at lower doses has not been investigated. Role of BDNF in the hippocampus has also been emphasized. Zanos et al. (2016) showed that (2*R*,6*R*)-HNK (10mg/kg, i.p.) increased mature BDNF levels in the hippocampus but not in the PFC, indicating that the BDNF in the hippocampus plays a more important role than the BDNF in the PFC. In terms of the role of downstream of BDNF–TrkB signaling, Zanos et al. (2016) did not observe any increase in mTORC1 phosphorylation in either the hippocampus or PFC in response to (2*R*,6*R*)-HNK. Consistent with this finding, Hashimoto's group reported observing that (*R*)-ketamine did not stimulate mTORC1 signaling, but (*S*)-ketamine did (Yang et al., 2018). Thus, the role of mTORC1 signaling in the actions of (2*R*,6*R*)-HNK requires further investigation.

## Role of (2*R*,6*R*)-HNK in the antidepressant effects of (*R*)-ketamine

As described above, there are several lines of evidence for the exertion of antidepressant effects by (2*R*,6*R*)-HNK in rodent models. There are, however, some reports to oppose these findings. In this section, the following two questions will be discussed in a context of exposure levels across the studies.

### Does (2*R*,6*R*)-HNK have antidepressant potential?

After Zanos et al. (2016) first reported the antidepressant-like effects of (2*R*,6*R*)-HNK in rodent models, its antidepressant effects were confirmed by several other research groups (Tables 1 and 2). As a whole, the research group has reported finding that (2*R*,6*R*)-HNK exhibits efficacy at doses of 10 mg/kg, i.p. or more, but at a dose of 3 mg/kg, i.p. (2*R*,6*R*)-HNK did not exert antidepressant effects in the models tested. By contrast, no antidepressant effects of (2*R*,6*R*)-HNK were observed in some rat and mouse models at 10 mg/kg or 20 mg/kg, i.p., however, (*R*)-ketamine consistently exhibited the antidepressant effects in animal models at 10 mg/kg, i.p. The precise reason for this discrepancy is unknown, but (2*R*,6*R*)-HNK is likely to have antidepressant potential, at least in animal models. However, the matters described below need to be considered.

First, according to our PK analyses, i.p. administration of (2*R*,6*R*)-HNK yields (2*R*,6*R*)-HNK concentrations in both plasma and cerebrospinal fluid (CSF) that are approximately 10 times higher than i.p. administration of the same dose of (*R*)-ketamine (Yamaguchi et al., 2018). If (2*R*,6*R*)-HNK is an active substance of (*R*,*S*)-ketamine, it should exert effects at doses as low as 1 mg/kg, i.p. Since (2*R*,6*R*)-HNK did not exhibit antidepressant effects at 3 mg/kg, i.p. (2*R*,6*R*)-HNK contributes little to the antidepressant effects of (*R*,*S*)-ketamine or (*R*)-ketamine. Second, (2*R*,6*R*)-HNK exerted antidepressant effects at a wide range of doses (5–125 mg/kg or 5–75 mg/kg, i.p.) in mice (Table 1), which was in stark contrast to the antidepressant effects of (*R*,*S*)-ketamine whose effective dose range is rather narrow (Chowdhury et al., 2017). Thus, only the 30 mg/kg dose of (*R*,*S*)-ketamine exerted antidepressant effects in the rat forced swimming test, and the other doses used, 3 and 80 mg/kg, did not exert significant effects. On the other hand, the range of effective doses administered by i.c.v. injection is rather limited (Table 2). Although the linearity of plasma exposure according to the dose tested needs to be confirmed, and differences in test conditions used need to be taken into account, some data regarding the antidepressant effects of (2*R*,6*R*)-HNK have been inconsistent. Moreover, even the doses administered by intra-mPFC injection have differed considerably from study to study (Fukumoto et al., 2019; Pham et al., 2018). Thus, while several findings regarding the antidepressant effects of (2*R*,6*R*)-HNK have been reported by different laboratories, the data obtained in regard to the antidepressant effects of (2*R*,6*R*)-HNK need to be carefully reviewed.

### Is formation of (2*R*,6*R*)-HNK essential for (*R*)-ketamine to exert its antidepressant effects?

Hashimoto's group found that i.c.v. injection of (2*R*,6*R*)-HNK (20 μg, 83 nmol) did not exert antidepressant effects in the chronic social defeat stress model, but that i.c.v. injection of the same dose of (*R*)-ketamine exerted antidepressant effects (Zhang et al., 2018). By contrast, Zanos et al. (2019) observed that i.c.v. injection of lower doses (1–10 nmol) of (2*R*,6*R*)-HNK exerted antidepressant effects in animal models, but that no antidepressant effects were observed at a higher dose (30 nmol), suggesting that the dose Hashimoto used was too high to exert antidepressant effects. Hashimoto also detected a small amount of (2*R*,6*R*)-HNK in the brain after i.c.v. injection of (*R*)-ketamine [approximately 1/6 after i.c.v. injection of (2*R*,6*R*)-HNK], meaning that a role of (2*R*,6*R*)-HNK in the antidepressant effects of i.c.v. injected (*R*)-ketamine cannot be fully ruled out. Fukumoto et al. (2019) used a 10 ng (0.042 nmol)/side microinjection (intra-mPFC), while Pham et al. (2018) used a 1 nmol/side dose, and both groups observed antidepressant effects. As described above, it needs to be carefully analyzed whether there is not a discrepancy that both doses (approximately 24 times differences) can be effective despite inverted U-shaped nature of the effects of (2*R*,6*R*)-HNK. However, the results of brain injection studies themselves may not clarify whether (2*R*,6*R*)-HNK is essential for (*R*)-ketamine to exert antidepressant effects.

Blockade of the metabolism of (*R*)-ketamine to (2*R*,6*R*)-HNK would provide a clear answer to the above question. To do so, both Hashimoto's group (Zhang et al., 2018) and Gould's group (Zanos et al., 2019) synthesized deuterium substitution at the C6 position of (*R*)-ketamine to prevent its metabolism to (2*R*,6*R*)-HNK. The deuterium substitution partly prevented its metabolism to (2*R*,6*R*)-HNK in both studies, and the brain concentration of (2*R*,6*R*)-HNK after (*R*)-$d_2$-ketamine administration was more than 4 times lower than after (*R*)-ketamine administration. Moreover, the 10 mg/kg dose of (*R*)-$d_2$-ketamine in both studies exerted practically the same antidepressant effects as (*R*)-ketamine. Zanos et al. (2019) also evaluated the effect of lower doses (2.5, 5 mg/kg), and the relationship between PK and PD assessed on the basis of their data is shown in Fig. 2. At 5 mg/kg, (*R*)-ketamine was effective, but (*R*)-$d_2$-ketamine was ineffective. In this case, similar exposures to the unchanged drug are predicted, but (2*R*,6*R*)-HNK

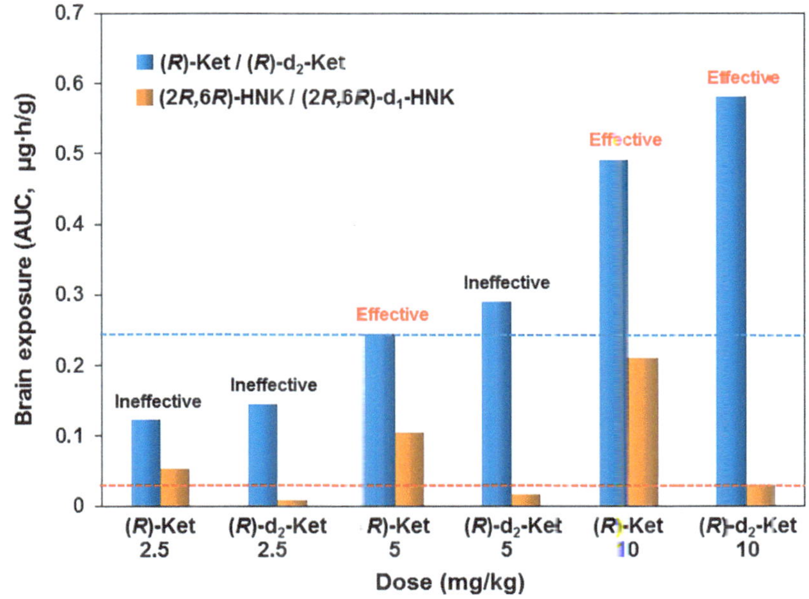


**FIG. 2** Relationship between antidepressant effects and brain exposures of (*R*)-ketamine (Ket) or (2*R*,6*R*)-HNK. This figure was constructed by referring the data from a comparative study on antidepressant effects after intraperitoneal administration of (*R*)-Ket or its C6-deuterium labeled compound ((*R*)-$d_2$-Ket) to mice reported by Zanos et al. (2019). The brain exposures postdoses of 2.5 and 5 mg/kg were calculated from the measured values postdose of 10 mg/kg. The blue dash line is minimum exposure of (*R*)-Ket/(*R*)-$d_2$-Ket at the effective doses. The red dash line is minimum exposure of (2*R*,6*R*)-HNK/(2*R*,6*R*)-$d_1$-HNK at effective doses.

exposures postdose of (*R*)-ketamine are predicted to be higher than after (*R*)-$d_2$-ketamine. Based on these findings, Zanos et al. (2019) concluded that metabolism to (2*R*,6*R*)-HNK increases the potency of (*R*)-ketamine to exert antidepressant actions in mice. However, it should be noted that (*R*)-ketamine did not exert any effects at the 2.5 mg/kg dose. In this case, a higher (2*R*,6*R*)-HNK exposure than those observed posteffective dose of (*R*)-$d_2$-ketamine is predicted. This indicates that the metabolism to (2*R*,6*R*)-HNK is not essential for exertion of the antidepressant effects of (*R*)-ketamine.

To clarify the role of metabolism to (2*R*,6*R*)-HNK in the antidepressant-like effects of (*R*)-ketamine, we established conditions in which metabolism to (2*R*,6*R*)-HNK was totally blocked (Yamaguchi et al., 2018). Interestingly, the concentration of (*R*)-ketamine increased due to inhibition of its metabolism. Under these conditions, (*R*)-ketamine exerted antidepressant effects that were dependent on the concentration of (*R*)-ketamine in the brain and CSF. This finding clearly shows that (*R*)-ketamine itself exerts antidepressant effects, and that metabolism to (2*R*,6*R*)-HNK is unnecessary for (*R*)-ketamine to exert them. This conclusion is strongly supported by results showing that (2*R*,6*R*)-HNK does not exert antidepressant effects at exposure levels predicted by the PK/PD relationship of (*R*)-ketamine, as described above. Nonetheless, the possibility that (2*R*,6*R*)-HNK potentiates the effects of (*R*)-ketamine cannot be completely ruled out. However, it may be impossible to explain simply based on the level of exposure to (2*R*,6*R*)-HNK, and how (2*R*,6*R*)-HNK is able to potentiate the antidepressant effects of (*R*)-ketamine needs to be elucidated.

## Conclusion

The proposal that formation of (2*R*,6*R*)-HNK is necessary for (*R*,*S*)-ketamine (or (*R*)-ketamine) to exert its antidepressant-like effects is interesting because (2*R*,6*R*)-HNK can be developed as an alternative to ketamine without ketamine-like side effects. However, PK/PD analyses of both (*R*)-ketamine and (2*R*,6*R*)-HNK, as well as the results of a study in which the metabolism of (*R*)-ketamine to (2*R*,6*R*)-HNK was completely blocked clearly indicate that metabolism to (2*R*,6*R*)-HNK contributes very little to the antidepressant-like effects of (*R*)-ketamine.

However, the antidepressant-like effects of (2*R*,6*R*)-HNK itself should be discussed separately from the role of (2*R*,6*R*)-HNK in the antidepressant-like effects of (*R*)-ketamine. Based on the results reported to date, it is likely that (2*R*,6*R*)-HNK exerts antidepressant effects, but that its effects occur at a much higher exposure level than produced by the antidepressant doses of (*R*)-ketamine. Moreover, it is premature to conclude that (2*R*,6*R*)-HNK exerts ketamine-like antidepressant effects in rodent models; rapid acting and efficacy in treatment-resistant models. Chou et al. (2018) observed that (2*R*,6*R*)-HNK exerted antidepressant effects in a modified learned helplessness paradigm, while Yokoyama et al. (2020) reported observing no effects of (2*R*,6*R*)-HNK in a chronic corticosterone-repeated model, both of which are regarded as treatment-resistant models. Therefore, it has not been consistently demonstrated that (2*R*,6*R*)-HNK exerts ketamine-like antidepressant effects in rodents.

Collectively, the antidepressant effects of (2*R*,6*R*)-HNK reported to date are insufficient to predict the ability of

(2*R*,6*R*)-HNK to serve as an alternative to ketamine, and it is necessary to conduct additional, more detailed experiments on the antidepressant profile of (2*R*,6*R*)-HNK in comparison with that of (*R*,*S*)-ketamine or (*R*)-ketamine. Of course, clinical trials will be necessary to provide a clearer picture of the antidepressant profiles of (2*R*,6*R*)-HNK. However, it is important to note that the exposure levels at which (2*R*,6*R*)-HNK exerts antidepressant effects should be taken into account when judging whether metabolism to (2*R*,6*R*)-HNK is essential for (*R*,*S*)-ketamine to exert its antidepressant effects. Because (2*R*,6*R*)-HNK may not have ketamine-like side effects, it has the potential to become a very useful antidepressant. Therefore, studies of (2*R*,6*R*)-HNK should be carefully evaluated.

## Key facts of major depressive disorder

- MDD is among the most prevalent forms of mental disorders, which affect more than 300 million people worldwide.
- There are significant amount of patients who do not respond to current medications.
- There are unmet medical needs for current medications in terms of onset of action and efficacy.
- Ketamine has been a breakthrough in depression therapy in over 60 years.
- Elucidation of underling mechanisms of ketamine is a key to develop novel antidepressants.

## Summary points

- This chapter focuses on antidepressant potential of (2*R*,6*R*)-HNK.
- Among enantiomers of racemic ketamine, (*R*)-ketamine may have more potent antidepressant effects than (*S*)-ketamine.
- (2*R*,6*R*)-HNK is one of the major metabolites formed from (*R*)-ketamine.
- The antidepressant effects of (2*R*,6*R*)-HNK have been reported in rodents.
- However, some data show lack of the antidepressant effects of (2*R*,6*R*)-HNK.
- Therefore, in this chapter, antidepressant potential of (2*R*,6*R*)-HNK is reviewed from data reported to date, and its potential as a novel antidepressant is discussed.

## Mini-directory of terms

**Major depressive disorder** A mental disorder characterized by depressive mood and anhedonia.

**Ketamine** A dissociative anesthetic which can also treat pain and depression at lower dose.

**Drug metabolism** A process through which a drug administered modified its structure mainly by liver enzymes.

## References

Autry, A. E., Adachi, M., Nosyreva, E., Na, E. S., Los, M. F., Cheng, P. F., et al. (2011). NMDA receptor blockade at rest triggers rapid behavioural antidepressant responses. *Nature*, *475*, 91–95.

Berman, R. M., Cappiello, A., Anand, A., Oren, D. A., Heninger, G. R., Charney, D. S., et al. (2000). Antidepressant effects of ketamine in depressed patients. *Biological Psychiatry*, *47*, 351–354.

Chang, T., Savory, A., Albin, M., Goulet, R., & Glazko, A. J. (1970). Metabolic disposition of tritium-labelled ketamine (Ketalar®: CI-581) in normal human subjects. *Clinical Research*, *18*, 597–601.

Chou, D., Peng, H. Y., Lin, T. B., Lai, C. Y., Hsieh, M. C., Wen, Y. C., et al. (2018). (2R,6R)-Hydroxynorketamine rescues chronic stress-induced depression-like behavior through its actions in the midbrain periaqueductal gray. *Neuropharmacology*, *139*, 1–12.

Chowdhury, G. M., Zhang, J., Thomas, M., Banasr, M., Ma, X., Pittman, B., et al. (2017). Transiently increased glutamate cycling in rat PFC is associated with rapid onset of antidepressant-like effects. *Molecular Psychiatry*, *22*, 120–126.

Clements, J. A., & Nimmo, W. S. (1981). Pharmacokinetics and analgesic effect of ketamine in man. *British Journal of Anaesthesia*, *53*, 27–30.

Clements, J. A., Nimmo, W. S., & Grant, I. S. (1982). Bioavailability, pharmacokinetics, and analgesic activity of ketamine in humans. *Journal of Pharmaceutical Sciences*, *71*, 539–542.

Dayton, P. G., Stiller, R. L., Cook, D. R., & Perel, J. M. (1983). The binding of ketamine to plasma proteins: Emphasis on human plasma. *European Journal of Clinical Pharmacology*, *24*, 825–831.

Desta, Z., Moaddel, R., Ogburn, E. T., Xu, C., Ramamoorthy, A., Venkata, S. L., et al. (2012). Stereoselective and regiospecific hydroxylation of ketamine and norketamine. *Xenobiotica*, *42*, 1076–1087.

Fukumoto, K., Fogaça, M. V., Liu, R. J., Duman, C., Kato, T., Li, X. Y., et al. (2019). Activity-dependent brain-derived neurotrophic factor signaling is required for the antidepressant actions of (2R,6R)-hydroxynorketamine. *Proceedings of the National Academy of Sciences of the United States of America*, *116*, 297–302.

Fukumoto, K., Iijima, M., & Chaki, S. (2016). The antidepressant effects of an mGlu2/3 receptor antagonist and ketamine require AMPA receptor stimulation in the mPFC and subsequent activation of the 5-HT neurons in the DRN. *Neuropsychopharmacology*, *41*, 1046–1056.

Fukumoto, K., Toki, H., Iijima, M., Hashihayata, T., Yamaguchi, J. I., Hashimoto, K., et al. (2017). Antidepressant potential of (R)-ketamine in rodent models: Comparison with (S)-ketamine. *The Journal of Pharmacology and Experimental Therapeutics*, *361*, 9–16.

Gerhard, D. M., Pothula, S., Liu, R. J., Wu, M., Li, X. Y., Girgenti, M. J., et al. (2020). GABA interneurons are the cellular trigger for ketamine's rapid antidepressant actions. *Journal of Clinical Investigation*. https://doi.org/10.1172/JCI130808.

Hashimoto, K. (2016). Ketamine's antidepressant action: Beyond NMDA receptor inhibition. *Expert Opinion on Therapeutic Targets*, *20*, 1389–1392.

Ho, M. F., Correia, C., Ingle, J. N., Kaddurah-Daouk, R., Wang, L., Kaufmann, S. H., et al. (2018). Ketamine and ketamine metabolites as novel estrogen receptor ligands: Induction of cytochrome P450 and AMPA glutamate receptor gene expression. *Biochemical Pharmacology*, *152*, 279–292.

Leal, G. C., Bandeira, I. D., Correia-Melo, F. S., Telles, M., Mello, R. P., Vieira, F., et al. (2020). Intravenous arketamine for treatment-resistant depression: Open-label pilot study. *European Archives of Psychiatry and Clinical Neuroscience*. https://doi.org/10.1007/s00406-020-01110-5.

Lumsden, E. W., Troppoli, T. A., Myers, S. J., Zanos, P., Aracava, Y., Kehr, J., et al. (2019). Antidepressant-relevant concentrations of the ketamine metabolite (2R,6R)-hydroxynorketamine do not block NMDA receptor function. *Proceedings of the National Academy of Sciences of the United States of America*, *116*, 5160–5169.

Moaddel, R., Sanghvi, M., Ramamoorthy, A., Jozwiak, K., Singh, N., Green, C., et al. (2016). Subchronic administration of (R,S)-ketamine induces ketamine ring hydroxylation in Wistar rats. *Journal of Pharmaceutical and Biomedical Analysis*, *127*, 3–8.

Moaddel, R., Venkata, S. L., Tanga, M. J., Bupp, J. E., Green, C. E., Iyer, L., et al. (2010). A parallel chiral–achiral liquid chromatographic method for the determination of the stereoisomers of ketamine and ketamine metabolites in the plasma and urine of patients with complex regional pain syndrome. *Talanta*, *82*, 1892–1904.

Pham, T. H., Defaix, C., Xu, X., Deng, S. X., Fabresse, N., Alvarez, J. C., et al. (2018). Common neurotransmission recruited in (R,S)-ketamine and (2R,6R)-hydroxynorketamine-induced sustained antidepressant-like effects. *Biological Psychiatry*, *84*, e3–e6.

Portmann, S., Kwan, H. Y., Theurillat, R., Schmitz, A., Mevissen, M., & Thormann, W. (2010). Enantioselective capillary electrophoresis for identification and characterization of human cytochrome P450 enzymes which metabolize ketamine and norketamine in vitro. *Journal of Chromatography A*, *1217*, 7942–7948.

Riggs, L. M., Aracava, Y., Zanos, P., Fischell, J., Albuquerque, E. X., Pereira, E. F. R., et al. (2020). (2R,6R)-Hydroxynorketamine rapidly potentiates hippocampal glutamatergic transmission through a synapse-specific presynaptic mechanism. *Neuropsychopharmacology*, *45*, 426–436.

Shirayama, Y., & Hashimoto, K. (2017). Effects of a single bilateral infusion of R-ketamine in the rat brain regions of a learned helplessness model of depression. *European Archives of Psychiatry and Clinical Neuroscience*, *267*, 177–182.

Shirayama, Y., & Hashimoto, K. (2018). Lack of antidepressant effects of (2R,6R)-hydroxynorketamine in a rat learned helplessness model: Comparison with (R)-ketamine. *International Journal of Neuropsychopharmacology*, *21*, 84–88.

Suzuki, K., Nosyreva, E., Hunt, K. W., Kavalali, E. T., & Monteggia, L. M. (2017). Effects of a ketamine metabolite on synaptic NMDAR function. *Nature*, *546*, E1–E3.

Turfus, S. C., Parkin, M. C., Cowan, D. A., Halket, J. M., Smith, N. W., Braithwaite, R. A., et al. (2009). Use of human microsomes and deuterated substrates: An alternative approach for the identification of novel metabolites of ketamine by mass spectrometry. *Drug Metabolism & Disposition*, *37*, 1769–1778.

Yamaguchi, J. I., Toki, H., Qu, Y., Yang, C., Koike, H., Hashimoto, K., et al. (2018). (2R,6R)-Hydroxynorketamine is not essential for the antidepressant actions of (R)-ketamine in mice. *Neuropsychopharmacology*, *43*, 1900–1907.

Yang, C., Qu, Y., Abe, M., Nozawa, D., Chaki, S., & Hashimoto, K. (2017). (R)-Ketamine shows greater potency and longer lasting antidepressant effects than its metabolite (2R,6R)-hydroxynorketamine. *Biological Psychiatry*, *82*, e43–e44.

Yang, C., Ren, Q., Qu, Y., Zhang, J. C., Ma, M., Dong, C., et al. (2018). Mechanistic target of rapamycin-independent antidepressant effects of (R)-ketamine in a social defeat stress model. *Biological Psychiatry*, *83*, 18–28.

Yang, C., Shirayama, Y., Zhang, J. C., Ren, Q., Yao, W., Ma, M., et al. (2015). R-ketamine: A rapid-onset and sustained antidepressant without psychotomimetic side effects. *Translational Psychiatry*, *5*, e632.

Yokoyama, R., Higuchi, M., Tanabe, W., Tsukada, S., Naito, M., Yamaguchi, T., et al. (2020). (S)-Norketamine and (2S,6S)-hydroxynorketamine exert potent antidepressant-like effects in a chronic corticosterone-induced mouse model of depression. *Pharmacology, Biochemistry & Behavior*, *191*, 172876.

Zanos, P., Highland, J. N., Liu, X., Troppoli, T. A., Georgiou, P., Lovett, J., et al. (2019). (R)-Ketamine exerts antidepressant actions partly via conversion to (2R,6R)-hydroxynorketamine, while causing adverse effects at sub-anaesthetic doses. *British Journal of Pharmacology*, *176*, 2573–2592.

Zanos, P., Moaddel, R., Morris, P. J., Georgiou, P., Fischell, J., Elmer, G. I., et al. (2016). NMDAR inhibition-independent antidepressant actions of ketamine metabolites. *Nature*, *533*, 481–486.

Zarate, C. A., Jr., Brutsche, N., Laje, G., Luckenbaugh, D. A., Venkata, S. L., Ramamoorthy, A., et al. (2012). Relationship of ketamine's plasma metabolites with response, diagnosis, and side effects in major depression. *Biological Psychiatry*, *72*, 331–338.

Zarate, C. A., Jr., Singh, J. B., Carlson, P. J., Brutsche, N. E., Ameli, R., Luckenbaugh, D. A., et al. (2006). A randomized trial of an N-methyl-D-aspartate antagonist in treatment-resistant major depression. *Archives of General Psychiatry*, *63*, 856–864.

Zhang, K., Fujita, Y., & Hashimoto, K. (2018). Lack of metabolism in (R)-ketamine's antidepressant actions in a chronic social defeat stress model. *Scientific Reports*, *8*, 4007.

Zhao, X., Venkata, S. L., Moaddel, R., Luckenbaugh, D. A., Brutsche, N. E., Ibrahim, L., et al. (2012). Simultaneous population pharmacokinetic modelling of ketamine and three major metabolites in patients with treatment-resistant bipolar depression. *British Journal of Clinical Pharmacology*, *74*, 304–314.

Chapter 38

# Linking 5-hydroxytryptamine to antidepressant actions of (*R*)-ketamine and social stress model

**Kai Zhang**[a] **and Kenji Hashimoto**[b]

[a]*Department of Psychiatry, Chaohu Hospital, Anhui Medical University, Hefei, China;* [b]*Division of Clinical Neuroscience, Chiba University Center for Forensic Mental Health, Chiba, Japan*

## List of abbreviations

| | |
|---|---|
| **5-HIAA** | 5-hydroxyindole acetic acid |
| **5-HT** | 5-hydroxytryptamine |
| **5HTP** | 5-hydroxytryptophan |
| **CNS** | central nervous system |
| **CSDS** | chronic social defeat stress |
| **CUMS** | chronic unpredictable mild stress |
| **FDA** | Food and Drug Administration |
| **FSL** | Flinders sensitive line |
| **FST** | forced swimming test |
| **LH** | learned helplessness |
| **LPS** | lipopolysaccharide |
| **MAO** | monoamine oxidase |
| **MDD** | major depressive disorder |
| **NMDAR** | *N*-methyl-D-aspartate receptor |
| **PCP** | phencyclidine |
| **PCPA** | *para*-chlorophenylalanine methyl ester hydrochloride |
| **PFC** | prefrontal cortex |
| **PPI** | prepulse inhibition |
| **SSRIs** | selective serotonin reuptake inhibitors |
| **TRD** | treatment-resistant depression |

## Introduction

### 5-Hydroxytryptamine, synthesis, and metabolism

5-Hydroxytryptamine (5-HT), also named serotonin, is a biogenic amine that is used as a neurotransmitter in the central nervous system (CNS). In 1948, Maurice Rapport and colleagues isolated and identified crystalline 5-HT from beef serum (Rapport, Green, & Page, 1948).

5-HT is processed through a two-step mechanism: hydroxylation and decarboxylation (Clark, Weissbach, & Udenfriend, 1954). First, tryptophan hydroxylase hydroxylates the essential amino acid tryptophan to 5-hydroxytryptophan (5-HTP). And then, 5-HT is decarboxylated from 5-HTP. In the presence of tryptophan, the hydroxylation of tryptophan and the decarboxylation of 5-HTP occur almost at the same time. 5-HT is synthesized and stored in presynaptic neurons in the CNS, such as serotonergic and catecholaminergic neurons (Darhstrom, 1964).

Similar to other biogenic amines like norepinephrine, dopamine, and histamine, metabolism by monoamine oxidase (MAO) is the primary metabolic pathway for serotonin (McIsaac & Page, 1959). MAO is a ubiquitous enzyme that exists mainly in the form of MAO-A and MAO-B. While MAO-A is responsible for the majority of 5-HT metabolism, only a small portion of 5-HT is metabolized by MAO-B. McIsaac and Page (1959) proved that 5-hydroxyindoleacetic acid (5-HIAA) is the main metabolite produced by the MAO metabolism of 5-HT. Their results also showed that excretion of 5-HIAA occurred mostly within 24 h after the action of exogenous 5-HT

## 5-HT processing

In the CNS, 5-HT is processed in many ways. 5-HT is released into the synaptic cleft after neuronal depolarization. On the presynaptic membrane, serotonin binds to postsynaptic 5-HT receptors or serotonin autoreceptors (Cerrito & Raiteri, 1979). The binding of 5-HT to its autoreceptor prevents the further release of 5-HT into the synaptic cleft and acts as a negative feedback mechanism. The highly selective serotonin transporter, located on the presynaptic membrane, is responsible for clearing serotonin from the synaptic cleft. 5-HT is recovered into presynaptic vesicles after transport to presynaptic neurons. Presynaptic vesicles protect 5-HT from metabolism.

Selective serotonin reuptake inhibitors (SSRIs) have an antidepressant effect through inhibition of the reuptake of

The Neuroscience of Depression. https://doi.org/10.1016/B978-0-12-817933-8.00056-6

5-HT, which improves depressive symptoms of patients with major depressive disorder (MDD). There are several obvious deficiencies of SSRIs (Baker, Tweedie, Duval, et al., 2003). First, the effective rate of SSRIs treatment is low. Only 33% of the patients were affected after a sufficient course of treatment with SSRIs. The second deficiency is the time lag between the initiation of the drug and the antidepressant effect: it usually takes weeks to months to have an antidepressant effect. Consequently, there is an urgent need to develop antidepressants that act quickly and can effectively improve patients' emotional symptoms and promote their social function. New antidepressants are required to meet these unmet needs. Since 2000, because of its robust antidepressant effects, (*R,S*)-ketamine has received extensive attention in the field of psychiatry worldwide (Hashimoto, 2019; Wei, Chang, & Hashimoto, 2020; Yang, Yang, Luo, & Hashimoto, 2019; Zhang & Hashimoto, 2019).

## (*R,S*)-Ketamine

### Brief history of (*R,S*)-ketamine

In 1956, Parker Davis Pharmaceutical Company first synthesized the general anesthetic phencyclidine (PCP) (Maddox, Godefroi, & Parcell, 1965). Pharmacological studies showed that PCP exerts its anesthetic effect by antagonizing the *N*-methyl-D-aspartate receptor (NMDAR) (Domino & Luby, 2012; Mion, 2017). However, subsequent clinical studies have shown that PCP also caused many severe side effects, such as abuse, hallucinations, and insanity (Domino & Luby, 2012). In 1965, PCP clinical used was ceased (Morris & Wallach, 2014). Therefore, it is necessary to find an anesthetic drug substitute for PCP without the serious psychotomimetic side effects (Mion, 2017). And then, the original form of (*R,S*)-ketamine (CI-581) was synthesized by Parker Davis Pharmaceutical Company (Domino, 2010). In 1964, Dr. Edward Domino and colleagues performed the first pharmacological study of (*R,S*)-ketamine in healthy people (Domino, 2010). In 1970, the US Food and Drug Administration (FDA) approved (*R,S*)-ketamine as an anesthetic drug (Mion, 2017). However, in subsequent decades, ketamine became a popular recreational drug worldwide due to its euphoric effects (Li & Vlisides, 2016; Wei et al., 2020).

### Antidepressant-like effect of (*R,S*)-ketamine in rodents

In 1975, Sofia and Harakal were the first to report that (*R,S*)-ketamine showed antidepressant-like effects in rodents (Sofia & Harakal, 1975). For example, (*R,S*)-ketamine enhanced yohimbine toxicity, reversed tetrabenazine-induced ptosis, and inhibited oxotremorine-induced tremors in a mouse model, and also reversed reserpine-induced hypothermia in a rat model. However, the antidepressant-like activity of (*R,S*)-ketamine was substantially lower than that of imipramine in these models (Sofia & Harakal, 1975; Wei et al., 2020).

(*R,S*)-ketamine also exerted rapid and sustained antidepressant-like effects in rat learned helplessness (LH), lipopolysaccharide (LPS), and chronic unpredictable mild stress (CUMS)-induced depression model (Koike, Iijima, & Chaki, 2011; Ma, Dang, Jia, et al., 2013; Yamaguchi, Toki, Qu, et al., 2018). In addition, comparable to tropomyosin receptor kinase B (TrkB) agonists and antagonists (7,8-dihydroxyflavone and ANA-12, respectively), (*R,S*)-ketamine showed rapid-acting and long-lasting antidepressant-like effects in mice after chronic social defeat stress (CSDS) (Zhang, Yao, Dong, et al., 2015).

Unfortunately, the potential psychotomimetic and other unwanted side effects of (*R,S*)-ketamine have been well documented in rodents and monkeys. These side effects include hyperlocomotion, schizophrenia-like psychotic symptoms, cognitive deficits, conditioned place preference, and prepulse inhibition (PPI) deficits (Chan, Lee, Siu, et al., 2013; Chang, Zhang, Pu, et al., 2019; Giorgetti, Marcotulli, Tagliabracci, et al., 2015; Tan & Hashimoto, 2020; Yang, Shirayama, Zhang, et al., 2015).

### Antidepressant effect of (*R,S*)-ketamine in humans

In 2000, Berman, Cappiello, Anand, et al. (2000) reported the antidepressant effects of (*R,S*)-ketamine in patients with MDD. To our knowledge, this was the first double-blind, placebo-controlled study of (*R,S*)-ketamine in patients with MDD. Subsequently, the antidepressant effect of (*R,S*)-ketamine was replicated in patients with treatment-resistant depression (TRD) (Zarate, Singh, Carlson, et al., 2006). Interestingly, (*R,S*)-ketamine is effective for the treatment of bipolar depression and suicidal ideation (Grunebaum, Galfalvy, Choo, et al., 2018; Wilkinson, Ballard, Bloch, et al., 2018). These results supported the validity of single-dose therapy at either 0.5 or 1.0 mg/kg in patients with MDD. Repeated infusion maintains the curative effect of (*R,S*)-ketamine in patients with MDD (Phillips, Norris, Talbot, et al., 2019; Strong & Kabbaj, 2018). In a double-blind, randomized study, Singh, Fedgchin, Daly, et al. (2016) reported that twice or thrice per week infusion of (*R,S*)-ketamine (0.5 mg/kg) was sufficient to maintain antidepressant efficacy in patients with TRD over 15 days. However, there is a lack of data on the long-term safety and efficacy of repeated administration of (*R,S*)-ketamine for patients with TRD. This is important information that is missing given the potential for severe side effects associated with (*R,S*)-ketamine use (Hashimoto, 2019; Zhang & Hashimoto, 2019).

## (*R*)-Ketamine

(*R*,*S*)-Ketamine is a racemic mixture that consists of two optical enantiomers, (*S*)- and (*R*)-ketamine (Domino, 2010). The affinity of (*S*)-ketamine at the NMDAR is threefold to fourfold greater than that of (*R*)-ketamine (Ebert, Mikkelsen, Thorkildsen, et al., 1997). Accordingly, (*S*)-ketamine showed approximately threefold more anesthetic potency than (*R*)-ketamine, and a twofold greater potency than (*R*,*S*)-ketamine in the clinical setting.

In a neonatal dexamethasone-treated model of depression, our group reported that (*R*)-ketamine had a greater potency and longer-lasting antidepressant effect than (*S*)-ketamine (Zhang, Li, & Hashimoto, 2014). As far as we know, this is the first study to directly compare the antidepressant effects of (*R*)-ketamine and (*S*)-ketamine. In addition, in 2015, our group confirmed that the antidepressant effect of (*R*)-ketamine was greater than that of (*S*)-ketamine in CSDS and LH models of depression (Yang et al., 2015). Subsequently, the superiority of (*R*)-ketamine antidepressant activity over (*S*)-ketamine in rodents was confirmed by other groups (Fukumoto, Toki, Iijima, et al., 2017; Zanos, Moaddel, Morris, et al., 2016). In addition, compared with racemic (*R*,*S*)-ketamine or (*S*)-ketamine, (*R*)-ketamine has fewer detrimental side effects in rodents and monkeys (Chang et al., 2019; Fukumoto et al., 2017; Hashimoto, Kakiuchi, Ohba, et al., 2017; Tian, Dong, Fujita, et al., 2018; Yang et al., 2015; Yang, Han, Zhang, et al., 2016).

Recently, Masaki, Kashiwagi, Watabe, et al. (2019) investigated functional MRI (fMRI) responses in rats after a single injection of (*R*,*S*)-ketamine, (*S*)-ketamine, or (*R*)-ketamine as well as the selective NMDAR antagonist (+)-MK-801. The fMRI responses after (*R*)-ketamine injection differed from the other drugs. Collectively, it is unlikely that NMDAR inhibition plays a major role in the pharmacological effects of (*R*)-ketamine in rats (Hashimoto, 2019; Yang et al., 2019).

Just this year, a pilot study in Brazil demonstrated that one single intravenous injection of (*R*)-ketamine (0.5 mg/kg for 40 min) elicited rapid-acting and sustained antidepressant actions in patients with TRD (Leal, Bandeira, Correia-Melo, et al., 2020). The study also found that (*R*)-ketamine had fewer side effects compared to (*R*,*S*)-ketamine and (*S*)-ketamine. This is the first study to show the rapid-acting antidepressant effect of (*R*)-ketamine in TRD patients. Nonetheless, additional randomized, double-blinded studies will be needed to investigate the antidepressant effects of (*R*)-ketamine in patients with MDD and bipolar disorder.

Thus far, the cumulative preclinical and clinical data suggest that the antidepressant effect of (*R*)-ketamine is superior to (*R*,*S*)-ketamine and (*S*)-ketamine (Chang et al., 2019; Hashimoto, 2019; Hashimoto, 2020; Wei et al., 2020; Yang et al., 2019). However, there is scant knowledge about the precise molecular mechanisms underlying the different antidepressant effects of (*R*)-ketamine. Is the antidepressant effect of (*R*)-ketamine related to the 5-HT system, like SSRIs?

# Linking 5-HT and the antidepressant actions of (*R*)-ketamine in a chronic social stress model

## Chronic social defeat stress

CSDS is an important type of adversity, which is closely related to the development of human psychopathology, such as depression (Golden, Covington III, Berton, et al., 2011). The CSDS model has some drawbacks (Hollis & Kabbaj, 2014; Yan, Cao Das, et al., 2010). First, a glaring drawback to this model is that females cannot be used due to the unique nature of the model, which induces depression by repeatedly exposing male naive mice to aggressor mice so gender differences in depression cannot be studied. The other drawback of the model is that the mice must be adult mice, therefore, studies in juvenile mice with depression cannot be studied. However, chronic social defeat has been proven to be a reliable model for depression and can be used to study the molecular mechanisms and phenotype of depression.

It is known that the 5-HT system plays a role in the antidepressant effects of ketamine and other rapid-acting antidepressant candidate $mGlu_{2/3}$ receptor antagonists (Chaki, 2017; Chaki & Fukumoto, 2019). To elucidate the role of 5-HT in the antidepressant actions of (*R*,*S*)-ketamine, (*S*)-ketamine, and (*R*)-ketamine, researchers have performed comprehensive studies to examine whether 5-HT depletion in the brain, caused by *para*-chlorophenylalanine (PCPA) injection, could influence the antidepressant-like effects of (*R*,*S*)-ketamine, (*S*)-ketamine, and (*R*)-ketamine in mice and in a transgenic rat depression model (du Jardin et al., 2018; Fukumoto, Iijima, & Chaki, 2014 Fukumoto, Iijima, & Chaki, 2016; Gigliucci, O'Dowd, Casey, et al., 2013; Pham et al., 2017 Zhang, Dong, Fujita, et al., 2018).

## Role of 5-HT in antidepressant-like effects of ketamine and its enantiomers

To date, six studies have investigated whether 5-HT depletion affects the antidepressant effects of (*R*,*S*)-ketamine and its enantiomers (Table 1). As previously mentioned, PCPA, a tryptophan hydroxylase inhibitor, can inhibit 5-HT synthesis. Gigliucci et al. (2013) reported that injection of PCPA affected 5-HT synthesis in control Sprague-Dawley rats, but did not eliminate the acute antidepressant effect of (*R*,*S*)-ketamine (25 mg/kg, 1 h prior to FST). However, in control C57BL/6J mice, Fukumoto

**TABLE 1 Role of 5-HT in antidepressant-like effects of (*R,S*)-ketamine and its enantiomers.**

| Agents | Dosage | Administration route and time | Animals | Effects | Role of serotonergic system | References |
|---|---|---|---|---|---|---|
| (*R,S*)-Ketamine | 30 mg/kg | I.p., 30 min after PCPA injection | Mice | Antidepressant effect (NSFT) was blocked by PCPA (i.p.) | Serotonergic system is involved in acute antidepressant effects | Fukumoto et al. (2014) |
| (*R,S*)-Ketamine | 30 mg/kg | I.p., 30 min or 24 h after PCPA injection | Mice | Antidepressant effect (FST) was blocked by PCPA (i.p.) | Serotonergic system is involved in acute and sustained antidepressant effect | Fukumoto et al. (2016) |
| (*R,S*)-Ketamine | 25 mg/kg | I.p., 24 h after PCPA injection | Rat | Antidepressant effect (FST) was blocked by PCPA (i.p.) | Serotonergic system is involved in sustained antidepressant effect | Gigliucci et al. (2013) |
| (*R,S*)-Ketamine | 10 mg/kg | I.p., 24 h after PCPA injection | Mice | Antidepressant effect (FST) was blocked by PCPA (i.p.) | Serotonergic system is involved in sustained antidepressant effect | Pham et al., (2017) |
| (*S*)-Ketamine | 15 mg/kg | I.p., 1 h or 48 h after PCPA injection | 5-HT depleted Flinders sensitive line rats | Antidepressant effect (FST) was blocked by PCPA (i.p.) | Serotonergic system is involved in acute and sustained antidepressant effect | du Jardin et al., (2018) |
| (*R*)-Ketamine | 10 mg/kg | I.p., 1 h or 48 h after PCPA injection | Mice | Antidepressant effects (TST, SPT) in CSDS model were NOT blocked by PCPA (i.p.) | Serotonergic system is NOT involved in acute and sustained antidepressant effect | Zhang et al. (2018) |

This table is a modification of the table from Chaki and Fukumoto (2019).
*CSDS*, chronic social defeat stress; *FST*, forced swimming test; *NSFT*, novelty-suppressed feeding test; *PCPA*, *para*-chlorophenylalanine; *SPT*, sucrose preference test; *TST*, tail suspension test.

and colleagues showed that (*R,S*)-ketamine exhibited antidepressant effects, but 5-HT depletion by injection of PCPA eliminated the acute antidepressant effects of (*R,S*)-ketamine (30 mg/kg, 30 min before FST) (Fukumoto et al., 2016).

Currently, the exact reasons underlying this discrepancy are unclear, but it may be related to the following factors. First, the dose of (*R,S*)-ketamine was different (Table 1), therefore, the dosage may contribute to this discrepancy, because single injections of high doses of (*R,S*)-ketamine increased locomotor activity in rodents (Fukumoto et al., 2017; Yang et al., 2015). Second, the use of different rodent species may also contribute to this discrepancy (Table 1). Third, the use of rodents with or without depression-like phenotypes may also play a role in the differences in the study findings. Therefore, the acute hyperactivity induced by ketamine and the rodent models with the depression-like phenotype should be taken into consideration when examining the acute antidepressant effects of (*R,S*)-ketamine and its enantiomers.

du Jardin, Liebenberg, Müller, et al. (2016) detected the link between 5-HT and the antidepressant-like effects of (*S*)-ketamine (15 mg/kg). Different from the above studies, du Jardin and colleagues used Flinders sensitive line (FSL) rats. The FSL rat is a genetic model of depression that has been validated by many different studies. FSL rats exhibited depression-like behavior and memory deficits in the novel object recognition and the FST, respectively. There was no direct correlation between depression-like behavior of FSL rats and 5-HT depletion. Notably, the acute and sustained antidepressant-like effects of (*S*)-ketamine were eliminated by 5-HT depletion in the FSL rats. Based on these results, du Jardin et al. (2016) concluded that (*S*)-ketamine's antidepressant effects depend on the endogenous concentration of 5-HT in this genetic model of depression.

Using a CSDS model, we examined whether the antidepressant effects of (*R*)-ketamine were related to the 5-HT system (Zhang et al., 2018). In the first part of our study, PCPA or vehicle was administered to control and CSDS-susceptible mice. We injected 300 mg/kg of PCPA into the abdominal cavity of CSDS-susceptible mice twice a day for 3 days, and then measured the levels of 5-HT and 5-HIAA using high-performance liquid chromatography in different regions of the mouse brain, such as prefrontal cortex (PFC), hippocampus, and striatum. In the second part of the study, we detected the antidepressant-like effects of (*R*)-ketamine (10 mg/kg) in the vehicle- and PCPA-treated CSDS-susceptible mice using behavioral tests, including tail suspension test and 1% sucrose preference test. We found that the levels of 5-HT and 5-HIAA in the PFC, hippocampus, and striatum were markedly reduced after the administration of PCPA. More importantly, PCPA injection did not affect the depression-like phenotype of the CSDS-susceptible mice. Interestingly, PCPA-induced 5-HT depletion did not abolish the antidepressant-like effects of (*R*)-ketamine in the CSDS-susceptible mice. Taken together, these results suggest that 5-HT may not play a major role in the antidepressant-like actions of (*R*)-ketamine in CSDS-susceptible mice. To our knowledge, this is the first study to show that 5-HT does not play a significant role in the antidepressant-like effect of (*R*)-ketamine in mice with a depression-like phenotype. Thus, there is no link between 5-HT and the antidepressant-like actions of (*R*)-ketamine in the CSDS model. Further studies investigating the role of ketamine enantiomers in rodents with depression-like phenotypes are needed.

## Conclusion

In conclusion, 5-HT depletion by PCPA does exert influences on the antidepressant effects of (*R*,*S*)-ketamine and (*S*)-ketamine, but it does not affect the acute and long-lasting antidepressant effects of (*R*)-ketamine in a CSDS model. Based on the above results, we conclude that 5-HT may not play an important role in the antidepressant effect of (*R*)-ketamine, although further study is needed.

## Key facts

- Antidepressant actions of (*R*,*S*)-ketamine are of great interest in the psychiatry.
- (*S*)-ketamine nasal spray for treatment-resistant depression was approved in the United States and Europe.
- Clinical study of (*R*)-ketamine in humans is underway.

## Summary points

- This chapter focuses on 5-HT system in the antidepressant-like effects of ketamine and its enantiomers.
- (*R*,*S*)-ketamine produces rapid-acting and sustained antidepressant effects in depressed patients.
- (*R*)-ketamine has greater potency and longer-lasting antidepressant effects than (*S*)-ketamine in rodents.
- Depletion of 5-HT by PCPA did not alter antidepressant-like effects of (*R*)-ketamine in mice with depression-like phenotype.
- Further study on the 5-HT system in the antidepressant effects of ketamine enantiomers in rodents with depression-like phenotype is needed.

## References


Baker, C. B., Tweedie, R., Duval, S., et al. (2003). Evidence that the SSRI dose response in treating major depression should be reassessed: A meta-analysis. *Depression and Anxiety*, *17*(1), 1–9.

Berman, R. M., Cappiello, A., Anand, A., et al. (2000). Antidepressant effects of ketamine in depressed patients. *Biological Psychiatry*, *47*(4), 351–354.

Cerrito, F., & Raiteri, M. (1979). Serotonin release is modulated by presynaptic autoreceptors. *European Journal of Pharmacology*, *57*(4), 427–430.

Chaki, S. (2017). mGlu$_{2/3}$ receptor antagonists as novel antidepressants. *Trends in Pharmacological Sciences*, *38*(6), 569–580.

Chaki, S., & Fukumoto, K. (2019). Role of serotonergic system in the antidepressant actions of mGlu$_{2/3}$ receptor antagonists: Similarity to ketamine. *International Journal of Molecular Medicine*. *20*(6), 1270.

Chan, K. W. S., Lee, T. M. C., Siu, A. M. H., et al. (2013). Effects of chronic ketamine use on frontal and medial temporal cognition. *Addictive Behaviors*, *38*(5), 2128–2132.

Chang, L., Zhang, K., Fu, Y., et al. (2019). Comparison of antidepressant and side effects in mice after intranasal administration of (*R*,*S*)-ketamine, (*R*)-ketamine, and (*S*)-ketamine. *Pharmacology Biochemistry and Behavior*, *181*, 53–59.

Clark, C. T., Weissbach, H., & Udenfriend, S. (1954). 5-Hydroxytryptophan decarboxylase: Preparation and properties. *Journal of Biological Chemistry*, *210*(1), 139–148.

Darhstrom, A. (1964). Evidence for the existence of monoamine-containing neurons in the central nervous system. *Acta Physiologica Scandinavica*, *62*(232), 5–55.

Domino, E. F. (2010). Taming the ketamine tiger. 1965. *Anesthesiology*, *113*(3), 678–684.

Domino, E. F., & Luby, E. D. (2012). Phencyclidine/schizophrenia: One view toward the past, the other to the future. *Schizophrenia Bulletin*, *38*(5), 914–919.

du Jardin, K. G., Liebenberg, N., Cajina, M., Muller, H. K. Elfving, B., Sanchez, C., et al. (2018). *S*-Ketamine mediates its acute and sustained antidepressant-like activity through a 5-HT $_{1B}$ receptor dependent mechanism in a genetic rat model of depression. *Frontiers in Pharmacology*, *8*, 978. doi:10.3389/fphar.2017.00978.

du Jardin, K. G., Liebenberg, N., Müller, H. K., et al. (2016). Differential interaction with the serotonin system by S-ketamine, vortioxetine, and fluoxetine in a genetic rat model of depression. *Psychopharmacology*, *233*(14), 2813–2825.

Ebert, B., Mikkelsen, S., Thorkildsen, C., et al. (1997). Norketamine, the main metabolite of ketamine, is a non-competitive NMDA receptor antagonist in the rat cortex and spinal cord. *European Journal of Pharmacology*, *333*(1), 99–104.

Fukumoto, K., Iijima, M., & Chaki, S. (2014). Serotonin-1A receptor stimulation mediates effects of a metabotropic glutamate 2/3 receptor antagonist, 2S-2-amino-2-(1S,2S-2-carboxycycloprop-1-yl)-3-(xanth-9-yl)propanoic acid (LY341495), and an N-methyl-D-aspartate receptor antagonist, ketamine, in the novelty-suppressed feeding test. *Psychopharmacology*, *231*(11), 2291–2298. https://doi.org/10.1007/s00213-013-3378-0.

Fukumoto, K., Iijima, M., & Chaki, S. (2016). The antidepressant effects of an mGlu2/3 receptor antagonist and ketamine require AMPA receptor stimulation in the mPFC and subsequent activation of the 5-HT neurons in the DRN. *Neuropsychopharmacology*, *41*(4), 1046–1056.

Fukumoto, K., Toki, H., Iijima, M., et al. (2017). Antidepressant potential of (*R*)-ketamine in rodent models: Comparison with (*S*)-ketamine. *Journal of Pharmacology and Experimental Therapeutics*, *361*(1), 9–16.

Gigliucci, V., O'Dowd, G., Casey, S., et al. (2013). Ketamine elicits sustained antidepressant-like activity via a serotonin-dependent mechanism. *Psychopharmacology*, *228*(1), 157–166.

Giorgetti, R., Marcotulli, D., Tagliabracci, A., et al. (2015). Effects of ketamine on psychomotor, sensory and cognitive functions relevant for driving ability. *Forensic Science International*, *252*, 127–142.

Golden, S. A., Covington, H. E., III, Berton, O., et al. (2011). A standardized protocol for repeated social defeat stress in mice. *Nature Protocols*, *6*(8), 1 83.

Grunebaum, M. F., Galfalvy, H. C., Choo, T. H., et al. (2018). Ketamine for rapid reduction of suicidal thoughts in major depression: A midazolam-controlled randomized clinical trial. *American Journal of Psychiatry*, *175*(4), 327–335.

Hashimoto, K. (2019). Rapid-acting antidepressant ketamine, its metabolites and other candidates: A historical overview and future perspective. *Psychiatry and Clinical Neurosciences*, *73*(10), 613–627.

Hashimoto, K. (2020). Molecular mechanisms of the rapid-acting and long-lasting antidepressant actions of (R)-ketamine. *Biochemical Pharmacology*, *177*, 113935. https://doi.org/10.1016/j.bcp.2020.113935.

Hashimoto, K., Kakiuchi, T., Ohba, H., et al. (2017). Reduction of dopamine $D_{2/3}$ receptor binding in the striatum after a single administration of esketamine, but not *R*-ketamine: A PET study in conscious monkeys. *European Archives of Psychiatry and Clinical Neuroscience*, *267*(2), 173–176.

Hollis, F., & Kabbaj, M. (2014). Social defeat as an animal model for depression. *ILAR Journal*, *55*(2), 221–232.

Koike, H., Iijima, M., & Chaki, S. (2011). Involvement of AMPA receptor in both the rapid and sustained antidepressant-like effects of ketamine in animal models of depression. *Behavioural Brain Research*, *224*(1), 107–111.

Leal, G. C., Bandeira, I. D., Correia-Melo, F. S., et al. (2020). Intravenous arketamine for treatment-resistant depression: open-label pilot study. *European Archives of Psychiatry and Clinical Neuroscience*, . https://doi.org/10.1007/s00406-020-01110-5. in press.

Li, L., & Vlisides, P. E. (2016). Ketamine: 50 years of modulating the mind. *Frontiers in Human Neuroscience*, *10*, 612.

Ma, X. C., Dang, Y. H., Jia, M., et al. (2013). Long-lasting antidepressant action of ketamine, but not glycogen synthase kinase-3 inhibitor SB216763, in the chronic mild stress model of mice. *PLoS One*, *8*(2), e56053.

Maddox, V. H., Godefroi, E. F., & Parcell, R. F. (1965). The synthesis of phencyclidine and other 1-arylcyclohexylamines. *Journal of Medicinal Chemistry*, *8*(2), 230–235.

Masaki, Y., Kashiwagi, Y., Watabe, H., et al. (2019). (*R*)-and (*S*)-ketamine induce differential fMRI responses in conscious rats. *Synapse*, *73*(12), e22126.

McIsaac, W. M., & Page, I. H. (1959). The metabolism of serotonin (5-hydroxytryptamine). *The Journal of Biological Chemistry*, *234*(4), 858–864.

Mion, G. (2017). History of anaesthesia: The ketamine story–past, present and future. *European Journal of Anaesthesiology (EJA)*, *34*(9), 571–575.

Morris, H., & Wallach, J. (2014). From PCP to MXE: A comprehensive review of the non-medical use of dissociative drugs. *Drug Testing and Analysis*, *6*(7–8), 614–632.

Pham, T. H., Mendez-David, I., Defaix, C., Guiard, B. P., Tritschler, L., David, D. J., et al. (2017). Ketamine treatment involves medial prefrontal cortex serotonin to induce a rapid antidepressant-like activity in BALB/cJ mice. *Neuropharmacology*, *112*(Pt A), 198–209. https://doi.org/10.1016/j.neuropharm.2016.05.010.

Phillips, J. L., Norris, S., Talbot, J., et al. (2019). Single, repeated, and maintenance ketamine infusions for treatment-resistant depression: A randomized controlled trial. *American Journal of Psychiatry*, *176*(5), 401–409.

Rapport, M. M., Green, A. A., & Page, I. H. (1948). Crystalline serotonin. *Science*, *108*(2804), 329–330.

Singh, J. B., Fedgchin, M., Daly, E., et al. (2016). A double-blind, randomized, placebo-controlled, dose-frequency study of intravenous ketamine in patients with treatment-resistant depression. *American Journal of Psychiatry*, *173*(8), 816–826.

Sofia, R. D., & Harakal, J. J. (1975). Evaluation of ketamine HCl for antidepressant activity. *Archives Internationales de Pharmacodynamie et de Thérapie*, *214*(1), 68–74.

Strong, C. E., & Kabbaj, M. (2018). On the safety of repeated ketamine infusions for the treatment of depression: Effects of sex and developmental periods. *Neurobiology of Stress*, *9*, 166–175.

Tan, Y., & Hashimoto, K. (2020). Risk of psychosis after repeated intermittent administration of (*S*)-ketamine, but not (*R*)-ketamine, in mice. *Journal of Affective Disorders*, *269*, 198–200.

Tian, Z., Dong, C., Fujita, A., et al. (2018). Expression of heat shock protein HSP-70 in the retrosplenial cortex of rat brain after administration of (*R,S*)-ketamine and (*S*)-ketamine, but not (*R*)-ketamine. *Pharmacology Biochemistry and Behavior*, *172*, 17–21.

Wei, Y., Chang, L., & Hashimoto, K. (2020). A historical review of antidepressant effects of ketamine and its enantiomers. *Pharmacology, Biochemistry and Behavior*, *190*, 172870.

Wilkinson, S. T., Ballard, E. D., Bloch, M. H., et al. (2018). The effect of a single dose of intravenous ketamine on suicidal ideation: A systematic review and individual participant data meta-analysis. *American Journal of Psychiatry*, *175*(2), 150–158.

Yamaguchi, J., Toki, H., Qu, Y., et al. (2018). (2*R*,6*R*)-Hydroxynorketamine is not essential for the antidepressant actions of (*R*)-ketamine in mice. *Neuropsychopharmacology*, *43*(9), 1900–1907.

Yan, H. C., Cao, X., Das, M., et al. (2010). Behavioral animal models of depression. *Neuroscience Bulletin*, *26*(4), 327–337.

Yang, C., Han, M., Zhang, J., et al. (2016). Loss of parvalbumin-immunoreactivity in mouse brain regions after repeated intermittent administration of esketamine, but not *R*-ketamine. *Psychiatry Research*, *239*, 281–283.

Yang, C., Shirayama, Y., Zhang, J. C., et al. (2015). *R*-ketamine: A rapid-onset and sustained antidepressant without psychotomimetic side effects. *Translational Psychiatry*, *5*(9), e632.

Yang, C., Yang, J., Luo, A., & Hashimoto, K. (2019). Molecular and cellular mechanisms underlying the antidepressant effects of ketamine enantiomers and its metabolites. *Translational Psychiatry*, *9*(1), 280.

Zanos, P., Moaddel, R., Morris, P. J., et al. (2016). NMDAR inhibition-independent antidepressant actions of ketamine metabolites. *Nature*, *533*(7604), 481–486.

Zarate, C. A., Singh, J. B., Carlson, P. J., et al. (2006). A randomized trial of an N-methyl-D-aspartate antagonist in treatment-resistant major depression. *Archives of General Psychiatry*, *63*(8), 856–864.

Zhang, K., Dong, C., Fujita, Y., et al. (2018). 5-Hydroxytryptamine-independent antidepressant actions of (*R*)-ketamine in a chronic social defeat stress model. *International Journal of Neuropsychopharmacology*, *21*(2), 157–163.

Zhang, K., & Hashimoto, K. (2019). An update on ketamine and its two enantiomers as rapid-acting antidepressants. *Expert Review of Neurotherapeutics*, *19*(1), 83–92.

Zhang, J., Li, S., & Hashimoto, K. (2014). *R*(−)-ketamine shows greater potency and longer lasting antidepressant effects than *S*(+)-ketamine. *Pharmacology Biochemistry and Behavior*, *116*, 137–141.

Zhang, J., Yao, W., Dong, C., et al. (2015). Comparison of ketamine, 7, 8-dihydroxyflavone, and ANA-12 antidepressant effects in the social defeat stress model of depression. *Psychopharmacology*, *232*(23), 4325–4335.

Chapter 39

# Mirtazapine: Multitarget strategies for treating substance use disorder and depression

Susana Barbosa-Méndez[a], Luis Enrique Becerril-Villanueva[b], María Dolores Ponce-Regalado[c], and Alberto Salazar-Juárez[a]

[a]Laboratorio de Neurofarmacología Conductual, Microcirugía y Terapéutica Experimental, Subdirección de Investigaciones Clínicas, Instituto Nacional de Psiquiatría "Ramón de la Fuente", México DF, México, [b]Laboratorio de Psicoinmunología, Dirección de Neurociencias, Instituto Nacional de Psiquiatría "Ramón de la Fuente", México DF, México, [c]Laboratorio de investigaciones en Biociencias, Departamento de Ciencias de la Salud, Centro Universitario de los Altos, Universidad de Guadalajara, Guadalajara, Mexico

## List of abbreviations

| | |
|---|---|
| **α1-NER, α2-NER** | norepinephrine receptor |
| **$\alpha_4\beta_2$, $\alpha_7$, $\beta_3$ nAChR** | nicotine receptor |
| **5-HT** | serotonergic system |
| **5-HT$_{2A}$R, 5-HT$_{2c}$R, 5-HT$_3$R** | serotonin receptors |
| **ACh** | holinergic system |
| **CB$_1$R** | cannabinoids receptor |
| **CPP** | conditioned place preference |
| **D$_1$R, D$_2$R** | dopamine receptors |
| **DA** | dopamine neurons |
| **DAT** | reuptake dopamine transporter |
| **DD** | dual diagnosis |
| **DRN** | dorsal raphe nucleus |
| **GABA** | GABAergic system |
| **GABA$_A$R** | GABA receptor |
| **GLU** | glutamatergic system |
| **GLUR** | glutamate receptor |
| **GLUT** | glutamate transporter |
| **H$_1$R** | histamine receptor |
| **i.p.** | intraperitoneal administration |
| **LC** | locus coeruleus |
| **M$_1$mAchR** | muscarinic receptor |
| **MDD** | major depressive disorder |
| **MLCS** | mesocorticolimbic dopaminergic system |
| **MOR, DOR, KOR** | opioid receptors |
| **NAcc** | nucleus accumbens |
| **NE** | noradrenergic system |
| **NET** | reuptake norepinephrine transporter |
| **PFC** | prefrontal cortex |
| **SERT** | reuptake serotonin transporter |
| **SUD** | substance use disorder |
| **VTA** | ventral-tegmental area |
| **WHO** | World Health Organization |

## Introduction

Major depressive disorder (MDD) is considered a serious health problem by the World Health Organization (WHO), especially when its duration and intensity affect the work, family, and/or school activity of the sufferer. That is why MDD is one of the main causes of morbicity worldwide (Walker, McGee, & Druss, 2015). The main characteristics of MDD in adults and adolescents are: a recurrent and/or long-lasting depressed and/or an hedonic mood, a decrease in interest in everyday activities, significant weight loss or gain, without dieting, insomnia or hypersomnia, agitation, psychomotor retardation, fatigue, loss of energy, feeling of worthlessness, excessive or inappropriate guilt, decreased ability to think and concentrate, decreased decision-making, thoughts of death and suicidal thoughts without a specific plan, or structured suicide attempt (Table 1).

## Substance use disorder (SUD)

Individuals with substance use disorder (SUD) show symptoms that are common in some mental illnesses, such as anxiety states, impulsiveness, psychotic episodes and/or mania, and obsessive-compulsive disorder as well as those related to mood. SUD is characterized by an excessive desire for the consumption of psychoactive substances, and the consequent withdrawal symptoms, a progressive loss of interests or pleasures not associated to substance use, and an increase in tolerance and craving for consumption (Santucci, 2012).

The Neuroscience of Depression. https://doi.org/10.1016/B978-0-12-817933-8.00043-8

**TABLE 1 Antidepressants used in the treatment of comorbid depression with substance abuse.**

| Antidepressant (dose) | Mood disorder diagnosis | Drug use disorder |
|---|---|---|
| Citalopram (40 mg/day) | MDD | Alcohol dependence or abuse |
| Fluoxetine (20–60 mg/kg) | MDD | Alcohol, cocaine, nicotine or opioid dependence, or abuse |
| Desipramine (100–300 mg/kg) | MDD | Alcohol or cocaine dependence or abuse |
| Nefazodone (300–500 mg/kg) | MDD | Alcohol dependence or abuse |
| Sertraline (100–200 mg/kg) | MDD | Nicotine or opioid dependence or abuse |
| Viloxazine (400 mg/kg) | MDD | Alcohol dependence or abuse |
| Imipramine (150–300 mg/kg) | MDD | Cocaine or opioid dependence or abuse |
| Gepirone (116–25 mg/kg) | MDD | Alcohol or cocaine dependence or abuse |
| Ritanserin (10 mg/kg) | MDD | Alcohol or cocaine dependence or abuse |
| Bupropion (300 mg/kg) | MDD | Cocaine or nicotine dependence or abuse |
| Mirtazapine (30–60 mg/kg) | MDD | Alcohol or cocaine dependence or abuse |
| Doxepin (100–200 mg/kg) | MDD | Nicotine or opioid dependence or abuse |
| Nortriptyline (100 mg/kg) | MDD | Nicotine dependence or abuse |
| Paroxetine (20–40 mg/kg) | MDD | Nicotine dependence or abuse |

## MDD-SUD comorbidity

Comorbidity is a fundamental characteristic of MDD. While several symptoms are considered diagnostic, others are frequently a sequel to MDD or part of other psychiatric disorders, as SUD.

Currently, dual diagnosis (DD) has been established between SUD and other mental illnesses (Ng, Browne, Samsom, & Wong, 2017) as MDD.

## Neurobiological mechanisms of DD (MDD-SUD)

The neural substrate associated to DD is the mesocorticolimbic dopaminergic system (MCLD). This system is directly interconnected to the serotonergic (5-HT), noradrenergic (NE), GABAergic (GABA), cholinergic (ACh), glutamatergic (GLU), and endogenous opioid systems in the brain, among others. MCLD consists of a series of interconnected neural structures, in close communication with each other: ventral-tegmental area (VTA), nucleus accumbens (NAcc) in its two portions (core and shell), dorsal striatum, prefrontal cortex (PFC), and axis structures through which the reinforcing effect of a psychoactive drug is processed.

From the molecular point of view, in the dopaminergic (DA) neurons located in these brain structures, different types of membrane receptors associated with psychoactive drugs are colocated: nicotine ($\alpha_4\beta_2$, $\alpha_7$, $\beta_3$ nAChR), cannabinoids ($CB_1R$), opioids (MOR, DOR, KOR), and alcohol ($GABA_AR$), among others. In addition, a high density of the dopamine transporter (DAT) and different subtypes of dopamine receptors ($D_1R$ and $D_2R$) are located at the terminals of these DA neurons. Due to the interconnections that MLCS shows with other brain structures, receptors and/or transporters for serotonin (5-$HT_{2A}R$, 5-$HT_{2c}R$, 5-$HT_3R$, and SET), norepinephrine (α1-NER, α2-NER, and NET), acetylcholine ($\alpha_4\beta_2$, $\alpha_7$, and $\beta_3$ nAChR), and glutamate (GLUR and GLUT), among others are also colocated in DA neurons, which are neurotransmission systems responsible for modulating by the drug reinforcing effect.

## Mirtazapine

Mirtazapine is an atypical antidepressant approved by the Food and Drug Administration (FDA) for the treatment of moderate to severe MDD and a wide range of symptoms associated with it (Croom, Perry, & Plosker, 2009).

This drug is a part of the group of tetracyclic antidepressants carrying out effects through the combination of noradrenergic and serotonergic mechanisms (Anttila & Leinonen, 2001). Its mechanism of action is complex since it involves the simultaneous interaction of mirtazapine with different subtypes of adrenergic ($\alpha_1$-NER and $\alpha_2$-NER) and serotonergic (5-$HT_{1A}R$, 5-$HT_{2A/C}R$, 5-$HT_{2C}R$, and 5-$HT_3R$) receptors; although it is also capable of interacting with histamine ($H_1R$) and acetylcholine ($M_1$mAchR) receptors (de Boer, 1996).

Mirtazapine potentiates serotonergic neurotransmission through the postsynaptic antagonism of 5-$HT_{2A}$ and 5-$HT_3$ receptors, resulting in increased excitation of dorsal raphe nucleus (DRN) neurons. Furthermore, the blockade of presynaptic $\alpha_2$-NERs, which normally inhibit serotonergic and noradrenergic terminals in the locus coeruleus (LC), promotes the release of NE and leads to increased adrenergic activation in the $\alpha_1$-NERs in the DRN, which enhances the release of 5-HT. This ability to stimulate DRN is complemented by antagonism in $\alpha_2$-NERs located in serotonergic neurons, which also leads to intensified DRN activity and a surge in 5-HT flow (Haddjeri, Blier, & de-Montigny, 1998).

On the other hand, mirtazapine is not able to block the reuptake systems of dopamine (DAT), serotonin (SERT), and norepinephrine (NET) in the brain and has a low affinity for dopamine receptors ($D_1R$ and $D_2R$), while it is a potent antagonist of histaminergic ($H_1$) receptors and a moderate antagonist of $M_1$mAchR muscarinic cholinergic receptors and $\alpha_1$-adrenergic receptors (Anttila & Leinonen, 2001).

## Mirtazapine-SUD

For the past decades, several antidepressants have been explored as possible therapeutic drugs to relieve SUD (Torrens, Fonseca, Mateu, & Farré, 2005). In fact, several clinical trials have evaluated the efficacy of various antidepressants to reduce withdrawal symptoms and drug craving, and/or prevent relapse during early withdrawal. However, the results of these studies have not been conclusive since they have reported that only some of the patients who received the medications during withdrawal from psychoactive drugs showed improvement and/or relief of any withdrawal symptoms (Lima, Reisser, Soares, & Farrell, 2010).

In general, these studies suggest that the symptomatology of SUD is heterogeneous and that one of the main limitations of antidepressants used in the SUD treatment is that they generally act selectively on a receptor subtype and only alleviate some symptoms of SUD. Therefore, molecules capable of acting simultaneously on several receptor subtypes (multitarget drugs), in different brain nuclei are required to simultaneously reduce different withdrawal symptoms and also prevent relapses.

## Preclinical studies

Pioneer studies in rodents show that mirtazapine administration to Sprague-Dawley rats reduced morphine-induced reward effects and the acquisition of morphine dependence (Kang et al., 2008). In addition, in a series of studies, Graves et al. demonstrated that Sprague-Dawley rats treated with mirtazapine showed an attenuation in the establishment and expression of the conditioned place preference (CPP) to morphine and methamphetamine. Other studies described that rats dosed with mirtazapine showed a reduction in the reacquisition of methamphetamine self-administration and the expression of methamphetamine-induced locomotor sensitization (Graves & Napier, 2012; Graves, Rafeyan, Watts, & Napier, 2012; Herrold et al., 2009; McDaid et al., 2007; Voigt, Mickiewicz, & Napier, 2011; Voigt & Napier, 2012).

In addition, it was reported that Wistar rats dosed with mirtazapine (30 mg/kg, i.p., daily, 30 days) during withdrawal showed a significant decrease in induction and expression of locomotor sensitization, CPP expression, duration of locomotor effect, and self-administration induced by different doses of cocaine, heroin, or nicotine (Barbosa-Méndez, Jurado, et al., 2017; Barbosa-Méndez, Leff, et al., 2017; Barbosa-Méndez, Matus-Ortega, Jacinto-Gutiérrez, & Salazar-Juárez, 2018; Barbosa-Méndez & Salazar-Juárez, 2018a; Fig. 1). These results support the efficacy of mirtazapine in reducing the behavior reinforcement effect induced by a wide variety of drugs.

In addition, these studies demonstrated that (1) the dose, (2) the duration of the dosage, and (3) the stage of drug withdrawal in which mirtazapine is administered are key parameters (dosing schedule) on which the efficacy of

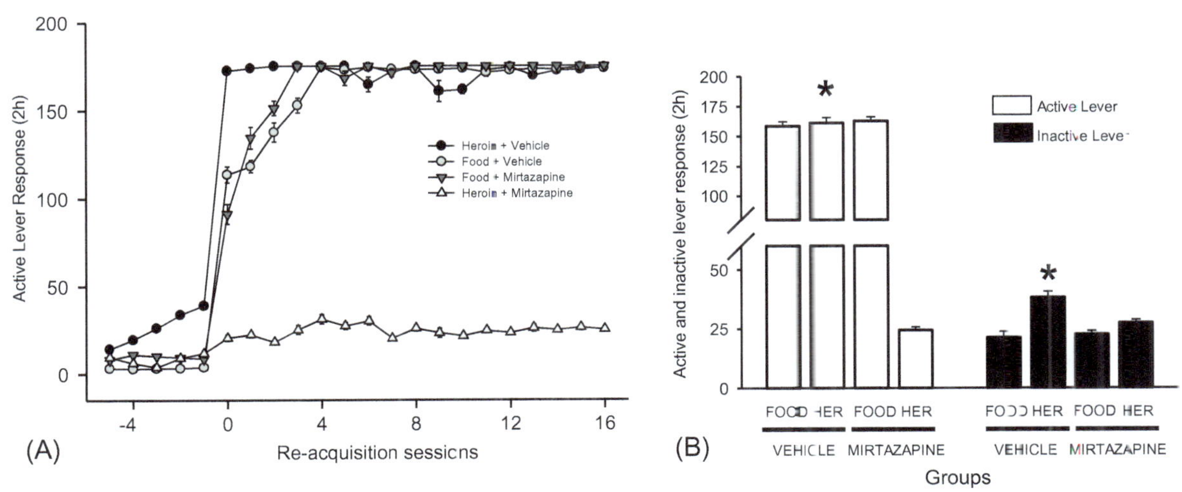


FIG. 1 Mirtazapine decreases heroin self-administration. The heroin + vehicle and heroin + mirtazapine groups received saline and mirtazapine, respectively, during extinction (30 days) and were then exposed to the reacquisition of heroin self-administration. The heroin + mirtazapine group also received mirtazapine during reacquisition (10 days). (A) Cocaine seeking during reacquisition in which a decrease in active lever pressures in the animals of the heroin + mirtazapine group is observed. (B) Mean active lever pressures during the reacquisition phase. Two-way ANOVA followed by a post hoc Tukey's test. (*) indicates that significant differences ($P < 0.01$) were found in the heroin + mirtazapine group, compared to the heroin + vehicle group.

mirtazapine depends as a therapeutic agent for the treatment of SUD (Barbosa-Méndez, Matus-Ortega, Flores-Zamora, Jurado, & Salazar-Juárez, 2017).

(1) *Dose*. Studies in rats showed that the administration of low doses of mirtazapine (0.5, 1, or 5 mg/kg) produced a significant attenuation in the maintenance of methamphetamine-induced CPP (Graves & Napier, 2011; Graves, Persons, Riddle, & Napier, 2012; Herrold et al., 2009), as well as a decrease in methamphetamine-seeking behavior (Graves & Napier, 2012; Graves, Rafeyan, et al., 2012). Other authors reported that the effectiveness of 5 mg/kg mirtazapine to attenuate methamphetamine-induced CPP in rats depended on the contextual signals associated with the administration of mirtazapine (Voigt et al., 2011; Voigt & Napier, 2012) rather than the dose used. Other studies reported that a dose of 10 or 15 mg/kg of mirtazapine not attenuate the expression of methamphetamine-, cocaine-, or nicotine-induced locomotor sensitization in rats (Barbosa-Méndez, Matus-Ortega, Flores-Zamora, et al., 2017; Bhatia, Szabo, Fowler, Wetsel, & Lee, 2011).

In 2008, Kang et al. reported that mirtazapine (30 mg/kg i.p.) attenuated morphine-induced CPP in rats. In addition, we have reported that 30 mg/kg of mirtazapine administered to rats previously treated with cocaine or nicotine decreased the expression of locomotor sensitization (Barbosa-Méndez, Jurado, Matus-Ortega, et al., 2017; Salazar-Juárez et al., 2016), CPP (Barbosa-Méndez et al., 2018), and reacquisition of self-administration to cocaine, heroin, or nicotine (Barbosa-Méndez, Leff, et al., 2017; Barbosa-Méndez & Salazar-Juárez, 2018a). The daily dosage (30 mg/kg) of mirtazapine reduced the toxic effects generated by a sublethal cocaine dose. However, when increasing the dose to 60 mg/kg, no behavioral effects were significantly different from those generated by a 30 mg/kg dose (Barbosa-Méndez, Matus-Ortega, Flores-Zamora, et al., 2017).

(2) Dosage duration. On the other hand, clinical studies have described that the pharmacological action of antidepressants is slow and their therapeutic effects become evident after weeks of treatment.

Studies in rodents have described the effects of acute administration of mirtazapine and found that rats dosed 15 min before cocaine administration showed a decrease in reward effects and the acquisition of morphine-induced dependence (Kang et al., 2008). In addition, rodents administered with mirtazapine 24 h before cocaine showed attenuation in the establishment of morphine- or methamphetamine-induced CPP (Graves, Persons, et al., 2012; Voigt & Napier, 2012).

Other studies reported that rats administered with mirtazapine for periods from 3 to 15 days showed a decrease in self-administration, locomotor sensitization, and maintenance of CPP induced by methamphetamine (Graves & Napier, 2011; Herrold et al., 2009; Voigt et al., 2011).

However, a recent study described the effects of chronic administration of mirtazapine in rodents and found that rats administered with mirtazapine for 15 days did not show attenuation of the expression of cocaine- or nicotine-induced locomotor sensitization (Barbosa-Méndez, Matus-Ortega, Flores-Zamora, et al., 2017). In contrast, animals pretreated with cocaine and dosed with mirtazapine for 30 days showed a significant decrease in cocaine self-administration, the expression of cocaine-induced locomotor sensitization, and CPP (Barbosa-Méndez et al., 2018; Salazar-Juárez et al., 2016). In addition, the study reported that a longer treatment with mirtazapine (60 days) significantly improved the decrease in the expression of cocaine-induced locomotor sensitization, compared to that generated by the administration of mirtazapine for 30 days (Barbosa-Méndez, Matus-Ortega, Flores-Zamora, et al., 2017).

(3) *Drug-withdrawal stage*. Some clinical studies have suggested that the early withdrawal phase is an important therapeutic window. Animal studies have reported that the administration of mirtazapine in the first days of drug withdrawal (10 days) reduced methamphetamine-induced CPP (Graves & Napier, 2011; Herrold et al., 2009; Voigt et al., 2011). However, an important limitation of these studies was the lack of follow-up on the effect of mirtazapine, even after withdrawing the medication.

We have previously (2016) described that (1) mirtazapine and cocaine showed an enhancement of cocaine-induced locomotor activity in dosed simultaneously during the active consumption phase; contrastingly, (2) the treatment with mirtazapine for $\geq$30 days during drug withdrawal caused a permanent decrease in the expression of cocaine-induced locomotor sensitization, even in the absence of mirtazapine (Salazar-Juárez et al., 2016). These findings suggest that the first 30 days of drug withdrawal are essential for mirtazapine to exert permanent therapeutic effects.

*Sedation*: Clinical and preclinical studies have shown that sedation generated during treatment with mirtazapine is dose-time dependent (Salazar-Juárez et al., 2016; (Salazar-Juárez et al., 2017)Sasada et al., 2013). Pioneer studies showed that $\leq$15 mg/kg doses of mirtazapine produced hyperphagia and weight gain, while doses of $\geq$60 mg/kg produced hypoactivity, drowsiness, and reduced exploratory activity as well as altered behavioral responses (de Boer, 1996(Iwamoto et al., 2013)).

Several clinical studies have shown that an initial dose of ≤15 mg/kg of mirtazapine acts as a potent histamine block that induces clear sedation and drowsiness (Grasmäder et al., 2005), while a dose of ≥30 mg/kg is associated with a decrease in sedative antihistamine activity due to enhanced noradrenergic transmission. In addition, the acute treatment with mirtazapine has been reported to produce temporary sedative effects.

Sasada et al. (2013) reported that the sedative effects induced by mirtazapine (7.5 or 15 mg/kg) disappeared after repeated administrations to healthy volunteers. Other studies indicated that sedation induced by 15 mg/kg mirtazapine decreased overtime (Grasmäder et al., 2005).

A recent study revealed that a single dose of ≤15 mg/day of mirtazapine administered to rats decreased spontaneous locomotor activity, balance, and motor coordination for several days. In contrast, animals dosed with ≥30 mg/day of mirtazapine showed sedation within 20 min after administration. The effect gradually decreased and reached reference levels at 60 min to 2 days until the animals no longer showed signs of sedation. This suggests that the chronic administration of ≥30 mg/kg of mirtazapine doses does not cause sedation (Salazar-Juárez et al., 2016).

*Drug withdrawal*: Preclinical studies have described that rodents exposed to different withdrawal periods showed an increase in anxiety- and depressive-like behavior. Kang et al. (2008) reported that mirtazapine decreased the severity of symptoms during morphine withdrawal in rats.

In a recent study, Wistar rats pretreated with cocaine underwent a 60-day cocaine-withdrawal period during which depression- and anxiety-like behaviors were evaluated in open field, elevated plus maze, light-dark box, forced swimming tests, and spontaneous locomotor activity. This study found that the chronic treatment with 30 or 60 mg/kg of mirtazapine decreased the depression- and anxiety-like behaviors induced by different doses of cocaine (10, 20, and 40 mg/kg) during the 60 days of cocaine withdrawal (Barbosa-Méndez & Salazar-Juárez, 2019a).

On the other hand, in a double-blind, placebo-controlled clinical trial whose participants were DD in patients with alcohol or cocaine dependence, the treatment with mirtazapine produced a significant decrease in depressive symptoms at week 2 and in all subsequent evaluations (weeks 3, 4, 6, 8, 10, and 12) during a 12-week study (Afshar et al., 2012; Cornelius et al., 2012, 2013, 2016b). Other double-blind, placebo-controlled clinical trials have shown that mirtazapine significantly improved the symptoms of depression, anxiety, insomnia, and dysphoria that appear during benzodiazepine, methamphetamine, and cocaine withdrawal (Afshar et al., 2012; Cruickshank et al., 2008; Kongsakon, Papadopoulos, & Saguansiritham, 2005).

## Clinical trials

A pioneering study in patients with benzodiazepine abuse showed that the administration of mirtazapine (60 mg/day) stopped benzodiazepine abuse in the streets, including the reduction of cocaine abuse (Zueco-Pérez, 2002).

Subsequently, other randomized, double-blind, placebo-controlled trials reported that 50% of the subjects treated with mirtazapine (30 mg/day) for 12 weeks showed a decrease in methamphetamine use (Colfax et al., 2011; de Bejczy & Söderpalm, 2015; McGregor, Srisurapanont, Mitchell, Wickes, & White, 2008). Studies in patients with alcohol use disorder showed that the dose of 30 or 45 mg/day of mirtazapine for 8–11 weeks reduced alcohol consumption (Brunette, Akerman, Dawson, O'Keefe, & Green, 2016; Liappas, Paparrigopoulos, Tzavellas, & Christodoulou, 2003; Liappas, Paparrigopoulos, Tzavellas, & Rabavilas, 2005).

However, other studies found no effect of mirtazapine treatment on cocaine or alcohol use (Afshar et al., 2012; Cornelius et al., 2016a). Afshar et al. (2012) found no significant differences in the urine positive cocaine tests between the placebo and mirtazapine group at week 11. In an open trial, Cornelius et al. (2012, 2016a, 2016b) found no evidence of the efficacy of mirtazapine in reducing the level of alcohol consumption in patients with alcohol use disorder.

Furthermore, double-blind, placebo-controlled clinical trials in which patients dependent on cocaine or alcohol were treated with 30 or 45 mg/day of mirtazapine for 4–14 weeks showed a decrease in drug craving as compared to patients treated with placebo (Brunette et al., 2016; Cornelius et al., 2016a; Graves, Rafeyan, et al., 2012; Yoon et al., 2006; Zueco-Pérez, 2002). However, Afshar et al. (2012) reported that the daily treatment with 45 mg of mirtazapine for 12 weeks failed to reduce cocaine craving.

On the other hand, clinical studies have described that mirtazapine dosed for several weeks is well tolerated and does not cause serious side effects (Afshar et al., 2012; Brunette et al., 2016; Kongsakon et al., 2005; McGregor et al., 2008; Yoon et al., 2006). These studies demonstrated that an initial dose of 15 mg/day followed by 30 mg/day of mirtazapine to treat methamphetamine-dependent withdrawal symptoms resulted in a safe and well-tolerated dose, which produced a significant decrease in the frequency of methamphetamine use (Colfax et al., 2011; McGregor et al., 2008).

As mentioned above, emotional and cognitive adverse symptoms are considered as the main factors that trigger the relapse to chronic drug abuse. Thus, effectively dealing these symptoms is one of the main objectives that a treatment against SUD must meet.

In an open trial study, McGregor et al. (2008) reported the efficacy of mirtazapine (60 mg/day for 10 days) in hospitalized patients. It significantly reduced the symptoms of methamphetamine withdrawal, such as agitation, fatigue, irritability, paranoid and suicidal ideation, anhedonia, and vivid dreams; it also decreased symptoms of depression and anxiety. Other double-blind, placebo-controlled clinical studies showed that mirtazapine significantly improved depression, anxiety, insomnia, minimized physical and subjective discomfort, and dysphoric symptoms that appear during benzodiazepine and alcohol withdrawal (Altintoprak, Zorlu, Coskunol, Akdeniz, & Kitapcioglu, 2008; Chandrasekaran, 2008; Cruickshank et al., 2008; Kongsakon et al., 2005; Liappas, Paparrigopoulos, Malitas, Tzavellas, & Christodoulou, 2004). In addition, the treatment with 45 mg/day of mirtazapine, administered for 12 weeks in outpatients, increased the amount of sleep and decreased depression in patients with depression and cocaine dependence (Afshar et al., 2012).

## Preclinical studies in models of polydrug

Epidemiological studies have indicated that a common practice among drug users is the simultaneous mixing or use of two or more drugs. These studies state that nicotine is a drug commonly used by cocaine dependents. In animals, the concurrent self-administration of cocaine and nicotine has been reported to increase the response rate in rhesus macaques (Mello & Newman, 2011) and increase dopamine release in rats (Gerasimov et al., 2000). We have reported that the combination of cocaine and nicotine increases the self-administration and consumption pattern of both drugs (Fig. 2) and enhances locomotor activity and the induction and expression of locomotor sensitization in rats (Barbosa-Méndez & Salazar-Juárez, 2018b).

In addition, we found that mirtazapine decreases the induction and expression of locomotor sensitization induced by the concurrent administration of cocaine and nicotine (Barbosa-Méndez & Salazar-Juárez, 2019b).

## Selective agent therapies

The role of adrenergic and serotonergic systems in the development and consolidation of SUD has been well documented.

Several preclinical studies have shown that the individual treatment with 5-$HT_2$, 5-$HT_3$, or $\alpha_2$NE receptor antagonists has not generated consistent and permanent results. Occasionally, the administration of 5-$HT_{2A}$ receptor antagonists such as ketanserin or M100907, 5-$HT_{2C}$ receptor agonists, (Ro60-0175), 5-$HT_3$ receptor antagonists such as tropisetron, and ondansetron or $\alpha_2$NE receptor antagonists (yohimbine) for several days during drug withdrawal caused a decrease in dopamine release and an attenuation in the induction and expression of locomotor sensitization, in self-administration and reacquisition of CPP induced by drugs like as cocaine and nicotine (Zayara et al., 2011). However, the treatment with these antagonists or agonists has also failed to decrease self-administration and drug-generated hyperactivity (Schenk et al., 2016).

## Multitarget therapy

Drugs such as cocaine simultaneously affect several neurotransmission systems and alter the functioning of different subtypes of monoaminergic and non-monoaminergic receptors. Mirtazapine decreases many behavioral effects of drugs by simultaneously antagonizing the 5-$HT_{2A}$, 5-$HT_{2C}$, and 5-$HT_3$ receptors, and even the $\alpha_2$NE. However, its limitations include the low affinity shown by the $\alpha_1$NE and $\alpha4\beta2$nAch cholinergic receptors, which have been reported to be involved in the development and establishment of the reinforcement effects induced by drugs such as cocaine, nicotine, and even opioids.

To compensate for this pharmacological shortcoming, different strategies have been carried out: (1) coadministration of mirtazapine with some $\alpha_1$NE or $\alpha4\beta2$nAch cholinergic receptor antagonists (Barbosa-Méndez, Matus-Ortega, & Salazar-Juárez, 2017; Barbosa-Méndez & Salazar-Juárez, 2019b), (2) development of a new multitarget drug capable of acting on the different receptor subtypes (Millan, 2006), and (3) development of a new pharmaceutical formulation containing mirtazapine and another/other antagonist(s) (Ranjan et al., 2011; Vysloužil et al., 2016).

In 2017, we reported that the coadministration of mirtazapine and prazosin ($\alpha_1$NER antagonist) for 30 days during drug withdrawal significantly reduced the expression of cocaine sensitization, compared with the individual treatment with mirtazapine or prazosin (Barbosa-Méndez, Matus-Ortega, & Salazar-Juárez, 2017). In addition, we have found that the coadministration of mirtazapine and mecamylamine potentiated the effect of mirtazapine on the expression of locomotor sensitization induced by cocaine, nicotine, and even the combination of both drugs (Barbosa-Méndez & Salazar-Juárez, 2019b).

These results suggest that coadministration of mirtazapine and other monoamine and non-monoamine antagonists generated promising results, especially in the absence of new selective or multitarget drugs specifically designed to treat SUD.

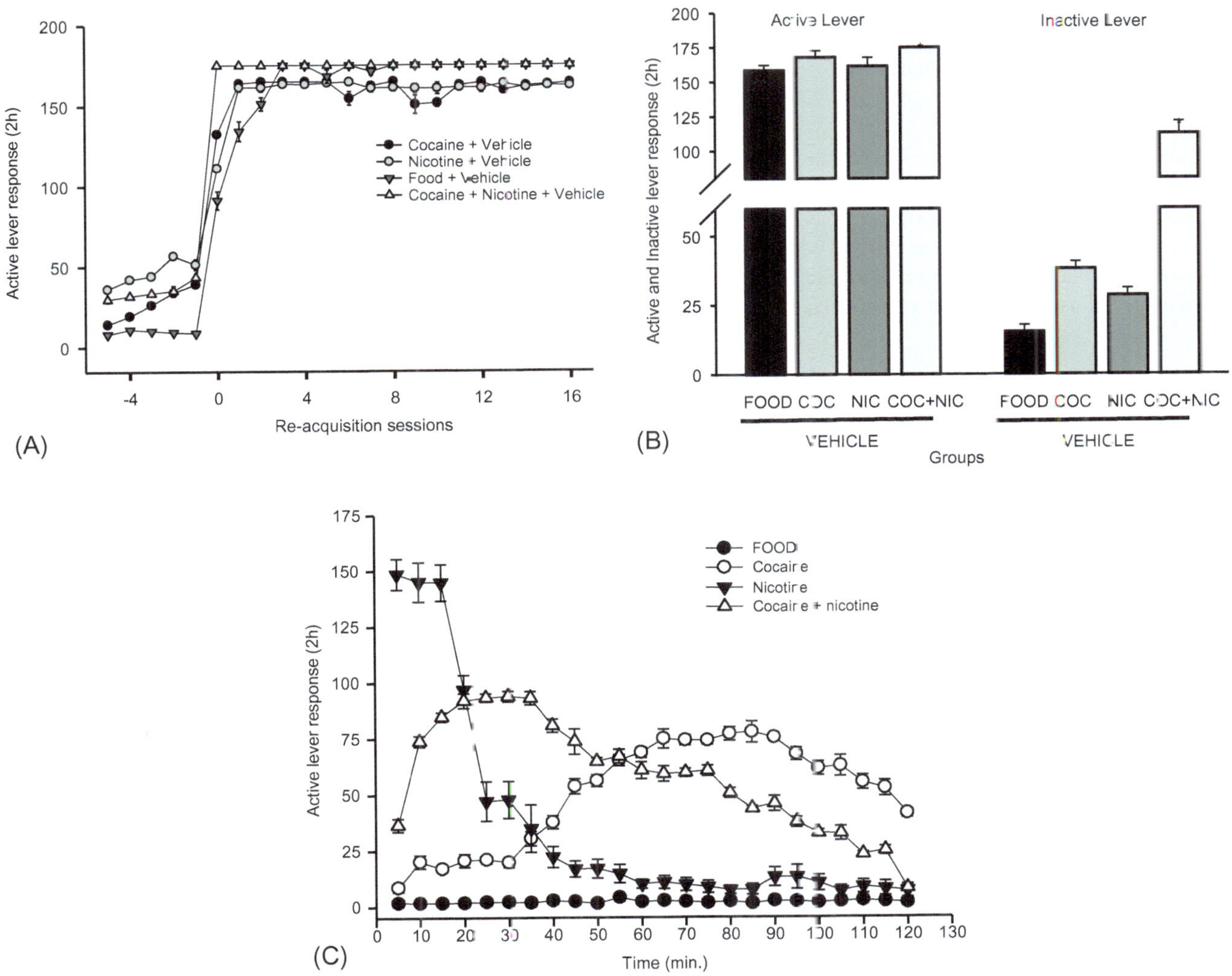


FIG. 2 The cocaine+nicotine mixture increases self-administration. (A) Concurrent administration of cocaine and nicotine increases the number of active lever pressures. (B) Mean active and inactive lever pressures. Two-way ANOVA followed by a post hoc Tukey's test. (*) indicates that significant differences ($P < 0.01$) were found in the cocaine+nicotine+vehicle group, compared to the cocaine+vehicle and nicotine+vehicle group (C) The cocaine+nicotine mixture modifies the consumption pattern of the drug. Nicotine is consumed in the first 20min of the test and cocaine at 40min, while the cocaine+nicotine mixture was consumed during the 120-min test.

## Conclusion and future perspectives

In view of the high prevalence of DD and the absence of effective medications for its treatment, there remains an urgent need to discover new and improved medications, whether they be highly selective or multitarget agents.

The use of drugs that selectively interact with specific targets involved in the etiology of DD has not been very successful; however, drugs that have multitarget mechanisms of action can incorporate monoaminergic (DA, 5-HT, and NE) and non-monoaminergic action mechanisms (ACH, GLU), which opens a range of possibilities to improve the treatment of DD and other complex CNS disorders.

Mirtazapine has a complex action mechanism that covers the characteristics of a multitarget drug, which as evidenced by coadministration studies with monoaminergic agents (prazosin, $\alpha_1$NER) and non-monoaminergic agents ($\alpha4\beta2$nAChR) can be used as a key molecule to design new drugs.

## Key facts

- MDD is a highly prevalent disorder, affecting more than 350 million people worldwide.
- SUD affects 330 million users around the world, leading to 166,000 direct deaths per year; approximately 5%–

6% of the population between ages 15 and 64 has consumed an illegal substance on at least one occasion.
- Dual diagnosis is considered as a serious public health problem, with a prevalence of 11%–27% of the world's population.
- Several preclinical and clinical studies have described the efficacy of mirtazapine in mitigating various behavioral alterations characteristic of SUD.
- Mirtazapine is a multitarget drug, which has demonstrated efficacy in preclinical and clinical studies, supporting its use in different clinical trials.

## Summary points

- Several clinical studies have shown that MDD is a strong risk factor for the development of other health problems such as SUD.
- Mirtazapine is an atypical antidepressant approved by the FDA for the treatment of moderate to severe MDD; due to its sedative, antiemetic, anxiolytic, and appetite-stimulating effects, mirtazapine has also been used in the treatment of other psychiatric disorders.
- Mirtazapine dosed for 30 days during drug withdrawal significantly reduced the induction and expression of locomotor sensitization and CPP, the duration of locomotor effect, and self-administration induced by different doses of cocaine, heroin, methamphetamines, or nicotine.
- Several clinical trials show that mirtazapine decreased withdrawal symptoms, use, and drug craving.
- Mirtazapine is a multitarget molecule, which not only is able to reduce the reinforcing effect of various psychoactive substances through its complex mechanism of action but can also simultaneously decrease withdrawal symptoms; these evidences support the use of mirtazapine in future clinical trials.

## Mini-dictionary of terms

**Substance use disorder** Substance use disorder is a psychiatric disorder characterized by the intense desire and uncontrolled consumption of large amounts of a substance, leading to noncompliance and/or abandonment of important work, social, academic, or domestic activities.

**Dual diagnosis** Dual diagnosis is characterized by the comorbidity of a conventional psychiatric disorder with the use of psychoactive substances.

**Self-administration** Self-administration is a pharmacological procedure that models various aspects of SUD under controlled laboratory conditions in which the animal self-administers psychoactive substances based on drug functioning as a reinforcer that maintains the seeking and intake of the psychoactive substance.

**Behavioral sensitization** Behavioral sensitization is the process whereby repeated, intermittent stimulant administration produces a progressively greater and enduring-behavioral response.

**CPP** CPP is a pharmacological procedure used to evaluate the efficacy of the reinforcement of psychoactive substances using classical or Pavlovian conditioning procedures.

**Polydrug abuse** Polydrug abuse is the simultaneous use of the effects of more than one psychoactive substance in a short period of time.

**Drug withdrawal** Drug withdrawal is the abrupt discontinuation or decrease in intake of psychoactive substances and the symptoms that characterize this stage.

**Dosing program** Dosing program is the set of pharmacological parameters (dose, route administration, days, etc.) under which a drug or medication will be administered.

**Multitarget drug** Multitarget drug is a drug that can act on multiple pharmacologically relevant targets for the generation of a therapeutic effect.

## References


Afshar, M., Knapp, C. M., Sarid-Segal, O., Devine, E., Colaneri, L. S., Tozier, L., et al. (2012). The efficacy of mirtazapine in the treatment of cocaine dependence with comorbid depression. *The American Journal of Drug and Alcohol Abuse*, *38*, 181–186.

Altintoprak, A. E., Zorlu, N., Coskunol, H., Akdeniz, F., & Kitapcioglu, G. (2008). Effectiveness and tolerability of mirtazapine and amitriptyline in alcoholic patients with co-morbid depressive disorder: A randomized, double-blind study. *Human Psychopharmacology*, *23*, 313–319.

Anttila, S. A., & Leinonen, E. V. (2001). A review of the pharmacological and clinical profile of mirtazapine. *CNS Drug Reviews*, *7*, 249–264.

Barbosa-Méndez, S., Jurado, N., Matus-Ortega, M., Martiñon, S., Heinze, G., & Salazar-Juárez, A. (2017). Mirtazapine attenuates the expression of nicotine-induced locomotor sensitization in rats. *European Journal of Pharmacology*, *812*, 28–37.

Barbosa-Méndez, S., Leff, P., Arías-Caballero, A., Hernández-Miramontes, R., Heinze, G., & Salazar-Juárez, A. (2017). Mirtazapine attenuates cocaine seeking in rats. *Journal of Psychiatric Research*, *92*, 38–46.

Barbosa-Méndez, S., Matus-Ortega, M., Flores-Zamora, A., Jurado, N., & Salazar-Juárez, A. (2017). Dose- and time-dependent effects of mirtazapine on the expression of cocaine-induced behavioral sensitization in rats. *Psychiatry Research*, *254*, 301–310.

Barbosa-Méndez, S., Matus-Ortega, M., Jacinto-Gutiérrez, S., & Salazar-Juárez, A. (2018). Mirtazapine impairs acquisition and reinstatement of cocaine-induced place preference in rats. *European Journal of Pharmacology*, *820*, 183–190.

Barbosa-Méndez, S., Matus-Ortega, M., & Salazar-Juárez, A. (2017). Synergistic interactions between mirtazapine and prazosin prevent the induction and expression of behavioral sensitization to cocaine in rats. *Physiology & Behavior*, *180*, 137–145.

Barbosa-Méndez, S., & Salazar-Juárez, A. (2018a). Mirtazapine attenuates nicotine-seeking behavior in rats. *Journal of Psychopharmacology*, *32*, 1010–1017.

Barbosa-Méndez, S., & Salazar-Juárez, A. (2018b). Cocaine + nicotine mixture enhances induction and expression of behavioral sensitization in rats. *Journal of Psychiatric Research*, *100*, 88–98.

Barbosa-Méndez, S., & Salazar-Juárez, A. (2019a). Mirtazapine attenuates anxiety- and depression-like behaviors in rats during cocaine withdrawal. *Journal of Psychopharmacology*, *33*, 589–605.

Barbosa-Méndez, S., & Salazar-Juárez, A. (2019b). Mirtazapine decreased induction and expression of cocaine + nicotine-induced locomotor sensitization in rats. *The World Journal of Biological Psychiatry*, *18*, 1–17.

Bhatia, K. S., Szabo, S. T., Fowler, J. C., Wetsel, W. C., & Lee, T. H. (2011). Reversal of long-term methamphetamine sensitization by combination of pergolide with ondansetron or ketanserin, but not mirtazapine. *Behavioural Brain Research*, *223*, 227–232.

Brunette, M. F., Akerman, S. C., Dawson, R., O'Keefe, C. D., & Green, A. I. (2016). An open-label pilot study of quetiapine plus mirtazapine for heavy drinkers with alcohol use disorder. *Alcohol*, *53*, 45–50.

Chandrasekaran, P. K. (2008). Employing mirtazapine to aid benzodiazepine withdrawal. *Singapore Medical Journal*, *49*, e166–e167.

Colfax, G. N., Santos, G. M., Das, M., Santos, D. M., Matheson, T., Gasper, J., et al. (2011). Mirtazapine to reduce methamphetamine use: A randomized controlled trial. *Archives of General Psychiatry*, *68*, 1168–1175.

Cornelius, J. R., Chung, T., Douaihy, A. B., Kirisci, L., Glance, J., Kmiec, J., et al. (2016a). Mirtazapine in comorbid major depression and an alcohol use disorder: A double-blind placebo-controlled pilot trial. *Psychiatry Research*, *242*, 326–330.

Cornelius, J. R., Chung, T. A., Douaihy, A. B., Kirisci, L., Glance, J., Kmiec, J., et al. (2016b). A review of the literature of mirtazapine in co-occurring depression and an alcohol use disorder. *Journal of Addictive Behaviors, Therapy & Rehabilitation*, *5*. pii:159.

Cornelius, J. R., Douaihy, A. B., Clark, D. B., Chung, T., Wood, D. S., & Daley, D. (2012). Mirtazapine in comorbid major depression and alcohol dependence: An open-label trial. *Journal of Dual Diagnosis*, *8*, 200–204.

Cornelius, J. R., Douaihy, A. B., Clark, D. B., Daley, D. C., Chung, T. A., Wesesky, M. A., et al. (2013). Mirtazapine in comorbid major depression and alcohol use disorder: A long-term follow-up study. *Journal of Addictive Behaviors, Therapy & Rehabilitation*, *3*(1). pii:1560.

Croom, K. F., Perry, C. M., & Plosker, G. L. (2009). Mirtazapine: A review of its use in major depression and other psychiatric disorders. *CNS Drugs*, *23*, 427–452.

Cruickshank, C. C., Montebello, M. E., Dyer, K. R., Quigley, A., Blaszczyk, J., Tomkins, S., et al. (2008). A placebo-controlled trial of mirtazapine for the management of methamphetamine withdrawal. *Drug and Alcohol Review*, *27*, 326–333.

de Bejczy, A., & Söderpalm, B. (2015). The effects of mirtazapine versus placebo on alcohol consumption in male high consumers of alcohol: A randomized, controlled trial. *Journal of Clinical Psychopharmacology*, *35*, 43–50.

de Boer, T. (1996). The pharmacologic profile of mirtazapine. *The Journal of Clinical Psychiatry*, *57*(4), 19–25.

Gerasimov, M. R., Franceschi, M., Volkow, N. D., Rice, O., Schiffer, W. K., & Dewey, S. L. (2000). Synergistic interactions between nicotine and cocaine or methylphenidate depend on the dose of dopamine transporter inhibitor. *Synapse*, *38*, 432–437.

Grasmäder, K., Verwohlt, P. L., Kühn, K. U., Frahnert, C., Hiemke, C., Dragicevic, A., et al. (2005). Relationship between mirtazapine dose, plasma concentration, response, and side effects in clinical practice. *Pharmacopsychiatry*, *38*, 13–17.

Graves, S. M., & Napier, T. C. (2011). Mirtazapine alters cue-associated methamphetamine seeking in rats. *Biological Psychiatry*, *69*, 275–281.

Graves, S. M., & Napier, T. C. (2012). SB 206553, a putative 5-HT2C inverse agonist, attenuates methamphetamine-seeking in rats. *BMC Neuroscience*, *13*, 65. 14.

Graves, S. M., Persons, A. L., Riddle, J. L., & Napier, T. C. (2012). The atypical antidepressant mirtazapine attenuates expression of morphine-induced place preference and motor sensitization. *Brain Research*, *1472*, 45–53.

Graves, S. M., Rafeyan, R., Watts, J., & Napier, T. C. (2012). Mirtazapine, and mirtazapine-like compounds as possible pharmacotherapy for substance abuse disorders: Evidence from the bench and the bedside. *Pharmacology & Therapeutics*, *136*, 343–353.

Haddjeri, N., Blier, P., & de-Montigny, C. (1998). Acute and long-term actions of the antidepressant drug mirtazapine on central 5-HT neurotransmission. *Journal of Affective Disorders*, *51*, 255–256.

Herrold, A. A., Shen, F., Graham, M. P., Harper, L. K., Specio, S. E., Tedford, C. E., et al. (2009). Mirtazapine treatment after conditioning with methamphetamine alters subsequent expression of place preference. *Drug and Alcohol Dependence*, *99*, 231–239.

Iwamoto, K., Kawano, N., Sasada, K., Kohmura, K., Yamamoto, M., Ebe, K., et al. (2013). Effects of low-dose mirtazapine on driving performance in healthy volunteers. *Human Psychopharmacology*, *28*, 523–528.

Kang, L., Wang, D., Li, B., Hu, M., Zhang, P., & Li, J. (2008). Mirtazapine, a noradrenergic and specific serotonergic antidepressant, attenuates morphine dependence and withdrawal in Sprague-Dawley rats. *The American Journal of Drug and Alcohol Abuse*, *34*, 541–552.

Kongsakon, R., Papadopoulos, K. I., & Saguansiritham, R. (2005). Mirtazapine in amphetamine detoxification: A placebo-controlled pilot study. *International Clinical Psychopharmacology*, *20*, 253–256.

Liappas, J., Paparrigopoulos, T., Malitas, P., Tzavellas, E., & Christodoulou, G. (2004). Mirtazapine improves alcohol detoxification. *Journal of Psychopharmacology*, *18*, 88–93.

Liappas, J., Paparrigopoulos, T., Tzavellas, E., & Christodoulou, G. (2003). Alcohol detoxification and social anxiety symptoms: A preliminary study of the impact of mirtazapine administration. *Journal of Affective Disorders*, *76*, 279–284.

Liappas, J., Paparrigopoulos, T., Tzavellas, E., & Rabavilas, A. (2005). Mirtazapine and venlafaxine in the management of collateral psychopathology during alcohol detoxification. *Progress in Neuro-Psychopharmacology & Biological Psychiatry*, *29*, 55–60.

Lima, M. S., Reisser, A. A., Soares, B. G., & Farrell, M. (2010). Antidepressants for cocaine dependence. *Cochrane Database of Systematic Reviews*, *2*, CD002950.

McDaid, J., Tedford, C. E., Mackie, A. R., Dallimore, J. E., Mickiewicz, A. L., Shen, F., et al. (2007). Nullifying drug-induced sensitization: Behavioral and electrophysiological evaluations of dopaminergic and serotonergic ligands in methamphetamine-sensitized rats. *Drug and Alcohol Dependence*, *86*, 55–66.

McGregor, C., Srisurapanont, M., Mitchell, A., Wickes, W., & White, J. M. (2008). Symptoms and sleep patterns during inpatient treatment of methamphetamine withdrawal: A comparison of mirtazapine and modafinil with treatment as usual. *Journal of Substance Abuse Treatment*, *35*, 334–342.

Mello, N. K., & Newman, J. L. (2011). Discriminative and reinforcing stimulus effects of nicotine, cocaine, and cocaine + nicotine combinations in rhesus monkeys. *Experimental and Clinical Psychopharmacology*, *19*, 203–214.

Millan, M. J. (2006). Multi-target strategies for the improved treatment of depressive states: Conceptual foundations and neuronal substrates, drug discovery and therapeutic application. *Pharmacology & Therapeutics*, *110*, 135–370.

Ng, E., Browne, C. J., Samsom, J. N., & Wong, A. H. C. (2017). Depression and substance use comorbidity: What we have learned from animal studies. *The American Journal of Drug and Alcohol Abuse*, *43*, 456–474.

Ranjan, O. P., Shavi, G. V., Nayak, U. Y., Arumugam, K., Averineni, R. K., Meka, S. R., et al. (2011). Controlled release chitosan microspheres of mirtazapine: In vitro and in vivo evaluation. *Archives of Pharmacal Research*, *34*, 1919–1929.

Salazar-Juárez, A., Barbosa-Méndez, S., Jurado, N., Hernández-Miramontes, R., Leff, P., & Antón, B. (2016). Mirtazapine prevents induction and expression of cocaine-induced behavioral sensitization in rats. *Progress in Neuro-Psychopharmacology & Biological Psychiatry*, *68*, 15–24.

Salazar-Juárez, A., Barbosa-Méndez, S., Merino-Reyes, P., Matus-Ortega, M., Hernández-Calderón, J. A., & Antón, B. (2017). Chronic dosing with mirtazapine does not produce sedation in rats. *Brazilian Journal of Psychiatry*, *39*, 228–236.

Santucci, K. (2012). Psychiatric disease and drug abuse. *Current Opinion in Pediatrics*, *24*, 233–237.

Sasada, K., Iwamoto, K., Kawano, N., Kohmura, K., Yamamoto, M., Aleksic, B., et al. (2013). Effects of repeated dosing with mirtazapine, trazodone, or placebo on driving performance and cognitive function in healthy volunteers. *Human Psychopharmacology*, *28*, 281–286.

Schenk, S., Foote, J., Aronsen, D., Bukholt, N., Highgate, Q., Van de Wetering, R., et al. (2016). Serotonin antagonists fail to alter MDMA self-administration in rats. *Pharmacology, Biochemistry, and Behavior*, *148*, 38–45.

Torrens, M., Fonseca, F., Mateu, G., & Farré, M. (2005). Efficacy of antidepressants in substance use disorders with and without comorbid depression. A systematic review and meta-analysis. *Drug and Alcohol Dependence*, *78*, 1–22.

Voigt, R. M., Mickiewicz, A. L., & Napier, T. C. (2011). Repeated mirtazapine nullifies the maintenance of previously established methamphetamine-induced conditioned place preference in rats. *Behavioural Brain Research*, *225*, 91–96.

Voigt, R. M., & Napier, T. C. (2012). Context-dependent effects of a single administration of mirtazapine on the expression of methamphetamine-induced conditioned place preference. *Frontiers in Behavioral Neuroscience*, *5*, 92. 13.

Vysloužil, J., Doležel, P., Kejdušová, M., Košťál, V., Beneš, L., & Dvořáčková, K. (2016). Long-term controlled release of PLGA microparticles containing antidepressant mirtazapine. *Pharmaceutical Development and Technology*, *21*, 214–221.

Walker, E. R., McGee, R. E., & Druss, B. G. (2015). Mortality in mental disorders and global disease burden implications a systematic review and meta-analysis. *JAMA Psychiatry*, *72*, 334–341.

Yoon, S. J., Pae, C. U., Kim, D. J., Namkoong, K., Lee, E., Oh, D. Y., et al. (2006). Mirtazapine for patients with alcohol dependence and comorbid depressive disorders: A multicentre, open label study. *Progress in Neuro-Psychopharmacology & Biological Psychiatry*, *30*, 1196–1201.

Zayara, A. E., McIver, G., Valdivia, P. N., Lominac, K. D., McCreary, A. C., & Szumlinski, K. K. (2011). Blockade of nucleus accumbens 5-HT2A and 5-HT2C receptors prevents the expression of cocaine-induced behavioral and neurochemical sensitization in rats. *Psychopharmacology*, *213*, 321–335.

Zueco-Pérez, P. L. (2002). Mirtazapine in the treatment of cocaine-dependence in patients with methadone. *Actas Españolas de Psiquiatría*, *30*, 337–342.

# Part IV

# Counselling, psychotherapy and behavioural treatments for depression

Chapter 40

# Mindfulness-based cognitive therapy and depression

Tatjana Ewais[a,b]

[a]Clinical Medicine, School of Medicine, Griffith University, Gold Coast, QLD, Australia; [b]Mater Clinical School, Faculty of Medicine, University of Queensland, Brisbane, QLD, Australia

## List of abbreviations

| | |
|---|---|
| **CNS** | central nervous system |
| **HPA axis** | hypothalamus-pituitary-adrenal axis |
| **MBIs** | mindfulness-based interventions |
| **MBCT** | mindfulness-based cognitive therapy |
| **MBSR** | mindfulness-based stress reduction-MBSR |
| **3MBS** | 3-min breathing space |
| **TRD** | treatment-resistant depression |

## Overview

Depression is one of the leading causes of morbidity and disability worldwide with high individual, health care, and societal costs (Vos et al., 2017). Despite major advances in the understanding of etiology and neurobiology of depression, only a third of those treated achieve remission after the first-line treatment, and a third fail to respond to multiple antidepressant trials and are identified as treatment resistant (Gaynes et al., 2009). Depression is a recurring and relapsing condition, with probability of depressive relapse increasing with every subsequent depressive episode (Eaton et al., 2008). Depression recurrence and relapse are the key factors contributing to its high prevalence and costs, disability, and disease burden.

In this context, there has been an ongoing interest in mindfulness-based therapies as novel interventions able to target different pathways involved in the pathogenesis of depression and depressive relapse (Kuyken et al., 2016; Strauss et al., 2014). Mindfulness-based interventions are increasingly used in the treatment of depressive disorders in various clinical populations and in a variety of settings (Abbott et al., 2014; Chi et al., 2018; Goldberg et al., 2018). This chapter examines the evidence regarding the applications and effectiveness of mindfulness interventions in the treatment of depressive disorders, with focus on mindfulness-based cognitive therapy (MBCT).

## Mindfulness-based interventions and depression

Mindfulness-based interventions (MBIs) are therapeutic interventions, which contain and are based on core mindfulness principles and practices. Broadly defined, MBIs include mindfulness-based programs, third-wave mindfulness therapies, mindfulness-informed interventions, and yoga. Mindfulness-based interventions place emphasis on noticing and accepting our thoughts, feelings, bodily sensations, and other experiences as they are and the resulting shift in our relationship with them. This emphasis on acceptance and change in our relationship with experiences, rather than on trying to change or avoid those experiences, is a key feature distinguishing MBIs from other psychological therapies.

The use of MBIs in mainstream health care has burgeoned since the introduction of the first mindfulness-based program called mindfulness-based stress reduction (MBSR) by Jon Kabat-Zinn in the late 1970s (Kabat-Zinn, 1991). Although MBSR was initially designed as a complementary therapy for medically ill, it has since been widely implemented in diverse clinical and nonclinical settings (Crowe et al., 2015; Jamieson & Tuckey, 2017; Khoury et al., 2015). Mindfulness-based cognitive therapy was developed by Segal, Williams, and Teasdale, who adapted the MBSR curriculum through addition of cognitive therapy components aimed at preventing relapse of depression (Segal, Williams, & Teasdale, 2002). Both MBSR and MBCT are evidence-based, 8-week group, mindfulness-based programs, which contain a mixture of mindfulness practices and psychoeducation.

The use of MBIs in the treatment of depression has evolved from research and clinical evidence of the effectiveness of mindfulness training in ameliorating psychological triggers of depression onset and relapse such as stress and rumination, as well as modifying some of the

The Neuroscience of Depression. https://doi.org/10.1016/B978-0-12-817933-8.00042-6

biological mechanisms underlying depression, such as inflammation and brain connectedness (van der Velden et al., 2015). Recent systematic reviews have confirmed the effectiveness of MBIs in treating depression in different clinical cohorts including youth (Chi et al., 2018; Zoogman et al., 2015), women during perinatal period (Dhillon, Sparkes, & Duarte, 2017; Taylor, Cavanagh, & Strauss, 2016), those with recurrent (Kuyken et al., 2016) and treatment-resistant depression (TRD) (Kenny & Williams, 2007), and comorbid mental health (Goldberg et al., 2018) and physical disorders (Abbott et al., 2014; Ewais et al., 2019a; Goldberg et al., 2018).

## MBCT for prevention of depressive relapse

MBCT is an 8-week, manualized, mindfulness-based group program that integrates mindfulness practices with elements of cognitive-behavioral therapy for depression. The program contains weekly group sessions of about 2h duration, an all-day silent retreat, and between-session individual homework practices of up to 45min a day. Segal, Williams, and Teasdale originally designed MBCT as a relapse prevention intervention for people in remission who had a history of three or more depressive episodes and were at high risk of relapse (Segal, Williams, & Teasdale, 2013). They recognized the need to create a therapeutic intervention combining mindfulness and cognitive approaches as relapse rates with maintenance antidepressants were high and there was a sufficient evidence of efficacy of both cognitive therapy and mindfulness training in targeting processes involved in relapse of depression (Segal et al., 2013).

Theoretical model of depressive relapse underpinning MBCT is based on the evidence that individuals with history of recurrent depression have heightened cognitive vulnerability that causes automatic activation of negative thinking patterns in response to relatively small changes in mood. These negative thoughts persist through rumination, and lead to deepening of the initial mood disturbance and depressive relapse (Williams & Kuyken, 2012). Rumination is a repetitive, preoccupied thinking pattern, triggered and maintained through "discrepancy monitoring" or looking for a mismatch between the reality and one's desires and expectations. Discrepancy monitoring and rumination are hallmarks of the "driven-doing mode" state of mind, characterized by recurrent, ineffective attempts to change one's thoughts and feelings (Crane, 2017). MBCT helps those at risk of depression recognize the automatic activation of maladaptive cognitive patterns early and disengage from them through the use of mindfulness skills and shifting of their relationship to their thoughts and feelings. Fig. 40.1 shows the role of driven mode, rumination, and discrepancy monitoring in triggering depression and the impact of MBCT on preventing

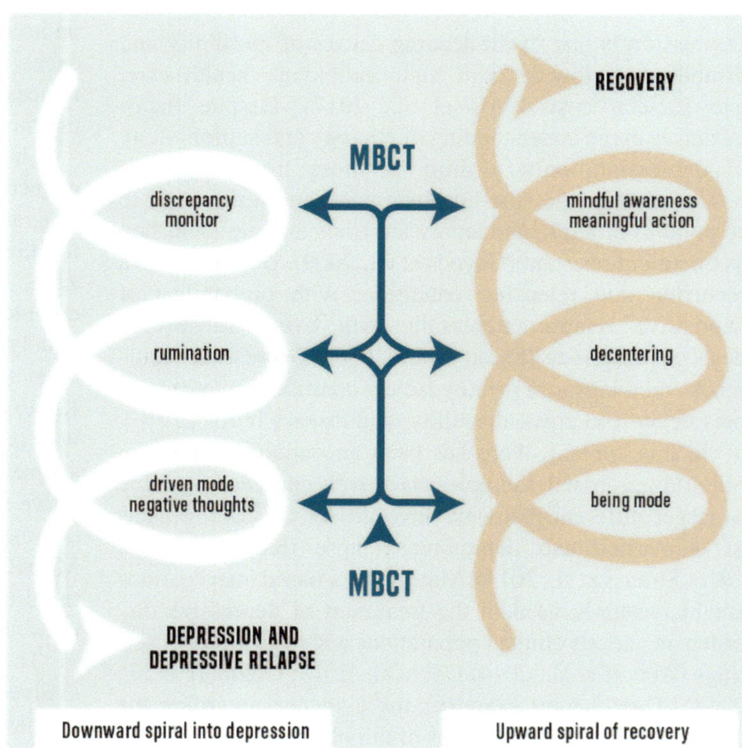


FIG. 40.1 Impact of MBCT on preventing a depressive relapse by decreasing rumination, driven mode, and discrepancy monitoring, and leading to recovery through use of decentering, mindful awareness, and value-based, meaningful action.

depressive relapse through the use of decentering, mindful awareness, and staying present in the being mode, focused on moment-by-moment experience.

Creators of MBCT adapted the existing MBSR curriculum by adding cognitive therapy components targeting processes involved in depressive relapse such as rumination, and included metacognitive and attentional practices of "decentering" or observing one's thoughts as passing mental phenomena rather than absolute truths. They modified the psychoeducation components of MBSR by adding educational content specific for depression and introduced shorter mindfulness practices such as a "3-min breathing space," which has become one of the signature practices of the MBCT program (Segal et al., 2013).

Segal, Williams, and Teasdale describe the 3-min breathing space (3MBS) as an initial, brief, and easily accessible practice that could assist participants in dealing with difficult situations, thoughts, and emotions in their daily lives, as well as bring the perspectives of the longer formal mindfulness practices (Segal et al., 2013).

The practice during the 3MBS involves open or wide-angle attention and focused or narrow-angle attention and their flow through three consecutive steps. The practice begins with step 1 or "awareness" during which participants are invited to direct their attention to their current inner experiences, thoughts, feelings, and body sensations, and acknowledge them as they are. Step 2 or "gathering" emphasizes the gathering or focusing one's attention to the physical sensation of the breath in a particular part of the body. Step 3 or "expanding" is about broadening the attention to the body as a whole and its surroundings.

The movement of the attention during the 3MBS is described using the metaphor of the sand moving through an hourglass as illustrated in Fig. 40.2.

In addition to attending the weekly MBCT sessions, participants in the MBCT program are encouraged to engage in regular formal and informal mindfulness practices between the weekly group sessions. These between-session practices are an integral component of the MBCT curriculum and crucial in achieving the beneficial "dose" of mindfulness training. Parsons and colleagues conducted a systematic review and meta-analysis of participants' between-session formal mindfulness practices, which showed completion rates of 64% of the assigned home practices and a small but significant association of completion of formal practices with improved treatment outcomes (Parsons et al., 2017).

Initial trials of MBCT confirmed its efficacy in preventing depressive relapse in currently well individuals with history of three or more depressive episodes, and found that MBCT halved the risk of relapse when compared to standard care in this cohort (Teasdale et al., 2000). Subsequent trials demonstrated that MBCT was cost effective and comparable to antidepressants in its ability to reduce depressive relapse (Kuyken et al., 2008, 2015). A recent meta-analysis of individual patient data from RCTs provided further evidence regarding the long-term efficacy of MBCT, reporting that patients receiving MBCT had significantly reduced risk of depressive relapse during a 60-week follow-up period compared with those who received treatment as usual (Kuyken et al., 2016). The same meta-analysis found that patients who received MBCT had comparable outcomes to those who received other active treatments and that those who were more depressed at baseline experienced greater improvement, indicating that MBCT may be particularly helpful for individuals with significant residual depressive symptoms (Kuyken et al., 2016).

FIG. 40.2 The hourglass model of 3-min breathing space.

Overall, the findings from several systematic reviews and meta-analyses have shown that MBCT is an effective treatment for reducing the rate of depressive relapse in individuals with recurrent depression who are in remission. As a result of these findings and ongoing research and training, MBCT is currently recommended in several national clinical guidelines for maintenance and preventing relapse of depression in individuals suffering from recurrent major depressive disorder (Malhi et al., 2018).

## MBCT in the treatment of current depression

Strong empirical support for the efficacy of MBCT in preventing depressive relapse, combined with improved understanding of its mechanisms of action, led to MBCT being trialed in people with current depression and with less than three past depressive episodes. Studies of MBCT for treatment of currently depressed individuals applied theoretical model of MBCT-mediated prevention of depressive relapse and showed it to be effective in the treatment of current depression through decrease in rumination and cognitive and emotional reactivity (Aalderen et al., 2012). Fig. 40.1 shows the cognitive processes underlying vulnerability to depressive relapse and perpetuating established depression, and how MBCT can both prevent depressive relapse and treat current depressive symptoms.

Meta-analysis of MBIs for people diagnosed with a current episode of an anxiety or depressive disorder found them to be effective in treating depressive symptoms with a large effect size in favor of MBIs on primary symptom severity (Strauss et al., 2014). The authors also conducted subgroup analyses, which showed that effects on primary depressive symptoms severity were significant for MBCT but not for MBSR and recommended the use of MBCT for treatment of current depression (Strauss et al., 2014). Further studies also tested the long-term effects of MBCT in individuals in remission vs those who were currently depressed and showed that long-term outcomes after MBCT did not differ between patients with current depression and those in remission (Aalderen et al., 2015). Moreover, studies with currently depressed individuals showed that those with current depression could participate in and benefit from MBCT, in contrast to previous suggestions that actively depressed patients could not concentrate well enough to utilize MBCT.

These research findings, when taken in conjunction with the theoretical underpinnings of MBCT, support the use of MBCT in individuals with a current diagnosis of depression and extend the evidence for MBCT for use in current depression as well as for reducing the risk of relapse in those with history of three or more depressive episodes. Future research will continue exploring specific modifications of MBCT curriculum, which are likely to facilitate engagement and therapeutic response in currently depressed individuals and identify subgroups which may be particularly responsive to MBCT.

## MBCT for TRD

TRD, defined as a failure to achieve remission from depressive symptoms after two or more antidepressant medications trials, affects at least a third of all depressed individuals and is a major cause of depression-related burden and disability (Gaynes et al., 2009). Since people with TRD are likely to maintain their depressive symptoms via the same mechanisms underlying depressive relapse and current depressive symptoms, the use of MBCT may help achieve remission in TRD, however, multifactorial nature of TRD may necessitate the use of MBCT in addition to antidepressant medications and other psychosocial interventions in TRD (Eisendrath, Chartier, & McLane, 2011).

Early MBCT trials showed promising results in chronically depressed individuals with TRD and acute depressive symptoms including suicidal thoughts (Kenny & Williams, 2007). They showed that the severity of depressive symptoms and suicidal ideation did not interfere with improvement and that MBCT appeared to be a safe and effective treatment in this patient cohort. A recent RCT of MBCT as an adjunctive treatment to antidepressant medications, in individuals with TRD who were currently depressed, showed MBCT to be significantly more effective in reducing depressive symptoms and to have significantly higher response rates than the control, Health Enhancement program, at the 8-week end point (Eisendrath et al., 2016). Although MBCT also showed greater remission rates than the control intervention, they did not reach statistical significance (Eisendrath et al., 2016). However, remission rates were still superior to many previous psychotherapy and medication trials in TRD, and MBCT was well accepted and feasible, as evidenced by high rates of compliance and low dropout rates. Overall, the results of research to date support the use of MBCT as a viable adjunct in the management of TRD. Future avenues of research may explore mechanisms of change of MBCT in TRD and specific MBCT adaptations in this population.

## Mechanisms of change in MBCT

Research into the mechanisms of therapeutic change in MBCT has been informed by the growing evidence of the efficacy of MBCT and advances in understanding of the neurobiology of depression. There have been two broad areas of research into the MBCT mechanism of change, one exploring the psychological or cognitive mechanisms and one investigating the mechanisms of MBCT impact on neurobiological pathways.

Early hypotheses into the mechanisms of change in mindfulness emerged from Kabat-Zinn's definition of mindfulness as "paying attention in a particular way: on purpose, in the present moment, and nonjudgmentally" (p. 4) (Kabat-Zinn, 1994). Shapiro and colleagues used this definition of mindfulness to propose three fundamental building blocks of mindfulness, intention, attention, and attitude (Shapiro et al., 2006). Their research was expanded by three recent studies, a systematic review and meta-analysis of studies of mediators of MBIs effect, a systematic review of mechanisms of change in MBCT in the treatment of recurrent depression, and a meta-synthesis of qualitative studies on MBCT mechanism of change.

Systematic review and meta-analysis of studies investigating psychological mechanisms mediating the impact of MBIs identified strong evidence for cognitive and emotional reactivity, moderate and consistent evidence for mindfulness, rumination, and worry, and preliminary but insufficient evidence for self-compassion and psychological flexibility (Gu et al., 2015). Systematic review of mechanisms of change in MBCT in recurrent depression found that changes in mindfulness, rumination worry, self-compassion, decentering, or meta-awareness mediated the effect of MBCT on treatment outcome, and in line with the theoretical predicted mechanisms (van der Velden et al., 2015). The same review also reported preliminary evidence regarding the role of changes in attention, memory specificity, self-discrepancy, and emotional reactivity in MBCT-mediated therapeutic change (van der Velden et al., 2015).

Meta-synthesis of qualitative studies identified five key themes underlying the process of therapeutic change in MBCT including translation of skills in everyday life, taking control through understanding, awareness, and acceptance, feelings toward self, expectations, and group impact (Cairns & Murray, 2015). It expanded on the previous work by adding another key component of translating mindfulness skills into everyday life and placed them in the context of individual expectations and group impact (Cairns & Murray, 2015). Fig. 40.3 depicts the key components and key mediators of therapeutic change in MBCT in the treatment of depression.

Research into neurobiological mechanisms of change in MBCT has been informed by advances in understanding of neurobiological pathways involved in depression and the impact of mindfulness on those pathways. The pathways that are most commonly identified as involved in MBCT-mediated therapeutic change are inflammatory and other immune system pathways, hypothalamus-pituitary-adrenal (HPA) axis, microbiome-gut-brain axis, and brain structural and functional neuroconnectivity changes.

Several studies have reported that inflammation plays a key role in the pathophysiology of depression and that the relationship between depression and proinflammatory cytokines is bidirectional (Berk et al., 2013; Mac Giollabhui et al., 2020). There is also increasing evidence regarding the beneficial impact of mindfulness training on inflammation and other components of immune system (Black & Slavich, 2016) Taken together, these findings provide support for the mechanism of change in MBCT through modifying inflammation and other immune system parameters. Furthermore, MBIs have been shown to reduce stress and enhance recovery from stress in chronic illness (Abbott et al., 2014; Ewais et al., 2019a; Sanada et al., 2017) and stress plays an important role in precipitating the onset of depression, both through central nervous system (CNS) changes and peripheral changes in gut microbiota (Bailey et al., 2011; Cryan, 2016). Some of the CNS changes in response to stress involve microglial activation and HPA-axis activation, while the peripheral changes involve dysbiosis in the gut microbiota (Househam et al., 2017). Fig. 40.4 shows the role of stress in microbiome-gut-brain

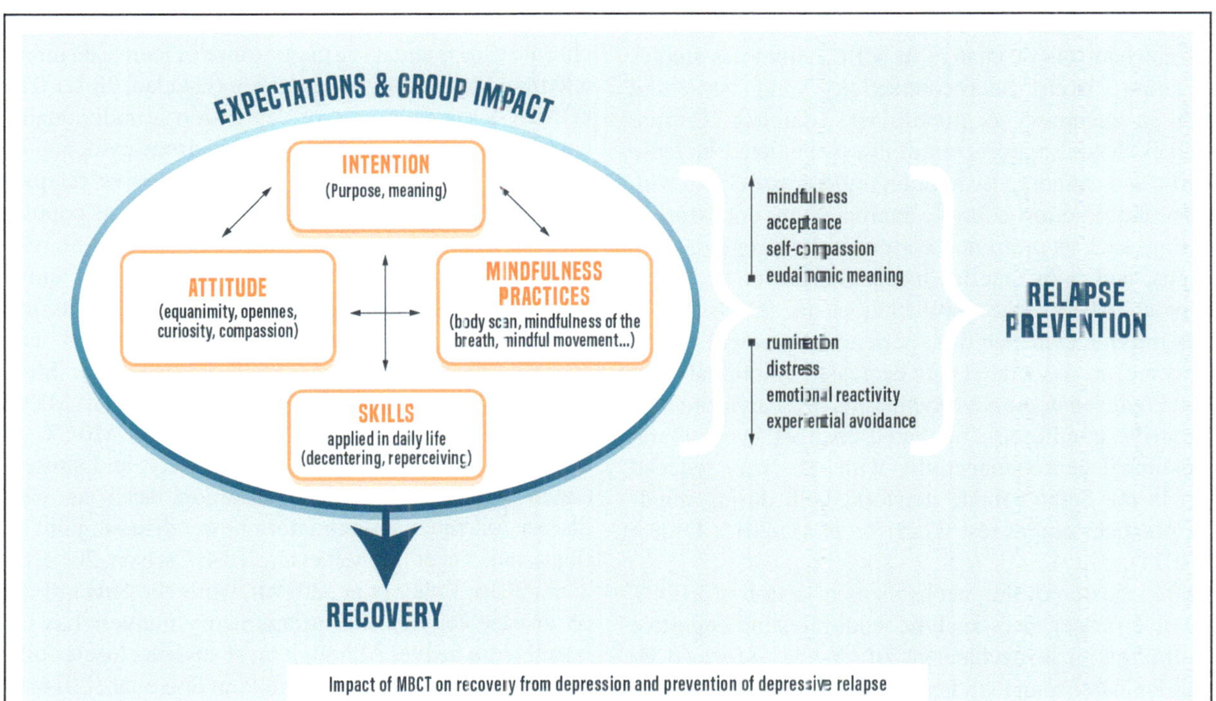


FIG. 40.3 The mechanisms of change in MBCT leading to recovery from depression and prevention of depressive relapse.

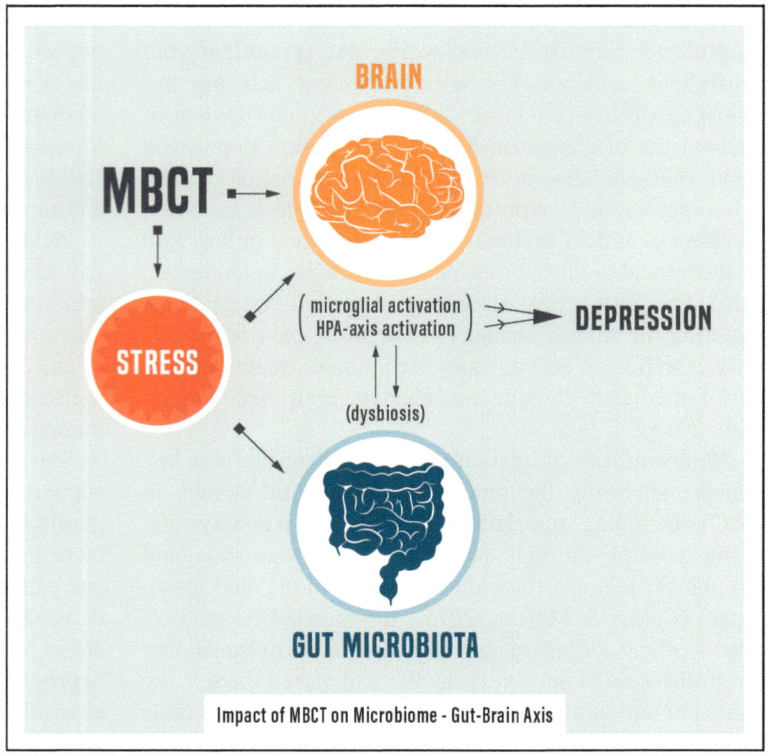


FIG. 40.4 Impact of MBCT on stress and microbiome-gut-brain axis and its role in treating depression.

axis dysregulation and onset of depression as well as the impact of MBCT on attenuating these stress-mediated changes.

Another important area in research exploring neurobiological mechanisms of change in MBCT involves studies investigating brain neuroconnectivity and structural changes in response to mindfulness training. Current research evidence supports mindfulness-mediated increase in volume and connectivity in brain regions associated with attention and emotion control, interoception, and sensory processing, such as prefrontal cortex, parietal regions, hippocampus, and right anterior insula (Hölzel et al., 2011; Lazar et al., 2005; Mac Giollabhui et al., 2020). These changes provide evidence of experience-dependent neuroplasticity and are associated with decreased emotional reactivity and reduced depressive symptoms. Research findings also describe mindfulness-mediated changes in structural and functional neuroconnectivity with relatively reduced activity in the default mode network, both during mindfulness practices and at rest (Garrison et al., 2015; Luders et al., 2011).

Future research of the mechanisms of action of MBCT will involve further meta-analytic studies testing cognitive and neurobiological mechanisms of change, informed by methodologically sound studies and supported by the latest advances in understanding of neurobiology of depression and mindfulness.

## MBCT for depression associated with chronic illness

Depression in medically ill worsens the course of the chronic illness and it is a major cause of increased morbidity and mortality in this patient group (Haddad, 2009). The use of MBCT for treatment of depression in individuals with comorbid physical illness emerged from evidence of the efficacy of MBCT in preventing depressive relapse and treating current depression in various clinical populations and settings. It was also a natural extension of research showing bidirectional links between depression and many chronic physical illnesses and shared common inflammatory pathways underlying depression and chronic physical illness (Mikocka-Walus et al., 2016; Moulton, Pickup, & Ismail, 2015). Recent reviews of MBCT in chronic physical disorders have shown MBCT to be effective in reducing depression, anxiety, and stress associated with many chronic disorders such as vascular disease, diabetes, inflammatory bowel disease, joint conditions, and cancer (Abbott et al., 2014; Carlson, 2012; Crowe et al., 2015; Ewais et al., 2019a), while the effect of MBCT on disease activity and inflammatory markers has largely been inconclusive. Although most studies closely followed the original MBCT curriculum (van Son et al., 2014), some introduced adaptations to both educational component and mindfulness practices aimed at providing disease-specific education and accommodating disease-related limitations such as pain and fatigue (Ewais et al., 2019b). Future research avenues in area of MBCT in individuals with chronic illness will further the exploration of biological markers, MBCT adaptations and innovative modes of delivery such as distance-delivered MBCT in group and individual format, and synchronous and asynchronous participation.

## Summary and future directions in the use of MBCT

There is a strong research evidence supporting the effectiveness of MBCT in both preventing depressive relapse in individuals with recurrent depression and treating current depression. Research findings have also shown positive impact of between-session formal mindfulness practices on depression and other outcomes.

In addition to evidence of MBCT effectiveness in adults with depression, there is a growing body of evidence demonstrating benefits of MBCT in different clinical populations including youth, women during the peripartum period, individuals with TRD, those with other mental health conditions, and individuals with chronic physical illness.

Research on the mechanisms of change in MBCT in depression has identified the key psychological and neurobiological components and mediators of the process of therapeutic change in MBCT, and it progresses with deepening of our understanding of cognitive and neurobiological underpinnings of depression.

Future research of MBCT in prevention and treatment of depression would benefit from addressing the remaining methodological and conceptual issues, exploring novel adaptations and mechanisms of action, and examining the quality and composition of mindfulness practices and their continuation rates.

## Key facts of mindfulness-based cognitive therapy and depression

- Depression is one of the leading causes of morbidity and disability worldwide with high individual, health care, and societal costs.
- Recurrence and relapse of depression are the key factors contributing to its high prevalence, costs, and disease burden.
- Recent systematic reviews have confirmed the effectiveness of mindfulness-based interventions in treating depression in different clinical populations.
- Mindfulness-based cognitive therapy (MBCT) is an 8-week, manualized, mindfulness-based program, which was designed for the prevention of depressive relapse in individuals with three or more depressive episodes.
- Studies have shown that MBCT is also effective in treating current depression as well as those with history of less than three depressive episodes.
- Changes in mindfulness, rumination, self-compassion, and decentering or meta-awareness have been shown to mediate the effect of MBCT
- MBCT is effective in reducing depression, anxiety, and stress associated with many chronic disorders such as vascular disease, diabetes, inflammatory bowel disease, joint conditions, and cancer.
- Future research of MBCT would benefit from exploring its novel adaptations and mechanisms of action and examining composition of mindfulness practices and their continuation rates.

## Summary points

- This chapter examines the evidence regarding the applications and effectiveness of mindfulness interventions in the treatment of depressive disorders, with focus on mindfulness-based cognitive therapy (MBCT).
- MBCT is effective in both preventing depressive relapse and treating those with current depressive symptoms.
- MBCT is a viable adjunct in the management of treatment-resistant depression (TRD).
- The effect of MBCT in recurrent depression is mediated by changes in mindfulness, rumination, self-compassion, and decentering or meta-awareness.
- Meta-synthesis of qualitative studies identified five key themes underlying therapeutic change in MBCT: translation of skills in everyday life, taking control through understanding, awareness and acceptance, feelings toward self, expectations and group impact.
- Neurobiological pathways underpinning therapeutic change in MBCT include inflammatory and other immune system pathways, hypothalamus-pituitary-adrenal (HPA) axis, microbiome-gut-brain axis, and brain structural and functional neuroconnectivity changes.
- MBCT is effective in reducing depression, anxiety, and stress associated with many chronic disorders.
- Future research in the use of MBCT in prevention and treatment of depression would benefit from further exploration of its mechanisms of action, adaptations and innovative modes of delivery such as distance-delivered MBCT in group and individual format, and synchronous and asynchronous participation.

## Mini-dictionary of terms

**Depression remission** The absence of depressive symptoms associated with return to normal functioning.

**Recovery from depression** The absence of depressive symptoms for at least 4 months following the onset of remission.

**Depression relapse** Return of depressive symptoms after remission but before full recovery was achieved.

**Depression recurrence** Return of depressive symptoms after recovery from depression.

**Mindfulness-based interventions (MBIs)** Therapeutic interventions which contain and are based on core mindfulness principles and practices.

**Mindfulness-based cognitive therapy (MBCT)** Eight-week, manualized, mindfulness-based group program that integrates mindfulness practices with elements of cognitive-behavioral therapy for depression.

**Mindfulness-based stress reduction (MBSR)** Eight-week group, manualized, mindfulness-based group program, which contain a mixture of various formal and informal mindfulness practices and psychoeducation.

**Treatment-resistant depression (TRD)** Depression characterized by failure to achieve remission after two or more trials of antidepressant medications.

**Decentering** Observing one's thoughts and emotions as transient mental phenomena rather than identifying or fusing with them or considering them as absolute truths.

**Metacognitive awareness** The process of being aware of one's thoughts, sometimes used interchangeably with decentering.

## References

Aalderen, J. R.v., et al. (2012). The efficacy of mindfulness-based cognitive therapy in recurrent depressed patients with and without a current depressive episode: A randomized controlled trial. *Psychological Medicine*, *42*, 989–1001.

Aalderen, J. R., et al. (2015). Long-term outcome of mindfulness-based cognitive therapy in recurrently depressed patients with and without a depressive episode at baseline. *Depression and Anxiety*, *32*(8), 563–569.

Abbott, R. A., et al. (2014). Effectiveness of mindfulness-based stress reduction and mindfulness based cognitive therapy in vascular disease: A systematic review and meta-analysis of randomised controlled trials. *Journal of Psychosomatic Research*, *76*(5), 341–351.

Bailey, M. T., et al. (2011). Exposure to a social stressor alters the structure of the intestinal microbiota: Implications for stressor-induced immunomodulation. *Brain, Behavior, and Immunity*, *25*(3), 397–407.

Berk, M., et al. (2013). So depression is an inflammatory disease, but where does the inflammation come from? *BMC Medicine*, *11*(1), 1. 200.

Black, D. S., & Slavich, G. M. (2016). Mindfulness meditation and the immune system: A systematic review of randomized controlled trials. *Annals of the New York Academy of Sciences*, *13731*(1), 13–24.

Cairns, V., & Murray, C. (2015). How do the features of mindfulness-based cognitive therapy contribute to positive therapeutic change? A meta-synthesis of qualitative studies. *Behavioural and Cognitive Psychotherapy*, *43*(3), 342–359.

Carlson, L. E. (2012). Mindfulness-based interventions for physical conditions: A narrative review evaluating levels of evidence. *International Scholarly Research Network Psychiatry*, *2012*, 6. 651583.

Chi, X. L., et al. (2018). Effects of mindfulness-based stress reduction on depression in adolescents and young adults: A systematic review and meta-analysis. *Frontiers in Psychology*, *9*, 1. 1034.

Crane, R. (2017). *Mindfulness-based cognitive therapy: Distinctive features* (2nd ed.). Abingdon, Oxon, New York, NY: Routledge, an imprint of the Taylor & Francis Group. ProQuest.

Crowe, M., et al. (2015). Mindfulness-based stress reduction for long-term physical conditions: A systematic review. *The Australian and New Zealand Journal of Psychiatry*, *50*(1), 21–32.

Cryan, J. F. (2016). Stress and the microbiota-gut-brain axis: An evolving concept in psychiatry. *The Canadian Journal of Psychiatry*, *61*(4), 201–203.

Dhillon, A., Sparkes, E., & Duarte, R. (2017). Mindfulness-based interventions during pregnancy: A systematic review and meta-analysis. *Mindfulness*, *8*(6), 1421–1437.

Eaton, W. W., et al. (2008). Population-based study of first onset and chronicity in major depressive disorder. *Archives of General Psychiatry*, *65* (5), 513–520.

Eisendrath, S., Chartier, M., & McLane, M. (2011). Adapting mindfulness-based cognitive therapy for treatment-resistant depression. *Cognitive and Behavioral Practice*, *18*(3), 362–370.

Eisendrath, S. J., et al. (2016). A randomized controlled trial of mindfulness-based cognitive therapy for treatment-resistant depression. *Psychotherapy and Psychosomatics*, *85*(2), 99–110.

Ewais, T., et al. (2019a). A systematic review and meta-analysis of mindfulness based interventions and yoga in inflammatory bowel disease. *Journal of Psychosomatic Research*, *116*, 44–53.

Ewais, T., et al. (2019b). Protocol for a pilot randomised controlled trial of mindfulness-based cognitive therapy in youth with inflammatory bowel disease and depression. *BMJ Open*, *9*(4), e025568.

Garrison, K., et al. (2015). Meditation leads to reduced default mode network activity beyond an active task. *Cognitive, Affective, & Behavioral Neuroscience*, *15*(3), 712–720.

Gaynes, B. N., et al. (2009). What did STARD teach us? Results from a large-scale, practical, clinical trial for patients with depression. *Psychiatric Services*, *60*(11), 1439–1445.

Goldberg, S. B., et al. (2018). Mindfulness-based interventions for psychiatric disorders: A systematic review and meta-analysis. *Clinical Psychology Review*, *59*, 52–60.

Gu, J., et al. (2015). How do mindfulness-based cognitive therapy and mindfulness-based stress reduction improve mental health and well-being? A systematic review and meta-analysis of mediation studies. *Clinical Psychology Review*, *37*, 1–12.

Haddad, M. (2009). Depression in adults with a chronic physical health problem: Treatment and management. *International Journal of Nursing Studies*, *46*(11), 1411–1414.

Hölzel, B. K., et al. (2011). How does mindfulness meditation work? Proposing mechanisms of action from a conceptual and neural perspective. *Perspectives on Psychological Science*, *6*(6), 537–559.

Househam, A. M., et al. (2017). The effects of stress and meditation on the immune system, human microbiota, and epigenetics. *Advances in Mind-Body Medicine*, *31*(4), 10.

Jamieson, S. D., & Tuckey, M. R. (2017). Mindfulness interventions in the workplace: A critique of the current state of the literature. *Journal of Occupational Health Psychology*, *22*(2), 180–193.

Kabat-Zinn, J. (1991). Full catastrophe living: Using the wisdom of your body and mind to face stress, pain, and illness. University of Massachusetts Medical Center, Worcester In J. Kabat-Zinn (Ed.), *Stress reduction* (pp. 1–14). New York: New York: Dell Publishing.

Kabat-Zinn, J. (1994). *Wherever you go, there you are: Mindfulness meditation in everyday life*. New York: Hyperion.

Kenny, M. A., & Williams, J. M. G. (2007). Treatment-resistant depressed patients show a good response to mindfulness-based cognitive therapy. *Behaviour Research and Therapy*, *45*(3), 617–625.

Khoury, B., et al. (2015). Mindfulness-based stress reduction for healthy individuals: A meta-analysis. *Journal of Psychosomatic Research*, *78*(6), 519–528.

Kuyken, W., et al. (2008). Mindfulness-based cognitive therapy to prevent relapse in recurrent depression. *Journal of Consulting and Clinical Psychology*, *76*(6), 966–978.

Kuyken, W., et al. (2015). Effectiveness and cost-effectiveness of mindfulness-based cognitive therapy compared with maintenance antidepressant treatment in the prevention of depressive relapse or recurrence (PREVENT): A randomised controlled trial. *The Lancet*, *386*(9988), 63–73.

Kuyken, W., et al. (2016). Efficacy of mindfulness-based cognitive therapy in prevention of depressive relapse: An individual patient data meta-analysis from randomized trials. *JAMA Psychiatry*, *73*(6), 565–574.

Lazar, W. S., et al. (2005). Meditation experience is associated with increased cortical thickness. *Neuroreport*, *16*(17), 1893–1897.

Luders, E., et al. (2011). Enhanced brain connectivity in long-term meditation practitioners. *NeuroImage*, *57*(4), 1308–1316.

Mac Giollabhui, N., Swistun, D., Murray, S., Moriarity, D. P., Kautz, M. M., Ellman, L. M., et al. (2020). Executive dysfunction in depression in adolescence: The role of inflammation and higher body mass. *Psychological Medicine*, *50*(4), 683–691.

Malhi, G. S., et al. (2018). Royal Australian and New Zealand College of Psychiatrists clinical practice guidelines for mood disorders: Major depression summary. *Medical Journal of Australia*, *208*(4), 175–180.

Mikocka-Walus, A., et al. (2016). Symptoms of depression and anxiety are independently associated with clinical recurrence of inflammatory bowel disease. *Clinical Gastroenterology and Hepatology*, *14*(6). 829-835.e1.

Moulton, C. D., Pickup, J. C., & Ismail, K. (2015). The link between depression and diabetes: The search for shared mechanisms. *The Lancet Diabetes and Endocrinology*, *3*(6), 461–471.

Parsons, C. E., et al. (2017). Home practice in mindfulness-based cognitive therapy and mindfulness-based stress reduction: A systematic review and meta-analysis of participants' mindfulness practice and its association with outcomes. *Behaviour Research and Therapy*, *95*(Suppl. C), 29–41.

Sanada, K., et al. (2017). Effects of mindfulness-based interventions on biomarkers in healthy and cancer populations: A systematic review. Report *BMC Complementary and Alternative Medicine*, *17*(1) 1. 125.

Segal, Z. V., Williams, J. M. G., & Teasdale, J. D. (2002). *Mindfulness-based cognitive therapy for depression: A new approach to preventing relapse. Mindfulness-based cognitive therapy for depression: A new approach to preventing relapse. Vol. xiv* (p. 351). New York, NY: Guilford Press.

Segal, Z. V., Williams, J. M. G., & Teasdale, J. D. (2013). *Mindfulness-based cognitive therapy for depression* (2nd ed.). The Guilford Press.

Shapiro, S. L., et al. (2006). *Mechanisms of mindfulness* (pp. 373–386). Hoboken: Wiley Subscription Services, Inc., A Wiley Company.

Strauss, C., et al. (2014). Mindfulness-based interventions for people diagnosed with a current episode of an anxiety or depressive disorder: A meta-analysis of randomised controlled trials. Report *PLoS One*, *9*, 4.

Taylor, B., Cavanagh, K., & Strauss, C. (2016). The effectiveness of mindfulness-based interventions in the perinatal period: A systematic review and meta-analysis. *PLoS One*, *11*(5), 1. e0155720.

Teasdale, J. D., et al. (2000). Prevention of relapse/recurrence in major depression by mindfulness-based cognitive therapy. *Journal of Consulting and Clinical Psychology*, *68*(4), 615–623.

van der Velden, A. M., et al. (2015). A systematic review of mechanisms of change in mindfulness-based cognitive therapy in the treatment of recurrent major depressive disorder. *Clinical Psychology Review*, *37*, 26–39.

van Son, J., et al. (2014). Mindfulness-based cognitive therapy for people with diabetes and emotional problems: Long-term follow-up findings from the DiaMind randomized controlled trial. *Journal of Psychosomatic Research*, *77*(1), 81–84.

Vos, T., et al. (2017). Global, regional, and national incidence, prevalence, and years lived with disability for 328 diseases and injuries for 195 countries, 1990–2016: A systematic analysis for the Global Burden of Disease Study 2016. *The Lancet*, *390*(10100), 1211–1259.

Williams, J. M. G., & Kuyken, W. (2012). Mindfulness-based cognitive therapy: A promising new approach to preventing depressive relapse. *The British Journal of Psychiatry*, *200*(5), 359–360. https://doi.org/10.1192/bjp.bp.111.104745.

Zoogman, S., et al. (2015). Mindfulness interventions with youth: A meta-analysis. *Mindfulness*, *6*(2), 290–302.

Chapter 41

# Online programs for depression

Philip J. Batterham, Alison L. Calear, and Ella Kurz
*Centre for Mental Health Research, Research School of Population Health, The Australian National University, Canberra, ACT, Australia*

## List of abbreviations

**Apps** applications
**CBT** cognitive behavioral therapy
**iCBT** internet cognitive behavioral therapy
**QALY** quality-adjusted life year
**RCT** randomized controlled trial
**WTP** willingness-to-pay

## Introduction

Depression is the leading cause of disability worldwide and a major contributor to the overall global burden of disease, affecting more than 300 million people (WHO, 2018). However, a minority of people experiencing clinically relevant depression symptoms seek professional help, with estimates that up to 56% of people with depression remain untreated (Wang et al., 2007). Treatment delays for depression are also common (Wang et al., 2005). There are myriad reasons why people experiencing depression symptoms do not seek professional help, including structural barriers (cost, time, and accessibility), attitudinal barriers (stigma, negative attitudes toward help seeking, negative past experiences, and self-reliance), knowledge barriers (mental health literacy and low perceived need), and contextual barriers (social and demographic factors) (Gulliver, Griffiths, & Christensen, 2010; Magaard, Seeralan, Schulz, & Brütt, 2017). Online treatment programs for depression, also referred to as internet interventions, can overcome many of these barriers to care by providing accessible evidence-based interventions at low to no cost from the individual's desired location (e.g., home, workplace, and library) at any time of the day or night. The automation of online programs ensures high fidelity and consistency in intervention delivery and allows for efficient population reach and scalability, particularly in settings where mental health expertise may be limited or not readily available (e.g., rural or remote areas, schools, and workplaces) (Batterham et al., 2015). The aim of this chapter is to provide an overview of the role and effectiveness of online programs for depression in the community, the facilitators and barriers to the dissemination and uptake of these interventions, and the future directions for digital mental health interventions for depression.

## The structure and function of online programs

Online programs for depression come in a variety of packages. Many of the original programs were developed as websites, although mobile applications (apps) are growing in popularity. Online programs typically adapt the principles and strategies from cognitive behavioral therapy (CBT) and related evidence-based forms of therapy into a package that can be completed in either a self-guided or clinician-supported format. A range of other forms of therapy for depression have been evaluated in the online context, including acceptance and commitment therapy, behavioral activation, CBT for insomnia, cognitive bias modification, problem-solving therapy, mindfulness-based therapy, and psychodynamic therapy, although CBT programs have the greatest evidence of efficacy (Koenigbauer, Letsch, Doebler, Ebert, & Baumeister, 2017; Richards & Richardson, 2012; Spijkerman, Pots, & Bohlmeijer, 2016; Ye et al., 2015). In many cases, the process of adaptation takes the 50-min psychologist consultation format and converts it directly into an online delivery format. However, with the development of technology and better understanding of how people interact with web-based technologies, considerable variations in delivery formats have been evaluated. Breaking therapy into briefer modules that focus on the core principles of CBT may increase user uptake and engagement with online programs (Batterham, Calear, Farrer, McCallum, & Cheng, 2018). Apps that focus on single therapeutic aspects, such as mood monitoring, behavioral activation, mindfulness, or cognitive restructuring are becoming more common and may better fit with people's mental health needs, availability of time, and how they use apps in everyday situations (Huguet et al., 2018; Lattie et al., 2016). Programs that incorporate greater interactivity (Richards, Dowling, O'Brien, Viganò, & Timulak, 2018), user-centered design features (Stawarz, Preist, Tallon, Wiles, & Coyle, 2018), gamification (inclusion of

The Neuroscience of Depression. https://doi.org/10.1016/B978-0-12-817933-8.00012-8

gaming elements) (Dias, Barbosa, & Vianna, 2018), and tailoring of intervention content to user needs and preferences (Batterham et al., 2018) may also lead to better participant adherence and consequently improved outcomes.

Online depression programs have typically been designed to treat symptoms of depression, and are usually aimed at individuals experiencing mild-to-moderate clinical symptoms. Nevertheless, there is evidence that many of these programs can be effective in the prevention of depression onset (Sander, Rausch, & Baumeister, 2016), supporting individuals with more severe depression (Meyer et al., 2015), and in relapse prevention (Holländare et al., 2013). Delivering prevention programs, including relapse prevention programs, may have different challenges to the delivery of treatment programs. In particular, consideration of the most appropriate settings for implementation should be a priority—for example, prevention programs may be more effective in younger populations and in settings where people spend most of their time, so delivery of prevention programs within schools or workplaces may lead to greater uptake and impact (Carolan, Harris, & Cavanagh, 2017; Deady et al., 2017; Ebert, Cuijpers, Munoz, & Baumeister, 2017). In contrast, delivery of treatment programs within clinical settings may be more appropriate. Nevertheless, barriers to implementation in clinical services remain (Andersson & Titov, 2014), and many people with depression symptoms are reluctant to engage with health services so would not be captured in this setting (Wang et al., 2005).

Some existing programs have been designed specifically for target populations, such as children and adolescents (Calear & Christensen, 2010), people from culturally diverse backgrounds (Shehadeh, Heim, Chowdhary, Maercker, & Albanese, 2016) and women in the perinatal period (Ashford, Olander, & Ayers, 2016). Comorbidity, which is not uncommon with depression, can also be addressed using online programs, by delivering transdiagnostic therapeutic elements, combinations of treatment approaches, or treatment that is tailored to individual symptom patterns (Batterham et al., 2018). There is also evidence that programs targeted at related mental disorders or risk factors for depression, such as anxiety, insomnia, and chronic physical health problems (e.g., diabetes), may also be effective in the treatment and prevention of depression (Boettcher et al., 2014; Christensen et al., 2016; Trompetter, Bohlmeijer, Veehof, & Schreurs, 2015).

## Self-guided vs clinician-supported programs

Online programs may be self-guided or clinician-supported. The former are automated and the latter augmented by a clinician, who may provide either real-time or delayed support and feedback via telephone, video, or messenger services (Andersson & Titov, 2014). While reviews of the literature show that treatments which include human support lead to better outcomes than unguided treatments (Palmqvist, Carlbring, & Andersson, 2007; Richards & Richardson, 2012; Spek et al., 2007), self-guided programs are also effective in reducing depressive symptoms (Berger, Hämmerli, Gubser, Andersson, & Caspar, 2011; Cuijpers et al., 2011). Advantages of clinician-guided programs are that clinicians may be able to offer a diagnosis to determine the suitability of the treatment (self-assessments are currently unable to replace structured diagnostic interviews), they may be tailored to accommodate comorbidities (for example, cooccurrence of mental disorders or cooccurrence of a mental and physical disorder) and clinicians can be of assistance in monitoring symptoms for deterioration and making referrals to other services that may be required, such as social, health, and crisis services (Andersson & Titov, 2014). Perhaps most importantly, clinician support may increase adherence to and prevent dropout of the program, as someone is monitoring intervention progress and can provide reminders to complete the intervention (Andersson & Titov, 2014). Advantages of self-guided programs are that they are usually less costly, more scalable, and more accessible than clinician-guided programs, as they don't require the workforce and paid services of clinicians (Batterham et al., 2015). Self-guided programs also meet the needs of people who, due to stigma or previous negative experiences, may not want to engage with a clinician (Batterham et al., 2015; Karyotaki et al., 2017). The continuing development of transdiagnostic and tailored capacities, automatic reminders, and other features that facilitate completion rates and outcomes in self-guided programs will likely make them an increasingly viable option.

## Evidence for the effectiveness of online programs

A plethora of reviews and meta-analyses have shown online programs for depression to be effective and cost effective. For example, Ahern, Kinsella, and Semkovska (2018) identified 29 randomized controlled trials (RCTs) of online programs for depression, demonstrating the efficacy of online treatment of depression to be equivalent to face-to-face treatment with an effect size of 0.44 (95% CI: 0.31, 0.57, $P < 0.001$). Wright et al. (2019) reported that among 40 RCTs of online depression programs, effect sizes ranged from 0.239 (95% CI: 0.115–0.364, $P < 0.001$) to 0.673 (95% CI: 0.546–0.801, $P < 0.001$) depending on the degree of support provided. Koenigbauer et al. (2017) reported a pooled effect size of −0.90 (95% CI: −1.07, −0.73), favoring online programs over no treatment, while Firth et al. (2017) reported pooled effect sizes ranging from

0.216 (95% CI: 0.10, 0.33, $P<0.01$) to 0.558 (95% CI: 0.38, 0.74, $P<0.01$). Sztein, Koransky, Fegan, and Himelhoch (2018) found online programs provided a large sustained effect in maintaining reduction of depressive symptoms in follow-up measures 3–6 months after the conclusion of the therapy [standardized mean difference: 0.83 (95% CI: 0.69, 0.99, $P<0.001$)]. A meta-analysis specifically examining the efficacy of self-guided internet treatment programs for depression also found these programs to be effective in reducing symptom severity (Hedges $g=0.27$) with number needed to treat of 8 (Karyotaki et al., 2017).

Systematic reviews and meta-analyses have also demonstrated the efficacy of online programs for depression in target populations. For example, Xiang et al. (2019) concluded online programs are a promising approach for reducing symptoms of depression among older adults (mean age: 66 years), with a within-group effect size of 1.27 (95% CI: 1.09, 1.45, $P<0.05$) and mean between-group effect size of 1.18 (95% CI: 0.63, 1.73, $P<0.05$). Loughnan, Joubert, Grierson, Andrews, & Newby, (2019) found online programs demonstrated large and significant improvements in depression in perinatal women pretreatment to post-treatment, with a large within-group effect size of 1.67 (95% CI 1.38, 1.96, $P<0.01$) and Hollis et al. (2017) demonstrated the effectiveness of online programs in children and young people with effect sizes ranging from 0.16 to 0.62.

The economic evaluation of online programs for depression by Ahern et al. (2018) in their systematic review and meta-analysis found online programs to be more cost effective compared to treatment as usual and no treatment, with average likelihood of cost effectiveness ranging from 56% [willingness-to-pay (WTP) threshold £20,000/QALY] in the United Kingdom to 52% (WTP threshold of €30,000/QALY) in Holland to >95% (WTP threshold of $95,000/QALY) in Australia. Paganini, Teigelkoetter, Buntrock, and Baumeister (2018) reviewed seven economic evaluations of online programs and found they were likely to be cost effective with an incremental cost-utility ratio between €3088 and €22,609 in six of the seven studies. Lokkerbol et al. (2014) reported that online programs are likely to increase the overall cost effectiveness of the health-care system, with health gains worth €1.45–€1.77 for every euro invested and Hedman, Ljótsson, and Lindefors (2012) found online programs to be significantly more cost effective compared to no treatment, with an average probability of cost effectiveness at 57% at a willingness to pay of zero. Similarly, evidence for the cost effectiveness of school-based internet programs for depression has also been found, with a cost of AU$7350 per disability-adjusted life year averted for universal prevention programs (Lee et al., 2017).

Much of the evidence for the effectiveness of online programs for depression comes from RCTs, which are widely recognized and accepted as the gold standard for program evaluation. However, other online clinical service outcomes, such as adherence, clinical disorder remission rates, and consumer satisfaction, may also be relevant to the assessment of intervention effectiveness if these data are available (Batterham, Calear, O'Dea, et al., 2019).

## Examples of effective programs

Examples of effective online programs include *ThisWayUp*, *GET.ON Mood Enhancer*, *MindSpot Clinic*, *myCompass*, *SPARX*, and *SHUTi*. *ThisWayUp* incorporates several cognitive behavioral components based on psychoeducation and management of thoughts, emotions, and behaviors. Participants receive homework assignments as well as computer-administered questionnaires relevant to their presenting problems, which allow for a therapist to monitor their progress, safety, and outcomes (Wong, Kady, Mewton, Sunderland, & Andrews, 2014). The *GET.ON Mood Enhancer* program focuses on psychoeducation, behavioral activation, problem solving, and relapse prevention, offering six lessons of text-based exercises, which include testimonials and interactive elements such as audio and video clips. Participants are offered four additional modules, which are chosen based on individual need and/or preference and are directed at sleep problems, relaxation techniques, and dealing with worrying thoughts (Ebert et al., 2014). *MindSpot Clinic* is structured on the principles of CBT and interpersonal therapy, with every participant working through the course with weekly contact from a therapist (Titov et al., 2015). The *myCompass* program is an automatic, self-guided, public health program delivered via mobile phone and computer (Proudfoot et al., 2013). *SPARX* is a gamified online program targeting and preventing depression in youth (Perry et al., 2017) and *SHUTi* treats insomnia but has also been shown to be effective in reducing depressive symptoms (Christensen et al., 2016).

## Opportunities and challenges in delivering online programs

Online programs overcome many of the barriers of face-to-face care. Accessibility is typically high for online programs, as they are available at any time of day, in virtually any location, and at little cost to the consumer (Andersson & Titov, 2014). By delivering evidence-based care at scale to a large section of the population in need, it may be possible to reduce the burden of depression. The potential for better access to evidence-based care, particularly for people living in rural or remote locations or those experiencing mild-moderate symptoms, is considerable but has yet to be realized. There are a number of barriers to reaching the population at need of care.

First, there is a considerable implementation gap between the evidence for online programs and the delivery of these programs into the community. While the best established programs such as *MindSpot* and *myCompass* reach thousands of users monthly, there is often reluctance from clinicians and consumers to engage with online programs. Pathways into care for people with mental health problems can be complex, further hindering efforts to increase uptake of online programs (Batterham et al., 2015). Clinical providers may be resistant to changes in practice, and appropriately concerned about patient safety when offering new forms of interventions (Andersson & Titov, 2014).

A lack of awareness of the existence of online programs is another barrier to their uptake. Clinicians and community members may be unaware of these programs or their effectiveness in treating and preventing depression (Batterham et al., 2015). Clinicians and community members who do have knowledge of online programs may encounter difficulties in selecting a program or discerning its quality or clinical authority. Concerns about data privacy and security may also be a barrier to uptake, with people worried about their or their patients' sensitive data being shared or accessed without permission. Implementation trials (e.g., the ImpleMentAll trial; Finch et al., 2018) are currently being conducted to better overcome the barriers to normalization of delivering online programs within clinical settings.

Nevertheless, online programs can bypass clinical services by directly marketing to consumers. There remains a need however to ensure that treatments are effective, safe, and acceptable. Despite considerable level 1 evidence that online programs can be highly effective in treating depression, the marketplace for such programs is dominated by commercial enterprises that may lack evidence of safety or effectiveness. Accreditation processes for online programs, such as NHS Choices and Beacon (Christensen et al., 2010), which ensure consumers and clinicians can make informed choices about which online services are appropriate and likely to be effective, are nascent, and come with considerable challenges (Batterham, Calear, O'Dea, et al., 2019).

There are many reasons why people experiencing depression symptoms do not seek professional help, summarized in Table 1. These include structural barriers, attitudinal barriers, knowledge barriers, and contextual barriers (Gulliver et al., 2010; Magaard et al., 2017). Online programs overcome many of these barriers to care, while providing high-fidelity, automated, and low-cost evidence-based treatments (Andersson & Titov, 2014). Nevertheless, as noted in Table 1, there are other barriers to accessing internet treatments, primarily related to public awareness of the availability and effectiveness of these services.

There is also considerable heterogeneity in the programs that are currently available online. Many programs that have strong research evidence are not subsequently delivered as a clinical service, typically due to resource constraints on the researchers who develop and trial such programs. Online programs require longer-term funding to cover storage and data security costs and to allow programs to remain relevant and responsive to technological advances. Some available programs may only be effective within particular settings or for particular audiences. Consumer engagement with online programs is highly variable, with low adherence a challenging shortcoming of many programs (Van Ballegooijen et al., 2014), as with face-to-face therapy. In the recent years, there has been a greater movement toward the involvement of end users in the design and development of online programs. The use of codesign methods (e.g., prototyping, focus groups) has allowed the needs and preferences of end users to be better captured and incorporated into interventions to promote uptake, engagement, and adherence (Batalden et al., 2016; Eyles et al., 2016). Understanding and challenging the barriers to engagement with therapy, both in online and traditional settings, may increase the impact of these programs in the community (Batterham, Calear, Sunderland, et al., 2019). Nonetheless, keeping up with new technologies and changes in how people engage with technology will continue to be a challenge for optimizing engagement with online programs. Disparities between the time frames of implementation processes (typically spanning 10–20 years (Morris, Wooding, & Grant, 2011)), research trials (typically 2–5 years to develop and measure long-term outcomes), and changes in technology use (typically over months or years) are also a clear challenge for ensuring that online programs remain safe, effective, engaging and relevant, and for ensuring that clinical practice keeps up with evolving technology and evidence.

## The future of online programs

As technology evolves, so does the nature and delivery of online programs for depression. Agile software development has become increasingly common in the field of e-mental health, and involves the use of user perspectives to continuously update and fine-tune existing programs (Brhel, Meth, Maedche, & Werder, 2015). There is also evidence that social connectedness is strongly protective against the development of depression. Incorporating technological features that capitalize on the social network of the individual may increase efficacy and adherence to online programs by ensuring the individual is supported in engaging with online programs (Mohr, Cuijpers, & Lehman, 2011). Network interventions have been shown to increase efficacy by targeting the most influential or most connected individuals within a social network to elicit change within the community (Griffiths et al., 2012). Use of social networks to support the dissemination of online programs also has promise, particularly given evidence that depression may propagate through social networks (Rosenquist, Fowler, & Christakis, 2011) and peers can be a positive influence on help seeking (Naslund, Aschbrenner, Marsch, & Bartels, 2016).

The use of passive sensing or "digital phenotyping" on mobile devices is another area showing promise. The use of real-time data and machine learning approaches to better understand patterns of behavior may lead to insights into how behavior changes in response to variations in mood (Insel, 2017). There are many sensors on a phone that can be used to infer factors such as physical activity, social connectedness, and sleep cycles. The combination of such data, the "digital phenotype" of the individual, with online psychosocial interventions may lead to personalized interventions that respond dynamically to the needs and preferences of the individual (Insel, 2017). Ongoing passive monitoring could be used to alert the individual and potentially their clinician or others in their social network to changes in mood or behavior, and deliver ultrabrief psychosocial interventions that are optimized to time and place. Such ecological momentary interventions, that are highly adaptive to the needs of the individual, may revolutionize the treatment of depression.

There is also considerable scope for implementing online programs more seamlessly into health-care systems. Stepped care programs that incorporate internet interventions may increase efficiency in the delivery of depression treatment, with early trials providing positive outcomes in terms of efficacy and cost effectiveness (Mohr et al., 2019). Further research to identify appropriate clinical pathways for increasing uptake of online programs is also warranted (Batterham, Calear, Gulliver, & Farrer, 2019), given the suboptimal implementation of these programs noted above.

## Conclusion

Online programs for depression provide high-fidelity, evidence-based treatments and prevention to people who can't, or do not want to, access traditional services. Self-guided programs have the potential for wide dissemination and reach, while clinician-guided programs can improve participant adherence and engagement. The evidence for online depression programs is strong, with both general and tailored programs demonstrating positive effects in a range of settings and populations. Current research is focusing on addressing the barriers to the uptake and completion of online programs, as well as the use of new digital technologies to further understand, monitor, treat, and prevent depression.

## Key facts

- Online programs for depression are effective for both adults and young people, with typical effect sizes from meta-analyses between 0.4 and 0.9 standard deviations.
- There is evidence that online programs are cost-effective, with significant health gains.
- Online programs address many of the barriers to accessing face-to-face health services.
- There is emerging evidence that social networks and "digital phenotypes" from mobile devices may be used to deliver interventions to treat depression symptoms in real time.
- Further research is needed to bridge the gap between the extensive evidence for the effectiveness of online programs and their implementation in the population.

**TABLE 1** Barriers to use of traditional services (face-to-face care) vs barriers to use of online programs.

| Domain | Traditional service barriers | Online program barriers |
|---|---|---|
| Structural | – Cost<br>– Time<br>– Accessibility | – Reliable internet access<br>– Accreditation processes<br>– Availability of digital apothecaries (online directories) and search engine optimization |
| Attitudinal | – Stigma<br>– Perceived lack of cultural safety<br>– Negative attitudes toward help seeking<br>– Negative past experiences<br>– Preference for self-reliance | – Perceived lack of evidence base<br>– Concerns about privacy of data<br>– Preference for face-to-face therapy<br>– Resistance to change |
| Knowledge | – Mental health literacy | – Mental health literacy<br>– Limited exposure to and experience of online programs<br>– Digital literacy/fluency |
| Contextual | – Social and demographic factors | – Social and demographic factors |

## Summary points

- Online programs for depression provide high-fidelity, evidence-based treatments and prevention to people who can't, or do not want to, access traditional services.
- Online programs typically adapt the principles and strategies from cognitive behavioral therapy (CBT) and related evidence-based forms of therapy into a package that can be completed in either a self-guided or clinician-supported format.
- Originally, online programs were developed as websites, while many emerging programs are developed as mobile applications.
- There is strong evidence for the efficacy and cost effectiveness of online programs, with both general and tailored programs demonstrating positive effects in a range of settings and populations.
- Barriers to the uptake and completion of online programs include an implementation gap between the evidence for online programs and their delivery into the community, reluctance from clinicians and consumers to engage with online programs, a lack of awareness of the existence of online programs, as well as a lack of adequate program accreditation processes.
- As technology like agile software and digital phenotyping evolves, so does the scope and delivery of online programs for depression.

## Mini-dictionary of terms

**Online programs** The adaption of the principles and strategies from cognitive behavioral therapy (CBT) and/or related evidence-based forms of therapy for depression into a package that can be completed online in either a self-guided or clinician-supported format. Also known as internet interventions.

**Self-guided program** Online program which is automated and accessed without clinician support.

**Clinician-supported program** Online program which is augmented by a clinician, who may provide either real-time or delayed support and feedback via telephone, video, or messenger services.

**Transdiagnostic program** Treatments that apply the same underlying treatment principles across mental disorders, without tailoring the protocol to specific diagnoses.

**Stepped care** An evidence-based hierarchy of interventions, which are matched to the individual's needs in order to provide the right level of care at the right time.

## References

Ahern, E., Kinsella, S., & Semkovska, M. (2018). Clinical efficacy and economic evaluation of online cognitive behavioral therapy for major depressive disorder: A systematic review and meta-analysis. *Expert Review of Pharmacoeconomics & Outcomes Research*, *18*(1), 25–41.

Andersson, G., & Titov, N. (2014). Advantages and limitations of internet-based interventions for common mental disorders. *World Psychiatry*, *13*(1), 4–11. https://doi.org/10.1002/wps.20083.

Ashford, M. T., Olander, E. K., & Ayers, S. (2016). Computer-or web-based interventions for perinatal mental health: A systematic review. *Journal of Affective Disorders*, *197*, 134–146.

Batalden, M., Batalden, P., Margolis, P., Seid, M., Armstrong, G., Opipari-Arrigan, L., et al. (2016). Coproduction of healthcare service. *BMJ Quality and Safety*, *25*(7), 509–517.

Batterham, P. J., Calear, A. L., Farrer, L., McCallum, S. M., & Cheng, V. W. S. (2018). FitMindKit: Randomised controlled trial of an automatically tailored online program for mood, anxiety, substance use and suicidality. *Internet Interventions*, *12*, 91–99. https://doi.org/10.1016/j.invent.2017.08.002.

Batterham, P. J., Calear, A. L., Gulliver, A., & Farrer, L. M. (2019). Efficacy of a transdiagnostic, video-based online program for reducing depression, anxiety, and suicidal ideation in adults: Protocol for a randomised controlled trial. *Contemporary Clinical Trials Communications*, *14*, 100341.

Batterham, P. J., Calear, A. L., O'Dea, B., Larsen, M. E., Kavanagh, D., Titov, D., et al. (2019). Stakeholder perspectives on evidence for digital mental health interventions: Implications for accreditation systems. *Digital Health*, *5*. https://doi.org/10.1177/2055207619878069.

Batterham, P. J., Calear, A. L., Sunderland, M., Kay-Lambkin, F., Farrer, L. M., & Gulliver, A. (2019). A brief intervention to increase uptake and adherence of an online program for depression and anxiety: Protocol for the enhancing engagement with psychosocial interventions (EEPI) randomized controlled trial. *Contemporary Clinical Trials*, *78*, 107–115.

Batterham, P. J., Sunderland, M., Calear, A. L., Davey, C. G., Christensen, H., Teesson, M., et al. (2015). Developing a roadmap for the translation of e-mental health services for depression. *The Australian and New Zealand Journal of Psychiatry*, *49*(9), 776–784.

Berger, T., Hämmerli, K., Gubser, N., Andersson, G., & Caspar, F. (2011). Internet-based treatment of depression: A randomized controlled trial comparing guided with unguided self-help. *Cognitive Behaviour Therapy*, *40*(4), 251–266.

Boettcher, J., Åström, V., Påhlsson, D., Schenström, O., Andersson, G., & Carlbring, P. (2014). Internet-based mindfulness treatment for anxiety disorders: A randomized controlled trial. *Behavior Therapy*, *45*(2), 241–253.

Brhel, M., Meth, H., Maedche, A., & Werder, K. (2015). Exploring principles of user-centered agile software development: A literature review. *Information and Software Technology*, *61*, 163–181.

Calear, A. L., & Christensen, H. (2010). Review of internet-based prevention and treatment programs for anxiety and depression in children and adolescents. *Medical Journal of Australia*, *192*, S12–S14.

Carolan, S., Harris, P. R., & Cavanagh, K. (2017). Improving employee well-being and effectiveness: Systematic review and meta-analysis of web-based psychological interventions delivered in the workplace. *Journal of Medical Internet Research*, *19*(7), e271. https://doi.org/10.2196/jmir.7583.

Christensen, H., Batterham, P. J., Gosling, J. A., Ritterband, L. M., Griffiths, K. M., Thorndike, F. P., et al. (2016). Effectiveness of an online insomnia program (SHUTi) for prevention of depressive episodes (the GoodNight Study): A randomised controlled trial. *Lancet Psychiatry*, *3*(4), 333–341. https://doi.org/10.1016/S2215-0366(15)00536-2.

Christensen, H., Murray, K., Calear, A. L., Bennett, K., Bennett, A., & Griffiths, K. M. (2010). Beacon: A web portal to high-quality mental health websites for use by health professionals and the public. *Medical Journal of Australia*, *192*, S40–S44.

Cuijpers, P., Donker, T., Johansson, R., Mohr, D. C., van Straten, A., & Andersson, G. (2011). Self-guided psychological treatment for depressive symptoms: A meta-analysis. *PLoS One*, *6*(6), e21274.

Deady, M., Choi, I., Calvo, R. A., Glozier, N., Christensen, H., & Harvey, S. B. (2017). eHealth interventions for the prevention of depression and anxiety in the general population: A systematic review and meta-analysis. *BMC Psychiatry*, *17*(1), 310. https://doi.org/10.1186/s12888-017-1473-1.

Dias, L. P. S., Barbosa, J. L. V., & Vianna, H. D. (2018). Gamification and serious games in depression care: A systematic mapping study. *Telematics and Informatics*, *35*(1), 213–224.

Ebert, D. D., Cuijpers, P., Munoz, R. F., & Baumeister, H. (2017). Prevention of mental health disorders using internet- and mobile-based interventions: A narrative review and recommendations for future research. *Frontiers in Psychiatry*, *8*, 116. https://doi.org/10.3389/fpsyt.2017.00116.

Ebert, D. D., Lehr, D., Baumeister, H., Boß, L., Riper, H., Cuijpers, P., et al. (2014). GET. ON mood enhancer: Efficacy of internet-based guided self-help compared to psychoeducation for depression: An investigator-blinded randomised controlled trial. *Trials*, *15*(1), 39.

Eyles, H., Jull, A., Dobson, R., Firestone, R., Whittaker, R., Te Morenga, L., et al. (2016). Co-design of mHealth delivered interventions: A systematic review to assess key methods and processes. *Current Nutrition Reports*, *5*(3), 160–167.

Finch, T. L., Girling, M., May, C. R., Mair, F. S., Murray, E., Treweek, S., et al. (2018). Improving the normalization of complex interventions: Part 2-validation of the NoMAD instrument for assessing implementation work based on normalization process theory (NPT). *BMC Medical Research Methodology*, *18*(1), 135.

Firth, J., Torous, J., Nicholas, J., Carney, R., Pratap, A., Rosenbaum, S., et al. (2017). The efficacy of smartphone-based mental health interventions for depressive symptoms: A meta-analysis of randomized controlled trials. *World Psychiatry*, *16*(3), 287–298.

Griffiths, F., Cave, J., Boardman, F., Ren, J., Pawlikowska, T., Ball, R., et al. (2012). Social networks—The future for health care delivery. *Social Science & Medicine*, *75*(12), 2233–2241.

Gulliver, A., Griffiths, K. M., & Christensen, H. (2010). Perceived barriers and facilitators to mental health help-seeking in young people: A systematic review. *BMC Psychiatry*, *10*, 113. https://doi.org/10.1186/1471-244X-10-113.

Hedman, E., Ljótsson, B., & Lindefors, N. (2012). Cognitive behavior therapy via the Internet: A systematic review of applications, clinical efficacy and cost-effectiveness. *Expert Review of Pharmacoeconomics & Outcomes Research*, *12*(6), 745–764.

Holländare, F., Anthony, S. A., Randestad, M., Tillfors, M., Carlbring, P., Andersson, G., et al. (2013). Two-year outcome of internet-based relapse prevention for partially remitted depression. *Behaviour Research and Therapy*, *51*(11), 719–722.

Hollis, C., Falconer, C. J., Martin, J. L., Whittington, C., Stockton, S., Glazebrook, C., et al. (2017). Annual research review: Digital health interventions for children and young people with mental health problems—A systematic and meta-review. *Journal of Child Psychology and Psychiatry*, *58*(4), 474–503.

Huguet, A., Miller, A., Kisely, S., Rao, S., Saadat, N., & McGrath, P. J. (2018). A systematic review and meta-analysis on the efficacy of internet-delivered behavioral activation. *Journal of Affective Disorders*, *235*, 27–38.

Insel, T. R. (2017). Digital phenotyping: Technology for a new science of behavior. *JAMA*, *318*(13), 1215–1216.

Karyotaki, E., Riper, H., Twisk, J., Hoogendoorn, A., Kleiboer, A., Mira, A., et al. (2017). Efficacy of self-guided internet-based cognitive behavioral therapy in the treatment of depressive symptoms: A meta-analysis of individual participant data. *JAMA Psychiatry*, *74*(4), 351–359. https://doi.org/10.1001/jamapsychiatry.2017.0044.

Koenigbauer, K., Letsch, J., Doebler, P., Ebert, D., & Baumeister, H. (2017). Internet- and mobile-based depression interventions for people with diagnosed depression: A systematic review and meta-analysis. *Journal of Affective Disorders*, *223*, 28–40.

Lattie, E. G., Schueller, S. M., Sargent, E., Stiles-Shields, C., Tomasino K. N., Corden, M. E., et al. (2016). Uptake and usage of IntelliCare: A publicly available suite of mental health and well-being apps. *Internet Interventions*, *4*. 152–158.

Lee, Y. Y., Barendregt, J. J., Stockings, E. A. Ferrari, A. J., Whiteford, H. A., Patton, G. A., et al. (2017). The population cost-effectiveness of delivering universal and indicated school-based interventions to prevent the onset of major depression among youth in Australia. *Epidemiology and Psychiatric Sciences*, *26*(5), 545–564. https://doi.org/10.1017/S2045796016000469.

Lokkerbol, J., Adema, D., Cuijpers, P., Reynolds, C. F., III, Schulz, R., Weehuizen, R., et al. (2014). Improving the cost-effectiveness of a healthcare system for depressive disorders by implementing telemedicine: A health economic modeling study. *The American Journal of Geriatric Psychiatry*, *22*(3), 253–262.

Loughnan, S. A., Joubert, A. E., Grierson, A., Andrews, G., & Newby, J. M. (2019). Internet-delivered psychological interventions for clinical anxiety and depression in perinatal women: A systematic review and meta-analysis. *Archives of Women's Mental Health*, *22*, 737–750.

Magaard, J. L., Seeralan, T., Schulz, H., & Brütt, A. L. (2017). Factors associated with help-seeking behaviour among individuals with major depression: A systematic review. *PLoS One*, *12*(5), e0176730.

Meyer, B., Bierbrodt, J., Schröder, J., Berger, T., Beevers, C. G., Weiss, M., et al. (2015). Effects of an internet intervention (deprexis) on severe depression symptoms: Randomized controlled trial. *Internet Interventions*, *2*(1), 48–59.

Mohr, D., Cuijpers, P., & Lehman, K. (2011). Supportive accountability A model for providing human support to enhance adherence to eHealth interventions. *Journal of Medical Internet Research*, *13*(1), e30.

Mohr, D. C., Lattie, E. G., Tomasino, K. N., Kwasny, M. J., Kaiser, S. M., Gray, E. L., et al (2019). A randomized noninferiority trial evaluating remotely-delivered stepped care for depression using internet cognitive behavioral therapy (CBT) and telephone CBT. *Behaviour Research and Therapy*, *123*, 103485.

Morris, Z. S., Wooding, S., & Grant, J. (2011). The answer is 17 years, what is the question: Understanding time lags in translational research. *Journal of the Royal Society of Medicine*, *104*(12), 510–520.

Naslund, J., Aschbrenner, K., Marsch, L., & Bartels, S. (2016). The future of mental health care: Peer-to-peer support and social media. *Epidemiology and Psychiatric Sciences*, *25*(2), 113–122.

Paganini, S., Teigelkoetter, W., Buntrock, C., & Baumeister, H. (2018). Economic evaluations of internet-and mobile-based interventions for the treatment and prevention of depression: A systematic review. *Journal of Affective Disorders*, *225*, 733–755.

Palmqvist, B., Carlbring, P., & Andersson, G. (2007). Internet-delivered treatments with or without therapist input. Does the therapist factor have implications for efficacy and cost? *Expert Review of Pharmacoeconomics & Outcomes Research*, *7*(3), 291–297.

Perry, Y., Werner-Seidler, A., Calear, A., Mackinnon, A., King, C., Scott, J., et al. (2017). Preventing depression in final year secondary students:

School-based randomized controlled trial. *Journal of Medical Internet Research*, *19*(11), e369.

Proudfoot, J., Clarke, J., Birch, M.-R., Whitton, A. E., Parker, G., Manicavasagar, V., et al. (2013). Impact of a mobile phone and web program on symptom and functional outcomes for people with mild-to-moderate depression, anxiety and stress: A randomised controlled trial. *BMC Psychiatry*, *13*(1), 312.

Richards, D., Dowling, M., O'Brien, E., Viganò, N., & Timulak, L. (2018). Significant events in an internet-delivered (space from depression) intervention for depression. *Counselling and Psychotherapy Research*, *18*(1), 35–48.

Richards, D., & Richardson, T. (2012). Computer-based psychological treatments for depression: A systematic review and meta-analysis. *Clinical Psychology Review*, *32*(4), 329–342.

Rosenquist, J. N., Fowler, J. H., & Christakis, N. A. (2011). Social network determinants of depression. *Molecular Psychiatry*, *16*(3), 273.

Sander, L., Rausch, L., & Baumeister, H. (2016). Effectiveness of internet-based interventions for the prevention of mental disorders: A systematic review and meta-analysis. *JMIR Mental Health*, *3*(3), e38.

Shehadeh, M. H., Heim, E., Chowdhary, N., Maercker, A., & Albanese, E. (2016). Cultural adaptation of minimally guided interventions for common mental disorders: A systematic review and meta-analysis. *JMIR Mental Health*, *3*(3), e44.

Spek, V., Cuijpers, P., Nyklíček, I., Riper, H., Keyzer, J., & Pop, V. (2007). Internet-based cognitive behaviour therapy for symptoms of depression and anxiety: A meta-analysis. *Psychological Medicine*, *37*(3), 319–328.

Spijkerman, M. P., Pots, W. T., & Bohlmeijer, E. T. (2016). Effectiveness of online mindfulness-based interventions in improving mental health: A review and meta-analysis of randomised controlled trials. *Clinical Psychology Review*, *45*, 102–114. https://doi.org/10.1016/j.cpr.2016.03.009.

Stawarz, K., Preist, C., Tallon, D., Wiles, N., & Coyle, D. (2018). User experience of cognitive behavioral therapy apps for depression: An analysis of app functionality and user reviews. *Journal of Medical Internet Research*, *20*(6), e10120.

Sztein, D. M., Koransky, C. E., Fegan, L., & Himelhoch, S. (2018). Efficacy of cognitive behavioural therapy delivered over the internet for depressive symptoms: A systematic review and meta-analysis. *Journal of Telemedicine and Telecare*, *24*(8), 527–539.

Titov, N., Dear, B. F., Staples, L. G., Bennett-Levy, J., Klein, B., Rapee, R. M., et al. (2015). MindSpot clinic: An accessible, efficient, and effective online treatment service for anxiety and depression. *Psychiatric Services*, *66*(10), 1043–1050.

Trompetter, H. R., Bohlmeijer, E. T., Veehof, M. M., & Schreurs, K. M. (2015). Internet-based guided self-help intervention for chronic pain based on acceptance and commitment therapy: A randomized controlled trial. *Journal of Behavioral Medicine*, *38*(1), 66–80.

Van Ballegooijen, W., Cuijpers, P., Van Straten, A., Karyotaki, E., Andersson, G., Smit, J. H., et al. (2014). Adherence to internet-based and face-to-face cognitive behavioural therapy for depression: A meta-analysis. *PLoS One*, *9*(7), e100674.

Wang, P. S., Aguilar-Gaxiola, S., Alonso, J., Angermeyer, M. C., Borges, G., Bromet, E. J., et al. (2007). Use of mental health services for anxiety, mood, and substance disorders in 17 countries in the WHO world mental health surveys. *The Lancet*, *370*(9590), 841–850.

Wang, P. S., Berglund, P., Olfson, M., Pincus, H. A., Wells, K. B., & Kessler, R. C. (2005). Failure and delay in initial treatment contact after first onset of mental disorders in the National Comorbidity Survey Replication. *Archives of General Psychiatry*, *62*(6), 603–613.

WHO. (2018). *Depression*. Fact Sheets. Retrieved from https://www.who.int/news-room/fact-sheets/detail/depression.

Wong, N., Kady, L., Mewton, L., Sunderland, M., & Andrews, G. (2014). Preventing anxiety and depression in adolescents: A randomised controlled trial of two school based internet-delivered cognitive behavioural therapy programmes. *Internet Interventions*, *1*(2), 90–94.

Wright, J. H., Owen, J. J., Richards, D., Eells, T. D., Richardson, T., Brown, G. K., et al. (2019). Computer-assisted cognitive-behavior therapy for depression: A systematic review and meta-analysis. *The Journal of Clinical Psychiatry*, *80*(2), 18r12188.

Xiang, X., Wu, S., Zuverink, A., Tomasino, K. N., An, R., & Himle, J. A. (2019). Internet-delivered cognitive behavioral therapies for late-life depressive symptoms: A systematic review and meta-analysis. *Aging & Mental Health*, *24*(8), 1196–1206.

Ye, Y. Y., Zhang, Y. F., Chen, J., Liu, J., Li, X. J., Liu, Y. Z., et al. (2015). Internet-based cognitive behavioral therapy for insomnia (ICBT-i) improves comorbid anxiety and depression-a meta-analysis of randomized controlled trials. *PLoS One*, *10*(11). https://doi.org/10.1371/journal.pone.0142258.

Chapter 42

# Clay art therapy on emotion regulation: Research, theoretical underpinnings, and treatment mechanisms

Joshua K.M. Nan[a], Lisa D. Hinz[b], and Vija B. Lusebrink[c]

[a]*Department of Social Work, Faculty of Social Science, Hong Kong Baptist University, Kowloon Tong, Kowloon, Hong Kong;* [b]*Art Therapy Doctoral Program, Norte Dame de Namur University, Belmont, CA, United States;* [c]*University of Louisville, Louisville, KY, United States*

## List of abbreviations

| | |
|---|---|
| **2D** | two-dimensional |
| **3D** | three-dimensional |
| **ANS** | autonomic nervous system |
| **C/Sy** | cognitive/symbolic |
| **CNS** | central nervous system |
| **CAT** | clay art therapy |
| **EEG** | electroencephalography |
| **ER** | emotional regulation |
| **ETC** | expressive therapies continuum |
| **FEATS** | formal elements art therapy scales |
| **K/S** | kinesthetic/sensory |
| **P/A** | perceptual/affective |
| **PD** | personality disordered |
| **RCT** | randomized controlled trial |

## Introduction

Although visual art was once overlooked by the scientific realm, many studies have related art activities to various cortical processes in recent decades (Liu & Miller, 2008). Scientists and health workers have paid increasing attention on how art activities can foster health at various levels; for instance, the division of work between the left and right hemispheres of the brain when creating or viewing artwork, the relationship between art activities and functions of the two hemispheres of the brain, and the benefits of effective emotion regulation (ER) in individuals experiencing depression (de Schotten et al., 2011; Friederici, 2011).

## Section one: Emotion regulation: Research and theories relating visual art to neuroscience

Emotion is a dynamic mechanism involving the interaction between the neurological system, the mind, and social relationships. Thus, basic emotions are closely associated with physical arousal systems (high/low arousal) and corporal states and according to the context, stimulate an individual's behavior and secondary or tertiary feelings (Schore, 2009). Psychophysiological problems emerge during periods of emotional dysregulation (Panksepp, 2009), and prolonged periods of this state may arise as forms of depression and anxiety.

### Constructs of emotion regulation

ER is the automatic adoption of cognitive strategies to balance the expression, tension, frequency, and duration of an emotional response (Gross, 2014). ER consists of three interlacing compartments. The first indicates personal positive and negative affective states (Joormann & Stanton, 2016; Nan, 2019). By investigating the cause of undermining negative affects and the methods of inducing positive affective experiences, an extensive understanding and treatments for depressive moods can be discovered (Joormann & Stanton, 2016; Werner-Seidler, Banks, Dunn, & Moulds, 2013).

The second constituent underlines cognitive-emotional interaction. Typical cognitive disabilities among individuals with depression include inelastic thinking, diminishing memorizing ability, hesitation during decision-making (Trivedi & Greer, 2014), and alexithymia, which is difficulty in differentiating emotions from sensations (Karlsson, Naatanen, & Stenman, 2008; Joormann & Stanton, 2016). Negative emotions have preceded biases and irrational cognitive strategies that cause failure in regulating emotions (Joormann & Stanton, 2016).

The third constituent links the physiological arousal systems of the human brain to neural mechanisms that process emotions (Cooney, Joormann, Atlas, Eugene,

The Neuroscience of Depression. https://doi.org/10.1016/B978-0-12-817933-8.00009-8

& Gotlib, 2007). Emotional processes that involve autonomic nervous activities and hormonal and cortical responses are integrated into the amygdala through the thalamus (Fuster, 2003). When the emotional arousal system is operating optimally, the parasympathetic ventral vagal system in the autonomic nervous system (ANS) induces a sense of safety and soothes physiological responses (e.g., sweating, muscular tension, and secreting hormones such as cortisol). In return, the unwind physiological state encourages effective emotional communication (Schore, 2009). Dysregulation can be exhibited in the physiological symptoms of hypoarousal or hyperarousal during prolonged stress (Panksepp & Watt, 2011; Schore, 2002). Therefore, ER was developed with transdiagnostic functions to treat depression and anxiety disorders (Sloan et al., 2017).

## Emotion regulation in relation to the left/right hemispheric brain

Neuroimaging studies have indicated that the left hemispheric brain helps process verbal language and fine motor control, whereas the right hemispheric brain controls the processing of nonverbal, visuospatial information, and gross motor skills (Gotts et al., 2013). The right hemispheric brain is more active and sensitive in displaying emotional stimuli (Kret & De Gelder, 2012) and predominantly processes the recognition, experience, and nonverbal expression of emotions. The left hemispheric brain is engaged when the verbal articulation of emotions is required (Nan, 2019).

Distinct differences between artists' and nonartists' brain activity have been discovered from electroencephalography (EEG) research. Major differences were noted in the mental drawings (i.e., art-making activity via memory and imagination) of trained and untrained artists in which a higher level of synchronization occurred in the right hemispheric brain of the artists (Bhattacharya & Petsche, 2005). In another study researching brain processes when viewing paintings, higher levels of synchronization in the right brain were detected in the artist group than in the nonartist group (Bhattacharya & Petsche, 2002).

Furthermore, combined fMRI and behavior analysis by Vessel, Starr, and Rubin (2012) revealed that individuals showed various degrees of emotional response to viewing art (i.e., viewing personally pleasing artwork), as indicated from differences in activity in the prefrontal cortex and amygdala (Vessel et al., 2012). Neuroimaging studies relating visual art to neuroscience have noted that long-term art activities may aid the activation and synchronization of neuronal groupings in various parts of the brain, which leads to the growth of integrative neuronal fibers (Siegel, 2009) that benefit mental health. [Note: More research records are outlined by Nan, 2015 regarding the distinctive functions of parts of the brain for processing emotions during art activities.]

## Treatment efficacy of art therapy on emotion regulation

Compared with pharmacological treatment, verbal psychotherapy results in a high adherence rate and a high rate of initiating patients to share personal feelings and experiences with psychotherapists (Lee, Tsang, & Kwok, 2007). Other prominent effects of psychotherapy include reduced treatment relapses (Thase, 2009), higher levels of the emotional regulatory neurological chemical serotonin (Karlsson, 2012; Karlsson, Hirvonen, Salminen, & Hietala, 2013), increased momentary positive affects, reduced negative affective states, and lowered emotional awareness (van der Velden et al., 2015).

The literature reports that art therapy helps assuage depressive moods and their negative affects (Bar-Sela, Atid, Danos, Gabay, & Epelbaum, 2007; Gussak, 2009), which alleviates physiological stress symptoms (Curl, 2008) and enhances emotional expression and mental well-being (Puig, Lee, Goodwin, & Sherrard, 2006) among populations. Elkis-Abuhoff, Goldblatt, Gaydos, and Corrato (2008) revealed the significant effects of clay work art therapy in improving the somatic functions and emotional status of patients with Parkinson's disease. In addition, de Morais et al. (2014) reported signs of reduced depression and anxiety symptoms from applying clay work to psychiatric patients in a day hospital.

In a randomized controlled trial, Nan and Ho (2017) adopted a specific approach to using clay, clay art therapy (CAT), on adult outpatients with major depressive disorders. Adults with depression were guided through physical clay activities and visual and cognitive engagements while producing clay products. During the various clay work processes, parts of the neurological cortex (Lusebrink, 2004) and the ANS are theorized to be activated simultaneously during sophisticated coordination, which may enhance cognitive abilities and mutate emotional learning (Bastos, Guimaraes, & Trentini, 2013), which fosters a positive affective experience through the active dialogue between cognition and emotion. Aside from improving depressive signs, the treatment results alleviate difficulties arising from alexithymia. CAT adopts a novel art therapy theory known as the expressive therapies continuum (ETC) as its intervention model. Research has supported the notion that the sensuous quality of art materials benefits the structure of ETC-aided emotional expression and problem-solving skills. Symbolic understanding from previous problems among individuals with personality disorders was reported to enhance participants' ER tactics (Haeyen, Kleijberg, & Hinz, 2018).

## Section two: Therapeutic functions of clay on emotion regulation

Clay is one of the most primeval materials and has been widely used since early in human history. Ancient myths, legends, and religions from around the world mention earth or clay (Nan, 2020). Records reveal that clay products have been created since prehistoric times. Firing clay into pottery products has been common in various contexts, such as cultivation and farming (beginning approximately 10,000–8000 BC; Mattison, 2003). With its copiousness, clay products can embody rich values from life (McGriffin, 2002).

Clay can be a supple and flexible art material when kneaded evenly with water for pottery hand building and wheel throwing. However, it can become extremely hard when parched and fired. Clay can be transformed into many forms of ceramic artwork, two-dimensional or three-dimensional (3D). The litheness and expressive disposition of clay provide numerous therapeutic benefits and functions. The sophisticated clay art creation process involves integrating actions such as clay slump molding (e.g., smashing, pressing, rolling, and kneading) and various pottery techniques (e.g., hand building, wheel throwing, coiling, pinching, or glazing). To prevent the clay from drying quickly during work, the artist must keep it hydrated and attentively plan the various steps of clay work processes. Physical strength and patience are required for the artist to maintain the workability of the clay throughout the repetitive process of clay molding and kneading.

### Haptic and proprioceptive sensations in clay work helps build mindful awareness of the physical environment

The "haptic" sense of touch involved when creating clay art is rich. It includes the cutaneous sense responding to physical experiences an object creates (e.g., pressure, vibration, and temperature) (Carr, 2008); the spatial information of holding/touching an object (e.g., shape, size, and position); different textures (e.g., roughness, hardness, and comfortability); and the pain induced (Elbrecht, 2013; Grunwald, 2008). Haptic perception is a unique means of communication that facilitates people's awareness of their spatial position through the body's "built-in" sensors in the muscles, joints, tendons, ligaments, and skin (Grunwald, 2008; Wilson, 1998). This is also known as "proprioception" or "proprioceptive sensation" (Elbrecht, 2013). Proprioceptive sensations help the artist adopt corresponding actions to interact with the surrounding physical environment. For instance, when creating clay art, the artist often must handle pieces of clay slumps or clay slabs simultaneously and preserve a similar level of moisture in the clay pieces. Thus, it is vital for the artist to be mindful of where, how, and when to use these pieces of clay. The habit of using the hands (including the palms, fingers, skin, and joints) to handle, organize, feel, and maintain clay is essential to help the artist recognize and manipulate objects through touch. This major pathway of communication during clay work processes facilitates somatic mindfulness in the artist (Nan, 2020).

### Sensational processes of touch from creating clay art attunes the psychobiological arousal system

Before developing verbal communication, an infant learns to develop a secure relationship with the mother through nonverbal communication. Such nonverbal communication includes at least physical, visual (reading facial expressions), auditory (recognizing amplitude of sound, tone, prosody, and tempo), and body (gestures, posture, and movement) cues (Schore, 2009). These nonverbal means operate as body-based affective communication patterns (Dael, Mortillaro, & Scherer, 2012; Panksepp & Watt, 2011).

The tactile quality of clay plays a role similar to nonverbal communication because touch can stimulate intense haptic sensational processes. Skillful manipulation of clay can foster an optimal arousal condition and effective communication that allows the artist to adjust the sympathetic and parasympathetic systems so that he or she can feel secure, relaxed, and immersed in the creative process (Bratton & Ferebee, 1999; Nan & Ho, 2014). This alternative body-based communication pattern imbedded in creating clay art can benefit the cortical and limbic-autonomic circuits of the brain (Cozolino, 2002; Schore, 2009; Siegel, 1999).

Meanwhile, creating clay art requires the active participation and rhythmic movement of the body. For instance, kneading a clay slump requires an artist to use the forearms, upper arms, or even the whole upper body to repeatedly press the clay slump. Pinching clay balls, rolling, and coiling the clay also involve rhythmic movements from the palms, hands, and fingers for fine motor activities. During the clay work process, the kinesthetic sensation and rhythmic engagement of the body can stimulate a feeling of inner rhythm, which can incite positive emotions and encourage self-exposure (Kronholm, Partonen, Salminen, Mattila, & Joukamaa, 2007). This constant facilitation of psychophysiological attunement may improve psychosomatic conditions, such as dysrhythmias (Hinz, 2019).

### Facilitation of soothing and expressing difficult emotions

The constant and repetitive manipulation of the clay involving body movement and intense visual processes can help integrate various sensory processes. These processes potentially can activate neural pathways associated with the primary somatosensory cortex of the brain (Carr, 2008). The integrated sensory processes also can form an emotion-calming effect. Thus, the constant creation of clay art is particularly beneficial to those who are oversensitive to or hyperaroused by external stimulation.

Integrated sensory processes from creating clay can strengthen interoception or awareness of bodily sensations as a basic component of emotional awareness and strengthen the differentiation between sensational processes and emotions (Barrett, 2017; Gu, Hof, Friston, & Fan, 2013). Elkis-Abuhoff et al. (2008) addressed the effect of clay work on improving the somatic functions and emotional status of patients with Parkinson's disease (Elkis-Abuhoff et al. 2008). Kimport and Robbins (2012) reported that tactile sensations associated with clay work promote emotional expression, and malleable clay molding actions induce creativity.

The distinctive ability of clay to "remember" the artist's movements on the clay surface can create profound expressions of inner processes (Nan, 2020; Sholt & Gavron, 2006). For individuals who are emotionally traumatized, both clay work processes and finished pottery works can create powerful therapeutic effects on reorganizing and integrating traumatic memories; mastering personal feelings; and instilling new meanings to struggles, conflicts, and problems in life. The transformative effects are similar to some of the effects of other art media reported in the art therapy literature (Appleton, 2001; Chapman, Morabito, Ladakakos, Schreier, & Knudson, 2001; C. Elbrecht & Antcliff, 2014).

### Increasing cognitive abilities and expressing abstract ideas by creating three-dimensional objects

Clay work processes can be complex because they involve various steps and skills. For instance, creating 3D clay products requires a combination of techniques, including kneading the clay, pinching, combining pinched pieces, sculpting the form, and glazing the final sculpture. These processes require different parts of the brain to cooperate to execute complex processes, such as intensely concentrating, deploying problem-solving skills, planning, organizing work steps, and thinking analytically. The continual stimulation of haptic perception, physical movements during clay work, and the involvement of visual sensations activate processes in the primary somatosensory cortex and foster activities of the orbitofrontal cortex (Carlson, 2001; Lusebrink, 2004).

Creating 3D clay products facilitates the free association and visualization of significant representational objects that may not be easily recalled by artists from imagination or verbal conversation (Keyes, 1984; Mattes & Robbins, 1981). Both the creative process and products foster the expression of abstract ideas and enhanced cognition to create emotional interaction. The beautiful glazed ceramic products also enrich aesthetic expressions and enhance a sense of ownership (Graziano, 1999). The term "concretization" best describes the embodiment of inner representations in clay products (Mattes & Robbins, 1981; Simon, 1996; Winship & Haigh, 1998).

## Section three: Treatment mechanisms of clay art therapy for emotion regulation

### Theoretical underpinnings: Expressive therapies continuum

The ETC is a theoretic model explaining how people absorb, process, and express information when interacting with art media. The ETC can be a guide for the therapeutic use of art media to aid the expression of sensations, feelings, images, and thoughts and render them accessible for cognitive processing (Lusebrink, 2016). As a developmentally and hierarchically based system, the ETC proposes that information processing begins with basic sensorimotor input on the kinesthetic/sensory (K/S) level, which represents activity in the sensorimotor cortices. Emotional value is attached to the second level, the perceptual/affective (P/A) level of the model, which represents the influence of the limbic system. Finally, visual input is forwarded to areas of the cerebral cortex where sophisticated cognitive operations occur on the cognitive/symbolic (C/Sy) level. When treating depression, K/S work can physically enliven people with depression; P/A work can stimulate blocked emotions; and C/S work can transform negative views of the self, world, and future (Hinz, 2019).

Applying the ETC in therapy is based on its system characteristics. Activities and expressions involving the three distinct levels of the ETC include information intrinsic to each level when all levels are interrelated (Lusebrink, 2016). The progressive differentiation of visual and tactile expression occurs when information processing progresses from the K/S level to higher levels of the ETC, which lead to an individual's enhanced emotional and cognitive functioning. Through media choices (e.g., clay, pencil, and poster color), task instructions, and directed questions, art therapists can emphasize certain levels of expression and thereby strengthen information processing flows (Hinz, 2019).

### Bottom-up approach of clay art therapy for emotion regulation

As an art medium, clay is widely used for different psychotherapeutic approaches. This chapter illustrates a specific clay work approach, CAT, as a body-mind approach to intervene in human cognitive, emotional, and physiological mechanisms. The emphasis on rationally treating cognitive processes and functioning to alter affects is a "top-down" model, whereas the CAT approach targets the mind and body to equally incorporate aspects of cognition, emotions, and physiology into treatment considerations. By adopting the ETC as its fundamental mechanism, CAT commences from bodily stimulation and experiences at the K/S level, and progressively, such experiences become attached with subjective feelings at the P/A level. These eventually transform into thoughts and personal representations at the C/Sy level. Every level of CAT activities embeds the creative level in its process. Thus, the body-mind CAT model endorses a "bottom-up" model (Fig. 1) that commences through an individual's direct contact with clay instead of eloquently talking through a client's problem. Both psychotherapeutic approaches complement one another, and both top-down and bottom-up models adapt to explain the human brain's neurological mechanisms (McRae et al., 2012).

Visual and tactile expressions through clay work can provide valuable information on an individual's strengths and weaknesses in ETC component functioning (e.g., kinesthetic, sensory, and affective). An art therapist can typically assess the strengths and weaknesses of ETC functioning from the first few artistic expressions of an individual. The art therapist can then utilize the manifested strengths of some ETC components to compensate for impaired components by gradually changing ETC components (Lusebrink, 2016; Lusebrink & Hinz, 2016). In a group work setting, an art therapist must provide ongoing assessments and guide each individual through the stages to observe how they interact with the clay and express the inner processes in their clay work. Research indicates that individuals with depression may exhibit similar signs and problems during artistic expression. For instance, research with the Formal Elements Art Therapy Scales revealed that drawings created by individuals with depression commonly exhibit reduced energy, few details, and poor problem-solving ability, which imply impairments of the kinesthetic, perceptual, and cognitive components of the ETC (Gantt, 2016; Manickam & Sajani, 2015). Therefore, with the ETC as the treatment protocol (Blomdahl, Guregård, Rusner, & Wijk, 2018), the art therapist can accordingly design and guide the CAT treatment processes.

## Goals and stages of clay art therapy

### *Stage one*

From initially touching the clay slump, participants of a CAT group are invited to use their hands to freely interact with the clay in different manners, such as squeezing, pinching, and rolling. Some rigorous physical movements such as pounding, kneading, and throwing the clay require the artists to invest intense physical strength involving the whole body. The participants are gradually led to form the clay into simple shapes, such as a snowball or brick. Through these somatosensory processes (Nan & Ho, 2017; Fig. 2), the art therapist can observe how an individual interacts with the clay and determine whether the individual should use kinesthetic or sensory components more or less. For the ETC, the clay work processes at this level aim to balance activities on the two poles of the K/S movement and achieve the creative integration of clay art.

Repetitive sensory-rich clay work processes can enhance the optimal condition of the ANS, which benefits the production of positive affective states. Haptic sensation also leads to somatic consciousness to benefit people with sensitive haptic perception. (Please refer to Section Two for details.) Therefore, clay activities at the K/S level gradually

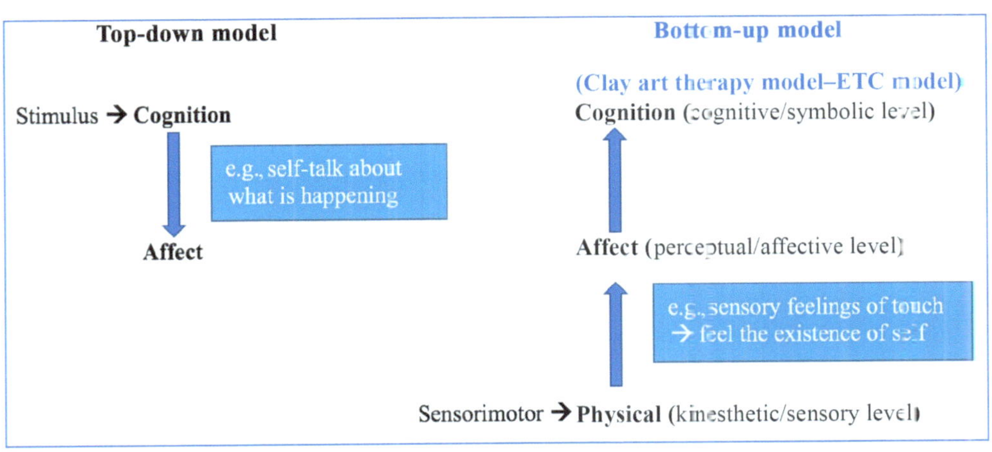


FIG. 1 Top-down model and bottom-up model. Verbal psychotherapy typically adopts the top-down approach (left) in which guided or self-talking is used to allow clients to understand their affective states. The clay art therapy (CAT) model adopts a bottom-up model in which kinesthetic movements and sensory stimulations are used to produce perceptual and affective experiences to alter the cognitive understanding of emotions.

FIG. 2 Clay art creation activities (kneading and pinching). Activities involving somatosensory processes and haptic perception during clay art therapy (CAT). *Adapted from Nan, J. K. M., & Ho, R. T. C. (2017). Effects of clay art therapy on adults outpatients with major depressive disorder: A randomized controlled trial.* Journal of Affective Disorders, 217, *237–245.*

lead to the emerging functions of the affective and perceptual components (Fig. 2). As the art therapist guides the participants to create clay products by building work with their hands and glazing colors, the corresponding psychological processes from making judgments on various dimensions (e.g., lines, shapes, height, and thickness) of the semiproducts can enhance perceptual skills and train objective judgments. These positive effects alleviate the typical features of indecisiveness and disorganized thoughts exhibited in individuals with depression. The visual process of coloring or glazing clay products facilitates the expression of inner affective states. This is a new method of shaping, reexperiencing, and harmonizing emotions. All aforementioned processes generally foster effective affect regulation strategies.

The number of clay sessions at this stage depends on the objectives of the group; however, conducting at least three to four sessions is typically beneficial. Despite clay work being similar to an ordinary clay workshop, the composition of various psychotherapeutic concepts, techniques, and treatment methods go beyond only art training.

## Stage two

To enhance the C/S components at stage two, the work steps for creating clay art become complicated. At this stage, the art therapist leads the participants to learn to integrate several clay products together to create a clay sculpture. The clay work processes use the prefrontal cortex to execute high levels of cognitive functioning, such as assimilating previously learned clay work techniques (e.g., kneading, pinching, and glazing clay) with new ones (integrating two pieces of clay work together; Piaget, 1983). Sculpture creation processes are more complex than the processes of stage one and require more intensive cortical-subcortical cooperation (e.g., perceptual skills, affective expression, good organization, and problem-solving skills), high energy input, and a tolerance for frustration (Fig. 3). Therefore, the creative processes become more intense at this stage.

The participants then design the sculpture features by themselves. Creating sculptures and products can thus embody rich personal meanings to the artists (Nan, 2020; Fig. 4). Focusing on the symbolism of creating a personally meaningful miniature simultaneously facilitates the integration of all ETC components and allows the achievement of a sense of flow (Csikszentmihalyi, 2008) or an enhanced sense of self-efficacy. To facilitate the creative process, group sessions in stage two are less structured than those of stage one, and the participants are given more freedom regarding their expressive style in the creative artwork. Fig. 5 summarizes the coordination of the various dimensions of task instructions for CAT intervention. Fig. 6 lists the basic structure of the CAT intervention framework.

In both stages one and two, CAT consists of repetitive physical interaction with the clay. These processes create

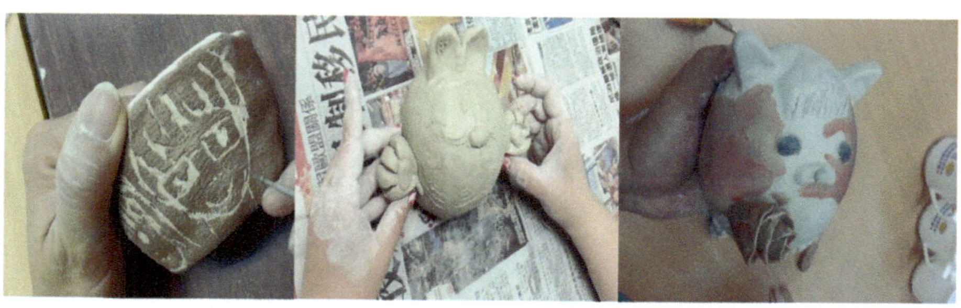

FIG. 3 Clay art creating activities (surface treatment and decorating personal sculptures). Activities involving perceptual skills, cognitive functions, and affective expression during clay art therapy (CAT). *Adapted from Nan, J. K. M., & Ho, R. T. C. (2017). Effects of clay art therapy on adults outpatients with major depressive disorder: A randomized controlled trial.* Journal of Affective Disorders, 217, *237–245.*

FIG. 4 Participants' clay artwork (minisculptures). Integration of various inner processes to create personally meaningful ceramic products.

| | STAGE 1 | STAGE 2 |
|---|---|---|
| **Task structure** | Highly structured | Semi structured |
| **Task difficulty/required skill level** | Simple<br>Low skill level | Complex/creative<br>High skill level |
| **Energy input** | Low input | High input, more intense attention |
| **Expressive style and boundary/limit** | Constricted<br>Create similar products | More freedom to create personally meaningful sculptures |

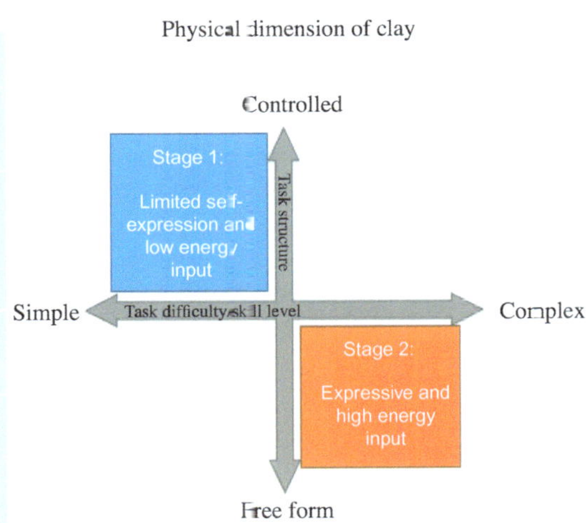


FIG. 5 Design of the two-stage intervention model. The two-stage intervention model carefully monitors the physical dimensions of clay work throughout the group process. These dimensions included (A) task structure (controlling the interaction with clay and style of expression), (B) task complexity and required skills level, (C) energy input, and (D) personal expressive style and boundaries/limits.

a psychophysiological feedback loop (Hinz, 2019) from participants to produce positive affective experiences. Therefore, art therapists must enable participants to experience, reexperience, conceptualize, and express these positive affective experiences through psychosomatic feedback. These new emotional experiences can aid the transformation of negative affects (Bidabadi & Mehryar, 2015; Nan, 2019, 2020). CAT treatment is a nonverbal process and is essential for art therapists to help participants make sense of their new affective experiences and provide

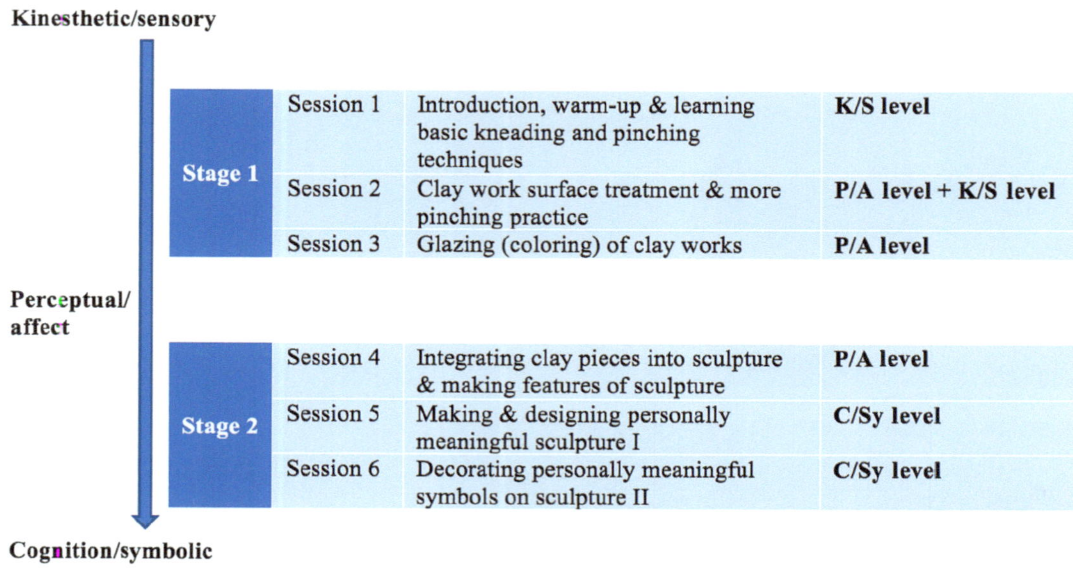


FIG. 6 Intervention framework of clay art therapy. Session plans of the two-stage intervention model, which adopts the bottom-up ETC model, progressively work from K/S-level activities to C/Sy-level activities.

psychoeducation during treatment sessions. Verbal interaction is thus indispensable during treatment.

## Concluding comments

### Future research direction

Research and clinical application of clay art therapy point toward the essential need to continually investigate methods on reducing negative emotional experience (Seidel et al., 2018; Menninghaus et al., 2017; Johnsen & Friborg, 2015) and promoting the positive affective experiences (Joormann & Stanton, 2016) via creative arts intervention. There are many aspects that warrant research rigor: the combined use of verbal and nonverbal processes in art therapy, long-term intervention effects of CAT, and the therapeutic effects of other visual art media, such as painting and drawing, and more empirical studies.

## Key facts

- Key facts of emotional regulation
  - Emotional regulation (ER) is the automatic adoption of cognitive strategies to balance the expression, tension, frequency, and duration of an emotional response (Gross, 2014).
  - ER consists of three interlacing elements: (1) levels of positive and negative affective mood, (2) interdepending cognitive and emotional experiences, and (3) mutual influences between physical arousal and emotions (Nan, 2019).
  - Depression is a product of ER difficulty. Sustained experience of negative mood and continuous decline in positive affect are distinct symptoms for depressive disorder. Understanding individual's cognitive mechanism for regulating emotional responses is essential in learning depression (Joormann & Stanton, 2016).
  - A prolonged period of depressed mood can disrupt the limbic system and cause regulatory impairment in the ANS. The disruption could lead to an extended negative affective state and physical-somatic symptoms in depressive individuals (Schore, 2009; Werner-Seidler et al., 2013).
  - Nonverbal psychotherapy, such as art therapy (de Morais et al., 2014), was found effective in improving ER. Nan and Ho's (2017) research on depression among outpatients has discovered the effects of clay art therapy (CAT), including physical interplay with clay, visual and cognitive processes of viewing, and creating clay art products, in alleviating depressive symptoms. The sophisticated coordination of different cortical areas (Lusebrink, 2004) and autonomic nervous system engaged during clay-work activities can strengthen cognitive competence and mutate emotional learning (Bastos et al., 2013; Carr, 2008).
- Key facts of art therapy
  - Art therapy is a type of nonverbal expressive therapy that stimulates sensory experience and utilizes the creative assets of the individual through engaging in art activities, which can benefit a person's physical and emotional well-being (Lusebrink, 2016; Hinz, 2019).

- Engaging in an art creative process vitalizes one's cognitive system that assists the individual develop, reorganize, express, and resolve own behaviors and emotions (Lusebrink, 2016; Hinz, 2019).
- Art therapy was found with effects in assuaging depressive moods and their negative emotions (Bar-Sela et al., 2007; Gussak, 2009), in consequence, alleviates physiological stress symptoms (Curl, 2008) and enhances emotional expression and mental well-being (Puig et al., 2006).
- In every psychotherapy, the practitioner should work toward a form of "attunement" with their clients, which he/she is capable to understand and identify the nonverbal rhythms and responses of others (Malchiodi, 2013). Whereas in art therapy, such attunement adopts a "bottom-up" approach that is nonverbal and sensory-based, the therapeutic process is less reliant on words (Lusebrink, 2016; Hinz, 2019).
- Every art form and art media has its distinctive physical properties and characteristics that advocate different sensory forms of attunement and builds unique practitioner-client relational dynamics. The art therapy process accentuates on the practitioner-client interaction in their experiential and sensory exchanges through art activities, which aims to attain a sense of emotional attunement and understanding (Malchiodi, 2013).
- Clay is a distinct art media that provides tactile quality similar to nonverbal communication, as the sense of touch can stimulate intense haptic sensational processes. Skillful manipulation of clay can foster an optimal arousal condition and effective communication that allows the artist to adjust the sympathetic and parasympathetic systems so that he or she can feel secure, relaxed, and immersed in the creative process (Bratton & Ferebee, 1999; Nan & Ho, 2014).

## Summary points

- Emotion regulation consists of three main intertwining compartments: (i) personal positive and negative affective states, (ii) cognitive-emotional interdependence, and (iii) neurological mechanisms of emotional processing.
- Research has indicated the treatment efficacy of art therapy in reducing depression and physiological stress symptoms, enhancing emotional expression, and improving holistic well-being. CAT provides benefits to patients with major depressive disorders to improve their depressive signs, mutate emotional learning, and alleviate difficulties arising from alexithymia.
- Because clay can be very malleable and fragile, the haptic and proprioceptive sensations are critical during clay work processes, which allow artists to be mindfully aware of their somatic experiences.
- The tactile property of clay stimulates intense haptic sensational processes that can optimally arouse the cognitive system. Through rhythmic repetition of kinesthetic movements, the artist can sense an attuned inner rhythm that produces a sense of pleasure and promotes emotional expression.
- Clay work can be a profound representation of inner processes to allow the artist to transform personal values and feelings into concrete outputs.
- The CAT model was developed from the ETC model, which adopts a bottom-up approach, proposes that information processing begins from sensorimotor stimulation at the K/S level, and gradually attaching emotional values at the P/A level before finally transforming into cognitive representations at the C/Sy level.
- The CAT protocol adopts a two-stage intervention design that progressively complies with the levels of the ETC model: (i) Stage 1: Clients are introduced to basic techniques of clay making, and tasks are highly structured, repetitive, and simple (ii) Stage 2: Clients learn more complex techniques and have the freedom to produce personally meaningful artwork via their creative abilities.

## Mini-dictionary of terms

**Alexithymia** The cognitive inability to encode, identify, and describe one's own and another person's emotions.

**Art therapy** A form of nonverbal psychotherapy that uses creative art-making processes to allow clients to express personal feelings and experience a sense of healing. This allows art psychotherapists to understand and analyze clients' mental state through their artwork.

**Bottom-up model** An explanation of how human perception processing begins from physical/sensory stimulation and gradually becomes cognitive representation in the mind.

**Clay art therapy** A type of art therapy utilizing clay as media. The tactile property of clay can induce intense haptic sensations that, along with visual stimulation, is believed to soothe emotions and promote emotion regulation.

**Concretization** The process of turning inner abstract values, emotions, and thoughts into concrete clay representations.

**Emotion regulation** The autonomic use of cognitive strategies to rectify optimal and adaptive expression, tension, frequency, and duration of an emotional response.

**Expressive therapies continuum** The theoretical model describing how people engage in, process, and express information when they interact with art media.

**Proprioceptive sensation** The haptic perception of one's spatial position within the body's "built-in" sensors in the muscles, joints, tendons, ligaments, and skin.

**Psychophysiological attunement** The sense of optimal coordination between one's bodily state and one's mental state, which produces a sense of pleasure.

## References


Appleton, V. (2001). Avenues of hope: Art therapy and the resolution of trauma. *Art Therapy: Journal of the American Art Therapy Association*, *18*(1), 6–13.

Barrett, L. F. (2017). The theory of constructed emotion: An active inference account of interoception and categorization. *Social Cognitive and Affective Neuroscience*, *12*(1), 1–23.

Bar-Sela, G., Atid, L., Danos, S., Gabay, N., & Epelbaum, R. (2007). Art therapy improved depression and influenced fatigue levels in cancer patients on chemotherapy. *Psycho-Oncology, Journal of the Psychological, Social and Behavioral Dimensions of Cancer*, *16*(11), 980–984. https://doi.org/10.1002/pon.1175.

Bastos, A. G., Guimaraes, L. S. P., & Trentini, C. M. (2013). Neurocognitive changes in depressed patients in psychodynamic psychotherapy, therapy with fluoxetine and combination therapy. *Journal of Affective Disorders*, *151*, 1066–1075.

Bhattacharya, J., & Petsche, H. (2002). Shadows of artistry: Cortical synchrony during perception and imagery of visual art. *Cognitive Brain Research*, *13*(2), 179–186.

Bhattacharya, J., & Petsche, H. (2005). Drawing on mind's canvas: Differences in cortical integration patterns between artists and non-artists. *Human Brain Mapping*, *26*(1), 1–14.

Bidabadi, S. S., & Mehryar, A. (2015). Music therapy as an adjunct to standard treatment for obsessive compulsive disorder and co-morbid anxiety and depression: A randomized clinical trial. *Journal of Affective Disorders.*, *184*, 13–17.

Blomdahl, C., Guregård, S., Rusner, M., & Wijk, H. (2018). A manual-based phenomenological art therapy for individuals diagnosed with moderate to severe depression (PATd): A randomized controlled study. *Psychiatric Rehabilitation Journal*, *41*(3), 169.

Bratton, S. C., & Ferebee, K. W. (1999). The use of structured expressive art activities in group activity therapy with preadolescents. In D. S. Sweeney, & L. E. Homeyer (Eds.), *The handbook of group play therapy: How to do it, how it works, whom it's best for* (pp. 192–214). San Francisco: Jossey-Bass.

Carlson, N. R. (2001). *Physiology of behavior* (7th ed., pp. 43–61). Boston: Allyn & Bacon.

Carr, R. (2008). Sensory processes and responses. In N. Hass-Cohen, & R. Carr (Eds.), *Art therapy and clinical neuroscience*. London & Philadelphia: Jessica Kingsley Publishers.

Chapman, L., Morabito, D., Ladakakos, C., Schreier, H., & Knudson, M. M. (2001). The effectiveness of art therapy interventions in reducing post traumatic stress disorder (PTSD) symptoms in pediatric trauma patients. *Art Therapy: Journal of the American Art Therapy Association*, *18*(2), 100–104. https://doi.org/10.1080/07421656.2001.10129750.

Cooney, R. E., Joormann, J., Atlas, L. Y., Eugene, F., & Gotlib, I. H. (2007). Remembering the good times: Neural correlates of affect regulation. *Neuroreport*, *18*(17), 1771–1774.

Cozolino, L. L. (2002). *The neuroscience of psychotherapy*. New York: Norton.

Csikszentmihalyi, M. (2008). *Flow: The psychology of optimal experience*. USA: HarperPerennial.

Curl, K. (2008). Assessing stress reduction as a function of artistic creation and cognitive focus. *Art Therapy*, *25*(4), 164–169.

Dael, N., Mortillaro, M., & Scherer, K. R. (2012). Emotion expression in body action and posture. *Emotion*, *12*(5), 1085–1101.

de Morais, A. H., Dalecio, M. A. N., Vizmann, S., Bueno, V. L. R.d. C., Roecker, S., Salvagioni, D. A. J., et al. (2014). Effect on scores of dep and anx in psychiatric patients after clay work in a day hospital. *The Arts in Psychotherapy*, *41*(2), 205–210.

de Schotten, M. T., Dell'Acqua, F., Forkel, S., Simmons, A., Vergani, F., Murphy, D. G., et al. (2011). A lateralized brain network for visuospatial attention. *Nature Precedings*, 1.

Elbrecht, C. (2013). *Trauma healing at the clay field*. London and Philadelphia: Kessica Kingsley Publishers.

Elbrecht, C., & Antcliff, L. R. (2014). Being touched through touch. Trauma treatment through haptic perception at the clay field: A sensorimotor art therapy. *International Journal of Art Therapy*, *19*(1), 19–30.

Elkis-Abuhoff, D. L., Goldblatt, R. B., Gaydos, M., & Corrato, S. (2008). Effects of clay manipulation on somatic dysfunction and emotional distress in patients with Parkinson's disease. *Art Therapy: Journal of the American Art Therapy Association*, *25*(3), 122–128.

Friederici, A. D. (2011). The brain basis of language processing: From structure to function. *Physiological Reviews*, *91*(4), 1357–1392.

Fuster, J. M. (2003). *Cortex and mind: Unifying cognition*. New York: Oxford University Press.

Gantt, L. (2016). The formal elements art therapy scale (FEATS). In D. Gussac, & M. L. Rosal (Eds.), *The Wiley handbook of Art Therapy* (pp. 569–578). John Wiley & Sons Inc.

Gotts, S. J., Jo, H. J., Wallace, G. L., Saad, Z. S., Cox, R. W., & Martin, A. (2013). Two distinct forms of functional lateralization in the human brain. *Proceedings of the National Academy of Sciences of the United States of America*, *110*(36), E3435–E3444.

Graziano, J. E. (1999). *Narrative and clay: The storied forms of early adolescents ten to fourteen* (doctoral). Colombia University. Unpublished.

Gross, J. J. (2014). Emotion regulation: Conceptual and empirical foundations. *Handbook of emotion regulation* (2nd, pp. 3–20). The Guilford Press.

Grunwald, M. (Ed.). (2008). *Human haptic perception: Basics and application*. Basel/Bosston, MA/Berlin: Birkhauser.

Gu, X., Hof, P. R., Friston, K. J., & Fan, J. (2013). Anterior insular cortex and emotional awareness. *Journal of Comparative Neurology*, *521*(15), 3371–3388.

Gussak, D. (2009). Comparing the effectiveness of art therapy on depression and locus of control of male and female inmates. *The Arts in Psychotherapy*, *36*, 202–207.

Haeyen, S., Kleijberg, M., & Hinz, L. (2018). Art therapy for patients diagnosed with personality disorders cluster B/C: A thematic analysis of emotion regulation from patient and art therapist perspectives. *International Journal of Art Therapy*, *23*(4), 156–168.

Hinz, L. D. (2019). *Expressive therapies continuum: A framework for using art in therapy*. Routledge.

Johnsen, T. J., & Friborg, O. (2015). The effects of cognitive behavioral therapy as an anti-depressive treatment is falling: A meta-analysis. *Psychological Bulletin*, *141*(4), 747.

Joormann, J., & Stanton, C. H. (2016). Examining emotion regulation in depression: A review and future directions. *Behaviour Research and Therapy*, *86*, 35–49. https://doi.org/10.1016/j.brat.2016.07.007.

Karlsson, H. (2012). Psychotherapy increases the amount of serotonin receptors in the brains of patients with major depressive disorder. In R. Levy, J. Ablon, & H. Kächele (Eds.), *Psychodynamic psychotherapy research: Evidence-based practice and practice-based evidence* (pp. 233–238). New York, NY: Humana Press.

Karlsson, H., Naatanen, P., & Stenman, H. (2008). Cortical activation in alexithymia as a response to emotional stimuli. *The British Journal of Psychiatry*, *192*(1), 32–38.

Karlsson, H., Hirvonen, J., Salminen, J., & Hietala, J. (2013). Increased serotonin receptor 1A binding in major depressive disorder after psychotherapy, but not after SSRI pharmacotherapy, is related to improved social functioning capacity. *Psychotherapy and Psychosomatics*, *82*(4), 260–261. https://doi.org/10.1159/000346143.

Keyes, M. F. (1984). The family clay sculpture. *The Arts in Psychotherapy*, *11*(1), 25–28. https://doi.org/10.1016/0197-4556(84)90015-7.

Kimport, R. E., & Robbins, J. S. (2012). Efficacy of creative clay work for reducing negative mood: A randomized controlled trial. *Art Therapy: Journal of the American Art Therapy Association*, *29*(2), 74–79.

Kret, M. E., & De Gelder, B. (2012). A review on sex differences in processing emotional signals. *Neuropsychologia*, *50*(7), 1211–1221.

Kronholm, E., Partonen, T., Salminen, J. K., Mattila, A. K., & Joukamaa, M. (2007). Alexithymia, depression and sleep disturbance symptoms. *Psychotherapy and Psychosomatics*, *77*(1), 63–65.

Lee, S., Tsang, A., & Kwok, K. (2007). Twelve-month prevalence, correlates, and treatment preference of adults with DSM-IV major depressive episode in Hong Kong. *Journal of Affective Disorders*, *98*(1–2), 129–136.

Liu, A., & Miller, B. (2008). Visual art and brain. In *Vol. 88. Handbook of clinical neurology* (pp. 471–488). SFO, USA: Memory and Aging Center, University of California.

Lusebrink, V. B. (2004). Art therapy and the brain: An attempt to understand the underlying processes of art expression in therapy. *Art Therapy: Journal of the American Art Therapy Association*, *21*(3), 125–135.

Lusebrink, V. B. (2016). Expressive therapies continuum In *The Wiley handbook of Art Therapy* (2nd, pp. 57–67). John Wiley & Sons Inc.

Lusebrink, V. B., & Hinz, L. D. (2016). The expressive therapies continuum as a framework in the treatment of trauma. In J. L. King (Ed.), *Art Therapy, trauma, and neuroscience: Theoretical and practical perspectives* (pp. 42–66). New York, NY: Routledge.

Malchiodi, C. A. (Ed.). (2013). *Expressive therapies* Guilford Publications.

Manickam, L., & Sajani, V. (2016). Elements of drawing and depression among adolescents in Kerala, South India: An explorative study using a person picking a mango from a tree. *SIS Journal of Projective Psychology & Mental Health*, *23*(1), 41.

Mattes, J., & Robbins, A. (1981). Creative interventions in psychoanalytic psychotherapy. *The Psychoanalytic Review*, *68*, 386–394.

Mattison, S. (2003). *The complete potter: The complete reference to tools, materials and techniques for all potters and ceramicists*. Apple Press.

McGriffin, M. S. P. (2002). *Reflections in clay: Mirror of truth*. Bloomington, IN: 1st Books Library.

McRae, K., Gross, J. J., Weber, J., Robertson, E. R., Sokol-Hessner, P., Ray, R. D., et al. (2012). The development of emotion regulation: An fMRI study of cognitive reappraisal in children, adolescents and young adults. *Social Cognitive and Affective Neuroscience*, *7*(1), 11–22.

Menninghaus, W., Wagner, V., Hanich, J., Wassiliwizky, E., Jacobsen, T., & Koelsch, S. (2017). The distancing-embracing model of the enjoyment of negative emotions in art reception. *Behavioral and Brain Sciences*, *40*, 1–15.

Nan, J. K. M. (2015). *Therapeutic effects of clay art therapy for patients with depression* (doctoral dissertation). The University of Hong Kong.

Nan, J. K. M. (2019). Affect regulation. In D. Gu, & M. E. Dupre (Eds.), *Encyclopedia of gerontology and public aging* Springer Pub. Co.

Nan, J. K. M. (2020). From clay to ceramic: An alchemical process of self-transformation. In L. Leone (Ed.), *Craft in Art Therapy* Routledge.

Nan, J. K. M., & Ho, R. T. H. (2014). Affect regulation and treatment for depression and anxiety through art: Theoretical ground and clinical issues. *Annals of Depression and Anxiety*, *1*(2), 1008.

Nan, J. K. M., & Ho, R. T. H. (2017). Effects of clay art therapy on adults outpatients with major depressive disorder: A randomized controlled trial. *Journal of Affective Disorders*, *217*, 237–245.

Panksepp, J. (2009). Brain emotion systems and qualities of mental life—From animal models of affect to implications for psychotherapeutics. In D. Fosha, D. J. Siegal, & M. F. Solomon (Eds.), *The healing power of emotion: Affective neuroscience, development, and clinical practice* W. W. Norton & Company.

Panksepp, J., & Watt, D. (2011). Why does depression hurt? Ancestral primary-process separation-distress (PANIC/GRIEF) and diminished brain reward (SEEKING) processes in the genesis of depressive affect. *Psychiatry: Interpersonal & Biological Processes*, *74*(1), 5–13. https://doi.org/10.1521/psyc.2011.74.1.5.

Piaget, J. (1983). Piaget's theory. In P. Mussen (Ed.), *Vol. 1. Handbook of child psychology* (4th ed.). New York: Wiley.

Puig, A., Lee, S. M., Goodwin, L., & Sherrard P. (2006). The efficacy of creative arts therapies to enhance emotional expression, spirituality, and psychological well-being of newly diagnosed Stage I and Stage II breast cancer patients: A preliminary study. *The Arts in Psychotherapy*, *33*(3), 218–228.

Schore, A. N. (2002). Dysregulation of the right brain: A fundamental mechanism of traumatic attachment and the psychopathogenesis of posttraumatic stress disorder. *Australian & New Zealand Journal of Psychiatry*, *36*(1), 9–30.

Schore, A. N. (2009). Right-brain affect regulation—An essential mechanism of development, trauma, dissociation, and psychotherapy. In D. Fosha, D. J. Siegal, & M. F. Solomon (Eds.) *The healing power of emotion: Affective neuroscience, development & clinical practice* (pp. 112–144). W.W. Norton & Company.

Seidel, M., King, J. A., Ritschel, F., Boehm, I. Geisler. D., Bernardoni, F., et al. (2018). Processing and regulation of negative emotions in anorexia nervosa: An fMRI study. *NeuroImage: Clinical*, *18*. 1–8.

Siegel, D. J. (2009). Emotion as integration: A possible answer to the question, what is emotion? In D. Fosha, D. J. Siegel, & M. Solomon (Eds.), *The healing power of emotion: Affective neuroscience, development, and clinical practice*. New York & London: W. W. Norton & Company.

Sholt, M., & Gavron. T. (2006). Therapeutic qualities of clay-work in art therapy and psychotherapy: A review. *Art Therapy: Journal of the American Art Therapy Association*, *23*(2), 66–72.

Siegel, D. J. (1999). *The developing mind: Toward a neurobiology of interpersonal experience*. New York: Guilford Press.

Simon, R. M. (1996) *Symbolic images in art as therapy*. London: Routledge.

Sloan, E., Hall, K., Moulding, R., Bryce, S., Mildred, H., & Staiger, P. K. (2017). Emotion regulation as a transdiagnostic treatment construct across anxiety, depression, substance, eating and borderline personality disorders: A systematic review. *Clinical Psychology Review*, *57*, 141–163.

Thase, M. (2009). Psychotherapies for depression in adults: A review of recent developments. In H. Herrman, M. Maj, & N. Sartorius (Eds.), *WPA series in evidence & experience in psychiatry. Depressive disorders* (3rd ed.). UK: Wiley-Blackwell.

Trivedi, M. H., & Greer, T. L. (2014). Cognitive dysfunction in unipolar depression: Implications for treatment. *Journal of Affective Disorders*, *152–154*, 19–27.

van der Velden, A. M., Kuyken, W., Wattar, U., Crane, C., Pallesen, K. J., Dahlgaard, J., et al. (2015). A systematic review of mechanisms of change in mindfulness-based cognitive therapy in the treatment of recurrent major depressive disorder. *Clinical Psychology Review*, *37*, 26–39.

Vessel, E. A., Starr, G. G., & Rubin, N. (2012). The brain on art: Intense aesthetic experience activates the default mode network. *Frontiers in Human Neuroscience*, *6*(66). https://doi.org/10.3389/fnhum.2012.00066.

Werner-Seidler, A., Banks, R., Dunn, B. D., & Moulds, M. L. (2013). An investigation of the relationship between positive affect regulation and depression. *Behaviour Research and Therapy*, *51*(1), 46–56.

Wilson, F. R. (1998). *The hand: How its use shapes the brain, language, and human culture* (1st). New York, NY: Vintage Books.

Winship, G., & Haigh, R. (1998). The formation of objects in the group matrix: Reflections on creative therapy with clay. *Group Analysis*, *31*, 71–81.

## Further Reading

Ehrich, E., Turncliff, R., Du, Y., Leigh-Pemberton, R., Fernandez, E., Jones, R., et al. (2015). Evaluation of opioid modulation in major depressive disorder. *Neuropsychopharmacology*, *40*(6), 1448–1455.

Ziadni, M. S., Jasinski, M. J., Labouvie-Vief, G., & Lumley, M. A. (2016). Alexithymia, defenses, and ego strength: Cross-sectional and longitudinal relationships with psychological well-being and depression. *Journal of Happiness Studies*, *18*(6), 1799–1813.

Chapter 43

# Solution-focused counseling and its use in postpartum depression

Seyed Abbas Mousavi[a], Somayeh Ramezani[b], and Ahmad Khosravi[c]

[a]*Department of Psychiatry, Mazandaran University of Medical Sciences, Sari, Iran;* [b]*Sexual Health and Fertility Research Center, Bahr Hospital, Shahroud University of Medical Sciences, Shahroud, Iran;* [c]*Center for Health Related Social and Behavioural Sciences Research, Department of Epidemiology, Shahroud University of Medical Sciences, Shahroud, Iran*

## List of abbreviations

SFBC solution-focused brief counseling
SFBT solution-focused brief therapy

## Introduction

To treat psychological problems such as depression, researchers have used various approaches including drug therapies, psychotherapies, and combination therapies (Korey, 2005). Mindfulness-based cognitive therapy, meta-cognitive therapy, and solution-focused brief therapy (Papageorgio & Wells, 2004) are among the psychological therapies for depression that can be combined even with medication (Habibi, 2015). Solution-focused brief therapy or Counseling has also had positive effects in the areas of marital problems, sexual abuse, stress reduction, and improved attitude and behavior (Sharifzadeh, Bolbol Haghighi, Keramat, Goli, & Motaghi, 2018). The findings also indicate the efficacy of solution-focused brief therapy (SFBT) on symptoms of depression among women (Dashtizadeh, Sajedi, Nazari, Davarniya, & Shakaram, 2015). Solution-focused brief therapy is a way of enabling mothers to build change in their lives within a short space of time. A review study of SFBT showed that 74% of studies reported significant effects of this kind of therapy. Because of its cost-effectiveness as well as its time-effectiveness, this method can be used as a convenient method for the treatment of many behavioral problems (Gingerich & Peterson, 2013).

In postpartum depression, preventive interventions are of paramount importance at the three levels of prevention. At the first level of prevention, health professionals' main concern is to reduce the vulnerability, eliminate, or minimize the behaviors and environmental conditions that lead to an increased risk. The second-level prevention aims at early diagnosis of depression, reduction of its progression, or delaying the subsequent disability of the early affliction, and the third-level prevention focuses on confining the problems and symptoms of the disease (Hanusa, Scholle, Haskett, Spadaro, & Wisner, 2008).

Solution-focused brief therapy/counseling are an effective solution to prevent and treat depression. This counseling model is also effective in treating behavioral disorders.

## Solution-focused brief therapy and counseling

Solution-focused brief therapy was developed by Steve De Shazer and his colleagues at the Milwaukee brief family therapy center. The way people deal with problems is different, and most of the problems that happen in life are beyond the control of the individual, but we become preoccupied with these problems, which is natural to some extent because it originates from human nature. However, the noteworthy point is that we often are obsessed with problems without thinking that there may be solutions to the problems. In this approach, the changes come from encouraging people to describe their preferred future and from detailing the skills (Ratner, George, & Iveson, 2012).

In SFBT, the goals and techniques of therapy are determined by the strengths, resources, capabilities, and attitudes of the client, and unlike problem-focused approaches, the solutions are of paramount importance. For this reason, SFBT is also called *hope counseling*. Within SFBT, the therapist as a researcher focuses on aspects of the client's specific conditions and leads the dialogs so that they can be helpful and related to the client's life. This process can be continued with the collaboration of the client and the therapist. This therapy method does not deny the possibility of learning from the past as valuable experiences. The difference between this approach and other approaches is

The Neuroscience of Depression. https://doi.org/10.1016/B978-0-12-817933-8.00047-5

that it encourages clients to share their past experiences. This approach emphasizes aspects of clients' past experiences that are effective, useful, and repeatable in their future lives. The purpose of having clients tell their past experiences is to discover their capabilities which can help them and make them hopeful about the future. In solution-focused brief therapy, in addition to the client, the therapist should also avoid thinking about how problems may arise. It should be remembered that thinking about the "problem" keeps people away from recognizing the effective solutions they have used in the past or the new solutions they can find. The task of the solution-focused therapist is to help develop effective solutions. The therapist attempts to look at the problem through the client's eye and look for behavior change with the same understanding the client has (Mofid, Ahmadi, & Etemadi, 2014). In this approach, nothing is fixed and the emphasis is only on solutions and the power of language. In this method, clients are helped to form their mental perceptions of solutions and to replicate solution-oriented behaviors that allow them to achieve their goals (Samaras, Rossi, Giannakopoulos, & Gold, 2010).

The basic principles on which the solution-focused brief therapy and counseling is based include the following:

1. Problems do not always occur and solutions are discovered through exceptions.
2. People rely on their own abilities to perform successful behaviors rather than adapting themselves to strategies that are beneficial to others.
3. Small and trivial achievements also give a great deal of hope to one's own life.
4. Solutions that the individual has yet to find have been created by others.
5. There is no perfect relationship between the problem and the solution, and each solution is unique.

All these principles are intertwined and integral to the solution-focused brief therapy (Nazari, 2014).

Solution-focused brief therapy and counseling is based on the assumption that all humans are capable of creating solutions in their lives, but only because of their life's concerns and failures, they need help finding such solutions. Solution-focused therapists and counselors talk to the clients to help them develop exceptions in their lives rather than looking for and focusing on problems.

## Solution-focused brief therapy principles

The core philosophy of SFBT is that change is constant and inevitable. Therapy is performed based on possible and changeable affairs, and this model of counseling emphasizes taking small steps to initiate change. The model also emphasizes capabilities and solutions rather than problems.

### How are the solutions made?

The underlying assumptions that help the therapist and the client to build a solution are:

1. Solutions do exist.
2. There is more than one solution.
3. Solutions can be made.
4. We can build solutions.
5. Instead of discovering solutions, we create and invent them.
6. There are some processes that can be modeled.

The best solutions are built if we pay attention to what the client wants and what is effective with him or her and replace what he does and is not effective with another task.

## Solution-focused brief therapy methods and techniques

Solution-focused brief therapy and counseling incorporates innovative techniques. Some of these techniques sometimes require creativity (Gingerich & Eisengart, 2000; Nazari, 2014). Among these techniques are:

### Highlighting strengths and resources

One of the important points in SFBT which is achieved by close and initiate talking with the client is highlighting. Milton Ericsson has devised this method and defined it as the concept of intimately engaging with the client and accepting what he has to offer. That is, to acknowledge the client as he or she is, to approve what is brought to the counseling session, and to enter it consciously into the counseling process.

The highlighting process is primarily about identifying and cataloging the client's abilities and talents. This practice is to create a new framework by which we can use what is initially seen as a weak point and drawback, as a positive element and an asset.

### Admiration

There's an old saying that if you can't say good things about others, don't talk at all. This sentence means that we should be highly respectful to others and emphasize their strengths and ignore their weaknesses, and in interacting with them we need to be honest and committed.

### Using future language or taking a presuppositional position

Since the view we have toward ourselves influences the language we use, the therapist should use promising statements. The therapist can say the mothers, "When you will be recovered, how it can affect your baby?" This embedded question has hope background.

### Changing the attitude

It is a process by which one looks at other aspects of a phenomenon that previously seemed negative. By reframing the view of clients, they are able to pick out a new meaning for an essential change.

### Finding and highlighting exceptions

Exceptions are times when the problem has not been a problem. The exceptions are the principles of solutions. For example, in a depressed mother, finding a good and happy day in their life for attention and focusing. In the solution-focused brief therapy and counseling approach, there are four exception types.

1. Past Exceptions: Experiences that go back to the time of the problem.
2. New Exceptions: Nonproblematic situations that have occurred recently.
3. Recurrent Exceptions: Cases that occur periodically and often without prior knowledge.
4. Future Exceptions: Indicate a time when the reported problem will be reduced or eliminated.

### Miracle questions

A miracle question helps one better understand the different aspects of the solutions that have happened before.

### Using the important word "instead"

This term helps clients try to see their success and status from different angles. This method helps mothers to find new ways of thinking, feeling, and behavior instead of doing now.

## The structure of solution-focused brief therapy and counseling

The basic structure of a solution-focused therapy involves explaining the problem precisely + skillfully building the solution + developing effective strategies + stopping ineffective strategies. Clarifying the problem and asking about it can be therapeutic and make clear what was unclear. If a client makes a general complaint, the therapist will ask him or her to make the case clear because what is clear and lucid is potentially controllable. The information you gain at this stage is useful for your conversations about goals and exceptions. Having a baseline of the problem makes it easier to measure progress in later stages.

## Solution-focused brief therapy as a practical skill in preventing postpartum blues and depression

Postpartum mood disorders including blues and depression are among the most common postpartum disorders. Postpartum blues has an incidence of 50%–80% and occurs in the first ten days after childbirth in mothers. Usually, many of the early symptoms including insomnia, anxiety, irritability, fatigue, unconsciousness, decreased libido, mood swings, and unreasonable crying at different times last up to 2 weeks after the delivery and then they disappear. In case, they continue, the mother should be checked for further evaluation (Ramezani, Khosravi, Motaghi, Hamidzadeh, & Mousavi, 2017; Wisner, Parry, & Piontek, 2002). Mothers who experience blues are usually at higher risks for depression (Gavin et al., 2005).

Therefore, identifying problems and learning how to resolve these problems can be effective in reducing anxiety and increasing mothers' coping power. This section describes, in brief, the SFBT model for the prevention of postpartum blues and depression. This model consists of three sessions that are offered to pregnant mothers during pregnancy. The effectiveness of this model has also been confirmed in a study conducted by the authors. Table 1 summarizes the material presented in these three sessions for preventing the depression during the pregnancy.

## Mini-dictionary of terms

Solution-focused brief therapy/counseling: This psychological approach encourages clients to share their past experiences. This approach emphasizes aspects of clients' past experiences that are effective, useful, and repeatable in their future lives.

## Key facts of solution-focused brief therapy

- Solution-focused therapy/counseling is an effective solution to prevent and treat depression and anxiety.
- Solution-focused counseling in prenatal care can be effective for improving the mental health of pregnant women.
- Solution-focused counseling encourages clients to express past beneficial experiences. It helps them to

**TABLE 1** Structure of solution-focused counseling for preventing the postpartum depression during the pregnancy

| | Issues |
|---|---|
| Session 1 | 1. Welcome and introduction of group members and description of the number of sessions.<br>2. Explaining the purpose of counseling sessions.<br>3. Materials about depression and blue: Definition of postpartum blues and depression and explaining their causes and symptoms. |
| Session 2 | 1. Precise defining of the problem by the client in the form of a sentence and a word.<br>2. Setting objective and operational goals.<br>3. Reviewing the solutions to the problem using SFBT techniques.<br>4. Providing homework. |
| Session 3 | 1. Checking previous homework and discussion.<br>2. Finding exceptions in their lives.<br>3. Summing up meeting. |

determine their ability to benefit from their past experiences that have repeatability in their future life.

## Summary points

- Solution-focused brief therapy/counseling are an effective solution to prevent and treat depression.
- Solution-focused brief therapy and counseling is based on the assumption that all humans are capable of creating solutions in their lives, but only because of their life's concerns and failures, they need help finding such solutions.
- Solution-focused therapists and counselors talk to the clients to help them develop exceptions in their lives rather than looking for and focusing on problems.
- The best solutions are built if we pay attention to what the client wants and what is effective with him or her and replace what he does and is not effective with another task.
- Clarifying the problem and asking about it can be therapeutic and make clear what was unclear. If a client makes a general complaint, the therapist will ask him or her to make the case clear because what is clear and lucid is potentially controllable. The information you gain at this stage is useful for your conversations about goals and exceptions. Having a baseline of the problem makes it easier to measure progress in later stages.

## References

Dashtizadeh, N., Sajedi, H., Nazari, A., Davarniya, R., & Shakaram, M. (2015). Effectiveness of solution-focused brief therapy (SFBT) on reducing symptoms of depression in women. *Journal of Clinical Nursing and Midwifery*, *4*(3), 67–78.

Gavin, N. I., Gaynes, B. N., Lohr, K. N., Meltzer-Brody, S., Gartlehner, G., & Swinson, T. (2005). Perinatal depression: A systematic review of prevalence and incidence. *Obstetrics & Gynecology*, *106*(5), 1071–1083. https://doi.org/10.1097/01.AOG.0000183597.31630.db.

Gingerich, W. J., & Eisengart, S. (2000). Solution-focused brief therapy: A review of the outcome research. *Family Process*, *39*(4), 477–498.

Gingerich, W. J., & Peterson, L. T. (2013). Effectiveness of solution-focused brief therapy a systematic qualitative review of controlled outcome studies. *Research on Social Work Practice*, *23*(3), 266–283.

Habibi, M. (2015). Solution-orientedtherapyfordepression: Atheoreticalorientationand introduction to therapeutic approach. *J Res Behave Sci*, *13*(2), 166–175.

Hanusa, B. H., Scholle, S. H., Haskett, R. F., Spadaro, K., & Wisner, K. L. (2008). Screening for depression in the postpartum period: A comparison of three instruments. *Journal of Women's Health*, *17*(4), 585–596. https://doi.org/10.1089/jwh.2006.0248.

Korey, J. (2005). *Theory and application of counseling and psychotherapy* (3rd ed.). Tehran: Arasbaran.

Mofid, W., Ahmadi, A., & Etemadi, O. (2014). The Comparison of cognitive-behavioral counseling and solution-oriented counseling on women's sexual satisfaction in Isfahan. *Women and Community*, *5* (19), 67–83.

Nazari, A. M. (2014). *Principles couples therapy and family therapy*. Tehran.

Papageorgio, C., & Wells, A. (2004). *Depressiverumination:Nature,theoryandtreatmentchiChester* (5th ed.). UK: Wiley.

Ramezani, S., Khosravi, A., Motaghi, Z., Hamidzadeh, A., & Mousavi, S. A. (2017). The effect of cognitive-behavioural and solution-focused counselling on prevention of postpartum depression in nulliparous pregnant women. *Journal of Reproductive and Infant Psychology*, *35*(2), 172–182. https://doi.org/10.1080/02646838.2016.1266470.

Ratner, H., George, E., & Iveson, C. (2012). *Solution focused brief therapy (100 key points)* (1 ed.). New York: Routledge.

Samaras, N., Rossi, G., Giannakopoulos, P., & Gold, G. (2010). Vascular depression. An age-related mood disorder. *European Geriatric Medicine*, *1*(4), 220–225.

Sharifzadeh, M., Bolbol Haghighi, N., Keramat, A., Goli, M., & Motaghi, Z. (2018). Effectiveness of midwifery counseling based on solution-focused approaches on fear of childbirth. *Koomesh*, *20*(2), 375–383.

Wisner, K. L., Parry, B. L., & Piontek, C. M. (2002). Postpartum depression. *New England Journal of Medicine*, *347*(3), 194–199. https://doi.org/10.1056/NEJMcp011542.

Chapter 44

# Transcranial direct current stimulation (tDCS) combined with cognitive emotional training (CET) as a novel treatment for depression

Donel Martin and Stevan Nikolin

*Black Dog Institute, School of Psychiatry, University of New South Wales, Randwick, NSW, Australia*

## Abbreviations

| | |
|---|---|
| **CET** | cognitive emotional training |
| **CT** | cognitive training |
| **DLPFC** | dorsolateral prefrontal cortex |
| **EEG** | electroencephalography |
| **EFMT** | emotional faces memory task |
| **ERP** | event-related potential |
| **IPD MA** | individual patient data meta-analysis |
| **MDD** | major depressive disorder |
| **RCT** | randomized controlled trial |
| **tDCS** | transcranial direct current stimulation |
| **TMS** | transcranial magnetic stimulation |

## Introduction

Over the last 20 years, there has been renewed interest in novel therapeutics for major depressive disorder (MDD). In particular, the application of noninvasive brain stimulation methods, including transcranial magnetic stimulation (TMS), and more recently, transcranial direct current stimulation (tDCS), have facilitated alternative therapies and strategies for treatment. While these methods have been demonstrated to produce antidepressant effects when given as a monotherapy or with concomitant pharmacotherapy, the clinical effects have been found to be highly variable between patients, with reliable predictors of response lacking. In this context, research has focused on novel methods for improving the antidepressant efficacy of these treatments, including their combination with behavioral interventions (Table 44.1).

Behavioural treatments for MDD are considered first-line treatments, particularly for mild or moderate symptoms. While these interventions have historically been implemented by therapists in person, increasingly aspects have been translated into the digital space (e.g., with smartphones), with controlled clinical trials showing similar efficacy to face-to-face interventions (Firth et al., 2017). While with TMS patients need to attend a clinic regularly to receive treatment, the combination of tDCS with a digitally administered behavioral intervention has enhanced translational appeal, as both interventions can potentially be administered by patients at home under remote medical supervision. In this chapter, we review the clinical evidence for these interventions given alone, as well as preliminary clinical and neurophysiological evidence for these treatments combined as a novel intervention for MDD.

## Transcranial direct current stimulation (tDCS)

tDCS involves the administration of a constant unidirectional electrical current of between 0 and 3 mA (milliamperes) to the brain to modulate ongoing brain activity. The current is delivered using a small, battery-powered device, and has advantages of portability and reduced costs compared to other noninvasive brain stimulation methods. The current is typically administered to the brain via two rubber electrodes, a positively charged anode and a negatively charged cathode that are placed upon the scalp. The electrodes drive a small electrical current of negatively charged ions from the cathode to the anode and positively charged ions toward the cathode. A sponge soaked in saline solution (a mixture of water and sodium chloride) is commonly used to facilitate current flow through the skin and prevent localized areas of increased current density, which can increase the risk of skin lesions. tDCS, when used therapeutically for the treatment of depression, is typically given for 20–40 min on a daily basis over several weeks

The Neuroscience of Depression. https://doi.org/10.1016/B978-0-12-817933-8.00054-2

**TABLE 44.1** Examples of available treatments for major depressive disorder.

| Pharmacological | Behavioral/psychological | Interventional |
|---|---|---|
| Selective serotonin reuptake inhibitor (SSRI)<br>Melatonergic agonist<br>Serotonin and noradrenaline reuptake inhibitor (SNRI)<br>Tricyclic antidepressant (TCA)<br>Monoamine oxidaze inhibitor (MAOI)<br>Ketamine | Cognitive behavioral therapy (CBT)<br>Interpersonal therapy<br>Acceptance and commitment therapy<br>Mindfulness-based cognitive therapy | Electroconvulsive therapy (ECT)<br>Transcranial magnetic stimulation (TMS)<br>Vagal nerve stimulation (VNS)<br>Transcranial direct current stimulation (tDCS) |

(Borrione, Moffa, Martin, Loo, & Brunoni, 2018). With the use of modern stimulation parameters and appropriate safety precautions, tDCS has been found to be safe and associated with only minimal and transient side effects (Bikson et al., 2016) (Fig. 44.1).

## Clinical effects of tDCS in MDD

The first controlled clinical studies of tDCS were conducted in the 1960s and 1970s and showed mixed efficacy results (Arul-Anandam & Loo, 2009). It was not until 2006 that the technique was revisited in two small sham-controlled studies, where significant mood and cognitive improvement (working memory) was observed with active tDCS compared to sham stimulation (Fregni et al., 2006; Fregni, Boggio, Nitsche, Rigonatti, & Pascual-Leone, 2006). Since those first modern trials, multiple other sham-controlled studies have investigated the antidepressant and cognitive effects of tDCS given either as a monotherapy or in combination with concomitant pharmacotherapy. In the majority of these studies, participants have sat at rest during tDCS administration and not engaged in any specified activity while receiving stimulation. The stimulation has targeted the left dorsolateral prefrontal cortex (L-DLPFC), with the anode or positively charged electrode placed over this region, and the cathode or negatively charged electrode placed over the contralateral side of the head. These studies have varied in the stimulation parameters used, including current intensity (1–2.5 mA), stimulus duration (20–30 min), the total number of treatments (5–22 sessions), as well as in sample characteristics (e.g., treatment resistance). Overall, the clinical effects of tDCS given either alone as a monotherapy or with concomitant medication in these studies have been mixed (Borrione et al., 2018).

FIG. 44.1 Transcranial direct current stimulation (tDCS).

To address this variability in clinical outcomes between studies and to examine potential predictors of clinical response, Moffa et al. (2020) conducted an individual patient data meta-analysis (IPD MA) which included data from 9 double-blind randomized controlled trials (RCTs) and 572 MDD participants. The results confirmed the findings from the previous IPD MA (Brunoni et al., 2016) in showing that tDCS had significant antidepressant effects compared to sham. While these results were important in confirming tDCS clinical effects, the overall magnitude of this effect [i.e., number needed to treat (NNT)] was modest, with NNT = 9 for a clinical response (i.e., $\geq$50% reduction in symptoms compared to pretreatment) and NNT = 13 for remission (i.e., near-complete symptom amelioration). Interestingly, none of the clinical, demographic, or treatment variables examined significantly predicted mood improvement with active tDCS.

The same methodological approach was also used to examine the cognitive effects of tDCS treatment for MDD. Martin, Moffa, et al. (2018) conducted an IPD MA which included data from 7 RCTs and 478 MDD participants. In each included study, cognition was assessed at both pretreatment and posttreatment. Across the 12 different cognitive outcomes examined, no evidence for cognitive improvement was found with active compared to sham tDCS after controlling for mood effects. Similar to the IPD MA for clinical outcomes above, no significant moderating effects were found for predicting cognitive improvement with active tDCS. The study concluded that there was little evidence for cognitive benefit for tDCS given alone as a monotherapy or with medication for MDD.

## Rationale for combining tDCS and CET in MDD

Based on the findings described above, research interest has since focused on potential methods for improving the clinical efficacy of tDCS for MDD. One promising approach is combining tDCS with a behavioral intervention instead of giving tDCS alone as reviewed above. The rationale for the combined approach is based on preclinical neuroscience research into the mechanisms for tDCS neuromodulatory effects. This research has found that concurrent brain activity during tDCS administration (e.g., from a task) is necessary to produce long-term neuroplastic changes (Fritsch et al., 2010), consistent with the "activation selectivity" model of tDCS effects (Bikson & Rahman, 2013).

## Cognitive emotional training (CET)

CET is a digitally administered behavioral intervention designed to treat depression. It requires participants to identify and remember emotions in an adaptive N-back working memory cognitive training paradigm. The rationale was based on neuroimaging research which showed task-induced increased metabolic activity in the DLPFC and amygdala during task performance (Neta & Whalen, 2011). These regions have been implicated in reduced cognitive control and emotion dysregulation (Dolcos, Iordan, & Dolcos, 2011), characteristics considered associated with core depressive symptoms (e.g., low mood, reduced concentration, and rumination). CET is administered in an adaptive cognitive training paradigm using the emotional faces Memory task (EMFT: Fig. 44 2). As the EMFT simultaneously activates the DLPFC and amygdala, it was hypothesized that repeated training on the EMFT would inhibit abnormal limbic activity associated with reduced cognitive control and emotion dysregulation and produce antidepressant effects (Iacoviello et al., 2014).

## Clinical effects of CET

The antidepressant effects of CET using the EMFT were first investigated in a small proof of concept RCT, which involved 21 participants with MDD (Iacoviello et al., 2014). The study was conducted in participants with moderate level symptoms and excluded those with a history of nonresponse to pharmacotherapy or chronic nonepisodic MDD. Participants were randomized to receive 8 sessions of CET or a control cognitive training (CT) intervention over 4 weeks (2 sessions per week). The CET group showed significantly greater mood improvement compared to the control CT group at post 4 weeks, with 6/11 CET participants meeting criteria for a clinical response. There was no significant effect on cognition in either the CET or control CT group.

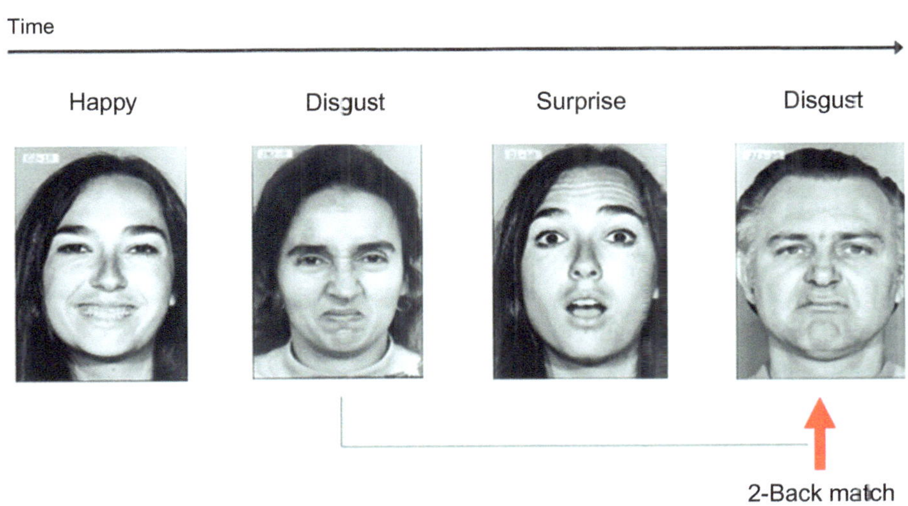


FIG. 44.2 Emotional faces memory task. An example of a 2-back match. Participants are required to match the emotion of the face shown according to n positions back in the sequence.

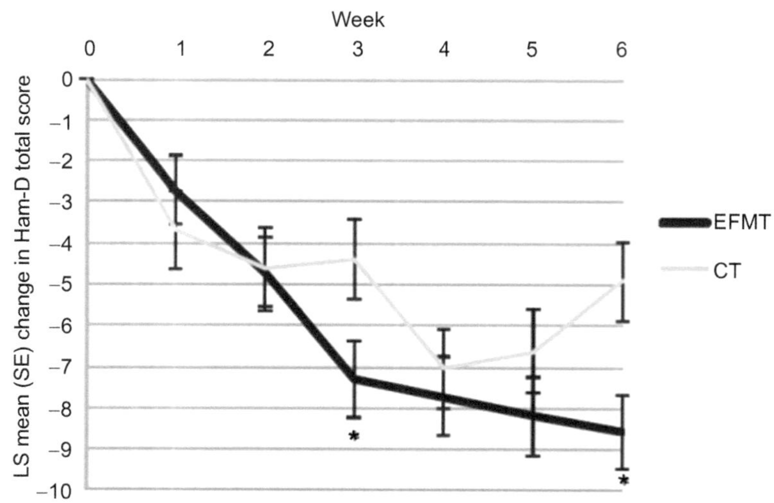


FIG. 44.3 Change in depression severity over time for participants randomized to complete the emotional faces memory task (EFMT) and Cognitive Training (CT) from Iacoviello et al. (2018). The figure is reproduced with permission.

The antidepressant effects of CET were then replicated by the same group in a larger RCT which included 51 participants with MDD (Iacoviello et al., 2018). Similarly to the earlier study, only participants with moderate level symptoms were included. Participants were excluded if they had taken any antidepressant medication during their current major depressive episode or had a history of nonresponse to pharmacotherapy. Participants were randomized to receive 18 sessions of CET or control CT over 6 weeks (3 sessions per week). The efficacy results are shown in Fig. 44.3. Participants who received CET had significantly better mood scores at both weeks 3 and 6. In the CET group, 10/28 (36%) of participants met criteria for a clinical response. Both groups showed significant improvement in working memory at posttreatment.

## Preliminary evidence for clinical effects of tDCS combined with CET

To date, only one study has investigated the clinical effects of tDCS combined with CET. In an open-label pilot study, 20 participants received a total of 18 combined tDCS and CET sessions three times a week for a total of 18 sessions. Unlike the studies of CET given as a monotherapy described above, only participants with a history of medication resistance were recruited (i.e., with treatment-resistant depression). Participants with MDD were required to have failed at least two different antidepressant medications at adequate doses for an adequate duration in the current or in previous episodes. tDCS was administered concurrently during CET for 30 min in 9 participants and for 40 min in 11 participants. Similarly to previous studies of tDCS given as a monotherapy in MDD, the anode was placed over the L-DLPFC, while the cathode was instead placed extracephalically over the right upper arm. The choice of montage was based on computational modeling of the tDCS stimulus which suggested greater stimulation of limbic regions relative to the standard bifrontal montage (Bai, Dokos, Ho, & Loo, 2014). The rationale was that concurrent tDCS stimulation of the L-DLPFC and limbic regions simultaneously targeted by CET would be beneficial for clinical efficacy.

The antidepressant effects of tDCS combined with CET are shown in Fig. 44.4. In this study, 7/17 study completers (46%) showed a clinical response. Participants additionally demonstrated significant improvements in working memory and emotion recognition. A limitation of the study was that it was not sham-controlled, so the potential for placebo and practice effects on the cognitive tasks cannot be ruled out.

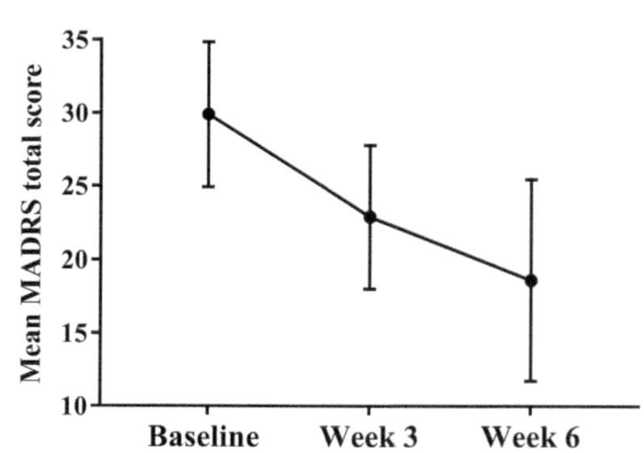


FIG. 44.4 Mean depression scores of tDCS combined with CET from a clinical pilot study (Martin, Teng, et al., 2018).

## Neurophysiological effects of tDCS combined with CET

### Electroencephalography (EEG)

Cognitive training is presumed to modulate underlying neurophysiological functioning through repetitive activation and training of the neural circuits that subserve cognitive functions. Electroencephalography (EEG) can be used to detect electrical activity generated by large populations of neurons acting in synchrony. Specifically, EEG detects electrical potential differences generated at the synaptic terminals of dendrites from cortical pyramidal neurons (Buzsáki, Anastassiou, & Koch, 2012). Compared to other neuroimaging modalities, EEG allows for noninvasive assessment of brain activity and has a high temporal resolution, making it well-suited to investigate neurocognitive processes, which involve rapid communication between distal and functionally connected brain regions. It has been used to gain further insights into the functional changes associated with antidepressant response and changes associated with CT. Broadly speaking, EEG can be used to detect brain activity under two distinct states, while individuals are at rest and during a task. Measures acquired at rest can reveal local and/or global changes in neuronal oscillations at specific frequency bands (Pizzagalli, Oakes, & Davidson, 2003), and those acquired during a task can be used to investigate event-related subprocesses that, in aggregate, form the cognitive functions being investigated (Keeser et al., 2011).

To date, there has only been one trial examining the neurophysiological effects of tDCS combined with CET in MDD using EEG. Nikolin, Martin, Loo, Iacoviello, and Boonstra (2020) investigated the effects of 18 sessions of the combined intervention delivered over 6 weeks in an open-label study. Totally 20 participants with treatment-resistant depression received 40 min of concurrent prefrontal tDCS with a current intensity of 2 mA. They were tested at baseline and immediately following the training course using resting-state and task-related EEG measures. The following sections describe the EEG markers of depression and cognitive processes selected for analyses and the rationale for their inclusion.

#### *Frontal alpha asymmetry*

Neuronal oscillations occurring in the alpha frequency band (8–12 Hz) were one of the first features of brain activity discovered using EEG. They are thought to reflect inhibitory processes necessary for task-relevant regions of the brain to operate unencumbered by neural background noise (Cooper, Croft, Dominey, Burgess, & Gruzelier, 2003). Frontal alpha asymmetry in which alpha power is disproportionately lateralized to the left prefrontal cortex, has been reported in multiple studies of depression (Debener et al., 2000; Stewart, Bismark, Towers, Coan, & Allen, 2010). Models linking alpha asymmetry to depressive symptomatology have posited that increased left prefrontal alpha is associated with decreased approach behaviors (e.g., goal-seeking) and positive information processing. Conversely, lower right-sided prefrontal alpha is linked to increased negative affect (Jaworska, Blier, Fusee, & Knott, 2012). These results broadly agree with findings that show alpha asymmetry to be inversely correlated with depression scores (Diego, Field, & Hernandez-Reif, 2001), and positively associated with vulnerability of depression relapse (Keune, Bostanov, Hautzinger, & Kotchoubey, 2011). Studies investigating the effects of behavioral therapies for MDD have shown that although frontal alpha asymmetry is higher in depressed participants compared to healthy individuals at baseline (Gollan et al., 2014), it does not change over the course treatment (Keune et al., 2011). These findings imply that frontal alpha asymmetry may be a stable trait of depression, rather than a marker of the state of depression. Notwithstanding, given that both tDCS and CET involve concomitant stimulation and recruitment of left prefrontal regions with the aim to enhance functioning, the potential for frontal asymmetry to be modified after the combined intervention in participants with MDD was considered worthy of investigation.

#### *Frontal theta*

Another potential biomarker for depression is resting-state frontal theta (4–8 Hz) power (Bailey et al., 2018). Source localization techniques have identified the anterior cingulate cortex as the generator of frontal theta activity (Mulert et al., 2007; Pizzagalli et al., 2003). This region has been implicated in emotion processing (Etkin, Egner, & Kalisch, 2011) and often displays abnormal activity in neuroimaging studies in depression (Yoshimura et al., 2010). Interestingly, there are several reports of heightened frontal theta as a positive predictor for response to depression treatment (Arns et al., 2015; Rentzsch, Adli, Wiethoff, de Castro, & Gallinat, 2014). Additionally, frontal theta has been linked to a network consisting of the hippocampus, prefrontal cortex, and anterior cingulate cortex that activates during executive functions such as working memory and cognitive control processes (Sauseng, Griesmayr, Freunberger, & Klimesch, 2010). In support of these correlational findings, midline frontal theta has been found to increase in response to 4 weeks of working memory cognitive training (Langer, von Bastian, Wirz, Oberauer, & Jäncke, 2013), as well as following a course of CT paired with concurrent anodal high-definition tDCS (HD-tDCS) delivered to the DLPFC (Choe, Coffman, Bergstedt, Ziegler, & Phillips, 2016). Best, Gale, Tran, Haque, and Bowie (2019) observed similar increases in theta in participants with severe mental illness, including depression, after 2 weeks of CT of the executive functions spatial working memory, set-shifting, and spatial manipulation. Based on these positive findings of frontal theta as a biomarker in MDD and evidence for change after CT interventions, the potential for the moderation of frontal theta after tDCS combined with CET was investigated.

## Task-related EEG

### *Event-related potentials*

ERPs reflect the average activity of a population of neurons time-locked to the onset of an event, such as the presentation of task stimuli (Sur & Sinha, 2009). Two ERP components, in particular, have consistently been reported in the EEG literature across several cognitive tasks. The anterior centro-frontal positive ERP occurring 150–200ms following stimulus onset, called P2, has been linked to background sustained attention modulation and context updating processes (Luu et al., 2014; Vilà-Balló et al., 2018). Its amplitude has been observed to be positively correlated to performance in working memory tasks (Han, Liu, Zhang, Jin, & Luo, 2013). Similarly, another positive frontal ERP component called P3, occurring approximately 300ms following stimulus onset, is positively correlated with working memory performance (Gajewski & Falkenstein, 2018; Gevins & Smith, 2000). The P3 component may reflect higher-order mental operations such as task switching and cognitive control (Brydges & Barceló, 2018; Polich, 2007). It is thought to be generated by multiple sources, one of which is the DLPFC, a common target for noninvasive brain stimulation in depression (Nakao, Kodabashi, Yarita, Fujimoto, & Tamura, 2012). The literature regarding ERP components in depression is relatively sparse, but there is some evidence to suggest that amplitudes are reduced during a working memory task for P2 (Kemp et al., 2006), and P3 (Bruder et al., 2009; lv, Zhao, Gong, Chen, & Miao, 2010), compared to healthy controls. These results suggest that working memory impairments reported in depression may be partially a result of deficits in early attentional processes (Snyder, 2013). As CET involves an adaptive n-back working memory paradigm, the potential for change in ERPs during the performance of a 2-back working memory task following combined tDCS and CET treatment was examined.

### *Event-related synchronization and desynchronization*

Time-frequency analyses track changes in the synchronous activity of neuronal populations during a task through spectral decomposition of the EEG signal. Frequency-specific increases in time-locked neural activity are referred to as event-related synchronization (ERS), and decreases in amplitude are referred to as event-related desynchronization (ERD) (Pfurtscheller & Da Silva, 1999). Two frequency-specific oscillations have repeatedly been shown to be relevant for executive functioning; alpha ERD, and theta ERS. Alpha desynchronization is thought to reflect selective inhibition of task-irrelevant brain regions to maintain functional engagement of regions needed for task performance (Jensen & Mazaheri, 2010). It becomes more pronounced with increasing cognitive demands (Klimesch, Doppelmayr, & Hanslmayr, 2006; Stipacek, Grabner, Neuper, Fink, & Neubauer, 2003). Interestingly, depressed patients show weakened alpha ERD during working memory, which may indicate impaired inhibitory processes necessary for cognitive control processes (Segrave et al., 2010). Furthermore, Schoenberg and Speckens (2015) found that changes in alpha ERD while viewing negative emotional stimuli were correlated with improvements in mood scores.

Theta ERS may reflect a gating mechanism and has been linked to the allocation of attention to task-relevant stimuli (Missonnier et al., 2006). It is also thought to reflect the integration and coordination of activity between multiple brain regions necessary for working memory processes (Sauseng et al., 2010). Despite a solid evidence base for theta ERS as an important task-related EEG feature, few studies have investigated theta ERS in depression and potential changes following treatment. Schoenberg and Speckens (2014) found increased theta ERS with mindfulness-based cognitive therapy for depression and found a correlation with symptom improvement.

## Preliminary EEG results of tDCS combined with CET

Nikolin et al. (2020) investigated changes in the EEG measures outlined above following 18 sessions of tDCS with concurrent CET in MDD. Somewhat surprisingly, results showed no significant changes for investigated EEG measures, following statistical correction for multiple comparisons. Equivalence testing, used to determine whether null findings can be ascribed to a lack of effect or to lack of statistical power to detect an effect, was inconclusive. This suggested that there was sufficient variance in outcomes that results may not be indicative of a "true negative." A larger sample may, therefore, be needed to combat interindividual heterogeneity in EEG outcomes and improve measurement precision.

Interestingly, there was some support for a moderate (Cohen's $d = 0.46$) increase in P2 amplitude (see Fig. 44.5), although this may be a chance finding as the result did not survive statistical correction. If confirmed, the findings could suggest a general effect of the intervention on sensory-attentional processes, but not higher-order cognitive functions. This would agree with similar findings obtained following treatment for depression using

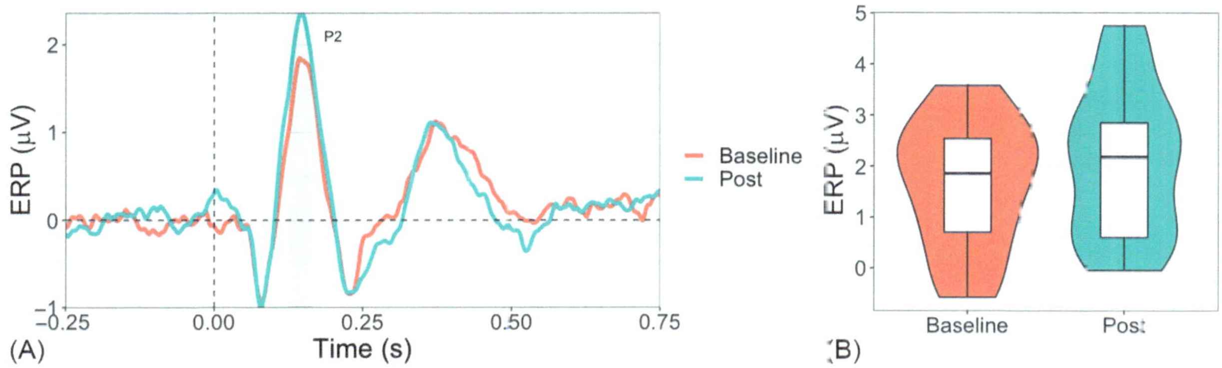


**FIG. 44.5** Event-related potentials (ERPs) during a working memory task. (A) Traces at Baseline and immediately post a course of tDCS combined with cognitive emotional training obtained from anterior midline channel Fz (according to the 10–20 International System), (B) A violin and box plot for the P2 component. Box plots display the interquartile range (50% of the data distribution), and centre lines mark the median values.

transcranial magnetic stimulation, another form of noninvasive brain stimulation that showed increased P2 amplitude in frontal regions (Choi et al., 2014; Spronk, Arns, Bootsma, van Ruth, & Fitzgerald, 2008).

## Conclusions and future research

Despite the availability of effective treatments, pharmacoresistant or treatment-resistant MDD remains a growing health and economic burden and the leading cause of global disability (Friedrich, 2017). The combination of tDCS with a concurrent digitally administered behavioral intervention has the potential to improve the therapeutic effects of tDCS and facilitate future broader clinical translation. In this chapter, we reviewed current clinical evidence for therapeutic effects of tDCS and CET given alone as monotherapies, and also preliminary clinical and neurophysiological findings of the novel combined intervention from recent open-label studies. Sham-controlled RCTs of tDCS given alone have demonstrated modest antidepressant effects. For CET as a monotherapy, two small RCTs found significant antidepressant effects in patients with moderate symptoms and no pharmacological resistance. There was limited evidence for either intervention producing cognitive enhancement as a monotherapy.

Preliminary clinical data from tDCS combined with CET showed promising therapeutic effects in patients with treatment-resistant MDD. Indeed, the clinical results from the pilot study (46% of patients were clinical responders) suggested that the combined treatment approach was more efficacious than tDCS given alone in treatment-resistant MDD patients (i.e., Bennabi et al., 2015; Blumberger, Tran, Fitzgerald, Hoy, & Daskalakis, 2012). Preliminary results further suggested the potential for improved cognitive functioning, which is important considering that MDD is commonly associated with neurocognitive dysfunction (Semkovska et al., 2019). These results support the need for future controlled studies of tDCS combined with CET to establish therapeutic efficacy. Based on the recent development of remote home-based treatment with tDCS using smaller handheld devices (Alonzo et al., 2019), and the availability of CET on smartphones, such future studies could be conducted primarily off-site at participant's homes. If proven effective, tDCS combined with CET could have important translational implications, for example, in MDD patients who are unable to regularly commute to a clinic to receive treatment (i.e., as is required for TMS, an approved noninvasive brain stimulation intervention for treatment-resistant depression).

In this chapter, we additionally reviewed preliminary evidence for neurophysiological changes following tDCS combined with CET. The rationale for examining EEG-specific biomarkers of depression and candidate targets for assessing changes following treatment was provided. While the majority of candidate neurophysiological measures did not show evidence for changes following treatment, there was some indication for potential improvement in a marker of general sensory attentional processing, although this effect did not survive statistical correction. Although generally negative, these preliminary results may assist in guiding future neuroscientific research into the neurophysiological mechanisms for antidepressant and cognitive changes following treatment in MDD. This research is important for the future development of novel neuroscientific based interventions, such as tDCS combined with CET, as they can help to inform improved targeting and individualization of treatment, for example, in the parameter space (i.e., by modifying and enhancing stimulation and behavioral targeting). Future studies into the neurophysiological changes in MDD following tDCS combined with CET, or other novel interventions, may benefit from including larger or more restricted sample sizes (e.g., including only patients who show clinical response or remission) and appropriate control conditions.

## Key facts of tDCS and CET

- tDCS as a monotherapy for MDD has demonstrated antidepressant effects and limited cognitive effects.
- CET as a monotherapy has significant antidepressant effects in patients with moderate and nonpharmacoresistant depression.
- tDCS combined with CET has significant antidepressant effects in treatment-resistant MDD patients.

## Summary points

- tDCS combined with CET in MDD may have greater efficacy than either intervention given alone.
- tDCS combined with CET may potentially be administered by patients at home under remote medical supervision.
- Future controlled trials are required to establish the therapeutic and neurophysiological effects of tDCS combined with CET in MDD.

## References


Alonzo, A., Fong, J., Ball, N., Martin, D., Chand, N., & Loo, C. (2019). Pilot trial of home-administered transcranial direct current stimulation for the treatment of depression. *Journal of Affective Disorders*, *252*, 475–483.

Arns, M., Etkin, A., Hegerl, U., Williams, L. M., DeBattista, C., Palmer, D. M., et al. (2015). Frontal and rostral anterior cingulate (rACC) theta EEG in depression: Implications for treatment outcome? *European Neuropsychopharmacology*, *25*(8), 1190–1200.

Arul-Anandam, A. P., & Loo, C. (2009). Transcranial direct current stimulation: A new tool for the treatment of depression? *Journal of Affective Disorders*, *117*(3), 137–145.

Bai, S., Dokos, S., Ho, K. A., & Loo, C. (2014). A computational modelling study of transcranial direct current stimulation montages used in depression. *NeuroImage*, *87*, 332–344.

Bailey, N., Hoy, K., Rogasch, N., Thomson, R., McQueen, S., Elliot, D., et al. (2018). Responders to rTMS for depression show increased fronto-midline theta and theta connectivity compared to non-responders. *Brain Stimulation*, *11*(1), 190–203.

Bennabi, D, Nikolier, M, Monnin, J, Tio, G, Pazart, L, Vandel, P, et al. (2015). Pilot study of feasibility of the effect of treatment with tDCS in patients suffering treatment-resistant depression with escitalopram. *Clinical Neurophysiology*, *126*, 1185–1189.

Best, M. W., Gale, D., Tran, T., Haque, M. K., & Bowie, C. R. (2019). Brief executive function training for individuals with severe mental illness: Effects on EEG synchronization and executive functioning. *Schizophrenia Research*, *203*, 32–40.

Bikson, M., Grossman, P., Thomas, C., Zannou, A. L., Jiang, J., Adnan, T., et al. (2016). Safety of transcranial direct current stimulation: Evidence based update 2016. *Brain Stimulation*, *9*(5), 641–661.

Bikson, M., & Rahman, A. (2013). Origins of specificity during tDCS: Anatomical, activity-selective, and input-bias mechanisms. *Frontiers in Human Neuroscience*, *7*, 688.

Blumberger, D. M., Tran, L. C., Fitzgerald, P. B., Hoy, K. E., & Daskalakis, Z. J. (2012). A randomized double-blind sham-controlled study of transcranial direct current stimulation for treatment-resistant major depression. *Front Psychiatry*, *3*, 74.

Borrione, L., Moffa, A. H., Martin, D., Loo, C. K., & Brunoni, A. R. (2018). Transcranial direct current stimulation in the acute depressive episode: A systematic review of current knowledge. *The Journal of ECT*, *34*(3), 153–163.

Bruder, G. E., Kroppmann, C. J., Kayser, J., Stewart, J. W., McGrath, P. J., & Tenke, C. E. (2009). Reduced brain responses to novel sounds in depression: P3 findings in a novelty oddball task. *Psychiatry Research*, *170*(2–3), 218–223.

Brunoni, A. R., Moffa, A. H., Fregni, F., Palm, U., Padberg, F., Blumberger, D. M., et al. (2016). Transcranial direct current stimulation for acute major depressive episodes: Meta-analysis of individual patient data. *The British Journal of Psychiatry*, *208*(6), 522–531.

Brydges, C. R., & Barceló, F. (2018). Functional dissociation of latency-variable, stimulus-and response-locked target P3 sub-components in task-switching. *Frontiers in Human Neuroscience*, *12*, 60.

Buzsáki, G., Anastassiou, C. A., & Koch, C. (2012). The origin of extracellular fields and currents—EEG, ECoG, LFP and spikes. *Nature Reviews. Neuroscience*, *13*(6), 407.

Choe, J., Coffman, B. A., Bergstedt, D. T., Ziegler, M. D., & Phillips, M. E. (2016). Transcranial direct current stimulation modulates neuronal activity and learning in pilot training. *Frontiers in Human Neuroscience*, *10*, 34.

Choi, K. M., Jang, K.-M., Jang, K. I., Um, Y. H., Kim, M.-S., Kim, D.-W., et al. (2014). The effects of 3 weeks of rTMS treatment on P200 amplitude in patients with depression. *Neuroscience Letters*, *577*, 22–27.

Cooper, N. R., Croft, R. J., Dominey, S. J., Burgess, A. P., & Gruzelier, J. H. (2003). Paradox lost?Exploring the role of alpha oscillations during externally vs. internally directed attention and the implications for idling and inhibition hypotheses. *International Journal of Psychophysiology*, *47*(1), 65–74.

Debener, S., Beauducel, A., Nessler, D., Brocke, B., Heilemann, H., & Kayser, J. (2000). Is resting anterior EEG alpha asymmetry a trait marker for depression? *Neuropsychobiology*, *41*(1), 31–37.

Diego, M. A., Field, T., & Hernandez-Reif, M. (2001). CES-D depression scores are correlated with frontal EEG alpha asymmetry. *Depression and Anxiety*, *13*(1), 32–37.

Dolcos, F., Iordan, A. D., & Dolcos, S. (2011). Neural correlates of emotion-cognition interactions: A review of evidence from brain imaging investigations. *Journal of Cognitive Psychology (Hove, England)*, *23*(6), 669–694.

Etkin, A., Egner, T., & Kalisch, R. (2011). Emotional processing in anterior cingulate and medial prefrontal cortex. *Trends in Cognitive Sciences*, *15*(2), 85–93.

Firth, J., Torous, J., Nicholas, J., Carney, R., Pratap, A., Rosenbaum, S., et al. (2017). The efficacy of smartphone-based mental health interventions for depressive symptoms: A meta-analysis of randomized controlled trials. *World Psychiatry*, *16*(3), 287–298.

Fregni, F., Boggio, P. S., Nitsche, M. A., Marcolin, M. A., Rigonatti, S. P., & Pascual-Leone, A. (2006). Treatment of major depression with transcranial direct current stimulation. *Bipolar Disorders*, *8*(2), 203–204.

Fregni, F., Boggio, P. S., Nitsche, M. A., Rigonatti, S. P., & Pascual-Leone, A. (2006). Cognitive effects of repeated sessions of transcranial direct current stimulation in patients with depression. *Depression and Anxiety*, *23*(8), 482–484.

Friedrich, M. J. (2017). Depression is the leading cause of disability around the world. *JAMA*, *317*(15), 1517.

Fritsch, B., Reis, J., Martinowich, K., Schambra, H. M., J., Y., Cohen, L. G., et al. (2010). Direct current stimulation promotes BDNF-dependent synaptic plasticity: Potential implications for motor learning. *Neuron*, *66*(2), 198–204.

Gajewski, P. D., & Falkenstein, M. (2018). ERP and behavioral effects of physical and cognitive training on working memory in aging: A randomized controlled study. *Neural Plasticity*, *2018*, 12.

Gevins, A., & Smith, M. E. (2000). Neurophysiological measures of working memory and individual differences in cognitive ability and cognitive style. *Cerebral Cortex*, *10*(9), 829–839.

Gollan, J. K., Hoxha, D., Chihade, D., Pflieger, M. E., Rosebrock, L., & Cacioppo, J. (2014). Frontal alpha EEG asymmetry before and after behavioral activation treatment for depression. *Biological Psychology*, *99*, 198–208.

Han, L., Liu, Y., Zhang, D., Jin, Y., & Luo, Y. (2013). Low-arousal speech noise improves performance in N-back task: An ERP study. *PLoS One*, *8*(10), e76261.

Iacoviello, B. M., Murrough, J. W., Hoch, M. M., Huryk, K. M., Collins, K. A., Cutter, G. R., et al. (2018). A randomized, controlled pilot trial of the emotional faces memory task: A digital therapeutic for depression. *NPJ Digital Medicine*, *1*.

Iacoviello, B. M., Wu, G., Alvarez, E., Huryk, K., Collins, K. A., Murrough, J. W., et al. (2014). Cognitive-emotional training as an intervention for major depressive disorder. *Depression and Anxiety*, *31*(8), 699–706.

Jaworska, N., Blier, P., Fusee, W., & Knott, V. (2012). Alpha power, alpha asymmetry and anterior cingulate cortex activity in depressed males and females. *Journal of Psychiatric Research*, *46*(11), 1483–1491.

Jensen, O., & Mazaheri, A. (2010). Shaping functional architecture by oscillatory alpha activity: Gating by inhibition. *Frontiers in Human Neuroscience*, *4*, 186.

Keeser, D., Padberg, F., Reisinger, E., Pogarell, O., Kirsch, V., Palm, U., et al. (2011). Prefrontal direct current stimulation modulates resting EEG and event-related potentials in healthy subjects: A standardized low resolution tomography (sLORETA) study. *NeuroImage*, *55*(2), 644–657.

Kemp, A. H., Hopkinson, P. J., Stephan, B. C., Clark, C. R., Gordon, E., Bryant, R. A., et al. (2006). Predicting severity of non-clinical depression: Preliminary findings using an integrated approach. *Journal of Integrative Neuroscience*, *5*(1), 89–110.

Keune, P. M., Bostanov, V., Hautzinger, M., & Kotchoubey, B. (2011). Mindfulness-based cognitive therapy (MBCT), cognitive style, and the temporal dynamics of frontal EEG alpha asymmetry in recurrently depressed patients. *Biological Psychology*, *88*(2–3), 243–252.

Klimesch, W., Doppelmayr, M., & Hanslmayr, S. (2006). Upper alpha ERD and absolute power: Their meaning for memory performance. *Progress in Brain Research*, *159*, 151–165.

Langer, N., von Bastian, C. C., Wirz, H., Oberauer, K., & Jäncke, L. (2013). The effects of working memory training on functional brain network efficiency. *Cortex*, *49*(9), 2424–2438.

Luu, P., Caggiano, D., Geyer, A., Lewis, J., Cohn, J., & Tucker, D. (2014). Time-course of cortical networks involved in working memory. *Frontiers in Human Neuroscience*, *8*, 4.

lv, J., Zhao, L., Gong, J., Chen, C., & Miao, D. (2010). Event-related potential based evidence of cognitive dysfunction in patients during the first episode of depression using a novelty oddball task. *Psychiatry Research*, *182*(1), 58–66.

Martin, D. M., Moffa, A., Nikolin, S., Bennabi, D., Brunoni, A. R., Flannery, W., et al. (2018). Cognitive effects of transcranial direct current stimulation treatment in patients with major depressive disorder: An individual patient data meta-analysis of randomised, sham-controlled trials. *Neuroscience and Biobehavioral Reviews*, *90*, 137–145.

Martin, D. M., Teng, J. Z., Lo, T. Y., Alonzo, A., Goh, T., Iacoviello, B. M., et al. (2018). Clinical pilot study of transcranial direct current stimulation combined with cognitive emotional training for medication resistant depression. *Journal of Affective Disorders*, *232*, 89–95.

Missonnier, P., Deiber, M.-P., Gold, G., Millet, P., Pun, M. G.-F., Fazio-Costa, L., et al. (2006). Frontal theta event-related synchronization: Comparison of directed attention and working memory load effects. *Journal of Neural Transmission*, *113*(10), 1477–1486.

Moffa, A. H., Martin, D., Alonzo, A., Bennabi, D., Blumberger, D. M., Bensonor, I. M., et al. (2020). Efficacy and acceptability of transcranial direct current stimulation (tDCS) for major depressive disorder: An individual patient data meta-analysis. *Progress in Neuro-Psychopharmacology & Biological Psychiatry*, *99*, 109836.

Mulert, C., Juckel, G., Brunnmeier, M., Karch, S., Leicht, G., Mergl, R., et al. (2007). Rostral anterior cingulate cortex activity in the theta band predicts response to antidepressive medication. *Clinical EEG and Neuroscience*, *38*(2), 78–81.

Nakao, Y., Kodabashi, A., Yarita, M., Fujimoto, T., & Tamura, T. (2012). Temporal activities during P3 components on the working memory-related brain regions: N-back ERP study. In *Paper presented at the proceedings of 2012 IEEE-EMBS international conference on biomedical and health informatics*.

Neta, M., & Whalen, P. J. (2011). Individual differences in neural activity during a facial expression vs. identity working memory task. *NeuroImage*, *56*(3), 1685–1692.

Nikolin, S., Martin, D., Loo, C. K., Iacoviello, B. M., & Boonstra, T. W. (2020). Assessing neurophysiological changes associated with combined transcranial direct current stimulation and cognitive emotional training for treatment-resistant depression. *European Journal of Neurophysiology*, *51*(10), 2119–2133.

Pfurtscheller, G., & Da Silva, F. L. (1999). Event-related EEG/MEG synchronization and desynchronization: Basic principles. *Clinical Neurophysiology*, *110*(11), 1842–1857.

Pizzagalli, D. A., Oakes, T. R., & Davidson, R. J. (2003). Coupling of theta activity and glucose metabolism in the human rostral anterior cingulate cortex: An EEG/PET study of normal and depressed subjects. *Psychophysiology*, *40*(6), 939–949.

Polich, J. (2007). Updating P300: An integrative theory of P3a and P3b. *Clinical Neurophysiology*, *118*(10), 2128–2148.

Rentzsch, J., Adli, M., Wiethoff, K., de Castro, A. G.-C., & Gallinat, J. (2014). Pretreatment anterior cingulate activity predicts antidepressant treatment response in major depressive episodes. *European Archives of Psychiatry and Clinical Neuroscience*, *264*(3), 213–223.

Sauseng, P., Griesmayr, B., Freunberger, R., & Klimesch, W. (2010). Control mechanisms in working memory: A possible function of EEG theta oscillations. *Neuroscience and Biobehavioral Reviews*, *34*(7), 1015–1022.

Schoenberg, P. L., & Speckens, A. E. (2014). Modulation of induced frontocentral theta (Fm-θ) event-related (de-) synchronisation dynamics following mindfulness-based cognitive therapy in major depressive disorder. *Cognitive Neurodynamics*, *8*(5), 373–388.

Schoenberg, P. L., & Speckens, A. E. (2015). Multi-dimensional modulations of α and γ cortical dynamics following mindfulness-based cognitive therapy in major depressive disorder. *Cognitive Neurodynamics*, *9*(1), 13–29.

Segrave, R., Thomson, R., Cooper, N. R., Croft, R. J., Sheppard, D., & Fitzgerald, P. (2010). Upper alpha activity during working memory processing reflects abnormal inhibition in major depression. *Journal of Affective Disorders*, *127*(1–3), 191–198.

Semkovska, M, Quinlivan, L, O'Grady, T, Johnson, R., Collins, A, O'Connor, J, et al. (2019). Cognitive function following a major depressive episode: A systematic review and meta-analysis. *The Lancet Psychiatry*, *6*(10), 851–861.

Snyder, H. R. (2013). Major depressive disorder is associated with broad impairments on neuropsychological measures of executive function: A meta-analysis and review. *Psychological Bulletin*, *139*(1), 81.

Spronk, D., Arns, M., Bootsma, A., van Ruth, R., & Fitzgerald, P. B. (2008). Long term effects of left frontal rTMS on EEG and ERPs in patients with depression. *Clinical EEG and Neuroscience*, *39*(3), 118–124.

Stewart, J. L., Bismark, A. W., Towers, D. N., Coan, J. A., & Allen, J. J. (2010). Resting frontal EEG asymmetry as an endophenotype for depression risk: Sex-specific patterns of frontal brain asymmetry. *Journal of Abnormal Psychology*, *119*(3), 502.

Stipacek, A., Grabner, R., Neuper, C., Fink, A., & Neubauer, A. (2003). Sensitivity of human EEG alpha band desynchronization to different working memory components and increasing levels of memory load. *Neuroscience Letters*, *353*(3), 193–196.

Sur, S., & Sinha, V. (2009). Event-related potential: An overview. *Industrial Psychiatry Journal*, *18*(1), 70.

Vilà-Balló, A., Salmi, J., Soveri, A., Rodríguez-Fornells, A., Lehtonen, M., & Laine, M. (2018). Neural signatures for active maintenance and interference during working memory updating. *Biological Psychology*, *132*, 233–243.

Yoshimura, S., Okamoto, Y., Onoda, K., Matsunaga, M., Ueda, K., & Suzuki, S.-i. (2010). Rostral anterior cingulate cortex activity mediates the relationship between the depressive symptoms and the medial prefrontal cortex activity. *Journal of Affective Disorders*, *122*(1–2), 76–85.

Part V

# Other aspects of treatment: Specific groups, monitoring and novel regimens

Chapter 45

# Putative effects of cannabidiol in depression and synaptic plasticity

Samia Joca[a,d], Gabriela P. Silote[a,b], Ariandra Sartim[a], Amanda Sales[c], Francisco Guimarães[c], and Gregers Wegener[b]

[a]*Department of BioMolecular Sciences, School of Pharmaceutical Sciences of Ribeirao Preto (FCFRP), University of São Paulo (USP), Ribeirão Preto, Brazil;* [b]*Department of Clinical Medicine, Translational Neuropsychiatry Unit (TNU), Aarhus University, Aarhus, Denmark;* [c]*Department of Pharmacology, School of Medicine of Ribeirão Preto (FMRP), University of São Paulo (USP), Ribeirão Preto, Brazil;* [d]*Aarhus Institute of Advanced Studies (AIAS), Aarhus University, Aarhus, Denmark*

## Abbreviations

| | |
|---|---|
| **2-AG** | 2-arachidonoyl glycerol |
| **5HT1A** | serotonin receptor type 1A |
| **$\Delta^9$-THC** | $\Delta^9$-tetrahydrocannabinol |
| **AEA** | anandamide |
| **BDNF** | brain-derived neurotrophic factor |
| **CB1** | cannabinoid receptor type 1 |
| **CB2** | cannabinoid receptor type 2 |
| **CBD** | cannabidiol |
| **CMS** | chronic mild stress model |
| **DRN** | dorsal raphe nucleus |
| **eCB** | endocannabinoid |
| **FST** | forced swimming test |
| **HPC** | hippocampus |
| **LH** | learned helplessness |
| **MDD** | major depressive disorder |
| **OBX** | olfactory bulbectomized rodent model |
| **PFC** | prefrontal cortex |
| **PPARγ** | peroxisome proliferator-activated receptor |
| **SSRI** | selective serotonin reuptake inhibitors |
| **TRPV1** | transient receptor potential subfamily V member 1 cation channel |
| **TST** | tail suspension test |

## Introduction

Major depressive disorder (MDD) is a chronic debilitating and life-threathning psychiatric disorder that affects more than 300 million people worldwide (WHO, 2017) with a lifetime prevalence of about 20% (Hasin et al., 2018; Kessler & Bromet, 2013). MDD is a leading cause of disability worldwide, contributing to 2.5% of the global burden of disease, with more than 70 million disability-adjusted life years ("GBD 2016 Disease and Injury Incidence and Prevalence Collaborators, 2017). The socioeconomic impact is very large, as MDD is associated with an increased risk of all-cause mortality and reduced life expectancy (Chang et al., 2011; Gustavsson et al., 2011; Laursen et al., 2016). Indeed, a recent study estimated an annual loss in the global economy of approximately $1 trillion, due to decreased productivity associated with MDD and anxiety, and that appropriate treatment of the conditions would result in an economic gain of $230 billion (Chisholm et al., 2016).

For many decades, MDD has been primarily treated with drugs targeting brain monoaminergic signaling systems, e.g., the tricyclic antidepressants and the selective serotonin reuptake inhibitors (SSRIs). Although the antidepressants available are overall effective (Cipriani et al., 2018), several weeks of continuous treatment are required to achieve significant mood-improving effects, and around 30%–40% of the patients remains depressed even after months of treatment (Cipriani et al., 2018; Keks et al., 2007). These substantial limitations makes the discovery of new and more effective antidepressants, primarily focused on nonmonoaminergic mechanisms, necessary. In this context, the discovery that ketamine, a dissociative anesthetic, can induce a rapid (within hours) and relatively sustained antidepressant effects following a single injection of subanesthetic doses (reviewed in Gould, Zarate, & Thompson, 2019; Zanos & Gould, 2018; Zanos et al., 2018), is of crucial importance. Additional findings showing that the effects of ketamine are associated with synaptogenesis in cortical and limbic brain regions paralleled with rapid elevations in protein translation mechanisms highlighted new avenues for finding different compounds with multimodal mechanisms of action, which also could be fast-acting antidepressants (Ceskova & Silhan, 2018).

The Neuroscience of Depression. https://doi.org/10.1016/B978-0-12-817933-8.00052-9

In 2010, our groups published the first evidence that Cannabidiol (CBD), a nonpsychoactive constituent extract from the *Cannabis sativa* plant, induces antidepressant-like effects in animal models (Zanelati et al., 2010). Although clinical evidence from randomized controlled trials is still lacking, CBD has received considerable attention in the past decade as a potential new antidepressant. Several preclinical studies have brought forward supportive evidence for the antidepressant properties of CBD (Russo & Marcu, 2017; Silote et al., 2019). The complex pharmacology of CBD, which may be associated to its potential beneficial broad therapeutic profile (antidepressant, antipsychotic, and anxiolytic) has prompted research in further characterizing its mechanisms and efficacy as a novel treatment option for depression (Mandolini et al., 2018).

## Cannabidiol pharmacology and therapeutic potential

*Cannabis sativa* (cannabis) is one of the world's oldest cultivated plants. Medicinal preparations from the flowers and resin of cannabis have been used in China since ~2700 BC (as reported in the world's oldest pharmacopeia, the *pen-ts'ao ching)* and cannabis was also one of the main drugs in the Assyrian pharmacopeia 3000 years ago (Russo, 2014; Zuardi, 2006). In India, the ancient Hindu scripture "Atharva Veda" mentions cannabis as one of five sacred plants, referring to it as a source of "happiness, donator of joy and bringer of freedom" (India Hemp Drugs Commission, 1894; Russo, 2014; Zuardi, 2006). Cannabis was brought to the Western medicine in the 19th century, through the works of O'Shaughnessy (1843)) and Moreau, (1845), but its use started to decrease in the first decades of the 20th century, due to the difficulty of obtaining consistent effects and the increased fear that recreational use of cannabis may be related to violence, crime, and other socially deviant behaviors (Russo, 2014; Zuardi, 2006). The result was legal restrictions in cannabis use, both medical and recreational, which happened in 1937 in the United States (Musto, 1972). As of today, cannabis remains the most common illicit drug used in the world (Volkow et al., 2014).

Despite use and knowledge about cannabis, little was known about the phytochemistry and pharmacology before 1964, when development of better chromatographic techniques allowed isolation of several different constituents by the group of Raphael Mechoulam, including $\Delta^9$-tetrahydrocannabinol (THC) and cannabidiol (CBD) (Gaoni & Mechoulam, 1964; Mechoulam et al., 1970). In these seminal works, only THC showed potent activity, and CBD was therefore considered a nonpsychoactive and less relevant constituent of cannabis. Today, it is known that the cannabis plant produces more than 100 different terpenophenolic compounds, termed (phyto)cannabinoids, all present in varying relative proportions depending on the strain of plant (Small, 2015).

Following isolation and identification of the constituents of cannabis, the scientific interest in the plant increased, culminating with the identification of cannabinoid receptors (Devane et al., 1992; Matsuda et al., 1990) and, later, endogenous cannabinoid ligands and a signaling system in the brain (Di Marzo et al., 1994; Munro, Thomas, & Abu-Shaar, 1993; Zygmunt et al., 1999). The endocannabinoid system consists of the cannabinoid receptors (CB1 and CB2), endogenous ligands [mainly anandamide (AEA) and 2-arachidonoyl glycerol (2-AG)], and processes responsible for endocannabinoid biosynthesis, cellular uptake, and metabolism (Pertwee, 2015; Piomelli, 2013). Therefore, the word cannabinoids usually refer to a heterogeneous group of compounds including endogenous, synthetic, and phytocannabinoids (Fig. 1).

While $\Delta^9$-THC was recognized as the primary psychoactive compound responsible for marijuana effects, such as euphoria, altered sensory perception, and relaxation, CBD was shown to be nonpsychostimulant, initially with the suggestion that it was an inactive substance (Zuardi, 2008). However, it was later discovered that CBD has multiple effects in the central nervous system, ranging from attenuating the subjective rates of intoxication induced by $\Delta^9$-THC in animals and humans (Martin-Santos et al., 2012; Robson, 2011; Schoedel et al., 2011; Zuardi et al., 1982), to showing promising therapeutic effects on several psychiatric conditions, such as anxiety (Crippa et al., 2011; Guimarães et al., 1990; Moreira, Aguiar, & Guimarães, 2006; Zuardi et al., 2017), psychosis (Iseger & Bossong, 2015; Moreira & Guimarães, 2005), epilepsy (Crippa et al., 2016; Do Val-da Silva et al., 2017; Gobira et al., 2015), and movement disorders (Peres et al., 2018). The therapeutic potential of CBD in these conditions is excellently reviewed elsewhere and lies beyond the scope of the present book chapter (Crippa et al., 2018; Mandolini et al., 2018).

The proposed broad range of effects induced by CBD stimulated an intense investigation about its pharmacological properties, which soon appeared to be complex, involving multiple receptors, transporters, ions channels, and enzymes (Ibeas Bih et al., 2015; Izzo et al., 2009)—Fig. 1). To date, only a relatively limited number of the possible targets have been explored as mediators of the effects of CBD in psychiatric and neurological disorders.

In contrast to THC, CBD has low affinity for the endocannabinoid receptors, CB1 and CB2, but may antagonize the former in the presence of THC (Thomas et al., 2007). The pharmacological rationale underlying this remarkable effects is not fully understood but has been proposed to involve the ability of CBD to act as a negative allosteric modulator at CB1 receptors (Laprairie et al., 2015) although it has also been argued that CBD can affect the activity of the CB1 in vivo through indirect mechanisms

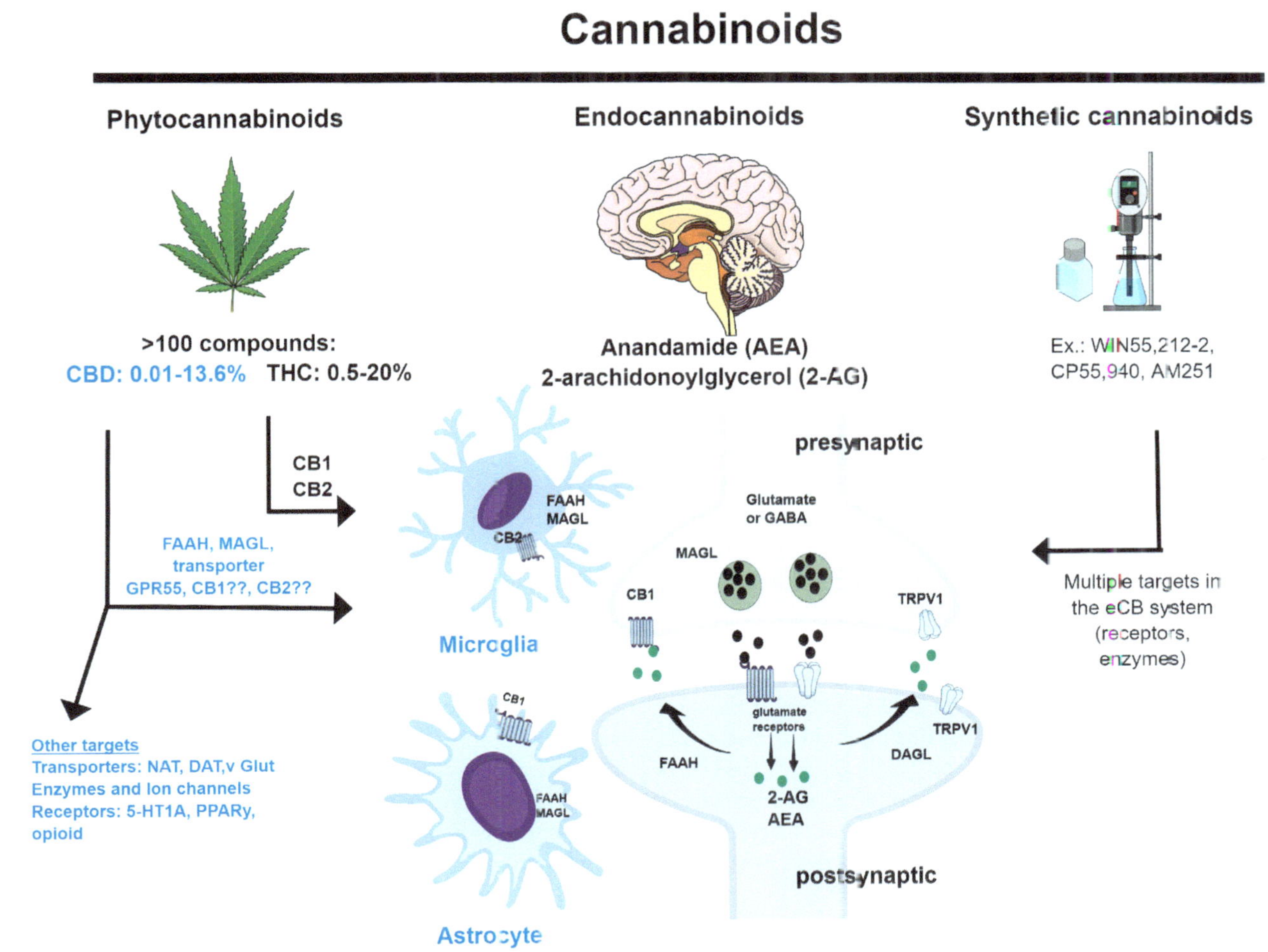


FIG. 1 Cannabinoids: heterogeneous group of compounds including endogenous, synthetic and phytocannabinoids. *Cannabis sativa* produces more than 100 compounds named phytocannabinoids, which are present in varying relative proportions depending on the plant strain. The most abundant cannabinoids in the plant are $\Delta^9$-tetrahidrocannabinol (THC) and cannabidiol (CBD). Anandamide (AEA) and 2-aracdonoylglycerol are the main endogenous cannabinoids (endocannabinoids, eCB), which are synthesized upon increased postsynaptic activity. The eCB signaling system comprises the cannabinoid type 1 and type 2 receptors (CB1 and CB2), to which AEA and 2-AG bind; synthetic and degradative enzymes (FAAH, MAGL); and transporters that regulate eCB levels at the synapse. Whereas THC activates CB1 and CB2, CBD has very low affinity for these receptors, but can increase eCB levels due to the inhibition of degrading enzymes. Moreover, CBD binds to a variety of other targets outside the eCB system, including monoamine transporters and receptors (ex. 5-HT1A, PPARy). *Source: Data from Russo, E.B., & J.Marcu (2017). Cannabis pharmacology: The usual suspects and a few promising leads.* Advances in Pharmacology, 80, *67–134; Small, E. (2015b). Evolution and classification of Cannabis sativa (Marijuana, Hemp) in relation to human utilization.* Botanical Review, 81*(3), 189–294; Pertwee, R.G. (2015b). Endocannabinoids and their pharmacological actions. In: Handbook of experimental pharmacology. pp. 1–37.; Lucas, C.J., Galettis, P., & Schneider, J. (2018). The pharmacokinetics and the pharmacodynamics of cannabinoids.* British Journal of Clinical Pharmacology, 84*(11), 2477–2482.; Castillo, Pablo E. et al. (2012). Endocannabinoid signaling and synaptic function.* Neuron, 76*(1), 70–81.*

(McPartland et al., 2015). Interestingly, CBD can increase the endogenous levels of AEA by inhibiting the membrane transporters or the enzyme responsible for its metabolism, Fatty Acid Amide Hydrolase (FAAH), which in turn could explain its indirect modulatory activity on CB1 and CB2 signaling (Ibeas Bih et al., 2015). Moreover, CBD has been reported to modulate the activity of many other neurotransmitter transporters, such as transporters for dopamine, glutamate, noradrenaline, and adenosine, in addition to AEA (Ibeas Bih et al., 2015).

Although CBD seemingly does not directly affect the CB1 and CB2 receptors, it may be an agonist at the transient receptor potential subfamily V member 1 cation channel (TRPV1, Bisogno et al., 2001), 5-HT1A receptors (Russo et al., 2005), and peroxisome proliferator-activated receptor (PPARy, Ramer et al., 2013). Similarly, CBD may also negatively modulate the orphan G protein-coupled receptor 55, GPR55, which is considered part of the eCB system with lysophosphatidylinositol as an endogenous ligand, as well as μ- and δ-opioid receptors (Ryberg et al., 2007). Finally, CBD can modulate the activity of many different ion channels and enzymes (Ibeas Bih et al., 2015).

It is unclear, however, how these complex neurochemical mechanisms can contribute to the effects of

CBD in vivo. Indeed, some of the proposed CBD targets are only activated at high concentrations of CBD using in vitro assays, and may, therefore, be irrelevant in vivo, especially when considering the relatively poor bioavailability of CBD. On the other hand, the diversity of potential pharmacological targets may help to explain the wide range of potential therapeutic effects, such as anticonvulsant, antiinflammatory, antipsychotic, anxiolytic, and neuroprotective effects (Elsaid, Kloiber, & Le Foll, 2019).

## The effects of CBD in animal models of depression

The first demonstration that CBD could be effective in alleviating depressive-like symptoms in animals was reported in 2010 when it was shown that CBD reduces the immobility time in Swiss mice exposed to the forced swimming test, a finding similar to what can be observed following administration of the established antidepressant imipramine (Zanelati et al., 2010). Interestingly, pretreatment with the 5-HT1A receptor antagonist (WAY100635) counteracted the effects of CBD, suggesting that they depend on intact 5HT1A receptor function (Zanelati et al., 2010). The finding that the antidepressant-like effects of CBD are mediated by facilitation of 5-HT1A-mediated signaling agrees with the proposed involvement of these receptors in the pathogenesis of depression and responsiveness to serotoninergic antidepressant drugs (Blier & El-Mansari, 2013; Kaufman et al., 2016).

The initial finding that CBD has antidepressant potential has been confirmed by other groups although at quite variable doses. In addition to the acute effect observed in animals exposed to the FST and Tail Suspension Test (TST), CBD is also effective following repeated treatment, as shown in Swiss mice subjected to the TST (Schiavon et al., 2016) and in Wistar rats subjected to the FST (Réus et al., 2011). CBD has also shown promising effects in animal models with better face and construct validity, such as the rat learned helplessness (LH; Sales et al., 2018), the chronic mild stress model (CMS; Xu et al., 2019), the olfactory bulbectomized rodent model (OBX; Linge et al., 2016, see later), and in the rat lines that are genetically vulnerable for the depressive phenotype, the Flinders Sensitive Line (FSL; Sales et al., 2019; Shbiro et al., 2019), and the Wistar-Kyoto (Shbiro et al., 2019; Shoval et al., 2016). CBD was shown to be effective in both male and female Wistar-Kyoto, but only in male FSL rats (Shbiro et al., 2019), thus bringing the question forward that the effects of CBD may be gender-dependent, and might be less effective in some female strains. However, the listed study only examined one dose of CBD ($30\,mg\,kg^{-1}$), making it difficult to conclude about its effectiveness in females. Additional investigation is necessary to answer this question.

The potential brain sites of CBD effects have been investigated by direct injections into the ventromedial PFC (Sartim, Guimarães, & Joca, 2016) and dorsal hippocampus (Sartim et al., 2018). In both regions, CBD promoted antidepressant-like effects. These findings indicate that it is CBD itself, and not a product of its metabolism, the responsible for inducing its behavioral effects. Moreover, the study also highlights that both brain regions may mediate the antidepressant-like effects of CBD. In line with the earlier studies demonstrating the involvement of the 5HT1A receptors, the behavioral effect produced by intra-mPFC injection of CBD could be counteracted by pretreatment with antagonists of not only 5-HT1A but also CB1 receptors, thereby implicating the participation of the local serotonergic and eCB system in CBD effects (Sartim et al., 2016). Given that CBD does not directly bind to CB1 receptors, it can be hypothesized that the observed effect is due to indirect modulation of eCB levels in the PFC, which subsequently would activate CB1 receptors, disinhibiting the neuronal projections to the Dorsal Raphe Nucleus (DRN) and thereby promoting 5-HT1A activation. As a consequence, the antidepressant effects would depend on increased serotonin availability in the ventromedial PFC, rather than direct activation of 5-HT1A receptors. This hypothesis, however, warrants further investigation.

Interestingly, in contrast to conventional monoaminergic antidepressants, CBD has been described to promote rapid and sustained antidepressant effects in the OBX model (Linge et al., 2016). Pretreatment with a 5-HT1A receptor antagonist, but not a CB1 antagonist, blocked CBD rapid effects in OBX mice, thus indicating a serotonergic-dependent mechanism. Increased glutamate and 5-HT availability were observed in the ventromedial PFC of OBX mice treated with CBD, particularly after chronic treatment, along with normalization of 5-HT1A functionality in the PFC and hippocampus (Linge et al., 2016). Based on these findings, it was suggested that CBD could induce antidepressant-like effects involving increased 5-HT signaling in cortical and limbic brain regions. In line with this finding, Sales and colleagues demonstrated that the effects of CBD in the FST are not observed in animals which have been depleted of brain serotonin following the treatment with PCPA, a serotonin synthesis inhibitor (Sales et al., 2018). Furthermore, the combination of subeffective doses of CBD and fluoxetine (SSRI) produced synergistic effects in Swiss mice tested in the FST (Sales et al., 2018). Altogether, these results indicate that the effects of CBD depend on intact serotonin signaling in cortical and limbic brain regions although the mechanism behind this effect is not yet clear.

Chronic treatment with CBD is also known to induce behavioral effects associated with changes in neuroplasticity, primarily neurogenesis, and synaptogenesis (Campos et al., 2017). CBD has been shown to increase neurogenesis both in vitro and in vivo through CB1-dependent mechanisms (Campos et al., 2013; Fogaça et al., 2018; Wolf et al., 2010). Indeed, in mice submitted to chronic mild stress, treatment with CBD during 14 days promoted behavioral adaptation and counteracted stress-induced impairments in neurogenesis and dendritic remodeling in the hippocampus (Fogaça et al., 2018). Interestingly, both effects induced by CBD on behavior and neuroplasticity were attenuated by concomitant treatment with CB1 and CB2 antagonists, but not by 5-HT1A antagonists, indicating an endocannabinoid-dependent mechanism in synaptic plasticity following repeated treatment with CBD (Campos et al., 2013; Fogaça et al., 2018). This finding agrees with the observation that chronic CBD treatment increases AEA signaling in the hippocampus (Campos et al., 2013; Fogaça et al., 2018) and that increases in eCB signaling promote neurogenesis through CB1 and CB2 receptors (Bambico et al., 2016; Bambico & Gobbi, 2008; Zhang et al., 2015).

On the other hand, it has also been reported that the behavioral effects induced by CBD treatment are independent of increased hippocampal neurogenesis (Schiavon et al., 2016). In this research, both acute and chronic treatments with CBD induced antidepressant-like effect in the TST. However, CBD increased neurogenesis at the lowest dose, but decreased it at the highest dose, indicating that CBD might be able to induce antidepressant-like effects even under conditions of decreased hippocampal neurogenesis. Indeed, such findings are of relevance, due to similar findings following chronic administration of antidepressant drugs, where increased neurogenesis, and promotion of integration of newborn neurons into the dentate gyrus, thus favoring behavioral adaptation to stress and relieve of depressive symptoms, have been observed (Park, 2019; Surget et al., 2011). However, contrasting results have also been shown, and it has been argued that not all antidepressant effects are dependent on hippocampal neurogenesis (Santarelli et al., 2003). However, it seems that neurogenesis may be relevant to the behavioral effects of monoaminergic antidepressants in counteracting hippocampal deficits produced by chronic stress exposure (Eliwa, Belzung, & Surget, 2017). In fact, the acute effects of fast-acting antidepressants, such as ketamine, cannot be explained by increased neurogenesis, as these effects do not correlate with the time required for the birth of new neurons (Eliwa et al., 2017). The rapid-acting effect of ketamine and similar drugs have been attributed to glutamate-mediated changes in the PFC, which is thought to activate a cascade of events resulting in increased synaptogenesis and attenuations of the negative effects of chronic stress and depression (Abdallah et al., 2016).

Indeed, in mice submitted to the FST, the acute antidepressant effect induced by CBD (Sales et al., 2019) was accompanied by increased BDNF and TrkB levels in the PFC, 30 min following injection, suggesting increased synaptic plasticity. Corroborating this idea, increased levels of the synaptic protein synaptophysin and increased dendritic branching were observed in the PFC, but not the hippocampus, of mice after CBD acute administration (Sales et al., 2019). Curiously, no effects were observed 7 days later, when an antidepressant effect could still be observed after a single CBD administration. Therefore, although the acute effect of CBD seems to involve a fast increase in synaptogenesis in the PFC, little is known about the molecular mechanisms mediating these effects, as well as it is unknown which mechanisms trigger CBD sustained effects.

Altogether, the findings suggest that CBD produces antidepressant effects after acute or repeated administration, independently of rodent species employed and the behavioral readout used to assess the antidepressant-like profile. Furthermore, the effects of CBD seem to involve changes in synaptic plasticity characterized by increased synaptogenesis in the PFC (acutely) and neurogenesis in the HPC (chronically). However, the molecular mechanism involved in these effects remains to be investigated and further characterized.

## CBD effects on depressed patients

*Cannabis sativa* has been used for centuries as a natural alternative for treating mood-related mental changes (Turna, Patterson, & Van Ameringen, 2017) and CBD could be involved in some of these mood-improving effects (Ashton et al., 2005) although THC also demonstrates antidepressant effect at low doses (Ashton et al., 2005). However, the available evidence and literature relating to the effects of CBD on mood, especially in depressed patients, are very sparse.

Hindocha and colleagues conducted in 2015 a study on cannabis users, in order to evaluate the acute effects of THC and CBD on the emotional facial recognition test. Inhalation of CBD (16 mg) improved the subjects' performance on the test, while THC (8 mg) impaired it (Hindocha et al., 2015). These results suggest that CBD possess the potential to enhance the emotional process assessed in recognition of emotional face test, while THC does the opposite. It is important to emphasize that the study evaluated only one dose of each substance and that the drug-response profile could change using different doses of cannabinoids. In another recent study, a questionnaire was applied to CBD users to identify the main reasons for use (Corroon & Phillips, 2018). From a total of 2409 people who declared themselves to be CBD users, 1483 reported a therapeutic use. Of these, more than 400 said they used CBD to treat depression-related symptoms, and of these, 250 patients rated the effect of CBD as either "very well by itself" or "moderately well by itself") (Corroon & Phillips, 2018).

Although the evidence described above, coupled with basic research data, suggests that CBD has mood-stabilizing properties, none of the studies fulfills the necessary proper scientific rigor, and the scientific literature still lacks controlled clinical studies in patients diagnosed with major depression.

In a recent systematic review and meta-analysis about the medical use of cannabinoids, including CBD, none of the studies evaluated cannabinoids for the treatment of depression (Whiting et al., 2015). Another recent review on human studies assessing CBD therapeutic potential described no evidence for RTC in humans that support CBD effects in depression (White, 2019). Altogether, there is, to date, no convincing evidence from human studies that support CBD use in depression treatment.

Currently, two randomized clinical trials are being conducted in patients with bipolar depression and are soon expected to be published (EUCTR2015-000465-31-DE (2015-000465-31, 2015) and NCT03310593 (NCT03310593, 2017)). Although the diagnostic entity—bipolar depression—is likely to be pathophysiologically distinct from MDD, the results of these studies, ideally associated with patients with unipolar depression, will help to understand the therapeutic potential of cannabidiol in these disorders.

Although relatively safe with either acute or chronic administration, CBD is not risk-free, as it can induce significant side effects (somnolence, decreased appetite, and diarrhea) and drug interactions due to interference in liver function (Elsaid et al., 2019; Iffland & Grotenhermen, 2017; VanDolah, Bauer, & Mauck, 2019). The half-life (t1/2) of CBD largely depends on the dose and the route of administration and varies considerably between humans and rodents, thus making it difficult to predict effective treatment regimes at this point (Iffland & Grotenhermen, 2017). Therefore, double-blind, placebo-controlled trials are still required to elucidate the potential antidepressant effect in humans. If confirmed, the study of CBD antidepressant properties could pave the way for new antidepressant therapy and help understanding depression neurobiology given its complex multitargeted pharmacology.

## Final considerations

Several preclinical studies reviewed in this chapter have consistently reported an antidepressant-like profile for CBD in different behavioral readouts, thus reinforcing the idea that it could be beneficial in depression treatment. It is important to notice, however, that there is still insufficient evidence from human studies to support this use.

## Key facts

- CBD has shown antidepressant-like effects in different animal models
- CBD has a complex pharmacology, targeting different receptors (e.g., 5-HT1A, cannabinoids, PPARγ, and GPR55), enzymes (FAAH and MAGL), and transporters (noradrenaline, dopamine, glutamate)
- CBD promotes rapid changes in brain neurochemistry and synaptic plasticity
- Evidence from human studies about CBD effects in depression are still inconclusive or insufficient

## Summary points

- CBD and THC are phytocannabinoids extracted from *Cannabis sativa*
- Contrary to THC, CBD is nonpsychostimulant, but induces a wide range of pharmacological effects
- CBD has shown promising therapeutic effects in different animal models of depression
- The antidepressant effect of CBD might involve the modulation of serotonergic neurotransmission as well as the endocannabinoid system. However, due to CBD complex pharmacology, its precise mechanism is not yet known
- CBD promotes rapid changes in brain neurochemistry and synaptic plasticity in cortical and subcortical brain regions with relevance to depression
- Evidence from human studies about CBD effects in depression are still inconclusive or insufficient

## Mini-dictionary of terms

**Negative allosteric modulator** A group of substances that bind to a receptor and decrease the agonist affinity and/or efficacy.

**Neuroplasticity** The ability of the brain to reorganize itself throughout life in response to new information, sensory stimulation, development, damage, or dysfunction.

**Psychoactive compound** "Substances that, when taken in or administered into one's system, affect mental processes, e.g., perception, consciousness, cognition or affect" (WHO).

**Synaptic plasticity** Strengthening or weakening of synapses over time, in response to increased or decreased activity, respectively.

## Acknowledgments

This study was supported by grants from the National Institute of Science and Translational Medicine, CNPq (465458/2014-9), Coordenação de Aperfeiçoamento de Pessoal de Nível Superior—Brazil (CAPES)—Finance Code 001, FAPESP (2012/17626-7, 2017/24304-0), Independent Research Fund Denmark (grant 8020-00310B to GW); Aarhus University Research Foundation, Denmark (AU-IDEAS initiative (eMOOD) to GW); and EU Horizon 2020 (ExEDE to GW). S.J. is thankful to AIAS-Cofund (Marie Curie) Fellowship Program and CNPq (306648/2014-8). G.S. and A.S. are thankful to FAPESP (Grant#2018/12119-6; #2017/26815-1) and CNPq (Grant#141.585/2016-1) for research fellowship.

## Disclosures

G.W. declares having received lecture/consultancy fees from H. Lundbeck A/S, Servier SA, Astra Zeneca AB, Eli Lilly A/S, Sun Pharma Pty Ltd., Pfizer Inc., Shire A/S, HB Pharma A/S, Arla Foods A.m.b.A., Alkermes Inc., Johnson & Johnson Inc., and Mundipharma International Ltd. F.G. is a coinventor (Mechoulam R, Crippa JA, Guimaraes FS, Zuardi A, Hallak, JE, Breuer A) of the patent "Fluorinated CBD compounds, compositions and uses thereof. Pub. No.: WO/2014/108899. International Application No.: PCT/IL2014/050023" Def. US no. Reg. 62193296; 29/07/2015; INPI on 19/08/2015 (BR1120150164927). The University of São Paulo has licensed the patent to Phytecs Pharm (USP Resolution No. 15.1.130002.1.1). The University of São Paulo has an agreement with Prati-Donaduzzi (Toledo, Brazil) to "develop a pharmaceutical product containing synthetic cannabidiol and prove its safety and therapeutic efficacy in the treatment of epilepsy, schizophrenia, Parkinson's disease, and anxiety disorders." All other authors declare no conflict of interest.

## References

2015-000465-31. (2015). *Cannabidiol in Bipolar Depression—CannaBiD-Study: An 8-week randomized, double-blind, placebo-controlled clinical trial of Cannabidiol as add-on therapy in bipolar depression.* [cited 2019 November 25]; Available from: https://www.clinicaltrialsregister.eu/ctr-search/search?query=2015-000465-31.

Abdallah, C. G., et al. (2016). Ketamine's mechanism of action: A path torapid-acting antidepressants. *Depression and Anxiety*, *33*(8), 689–697.

Ashton, C. H., et al. (2005). Cannabinoids in bipolar affective disorder: A review and discussion of their therapeutic potential. *Journal of Psychopharmacology*, *19*(3), 293–300.

Bambico, F. R., & Gobbi, G. (2008). The cannabinoid CB1 receptor and the endocannabinoid anandamide: Possible antidepressant targets. *Expert Opinion on Therapeutic Targets*, *12*(11), 1347–1366.

Bambico, F. R., et al. (2016). The fatty acid amide hydrolase inhibitor URB597 modulates serotonin-dependent emotional behaviour, and serotonin1A and serotonin2A/C activity in the hippocampus. *European Neuropsychopharmacology*, *26*(3), 578–590.

Bisogno, T., et al. (2001). Molecular targets for cannabidiol and its synthetic analogues: Effect on vanilloid VR1 receptors and on the cellular uptake and enzymatic hydrolysis of anandamide. *British Journal of Pharmacology*, *134*(4), 845–852.

Blier, P., & El-Mansari, M. (2013). Serotonin and beyond: Therapeutics for major depression. *Philosophical Transactions of the Royal Society B: Biological Sciences*, *368*(1615), 1–7. https://doi.org/10.1098/rstb.2012.0536.

Campos, A. C., et al. (2013). The anxiolytic effect of cannabidiol on chronically stressed mice depends on hippocampal neurogenesis: Involvement of the endocannabinoid system. *International Journal of Neuropsychopharmacology*, *16*(6), 1407–1419.

Campos, A. C., et al. (2017). Plastic and neuroprotective mechanisms involved in the therapeutic effects of cannabidiol in psychiatric disorders. *Frontiers in Pharmacology*, *8*, 1–18.

Ceskova, E., & Silhan, P. (2018). Novel treatment options in depression and psychosis. *Neuropsychiatric Disease and Treatment*, *14*, 741–747.

Chang, C. K., et al. (2011). Life expectancy at birth for people with serious mental illness and other major disorders from a secondary mental health care case register in London. *PLoS One*, *6*(5), e19590.

Chisholm, D., et al. (2016). Scaling-up treatment of depression and anxiety: A global return on investment analysis. *The Lancet Psychiatry*, *3*(5), 415–424.

Cipriani, A., et al. (2018). Comparative efficacy and acceptability of 21 antidepressant drugs for the acute treatment of adults with major depressive disorder: A systematic review and network meta-analysis. *The Lancet*, *391*(10128), 1357–1366.

Corroon, J., & Phillips, J. A. (2018). A cross-sectional study of Cannabidiol users. *Cannabis and Cannabinoid Research*, *3*(1), 152–161.

Crippa, J. A., et al. (2018). Translational investigation of the therapeutic potential of cannabidiol (CBD): Toward a new age. *Frontiers in Immunology*, *9*(2009), 1–16.

Crippa, J. A. S., et al. (2011). Neural basis of anxiolytic effects of cannabidiol (CBD) in generalized social anxiety disorder: A preliminary report. *Journal of Psychopharmacology*, *25*(1), 121–130.

Crippa, J. A. S., et al. (2016). Δ9-THC intoxication by cannabidiol-enriched cannabis extract in two children with refractory epilepsy: Full remission after switching to purified cannabidiol. *Frontiers in Pharmacology*, *7*, 359.

Devane, W. A., et al. (1992). Isolation and structure of a brain constituent that binds to the cannabinoid receptor. *Science*, *258*(5090), 1946–1949.

Di Marzo, V., et al. (1994). Formation and inactivation of endogenous cannabinoid anandamide in central neurons. *Nature*, *372*(6507), 686–691.

Do Val-da Silva, R. A., et al. (2017). Protective effects of cannabidiol against seizures and neuronal death in a rat model of mesial temporal lobe epilepsy. *Frontiers in Pharmacology*, *8*, 131.

Eliwa, H., Belzung, C., & Surget, A. (2017). Adult hippocampal neurogenesis: Is it the alpha and omega of antidepressant action? *Biochemical Pharmacology*, *141*, 86–99.

Elsaid, S., Kloiber, S., & Le Foll, B. (2019). Effects of cannabidiol (CBD) in neuropsychiatric disorders: A review of pre-clinical and clinical findings. *Progress in Molecular Biology and Translational Science*, *167*, 25–75.

Fogaça, M. V., et al. (2018). The anxiolytic effects of cannabidiol in chronically stressed mice are mediated by the endocannabinoid system: Role of neurogenesis and dendritic remodeling. *Neuropharmacology*, *135*, 22–33.

Gaoni, Y., & Mechoulam, R. (1964). Isolation, structure, and partial synthesis of an active constituent of hashish. *Journal of the American Chemical Society*, *86*(8), 1646–1647.

GBD 2016 Disease and Injury Incidence and Prevalence Collaborators. (2017). Global, regional, and national incidence, prevalence, and years lived with disability for 328 diseases and injuries for 195 countries, 1990–2016: A systematic analysis for the Global Burden of Disease Study 2016 [published correction appears in Lancet. 2017 Oct 28;390(10106) e38]. *The Lancet*, *390*(10100), 1211–1259. https://doi.org/10.1016/S0140-6736(17)32154-2.

Gobira, P. H., et al. (2015). Cannabidiol, a Cannabis sativa constituent, inhibits cocaine-induced seizures in mice: Possible role of the mTOR pathway and reduction in glutamate release. *Neurotoxicology*, *50*, 116–121.

Gould, T. D., Zarate, C. A., Jr., & Thompson, S. M. (2019). Molecular pharmacology and neurobiology of rapid-acting antidepressants. *Annual Review of Pharmacology and Toxicology*, *59*, 213–236.

Guimarães, F. S., et al. (1990). Antianxiety effect of cannabidiol in the elevated plus-maze. *Psychopharmacology*, *100*(4), 558–559.

Gustavsson, A., et al. (2011). Cost of disorders of the brain in Europe 2010. *European Neuropsychopharmacology*, *21*(10), 718–779.

Hasin, D. S., et al. (2018). Epidemiology of adult DSM-5 major depressive disorder and its specifiers in the United States. *JAMA Psychiatry*, *75* (4), 336–346.

Hindocha, C., et al. (2015). Acute effects of delta-9-tetrahydrocannabinol, cannabidiol and their combination on facial emotion recognition: A randomised, double-blind, placebo-controlled study in cannabis users. *European Neuropsychopharmacology*, *25*(3), 325–334.

Ibeas Bih, C., et al. (2015). Molecular targets of Cannabidiol in neurological disorders. *Neurotherapeutics*, *12*(4), 699–730.

Iffland, K., & Grotenhermen, F. (2017). An update on safety and side effects of Cannabidiol: A review of clinical data and relevant animal studies. *Cannabis and Cannabinoid Research*, *2*(1), 139–154.

Iseger, T. A., & Bossong, M. G. (2015). A systematic review of the antipsychotic properties of cannabidiol in humans. *Schizophrenia Research*, *162*(1–3), 153–161.

Izzo, A. A., et al. (2009). Non-psychotropic plant cannabinoids: New therapeutic opportunities from an ancient herb. *Trends in Pharmacological Sciences*, *30*(10), 515–527.

Kaufman, J., et al. (2016). The 5-HT1A receptor in major depressive disorder. *European Neuropsychopharmacology*, *26*(3), 397–410.

Keks, N. A., et al. (2007). Beyond the evidence: Is there a place for antidepressant combinations in the pharmacotherapy of depression? *The Medical Journal of Australia*, *186*(3), 142–144.

Kessler, R. C., & Bromet, E. J. (2013). The epidemiology of depression across cultures. *Annual Review of Public Health*, *34*(1), 119–138.

Laprairie, R. B., et al. (2015). Cannabidiol is a negative allosteric modulator of the cannabinoid CB1 receptor. *British Journal of Pharmacology*, *172*(20), 4790–4805.

Laursen, T. M., et al. (2016). Mortality and life expectancy in persons with severe unipolar depression. *Journal of Affective Disorders*, *193*, 203–207.

Linge, R., et al. (2016). Cannabidiol induces rapid-acting antidepressant-like effects and enhances cortical 5-HT/glutamate neurotransmission: Role of 5-HT1A receptors. *Neuropharmacology*, *103*, 16–26.

Mandolini, G. M., et al. (2018). Pharmacological properties of cannabidiol in the treatment of psychiatric disorders: A critical overview. *Epidemiology and Psychiatric Sciences*, *27*(4), 327–335.

Martin-Santos, R., et al. (2012). Acute effects of a single, oral dose of d9-tetrahydrocannabinol (THC) and cannabidiol (CBD) administration in healthy volunteers. *Current Pharmaceutical Design*, *18*(32), 4966–4979.

Matsuda, L. A., et al. (1990). Structure of a cannabinoid receptor and functional expression of the cloned cDNA. *Nature*, *346*(6284), 561–564.

McPartland, J. M. et al. (2015). Are cannabidiol and Delta(9)-tetrahydrocannabivarin negative modulators of the endocannabinoid system? A systematic review. *British Journal of Pharmacology*, *172* (3), 737–753.

Mechoulam, R., et al. (1970). Chemical basis of hashish activity. *Science*, *169*(3945), 611–612.

Moreau, J. J. (1845). *Du hachisch et de l'aliénation mentale: études psychologiques*. Masson: Fortin.

Moreira, F. A., Aguiar, D. C., & Guimarães, F. S. (2006). Anxiolytic-like effect of cannabidiol in the rat Vogel conflict test. *Progress in Neuro-Psychopharmacology and Biological Psychiatry*, *30*(8), 1466–1471.

Moreira, F. A., & Guimarães, F. S. (2005). Cannabidiol inhibits the hyperlocomotion induced by psychotomimetic drugs in mice. *European Journal of Pharmacology*, *512*(2–3), 199–205.

Munro, S., Thomas, K. L., & Abu-Shaar, M. (1993). Molecular characterization of a peripheral receptor for cannabinoids. *Nature*, *365*(6441), 61–65.

Musto, D. F. (1972). The marihuana tax act of 1937. *Archives of General Psychiatry*, *26*(2), 101–108.

NCT03310593. (2017). *A double-blind, randomized, placebo-controlled clinical trial of adjunctive cannabidiol for bipolar depression.* [cited 2019 November 25]; Available from: https://clinicaltrials.gov/ct2/show/NCT03310593.

O'Shaughnessy, W. B. (1843). On the preparations of the Indian Hemp, or Gunjah: Cannabis indica their effects on the animal system in health, and their utility in the treatment of tetanus and other convulsive diseases. *Provincial Medical Journal and Retrospect of the Medical Sciences*, *5*(123), 363–369.

Park, S. C. (2019). Neurogenesis and antidepressant action. *Cell and Tissue Research*, *377*(1), 95–106.

Peres, F. F., et al. (2018). Cannabidiol as a promising strategy to treat and prevent movement disorders? *Frontiers in Pharmacology*, *9*(482), 1–12.

Pertwee, R. G. (2015). Endocannabinoids and their pharmacological actions. In R. G. Pertwee (Ed.), *Endocannabinoids* (pp. 1–37). Cham: Springer International Publishing.

Piomelli, D. (2013). Endocannabinoids. In W. J. Lennarz, & M. D. Lane (Eds.), *Encyclopedia of biological chemistry* (2nd ed., pp. 194–196). Waltham: Academic Press.

Ramer, R., et al. (2013). COX-2 and PPAR-γ confer cannabidiol-induced apoptosis of human lung cancer cells. *Molecular Cancer Therapeutics*, *12*(1), 69–82.

Réus, G. Z., et al. (2011). Administration of cannabidiol and imipramine induces antidepressant-like effects in the forced swimming test and increases brain-derived neurotrophic factor levels in the rat amygdala. *Acta Neuropsychiatrica*, *23*(5), 241–248.

Robson, P. (2011). Abuse potential and psychoactive effects of δ-9-tetrahydrocannabinol and cannabidiol oromucosal spray (Sativex), a new cannabinoid medicine. *Expert Opinion on Drug Safety*, *10*(5), 675–685.

Russo, E. (2014). The pharmacological history of cannabis. In R. G. Pertwee (Ed.), *Handbook of Cannabis* (pp. 23–43). Oxford: Oxford University Press.

Russo, E. B., & Marcu, J. (2017). Cannabis pharmacology: The usual suspects and a few promising leads. *Advances in Pharmacology*, *80*, 67–134.

Russo, E. B., et al. (2005). Agonistic properties of cannabidiol at 5-HT1a receptors. *Neurochemical Research*, *30*(8), 1037–1043.

Ryberg, E., et al. (2007). The orphan receptor GPR55 is a novel cannabinoid receptor. *British Journal of Pharmacology*, *152*(7), 1092–1101.

Sales, A. J., et al. (2018). Antidepressant-like effect induced by Cannabidiol is dependent on brain serotonin levels. *Progress in Neuro-Psychopharmacology and Biological Psychiatry*, *86*, 255–261.

Sales, A. J., et al. (2019). Cannabidiol induces rapid and sustained antidepressant-like effects through increased BDNF signaling and synaptogenesis in the prefrontal cortex. *Molecular Neurobiology*, *56* (2), 1070–1081.

Santarelli, L., et al. (2003). Requirement of hippocampal neurogenesis for the behavioral effects of antidepressants. *Science*, *301*(5634), 805–809.

Sartim, A. G., Guimarães, F. S., & Joca, S. R. L. (2016). Antidepressant-like effect of cannabidiol injection into the ventral medial prefrontal cortex-possible involvement of 5-HT1A and CB1 receptors. *Behavioural Brain Research*, *303*, 218–227.

Sartim, A. G., et al. (2018). Hippocampal mammalian target of rapamycin is implicated in stress-coping behavior induced by cannabidiol in the forced swim test. *Journal of Psychopharmacology*, *32*(8), 922–931.

Schiavon, A. P., et al. (2016). Influence of single and repeated cannabidiol administration on emotional behavior and markers of cell proliferation and neurogenesis in non-stressed mice. *Progress in Neuro-Psychopharmacology & Biological Psychiatry*, *64*, 27–34.

Schoedel, K. A., et al. (2011). A randomized, double-blind, placebo-controlled, crossover study to evaluate the subjective abuse potential and cognitive effects of nabiximols oromucosal spray in subjects with a history of recreational cannabis use. *Human Psychopharmacology*, *26*(3), 224–236.

Shbiro, L., et al. (2019). Effects of cannabidiol in males and females in two different rat models of depression. *Physiology and Behavior*, *201*, 59–63.

Shoval, G., et al. (2016). Prohedonic effect of cannabidiol in a rat model of depression. *Neuropsychobiology*, *73*(2), 123–129.

Silote, G. P., et al. (2019). Emerging evidence for the antidepressant effect of cannabidiol and the underlying molecular mechanisms. *Journal of Chemical Neuroanatomy*, *98*, 104–116.

Small, E. (2015). Evolution and classification of Cannabis sativa (marijuana, hemp) in relation to human utilization. *The Botanical Review*, *81*(3), 189–294.

Surget, A., et al. (2011). Antidepressants recruit new neurons to improve stress response regulation. *Molecular Psychiatry*, *16*(12), 1177–1188.

The Indian Hemp Drugs Commission. (1894). In W. M. Young (Ed.), *Report of the Indian Hemp Drugs Commission, 1893–1894*. Simla: Government Press.

Thomas, A., et al. (2007). Cannabidiol displays unexpectedly high potency as an antagonist of CB1 and CB2receptor agonists in vitro. *British Journal of Pharmacology*, *150*(5), 613–623.

Turna, J., Patterson, B., & Van Ameringen, M. (2017). Is cannabis treatment for anxiety, mood, and related disorders ready for prime time? *Depression and Anxiety*, *34*(11), 1006–1017.

VanDolah, H. J., Bauer, B. A., & Mauck, K. F. (2019). Clinicians' guide to Cannabidiol and hempoils. *Mayo Clinic Proceedings*, *94*(9), 1840–1851.

Volkow, N. D., et al. (2014). Adverse health effects of marijuana use. *New England Journal of Medicine*, *370*(23), 2219–2227

White, C. M. (2019). A review of human studies assessing Cannabidiol's (CBD) therapeutic actions and potential. *Journal of Clinical Pharmacology*, *59*(7), 923–934.

Whiting, P. F., et al. (2015). Cannabinoids for medical use: A systematic review and meta-analysis. *JAMA*, *313*(24), 2456–2473.

WHO. (2017). *WHO investing in treatment for depression and anxiety leads to fourfold return*. WHO.

Wolf, S. A., et al. (2010). Cannabinoid receptor CB1 mediates baseline and activity-induced survival of new neurons in adult hippocampal neurogenesis. *Cell Communication and Signaling*, *8*(12), 1–14.

Xu, C., et al. (2019). Pharmacokinetics of oral and intravenous cannabidiol and its antidepressant-like effects in chronic mild stress mouse model. *Environmental Toxicology and Pharmacology*, *70*, 103202.

Zanelati, T. V., et al. (2010). Antidepressant-like effects of cannabidiol in mice: Possible involvement of 5-HT(1A) receptors. *British Journal of Pharmacology*, *159*(1), 122–128.

Zanos, P., & Gould, T. D. (2018). Mechanisms of ketamine action as an antidepressant. *Molecular Psychiatry*, *23*(4), 801–811.

Zanos, P., et al. (2018). Ketamine and ketamine metabolite pharmacology Insights into therapeutic mechanisms. *Pharmacological Reviews*, *70*(3), 621–660.

Zhang, Z., et al. (2015). Blockade of 2-arachidonoylglycerol hydrolysis produces antidepressant-like effects and enhances adult hippocampal neurogenesis and synaptic plasticity. *Hippocampus*, *25*(1), 16–26.

Zuardi, A. W. (2006). History of cannabis as a medicine: A review. *Brazilian Journal of Psychiatry*, *28*, 153–157.

Zuardi, A. W. (2008). Cannabidiol: From an inactive cannabinoid to a drug with wide spectrum of action. *Revista Brasileira de Psiquiatria*, *30*(3), 271–280.

Zuardi, A. W., et al. (1982). Action of cannabidiol on the anxiety and other effects produced by δ9-THC in normal subjects. *Psychopharmacology*, *76*(3), 245–250.

Zuardi, A. W., et al. (2017). Inverted U-shaped dose-response curve of the anxiolytic effect of cannabidiol during public speaking in real life. *Frontiers in Pharmacology*, *8*(259), 1–14.

Zygmunt, P. M., et al. (1999). Vanilloid receptors on sensory nerves mediate the vasodilator action of anandamide. *Nature*, *400*(6743), 452–457.

Chapter 46

# Transcutaneous auricular vagus nerve stimulation in the treatment of depression

Jian Kong[a], Georgia Wilson[a], and Peijing Rong[b]

[a]*Department of Psychiatry, Massachusetts General Hospital, Harvard Medical School, Charlestown, MA, United States;* [b]*Institute of Acupuncture and Moxibustion, China Academy of Chinese Medical Sciences, Beijing, China*

## List of abbreviations

**ABVN** auricular branch of the vagus nerve
**ACC** anterior cingulate cortex
**BDI** Beck depression inventory
**DMN** default mode network
**FC** functional connectivity
**HAMD** Hamilton depression rating scale
**LH** lateral hypothalamus
**LPFC** lateral prefrontal cortex
**MDD** major depressive disorder
**MH** medial hypothalamus
**NTS** nucleus tractus solitarii
**NAc** nucleus accumbens
**OFC** orbitofrontal cortex
**PAG** periaqueductal gray
**rsFC** resting-state functional connectivity
**taVNS** transcutaneous auricular vagus nerve stimulation

## Introduction

As a control center that integrates interoceptive information and responds with appropriate adaptive modulatory feedback (Yuan & Silberstein, 2016), the vagus nerve (VN) consists of a complex network that regulates mood, the neuro-endocrine-immune axis, and memory (Kong, Fang, Park, Li, & Rong, 2018; Yuan, Li, Sun, Arias-Carrion, & Machado, 2016). Due to its wide influence on multiple systems and important role in maintaining homeostasis, stimulating the vagus nerve to relieve clinical symptoms has drawn the attention of clinicians and investigators (Yuan et al., 2016).

Vagus nerve stimulation (VNS) is an FDA-approved intervention for treatment-resistant depression that can produce clinically significant antidepressant effects (Daban, Martinez-Aran, Cruz, & Vieta, 2008; Sackeim & Lisanby, 2001). However, the side effects, cost, and limited accessibility associated with implantable VNS limit its application. To overcome the barriers of applying invasive VNS (iVNS), noninvasive transcutaneous vagus nerve stimulation (tVNS) methods have been developed. Currently, there are two main ways to apply tVNS. One method is to superficially apply stimulation on the cervical nerve using a specially designed device such as GammaCore, while the other method is to apply stimulation on the ear via transcutaneous auricular vagus nerve stimulation (taVNS). In this chapter, we will focus on taVNS, which has been applied to treat depression in several studies.

The rationale of taVNS is based on anatomical studies suggesting the presence of afferent VN distribution at the ear (Peuker & Filler, 2002). Reciprocal interactions between the brain and peripheral nerves raise the possibility that propagation of electrical impulses may follow an inverse path from the peripheral nerves (Shiozawa et al., 2014). Thus, stimulation of VN fibers on the ear may produce an effect similar to classic VNS in reducing depression symptoms. This possibility has led to the development of taVNS, a noninvasive alternative to VNS that can be self-administered (Kong et al., 2018). In this chapter, we will first introduce findings of clinical studies on taVNS treatment effects on depression. Then, we will discuss the potential underlying mechanism, limitations, and future directions of taVNS.

## Clinical trials on taVNS treatment of depression

Hein et al. (2013) performed the first taVNS clinical trial on individuals with MDD using an add-on design (antidepressant therapy with real or sham taVNS). They found that compared to the sham group, there was no significant difference between the two treatment groups on the Hamilton

The Neuroscience of Depression. https://doi.org/10.1016/B978-0-12-817933-8.00031-1

Depression Rating Scale (HAMD) after a 2-week treatment (5 times per week). Nevertheless, they found that the real taVNS group showed significant improvement on the Beck Depression Inventory (BDI). A potential limitation of this study was its relatively small sample size (real taVNS = 18, sham taVNS = 19). In addition, this study used a sub-threshold stimulation intensity (i.e., below the threshold of perception), raising concern that the vagus nerve may not have been activated. Finally, treatment only lasted for 2 weeks.

In a subsequent nonrandomized clinical study with a relatively large sample size (160 MDD patients) (Rong et al., 2016), Rong and colleagues investigated the taVNS treatment effect by training patients to apply taVNS at home. Due to ethical concerns, they first recruited 91 patients and trained them to administer the treatment 2 times a day for 12 weeks. After demonstrating the effect of taVNS in this first cohort, they recruited 69 patients who received 4 weeks of sham taVNS followed by 8 weeks of real taVNS. The results showed that after the fourth week, the taVNS group had greater decreases in the 24-item HAMD score and higher rates of good responders (as determined by a 50% reduction in HAMD score) than the sham taVNS group. The clinical improvements continued until week 12. The main limitation of this study was its nonrandomized design.

By reanalyzing the data from the above study (Rong et al., 2016) using a repeated measurement analysis, we also explored how taVNS can modulate the specific symptoms of MDD. Results showed that 1 month of taVNS can significantly reduce multiple symptoms of MDD, including anxiety, retardation, sleep disturbance, and hopelessness, with medium effect sizes, as well as a downward trend in cognitive disturbance and diurnal variation (Kong et al., 2018).

In a more recent study (Trevizol et al., 2016), Trevizol and colleagues tested the effect of taVNS on the bilateral mastoid process in 12 patients with MDD using a single-arm design. All patients underwent 10 sessions of taVNS treatment over 2 weeks. Results showed that 17-item HAMD scores were significantly reduced after the 2-week treatment. All patients exhibited a clinical response (a reduction of HAMD scores of at least 50%), and the effect remained for 1 month after treatment. The main limitation of this study was its lack of a control group and a small sample size.

In summary, taVNS has shown promising potential for treating depression. It can significantly reduce multiple symptoms of MDD, implying that it may also hold the potential to treat other disorders with these symptoms. Nevertheless, early investigations suffer from important methodology limitations and none have used follow-up assessments.

## Potential side effects of taVNS

Literature has reported that taVNS is associated with adverse effects such as tinnitus or exacerbation of preexisting tinnitus, pain, paresthesia, or pruritus during or after stimulation (Rong et al., 2016; Straube, Ellrich, Eren, Blum, & Ruscheweyh, 2015). Since there are no direct nerve fibers connecting the auricular vagus nerve to the heart, both the left and right ear may be safe locations for taVNS. In support of this, Kreuzer et al. (2012) found that taVNS has no arrhythmic effects on cardiac function in tinnitus patients with no known preexisting cardiac pathology. Hein et al. (2013) also found that heart rate, blood pressure, and blood test values did not change over the 2-week treatment period in patients with MDD.

It is worth noting that the method of applying taVNS may be associated with side effects produced. Side effects may be more severe when electrical current flows across the whole brain during bilateral stimulation. For instance, in the previous study applying taVNS on the bilateral mastoid process (Trevizol et al., 2016), investigators found that of the 12 patients included in the study, 10 reported mild to moderate diurnal sleepiness after stimulation, 6 reported mild to moderate tension headaches with no need for medication, and 4 reported mild to moderate nausea.

Taken together, taVNS is a safe and well-tolerated therapeutic method for MDD, particularly when taVNS is applied on the conchae and electrical current does not flow across the whole brain.

## Potential mechanisms underlying taVNS treatment of depression

### Modulating the brain network associated with the pathophysiology of depression

Neural anatomy has shown that the auricular branch of the vagus nerve (ABVN) projects to the nucleus tractus solitarii (NTS), which is further connected with other brain regions such as the locus coeruleus, parabrachial nucleus, periaqueductal gray (PAG), hypothalamus, thalamus, nucleus accumbens (NAc), amygdala, hippocampus, anterior cingulate cortex (ACC), anterior insula, and lateral prefrontal cortex (LPFC). Neuroimaging studies (Dietrich et al., 2008; Fang et al., 2009; Fang et al., 2014; Frangos, Ellrich, & Komisaruk, 2015; Kraus et al., 2007; Kraus et al., 2013; Zhang et al., 2019) have found that, compared to a control condition, taVNS stimulation can activate or deactivate "classical" central vagal projections. Thus, taVNS may directly and indirectly modulate depression-related cortical-limbic-thalamic-striatal neural circuits, influencing the activity of these regions and therefore providing a

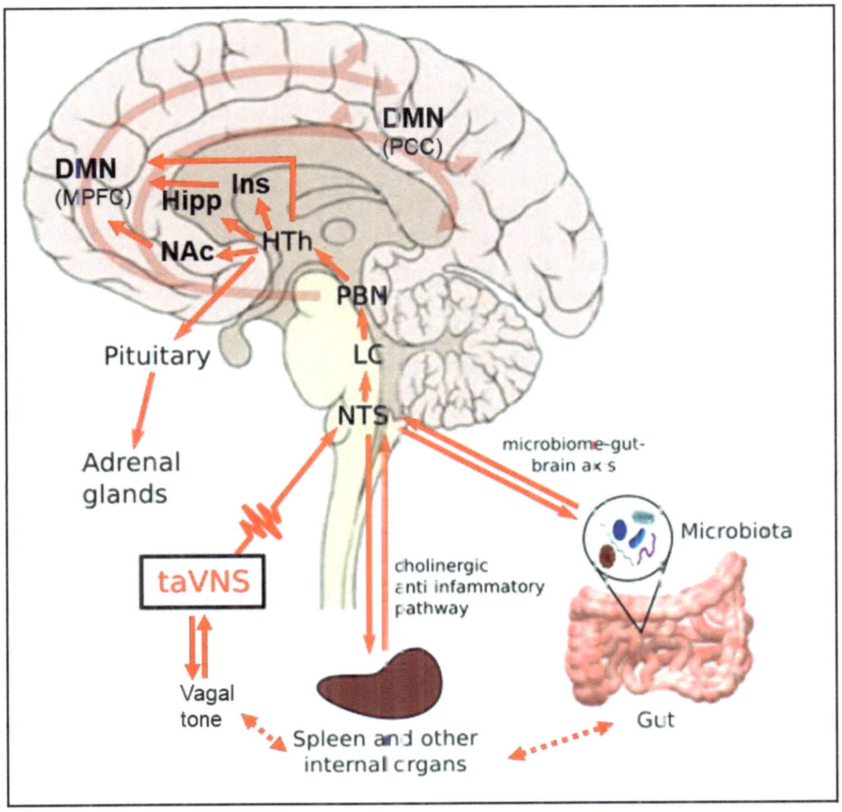


**FIG. 1** Hypothetical schematic illustration of the pathways/mechanism of taVNS treatment of depression. taVNS can modulate the brain network associated with the neuropathology of depression, inflammation response, microbiome-brain-gut axis, and vagal tone. Figure shows stimulation of the auricular branch of the vagus nerve (indicated in *red*), which projects to the nucleus tractus solitarii and continues to the locus coeruleus, parabrachial nucleus, other brainstem, basal ganglia, and cerebral cortex. *DMN*, default mode network (medial prefrontal cortex, posterior cingulate cortex); *Hipp*, hippocampus; *HTh*, hypothalamus; *Ins*, insula; *LC*, locus coeruleus; *NAc*, nucleus accumbens; *NTS*, nucleus tractus solitarii; *PBN*, parabrachial nucleus.

foundation for taVNS treatment of depression. Specifically, we believe that taVNS may work through multiple mechanisms (Fig. 1).

## *Modulating default mode network*

Accumulating evidence has demonstrated that major depressive disorder (MDD) is associated with structural and functional abnormalities in brain circuits involved in emotional processing, self-representation, reward, and external stimulus interactions (stress and distress) (Pizzagalli, 2011; Wang et al., 2012). These brain regions include the medial prefrontal cortex, posterior cingulate cortex/precuneus, and hippocampus. Interestingly, these regions also fall within the default mode network (DMN), a brain network believed to be involved in self-referential processing (Buckner, Andrews-Hanna, & Schacter, 2008). Additional studies (Wang, Hermens, Hickies, & Lagopoulos, 2012) have found DMN functional connectivity changes in MDD patients.

In a recent study, we investigated resting-state functional connectivity changes before and after 1 month of taVNS as compared to 1 month of sham taVNS. We found that the rsFC between the DMN and dorsal anterior cingulate cortex, anterior insula, and parahippocampus decreased, while FC between the DMN, precuneus, and orbital prefrontal cortex increased in patients with depression after 1 month of taVNS compared to 1 month of sham taVNS. These rsFC increases/decreases are associated with HAMD-24 score reduction (Fang et al., 2016). These results suggest that taVNS can significantly modulate the DMN in patients with depression.

## *Modulating reward/motivation network*

A core symptom of depression is anhedonia, the loss of interest in pleasurable activities and impairment in multiple dimensions of well-being (Bogdan, Nikolova, & Pizzagalli, 2013; Hwang et al., 2016). Thus, the reward system may play an important role in depression (Phillips et al., 2015). The nucleus accumbens (NAc) is a key region for reward, pleasure, and motivation (Floresco, 2015). The NAc also plays a key role in the integration of cognitive and affective information to augment the efficiency and vigor of motivated behaviors (Floresco, 2015). In a previous study, we found that during continuous 20 Hz taVNS, the FC between the nucleus accumbens and rACC was significantly increased in patients with depression (Wang et al., 2018). This result indicates that taVNS may also modulate the reward/motivation network to reduce depression symptoms.

### *Modulating hypothalamic-pituitary-adrenal axis*

The hypothalamus is known to mediate many neuroendocrine and neurovegetative functions (Herman, Ostrander, Mueller, & Figueiredo, 2005; Nestler et al., 2002). Through the hypothalamic-pituitary-adrenal (HPA) axis, the hypothalamus modulates the acute cortisol response to stress and variations in cortisol secretion over the circadian rhythm (Pariante & Lightman, 2008). Dysregulation or hyperactivity of the HPA axis is well characterized in major depression (Sudheimer et al., 2015). A recent brain imaging study showed that both lateral hypothalamus (LH) and medial hypothalamus (MH) circuitries showed significant connections to the striatum, thalamus, brainstem, orbitofrontal cortex (OFC), cingulate cortex, and temporal regions (Kullmann et al., 2014). We recently found that 6 min of 20 Hz taVNS significantly decreased functional connectivity between the medial hypothalamus and rACC in patients with depression. Furthermore, the strength of the MH—rACC connection during taVNS could predict the treatment effect after 4 weeks, suggesting that MH—rACC FC may be a biomarker to predict taVNS treatment response (Tu et al., 2018).

### *Modulating interoception*

Interoception is contemporarily defined as the sense of the internal state of the body. It encompasses the brain's process of integrating signals relayed from the body into specific subregions to form a representation of one's physiological state. It is believed that the insula plays an important role in interoception (Strigo & Craig, 2016), and disrupted interoception is a prominent feature of the diagnostic classification of psychiatric disorders (Khalsa & Lapidus, 2016). In a recent study (Fang et al., 2017), we found that taVNS produced significant fMRI signal increases in the anterior insula compared to sham stimulation in patients with MDD. The insula activation level during the first stimulation session in the taVNS group was significantly associated with clinical improvement after 4 weeks, as shown by the reduction in Hamilton Depression Rating Scale (HAMD) scores. This result is consistent with a previous study (McGrath et al., 2013) in which the authors found that metabolism at the anterior insula at baseline can predict treatment response.

## Modulating the inflammation system

Literature suggests that stress initiates cognitive, affective, and possibly biological processes that increase one's risk for depression, and neuroinflammatory sensitization may play an important role in this process (Slavich & Irwin, 2014).

Studies have suggested that the vagus nerve plays a crucial role in bidirectionally connecting the brain and immune system, thereby reducing exacerbated inflammation processes outside the central nervous system (CNS) (Pereira & Leite, 2016). Specifically, the vagus nerve may modulate the inflammation system by (a) activating the hypothalamic-pituitary-adrenal (HPA) axis and suppressing peripheral inflammation via glucocorticoids (Bellavance & Rivest, 2014), and (b) through the "inflammatory reflex" mechanism (Browning, Verheijden, & Boeckxstaens, 2016).

Studies on vagal tone, as measured by heart rate variability, also support the hypothesis that taVNS may relieve symptoms of depression through the antiinflammation mechanism. For instance, studies have identified an association between depression and reduced vagal activity, as indexed by high frequency heart rate variability (HF-HRV), which has been shown to inversely correlate with depression severity in adults (Koenig, Kemp, Beauchaine, Thayer, & Kaess, 2016), children, and adolescents (Koenig et al., 2016). Studies have also shown that reduced heart rate variability indices (low vagal tone) are independently associated with increased serum C-reactive protein and IL-6 levels and increased TNF and IL-6 production (Huston & Tracey, 2011; Jarczok, Koenig, Mauss, Fischer, & Thayer, 2014).

Taken together, these studies suggest that taVNS may at least partly work by modulating inflammation response.

## Other potential mechanisms

The vagus nerve can significantly modulate the gastrointestinal, immune, and endocrine systems (Yuan & Silberstein, 2016). Accumulating evidence has demonstrated that microbe interactions are crucial in maintaining homeostasis in humans. Studies have also shown that the microbiota is essential in modulating the stress response and stress-related conditions such as depression and anxiety (Forsythe, Bienenstock, & Kunze, 2014). Thus, taVNS may also regulate the functions of the above systems and achieve a treatment effect in depression by adjusting the microbiome-brain-gut axis (Forsythe et al., 2014).

VNS can also alter the transmission of neurotransmitters such as serotonin and norepinephrine, which can modulate hippocampal cell proliferation (Yuan et al., 2016). Based on the neurogenic theory of depression (Miller & Hen, 2015), depression is a consequence of impaired adult hippocampal neurogenesis, and correcting this impairment can be an effective treatment for the disorder. Thus, taVNS may also relieve depression symptoms by modulating hippocampal neurogenesis (Yuan et al., 2016).

## Challenges and future directions

### Locations

Neuroanatomy indicates that the inner tragus, inferoposterior wall of the ear canal, and cymba conchae all have VN distribution (Peuker & Filler, 2002). In a recent imaging study, investigators found that only stimulation of the cymba conchae produced a significantly stronger activation in both the NTS and LC compared to sham stimulation in healthy subjects (Yakunina, Kim, & Nam, 2017). We also found that stimulating at the conchae can change the activity of the NTS and LC in patients with migraine (Zhang et al., 2019).

In addition, VN distribution on the inner tragus is under debate (Burger & Verkuil, 2018) due to inconsistent descriptions of nerve distribution on the ear (Peuker & Filler, 2002). Previous studies have found that stimulation of the cymba conchae and cavity of concha can produce significant findings (Fang et al., 2014, 2016, 2017; Liu et al., 2016; Rong et al., 2016; Tu et al., 2018; Wang et al., 2018). Thus, these areas may be preferential locations for taVNS treatment of depression.

Nevertheless, taVNS at different locations of the ear with VN innervation may modulate different brain pathways and be associated with different modulation effects. Other locations should thus be considered if the conchae area cannot achieve significant results.

### Stimulation frequency

Previous studies have suggested that different stimulation frequencies could produce different brain changes and neurotransmitter releases (Han, 2003; Zhang et al., 2003). Investigators have used different frequencies [1.5 Hz (Hein et al., 2013), 20 Hz (Rong et al., 2016), and 120 Hz (Trevizol et al., 2016)] in previous taVNS depression studies. In a recent study of healthy subjects, investigators compared the effect of 2, 10, and 20 Hz on heart rate and found that 20 and 10 Hz significantly decreased heart rate (Badran et al., 2018). A clinical study (Bauer et al., 2016) of taVNS treatment for medication-resistant epilepsy observed a significant reduction in seizure frequency for patients receiving 25 Hz, compared to 1 Hz, stimulation frequency. However, in a study of patients with chronic migraine headaches (Straube et al., 2015), investigators found that both 1 and 25 Hz taVNS improved clinical outcomes, but 1 Hz taVNS produced the greatest improvement. These results suggest that disorders may respond differently to different taVNS frequencies. Comparative studies are needed to identify the optimal stimulation frequency.

### Dose effect

The dose effect of taVNS remains unclear. We thus do not know how long and how frequently we should apply taVNS to achieve optimal results. In the current study, the duration of stimulation ranges from 30-min 2 times per day (Rong et al., 2016) to 15-min 5 times per week (Hein et al., 2013). Further studies are needed to explore the optimal dose of taVNS for the treatment of depression.

### Future directions

(1) Confirmatory clinical study to investigate the efficacy of taVNS on MDD and underlying mechanisms of taVNS used to treat depression
(2) Comparative studies to elucidate the optimal parameters and doses of taVNS.
(3) Investigate the effect of taVNS on disorders comorbid with depression. Depression can be comorbid with disorders such as chronic pain (Dersh, Gatchel, Mayer, Polatin, & Temple, 2006), inflammatory bowel disease (Abautret-Daly, Dempsey, Parra-Blanco, Medina, & Harkin, 2018), and autism (Jin & Kong, 2016). Thus, taVNS may also provide a new "two-in-one" treatment option for patients with disorders comorbid with depression.

In summary, taVNS is a safe, noninvasive, cost-effective and easy-to-apply therapeutic method that has great potential for the treatment of MDD. Nevertheless, further research is needed to test the efficacy of taVNS and elucidate its underlying mechanism.

## Key facts of taVNS

- taVNS can significantly relieve symptoms in patients with depression.
- taVNS can modulate the resting-state functional connectivity of the DMN and extended amygdala network.
- Continuous taVNS can decrease functional connectivity between the medial hypothalamus and anterior cingulate cortex.
- Continuous taVNS can increase functional connectivity between the nucleus accumbens and anterior cingulate cortex.
- taVNS may modulate inflammation response.

## Summary points

- taVNS is a safe, noninvasive, cost-effective and easy-to-apply therapeutic method.
- taVNS has demonstrated great potential for the treatment of depression.

- taVNS may work through multiple mechanisms including modulating brain networks associated with depression and mood regulation, inflammation, hippocampal neurogenesis, and the microbiome-brain-gut axis.
- A confirmative study with a large sample size is needed to test the efficacy of taVNS.
- Comparative studies are needed to elucidate the optimal parameters and doses of taVNS.

## Mini-dictionary of terms

**Vagus nerve** The tenth cranial nerve (CN X); interfaces with the parasympathetic control of the heart, lungs, digestive tract, and other internal organs.

**Vagus nerve stimulation** A therapeutic treatment method of applying electrical stimulation to the vagus nerve.

**Transcutaneous auricular vagus nerve stimulation** A vagus nerve stimulation method that applies stimulation on an area of the ear with vagus nerve distribution.

**Default mode network** A series of brain areas that are active when a person is not focused on the outside world and the brain is at wakeful rest.

**Resting-state functional connectivity** A brain imaging data analysis method used to investigate the association (connectivity) between different brain areas during wakeful rest using correlation or other related analysis.

## References

Abautret-Daly, A., Dempsey, E., Parra-Blanco, A., Medina, C., & Harkin, A. (2018). Gut-brain actions underlying comorbid anxiety and depression associated with inflammatory bowel disease. *Acta Neuropsychiatrica*, *30*, 275–296. https://doi.org/10.1017/neu.2017.3.

Badran, B. W., Mithoefer, O. J., Summer, C. E., LaBate, N. T., Glusman, C. E., Badran, A. W., et al. (2018). Short trains of transcutaneous auricular vagus nerve stimulation (taVNS) have parameter-specific effects on heart rate. *Brain Stimulation*, *11*(4), 699–708. https://doi.org/10.1016/j.brs.2018.04.004.

Bauer, S., Baier, H., Baumgartner, C., Bohlmann, K., Fauser, S., Graf, W., et al. (2016). Transcutaneous vagus nerve stimulation (tVNS) for treatment of drug-resistant epilepsy: A randomized, double-blind clinical trial (cMPsE02). *Brain Stimulation*, *9*(3), 356–363. https://doi.org/10.1016/j.brs.2015.11.003.

Bellavance, M. A., & Rivest, S. (2014). The HPA—Immune axis and the immunomodulatory actions of glucocorticoids in the brain. *Frontiers in Immunology*, *5*, 136. https://doi.org/10.3389/fimmu.2014.00136.

Bogdan, R., Nikolova, Y. S., & Pizzagalli, D. A. (2013). Neurogenetics of depression: A focus on reward processing and stress sensitivity. *Neurobiology of Disease*, *52*, 12–23. https://doi.org/10.1016/j.nbd.2012.05.007.

Browning, K. N., Verheijden, S., & Boeckxstaens, G. E. (2016). The vagus nerve in appetite regulation, mood, and intestinal inflammation. *Gastroenterology*, . https://doi.org/10.1053/j.gastro.2016.10.046.

Buckner, R. L., Andrews-Hanna, J. R., & Schacter, D. L. (2008). The brain's default network: Anatomy, function, and relevance to disease. *Annals of the New York Academy of Sciences*, *1124*, 1–38.

Burger, A. M., & Verkuil, B. (2018). Transcutaneous nerve stimulation via the tragus: Are we really stimulating the vagus nerve? *Brain Stimulation*, *11*(4), 945–946. https://doi.org/10.1016/j.brs.2018.03.018.

Daban, C., Martinez-Aran, A., Cruz, N., & Vieta, E. (2008). Safety and efficacy of vagus nerve stimulation in treatment-resistant depression. A systematic review. *Journal of Affective Disorders*, *110*(1–2), 1–15.

Dersh, J., Gatchel, R. J., Mayer, T., Polatin, P., & Temple, O. R. (2006). Prevalence of psychiatric disorders in patients with chronic disabling occupational spinal disorders. *Spine (Phila Pa 1976)*, *31*(10), 1156–1162. https://doi.org/10.1097/01.brs.0000216441.83135.6f.

Dietrich, S., Smith, J., Scherzinger, C., Hofmann-Preiss, K., Freitag, T., Eisenkolb, A., et al. (2008). A novel transcutaneous vagus nerve stimulation leads to brainstem and cerebral activations measured by functional MRI. *Biomedizinische Technik. Biomedical Engineering*, *53*(3), 104–111. https://doi.org/10.1515/BMT.2008.022.

Fang, J., Egorova, N., Rong, P., Liu, J., Hong, Y., Fan, Y., et al. (2017). Early cortical biomarkers of longitudinal transcutaneous vagus nerve stimulation treatment success in depression. *Neuroimage. Clinical*, *14*, 105–111. https://doi.org/10.1016/j.nicl.2016.12.016.

Fang, J. L., Hong, Y., Fan, Y. Y., Liu, J., Ma, Y. Y., Xiu, C. H., et al. (2014). Brain response to transcutaneous electronical stimulation on auricular concha of the healthy subjects using fMRI. *Chinese Journal of Magnetic Resonance Imaging*, *5*(6), 416–422.

Fang, J., Jin, Z., Wang, Y., Li, K., Kong, J., Nixon, E. E., et al. (2009). The salient characteristics of the central effects of acupuncture needling: Limbic-paralimbic-neocortical network modulation. *Human Brain Mapping*, *30*(4), 1196–1206. 18571795.

Fang, J., Rong, P., Hong, Y., Fan, Y., Liu, J., Wang, H., et al. (2016). Transcutaneous vagus nerve stimulation modulates default mode network in major depressive disorder. *Biological Psychiatry*, *79*(4), 266–273. https://doi.org/10.1016/j.biopsych.2015.03.025.

Floresco, S. B. (2015). The nucleus accumbens: An interface between cognition, emotion, and action. *Annual Review of Psychology*, *66*, 25–52. https://doi.org/10.1146/annurev-psych-010213-115159.

Forsythe, P., Bienenstock, J., & Kunze, W. A. (2014). Vagal pathways for microbiome-brain-gut axis communication. *Advances in Experimental Medicine and Biology*, *817*, 115–133. https://doi.org/10.1007/978-1-4939-0897-4_5.

Frangos, E., Ellrich, J., & Komisaruk, B. R. (2015). Non-invasive access to the vagus nerve central projections via electrical stimulation of the external ear: fMRI evidence in humans. *Brain Stimulation*, *8*(3), 624–636. https://doi.org/10.1016/j.brs.2014.11.018.

Han, J. S. (2003). Acupuncture: Neuropeptide release produced by electrical stimulation of different frequencies. *Trends in Neurosciences*, *26*, 17–22.

Hein, E., Nowak, M., Kiess, O., Biermann, T., Bayerlein, K., Kornhuber, J., et al. (2013). Auricular transcutaneous electrical nerve stimulation in depressed patients: A randomized controlled pilot study. *Journal of Neural Transmission*, *120*(5), 821–827. https://doi.org/10.1007/s00702-012-0908-6.

Herman, J. P., Ostrander, M. M., Mueller, N. K., & Figueiredo, H. (2005). Limbic system mechanisms of stress regulation: Hypothalamo-pituitary-adrenocortical axis. *Progress in Neuro-Psychopharmacology & Biological Psychiatry*, *29*(8), 1201–1213. https://doi.org/10.1016/j.pnpbp.2005.08.006.

Huston, J. M., & Tracey, K. J. (2011). The pulse of inflammation: Heart rate variability, the cholinergic anti-inflammatory pathway and implications for therapy. *Journal of Internal Medicine*, *269*(1), 45–53. https://doi.org/10.1111/j.1365-2796.2010.02321.x.

Hwang, J. W., Xin, S. C., Ou, Y. M., Zhang, W. Y., Liang, Y. L., Chen, J., et al. (2016). Enhanced default mode network connectivity with ventral striatum in subthreshold depression individuals. *Journal of Psychiatric Research*, *76*, 111–120. https://doi.org/10.1016/j.jpsychires.2016.02.005.

Jarczok, M. N., Koenig, J., Mauss, D., Fischer, J. E., & Thayer, J. F. (2014). Lower heart rate variability predicts increased level of C-reactive protein 4 years later in healthy, nonsmoking adults. *Journal of Internal Medicine*, *276*(6), 667–671. https://doi.org/10.1111/joim.12295.

Jin, Y., & Kong, J. (2016). Transcutaneous vagus nerve stimulation: A promising method for treatment of autism spectrum disorders. *Frontiers in Neuroscience*, *10*, 609. https://doi.org/10.3389/fnins.2016.00609.

Khalsa, S. S., & Lapidus, R. C. (2016). Can interoception improve the pragmatic search for biomarkers in psychiatry? *Frontiers in Psychiatry*, *7*, 121. https://doi.org/10.3389/fpsyt.2016.00121.

Koenig, J., Kemp, A. H., Beauchaine, T. P., Thayer, J. F., & Kaess, M. (2016). Depression and resting state heart rate variability in children and adolescents—A systematic review and meta-analysis. *Clinical Psychology Review*, *46*, 136–150. https://doi.org/10.1016/j.cpr.2016.04.013.

Kong, J., Fang, J., Park, J., Li, S., & Rong, P. (2018). Treating depression with transcutaneous auricular vagus nerve stimulation: State of the art and future perspectives. *Frontiers in Psychiatry*, *9*, 20. https://doi.org/10.3389/fpsyt.2018.00020.

Kraus, T., Hosl, K., Kiess, O., Schanze, A., Kornhuber, J., & Forster, C. (2007). BOLD fMRI deactivation of limbic and temporal brain structures and mood enhancing effect by transcutaneous vagus nerve stimulation. *Journal of Neural Transmission*, *114*(11), 1485–1493.

Kraus, T., Kiess, O., Hosl, K., Terekhin, P., Kornhuber, J., & Forster, C. (2013). CNS BOLD fMRI effects of sham-controlled transcutaneous electrical nerve stimulation in the left outer auditory canal—A pilot study. *Brain Stimulation*, *6*(5), 798–804. https://doi.org/10.1016/j.brs.2013.01.011.

Kreuzer, P. M., Landgrebe, M., Husser, O., Resch, M., Schecklmann, M., Geisreiter, F., et al. (2012). Transcutaneous vagus nerve stimulation: Retrospective assessment of cardiac safety in a pilot study. *Frontiers in Psychiatry*, *3*, 70. https://doi.org/10.3389/fpsyt.2012.00070.

Kullmann, S., Heni, M., Linder, K., Zipfel, S., Haring, H. U., Veit, R., et al. (2014). Resting-state functional connectivity of the human hypothalamus. *Human Brain Mapping*, *35*(12), 6088–6096. https://doi.org/10.1002/hbm.22607.

Liu, J., Fang, J., Wang, Z., Rong, P., Hong, Y., Fan, Y., et al. (2016). Transcutaneous vagus nerve stimulation modulates amygdala functional connectivity in patients with depression. *Journal of Affective Disorders*, *205*, 319–326. https://doi.org/10.1016/j.jad.2016.08.003.

McGrath, C. L., Kelley, M. E., Holtzheimer, P. E., Dunlop, B. W., Craighead, W. E., Franco, A. R., et al. (2013). Toward a neuroimaging treatment selection biomarker for major depressive disorder. *JAMA Psychiatry*, *70*(8), 821–829. https://doi.org/10.1001/jamapsychiatry.2013.143.

Miller, B. R., & Hen, R. (2015). The current state of the neurogenic theory of depression and anxiety. *Current Opinion in Neurobiology*, *30*, 51–58. https://doi.org/10.1016/j.conb.2014.08.012.

Nestler, E. J., Barrot, M., DiLeone, R. J., Eisch, A. J., Gold, S. J., & Monteggia, L. M. (2002). Neurobiology of depression. *Neuron*, *34*(1), 13–25.

Pariante, C. M., & Lightman, S. L. (2008). The HPA axis in major depression: Classical theories and new developments. *Trends in Neurosciences*, *31*(9), 464–468.

Pereira, M. R., & Leite, P. E. (2016). The involvement of parasympathetic and sympathetic nerve in the inflammatory reflex. *Journal of Cellular Physiology*, *231*(9), 1862–1869. https://doi.org/10.1002/jcp.25307.

Peuker, E. T., & Filler, T. J. (2002). The nerve supply of the human auricle. *Clinical Anatomy*, *15*(1), 35–37.

Phillips, M. L., Chase, H. W., Sheline, Y. I., Etkin, A., Almeida, J. R., Deckersbach, T., et al. (2015). Identifying predictors, moderators, and mediators of antidepressant response in major depressive disorder: Neuroimaging approaches. *The American Journal of Psychiatry*, *172*(2), 124–138. https://doi.org/10.1176/appi.ajp.2014.14010076.

Pizzagalli, D. A. (2011). Frontocingulate dysfunction in depression: Toward biomarkers of treatment response. *Neuropsychopharmacology*, *36*(1), 183–206. https://doi.org/10.1038/npp.2010.166.

Rong, P., Liu, J., Wang, L., Liu, R., Fang, J., Zhao, J., et al. (2016). Effect of transcutaneous auricular vagus nerve stimulation on major depressive disorder: A nonrandomized controlled pilot study. *Journal of Affective Disorders*, *195*, 172–179. https://doi.org/10.1016/j.jad.2016.02.031.

Sackeim, H. A., & Lisanby, S. H. (2001). Physical treatments in psychiatry. In M. M. Weissman (Ed.), *Treatment of depression: Bridging the 21st century* (pp. 151–172). Washington, DC: American Psychiatric Press.

Shiozawa, P., Silva, M. E., Carvalho, T. C., Cordeiro, Q., Brunoni, A. R., & Fregni, F. (2014). Transcutaneous vagus and trigeminal nerve stimulation for neuropsychiatric disorders: A systematic review. *Arquivos de Neuro-Psiquiatria*, *72*(7), 542–547.

Slavich, G. M., & Irwin, M. R. (2014). From stress to inflammation and major depressive disorder: A social signal transduction theory of depression. *Psychological Bulletin*, *140*(3), 774–815. https://doi.org/10.1037/a0035302

Straube, A., Ellrich, J., Eren, O., Blum, B., & Ruscheweyh, R. (2015). Treatment of chronic migraine with transcutaneous stimulation of the auricular branch of the vagal nerve (auricular t-VNS): A randomized, monocentric clinical trial. *The Journal of Headache and Pain*, *16*, 543. https://doi.org/10.1186/s10194-015-0543-3.

Strigo, I. A., & Craig, A. D. (2016). Interoception, homeostatic emotions and sympathovagal balance. *Philosophical Transactions of the Royal Society of London. Series B, Biological Sciences*, *371*(1708), 20160010. https://doi.org/10.1098/rstb.2016.0010.

Sudheimer, K., Keller, J., Gomez, R., Tennakoon, L., Reiss, A., Garrett, A., et al. (2015). Decreased hypothalamic functional connectivity with subgenual cortex in psychotic major depression. *Neuropsychopharmacology*, *40*(4), 849–860. https://doi.org/10.1038/npp.2014.259.

Trevizol, A. P., Shiozawa, P., Taiar, I., Soares, A., Gomes, J. S., Barros, M. D., et al. (2016). Transcutaneous vagus nerve stimulation (taVNS) for major depressive disorder: An open label proof-of-concept trial. *Brain Stimulation*, *9*(3), 453–454. https://doi.org/10.1016/j.brs.2016.02.001.

Tu, Y., Fang, J., Cao, J., Wang, Z., Park, J., Jorgenson, K., et al. (2018). A distinct biomarker of continuous transcutaneous vagus nerve stimulation treatment in major depressive disorder. *Brain Stimulation*, *11*(3), 501–508. https://doi.org/10.1016/j.brs.2018.01.006.

Wang, Z., Fang, J., Liu, J., Rong, P., Jorgenson, K., Park, J., et al. (2018). Frequency-dependent functional connectivity of the nucleus accumbens during continuous transcutaneous vagus nerve stimulation

in major depressive disorder. *Journal of Psychiatric Research*, *102*, 123–131. https://doi.org/10.1016/j.jpsychires.2017.12.018.
Wang, L., Hermens, D. F., Hickies, I. B., & Lagopoulos, J. (2012). A systematic review of resting-state functional-MRI studies in major depression. *Journal of Affective Disorders*, *142*(1–3), 6–12. https://doi.org/10.1016/j.jad.2012.04.013.
Wang, Z., Liu, J., Zhong, N., Qin, Y., Zhou, H., & Li, K. (2012). Changes in the brain intrinsic organization in both on-task state and post-task resting state. *NeuroImage*, *62*(1), 394–407.
Yakunina, N., Kim, S. S., & Nam, E. C. (2017). Optimization of transcutaneous vagus nerve stimulation using functional MRI. *Neuromodulation*, *20*(3), 290–300. https://doi.org/10.1111/ner.12541.
Yuan, T. F., Li, A., Sun, X., Arias-Carrion, O., & Machado, S. (2016). Vagus nerve stimulation in treating depression: A tale of two stories. *Current Molecular Medicine*, *16*(1), 33–39.
Yuan, H., & Silberstein, S. D. (2016). Vagus nerve and vagus nerve stimulation, a comprehensive review: Part I. *Headache*, *56*(1), 71–78. https://doi.org/10.1111/head.12647.
Zhang, W. T., Jin, Z., Cui, G. H., Zhang, K. L., Zhang, L., Zeng, Y. W., et al. (2003). Relations between brain network activation and analgesic effect induced by low vs. high frequency electrical acupoint stimulation in different subjects: A functional magnetic resonance imaging study. *Brain Research*, *982*, 168–178.
Zhang, Y., Liu, J., Li, H., Yan, Z., Liu, X., Cao, J., et al. (2019). Transcutaneous auricular vagus nerve stimulation at 1Hz modulates locus coeruleus activity and resting state functional connectivity in patients with migraine: An fMRI study. *Neuroimage. Clinical*, *24*, 101971. https://doi.org/10.1016/j.nicl.2019.101971.

Chapter 47

# Exercise for depression as a primary and comorbid with obesity disorder: A narrative

Ioannis D. Morres, Antonis Hatzigeorgiadis, and Yannis Theodorakis
*Department of Physical Education and Sport Science, University of Thessaly, Trikala, Greece*

## List of abbreviations

| | |
|---|---|
| **AE** | aerobic exercise |
| **BDI** | beck depression inventory |
| **BMI** | body mass index |
| **ES** | effect-size |
| **GWBS** | general well-being scale |
| **MDD** | major depressive disorder |
| **MHR** | maximum heart rate |
| **MMPI** | Minnesota Multiphasic Personality Inventory |
| **RCTs** | randomized controlled trials |

## Introduction

Physical activity is widely recommended as a treatment modality for various physical or mental health disorders. Although the term physical activity and physical exercise are often used interchangeably, there are distinct differences. Physical activity is defined as "any bodily movement produced by skeletal muscles that result in energy expenditure" (Caspersen, Powell, & Christenson, 1985; pp. 126). It is naturally embedded in daily life via various subsets (e.g., long hikes, transport to work, household activities, etc.), settings (home, work, leisure, etc.), or time-frames (any day or time of the day) and linked to lower depression in outpatients with MDD (Morres, Hatzigeorgiadis, Krommidas, et al., 2019).

Physical exercise is a physical activity subset, but it is "structured, planned, repetitive, and purposive toward fitness gains" (Caspersen et al., 1985; pp.128). Physical exercise is the most widely recommended PA subset for depression treatment. Guidelines for mild/moderate depression treatment recommend supervised, group-based, and moderate-intensity physical exercise of three times per week for a period of 10–14 weeks (National Institute for Health and Care Excellence [NICE], 2009). MDD, also referred to as clinical depression, is a disabling and life-threatening mental illness and represents the most common type of depression. The prevalence of MDD is increasing at an alarming rate (affects 7% of the US population) and is threefold higher in adults aged 18–29 years compared to adults +60 years.

## Clinical evidence for exercise for MDD as a primary disorder

Clinical RCTs are considered the "golden standard" for clinical evidence, in particular for evidence-based conclusions on the causal relationship between treatment effect and disease improvement. These RCTs are implemented in well-organized settings by specialized staff and are structured upon certain methodological standards ensuring increased internal validity and thus, low risk of bias. For example, clinical RCTs employ narrow/specific criteria for subject participation and prioritize their goal toward symptom disease improvement. Clinical RCTs, in return, are the input of systematic reviews evaluating treatment effects via group-based statistical and practical significance analyses including meta-analytic techniques. These reviews, i.e., meta-analyses, stand at the top of evidence based-pyramid representing the most vital recourses for extracting firm causal conclusions.

This narrative will focus on meta-analyses that reviewed RCTs examining the antidepressant efficacy (ES) of exercise excluding mindful-activities (yoga, qigong, tai chi, etc.) vs controls in adults aged 18–65 years with a referral or a clinical diagnosis of MDD as a primary disorder. High risk of bias will be discussed as it is linked to overestimation of treatment effects. Readers may interpret meta-analytic findings in Tables 1 and 2 via a twofold consideration. First, the widely used Cohen's d criterion reports that ESs of 0.20, 0.50, and 0.80 indicate a small, moderate, and large treatment effect, respectively

The Neuroscience of Depression. https://doi.org/10.1016/B978-0-12-817933-8.00028-1

**TABLE 1** Meta-analytic findings for exercise and major depression in randomized controlled trials (RCTs).

| Meta-analysis | Control condition | ES (CI 95%; *N* = arms/participants) |
|---|---|---|
| Craft and Landers (1998) | Depression treatments[a] | −0.75 (−1.19, −0.98; 23 = 23/2.110)[b] |
| Lawlor and Hopker (2001) | Depression treatments[a] | −1.10 (−1.50, −0.60; *N* = 10/432) |
| Mead, Morley, Campbell, and Greig (2009) | Depression treatments[a] | −0.82 (−1.12, −0.51; *N* = 23/907) |
| Rethorst et al. (2009) | Depression treatments[a] | −1.03 (−0.85, −1.21; *N* = 17/842)[b] |
| Krogh et al. (2011) | Depression treatments[a] | −0.40 (−0.66, −0.14; *N* = 13/736)[b] |
| Rimer, Dwan, Lawlor, and Greig (2012) | Depression treatments[a] | −0.67 (−0.90, −0.43; *N* = 28/1.101) |
| Robertson et al. (2012) | Depression treatments[a] | −0.86 (−1.12, −0.61; *N* = 8/295) |
| Cooney, Dwan, Greig, and Lawlor (2013) | Depression treatments[a] | −0.57 (−0.81, −0.32; N = 23/967)[b] |
| Danielsson, Noras, Waern, and Carlsson (2013) | Depression treatments[a] | −0.44 (−0.79, −0.09; *N* = 4/209) |
| Silveira et al. (2013) | Depression treatments[a] | −0.61 (−0.88, −0.33; *N* = 18/1.759)[b] |
| Schuch et al. (2016) | Depression treatments[a] | −1.14 (0.46, 1.81; *N* = 9/355)[b] |
| Kvam et al. (2016) | Depression treatments[a] | −0.68 (−0.92, −0.44; *N* = 23/977)[b] |
| Krogh et al. (2017) | Depression treatments[a] | −0.75 (−0.98, −0.52; *N* = 31/1.968) |
| Nebiker et al. (2018) | Depression treatments[a] | −0.93 (−1.17, −0.70; *N* = 29/1.421) |
| Morres, Hatzigeorgiadis, Stathi, et al. (2019) | Depression treatments[c] | −0.79 (−1.01, −0.57; N = 13/455)[b] |

[a] *Conventional treatment conditions for depression (e.g., psychotherapy, medication, waiting list, etc.) and/or exercise conditions for depression.*
[b] *defined by studies as clinically depressed.*
[c] *Conventional treatment conditions for depression (e.g., psychotherapy, medication, waiting list, etc.) excluding exercise conditions for depression.*
Table 1 shows an overview of the antidepressant effect of exercise (unpublished).
CI: confidence interval; ES: effect-size; N: number.

(Cohen, 1992). Second, comparisons involving <5 arms may record imprecise results (Borenstein, Hedges, Higgins, & Rothstein, 2009).

## Meta-analytic studies for exercise and MDD

The majority of meta-analyses have provided evidence for the clinical efficacy of exercise on depression among MDD adults (Tables 1 and 2). In an early meta-analysis,Craft and Landers (1998) exercise was found to beneficial in MDD as subgroup analyses showed a statistically significant large antidepressant effect of exercise. Ever since, various meta-analyses have also found comparably optimistic findings for exercise. Lawlor and Hopker (2001) reported a significantly large antidepressant ES for exercise. Analysis of the effects of exercise on depression among the low risk of bias trials was not computed. In Rethorst, Wipfli, and Landers (2009) the effect of exercise on depression was significantly large and comparably large in depressed samples vs nonclinically depressed samples. Moreover, the large clinical effect was significantly higher in trials satisfying rather than the dissatisfying risk of bias criteria including intention-to-treat analysis or concealed allocation. In Schuch et al. (2016), subgroup analysis for samples diagnosed with MDD recorded significantly large effects for exercise on depression, which remained large but marginally nonsignificant in low risk of bias RCTs. The meta-analytic studies of Nebiker et al. (2018) and Robertson, Robertson, Jepson, and Maxwell (2012) recorded also significantly large antidepressant effects for exercise, which remained moderate to large in low risk of bias trials. In the Kvam, Kleppe, Nordhus, and Hovland (2016) meta-analysis, the exercise showed a large antidepressant ES. This ES remained large across RCTs satisfying the risk of bias criteria of intention-to-treat or blinding outcomes, but it was reduced to small and nonsignificant in RCTs satisfying the above two criteria + concealed allocation. Danielsson et al. (2013) performed a subgroup analysis for exercise vs treatment as usual and recorded a significantly moderate effect of exercise on depression.

In the Krogh, Nordentoft, Sterne, and Lawlor (2011) meta-analysis, the exercise demonstrated a statistically significant moderate ES on depression in clinically diagnosed MDD samples. This effect was reduced to small/nonsignificant in low risk of bias trials. Similarly, the Krogh, Hjorthøj, Speyer, Gluud, and Nordentoft (2017)

**TABLE 2** **Meta-analytic findings for exercise and depression in low risk of bias randomized controlled trials (RCTs).**

| Meta-analysis | ES (CI 95%; $N$ = arms) |
|---|---|
| Craft and Landers (1998) | NR |
| Lawlor and Hopker (2001) | NR |
| Mead et al. (2009) | −0.77 (−1.12, −0.42; $N$=8) concealed allocation<br>−0.63 (−0.1,16–0.10; $N$=9) intention to treat<br>−0.39 (−0.75, −0.03; $N$=7) blind outcome<br>−0.42 (−0.88, 0.03; $N$=3) [a] |
| Rethorst et al. (2009) | −1.72 (−1.27, −2.17; $N$=3) intention to treat<br>−1.36 (−1.03, −1.69; $N$=5) concealed allocation |
| Krogh et al. (2011) | −0.19 (−0.70, 0.31; $N$=3) [a] |
| Rimer et al. (2012) | −0.64 (−0.91, −0.36; $N$=11) concealed allocation<br>−0.69 (−1.09–0.28; $N$=10) intention to treat<br>−0.40 (−0.66, −0.13; N=9) blind outcome<br>−0.31 (−0.63, 0.01; $N$=4) [a] |
| Robertson et al. (2012) | −0.69 (−0.99, −0.39 $N$=4) [a] and statistical power analysis |
| Cooney et al. (2013) | −0.49 (−0.75, −0.24 $N$=14) concealed allocation<br>−0.61 (−1.00, −0.22; $N$=11) intention to treat<br>−0.36 (−0.60, −0.12; $N$=12) blind outcome<br>−0.18 (−0.47, 0.11; $N$=6) [a] |
| Danielsson, Noras, Waern, & Carlsson, 2013 | NR |
| Silveira et al. (2013) | NR |
| Schuch et al. (2016) | 1.13 (−0.14, 2.40; N=3)[b] |
| Kvam et al. (2016) | −0.56 (−0.87, −0.25 N=12) intention to treat<br>−0.40 (−0.70, −0.11; N=10) blind outcome<br>−0.26 (−0.61, 0.08; N=6)[a] |
| Krogh et al. (2017) | −0.11 (−0.41, 0.18 N=4) Cochrane |
| Nebiker et al. (2018) | −0.83 (−1.13, −0.54 $N$=22) PEDro |
| Morres, Hatzigeorgiadis, Stathi, et al. (2019) | −0.70 (−0.94, −0.45 N=7) bPEDro |

[a]*Intention to treat, concealed allocation and blind outcome.*
[b]*Studies as clinically depressed.*
Table 2 shows an overview of the antidepressant effect of exercise in low risk of bias trials (unpublished).
CI: confidence interval; ES: effect-size; N: number; NR: not reported.
Cochrane: Handbook Cochrane criteria for Systematic Reviews of Interventions; PEDro: Physiotherapy Evidence-Based Scale.

meta-analysis recorded a significantly large effect for exercise on depression, but this effect was reduced to small/nonsignificant when the four low risk of bias trials were synthesized. This is unsurprising, as two of the four trials compared aerobic exercise (AE) vs other types of exercise, which all proved equally effective. Indeed, the first RCT (Krogh, Videbech, Thomsen, Gluud, & Nordentoft, 2012) compared AE to stretching and within-arm analyses showed large antidepressant ESs (AE: ES = −1.37 CI 95% −1.78 to −0.96; stretching: ES = −1.51 CI 95% −1.92 to −1.10). Thus, between-arm comparisons were apparently unable to record differences. The second RCT (Krogh, Saltin, Gluud, & Nordentoft, 2009) compared AE vs strength vs low intensity (relaxation) exercise. Again, within-arm analyses showed large antidepressant effects, indicating that between-arm comparisons can record no differences (AE: ES = −1.57 CI 95% −2.02 to −1.13; strength; ES = −1.60 CI 95% −2.05 to −1.15; low intensity: ES = −1.30 CI 95% −1.74 to −0.86). Hence, the reduced ES of exercise in the low risk of bias analysis of Krogh et al. (2017) is possibly due to the comparison of exercise vs exercise arms, which all proved to be antidepressant and thus, unable to record between-arm differences. Comparing the antidepressant effects of exercise vs other exercise arms reveals no between-arm differences (Danielsson et al., 2013;

Schuch, Morres, Ekkekakis, Rosenbaum, & Stubbs, 2017) and misleads conclusions on the reductive effects of exercise interventions on depression (Schuch et al., 2017).

## Clinical attributes of trials reviewed by meta-analyses

Despite the positive evidence by these meta-analyses, clinical attributes warrant consideration. In particular, the above reviews included a number of exercise trials with samples comprising community volunteers with a psychometric evaluation of depression. However, community volunteers may have well-established expectations and determination toward change. Blumenthal and Ong (2009) report that community volunteers to exercise trials are motivated toward exercise. Instead, depressed patients have shown persistent depression and psychosocial impairment that led to a health service, which, in turn, causes frustration and disappointment, as it reveals the disease complexity/severity and the need for systematic care. Also, a number of exercise trials in these reviews included mixed adult (18–65 years) and elderly adult (+65 years) depressed samples. However, older adults show distinct differences in depression, and higher depression relief via exercise (Silveira et al., 2013) and are more likely to complete exercise on prescription programs than adult peers ( James et al., 2008). In addition, a number of exercise trials in the above meta-analyses included comparisons of exercise vs other exercise arms. As above, such comparisons may mask the true antidepressant effect of exercise interventions because within-arm improvements reveal no between-arm differences. Also, some of the above meta-analyses included a number of non-randomized CTs or RCTs with patients with depression not as a primary disorder but as a result of a medical condition (e.g., postnatal depression). Finally, none of the above meta-analyses has exclusively reviewed the effects of AE on depression, although AE is the most frequently preferred type of exercise in mental health services.

## Meta-analysis for AE in adult MDD patients in mental health services

In light of the clinical attributes seen in meta-analytic studies for exercise and MDD, Morres et al. (2019) explored the antidepressant ES of supervised AE vs non-exercise arms in mental health services for patients 18–65 years with a referral or clinical diagnosis of MDD as a primary disorder. In detail, Morres et al. reviewed 11 RCTs involving 455 MDD patients and reported a statistically significant large antidepressant effect ($ES = -0.79$ CI $-1.01$ to $-0.57$) for supervised AE vs psychological treatments, treatment as usual or psychotropic agents. Heterogeneity was nonsignificant and low ($I^2 = 21\%$) and dropouts were $<15\%$. The average AE was 45 min of moderate intensity, three times per week, for 9.2 weeks. Subgroup analyses revealed comparably large effects for AE delivered in equipment-free or equipment-based formats, in groups or individually, in an outdoor, indoor, hospital or nonhospital settings and in out- or in-patients with mild–moderate or moderate–severe depression. Based on sensitivity analyses, the large effect of AE on depression did not change in RCTs with low risk of bias (comparisons included $>5$ arms) or with short-term or preference-based AE interventions. Notwithstanding the limited number of RCTs reviewed, findings indicate that moderate intensity AE as an add-on intervention in mental health services showed higher antidepressant effects vs conventional depression treatments. The employment of AE as an add-on intervention in mental health services has been previously suggested (Morres, Stathi, Martinsen, & Sørensen, 2014).

## Pragmatic evidence for exercise and depression

According to meta-analytic clinical evidence, exercise shows antidepressant efficacy with relatively low attrition among MDD patients in clinical settings. The established antidepressant efficacy of exercise, however, needs to be replicated by pragmatic RCTs in routine practice to provide "real life" evidence. Pragmatic RCTs measure treatment effectiveness which is defined as the benefit the treatment produces in routine practice (Roland & Torgerson, 1998). These trials rely on externally valid criteria including recruitment via wide inclusion criteria and via patient's presentation, as opposed to diagnosis. Additional criteria include functional health outcomes (e.g., quality of life) (Hotopf, 2002), acceptance of non-blinded patients/therapists (Ford & Norrie, 2016), and comparison of usual to new treatments while varying the intervention to patients' key problems (MacPherson, 2004). Also, in pragmatic trials, collateral interventions supporting compliance are not a methodological drawback. Collateral interventions integrated into a broader "complex" intervention is a typical phenomenon in routine practice within a stepped collaborative care model for treatment disease (Hotopf, 2002).

Collectively, pragmatic trials are designed to provide routine practice with direct real-life evidence attributable to increased external validity. Field studies from pragmatic settings point that community-based exercise on prescription (or referral) schemes for depressed outpatients are linked to high attrition; depressed outpatients show up to 60% attrition due to various psychosocial factors including lack of motivation, time or recourses (Hanson, Allin, Ellis, & Dodd-Reynolds, 2013; James et al., 2009). However, depressed patients appear to be motivated in the initial stage from the referral endorsement through to the first appointment of the exercise on the prescription scheme (Crone, Johnston, Gidlow, Henley, & James, 2008; Harrison, McNair, & Dugdill, 2005). Hence, prescribed intensity itself is also correlating to high attrition (Morgan et al., 2016).

## Ideographic vs nomothetic exercise

Prescribed intensity exercise represents the nomothetic exercise model (Morgan, 2001). In this model, training is exclusively based on fixed intensities and thus, possibly untailored to the challenging profile of depression. Indeed, depressed people on prescribed intensity exercise may often show disturbed exertion; they experience disturbed processing of information to the prescribed intensity with negative cognitive, behavioral, and affective responses (Borg, 1998; Morgan, 1973, 1994). Thus, exercise has to consider patient's personal grounds, especially since guidelines highlight individual variability in exercise. Particularly, physical activity guidelines for depression treatment (NICE, 2009) or for promotion in primary care (NICE, 2013) in the United Kingdom highlight the satisfaction of personal needs and preferences. Similarly, the Canadian Network for Mood and Anxiety Treatments (CANMAT) recommends exercise for depression prioritizes self-directed over practitioner-directed approaches, as the satisfaction of individual variability components, such as personal preferences, may maximize outcome improvement (Ravindran et al., 2016).

Given the link of the nomothetic model of prescribed intensity exercise to disturbed exertion and untailored individual variability, alternative exercise models are essential in order to reduce depression and attrition in pragmatic settings. The ideographic exercise model of preferred intensity exercise is a promising way out; it excludes fixed intensities by promoting self-selected intensities or preferred levels of exertion within a flexible context to satisfy vital individual variability components (e.g., needs and preferences) (Morgan, 2001). Studies have favored preferred over prescribed intensity exercise with respect to affective responses (e.g., Oliveira, Deslandes, & Santos, 2015; Parfitt, Rose, & Burgess, 2006; Rose & Parfitt, 2007).

## Pragmatic trial ideographic vs nomothetic exercise for depression

To our knowledge, Callaghan, Khalil, Morres, and Carter (2011) is the only currently available pragmatic RCT of community-based exercise for depressed outpatients recruited via health services. Of the initially eligible sample, 65% gave a consent form and was randomly assigned to preferred or prescribed intensity. Participants were women. The preferred intensity group exercised via preferred physical exertion and the prescribed intensity group via fixed intensities based on guidelines suggesting moderate intensity (60%–80% of the age predicted MHR = 220-age). All women exercised three times per week over 4 weeks on treadmill equipment, and in groups of up to five in public gyms. They also received social support for exercise by a psychologist and were supervised by an exercise therapist, who recorded MHR and perceived effort during training. Mean-based statistical analysis of postintervention outcomes revealed larger depression and quality of life improvement, and comparable MHR but lower perceived effort at peak exercise for preferred vs prescribed intensity groups. Both groups showed $<15\%$ dropouts. In the qualitative part of this trial, patients reported the benefits of preferred intensity exercise (Khalil, Callaghan, Carter, & Morres, 2012). Although mean-based statistical/practical significance analyses are vital, individual and clinical significant analysis is also needed to help toward personalized treatment given the patients' diverse needs/preferences. The individual and clinical significant analysis demonstrates if a patient shows treatment change and, subsequently, if this change has moved from a dysfunctional to a normative range.

## Individual clinical significant analysis

The Jacobson and Truax (JT) (1991) method of individual and clinical significant analysis demonstrates if greater clinically significant individual treatment change is observed in the intervention vs the comparator postintervention. The JT employs the pre- and postintervention variances and the test–retest reliability coefficient from the comparator to explore individual treatment change differences of the intervention group in excess of what would be projected under the comparator. The JT also explores if a postintervention score (treatment change) is likely to fall within the desired or normal (general) population range. This method places the cut-off ($C_{cutoff}$) midway between the dysfunctional and general population distributions.

Based on the JT method, Morres, Hinton-Bayre, Motakis, Carter, and Callaghan (2019) run a secondary analysis of the RCT of Callaghan et al. (2011). In detail, Morres, Hinton-Bayre, Motakis, et al. (2019) investigated if the higher antidepressant effects of the preferred intensity group ($n = 19$) vs the prescribed intensity group ($n = 19$) seen in Callaghan et al. corresponded also to individual and clinically significant depression relief. At postintervention, results showed that preferred intensity exercise led 32% of the patients ($n = 6$) to significantly greater depression relief (four patients were not prescribed antidepressants). This relief indicated recovery from depression because the statistically significant decreases in postintervention scores in depression also fell within the normal population range. Further, 42% of the preferred intensity patients ($n = 8$) showed a trend-level improvement in depression scores postintervention (six patients were prescribed antidepressants). No patient experienced deterioration in depression. Finally, Morres et al. reversed the data set in order to compare the individual and clinically significant antidepressant effects of prescribed vs preferred intensity. Only one patient in the prescribed intensity group showed significantly greater depression alleviation, consistent with chance. No patient experienced deterioration in depression.

The findings by Morres, Hinton-Bayre, Motakis, et al. (2019) are promising, especially since the use/accuracy of the JT method is supported by the fact that prescribed intensity showed no depression relief postintervention in Callaghan et al. (2011). Due to a large number of medicated patients ($n = 15$) in a small sample size ($n = 19$), however, Morres et al. did not clarify if less medicated than medicated-free patients in the preferred intensity group showed recovery from depression. Larger studies are needed to examine the individual and clinically significant effects of preferred intensity exercise in depressed women.

## Exercise for depression as a comorbid with obesity disorder

To our best knowledge, there appear to be only a few exercise-depression RCTs for obese adults aged 18–65 (BMI: $\geq 30\,kg/m^2$), despite the causal link of obesity to depression development regardless the presence of weight inducing metabolic side effects (Tyrrell et al., 2018).

Particularly, Baillot et al. (2018) in their recent meta-analysis recorded only four RCTs examining the effects of exercise on weight loss and depression in obese adults excluding menopausal women; two trials focused on elderly adults (+65) and two trials on adults aged 18–65 years. Of the latter two RCTs, the Sarsan, Ardiç, Özgen, Topuz and Sermez (2006) trial recruited obese women admitted to endocrinology and metabolic disease unit with main inclusion criteria referring to the sedentary lifestyle and eating as usual (no energy-restricted diets). The authors compared the effects of 12 week AE vs strength exercise vs control condition-(not defined) on weight loss. They also measured depression as a secondary outcome via the BDI, which is a widely used tool for depression. Participants' baseline depression severity was minimal to mild. All but the control group showed significant weight loss and depression relief postintervention. Only the AE group showed significantly higher depression relief vs the control condition.

In the second RCT, Nieman, Custer, Butterworth, Utter, and Henson (2000) employed a 12-week supervised program to compare the weight loss and antidepressant effects of AE vs AE+diet vs diet vs control condition (stretching+ calisthenics) in obese adults recruited via media advertisements. Main inclusion criteria were eating, as usual, no history of smoking or abusing alcohol, sedentary lifestyle, and, good health (no mood problems). An additional group of participants meeting the above inclusion criteria but with a physically active lifestyle and non-overweight profile ($BMI \leq 25\,kg/m^2$) were also recruited for prestudy cross-sectional comparisons. Depression was measured as a secondary outcome with the subscale of the GWBS. Prestudy results showed more severe depression (and higher BMI) for obese vs nonoverweight participants. Poststudy results showed that the AE and the AE+diet groups brought about a significantly lower body mass/fat and depression than the control group.

## Additional exercise trials for depression as a comorbid with obesity disorder

Four additional trials for obese media respondents have also examined if exercise can reduce weight and depression. First, Geliebter et al. (1997) recruited a mixed-gender sample. Main inclusion criteria were sedentary lifestyle, no smoking, no history of drug abuse, and women on the premenopausal stage. This RCT compared the effects of an 8-week supervised AE+diet vs strength+diet vs diet on weight loss and depression. At postintervention, all groups exhibited statistically significant comparable weight loss, whereas the AE+diet and the strength+diet groups exhibited significantly lower depression vs the diet group. This is supportive evidence for the add-on antidepressant value of exercise combined with diet vs diet alone. Accordingly, embedding exercise to diet programs enables obese women to experience weight loss and depression alleviation.

Second, Imayama et al. (2011) examined the antidepressant effects of a long-term (48 weeks) supervised program of AE+dietary weight loss vs AE vs dietary weight loss vs control group (on a waiting list to receive four diet sessions and 8 weeks of exercise postintervention). Main participation criteria referred to adult and elderly adult postmenopausal women (50–75 years), nonsmoking, alcohol intake of <2 drinks/day, and $BMI = \geq 25\,kg/m^2$. At baseline, the mean BMI of all groups was $>30\,kg/m^2$. At postintervention, the AE+dietary weight loss group exhibited significantly higher improvement in depression measured with the BSI-18. This improvement was significantly different from the control group.

Third, the Koeppl et al. (1992) RCT compared the weight loss and antidepressant effects of a long-term (55 weeks) supervised AE or dietary weight loss program in overweight/obese adult and elderly adult men vs general population healthy males comprising the control group. Main inclusion criteria were a healthy profile including no physical/psychiatric disorders or a history of alcohol or drug abuse. Depression was measured with a subscale of the MMPI. No within-or between-groups differences in weight loss postintervention were found. However, both the AE and the dietary weight loss arms showed a favorable comparison over controls in depression. According to Koeppl et al. (1992), this favorable comparison emerged from depression deterioration in the controls rather than from depression relief in the intervention groups. Hence, the authors suggested that the exercise and the weight loss programs recorded a "neutralizing" effect on the often-described increase in depression among elderly adults. However, it should be noted that controls' outcomes were administrated a year after the completion of the trial.

Finally, the multiarm trial of Wadden et al. (1997) recruited middle-aged adult (41.1 ± 8.6 years) obese women (Mean BMI = 36.5 ± 5.1 kg/m$^2$) for a 48-week supervised weight loss program with main inclusion criteria being no bulimia or other mental health disorders including severe depression. Depression was measured with the BDI. The authors compared the weight loss and antidepressant effects of diet vs diet + AE vs diet + strength exercise vs diet combined with AE and strength exercise. At the end of the program (week-48), all interventions proved comparably effective in reducing weight. Also, all interventions significantly reduced depression scores at program time point-1 (week-4). Reduced depression scores remained stably low by week-48. Trials are presented in Table 3.

**TABLE 3** Randomized controlled trials (RCTs) examining the effects of exercise on depressive symptoms in obese people.

| RCTs | Sample | Baseline depression tool/severity | Interventions | Control | Duration | Discharge depression |
|---|---|---|---|---|---|---|
| Sarsan et al. (2006) | Women 20–60 years $N=76$ | BDI/minimal | 1. Aerobic epsxercise (AE) 2. Strength exercise (SE) | Condition was not defined | 12 weeks | AE>C |
| Nieman et al. (2000) | Women 25–70 years $N=91$ | GWBS subscale/NR | 1. Exercise (E) 2. Exercise + diet (ED) 3. Diet + stretching/ calisthenics (DSC) | Stretching/ calisthenics | 12 weeks | ED>C |
| Geliebter et al. (1997) | Mixed gender 19–48 years $N=65$ | BDI/minimal-mild | 1. Aerobic exercise + diet (AED) 2. Strength exercise + diet (SED) | Diet | 8 weeks | AED>C SED>C |
| Imayama et al. (2011) | Postmenopausal women 50–75 years $N=439$ | BSI-18/NR | 1. Aerobic exercise (AE) 2. Aerobic exercise + diet (AED) 3. Diet | Four diet sessions and 8 weeks of exercise were delivered a year postintervention | 48 weeks | AED>C |
| Koeppl et al. (1992) | Men 46–78 years $N=145$ | MMPI subscale/ scores were within the "normal" range | 1. Aerobic exercise (AE) 2. Diet | General population participants | 55 weeks | AE>C D>C |
| Wadden et al. (1997)[M] | Women Mean = 41.1 ±8.6 years $N=128$ | BDI/minimal-mild | 1. Diet 2. Strength exercise + diet (SED) 3. Aerobic exercise + diet (AED) 4. Aerobic/strength Exercise + diet (ASED) | | 48 weeks | Comparable improvement across all groups |

Table3 shows a descriptive analysis of exercise trials for depression as a comorbid with obesity disorder (unpublished).
>: More effective; *BDI*: beck depression inventory; *BSI*: brief symptom inventory-18; *C*: control; *GWBS*: general well-being scale; *M*: multiarm-randomized trial. *MMPI*: Minnesota Multiphasic Personality Inventory; *N*: number; *NR*: not reported.

## Collective evidence

Supervised exercise activities seem to be a promising strategy in reducing depression (and weight) in obesity, as they often compare favorably over dietary weight loss programs. However, current evidence is based on a few RCTs employing: (1) depression as a secondary outcome; (2) samples often consisted of mixed adults/elderly adults with minimal depression and recruited via media advertisements and not via health services; (3) depression subscales rather than depression-specific scales comprehensive enough to assess depressive symptom/severity and sensitive enough to capture treatment change; (4) design flaws and poor reporting. Hence, no RCT has primarily examined the effects of exercise on depression as comorbid with obesity disorder in adult depressed samples (18–65 years) in mental health services. Relevant RCTs are, therefore, essential, especially for MDD, which is the most common type of depression. Also, RCTs with a longitudinal design may help clarify if depression and obesity have reciprocal associations within an exercise context. To this extent, further research is needed to examine the effects of exercise on depression as a comorbid with obesity disorder in adults (18–65) in health services.

## Key facts of the safety of exercise for depressed patients

- Based on individual clinical significance analysis, preferred or prescribed intensity exercise leads no patient to depression deterioration.
- The majority of depressed patients show a sedentary lifestyle.
- Exercise may cause occasional mild aches or fatigue from using dormant muscles.
- Depressed patients should build their exercise programs gradually; initial intensity should be low and gradually increase during the first weeks.
- Depressed patients should be assessed for cardiopulmonary risk before exercise, and moderate- or high-risk patients should undergo careful clinical testing before exercise participation.

## Summary points

- Clinical evidence shows a causal relationship between exercise participation and improved depression.
- Moderate intensity exercise is widely used and broadly effective in improving depression severity.
- Pragmatic evidence suggests that the ideographic model of preferred intensity exercise brings about higher antidepressant effects vs the nomothetic model of prescribed intensity exercise.
- Preferred vs prescribed intensity exercise in pragmatic settings leads 32% of depressed patients to clinically significant improvement in depression with low attrition rates.
- Based on a few exercise trials for depression as a comorbid with obesity disorder, exercise seems to be a promising antidepressant intervention.
- Additional exercise trials for depression as a comorbid with obesity disorder are essential.

## References


Baillot, A., Saunders, S., Brunet, J., Romain, A. J., Trottier, A., & Bernard, P. (2018). A systematic review and meta-analysis of the effect of exercise on psychosocial outcomes in adults with obesity: A call for more research. *Mental Health and Physical Activity*, *14*, 1–10. https://doi.org/10.1016/j.mhpa.2017.12.004.

Blumenthal, J. A., & Ong, L. (2009). A commentary on 'exercise and depression' (Mead et al., 2009): And the verdict is…. *Mental Health and Physical Activity*, *2*(2), 97–99. https://doi.org/10.1016/j.mhpa.2009.08.001.

Borenstein, M., Hedges, L. V., Higgins, J. P. T., & Rothstein, H. R. (2009). *Introduction to meta-analysis*. John Wiley & Sons(Ltd.).

Borg, G. (1998). *Borg's perceived exertion and pain scales*. Champaign: Human Kinetics.

Callaghan, P., Khalil, E., Morres, I., & Carter, T. (2011). Pragmatic randomised controlled trial of preferred intensity exercise in women living with depression. *BMC Public Health*, *11*(1), 465. https://doi.org/10.1186/1471-2458-11-465.

Caspersen, C. J., Powell, K. E., & Christenson, G. M. (1985). Physical activity, exercise, and physical fitness: Definitions and distinctions for health-related research. *Public Health Reports*, *100*(2), 126–131.

Cohen, J. (1992). Statistical power analysis. *Current Directions in Psychological Science*, *1*(3), 98–101. https://doi.org/10.1111/1467-8721.ep10768783.

Cooney, G., Dwan, K., Greig, C. A., Lawlor, D. A., et al. (2013). Exercise for depression. *Cochrane Database Systematic Reviews*, *9*, CD004366. https://doi.org/10.1002/14651858.CD004366.pub6.

Craft, L. L., & Landers, D. M. (1998). The effect of exercise on clinical depression and depression resulting from mental illness: A meta-analysis. *Journal of Sport & Exercise Psychology*, *20*, 339–357. https://doi.org/10.1123/jsep.20.4.339.

Crone, D., Johnston, L. H., Gidlow, C., Henley, C., & James, D. V. B. (2008). Uptake and participation in physical activity referral schemes in the UK: An investigation of patients referred with mental health problems. *Issues in Mental Health Nursing*, *29*(10), 1088–1097. https://doi.org/10.1080/01612840802319837.

Danielsson, L., Noras, A. M., Waern, M., & Carlsson, J. (2013). Exercise in the treatment of major depression: A systematic review grading the quality of evidence. *Physiotherapy Theory and Practice*, *29*(8), 573–585. https://doi.org/10.3109/09593985.2013.774452.

Ford, I., & Norrie, J. (2016). Pragmatic trials. *New England Journal of Medicine*, *375*(5), 454–463. https://doi.org/10.1056/NEJMra1510059.

Geliebter, A., Maher, M. M., Gerace, L., Gutin, B., Heymsfield, S. B., & Hashim, S. A. (1997). Effects of strength or aerobic training on body composition, resting metabolic rate, and peak oxygen consumption in obese dieting subjects. *The American Journal of Clinical Nutrition*, *66* (3), 557–563. https://doi.org/10.1093/ajcn/66.3.557.

Hanson, C. L., Allin, L. J., Ellis, J. G., & Dodd-Reynolds, C. J. (2013). An evaluation of the efficacy of the exercise on referral scheme in Northumberland, UK: Association with physical activity and predictors of engagement. A naturalistic observation study. *BMJ Open*, *3*(8). https://doi.org/10.1136/bmjopen-2013-002849.

Harrison, R., McNair, F., & Dugdill, L. (2005). Access to exercise referral schemes—A population based analysis. *Journal of Public Health*, *27* (4), 326–330. https://doi.org/10.1093/pubmed/fdi048

Hotopf, M. (2002). The pragmatic randomised controlled trial. *Advances in Psychiatric Treatment*, *8*, 326–333. https://doi.org/10.1192/apt.8.5.326.

Imayama, I., Alfano, C. M., Kong, A., Foster-Schubert, K. E., Bain, C. E., Xiao, L., et al. (2011). Dietary weight loss and exercise interventions effects on quality of life in overweight/obese postmenopausal women: A randomized controlled trial. *The International Journal of Behavioral Nutrition and Physical Activity*, *8*, 118. https://doi.org/10.1186/1479-5868-8-118.

Jacobson, N. S., & Truax, P. (1991). Clinical significance: A statistical approach to defining meaningful change in psychotherapy research. *Journal of Consulting and Clinical Psychology*, *59*(1), 12–19. https://doi.org/10.1037/0022-006X.59.1.12.

James, D. V., Johnston, L. H., Crone, D., Sidford, A. H., Gidlow, C., Morris, C., et al. (2008). Factors associated with physical activity referral uptake and participation. *Journal of Sports Sciences*, *26*(2), 217–224. https://doi.org/10.1080/02640410701468863.

James, D., Mills, H., Crone, D., Johnston, L. H., Morris, C., & Gidlow, C. J. (2009). Factors associated with physical activity referral completion and health outcomes. *Journal of Sports Sciences*, *27*(10), 1007–1017.

Khalil, E., Callaghan, P., Carter, T., & Morres, I. (2012). Pragmatic randomised controlled trial of an exercise programme to improve wellbeing outcomes in women with depression: Findings from the qualitative component. *Psychology (Savannah, Ga.)*, *3*(11), 979–986.

Koeppl, P. M., Heller, J., Bleecker, E. R., Meyers, D. A., Goldberg, A. P., & Bleecker, M. L. (1992). The influence of weight reduction and exercise regimes upon the personality profiles of overweight males. *Journal of Clinical Psychology*, *48*(4), 463–471.

Krogh, J., Hjorthøj, C., Speyer, H., Gluud, C., & Nordentoft, M. (2017). Exercise for patients with major depression: A systematic review with meta-analysis and trial sequential analysis. *BMJ Open*, *7*(9). https://doi.org/10.1136/bmjopen-2016-014820.

Krogh, J., Nordentoft, M., Sterne, J. A., & Lawlor, D. A. (2011). The effect of exercise in clinically depressed adults: Systematic review and meta-analysis of randomized controlled trials. *The Journal of Clinical Psychiatry*, *72*(4), 529–538. https://doi.org/10.4088/JCP.08r04913blu.

Krogh, J., Saltin, B., Gluud, C., & Nordentoft, M. (2009). The DEMO trial: A randomized, parallel-group, observer-blinded clinical trial of strength versus aerobic versus relaxation training for patients with mild to moderate depression. *The Journal of Clinical Psychiatry*, *70* (6), 790–800. https://doi.org/10.4088/JCP.08m04241.

Krogh, J., Videbech, P., Thomsen, C., Gluud, C., & Nordentoft, M. (2012). DEMO-II trial. Aerobic exercise versus stretching exercise in patients with major depression-a randomised clinical trial. *PLoS One*, *7*(10). https://doi.org/10.1371/journal.pone.0048316, e48316.

Kvam, S., Kleppe, C. L., Nordhus, I. H., & Hovland, A. (2016). Exercise as a treatment for depression: A meta-analysis. *Journal of Affective Disorders*, *202*, 67–86. https://doi.org/10.1016/j.jad.2016.03.063.

Lawlor, D. A., & Hopker, S. W. (2001). The effectiveness of exercise as an intervention in the management of depression: Systematic review and meta-regression analysis of randomised controlled trials. *BMJ*, *322* (7289), 763–767. https://doi.org/10.1136/bmj.322.7289.763

MacPherson, H. (2004). Pragmatic clinical trials. *Complementary Therapies in Medicine*, *12*(2–3), 136–140. https://doi.org/10.1016/j.ctim.2004.07.043.

Mead, G. E., Morley, W., Campbell, P., Greig, C. A., et al. (2009). Exercise for depression. *Cochrane Database of Systematic Reviews*, *8*(3), CD004366. https://doi.org/10.1002/14651858.CD004366.pub4.

Morgan, W. P. (1973). Psychological factors influencing perceived exertion. *Medicine and Science in Sports*, *5*(2), 97–103. https://doi.org/10.1249/00005768-197300520-00019.

Morgan, W. P. (1994). Psychological components of effort sense. *Medicine and Science in Sports and Exercise*, *26*(9), 1071–1077. https://doi.org/10.1249/00005768-199409000-00001.

Morgan, W. P. (2001). Prescription of physical activity: A paradigm shift. *Quest*, *53*(3), 366–382. https://doi.org/10.1080/00336297.2001.10491754.

Morgan, F., Battersby, A., Weightman, A. L., Searchfield, L., Turley, R., Morgan, H., et al. (2016). Adherence to exercise referral schemes by participants—What do providers and commissioners need to know? A systematic review of barriers and facilitators. *BMC Public Health*, *16* (1), 227. https://doi.org/10.1186/s12889-016-2882-7.

Morres, I. D., Hatzigeorgiadis, A., Krommidas, C., Comoutos, N., Sideri, E., Ploumpidis, D., et al. (2019). Objectively measured physical activity and depressive symptoms in adult outpatients diagnosed with major depression. Clinical perspectives. *Psychiatry Research*, *280*, 112489. https://doi.org/10.1016/j.psychres.2019.112489.

Morres, I. D., Hatzigeorgiadis, A., Stathi, A., Comoutos, N., Arpin-Cribbie, C., Krommidas, C., et al. (2019). Aerobic exercise for adult patients with major depressive disorder in mental health services: A systematic review and meta-analysis. *Depression and Anxiety*, *36*(1), 39–53. https://doi.org/10.1002/da.22842.

Morres, I. D., Hinton-Bayre, A., Motakis, E., Carter, T., & Callaghan, P. (2019). A pragmatic randomised controlled trial of preferred intensity exercise in depressed adult women in the United Kingdom: Secondary analysis of individual variability of depression. *BMC Public Health*, *19*(1), 941. https://doi.org/10.1186/s12889-019-7238-7.

Morres, I., Stathi, A., Martinsen, E. W., & Sørensen, M. (2014). Physical exercise and major depressive disorder in adult patients. In A. G. Papaioannou, & D. Hackfort (Eds.), *Routledge companion to sport and exercise psychology: Global perspectives and fundamental concepts* (pp. 823–834). London: Taylor & Francis.

National Institute for Health and Care Excellence. (2009). Depression: The treatment and management of depression in adults (update). In *Clinical guidelines, [CG90]*. London, UK.

National Institute for Health and Care Excellence. (2013). Physical activity: Brief advice for adults in primary care. In *NICE guidelines [PH44]*. London, UK.

Nebiker, L., Lichtenstein, E., Minghetti, A., Zahner, L., Gerber, M., Faude, O., et al. (2018). Moderating effects of exercise duration and intensity in neuromuscular vs endurance exercise interventions for the treatment of depression: A meta-analytical review. *Frontiers in Psychiatry*, *9* (305). https://doi.org/10.3389/fpsyt.2018.00305.

Nieman, D. C., Custer, W. F., Butterworth, D. E., Utter, A. C., & Henson, D. A. (2000). Psychological response to exercise training and/or energy restriction in obese women. *Journal of Psychosomatic*

*Research*, *48*(1), 23–29. https://doi.org/10.1016/S0022-3999(99)00066-5.

Oliveira, B., Deslandes, A., & Santos, T. (2015). Differences in exercise intensity seems to influence the affective responses in self-selected and imposed exercise: A meta-analysis. *Frontiers in Psychology*, *6* (1105). https://doi.org/10.3389/fpsyg.2015.01105.

Parfitt, G., Rose, E. A., & Burgess, W. M. (2006). The psychological and physiological responses of sedentary individuals to prescribed and preferred intensity exercise. *British Journal of Health Psychology*, *11*(1), 39–53. https://doi.org/10.1348/135910705X43606.

Ravindran, A. V., Balneaves, L. G., Faulkner, G., Ortiz, A., McIntosh, D., Morehouse, R. L., et al. (2016). Canadian network for mood and anxiety treatments (CANMAT) 2016 clinical guidelines for the management of adults with major depressive disorder:Section 5. Complementary and alternative medicine treatments. *The Canadian Journal of Psychiatry*, *61*(9), 576–587. https://doi.org/10.1177/0706743716660290.

Rethorst, C. D., Wipfli, B. M., & Landers, D. M. (2009). The antidepressive effects of exercise: A meta-analysis of randomized trials. *Sports Medicine*, *39*(6), 491–511. https://doi.org/10.2165/00007256-200939060-00004.

Rimer, J., Dwan, K., Lawlor, D. A., Greig, C. A., et al. (2012). Exercise for depression. *Cochrane Database Systematic Reviews*, *7*, CD004366. https://doi.org/10.1002/14651858.CD004366.pub5.

Robertson, R., Robertson, A., Jepson, R., & Maxwell, M. (2012). Walking for depression or depressive symptoms: A systematic review and meta-analysis. *Mental Health and Physical Activity*, *5*(1), 66–75. https://doi.org/10.1016/j.mhpa.2012.03.002.

Roland, M., & Torgerson, D. J. (1998). Understanding controlled trials: What are pragmatic trials? *BMJ*, *316*(7127), 285. https://doi.org/10.1136/bmj.316.7127.285.

Rose, E. E., & Parfitt, G. (2007). A quantitative analysis and qualitative explanation of the individual differences in affective responses to prescribed and self-selected exercise intensities. *Journal of Sport and Exercise Psychology*, *29*(3), 281–309. https://doi.org/10.1123/jsep.29.3.281.

Sarsan, A., Ardiç, F., Özgen, M., Topuz, O., & Sermez, Y. (2006). The effects of aerobic and resistance exercises in obese women. *Clinical Rehabilitation*, *20*(9), 773–782. https://doi.org/10.1177/0269215506070795.

Schuch, F. B., Morres, Ekkekakis, P., Rosenbaum, S., & Stubbs, B. (2017). A critical review of exercise as a treatment for clinically depressed adults: Time to get pragmatic. *Acta Neuropsychiatrica*, *29*(2), 65–71. https://doi.org/10.1017/neu.2016.21.

Schuch, F. B., Vancampfort, D., Richards, J., Rosenbaum, S., Ward, P. B., & Stubbs, B. (2016). Exercise as a treatment for depression: A meta-analysis adjusting for publication bias. *Journal of Psychiatric Research*, *77*, 42–51. https://doi.org/10.1016/j.jpsychires.2016.02.023.

Silveira, H., Moraes, H., Oliveira, N., Coutinho, E. S., Laks, J., & Deslandes, A. (2013). Physical exercise and clinically depressed patients: A systematic review and meta-analysis. *Neuropsychobiology*, *67*(2), 61–68. https://doi.org/10.1159/000345160.

Tyrrell, J., Mulugeta, A., Wood, A. R., Zhou, A., Beaumont, R. N., Tuke, M. A., et al. (2018). Using genetics to understand the causal influence of higher BMI on depression. *International Journal of Epidemiology*, *48*(3), 834–848. https://doi.org/10.1093/ije/dyy223.

Wadden, T. A., Vogt, R. A., Andersen, R. E., Bartlett, S. J., Foster, G. D., Kuehnel, R. H., et al. (1997). Exercise in the treatment of obesity: Effects of four interventions on body composition, resting energy expenditure, appetite, and mood. *Journal of Consulting and Clinical Psychology*, *65*(2), 269–277.

## Examples of mini-dictionary of terms

**Aerobic exercise** A method of exercise training in which oxygen is the main "fuel" for energy production.

**Moderate intensity exercise** The heart beats faster and the breath becomes harder than normal, but the exerciser is still able to talk.

**Obesity** The accumulation of excessive amount of fat that presents a risk to health.

**Exercise on prescription (or referral) schemes** Health professionals prescribe physical exercise to people with health disorders including depression.

**Disturbed physical exertion** Depressed people during exercise may often show excessive fear or worry, and exaggerated or inaccurate intensity sensing.

Chapter 48

# Acupressure and depression: A scientific narrative

Nant Thin Thin Hmwe[a] and Sally Wai-Chi Chan[b,c]

[a]School of Nursing and Midwifery, The University of Newcastle, Callaghan, NSW, Australia, [b]Division of Global and Engagement and Partnership, The University of Newcastle, Callaghan, NSW, Australia, [c]Centre for Brain and Mental Health, The University of Newcastle Callaghan, NSW, Australia

## List of abbreviations

| | |
|---|---|
| **HT 7** | heart 7 |
| **KI 1** | kidney 1 |
| **KI 3** | kidney 3 |
| **LI 4** | liver 4 |
| **PPS** | pressure pain sensitivity |
| **RCT** | randomized controlled trial |
| **SP 6** | spleen 6 |
| **ST 36** | stomach 36 |
| **TCM** | traditional Chinese medicine |
| **TEAS** | trans-electrical acupoint stimulation |

## Introduction

Acupressure is a practice in traditional Chinese medicine. It refers to the stimulation of acupuncture points (acupoints) using fingers, hands, or devices (Tsay, 2004). The purpose of acupressure is to regulate vital energy in the human body, which is essential to maintain good health, prevent illness, and promote psychological well-being (Freeman, 2004). Acupressure has been suggested as a therapeutic modality for depression in a variety of study population (Lee & Frazier, 2011; Robinson, Lorenc, & Liao, 2011). It is believed that acupressure regulates the endocrine and nervous system, thus maintaining normal physiological functions of human body. Stimulation of acupoints by giving fingertip pressure releases endorphin and serotonin, which helps the patients relax, alleviate anxiety, and reduce the symptoms of depression (Cho & Tsay, 2004; Mehta, Dhapte, Kadam, & Dhapte, 2017). Acupressure is recommended as a technique for stress management and promotes emotional well-being (Gach & Henning, 2004; Lane, 2009). Acupressure is also suggested as an adjuvant treatment for depression together with psychotherapy (Lane, 2009).

Depression is one of the common mental disorders. It is a leading cause of disability worldwide and a major contributor to the global burden of disease (World Health Organization, 2018). Depression is highly prevalent in patients with chronic physical diseases, which significantly worsen their health status (National Collaborating Centre for Mental Health (UK), 2010). Thus, prevention and management of depression is an important area that needs to be addressed with effective intervention strategies. Conventional medical treatment could be integrated with complementary therapy such as acupressure, in the treatment of depression. This chapter discusses the basic concepts of acupressure, application of acupressure, scientific evidences for therapeutic effect of acupressure on depression, and implications for research and clinical practice.

## Basic concepts of acupressure

### Traditional Chinese medicine perspectives

Theoretical basis of acupressure is derived from Traditional Chinese Medicine (TCM) theory. In TCM concept, the vital energy known as *Qi* is flowing through the human body in channels, which is called meridians (Beal, 1999). These meridians are distributed throughout the body, connecting the visceral organs, extremities, superficial organs, and tissues (Hsieh, 2013; World Health Organization (WHO), 1993). The meridian system is a transport pathway of *Qi* and blood that links together as *Yin* and *Yang* (Hsieh, 2013). There are specific acupoints located along the meridians, and these acupoints are able to receive stimulation to regulate the *Qi* deficiency and excess, *Yin* and *Yang* balance, and functions of the internal organs (Zhang, 2014).

In TCM concept, health seems as the maintenance of smooth flow of *Qi* in the body (Hsieh, 2013). Physical and mental illnesses are caused by *Qi* irregularities; *Qi* deficiency or excess. When *Qi* flows freely, the person is healthy and well balanced. If *Qi* flow is disrupted or deficient, dysfunction occurs and then physical and mental illnesses are developed (Liu & Liu, 2010). Stimulation of acupoints regulates the smooth flow of *Qi* that improves blood circulation, maintains normal function of internal

The Neuroscience of Depression. https://doi.org/10.1016/B978-0-12-817933-8.00049-9

organs, and contributes to good health (Freeman, 2004; Zeng et al., 2016). Each acupoint has specific therapeutic properties, but stimulation on certain acupoints can elicit different beneficial responses under different conditions (Huang, Kutner, & Bliwise, 2009; Liu & Liu, 2010). There are specific acupoints that can be used for emotional imbalances and depression.

### Biomedical perspectives

In biomedical perspectives, mechanism of actions regarding acupressure is not clearly understood. There is a growing body of researches and reviews, exploring the physiological mechanism and treatment effects of acupressure on a variety of health conditions. Acupoint stimulation regulates immune, endocrine, and nervous system functions (Ding et al., 2014; Huang, Kutner, & Bliwise, 2011). Stimulation of acupoints using fingers or thumb pressure regulates the release of neurotransmitters and neurohormones, contributing to maintain normal physiological functions (Weiss-Farnan, 2010). Lane (2009) suggested that manual stimulation of acupoints regulates cortisol and releases serotonin, which helps the persons relax and induce calmness.

Acupoints are shown to have high electrical conductance and become sensitive under pathological conditions of the internal organs (Kim et al., 2017). Therefore, stimulating acupoints can relieve the symptoms associated with the visceral organs. Another possible mechanism is that acupressure mediates the nitric oxide signal that increases microcirculation, regulates central nervous system, and cardiovascular functions (Hsiao, 2011). These regulatory effects of acupressure have positive impact on a person's moods and emotional well-being.

## Application of acupressure

Acupressure is a noninvasive technique, which is easy to perform. Acupressure applied manually using fingertip does not require the use of needle or any other devices, thus adverse effects are less likely to occur compared to invasive needle acupuncture. Nurses and therapists can learn acupressure through short-term training on basic concepts and techniques of acupressure. Patients can also learn simple acupressure technique to perform self-administered acupressure for emotional well-being. The following section describes common acupoints for depression, acupressure technique, general guidelines, and safety and precautions in application of acupressure.

### Acupressure techniques

The fundamental principle of acupressure is stimulation of acupoints to regulate the flow of *Qi*. There are many different techniques used to apply pressure on acupoints. Application of thumb or finger pressure is the most common technique (Gach, 2014; Kolster & Waskowiak, 2007). Finger pressure can be applied using firm pressure or circular pressure (rotational movement).

*Static/firm pressure*: Use the steady or stationary pressure with thumb or fingers; apply pressure gradually and hold without any movement for 1–2 min. One minute of steady pressure calms and relaxes the nervous system, promoting greater healing (Gach, 2014).

*Circular pressure*: Begin with firm pressure and then make small circular movements with rhythmic manners for 1–2 min. The circular movement has a relaxing and balancing effect (Kolster & Waskowiak, 2007).

The pressure can be applied unilateral or simultaneous bilateral pressure (Kolster & Waskowiak, 2007; Wagner, 2015). The amount of pressure applied depends on the thickness of skin, muscle, and fatty tissue at the point (Gach & Henning, 2004; Maa et al., 2003). The intensity of pressure is to be adjusted according to the person's level of tolerance (Shen et al., 2017; Wagner, 2015).

### General guidelines

The following are the guiding principles when conducting acupressure:

- Perform acupressure in a pleasant and comfortable environment to promote relaxation.
- Keep the fingernails short to avoid making scrapes on the patient's skin.
- The hand and fingers should be warm and relax.
- Use a comfortable position.
- Encourage client to relax while administering acupressure.
- Concentrate on the points that are treating.
- Apply finger pressure in a slow, rhythmic manner to enable layers of tissues and the internal organs to respond (Gach, 2014).

### Safety and precautions

There are certain precautions and considerations required for the application of acupressure:

- Avoid pressing any area in an abrupt, forceful, or jarring way (Gach & Henning, 2004; Wagner, 2015).
- Acupressure should be not be performed immediately after meals, or alcohol consumption (Gach & Henning, 2004; Wagner, 2015).
- Acupressure is not suitable to apply on the skin that is inflamed, injured, newly formed scarred or having rashes, lesions, bruises, or open wound (Gach & Henning, 2004).
- Cautions should be taken for people with pregnancy, acute illness or infection, serious cardiac problems, and blood clotting disorders.

## Research evidences for the effect of acupressure on depression

Research studies investigating the treatment effect of acupressure on depression have been conducted in many different settings. Study populations include patients with hemodialysis, older people, college students majoring in acupuncture and moxibustion, patients with stroke, patients with ischemic heart disease, patients with cancer, and patients with chronic obstructive pulmonary disease (COPD). Summary of the studies is shown in Table 1.

There are three studies conducted in patients with hemodialysis. The studies used 4-week acupressure intervention (three times per week) when the patients were undergoing hemodailysis treatment. In Cho and Tsay (2004), hemodialysis patients ($n = 62$) were randomly assigned into acupressure group and control group. The results indicated that depression scores in acupressure group significantly reduced after intervention, when compared to the control group. Tsay et al. (2004) compared the effect of acupressure and transcutaneous electrical acupoint stimulation (TEAS) with a control group ($n = 108$). The result showed that depression scores were significantly lower in the acupressure and TEAS groups when compared to the control group. There was no difference between acupressure and the TEAS groups, suggesting that acupressure and TEAS were equally effective in reducing depression in hemodialysis patients. Another randomized controlled trial (RCT, $n = 108$) indicated that depression was significantly reduced in acupressure group compared to control group (Hmwe, Subramanian, Tan, & Chong, 2015). The consistent findings from the three reviewed studies conducted in hemodialysis population suggested that acupressure had a positive treatment effect on depression for patients undergoing hemodialysis.

In older adult population, four studies have been conducted in participants with different health conditions; older person with chronic knee pain, older women living alone, frail elderly living in community, and older people diagnosed with depression. Tse and Au (2010) conducted a quasi-experimental study among older adults with chronic knee pain ($n = 62$). The results showed that depression was significantly reduced after 4 weeks in the acupressure group when compared to the control group. In study conducted by Tse, Tang, Wan, and Vong (2014), a 8-week physical exercise program (PEP) included with self-administered acupressure was conducted among older persons with chronic pain. The results found that the PEP significantly reduced the intensity of pain and improved psychological well-being (happiness, loneliness, life

**TABLE 1** Summary of studies investigating the effect of acupressure on depression.

| Authors (year) and country | Study design | Population and sample | Study groups | Findings |
|---|---|---|---|---|
| Tsay, Cho, and Chen (2004)<br><br>Taiwan | Experimental | Hemodialysis patients<br>$N = 106$ | i. Acupressure<br>ii. TEAS<br>iii. Control; usual care | Acupressure and TEAS group had lower levels of depression than control group<br>No difference between acupressure and TEAS |
| Cho and Tsay (2004)<br><br>Taiwan | Experimental | Hemodialysis patients<br>$N = 58$ | i. Acupressure<br>ii. Control | Acupressure significantly improved depression, compared to control group |
| Wu, Lin, Wu, and Lin (2007)<br><br>Taiwan | Randomized block experimental | Patient with COPD<br>$N = 22$ | i. Acupressure<br>ii. Sham acupressure | Reduction of depression scores in acupressure significantly greater than sham acupressure |
| Kang, Sok, and Kang (2009)<br><br>Korea | Randomized clinical trial | Stroke patients with hemiplegia<br>$N = 56$ | i. Acupressure<br>ii. Control | Depression in acupressure group was reduced significantly, compared control group |
| Tse and Au (2010)<br><br>Hong Kong | Quasi-experimental | Older persons with chronic knee pain<br>$N = 62$ | i. Acupressure<br>ii. Control | Depression level decreased significantly after acupressure therapy |

*Continued*

**TABLE 1 Summary of studies investigating the effect of acupressure on depression—cont'd**

| Authors (year) and country | Study design | Population and sample | Study groups | Findings |
|---|---|---|---|---|
| Honda, Tsuda, and Horiuchi (2012)<br><br>Japan | Study design is not reported | College students majoring in acupuncture and moxibustion, $N=25$ | i. Acupressure<br>ii. Control | Depression score in self-acupressure group decreased significantly from baseline |
| Bergmann et al. (2014)<br><br>Denmark | Randomized clinical trial | Patients with ischemic heart disease<br>$N=213$ | i. Active group; PPS measurement followed by acupressure<br>i. Control | Reduction of depression was significantly greater in active group compared to control group |
| Tang et al. (2014)<br><br>Taiwan | Experimental | Patients with lung cancer<br>$N=57$ | i. Acupressure with essential oil<br>ii. Acupressure only<br>iii. Sham acupressure | No significant difference among the three groups on depression |
| Tse et al. (2014)<br><br>Hong Kong | Randomized controlled trial | Nursing home residents with pain<br>$N=396$ | i. PEP included with acupressure<br>ii. Control | PEP with acupressure had significantly improved psychological well-being (happiness, loneliness, life satisfaction, and depression) |
| Sok (2016)<br><br>Korea | Quasi-experimental | Older women living alone<br>$N=76$ | i. Acupressure<br>i. Control | Depression was significantly reduced in experimental group compared to control group |
| Lan et al. (2015)<br><br>Taiwan | Quasi-experimental | Patients with HCP<br>$N=62$ | i. Acupressure<br>ii. Control | No significant difference between groups on depression |
| Hmwe et al. (2015)<br><br>Malaysia | Randomized controlled trial | Hemodialysis patients<br>$N=108$ | i. Acupressure<br>ii. Control | Depression scores were significantly reduced in acupressure group, compared to control group |
| Chan et al. (2017)<br><br>Hong Kong | Randomized controlled trial | Frail elderly living in community<br>$N=101$ | i. Acupressure<br>ii. Control; waiting list | Acupressure significantly reduced depression scores |
| Molassiotis et al. (2020)<br><br>Hong Kong | Randomized controlled trial | Older people diagnosed with depression<br>$N=118$ | i. Acupressure<br>ii. Sham acupressure<br>iii. Control | Depression scores reduced significantly in both acupressure and sham acupressure group when compared to control group |

*COPD*, chronic obstructive pulmonary disease; *HCP*, hepatocellular carcinoma; *PEP*, physical exercise program; *PPS*, pressure pain sensitivity; *TEAS*, transcutaneous electrical acupoint stimulation.
Summary table created by authors.

satisfaction, and depression), compared to the control group who did not receive PEP. Sok (2016) conducted a quasi-experimental with control group among older women living alone ($n=76$). Acupressure was provided to the study group twice weekly for 8 weeks. The results indicated that depression level decreased significantly in the experimental group, when compared to the control group.

(Chan et al. 2017) conducted an RCT among frail elderly ($n=101$), given acupressure 4 times per week for 12 weeks. The results indicated that acupressure significantly reduced depression, however, the completion rate was lower than 80% (79 out of 101). It limited the validity of the results. In another RCT conducted among older people with depression ($n = 118$), acupressure was compared with sham acupressure and usual care control group (Molassiotis et al., 2020). The course of acupressure was 4 times a week for 12 weeks. The findings indicated that depression scores reduced significantly in both acupressure and sham acupressure group. In this study, the finding was inconclusive due to high attrition rate (28.8%, 34 out of 118), and significant positive effect shown in the sham acupressure group.

Honda et al. (2012) conducted their study among healthy volunteers, who were college students ($n = 25$) majoring in acupuncture and moxibustion. In all, 30 participants were randomly selected and assigned to self-administered acupressure group for 4 weeks, and the remaining participants ($n = 12$) were assigned to the control group having daily routine activities only. The finding indicated depression scores in the self-acupressure group decreased significantly from baseline, but no change was found in the control group. The finding suggested that acupressure might be a method of choice for self-management of depressive mood. However, the findings may not be generalizable because of small sample size. To have a more conclusive results, there is a need to replicate the study with a larger sample size in other group of students or healthy volunteers.

Wu et al. (2007) investigated the effect of acupressure on depression in patients with COPD ($n = 44$), which included an acupressure and sham acupressure group. Acupressure was given five sessions per week for 4 weeks. The results showed that depression was significantly reduced in true acupressure group compared to the sham acupressure group. In Kang et al. (2009), 14-point meridian acupressure was applied daily for 2 weeks among stroke patients having hemiplegia ($n = 56$). The results were similar to other studies, indicating that the experimental group had significant improvement in their depression when compared with the control group. In Bergmann et al. (2014) study, self-measurement of pressure pain sensitivity (PPS) followed by acupressure was employed among patients with ischemic heart disease ($n = 213$). The results showed that reduction of depression score in intervention group was significantly greater, compared to the control group.

There are two studies conducted in patients with cancer; lung cancer undergoing chemotherapy (Tang et al., 2014), and hepatocellular carcinoma treated with transcatheter arterial chemoembolization (Lan, Lin, Chen, Lin, & Wang, 2015). In Tang et al.'s (2014) study, the participants performed self-acupressure once daily for 5 months. The results showed that there was no significant difference in depression level between acupressure and sham acupressure groups. In Lan et al.'s (2015) study, the participants received acupressure for 5 days, and the results showed that acupressure did not alleviate depression. The 5-day course of acupressure might not be adequate to produce therapeutic effect on depression. Other possible reason for no significant effect would be the severity of depression in the cancer patients, which could be affected by multiple factors. Acupressure combined with other therapeutic approaches might be more effective in reducing the level of depression.

Overall study findings showed promising results. However, there were two studies conducted in patients with cancer showed no significant effect (Lan et al., 2015; Tang et al., 2014). And, one study showed that acupressure and sham acupressure had similar treatment effect on depression (Molassiotis et al., 2020). Further studies are required to produce strong evidences, as well to develop standard protocol for acupressure practice.

## Acupressure techniques, frequency, and duration

Most of the study used finger pressure on the selected acupoints. Some study began the session with massage, and then applied pressure on the acupoints. One study incorporated self-acupressure in physical exercise program that involved different type of exercises (Tse et al., 2014). The summary of acupressure interventions applied in the reviewed studies is shown in Table 2.

The duration of pressure applied on each acupoint ranged from 1 to 3 min. The duration of each acupressure session lasted for 4–16 min, 15-min duration was most commonly applied. The frequency of acupressure session also varied: three times daily, twice daily, once daily, five times per week, four times a week, three times a week, and twice a week. The overall duration of acupressure intervention program ranged from 5 days to 5 months, 4-week intervention program was most common.

## Common acupoints for depression

The acupoints used for depression varied across the studies (Table 2). Some of the commonly used acupoints in the reviewed studies are SP 6, ST 36, KI 1, KI 3, and HT 7. In a quantitative synthesis of acupoint stimulation for anxiety and depression, the commonly used appoints were SP 6, ST 36, and LI 4 (Wang, Deng, Tan, & Guan, 2016). Fig. 1 shows the location of these common acupoints. A combination of three and more acupoints was used in all the research studies. More researches are needed that focus on specific acupoints for depression, and the dosage (duration and frequency) of acupressure treatment for depression.

**TABLE 2 Summary acupressure interventions: acupoints, techniques, frequency, and duration.**

| Authors (year) | Acupoints applied | Acupressure technique | Frequency and duration |
|---|---|---|---|
| Tsay et al. (2004) | KI1, ST36, GB34, SP6 | Consistent pressure on correct acupoints, using finger (3 min per point)<br>The force of finger pressure was between 3 and 4 kg<br>Begin with 3 min of massage to relax the person and 12 min of acupressure | 15 min per session<br>3 times/week for 4 weeks |
| Cho and Tsay (2004) | ST6, SP6, KI3, KI1 | Pressing and rubbing on acupoints using the finger pads with a force of 3–5 kg, once per second and after 5 s release for 1 s<br>Each acupoint was pressed for 3 min and then the lower limbs were massaged for 3 min<br>Patients were scheduled to undergo acupressure therapy while having hemodialysis treatment | 15-min session: 12 min acupressure and 3 min massage<br>3 times/week for 4 weeks |
| Wu et al. (2007) | GV14, CV22, B13, B23, L10 | Effleurage (massage) neck and shoulder, and then press and rub each acupoint | 16-min session: 5 times per week for 4 weeks |
| Kang et al. (2009) | 14 Meridian points in qi flow | A finger acupressure, pressing on the 14 Meridian points in qi flow<br>Applied after general physical therapy | 10 min for each session, every day for 2 weeks |
| Tse and Au (2010) | ST35, ST36, and Xiyan (located around knee joint) | 6 Cycles in total; 15-s pressure and 15-s rest for each point, and repeat<br>Arm and thumb forming a vertical line on the knee, using 5 pound force of pressure<br>Patient position; lying on bed with knee flex to 30 degrees | 9 min for each knee<br>Twice a week for 4 weeks |
| Honda et al. (2012) | Six acupoints on the neck (not specify the name of acupoints used) | Self-acupressure in the morning, midday, and at night<br>Pressing on six acupressure points on the neck (three points on the left and right sides each) for 5 s, with sufficient pressure without causing discomfort | 3 Times daily for 4 weeks |
| Bergmann et al. (2014) | Two points:<br>– at sternum at the level of fourth intercostal space<br>– on the back 1.5″ lateral to fourth and fifth thoracic vertebra | Patients performed PPS measurements twice daily in the morning and evening<br>If the stress level was high, acupressure was performed by applying finger pressure for 1 min sufficiently enough to feel the pressure without causing pain | 3 months |
| Tang et al. (2014) | LI4, ST36, PC6 | Acupressure skills included pointing, pressing, kneading, and pushing<br>*Pointing:* finding the acupoint<br>*Pressing*: applying enough pressure to release the flow of qi, which the patient experiences as a dull ache, tingling, and soreness<br>*Kneading:* rotating the finger around the acupoint to stimulate the curing effect<br>*Pushing:* Pushing the thumb from the acupoint to the limb extremity to relax the patient's muscles and enhance blood circulation<br>1-min acupressure for each point | Once daily in the morning, continued for 5 months |

**TABLE 2 Summary acupressure interventions: acupoints, techniques, frequency, and duration—cont'd**

| Authors (year) | Acupoints applied | Acupressure technique | Frequency and duration |
|---|---|---|---|
| Tse et al. (2014) | Acupoints not specified | 1-h PEP: exercises and acupressure<br>15-min warm-up exercise, followed by 45-min exercise training<br>Different types of exercises were offered in each session; muscle strengthening using an elastic band, stretching, balancing, towel dance, and *self-administered acupressure* and massage<br>Not reported about self-acupressure technique | Twice a week for 8 weeks |
| Sok (2016) | In Joktaeyang Bangkwang Kyeong in qi flow | Finger acupressure pressing on the Joktaeyang Bangkwang Kyeong in qi flow; half of the total body surface | 15 min, twice per week, for 8 weeks |
| Lan et al. (2015) | Eight acupoints on ears: Yintang, Shenting, Cuanzhu, Taiyang, Jingming, Yangbai, Baihui, Fenchi | The duration of each acupoint massage was 4 min (not clear about duration per point and total duration) | 2 Times daily for 5 days |
| Hmwe et al. (2015) | Ying Tang, HT7, KI3 | Consistent finger pressure with rotational movement, applied 3 min on each point (did not apply finger pressure on the hand with arteriovenous fistula or graft) | 15-min; 3-min light massage; and 12-min acupressure<br>3 Times a week for 4 weeks |
| Chan et al. (2017) | Acupoints not specified | Not reported | 15-min sessions; 4 times/week for 12 weeks |
| Molassiotis et al. (2020) | ST36, LI3, HT7, LU1, GV 20 | Using the thumb, the therapist/caregiver applied and released pressure slowly to each point, holding for about 1 min in each point. Without massaging the area, the therapist/caregiver held firm pressure until it "hurt good"<br>A good rule of thumb was to apply slow, even but firm pressure with the finger at a 90-degree angle<br>The therapist/caregiver should begin to feel a pulsing beneath the finger | 4 Times a week for 12 weeks (research associated performed twice a week, and then caregiver or patient themselves performed twice a week) |

*PEP*; physical exercise program; *PPS*, pressure pain sensitivity.
Summary table created by authors.

## Implications for clinical practice and research

There are some emerging evidences suggesting that acupressure may have potential to be applied in clinical practice to promote patients' well-being. The current review shows a promising effect of acupressure on depression, in different groups of older people and other clinical populations. The results of the current review concurred with a recent integrative review of acupressure interventions for older people that indicated positive effect of acupressure on depression and psychological well-being (Hmwe, Browne, Mollart, Allanson, & Chan, 2019). However, the results are inconclusive due to study limitations such as small sample size, lack of blinding and randomization, high attrition rate, placebo effect shown in sham acupressure. Future well-designed trials with larger sample size would provide good evidence. Studies are required to focus on specific acupoints and standardized acupressure protocol for the management of depression. Using mix-method approach will obtain in-depth understanding of patients' perspectives about the acupressure intervention. Based on the research evidences, clinical practice guidelines should be developed for application of acupressure for managing depression and psychological symptoms.

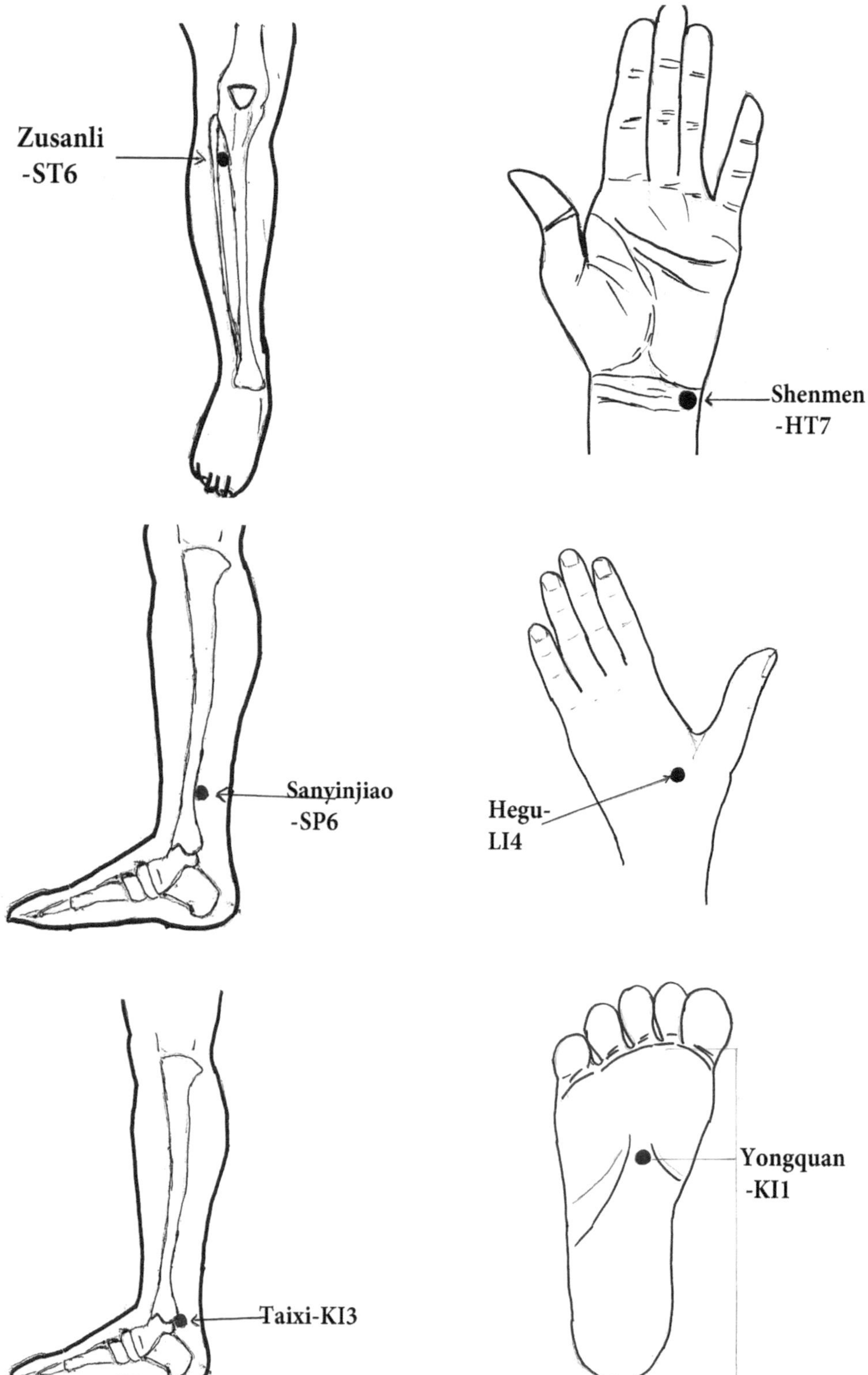


FIG. 1 Commonly used acupoints. *Acupoint images created by authors.*

## Key facts

- Acupressure has been recommended as a therapeutic modality for symptom management and promoting psychological well-being.
- The fundamental principle of acupressure is stimulation of acupoints to regulate the flow of vital energy (*Qi*) within the body.
- In biomedical perspectives, acupoint stimulation regulates endocrine and nervous systems.
- Acupressure technique is simple, noninvasive, and easy to learn.
- Application of thumb or finger pressure is the most common acupressure technique.

## Summary points

- This chapter focuses on the potential benefits of acupressure intervention for the management of depression.
- Research findings suggest that acupressure had a positive effect to reduce level of depression.
- Further studies are required to produce strong evidences, as well to develop standard protocol for acupressure practice.
- Studies are required to focus on specific acupoints and standardized acupressure protocol for the management of depression.

## Mini-dictionary of terms

**Acupressure** Stimulation of acupoints using hand, fingers, or devices.

**Acupoints** Acupoints are the specific points located along the meridians, and able to receive stimulation to regulate the Qi flow.

**Meridians** The system of vessels or pathways for *Qi* flow, connecting the visceral organs, extremities, superficial organs, and tissues.

**Traditional Chinese Medicine theory** Theoretical system based on ancient Chinese Philosophical thoughts.

**Vital energy (*Qi*)** The basic life force that supports the functions of body and mind.

**Yin and Yang** The opposite, interrelated cosmic forces found in all matter in nature. They work together to maintain dynamic equilibrium.

## References

Beal, M. W. (1999). Acupuncture and acupressure. *Journal of Midwifery & Women's Health*, *44*(3), 217–230.

Bergmann, N., Ballegaard, S., Bech, P., Hjalmarson, A., Krogh, J., Gyntelberg, F., et al. (2014). The effect of daily self-measurement of pressure pain sensitivity followed by acupressure on depression and quality of life versus treatment as usual in ischemic heart disease: A randomized clinical trial. *PLoS One [Electronic Resource]*, *9*(5), e97553. https://doi.org/10.1371/journal.pone.0097553.

Chan, C. W. C., Pui Hing, C., Leung, A. Y. M., Kwai Ching, L. O., Huafeng, S. H. I., Tin Pui, Y. U. M., et al. (2017). Acupressure for frail older people in community dwellings: A randomised controlled trial. *Age & Ageing*, *46*(6), 957–964. https://doi.org/10.1093/ageing/afx050.

Cho, Y. C., & Tsay, S. L. (2004). The effect of acupressure with massage on fatigue and depression in patients with end-stage renal disease. *Journal of Nursing Research*, *12*(1), 51–59.

Ding, S. S., Hong, S. H., Wang, C., Guo, Y., Wang, Z. K., & Xu, Y (2014). Acupuncture modulates the neuro–endocrine–immune network. *QJM: An International Journal of Medicine*, *107*(5), 341–345. https://doi.org/10.1093/qjmed/hct196.

Freeman, L. (2004). *Mosby's Complementary & Alternative Medicine A research-based approach* (2nd ed.). USA: Mosby.

Gach, M. R. (2014). *Acupressure massage: Tuina Chinese massage techniques*. Retrieved from http://www.acupressure.com/articles/acupressure_massage_techniques.htm.

Gach, M. R., & Henning, B. A. (2004). *Acupressure for emotional healing: A self-care guide for trauma, stress, & common emotional imbalances*: Bantam.

Hmwe, N. T. T., Browne, G., Mollart, L., Allanson, V., & Chan, S. W. (2019). An integrative review of acupressure interventions for older people: A focus on sleep quality, depression, anxiety, and agitation. *International Journal of Geriatric Psychiatry*, *34*(3), 381–396. https://doi.org/10.1002/gps.5031.

Hmwe, N. T. T., Subramanian, P., Tan, L. P., & Chong, W. K. (2015). The effects of acupressure on depression, anxiety and stress in patients with hemodialysis: A randomized controlled trial. *International Journal of Nursing Studies*, *52*(2), 509–518.

Honda, Y., Tsuda, A., & Horiuchi, S. (2012). Four-week self-administered acupressure improves depressive mood. *Psychology*, *03*, 802–804. https://doi.org/10.4236/psych.2012.329121.

Hsiao, S. H. (2011). A neurovascular blood-flow modulation model via acupuncture induced nitric oxide. In M. Saad (Ed.), *Tech Open Access. Acupuncture—Concepts and physiology*. Retrieved from https://www.intechopen.com/books/acupuncture-concepts-and-physiology/a-neurovascular-blood-flow-modulation-model-via-acupuncture-induced-nitric-oxide.

Hsieh, C. L. (2013). Modern scientific explanation of traditional acupuncture theory. In H. Hong (Ed.), *Acupuncture: Theory and evidence* (pp. 53–62). World Scientific Publishing

Huang, W., Kutner, N., & Bliwise, D. L. (2009). A systematic review of the effects of acupuncture in treating insomnia. *Sleep Medicine Reviews*, *13*(1), 73–104. https://doi.org/10.1016/j.smrv.2008.04.002.

Huang, W., Kutner, N. & Bliwise, D. L. (2011). Autonomic activation in insomnia: The case for acupuncture. *Journal of clinical sleep medicine: JCSM: Official Publication of the American Academy of Sleep Medicine*, *7*(1), 95.

Kang, H. S., Sok, S. R., & Kang, J. S. (2009). Effects of Meridian acupressure for stroke patients in Korea. *Journal of Clinical Nursing*, *18*(15), 2145–2152. https://doi.org/10.1111/j.1365-2702.2008.02522.x.

Kim, D. H., Ryu, Y., Hahm, D. H., Sohn, B. Y., Shim, I., Kwon, O. S., et al. (2017). Acupuncture points can be identified as cutaneous neurogenic inflammatory spots. *Scientific Reports*, *7*(1), 15214. https://doi.org/10.1038/s41598-017-14359-z.

Kolster, B. C., & Waskowiak, A. (2007). *The acupressure atlas (N. Win-Myint, Trans.)*. Healing Arts Press, Inner Traditions International.

Lan, S.-C., Lin, Y.-E. Chen, S.-C., Lin, Y.-F., & Wang, Y.-J. (2015). Effects of acupressure on fatigue and depression in hepatocellular carcinoma patients treated with transcatheter arterial chemoembolization: A quasi-experimental study. *Evidence-Based Complementary and*

*Alternative Medicine : eCAM, 2015*, 496485. https://doi.org/10.1155/2015/496485.

Lane, J. (2009). The neurochemistry of counterconditioning: Acupressure desensitization in psychotherapy. *Energy Psychology: Theory, Research, and Treatment*, *1*(1), 31–44.

Lee, E. J., & Frazier S. K. (2011). The efficacy of acupressure for symptom management: A systematic review. *Journal of Pain and Symptom Management*, *42*, 589–603. https://doi.org/10.1016/j.jpainsymman.2011.01.007.

Liu, Z. W., & Liu, L. (Eds.). (2010). *Essentials of Chinese medicine (Vol. 1)*. London: Springer: Verlag London Limited.

Maa, S. H., Sun, M. F., Hsu, K. H., Hung, T. J., Chen, H. C., Yu, C. T., et al. (2003). Effect of acupuncture or acupressure on quality of life of patients with chronic obstructive asthma: A pilot study. *Journal of Alternative & Complementary Medicine*, *9*(5), 659–670. https://doi.org/10.1089/107555303322524517.

Mehta, P., Dhapte, V., Kadam, S., & Dhapte, V. (2017). Contemporary acupressure therapy: Adroit cure for painless recovery of therapeutic ailments. *Journal of Traditional and Complementary Medicine*, *7*, 251–263. https://doi.org/10.1016/j.jtcme.2016.06.004.

Molassiotis, A., Suen, L., Lai, C., Chan, B., Wat, K. H. Y., Tang, J., et al. (2020). The effectiveness of acupressure in the management of depressive symptoms and in improving quality of life in older people living in the community: A randomised sham-controlled trial. *Aging and Mental Health*, *24*(6), 1001–1009. https://doi.org/10.1080/13607863.2019.1584789.

National Collaborating Centre for Mental Health (UK). (2010). Depression in adults with a chronic physical Health problem: Treatment and management. In *NICE clinical guidelines No. 91*. Retrieved from https://www.ncbi.nlm.nih.gov/books/NBK82930/.

Robinson, N., Lorenc, A., & Liao, X. (2011). The evidence for shiatsu: A systematic review of shiatsu and acupressure. *BMC Complementary and Alternative Medicine*, *11*(1), 88. https://doi.org/10.1186/1472-6882-11-88.

Shen, K., Cho, Y., Pascoe, E. M., Hawley, C. M., Oliver, V., Hughes, K. M., et al. (2017). The SIESTA trial: A randomized study investigating the efficacy, safety, and tolerability of acupressure versus sham therapy for improving sleep quality in patients with end-stage kidney disease on hemodialysis. *Evidence-Based Complementary & Alternative Medicine (eCAM)*, 1–10. https://doi.org/10.1155/2017/7570352.

Sok, S. R. (2016). Effects of meridian acupressure in Joktaeyang Bangkwang Kyeong on the physical health state, depression, and life satisfaction of Korean older adult women living alone. *Journal of Gerontological Nursing*, *42*(2), 28–35. https://doi.org/10.3928/00989134-20151218-01.

Tang, W. R., Chen, W. J., Yu, C. T., Chang, Y. C., Chen, C. M., Wang, C. H., et al. (2014). Effects of acupressure on fatigue of lung cancer patients undergoing chemotherapy: An experimental pilot study. *Complementary Therapies in Medicine*, *22*(4), 581–591.

Tsay, S. L. (2004). Acupressure and fatigue in patients with end-stage renal disease-a randomized controlled trial. *International Journal of Nursing Studies*, *41*(1), 99–106.

Tsay, S. L., Cho, Y. C., & Chen, M. L. (2004). Acupressure and transcutaneous electrical acupoint stimulation in improving fatigue, sleep quality and depression in hemodialysis patients. *American Journal of Chinese Medicine*, *32*(3), 407–416. https://doi.org/10.1142/s0192415x04002065.

Tse, M., & Au, J. (2010). The effects of acupressure in older adults with chronic knee pain: Depression, pain, activities of daily living and mobility. *Journal of Pain Management*, *3*(4), 399–410.

Tse, M. M., Tang, S. K., Wan, V. T., & Vong, S. K. (2014). The effectiveness of physical exercise training in pain, mobility, and psychological well-being of older persons living in nursing homes. *Pain Management Nursing*, *15*(4), 778–788. https://doi.org/10.1016/j.pmn.2013.08.003.

Wagner, J. (2015). Incorporating acupressure into nursing practice. *American Journal of Nursing*, *115*(12), 40–45. https://doi.org/10.1097/01.Naj.0000475290.20362.77.

Wang, T., Deng, R., Tan, J. Y., & Guan, F. G. (2016). Acupoints stimulation for anxiety and depression in cancer patients: A quantitative synthesis of randomized controlled trials. *Evidence-based Complementary and Alternative Medicine*, *2016*, 15. https://doi.org/10.1155/2016/5645632.

Weiss-Farnan, P. (2010). Acupressure. In M. Snyder, & R. Lindquist (Eds.), *Complementary and alternative therapies in nursing* (6th ed.). NewYork: Springer Publishing Company, LLC.

World Health Organization. (2018). *Depression fact sheet*. Retrieved from https://www.who.int/en/news-room/fact-sheets/detail/depression.

World Health Organization (WHO). (1993). *Standard acupuncture nomenclature: A brief explanation of 361 classical acupuncture point names and their multilingual comparative list*. Manila: WHO Regional Office for the Western Pacific.

Wu, H. S., Lin, L. C., Wu, S. C., & Lin, J. G. (2007). The psychologic consequences of chronic dyspnea in chronic pulmonary obstruction disease: The effects of acupressure on depression. *The Journal of Alternative and Complementary Medicine*, *13*(2), 253–261.

Zeng, H., Liu, M., Wang, P., Kang, J., Lu, F., & Pan, L. (2016). The effects of acupressure training on sleep quality and cognitive function of older adults: A 1-year randomized controlled trial. *Research in Nursing & Health*, *39*(5), 328–336. https://doi.org/10.1002/nur.21738.

Zhang, R. (2014). *Introduction to acupuncture and moxibustion. (X. Wang, Trans.)*. Singapore: World Century Publishing Corporation. Original work published 2013.

Chapter 49

# Potential beneficial effects of *Bifidobacterium breve* A1 on cognitive impairment and psychiatric disorders

Ryo Okubo[a], Jinzhong Xiao[b], and Yutaka J. Matsuoka[c]

[a]*Department of Clinical Epidemiology, Translational Medical Center, National Center of Neurology and Psychiatry, Kodaira, Tokyo, Japan*, [b]*Next Generation Science Institute, Morinaga Milk Industry Co., Ltd., Zama City, Kanagawa, Japan*, [c]*Division of Health Care Research, Behavioral Science and Survivorship Research Group, Center for Public Health Sciences, National Cancer Center Japan, Tokyo, Japan*

## List of abbreviations

| | |
|---|---|
| **AD** | Alzheimer's disease |
| **Aβ** | β-amyloid peptide |
| **IL** | interleukin |
| **MCI** | mild cognitive impairment |
| **MGB** | microbiota-gut-brain |
| **MMSE** | mini-mental state examination |
| **PANSS** | positive and negative syndrome scale |
| **RBANS** | repeatable battery for the assessment of neuropsychological status |
| **TRANCE** | tumor necrosis factor-related activation-induced cytokine |

## Microbiota-gut-brain axis

Tens of trillions of microorganisms including bacteria live in our digestive tracts and are known as the gut microbiota. They play profound roles in our health. Importantly, the gut environment has been shown to influence the brain and vice versa, while the brain also affects the gut environment. The gut and the brain are linked bidirectionally and this is now widely known as the microbiota-gut-brain (MGB) axis (Fleshner, 2011; Mayer, 2011; Montiel-Castro, Gonzalez-Cervantes, Bravo-Ruiseco, & Pacheco-Lopez, 2013; Rieder, Wisniewski, Alderman, & Campbell, 2017).

Studies in rats and humans suggest that the gut microbiota composition affects brain functions through at least three pathways: an endocrine pathway related to the hypothalamic–pituitary–adrenal axis; an immune pathway that can pass the brain–blood barrier and affect microglia in the brain; and a neural pathway via stimulating the vagus nerve and affecting neurochemicals such as 5-hydroxytryptamine, dopamine, gamma-aminobutyric acid, and brain-derived neurotrophic factor (Pirbaglou et al., 2016; Romijn & Rucklidge, 2015; Wang, Lee, Braun, & Enck, 2016). Recent studies have indicated the strong influence of these pathways on the pathophysiology of psychiatric disorders (Okubo, Chen, Sekiguchi, Hamazaki, & Matsuoka, 2018; Petra et al., 2015).

Here, we present an overview of the effects of the gut microbiota in patients with neurological and cognitive impairments, including the findings of our work to date on the effects of the probiotic *Bifidobacterium breve* A1.

## MGB axis in Alzheimer's disease

A recent age- and sex-matched case–control study showed that there was a decrease in microbial diversity and changes in bacterial abundance in patients with Alzheimer's disease (AD) compared with controls. These changes included a decrease in Firmicutes and Bifidobacterium levels and an increase in Bacteroidetes levels (Vogt et al., 2017). These changes in patients with AD could influence the pathophysiology of AD.

A cross-sectional study simultaneously assessed the gut microbiome, inflammatory molecules in blood, and brain amyloid and suggested that changes in the gut microbiome were related to systematic inflammation and brain amyloidosis in patients with cognitive impairment (Cattaneo et al., 2017). Interestingly, gram-negative bacterial molecules such as lipopolysaccharide colocalized with β-amyloid peptide (Aβ) in amyloid plaques and around vessels in the AD brain (Zhan et al., 2016). These results imply that changes in the gut microbiome could influence translocation of gram-negative bacterial molecules and amyloid formation, leading to cognitive impairment in patients with AD (Zhao & Lukiw, 2015).

The Neuroscience of Depression. https://doi.org/10.1016/B978-0-12-817933-8.00040-2

## MGB axis in schizophrenia

A recent systematic review revealed that the gut microbiome of patients with schizophrenia was consistently reported to be distinct from that of controls in terms of beta diversity (Nguyen, Hathaway, Kosciolek, Knight, & Jeste, 2019). Furthermore, changes in gut microbiota composition have been shown to strongly affect immunity and inflammation (Belkaid & Hand, 2014). Therefore, gut microbiota alterations in patients with schizophrenia could influence the pathophysiology of their schizophrenia.

Previous studies have suggested that levels of proinflammatory cytokines in the blood are elevated in patients with schizophrenia (Miller, Buckley, Seabolt, Mellor, & Kirkpatrick, 2011; Na, Jung, & Kim, 2014). Translocation of the gut bacteria into the body may contribute to this elevation and thus produce an immune response (Severance, Yolken, & Eaton, 2016). The translocation is elevated in drug-naïve patients with schizophrenia (Severance et al., 2013). Furthermore, defects in genes related to gut epithelial barrier function that controls the translocation have been found in patients with schizophrenia (Sun et al., 2004; Wu et al., 2010). These findings together suggest that the gut microbiota composition and compromised gut epithelial barrier function affect the immune system, which then may contribute to the pathophysiology of schizophrenia (Desbonnet et al., 2010; O'Mahony et al., 2009; Okubo et al., 2019).

## *Bifidobacterium breve* A1 as probiotics

Probiotics are ' live microorganisms which when administered in adequate amounts confer a health benefit on the host" (FAO/WHO Joint Working Group, 2002) . Numerous studies have observed improvement of symptoms in patients with psychiatric disorders who receive probiotic interventions (Nishimura et al., 2016; Pirbaglou et al., 2016; Romijn & Rucklidge, 2015; Wang et al., 2016; Yadav, Chatterjee, Majeed, & Kumar, 2016). Probiotics are most commonly provided in the form of food or drink containing *Lactobacillus* or *Bifidobacterium* (Ng, Peters, Ho, Lim, & Yeo, 2018). In the colon of human infants, *Bifidobacterium* is the predominant microorganism and contributes to healthy growth by breaking down oligosaccharides in breast milk. The proportion of *Bifidobacterium* decreases with age as the gut microbiota becomes more diverse (Odamaki et al., 2016).

*B. breve* A1 was first isolated from the feces of human infants, which was stocked as MCC1274 in the culture collection of Morinaga Milk Industry Co., Ltd., Kanagawa, Japan, and has demonstrated anti-inflammatory effects in an animal model using Aβ (Kobayashi et al., 2017). Also, beneficial effects have been found for psychiatric disorders such as cognitive impairment in AD (Kobayashi, Kinoshita et al., 2019; Kobayashi, Kuhara, Oki, & Xiao, 2019) and anxiety and depressive symptoms in schizophrenia (Okubo et al., 2019).

## Administration of *B. breve* A1 to subjects with mild cognitive impairment

Based on the antiinflammatory effects of *B. breve* A1, we previously conducted an open-label single-arm study and a double-blind randomized control study in subjects with mild cognitive impairment (MCI) (Kobayashi, Kinoshita, et al., 2019; Kobayashi, Kuhara, et al., 2019). Neuronal inflammation has a critical role in the development of neurodegenerative diseases such as AD (Yasuno et al., 2012). We hypothesized that the inflammatory property of *B. breve* A1 would have therapeutic effects to alleviate cognitive impairment in subjects with MCI.

The preliminary single-arm study (Kobayashi, Kinoshita, et al., 2019) suggested that consumption of *B. breve* A1 at $2.0 \times 10^{10}$ colony-forming units for 24 weeks could improve not only cognitive function, but also mental health in patients with MCI. Although a significant improvement in mood was not shown by assessment using the Profile of Mood States second edition, the average scores of total mood disturbance tended to improve. Moreover, secondary analyses showed that *B. breve* A1 significantly improved anxious mood. Further studies using larger samples are needed to clarify the effects of *B. breve* A1 on mental health in patients with MCI.

To clarify the effects of consumption of *B. breve* A1 on cognitive impairment, we conducted a double-blind randomized control study for 12 weeks in older adults with memory complaints (Kobayashi, Kuhara, et al., 2019). In both placebo and *B. breve* A1 groups, there was a significant improvement in cognitive impairment, as shown by neuropsychological tests such as the Repeatable Battery for the Assessment of Neuropsychological Status (RBANS) and the Mini-Mental State Examination (MMSE). There was no significant difference in the neuropsychological tests between the placebo and *B. breve* A1 groups in the whole subjects, however, significant improvement was found in the changes from baseline of "immediate memory" subscale of the RBANS and MMSE total score, in a subgroup of subjects with worse cognitive function (RBANS score < 41) at baseline (Fig. 1).

## Effect of *B. breve* A1 on anxiety and depressive symptoms in schizophrenia

The beneficial effects of probiotics have also been investigated in patients with schizophrenia. One study

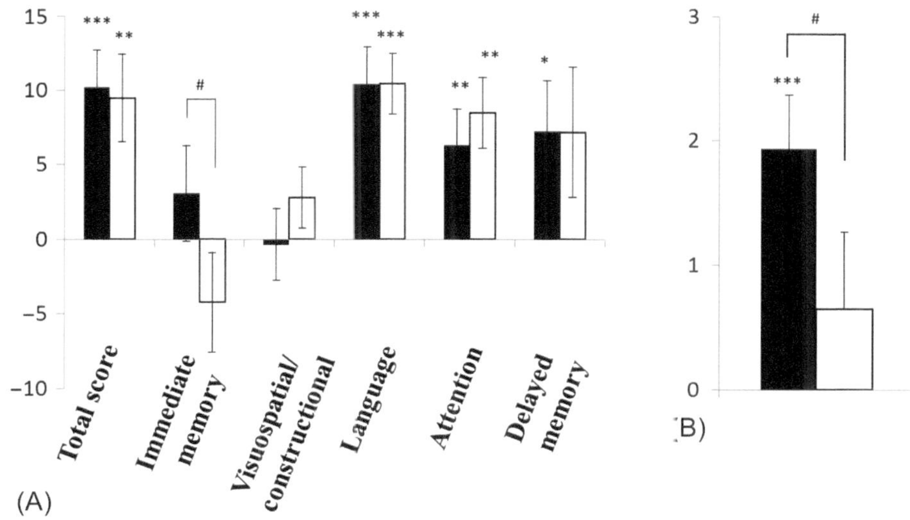


**FIG. 1** Results of the neuropsychological tests after supplementation of *Bifidobacterium breve* A1 for 12 weeks in a subgroup of patient with cognitive impairment who had a low RBANS score (<41). Cognitive function in older adults who complained of memory problems was assessed at baseline and after 12 weeks of *B. breve* A1 supplementation using the Japanese version of the Repeatable Battery for the Assessment of Neuropsychological Status (RBANS) and the Mini-Mental State Examination (MMSE). (A) Changes in RBANS score. (B) Changes in MMSE total score. Filled column: *B. breve* A1; open column: placebo. Values are shown as means ± standard deviation. *$P < 0.05$, **$P < 0.01$, ***$P < 0.001$, Wilcoxon signed-rank test within the group from baseline to 12-weeks examination; #$P < 0.05$, Wilcoxon rank-sum test between the treatment and placebo groups.

suggested administration of probiotics in patients with schizophrenia changed their inflammatory cytokines levels, although there were no positive or negative effects on symptoms (Dickerson et al., 2014). Based on the aforementioned studies, we hypothesized that consumption of *B. breve* A1 could change the gut microbiota composition and compromised gut epithelial barrier function in patients with schizophrenia, which would then affect the immune system and in turn contribute to improving anxiety and depressive symptoms in schizophrenia (Fig. 2A).

To examine our hypothesis, we conducted a proof-of-concept study examining the effects of consumption of *B. breve* A1 daily over 4 weeks on anxiety and depressive symptoms, as reported previously (Okubo et al., 2019). We assessed not only the effect of *B. breve* A1 on anxiety and depressive symptoms in patients with schizophrenia, but also the composition of the gut microbiota and immune products such as cytokines and chemokines before and after consumption of *B. breve* A1 (Fig. 2B).

All participants consumed freeze-dried *B. breve* A1 at $5.0 \times 10^{10}$ colony-forming units daily. Subjective and objective anxiety and depressive symptoms improved from baseline in the participants according to scores on the Hospital Anxiety and Depression Scale (Kugaya, Akechi, Okuyama, Okamura, & Uchitomi, 1998; Zigmond & Snaith, 1983) and scores for anxiety and depression on the Positive and Negative Syndrome Scale (Citrome, Meng, & Hochfeld, 2011) (Fig. 3).

## No significant change in the gut microbiota was observed, but *B. breve* A1 may have affected gut epithelial barrier function

Contrary to our hypothesis, after 4 weeks of consumption of *B. breve* A1 daily, the gut microbiota composition had not significantly change at 4 and 8 weeks (Okubo et al., 2019). These findings imply that there is no relationship between direct change in the gut microbiota and the effect of *B. breve* A1 on anxiety and depressive symptoms in patients with schizophrenia.

Our findings do, however, indicate that the effect of *B. breve* A1 on anxiety and depressive symptoms might be related to enhanced gut epithelial barrier function. We analyzed the Th1, Th2, and Th17 cytokines and after applying the false discovery rate for multiple comparisons (Benjamini, Drai, Elmer, Kafkafi, & Golani, 2001), we found a significant upregulation of expression levels of interleukin (IL)-22 and tumor necrosis factor-related activation-induced cytokine (TRANCE) in patients with schizophrenia who responded to *B. breve* A1. The role of these cytokines is important in gut epithelial barrier function. IL-22 deficiency has been found to change the expression of antimicrobial peptides and mucins a mouse model of colitis, resulting in greater disease severity than in wild-type controls (Sugimoto et al., 2008; Zenewicz et al., 2008; Zenewicz et al., 2013). TRANCE deficiency has been shown to induce severely impaired IgA production

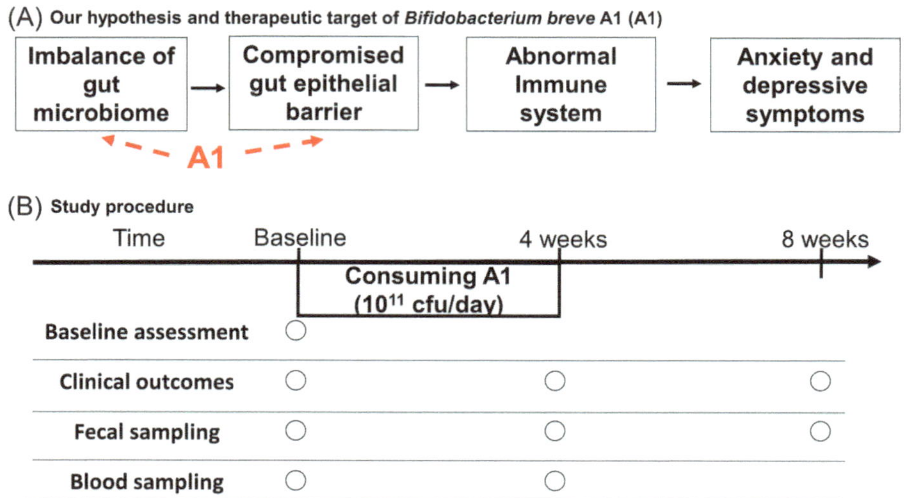


FIG. 2 Proof-of-concept study. (A) Our hypothesis and therapeutic target of *Bifidobacterium breve* A1 (A1): we hypothesized that A1 could change the composition of the gut microbiota and compromised gut epithelial barrier function in patients with schizophrenia and then affect the immune system, which in turn could affect anxiety and depressive symptoms in schizophrenia. (B) Study procedure: all participants consumed the freeze-dried form of A1 daily. *(From Okubo, R., Koga, M., Katsumata, N., Odamaki, T., Matsuyama, S., Oka, M., et al. (2019). Effect of Bifidobacterium breve A-1 on anxiety and depressive symptoms in schizophrenia: A proof-of-concept study.* Journal of Affective Disorders, 245, *377–385. https://doi.org/10.1016/j.jad.2018.11.011, with permission.)*

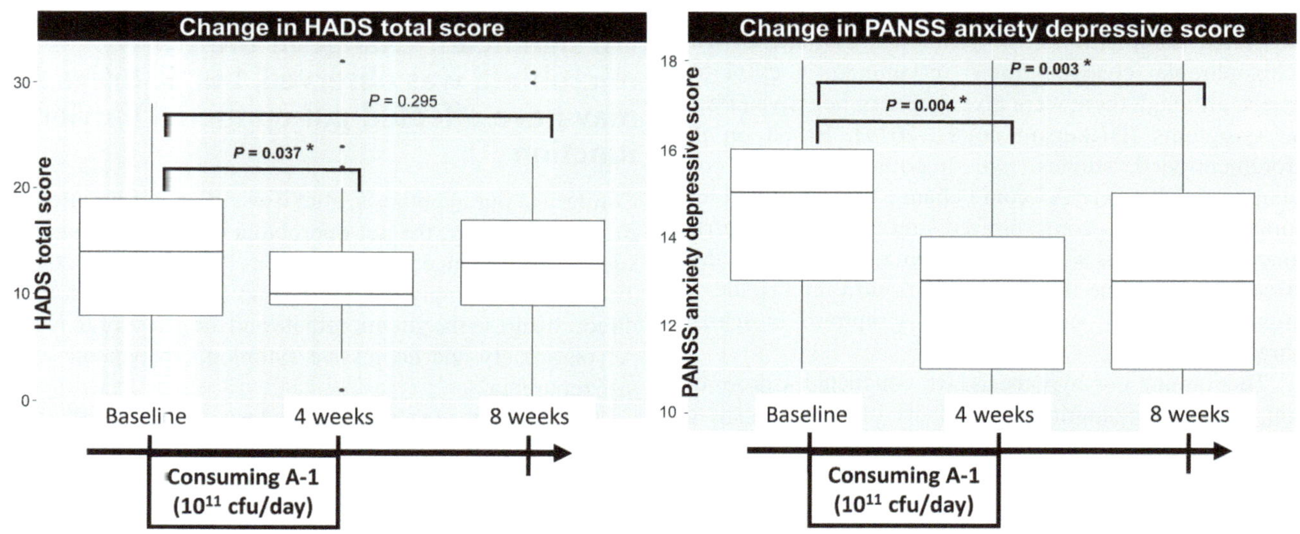


FIG. 3 Change in anxiety and depressive symptoms after patients with schizophrenia consumed *Bifidobacterium breve* A1 daily over 4 weeks. Significant differences are indicated by asterisks (Wilcoxon's signed-rank test). Significant *P*-values are shown in bold. *Abbreviations*: HADS, Hospital Anxiety and Depression Scale; PANSS, Positive and Negative Syndrome Scale. *(From Okubo, R., Koga, M., Katsumata, N., Odamaki, T., Matsuyama, S., Oka, M., et al. (2019). Effect of Bifidobacterium breve A-1 on anxiety and depressive symptoms in schizophrenia: A proof-of-concept study.* Journal of Affective Disorders, 245, *377–385. https://doi.org/10.1016/j.jad.2018.11.011, with permission.)*

in mice and to downregulate the expression of the epithelial chemokine CCL20 (Nagashima et al., 2017). Taken together, the findings of these studies suggest that consumption of *B. breve* A1 for 4 weeks may change gut epithelial barrier function related to TRANCE and IL-22 and thus improve the anxiety and depressive symptoms in patients with schizophrenia.

## Dietary habits and baseline gut microbiota could influence the effect of *B. breve* A1 on anxiety and depressive symptoms

Because no studies on probiotics had evaluated dietary habits and gut microbiota composition, we did so to assess differences from baseline between patients with

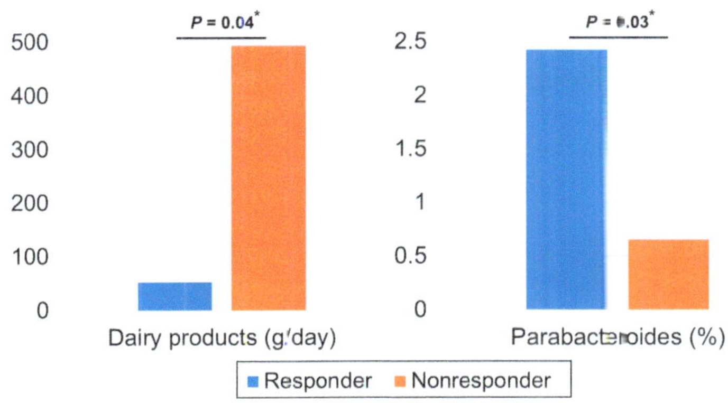


**FIG. 4** Baseline characteristics compared between patients with schizophrenia who responded to daily consumption of *Bifidobacterium breve* A1 over 4 weeks and nonresponders. Parabacteroides (%) represented relative abundance in gut microbiota. Wilcoxon's rank-sum test was used for statistical analyses.

schizophrenia who responded to *B. breve* A1 (responders) and those who did not (nonresponders) (Okubo et al., 2019).

The likelihood that responders would consume dairy products at baseline was less than that of nonresponders (Fig. 4). In one cohort study involving 14,539 adults, high yogurt consumption was found to be associated with a lower risk of depression in women (Perez-Cornago et al., 2016). Studies in Japan also revealed that a lower risk of depressive symptoms was associated with high yogurt consumption during pregnancy (Miyake, Tanaka, Okubo, Sasaki, & Arakawa, 2015) and low-fat dairy consumption in adults (Cui et al., 2017). The results of these studies indicate that dietary habits including dairy consumption may decrease depressive symptoms via the MGB axis and are consistent with our results that dietary habits including dairy consumption improve responsiveness to probiotics treatment, thereby improving anxiety and depressive symptoms in patients with schizophrenia.

Furthermore, *Parabacteroides* at baseline had a significantly greater relative abundance in responders than in nonresponders, but no significant difference was found in diversity analyses between responders and nonresponders (Fig. 4). This finding indicates that consumption of *B. breve* A1 may improve anxiety and depressive symptoms in schizophrenia patients, especially those who have significantly more *Parabacteroides* in the gut microbiota. A case–control study recently showed that high levels of *Parabacteroides* were associated with major depressive disorder (Jiang et al., 2015). This finding indicates that *Parabacteroides* can affect depressive symptoms regardless of the type of mental disorder. Further evaluation of dietary habits and the gut microbiota may allow for the realization of individualized probiotics treatment regimens for improving anxiety and depressive symptoms.

## Conclusions and perspectives

Clinical studies have shown that *B. breve* A1, a probiotic strain that has demonstrated antiinflammatory properties, could have beneficial effects on brain function and mental health as improving cognitive impairment and anxiety and depressive symptoms. Our preliminary study in 2019 suggested the beneficial effects on not only cognitive function, but also mental health in patients with MCI. Our double-blind randomized control study then confirmed the beneficial effects on cognitive impairment as evaluated by neuropsychological tests, especially in patients with worse cognitive function at baseline. Our recent 2019 proof-of-concept study also suggested that consumption of *B. breve* A1 improves gut epithelial barrier function, thereby improving anxiety and depressive symptoms in patients with schizophrenia. These beneficial effects on mental health across psychiatric diagnoses might be associated with enhanced gut epithelial barrier function. However, future studies with larger sample sizes are necessary to validate our findings. Moreover, further studies should evaluate dietary habits and the gut microbiota to allow for the realization of individualized probiotic treatment regimens for psychiatric disorders.

## Key facts

- Probiotics are "live microorganisms which when administered in adequate amounts confer a health benefit on the host". The beneficial effect of probiotics on depression has been confirmed by a recent meta-analysis, although the effect size was small. However, for those who had any mental disorders or physical illness, the effect size was moderate (Liu, Walsh, & Sheehan, 2019).

## Summary points

- *B. breve* A1 was first isolated from the feces of human infants and exhibited antiinflammatory effects in an animal model.

- Consumption of *B. breve* A1 could improve cognitive impairment and mental health in patients with MCI and anxiety and depressive symptoms in schizophrenia.
- Consumption of *B. breve* A1 could change gut epithelial barrier function relating to TRANCE and IL-22, which would then improve anxiety and depressive symptoms in patients with schizophrenia.
- Patients with schizophrenia who responded to *B. breve* A1 consumed dairy products less than nonresponders and had a greater relative abundance of *Parabacteroides* at baseline.
- *B. breve* A1 could have beneficial effects on mental health conditions such as anxiety and depression across psychiatric diagnoses.
- Evaluation of dietary habits and the gut microbiota could allow for the realization of individualized probiotics treatment regimens.

## Mini-dictionary of terms

**Gut microbiota** Microbiota is defined as a community of microorganisms (such as bacteria, fungi, and viruses). The colon has the densest community in the human body. The number of microorganisms in the human body is 10 times greater than the number of human cells. The gut microbiota is now recognized as an organ of the human body.

**Microbiota-gut-brain (MGB) axis** Gut microbiota and its relationship with brain function has received increased attention in the recent literature and is now known as the MGB axis (Dinan, Stanton, & Cryan, 2013; Sudo, 2014). Germ-free animal studies have been conducted to examine the effect of the absence of microbiota on host brain function. Results showed impacts of the MGB axis on behaviors and development of neuronal circuits in the brain (Diaz Heijtz et al., 2011).

**Gut epithelial barrier function** The composition of gut microbiota influences the permeability of the gut epithelial barrier that modulates the flow of substances between the gastrointestinal tract and bloodstream (Hollander, 1999). A compromised epithelial barrier (leaky gut) results in increased plasma lipopolysaccharides (endotoxemia) (Cani et al., 2007) and increases inflammatory cytokine expression by binding to Toll-like receptor 4 (Jia et al., 2014).

**Proof-of-concept study** Proof-of-concept studies provide first evidence for use of a candidate drug or intervention in a specific population and contribute to identifying the most appropriate treatment strategy. Such studies enable researchers to avoid wasting resources on a candidate drug or intervention that will probably fail (Campbell, Gilron, Doshi, & Raja, 2019).

## References


Belkaid, Y., & Hand, T. W. (2014). Role of the microbiota in immunity and inflammation. *Cell*, *157*(1), 121–141. https://doi.org/10.1016/j.cell.2014.03.011.

Benjamini, Y., Drai, D., Elmer, G., Kafkafi, N., & Golani, I. (2001). Controlling the false discovery rate in behavior genetics research. *Behavioural Brain Research*, *125*(1–2), 279–284.

Campbell, C. M., Gilron, I., Doshi, T., & Raja, S. (2019). Designing and conducting proof-of-concept chronic pain analgesic clinical trials. *Pain Report*, *4*(3). https://doi.org/10.1097/PR9.0000000000000697.

Cani, P. D., Amar, J., Iglesias, M. A., Poggi, M., Knauf, C., Bastelica, D., et al. (2007). Metabolic endotoxemia initiates obesity and insulin resistance. *Diabetes*, *56*(7), 1761–1772. https://doi.org/10.2337/db06-1491.

Cattaneo, A., Cattane, N., Galluzzi, S., Provasi, S., Lopizzo, N., Festari, C., et al. (2017). Association of brain amyloidosis with pro-inflammatory gut bacterial taxa and peripheral inflammation markers in cognitively impaired elderly. *Neurobiology of Aging*, *49*, 60–68. https://doi.org/10.1016/j.neurobiolaging.2016.08.019.

Citrome, L., Meng, X., & Hochfeld, M. (2011). Efficacy of iloperidone in schizophrenia: A PANSS five-factor analysis. *Schizophrenia Research*, *131*(1–3), 75–81. https://doi.org/10.1016/j.schres.2011.05.018.

Cui, Y., Huang, C., Momma, H., Ren, Z., Sugiyama, S., Guan, L., et al. (2017). Consumption of low-fat dairy, but not whole-fat dairy, is inversely associated with depressive symptoms in Japanese adults. *Social Psychiatry and Psychiatric Epidemiology*, *52*(7), 847–853. https://doi.org/10.1007/s00127-016-1333-1.

Desbonnet, L., Garrett, L., Clarke, G., Kiely, B., Cryan, J. F., & Dinan, T. G. (2010). Effects of the probiotic Bifidobacterium infantis in the maternal separation model of depression. *Neuroscience*, *170*(4), 1179–1188. https://doi.org/10.1016/j.neuroscience.2010.08.005.

Diaz Heijtz, R., Wang, S., Anuar, F., Qian, Y., Bjorkholm, B., Samuelsson, A., et al. (2011). Normal gut microbiota modulates brain development and behavior. *Proceedings of the National Academy of Sciences of the United States of America*, *108*(7), 3047–3052. https://doi.org/10.1073/pnas.1010529108.

Dickerson, F. B., Stallings, C., Origoni, A., Katsafanas, E., Savage, C. L., Schweinfurth, L. A., et al. (2014). Effect of probiotic supplementation on schizophrenia symptoms and association with gastrointestinal functioning: A randomized, placebo-controlled trial. *Prim Care Companion CNS Disord*, *16*(1). https://doi.org/10.4088/PCC.13m01579.

Dinan, T. G., Stanton, C., & Cryan, J. F. (2013). Psychobiotics: A novel class of psychotropic. *Biological Psychiatry*, *74*(10), 720–726. https://doi.org/10.1016/j.biopsych.2013.05.001.

FAO/WHO Joint Working Group. (2002). Report on Drafting Guidelines for the Evaluation of Probiotics in Food, London, Ontario, Canada, April 30 and May 1. 2002 (Accessed 05 September 2020). https://www.who.int/foodsafety/fs_management/en/probiotic_guidelines.pdf.

Fleshner, M. (2011). The gut microbiota: A new player in the innate immune stress response? *Brain, Behavior, and Immunity*, *25*(3), 395–396. https://doi.org/10.1016/j.bbi.2010.12.007.

Hollander, D. (1999). Intestinal permeability, leaky gut, and intestinal disorders. *Current Gastroenterology Reports*, *1*(5), 410–416.

Jia, L., Vianna, C. R., Fukuda, M., Berglund, E. D., Liu, C., Tao, C., et al. (2014). Hepatocyte toll-like receptor 4 regulates obesity-induced inflammation and insulin resistance. *Nature Communications*, *5*, 3878. https://doi.org/10.1038/ncomms4878.

Jiang, H., Ling, Z., Zhang, Y., Mao, H., Ma, Z., Yin, Y., et al. (2015). Altered fecal microbiota composition in patients with major depressive disorder. *Brain, Behavior, and Immunity*, *48*, 186–194. https://doi.org/10.1016/j.bbi.2015.03.016.

Kobayashi, Y., Kinoshita, T., Matsumoto, A., Yoshino, K., Saito, I., & Xiao, J. Z. (2019). Bifidobacterium breve A1 supplementation improved cognitive decline in older adults with mild cognitive impairment: An open-label, single-arm study. *Journal of Prevention of Alzheimer' Disease*, *6*(1), 70–75. https://doi.org/10.14283/jpad.2018.32.

Kobayashi, Y., Kuhara, T., Oki, M., & Xiao, J. Z. (2019). Effects of Bifidobacterium breve A1 on the cognitive function of older adults with memory complaints: A randomised, double-blind, placebo-controlled trial. *Beneficial Microbes*, *10*(5), 511–520. https://doi.org/10.3920/BM2018.0170.

Kobayashi, Y., Sugahara, H., Shimada, K., Mitsuyama, E., Kuhara, T., Yasuoka, A., et al. (2017). Therapeutic potential of Bifidobacterium breve strain A1 for preventing cognitive impairment in Alzheimer's disease. *Scientific Reports*, *7*(1), 13510. https://doi.org/10.1038/s41598-017-13368-2.

Kugaya, A., Akechi, T., Okuyama, T., Okamura, H., & Uchitomi, Y. (1998). Screening for psychological distress in Japanese cancer patients. *Japanese Journal of Clinical Oncology*, *28*(5), 333–338.

Liu, R. T., Walsh, R. F. L., & Sheehan, A. E. (2019). Prebiotics and probiotics for depression and anxiety: A systematic review and meta-analysis of controlled clinical trials. *Neuroscience and Biobehavioral Reviews*, *102*, 13–23. https://doi.org/10.1016/j.neubiorev.2019.03.023.

Mayer, E. A. (2011). Gut feelings: The emerging biology of gut-brain communication. *Nature Reviews. Neuroscience*, *12*(8), 453–466. https://doi.org/10.1038/nrn3071.

Miller, B. J., Buckley, P., Seabolt, W., Mellor, A., & Kirkpatrick, B. (2011). Meta-analysis of cytokine alterations in schizophrenia: Clinical status and antipsychotic effects. *Biological Psychiatry*, *70*(7), 663–671. https://doi.org/10.1016/j.biopsych.2011.04.013.

Miyake, Y., Tanaka, K., Okubo, H., Sasaki, S., & Arakawa, M. (2015). Intake of dairy products and calcium and prevalence of depressive symptoms during pregnancy in Japan: A cross-sectional study. *BJOG*, *122*(3), 336–343. https://doi.org/10.1111/1471-0528.12972.

Montiel-Castro, A. J., Gonzalez-Cervantes, R. M., Bravo-Ruiseco, G., & Pacheco-Lopez, G. (2013). The microbiota-gut-brain axis: Neurobehavioral correlates, health and sociality. *Frontiers in Integrative Neuroscience*, *7*, 70. https://doi.org/10.3389/fnint.2013.00070.

Na, K. S., Jung, H. Y., & Kim, Y. K. (2014). The role of pro-inflammatory cytokines in the neuroinflammation and neurogenesis of schizophrenia. *Progress in Neuro-Psychopharmacology & Biological Psychiatry*, *48*, 277–286. https://doi.org/10.1016/j.pnpbp.2012.10.022.

Nagashima, K., Sawa, S., Nitta, T., Tsutsumi, M., Okamura, T., Penninger, J. M., et al. (2017). Identification of subepithelial mesenchymal cells that induce IgA and diversify gut microbiota. *Nature Immunology*, *18*(6), 675–682. https://doi.org/10.1038/ni.3732.

Ng, Q. X., Peters, C., Ho, C. Y. X., Lim, D. Y., & Yeo, W.-S. (2018). A meta-analysis of the use of probiotics to alleviate depressive symptoms. *Journal of Affective Disorders*, *228*, 13–19. https://doi.org/10.1016/j.jad.2017.11.063.

Nguyen, T. T., Hathaway, H., Kosciolek, T., Knight, R., & Jeste, D. V. (2019). Gut microbiome in serious mental illnesses: A systematic review and critical evaluation. *Schizophrenia Research*. https://doi.org/10.1016/j.schres.2019.08.026.

Nishimura, M., Ohkawara, T., Tetsuka, K., Kawasaki, Y., Nakagawa, R., Satoh, H., et al. (2016). Effects of yogurt containing Lactobacillus plantarum HOKKAIDO on immune function and stress markers. *Journal of Traditional and Complementary Medicine*, *6*(3), 275–280. https://doi.org/10.1016/j.jtcme.2015.07.003.

Odamaki, T., Kato, K., Sugahara, H., Hashikura, N., Takahashi, S., Xiao, J. Z., et al. (2016). Age-related changes in gut microbiota composition from newborn to centenarian: A cross-sectional study. *BMC Microbiology*, *16*, 90. https://doi.org/10.1186/s12866-016-0708-5.

Okubo, R., Chen, C., Sekiguchi, M., Hamazaki, K., & Matsuoka, Y. J. (2018). Mechanisms underlying the effects of n-3 polyunsaturated fatty acids on fear memory processing and their hypothetical effects on fear of cancer recurrence in cancer survivors. *Prostaglandins, Leukotrienes, and Essential Fatty Acids*, *131*, 14–23. https://doi.org/10.1016/j.plefa.2018.03.006.

Okubo, R., Koga, M., Katsumata, N., Odamaki, T., Matsuyama, S., Oka, M., et al. (2019). Effect of Bifidobacterium breve A-1 on anxiety and depressive symptoms in schizophrenia: A proof-of-concept study. *Journal of Affective Disorders*, *245*, 377–385. https://doi.org/10.1016/j.jad.2018.11.011.

O'Mahony, S. M., Marchesi, J. R., Scully, P., Codling, C., Ceolho, A. M., Quigley, E. M., et al. (2009). Early life stress alters behavior, immunity, and microbiota in rats: Implications for irritable bowel syndrome and psychiatric illnesses. *Biological Psychiatry*, *65*(3), 263–267. https://doi.org/10.1016/j.biopsych.2008.06.026.

Perez-Cornago, A., Sanchez-Villegas, A., Bes-Rastrollo, M., Gea, A., Molero, P., Lahortiga-Ramos, F., et al. (2016). Intake of high-fat yogurt, but not of low-fat yogurt or prebiotics, is related to lower risk of depression in women of the SUN cohort study. *The Journal of Nutrition*, *146*(9), 1731–1739. https://doi.org/10.3945/jn.116.233858.

Petra, A. I., Panagiotidou, S., Hatziagelaki, E., Stewart, J. M., Conti, P., & Theoharides, T. C. (2015). Gut-microbiota-brain Axis and its effect on neuropsychiatric disorders with suspected immune dysregulation. *Clinical Therapeutics*, *37*(5), 984–995. https://doi.org/10.1016/j.clinthera.2015.04.002.

Pirbaglou, M., Katz, J., de Souza, R. J., Stearns, J. C., Motamed, M., & Ritvo, P. (2016). Probiotic supplementation can positively affect anxiety and depressive symptoms: A systematic review of randomized controlled trials. *Nutrition Research*, *36*(9), 889–898. https://doi.org/10.1016/j.nutres.2016.06.009.

Rieder, R., Wisniewski, P. J., Alderman, B. L., & Campbell, S. C. (2017). Microbes and mental health: A review. *Brain, Behavior, and Immunity*. https://doi.org/10.1016/j.bbi.2017.01.016.

Romijn, A. R., & Rucklidge, J. J. (2015). Systematic review of evidence to support the theory of psychobiotics. *Nutrition Reviews*, *73*(10), 675–693. https://doi.org/10.1093/nutrit/nuv025.

Severance, E. G., Gressitt, K. L., Stallings, C. R., Origoni, A. E., Khushalani, S., Leweke, F. M., et al. (2013). Discordant patterns of bacterial translocation markers and implications for innate immune imbalances in schizophrenia. *Schizophrenia Research*, *148*(1–3), 130–137. https://doi.org/10.1016/j.schres.2013.05.018.

Severance, E. G., Yolken, R. H., & Eaton, W. W. (2016). Autoimmune diseases, gastrointestinal disorders and the microbiome in schizophrenia: More than a gut feeling. *Schizophrenia Research*, *176*(1), 23–35. https://doi.org/10.1016/j.schres.2014.06.027.

Sudo, N. (2014). Microbiome, HPA axis and production of endocrine hormones in the gut. *Advances in Experimental Medicine and Biology*, *817*, 177–194. https://doi.org/10.1007/978-1-4939-0897-4_8.

Sugimoto, K., Ogawa, A., Mizoguchi, E., Shimomura, Y., Andoh, A., Bhan, A. K., et al. (2008). IL-22 ameliorates intestinal inflammation in a mouse model of ulcerative colitis. *The Journal of Clinical Investigation*, *118*(2), 534–544. https://doi.org/10.1172/JCI33194.

Sun, Z. Y., Wei, J., Xie, L., Shen, Y., Liu, S. Z., Ju, G. Z., et al. (2004). The CLDN5 locus may be involved in the vulnerability to schizophrenia. *European Psychiatry*, *19*(6), 354–357. https://doi.org/10.1016/j.eurpsy.2004.06.007.

Vogt, N. M., Kerby, R. L., Dill-McFarland, K. A., Harding, S. J., Merluzzi, A. P., Johnson, S. C., et al. (2017). Gut microbiome alterations in

Alzheimer's disease. *Scientific Reports*, *7*(1), 13537. https://doi.org/10.1038/s41598-017-13601-y.

Wang, H., Lee, I. S., Braun, C., & Enck, P. (2016). Effect of probiotics on central nervous system functions in animals and humans: A systematic review. *Journal of Neurogastroenterology and Motility*, *22*(4), 589–605. https://doi.org/10.5056/jnm16018.

Wu, N., Zhang, X., Jin, S., Liu, S., Ju, G., Wang, Z., et al. (2010). A weak association of the CLDN5 locus with schizophrenia in Chinese case-control samples. *Psychiatry Research*, *178*(1), 223. https://doi.org/10.1016/j.psychres.2009.11.019.

Yadav, V., Chatterjee, S. S., Majeed, M., & Kumar, V. (2016). Preventive potentials of piperlongumine and a Piper longum extract against stress responses and pain. *Journal of Traditional and Complementary Medicine*, *6*(4), 413–423. https://doi.org/10.1016/j.jtcme.2015.11.001.

Yasuno, F., Kosaka, J., Ota, M., Higuchi, M., Ito, H., Fujimura, Y., et al. (2012). Increased binding of peripheral benzodiazepine receptor in mild cognitive impairment-dementia converters measured by positron emission tomography with [(1)(1)C]DAA1106. *Psychiatry Research*, *203*(1), 67–74. https://doi.org/10.1016/j.pscychresns.2011.08.013.

Zenewicz, L. A., Yancopoulos, G. D., Valenzuela, D. M., Murphy, A. J., Stevens, S., & Flavell, R. A. (2008). Innate and adaptive interleukin-22 protects mice from inflammatory bowel disease. *Immunity*, *29*(6), 947–957. https://doi.org/10.1016/j.immuni.2008.11.003.

Zenewicz, L. A., Yin, X., Wang, G., Elinav, E., Hao, L., Zhao, L., et al. (2013). IL-22 deficiency alters colonic microbiota to be transmissible and colitogenic. *Journal of Immunology*, *190*(10), 5306–5312. https://doi.org/10.4049/jimmunol.1300016.

Zhan, X., Stamova, B., Jin, L. W., DeCarli, C., Phinney, B., & Sharp, F. R. (2016). Gram-negative bacterial molecules associate with Alzheimer disease pathology. *Neurology*, *87*(22), 2324–2332. https://doi.org/10.1212/WNL.0000000000003391.

Zhao, Y., & Lukiw, W. J. (2015). Microbiome-generated amyloid and potential impact on amyloidogenesis in Alzheimer's disease (AD). *Journal of Nature and Science*, *1*(7).

Zigmond, A. S., & Snaith, R. P. (1983). The hospital anxiety and depression scale. *Acta Psychiatrica Scandinavica*, *67*(6), 361–370.

Chapter 50

# Coenzyme $Q_{10}$ and depression

Amir Sasan Bayani Ershadi[a,b] and Mir-Jamal Hosseini[a,b]

[a]Departments of Pharmacology and Toxicology, School of Pharmacy, Zanjan University of Medical Sciences, Zanjan, Iran, [b]Applied Pharmacology Research Center, Zanjan University of Medical sciences, Zanjan, Iran

## List of abbreviation

| | |
|---|---|
| **$CoQ_{10}$** | coenzyme $Q_{10}$ |
| **HPA axis** | hypothalamus, pituitary gland, and adrenal axis |
| **GABA** | gamma-aminobutyric acid |
| **cAMP** | cyclic adenosine monophosphate |
| **ATP** | adenosine triphosphate |
| **HMG-CoA** | 3-hydroxy-3-methylglutaryl-coenzyme A |
| **PD** | Parkinson's disease |
| **MS** | multiple sclerosis |
| **ETC** | electron transfer chain |
| **MPT** | mitochondrial permeability transition |
| **FST** | forced swimming test |
| **TNF-α** | tumor necrosis factor alpha |

## A brief introduction to depression treatment regimens

Understanding depression etiology has been one of the challenging topics for scientists and clinicians for decades. Exploring the facts underlying the disease can lighten the path to find a comprehensive cure for depression. Besides accidental discovery of mood-boosting properties of some compounds leading to synaptic monoamine-deficiency theory, hyperactivity of HPA axis, activity reduction of GABA-related systems, hereditary factors underlying the disease, hypothalamic volume loss, or even circadian impairments are among the etiologies proposed to answer main questions about this multifactorial disease (Hasler, 2010). Some other etiological studies emphasize on the relationship between neuronal-endocrine systems and immune system. Pro-inflammatory cytokines can disturb usual pathways leading to disorders such as major depressive disorder and inhibit the production of serotonin precursors or even the abovementioned HPA hyperactivity (Schiepers, Wichers, & Maes, 2005). According to the abovementioned etiologies, adverse treatments are available, which are summarized in Tables 1 and 2.

Most of the patients diagnosed with depression ameliorate by the abovementioned therapies but a minority remains unresponsive. This population disease severity could be classified into a spectrum from minor symptoms to the severe ones. Lots of studies have been designed on this population in order to surpass this obstacle and some of them were promising, however, an important number of patients cannot be cured with these manipulations. In this table, treatments for the patients who cannot utilize conventional ones as a cure are also provided.

There are some minor to major interventions in this group, of which some of them are not easily accepted by patients. And even if the patient is convinced to accept the treatment, there is an important percentage of patients who do not respond to these interventions.

Over 90% of patients diagnosed with major depression experience fatigue in different levels. Fatigue, anhedonia, and impairment in energy levels are prominent features, which are commonly observed in human and animals in this disease. Recent studies highlighted involvement of mitochondrial dysfunction comes from even early stages of stressful environments and plays a newly discovered role in etiology of depression (Amiri et al., 2017; Demyttenaere, De Fruyt, & Stahl, 2005; Sonei et al., 2017). It supposed that the brain could be particularly susceptible to oxidative damage and energy deficits due to its high-energy demand, low amounts of antioxidants, and high ratio of immature cells during its development (Sonei et al., 2017). Oxidative and nitrosative stresses can reregulate tryptophan catabolites pathways in response to the damage to lipid membranes and proteins, which can affect cAMP regulation and alter circadian genes expression leading to permeability changes in the blood-brain barrier as well as decreased neurogenesis and neuroplasticity. None of the abovementioned hypotheses are comprehensive or helpful for the complete recovery of all patients. In fact, evidences like nearly 20% of the depressed patients remained unresponsive to the current medications or recurrence of the disease after medication in some of the patients indicate that further studies are needed in this area (Crown et al., 2002). In this regard, $CoQ_{10}$ level is introduced as a hallmark of depression, which is related to oxidative stress pathways.

The Neuroscience of Depression. https://doi.org/10.1016/B978-0-12-817933-8.00036-0

**TABLE 1** Unipolar depression treatments (Kupfer & Regier, 2013).

| | |
|---|---|
| Psychotherapy | Cognitive-behavioral therapy (CBT) |
| | Interpersonal psychotherapy (IPT) |
| Pharmacotherapy | Selective serotonin reuptake inhibitors (SSRIs) |
| | Serotonin-norepinephrine reuptake inhibitors (SNRIs) |
| | Atypical agents |
| | Tricyclic and tetracyclic antidepressants (TCAs) |
| | Monoamine oxidase inhibitors (MAOIs) |
| Psychotherapy combined with pharmacotherapy | |

**TABLE 2** Treatment-resistant cases (McIntyre et al., 2014; Rizvi et al., 2011).

| | |
|---|---|
| Pharmacotherapy | Ketamine, antipsychotics, lithium, levothyroxin, NSAIDs, etc. |
| Psychotherapy | Cognitive-behavioral therapy (CBT) |
| | Interpersonal psychotherapy (IPT) |
| Adjunctive therapy | |
| Neuromodulatory | Repetitive transcranial magnetic stimulation (rTMS) |
| | Electroconvulsive therapy (ECT) |
| | Deep brain stimulation (DBS) |
| | Vagus nerve stimulation (VNS) |
| Aerobic exercise | |

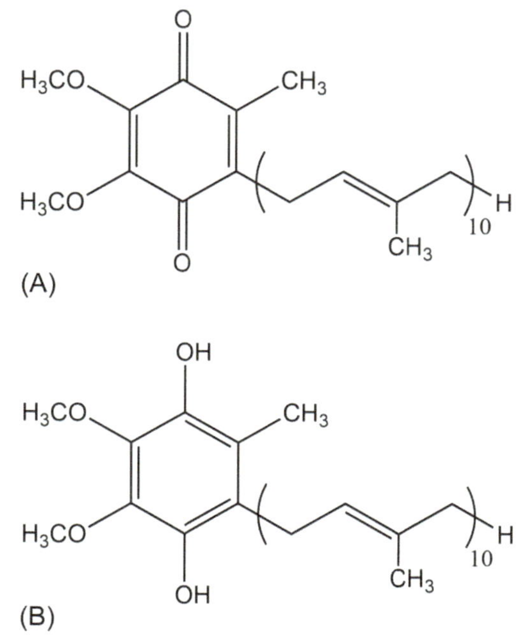


**FIG. 1** Coenzyme $Q_{10}$ structure: (A) oxidized form and (B) fully reduced form.

## $CoQ_{10}$ and its pharmacological application

Coenzyme $Q_{10}$ is a lipid-soluble molecule distributed ubiquitously in our body with a quinone structure summed up as "ubiquinone," which refers to the oxidized form of this coenzyme and can be reduced gradually from ubisemiquinone to ubiquinol (Littarru & Tiano, 2007) (Fig. 1). Due to its de novo biosynthesis in body, coenzyme $Q_{10}$ is not considered as a vitamin. In fact, each cell produces enough amount of it, if needed (Shukla & Dubey, 2018). Although $CoQ_{10}$ is found in different organelles of body, the highest concentrations are observed in the mitochondria of the heart, kidneys, liver, and brain (Turunen, Olsson, & Dallner, 2004). Therefore, brain as one of the highest energy consuming parts of our body requires approximately 10 times more energy than others. This energy is provided by the energy factories located inside each cell called mitochondria. The more cells need energy, the more they have mitochondria (Liesa, Palacín, & Zorzano, 2009).

Oxidative phosphorylation is the major pathway in mitochondria, which produces 92% of ATP required for our everyday life. This pathway consists of various components including coenzyme $Q_{10}$. $CoQ_{10}$ is placed in the inner mitochondrial membrane of human cells and transfers—accepts and donates—electrons from membrane-bound dehydrogenases to complex III of electron transfer chain (ETC). ETC is one of the pivotal pathways in ATP formation and energy reproduction (Lenaz, Fato, Formiggini, & Genova, 2007). ETC has five complexes working along each other, from which complexes I, III, and IV transfer $H^+$ to mitochondrial intermembrane to produce the gradient needed for ATP production (Fig. 2).

Complexes I and III make superoxide onions $(O_2^-)$ as a by-product of ATP synthesis pathway. Superoxide dismutase (SOD) is the first enzyme responsible for converting $O_2^-$ into $H_2O_2$ and $O_2$, which are much safer. Then, glutathione peroxidase and catalase converts $H_2O_2$ into water. If $H_2O_2$ reacts with $Cu^{2+}$ or $Fe^{2+}$, it produces $OH^-$, a highly reactive free radical, which causes oxidative damages (Bansal & Kuhad, 2016). In the presence of free radicals,

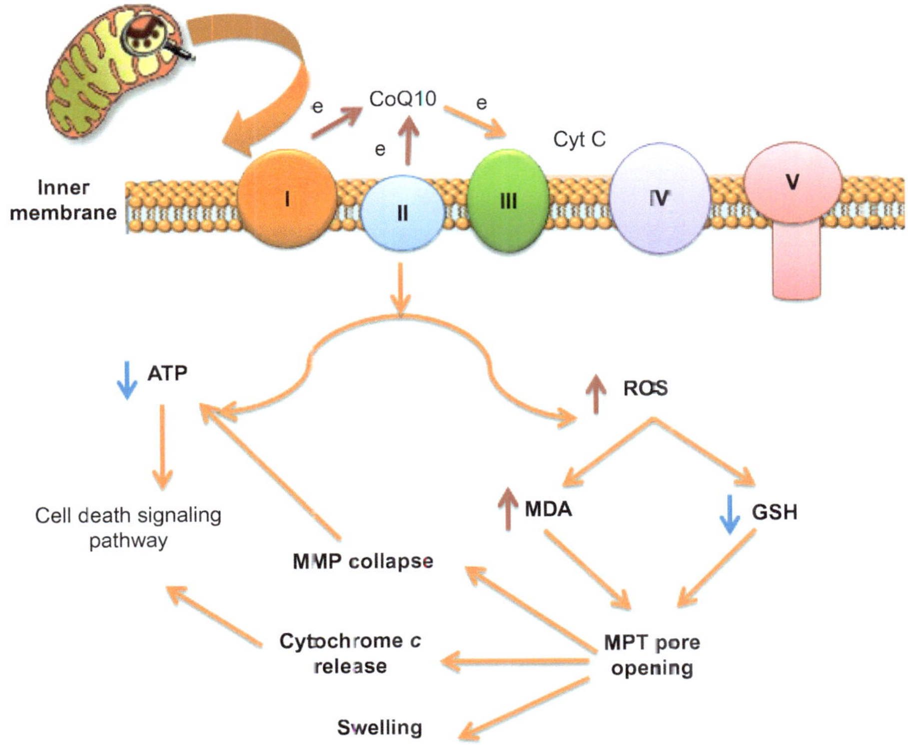


FIG. 2 Scheme of electron transfer chain (ETC).

monoamine oxidase A and B which are located on outer membrane of mitochondria, are metabolized to serotonin, dopamine, and noradrenaline. Production of ROS is needed in some signaling pathways but its overproduction can damage DNA, proteins, and lipids. $CoQ_{10}$, as an electron carrier, is a strong antioxidant, which plays a key role in the production of ATP and prevention of oxidative damages (Mattson, Gleichmann, & Cheng, 2008) (Fig. 3).

Coenzyme Q molecules play their roles in different shapes during evolution. In fact, CoQs are slightly different in different creatures. CoQ molecules are distinguished by the difference in the number of isoprenoids in their side chain.

For instance, major CoQ molecule in *Saccharomyces cervisiae*, *Escherichia coli*, rodents, and humans are $CoQ_6$, $CoQ_8$, $CoQ_9$, and $CoQ_{10}$, respectively (Fig. 4), all of which are vital for oxidative respiration. Moreover, $CoQ_{10}$ can interact as a radical scavenger and halt any oxidative or nitrosative stress occurring for lipids, proteins, and DNA (Shukla & Dubey, 2018). $CoQ_{10}$ can directly extinguish lipid peroxyl radicals and help improve other antioxidants like alpha-tocopherol (Ernster, Forsmark, & Nordenbrand, 1992; Kagan, Serbinova, & Packer, 1990).

Recent studies have clarified the effect of $CoQ_{10}$ on gene expression involved in oxidation/reduction of control on cell signaling. They also indicate that ROS are able to activate the expression of pro-inflammatory genes such as nuclear transcription factor (NF-kß). Through suppressing NFkß1 gene expression, $CoQ_{10}$ displays its antiinflammatory properties as well (Pierce, Shen, Peltzer, Thimmesch, & Hiebert, 2017; Schmelzer et al., 2008). Furthermore, the expression of another pro-inflammatory gene (C-X-C motif) ligand 2 gene (CXCL2) reduces through $CoQ_{10}$ administration (150mg/day for 14days). This feature is achieved by region-specific demethylation of $CoQ_{10}$. Also, PMAIP1 and MMD are other pro-inflammatory genes, which have been suppressed by $CoQ_{10}$ (Fischer, Onur, Schmelzer, & Döring, 2012).

Inner membrane of mitochondria has unique properties that make mitochondrial energy conservation possible by generation of electron and PH gradient over its both sides. This gradient is made by specific ion channels and complex transporters called mitochondrial permeability transition (MPT) pores. MTP pore opening results in ion imbalance, pyridine nucleotide loss, and finally hydrolysis of ATP. There are more than 40 factors, which modulate the opening of MPT pores. $CoQ_{10}$ prevents MPT pore opening through binding to a site on the pore (Walter et al., 2000). Moreover, inner mitochondrial membrane contains uncoupling proteins (UCPs) that can transport $H^+$ into the mitochondria, counteract the oxidative phosphorylation, and produce heat rather than ATP. UCPs have different subtypes that are distributed in various tissues. In addition to thermogenesis, they play the role of an oxygen radical scavenger that can cause some diseases such as

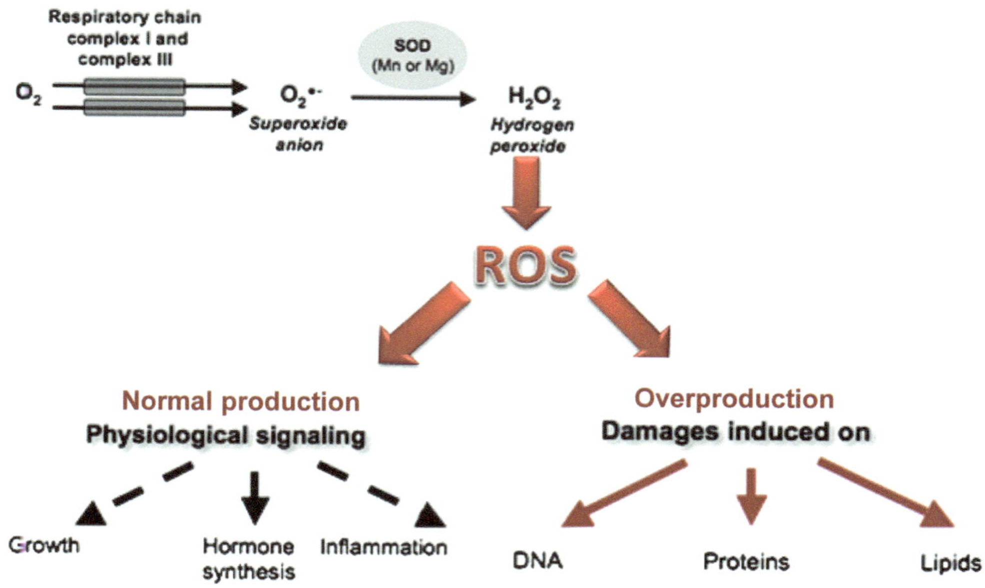


FIG. 3 Normal and overproduction of ROS during oxidative phosphorylation.

diabetes and obesity with its abnormalities. The elimination of $CoQ_{10}$ as a fundamental cofactor of UCP can make UCPs unfunctional while transporting $H^+$ (Echtay, Winkler, & Klingenberg, 2000).

$CoQ_{10}$ administration can also improve arterial endothelial dysfunction through increasing endothelial release of nitric oxide results in its dilation. Other functions of $CoQ_{10}$ in cells are transport of extramitochondrial electron to plasma membranes and lysosomes, alterations on physicochemical properties of membranes, and regulation of surface $\beta_2$-integrin amount of blood monocytes (Littarru & Tiano, 2007; Turunen et al., 2004). $CoQ_{10}$ deficiency may be primary, caused by mutations in $CoQ_{10}$ biosynthesis genes, or secondary, resulted from other diseases including mitochondrial abnormalities. Vitamin $B_6$ is another fundamental element in $CoQ_{10}$ biosynthesis whose deficiency can lead to $CoQ_{10}$ deficiency.

Moreover, statin therapy for the treatment of hyperlipidemia is developed through the inhibition of HMG-CoA reductase enzyme. Following inhibition of enzyme, mevalonate production is suppressed and consequently the cholesterol production decreases. Although these drugs are widely prescribed, long-term uses can provoke secondary $CoQ_{10}$ deficiency to the consumption of HMG-CoA reductase inhibitors. That is why the muscular complications occur after using statin. Myalgia, myositis, and rhabdomyolysis with or without increase in creatine kinase serum levels are the main symptoms of this adverse effect. Beneficial effect of $CoQ_{10}$ administration on this disorder remains controversial due to lack of sufficient evidence supporting them (Potgieter, Pretorius, & Pepper, 2013).

## Toxicity

Available evidences have shown that acute and chronic use of $CoQ_{10}$ do not cause a serious adverse effect on human or experimental animals. $CoQ_{10}$ is tolerated very well in all human population treated with doses of 200–3000 mg/day. However, gastrointestinal disturbances such as abdominal discomfort, nausea, vomiting, diarrhea, and anorexia as well as allergic rash and headache have been reported only in less than 1% of the patients participating in clinical trials (Bonakdar & Guarneri, 2005). The total pool of $CoQ_{10}$ is estimated to be 0.5–1.5 g in healthy adults. There is limited data on the metabolism of $CoQ_{10}$ in animals and humans and it is supposed that $CoQ_{10}$ is metabolized in all tissues, transported into the blood, and mainly excreted from the biliary and fecal pathway and slightly from urine pathway (Bhagavan & Chopra, 2006).

## Therapeutic uses of $CoQ_{10}$

In blood circulation in humans, about 95% of $CoQ_{10}$ exists in its reduced form as ubiquinol. The limited data on plasma $CoQ_{10}$ level indicate not only the amount of absorbed $CoQ_{10}$, but also the tissue $CoQ_{10}$ content. Although plasma $CoQ_{10}$ level does not necessarily reflect the tissue status (Bhagavan & Chopra, 2006).

$CoQ_{10}$ antiplatelet properties should be considered with more attention to rule out any risk of bleeding. Its dosage should be adjusted when prescribed with anticoagulant agents like coumarin in regard of reaching optimum prothrombin time in concomitant usage (Singh, Devaraj, & Jialal, 2007). $CoQ_{10}$ positive inotropic effect and blood

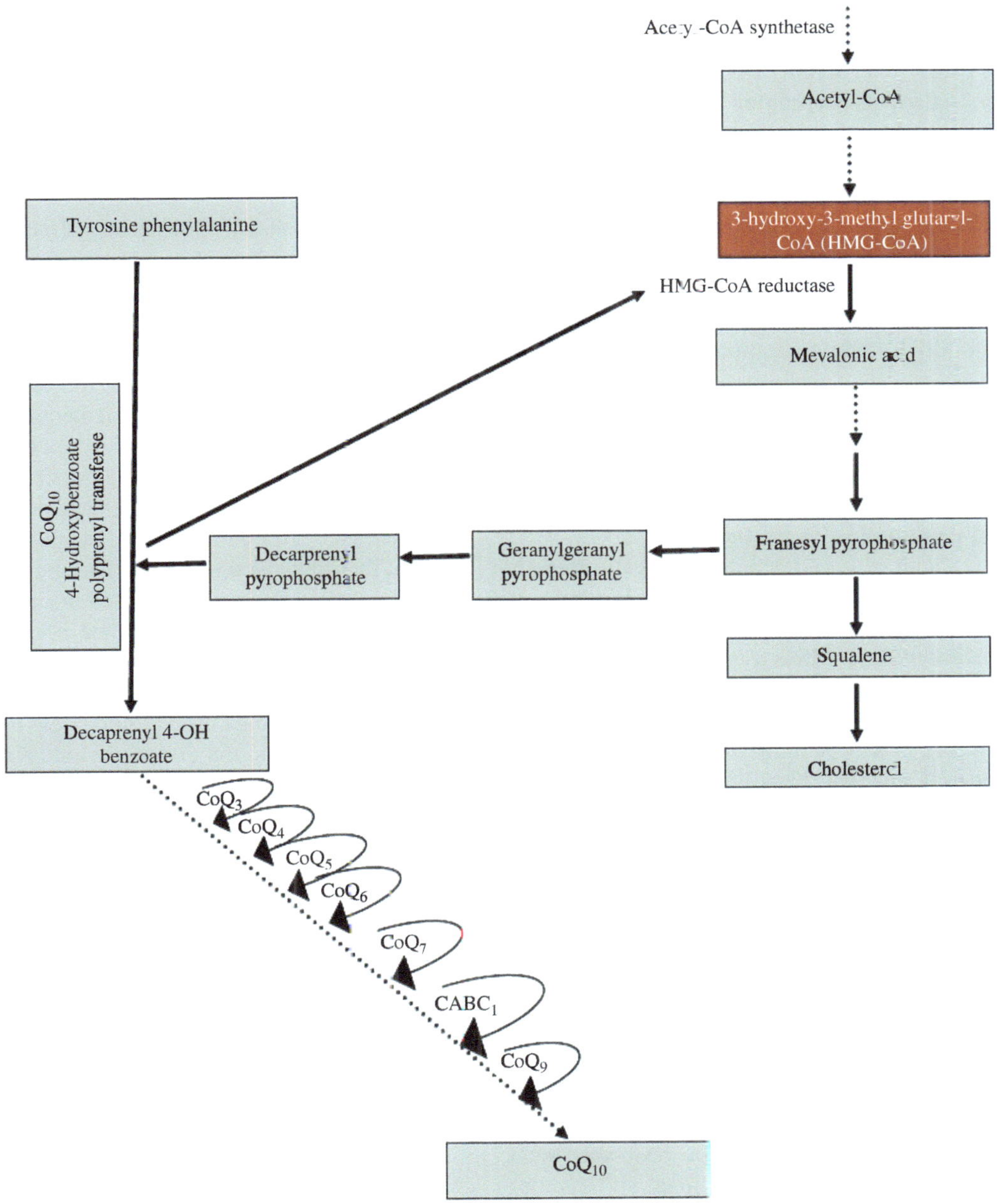


**FIG. 4** Scheme of $CoQ_{10}$ biosynthesis pathway.

pressure reduction are among causes of $CoQ_{10}$ interaction with beta-blockers. Furthermore, HMG-CoA reductase inhibitors lower $CoQ_{10}$ serum levels; thus, higher doses of $CoQ_{10}$ are suggested to attain the favorable outcome of $CoQ_{10}$ supplementation in patients that take these medications at the same time (Young, Johnson, Steffens, & Doraiswamy, 2007).

Different diseases can alter production/consumption ratio of $CoQ_{10}$. Aging is another factor that declines $CoQ_{10}$ level in tissues (Barcelos & Haas, 2019). Mitochondrial dysfunction resulted from $CoQ_{10}$ deficiency is involved in pathophysiology of several diseases including Parkinson's disease, Huntington's disease, Alzheimer's disease, fibromyalgia, atherosclerosis, hypertension, diabetes, male infertility, and migraine (Mohammadi-Bardbori & Hosseini, 2015). Clinical trials have indicated that $CoQ_{10}$ supplementation ameliorates mitochondrial abnormalities in the abovementioned diseases:

*Parkinson's disease (PD):* Mitochondrial dysfunction in dopaminergic pathway is one of the key factors leading to pathophysiology of disease (Schapira, 2008). According to the researchers, PD cellular dysfunction is correlated with plasma $CoQ_{10}$ deficiency. $CoQ_{10}$ neuroprotection studies on PD patients also revealed nigrostriatal dopaminergic system protection (Sohmiya et al., 2004).

*Huntington's disease:* Huntington's is a genetic neurodegenerative disease causing striatal neurons' death. One of the most important mechanisms reported for these neurodegenerations is mitochondrial dysfunction, accompanied with oxidative and nitrosative stresses and a marked decline in antioxidant ability of brain tissues. Several studies referred to the reduction of striatal lesions, survival extension, and some developments in motor disabilities as responses to $CoQ_{10}$ administration (Naia, Ribeiro, & Rego, 2012).

*Alzheimer's disease:* Oxidative stresses play a key role in Alzheimer's process as a neurodegenerative disease, which was consequently suggested to be treated by mitochondrial antioxidants as a potential treatment. Beneficial behavioral and neuroprotective properties of $CoQ_{10}$ were confirmed in the studies on Alzheimer's in both animals and humans (Wadsworth, Bishop, Pappu, Woltjer, & Quinn, 2008; Young et al., 2007).

*Multiple sclerosis (MS):* MS is a central nervous system disorder accompanied with demyelination and lasting inflammation. Depression and fatigue are reported as most common symptoms in MS patients (Janardhan & Bakshi, 2002). Sanoobar, Dehghan, Khalili, Azimi, and Seifar (2016) designed a randomized, double-blind trial with placebo to determine the effect of $CoQ_{10}$ supplementation (500 mg/day) on MS patients for a 12-week period. They measured fatigue and depression by Fatigue Severity Scale (FSS) and Beck Depression Inventory (BDI). The results demonstrated the significant improvement of these two areas by $CoQ_{10}$ supplementation (Sanoobar et al., 2016).

## Correlation of $CoQ_{10}$ with depression

Although the way $CoQ_{10}$ alleviates depressive symptoms in large-sized double-blind clinical trials has not been completely investigated, nearly all human and animal studies confirmed its beneficial properties in the treatment of depression. There are a lot of evidences for the immune-inflammatory response-mediated theory of pathobiology of depression:

*Animal studies*: Aboul-Fotouh (2013a, 2013b) demonstrated a significant antidepressant effect of the chronic treatment (25, 50, 100, and 150 mg/kg/day i.p. for 3 weeks) of $CoQ_{10}$ on chronic restraint stress (CRS) rat models of depression. In addition to behavioral ameliorations, hippocampal catalase and glutathione peroxidase were restored to normal condition in dose-dependent manner. Furthermore, malondialdehyde (lipid peroxidation biomarker) nitric oxide and 8-hydroxy-2′-deoxyguanosine levels (DNA oxidation biomarker) decreased following $CoQ_{10}$ administration in hippocampus (Aboul-Fotouh, 2013b). In another study on CRS, Aboul-Fotouh (2013a, 2013b) found that rats adrenal gland weight decreased in a dose-dependent therapy with $CoQ_{10}$. On the other hand, the treatment led to an increase in hippocampal and prefrontal cortex creatine kinase activity, which supported brain energy impairment role in depression (Aboul-Fotouh, 2013a).

Andalib, Mashhadi-Mousapour, Bijani, and Hosseini (2019) investigated brain $CoQ_{10}$ levels and behavioral properties of depressive mice using high-performance liquid chromatography (HPLC) method. The results demonstrated decreased brain $CoQ_{10}$ after streptozotocin administration and behavioral deterioration in forced swimming test (FST) and splash test. $CoQ_{10}$ supplementation compensated all of the abovementioned defects including behavioral declines and brain $CoQ_{10}$ levels (Andalib et al., 2019). Fu, Ji, and Dam (2010) discovered that $CoQ_{10}$-supplemented ICR mice had extra endurance in swimming-like activities and that liver glycogen reservoirs had greater content when serum urea nitrogen levels decreased in animals. The results supported antifatigue and endurance boosting property of coenzyme $Q_{10}$ (Fu et al., 2010).

Antidepressant effects of $CoQ_{10}$ were investigated in male Wistar rats using chronic mild stress (CMS) protocol. Based on the results, the effects include restoration of the balance of tryptophan catabolites pathways (kynurenine and serotonin) into serotonin was reported by downregulating hippocampal indoleamine 2,3-dioxygenase 1 (IDO-1), amelioration of depressive behaviors and glutathione and catalase levels, normalization of the elevated malondialdehyde and 4-hydroxynonenal, as well as reduction in pro-inflammatory cytokines (Abuelezz, Hendawy, & Magdy, 2017). Studies demonstrated that TNF-α, IL-5, thrombin, C reactive protein, vitronectin, and vitronectin receptor are modulated by $CoQ_{10}$ through NFkB1. It supposed that $CoQ_{10}$ can significantly reduce the pro-inflammatory cytokines (Schmelzer et al., 2008).

*Human studies*: Maes et al. (2009) investigated plasma concentrations of $CoQ_{10}$ in 35 depressed patients and 22 normal people and demonstrated the decreased $CoQ_{10}$ levels in depressed patients compared with the normal people. Since treatment-resistant depression patients had the lowest $CoQ_{10}$ levels, further studies are suggested to evaluate possible beneficial role of $CoQ_{10}$ supplementation in their treatment (Maes et al., 2009). Mehrpooya, Yasrebifar, Haghighi, Mohammadi, and Jahangard (2018) designed a double-blind study with placebo on 69 bipolar patients in their depressive episodes. They evaluated their participants at the beginning of the study, 4 weeks and finally 8 weeks after treatment with $CoQ_{10}$ (200 mg/day). $CoQ_{10}$ had minimal adverse effects and significantly decreased the symptoms of depression in comparison with placebo-treated patients (Mehrpooya et al., 2018). Along with this study, Jahangard, Yasrebifar, Haghighi, Ranjbar, and Mehrpooya

(2019) proclaimed that same 69 patients with abovementioned treatment had significantly higher serum levels of total antioxidant capacity (TAC) and total thiol group (TTG). TNF-α, IL-10, and NO were also decreased following $CoQ_{10}$ administration (Jahangard et al., 2019). Forester et al. (2015) indicated that $CoQ_{10}$ supplementation (800 mg per day) for 4 weeks in depressed bipolar people improved their depressive behavior (Forester et al., 2015). Several case studies confirmed the amelioration of depressive symptom in different psychiatric disorders due to $CoQ_{10}$ supplementation (60–200 mg/kg) (Onishi et al., 1997; Plotnikoff & Barber, 2016; Shinkai, Nakashima, Ohmori, Terao, & Nakamura, 2000; Yamazaki, Igarashi, Hamamoto, Miyazaki, & Nonaka, 1991).

## Conclusion

During the recent decades, decline in $CoQ_{10}$ level in human plasma and animal tissues is confirmed in depression and psychiatric diseases. In this regard, numerous clinical and preclinical studies have shown the administration of $CoQ_{10}$ relatively reversed mitochondrial dysfunction, behavioral abnormality, and overexpression of inflammatory responses in depression. Furthermore, according to the studies, no serious adverse effects reported in acute or chronic administration of $CoQ_{10}$ on human or experimental animals and it is tolerated very well in all human population. Therefore, it is supposed that antioxidant activity of $CoQ_{10}$ is accompanied with decreasing mitochondrial ROS in ETC, enhancing of mitochondrial bioenergetics, and produces neuroprotective effects via rising in mitochondrial and tissue $CoQ_{10}$ levels in depression. It is assumed that $CoQ_{10}$ supplementation may be useful in the treatment of neurodegenerative diseases.

## Key facts of mice behavioral tests

- They are usually performed for investigation of different diseases and medications.
- Depression is a widely studied disease that has known behavioral tests to assess its status after and before certain alterations.
- They usually induce stressful situations to animals and assess its certain parameters. These tests are based on tendency of animal to escape distress or induction of their reward centers by sweet solution consumption, etc.
- Forced swimming test is a well-known behavioral test and decrease in mobility time of animal during its standard swimming condition is a sign of deteriorating depression condition.
- Splash test is another test based on tendency of animal on self-grooming and sweet consumption. The more seconds animal groom itself, the better its mental condition is.

## Key facts of human depression studies

- Most of depression studies on humans contain questionnaires that evaluate changes on participants.
- Some other blood biomarkers like BDNF, SOD, and MDA are other factors that can be measured in these studies.
- Genetic and epigenetic studies and also different imaging techniques to evaluate volume of various parts of brain including hippocampus are other aspects that paid attention to in such studies.

## Key facts of oxidative stress and antioxidants

- Oxidation is a reaction with oxygen that can be harmful in certain circumstances.
- Oxidative phosphorylation has a by-product that is known as free radicals. These molecules have uneven electrons on their oxygen molecule.
- Free radicals can react very fast and produce chain reactions that oxidize anything around them.
- Antioxidants are usually electron donor compounds that can stabilize excessive free radicals and prevent these harmful chains continuance.

## Summary points

- Conventional theories of depression do not lead to a proper treatment for all patients and in this regard a comprehensive theory is needed.
- Mitochondrial dysfunction as a depression etiology highlights whole other aspects of this disease and compounds like $CoQ_{10}$ has promising abilities to help patients improve their symptoms.
- Nearly, all animal and human studies of $CoQ_{10}$ supplementation are in favor of alleviation of depressed animal models and patients. Behavioral improvements are obvious and novel precise biomarker measurements using HPLC demonstrated even more precise information on $CoQ_{10}$ mechanism on depression treatment.
- Clinical trials with wide range of participants could help to attain the best guidelines in depression treatment.

## Mini-dictionary of terms

**HPA axis** It is one of the most important systems in stress induction. Hypothalamic-pituitary-adrenal axis can over-activate or suppressed by various compounds and situations.

**Pro-inflammatory cytokine** TNF-α, IL-1β, and IL-6, IL-10, and IL-13 are classified as pro-inflammatory cytokines and upregulate inflammatory responses and secreted from certain immune cells. Some diseases can dysregulate their production and release.

**Free radicals** Free radicals contain unpaired electrons and can easily react with any other molecule. They mostly produce as a by-product of oxygen metabolism and can damage any compound interact with.

**Electron transfer chain** There are a series of subunits in inner membrane of mitochondria that shuttles electron to create $H^+$ gradient in order to produce ATP.

**Quinone** It is an organic oxidized compound containing double binds of carbon and oxygen. They have electrophilic properties and stabilize by receiving electrons.

**HPLC** It is a kind of column chromatography with high sensitivity. The sample usually injected into the apparatus and goes through a column with specific properties with high pressure in order to identify and quantify different compounds.

**Myalgia** It is a mild to severe pain in muscles and can be caused by various origins including any muscle injuries, some medications, or even dehydration.

**Rhabdomyolysis** Muscles contains myoglobin for their oxygen reservoir and any breakdown in muscle cells can end in its release into blood and induce life-threatening kidney damages. Injuries and some medications are listed as one of their causes.

**GABA** Gamma-aminobutyric acid is neurotransmitter in central nervous system that acts as an inhibitor in mature brains and decreases some activities of neural complexes.

**Placebo** It could be any kind of drug dosage form with no amount of active substance in order to compare it with groups of participants that receive the actual drugs in clinical trials.

**Nigrostriatal** It is a dopamine brain pathway that comes from substantia nigra in the midbrain to dorsal striatum in forebrain. It has pivotal role in voluntary movement, cognition, and reward functions.

**Anticoagulant** It is a class of antithrombotic agents reducing fibrin formation in order to prevent blood clotting. A blood cloth can clump vessels and this blockage dysregulates nutrition supply of tissues leading to probable fatal situations.

## References


Aboul-Fotouh, S. (2013a). Chronic treatment with coenzyme Q10 reverses restraint stress-induced anhedonia and enhances brain mitochondrial respiratory chain and creatine kinase activities in rats. *Behavioural Pharmacology*, *24*(7), 552–560. https://doi.org/10.1097/FBP.0b013e3283654029.

Aboul-Fotouh, S. (2013b). Coenzyme Q10 displays antidepressant-like activity with reduction of hippocampal oxidative/nitrosative DNA damage in chronically stressed rats. *Pharmacology, Biochemistry, and Behavior*, *104*, 105–112. https://doi.org/10.1016/j.pbb.2012.12.027.

Abuelezz, S. A., Hendawy, N., & Magdy, Y. (2017). Targeting oxidative stress, cytokines and serotonin interactions via Indoleamine 2, 3 dioxygenase by coenzyme Q10: Role in suppressing depressive like behavior in rats. *Journal of Neuroimmune Pharmacology*, *12*(2), 277–291. https://doi.org/10.1007/s11481-016-9712-7.

Amiri, S., Haj-Mirzaian, A., Momeny, M., Amini-Khoei, H., Rahimi-Balaei, M., Poursaman, S., et al. (2017). Streptozotocin induced oxidative stress, innate immune system responses and behavioral abnormalities in male mice. *Neuroscience*, *340*, 373–383.

Andalib, S., Mashhadi-Mousapour, M., Bijani, S., & Hosseini, M. J. (2019). Coenzyme Q10 alleviated behavioral dysfunction and bioenergetic function in an animal model of depression. *Neurochemical Research*, *44*(5), 1182–1191. https://doi.org/10.1007/s11064-019-02761-0.

Bansal, Y., & Kuhad, A. (2016). Mitochondrial dysfunction in depression. *Current Neuropharmacology*, *14*(6), 610–618.

Barcelos, I. P.d., & Haas, R. H. (2019). Coq10 and aging. *Biology*, *8*(2), 28.

Bhagavan, H. N., & Chopra, R. K. (2006). Coenzyme Q10: Absorption, tissue uptake, metabolism and pharmacokinetics. *Free Radical Research*, *40*(5), 445–453.

Bonakdar, R. A., & Guarneri, E. (2005). Coenzyme Q10. *American Family Physician*, *72*(6), 1065–1070.

Crown, W. H., Finkelstein, S., Berndt, E. R., Ling, D., Poret, A. W., Rush, A. J., et al. (2002). The impact of treatment-resistant depression on health care utilization and costs. *The Journal of Clinical Psychiatry*, *63*(11), 963–971.

Demyttenaere, K., De Fruyt, J., & Stahl, S. M. (2005). The many faces of fatigue in major depressive disorder. *International Journal of Neuropsychopharmacology*, *8*(1), 93–105.

Echtay, K. S., Winkler, E., & Klingenberg, M. (2000). Coenzyme Q is an obligatory cofactor for uncoupling protein function. *Nature*, *408*(6812), 609.

Ernster, L., Forsmark, P., & Nordenbrand, K. (1992). The mode of action of lipid-soluble antioxidants in biological membranes. Relationship between the effects of ubiquinol and vitamin E as inhibitors of lipid peroxidation in submitochondrial particles. *Journal of Nutritional Science and Vitaminology*, *38*(Special), 548–551.

Fischer, A., Onur, S., Schmelzer, C., & Döring, F. (2012). Ubiquinol decreases monocytic expression and DNA methylation of the pro-inflammatory chemokine ligand 2 gene in humans. *BMC Research Notes*, *5*(1), 540.

Forester, B. P., Harper, D. G., Georgakas, J., Ravichandran, C., Madurai, N., & Cohen, B. M. (2015). Antidepressant effects of open label treatment with coenzyme Q10 in geriatric bipolar depression. *Journal of Clinical Psychopharmacology*, *35*(3), 338.

Fu, X., Ji, R., & Dam, J. (2010). Antifatigue effect of coenzyme Q10 in mice. *Journal of Medicinal Food*, *13*(1), 211–215.

Hasler, G. (2010). Pathophysiology of depression: Do we have any solid evidence of interest to clinicians? *World Psychiatry*, *9*(3), 155–161.

Jahangard, L., Yasrebifar, F., Haghighi, M., Ranjbar, A., & Mehrpooya, M. (2019). Influence of adjuvant coenzyme Q10 on inflammatory and oxidative stress biomarkers in patients with bipolar disorders during the depressive episode. *Molecular Biology Reports*, *46*(5), 5333–5343.

Janardhan, V., & Bakshi, R. (2002). Quality of life in patients with multiple sclerosis: The impact of fatigue and depression. *Journal of the Neurological Sciences*, *205*(1), 51–58.

Kagan, V., Serbinova, E., & Packer, L. (1990). Antioxidant effects of ubiquinones in microsomes and mitochondria are mediated by tocopherol recycling. *Biochemical and Biophysical Research Communications*, *169*(3), 851–857.

Kupfer, D. J., & Regier, D. A. (2013). *Diagnostic and statistical manual of mental disorders (DSM-5®)* (5th ed.). Washington: American Psychiatric Association.

Lenaz, G., Fato, R., Formiggini, G., & Genova, M. L. (2007). The role of coenzyme Q in mitochondrial electron transport. *Mitochondrion*, *7*, 8–33. https://doi.org/10.1016/j.mito.2007.03.009.

Liesa, M., Palacín, M., & Zorzano, A. (2009). Mitochondrial dynamics in mammalian health and disease. *Physiological Reviews*, *89*(3), 799–845.

Littarru, G. P., & Tiano, L. (2007). Bioenergetic and antioxidant properties of coenzyme Q 10: Recent developments. *Molecular Biotechnology*, *37*(1), 31–37.

Maes, M., Mihaylova, I., Kubera, M., Uytterhoeven, M., Vrydags, N., & Bosmans, E. (2009). Lower plasma coenzyme Q 10 in depression: A marker for treatment resistance and chronic fatigue in depression and a risk factor to cardiovascular disorder in that illness. *Neuroendocrinology Letters*, *30*(4), 462–469.

Mattson, M. P., Gleichmann, M., & Cheng, A. (2008). Mitochondria in neuroplasticity and neurological disorders. *Neuron*, *60*(5), 748–766.

McIntyre, R. S., Filteau, M.-J., Martin, L., Patry, S., Carvalho, A., Cha, D. S., et al. (2014). Treatment-resistant depression: Definitions, review of the evidence, and algorithmic approach. *Journal of Affective Disorders*, *156*, 1–7.

Mehrpooya, M., Yasrebifar, F., Haghighi, M., Mohammadi, Y., & Jahangard, L. (2018). Evaluating the effect of coenzyme Q10 augmentation on treatment of bipolar depression: A double-blind controlled clinical trial. *Journal of Clinical Psychopharmacology*, *38*(5), 460–466. https://doi.org/10.1097/JCP.0000000000000938.

Mohammadi-Bardbori, A., & Hosseini, M.-J. (2015). Therapeutic implication of coenzyme Q10 during statin therapy: Pros and cons. *Trends in Pharmaceutical Sciences*, *1*(3), 119–128.

Naia, L., Ribeiro, M. J., & Rego, A. C. (2012). Mitochondrial and metabolic-based protective strategies in Huntington's disease: The case of creatine and coenzyme Q. *Reviews in the Neurosciences*, *23*(1), 13–28.

Onishi, H., Kawanishi, C., Iwasawa, T., Osaka, H., Hanihara, T., Inoue, K., et al. (1997). Depressive disorder due to mitochondrial transfer RNALeu (UUR) mutation. *Biological Psychiatry*, *41*(11), 1137–1139.

Pierce, J. D., Shen, Q., Peltzer, J., Thimmesch, A., & Hiebert, J. B. (2017). A pilot study exploring the effects of ubiquinol on brain genomics after traumatic brain injury. *Nursing Outlook*, *65*(5), S44–S52.

Plotnikoff, G., & Barber, M. (2016). Refractory depression, fatigue, irritable bowel syndrome, and chronic pain: A functional medicine case report. *The Permanente Journal*, *20*(4), 104.

Potgieter, M., Pretorius, E., & Pepper, M. S. (2013). Primary and secondary coenzyme Q10 deficiency: The role of therapeutic supplementation. *Nutrition Reviews*, *71*(3), 180–188.

Rizvi, S. J., Donovan, M., Giacobbe, P., Placenza, F., Rotzinger, S., & Kennedy, S. H. (2011). *Neurostimulation therapies for treatment resistant depression: A focus on vagus nerve stimulation and deep brain stimulation*. Taylor & Francis.

Sanoobar, M., Dehghan, P., Khalili, M., Azimi, A., & Seifar, F. (2016). Coenzyme Q10 as a treatment for fatigue and depression in multiple sclerosis patients: A double blind randomized clinical trial. *Nutritional Neuroscience*, *19*(3), 138–143. https://doi.org/10.1179/1476830515Y.0000000002.

Schapira, A. H. (2008). Mitochondria in the aetiology and pathogenesis of Parkinson's disease. *The Lancet Neurology*, *7*(1), 97–109.

Schiepers, O. J., Wichers, M. C., & Maes, M. (2005). Cytokines and major depression. *Progress in Neuro-Psychopharmacology and Biological Psychiatry*, *29*(2), 201–217.

Schmelzer, C., Lindner, I., Rimbach, G., Niklowitz, P., Menke, T., & Döring, F. (2008). Functions of coenzyme Q10 in inflammation and gene expression. *BioFactors*, *32*(1–4), 179–183.

Shinkai, T., Nakashima, M., Ohmori, O., Terao, T., & Nakamura, J. (2000). Coenzyme Q10 improves psychiatric symptoms in adult-onset mitochondrial myopathy, encephalopathy, lactic acidosis and stroke-like episodes: A case report. *Australian & New Zealand Journal of Psychiatry*, *34*(6), 1034–1035.

Shukla, S., & Dubey, K. K. (2018). CoQ10 a super-vitamin: Review on application and biosynthesis. *3 Biotech*, *8*(5), 249. https://doi.org/10.1007/s13205-018-1271-6.

Singh, U., Devaraj, S., & Jialal, I. (2007). Coenzyme Q10 supplementation and heart failure. *Nutrition Reviews*, *65*(6), 286–293.

Sohmiya, M., Tanaka, M., Tak, N. W., Yamagisawa, M., Tanino, Y., Suzuki, Y., et al. (2004). Redox status of plasma coenzyme Q10 indicates elevated systemic oxidative stress in Parkinson's disease. *Journal of the Neurological Sciences*, *223*(2), 161–166.

Sonei, N., Amiri, S., Jafarian, I., Anoush, M., Rahimi-Balaei, M., Bergen, H., et al. (2017). Mitochondrial dysfunction bridges negative affective disorders and cardiomyopathy in socially isolated rats: Pros and cons of fluoxetine. *The World Journal of Biological Psychiatry*, *18*(1), 39–53.

Turunen, M., Olsson, J., & Dallner, G. (2004). Metabolism and function of coenzyme Q. *Biochimica et Biophysica Acta (BBA)-Biomembranes*, *1660*(1–2), 171–199.

Wadsworth, T. L., Bishop, J. A., Pappu, A. S., Woltjer, R. L., & Quinn, J. F. (2008). Evaluation of coenzyme Q as an antioxidant strategy for Alzheimer's disease. *Journal of Alzheimer's Disease*, *14*(2), 225–234.

Walter, L., Nogueira, V., Leverve, X., Heitz, M.-P., Bernardi, P., & Fontaine, E. (2000). Three classes of ubiquinone analogs regulate the mitochondrial permeability transition pore through a common site. *Journal of Biological Chemistry*, *275*(38), 29521–29527.

Yamazaki, M., Igarashi, H., Hamamoto, M., Miyazaki, T., & Nonaka, I. (1991). A case of mitochondrial encephalomyopathy with schizophrenic psychosis, dementia and neuroleptic malignant syndrome. *Rinsho shinkeigaku (Clinical Neurology)*, *31*(11), 1219–1223.

Young, A. J., Johnson, S., Steffens, D. C., & Doraiswamy, P. M. (2007). Coenzyme Q10: A review of its promise as a neuroprotectant. *CNS Spectrums*, *12*(1), 62–68.

Chapter 51

# Gene expression in major depressive disorder: Peripheral tissue and brain-based studies

Kristin Mignogna[a,b] and Fernando S. Goes[b]

[a]Department of Mental Health, Bloomberg School of Public Health, Johns Hopkins University, Baltimore, MA, United States; [b]Department of Psychiatry and Behavioral Sciences, Johns Hopkins University School of Medicine, Baltimore, MD, United States

## List of abbreviations

| | |
|---|---|
| **ACC** | anterior cingulate cortex |
| **AMYG** | amygdala |
| **cDNA** | complementary DNA |
| **DE** | differential expression |
| **DLPFC** | dorsolateral prefrontal cortex |
| **DNA** | deoxyribonucleic acid |
| **eQTL** | expression quantitative trait locus |
| **FDR** | false discovery rate |
| **GO** | gene ontology |
| **GWAS** | genome-wide association study |
| **HC** | hippocampus |
| **MDD** | major depressive disorder |
| **mRNA** | messenger RNA |
| **NAc** | nucleus accumbens |
| **NESDA** | Netherlands Study of Depression and Anxiety |
| **OFC** | orbitofrontal cortex |
| **PBMC** | peripheral blood mononuclear cells |
| **PFC** | prefrontal cortex |
| **PGC** | Psychiatric Genomics Consortium |
| **RNA** | ribonucleic acid |
| **RNAseq** | RNA sequencing |
| **SMRI** | Stanley Medical Research Institute |
| **TWAS** | transcriptome-wide association study |
| **WB** | whole blood |

## Introduction

Major depressive disorder (MDD) is a common, complex disorder that ranks among the most disabling and costly illnesses across the globe (Global Burden of Disease Study 2013 Collaborators, 2015). When severe, persistent, or comorbid with other conditions, it can be associated with high overall disability, increased rates of suicide, and premature mortality (Laursen, Musliner, Benros, Vestergaard, & Munk-Olsen, 2016). Pharmacological and psychotherapeutic treatments are moderately effective, although approximately a third of treated subjects suffer from treatment-resistant symptoms (Gaynes et al., 2009), highlighting the need to develop novel therapies that specifically target the underlying pathophysiology of MDD. However, coherent models of the pathophysiology of MDD remain elusive, necessitating novel, robust discoveries to usher a new era of more targeted therapies. Fortunately, rapid advances in technology have provided an unprecedented ability to measure biological markers that could provide these novel insights. While a variety of approaches may ultimately need to be tested and combined to best understand the underlying biology of MDD, in this chapter, we focus on the potential for gene expression, a fundamental molecular feature that can index both genetic and environmental risk, to provide key insights into the pathophysiology of MDD.

## Gene expression and its measurement

For the purposes of this chapter, we define "gene expression" as the amount of messenger RNA (mRNA) measured in a specific cell or tissue. mRNA is the intermediate molecule transcribed from DNA coding regions or exons, that is translated into a protein. Other forms of transcribed but noncoding RNA (Morris & Mattick, 2014), such as long noncoding RNAs or microRNAs, are also widely expressed and have increasingly been shown to have important regulatory function. An important level of variation arises from splicing (or cleaving of RNA molecules) during RNA processing, which produces different combinations of exons, called isoforms, that can yield different forms of proteins following translation (Fig. 1).

The two most common methods of measuring genome-wide RNA abundance are RNA microarrays and RNA sequencing (RNAseq). Both involve an initial reverse

The Neuroscience of Depression. https://doi.org/10.1016/B978-0-12-817933-8.00095-5

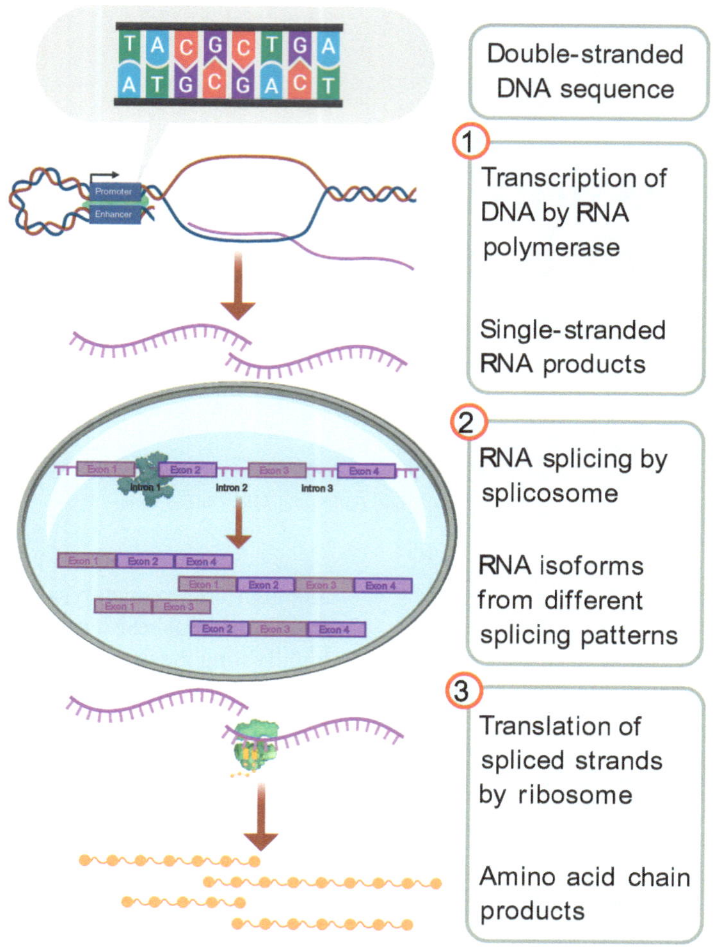


FIG. 1 Gene expression—an overview.

transcription step to convert RNA to complimentary DNA (cDNA), which is more stable and easier to quantify. Microarray technology involves labeling the resulting cDNA before annealing to synthetic DNA sequences anchored across the surface of a glass plate, or "chip." However, in the last decade, microarray technology has been supplanted by RNAseq, which provides a more comprehensive survey of RNA molecules, including information on specific combinations of exons, or isoforms. RNAseq usually employs one of two methods to enrich the yield of functionally relevant RNA: poly (A) capture, or ribosomal RNA depletion. The former targets the poly-adenosine tail found in mRNA (and some noncoding RNA) molecules, while the latter aims to reduce the amount of ribosomal RNA, thereby increasing the proportion of mRNA and long noncoding RNAs sequenced.

As sequencing generally requires an initial fragmentation step and a typical fragment usually consists of no more than 1–2 exons (or 300–500 base pairs), statistical tools are required to reconstruct contiguous reads into an isoform, albeit with some uncertainty. While emerging long-read sequencing technologies can provide better isoform resolution, all genome-wide studies currently published have used "short-read" technology, with limited isoform quantification. Once sequencing is performed, analytical methods are applied to normalize the data and reduce the effect of confounders such as varying RNA quality and experimental batch effects. Such procedures are important for reducing false positives and are well described in a number of recent reviews (Conesa et al., 2016; Van den Berge et al., 2019).

## Candidate gene vs whole-genome approaches

Hypothesis-driven candidate gene studies focusing on a small number of prespecified genes have been shown to have a high rate of false positives (Border et al., 2019). Potential reasons for this include failure to account for potential confounders, including genetic ancestry, familial relatedness, and batch effects (Johnson, Li, & Rabinovic, 2007). In contrast, whole-genome or transcriptome approaches more readily allow for these potentials confounds to be detected, and provide a more explicit framework for correction of multiple testing. As a result, genome-wide significant results have a much higher rate of replication. For this reason, we restrict our discussion of the literature to studies using whole-genome approaches.

## Confounding, expression, and causality

Gene expression is dynamic and influenced by both causal and noncausal influences, it can be difficult to establish the direction of causation. Therefore, as shown in Fig. 2, one should consider several causal possibilities: gene expression levels contribute to the risk for developing MDD (causal pathway); the presence of MDD changes gene expression levels (reverse causality); and the same etiologic pathways (environmental and/or biological) contribute both to risk for MDD and to gene expression changes, independently (shared causality). Longitudinal studies of presymptomatic individuals can help to clarify these relationships, but they have, to our knowledge, not yet been performed with comprehensive genome-wide approaches. Moreover, human studies are often also very likely to be affected by residual confounding from unmeasured covariates that can differ between cases and controls. Controlling for metrics such as postmortem interval, RNA quality, and medications can reduce, but not fully eliminate such confounding (Jaffe et al., 2017), highlighting the importance of further validation and the use of independent evidence to strengthen causal claims.

## Differential expression (DE) studies

The most common method for testing the association between gene expression and a specific phenotype is to compare normalized and batch-corrected RNA levels between two groups (e.g., case and control) to identify molecules with significant DE. These analyses typically focus on one or more of several molecular "units," such as all transcripts mapping to a gene, an exon, or an isoform. More broadly, they can also be network or pathway based, focusing on how patterns of co-expressed genes, or genes hypothesized to be in previously annotated pathways, may be associated with the disorder in question. Such analyses may have greater power to distinguish cases and controls since they may reflect the cumulative effect of several, sometimes hundreds, constituent genes.

## Gene expression as mediators of genetic risk

Gene expression studies are often used to characterize the association between variants in the DNA sequence and the abundance of specific RNA transcripts. Since DNA composition is generally unaffected by reverse causation, such studies, known as expression quantitative trait (eQTL) studies, can help provide causal evidence that specific expression features mediate inherited risk for a specific disorder. eQTLs can be divided into those near the expressed gene (cis-eQTLs) and those further away, potentially even residing in different chromosomes (trans-eQTLs). Cis-eQTLs have been discovered for most (86.1%) protein-coding genes and usually represent DNA variants in promoter regions that have effects on expression across tissues (GTEx Consortium et al., 2017). On the other hand, trans-eQTLs are more likely to be tissue specific, but are more challenging to detect due to smaller effect sizes, and greater susceptibility to confounding factors (Saha & Battle, 2018). eQTLs can be particularly informative if they colocalize to regions of the genome associated with a phenotype of interest by prior GWAS, since they can provide a mechanistic link between genetic susceptibility, gene expression, and disease association. However, such causal interpretations need to be approached cautiously, as

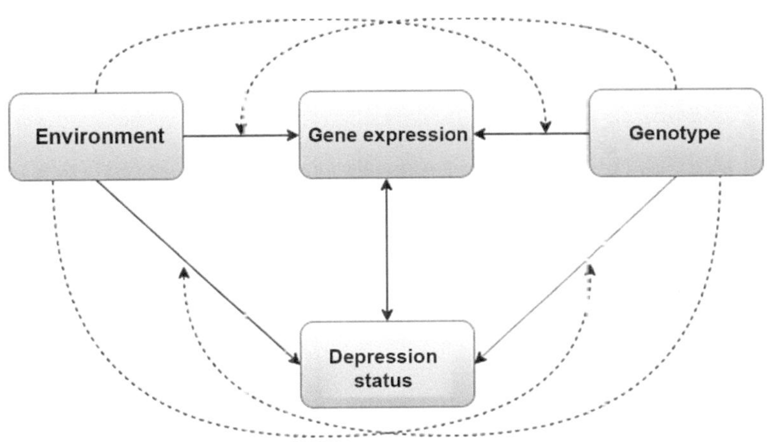


FIG. 2 Causality and direction of effects gene expression studies. *Solid lines* represent direct relationships, whereas *dotted lines* represent mediating effects.

physical overlap between a GWAS-significant locus and an eQTL may be coincidental and noncausal (Neumeyer, Hemani, & Zeggini, 2019).

## Tissue specificity and the use of surrogate markers

Gene expression patterns can be tissue and cell-type specific, highlighting the importance of studying disorder-relevant cells or tissues, such as specific regions of the brain (GTEx Consortium et al., 2017). However, postmortem tissue is limited and cannot, for obvious reasons, be collected from living patients. Therefore, several studies have focused on expression in surrogate tissues such as blood, which is reported to have expression levels that display correlation estimates between 50% and 81.9% with those in the brain (Liew, Ma, Tang, Zheng, & Dempsey, 2006; Sullivan, Fan, & Perou, 2006). Although gene expression from blood is unlikely to predict expression in the brain, they may reveal noncausal but potentially useful biomarkers that may correlate with clinically useful end points.

In this chapter, we limit our literature review to studies utilizing genome-wide approaches (Table 1); moreover, because of concerns of statistical power and reproducibility, our primary focus will be on studies with larger sample sizes when available and those that employ more recent RNAseq assay. We discuss the implications of their results regarding the biological mechanisms involved in MDD and look toward future directions as sample sizes increase and novel technologies become more widespread.

## Peripheral tissue studies in MDD

As listed in Table 1, most peripheral tissue studies have analyzed gene expression using whole blood, although some studies have focused on isolated lymphocytes. Given the difficulties of interpreting studies with small sample sizes due to high rates of false negatives and false positives,

**TABLE 1** Summary of genome-wide gene expression studies on major depressive disorder.

| Study | Pubmed ID | Year | Tissue | Sample | Assay | *N* per group (MDD/controls) |
|---|---|---|---|---|---|---|
| ***Blood-based studies*** | | | | | | |
| Belzeaux et al. | 23149449 | 2012 | PBMC | Aix Marseille Université | Microarray (SurePrint G3) | 16/12 |
| Savitz et al. | 23064081 | 2013 | PBMC | University of Oklahoma | Microarray (HumanHT-12) | 21/24 |
| Liu et al. | 24676134 | 2014 | PBMC | Shanxi Medical University | Microarray (GG-HTA) | 10/10 |
| Mostafavi et al. | 24296977 | 2014 | WB | Depression Genes Network (DGN) | RNAseq | 463/459 |
| Jansen et al. | 26008736 | 2016 | WB | Netherlands Study of Depression and Anxiety | Microarray (U219) | 1517/331 |
| Leday et al. | 28688579 | 2018 | WB | GSK-HiTDiP and Janssen-BRC | Microarray (U133) | 113/57<br>94/100 |
| Miyata et al. | 26926397 | 2016 | WB | Gunma University Hospital | Microarray (SurePrint G3) | 10/12 |
| Woo et al. | 30076325 | 2018 | WB | Samsung Medical Center | Microarray (Human Gene 1.0 ST) | 38/14 |
| ***Brain-based studies*** | | | | | | |
| Iwamoto et al. | 14743183 | 2004 | PFC | Stanley Brain Collection | Microarray (HU95A) | 11/15 |
| Sibille et al. | 14603265 | 2004 | PFC | Human Neurobiology Core, Columbia | Microarray (U133A) | 19/19 |

**TABLE 1 Summary of genome-wide gene expression studies on major depressive disorder—cont'd**

| Study | Pubmed ID | Year | Tissue | Sample | Assay | *N* per group (MDD/ controls) |
|---|---|---|---|---|---|---|
| Choudary et al. | 16230605 | 2005 | ACC, DLPFC | University of California Irvine | Microarray (U95Av2) | 9/7 |
| Aston et al. | 15303102 | 2005 | Temporal cortex | Stanley Brain Collection | Microarray (HU95A) | 12/14 |
| Sequeira et al. | 16389195 | 2006 | DLPFC, OFC, motor cortex | Quebec Suicide Brain Bank | Microarray (U133A) | 16/12 |
| Kang et al. | 18045927 | 2007 | DLPFC | Cuyahoga County Medical Examiner's Office | Microarray (Human 1A Oligo) | 14/14 |
| Sequeira et al. | 16389195 | 2007 | AMYG, HC, ACC, PCC | Quebec Suicide Brain Bank | Microarray (U133A) | 9–14/6–3 |
| Tochigi et al. | 18068248 | 2008 | PFC | Stanley Brain Collection | Microarray (HU95Av2) | 15/15 |
| Klempan et al. | 17938633 | 2009 | PFC | Quebec Suicide Brain Bank | Microarray (U133) | 16/13 |
| Sequeira et al. | 16389195 | 2009 | AMY, ACC HC, NAc, PFC | Quebec Suicide Brain Bank | Microarray (U133) | 16/13 |
| Sibille et al. | 14603265 | 2009 | AMY, ACC | Allegheny County Medical Examiner's Office | Microarray (U133) | 14–16/ 14–16 |
| Duric et al. | 20953200 | 2010/ 12 | HC | Cuyahoga County Medical Examiner's Office | Microarray (MI Ready) | 21/18 |
| Bernard et al. | 20386568 | 2011 | Locus coeruleus | University of California Irvine | Microarray (U133) | 12/9 |
| Kohen et al. | 24594777 | 2014 | HC | Stanley Brain Collection | RNAseq | 17/29 |
| Kim et al. | 26077692 | 2016 | HC | Stanley Brain Collection | RNAseq | 15/15 |
| Darby et al. | 27622934 | 2016 | OFC | Stanley Brain Collection | RNAseq | 15/15 |
| Labonte et al. | 28825715 | 2017 | DLPFC, NAc, OFC, anterior insula, ACC, ventral subiculum | Quebec Suicide Brain Bank UT Southwestern Brain | RNAseq | 26/22 |
| Pantazatos et al. | 27528462 | 2017 | DLPFC | New York State Psychiatric Institute | RNAseq | 30/29 |
| Ramaker et al. | 28754123 | 2017 | DLPFC, ACC, NAc | University of California Irvine | RNAseq | 24/24 |
| Mahajan et al. | 29175309 | 2018 | HC | Cuyahoga County Medical Examiner's Office | RNAseq | 23/23 |
| Gandal et al. | 29439242 | 2018 | ACC, PFC | Numerous | Microarray | 87/293 |

*Continued*

**TABLE 1 Summary of genome-wide gene expression studies on major depressive disorder—cont'd**

| Study | Pubmed ID | Year | Tissue | Sample | Assay | *N* per group (MDD/controls) |
|---|---|---|---|---|---|---|
| Lanz et al. | 31123247 | 2019 | DLPFC, HC, striatum | Allegheny County Medical Examiner's Office | Microarray (U133) | 19/19 |
| Scarr et al. | 31238704 | 2019 | PFC | Victorian Institute of Forensic Medicine | Microarray (Human Exon 1.0 ST) | 15/15 |

The following abbreviations are used in the table: *AMYG*=amygdala; *ACC*=anterior cingulate cortex; *DLPFC*=dorsolateral prefrontal cortex; *HC*=hippocampus; *OFC*=orbitofrontal cortex; *PBMC*=peripheral blood mononuclear cells; *PFC*=prefrontal cortex; *NAc*=nucleus accumbens; *WB*=whole blood.

we focus our discussion on three studies with relatively large samples sizes (Jansen et al., 2016; Leday et al., 2018; Mostafavi et al., 2014).

The first of these studies (Mostafavi et al., 2014) used RNAseq of whole blood to assess expression differences between 463 individuals with a lifetime diagnosis of recurrent MDD and 459 controls. The investigators accounted for a number of potential technical and physiological covariates and used permutation to control for multiple testing. While they did not identify any significantly differentially expressed genes, they found an enrichment in the genes with the largest expression differences for interferon α/β signaling, which is involved in immune response to viral infections, which has also been associated with the pathophysiology of depression (Miller & Raison, 2016).

The second large-scale whole blood-based study (Jansen et al., 2016) performed microarray-based analysis on samples from the Netherlands Study of Depression and Anxiety (NESDA) cohort, a longitudinal study of 1848 subjects that allowed the investigators to separate their cases into those in remission versus those with current symptoms of MDD at the time of the assay. Analyzing their 882 current MDD cases vs 331 controls, Jansen et al. found 18 significant associations at a genome-wide false discovery rate (FDR) threshold of less than 5%, compared to seven significant genes in a larger analysis of all 1517 MDD cases (remitted and current). Based on their greater number of significant findings and a pattern of smaller effects in the remitted cases, Jansen et al. hypothesized that subjects with active MDD symptoms have more significant expression changes, potentially reflecting the importance of the underlying mood state as a driver of expression changes. Subsequently, the authors performed pathway and gene co-expression analyses of the current MDD comparison and identified enrichment of the "natural killer cell-mediated cytotoxicity" pathway as well as co-expression modules enriched for IL-12, IL-6-mediated signaling events, and the previous natural killer cell pathway.

The most recent blood-based differential expression study (Leday et al., 2018) was performed using microarray assays of whole blood in two independent samples, totaling to 212 MDD cases and 175 controls. DE analyses were conducted in each sample separately, and genes with the strongest association across both studies were identified using a Bayesian analysis. The authors subsequently focused on this set of 165 genes, finding enrichment for gene ontology (GO) categories representing the "innate immune system" and "response to infection."

To compare the findings across these three major whole blood-based studies, we meta-analyzed the available summary statistics weighting each study by the square root of their total sample size. After combining the summary data at the gene level (selecting the most associated probeset per gene), there were 150 genes with data across all three studies, 30 of which were nominally significant (meta $P$-value $<0.05$) and had the same direction of effect. As shown in Table 2, the SRSF5 gene displayed the strongest consistent evidence for association, with a meta-analysis $P$-value of $3.4 \times 10^{-7}$, which would withstand multiple testing correction for the number of protein-coding genes. *SRSF5* (serine and arginine-rich splicing factor 5) encodes for a brain-expressed RNA-binding protein associated with alternative splicing. Although there is limited knowledge about *SRSF5* and the pathophysiology of depression, it is notable that GWAS studies of MDD have shown a strong association with *RBFOX1* (Wray et al., 2018), which is also an RNA-binding protein involved in alternative splicing. Testing for functional enrichment of all 30 nominally significant genes found several significant ($FDR < 0.05$) pathways related to "axon guidance related to netrin" and a number of broad immune response categories. Although

**TABLE 2** Meta-analysis of differentially expressed genes in blood-based studies.

| | Leday et al. (2018) (GSK sample) | | Leday et al. (2018) (Janssen sample) | | Mostafavi et al. (2014) (DGN sample) | | Jansen et al. (2016) (NESDA sample) | | | |
|---|---|---|---|---|---|---|---|---|---|---|
| Gene | Direction (OR) | *P*-value | Direction (OR) | *P*-value | Direction | *P*-value | Direction | *P*-value | Meta direction | Meta *P*-value |
| SRSF5 | – (0.90) | 0.013 | – *(0.93)* | 0.011 | – | 0.00016 | – | 0.006 | ---- | $3.40 \times 10^{-7}$ |
| SAMD3 | – (0.81) | 0.049 | – *(0.87)* | 0.018 | – | 0.34 | – | 0.0007 | [illegible] | $6.90 \times 10^{-5}$ |
| PIK3R1 | – (0.84) | 0.016 | – *(0.91)* | 0.0065 | – | 0.065 | – | 0.022 | ---- | $1.22 \times 10^{-4}$ |
| ST6GAL1 | – (0.85) | 0.0017 | – *(0.91)* | 0.011 | – | 0.41 | – | 0.0022 | ---- | $1.77 \times 10^{-4}$ |
| MTSS1 | – (0.81) | 0.015 | – *(0.87)* | 0.024 | – | 0.34 | – | 0.0012 | ---- | $1.86 \times 10^{-4}$ |
| RPL22 | – (0.86) | 0.0028 | – *(0.91)* | 0.014 | – | 0.015 | – | 0.078 | ---- | $2.65 \times 10^{-4}$ |
| THOC1 | – (0.84) | 0.0012 | – *(0.86)* | 0.003 | | 0.13 | – | 0.054 | ---- | $3.26 \times 10^{-4}$ |
| USP47 | – (0.84) | 0.012 | – *(0.88)* | 0.0033 | – | 0.035 | – | 0.059 | ---- | $3.53 \times 10^{-4}$ |
| NFATC2 | – (0.80) | 0.0003 | – *(0.88)* | 0.0050 | – | 0.76 | – | 0.0021 | ---- | $5.54 \times 10^{-4}$ |
| SRSF6 | – (0.85) | 0.0017 | – *(0.91)* | 0.019 | – | 0.15 | – | 0.037 | [illegible] | $8.43 \times 10^{-4}$ |

seemingly separate, axon guidance and immune function have been previously linked, and may represent pleomorphic functions of similar genes across organ systems and developmental time points (Feinstein & Ramkhelawon, 2017).

## Brain-based studies in MDD

While several studies of postmortem brain tissues utilizing microarray technology were published in the last decade, most made use of early microarray technology and were limited by small sample sizes (Table 1). Gandal et al. combined datasets from several of these earlier studies to perform the largest published postmortem gene expression study of MDD, identifying 328 differentially expressed genes at an FDR < 5% (Gandal et al., 2018). Based on these results, the authors highlighted a differentially co-expressed modules enriched for several pathways of note, including G-protein-coupled receptors, cytokine signaling, and neuroactive ligand-receptor interactions (Gandal, Haney, et al., 2018). This study was also part of a cross-disorder investigation that found significant genome-wide correlations between differential expression patterns seen in MDD compared with schizophrenia and bipolar disorder.

More recent brain-based studies have used RNAseq to measure gene expression. To our knowledge, seven such studies have been published on brain tissue (Table 1), with MDD case group sample sizes ranging from 15 to 30 subjects. Kohen et al. published an early study utilizing 17 MDD cases and 29 controls from the Stanley Medical Research Institute (SMRI) Neuropathology Consortium that performed laser-captured microdissection of the dentate gyrus, the primary region of the hippocampus responsible for neurogenesis (Kohen, Dobra, Tracy, & Haugen, 2014). Although the focus on a more homogenous cell population within the dentate gyrus anticipated current efforts to study gene expression at the single-cell level, the authors used an earlier sequencing platform and nontraditional analytical methods that make a comparison with other studies challenging. In a more recent study of homogenate tissue but also from the dentate gyrus, Mahajan et al. performed RNAseq in 23 MDD cases and 23 controls in the Cuyahoga County Medical Examiner's office, identifying 30 differentially expressed genes (FDR < 5%), with pathway analyses again highlighting several inflammatory categories, including interferon and IL-17 signaling (Mahajan et al., 2018).

In contrast, Pantazatos et al. investigated gene expression from the dorsolateral prefrontal cortex (DLPFC) (Pantazatos et al., 2017). This study included 30 subjects with MDD and 29 controls from the New York State Psychiatric Institute, a collection notable for including only subjects free of the potential confounds of psychotropic medications at time of death. Performing multigroup omnibus tests to differentiate cases dying of suicide from those who died by other means, Pantazatos et al. identified four differentially expressed genes at FDR < 5%, all of which showed greater effect sizes in the cases with suicide. Pathway analyses revealed multiple, broad categories that included several potentially inflammation-related subsets such as "chemokine receptor binding" and "cellular response to lipopolysaccharide."

Two more recent studies were noteworthy for performing RNAseq on several brain regions in the same individuals, allowing the investigators to test for heterogeneity across varying regions of the brain (Labonte et al., 2017; Ramaker et al., 2017). The first of these performed RNAseq on tissue from the anterior cingulate cortex (ACC), the DLPFC, and the nucleus accumbens (NAc) in 24 cases and 24 controls in three disorder-control groups: schizophrenia, bipolar disorder, and MDD (Ramaker et al., 2017). This study primarily focused on DE in schizophrenia and found no genome-wide significant DE genes in bipolar disorder or MDD. In the second multi-region study, Labonte et al. performed RNAseq on 26 MDD samples and 22 controls in six brain regions: ventromedial and dorsolateral prefrontal cortex, orbitofrontal cortex, NAc, anterior insula, and ventral subiculum (Labonte et al., 2017). This study focused on sex-stratified analysis, reporting overall limited overlap of differentially expressed genes between males and females. The authors highlighted patterns of differential expression and co-expression across the genome and performed experimental validation of one of the most significant "hub genes" (a gene central to co-expression network structure), *DUSP6*, which encodes a stress-induced phosphatase linked to pyramidal cell hyperactivity. However, further comparison with other studies is unfortunately limited by the lack of sex specificity in other analyses.

## Single-cell sequencing

All the studies described above have sequenced RNA from bulk tissues, such as whole blood or homogenized brain sections, where measured expression levels reflect the overall combination of that tissue's constituent cells. However, emerging technologies now allow for the measurement of genomic features, such as gene expression, at single-cellular resolution, allowing the interrogation of cell-specific differential expression and eQTL analyses within the major types of cells found in the human brain. The application of single-cell RNAseq (scRNA-seq) to MDD has been limited to an initial preliminary study of the DLPFC in 17 cases with MDD and 17 controls (Nagy et al., 2020). This study identified 26 cellular clusters that were annotated to excitatory and inhibitory neurons, as well as, the major types of glial cells. The authors performed DE analyses within each cluster and identified 45 significant genes (FDR < 5%) in 12 clusters, representing several types of cells, including excitatory and inhibitory neurons, as well as in specific oligodendrocyte and astrocytic lineages. Pathway analyses highlighted synaptic pathways, which have not yet been implicated in the homogenate-based studies, but which have been linked to MDD in prior genetic studies (Pirooznia et al., 2016; Wray et al., 2018).

## Emerging themes

Among the brain-based RNAseq studies, there was little concordance across the genome-wide significant findings across the discussed brain tissue studies, with only one protein-coding gene being reported as significant in more than one study (*CCL2*), although with opposite directions of effect (Gandal, Haney, et al., 2018; Pantazatos et al., 2017). However, this lack of convergent findings at the level of the gene is perhaps unsurprising, since the relatively small sample sizes used in most postmortem brain studies are underpowered to detect DE of modest effects, and make the studies susceptible to the "winner's curse," a form of ascertainment bias that results in upwardly biased effect size estimates in the initial discovery sample. In addition, the studies reviewed have important methodological differences (including assay variability, variable correction for known and unknown confounds, differing normalization methods, and nonuniform means of correction for multiple testing) that further contribute to the lack of replication. Yet, despite the lack of replication of significant genes, pathway analyses have generally shown overrepresentation of similar biological functions, focusing broadly on the immune response, which has been hypothesized to play a role in pathophysiology of at least a subset of MDD cases (Miller & Raison, 2016).

## Integration with genetic risk

Gene expression datasets can be integrated with emerging results from much larger GWAS studies to specifically highlight the effect of genetically mediated expression changes on disease risk. When performed comprehensively across the genome, such studies are known as transcriptome-wide association studies (TWAS) (Gusev et al., 2018). While not susceptible to the typical confounding from reverse causality that hinders case–control DE studies, TWAS studies can have significant limitations due to widespread, but noncausal, correlation across makers that can mislead fine-mapping efforts. Hence, as with almost all investigative modalities, TWAS studies should be interpreted alongside other orthogonal forms of evidence before claims of causation are made.

Three such "integrative studies" have been performed in MDD (Gerring et al., 2019; Wray et al., 2018; Zhong et al., 2019) and while all use the PGC Study of MDD as their GWAS "backbone," they use different eQTL reference datasets, leading to some expected variability in statistical power and tissue specificity. When considered together, there were a total of 132 significant genes across all three studies, with 19 genes implicated by two studies, and four genes found to be significant in all three studies. The consistency of the results for the four genes (*DDAH2*, *BTN3A2*, *DENND1B*, *NEGR1*) provides strong impetus for further investigation and highlights again the potential role of axon regulation (*NEGR1*) and immune processes (*BTN3A2*, *DENND1B*).

## Future directions

Gene expression studies in complex psychiatric disorders such as MDD are in a promising but early phase of development, with most studies, particularly those utilizing postmortem brain being limited by small sample sizes, and lack of analytical standardization across studies. As a result, even when genome-wide significant, it is unclear whether differentially expressed genes represent generalizable associations. As shown in Fig. 3, all published postmortem studies have been underpowered to detect modest effect sizes (OR < 1.2), which often the rule in complex disorders such as MDD. Therefore, larger sample sizes will be necessary, along with more emphasis on analytical standardization and data sharing across studies. As larger studies are performed, an additional important consideration will be quality of the phenotypic metrics since "minimal phenotyping" efforts in depression research may lead to greater heterogeneity that becomes more apparent in larger sample collections. Indeed, sharing of phenotypic data will, therefore, be arguably as important as that of genomic data.

With these challenges in mind, there are ongoing international efforts to perform more comprehensive assays of large-scale postmortem brain collections (BrainSeq: A Human Brain Genomics Consortium, 2015; Gandal et al., 2018) that will facilitate harmonized comparisons in different brain regions and across the major psychiatric diagnoses. Not only will this produce more replicable results in gene-level studies, but it will also allow for a more comprehensive evaluation of isoforms and noncoding RNA on disease pathophysiology. Although it has received limited attention from the initial expression studies of MDD, isoform dysregulation may be particularly relevant for psychiatric disorders (Gandal, Zhang, et al., 2018) and for fine-mapping, or linking, genomic loci identified in GWAS studies with the relevant downstream transcript(s) (Clark et al., 2020). Analysis of noncoding RNA molecules, such as microRNAs or lncRNA (long noncoding RNAs), will also be important to consider in more comprehensive manner, since these molecular entities are important modulators of gene expression (Marchese, Raimondi, & Huarte, 2017).

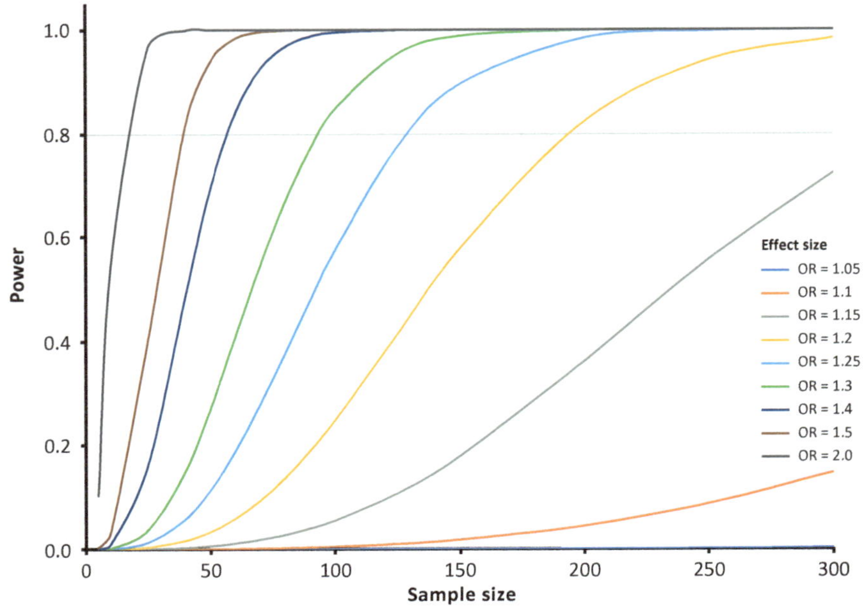


FIG. 3 Power estimates for gene expression studies. The X-axis refers to case sample sizes, assuming an equally sized control sample. Power estimates are shown for various plausible effect sizes (OR represents odds ratios).

In addition, single-cell analysis will also likely provide more nuanced insight into the specific mechanistic pathways involved in the pathophysiology of MDD. Although several encouraging proof-of-concept studies have been recently performed, the technology is relatively nascent and limited at present by cost, reliability, and ability to discern isoforms or obtain longer reads at the single-cell level. Hence, further technological advances need to be made before it is considered to be the assay of choice for measuring gene expression.

## Triangulation and strengthening of causal claims

While the increasing sample sizes and technological advances will lead to more robust, comprehensive, and granular insights into the molecular underpinnings of MDD, gene expression studies in humans should be seen as initial exploratory studies that begin the process of hypothesis testing for causal disease-related factors. As described above, the usual case–control confounds of gene expression studies are akin to the challenges encountered in observational studies, where the framework of triangulation, or integration of information from studies using different methodologies (each with its own independent biases) has been embraced to enhance reproducibility and strengthen causal claims (Lawlor, Tilling, & Davey Smith, 2016). Large-scale genetic studies such as GWAS, for example, can provide independent evidence for causality if an implicated locus has been linked to eQTL consistent with the same direction of effect seen in a gene expression differential expression analysis. However, linking a locus with a causal eQTL may be more arduous than initially assumed (Neumeyer et al., 2019) and DE studies remain underpowered to detect expected changes in gene expression given the small effects seen in GWAS studies (Fig. 3).

Ultimately, triangulation will require evidence from experimental studies using emerging gene-editing and optogenetic technologies to more faithfully test the effect of the presumed causal factor in model systems such as reprogrammed neurons, organoids, as well as more traditional animal-based models. Technological advances now allow for appropriate gene-dosage experiments to be designed, where the subtle effect seen in human studies can be recapitulated in the model system instead of the more traditional gene-knock out or knock in experiments that represent extreme molecular perturbations unlikely to be relevant in the study common complex disorders. These advances also provide the ability to target specific cells and circuits in animal models that can build upon the evidence provided by emerging single-cell studies.

## Summary

Gene expression represents the initial functional output of genetic risk and may arguably been one of the most informative study modalities to facilitate novel drug discovery relevant for various clinical traits. In psychiatric disorders such as MDD, expression studies present greater challenges due to the limited accessibility of the primary tissue of relevance. Although most published studies have been limited by small sample sizes, there are ongoing efforts, such as those carried out by the psychENCODE consortium, which are amassing much larger collections using harmonized pipelines and open access to data, which will provide a model for current and future collaborative work. Moreover, emerging technologies will soon permit a comprehensive characterization of gene expression at the cellular level, providing a more granular and nuanced framework for the understanding of causal etiological factors.

## Summary points

- Gene expression represents an important, easily measurable molecular feature that can reflect the influence of both inherited and/or environmental risk factors.
- Case–control studies need to be carefully assessed for potential confounders, including the long-term effects of treatment as well as the illness itself.
- When integrated with genetic studies, gene expression studies can provide functional insights into genetic risk that may be more causally interpretable.
- RNA-sequencing technology allows for a more comprehensive measurement of all gene expression-related features, such as isoform quantification.
- Emerging single-cell-based sequencing technology will soon be the likely method of choice for expression studies, providing a level of granularity more relevant to complex organs such as the brain.
- Most studies to date have been underpowered, highlighting the need for larger samples sizes and more open data sharing. However, initial results suggest an enrichment of axon guidance and immune-based pathways.

## Mini-dictionary of terms

**RNA microarray** An assay that measures gene expression via quantification of tagged sample RNA fragments bound to an array of sequence-specific oligonucleotides.

**RNA sequencing** An assay that uses next generation sequencing to measure expression levels via quantification of sequence fragments from a particular gene, transcript, or RNA molecule.

**MicroRNA** RNA molecules with roughly 22 nucleotides that are not translated into protein but can regulate the expression of genes by binding to messenger RNA molecules.

**Long noncoding RNA** RNA molecules with more than 200 nucleotides that are not translated into protein but can regulate the expression of messenger RNA molecules, largely by mediating chromatin modification.

**Exon** A region of RNA in a gene that can be translated into proteins.

**Isoform** One of many potential combinations of exons in the final RNA product and amino acid sequence.

**Differential expression** Significantly different gene expression levels between two comparison groups, typically case and control samples.

**Expression quantitative trait locus** A region of DNA sequence in which the genotype of linked (or co-inherited) sites across that sequence are associated with the expression levels of a given gene.

**Genome-wide association study** A study in which a trait is tested for association with single nucleotide polymorphisms across the entire genome, with significance of these association being corrected for multiple testing.

**Transcriptome-wide association study** A study in which a trait is tested for association with predicted expression levels of genes across the entire genome from genotypes of loci known to affect each gene's expression.

**Gene ontology** A comprehensive list of all known molecular and biological characteristics of a gene.

## References


Border, R., Johnson, E. C., Evans, L. M., Smolen, A., Berley, N., Sullivan, P. F., et al. (2019). No Support for Historical Candidate Gene or Candidate Gene-by-Interaction Hypotheses for Major Depression Across Multiple Large Samples. *The American Journal of Psychiatry*, *176*, 376–387.

BrainSeq: A Human Brain Genomics Consortium. (2015). *Neuron*, *88*, 1078–1083.

Clark, M. B., Wrzesinski, T., Garcia, A. B., Hall, N. A. L., Kleinman, J. E., Hyde, T., … Tunbridge, E. M. (2020). Long-read sequencing reveals the complex splicing profile of the psychiatric risk gene CACNA1C in human brain. *Molecular Psychiatry*, *25*, 37–47.

Conesa, A., Madrigal, P., Tarazona, S., Gomez-Cabrero, D., Cervera, A., McPherson, A., et al. (2016). A survey of best practices for RNA-seq data analysis. *Genome Biology*, *17*, 13–13.

Feinstein, J., & Ramkhelawon, B. (2017). Netrins & Semaphorins: Novel regulators of the immune response. *Biochimica Et Biophysica Acta. Molecular Basis of Disease*, *1863*, 3183–3189.

Gandal, M. J., Haney, J. R., Parikshak, N. N., Leppa, V., Ramaswami, G., Hartl, C., et al. (2018). Shared molecular neuropathology across major psychiatric disorders parallels polygenic overlap. *Science*, *359*, 693–697.

Gandal, M. J., Zhang, P., Hadjimichael, E., Walker, R. L., Chen, C., Liu, S., et al. (2018). Transcriptome-wide isoform-level dysregulation in ASD, schizophrenia, and bipolar disorder. *Science*, *362*. https://doi.org/10.1126/science.aat8127.

Gaynes, B. N., Warden, D., Trivedi, M. H., Wisniewski, S. R., Fava, M., & Rush, A. J. (2009). What did STAR*D teach us? results from a large-scale, practical, clinical trial for patients with depression. *Psychiatric Services*, *60*, 1439–1445.

Gerring, Z. F., Gamazon, E. R., Derks, E. M., & Major Depressive Disorder Working Group of the Psychiatric Genomics Consortium. (2019). A gene co-expression network-based analysis of multiple brain tissues reveals novel genes and molecular pathways underlying major depression. *PLoS Genetics*, *15*, e1008245.

Global Burden of Disease Study 2013 Collaborators. (2015). Global, regional, and national incidence, prevalence, and years lived with disability for 301 acute and chronic diseases and injuries in 188 countries, 1990-2013: A systematic analysis for the global burden of disease study 2013. *Lancet*, *386*, 743–800.

GTEx Consortium, Laboratory, Data Analysis & Coordinating Center (LDACC)-Analysis Working Group, Statistical Methods groups-Analysis Working Group, Enhancing GTEx (eGTEx) groups, NIH Common Fund, NIH/NCI, … Montgomery, S. B. (2017). Genetic effects on gene expression across human tissues. *Nature*, *550*, 204–213.

Gusev, A., Mancuso, N., Won, H., Kousi, M., Finucane, H. K., Reshef, Y., et al. (2018). Transcriptome-wide association study of schizophrenia and chromatin activity yields mechanistic disease insights. *Nature Genetics*, *50*, 538–548.

Jaffe, A. E., Tao, R., Norris, A. L., Kealhofer, M., Nellore, A., Shin, J. H., et al. (2017). qSVA framework for RNA quality correction in differential expression analysis. *Proceedings of the National Academy of Sciences of the United States of America*, *114*, 7130–7135.

Jansen, R., Penninx, B. W. J. H., Madar, V., Xia, K., Milaneschi, Y., Hottenga, J. J., et al. (2016). Gene expression in major depressive disorder. *Molecular Psychiatry*, *21*, 444.

Johnson, W. E., Li, C., & Rabinovic, A. (2007). Adjusting batch effects in microarray expression data using empirical bayes methods. *Biostatistics*, *8*, 118–127.

Kohen, R., Dobra, A., Tracy, J. H., & Haugen, E. (2014). Transcriptome profiling of human hippocampus dentate gyrus granule cells in mental illness. *Translational Psychiatry*, *4*, e366.

Labonte, B., Engmann, O., Purushothaman, I., Menard, C., Wang, J., Tan, C., et al. (2017). Sex-specific transcriptional signatures in human depression. *Nature Medicine*, *23*, 1102–1111.

Laursen, T. M., Musliner, K. L., Benros, M. E., Vestergaard, M., & Munk-Olsen, T. (2016). Mortality and life expectancy in persons with severe unipolar depression. *Journal of Affective Disorders*, *193*, 203–207.

Lawlor, D. A., Tilling, K., & Davey Smith, G. (2016). Triangulation in aetiological epidemiology. *International Journal of Epidemiology*, *45*, 1866–1886.

Leday, G. G. R., Vertes, P. E., Richardson, S., Greene, J. R., Regan, T., Khan, S., et al. (2018). Replicable and coupled changes in innate and adaptive immune gene expression in two case-control studies of blood microarrays in major depressive disorder. *Biological Psychiatry*, *83*, 70–80.

Liew, C. C., Ma, J., Tang, H. C., Zheng, R., & Dempsey, A. A. (2006). The peripheral blood transcriptome dynamically reflects system wide biology: A potential diagnostic tool. *The Journal of Laboratory and Clinical Medicine*, *147*, 126–132.

Mahajan, G. J., Vallender, E. J., Garrett, M. R., Challagundla, L., Overholser, J. C., Jurjus, G., et al. (2018). Altered neuro-inflammatory gene expression in hippocampus in major depressive disorder. *Progress in Neuro-Psychopharmacology & Biological Psychiatry*, *82*, 177–186.

Marchese, F. P., Raimondi, I., & Huarte, M. (2017). The multidimensional mechanisms of long noncoding RNA function. *Genome Biology*, *18*, 206–212.

Miller, A. H., & Raison, C. L. (2016). The role of inflammation in depression: From evolutionary imperative to modern treatment target. *Nature Reviews. Immunology*, *16*, 22–34.

Morris, K. V., & Mattick, J. S. (2014). The rise of regulatory RNA. *Nature Reviews. Genetics*, *15*, 423–437.

Mostafavi, S., Battle, A., Zhu, X., Potash, J. B., Weissman, M. M., Shi, J., et al. (2014). Type I interferon signaling genes in recurrent major depression: Increased expression detected by whole-blood RNA sequencing. *Molecular Psychiatry*, *19*, 1267–1274.

Nagy, C., Maitra, M., Tanti, A., Suderman, M., Théroux, J., Davoli, M., et al. (2020). Single-nucleus transcriptomics of the prefrontal cortex in major depressive disorder implicates oligodendrocyte precursor cells and excitatory neurons. *Nature Neuroscience*, *23*, 771–781. https://doi.org/10.1038/s41593-020-0621-y.

Neumeyer, S., Hemani, G., & Zeggini, E. (2019). Strengthening causal inference for complex disease using molecular quantitative trait loci. *Trends in Molecular Medicine*. S1471-4914(19)30266-7.

Pantazatos, S. P., Huang, Y., Rosoklija, G. B., Dwork, A. J., Arango, V., & Mann, J. J. (2017). Whole-transcriptome brain expression and exon-usage profiling in major depression and suicide: Evidence for altered glial, endothelial and ATPase activity. *Molecular Psychiatry*, *22*, 760–773.

Pirooznia, M., Wang, T., Avramopoulos, D., Potash, J. B., Zandi, P. P., & Goes, F. S. (2016). High-throughput sequencing of the synaptome in major depressive disorder. *Molecular Psychiatry*, *21*, 650–655.

Ramaker, R. C., Bowling, K. M., Lasseigne, B. N., Hagenauer, M. H., Hardigan, A. A., Davis, N. S., et al. (2017). Post-mortem molecular profiling of three psychiatric disorders. *Genome Medicine*, *9*, 72–75.

Saha, A., & Battle, A. (2018). False positives in trans-eQTL and co-expression analyses arising from RNA-sequencing alignment errors. *F1000Research*, *7*, 1860.

Sullivan, P. F., Fan, C., & Perou, C. M. (2006). Evaluating the comparability of gene expression in blood and brain. *American Journal of Medical Genetics. Part B, Neuropsychiatric Genetics*, *141B*, 261–268.

Van den Berge, K., Hembach, K. M., Soneson, C., Tiberi, S., Clement, L., Love, M. I., et al. (2019). RNA sequencing data: Hitchhiker's guide to expression analysis. *Annual Review of Biomedical Data Science*, *2*, 139–173.

Wray, N. R., Ripke, S., Mattheisen, M., Trzaskowski, M., Byrne, E. M., Abdellaoui, A., et al. (2018). Genome-wide association analyses identify 44 risk variants and refine the genetic architecture of major depression. *Nature Genetics*, *50*, 668–681.

Zhong, J., Li, S., Zeng, W., Li, X., Gu, C., Liu, J., et al. (2019). Integration of GWAS and brain eQTL identifies FLOT1 as a risk gene for major depressive disorder. *Neuropsychopharmacology*, *44*, 1542–1551.

Chapter 52

# Electroconvulsive therapy for depression: Effectiveness, cognitive side-effects, and mechanisms of action

Maria Semkovska
*Department of Psychology, University of Southern Denmark, Odense M, Denmark*

## Abbreviations

**CRF** cerebrospinal fluid
**ECT** electroconvulsive therapy
**MAOI** monoamine oxidase inhibitors
**RAA** retrograde autobiographical amnesia
**RUL** right-unilateral (electrode placement)
**ST** (individual patient's) seizure threshold
**TCA** tricyclic antidepressants

Electroconvulsive therapy (ECT) has been used for over 80 years to treat a variety of mental health disorders. Since then, it remains the most acutely effective treatment for depression (Ross, Zivin, & Maixner, 2018; UK ECT Review Group, 2003). ECT consists of the controlled and safe application, under anesthesia, of an electrical stimulus to the patient's head to produce a generalized seizure for therapeutical purposes. A series of ECT sessions are required to achieve treatment efficacy, with twice or thrice weekly administration being the most common prescription frequencies.

Both randomized controlled trials and meta-analyses consistently show ECT to be more effective than pharmacotherapy. Despite this strong evidence of efficacy, ECT's use in depression remains limited, mainly because of cognitive side-effects concerns. Several modifications of ECT technique have been introduced over the years, which have substantially reduced the intervention's cognitive effects. Lack of consensus regarding both the treatment's mechanism of action and the reversibility of associated cognitive side-effects have contributed to its suboptimal use. In that context, this review aims to (a) summarize established knowledge of ECT's effectiveness and associated cognitive side-effects; (b) appraise recent cognitive research with a particular focus on retrograde amnesia; and (c) examine evidence supporting the predominant explanatory models of ECT's mechanism of action.

## Summary of ECT effectiveness research

Dating from 1938, the first demonstration of ECT's clinical efficacy in schizophrenia rapidly led to the intervention's testing for other mental disorders. From the early 1940s, it became clear that ECT was even more effective for mood disorders than for schizophrenia. Since then and until today, the primary indication of ECT is severe depression. In the 1940s, 1950s, and early 1960s, ECT was the predominant first-line treatment for depression in most countries. With the advent of antidepressant medication, ECTs prescription declined progressively until the 1990s when use stabilized in most western countries (Weiner, 2015). Worldwide, ECT remains an important treatment option, especially for treatment-resistant depression.

Meta-analyses of randomized controlled trials have consistently demonstrated ECT to be an effective treatment for major depression. In pairs of between-treatments comparisons, ECT was found to be significantly more efficacious than simulated ECT, placebo, and antidepressant medication by both early (e.g., Janicak et al., 1985) and modern (e.g., UK ECT Review Group, 2003) meta-analyses. Regarding antidepressant medication, Janicak et al. (1985) found that ECT results in, respectively, 20% and 45% higher response rates than treatment with TCAs and MAOIs. These results were supported by a more recent meta-analysis (Pagnin, de Queiroz, Pini, & Cassano, 2004) where patients receiving real ECT were 3 times more likely to have a positive response than a patient treated with TCA and 6 times more likely to respond than those receiving MAOIs. Further, Kho, van Vreeswijk, Simpson, and Zwinderman (2003) meta-analysis found a mean effect size of medium strength—0.68—favoring ECT over antidepressant medication, whereas the UK ECT Review Group (2003), which evaluated a more comprehensive sample with stringent statistical analyses, demonstrated this

The Neuroscience of Depression. https://doi.org/10.1016/B978-0-12-817933-8.00046-3

advantage to be of a large size, with a pooled weighted effect of 0.80.

In an initial trial relative to SSRIs for nonresponders to two different antidepressants (Folkerts et al., 1997), ECT doubled the reduction in depressive symptomatology (59.8%) that was observed with paroxetine (29.5%). This ECT advantage was coupled with 2.5 times higher response rates than paroxetine's rates. Since this first trial, several studies in patients failing to respond to two different antidepressants have shown that ECT results in 6–12 times higher remission rates than continuing with a third or fourth pharmacological treatment attempt (see Sackeim, 2017 review). Contemporaneous research further indicates that, overall, ECT is associated with 3–5 times higher remission rates than pharmacotherapy for depression (Ross et al., 2018). In a recent study of the entire depression inpatient population from the general hospitals of nine American states, Slade, Jahn, Regenold, and Case (2017) found that ECT resulted in 50% fewer readmissions in the 30 days post-discharge than pharmacological treatment. Beyond its short-term clinical effectiveness, ECT is also associated with significantly reduced long-term suicidal risk in both unipolar and bipolar depression (Liang, Chung, Ho, Tsai, & Chien, 2018) and lower all-cause mortality than comparator antidepressant treatments (Fink, Kellner, & McCall, 2014).

Despite consistently strong evidence of efficacy, ECT is not prescribed as first-line treatment for severe depression and is principally reserved for treatment-resistant depression. Reasons explaining this limited use include the stigma historically linked to the intervention, fears around the necessary seizure induction, and, most significantly, cognitive side-effects concerns. From a clinical perspective, uncertainty about the occurrence and persistence of the latter represents the main factor limiting ECT prescription.

## Cognitive effects

### Scope: Use of brief-pulse ECT in severe depression

Most of ECT research in the last 40 years has focused on understanding and minimizing the intervention's cognitive side-effects as the main factor limiting its use. The consecutive refinements of ECT technique over the years aimed at reducing these effects (Semkovska, Keane, Babalola, & McLoughlin, 2011). These improvements included moving from sine-wave electrical stimulus to more efficient brief-pulse stimuli; using right-unilateral (RUL), instead of bitemporal, electrode placement; and optimizing the stimulus intensity by adjusting it to the individual patient's seizure threshold (ST). While efficacy was maintained and safety increased, all these modifications have resulted in progressively lower levels of cognitive side-effects.

Originally, the electrical stimulus used for ECT was sine wave with a long pulse width (8.3 ms). However, it progressively emerged that this form of electrical stimulation was not optimal for discharging neurons (Weiner, 2015) and, without an efficacy advantage, was leading to significantly more cognitive side-effects than square-wave brief-pulse (0.5–1.5 ms) ECT (Semkovska & McLoughlin, 2010). Consequently, brief-pulse ECT gradually replaced sine wave and is currently the predominant delivery method worldwide. Thus, the present cognitive side-effects section will focus on brief-pulse ECT. Ultra-brief pulse (<0.5 ms) ECT is a recent technique refinement that emerged at the beginning of our century. Current meta-analytical evidence indicates that ultra-brief stimulus is indeed associated with less cognitive side-effects but also has lower efficacy than 0.5–1.5 brief-pulse stimulus (Tor et al., 2015). Thus, presently, expert consensus advises on caution upon prescribing ultra-brief ECT for depression, while more research is needed to establish its parameters of efficacy (Loo, Katalinic, Martin, & Schweitzer, 2012; Tor et al., 2015).

Prior to evaluating the evidence-based ECT's cognitive side-effects, it is indispensable to consider that depression itself is associated with cognitive issues. Specifically, cognitive deficits are a well-established feature of major depression, with some of them persisting even after remission (Semkovska et al., 2019). Indeed, numerous meta-analyses have demonstrated deficits in alertness, psychomotor speed, sustained attention, memory, and executive functioning in depressed patients. Notably, issues around the ability to recall specific events from one's past have been consistently observed during a depressive episode through a variety of psychometric instruments (see van Vreeswijk & de Wilde, 2004 meta-analysis and Semkovska & McLoughlin, 2013 review). At the same time, retrograde autobiographical amnesia (RAA or losing memories from one's past) represents the chief patients' and clinicians' concern relative to ECT (Kellner, 2012). Evidence regarding the nature of ECT-associated RAA is detailed in the following section. However, the context of ECT prescription, that is providing therapy to patients who already present with diverse cognitive difficulties, needs to be heeded when appraising evidence relative to ECT's cognitive effects.

### Acute effects: within the first 3 hours after an ECT session end

Although not sufficient, the induction of generalized seizure is a necessary condition for ECT's effectiveness. Immediately after the seizure's end, disorientation occurs for a relatively short period, usually from 15 min to an hour, depending on the ECT technique used. Specifically, time to reorientation varies as a function of both *electrode placement*, with bitemporal ECT leading to a longer duration than RUL ECT (e.g., Sackeim et al., 2008; Semkovska et al., 2016), and *electrical dosage*, with higher stimulus leading to longer duration than lower charges (e.-g., Sackeim et al., 1993, 2000). Evidence suggests that ultra-brief ECT is also linked to quicker reorientation (Sackeim et al., 2008), but in the context of also being less efficacious than brief-pulse ECT (Loo et al., 2012). Time to reorientation has been consistently shown to predict the importance of ECT-associated RAA (Martin, Galvez, & Loo, 2015). Specifically, increasing longer disorientation duration is associated with an increasingly higher number of previously known events or specific information that the patient is no longer able to recall after ECT.

## Subacute effects—3 hours to 3 days after the end of an ECT treatment course

The strongest evidence regarding brief-pulse ECT-associated cognitive side-effects concerns the first few days following the intervention's last session. During that period, cognitive tests consistently show a decreased performance relative to pretreatment levels. If there were no side-effects, it would be expected for the improvement in mood to be associated with better posttreatment cognitive function. The subacute effects concern most cognitive domains, with the notable exception of auditory attention that appears unaffected by ECT (Semkovska & McLoughlin, 2010). Specifically, global cognitive status, processing speed, and spatial orientation show a small decrease relative to pre-treatment performance levels, while visual memory and verbal fluency—a medium decrease. Large reductions are observed in the executive ability to switch between tasks, and in verbal memory, with the delayed recall of information learned in the posttreatment period being especially affected. The latter is known as the characteristic post-ECT anterograde amnesia.

Our two meta-analyses on the topic (Semkovska et al., 2011; Semkovska & McLoughlin, 2010) evaluated the effect of various treatment variables on the extent of post-ECT subacute effects. Results showed the cognitive effect sizes to be independent of both the total number of ECT sessions received within a treatment course and the sessions' frequency (twice weekly vs thrice weekly). However, the severity of the cognitive effects varied as a function of electrode placement, electrical stimulus dose for the right unilateral placement, and patient's age.

Except for working memory and visual recognition abilities where no electrode placement effect is observed, bitemporal ECT is associated with significantly larger subacute effects than RUL ECT in all cognitive domains. The largest differences between the two placements appear in delayed verbal memory (anterograde amnesia) where bitemporal ECT is associated with large reductions while RUL ECT leads to medium decreases relative to pre-treatment levels. Increased electrical dosage with RUL ECT is associated with larger reductions in all verbal memory variables. This finding suggests that the significant differences between the two placements could be explained by variations in RUL ECT stimulus dosage. Indeed, in trials comparing high-dose RUL ECT (e.g., Semkovska et al., 2016 with 6 times the ST) to standard bitemporal ECT, differences between the two placements do not concern as many cognitive domains as in trials using lower RUL electrical dosage (e.g. Sackeim et al., 2000, with 1.5 times the ST). A recent meta-analysis focusing specifically on comparing these two treatment modalities supported the view that the high-dose RUL ECT cognitive advantage over bitemporal ECT is limited to autobiographical memory function (Kolshus, Jelovac, & McLoughlin, 2017). Finally, increased patient's age predicts smaller decreases in verbal learning and larger decreases in executive function. Thus, during the subacute period, younger patients present with more memory difficulties, but are less impaired in their ability to plan, organize, and switch between tasks.

## Short-term effects—Up to 2 weeks post-ECT

Meta-analytical evidence indicates that, within 2 weeks following ECT's final session, cognitive performance returns to pretreatment levels (Semkovska & McLoughlin, 2010). Measures of global cognitive status even show a small to medium improvement during that period. These results suggest that while there is a progressive cognitive improvement in the first week of post-ECT, cognitive function is probably still affected by the treatment. Indeed, as patients recover normal mood, one would expect their cognitive function to improve above baseline. The comparable pre-post performance levels indicate a persistent, albeit decreasing in size, effect of ECT on cognition in the 4–14 days following the treatment's end. This conclusion is supported by the significant meta-regression coefficient effects of the post-ECT assessment time point on memory performance (Semkovska et al., 2011). Analyzing data from 39 studies, with intervals between end of ECT course and cognitive assessment time ranging from 1 to 183 days, longer delays predicted better memory performance, concerning equally verbal, visual, and autobiographical memory.

Moderator analyses of treatment parameters and patient characteristics effects on cognitive function post-ECT are not significant in the short-term interval (Semkovska et al., 2011). Thus, the progressive cognitive improvement observed in the first 2 weeks after ECT is independent of treatment and patient variables after the subacute period. These meta-analytical findings on ECT's short-term cognitive effects have been consistently replicated by subsequent randomized trials (e.g., Dybedal, Bjolseth, Benth, & Tanum, 2016; Semkovska et al., 2016; Vasavada et al., 2017; Verwijk et al., 2014).

## Long-term effects—From 15 days on after the end of an ECT treatment course

In a meta-analysis of studies assessing cognitive function from 15 days and up to 2 years after ECT's end, all cognitive functions affected by ECT were shown to significantly improve above pretreatment levels (Semkovska & McLoughlin, 2010). This was not the case for autobiographical memory measures due to the nature of tests assessing this function that does not permit higher than baseline performance (see Section "RAA"). For the remaining cognitive domains, improvement above pre-ECT levels was small for processing speed, working memory and verbal memory, and medium for global cognitive status, visual memory, and some aspects of executive function. These results suggest that, apart from retrograde amnesia, the cognitive effects of ECT are limited to the first 2 weeks after a treatment course. As with the short-term effects, treatment parameters and individual characteristics do not influence significantly cognitive improvement levels at long-term follow-ups (Semkovska et al., 2011). After these meta-analytical findings, studies of the long-term cognitive effects of ECT continue to observe similar results of improved function relative to pretreatment levels up to 2 years after ECT's end (Anderson, McAllister-Williams, Downey, Elliott, & Loo, 2020; Dybedal et al., 2016; Mohn & Rund, 2016; Mohn & Rund, 2019; Verwijk et al., 2015).

## Retrograde autobiographical amnesia (RAA)

Losing memories about events and facts experienced or learned before starting the treatment represent the biggest patients' concern relative to ECT (Semkovska & O'Grady, 2017). RAA is also the most researched effect of ECT, with many technical refinements through the years aiming to show reduction of this side-effect (Semkovska & McLoughlin, 2013). Most studies in the domain are using post-ECT assessments based on percentage consistency with baseline performance that is always fixed at 100%, irrespective of how detailed or accurate was the pre-ECT autobiographical memory recall (Semkovska & McLoughlin, 2014). Consequently, in ECT research, RAA is predominantly defined as percentage details not recalled at posttreatment relative to the pretreatment description. However, never-depressed controls also show loss in autobiographical memory consistency over time, without suffering from amnesia (Semkovska, Noone, Carton, & McLoughlin, 2012). Further, depression itself is associated with difficulties in recalling specific autobiographical events, such as how one spent last New Year's Eve or what happened on the first day of one's last summer holidays (see Sumner, Griffith, & Mineka, 2010 meta-analysis on the topic). Indeed, during a depressive episode, the autobiographical memory recall has been consistently found to be overgeneral with nonspecific answers provided to the above-described specific event questions, such as "we ate and had fun on New Year's" or "we probably swam as we usually do" (Semkovska et al., 2012). These difficulties in recalling specific events persist after remitting from depression as shown by a meta-analysis of data from individuals who were *not* treated with ECT (Semkovska et al., 2019). In sum, distinguishing ECT-related RAA from, on one hand, the depression-associated autobiographical memory difficulties and, on the other hand, the normal changes in autobiographical memory consistency with the passage of time poses a methodological challenge (Anderson et al., 2020). Finally, everybody's autobiographical memories are fundamentally different, which further compounds the issue of adequately delineating ECT-related RAA.

Bearing in mind these measurement challenges, what do we know to date about ECT-associated RAA?

1. Differences in ECT technique result in differences in autobiographical memory outcomes. Indeed, during the subacute period, bitemporal ECT leads to a higher loss in autobiographical memory consistency than RUL ECT (Semkovska et al., 2011), while ultra-brief pulse is associated with less such loss than brief-pulse ECT (Tor et al., 2015).
2. Unlike with tests of anterograde memory described above, the degree of change pre- and post-ECT in autobiographical memory is not associated with electrical stimulus dose as shown by a meta-analysis of 39 studies (Semkovska et al., 2011). Consistent with this, even when delivered at high intensity, RUL ECT leads to better autobiographical performance than bitemporal ECT up to 6 months posttreatment (Semkovska et al., 2016; Verwijk et al., 2015).
3. A significant meta-regression has also shown that longer intervals between final ECT sessions and cognitive retesting predicted higher consistency between post-ECT autobiographical recall and pretreatment performance (Semkovska et al., 2011). Similar improvements of autobiographical memory outcomes with the passage of time post-ECT are observed by recent studies (Anderson et al., 2020; Sackeim, 2017).
4. Taken together, these findings clearly indicate that ECT does cause a certain degree of RAA that may decrease in severity with time. However, the methodological limitations around its assessment and the paucity of longitudinal comparisons with normal and depressed

controls prevent us to determine if ECT-associated RAA is reversible.

Three preliminary studies, using validated instruments to assess autobiographical memory, have sought to provide some clarity on this topic. All distinguished between semantic and episodic autobiographical memories (see definitions in Section "Mini-dictionary of terms"). Importantly, these instruments allow for possibly improved post-ECT performance, because scores are not calculated as a function of pretreatment, thus theoretically permitting the examination of RAA's duration.

Verwijk et al. (2015) used the validated Kopelman Autobiographical Memory Interview in patients treated with RUL ECT. They also followed-up only remitters ($n=33$), thus potentially isolating the mood-associated from the ECT-associated memory effects. After a slight decrease immediately after ECT, episodic personal memories improved significantly relative to pretreatment levels at both 3- and 6-months follow-ups, while semantic autobiographical memory remained unaffected by ECT. Although preliminary, these results suggest that RAA associated with RUL ECT concerns only episodic personal memories and may be reversible within the first 3 months after end of treatment.

Söderlund, Percy, and Levine (2012) used the validated Levine's Autobiographical Interview in 22 depressed patients. At 3 months post-ECT, all participants showed persistent autobiographical memory issues. Patients treated with bitemporal ECT showed difficulties with both semantic and episodic personal memories, with the latter being more impaired than the former. Episodic RAA was also observed in patients treated with RUL ECT, but to a lesser extent than following bitemporal ECT. The study did not distinguish between remitters and nonremitters. Thus, in this mixed RUL ECT group, some individuals probably also presented confounding mood-associated autobiographical memory issues (i.e., the lack of specificity in episodic memory recall characteristic of depression), which could explain the difference with Verwijk et al.'s (2015) team results.

Finally, using the validated scoring system of the Columbia Autobiographical Memory Interview—Short Form, Semkovska and O'Grady (2017) compared 19 severely depressed inpatients never treated with ECT who received pharmacotherapy with 19 individually matched patients treated with bitemporal ECT. At pretreatment, both groups were comparable in the amount of semantic and episodic personal memories that they generated. At posttreatment and 3-month follow-up, the ECT group showed decreased episodic memories, but comparable levels of semantic memories relative to the pharmacotherapy group. These results were obtained on groups means. When the performance of each patient was compared to normative data, it was found that, while the pharmacotherapy group improved their semantic score, the ECT's group remained comparable to baseline, suggesting a slight semantic RAA at posttreatment. This difference between the two groups disappeared at follow-up. Furthermore, on individual level as well, 37% of participants showed impaired episodic personal memories at posttreatment, while for 26%, this episodic RAA persisted at the 3-month follow-up. Together, these results indicated that a slight ECT-associated semantic RAA is reversible within 3 months, but an episodic RAA persists in some individuals for at least 3-month posttreatment. Thus, RAA following bitemporal ECT is limited to specific personal episodes, does not affect all patients, and is reversible for some.

## ECT's mechanism of action models

Since ECT's inception, large evidence has accumulated on its diverse neurobiological effects. Consequently, many mechanisms were proposed to explain its clinical effectiveness, involving changes at the cell, synapse, or neurocircuitry levels. Supported models postulate improved neurotransmission, normalization of the neuroendocrine overdrive, anticonvulsant outturn, and neuroplasticity enhancement as the probable mechanisms through which ECT exerts its effect.

### Improved neurotransmission

The *monoamine* hypothesis postulates that depression results from a deficiency in the norepinephrine, dopamine, and/or serotonin neurotransmission (Hirschfeld, 2000). Supported by the effectiveness of antidepressants acting on increasing monoamines' availability, this hypothesis has also been tested as an explanatory model for the therapeutic effect of ECT. Enhanced monoamine transmission following electroconvulsive shock has been supported by extensive animal research (Fosse & Read, 2013). However, evidence in humans is limited to the demonstration of increased levels of CRF dopamine metabolites following ECT (e.g., Nikisch & Mathe, 2008), whereas evidence of the serotoninergic system involvement is lacking (Singh & Kar, 2017). Further, even if ECT does indeed result in the "normalization" of dopamine (or other monoamines) function, it is unclear how these effects could contribute to ECT's efficacy.

Research has also suggested that depressive symptoms could be explained by deficits in *GABA*ergic inhibitory neurotransmission (Luscher, Shen, & Sahir, 2011). Supporting evidence of reduced GABA levels in CRF and prefrontal regions of acutely depressed patients has prompted an examination of ECT's effect on the neurotransmitter. Both animal and human studies have demonstrated increased

GABA levels following ECT (Kellner, 2012). Giving the central role of GABA in regulating neurogenesis and synaptogenesis (Heaney & Kinney, 2016), its posttreatment increase appears also consistent with the neuroplasticity model of ECT action described below.

## Normalization of the neuroendocrine overdrive

The neuroendocrine hypothesis of depression emerged from a wealth of research showing hypothalamic–pituitary–adrenal axis dysregulation and, particularly, increased cortisol levels as key characteristics of the disorder. Consequently, ECT's antidepressant properties are postulated to result from its effect on releasing hypothalamic and/or pituitary hormones. Both animal and human studies show that ECT leads indeed to the increased transient discharge of many hormones, including prolactin, thyroid-stimulating hormone, adrenocorticotropic hormone, cortisol, and endorphins (Fosse & Read, 2013). In depressed patients, these hormones return to baseline levels within 1 hour of ECT's end, followed by a further progressive normalization of cortisol levels (Kellner, 2012). In animal models, ECT has also been shown to reduce cortisol-induced inhibition of neuroplasticity (Singh & Kar, 2017). Thus, the neuroendocrine overdrive normalization model also supports the neuroplasticity explanatory mechanism of ECT action.

## The anticonvulsant model

This model is supported by the various ECT anticonvulsant properties, and especially the continuous increase in ST and decrease in seizure duration over the treatment course (Sackeim, 2004). In epilepsy, continuous seizure activity could be observed for hours and up to days, whereas the ECT-generated seizure is limited to less than a minute and rarely extends over 90 seconds (Wang, Lin, Chiu, & Tseng, 2013), further demonstrating the treatment's anticonvulsant effect. The generalized seizure following ECT leads to a progressive (i.e., larger after each ECT session) slowing of brain activity, most prominent in the frontal cortex (Sackeim, 2004). This inhibition is predominantly due to the above-described GABA release, but also to endogenous opioids emission. ECT's strong anticonvulsant properties suggest the antidepressant effect is exerted through promotion of inhibitory brain processes. In a systematic review of research exploring the relationship between these inhibitory processes and clinical response, Mayur (2006) found that all 18 studies on the topic observed a significant positive association between diverse cerebral slowing parameters and improved depressive symptoms.

## Neuroplasticity enhancement

The most recent and well-supported model of ECT's mode of action is based on its potent neuroplasticity enhancing effect. Neuroplasticity refers to the brain's ability to adjust its functions and connections in response to environmental change (Robertson & Murre, 1999). This is achieved through synaptogenesis, neurogenesis, and gliogenesis. Varied animal models consistently show such neuroplastic changes following ECT (Bouckaert et al., 2014), especially in the hippocampal area (Takamiya et al., 2018). In patients with depression, initial studies evaluated plasma levels of brain-derived neurotrophic factors as an estimate of neuroplasticity occurring. This early research gave little support for the neuroplasticity model, as correlations between changes in plasma levels following treatment and clinical response were inconsistent (Bouckaert et al., 2014). However, subsequent neuroimaging studies cumulated support for the morphological and functional cerebral changes associated with ECT's action.

Longitudinal *structural imaging* research provides strong evidence of brain plasticity in patients treated with ECT for depression. Three recent meta-analyses indicated that ECT induces a cerebral volume increase in the limbic area, predominantly in the hippocampus (Gbyl & Videbech, 2018; Takamiya et al., 2018; Wilkinson, Sanacora, & Bloch, 2017). All demonstrated a significant bilateral increase in hippocampal volume, which was also found to be more pronounced in younger patients by Takamiya et al.' analyses (2018). The bilateral amygdala volume increase was also observed post-ECT, albeit to a smaller extent than in the hippocampus. While some of the meta-analyzed studies found a significant relationship between the hippocampal increase and clinical response (e.g., Dukart et al., 2014; Joshi et al., 2016), all three meta-analyses indicated that this relationship was not significant overall. In addition to synaptogenesis, both neurogenesis and gliogenesis can be considered as possible factors in the hippocampal volume growth. Indeed, ECT increases the number of hippocampal neurons and induces proliferation of glial cells in both mice and nonhuman primates. Hippocampal neurogenesis has also been observed in adult humans (e.g., Spalding et al., 2013), thus, increased neurogenesis might contribute to the hippocampal growth post-ECT.

Decreased hippocampal function and smaller hippocampal volume are consistent findings of the neuroimaging research in depression (see Schmaal et al., 2017 meta-analysis). Consequently, remediating these cortical abnormalities can reasonably explain the ECT antidepressant effect, without the relationship between relative hippocampal increase and level of clinical improvement having to be necessarily linear. Indeed, patients who remitted from depression without receiving ECT show persistent memory difficulties as measured by neuropsychological tests (Semkovska et al., 2019). These deficits increase proportionately to the number of previous depressive episodes. Thus, variations in hippocampal volume could be explained by the interaction between number of previous episodes, levels of clinical response, and depression-associated hippocampal loss, as opposed to clinical response alone. Another possibility is that only volumetric changes in specific parts of the hippocampus are associated with the clinical response. Two studies completed after the publication of the above-described meta-analyses on hippocampal volume support this hypothesis. Cao et al. (2018) observed that neuroplasticity in the granule cell layers appears essential to achieve remission following ECT. Nuninga's team (Nuninga et al., 2020) found that initial dentate gyrus volume predicted clinical response following ECT, whereas increased volume in that gyrus post-ECT was associated with a lower level of depressive symptoms. These specific results need replications with larger sample sizes, but the overall studies' findings further strengthen existing support for the neuroplasticity model of ECT effectiveness.

*Functional imaging* research provides evidence of post-seizure, ECT associated, continuous slowing down of specific cortical regions connectivity. For example, in a systematic review of nine studies, Abbott, Gallegos, Rediske, Lemke, and Quinn (2014) found decreased frontal perfusion and reduced frontal functional connectivity post-ECT. In their review of 20 brain imaging studies, Fosse and Read (2013) showed robust evidence of an association between reduced depression scores in the week following ECT sessions' end and frontal deactivation relative to pretreatment levels. Recent studies have also consistently demonstrated a decrease in and normalization of cerebral resting-state functional connectivity following treatment (e.g., Cano et al., 2017; Leaver et al., 2016).

The default mode network is a brain system of interacting cerebral regions characterized by coordinated activations that are distinct from the activities of other brain networks. It appears activated mainly when the person is engaged in reflection and not actively occupied or involved in a task (Raichle et al., 2001). Systematic reviews indicate that the default mode network, and especially its anterior component centered on the medial prefrontal cortex, show increased activation during depression relative to levels observed in nonclinical controls (e.g., Mulders, van Eijndhoven, Schene, Beckmann, & Tendolkar, 2015). This finding is consistent with the characteristic of depression regular engagement in rumination and self-referential negative reflection (Disner, Beevers, Haigh, & Beck, 2011). Both the frontal activation decrease and the cerebral resting-state normalization following ECT support this neuroplasticity model of its antidepressant action.

## Future directions: Testing an integrated model

In conclusion, while the neuroplasticity model of ECT's therapeutical action has received the most empirical support, evidence from the remaining models appears consistent and can be combined for an integrated model. Specifically, the anticonvulsant effect of ECT involves an increase of GABAergic transmission, which leads to, on one hand, a reduction in prefrontal activation and normalization of the default mode network state and, on the other hand, promotion of cerebral neurogenesis and synaptogenesis. In parallel, the ECT regulatory effect on the hypothalamic–pituitary–adrenal axis leads to a decrease in the depression-associated excessive cortisol concentrations, thus lifting the hormone's inhibitory effect on neuroplasticity. Normalization of the prefrontal and default mode network activations can explain the reduction in depressive symptoms involving rumination and negative internal self-representations. The disrupted cognitive functions during the first weeks after treatment could also be the consequence of this suddenly decreased activation of the frontotemporal connectivity.

The neuroplastic changes promoted by the ECT effects on hormonal release and neurotransmission result in increased hippocampal volume, which probably contributes to the improved cognitive functions observed at long-term follow-ups, especially with regard to anterograde memory. ECT-associated neuroplastic changes can also explain the occurrence and, sometimes, persistence of RAA for certain events. Indeed, in the same way as other cognitive functions, semantic personal memories that are overly rehearsed pieces of information (e.g., the birthday date of one's sister; the name of one's primary school teacher) show a decrease immediately after ECT but consecutively recover over the following 3 months. Access to episodic memories, which are less rehearsed, can be fragile and thus prone to be disrupted by the early ECT-associated cerebral deactivation and by the subsequent synaptic reshaping that likely occurs during the hippocampal growth. To investigate the accuracy of the present integrated model and its components, future studies should use translational methodologies to assess concurrently neurotransmission, activation, connectivity, and cognitive function within longitudinal designs using multiple assessment time-points. Such studies would permit to further clarify ECT's mechanism of action by ascertaining critical processes for the response and sustained remission following ECT, along with potential predictors of nonresponse and relapse.

## Key facts for electroconvulsive therapy

- ECT is superior to pharmacotherapy for the treatment of depression.

- Except for retrograde memory, cognitive side-effects are temporary and limited to the first few days post-ECT. Over the following weeks, cognition progressively improves above pretreatment levels.
- Retrograde autobiographical memory can be affected up to 3 months and possibly longer with bitemporal ECT. However, this retrograde amnesia is limited to a small number of specific personal events, does not affect every patient, and is reversible for some individuals.
- Differences in ECT techniques and patients' age explain variations in clinical outcome and cognitive side-effects during the first few days after ECT but account little for long-term outcomes variations.
- ECT most likely exerts its mechanism of action by promoting both structural and functional brain plasticity.
- Decreased cerebral resting state and increased bilateral hippocampal volume may be central in explaining ECT's effectiveness and cognitive side-effects.

## Summary points

- ECT is currently the most effective treatment for depression.
- Bitemporal ECT is associated with more cognitive side-effects during the first 3 days after treatment and a higher likelihood of persistent retrograde memory issues than right unilateral ECT.
- Retrograde amnesia does not affect all patients and, when occurring, it is reversible for most during the 3 months post-ECT.
- ECT increases GABAergic neurotransmission and regulates the hypothalamic–pituitary–adrenal axis function.
- ECT is associated with hippocampal volume growth, slowing of frontotemporal connectivity, and normalization of the default mode network levels of activity.

## Mini-dictionary of terms

**Anterograde amnesia** The diminished recall or loss of information that has been recently acquired; in the ECT context: "recently" means after the end of the treatment course.

**Anterograde memory** The ability to learn new memories and recall them after a delay.

**Episodic autobiographical memory** Memories about events that one witnessed or experienced and that are situated in a specific time and place, such as the events that happened in the morning of one's first day at work in a new job.

**Gliogenesis** The formation of new glial cells, e.g., astrocytes or oligodendrocytes.

**Neurogenesis** The brain plasticity process of new neurons creation.

**Retrograde autobiographical amnesia** The diminished memory recall or memory loss of self-related information that was acquired prior to starting ECT.

**Seizure threshold** The minimum charge required to induce a generalized seizure that is needed for therapeutic effect.

**Semantic autobiographical memory** Decontextualized knowledge about one's past that are facts about the self not related to a specific event, such as the name of one's kindergarten teacher.

**Synaptogenesis** The brain plasticity process of formation, reorganization, and maintenance of the synaptic connections.

## References


Abbott, C. C., Gallegos, P., Rediske, N., Lemke, N. T., & Quinn, D. K. (2014). A review of longitudinal electroconvulsive therapy: Neuroimaging investigations. *Journal of Geriatric Psychiatry and Neurology*, *27*(1), 33–46.

Anderson, I. M., McAllister-Williams, R. H., Downey, D., Elliott, R., & Loo, C. (2020). Cognitive function after electroconvulsive therapy for depression: Relationship to clinical response. *Psychological Medicine*, 1–10.

Bouckaert, F., Sienaert, P., Obbels, J., Dols, A., Vandenbulcke, M., Stek, M., et al. (2014). ECT: Its brain enabling effects: A review of electroconvulsive therapy-induced structural brain plasticity. *Journal of ECT*, *30*(2), 143–151.

Cano, M., Martinez-Zalacain, I., Bernabeu-Sanz, A., Contreras-Rodriguez, O., Hernandez-Ribas, R., Via, E., et al. (2017). Brain volumetric and metabolic correlates of electroconvulsive therapy for treatment-resistant depression: A longitudinal neuroimaging study. *Translational Psychiatry*, *7*(2), e1023.

Cao, B., Luo, Q., Fu, Y., Du, L., Qiu, T., Yang, X., et al. (2018). Predicting individual responses to the electroconvulsive therapy with hippocampal subfield volumes in major depression disorder. *Scientific Reports*, *8*(1), 5434.

Disner, S. G., Beevers, C. G., Haigh, E. A., & Beck, A. T. (2011). Neural mechanisms of the cognitive model of depression. Nature reviews. *Neuroscience*, *12*(8), 467–477.

Dukart, J., Regen, F., Kherif, F., Colla, M., Bajbouj, M., Heuser, I., et al. (2014). Electroconvulsive therapy-induced brain plasticity determines therapeutic outcome in mood disorders. *Proceedings of the National Academy of Sciences of the USA*, *111*(3), 1156–1161.

Dybedal, G. S., Bjolseth, T. M., Benth, J. S., & Tanum, L. (2016). Cognitive effects of Bifrontal versus right unilateral electroconvulsive therapy in the treatment of major depression in elderly patients: A randomized, controlled trial. *Journal of ECT*, *32*(3), 151–158.

Fink, M., Kellner, C. H., & McCall, W. V. (2014). The role of ECT in suicide prevention. *Journal of ECT*, *30*(1), 5–9.

Folkerts, H. W., Michael, N., Tolle, R., Schonauer, K., Mucke, S., & Schulze-Monking, H. (1997). Electroconvulsive therapy vs. paroxetine in treatment-resistant depression—a randomized study. *Acta Psychiatrica Scandinavica*, *96*(5), 334–342.

Fosse, R., & Read, J. (2013). Electroconvulsive treatment: Hypotheses about mechanisms of action. *Frontiers in Psychiatry*, *4*, 94.

Gbyl, K., & Videbech, P. (2018). Electroconvulsive therapy increases brain volume in major depression: A systematic review and meta-analysis. *Acta Psychiatrica Scandinavica*, *138*(3), 180–195.

Heaney, C. F., & Kinney, J. W. (2016). Role of GABA(B) receptors in learning and memory and neurological disorders. *Neuroscience and Biobehavioral Reviews*, *63*, 1–28.

Hirschfeld, R. M. (2000). History and evolution of the monoamine hypothesis of depression. *Journal of Clinical Psychiatry*, *61*(Suppl 6), 4–6.

Janicak, P. G., Davis, J. M., Gibbons, R. D., Ericksen, S., Chang, S., & Gallagher, P. (1985). Efficacy of ECT: A meta-analysis. *American Journal of Psychiatry*, *142*(3), 297–302.

Joshi, S. H., Espinoza, R. T., Pirnia, T., Shi, J., Wang, Y., Ayers, B., et al. (2016). Structural plasticity of the Hippocampus and amygdala induced by electroconvulsive therapy in major depression. *Biological Psychiatry*, *79*(4), 282–292.

Kellner, C. H. (2012). Electroconvulsive therapy: Basic concepts. In C. H. Kellner (Ed.), *Brain stimulation in psychiatry: ECT, DBS TMS and other modalities* (pp. 3–16). Cambridge: Cambridge University Press.

Kho, K. H., van Vreeswijk, M. F., Simpson, S., & Zwinderman, A. H. (2003). A meta-analysis of electroconvulsive therapy efficacy in depression. *Journal of ECT*, *19*(3), 139–147.

Kolshus, E., Jelovac, A., & McLoughlin, D. M. (2017). Bitemporal v. high-dose right unilateral electroconvulsive therapy for depression: A systematic review and meta-analysis of randomized controlled trials. *Psychological Medicine*, *47*(3), 518–530.

Leaver, A. M., Espinoza, R., Pirnia, T., Joshi, S. H., Woods, R. P., & Narr, K. L. (2016). Modulation of intrinsic brain activity by electroconvulsive therapy in major depression. *Biological Psychiatry Cognitive Neuroscience and Neuroimaging*, *1*(1), 77–86.

Liang, C. S., Chung, C. H., Ho, P. S., Tsai, C. K., & Chien, W. C. (2018). Superior anti-suicidal effects of electroconvulsive therapy in unipolar disorder and bipolar depression. *Bipolar Disorders*, *20*(6), 539–546.

Loo, C. K., Katalinic, N., Martin, D., & Schweitzer, I. (2012). A review of ultrabrief pulse width electroconvulsive therapy. *Therapeutic Advances in Chronic Disease*, *3*(2), 69–85.

Luscher, B., Shen, Q., & Sahir, N. (2011). The GABAergic deficit hypothesis of major depressive disorder. *Molecular Psychiatry*, *16* (4), 383–406.

Martin, D. M., Galvez, V., & Loo, C. K. (2015). Predicting retrograde autobiographical memory changes following electroconvulsive therapy: Relationships between individual, treatment, and early clinical factors. *International Journal of Neuropsychopharmacology*, *18*(12), pyv067.

Mayur, P. (2006). Ictal electroencephalographic characteristics during electroconvulsive therapy: A review of determination and clinical relevance. *Journal of ECT*, *22*(3), 213–217.

Mohn, C., & Rund, B. R. (2016). Maintained improvement of neurocognitive function in major depressive disorders 6 months after ECT. *Frontiers in Psychiatry*, *7*, 200.

Mohn, C., & Rund, B. R. (2019). Neurognitive function and symptom remission 2 years after ECT in major depressive disorders. *Journal of Affective Disorders*, *246*, 368–375.

Mulders, P. C., van Eijndhoven, P. F., Schene, A. H., Beckmann, C. F., & Tendolkar, I. (2015). Resting-state functional connectivity in major depressive disorder: A review. *Neuroscience and Biobehavioral Reviews*, *56*, 330–344.

Nikisch, G., & Mathe, A. A. (2008). CSF monoamine metabolites and neuropeptides in depressed patients before and after electroconvulsive therapy. *European Psychiatry*, *23*(5), 356–359.

Nuninga, J. O., Mandl, R. C. W., Boks, M. P., Bakker S., Somers, M., Heringa, S. M., et al. (2020). Volume increase in the dentate gyrus after electroconvulsive therapy in depressed patients as measured with 7T. *Molecular Psychiatry*, *25*, 1559–1568.

Pagnin, D., de Queiroz, V., Pini, S., & Cassano, G. B. (2004). Efficacy of ECT in depression: A meta-analytic review. *Journal of ECT*, *20*(1), 13–20.

Raichle, M. E., MacLeod, A. M., Snyder, A. Z., Powers, W. J., Gusnard, D. A., & Shulman, G. L. (2001). A default mode of brain function. *Proceedings of the National Academy of Sciences of the USA*, *98*(2), 676–682.

Robertson, I. H., & Murre, J. M. (1999). Rehabilitation of brain damage: Brain plasticity and principles of guided recovery. *Psychological Bulletin*, *125*(5), 544–575.

Ross, E. L., Zivin, K., & Maixner, D. F. (2018). Cost-effectiveness of electroconvulsive therapy vs pharmacotherapy/psychotherapy for treatment-resistant depression in the United States. *JAMA Psychiatry*, *75*(7), 713–722.

Sackeim, H. A. (2004). Convulsant and anticonvulsant properties of electroconvulsive therapy: Towards a focal form of brain stimulation. *Clinical Neuroscience Research*, *4*, 39–57.

Sackeim, H. A. (2017). Modern electroconvulsive therapy: Vastly improved yet greatly underused. *JAMA Psychiatry*, *74*(8), 779–780.

Sackeim, H. A., Prudic, J., Devanand, D. P., Kiersky J. E., Fitzsimons, L., Moody, B. J., et al. (1993). Effects of stimulus intensity and electrode placement on the efficacy and cognitive effects of electroconvulsive therapy. *New England Journal of Medicine*, *328*(12), 839–846.

Sackeim, H. A., Prudic, J., Devanand, D. P., Nobler, M. S., Lisanby, S. H., Peyser, S., et al. (2000). A prospective, randomized, double-blind comparison of bilateral and right unilateral electroconvulsive therapy at different stimulus intensities. *Archives of General Psychiatry*, *57*(5), 425–434.

Sackeim, H. A., Prudic, J., Nobler, M. S., Fitzsimons, L., Lisanby, S. H., Payne, N., et al. (2008). Effects of pulse width and electrode placement on the efficacy and cognitive effects of electroconvulsive therapy. *Brain Stimulation*, *1*(2), 71–83.

Schmaal, L., Hibar, D. P., Samann, P. G., Hall, G. B., Baune, B. T., Jahanshad, N., et al. (2017). Cortical abnormalities in adults and adolescents with major depression based on brain scans from 20 cohorts worldwide in the ENIGMA major depressive disorder working group. *Molecular Psychiatry*, *22*(6), 900–909.

Semkovska, M., Keane, D., Babalola, O., & McLoughlin, D. M. (2011). Unilateral brief-pulse electroconvulsive therapy and cognition: Effects of electrode placement, stimulus dosage and time. *Journal of Psychiatric Research*, *45*(6), 770–780.

Semkovska, M., Landau, S., Dunne, R., Kolshus, E., Kavanagh, A., Jelovac, A., et al. (2016). Bitemporal versus high-dose unilateral twice-weekly electroconvulsive therapy for depression (EFFECT-dep): A pragmatic, randomized, non-inferiority trial. *American Journal of Psychiatry*, *173*(4), 408–417.

Semkovska, M., & McLoughlin, D. M. (2010). Objective cognitive performance associated with electroconvulsive therapy for depression: A systematic review and meta-analysis. *Biological Psychiatry*, *68*(6), 568–577.

Semkovska, M., & McLoughlin, D. M. (2013). Measuring retrograde autobiographical amnesia following electroconvulsive therapy: Historical perspective and current issues. *Journal of ECT*, *29*(2), 127–133.

Semkovska, M., & McLoughlin, D. M. (2014). Retrograde autobiographical amnesia after electroconvulsive therapy: On the difficulty of finding the baby and clearing murky bathwater. *Journal of ECT*, *30*(3), 187–188 (discussion 189–190).

Semkovska, M., Noone, M., Carton, M., & McLoughlin, D. M. (2012). Measuring consistency of autobiographical memory recall in depression. *Psychiatry Research*, *197*(1–2), 41–48.

Semkovska, M., & O'Grady, T. (2017). Unravelling autobiographical retrograde amnesia following Bitemporal electroconvulsive therapy: Effect of treatment versus effect of time. *Psychology*, *8*, 611–626.

Semkovska, M., Quinlivan, L., O'Grady, T., Johnson, R., Collins, A., O'Connor, J., et al. (2019). Cognitive function following a major depressive episode: A systematic review and meta-analysis. *Lancet Psychiatry*, *6*(10), 851–861.

Singh, A., & Kar, S. K. (2017). How electroconvulsive therapy works? *Understanding the Neurobiological Mechanisms Clinical Psychopharmacology and Neuroscience*, *15*(3), 210–221.

Slade, E. P., Jahn, D. R., Regenold, W. T., & Case, B. G. (2017). Association of electroconvulsive therapy with psychiatric readmissions in US hospitals. *JAMA Psychiatry*, *74*(8), 798–804.

Söderlund, H., Percy, A., & Levine, B. (2012). Electroconvulsive therapy for depression and autobiographical memory. In A. Zeman, N. Kapur, & M. Jones-Gotman (Eds.), *Epilepsy and memory* (pp. 244–258). Oxford, UK: Oxford University Press.

Spalding, K. L., Bergmann, O., Alkass, K., Bernard, S., Salehpour, M., Huttner, H. B., et al. (2013). Dynamics of hippocampal neurogenesis in adult humans. *Cell*, *153*(6), 1219–1227.

Sumner, J. A., Griffith, J. W., & Mineka, S. (2010). Overgeneral autobiographical memory as a predictor of the course of depression: A meta-analysis. *Behaviour Research and Therapy*, *48*(7), 614–625.

Takamiya, A., Chung, J. K., Liang, K. C., Graff-Guerrero, A., Mimura, M., & Kishimoto, T. (2018). Effect of electroconvulsive therapy on hippocampal and amygdala volumes: Systematic review and meta-analysis. *British Journal of Psychiatry*, *212*(1), 19–26.

Tor, P. C., Bautovich, A., Wang, M. J., Martin, D., Harvey, S. B., & Loo, C. (2015). A systematic review and meta-analysis of brief versus ultrabrief right unilateral electroconvulsive therapy for depression. *Journal of Clinical Psychiatry*, *76*(9), e1092–e1098.

UK ECT Review Group. (2003). Efficacy and safety of electroconvulsive therapy in depressive disorders: A systematic review and meta-analysis. *Lancet*, *361*(9360), 799–808.

Van Vreeswijk, M. F., & De Wilde, E. J. (2004). Autobiographical memory specificity, psychopathology, depressed mood and the use of the autobiographical memory test: A meta-analysis. *Behaviour Research and Therapy*, *42*(6), 731–743.

Vasavada, M. M., Leaver, A. M., Njau, S., Joshi, S. H., Ercoli, L., Hellemann, G., et al. (2017). Short- and long-term cognitive outcomes in patients with major depression treated with electroconvulsive therapy. *Journal of ECT*, *33*(4), 278–285.

Verwijk, E., Comijs, H. C., Kok, R. M., Spaans, H. P., Tielkes, C. E., Scherder, E. J., et al. (2014). Short- and long-term neurocognitive functioning after electroconvulsive therapy in depressed elderly: A prospective naturalistic study. *International Psychogeriatrics*, *26*(2), 315–324.

Verwijk, E., Spaans, H. P., Comijs, H. C., Kho, K. H., Sienaert, P., Bouckaert, F., et al. (2015). Relapse and long-term cognitive performance after brief pulse or ultrabrief pulse right unilateral electroconvulsive therapy: A multicenter naturalistic follow up. *Journal of Affective Disorders*, *184*, 137–144.

Wang, C. C., Lin, C. H., Chiu, Y. C., & Tseng, C. C. (2013). The Clonic phase of seizures in patients treated with electroconvulsive therapy is related to age and stimulus intensity. *Frontiers in Psychiatry*, *4*, 166.

Weiner, R. D. (2015). Introduction to convulsive therapy. In I. M. Reti (Ed.), *Brain stimulation: Methodologies and interventions* (pp. 61–82). New Jersey: Wiley Blackwell.

Wilkinson, S. T., Sanacora, G., & Bloch, M. H. (2017). Hippocampal volume changes following electroconvulsive therapy: A systematic review and meta-analysis. *Biological Psychiatry Cognitive Neuroscience and Neuroimaging*, *2*(4), 327–335.

Chapter 53

# Prenatal depression and offspring DNA methylation

Sabrina Faleschini[a] and Andres Cardenas[b]

[a]School of Psychology, Laval University, Quebec, QC, Canada; [b]Division of Environmental Health Sciences, University of California, Berkeley, Berkeley, CA, United States

## Abbreviations

| | |
|---|---|
| **CpGs** | cytosine-guanine dinucleotides sites |
| **DOHaD** | developmental origins of health and disease |
| **EWAS** | epigenome-wide association studies |
| **DNA** | deoxyribonucleic acid |
| **SSRIs** | selective serotonin reuptake inhibitors |

## Prenatal maternal depression

Pregnancy is a sensitive period during which women are at higher risk of experiencing stress and mental health issues (Andersson et al., 2003). Notably, higher levels of depressive symptoms are generally observed during pregnancy than during the postpartum period (Evans, Heron, Francomb, Oke, & Golding, 2001). Approximately 15% of women will report one major depressive episode during pregnancy and many more pregnant women will experience subsyndromal depressive symptoms (Bennett, Einarson, Taddio, Koren, & Einarson, 2004; Brown & Solchany, 2004). Given the impact of depression on daily functioning, up to 8% of pregnant women will use antidepressants such as selective serotonin reuptake inhibitors (SSRIs) in order to reduce their symptoms (Andrade et al., 2008). The occurrence of depressive symptoms during pregnancy is likely to have a significant influence on the mother's health and behavior and potentially influences the development of her unborn child.

During pregnancy, maternal depression is associated with changes in fetal health such as elevated motor activity as well as poor perinatal outcomes including impaired fetal weight gain and a higher risk of preterm delivery (Field, Diego, & Hernandez-Reif, 2006; Henrichs et al., 2010). The long-term effects of prenatal maternal depression on child development following birth are also extensively documented. Notably, a meta-analysis from 2018 by Madigan and colleagues observed that prenatal stress, defined by the presence of depression and anxiety, could negatively impact child socioemotional development and behavioral functioning later in life. Interestingly, they observed that the odds of having children with behavioral difficulties were almost 1.5–2 times greater for women who have experienced prenatal depression or anxiety in comparison to women who did not experience these symptoms (Madigan et al., 2018). Also, by comparing children's exposure to maternal depression during the prenatal and postnatal periods, many studies have observed greater risk with exposure to prenatal depression on child cognitive and behavioral development during childhood (Faleschini, Rifas-Shiman, Tiemeier, Oken, & Hivert, 2019; Lin et al., 2017; O'Donnell, Glover, Barker, & O'Connor, 2014). A potential pathway for the prenatal maternal environment to influence child development could be examined by the developmental origins of health and disease hypothesis (DOHaD).

## Developmental origins of health and disease (DOHaD)

The developmental origins of health and disease (DOHaD) hypothesis implies a relationship between early life environment, especially during fetal development, and later onset of diseases and health trajectories. The DOHaD hypothesis suggests that the fetus will adapt to the maternal intrauterine environment in order to shape and adjust its development to its prenatal environment (Swanson, Entringer, Buss, & Wadhwa, 2009). The concept of DOHaD was first developed by David Barker and colleagues who observed that low birth weight in neonates increases the risk of cardiovascular disease and type 2 diabetes during adulthood (Barker et al., 1993). In fact, during the prenatal period, several types of environmental cues may be perceived by the fetus in development. The adaptive or maladaptive response of the fetus to its prenatal

The Neuroscience of Depression. https://doi.org/10.1016/B978-0-12-817933-8.00006-2

environment could potentially lead to changes and alteration of its postnatal phenotype later on (Fleming, Velazquez, & Eckert, 2015). This alteration in fetal physiological development may ultimately increase the risk of developing diseases and influence health trajectories in adulthood. Alterations to epigenetic marks and fetal epigenetic programming could explain the long-term programming of early life and prenatal exposures.

## Epigenetics mechanisms

Epigenetic modifications are heritable changes in gene expression that do not depend on the DNA sequence. Four major epigenetic modifications commonly studied are DNA methylation, histone modification, chromatin remodeling, and noncoding RNA-mediated modifications (Babenko, Kovalchuk, & Metz, 2015). Methylation of the DNA commonly occurs at cytosine nucleotides in the context of cytosine-guanine dinucleotides (CpGs) (Rakyan, Down, Balding, & Beck, 2011). DNA methylation is the most widely and commonly investigated epigenetic modification among recent studies, particularly among humans. Epigenetic programming is the process of erasure and re-establishment of almost all DNA methylation marks across the genome in the offspring, occurring throughout pregnancy and early fetal development. In fact, the prenatal period is a highly sensitive time for the developing fetus that may be affected by adverse environmental exposures. Critical influences that can occur during the prenatal development associated with epigenetic effects on children include fetal exposure to undernutrition, medication, toxic metals, endocrine disrupting compounds, or maternal mental health problems (Cardenas et al., 2019, 2017; Heijmans et al., 2008; Non, Binder, Kubzansky, & Michels, 2014).

A prime illustration of an epigenetic study is *The Dutch Famine Cohort Study* which examines the long-term impact of the Dutch famine that took place in 1944–1945 in the German-occupied Netherlands. During this time, food and fuel blockade limited access to food supplies leading to severe food scarcity and famine in the region. The study of the long-term effects of prenatally exposed children to the Dutch famine has observed a higher risk of coronary heart disease and obesity among exposed individuals (Painter, Roseboom, & Bleker, 2005). Heijmans and colleagues have studied epigenetics modifications in people who were prenatally exposed to famine as a potential mechanism and they found that almost six decades after being exposed in utero, these individuals have less DNA methylation of the imprinted *IGF2* gene when compared to their unexposed same-sex siblings (Heijmans et al., 2008).

An efficient way to study variations in DNA methylation is the implementation of Epigenome-Wide Association Studies (EWAS). EWAS can be designed in multiple ways and epidemiological studies have mostly used four designs: retrospective (case–control) cohort, parent-offspring pairs, monozygotic twins, and longitudinal cohorts (Table 1) (Rakyan et al., 2011). A retrospective cohort implies unrelated individuals recruited on the basis of their phenotype (case–control samples) and allows the examination of the associations between epigenetics effects and a clinical phenotype. A major limitation of this design is the chance for reverse causation due to the timing of sample collection. Also, the study of parent-offspring pairs allows to identify intergenerational transmissions of various epigenetics alterations from parents to their child and the

**TABLE 1** Four types of commonly implemented epigenome-wide association study (EWAS) designs.

| Case–control cohort | Families cohort | Monozygotic twins cohort | Longitudinal cohort |
|---|---|---|---|
| • Composed of unrelated individuals recruited on the basis of their phenotype or health outcome.<br>• Associations between epigenetics effects and a clinical phenotype and outcomes.<br>• Difficulty to control for environmental and genetic confounders.<br>• Increased chance for reverse causation where disease leads to epigenetic variation. | • Composed of parent-offspring pairs.<br>• The possibility of identifying intergenerational transmissions of various epigenetics changes.<br>• Allows to understand the long-term development of diseases.<br>• Lowers the chance for reverse causation. | • Composed of monozygotic twins who are discordant for a disease.<br>• Possibility to control for genetics.<br>• Allows to identify epigenetics variations associated with a disease of interest present in one of the two twins. | • Composed of people with initially no illness.<br>• Used to follow people without disease over the years.<br>• Can examine disease episodes and associated epigenetics variations and marks the preceded disease onset.<br>• Better to established temporality for disease prediction and mechanisms. |

*Note.* Based on the work of Rakyan et al., 2011.

influence of the early-life environment. This second type of study is used in order to understand the long-term development of diseases across generations. In contrast to parent-offspring pairs, monozygotic twin studies are used to identify epigenetic variations associated with a disease of interest present in one of the two twins, thereby eliminating genetic and minimizing early-life environmental confounding. Finally, longitudinal cohorts are used to follow individuals without disease over the years to examine diseases and outcomes associated with epigenetics variations. A cohort study is ideal as it minimizes the chances for reverse causation. In summary, EWAS can be used to study epigenetic alterations associated with the presence of a disease or a disorder. Yet, study designs need to be carefully considered as epigenetic changes can vary over time, are influenced by environmental exposures, are tissue and cell-type specific, and could be a consequence rather than drivers of health conditions. Assessments of these studies can provide a better understanding of the mechanisms by which prenatal exposure to different risk factors, such as maternal depression in utero, can potentially adversely influence children's long-term development and well-being or alternatively provide biomarkers of disease progression, detection, and monitor interventions.

## Epigenetics studies of fetal exposure to maternal depression

Experiencing some levels of stress is common for a woman who is expecting a child. However, stress is associated with a biochemical reaction in women's body which have the potential to influence fetal development. In fact, experiencing high levels of stress during pregnancy may potentially represent a risk factor for the development of mental health problems in pregnant women during this sensitive developmental window. Recently, several studies have attempted to better understand the way in which children could be affected by prenatal exposure to mental health issues such as depression. Notably, the study of epigenetics alterations in neonates and children exposed to prenatal depression has been the subject of an increasing interest in recent years. Research on prenatal exposure to maternal depression and related epigenetics effects is a relatively new field where studies have been performed mainly during the last decade. Globally, studies in the field have observed epigenetic modifications of exposed offspring, generally testing DNA methylation during the neonatal and the childhood periods in accessible tissue like blood or buccal cells.

A review performed by Nemoda and Szyf (2017) examined epigenetics alterations in offspring exposed to prenatal maternal depression. This review highlights genes discovered in two specific types of studies: candidate gene association studies and epigenome-wide association studies. Tables 2 and 3 present major genes reported in these studies. Regarding candidate gene association studies, the review reported epigenetic associations of exposure to prenatal depression notably on the gene *NR3C1*, a glucocorticoid receptor encoding gene (Nemoda & Szyf, 2017). For example, a Canadian study by Oberlander et al. (2008) found greater methylation in cord blood of neonates at certain sites of the *NR3C1* gene in association with exposure to maternal depression during the third trimester of pregnancy. Similar results have been found in a study by Conradt et al. (2013) using placenta DNA samples, reinforcing the idea that changes might affect multiple tissues. Interestingly, Oberlander noted that DNA methylation

**TABLE 2** **Candidate-gene studies with altered DNA methylation in offspring following prenatal exposure to maternal depression.**

| Candidate-gene studies (Hypothesis driven) | | |
|---|---|---|
| **References** | **Genes with altered methylation** | **Tissue** |
| Conradt, Lester, Appleton, Armstrong, and Marsit (2013) | *NR3C1* | Placenta |
| Braithwaite, Kundakovic, Ramchandani, Murphy, and Champagne (2015) | *NR3C1, BDNF* | Buccal cells |
| Devlin, Brain, Austin, and Oberlander (2010) | *SLC6A4* | Cord blood |
| Liu et al. (2012) | *MEG3* | Cord blood |
| Murgatroyd, Quinn, Sharp, Pickles, and Hill (2015) | *NR3C1* | Buccal cells |
| Oberlander et al. (2008) | *NR3C1* | Cord blood |
| Stonawski (2019) | *NR3C1, NR3C2, SLC6A4* | Buccal cells |
| Unternaehrer et al. (2016) | *OXTR* | Cord blood |

**TABLE 3** Epigenome-wide association studies with altered DNA methylation in offspring following prenatal exposure to maternal depression.

| | Epigenome-wide association studies (Hypothesis free) | |
|---|---|---|
| References | Genes with altered methylation | Tissue |
| Cardenas et al. (2019) | *KCNJ9, SLC39A2* | Cord blood |
| Nemoda et al. (2015) | Several genes including *BLK, FCER2* and *MGAT4A* | Cord blood |
| Non et al. (2014) | Several genes including *COL7A1* and *ENC1* | Cord blood |
| Viuff et al. (2018) | Several genes including *NNAT* | Cord blood |

differences were related to an increase of HPA stress reactivity at 3 months of age in exposed infants. Higher DNA methylation on the *NR3C1* gene was also found in buccal swab samples of 2-month-old male infants in a study by Braithwaite et al. (2015) based in the United Kingdom and in the salivary sample of the 14-month-old infant in a study by Murgatroyd et al. (2015) using participants from the *Wirral Child Health and Development Study*. It is of interest to note that a sex-specific difference on the *NR3C1* gene was found in the study of Braithwaite et al. (2015) where DNA methylation alterations were found only in male infants. This result points out potential sex differences in the way newborns could be adversely affected by early-life environment.

Another chosen candidate gene studied is the *SLC6A4* gene which transports serotonin from synaptic spaces into presynaptic neurons and plays a crucial role in brain development (Booij, Tremblay, Szyf, & Benkelfat, 2015; Nemoda & Szyf, 2017). Notably, a study by Devlin et al. (2010) found lower methylation at some sites of the *SLC6A4* gene in cord blood of neonates exposed to maternal depression during pregnancy. Moreover, exposure to prenatal depression has been associated with lower *OXTR* DNA methylation in cord blood of neonates in a study based in Switzerland by Unternaehrer et al. (2016). The *BDNF* promoter-IV gene has also been chosen by two candidate gene studies where Braithwaite et al. (2015) found lower DNA methylation in buccal swab samples of 2-month-old infants while Devlin et al. (2010) found no associations with methylation in cord blood of neonates and exposure to depression during pregnancy. Genes responsible for fetal growth have also been targeted and a study found greater DNA methylation on the *MEG3* gene (Liu et al., 2012). Another study looking at the salivary DNA sample of mothers and children observed that mothers with persistent depression during the prenatal and postnatal periods had significantly higher *OXTR* methylation but lower methylation in the region closer to the *AVP* gene (AVP IGR) (King et al., 2017). However, no DNA methylation differences were observed in exposed children. Recently, a study among school-age children observed an association between exposure to prenatal depressive symptoms and altered DNA methylation in the *NR3C1*, *NR3C2*, and *SLC6A4* genes with sex-specific effects (Stonawski et al., 2019). In comparison, Mansell tested this association on the *NR3C1* gene but did not find any significant effect after taking into account the multiple comparisons (Mansell et al., 2016).

Fewer studies have been performed using epigenome-wide association designs. Among those that have been conducted, a study observed an association between exposure to prenatal depression and DNA methylation of multiple genes including *BLK*, *FCER2*, and *MGAT4A* in cord blood samples of neonates (Nemoda et al., 2015). Similarly, Non et al. (2014) noted several other genes affected in cord blood of neonates, notably the *Col7a1* which is a gene related to collagen production. Moreover, using data from the *Avon Longitudinal Study of Parents and Children*, Viuff et al. (2018) carried epigenome-wide association analyses and identified DNA methylation differences in several genes, including *NNAT*, but could not replicate their results using the *Generation R Study*. Recently, a study by Cardenas et al. (2019) observed an association between exposure to maternal depression during pregnancy and DNA methylation in cord blood for the *KCNJ9* and *SLC39A2* genes in the *Project Viva* birth cohort. However, these associations did not replicate in the *Generation R Study*. Finally, two studies using an epigenome-wide design observed no significant association between exposure to maternal depressive symptoms and DNA methylation in neonates or infants (Schroeder et al., 2012; Wikenius et al., 2019).

Thus, epigenetics associations have been increasingly observed in studies looking at DNA methylation in prenatally exposed offspring mostly in leukocyte DNA and placenta samples. In fact, many studies have noted relevant epigenetics alterations following exposure to prenatal depression. However, a major limitation is that potential alteration result from antidepressant exposure during pregnancy or confounding by indication. In fact, up to 8% of women use antidepressant medication during pregnancy, notably SSRIs, in order to better cope with their anxious or depressive symptoms (Andrade et al., 2008). While generally considered to be safe to use during pregnancy, medications such as SSRIs have been found to be related to abnormal fetal activity as well as long-term developmental problems in children (El Marroun, White, Verhulst, & Tiemeier, 2014; Glover & Clinton, 2016). Similar to depression, antidepressant use during pregnancy could possibly have lasting epigenetics effects in exposed offspring. Heterogeneity in treatment, dosing, the timing of fetal exposure, and tissue tested could explain the current challenges and differences observed in multiple studies.

A review by Viuff et al. (2016) pointed out that very few studies have investigated epigenetics alteration associated with antidepressants use during pregnancy. Based on six included studies, the authors highlighted the lack of convergence and the high level of heterogeneity among performed studies. Moreover, they found that only half of these studies observed a significant association between exposure to prenatal antidepressants and DNA methylation in cord blood of neonates (Gurnot et al., 2015; Schroeder et al., 2012; Soubry et al., 2011). Thus, studies by Devlin et al. (2010), Non et al. (2014), and Oberlander et al. (2008) did not find any significant results. More recently, the study by Cardenas et al. (2019) using an epigenome-wide approach observed that newborns prenatally exposed to antidepressants had 7% lower DNA methylation in the *ZNF575* gene in the *Project Viva* cohort, a result replicated in the *Generation R Study*. In the first cohort, authors observed the persistence of the DNA methylation association during early childhood (~3 years), but no persistence of these effects was found in mid-childhood (~7 years). This result suggests that DNA methylation alterations might be malleable during development. Results of studies on epigenetics alterations associated with prenatal exposure to antidepressants are summarized in Table 4.

Results on epigenetics effects of depression and antidepressant use in mothers during pregnancy highlighted the importance of early programming in fetuses exposed to adversity. There is variation in the genes identified and there is heterogeneity in the types of studies, populations, and technologies used in the field. In fact, it is important to note that this field of research is fairly new and faces some limitations regarding the sample size, the scarcity of longitudinal follow-up, or the lack of results replications that have been performed until now (Table 5). Thus, much more research remains to be done to better understand the epigenetic alterations associated with exposure to prenatal maternal depression.

## Recommendations for future research

Studies examining the epigenetic associations of fetal exposure to prenatal maternal depression are relatively new and have been conducted primarily over the past decade. Several research directions can guide the work that remains to be done in this area (Table 6). First, although the majority of studies have used candidate gene approaches to assess epigenetic effects of depression in exposed children, an increasing number of studies are using epigenome-wide analyses to explore epigenetic effects over a wide range of genes without limiting the investigation to a few genes chosen a priori. These epigenome-wide studies have the advantage of being able to provide information on new genes and entire pathways and network of genes that may be affected by a specific disease without having an initial hypothesis. Tissue specificity of epigenetic marks remains an issue in epigenetic studies, particularly for mental health conditions where leukocyte DNA might not be the optimal biomarker. Many studies to date have been performed in accessible tissue like blood or buccal cells. Collecting target tissue can be challenging so experimental animal models might help address this current gap as well as blood–brain reference epigenomes. Also, although this domain is relatively young, some studies have observed sex-specific differences in epigenetics alterations among exposed offspring that merits further investigation in the future (Braithwaite et al., 2015; Stonawski et al., 2019). In addition, it is crucial to better document the persistence over time of epigenetic effects in children exposed to prenatal maternal mental health problems. A large percentage of the studies conducted have examined epigenetic effects in the newborn, but very few have studied the persistence of this effect over time to better understand its long-term consequences in children as well as their effects on their development (Cardenas et al., 2019; Wikenius et al., 2019). Finally, it is important to note that some proportion of women with depressive symptoms will use antidepressants in order to be able to function normally. Research to date has been heterogeneous and presents mixed results on this topic due to complexities in timing and type of medications (Viuff et al., 2016). Thus, further studies on the epigenetic effects of antidepressants are greatly needed to better understand the effects of this medication, which is being offered to many women during pregnancy.

## Conclusion

Several observations can be drawn from existing studies and the results discussed in this chapter. In order to better understand the effects of maternal depression on later child development and the role of epigenetics as an explanatory mechanism underlying this relationship, or alternatively a biomarker of disease, it is important to highlight pregnancy as a sensitive period for child development congruently with epigenetic programming. The effects of prenatal exposure to maternal depression or mental health disorders are well documented. According to the DOHaD hypothesis,

**TABLE 4** Genes with altered DNA methylation following prenatal exposure to antidepressants.

| References | Genes with altered methylation | Tissue |
|---|---|---|
| *Candidate-gene study* | | |
| Soubry et al. (2011) | *IGF2* | Cord blood |
| *Epigenome-wide association studies* | | |
| Cardenas et al. (2019) | *ZNF575* | Cord blood |
| Gurnot et al. (2015) | *CYP2E1, EVA1, SLMAP* | Cord blood |
| Schroeder et al. (2012) | *TNFRSF21, CHRNA2* | Cord blood |

**TABLE 5** Limits of the current literature about prenatal depression and offspring DNA methylation.

| Limits of the current literature |
|---|
| • *Sample size*: The use of small sample sizes may limit the possibility to find associations and may hinder the generalization of the results to larger samples.<br>• *Longitudinal effects*: Few studies have examined the long-term effects of prenatal exposure to depression on DNA methylation, thus we can't currently know if these effects persist or not over time. The associated mechanisms are also unclear.<br>• *Tissue specificity*: Studies often use tissues such as cord blood or buccal cells which may limit our knowledge about target organs like the brain, related to certain diseases that could be affected by prenatal exposure to depression.<br>• *Replication*: A limited number of studies replicated their result in other cohorts which limits the possibility of applying the results to different populations. |

**TABLE 6** Summary of recommendations for future avenues of research.

| Recommendations for future research |
|---|
| • *Epigenome-wide association studies*: An increasing number of studies are using epigenome-wide approaches to test epigenetic effects over a wide range of genes without limiting the investigation to a few genes chosen a priori. These studies have the advantage of being able to provide information on novel genes and pathways that may be affected by a specific disease without having an initial hypothesis. However, there is a need to increase sample sizes and perform meta-analyses across different cohorts for power and generalizability.<br>• *Tissue specificity*: Although cord blood and other accessible tissues like buccal cells have been used, there is a need to investigate disease relevant cells and tissues of offspring exposed during fetal development. Experimental animal models or epidemiological studies that leverage already collected samples can address this gap.<br>• *Timing of exposure*: The fetus is sensitive to environmental exposures and there are susceptible windows of fetal development for certain organs. Designing experiments and epidemiological cohort studies to address windows of susceptibly for depression and its treatment could help address current gaps.<br>• *Sex difference*: Some studies have observed epigenetic sex differences between male and female exposure to prenatal maternal mental health issues that merit further investigation. The design of studies specifically to look at sex-differences is needed.<br>• *Longitudinal effects*: It is important to better document the persistence over time of epigenetic effects in children exposed to prenatal maternal mental health problems. A large percentage of the studies have examined epigenetic effects in the newborn, but very few have studied the persistence of this effect over time to better understand its long-term consequences in children and potentially malleability of epigenetic marks.<br>• *Antidepressants use*: Given that up to 8% of women with depressive or anxious symptoms will use antidepressant medication, future studies should continue to investigate the epigenetics effects of exposed children. Confounding by indication remains a major challenge among epidemiological studies looking at depression. |

prenatal exposure to adverse environments can result in an increased risk of developing diseases and influence health trajectories later in life. Epigenetic mechanisms can help us understand the relationship between prenatal exposure to adverse environments and their consequences on later child development. In addition, epigenetic alterations might serve as early disease biomarkers that could be used for diagnosing, monitoring treatment, and identifying children at risk. The investigation of the epigenetic effects of prenatal exposure to maternal depression is relatively new and studies on the subject have been conducted mostly over the past decade in accessible tissue. Thus, much work remains to be done to better understand alterations associated with maternal depression in the epigenome, including the genes affected in neonates and children as well as sensitive tissues and timing of exposure. Finally, given the recent emergence of data in the field of epigenetics, public health efforts should be implemented to promote mental health for women during pregnancy. Better screening and follow-up of at-risk women could reduce later developmental problems and psychopathologies in their children. Thus, epigenetic biomarkers can play a critical role to monitor, diagnose, and identify children at risk.

## Key facts

- Almost 15% of women will report a major depressive episode during pregnancy and many more will experience depressive symptoms.
- Given the impact of depression on their daily functioning, up to 8% of pregnant women will use antidepressants to reduce their symptoms.
- The prenatal period is a highly sensitive time for the developing fetus who may be affected by adverse exposures.
- The occurrence of depressive symptoms during pregnancy is likely to have a significant influence on offspring DNA methylation.
- The study of DNA methylation alterations is one of the most widely studied epigenetic modifications tested in recent studies.

## Summary points

- Pregnancy is a sensitive period during which women are at higher risk of experiencing stress and mental health issues. Higher levels of depressive symptoms are generally observed during pregnancy than during the postpartum period.
- The occurrence of depressive symptoms during pregnancy is likely to have a significant influence on the development of the unborn child by affecting its epigenome or serving as a biomarker of exposure.
- The literature on prenatal exposure to maternal depression and related epigenetics alterations is a relatively new field of interest where studies have been performed mainly during the last decade.
- Studies in the field have observed some epigenetics alterations in offspring exposed to prenatal maternal depression. Results on epigenetics effects of prenatal depression and antidepressants highlighted the importance of early life programming in fetuses exposed to adversity.
- Futures studies should focus on sex-specific differences in offspring exposed to prenatal maternal depression in addition to examine long-term consequences in children.

## Mini-dictionary of terms

**Developmental origins of health and disease (DOHaD)** The DOHaD hypothesis suggests that the fetus will adapt itself to the maternal intrauterine environment in order to shape and adjust its development to its surrounding prenatal environment.

**Epigenetics** The examination of heritable molecular changes in gene expression that do not depend on the DNA sequence.

**DNA methylation** Epigenetic modification of DNA by the addition of a methyl group at the fifth position of cytosine nucleotides in the context of cytosine-guanine dinucleotides (CpGs).

**Epigenome-wide association studies** The study of the epigenome-wide alterations associated with the presence of a disease or a disorder; often in the context of DNA methylation.

**Candidate gene association studies** The examination of chosen a priori genes potentially associated with the presence of a disease or a disorder.

## References

Andersson, L., Sundström-Poromaa, I., Bixo, M., Wulff, M., Bondestam, K., & Åström, M. (2003). Point prevalence of psychiatric disorders during the second trimester of pregnancy: A population-based study. *American Journal of Obstetrics and Gynecology*, *189*, 148–154. https://doi.org/10.1067/mob.2003.336.

Andrade, S. E., Raebel, M. A., Brown, J., Lane, K., Livingston, J., Boudreau, D., et al. (2008). Use of antidepressant medications during pregnancy: A multisite study. *American Journal of Obstetrics and Gynecology*, *198*(2), 194.e1–194.e5. https://doi.org/10.1016/j.ajog.2007.07.036.

Babenko, O., Kovalchuk, I., & Metz, G. A. S. (2015). Stress-induced perinatal and transgenerational epigenetic programming of brain development and mental health. *Neuroscience and Biobehavioral Reviews*, *48*, 70–91. https://doi.org/10.1016/j.neubiorev.2014.11.013.

Barker, D. J. P., Hales, C. N., Fall, C. H. D., Osmond, C., Phipps, K., & Clark, P. M. S. (1993). Type 2 (non-insulin-dependent) diabetes mellitus, hypertension and hyperlipidaemia (syndrome X): Relation to reduced fetal growth. *Diabetologia*, *36*(1), 62–67. https://doi.org/10.1007/BF00399095.

Bennett, H. A., Einarson, A., Taddio, A., Koren, G., & Einarson, T. R. (2004). Prevalence of depression during pregnancy: Systematic review. *Obstetrics & Gynecology*, *103*(4), 698–709. https://doi.org/10.1097/01.AOG.0000116689.75396.5f.

Booij, L., Tremblay, R., Szyf, M., & Benkelfat, C. (2015). Genetic and early environmental influences on the serotonin system: Consequences for brain development and risk for psychopathology. *Journal of Psychiatry and Neuroscience*, *40*(1), 5–18. https://doi.org/10.1503/jpn.140099.

Braithwaite, E. C., Kundakovic, M., Ramchandani, P. G., Murphy, S. E., & Champagne, F. A. (2015). Maternal prenatal depressive symptoms predict infant NR3C1 1F and BDNF IV DNA methylation. *Epigenetics*, *10*(5), 408–417. https://doi.org/10.1080/15592294.2015.1039221.

Brown, M. A., & Solchany, J. E. (2004). Two overlooked mood disorders in women: Subsyndromal depression and prenatal depression. *Nursing Clinics of North America*, *39*, 83–95. https://doi.org/10.1016/j.cnur.2003.11.005.

Cardenas, A., Faleschini, S., Cortes Hidalgo, A., Rifas-Shiman, S. L., Baccarelli, A. A., Demeo, D. L., et al. (2019). Prenatal maternal antidepressants, anxiety, and depression and offspring DNA methylation: Epigenome-wide associations at birth and persistence into early childhood. *Clinical Epigenetics*, *11*, 56. https://doi.org/10.1186/s13148-019-0653-x.

Cardenas, A., Rifas-Shiman, S. L., Agha, G., Hivert, M. F., Litonjua, A. A., DeMeo, D. L., et al. (2017). Persistent DNA methylation changes associated with prenatal mercury exposure and cognitive performance during childhood. *Scientific Reports*, *7*, 288. https://doi.org/10.1038/s41598-017-00384-5.

Conradt, E., Lester, B. M., Appleton, A. A., Armstrong, D. A., & Marsit, C. J. (2013). The roles of DNA methylation of NR3C1 and 11β-HSD2 and exposure to maternal mood disorder in utero on newborn neurobehavior. *Epigenetics*, *8*(12), 1321–1329. https://doi.org/10.4161/epi.26634.

Devlin, A. M., Brain, U., Austin, J., & Oberlander, T. F. (2010). Prenatal exposure to maternal depressed mood and the MTHFR C677T variant affect SLC6A4 methylation in infants at birth. *PLoS One*, *5*(8), e12201. https://doi.org/10.1371/journal.pone.0012201.

El Marroun, H., White, T., Verhulst, F. C., & Tiemeier, H. (2014). Maternal use of antidepressant or anxiolytic medication during pregnancy and childhood neurodevelopmental outcomes: A systematic review. *European Child and Adolescent Psychiatry*, *23*, 973–992. https://doi.org/10.1007/s00787-014-0558-3.

Evans, J., Heron, J., Francomb, H., Oke, S., & Golding, J. (2001). Cohort study of depressed mood during pregnancy and after childbirth. *British Medical Journal*, *323*, 257–260. https://doi.org/10.1136/bmj.323.7307.257.

Faleschini, S., Rifas-Shiman, S. L., Tiemeier, H., Oken, E., & Hivert, M. F. (2019). Associations of prenatal and postnatal maternal depressive symptoms with offspring cognition and behavior in mid-childhood: A prospective cohort study. *International Journal of Environmental Research and Public Health*, *16*, 1007. https://doi.org/10.3390/ijerph16061007.

Field, T., Diego, M., & Hernandez-Reif, M. (2006). Prenatal depression effects on the fetus and newborn: A review. *Infant Behavior and Development*, *29*, 445–455. https://doi.org/10.1016/j.infbeh.2006.03.003.

Fleming, T. P., Velazquez, M. A., & Eckert, J. J. (2015). Embryos, DOHaD and David Barker. *Journal of Developmental Origins of Health and Disease*, *6*(5), 377–383. https://doi.org/10.1017/S2040174415001105.

Glover, M. E., & Clinton, S. M. (2016). Of rodents and humans: A comparative review of the neurobehavioral effects of early life SSRI exposure in preclinical and clinical research. *International Journal of Developmental Neuroscience*, *51*, 50–72. https://doi.org/10.1016/j.ijdevneu.2016.04.008.

Gurnot, C., Martin-Subero, I., Mah, S. M., Weikum, W., Goodman, S. J., Brain, U., et al. (2015). Prenatal antidepressant exposure associated with CYP2E1 DNA methylation change in neonates. *Epigenetics*, *10*(5), 361–372. https://doi.org/10.1080/15592294.2015.1026031.

Heijmans, B. T., Tobi, E. W., Stein, A. D., Putter, H., Blauw, G. J., Susser, E. S., et al. (2008). Persistent epigenetic differences associated with prenatal exposure to famine in humans. *Proceedings of the National Academy of Sciences of the United States of America*, *105*(44), 17046–17049. https://doi.org/10.1073/pnas.0806560105.

Henrichs, J., Schenk, J. J., Roza, S. J., van den Berg, M. P., Schmidt, H. G., Steegers, E. A. P., et al. (2010). Maternal psychological distress and fetal growth trajectories: The generation R study. *Psychological Medicine*, *40*, 633–643. https://doi.org/10.1017/S0033291709990894.

King, L., Robins, S., Chen, G., Yerko, V., Zhou, Y., Nagy, C., et al. (2017). Perinatal depression and DNA methylation of oxytocin-related genes: A study of mothers and their children. *Hormones and Behavior*, *96*, 84–94. https://doi.org/10.1016/j.yhbeh.2017.09.006.

Lin, Y., Xu, J., Huang, J., Jia, Y., Zhang, J., Yan, C., et al. (2017). Effects of prenatal and postnatal maternal emotional stress on toddlers' cognitive and temperamental development. *Journal of Affective Disorders*, *207*, 9–17. https://doi.org/10.1016/j.jad.2016.09.010.

Liu, Y., Murphy, S. K., Murtha, A. P., Fuemmeler, B. F., Schildkraut, J., Huang, Z., et al. (2012). Depression in pregnancy, infant birth weight and DNA methylation of imprint regulatory elements. *Epigenetics*, *7*(7), 735–746. https://doi.org/10.4161/epi.20734.

Madigan, S., Oatley, H., Racine, N., Fearon, R. M. P., Schumacher, L., Akbari, E., et al. (2018). A meta-analysis of maternal prenatal depression and anxiety on child socioemotional development. *Journal of the American Academy of Child and Adolescent Psychiatry*, *57*(9), 645–657. https://doi.org/10.1016/j.jaac.2018.06.012.

Mansell, T., Vuillermin, P., Ponsonby, A. L., Collier, F., Saffery, R., Barwon Infant Study Investigator Team, et al. (2016). Maternal mental well-being during pregnancy and glucocorticoid receptor gene promoter methylation in the neonate. *Development and Psychopathology*, *28*, 1421–1430. https://doi.org/10.1017/S0954579416000183.

Murgatroyd, C., Quinn, J. P., Sharp, H. M., Pickles, A., & Hill, J. (2015). Effects of prenatal and postnatal depression, and maternal stroking, at the glucocorticoid receptor gene. *Translational Psychiatry*, *5*, e560. https://doi.org/10.1038/tp.2014.140.

Nemoda, Z., Massart, R., Suderman, M., Hallett, M., Li, T., Coote, M., et al. (2015). Maternal depression is associated with DNA methylation changes in cord blood T lymphocytes and adult hippocampi. *Translational Psychiatry*, *5*, e545. https://doi.org/10.1038/tp.2015.32.

Nemoda, Z., & Szyf, M. (2017). Epigenetic alterations and prenatal maternal depression. *Birth Defects Research*, *109*, 888–897. https://doi.org/10.1002/bdr2.1081.

Non, A. L., Binder, A. M., Kubzansky, L. D., & Michels, K. B. (2014). Genome-wide DNA methylation in neonates exposed to maternal depression, anxiety, or SSRI medication during pregnancy. *Epigenetics*, *9*(7), 964–972. https://doi.org/10.4161/epi.28853.

O'Donnell, K. J., Glover, V., Barker, E. D., & O'Connor, T. G. (2014). The persisting effect of maternal mood in pregnancy on childhood psychopathology. *Development and Psychopathology*, *26*, 393–403. https://doi.org/10.1017/S0954579414000029.

Oberlander, T. F., Weinberg, J., Papsdorf, M., Grunau, R., Misri, S., & Devlin, A. M. (2008). Prenatal exposure to maternal depression, neonatal methylation of human glucocorticoid receptor gene (NR3C1) and infant cortisol stress responses. *Epigenetics*, *3*(2), 97–106. https://doi.org/10.4161/epi.3.2.6034.

Painter, R. C., Roseboom, T. J., & Bleker, O. P. (2005). Prenatal exposure to the Dutch famine and disease in later life: An overview. *Reproductive Toxicology*, *20*, 345–352. https://doi.org/10.1016/j.reprotox.2005.04.005.

Rakyan, V. K., Down, T. A., Balding, D. J., & Beck, S. (2011). Epigenome-wide association studies for common human diseases. *Nature Reviews Genetics*, *12*, 529–541. https://doi.org/10.1038/nrg3000.

Schroeder, J. W., Smith, A. K., Brennan, P. A., Conneely, K. N., Kilaru, V., Knight, B. T., et al. (2012). DNA methylation in neonates born to women receiving psychiatric care. *Epigenetics*, *7*(4), 409–414. https://doi.org/10.4161/epi.19551.

Soubry, A., Murphy, S. K., Huang, Z., Murtha, A., Schildkraut, J. M., Jirtle, R. L., et al. (2011). The effects of depression and use of antidepressive medicines during pregnancy on the methylation status of the IGF2 imprinted control regions in the offspring. *Clinical Epigenetics*, *3*, 2. https://doi.org/10.1186/1868-7083-3-2.

Stonawski, V., Frey, S., Golub, Y., Rohleder, N., Kriebel, J., Goecke, T. W., et al. (2019). Associations of prenatal depressive symptoms with DNA methylation of HPA axis-related genes and diurnal cortisol profiles in primary school-aged children. *Development and Psychopathology*, *31*, 419–431. https://doi.org/10.1017/S0954579418000056.

Swanson, J. M., Entringer, S., Buss, C., & Wadhwa, P. D. (2009). Developmental origins of health and disease: Environmental exposures. *Seminars in Reproductive Medicine*, *27*(5), 391–402. https://doi.org/10.1055/s-0029-1237427.

Unternaehrer, E., Bolten, M., Nast, I., Staehli, S., Meyer, A. H., Dempster, E., et al. (2016). Maternal adversities during pregnancy and cord blood oxytocin receptor (OXTR) DNA methylation. *Social Cognitive and Affective Neuroscience*, *11*(9), 1460–1470. https://doi.org/10.1093/scan/nsw051.

Viuff, A. C., Pedersen, L. H., Kyng, K., Staunstrup, N. H., Børglum, A., & Henriksen, T. B. (2016). Antidepressant medication during pregnancy and epigenetic changes in umbilical cord blood: A systematic review. *Clinical Epigenetics*, *8*, 94. https://doi.org/10.1186/s13148-016-0262-x.

Viuff, A. C., Sharp, G. C., Rai, D., Henriksen, T. B., Pedersen, L. H., Kyng, K. J., et al. (2018). Maternal depression during pregnancy and cord blood DNA methylation: Findings from the Avon longitudinal study of parents and children. *Translational Psychiatry*, *8*, 244. https://doi.org/10.1038/s41398-018-0286-4.

Wikenius, E., Myhre, A. M., Page, C. M., Moe, V., Smith, L., Heiervang, E. R., et al. (2019). Prenatal maternal depressive symptoms and infant DNA methylation: A longitudinal epigenome-wide study. *Nordic Journal of Psychiatry*, *73*(4–5), 257–263. https://doi.org/10.1080/08039488.2019.1613446.

Chapter 54

# Treating depression with theta burst stimulation (TBS)

Ankita Chattopadhyay
*Department of Psychiatry, All India Institute of Medical Sciences, New Delhi, India*

## List of abbreviations

| | |
|---|---|
| **rTMS** | repetitive transcranial magnetic stimulation |
| **TBS** | theta burst stimulation |
| **cTBS** | continuous TBS |
| **iTBS** | intermittent TBS |
| **imTBS** | intermediate TBS |
| **MDD** | major depressive disorder |
| **DLPFC** | dorsolateral prefrontal cortex |
| **LTP** | long-term potentiation |
| **LTD** | long-term depression |
| **MEP** | motor evoked potential |
| **RMT** | resting motor threshold |
| **RCT** | randomized controlled trial |
| **PET** | positron emission tomography |
| **fMRI** | functional magnetic resonance imaging |
| **NIRS** | near-infrared spectroscopy |
| **US FDA** | United States Food and Drugs Administration |

## Introduction

Major depressive disorder is one of the leading causes of disability all over the world (Murray, Vos, Lozano, et al., 2012). Various pharmacotherapeutic and psychotherapeutic approaches are available for treating depression. But even with the available options, about one-third of the patients suffering from major depression show responses (Rush, Kraemer, Sackeim, et al., 2006). Inadequate efficacy, side effects, and poor tolerability of current treatment strategies indicate the need for more effective and tolerable treatment options (Chung et al., 2017).

For such patients who respond inadequately to drug trials, newer modalities of treatment are being tried. Of these, non-invasive brain stimulation techniques are being used widely. Repetitive transcranial magnetic stimulation (rTMS) has shown promising results. Multiple randomized controlled trials, including multicentric trials as well as meta-analyses over the last 2 decades have shown the role of rTMS of the left dorsolateral prefrontal cortex (DLPFC) in treatment-resistant depression (Blumberger et al., 2018).

One of the newer forms of rTMS is theta-burst stimulation (TBS) (Huang, Edwards, Rounis, Bhatia, & Rothwell, 2005; Suppa, Huang, Funke, et al., 2016). In theta-burst stimulation, the stimulation mimics the endogenous theta rhythm and ultimately leads to synaptic long-term potentiation (LTP) (Suppa et al., 2016). Clinical trials have already shown similar effects of TBS as compared to conventional rTMS in patients with depression. Theta-burst stimulation, therefore, seems to be a promising intervention in such patients with the added advantage of less time duration of the stimulation process as compared to conventional 10 Hz rTMS.

## Role of transcranial magnetic stimulation in depression

The antidepressant potential of rTMS is being studied since the last two decades (George, Wassermann, Williams, et al., 1995; Pascual-Leone, Rubio, Pallardo, et al., 1996). In 2008, the United States Food and Drugs Administration (US FDA) approved the use of rTMS in patients of major depression who have failed the trial of an antidepressant drug. This was subsequent to the findings of a randomized sham-controlled trial of 300 participants demonstrating the efficacy of active rTMS over sham stimulation (O'Reardon, Solvason, & Janicak, 2007). Overall the efficacy of conventional rTMS over sham rTMS in major depression has been established through numerous sham-controlled randomized trials and meta-analyses (Berlim, van den Eynde, Tovar-Perdomo, et al., 2014; Teng, Guo, Peng, et al., 2017; Chen, Chung, Hoy, & Fitzgerald, 2019). The response of conventional rTMS in the treatment-resistant major depressive disorder has been around 40–60% with about 30%–40% patients achieving full remission (Chen et al., 2019).

In depression, rTMS is applied over the left or right dorsolateral prefrontal cortex (DLPFC). Usually, the session is given on a daily basis for 20–30 days over 4–6 weeks.

The Neuroscience of Depression. https://doi.org/10.1016/B978-0-12-817933-8.00033-5

A standard treatment course involves 20–30 daily stimulation sessions applied over 4–6 weeks (Perera, George, Grammer, et al., 2016). Depending on the response, the further course is decided. It usually takes 35–45 min to complete daily sessions. The most studied and accepted protocol is of 10 Hz rTMS applied to the left DLPFC, though 1 Hz rTMS applied to the right DLPFC and sequential bilateral rTMS have also been tried (Chen, Zhou, Wu, et al., 2013).

The efficacy and minimal side effects of rTMS have shown it to be a promising modality for treating depression. But there are some concerns and inconveniences regarding its use. The onset of action of the antidepressant profile of the TMS is slow at times and limits its use in patients where immediate response is warranted. Over the last two decades, several modifications have been done in the protocol of TMS to be followed. But the conventional high-frequency rTMS requires about 35 min for each session to deliver the adequate number of pulses (1000 – 1200). Over the years of study, it has also been evident that an increase in stimulation dose with an increase in the total number of pulses is more effective. Therefore, accelerated regimes of rTMS are also being tried with the hope of faster onset of antidepressant action with similar efficacy as conventional TMS (Fitzgerald, Hoy, Elliot, et al., 2018). Similarly, different stimulation patterns of TMS have also been studied. Of these, theta burst stimulation seems to be promising with similar efficacy as well as with the advantage of less time of stimulation being needed (Chen et al., 2019).

### Theta-burst stimulation

It is a type of TMS in which bursts of pulses are applied at theta frequency. The repetition of electrical stimulation has been seen to modulate synaptic transmission. In TBS repeated electromagnetic stimulation is used to modulate the synaptic transmission. Stimulation at repeated bursts of theta frequency has been shown to be more effective than at other frequencies in terms of long-term potentiation of synaptic transmission.

## Mechanism of action of theta-burst stimulation and its types

The term theta-burst stimulation originates from the fact that the bursts of pulses given are similar to the electroencephalographic theta range (4–8 Hz) (Suppa et al., 2016). The effect of TBS on the human cortex was studied by Huang and colleagues in 2005. They found that pulses of TMS of 50 Hz repeated at bursts of 5 Hz produced cortical excitability for a longer time as compared to a single pulse of TMS applied at the same intensity (Huang et al., 2005). TBS mimics the endogenous theta rhythms of the brain and induces long-lasting synaptic changes (Suppa et al., 2016). Depending on the pattern of administration of the bursts of pulses clinically, TBS can be of two types. When it is given uninterrupted in the form of $3 \times 50$ Hz pulses in 5 Hz bursts for 20 s or 40 s making a total of 300 or 600 pulses, respectively, it is called continuous TBS (cTBS). This causes a reduction in the amplitude of motor-evoked potential (MEP) due to a reduction in neuronal excitability, ultimately leading to long-term depression (LTD) of the synaptic transmission (Huang et al., 2005). When TBS is given intermittently in the form of $3 \times 50$ Hz pulses in 5 Hz bursts over 2 s with a gap of 8 s, such cycles being continued for 192 s, it is known as intermittent TBS (iTBS). iTBS causes an increase in the amplitude of motor-evoked potential (MEP) due to stimulation in neuronal excitability, ultimately leading to the long-term potentiation (LTP) of the synaptic transmission (Huang et al., 2005). By the long-term potentiation, increased neurotransmission occurs across the synapse due to the repetitive electromagnetic stimulation. Though TBS takes only a few minutes to be administered. The cortical conditioning induced by TBS tends to last for about 60 min after the stimulation (Huang et al., 2005). Thus, the LTP and LTD caused by TBS tend to last almost double the duration of that caused by conventional rTMS (Thut & Pascual-Leone, 2010). Apart from these two patterns, when TBS is given as $3 \times 50$ Hz pulses in 5 Hz bursts over 5 s and then followed by 15 s of no pulses, such cycles being repeated for 110 s, it is known as intermediate TBS (imTBS). This pattern however induces no apparent response (Chen et al., 2019).

These findings have been explored with TMS-EEG experiments on the prefrontal cortical reactivity before and after TBS. An increase in TMS-evoked theta oscillations and TMS-evoked potentials was seen following iTBS while cTBS led to a suppressive effect (Chung et al., 2017). The inability to demonstrate any changes following sham stimulation further strengthens the concept of cortical modulation by TBS. Similarly, researchers have tried to observe the changes following TBS using various neuroimaging techniques. cTBS has shown to affect the regional activation and stimulation profiles in fMRI studies (Agnew, Banissy, McGettigan, et al., 2018). Local cerebral perfusion has been seen to be changed in response to iTBS and the change has been evident in distal motor regions as well (Cardenas-Morales, Grön, & Kammer, 2010). Active TBS has shown changes in metabolic activity in the fronto-cingulo-temporal circuits in PET studies whereas sham TBS failed to cause any change (Li et al., 2018). Similar changes in deoxygenation-oxygenation patterns seen in near-infrared spectroscopy (NIRS) studies with TBS but without any change in sham stimulation (Tupak, Dresler, Badewien, et al., 2013). Thus, functional imaging also points to the action of TBS on the neuromodulation of the cortex, though more studies on the exact mechanism are warranted.

## Theta-burst stimulation in depression

Various case reports, chart reviews, and open-label trials have been published regarding the effectiveness of TBS in depression. Over the last one decade, there have been some randomized control trials supporting the evidence of TBS in managing depression. Some of these trials are listed in Table 1.

## Safety of theta-burst stimulation

Overall, transcranial magnetic stimulation being a non-invasive procedure is usually considered safe to use. As compared to conventional TMS, researchers who have used TBS have highlighted the use of paradigms that are short-lasting and use a lesser total number of pulses as compared to conventional rTMS (Oberman, Edwards, Eldaief, &

**TABLE 1** Trials of theta burst stimulation in management of depression.

| Authors | Sample size | Type of study | Duration of study | Parameters of stimulation | Study outcomes |
|---|---|---|---|---|---|
| Dhami et al. (2019) | 20 | Open label | 2 weeks | 800 iTBS pulses to left DLPFC, 1800 cTBS pulse to right DLPFC at 80% RMT | 4 out of 20 patients had >50% reduction in symptoms and 2 of them achieved remission, Significant reduction in HDRS-17 scores from baseline (22.4 ± 2.9) to session 5 (17.1 ± 5.73) ($P < 0.001$) and session 10 (13.5 ± 5.00) ($P < 0.0001$) |
| Blumberger et al. (2018) (THREE-D trial) | 414 participants, 205: rTMS 209: TBS | RCT | 4 weeks | rTMS group—3000 pulses/session at 120% RMT, 10 Hz, to left DLPFC, Time—37.5 min TBS group—600 pulses/ session at 120% RMT, triplet 50 Hz burst repeated at 5 Hz, 2 s on 8 s off, time = 3 min 9 s Total 20 session given in both groups with 5 session a week | HDRS scores improved from 23.5 (SD 4.4) to 13.4 (7.8) in rTMS group and from 23.6 (4.3) to 13.4 (7.9) in iTBS group ($p = 0.0011$) after treatment |
| Cheng et al. (2016) | 60 participants, 15 in each group Group 1: only cTBS; Group 2:only iTBS; Group 3: both iTBS and cTBS; and Group 4: sham TBS | RCT | Daily sessions for 10 days | 3-pulse 50-Hz bursts given every 200 ms (at 5 Hz) and an intensity of 80% active motor threshold In cTBS: a 120-s train of uninterrupted bursts given to the right DLPFC (1800 pulses) in each session per day In iTBS, a 2-s train of bursts repeated every 10 s for a total of 570 s (1800 pulses) to the left DLPFC in each session per day. | iTBS alone or in combination with cTBS was better than cTBS alone or sham |
| Duprat et al. (2016) | 50 participants | RCT | 4 days | 20 iTBS sessions, given as 5 session/d. intersession interval kept of 15 min, total 1620 pulses per session in 54 triplet bursts with train duration of 2 s and intertrain interval of 8 s | Accelerated iTBS led to clinical improvement (38% participants showed response) within 2-weeks post stimulation |

*Continued*

**TABLE 1** Trials of theta burst stimulation in management of depression.—cont'd

| Authors | Sample size | Type of study | Duration of study | Parameters of stimulation | Study outcomes |
|---|---|---|---|---|---|
| Chistyakov et al. (2015) | 29 participants, active cTBS = 15, sham cTBS = 14 | RCT | 20 sessions | cTBS given as triple-pulse 50 Hz bursts at rate of 5 Hz in uninterrupted trains | cTBS better than sham in those whose medication status remained unchanged or were medication-free |
| Bakker et al. (2015) | 185 subjects, 10 HzrTMS: 98, TBS: 87 | Naturalistic study, a retrospective case series | Mean number of sessions were 21.4 +/− 4.7 | rTMS: 6000 pulses (3000 to each hemisphere), 1 cycle-5 s on 10 s off, for 30 min<br>iTBS: 600 pulses per hemisphere (1200 pulses total), for 6 min | Effectiveness of 6 min iTBS and 30 min 10 Hz rTMS protocols comparable at 120% RMT |
| Prasser et al. (2015) | 56 participants | RCT | 1 session daily for 15 days | rTMS: Total 2000 stimuli/d<br>TBS: Total 2400 stimuli/d, with 1200 stimuli of cTBS followed by 1200 stimuli of iTBS to Left LPFC | TBS and conventional rTMS are better than sham |
| Li et al. (2014) | 60 participants, Group A: cTBS; Group B: iTBS; Group C: combined; Group Dsham | RCT with open label follow up | RCT: 2 weeks Open label : 12 weeks | 3-pulse 50-Hz bursts given every 200 ms (at 5 Hz) and at 80% of motor threshold for 10 sessions | iTBS and combined groups showed the best response |
| Plewnia et al. (2014) | 32 participants | RCT | 6 weeks | 2 trains of 600 stimuli applied in bursts of 3 pulses at 50 Hz given every 200 ms for 6 weeks (with 5 sessions/week) | cTBS better than sham |

cTBS: continuous TBS; iTBS: intermittent TBS; rTMS: repetitive transcranial magnetic stimulation; RMT: resting motor threshold; RCT: randomized controlled trial; TBS: theta burst stimulation.

Pascual-Leone, 2011). However, there is also a possible risk due to the use of high-frequency stimulation. The usual side effects reported are mild headache, tinnitus, mild discomfort due to cutaneous sensation and contraction of a neck muscle, nausea, light headedness, unilateral eye pain, and lacrimation. The risk of seizure is an important consideration while giving TBS. Though there is only one reported case of seizure in a person receiving continuous TBS with no previous history of seizures, sufficient caution should be ensured to prevent the risk while administering TBS (Oberman & Pascual-Leone, 2009). Also, there can be a risk of treatment-emergent switch to hypomania or mania which should be monitored while giving TBS (Chen et al., 2019). Currently, there are no guidelines regarding the safety of the use of theta-burst stimulation protocols (Rachid, 2017). Whether the stimulation parameters like frequency, duration, number of pulses can predict the risk of the adverse effect that still needs to be studied.

## Conclusion

The therapeutic potential of TMS in depression has been well documented and is being used in clinical practice. The effect on neuronal modulation along with changes in cerebral regional metabolism and electrophysiology is responsible for clinical improvement in depression. Theta burst stimulation is a novel technique of providing the transcranial magnetic stimulation. Recent trials have shown the noninferiority of the TBS protocol as compared to conventional rTMS. With the added benefit of delivering the pulses within 3–6 min increases the feasibility of catering to a larger number of patients within the same time. TBS is comparable to conventional TMS regarding tolerability and adverse effect profile. Recently FDA has approved the use of theta-burst stimulation in depression following the results of the THREE-D trial in August 2018 (Blumberger et al., 2018). Thus, TBS seems to be a promising intervention for treating depression, though further studies regarding the tolerability profile with appropriate guidelines for its use are needed.

## Key facts

- Major depressive disorder is one of the leading causes of disability worldwide.
- rTMS has been approved by US FDA in 2008 for treating patients with depression who have failed the trial of a single antidepressant.
- Theta burst stimulation is one of the types of patterned rTMS, which can be given intermittently (iTBS) or continuously (cTBS).
- iTBS is stimulatory, usually given at left DLPFC or cTBS is inhibitory, usually given to right DLPFC.
- US FDA has approved the use of TBS in depression in August 2018 based on the results of the THREE-D trial.
- The usual side effects of a mild headache, tinnitus, mild discomfort due to cutaneous sensation and contraction of a neck muscle, nausea, light headedness, unilateral eye pain, and lacrimation with a rare risk of seizure and switch to hypomania or mania.

## Summary

- A major depressive disorder is one of the leading causes of disability in the world with the emergent need for newer strategies of treatment.
- Repetitive transcranial magnetic stimulation (rTMS) with the new pattern of theta-burst stimulation (TBS) is one of the newer strategies to manage resistant depression.
- Theta burst stimulation (TBS) can be either stimulatory (intermittent TBS) or inhibitory (continuous TBS).
- Studies suggest that the effectiveness of theta-burst stimulation is noninferior to conventional rTMS with a comparable tolerability profile and less time of administration.
- In 2018, the FDA has approved the use of TBS in treating depression.
- There is a lack of proper guidelines for the administration, stimulation parameters, and safety profile of TBS.

## Mini-dictionary terms

**rTMS** A noninvasive form of brain stimulation in which a changing magnetic field is used to cause electric current at a specific area of the brain through electromagnetic induction.

**TBS** A type of patterned TMS in which the pulses are given in bursts corresponding to the theta frequency.

**iTBS** A type of TBS administered intermittently in the form of $3 \times 50$ Hz pulses in 5 Hz bursts over 2 s with a gap of 8 s, such cycles being continued for 192 s.

**cTBS** A type of TBS administered uninterruptedly in the form of $3 \times 50$ Hz pulses in 5 Hz bursts for 20 s or 40 s making a total of 300 or 600 pulses, respectively.

## References


Agnew, Z. K., Banissy, M., McGettigan, C., et al. (2018). Investigating the neural basis of theta burst stimulation to premotor cortex on emotional vocalization perception: A combined TMS-fMRI study. *Frontiers in Human Neuroscience*, *12*, 150.

Bakker, N., Shahab, S., Giacobbe, P., Blumberger, D. M., Daskalakis, Z. J., Kennedy, S. H., et al. (2015). rTMS of the dorsomedial prefrontal cortex for major depression: Safety, tolerability, effectiveness, and outcome predictors for 10 Hz versus intermittent theta-burst stimulation. *Brain Stimulation*, *8*(2), 208–215.

Berlim, M., van den Eynde, F., Tovar-Perdomo, S., et al. (2014). Response, remission and drop-out rates following high-frequency repetitive transcranial magnetic stimulation (rTMS) for treating major depression: A systematic review and meta-analysis of randomized, double-blind and sham-controlled trials. *Psychological Medicine*, *44*, 225–239.

Blumberger, D. M., Vila-Rodriguez, F., Thorpe, K. E., Feffer, K., Noda, Y., Giacobbe, P., et al. (2018). Effectiveness of theta burst versus high-frequency repetitive transcranial magnetic stimulation in patients with depression (THREE-D): A randomised non-inferiority trial. *The Lancet*, *391*(10131), 1683–1692.

Cardenas-Morales, L., Grön, G., & Kammer, T. (2010). Exploring the after-effects of theta burst magnetic stimulation on the human motor cortex: A functional imaging study. *Human Brain Mapping*, *32*, 1948–1960.

Chen, L., Chung, S. W., Hoy, K. E., & Fitzgerald, P. B. (2019). Is theta burst stimulation ready as a clinical treatment for depression? *Expert review of neurotherapeutics*, *19*(11), 1089–1102.

Chen, J., Zhou, C., Wu, B., et al. (2013). Left versus right repetitive transcranial magnetic stimulation in treating major depression: A meta-analysis of randomised controlled trials. *Psychiatry Research*, *210*(3), 1260–1264.

Cheng, C.-M., Juan, C.-H., Chen, M.-H., Chang, C.-F., Lu, H. J., Su, T.-P., et al. (2016). Different forms of prefrontal theta burst stimulation for executive function of medication- resistant depression: Evidence from a randomized sham-controlled study. *Progress in Neuro-Psychopharmacology and Biological Psychiatry*, *66*, 35–40.

Chistyakov, A. V., Kreinin, B., Marmor, S., Kaplan, B., Khatib, A., Darawsheh, N., et al. (2015). Preliminary assessment of the therapeutic efficacy of continuous theta-burst magnetic stimulation (cTBS) in major depression: A double-blind sham-controlled study. *Journal of Affective Disorders*, *170*, 225–229.

Chung, S. W., Lewis, B. P., Rogasch, N. C. et al. (2017 Jul). Demonstration of short-term plasticity in the dorsolateral prefrontal cortex with theta burst stimulation: A TMS-EEG study. *Clinical Neurophysiology*, *128*(7), 1117–1126.

Dhami, P., Knyahnytska, Y., Atluri, S., Lee, J., Courtney, D. B., Croarkin, P. E., et al. (2019). Feasibility and clinical effects of theta burst stimulation in youth with major depressive disorders: An open-label trial. *Journal of Affective Disorders*, *258*, 66–73.

Duprat, R., Desmyter, S., Raedt Rudi, D., van Heeringen, K., Van den Abbeele, D., Tandt, H., et al. (2016). Accelerated intermittent theta burst stimulation treatment in medication-resistant major depression: A fast road to remission? *Journal of Affective Disorders*, *200*, 6–14.

Fitzgerald, P. B., Hoy, K. E., Elliot, D., et al. (2018). Accelerated repetitive transcranial magnetic stimulation in the treatment of depression. *Neuropsychopharmacology*, *43*(7), 1565–1572.

George, M., Wassermann, E., Williams, W., et al. (1995). Daily repetitive transcranial magnetic stimulation (rTMS) improves mood in depression. *Neuroreport*, *6*, 1853–1856.

Huang, Y. Z., Edwards, M. J., Rounis, E., Bhatia, K. P., & Rothwell, J. C. (2005). Theta burst stimulation of the human motor cortex. *Neuron*, *45*, 201–206.

Li, C.-T., Chen, M.-H., Juan, C.-H., Huang, H.-H., Chen, L.-F., Hsieh, J.-C., et al. (2014). Efficacy of prefrontal theta-burst stimulation in refractory depression: A randomized sham-controlled study. *Brain: A Journal of Neurology*, *137*(Pt 7), 2088–2098.

Li, C. T., Chen, M. H., Juan, C. H., Liu, R. S., Lin, W. C., Bai, Y. M., et al. (2018). Effects of prefrontal theta-burst stimulation on brain function in treatment-resistant depression: A randomized sham-controlled neuroimaging study. *Brain Stimulation*, *11*(5), 1054–1062.

Murray, C. J., Vos, T., Lozano, R., et al. (2012). Disabilityadjusted life years (DALYs) for 291 diseases and injuries in 21 regions, 1990–2010: a systematic analysis for the Global Burden of Disease Study 2010. *Lancet*, *380*, 2197–2223.

O'Reardon, J., Solvason, H., & Janicak, P. (2007). Efficacy and safety of transcranial magnetic stimulation in the acute treatment of major depression: A muiltisite randomized contorlled trial. *Biological Psychiatry*, *62*(11), 1208–1216.

Oberman, L. M., & Pascual-Leone, A. (2009). Report of seizure induced by continuous theta burst stimulation. *Brain Stimulation*, *2*, 246–247.

Oberman, L., Edwards, D., Eldaief, M., & Pascual-Leone, A. (2011). Safety of theta burst transcranial magnetic stimulation: A systematic review of the literature. *The Journal of Clinical Neurophysiology*, *28*, 67–74.

Pascual-Leone, A., Rubio, B., Pallardo, F., et al. (1996). Rapid-rate transcranial magnetic stimulation of left dorsolateral prefrontal cortex in drug-resistant depression. *Lancet*, *348*, 233–237.

Perera, T., George, M. S., Grammer, G., et al. (2016). The clinical TMS society consensus review and Treatment Recommendations for TMS therapy for major depressive disorder. *Brain Stimulation*, *9*(3), 336–346.

Plewnia, C., Pasqualetti, P., Große, S., Schlipf, S., Wasserka, B., Zwissler, B., et al. (2014). Treatment of major depression with bilateral theta burst stimulation: A randomized controlled pilot trial. *Journal of Affective Disorders*, *156*, 219–223.

Prasser, J., Schecklmann, M., Poeppl, T. B., Frank, E., Kreuzer, P. M., Hajak, G., et al. (2015). Bilateral prefrontal rTMS and theta burst TMS as an add-on treatment for depression: A randomized placebo controlled trial. *The World Journal of Biological Psychiatry: The Official Journal of the World Federation of Societies of Biological Psychiatry*, *16*(1), 57–65.

Rachid, F. (2017). Safety and efficacy of theta-burst stimulation in the treatment of psychiatric disorders: A review of the literature. *The Journal of Nervous and Mental Disease*, *205*(11), 823–839.

Rush, A. J., Kraemer, H. C., Sackeim, H. A., et al. (2006). Report by the ACNP task force on response and remission in major depressive disorder. *Neuropsychopharmacology*, *31*, 1841–1853.

Suppa, A., Huang, Y. Z., Funke, K., et al. (2016). Ten years of theta burst stimulation in humans: Established knowledge, unknowns and prospects. *Brain Stimulation*, *9*, 323–335.

Teng, S., Guo, Z., Peng, H., et al. (2017 Mar). High-frequency repetitive transcranial magnetic stimulation over the left DLPFC for major depression: Session-dependent efficacy: A meta-analysis. *European Psychiatry*, *41*, 75–84.

Thut, G., & Pascual-Leone, A. (2010). A review of combined TMS-EEG studies to characterize lasting effects of repetitive TMS and assess their usefulness in cognitive and clinical neuroscience. *Brain Topography*, *22*, 219–232.

Tupak, S. V., Dresler, T., Badewien, M., et al. (2013 Jan). Inhibitory transcranial magnetic theta burst stimulation attenuates prefrontal cortex oxygenation. *Human Brain Mapping*, *34*(1), 150–157.

# Index

Note: Page numbers followed by *f* indicate figures, *t* indicate tables, and *b* indicate boxes.

## D

## F



## G



## H

## I



## J



## K



## L



## M

## N

## Q



## R



## S

## T